Conceptual Understanding

Students need to be equipped with both the methods and conceptual understanding of statistics. MyStatLab offers a full question library of over 1,000 conceptual-based questions to help tighten the comprehension of statistical concepts.

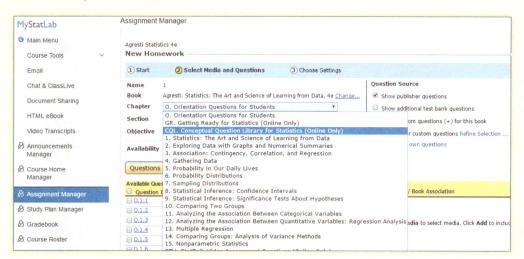

Real-World Statistics

MyStatLab video resources help foster conceptual understanding. StatTalk Videos, hosted by fun-loving statistician, Andrew Vickers, demonstrate important statistical concepts through interesting stories and real-life events. This series of 24 videos includes assignable questions built in MyStatLab and an instructor's guide.

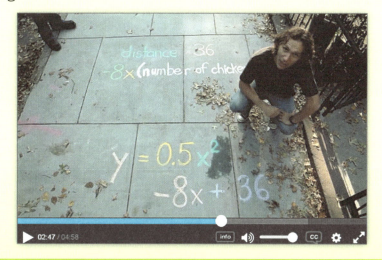

Visit www.mystatlab.com and click Get Trained to make sure you're getting the most out of MyStatLab.

ELEMENTARY STATISTICS

13th EDITION

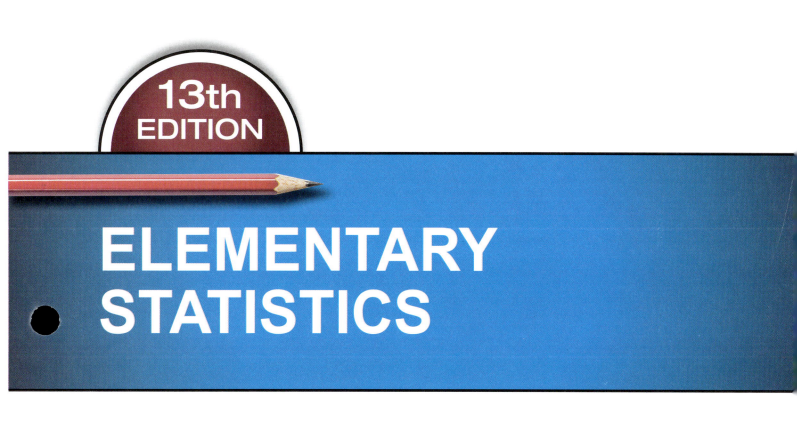

ELEMENTARY STATISTICS

MARIO F. TRIOLA

Special Contributions by Laura Iossi,
Broward College

Director, Portfolio Management Deirdre Lynch
Senior Portfolio Manager Suzy Bainbridge
Portfolio Management Assistant Justin Billing
Content Producer Peggy McMahon
Managing Producer Karen Wernholm
Manager, Courseware QA Mary Durnwald
Manager, Content Development Robert Carroll
Senior Producer Vicki Dreyfus
Product Marketing Manager Tiffany Bitzel
Field Marketing Manager Andrew Noble

Product Marketing Assistant Jennifer Myers
Senior Author Support/Technology Specialist Joe Vetere
Manager, Rights and Permissions Gina Cheselka
Text and Cover Design, Illustrations Production Coordination, Composition Cenveo Publisher Services
Cover Image: © Laura A. Watt/Getty Images
Field Marketing Assistant Erin Rush

Library of Congress Cataloging-in-Publication Data

Names: Triola, Mario F.
Title: Elementary statistics / Mario F. Triola, Dutchess Community College.
Description: 13th edition. | Boston : Pearson, [2018] | Includes bibliographical references and index.
Identifiers: LCCN 2016016750| ISBN 9780134462455 (hardcover) | ISBN 0134462459 (hardcover)
Subjects: LCSH: Statistics—Textbooks.
Classification: LCC QA276.12 .T76 2018 | DDC 519.5--dc23
LC record available at https://lccn.loc.gov/2016016750

4   18

Student Edition
ISBN 13: 978-0-13-446245-5
ISBN 10: 0-13-446245-9

To Ginny
Marc, Dushana, and Marisa
Scott, Anna, Siena, and Kaia

ABOUT THE AUTHOR

Mario F. Triola is a Professor Emeritus of Mathematics at Dutchess Community College, where he has taught statistics for over 30 years. Marty is the author of *Essentials of Statistics,* 5th edition, *Elementary Statistics Using Excel,* 6th edition, *Elementary Statistics Using the TI-83/84 Plus Calculator,* 4th edition, and he is a co-author of *Biostatistics for the Biological and Health Sciences,* 2nd edition, *Statistical Reasoning for Everyday Life,* 5th edition, and *Business Statistics. Elementary Statistics* is currently available as an International Edition, and it has been translated into several foreign languages. Marty designed the original Statdisk statistical software, and he has written several manuals and workbooks for technology supporting statistics education. He has been a speaker at many conferences and colleges. Marty's consulting work includes the design of casino slot machines and fishing rods. He has worked with attorneys in determining probabilities in paternity lawsuits, analyzing data in medical malpractice lawsuits, identifying salary inequities based on gender, and analyzing disputed election results. He has also used statistical methods in analyzing medical school surveys, and in analyzing survey results for the New York City Transit Authority. Marty has testified as an expert witness in the New York State Supreme Court. The Text and Academic Authors Association has awarded Marty a "Texty" for Excellence for his work on *Elementary Statistics.*

CONTENTS

1 INTRODUCTION TO STATISTICS   1

1-1 Statistical and Critical Thinking   3
1-2 Types of Data   13
1-3 Collecting Sample Data   25

2 EXPLORING DATA WITH TABLES AND GRAPHS   40

2-1 Frequency Distributions for Organizing and Summarizing Data   42
2-2 Histograms   51
2-3 Graphs That Enlighten and Graphs That Deceive   57
2-4 Scatterplots, Correlation, and Regression   67

3 DESCRIBING, EXPLORING, AND COMPARING DATA   80

3-1 Measures of Center   82
3-2 Measures of Variation   97
3-3 Measures of Relative Standing and Boxplots   112

4 PROBABILITY   131

4-1 Basic Concepts of Probability   133
4-2 Addition Rule and Multiplication Rule   147
4-3 Complements, Conditional Probability, and Bayes' Theorem   159
4-4 Counting   169
4-5 Probabilities Through Simulations (download only)   177

5 DISCRETE PROBABILITY DISTRIBUTIONS   184

5-1 Probability Distributions   186
5-2 Binomial Probability Distributions   199
5-3 Poisson Probability Distributions   214

6 NORMAL PROBABILITY DISTRIBUTIONS   226

6-1 The Standard Normal Distribution   228
6-2 Real Applications of Normal Distributions   242
6-3 Sampling Distributions and Estimators   254
6-4 The Central Limit Theorem   265
6-5 Assessing Normality   275
6-6 Normal as Approximation to Binomial   284

7 ESTIMATING PARAMETERS AND DETERMINING SAMPLE SIZES   297

7-1 Estimating a Population Proportion   299
7-2 Estimating a Population Mean   316
7-3 Estimating a Population Standard Deviation or Variance   332
7-4 Bootstrapping: Using Technology for Estimates   342

8 HYPOTHESIS TESTING   356

8-1 Basics of Hypothesis Testing   358
8-2 Testing a Claim About a Proportion   373
8-3 Testing a Claim About a Mean   387
8-4 Testing a Claim About a Standard Deviation or Variance   399

9 INFERENCES FROM TWO SAMPLES   414

9-1 Two Proportions   416
9-2 Two Means: Independent Samples   428
9-3 Two Dependent Samples (Matched Pairs)   442
9-4 Two Variances or Standard Deviations   452

10 CORRELATION AND REGRESSION 468

10-1 Correlation   470
10-2 Regression   489
10-3 Prediction Intervals and Variation   503
10-4 Multiple Regression   511
10-5 Nonlinear Regression   522

11 GOODNESS-OF-FIT AND CONTINGENCY TABLES 533

11-1 Goodness-of-Fit   535
11-2 Contingency Tables   546

12 ANALYSIS OF VARIANCE 566

12-1 One-Way ANOVA   568
12-2 Two-Way ANOVA   582

13 NONPARAMETRIC TESTS 597

13-1 Basics of Nonparametric Tests   599
13-2 Sign Test   601
13-3 Wilcoxon Signed-Ranks Test for Matched Pairs   612
13-4 Wilcoxon Rank-Sum Test for Two Independent Samples   619
13-5 Kruskal-Wallis Test for Three or More Samples   626
13-6 Rank Correlation   632
13-7 Runs Test for Randomness   640

14 STATISTICAL PROCESS CONTROL 654

14-1 Control Charts for Variation and Mean   656
14-2 Control Charts for Attributes   667

15 ETHICS IN STATISTICS 677

APPENDIX A TABLES 683

APPENDIX B DATA SETS 697

APPENDIX C WEBSITES AND BIBLIOGRAPHY OF BOOKS 709

APPENDIX D ANSWERS TO ODD-NUMBERED SECTION EXERCISES 710
(and all Quick Quizzes, all Review Exercises, and all Cumulative Review Exercises)

Credits   752
Index   756

PREFACE

Statistics permeates nearly every aspect of our lives. From opinion polls, to clinical trials in medicine, self-driving cars, drones, and biometric security, statistics influences and shapes the world around us. *Elementary Statistics* forges the relationship between statistics and our world through extensive use of a wide variety of real applications that bring life to theory and methods.

Goals of This Thirteenth Edition

- Foster personal growth of students through critical thinking, use of technology, collaborative work, and development of communication skills.

- Incorporate the latest and best methods used by professional statisticians.

- Include features that address all of the recommendations included in the *Guidelines for Assessment and Instruction in Statistics Education* (GAISE) as recommended by the American Statistical Association.

- Provide an abundance of new and interesting data sets, examples, and exercises, such as those involving biometric security, cybersecurity, drones, and smartphone data speeds.

- Enhance teaching and learning with the most extensive and best set of supplements and digital resources.

Audience/Prerequisites

Elementary Statistics is written for students majoring in any subject. Algebra is used minimally. It is recommended that students have completed at least an elementary algebra course or that students should learn the relevant algebra components through an integrated or co-requisite course available through MyStatLab. In many cases, underlying theory is included, but this book does not require the mathematical rigor more appropriate for mathematics majors.

Hallmark Features

Great care has been taken to ensure that each chapter of *Elementary Statistics* will help students understand the concepts presented. The following features are designed to help meet that objective of conceptual understanding.

Real Data

Hundreds of hours have been devoted to finding data that are real, meaningful, and interesting to students. 94% of the examples are based on real data, and 92% of the exercises are based on real data. Some exercises refer to the 32 data sets listed in Appendix B, and 12 of those data sets are new to this edition. Exercises requiring use of the Appendix B data sets are located toward the end of each exercise set and are marked with a special data set icon .ılı.

Real data sets are included throughout the book to provide relevant and interesting real-world statistical applications including biometric security, self-driving cars, smartphone data speeds and use of drones for delivery. Appendix B includes descriptions of the 32 data sets that can be downloaded from the companion website www.pearsonhighered.com/triola or www.TriolaStats.com.

The companion website and TriolaStats.com include downloadable data sets in formats for technologies including Excel, Minitab, JMP, SPSS, and TI-83/84 Plus calculators. The data sets are also included in the free Statdisk software, which is also available on the website.

Readability

Great care, enthusiasm, and passion have been devoted to creating a book that is readable, understandable, interesting, and relevant. Students pursuing any major are sure to find applications related to their future work.

Website

This textbook is supported by www.pearsonhighered.com/triola and the author's website www.TriolaStats.com which are continually updated to provide the latest digital resources for the *Triola Statistics Series,* including:

- Statdisk: A free robust statistical software package designed for this book.
- Downloadable Appendix B data sets in a variety of technology formats.
- Downloadable textbook supplements including Section 4-5 *Probabilities Through Simulations, Glossary of Statistical Terms* and *Formulas and Tables.*
- Online instructional videos created specifically for the 13th Edition that provide step-by-step technology instructions.
- Triola Blog which highlights current applications of statistics, statistics in the news and online resources.
- Contact link providing one-click access for instructors and students to contact the author, Marty Triola, with questions and comments.

Chapter Features

Chapter Opening Features

- Chapters begin with a *Chapter Problem* that uses real data and motivates the chapter material.
- *Chapter Objectives* provide a summary of key learning goals for each section in the chapter.

Exercises Many exercises require the *interpretation* of results. Great care has been taken to ensure their usefulness, relevance, and accuracy. Exercises are arranged in order of increasing difficulty and exercises are also divided into two groups: (1) *Basic Skills and Concepts* and (2) *Beyond the Basics. Beyond the Basics* exercises address more difficult concepts or require a stronger mathematical background. In a few cases, these exercises introduce a new concept.

End-of-Chapter Features

- *Chapter Quick Quiz* provides 10 review questions that require brief answers.
- *Review Exercises* offer practice on the chapter concepts and procedures.
- *Cumulative Review Exercises* reinforce earlier material.
- *Technology Project* provides an activity that can be used with a variety of technologies.
- *From Data to Decision* is a capstone problem that requires critical thinking and writing.
- *Cooperative Group Activities* encourage active learning in groups.

Other Features

Margin Essays There are 106 margin essays designed to highlight real-world topics and foster student interest. There are also many *Go Figure* items that briefly describe interesting numbers or statistics.

Flowcharts The text includes flowcharts that simplify and clarify more complex concepts and procedures. Animated versions of the text's flowcharts are available within MyStatLab and MathXL.

Detachable Formula and Table Card This insert, organized by chapter, gives students a quick reference for studying, or for use when taking tests (if allowed by the instructor). It also includes the most commonly used tables. This is also available for download at www.TriolaStats.com.

Technology Integration

As in the preceding edition, there are many displays of screens from technology throughout the book, and some exercises are based on displayed results from technology. Where appropriate, sections end with a new **Tech Center** subsection that includes new technology specific videos and detailed instructions for Statdisk, Minitab®, Excel®, StatCrunch, or a TI-83/84 Plus® calculator. (Throughout this text, "TI-83/84 Plus" is used to identify a TI-83 Plus or TI-84 Plus calculator). The end-of-chapter features include a *Technology Project*.

The Statdisk statistical software package is designed specifically for this textbook and contains all Appendix B data sets. Statdisk is free to users of this book and it can be downloaded at www.Statdisk.org.

Changes in This Edition

New Features

Chapter Objectives provide a summary of key learning goals for each section in the chapter.

Your Turn: Many examples include a new "your turn" feature that directs students to a relevant exercise so that they can immediately apply what they just learned from the example.

Tech Center: Improved technology instructions, supported by custom author created instructional videos and downloadable content available at www.TriolaStats.com.

Technology Videos. New, author-driven technology videos provide step-by-step details for key statistical procedures using Excel, TI-83/84 calculators, and Statdisk.

Larger Data Sets: Some of the data sets in Appendix B are much larger than in previous editions. It is no longer practical to print all of the Appendix B data sets in this book, so the data sets are *described* in Appendix B, and they can be downloaded at www.TriolaStats.com.

New Content: New examples, exercises and Chapter Problems provide relevant and interesting real-world statistical applications including biometric security, self-driving cars, smartphone data speeds, and use of drones for delivery.

	Number	New to This Edition	Use Real Data
Exercises	1756	81% (1427)	92% (1618)
Examples	211	73% (153)	94% (198)
Chapter Problems	14	93% (13)	100% (14)

Organization Changes

New Chapter Objectives: All chapters now begin with a list of key learning goals for that chapter. *Chapter Objectives* replaces the former *Review and Preview* numbered section. The first numbered section of each chapter now covers a major topic.

New Subsection 1-3, Part 2: Big Data and Missing Data: Too Much and Not Enough

New Section 2-4: Scatterplots, Correlation, and Regression

> The previous edition included scatterplots in Chapter 2, but this new section includes scatterplots in Part 1, the linear correlation coefficient *r* in Part 2, and linear regression in Part 3. These additions are intended to greatly facilitate coverage for those professors who prefer some early coverage of correlation and regression concepts. Chapter 10 continues to include these topics discussed with much greater detail.

New Subsection 4-3, Part 3: Bayes' Theorem

New Section 7-4: Bootstrapping: Using Technology for Estimates

Combined Sections:

- **4-2: Addition Rule and Multiplication Rule**
  Combines 12th edition Section 4-3 (*Addition Rule*) and Section 4-4 (*Multiplication Rule: Basics*).

- **5-2: Binomial Probability Distributions**
  Combines 12th edition Section 5-3 (*Binomial Probability Distributions*) and Section 5-4 (*Parameters for Binomial Distributions*)

Removed Sections:

Section 15-2 (*Projects*) has been changed to an insert in the Instructor's Edition and has been moved to accompany the first set of *Cooperative Group Activities* in Chapter 1. Section 15-3 (*Procedures*) and Section 15-4 (*Perspectives*) have been removed.

Changed Terminology

Significant: References in the previous edition to "unusual" outcomes are now described in terms of "significantly low" or "significantly high," so that the link to hypothesis testing is further reinforced.

Multiplication Counting Rule: References in Section 4-4 (*Counting*) to the "fundamental counting rule" now use "multiplication counting rule" so that the name of the rule better suggests how it is applied.

Flexible Syllabus

This book's organization reflects the preferences of most statistics instructors, but there are two common variations:

- *Early Coverage of Correlation and Regression:* Some instructors prefer to cover the basics of correlation and regression early in the course. Section 2-4 now includes basic concepts of scatterplots, correlation, and regression without the use of formulas and greater depth found in Sections 10-1 (*Correlation*) and 10-2 (*Regression*).

- *Minimum Probability:* Some instructors prefer extensive coverage of probability, while others prefer to include only basic concepts. Instructors preferring minimum coverage can include Section 4-1 while skipping the remaining sections of Chapter 4, as they are not essential for the chapters that follow. Many instructors prefer to cover the fundamentals of probability along with the basics of the addition rule and multiplication rule (Section 4-2).

GAISE This book reflects recommendations from the American Statistical Association and its *Guidelines for Assessment and Instruction in Statistics Education* (GAISE). Those guidelines suggest the following objectives and strategies.

1. *Emphasize statistical literacy and develop statistical thinking:* Each section exercise set begins with *Statistical Literacy and Critical Thinking* exercises. Many of the book's exercises are designed to encourage statistical thinking rather than the blind use of mechanical procedures.

2. *Use real data:* 94% of the examples and 92% of the exercises use real data.

3. *Stress conceptual understanding rather than mere knowledge of procedures:* Instead of seeking simple numerical answers, most exercises and examples involve conceptual understanding through questions that encourage practical interpretations of results. Also, each chapter includes a *From Data to Decision* project.

4. *Foster active learning in the classroom:* Each chapter ends with several *Cooperative Group Activities*.

5. *Use technology for developing conceptual understanding and analyzing data:* Computer software displays are included throughout the book. Special *Tech Center* subsections include instruction for using the software. Each chapter includes a *Technology Project*. When there are discrepancies between answers based on tables and answers based on technology, Appendix D provides *both* answers. The website www.TriolaStats.com includes free text-specific software (Statdisk), data sets formatted for several different technologies, and instructional videos for technologies.

6. *Use assessments to improve and evaluate student learning:* Assessment tools include an abundance of section exercises, *Chapter Quick Quizzes*, *Chapter Review Exercises*, *Cumulative Review Exercises*, *Technology Projects*, *From Data to Decision* projects, and *Cooperative Group Activities*.

Acknowledgments

I would like to thank the thousands of statistics professors and students who have contributed to the success of this book. I thank the reviewers for their suggestions for this thirteenth edition: Eric Gorenstein, Bunker Hill Community College; Rhonda Hatcher, Texas Christian University; Ladorian Latin, Franklin University; Joseph Pick, Palm Beach State College; and Lisa Whitaker, Keiser University. Special thanks to Laura Iossi of Broward College for her comprehensive work in reviewing and contributing to this 13th edition.

Other recent reviewers have included Raid W. Amin, University of West Florida; Robert Black, United States Air Force Academy; James Bryan, Merced College; Donald Burd, Monroe College; Keith Carroll, Benedictine University; Monte Cheney, Central Oregon Community College; Christopher Donnelly, Macomb Community College; Billy Edwards, University of Tennessee—Chattanooga; Marcos Enriquez, Moorpark College; Angela Everett, Chattanooga State Technical Community College; Joe Franko, Mount San Antonio College; Rob Fusco, Broward College; Sanford Geraci, Broward College; Laura Heath, Palm Beach State College; Richard Herbst, Montgomery County Community College; Richard Hertz; Diane Hollister, Reading Area Community College; Michael Huber, George Jahn, Palm Beach State College; Gary King, Ozarks Technical Community College; Kate Kozak, Coconino Community College; Dan Kumpf, Ventura College; Mickey Levendusky, Pima County Community College; Mitch Levy, Broward College; Tristan Londre, Blue River Community College; Alma

Lopez, South Plains College; Kim McHale, Heartland Community College; Carla Monticelli, Camden County Community College; Ken Mulzet, Florida State College at Jacksonville; Julia Norton, California State University Hayward; Michael Oriolo, Herkimer Community College; Jeanne Osborne, Middlesex Community College; Ali Saadat, University of California—Riverside; Radha Sankaran, Passaic County Community College; Steve Schwager, Cornell University; Pradipta Seal, Boston University; Kelly Smitch, Brevard College; Sandra Spain, Thomas Nelson Community College; Ellen G. Stutes, Louisiana State University, Eunice; Sharon Testone, Onondaga Community College; Chris Vertullo, Marist College; Dave Wallach, University of Findlay; Cheng Wang, Nova Southeastern University; Barbara Ward, Belmont University; Richard Weil, Brown College; Gail Wiltse, St. John River Community College; Claire Wladis, Borough of Manhattan Community College; Rick Woodmansee, Sacramento City College; Yong Zeng, University of Missouri at Kansas City; Jim Zimmer, Chattanooga State Technical Community College; Cathleen Zucco-Teveloff, Rowan University; Mark Z. Zuiker, Minnesota State University, Mankato.

This thirteenth edition of *Elementary Statistics* is truly a team effort, and I consider myself fortunate to work with the dedication and commitment of the Pearson team. I thank Suzy Bainbridge, Justin Billing, Deirdre Lynch, Peggy McMahon, Vicki Dreyfus, Christine O'Brien, Joe Vetere, and Rose Kernan of Cenveo Publisher Services.

I extend special thanks to Marc Triola, M.D., New York University School of Medicine, for his outstanding work on creating the new 13th edition of the Statdisk software. I thank Scott Triola for his very extensive help throughout the entire production process for this 13th edition.

I thank the following for their help in checking the accuracy of text and answers in this thirteenth edition: James Lapp, Paul Lorczak, and Dirk Tempelaar.

M.F.T.
Madison, Connecticut
September 2016

Resources for Success

MyStatLab® Online Course for Elementary Statistics, 13e by Mario F. Triola (access code required)

MyStatLab is available to accompany Pearson's market leading text offerings. To give students a consistent tone, voice, and teaching method each text's flavor and approach is tightly integrated throughout the accompanying MyStatLab course, making learning the material as seamless as possible.

Expanded objective-based MathXL coverage - MathXL is newly mapped to improve student learning outcomes. Homework reinforces and supports students' understanding of key statistics topics.

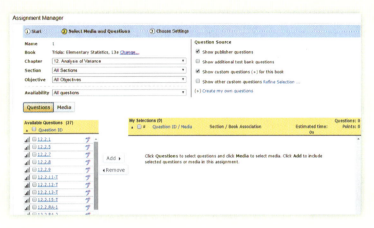

Enhanced video program to meet Introductory Statistics needs:

- **New! Tech-Specific Video Tutorials -** These short, topical videos address how to use varying technologies to complete exercises.
- **Updated! Chapter Review Exercise Videos -** Watch the Chapter Review Exercises come to life with new review videos that help students understand key chapter concepts.
- **Updated! Section Lecture Videos -** Watch author, Marty Triola, work through examples and elaborate on key objectives of the chapter.

Real-World Data Examples - Help understand how statistics applies to everyday life through the extensive current, real-world data examples and exercises provided throughout the text.

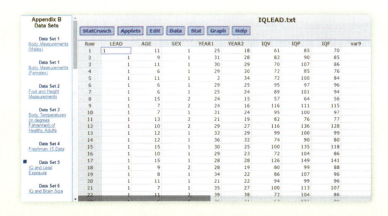

www.mystatlab.com

Resources for Success

Supplements

Student Resources

Student's Solutions Manual, by James Lapp (Colorado Mesa University), provides detailed, worked-out solutions to all odd-numbered text exercises.
(ISBN-13: 978-0-13-446429-9; ISBN-10: 0-13-446429-X)

Student Workbook for the Triola Statistics Series, by Laura Iossi (Broward College) offers additional examples, concept exercises, and vocabulary exercises for each chapter.
(ISBN- 13: 978-0-13-446423-7; ISBN 10: 0-13-446423-0)

The following technology manuals include instructions, examples from the main text, and interpretations to complement those given in the text.

Excel Student Laboratory Manual and Workbook, (Download Only) by Laurel Chiappetta (University of Pittsburgh).
(ISBN-13: 978-0-13-446427-5; ISBN-10: 0-13-446427-3)

MINITAB Student Laboratory Manual and Workbook, (Download Only) by Mario F. Triola.
(ISBN-13: 978-0-13-446418-3; ISBN-10: 0-13-446418-4)

Graphing Calculator Manual for the TI-83 Plus, TI-84 Plus, TI-84 Plus C and TI-84 Plus CE, (Download Only) by Kathleen McLaughlin (University of Connecticut) & Dorothy Wakefield (University of Connecticut Health Center).
(ISBN-13: 978-0-13-446422-0; ISBN 10: 0-13-446422-2)

Statdisk Student Laboratory Manual and Workbook (Download Only), by Mario F. Triola. These files are available to instructors and students through the Triola Statistics Series Web site, www.pearsonhighered.com/triola, and MyStatLab.

SPSS Student Laboratory Manual and Workbook (Download Only), by James J. Ball (Indiana State University). These files are available to instructors and students through the Triola Statistics Series Web site, www.pearsonhighered.com/triola, and MyStatLab.

Instructor Resources

Annotated Instructor's Edition, by Mario F. Triola, contains answers to exercises in the margin, plus recommended assignments, and teaching suggestions.
(ISBN-13: 978-0-13-446410-7; ISBN-10: 0-13-446410-9)

Instructor's Solutions Manual (Download Only), by James Lapp (Colorado Mesa University), contains solutions to all the exercises. These files are available to qualified instructors through Pearson Education's online catalog at www.pearsonhighered.com/irc or within MyStatLab.

Insider's Guide to Teaching with the Triola Statistics Series, (Download Only) by Mario F. Triola, contains sample syllabi and tips for incorporating projects, as well as lesson overviews, extra examples, minimum outcome objectives, and recommended assignments for each chapter.
(ISBN-13: 978-0-13-446425-1; ISBN-10: 0-13-446425-7)

TestGen® Computerized Test Bank (www.pearsoned.com/testgen), enables instructors to build, edit, print, and administer tests using a computerized bank of questions developed to cover all the objectives of the text. TestGen is algorithmically based, allowing instructors to create multiple but equivalent versions of the same question or test with the click of a button. Instructors can also modify test bank questions or add new questions. The software and testbank are available for download from Pearson Education's online catalog at www.pearsonhighered.com. Test Forms (Download Only) are also available from the online catalog.

PowerPoint® Lecture Slides: Free to qualified adopters, this classroom lecture presentation software is geared specifically to the sequence and philosophy of Elementary Statistics. Key graphics from the book are included to help bring the statistical concepts alive in the classroom. These files are available to qualified instructors through Pearson Education's online catalog at www.pearsonhighered.com/irc or within MyStatLab.

Learning Catalytics: Learning Catalytics is a web-based engagement and assessment tool. As a "bring-your-own-device" direct response system, Learning Catalytics offers a diverse library of dynamic question types that allow students to interact with and think critically about statistical concepts. As a real-time resource, instructors can take advantage of critical teaching moments both in the classroom or through assignable and gradeable homework.

Get the Most Out of
MyStatLab®

MyStatLab is the leading online homework, tutorial, and assessment program for teaching and learning statistics, built around Pearson's best-selling content. MyStatLab helps students and instructors improve results; it provides engaging experiences and personalized learning for each student so learning can happen in any environment. Plus, it offers flexible and time-saving course management features to allow instructors to easily manage their classes while remaining in complete control, regardless of course format.

Preparedness

One of the biggest challenges in many mathematics and statistics courses is making sure students are adequately prepared with the prerequisite skills needed to successfully complete their course work. Pearson offers a variety of content and course options to support students with just-in-time remediation and key-concept review.

- **MyStatLab with Integrated Review** ISBN 13: 978-0-13-449535-4, ISBN 10: 0-13-449535-7 can be used for just-in-time prerequisite review or co-requisite courses. These courses provide videos on review topics, along with pre-made, assignable skills-check quizzes and personalized review homework assignments.

- In recent years many new course models have emerged, as institutions "redesign" to help improve retention and results. At Pearson, we're focused on creating solutions tailored to support your plans and programs. In addition to the new Integrated Review courses, we offer new All-in-One solutions, non-STEM pathways, and STEM-track options.

Used by more than 37 million students worldwide, MyStatLab delivers consistent, measurable gains in student learning outcomes, retention, and subsequent course success.

www.mystatlab.com

Technology Resources

The following resources can be found at www.pearson highered.com/triola, the author maintained Triola Statistics Series Web site (http://www.triolastats.com), and MyStatLab

- Appendix B data sets formatted for Minitab, SPSS, SAS, Excel, JMP, and as text files. Additionally, these data sets are available as an APP for the TI-83/84 Plus calculators, and supplemental programs for the TI-83/84 Plus calculator are also available.

- Statdisk statistical software instructions for download. New features include the ability to directly use lists of data instead of requiring the use of their summary statistics.

- Extra data sets, *Probabilities Through Simulations, Bayes' Theorem,* an index of applications, and a symbols table.

Video Resources has been expanded, updated and now supplements most sections in the book, with many topics presented by the author. The videos aim to support both instructors and students through lecture, reinforcing statistical basics through technology, and applying concepts:

- **Section Lecture Videos**

- **Chapter Review Exercise Videos** walk students through the exercises and help them understand key chapter concepts.

- **New! Technology Video Tutorials** - These short, topical videos address how to use Excel, StatDisk, and the TI graphing calculator to complete exercises.

- **StatTalk Videos: 24 Conceptual Videos to Help You Actually Understand Statistics.** Fun-loving statistician Andrew Vickers takes to the streets of Brooklyn, NY, to demonstrate important statistical concepts through interesting stories and real-life events. These fun and engaging videos will help students actually understand statistical concepts. Available with an instructors user guide and assessment questions.

Videos also contain optional English and Spanish captioning. All videos are available through the MyStatLab online course.

MyStatLab™ Online Course (access code required)
MyStatLab is a course management system that delivers proven results in helping individual students succeed.

- MyStatLab can be successfully implemented in any environment—lab-based, hybrid, fully online, traditional—and demonstrates the quantifiable differ-

ence that integrated usage has on student retention, subsequent success, and overall achievement.

- MyStatLab's comprehensive online gradebook automatically tracks students' results on tests, quizzes, homework, and in the study plan. Instructors can use the gradebook to provide positive feedback or intervene if students have trouble. Gradebook data can be easily exported to a variety of spreadsheet programs, such as Microsoft Excel. You can determine which points of data you want to export, and then analyze the results to determine success.

MyStatLab provides engaging experiences that personalize, stimulate, and measure learning for each student. In addition to the resources below, each course includes a full interactive online version of the accompanying textbook.

- Tutorial Exercises with Multimedia Learning Aids: The homework and practice exercises in MyStatLab align with the exercises in the textbook, and they regenerate algorithmically to give students unlimited opportunity for practice and mastery. Exercises offer immediate helpful feedback, guided solutions, sample problems, animations, videos, and eText clips for extra help at point-of-use.

- Getting Ready for Statistics: A library of questions now appears within each MyStatLab course to offer the developmental math topics students need for the course. These can be assigned as a prerequisite to other assignments, if desired.

- Conceptual Question Library: In addition to algorithmically regenerated questions that are aligned with your textbook, there is a library of 1000 Conceptual Questions available in the assessment manager that require students to apply their statistical under-standing.

- StatCrunch™: MyStatLab integrates the Web-based statistical software, StatCrunch, within the online assessment platform so that students can easily analyze data sets from exercises and the text. In addition, MyStatLab includes access to www.StatCrunch.com, a Web site where users can access more than 15,000 shared data sets, conduct online surveys, perform complex analyses using the powerful statistical software, and generate compelling reports.

- Statistical Software Support: Knowing that students often use external statistical software, we make it easy to copy our data sets, both from the ebook and the MyStatLab questions, into software such as StatCrunch, Minitab, Excel, and more. Students have access to a variety of support tools—Technology

Tutorial Videos, Technology Study Cards, and Technology Manuals for select titles—to learn how to effectively use statistical software.

MathXL® for Statistics Online Course (access code required)

MathXL® is the homework and assessment engine that runs MyStatLab. (MyStatLab is MathXL plus a learning management system.)

With MathXL for Statistics, instructors can:

- Create, edit, and assign online homework and tests using algorithmically generated exercises correlated at the objective level to the textbook.

- Create and assign their own online exercises and import TestGen tests for added flexibility.

- Maintain records of all student work, tracked in MathXL's online gradebook.

With MathXL for Statistics, students can:

- Take chapter tests in MathXL and receive personalized study plans and/or personalized homework assignments based on their test results.

- Use the study plan and/or the homework to link directly to tutorial exercises for the objectives they need to study.

- Students can also access supplemental animations and video clips directly from selected exercises.

- Knowing that students often use external statistical software, we make it easy to copy our data sets, both from the ebook and the MyStatLab questions, into software like StatCrunch™, Minitab, Excel, and more.

MathXL for Statistics is available to qualified adopters. For more information, visit our web site at www.mathxl.com, or contact your Pearson representative.

StatCrunch™

StatCrunch is powerful, web-based statistical software that allows users to perform complex analyses, share data sets, and generate compelling reports. A vibrant online community offers more than 15,000 data sets for students to analyze.

- **Collect.** Users can upload their own data to StatCrunch or search a large library of publicly shared data sets, spanning almost any topic of interest. Also, an online survey tool allows users to quickly collect data via web-based surveys.

- **Crunch.** A full range of numerical and graphical methods allow users to analyze and gain insights from any data set. Interactive graphics help users understand statistical concepts and are available for export to enrich reports with visual representations of data.

- **Communicate.** Reporting options help users create a wide variety of visually appealing representations of their data.

Full access to StatCrunch is available with a MyStatLab kit, and StatCrunch is available by itself to qualified adopters. StatCrunch Mobile is now available to access from your mobile device. For more information, visit our Web site at www.StatCrunch.com, or contact your Pearson representative.

Minitab® 17 and Minitab Express™ make learning statistics easy and provide students with a skill-set that's in demand in today's data driven workforce. Bundling Minitab® software with educational materials ensures students have access to the software they need in the classroom, around campus, and at home. And having 12 months versions of Minitab 17 and Minitab Express available ensures students can use the software for the duration of their course.
ISBN 13: 978-0-13-445640-9
ISBN 10: 0-13-445640-8 (Access Card only; not sold as stand alone.)

JMP Student Edition, Version 12 is an easy-to-use, streamlined version of JMP desktop statistical discovery software from SAS Institute, Inc. and is available for bundling with the text.
ISBN-13: 978-0-13-467979-2
ISBN-10: 0-13-467979-2

13th EDITION

ELEMENTARY STATISTICS

1-1 Statistical and Critical Thinking

1-2 Types of Data

1-3 Collecting Sample Data

1

INTRODUCTION TO STATISTICS

CHAPTER PROBLEM

Survey Question: Do you prefer to read a *printed* book or an *electronic* book?

Surveys provide data that enable us to improve products or services. Surveys guide political candidates, shape business practices, influence social media, and affect many aspects of our lives. Surveys give us insight into the opinions and views of others. Let's consider one *USA Today* survey in which respondents were asked if they prefer to read a printed book or an electronic book. Among 281 respondents, 65% preferred a printed book and 35% preferred an electronic book. Figure 1-1 on the next page includes graphs that depict these results.

The survey results suggest that people overwhelmingly prefer reading printed books to reading ebooks. The graphs in Figure 1-1 visually depict the survey results, and they support a claim that people prefer printed books to ebooks by a wide margin. One of the most important objectives in this book is to encourage the use of critical thinking so that such results are not blindly accepted. We might question whether the survey results are valid. Who conducted the survey? How were respondents selected? Do the graphs in Figure 1-1

1

depict the results well, or are those graphs somehow misleading?

The survey results presented here have major flaws that are among the most commonly used, so they are especially important to recognize. Here are brief descriptions of each of the major flaws:

Flaw 1: Misleading Graphs

The bar chart in Figure 1-1(a) is very deceptive. By using a vertical scale that does not start at zero, the difference between the two percentages is grossly exaggerated. Figure 1-1(a) makes it appear that about eight times as many people choose a printed book over an ebook, but with response rates of 65% and 35%, that ratio is very roughly 2:1, not 8:1.

The illustration in Figure 1-1(b) is also deceptive. Again, the difference between the actual response rates of 65% for printed books and 35% for ebooks is a difference that is grossly distorted. The picture graph (or "pictograph") in Figure 1-1(b) makes it appear that people prefer printed books to ebooks by a ratio of roughly 4:1 instead of being the correct ratio of 65:35, or roughly 2:1. (Objects with area or volume can distort perceptions because they can be drawn to be disproportionately larger or smaller than the data indicate.)

Deceptive graphs are discussed in more detail in Section 2-3, but we see here that the illustrations in Figure 1-1 grossly exaggerate the preference for printed books.

Flaw 2: Bad Sampling Method

The aforementioned survey responses are from a *USA Today* survey of Internet users. The survey question was posted on a website and Internet users decided whether to respond. This is an example of a *voluntary response sample*—a sample in which respondents themselves decide whether to participate. With a voluntary response sample, it often happens that those with a strong interest in the topic are more likely to participate, so the results are very questionable. In this case, it is reasonable to suspect that Internet users might prefer ebooks at a rate higher than the rate in the general population. When using sample data to learn something about a population, it is *extremely* important to obtain sample data that are representative of the population from which the data are drawn. As we proceed

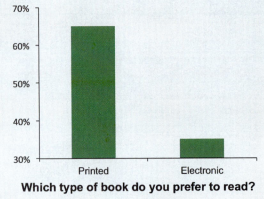

Which type of book do you prefer to read?

(a)

Readers Preferring Printed Books

Readers Preferring eBooks

(b)

FIGURE 1-1   Survey Results

through this chapter and discuss types of data and sampling methods, we should focus on these key concepts:

- **Sample data must be collected in an appropriate way, such as through a process of *random* selection.**
- **If sample data are not collected in an appropriate way, the data may be so completely useless that no amount of statistical torturing can salvage them.**

It would be easy to accept the preceding survey results and blindly proceed with calculations and statistical analyses, but we would miss the two critical flaws described above. We could then develop conclusions that are fundamentally wrong and misleading. Instead, we should develop skills in statistical thinking and critical thinking so that we can understand how the survey is so seriously flawed.

CHAPTER OBJECTIVES

Here is the single most important concept presented in this chapter: When using methods of statistics with sample data to form conclusions about a population, it is absolutely essential to collect sample data in a way that is appropriate. Here are the chapter objectives:

1-1 Statistical and Critical Thinking

- Analyze sample data relative to context, source, and sampling method.
- Understand the difference between *statistical significance* and *practical significance*.
- Define and identify a *voluntary response sample* and know that statistical conclusions based on data from such a sample are generally not valid.

1-2 Types of Data

- Distinguish between a *parameter* and a *statistic*.
- Distinguish between *quantitative data* and *categorical* (or *qualitative* or *attribute*) *data*.
- Distinguish between *discrete* data and *continuous* data.
- Determine whether basic statistical calculations are appropriate for a particular data set.

1-3 Collecting Sample Data

- Define and identify a *simple random sample*.
- Understand the importance of sound sampling methods and the importance of good design of experiments.

1-1   Statistical and Critical Thinking

Key Concept In this section we begin with a few very basic definitions, and then we consider an *overview* of the process involved in conducting a statistical study. This process consists of "prepare, analyze, and conclude." "Preparation" involves consideration of the *context*, the *source* of data, and *sampling method*. In future chapters we construct suitable graphs, explore the data, and execute computations required for the statistical method being used. In future chapters we also form conclusions by determining whether results have statistical significance and practical significance.

Statistical thinking involves critical thinking and the ability to make sense of results. Statistical thinking demands so much more than the ability to execute complicated calculations. Through numerous examples, exercises, and discussions, this text will help you develop the statistical thinking skills that are so important in today's world.

We begin with some very basic definitions.

DEFINITIONS

Data are collections of observations, such as measurements, genders, or survey responses. (A single data value is called a *datum*, a term rarely used. The term "data" is plural, so it is correct to say "data *are* . . ." not "data *is* . . .")

Statistics is the science of planning studies and experiments; obtaining data; and organizing, summarizing, presenting, analyzing, and interpreting those data and then drawing conclusions based on them.

A **population** is the complete collection of *all* measurements or data that are being considered. Typically, a population is the complete collection of data that we would like to make inferences about.

A **census** is the collection of data from *every* member of the population.

A **sample** is a *subcollection* of members selected from a population.

Because populations are often very large, a common objective of the use of statistics is to obtain data from a sample and then use those data to form a conclusion about the population.

EXAMPLE 1 Residential Carbon Monoxide Detectors

In the journal article "Residential Carbon Monoxide Detector Failure Rates in the United States" (by Ryan and Arnold, *American Journal of Public Health,* Vol. 101, No. 10), it was stated that there are 38 million carbon monoxide detectors installed in the United States. When 30 of them were randomly selected and tested, it was found that 12 of them failed to provide an alarm in hazardous carbon monoxide conditions. In this case, the population and sample are as follows:

Population: All 38 million carbon monoxide detectors in the United States

Sample: The 30 carbon monoxide detectors that were selected and tested

The objective is to use the sample data as a basis for drawing a conclusion about the population of all carbon monoxide detectors, and methods of statistics are helpful in drawing such conclusions.

> **YOUR TURN** Do part (a) of Exercise 2 "Reported Versus Measured."

We now proceed to consider the process involved in a statistical study. See Figure 1-2 for a summary of this process and note that the focus is on critical thinking, not mathematical calculations. Thanks to wonderful developments in technology, we have powerful tools that effectively do the number crunching so that we can focus on understanding and interpreting results.

Prepare

Context Figure 1-2 suggests that we begin our preparation by considering the *context* of the data, so let's start with context by considering the data in Table 1-1. Table 1-1 includes the numbers of registered pleasure boats in Florida (tens of thousands) and the numbers of manatee fatalities from encounters with boats in Florida for each of several recent years. The format of Table 1-1 suggests the following goal: Determine whether there is a *relationship* between numbers of boats and numbers of manatee deaths from

TABLE 1-1 Pleasure Boats and Manatee Fatalities from Boat Encounters

Pleasure Boats (tens of thousands)	99	99	97	95	90	90	87	90	90
Manatee Fatalities	92	73	90	97	83	88	81	73	68

boats. This goal suggests a reasonable hypothesis: As the numbers of boats increase, the numbers of manatee deaths increase.

Source of the Data The second step in our preparation is to consider the source (as indicated in Figure 1-2). The data in Table 1-1 are from the Florida Department of Highway Safety and Motor Vehicles and the Florida Marine Research Institute. The sources certainly appear to be reputable.

Sampling Method Figure 1-2 suggests that we conclude our preparation by considering the sampling method. The data in Table 1-1 were obtained from official government records known to be reliable. The sampling method appears to be sound.

Sampling methods and the use of randomization will be discussed in Section 1-3, but for now, we stress that a sound sampling method is absolutely essential for good results in a statistical study. It is generally a bad practice to use voluntary response (or self-selected) samples, even though their use is common.

Prepare

1. **Context**
   - What do the data represent?
   - What is the goal of study?
2. **Source of the Data**
   - Are the data from a source with a special interest so that there is pressure to obtain results that are favorable to the source?
3. **Sampling Method**
   - Were the data collected in a way that is unbiased, or were the data collected in a way that is biased (such as a procedure in which respondents volunteer to participate)?

↓

Analyze

1. **Graph the Data**
2. **Explore the Data**
   - Are there any outliers (numbers very far away from almost all of the other data)?
   - What important statistics summarize the data (such as the mean and standard deviation described in Chapter 3)?
   - How are the data distributed?
   - Are there missing data?
   - Did many selected subjects refuse to respond?
3. **Apply Statistical Methods**
   - Use technology to obtain results.

↓

Conclude

1. **Significance**
   - Do the results have statistical significance?
   - Do the results have practical significance?

FIGURE 1-2 Statistical and Critical Thinking

Survivorship Bias

In World War II, statistician Abraham Wald saved many lives with his work on the Applied Mathematics Panel. Military leaders asked the panel how they could improve the chances of aircraft bombers returning after missions. They wanted to add some armor for protection, and they recorded locations on the bombers where damaging holes were found. They reasoned that armor should be placed in locations with the most holes, but Wald said that strategy would be a big mistake. He said that armor should be placed where returning bombers were *not* damaged. His reasoning was this: The bombers that made it back with damage were *survivors*, so the damage they suffered could be survived. Locations on the aircraft that were not damaged were the most vulnerable, and aircraft suffering damage in those vulnerable areas were the ones that did not make it back. The military leaders would have made a big mistake with survivorship bias by studying the planes that survived instead of thinking about the planes that did not survive.

Origin of "Statistics"

The word *statistics* is derived from the Latin word *status* (meaning "state"). Early uses of statistics involved compilations of data and graphs describing various aspects of a state or country. In 1662, John Graunt published statistical information about births and deaths. Graunt's work was followed by studies of mortality and disease rates, population sizes, incomes, and unemployment rates. Households, governments, and businesses rely heavily on statistical data for guidance. For example, unemployment rates, inflation rates, consumer indexes, and birth and death rates are carefully compiled on a regular basis, and the resulting data are used by business leaders to make decisions affecting future hiring, production levels, and expansion into new markets.

> **DEFINITION**
>
> A **voluntary response sample** (or **self-selected sample**) is one in which the respondents themselves decide whether to be included.

The following types of polls are common examples of voluntary response samples. By their very nature, all are seriously flawed because we should not make conclusions about a population on the basis of samples with a strong possibility of bias:

- Internet polls, in which people online can decide whether to respond
- Mail-in polls, in which people can decide whether to reply
- Telephone call-in polls, in which newspaper, radio, or television announcements ask that you voluntarily call a special number to register your opinion

The Chapter Problem involves a *USA Today* survey with a voluntary response sample. See also the following Example 2.

EXAMPLE 2 Voluntary Response Sample

The ABC television show *Nightline* asked viewers to call with their opinion about whether the United Nations headquarters should remain in the United States. Viewers then decided themselves whether to call with their opinions, and 67% of 186,000 respondents said that the United Nations should be moved out of the United States. In a separate and independent survey, 500 respondents were randomly selected and surveyed, and 38% of this group wanted the United Nations to move out of the United States. The two polls produced dramatically different results. Even though the *Nightline* poll involved 186,000 volunteer respondents, the much smaller poll of 500 randomly selected respondents is more likely to provide better results because of the far superior sampling method.

YOUR TURN Do Exercise 1 "Online Medical Info."

Analyze

Figure 1-2 indicates that after completing our preparation by considering the context, source, and sampling method, we begin to *analyze* the data.

Graph and Explore An analysis should begin with appropriate graphs and explorations of the data. Graphs are discussed in Chapter 2, and important statistics are discussed in Chapter 3.

Apply Statistical Methods Later chapters describe important statistical methods, but application of these methods is often made easy with technology (calculators and/or statistical software packages). A good statistical analysis *does not* require strong computational skills. A good statistical analysis *does* require using common sense and paying careful attention to sound statistical methods.

Conclude

Figure 1-2 shows that the final step in our statistical process involves conclusions, and we should develop an ability to distinguish between statistical significance and practical significance.

Statistical Significance *Statistical significance* is achieved in a study when we get a result that is very unlikely to occur by chance. A common criterion is that we have statistical significance if the likelihood of an event occurring by chance is 5% or less.

- Getting 98 girls in 100 random births *is* statistically significant because such an extreme outcome is not likely to result from random chance.

- Getting 52 girls in 100 births *is not* statistically significant because that event could easily occur with random chance.

Practical Significance It is possible that some treatment or finding is effective, but common sense might suggest that the treatment or finding does not make enough of a difference to justify its use or to be practical, as illustrated in Example 3.

EXAMPLE 3 **Statistical Significance Versus Practical Significance**

ProCare Industries once supplied a product named Gender Choice that supposedly increased the chance of a couple having a baby with the gender that they desired. In the absence of any evidence of its effectiveness, the product was banned by the Food and Drug Administration (FDA) as a "gross deception of the consumer." But suppose that the product was tested with 10,000 couples who wanted to have baby girls, and the results consist of 5200 baby girls born in the 10,000 births. This result is statistically significant because the likelihood of it happening due to chance is only 0.003%, so chance doesn't seem like a feasible explanation. That 52% rate of girls is statistically significant, but it lacks practical significance because 52% is only slightly above 50%. Couples would not want to spend the time and money to increase the likelihood of a girl from 50% to 52%. (*Note*: In reality, the likelihood of a baby being a girl is about 48.8%, not 50%.)

YOUR TURN Do Exercise 15 "Gender Selection."

Analyzing Data: Potential Pitfalls

Here are a few more items that could cause problems when analyzing data.

Misleading Conclusions When forming a conclusion based on a statistical analysis, we should make statements that are clear even to those who have no understanding of statistics and its terminology. We should carefully avoid making statements not justified by the statistical analysis. For example, later in this book we introduce the concept of a correlation, or association between two variables, such as numbers of registered pleasure boats and numbers of manatee deaths from encounters with boats. A statistical analysis might justify the statement that there is a correlation between numbers of boats and numbers of manatee fatalities, but it would not justify a statement that an increase in the number of boats *causes* an increase in the number of manatee fatalities. Such a statement about causality can be justified by physical evidence, not by statistical analysis.

Correlation does not imply causation.

Sample Data Reported Instead of Measured When collecting data from people, it is better to take measurements yourself instead of asking subjects to *report* results. Ask people what they weigh and you are likely to get their *desired* weights, not their actual weights. People tend to round, usually down, sometimes *way* down. When asked, someone with a weight of 187 lb might respond that he or she weighs 160 lb. Accurate weights are collected by using a scale to *measure* weights, not by asking people what they weigh.

Publication Bias

There is a "publication bias" in professional journals. It is the tendency to publish positive results (such as showing that some treatment is effective) much more often than negative results (such as showing that some treatment has no effect). In the article "Registering Clinical Trials" (*Journal of the American Medical Association*, Vol. 290, No. 4), authors Kay Dickersin and Drummond Rennie state that "the result of not knowing who has performed what (clinical trial) is loss and distortion of the evidence, waste and duplication of trials, inability of funding agencies to plan, and a chaotic system from which only certain sponsors might benefit, and is invariably against the interest of those who offered to participate in trials and of patients in general." They support a process in which *all* clinical trials are registered in one central system, so that future researchers have access to all previous studies, not just the studies that were published.

Loaded Questions If survey questions are not worded carefully, the results of a study can be misleading. Survey questions can be "loaded," or intentionally worded to elicit a desired response. Here are the actual rates of "yes" responses for the two different wordings of a question:

97% yes: "Should the President have the line item veto to eliminate waste?"

57% yes: "Should the President have the line item veto, or not?"

Order of Questions Sometimes survey questions are unintentionally loaded by such factors as the order of the items being considered. See the following two questions from a poll conducted in Germany, along with the very different response rates:

"Would you say that traffic contributes more or less to air pollution than industry?" (45% blamed traffic; 27% blamed industry.)

"Would you say that industry contributes more or less to air pollution than traffic?" (24% blamed traffic; 57% blamed industry.)

In addition to the order of items within a question, as illustrated above, the order of separate questions could also affect responses.

Nonresponse A *nonresponse* occurs when someone either refuses to respond to a survey question or is unavailable. When people are asked survey questions, some firmly refuse to answer. The refusal rate has been growing in recent years, partly because many persistent telemarketers try to sell goods or services by beginning with a sales pitch that initially sounds as though it is part of an opinion poll. (This "selling under the guise" of a poll is called *sugging*.) In *Lies, Damn Lies, and Statistics*, author Michael Wheeler makes this very important observation:

People who refuse to talk to pollsters are likely to be different from those who do not. Some may be fearful of strangers and others jealous of their privacy, but their refusal to talk demonstrates that their view of the world around them is markedly different from that of those people who will let poll-takers into their homes.

Percentages Some studies cite misleading or unclear percentages. Note that 100% of some quantity is *all* of it, but if there are references made to percentages that exceed 100%, such references are often not justified. In an ad for The Club, a device used to discourage car thefts, it was stated that "The Club reduces your odds of car theft by 400%." If the Club eliminated *all* car thefts, it would reduce the odds of car theft by 100%, so the 400% figure is misleading and doesn't make sense.

The following list identifies some key principles to apply when dealing with percentages. These principles all use the basic concept that % or "percent" really means "divided by 100." The first principle that follows is used often in this book.

Percentage of: To find a percentage of an amount, replace the % symbol with division by 100, and then interpret "of" to be multiplication. This example shows that 6% of 1200 is 72:

$$6\% \text{ of } 1200 \text{ responses} = \frac{6}{100} \times 1200 = 72$$

Decimal → Percentage: To convert from a decimal to a percentage, multiply by 100%. This example shows that 0.25 is equivalent to 25%:

$$0.25 \rightarrow 0.25 \times 100\% = 25\%$$

Fraction → Percentage: *To convert from a fraction to a percentage,* divide the denominator into the numerator to get an equivalent decimal number; then multiply by 100%. This example shows that the fraction 3/4 is equivalent to 75%:

$$\frac{3}{4} = 0.75 \rightarrow 0.75 \times 100\% = 75\%$$

Percentage → Decimal: To convert from a percentage to a decimal number, replace the % symbol with division by 100. This example shows that 85% is equivalent to 0.85:

$$85\% = \frac{85}{100} = 0.85$$

1-1 Basic Skills and Concepts

Statistical Literacy and Critical Thinking

1. Online Medical Info *USA Today* posted this question on its website: "How often do you seek medical information online?" Of 1072 Internet users who chose to respond, 38% of them responded with "frequently." What term is used to describe this type of survey in which the people surveyed consist of those who decided to respond? What is wrong with this type of sampling method?

2. Reported Versus Measured In a survey of 1046 adults conducted by Bradley Corporation, subjects were asked how often they wash their hands when using a public restroom, and 70% of the respondents said "always."

a. Identify the sample and the population.

b. Why would better results be obtained by observing the hand washing instead of asking about it?

3. Statistical Significance Versus Practical Significance When testing a new treatment, what is the difference between statistical significance and practical significance? Can a treatment have statistical significance, but not practical significance?

4. Correlation One study showed that for a recent period of 11 years, there was a strong correlation (or association) between the numbers of people who drowned in swimming pools and the amounts of power generated by nuclear power plants (based on data from the Centers for Disease Control and Prevention and the Department of Energy). Does this imply that increasing power from nuclear power plants is the cause of more deaths in swimming pools? Why or why not?

Consider the Source. *In Exercises 5–8, determine whether the given source has the potential to create a bias in a statistical study.*

5. Physicians Committee for Responsible Medicine The Physicians Committee for Responsible Medicine tends to oppose the use of meat and dairy products in our diets, and that organization has received hundreds of thousands of dollars in funding from the Foundation to Support Animal Protection.

6. Arsenic in Rice Amounts of arsenic in samples of rice grown in Texas were measured by the Food and Drug Administration (FDA).

7. Brain Size A data set in Appendix B includes brain volumes from 10 pairs of monozygotic (identical) twins. The data were collected by researchers at Harvard University, Massachusetts General Hospital, Dartmouth College, and the University of California at Davis.

8. Chocolate An article in *Journal of Nutrition* (Vol. 130, No. 8) noted that chocolate is rich in flavonoids. The article notes "regular consumption of foods rich in flavonoids may reduce the risk of coronary heart disease." The study received funding from Mars, Inc., the candy company, and the Chocolate Manufacturers Association.

Sampling Method.   *In Exercises 9–12, determine whether the sampling method appears to be sound or is flawed.*

9. Nuclear Power Plants In a survey of 1368 subjects, the following question was posted on the *USA Today* website: "In your view, are nuclear plants safe?" The survey subjects were Internet users who chose to respond to the question posted on the electronic edition of *USA Today*.

10. Clinical Trials Researchers at Yale University conduct a wide variety of clinical trials by using subjects who volunteer after reading advertisements soliciting paid volunteers.

11. Credit Card Payments In an AARP, Inc. survey of 1019 randomly selected adults, each was asked how much credit card debt he or she pays off each month.

12. Smartphone Usage In a survey of smartphone ownership, the Pew Research Center randomly selected 1006 adults in the United States.

Statistical Significance and Practical Significance.   *In Exercises 13–16, determine whether the results appear to have statistical significance, and also determine whether the results appear to have practical significance.*

13. Diet and Exercise Program In a study of the Kingman diet and exercise program, 40 subjects lost an average of 22 pounds. There is about a 1% chance of getting such results with a program that has no effect.

14. MCAT The Medical College Admissions Test (MCAT) is commonly used as part of the decision-making process for determining which students to accept into medical schools. To test the effectiveness of the Siena MCAT preparation course, 16 students take the MCAT test, then they complete the preparatory course, and then they retake the MCAT test, with the result that the average (mean) score for this group rises from 25 to 30. There is a 0.3% chance of getting those results by chance. Does the course appear to be effective?

15. Gender Selection In a study of the Gender Aide method of gender selection used to increase the likelihood of a baby being born a girl, 2000 users of the method gave birth to 980 boys and 1020 girls. There is about a 19% chance of getting that many girls if the method had no effect.

16. IQ Scores Most people have IQ scores between 70 and 130. For $39.99, you can purchase a PC or Mac program from HighIQPro that is claimed to increase your IQ score by 10 to 20 points. The program claims to be "the only proven IQ increasing software in the brain training market," but the author of your text could find no data supporting that claim, so let's suppose that these results were obtained: In a study of 12 subjects using the program, the average increase in IQ score is 3 IQ points. There is a 25% chance of getting such results if the program has no effect.

In Exercises 17–20, refer to the sample of body temperatures (degrees Fahrenheit) in the table below. (The body temperatures are from a data set in Appendix B.)

	Subject				
	1	2	3	4	5
8 AM	97.0	98.5	97.6	97.7	98.7
12 AM	97.6	97.8	98.0	98.4	98.4

17. Context of the Data Refer to the table of body temperatures. Is there some meaningful way in which each body temperature recorded at 8 AM is matched with the 12 AM temperature?

18. Source The listed body temperatures were obtained from Dr. Steven Wasserman, Dr. Philip Mackowiak, and Dr. Myron Levine, who were researchers at the University of Maryland. Is the source of the data likely to be biased?

19. Conclusion Given the body temperatures in the table, what issue can be addressed by conducting a statistical analysis of the data?

20. Conclusion If we analyze the listed body temperatures with suitable methods of statistics, we conclude that when the differences are found between the 8 AM body temperatures and the 12 AM body temperatures, there is a 64% chance that the differences can be explained by random results obtained from populations that have the same 8 AM and 12 AM body temperatures. What should we conclude about the statistical significance of those differences?

In Exercises 21–24, refer to the data in the table below. The entries are white blood cell counts (1000 cells / μL) and red blood cell counts (million cells / μL) from male subjects examined as part of a large health study conducted by the National Center for Health Statistics. The data are matched, so that the first subject has a white blood cell count of 8.7 and a red blood cell count of 4.91, and so on.

	Subject				
	1	2	3	4	5
White	8.7	5.9	7.3	6.2	5.9
Red	4.91	5.59	4.44	4.80	5.17

21. Context Given that the data are matched and considering the units of the data, does it make sense to use the difference between each white blood cell count and the corresponding red blood cell count? Why or why not?

22. Analysis Given the context of the data in the table, what issue can be addressed by conducting a statistical analysis of the measurements?

23. Source of the Data Considering the source of the data, does that source appear to be biased in some way?

24. Conclusion If we analyze the sample data and conclude that there is a correlation between white blood cell counts and red blood cell counts, does it follow that higher white blood cell counts are the cause of higher red blood cell counts?

What's Wrong? *In Exercises 25–28, identify what is wrong.*

25. Potatoes In a poll sponsored by the Idaho Potato Commission, 1000 adults were asked to select their favorite vegetables, and the favorite choice was potatoes, which were selected by 26% of the respondents.

26. Healthy Water In a *USA Today* online poll, 951 Internet users chose to respond, and 57% of them said that they prefer drinking bottled water instead of tap water.

27. Motorcycles and Sour Cream In recent years, there has been a strong correlation between per capita consumption of sour cream and the numbers of motorcycle riders killed in noncollision accidents. Therefore, consumption of sour cream causes motorcycle fatalities.

28. Smokers The electronic cigarette maker V2 Cigs sponsored a poll showing that 55% of smokers surveyed say that they feel ostracized "sometimes," "often," or "always."

Percentages.   *In Exercises 29–36, answer the given questions, which are related to percentages.*

29. Workplace Attire In a survey conducted by Opinion Research Corporation, 1000 adults were asked to identify "what is inappropriate in the workplace." Of the 1000 subjects, 70% said that miniskirts were not appropriate in the workplace.

a. What is 70% of 1000?

b. Among the 1000 respondents, 550 said that shorts are unacceptable in the workplace. What percentage of respondents said that shorts are unacceptable in the workplace?

30. Checking Job Applicants In a study conducted by the Society for Human Resource Management, 347 human resource professionals were surveyed. Of those surveyed, 73% said that their companies conduct criminal background checks on all job applicants.

a. What is the exact value that is 73% of the 347 survey subjects?

b. Could the result from part (a) be the actual number of survey subjects who said that their companies conduct criminal background checks on all job applicants? Why or why not?

c. What is the actual number of survey subjects who said that their company conducts criminal background checks on all job applicants?

d. Assume that 112 of the survey subjects are females. What percentage of those surveyed are females?

31. Marriage Proposals In a survey conducted by TheKnot.com, 1165 engaged or married women were asked about the importance of a bended knee when making a marriage proposal. Among the 1165 respondents, 48% said that the bended knee was essential.

a. What is the exact value that is 48% of 1165 survey respondents?

b. Could the result from part (a) be the actual number of survey subjects who said that a bended knee is essential? Why or why not?

c. What is the actual number of survey respondents saying that the bended knee is essential?

d. Among the 1165 respondents, 93 said that a bended knee is corny and outdated. What percentage of respondents said that a bended knee is corny and outdated?

32. Chillax *USA Today* reported results from a Research Now for Keurig survey in which 1458 men and 1543 women were asked this: "In a typical week, how often can you kick back and relax?"

a. Among the women, 19% responded with "rarely, if ever." What is the exact value that is 19% of the number of women surveyed?

b. Could the result from part (a) be the actual number of women who responded with "rarely, if ever"? Why or why not?

c. What is the actual number of women who responded with "rarely, if ever"?

d. Among the men who responded, 219 responded with "rarely, if ever." What is the percentage of men who responded with "rarely, if ever."?

e. Consider the question that the subjects were asked. Is that question clear and unambiguous so that all respondents will interpret the question the same way? How might the survey be improved?

33. Percentages in Advertising An ad for Big Skinny wallets included the statement that one of their wallets "reduces your filled wallet size by 50%–200%." What is wrong with this statement?

34. Percentages in Advertising Continental Airlines ran ads claiming that lost baggage is "an area where we've already improved 100% in the past six months." What is wrong with this statement?

35. Percentages in Advertising A *New York Times* editorial criticized a chart caption that described a dental rinse as one that "reduces plaque on teeth by over 300%." What is wrong with this statement?

36. Percentages in Negotiations When the author was negotiating a contract for the faculty and administration at a college, a dean presented the argument that if faculty receive a 4% raise and administrators receive a 4% raise, that's an 8% raise and it would never be approved. What's wrong with that argument?

1-1   Beyond the Basics

37. What's Wrong with This Picture? The *Newport Chronicle* ran a survey by asking readers to call in their response to this question: "Do you support the development of atomic weapons that could kill millions of innocent people?" It was reported that 20 readers responded and that 87% said "no," while 13% said "yes." Identify four major flaws in this survey.

38. Falsifying Data A researcher at the Sloan-Kettering Cancer Research Center was once criticized for falsifying data. Among his data were figures obtained from 6 groups of mice, with 20 individual mice in each group. The following values were given for the percentage of successes in each group: 53%, 58%, 63%, 46%, 48%, 67%. What's wrong with those values?

1-2   Types of Data

Key Concept A major use of statistics is to collect and use sample data to make conclusions about populations. We should know and understand the meanings of the terms *statistic* and *parameter*, as defined below. In this section we describe a few different types of data. The type of data is one of the key factors that determine the statistical methods we use in our analysis.

In Part 1 of this section we describe the basics of different types of data, and then in Part 2 we consider "big data" and missing data.

PART 1   Basic Types of Data
Parameter/Statistic

DEFINITIONS

A **parameter** is a numerical measurement describing some characteristic of a *population*.

A **statistic** is a numerical measurement describing some characteristic of a *sample*.

HINT The alliteration in "population parameter" and "sample statistic" helps us remember the meanings of these terms.

If we have more than one statistic, we have "statistics." Another meaning of "statistics" was given in Section 1-1, where we defined *statistics* to be the science of planning studies and experiments; obtaining data; organizing, summarizing, presenting, analyzing, and interpreting those data; and then drawing conclusions based on them. We now have two different definitions of statistics, but we can determine which of these two definitions applies by considering the context in which the term *statistics* is used. The following example uses the first meaning of *statistics* as given on the previous page.

EXAMPLE 1   Parameter / Statistic

There are 17,246,372 high school students in the United States. In a study of 8505 U.S. high school students 16 years of age or older, 44.5% of them said that they texted while driving at least once during the previous 30 days (based on data in "Texting While Driving and Other Risky Motor Vehicle Behaviors Among US High School Students," by Olsen, Shults, Eaton, *Pediatrics*, Vol. 131, No. 6).

1. **Parameter:** The population size of 17,246,372 high school students is a parameter, because it is the entire population of all high school students in the United States. If we somehow knew the percentage of all 17,246,372 high school students who reported they had texted while driving, that percentage would also be a parameter.

2. **Statistic:** The sample size of 8505 surveyed high school students is a statistic, because it is based on a sample, not the entire population of all high school students in the United States. The value of 44.5% is another statistic, because it is also based on the sample, not on the entire population.

> YOUR TURN   Do Exercise 1 "Parameter and Statistic."

Quantitative / Categorical

Some data are numbers representing counts or measurements (such as an IQ score of 135), whereas others are attributes (such as eye color of green or brown) that are not counts or measurements. The terms quantitative data and categorical data distinguish between these types.

DEFINITIONS

Quantitative (or numerical) data consist of *numbers* representing counts or measurements.

Categorical (or qualitative or attribute) data consist of names or labels (not numbers that represent counts or measurements).

CAUTION Categorical data are sometimes coded with numbers, with those numbers replacing names. Although such numbers might appear to be quantitative, they are actually categorical data. See the third part of Example 2 that follows.

Include Units of Measurement With quantitative data, it is important to use the appropriate units of measurement, such as dollars, hours, feet, or meters. We should carefully observe information given about the units of measurement, such as "all amounts are in *thousands of dollars*" or "all units are in *kilograms*." Ignoring such units of measurement can be very costly. The National Aeronautics and Space Administration (NASA) lost its $125 million Mars Climate Orbiter when the orbiter crashed because the controlling software had acceleration data in *English* units, but they were incorrectly assumed to be in *metric* units.

EXAMPLE 2   **Quantitative / Categorical**

1. **Quantitative Data:** The ages (in years) of subjects enrolled in a clinical trial
2. **Categorical Data as Labels:** The genders (male / female) of subjects enrolled in a clinical trial
3. **Categorical Data as Numbers:** The identification numbers 1, 2, 3, . . . , 25 are assigned randomly to the 25 subjects in a clinical trial. Those numbers are substitutes for names. They don't measure or count anything, so they are categorical data.

> **YOUR TURN** Do Exercise 2 "Quantitative / Categorical Data."

Discrete / Continuous

Quantitative data can be further described by distinguishing between *discrete* and *continuous* types.

> **DEFINITIONS**
>
> **Discrete data** result when the data values are quantitative and the number of values is finite, or "countable." (If there are infinitely many values, the collection of values is countable if it is possible to count them individually, such as the number of tosses of a coin before getting tails.)
>
> **Continuous (numerical) data** result from infinitely many possible quantitative values, where the collection of values is not countable. (That is, it is impossible to count the individual items because at least some of them are on a continuous scale, such as the lengths of distances from 0 cm to 12 cm.)

CAUTION The concept of countable data plays a key role in the preceding definitions, but it is not a particularly easy concept to understand. Continuous data can be measured, but not counted. If you select a particular data value from continuous data, there is no "next" data value. See Example 3.

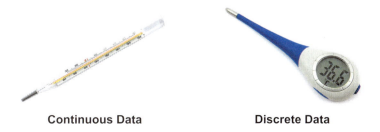

Continuous Data   **Discrete Data**

EXAMPLE 3 Discrete/Continuous

1. **Discrete Data of the Finite Type:** Each of several physicians plans to count the number of physical examinations given during the next full week. The data are discrete data because they are finite numbers, such as 27 and 46, that result from a counting process.

2. **Discrete Data of the Infinite Type:** Casino employees plan to roll a fair die until the number 5 turns up, and they count the number of rolls required to get a 5. It is possible that the rolls could go on forever without ever getting a 5, but the numbers of rolls can be counted, even though the counting might go on forever. The collection of the numbers of rolls is therefore countable.

3. **Continuous Data:** When the typical patient has blood drawn as part of a routine examination, the volume of blood drawn is between 0 mL and 50 mL. There are infinitely many values between 0 mL and 50 mL. Because it is impossible to count the number of different possible values on such a continuous scale, these amounts are continuous data.

YOUR TURN Do Exercise 3 "Discrete/Continuous Data."

GRAMMAR: FEWER VERSUS LESS When describing smaller amounts, it is correct grammar to use "fewer" for discrete amounts and "less" for continuous amounts. It is correct to say that we drank *fewer* cans of cola and that, in the process, we drank *less* cola. The numbers of cans of cola are discrete data, whereas the volume amounts of cola are continuous data.

Levels of Measurement

Another common way of classifying data is to use four levels of measurement: nominal, ordinal, interval, and ratio, all defined below. (Also see Table 1-2 for brief descriptions of the four levels of measurements.) When we are applying statistics to real problems, the level of measurement of the data helps us decide which procedure to use. There will be references to these levels of measurement in this book, but the important point here is based on common sense: *Don't do computations and don't use statistical methods that are not appropriate for the data.* For example, it would not make sense to compute an average (mean) of Social Security numbers, because those numbers are data that are used for identification, and they don't represent measurements or counts of anything.

TABLE 1-2 Levels of Measurement

Level of Measurement	Brief Description	Example
Ratio	There is a natural zero starting point and ratios make sense.	Heights, lengths, distances, volumes
Interval	Differences are meaningful, but there is no natural zero starting point and ratios are meaningless.	Body temperatures in degrees Fahrenheit or Celsius
Ordinal	Data can be arranged in order, but differences either can't be found or are meaningless.	Ranks of colleges in *U.S. News & World Report*
Nominal	Categories only. Data cannot be arranged in order.	Eye colors

DEFINITION

The **nominal level of measurement** is characterized by data that consist of names, labels, or categories only. The data cannot be arranged in some order (such as low to high).

EXAMPLE 4 Nominal Level

Here are examples of sample data at the nominal level of measurement.

1. **Yes/No/Undecided:** Survey responses of *yes*, *no*, and *undecided*
2. **Coded Survey Responses:** For an item on a survey, respondents are given a choice of possible answers, and they are coded as follows: "I agree" is coded as 1; "I disagree" is coded as 2; "I don't care" is coded as 3; "I refuse to answer" is coded as 4; "Go away and stop bothering me" is coded as 5. The numbers 1, 2, 3, 4, 5 don't measure or count anything.

> YOUR TURN Do Exercise 22 "Exit Poll."

Because nominal data lack any ordering or numerical significance, they should not be used for calculations. Numbers such as 1, 2, 3, and 4 are sometimes assigned to the different categories (especially when data are coded for computers), but these numbers have no real computational significance and any average (mean) calculated from them is meaningless and possibly misleading.

DEFINITION

Data are at the **ordinal level of measurement** if they can be arranged in some order, but differences (obtained by subtraction) between data values either cannot be determined or are meaningless.

EXAMPLE 5 Ordinal Level

Here is an example of sample data at the ordinal level of measurement.

Course Grades: A college professor assigns grades of A, B, C, D, or F. These grades can be arranged in order, but we can't determine differences between the grades. For example, we know that A is higher than B (so there is an ordering), but we cannot subtract B from A (so the difference cannot be found).

> YOUR TURN Do Exercise 21 "College Rankings."

Ordinal data provide information about relative comparisons, but not the *magnitudes* of the differences. Usually, ordinal data should not be used for calculations such as an average (mean), but this guideline is sometimes ignored (such as when we use letter grades to calculate a grade-point average).

DEFINITION

Data are at the **interval level of measurement** if they can be arranged in order, and differences between data values can be found and are meaningful. *Data at this level do not have a natural zero starting point at which none of the quantity is present.*

Measuring Disobedience

How are data collected about something that doesn't seem to be measurable, such as people's level of disobedience? Psychologist Stanley Milgram devised the following experiment: A researcher instructed a volunteer subject to operate a control board that gave increasingly painful "electrical shocks" to a third person. Actually, no real shocks were given, and the third person was an actor. The volunteer began with 15 volts and was instructed to increase the shocks by increments of 15 volts. The disobedience level was the point at which the subject refused to increase the voltage. Surprisingly, two-thirds of the subjects obeyed orders even when the actor screamed and faked a heart attack.

Six Degrees of Separation

Social psychologists, historians, political scientists, and communications specialists are interested in "The Small World Problem": Given any two people in the world, how many intermediate links are necessary to connect the two original people? In the 1950s and 1960s, social psychologist Stanley Milgram conducted an experiment in which subjects tried to contact other target people by mailing an information folder to an acquaintance who they thought would be closer to the target. Among 160 such chains that were initiated, only 44 were completed, so the failure rate was 73%. Among the successes, the number of intermediate acquaintances varied from 2 to 10, with a median of 6 (hence "six degrees of separation"). The experiment has been criticized for its high failure rate and its disproportionate inclusion of subjects with above-average incomes. A more recent study conducted by Microsoft researcher Eric Horvitz and Stanford Assistant Professor Jure Leskovec involved 30 billion instant messages and 240 million people. This study found that for instant messages that used Microsoft, the mean length of a path between two individuals is 6.6, suggesting "seven degrees of separation." Work continues in this important and interesting field.

EXAMPLE 6   Interval Level

These examples illustrate the interval level of measurement.

1. **Temperatures:** Body temperatures of 98.2°F and 98.8°F are examples of data at this interval level of measurement. Those values are ordered, and we can determine their difference of 0.6°F. However, there is no natural starting point. The value of 0°F might seem like a starting point, but it is arbitrary and does not represent the total absence of heat.

2. **Years:** The years 1492 and 1776 can be arranged in order, and the difference of 284 years can be found and is meaningful. However, time did not begin in the year 0, so the year 0 is arbitrary instead of being a natural zero starting point representing "no time."

YOUR TURN Do Exercise 25 "Baseball."

DEFINITION

Data are at the **ratio level of measurement** if they can be arranged in order, differences can be found and are meaningful, and *there is a natural zero starting point* (where zero indicates that none of the quantity is present). For data at this level, differences and ratios are both meaningful.

EXAMPLE 7   Ratio Level

The following are examples of data at the ratio level of measurement. Note the presence of the natural zero value, and also note the use of meaningful ratios of "twice" and "three times."

1. **Heights of Students:** Heights of 180 cm and 90 cm for a high school student and a preschool student (0 cm represents no height, and 180 cm is *twice* as tall as 90 cm.)

2. **Class Times:** The times of 50 min and 100 min for a statistics class (0 min represents no class time, and 100 min is *twice* as long as 50 min.)

YOUR TURN Do Exercise 24 "Fast Food Service Times."

HINT The distinction between the interval and ratio levels of measurement can be a bit tricky. Here are two tools to help with that distinction:

1. **Ratio Test** Focus on the term "ratio" and know that the term "twice" describes the ratio of one value to be double the other value. To distinguish between the interval and ratio levels of measurement, use a "ratio test" by asking this question: Does use of the term "twice" make sense? "Twice" makes sense for data at the ratio level of measurement, but it does not make sense for data at the interval level of measurement.

2. **True Zero** For ratios to make sense, there must be a value of "true zero," where the value of zero indicates that none of the quantity is present, and zero is not simply an arbitrary value on a scale. The temperature of 0°F is arbitrary and does not indicate that there is no heat, so temperatures on the Fahrenheit scale are at the interval level of measurement, not the ratio level.

EXAMPLE 8 **Distinguishing Between the Ratio Level and Interval Level**

For each of the following, determine whether the data are at the ratio level of measurement or the interval level of measurement:

  a. Times (minutes) it takes students to complete a statistics test.

  b. Body temperatures (Celsius) of statistics students.

SOLUTION

  a. Apply the "ratio test" described in the preceding hint. If one student completes the test in 40 minutes and another student completes the test in 20 minutes, does it make sense to say that the first student used *twice* as much time? Yes! So the times are at the ratio level of measurement. We could also apply the "true zero" test. A time of 0 minutes does represent "no time," so the value of 0 is a true zero indicating that no time was used.

  b. Apply the "ratio test" described in the preceding hint. If one student has a body temperature of 40°C and another student has a body temperature of 20°C, does it make sense to say that the first student is *twice* as hot as the second student? (Ignore subjective amounts of attractiveness and consider only science.) No! So the body temperatures are not at the ratio level of measurement. Because the difference between 40°C and 20°C is the same as the difference between 90°C and 70°C, the differences are meaningful, but because ratios do not make sense, the body temperatures are at the interval level of measurement. Also, the temperature of 0°C does not represent "no heat" so the value of 0 is not a true zero indicating that no heat is present.

Big Data Instead of a Clinical Trial

Nicholas Tatonetti of Columbia University searched Food and Drug Administration databases for adverse reactions in patients that resulted from different pairings of drugs. He discovered that the Paxil (paroxetine) drug for depression and the pravastatin drug for high cholesterol interacted to create increases in glucose (blood sugar) levels. When taken separately by patients, neither drug raised glucose levels, but the increase in glucose levels occurred when the two drugs were taken together. This finding resulted from a general database search of interactions from many pairings of drugs, not from a clinical trial involving patients using Paxil and pravastatin.

PART 2 **Big Data and Missing Data: Too Much and Not Enough**

When working with data, we might encounter some data sets that are excessively large, and we might also encounter some data sets with individual elements missing. Here in Part 2 we briefly discuss both cases.

Big Data

Some considered him to be a hero whistleblower while others thought of him as a traitor, but Edward Snowden used his employment at the NSA (National Security Agency) to reveal substantial top secret documents that led to the realization that the NSA was conducting telephone and Internet surveillance of U.S. citizens as well as world leaders. The NSA was collecting massive amounts of data that were analyzed in an attempt to prevent terrorism. Monitoring telephone calls and Internet communications is made possible with modern technology. The NSA can now compile *big data,* and such ginormous data sets have led to the birth of *data science*. There is not universal agreement on the following definitions, and various other definitions can be easily found elsewhere.

DEFINITIONS

Big data refers to data sets so large and so complex that their analysis is beyond the capabilities of traditional software tools. Analysis of big data may require software simultaneously running in parallel on many different computers.

Data science involves applications of statistics, computer science, and software engineering, along with some other relevant fields (such as sociology or finance).

Examples of Data Set Magnitudes We can see from the above definition of big data that there isn't a fixed number that serves as an exact boundary for determining whether a data set qualifies as being big data, but big data typically involves amounts of data such as the following:

- Terabytes (10^{12} or 1,000,000,000,000 bytes) of data
- Petabytes (10^{15} bytes) of data
- Exabytes (10^{18} bytes) of data
- Zettabytes (10^{21} bytes) of data
- Yottabytes (10^{24} bytes) of data

Examples of Applications of Big Data The following are a few examples involving big data:

- Google provides live traffic maps by recording and analyzing GPS (global positioning system) data collected from the smartphones of people traveling in their vehicles.
- Attempts to forecast flu epidemics by analyzing Internet searches of flu symptoms.
- The Sloan Digital Sky Survey started in the year 2000, and it quickly collected more astronomy data than in the history of mankind up to 2000. It now has more than 140 terabytes of astronomy data.
- Walmart has a sales database with more than 2.5 petabytes (2,500,000,000,000,000 bytes) of data. For online sales, Walmart developed the Polaris search engine that increased sales by 10% to 15%, worth billions of dollars.
- Amazon monitors and tracks 1.4 billion items in its store that are distributed across hundreds of fulfillment centers around the world.

Examples of Jobs According to Analytic Talent, there are 6000 companies hiring data scientists, and here are some job posting examples:

- Facebook: Data Scientist
- IBM: Data Scientist
- PayPal: Data Scientist
- The College Board: SAS Programmer/Data Scientist
- Netflix: Senior Data Engineer/Scientist

Statistics in Data Science The modern data scientist has a solid background in statistics and computer systems as well as expertise in fields that extend beyond statistics. The modern data scientist might be skilled with Hadoop software, which uses parallel processing on many computers for the analysis of big data. The modern data scientist might also have a strong background in some other field such as psychology, biology, medicine, chemistry, or economics. Because of the wide range of disciplines required, a data science project might typically involve a team of collaborating individuals with expertise in different fields. An introductory statistics course is a great first step in becoming a data scientist.

Missing Data

When collecting sample data, it is quite common to find that some values are missing. Ignoring missing data can sometimes create misleading results. If you make the mistake of skipping over a few different sample values when you are manually typing them into a statistics software program, the missing values are not likely to have a serious effect on the results. However, if a survey includes many missing salary entries because those with very low incomes are reluctant to reveal their salaries, those missing low values will have the serious effect of making salaries appear higher than they really are.

For an example of missing data, see the following table. The body temperature for Subject 2 at 12 AM on Day 2 is missing. (The table below includes the first three rows of data from Data Set 3 "Body Temperatures" in Appendix B.)

Body Temperatures (in degrees Fahrenheit) of Healthy Adults

Subject	Age	Sex	Smoke	Temperature Day 1 8 AM	Temperature Day 1 12 AM	Temperature Day 2 8 AM	Temperature Day 2 12 AM
1	22	M	Y	98.0	98.0	98.0	98.6
2	23	M	Y	97.0	97.6	97.4	----
3	22	M	Y	98.6	98.8	97.8	98.6

There are different categories of missing data. See the following definitions.

> **DEFINITION**
>
> A data value is **missing completely at random** if the likelihood of its being missing is independent of its value or any of the other values in the data set. That is, any data value is just as likely to be missing as any other data value.

(*NOTE:* More complete discussions of missing data will distinguish between *missing completely at random* and *missing at random*, which means that the likelihood of a value being missing is independent of its value after controlling for another variable. There is no need to know this distinction in this book.)

Example of Missing Data—Random: When using a keyboard to manually enter ages of survey respondents, the operator is distracted by a colleague singing "Daydream Believer" and makes the mistake of failing to enter the age of 37 years. This data value is missing completely at random.

> **DEFINITION**
>
> A data value is **missing not at random** if the missing value is related to the reason that it is missing.

Example of Missing Data—Not at Random A survey question asks each respondent to enter his or her annual income, but respondents with very low incomes skip this question because they find it embarrassing.

Biased Results? Based on the two definitions and examples from the previous page, it makes sense to conclude that if we ignore data *missing completely at random*, the remaining values are not likely to be biased and good results should be obtained. However, if we ignore data that are *missing not at random*, it is very possible that the remaining values are biased and results will be misleading.

Correcting for Missing Data There are different methods for dealing with missing data.

1. **Delete Cases:** One very common method for dealing with missing data is to delete all subjects having any missing values.

    - If the data are missing completely at random, the remaining values are not likely to be biased and good results can be obtained, but with a smaller sample size.

    - If the data are missing not at random, deleting subjects having any missing values can easily result in a bias among the remaining values, so results can be misleading.

2. **Impute Missing Values:** We "impute" missing data values when we substitute values for them. There are different methods of determining the replacement values, such as using the mean of the other values, or using a randomly selected value from other similar cases, or using a method based on regression analysis (which will make more sense after studying Chapter 10).

In this book we do not work much with missing data, but it is important to understand this:

> **When analyzing sample data with missing values, try to determine *why* they are missing, then decide whether it makes sense to treat the remaining values as being representative of the population. If it appears that there are missing values that are *missing not at random* (that is, their values are related to the reasons why they are missing), know that the remaining data may well be biased and any conclusions based on those remaining values may well be misleading.**

1-2 Basic Skills and Concepts

Statistical Literacy and Critical Thinking

1. Parameter and Statistic In a Harris Interactive survey of 2276 adults in the United States, it was found that 33% of those surveyed never travel using commercial airlines. Identify the population and sample. Is the value of 33% a statistic or a parameter?

2. Quantitative/Categorical Data Identify each of the following as quantitative data or categorical data.

a. The platelet counts in Data Set 1 "Body Data" in Appendix B

b. The cigarette brands in Data Set 13 "Cigarette Contents" in Appendix B

c. The colors of the M&M candies in Data Set 27 "M&M Weights" in Appendix B

d. The weights of the M&M candies in Data Set 27 "M&M Weights" in Appendix B

3. Discrete/Continuous Data Which of the following describe discrete data?

a. The numbers of people surveyed in each of the next several years for the National Health and Nutrition Examination Surveys

b. The exact foot lengths (measured in cm) of a random sample of statistics students

c. The exact times that randomly selected drivers spend texting while driving during the past 7 days

4. Health Survey In a survey of 1020 adults in the United States, 44% said that they wash their hands after riding public transportation (based on data from KRC Research).

a. Identify the sample and population.

b. Is the value of 44% a statistic or parameter?

c. What is the level of measurement of the value of 44%? (nominal, ordinal, interval, ratio)

d. Are the numbers of subjects in such surveys discrete or continuous?

In Exercises 5–12, identify whether the given value is a statistic or a parameter.

5. On-time Flights In a study of American Airlines flights from JFK in New York to LAX in Los Angeles, 48 flights are randomly selected and the average (mean) arrival time is 8.9 minutes late.

6. CHIS A recent California Health Interview Survey (CHIS) included 2799 adolescent residents of California.

7. Housing Units According to the Census Bureau, the total number of housing units in the United States is 132,802,859.

8. Triangle Fire Fatalities A deadly disaster in the United States was the Triangle Shirtwaist Factory Fire in New York City. A population of 146 garment workers died in that fire.

9. Birth Weight In a study of 400 babies born at four different hospitals in New York State, it was found that the average (mean) weight at birth was 3152.0 grams.

10. Birth Genders In the same study cited in the preceding exercise, 51% of the babies were girls.

11. *Titanic* A study was conducted of all 2223 passengers aboard the *Titanic* when it sank.

12. Periodic Table The average (mean) atomic weight of all elements in the periodic table is 134.355 unified atomic mass units.

In Exercises 13–20, determine whether the data are from a discrete or continuous data set.

13. Freshman 15 In a study of weight gains by college students in their freshman year, researchers record the amounts of weight gained by randomly selected students (as in Data Set 6 "Freshman 15" in Appendix B).

14. CHIS Among the subjects surveyed as part of the California Health Interview Survey (CHIS), several subjects are randomly selected and their heights are recorded.

15. McDonald's In a study of service times at a McDonald's drive-up window, the numbers of cars serviced each hour of several days are recorded.

16. House Attendance The Clerk of the U.S. House of Representatives records the number of representatives present at each session.

17. Corvettes A shift manager records the numbers of Corvettes manufactured during each day of production.

18. Criminal Forensics When studying the relationship between lengths of feet and heights so that footprint evidence at a crime scene can be used to estimate the height of the suspect, a researcher records the exact lengths of feet from a large sample of random subjects.

19. Smartphones Students in a statistics class record the exact lengths of times that they surreptitiously use their smartphones during class.

20. Texting Fatalities The Insurance Institute for Highway Safety collects data consisting of the numbers of motor vehicle fatalities caused by driving while texting.

In Exercises 21–28, determine which of the four levels of measurement (nominal, ordinal, interval, ratio) is most appropriate.

21. College Rankings *U.S. News & World Report* periodically provides its rankings of national universities, and in a recent year the ranks for Princeton, Harvard, and Yale were 1, 2, and 3, respectively.

22. Exit Poll For the presidential election of 2016, ABC News conducts an exit poll in which voters are asked to identify the political party (Democratic, Republican, and so on) that they registered with.

23. M&Ms Colors of M&Ms (red, orange, yellow, brown, blue, green) listed in Data Set 27 "M&M Weights" in Appendix B

24. Fast Food Service Times In a study of fast food service times, a researcher records the time intervals of drive-up customers beginning when they place their order and ending when they receive their order.

25. Baseball Baseball statistician Bill James records the years in which the baseball World Series is won by a team from the National League.

26. Movie Ratings The author rated the movie *Star Wars: The Force Awakens* with 5 stars on a scale of 5 stars.

27. Lead in Blood Blood lead levels of low, medium, and high used to describe the subjects in Data Set 7 "IQ and Lead" in Appendix B

28. Body Temperatures Body temperatures (in degrees Fahrenheit) listed in Data Set 3 "Body Temperatures" in Appendix B

In Exercises 29–32, identify the level of measurement of the data as nominal, ordinal, interval, or ratio. Also, explain what is wrong with the given calculation.

29. Super Bowl The first Super Bowl attended by the author was Super Bowl XLVIII. On the first play of the game, the Seattle defense scored on a safety. The defensive players wore jerseys numbered 31, 28, 41, 56, 25, 54, 69, 50, 91, 72, 29, and the average (mean) of those numbers is 49.6.

30. Social Security Numbers As part of a project in a statistics class, students report the last four digits of their Social Security numbers, and the average (mean) of those digits is computed to be 4.7.

31. Temperatures As this exercise is being written, it is 80°F at the author's home and it is 40°F in Auckland, New Zealand, so it is twice as warm at the author's home as it is in Auckland, New Zealand.

32. College Ranks As of this writing, *U.S. News & World Report* ranked national universities, including these results: Princeton (1), Harvard (2), Yale (3), and Columbia (4). The difference between Princeton and Harvard is the same as the difference between Yale and Columbia.

1-2 Beyond the Basics

33. Countable For each of the following, categorize the nature of the data using one of these three descriptions: (1) discrete because the number of possible values is finite; (2) discrete because the number of possible values is infinite but countable; (3) continuous because the number of possible values is infinite and not countable.

a. Exact lengths of the feet of members of the band the Monkees

b. Shoe sizes of members of the band the Monkees (such as 9, 9½, and so on)

c. The number of albums sold by the Monkees band

d. The numbers of monkeys sitting at keyboards before one of them randomly types the lyrics for the song "Daydream Believer."

1-3 Collecting Sample Data

Key Concept When using statistics in a study, planning is very important, and it is essential to use an appropriate method for collecting the sample data. This section includes comments about various methods and sampling procedures. Of particular importance is the method of using a *simple random sample*. We will make frequent use of this sampling method throughout the remainder of this book.

As you read this section, remember this:

> **If sample data are not collected in an appropriate way, the data may be so utterly useless that no amount of statistical torturing can salvage them.**

PART 1 Basics of Design of Experiments and Collecting Sample Data

The Gold Standard Randomization with placebo/treatment groups is sometimes called the "gold standard" because it is so effective. (A placebo such as a sugar pill has no medicinal effect.) The following example describes how the gold standard was used in the largest health experiment ever conducted.

 EXAMPLE 1 The Salk Vaccine Experiment

In 1954, an experiment was designed to test the effectiveness of the Salk vaccine in preventing polio, which had killed or paralyzed thousands of children. By random selection, 401,974 children were randomly assigned to two groups: (1) 200,745 children were given a *treatment* consisting of Salk vaccine injections; (2) 201,229 children were injected with a *placebo* that contained no drug. Children were assigned to the treatment or placebo group through a process of random selection, equivalent to flipping a coin. Among the children given the Salk vaccine, 33 later developed paralytic polio, and among the children given a placebo, 115 later developed paralytic polio.

Example 1 describes an *experiment* because subjects were given a treatment, but ethical, cost, time, and other considerations sometimes prohibit the use of an experiment. We would never want to conduct a driving/texting experiment in which we ask subjects to

Clinical Trials Versus Observational Studies

In a *New York Times* article about hormone therapy for women, reporter Denise Grady wrote about randomized clinical trials that involve subjects who were randomly assigned to a treatment group and another group not given the treatment. Such randomized clinical trials are often referred to as the "gold standard" for medical research. In contrast, observational studies can involve patients who decide themselves to undergo some treatment. Subjects who decide themselves to undergo treatments are often healthier than other subjects, so the treatment group might appear to be more successful simply because it involves healthier subjects, not necessarily because the treatment is effective. Researchers criticized observational studies of hormone therapy for women by saying that results might appear to make the treatment more effective than it really is.

text while driving—some of them could die. It would be far better to observe past crash results to understand the effects of driving while texting. See the following definitions.

> **DEFINITIONS**
>
> In an **experiment**, we apply some *treatment* and then proceed to observe its effects on the individuals. (The individuals in experiments are called **experimental units**, and they are often called **subjects** when they are people.)
>
> In an **observational study**, we observe and measure specific characteristics, but we don't attempt to *modify* the individuals being studied.

Experiments are often better than observational studies because well-planned experiments typically reduce the chance of having the results affected by some variable that is not part of a study. A *lurking variable* is one that affects the variables included in the study, but it is not included in the study.

EXAMPLE 2   Ice Cream and Drownings

Observational Study: Observe past data to conclude that ice cream causes drownings (based on data showing that increases in ice cream sales are associated with increases in drownings). The mistake is to miss the lurking variable of temperature and the failure to see that as the temperature increases, ice cream sales increase and drownings increase because more people swim.

Experiment: Conduct an *experiment* with one group treated with ice cream while another group gets no ice cream. We would see that the rate of drowning victims is about the same in both groups, so ice cream consumption has no effect on drownings.

Here, the experiment is clearly better than the observational study.

Design of Experiments

Good design of experiments includes *replication, blinding,* and *randomization.*

- **Replication** is the repetition of an experiment on more than one individual. Good use of replication requires sample sizes that are large enough so that we can see effects of treatments. In the Salk experiment in Example 1, the experiment used sufficiently large sample sizes, so the researchers could see that the Salk vaccine was effective.

- **Blinding** is used when the subject doesn't know whether he or she is receiving a treatment or a placebo. Blinding is a way to get around the **placebo effect,** which occurs when an untreated subject reports an improvement in symptoms. (The reported improvement in the placebo group may be real or imagined.) The Salk experiment in Example 1 was **double-blind,** which means that blinding occurred at two levels: (1) The children being injected didn't know whether they were getting the Salk vaccine or a placebo, and (2) the doctors who gave the injections and evaluated the results did not know either. Codes were used so that the researchers could objectively evaluate the effectiveness of the Salk vaccine.

- **Randomization** is used when individuals are assigned to different groups through a process of random selection, as in the Salk vaccine experiment in Example 1. The logic behind randomization is to use chance as a way to create two groups that are similar. The following definition refers to one common and effective way to collect sample data in a way that uses randomization.

DEFINITION

A **simple random sample** of *n* subjects is selected in such a way that every possible *sample of the same size n* has the same chance of being chosen. (A simple random sample is often called a random sample, but strictly speaking, a *random sample* has the weaker requirement that all members of the population have the same chance of being selected. That distinction is not so important in this text. See Exercise 37 "Simple Random Sample vs. Random Sample".)

Throughout, we will use various statistical procedures, and we often have a requirement that we have collected a *simple random sample*, as defined above.

Unlike careless or haphazard sampling, random sampling usually requires very careful planning and execution. Wayne Barber of Chemeketa Community College is quite correct when he tells his students that "randomness needs help."

Other Sampling Methods In addition to simple random sampling, here are some other sampling methods commonly used for surveys. Figure 1-3 illustrates these different sampling methods.

Simple Random Sample
A sample of *n* subjects is selected so that every sample of the same size *n* has the same chance of being selected.

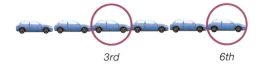

Systematic Sample
Select every *k*th subject.

Convenience Sample
Use data that are very easy to get.

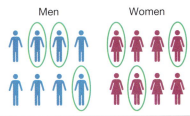

Stratified Sample
Subdivide population into strata (groups) with the same characteristics, then randomly sample within those strata.

Cluster Sample
Partition the population in clusters (groups), then randomly select some clusters, then select all members of the selected clusters.

FIGURE 1-3   Common Sampling Methods

Hawthorne and Experimenter Effects

The well-known placebo effect occurs when an untreated subject incorrectly believes that he or she is receiving a real treatment and reports an improvement in symptoms. The Hawthorne effect occurs when treated subjects somehow respond differently simply because they are part of an experiment. (This phenomenon was called the "Hawthorne effect" because it was first observed in a study of factory workers at Western Electric's Hawthorne plant.) An experimenter effect (sometimes called a Rosenthal effect) occurs when the researcher or experimenter unintentionally influences subjects through such factors as facial expression, tone of voice, or attitude.

DEFINITIONS

In **systematic sampling**, we select some starting point and then select every *k*th (such as every 50th) element in the population.

With **convenience sampling**, we simply use data that are very easy to get.

In **stratified sampling**, we subdivide the population into at least two different subgroups (or strata) so that subjects within the same subgroup share the same characteristics (such as gender). Then we draw a sample from each subgroup (or stratum).

In **cluster sampling**, we first divide the population area into sections (or clusters). Then we randomly select some of those clusters and choose *all* the members from those selected clusters.

Multistage Sampling Professional pollsters and government researchers often collect data by using some combination of the preceding sampling methods. In a multistage sample design, pollsters select a sample in different stages, and each stage might use different methods of sampling, as in the following example.

EXAMPLE 3 **Multistage Sample Design**

The U.S. government's unemployment statistics are based on surveys of households. It is impractical to personally survey each household in a simple random sample, because they would be scattered all over the country. Instead, the U.S. Census Bureau and the Bureau of Labor Statistics collaborate to conduct a survey called the Current Population Survey. A recent survey incorporates a multistage sample design, roughly following these steps:

1. The entire United States is partitioned into 2,007 different regions called *primary sampling units* (PSUs). The primary sampling units are metropolitan areas, large counties, or combinations of smaller counties. The 2,007 primary sampling units are then grouped into 824 different strata.

2. In each of the 824 different strata, one of the primary sampling units is selected so that the probability of selection is proportional to the size of the population in each primary sampling unit.

3. In each of the 824 selected primary sampling units, census data are used to identify a census *enumeration district*, with each containing about 300 households. Enumeration districts are then randomly selected.

4. In each of the selected enumeration districts, clusters of about four addresses (contiguous whenever possible) are randomly selected.

5. A responsible person in each of the 60,000 selected households is interviewed about the employment status of each household member of age 16 or older.

This multistage sample design includes a combination of random, stratified, and cluster sampling at different stages. The end result is a very complicated sampling design, but it is much more practical, less expensive, and faster than using a simpler design, such as a simple random sample.

PART 2 ## Beyond the Basics of Design of Experiments and Collecting Sample Data

In Part 2 of this section, we discuss different types of observational studies and different ways of designing experiments.

Observational Studies The following definitions identify the standard terminology used in professional journals for different types of observational studies. These definitions are illustrated in Figure 1-4.

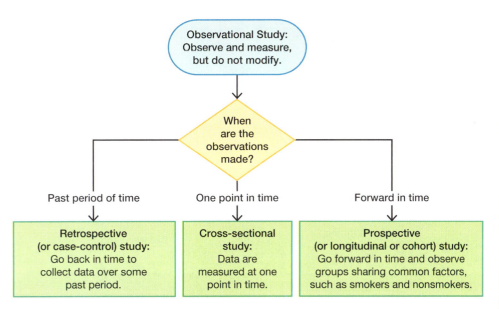

FIGURE 1-4 **Types of Observational Studies**

DEFINITIONS

In a **cross-sectional study,** data are observed, measured, and collected at one point in time, not over a period of time.

In a **retrospective (or case-control) study,** data are collected from a past time period by going back in time (through examination of records, interviews, and so on).

In a **prospective (or longitudinal or cohort) study,** data are collected in the future from groups that share common factors (such groups are called *cohorts*).

Experiments In an experiment, **confounding** occurs when we can see some effect, but we can't identify the specific factor that caused it, as in the ice cream and drowning observational study in Example 2. See also the bad experimental design illustrated in Figure 1-5(a), where confounding can occur when the treatment group of women shows strong positive results. Because the treatment group consists of women and the placebo group consists of men, confounding has occurred because we cannot determine whether the treatment or the gender of the subjects caused the positive results. The Salk vaccine experiment in Example 1 illustrates one method for controlling the effect of the treatment variable: Use a *completely randomized experimental design,*

Do Women Earn Less Than Men?

Evidence from the Census and the Bureau of Labor Statistics indicates that women earn about 77% as much as men. Jillian Berman reported in the *Huffington Post* that the PayScale company used salary data from its millions of website users to conclude that men and women earn about the same when they begin their careers, but men tend to earn more as they advance. She stated that according to the study, "women working in various non-managerial jobs earn about 98 percent of what men do on average." Berman notes that this conclusion is based on the Pay-Scale data, which website users report in online surveys, instead of on data obtained from government bureaus. The PayScale study accounts for factors such as education and job responsibilities. This study does appear to confirm that women hold disproportionately fewer high-level jobs and disproportionately more low-level jobs, so there is clearly a gender gap.

whereby randomness is used to assign subjects to the treatment group and the placebo group. A completely randomized experimental design is one of the following methods that are used to control effects of variables.

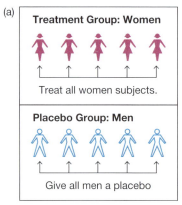

(a) **Bad experimental design:** Treat all women subjects and give the men a placebo. (Problem: We don't know if effects are due to sex or to treatment.)

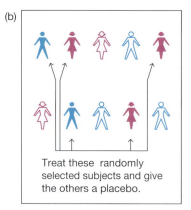

(b) **Completely randomized experimental design:** Use randomness to determine who gets the treatment and who gets the placebo.

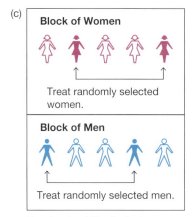

(c) **Randomized block design:**
1. Form a block of women and a block of men.
2. Within each block, randomly select subjects to be treated.

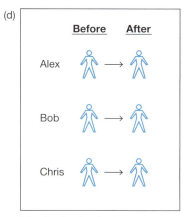

(d) **Matched pairs design:** Get measurements from the same subjects before and after some treatment.

FIGURE 1-5 **Designs of Experiments**

Completely Randomized Experimental Design: Assign subjects to different treatment groups through a process of *random selection*, as illustrated in Figure 1-5(b).

Randomized Block Design: See Figure 1-5c. A **block** is a group of subjects that are similar, but blocks differ in ways that might affect the outcome of the experiment. Use the following procedure, as illustrated in Figure 1-5(c):

1. Form blocks (or groups) of subjects with similar characteristics.

2. Randomly assign treatments to the subjects within each block.

For example, in designing an experiment to test the effectiveness of aspirin treatments on heart disease, we might form a block of men and a block of women, because it is known that the hearts of men and women can behave differently. By controlling for gender, this randomized block design eliminates gender as a possible source of confounding.

A randomized block design uses the same basic idea as stratified sampling, but randomized block designs are used when designing experiments, whereas stratified sampling is used for surveys.

Matched Pairs Design: Compare two treatment groups (such as treatment and placebo) by using subjects matched in pairs that are somehow related or have similar characteristics, as in the following cases.

- Before/After: Matched pairs might consist of measurements from subjects before and after some treatment, as illustrated in Figure 1-5(d). Each subject yields a "before" measurement and an "after" measurement, and each before/after pair of measurements is a matched pair.

- Twins: A test of Crest toothpaste used matched pairs of twins, where one twin used Crest and the other used another toothpaste.

Rigorously Controlled Design: Carefully assign subjects to different treatment groups, so that those given each treatment are similar in the ways that are important to the experiment. This can be extremely difficult to implement, and often we can never be sure that we have accounted for all of the relevant factors.

Sampling Errors

In statistics, you could use a good sampling method and do everything correctly, and yet it is possible to get wrong results. No matter how well you plan and execute the sample collection process, there is likely to be some error in the results. The different types of sampling errors are described here.

> **DEFINITIONS**
>
> A **sampling error** (or **random sampling error**) occurs when the sample has been selected with a random method, but there is a discrepancy between a sample result and the true population result; such an error results from chance sample fluctuations.
>
> A **nonsampling error** is the result of human error, including such factors as wrong data entries, computing errors, questions with biased wording, false data provided by respondents, forming biased conclusions, or applying statistical methods that are not appropriate for the circumstances.
>
> A **nonrandom sampling error** is the result of using a sampling method that is not random, such as using a convenience sample or a voluntary response sample.

Experimental design requires much more thought and care than we can describe in this relatively brief section. Taking a complete course in the design of experiments is a good start in learning so much more about this important topic.

1-3   Basic Skills and Concepts

Statistical Literacy and Critical Thinking

1. Back Pain Treatment In a study designed to test the effectiveness of paracetamol (or acetaminophen) as a treatment for lower back pain, 1643 patients were randomly assigned to one of three groups: (1) the 547 subjects in the placebo group were given pills containing no medication; (2) 550 subjects were in a group given pills with paracetamol taken at regular intervals; (3) 546 subjects were in a group given pills with paracetamol to be taken when needed for pain relief. (See "Efficacy of Paracetamol for Acute Low-Back Pain," by Williams, et al., *Lancet*, doi:10.1016/S0140-6736(14)60805-9.) Is this study an experiment or an observational study? Explain.

2. Blinding What does it mean when we say that the study cited in Exercise 1 was "double-blind"?

3. Replication In what specific way was replication applied in the study cited in Exercise 1?

4. Sampling Method The patients included in the study cited in Exercise 1 were those "who sought care for low-back pain directly or in response to a community advertisement." What type of sampling best describes the way in which the 1643 subjects were chosen: simple random sample, systematic sample, convenience sample, stratified sample, cluster sample? Does the method of sampling appear to adversely affect the quality of the results?

*Exercises 5–8 refer to the study of an association between which ear is used for cell phone calls and whether the subject is left-handed or right-handed. The study is reported in "Hemispheric Dominance and Cell Phone Use," by Seidman et al., **JAMA Otolaryngology— Head & Neck Surgery**, Vol. 139, No. 5. The study began with a survey e-mailed to 5000 people belonging to an otology online group, and 717 surveys were returned. (**Otology** relates to the ear and hearing.)*

5. Sampling Method What type of sampling best describes the way in which the 717 subjects were chosen: simple random sample, systematic sample, convenience sample, stratified sample, cluster sample? Does the method of sampling appear to adversely affect the quality of the results?

6. Experiment or Observational Study Is the study an experiment or an observational study? Explain.

7. Response Rate What percentage of the 5000 surveys were returned? Does that response rate appear to be low? In general, what is a problem with a very low response rate?

8. Sampling Method Assume that the population consists of all students currently in your statistics class. Describe how to obtain a sample of six students so that the result is a sample of the given type.

a. Simple random sample

b. Systematic sample

c. Stratified sample

d. Cluster sample

In Exercises 9–20, identify which of these types of sampling is used: random, systematic, convenience, stratified, or cluster.

9. Cormorant Density Cormorant bird population densities were studied by using the "line transect method" with aircraft observers flying along the shoreline of Lake Huron and collecting sample data at intervals of every 20 km (based on data from *Journal of Great Lakes Research*).

10. Sexuality of Women The sexuality of women was discussed in Shere Hite's book *Women and Love: A Cultural Revolution*. Her conclusions were based on sample data that consisted of 4500 mailed responses from 100,000 questionnaires that were sent to women.

11. UFO Poll In a Kelton Research poll, 1114 Americans 18 years of age or older were called after their telephone numbers were randomly generated by a computer, and 36% of the respondents said that they believe in the existence of UFOs.

12. Class Survey The author surveyed a sample from the population of his statistics class by identifying groups of males and females, then randomly selecting five students from each of those two groups.

13. Driving A student of the author conducted a survey on driving habits by randomly selecting three different classes and surveying all of the students as they left those classes.

14. Acupuncture Study In a study of treatments for back pain, 641 subjects were randomly assigned to the four different treatment groups of individualized acupuncture, standardized acupuncture, simulated acupuncture, and usual care (based on data from "A Randomized Trial Comparing Acupuncture, Simulated Acupuncture, and Usual Care for Chronic Low Back Pain," by Cherkin et al., *Archives of Internal Medicine*, Vol. 169, No. 9).

15. Dictionary The author collected sample data by randomly selecting five books from each of the categories of science, fiction, and history. The numbers of pages in the books were then identified.

16. Deforestation Rates Satellites are used to collect sample data for estimating deforestation rates. The Forest Resources Assessment of the United Nations (UN) Food and Agriculture Organization uses a method of selecting a sample of a 10-km-wide square at every 1° intersection of latitude and longitude.

17. Testing Lipitor In a clinical trial of the cholesterol drug Lipitor (atorvastatin), subjects were partitioned into groups given a placebo or Lipitor doses of 10 mg, 20 mg, 40 mg, or 80 mg. The subjects were randomly assigned to the different treatment groups (based on data from Pfizer, Inc.).

18. Exit Polls During the last presidential election, CNN conducted an exit poll in which specific polling stations were randomly selected and all voters were surveyed as they left the premises.

19. *Literary Digest* Poll In 1936, *Literary Digest* magazine mailed questionnaires to 10 million people and obtained 2,266,566 responses. The responses indicated that Alf Landon would win the presidential election. He didn't.

20. Highway Strength The New York State Department of Transportation evaluated the quality of the New York State Thruway by testing core samples collected at regular intervals of 1 mile.

Critical Thinking: What's Wrong? *In Exercises 21–28, determine whether the study is an experiment or an observational study, and then identify a major problem with the study.*

21. Online News In a survey conducted by *USA Today*, 1465 Internet users chose to respond to this question posted on the *USA Today* electronic edition: "Is news online as satisfying as print and TV news?" 52% of the respondents said "yes."

22. Physicians' Health Study The Physicians' Health Study involved 22,071 male physicians. Based on random selections, 11,037 of them were treated with aspirin and the other 11,034 were given placebos. The study was stopped early because it became clear that aspirin reduced the risk of myocardial infarctions by a substantial amount.

23. Drinking and Driving A researcher for a consortium of insurance companies plans to test for the effects of drinking on driving ability by randomly selecting 1000 drivers and then randomly assigning them to two groups: One group of 500 will drive in New York City after no alcohol consumption, and the second group will drive in New York City after consuming three shots of Jim Beam bourbon whiskey.

24. Blood Pressure A medical researcher tested for a difference in systolic blood pressure levels between male and female students who are 12 years of age. She randomly selected four males and four females for her study.

25. Driver Aggression In testing a treatment designed to reduce driver aggression in the United States, the original plan was to use a sample of 500 drivers randomly selected throughout the country. The program managers know that they would get a biased sample if they limit their study to drivers in New York City, so they planned to compensate for that bias by using a larger sample of 3000 drivers in New York City.

26. Atkins Weight Loss Program An independent researcher tested the effectiveness of the Atkins weight loss program by randomly selecting 1000 subjects using that program. Each of the subjects was called to report their weight before the diet and after the diet.

27. Crime Research A sociologist has created a brief survey to be given to 2000 adults randomly selected from the U.S. population. Here are her first two questions: (1) Have you ever been the victim of a felony crime? (2) Have you ever been convicted of a felony?

28. Medications The Pharmaceutical Research and Manufacturers of America wants information about the consumption of various medications. An independent researcher conducts a survey by mailing 10,000 questionnaires to randomly selected adults in the United States, and she receives 152 responses.

1-3   Beyond the Basics

In Exercises 29–32, indicate whether the observational study used is cross-sectional, retrospective, or prospective.

29. Nurses' Health Study II Phase II of the Nurses' Health Study was started in 1989 with 116,000 female registered nurses. The study is ongoing.

30. Heart Health Study Samples of subjects with and without heart disease were selected, then researchers looked back in time to determine whether they took aspirin on a regular basis.

31. Marijuana Study Researchers from the National Institutes of Health want to determine the current rates of marijuana consumption among adults living in states that have legalized the use of marijuana. They conduct a survey of 500 adults in those states.

32. Framingham Heart Study The Framingham Heart Study was started in 1948 and is ongoing. Its focus is on heart disease.

In Exercises 33–36, identify which of these designs is most appropriate for the given experiment: completely randomized design, randomized block design, or matched pairs design.

33. Lunesta Lunesta is a drug designed to treat insomnia. In a clinical trial of Lunesta, amounts of sleep each night are measured before and after subjects have been treated with the drug.

34. Lipitor A clinical trial of Lipitor treatments is being planned to determine whether its effects on diastolic blood pressure are different for men and women.

35. West Nile Vaccine Currently, there is no approved vaccine for the prevention of infection by West Nile virus. A clinical trial of a possible vaccine is being planned to include subjects treated with the vaccine while other subjects are given a placebo.

36. HIV Vaccine The HIV Trials Network is conducting a study to test the effectiveness of two different experimental HIV vaccines. Subjects will consist of 80 pairs of twins. For each pair of twins, one of the subjects will be treated with the DNA vaccine and the other twin will be treated with the adenoviral vector vaccine.

37. Simple Random Sample vs. Random Sample Refer to the definition of *simple random sample* on page 27 and its accompanying definition of *random sample* enclosed within parentheses. Determine whether each of the following is a simple random sample and a random sample.

a. A statistics class with 36 students is arranged so that there are 6 rows with 6 students in each row, and the rows are numbered from 1 through 6. A die is rolled and a sample consists of all students in the row corresponding to the outcome of the die.

b. For the same class described in part (a), the 36 student names are written on 36 individual index cards. The cards are shuffled and six names are drawn from the top.

c. For the same class described in part (a), the six youngest students are selected.

Chapter Quick Quiz

1. Hospitals In a study of births in New York State, data were collected from four hospitals coded as follows: (1) Albany Medical Center; (1438) Bellevue Hospital Center; (66) Olean General Hospital; (413) Strong Memorial Hospital. Does it make sense to calculate the average (mean) of the numbers 1, 1438, 66, and 413?

2. Hospitals Which of the following best describes the level of measurement of the numbers 1, 1438, 66, and 413 from Exercise 1: nominal, ordinal, interval, ratio?

3. Birth Weights In the same study cited in Exercise 1, birth weights of newborn babies are given in grams. Are these birth weights discrete data or continuous data?

4. Birth Weights Are the birth weights described in Exercise 3 quantitative data or categorical data?

5. Birth Weights Which of the following best describes the level of measurement of the birth weights described in Exercise 3: nominal, ordinal, interval, ratio?

6. Statistic/Parameter In an AARP survey of 1019 randomly selected adults, the respondents were asked to identify the number of credit cards they have, and 26% of them said they had no credit cards. Is the value of 26% a statistic or a parameter?

7. AARP Survey Refer to the survey described in Exercise 6. Because the 1019 subjects agreed to respond, do they constitute a voluntary response sample?

8. Observational Study or Experiment Are the data described in Exercise 6 the result of an observational study or an experiment?

9. Physicians' Health Study In the Physicians' Health Study, some of the subjects were treated with aspirin while others were given a placebo. For the subjects in this experiment, what is *blinding*?

10. Sampling In a statistical study, which of the following types of samples is generally best: convenience sample, voluntary response sample, simple random sample, biased sample?

Review Exercises

1. What's Wrong? In an American Optometric Association survey, 1009 adults were randomly selected and asked to identify what they worry most about losing. 51% of the respondents chose "vision." What's wrong here?

2. Paying for First Dates *USA Today* posted this question on the electronic version of its newspaper: "Should guys pay for the first date?" Of the 1148 subjects who decided to respond, 85% of them said "yes."

a. What is wrong with this survey?

b. Is the value of 85% a statistic or a parameter?

c. Does the survey constitute an experiment or an observational study?

3. Sample Design Literacy In "Cardiovascular Effects of Intravenous Triiodothyronine in Patients Undergoing Coronary Artery Bypass Graft Surgery" [*Journal of the American Medical Association (JAMA)*, Vol. 275, No. 9], the authors explain that patients were assigned to one of three groups: (1) a group treated with triidothyronine, (2) a group treated with normal saline bolus and dopamine, and (3) a placebo group given normal saline. The authors summarize the sample design as "randomized and double-blind." Describe the meaning of "randomized" and "double-blind" in the context of this study.

4. Divorces and Margarine One study showed that there is a very high correlation between the divorce rate in Maine and per capita consumption of margarine in the United States. Can we conclude that either one of those two variables is the cause of the other?

5. Simple Random Sample Which of the following is/are simple random samples?

a. As Lipitor pills are being manufactured, a quality control plan is to select every 500th pill and test it to confirm that it contains 80 mg of atorvastatin.

b. To test for a gender difference in the way that men and women make online purchases, Gallup surveys 500 randomly selected men and 500 randomly selected women.

c. A list of all 10,877 adults in Trinity County, California, is obtained; the list is numbered from 1 to 10,877; and then a computer is used to randomly generate 250 different numbers between 1 and 10,877. The sample consists of the adults corresponding to the selected numbers.

6. Defense of Marriage Act Both of the following questions are essentially the same. Does the difference in wording seem as though it could affect the way that people respond?

• Are you in favor of the "Defense of Marriage Act"?

• Are you in favor of an act that for federal and state aid, only heterosexual marriages should be recognized?

7. Colleges in United States Currently, there are 4612 colleges in the United States, and the number of full-time students is 13,203,477.

a. Are the numbers of full-time students at different colleges discrete or continuous?

b. What is the level of measurement for the numbers of full-time students at colleges? (nominal, ordinal, interval, ratio)

c. What is wrong with surveying college students by mailing questionnaires to 10,000 of them who are randomly selected?

d. If we randomly select 50 full-time college students in each of the 50 states, what type of sample is obtained? (random, systematic, convenience, stratified, cluster)

e. If we randomly select four colleges and survey all of their full-time students, what type of sample is obtained? (random, systematic, convenience, stratified, cluster)

8. Percentages

a. The labels on U-Turn protein energy bars include the statement that these bars contain "125% less fat than the leading chocolate candy brands" (based on data from *Consumer Reports* magazine). What is wrong with that claim?

b. In a Pew Research Center poll on driving, 58% of the 1182 respondents said that they like to drive. What is the actual number of respondents who said that they like to drive?

c. In a Pew Research Center poll on driving, 331 of the 1182 respondents said that driving is a chore. What percentage of respondents said that driving is a chore?

9. Types of Data In each of the following, identify the level of measurement of the sample data (nominal, ordinal, interval, ratio) and the type of sampling used to obtain the data (random, systematic, convenience, stratified, cluster).

a. At Albany Medical Center, every 10th newborn baby is selected and the body temperature is measured (degrees Fahrenheit).

b. In each of the 50 states, 50 voters are randomly selected and their political party affiliations are identified.

c. A pollster stops each person passing her office door and asks the person to rate the last movie that he or she saw (on a scale of 1 star to 4 stars).

10. Statistical Significance and Practical Significance The Technogene Research Group has developed a procedure designed to increase the likelihood that a baby will be born a girl. In a clinical trial of their procedure, 236 girls were born to 450 different couples. If the method has no effect, there is about a 15% chance that such extreme results would occur. Does the procedure appear to have statistical significance? Does the procedure appear to have practical significance?

Cumulative Review Exercises

For Chapter 2 through Chapter 14, the Cumulative Review Exercises include topics from preceding chapters. For this chapter, we present a few calculator warm-up exercises, with expressions similar to those found throughout this book. Use your calculator to find the indicated values.

1. Birth Weights Listed below are the weights (grams) of newborn babies from Albany Medical Center Hospital. What value is obtained when those weights are added and the total is divided by the number of weights? (This result, called the *mean*, is discussed in Chapter 3.) What is notable about these values, and what does it tell us about how the weights were measured?

$$3600 \quad 1700 \quad 4000 \quad 3900 \quad 3100 \quad 3800 \quad 2200 \quad 3000$$

2. Six Children Jule Cole is a founder of Mabel's Labels, and she is the mother of six children. The probability that six randomly selected children are all girls is found by evaluating 0.5^6. Find that value.

3. Tallest Person Robert Wadlow (1918–1940) is the tallest known person to have lived. The expression below converts his height of 272 cm to a standardized score. Find this value and round the result to two decimal places. Such standardized scores are considered to be significantly high if they are greater than 2 or 3. Is the result significantly high?

$$\frac{272 - 176}{6}$$

4. Body Temperature The given expression is used for determining the likelihood that the average (mean) human body temperature is different from the value of 98.6°F that is commonly used. Find the given value and round the result to two decimal places.

$$\frac{98.2 - 98.6}{\frac{0.62}{\sqrt{106}}}$$

5. Determining Sample Size The given expression is used to determine the size of the sample necessary to estimate the proportion of college students who have the profound wisdom to take a statistics course. Find the value and round the result to the nearest whole number.

$$\frac{1.96^2 \cdot 0.25}{0.03^2}$$

6. Standard Deviation One way to get a very rough approximation of the value of a standard deviation of sample data is to find the range, then divide it by 4. The range is the difference between the highest sample value and the lowest sample value. In using this approach, what value is obtained from the sample data listed in Exercise 1 "Birth Weights"?

7. Standard Deviation The standard deviation is an extremely important concept introduced in Chapter 3. Using the sample data from Exercise 1 "Birth Weights," part of the calculation of the standard deviation is shown in the expression below. Evaluate this expression. (Fortunately, calculators and software are designed to automatically execute such expressions, so our future work with standard deviations will not be burdened with cumbersome calculations.)

$$\frac{(3600 - 3162.5)^2}{7}$$

8. Standard Deviation The given expression is used to compute the standard deviation of three randomly selected body temperatures. Perform the calculation and round the result to two decimal places.

$$\sqrt{\frac{(98.4 - 98.6)^2 + (98.6 - 98.6)^2 + (98.8 - 98.6)^2}{3 - 1}}$$

Scientific Notation. *In Exercises 9–12, the given expressions are designed to yield results expressed in a form of scientific notation. For example, the calculator-displayed result of 1.23E5 can be expressed as 123,000, and the result of 1.23E-4 can be expressed as 0.000123. Perform the indicated operation and express the result as an ordinary number that is not in scientific notation.*

9. 0.4^8 **10.** 9^{11} **11.** 6^{14} **12.** 0.3^{12}

Technology Project

1. Missing Data The focus of this project is to download a data set and manipulate it to work around missing data.

a. First, download Data Set 3 "Body Temperatures" in Appendix B from TriolaStats.com. Choose the download format that matches your technology.

b. Some statistical procedures, such as those involved with correlation and regression (discussed in later chapters) require data that consist of matched pairs of values, and those procedures ignore pairs in which at least one of the data values in a matched pair is missing. Assume that we want to conduct analyses for correlation and regression on the last two columns of data in Data Set 3: body temperatures measured at 8 AM on day 2 and again at 12 AM on day 2. For those last two columns, identify the rows with at least one missing value. Note that in some technologies, such as TI-83/84 Plus calculators, missing data must be represented by a constant such as −9 or 999.

c. Here are two different strategies for reconfiguring the data set to work around the missing data in the last two columns (assuming that we need matched pairs of data with no missing values):

i. Manual Deletion Highlight rows with at least one missing value in the last two columns, then delete those rows. This can be tedious if there are many rows with missing data and those rows are interspersed throughout instead of being adjacent rows.

ii. Sort Most technologies have a Sort feature that allows you to rearrange all rows using one particular column as the basis for sorting (TI-83/84 Plus calculators *do not* have this type of sort feature). The result is that all rows remain the same but they are in a different order. First use the technology's Sort feature to rearrange all rows using the "8 AM day 2" column as the basis for sorting (so that all missing values in the "8 AM day 2" column are at the beginning); then highlight and delete all of those rows with missing values in the "8 AM day 2" column. Next, use the technology's Sort feature to rearrange all rows using the "12 AM day 2" column as the basis for sorting (so that all missing values in the "12 AM day 2" column are at the beginning); then highlight and delete all of those rows with missing values in the "12 AM day 2" column. The remaining rows will include matched pairs of body temperatures, and those rows will be suitable for analyses such as correlation and regression. Print the resulting reconfigured data set.

Critical Thinking:
Do Male Symphony Conductors Really Live Longer?

Several media reports made the interesting observation that male symphony conductors live longer than other males. John Amaral wrote in *Awaken* that orchestra conductors "live longer than almost any other group of people by three to seven years." Robert Levine wrote in Polyphonic.org that they live longer "because they stand up while working." Some provided other explanations for this phenomenon, often referring to cardiovascular activity. But do male symphony conductors really live longer than other groups of males? The Internet can be researched for possible answers. Let's also consider the following.

Analysis

1. Consider the statement that "male symphony conductors live longer." Identify the specific group that they supposedly live longer than. Does that other group consist of males randomly selected from the general population?

2. It is reasonable to assume that males do not become symphony conductors until they have reached at least the age of 40 years. When comparing life spans of male conductors, should we compare them to other males in the general population, or should we compare them to other males who lived until at least 40 years of age? Explain.

3. Without any disabilities, males qualify for Medicare if they are 65 or older and meet a few other requirements. If we compare life spans of males on Medicare to life spans of males randomly selected from the general population, why would we find that males on Medicare have longer life spans?

4. Explain in detail how to design a study for collecting data to determine whether it is misleading to state that male symphony conductors live longer. Should the study be an experiment or an observational study?

Cooperative Group Activities

1. In-class activity Working in groups of three or four, design an experiment to determine whether pulse rates of college students are the same while the students are standing and sitting. Conduct the experiment and collect the data. Save the data so that they can be analyzed with methods presented in the following chapters.

2. In-class activity Working in groups of three or four, construct a brief survey that includes only a few questions that can be quickly asked. Include some objective questions along with some that are biased, such as the first question below.

• Should your college force all students to pay a $100 activity fee?

• Should your college fund activities by collecting a $100 fee?

Conduct the survey and try to detect the effect that the biased wording has on the responses.

3. In-class activity Identify problems with a mailing from *Consumer Reports* magazine that included an annual questionnaire about cars and other consumer products. Also included were a request for a voluntary contribution of money and a voting ballot for the board of directors. Responses were to be mailed back in envelopes that required postage stamps.

4. Out-of-class activity Find a report of a survey that used a voluntary response sample. Describe how it is quite possible that the results do not accurately reflect the population.

5. Out-of-class activity Find a professional journal with an article that includes a statistical analysis of an experiment. Describe and comment on the design of the experiment. Identify one particular issue addressed by the study, and determine whether the results were found to be statistically significant. Determine whether those same results have practical significance.

2-1 Frequency Distributions for Organizing and Summarizing Data

2-2 Histograms

2-3 Graphs That Enlighten and Graphs That Deceive

2-4 Scatterplots, Correlation, and Regression

2

EXPLORING DATA WITH TABLES AND GRAPHS

CHAPTER PROBLEM

Fast food restaurants: Which one is fastest?

One attractive feature of fast food restaurants is that they are fast! To remain competitive, fast food restaurants must not only provide a good culinary experience but also must do it as quickly as their competitors. Data Set 25 "Fast Food" in Appendix B lists measured service times (seconds) obtained from samples of drive-through customers at different restau-

rants. Table 2-1 lists the 50 service time measurements from the first column of Data Set 25. It is an exceptionally rare person who can simply look at those data and form meaningful conclusions. We mere mortals must work at describing, exploring, and comparing the different lists of service times to gain meaningful insights. In this chapter we present methods that focus on organizing and summarizing the data and using graphs that enable us to understand important characteristics of the data, especially the *distribution* of the data. These methods will help us compare the restaurants.

TABLE 2-1 Drive-Through Service Times (seconds) for McDonald's Lunches

107	139	197	209	281	254	163	150	127	308	206	187	169	83	127	133	140
143	130	144	91	113	153	255	252	200	117	167	148	184	123	153	155	154
100	117	101	138	186	196	146	90	144	119	135	151	197	171	190	169	

CHAPTER OBJECTIVES

This chapter and the following chapter focus on important characteristics of data, including the following:

Characteristics of Data

1. **Center:** A representative value that shows us where the middle of the data set is located.

2. **Variation:** A measure of the amount that the data values vary.

3. **Distribution:** The nature or shape of the spread of the data over the range of values (such as bell-shaped).

4. **Outliers:** Sample values that lie very far away from the vast majority of the other sample values.

5. **Time:** Any change in the characteristics of the data over time.

This chapter provides tools that enable us to gain insight into data by organizing, summarizing, and representing them in ways that enable us to see important characteristics of the data. Here are the chapter objectives:

2-1 Frequency Distributions for Organizing and Summarizing Data

- Develop an ability to summarize data in the format of a frequency distribution and a relative frequency distribution.

- For a frequency distribution, identify values of class width, class midpoint, class limits, and class boundaries.

2-2 Histograms

- Develop the ability to picture the distribution of data in the format of a histogram or relative frequency histogram.

- Examine a histogram and identify common distributions, including a uniform distribution and a normal distribution.

2-3 Graphs That Enlighten and Graphs That Deceive

- Develop an ability to graph data using a dotplot, stemplot, time-series graph, Pareto chart, pie chart, and frequency polygon.

- Determine when a graph is deceptive through the use of a nonzero axis or a pictograph that uses an object of area or volume for one-dimensional data.

2-4 Scatterplots, Correlation, and Regression

- Develop an ability to construct a scatterplot of paired data.

- Analyze a scatterplot to determine whether there appears to be a correlation between two variables.

Frequency Distributions for Organizing and Summarizing Data

2-1

TABLE 2-2 McDonald's Lunch Drive-Through Service Times

Time (seconds)	Frequency
75–124	11
125–174	24
175–224	10
225–274	3
275–324	2

Key Concept When working with large data sets, a *frequency distribution* (or *frequency table*) is often helpful in organizing and summarizing data. A frequency distribution helps us to understand the nature of the *distribution* of a data set.

> **DEFINITION**
>
> A **frequency distribution** (or **frequency table**) shows how data are partitioned among several categories (or *classes*) by listing the categories along with the number (frequency) of data values in each of them.

Let's use the McDonald's lunch drive-through service times (in seconds) listed in Table 2-1. Table 2-2 is a frequency distribution summarizing those service times. The **frequency** for a particular class is the number of original values that fall into that class. For example, the first class in Table 2-2 has a frequency of 11, so 11 of the service times are between 75 seconds and 124 seconds, inclusive.

The following standard terms are often used in constructing frequency distributions and graphs.

> **DEFINITIONS**
>
> **Lower class limits** are the smallest numbers that can belong to each of the different classes. (Table 2-2 has lower class limits of 75, 125, 175, 225, and 275.)
>
> **Upper class limits** are the largest numbers that can belong to each of the different classes. (Table 2-2 has upper class limits of 124, 174, 224, 274, and 324.)
>
> **Class boundaries** are the numbers used to separate the classes, but without the gaps created by class limits. Figure 2-1 shows the gaps created by the class limits from Table 2-2. In Figure 2-1 we see that the values of 124.5, 174.5, 224.5, and 274.5 are in the centers of those gaps. Following the pattern of those class boundaries, we see that the lowest class boundary is 74.5 and the highest class boundary is 324.5. The complete list of class boundaries is 74.5, 124.5, 174.5, 224.5, 274.5, and 324.5.
>
> **Class midpoints** are the values in the middle of the classes. Table 2-2 has class midpoints of 99.5, 149.5, 199.5, 249.5, and 299.5. Each class midpoint can be found by adding the lower class limit to the upper class limit and dividing the sum by 2.
>
> **Class width** is the difference between two consecutive lower class limits (or two consecutive lower class boundaries) in a frequency distribution. Table 2-2 uses a class width of 50. (The first two lower class boundaries are 75 and 125, and their difference is 50.)

CAUTION Finding the correct class width can be tricky. For class width, don't make the most common mistake of using the difference between a lower class limit and an upper class limit. See Table 2-2 and note that the class width is 50, not 49.

CAUTION For class boundaries, remember that they split the difference between the end of one class and the beginning of the next class, as shown in Figure 2-1.

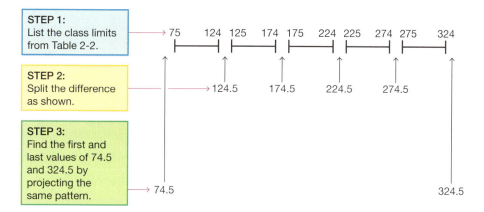

FIGURE 2-1 **Finding Class Boundaries from Class Limits in Table 2-2**

STEP 1:
List the class limits from Table 2-2.

STEP 2:
Split the difference as shown.

STEP 3:
Find the first and last values of 74.5 and 324.5 by projecting the same pattern.

No Phones or Bathtubs

Many statistical analyses must consider changing characteristics of populations over time. Here are some observations of life in the United States from 100 years ago:

- 8% of homes had a telephone.
- 14% of homes had a bathtub.
- The mean life expectancy was 47 years.
- The mean hourly wage was 22 cents.
- There were approximately 230 murders in the entire United States.

Although these observations from 100 years ago are in stark contrast to the United States of today, statistical analyses should always consider changing population characteristics that might have more subtle effects.

Procedure for Constructing a Frequency Distribution

We construct frequency distributions to (1) summarize large data sets, (2) see the distribution and identify outliers, and (3) have a basis for constructing graphs (such as *histograms,* introduced in Section 2-2). Technology can generate frequency distributions, but here are the steps for manually constructing them:

1. Select the number of classes, usually between 5 and 20. The number of classes might be affected by the convenience of using round numbers. (According to "Sturges' guideline," the ideal number of classes for a frequency distribution can be approximated by $1 + (\log n)/(\log 2)$ where n is the number of data values. We don't use this guideline in this book.)

2. Calculate the class width.

$$\text{Class width} \approx \frac{(\text{maximum data value}) - (\text{minimum data value})}{\text{number of classes}}$$

Round this result to get a convenient number. (It's usually best to round *up.*) Using a specific number of classes is not too important, and it's usually wise to change the number of classes so that they use convenient values for the class limits.

3. Choose the value for the first lower class limit by using either the minimum value or a convenient value below the minimum.

4. Using the first lower class limit and the class width, list the other lower class limits. (Do this by adding the class width to the first lower class limit to get the second lower class limit. Add the class width to the second lower class limit to get the third lower class limit, and so on.)

75–
125–
175–
225–
275–

5. List the lower class limits in a vertical column and then determine and enter the upper class limits.

6. Take each individual data value and put a tally mark in the appropriate class. Add the tally marks to find the total frequency for each class.

When constructing a frequency distribution, be sure the classes do not overlap. Each of the original values must belong to exactly one class. Include all classes, even those with a frequency of zero. Try to use the same width for all classes, although it is sometimes impossible to avoid open-ended intervals, such as "65 years or older."

EXAMPLE 1 McDonald's Lunch Service Times

Using the McDonald's lunch service times in Table 2-1, follow the above procedure to construct the frequency distribution shown in Table 2-2. Use five classes.

SOLUTION

Step 1: Select 5 as the number of desired classes.

Step 2: Calculate the class width as shown below. Note that we round 45 up to 50, which is a more convenient number.

$$\text{Class width} \approx \frac{(\text{maximum data value}) - (\text{minimum data value})}{\text{number of classes}}$$

$$= \frac{308 - 83}{5} = 45 \approx 50 \text{ (rounded up to a more convenient number)}$$

Step 3: The minimum data value is 83, which is not a very convenient starting point, so go to a value below 83 and select the more convenient value of 75 as the first lower class limit. (We could have used 80 or 50 instead.)

Step 4: Add the class width of 50 to the starting value of 75 to get the second lower class limit of 125. Continue to add the class width of 50 until we have five lower class limits. The lower class limits are therefore 75, 125, 175, 225, and 275.

Step 5: List the lower class limits vertically, as shown in the margin. From this list, we identify the corresponding upper class limits as 124, 174, 224, 274, and 324.

Step 6: Enter a tally mark for each data value in the appropriate class. Then add the tally marks to find the frequencies shown in Table 2-2.

 YOUR TURN Do Exercise 11 "Old Faithful."

Categorical Data So far we have discussed frequency distributions using only quantitative data sets, but frequency distributions can also be used to summarize categorical (or qualitative or attribute) data, as illustrated in Example 2.

EXAMPLE 2 Emergency Room Visits for Injuries from Sports and Recreation

Table 2-3 lists data for the highest seven sources of injuries resulting in a visit to a hospital emergency room (ER) in a recent year (based on data from the Centers for Disease Control). The activity names are categorical data at the nominal level of measurement, but we can create the frequency distribution as shown. It might be surprising to see that bicycling is at the top of this list, but this doesn't mean that bicycling is the most dangerous of these activities; many more people bicycle than play football or ride an all-terrain vehicle or do any of the other listed activities.

TABLE 2-3 Annual ER Visits for Injuries from Sports and Recreation

Activity	Frequency
Bicycling	26,212
Football	25,376
Playground	16,706
Basketball	13,987
Soccer	10,436
Baseball	9,634
All-terrain vehicle	6,337

Relative Frequency Distribution

A variation of the basic frequency distribution is a **relative frequency distribution** or **percentage frequency distribution**, in which each class frequency is replaced by a relative frequency (or proportion) or a percentage. In this text we use the term "relative frequency distribution" whether we use relative frequencies or percentages. Relative frequencies and percentages are calculated as follows.

$$\text{Relative frequency for a class} = \frac{\text{frequency for a class}}{\text{sum of all frequencies}}$$

$$\text{Percentage for a class} = \frac{\text{frequency for a class}}{\text{sum of all frequencies}} \times 100\%$$

Table 2-4 is an example of a relative frequency distribution. It is a variation of Table 2-2 in which each class frequency is replaced by the corresponding percentage value. Because there are 50 data values, divide each class frequency by 50, and then multiply by 100%. The first class of Table 2-2 has a frequency of 11, so divide 11 by 50 to get 0.22, and then multiply by 100% to get 22%. The sum of the percentages should be 100%, with a small discrepancy allowed for rounding errors, so a sum such as 99% or 101% is acceptable. The sum of the percentages in Table 2-4 is 100%.

The sum of the percentages in a relative frequency distribution must be very close to 100% (with a little wiggle room for rounding errors).

TABLE 2-4 Relative Frequency Distribution of McDonald's Lunch Service Times

Time (seconds)	Relative Frequency
75–124	22%
125–174	48%
175–224	20%
225–274	6%
275–324	4%

Cumulative Frequency Distribution

Another variation of a frequency distribution is a **cumulative frequency distribution** in which the frequency for each class is the sum of the frequencies for that class and all previous classes. Table 2-5 is the cumulative frequency distribution found from Table 2-2. Using the original frequencies of 11, 24, 10, 3, 2, we add 11 + 24 to get the second cumulative frequency of 35, then we add 11 + 24 + 10 to get the third, and so on. See Table 2-5, and note that in addition to the use of cumulative frequencies, the class limits are replaced by "less than" expressions that describe the new ranges of values.

Critical Thinking: Using Frequency Distributions to Understand Data

At the beginning of this section we noted that a frequency distribution can help us understand the *distribution* of a data set, which is the nature or shape of the spread of the data over the range of values (such as bell-shaped). In statistics we are often interested in determining whether the data have a *normal distribution*. (Normal distributions are

TABLE 2-5 Cumulative Frequency Distribution of McDonald's Lunch Service Times

Time (seconds)	Cumulative Frequency
Less than 125	11
Less than 175	35
Less than 225	45
Less than 275	48
Less than 325	50

Growth Charts Updated

Pediatricians typically use standardized growth charts to compare their patient's weight and height to a sample of other children. Children are considered to be in the normal range if their weight and height fall between the 5th and 95th percentiles. If they fall outside that range, they are often given tests to ensure that there are no serious medical problems. Pediatricians became increasingly aware of a major problem with the charts: Because they were based on children living between 1929 and 1975, the growth charts had become inaccurate. To rectify this problem, the charts were updated in 2000 to reflect the current measurements of millions of children. The weights and heights of children are good examples of populations that change over time. This is the reason for including changing characteristics of data over time as an important consideration for a population.

TABLE 2-7 Last Digits of Pulse Rates from the National Health and Examination Survey

Last Digit of Pulse Rate	Frequency
0	455
1	0
2	461
3	0
4	479
5	0
6	425
7	0
8	399
9	0

discussed extensively in Chapter 6.) Data that have an approximately normal distribution are characterized by a frequency distribution with the following features.

Normal Distribution

1. The frequencies start low, then increase to one or two high frequencies, and then decrease to a low frequency.
2. The distribution is approximately symmetric: Frequencies preceding the maximum frequency should be roughly a mirror image of those that follow the maximum frequency.

Table 2-6 satisfies these two conditions. The frequencies start low, increase to the maximum of 30, and then decrease to a low frequency. Also, the frequencies of 2 and 8 that precede the maximum are a mirror image of the frequencies 8 and 2 that follow the maximum. Real data sets are usually not so perfect as Table 2-6, and judgment must be used to determine whether the distribution comes "close enough" to satisfying those two conditions. (There are more objective procedures included later.)

TABLE 2-6 Frequency Distribution Showing a Normal Distribution

Time	Frequency	Normal Distribution
75–124	2	← Frequencies start low, . . .
125–174	8	
175–224	30	← Increase to this maximum, . . .
225–274	8	
275–324	2	← Decrease to become low again.

Analysis of Last Digits Example 3 illustrates this principle:

Frequencies of last digits sometimes reveal how the data were collected or measured.

 EXAMPLE 3 **Exploring Data: How Were the Pulse Rates Measured?**

Upon examination of measured pulse rates from 2219 adults included in the National Health and Examination Survey, the last digits of the recorded pulse rates are identified and the frequency distribution for those last digits is as shown in Table 2-7. Here is an important observation of those last digits: All of the last digits are *even* numbers. If the pulse rates were counted for 1 full minute, there would surely be a large number of them ending with an *odd* digit. So what happened?

One reasonable explanation is that even though the pulse rates are the number of heartbeats in 1 minute, they were likely counted for 30 seconds and the number of beats was doubled. (The original pulse rates are not all multiples of 4, so we can rule out a procedure of counting for 15 seconds and then multiplying by 4.) Analysis of these last digits reveals to us the method used to obtain these data.

In many surveys, we can determine that surveyed subjects were asked to *report* some values, such as their heights or weights, because disproportionately many values end in 0 or 5. This is a strong clue that the respondent is rounding instead of being physically measured. Fascinating stuff!

YOUR TURN Do Exercise 17 "Analysis of Last Digits."

Gaps Example 4 illustrates this principle:

> **The presence of gaps can suggest that the data are from two or more different populations.**

The converse of this principle is not true, because data from different populations do not necessarily result in gaps.

 EXAMPLE 4 Exploring Data: What Does a Gap Tell Us?

Table 2-8 is a frequency distribution of the weights (grams) of randomly selected pennies. Examination of the frequencies reveals a large *gap* between the lightest pennies and the heaviest pennies. This suggests that we have two different populations: Pennies made before 1983 are 95% copper and 5% zinc, but pennies made after 1983 are 2.5% copper and 97.5% zinc, which explains the large gap between the lightest pennies and the heaviest pennies represented in Table 2-8.

YOUR TURN Do Exercise 18 "Analysis of Last Digits" and determine whether there is a gap. If so, what is a reasonable explanation for it?

TABLE 2-8 Randomly Selected Pennies

Weight (grams) of Penny	Frequency
2.40–2.49	18
2.50–2.59	19
2.60–2.69	0
2.70–2.79	0
2.80–2.89	0
2.90–2.99	2
3.00–3.09	25
3.10–3.19	8

Comparisons Example 5 illustrates this principle:

> **Combining two or more relative frequency distributions in one table makes comparisons of data much easier.**

 EXAMPLE 5 Comparing McDonald's and Dunkin' Donuts

Table 2-9 shows the relative frequency distributions for the drive-through lunch service times (seconds) for McDonald's and Dunkin' Donuts. Because of the dramatic differences in their menus, we might expect the service times to be very different. By comparing the relative frequencies in Table 2-9, we see that there are major differences. The Dunkin' Donuts service times appear to be lower than those at McDonald's. This is not too surprising, given that many of the Dunkin' Donuts orders are probably for coffee and a donut.

TABLE 2-9 McDonald's and Dunkin' Donuts Lunch Service Times

Time (seconds)	McDonald's	Dunkin' Donuts
25–74		22%
75–124	22%	44%
125–174	48%	28%
175–224	20%	6%
225–274	6%	
275–324	4%	

YOUR TURN Do Exercise 19 "Oscar Winners."

TECH CENTER

 Frequency Distributions

Access tech supplements, videos, and data sets at **www.TriolaStats.com**

Frequency distributions are often easy to obtain after generating a histogram, as described in Section 2-2. With Statdisk, for example, we can generate a histogram with a desired starting point and class width, then check "Bar Labels" to see the frequency for each class. If histograms are not used, "sort" the data (arrange them in order) so that we can see the maximum data value and the minimum data value used for computing the class width. Once the class limits are established, it is easy to find the frequency for each class using sorted data. Every statistics software package includes a sort feature.

2-1   Basic Skills and Concepts

Statistical Literacy and Critical Thinking

Table for Exercise 1

McDonald's Dinner Service Times

Time (sec)	Frequency
60–119	7
120–179	22
180–239	14
240–299	2
300–359	5

Table for Exercise 4

Height (cm)	Relative Frequency
130–144	23%
145–159	25%
160–174	22%
175–189	27%
190–204	28%

1. McDonald's Dinner Service Times Refer to the accompanying table summarizing service times (seconds) of McDonald's dinners. How many individuals are included in the summary? Is it possible to identify the exact values of all of the original service times?

2. McDonald's Dinner Service Times Refer to the accompanying frequency distribution. What problem would be created by using classes of 60–120, 120–180, . . . , 300–360?

3. Relative Frequency Distribution Use percentages to construct the relative frequency distribution corresponding to the accompanying frequency distribution for McDonald's dinner service times.

4. What's Wrong? Heights of adult males are known to have a normal distribution, as described in this section. A researcher claims to have randomly selected adult males and measured their heights with the resulting relative frequency distribution as shown here. Identify two major flaws with these results.

In Exercises 5–8, identify the class width, class midpoints, and class boundaries for the given frequency distribution. Also identify the number of individuals included in the summary. The frequency distributions are based on real data from Appendix B.

5.

Age (yr) of Best Actress When Oscar Was Won	Frequency
20–29	29
30–39	34
40–49	14
50–59	3
60–69	5
70–79	1
80–89	1

6.

Age (yr) of Best Actor When Oscar Was Won	Frequency
20–29	1
30–39	28
40–49	36
50–59	15
60–69	6
70–79	1

7.

Blood Platelet Count of Males	Frequency
0–99	1
100–199	51
200–299	90
300–399	10
400–499	0
500–599	0
600–699	1

8.

Blood Platelet Count of Females	Frequency
100–199	25
200–299	92
300–399	28
400–499	0
500–599	2

Normal Distributions.   *In Exercises 9 and 10, using a loose interpretation of the criteria for determining whether a frequency distribution is approximately a normal distribution, determine whether the given frequency distribution is approximately a normal distribution. Give a brief explanation.*

9. Best Actresses Refer to the frequency distribution from Exercise 5.

10. Best Actors Refer to the frequency distribution from Exercise 6.

TABLE A-4 Chi-Square (χ^2) Distribution

Degrees of Freedom	Area to the *Right* of the Critical Value									
	0.995	0.99	0.975	0.95	0.90	0.10	0.05	0.025	0.01	0.005
1	—	—	0.001	0.004	0.016	2.706	3.841	5.024	6.635	7.879
2	0.010	0.020	0.051	0.103	0.211	4.605	5.991	7.378	9.210	10.597
3	0.072	0.115	0.216	0.352	0.584	6.251	7.815	9.348	11.345	12.838
4	0.207	0.297	0.484	0.711	1.064	7.779	9.488	11.143	13.277	14.860
5	0.412	0.554	0.831	1.145	1.610	9.236	11.071	12.833	15.086	16.750
6	0.676	0.872	1.237	1.635	2.204	10.645	12.592	14.449	16.812	18.548
7	0.989	1.239	1.690	2.167	2.833	12.017	14.067	16.013	18.475	20.278
8	1.344	1.646	2.180	2.733	3.490	13.362	15.507	17.535	20.090	21.955
9	1.735	2.088	2.700	3.325	4.168	14.684	16.919	19.023	21.666	23.589
10	2.156	2.558	3.247	3.940	4.865	15.987	18.307	20.483	23.209	25.188
11	2.603	3.053	3.816	4.575	5.578	17.275	19.675	21.920	24.725	26.757
12	3.074	3.571	4.404	5.226	6.304	18.549	21.026	23.337	26.217	28.299
13	3.565	4.107	5.009	5.892	7.042	19.812	22.362	24.736	27.688	29.819
14	4.075	4.660	5.629	6.571	7.790	21.064	23.685	26.119	29.141	31.319
15	4.601	5.229	6.262	7.261	8.547	22.307	24.996	27.488	30.578	32.801
16	5.142	5.812	6.908	7.962	9.312	23.542	26.296	28.845	32.000	34.267
17	5.697	6.408	7.564	8.672	10.085	24.769	27.587	30.191	33.409	35.718
18	6.265	7.015	8.231	9.390	10.865	25.989	28.869	31.526	34.805	37.156
19	6.844	7.633	8.907	10.117	11.651	27.204	30.144	32.852	36.191	38.582
20	7.434	8.260	9.591	10.851	12.443	28.412	31.410	34.170	37.566	39.997
21	8.034	8.897	10.283	11.591	13.240	29.615	32.671	35.479	38.932	41.401
22	8.643	9.542	10.982	12.338	14.042	30.813	33.924	36.781	40.289	42.796
23	9.260	10.196	11.689	13.091	14.848	32.007	35.172	38.076	41.638	44.181
24	9.886	10.856	12.401	13.848	15.659	33.196	36.415	39.364	42.980	45.559
25	10.520	11.524	13.120	14.611	16.473	34.382	37.652	40.646	44.314	46.928
26	11.160	12.198	13.844	15.379	17.292	35.563	38.885	41.923	45.642	48.290
27	11.808	12.879	14.573	16.151	18.114	36.741	40.113	43.194	46.963	49.645
28	12.461	13.565	15.308	16.928	18.939	37.916	41.337	44.461	48.278	50.993
29	13.121	14.257	16.047	17.708	19.768	39.087	42.557	45.722	49.588	52.336
30	13.787	14.954	16.791	18.493	20.599	40.256	43.773	46.979	50.892	53.672
40	20.707	22.164	24.433	26.509	29.051	51.805	55.758	59.342	63.691	66.766
50	27.991	29.707	32.357	34.764	37.689	63.167	67.505	71.420	76.154	79.490
60	35.534	37.485	40.482	43.188	46.459	74.397	79.082	83.298	88.379	91.952
70	43.275	45.442	48.758	51.739	55.329	85.527	90.531	95.023	100.425	104.215
80	51.172	53.540	57.153	60.391	64.278	96.578	101.879	106.629	112.329	116.321
90	59.196	61.754	65.647	69.126	73.291	107.565	113.145	118.136	124.116	128.299
100	67.328	70.065	74.222	77.929	82.358	118.498	124.342	129.561	135.807	140.169

Source: From Donald B. Owen, *Handbook of Statistical Tables.*

Degrees of Freedom

$n - 1$	Confidence Interval or Hypothesis Test for a standard deviation or variance
$k - 1$	Goodness-of-fit test with k different categories
$(r - 1)(c - 1)$	Contingency table test with r rows and c columns
$k - 1$	Kruskal-Wallis test with k different samples

Formulas and Tables by Mario F. Triola
Copyright 2018 Pearson Education, Inc.

Ch. 3: Descriptive Statistics

$$\bar{x} = \frac{\Sigma x}{n} \quad \text{Mean}$$

$$\bar{x} = \frac{\Sigma (f \cdot x)}{\Sigma f} \quad \text{Mean (frequency table)}$$

$$s = \sqrt{\frac{\Sigma (x - \bar{x})^2}{n - 1}} \quad \text{Standard deviation}$$

$$s = \sqrt{\frac{n(\Sigma x^2) - (\Sigma x)^2}{n(n-1)}} \quad \text{Standard deviation (shortcut)}$$

$$s = \sqrt{\frac{n[\Sigma (f \cdot x^2)] - [\Sigma (f \cdot x)]^2}{n(n-1)}} \quad \begin{array}{l}\text{Standard deviation}\\ \text{(frequency table)}\end{array}$$

$$\text{variance} = s^2$$

Ch. 4: Probability

$P(A \text{ or } B) = P(A) + P(B) \quad$ if A, B are mutually exclusive

$P(A \text{ or } B) = P(A) + P(B) - P(A \text{ and } B)$
  if A, B are not mutually exclusive

$P(A \text{ and } B) = P(A) \cdot P(B) \quad$ if A, B are independent

$P(A \text{ and } B) = P(A) \cdot P(B|A) \quad$ if A, B are dependent

$P(\bar{A}) = 1 - P(A) \quad$ Rule of complements

$${}_nP_r = \frac{n!}{(n-r)!} \quad \text{Permutations (no elements alike)}$$

$$\frac{n!}{n_1! \, n_2! \, \ldots \, n_k!} \quad \text{Permutations } (n_1 \text{ alike}, \ldots)$$

$${}_nC_r = \frac{n!}{(n-r)! \, r!} \quad \text{Combinations}$$

Ch. 5: Probability Distributions

$\mu = \Sigma [x \cdot P(x)] \quad$ Mean (prob. dist.)

$\sigma = \sqrt{\Sigma [x^2 \cdot P(x)] - \mu^2} \quad$ Standard deviation (prob. dist.)

$$P(x) = \frac{n!}{(n-x)! \, x!} \cdot p^x \cdot q^{n-x} \quad \text{Binomial probability}$$

$\mu = n \cdot p \qquad\qquad$ Mean (binomial)

$\sigma^2 = n \cdot p \cdot q \qquad\quad$ Variance (binomial)

$\sigma = \sqrt{n \cdot p \cdot q} \qquad$ Standard deviation (binomial)

$$P(x) = \frac{\mu^x \cdot e^{-\mu}}{x!} \quad \begin{array}{l}\text{Poisson distribution where}\\ e = 2.71828\end{array}$$

Ch. 6: Normal Distribution

$$z = \frac{x - \mu}{\sigma} \text{ or } \frac{x - \bar{x}}{s} \quad \text{Standard score}$$

$\mu_{\bar{x}} = \mu \quad$ Central limit theorem

$$\sigma_{\bar{x}} = \frac{\sigma}{\sqrt{n}} \quad \text{Central limit theorem (Standard error)}$$

Ch. 7: Confidence Intervals (one population)

$\hat{p} - E < p < \hat{p} + E \quad$ Proportion

$$\text{where } E = z_{\alpha/2}\sqrt{\frac{\hat{p}\hat{q}}{n}}$$

$\bar{x} - E < \mu < \bar{x} + E \quad$ Mean

$$\text{where } E = t_{\alpha/2}\frac{s}{\sqrt{n}} \quad (\sigma \text{ unknown})$$

$$\text{or } E = z_{\alpha/2}\frac{\sigma}{\sqrt{n}} \quad (\sigma \text{ known})$$

$$\frac{(n-1)s^2}{\chi_R^2} < \sigma^2 < \frac{(n-1)s^2}{\chi_L^2} \quad \text{Variance}$$

Ch. 7: Sample Size Determination

$$n = \frac{[z_{\alpha/2}]^2 0.25}{E^2} \quad \text{Proportion}$$

$$n = \frac{[z_{\alpha/2}]^2 \hat{p}\hat{q}}{E^2} \quad \text{Proportion } (\hat{p} \text{ and } \hat{q} \text{ are known})$$

$$n = \left[\frac{z_{\alpha/2}\sigma}{E}\right]^2 \quad \text{Mean}$$

Ch. 8: Test Statistics (one population)

$$z = \frac{\hat{p} - p}{\sqrt{\dfrac{pq}{n}}} \quad \text{Proportion—one population}$$

$$t = \frac{\bar{x} - \mu}{\dfrac{s}{\sqrt{n}}} \quad \text{Mean—one population } (\sigma \text{ unknown})$$

$$z = \frac{\bar{x} - \mu}{\dfrac{\sigma}{\sqrt{n}}} \quad \text{Mean—one population } (\sigma \text{ known})$$

$$\chi^2 = \frac{(n-1)s^2}{\sigma^2} \quad \begin{array}{l}\text{Standard deviation or variance—}\\ \text{one population}\end{array}$$

Procedure for Hypothesis Tests

1. Identify the Claim
Identify the claim to be tested and express it in symbolic form.

↓

2. Give Symbolic Form
Give the symbolic form that must be true when the original claim is false.

↓

3. Identify Null and Alternative Hypothesis
Consider the two symbolic expressions obtained so far:
- **Alternative hypothesis H_1** is the one *NOT* containing equality, so H_1 uses the symbol $>$ or $<$ or $\neq$.
- **Null hypothesis H_0** is the symbolic expression that the parameter equals the fixed value being considered.

↓

4. Select Significance Level
Select the **significance level** α based on the seriousness of a type I error. Make α small if the consequences of rejecting a true H_0 are severe.
- The values of 0.05 and 0.01 are very common.

↓

5. Identify the Test Statistic
Identify the test statistic that is relevant to the test and determine its sampling distribution (such as normal, t, chi-square).

↓

P-Value Method	Critical Value Method
6. Find Values Find the value of the **test statistic** and the **P-value** (see Figure 8-3). Draw a graph and show the test statistic and P-value.	**6. Find Values** Find the value of the **test statistic** and the **critical values**. Draw a graph showing the test statistic, critical value(s) and critical region.
↓	↓
7. Make a Decision • **Reject H_0** if P-value $\leq \alpha$. • **Fail to reject H_0** if P-value $> \alpha$.	**7. Make a Decision** • **Reject H_0** if the test statistic is in the critical region. • **Fail to reject H_0** if the test statistic is not in the critical region.

↓

8. Restate Decision in Nontechnical Terms
Restate this previous decision in simple nontechnical terms, and address the original claim.

Finding *P*-Values

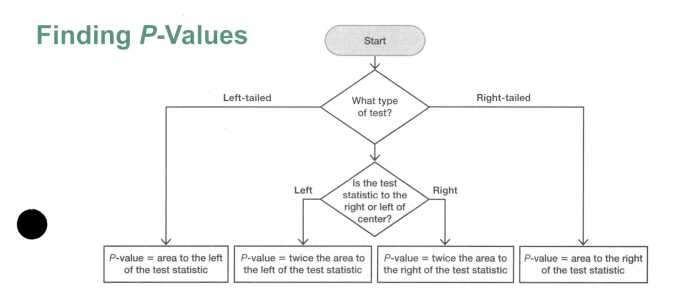

NEGATIVE z Scores

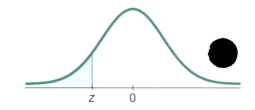

TABLE A-2 Standard Normal (z) Distribution: Cumulative Area from the LEFT

z	.00	.01	.02	.03	.04	.05	.06	.07	.08	.09
−3.50 and lower	.0001									
−3.4	.0003	.0003	.0003	.0003	.0003	.0003	.0003	.0003	.0003	.0002
−3.3	.0005	.0005	.0005	.0004	.0004	.0004	.0004	.0004	.0004	.0003
−3.2	.0007	.0007	.0006	.0006	.0006	.0006	.0006	.0005	.0005	.0005
−3.1	.0010	.0009	.0009	.0009	.0008	.0008	.0008	.0008	.0007	.0007
−3.0	.0013	.0013	.0013	.0012	.0012	.0011	.0011	.0011	.0010	.0010
−2.9	.0019	.0018	.0018	.0017	.0016	.0016	.0015	.0015	.0014	.0014
−2.8	.0026	.0025	.0024	.0023	.0023	.0022	.0021	.0021	.0020	.0019
−2.7	.0035	.0034	.0033	.0032	.0031	.0030	.0029	.0028	.0027	.0026
−2.6	.0047	.0045	.0044	.0043	.0041	.0040	.0039	.0038	.0037	.0036
−2.5	.0062	.0060	.0059	.0057	.0055	.0054	.0052	.0051 *	.0049	.0048
−2.4	.0082	.0080	.0078	.0075	.0073	.0071	.0069	.0068	.0066	.0064
−2.3	.0107	.0104	.0102	.0099	.0096	.0094	.0091	.0089	.0087	.0084
−2.2	.0139	.0136	.0132	.0129	.0125	.0122	.0119	.0116	.0113	.0110
−2.1	.0179	.0174	.0170	.0166	.0162	.0158	.0154	.0150	.0146	.0143
−2.0	.0228	.0222	.0217	.0212	.0207	.0202	.0197	.0192	.0188	.0183
−1.9	.0287	.0281	.0274	.0268	.0262	.0256	.0250	.0244	.0239	.0
−1.8	.0359	.0351	.0344	.0336	.0329	.0322	.0314	.0307	.0301	.02
−1.7	.0446	.0436	.0427	.0418	.0409	.0401	.0392	.0384	.0375	.0367
−1.6	.0548	.0537	.0526	.0516	.0505 *	.0495	.0485	.0475	.0465	.0455
−1.5	.0668	.0655	.0643	.0630	.0618	.0606	.0594	.0582	.0571	.0559
−1.4	.0808	.0793	.0778	.0764	.0749	.0735	.0721	.0708	.0694	.0681
−1.3	.0968	.0951	.0934	.0918	.0901	.0885	.0869	.0853	.0838	.0823
−1.2	.1151	.1131	.1112	.1093	.1075	.1056	.1038	.1020	.1003	.0985
−1.1	.1357	.1335	.1314	.1292	.1271	.1251	.1230	.1210	.1190	.1170
−1.0	.1587	.1562	.1539	.1515	.1492	.1469	.1446	.1423	.1401	.1379
−0.9	.1841	.1814	.1788	.1762	.1736	.1711	.1685	.1660	.1635	.1611
−0.8	.2119	.2090	.2061	.2033	.2005	.1977	.1949	.1922	.1894	.1867
−0.7	.2420	.2389	.2358	.2327	.2296	.2266	.2236	.2206	.2177	.2148
−0.6	.2743	.2709	.2676	.2643	.2611	.2578	.2546	.2514	.2483	.2451
−0.5	.3085	.3050	.3015	.2981	.2946	.2912	.2877	.2843	.2810	.2776
−0.4	.3446	.3409	.3372	.3336	.3300	.3264	.3228	.3192	.3156	.3121
−0.3	.3821	.3783	.3745	.3707	.3669	.3632	.3594	.3557	.3520	.3483
−0.2	.4207	.4168	.4129	.4090	.4052	.4013	.3974	.3936	.3897	.3859
−0.1	.4602	.4562	.4522	.4483	.4443	.4404	.4364	.4325	.4286	.4247
−0.0	.5000	.4960	.4920	.4880	.4840	.4801	.4761	.4721	.4681	.4641

NOTE: For values of z below −3.49, use 0.0001 for the area.
*Use these common values that result from interpolation:

z Score	Area
−1.645	0.0500
−2.575	0.0050

(*continued*)

Ch. 9: Confidence Intervals (two populations)

$$(\hat{p}_1 - \hat{p}_2) - E < (p_1 - p_2) < (\hat{p}_1 - \hat{p}_2) + E$$

where $E = z_{\alpha/2}\sqrt{\dfrac{\hat{p}_1 \hat{q}_1}{n_1} + \dfrac{\hat{p}_2 \hat{q}_2}{n_2}}$

$$(\bar{x}_1 - \bar{x}_2) - E < (\mu_1 - \mu_2) < (\bar{x}_1 - \bar{x}_2) + E \quad \text{(Indep.)}$$

where $E = t_{\alpha/2}\sqrt{\dfrac{s_1^2}{n_1} + \dfrac{s_2^2}{n_2}}$ (df = smaller of $n_1 - 1, n_2 - 1$)

(σ_1 and σ_2 unknown and not assumed equal)

$$E = t_{\alpha/2}\sqrt{\dfrac{s_p^2}{n_1} + \dfrac{s_p^2}{n_2}} \quad (\text{df} = n_1 + n_2 - 2)$$

$$s_p^2 = \dfrac{(n_1 - 1)s_1^2 + (n_2 - 1)s_2^2}{(n_1 - 1) + (n_2 - 1)}$$

(σ_1 and σ_2 unknown but assumed equal)

$$E = z_{\alpha/2}\sqrt{\dfrac{\sigma_1^2}{n_1} + \dfrac{\sigma_2^2}{n_2}}$$

(σ_1, σ_2 known)

$$\bar{d} - E < \mu_d < \bar{d} + E \quad \text{(Matched pairs)}$$

where $E = t_{\alpha/2}\dfrac{s_d}{\sqrt{n}}$ (df = $n - 1$)

Ch. 9: Test Statistics (two populations)

$$= \dfrac{(\hat{p}_1 - \hat{p}_2) - (p_1 - p_2)}{\sqrt{\dfrac{\bar{p}\,\bar{q}}{n_1} + \dfrac{\bar{p}\,\bar{q}}{n_2}}} \quad \begin{array}{l}\text{Two proportions}\\ \bar{p} = \dfrac{x_1 + x_2}{n_1 + n_2}\end{array}$$

$$t = \dfrac{(\bar{x}_1 - \bar{x}_2) - (\mu_1 - \mu_2)}{\sqrt{\dfrac{s_1^2}{n_1} + \dfrac{s_2^2}{n_2}}} \quad \begin{array}{l}\text{df = smaller of}\\ n_1 - 1, n_2 - 1\end{array}$$

Two means—independent; σ_1 and σ_2 unknown, and not assumed equal.

$$t = \dfrac{(\bar{x}_1 - \bar{x}_2) - (\mu_1 - \mu_2)}{\sqrt{\dfrac{s_p^2}{n_1} + \dfrac{s_p^2}{n_2}}} \quad \begin{array}{l}(\text{df} = n_1 + n_2 - 2)\\ s_p^2 = \dfrac{(n_1 - 1)s_1^2 + (n_2 - 1)s_2^2}{n_1 + n_2 - 2}\end{array}$$

Two means—independent; σ_1 and σ_2 unknown, but assumed equal.

$$z = \dfrac{(\bar{x}_1 - \bar{x}_2) - (\mu_1 - \mu_2)}{\sqrt{\dfrac{\sigma_1^2}{n_1} + \dfrac{\sigma_2^2}{n_2}}} \quad \begin{array}{l}\text{Two means—independent;}\\ \sigma_1, \sigma_2 \text{ known.}\end{array}$$

$$t = \dfrac{\bar{d} - \mu_d}{\dfrac{s_d}{\sqrt{n}}} \quad \text{Two means—matched pairs (df} = n - 1)$$

$$= \dfrac{s_1^2}{s_2^2} \quad \begin{array}{l}\text{Standard deviation or variance—}\\ \text{two populations (where } s_1^2 \geq s_2^2)\end{array}$$

Ch. 10: Linear Correlation/Regression

Correlation $r = \dfrac{n\Sigma xy - (\Sigma x)(\Sigma y)}{\sqrt{n(\Sigma x^2) - (\Sigma x)^2}\sqrt{n(\Sigma y^2) - (\Sigma y)^2}}$

or $r = \dfrac{\Sigma(z_x z_y)}{n - 1}$ where z_x = z score for x, z_y = z score for y

Slope: $b_1 = \dfrac{n\Sigma xy - (\Sigma x)(\Sigma y)}{n(\Sigma x^2) - (\Sigma x)^2}$ or $b_1 = r\dfrac{s_y}{s_x}$

y-Intercept:

$$b_0 = \bar{y} - b_1\bar{x} \text{ or } b_0 = \dfrac{(\Sigma y)(\Sigma x^2) - (\Sigma x)(\Sigma xy)}{n(\Sigma x^2) - (\Sigma x)^2}$$

$$\hat{y} = b_0 + b_1 x \quad \text{Estimated eq. of regression line}$$

$$r^2 = \dfrac{\text{explained variation}}{\text{total variation}}$$

$$s_e = \sqrt{\dfrac{\Sigma(y - \hat{y})^2}{n - 2}} \text{ or } \sqrt{\dfrac{\Sigma y^2 - b_0\Sigma y - b_1\Sigma xy}{n - 2}}$$

$$\hat{y} - E < y < \hat{y} + E \quad \text{Prediction interval}$$

where $E = t_{\alpha/2}s_e\sqrt{1 + \dfrac{1}{n} + \dfrac{n(x_0 - \bar{x})^2}{n(\Sigma x^2) - (\Sigma x)^2}}$

Ch. 11: Goodness-of-Fit and Contingency Tables

$$\chi^2 = \Sigma\dfrac{(O - E)^2}{E} \quad \text{Goodness-of-fit (df} = k - 1)$$

$$\chi^2 = \Sigma\dfrac{(O - E)^2}{E} \quad \text{Contingency table [df} = (r - 1)(c - 1)]$$

where $E = \dfrac{(\text{row total})(\text{column total})}{(\text{grand total})}$

$$\chi^2 = \dfrac{(|b - c| - 1)^2}{b + c} \quad \text{McNemar's test for matched pairs (df} = 1)$$

Ch. 12: One-Way Analysis of Variance

Procedure for testing $H_0: \mu_1 = \mu_2 = \mu_3 = \ldots$

1. Use software or calculator to obtain results.
2. Identify the P-value.
3. Form conclusion:

 If P-value $\leq \alpha$, reject the null hypothesis of equal means.
 If P-value $> \alpha$, fail to reject the null hypothesis of equal means.

Ch. 12: Two-Way Analysis of Variance

Procedure:

1. Use software or a calculator to obtain results.
2. Test H_0: There is no interaction between the row factor and column factor.
3. Stop if H_0 from Step 2 is rejected.

 If H_0 from Step 2 is not rejected (so there does not appear to be an interaction effect), proceed with these two tests:
 Test for effects from the row factor.
 Test for effects from the column factor.

Ch. 13: Nonparametric Tests

$$z = \dfrac{(x + 0.5) - (n/2)}{\dfrac{\sqrt{n}}{2}} \quad \text{Sign test for } n > 25$$

$$z = \dfrac{T - n(n + 1)/4}{\sqrt{\dfrac{n(n + 1)(2n + 1)}{24}}} \quad \begin{array}{l}\text{Wilcoxon signed ranks}\\ \text{(matched pairs and } n > 30)\end{array}$$

$$z = \dfrac{R - \mu_R}{\sigma_R} = \dfrac{R - \dfrac{n_1(n_1 + n_2 + 1)}{2}}{\sqrt{\dfrac{n_1 n_2(n_1 + n_2 + 1)}{12}}} \quad \begin{array}{l}\text{Wilcoxon rank-sum}\\ \text{(two independent}\\ \text{samples)}\end{array}$$

$$H = \dfrac{12}{N(N + 1)}\left(\dfrac{R_1^2}{n_1} + \dfrac{R_2^2}{n_2} + \cdots + \dfrac{R_k^2}{n_k}\right) - 3(N + 1)$$

Kruskal-Wallis (chi-square df = $k - 1$)

$$r_s = 1 - \dfrac{6\Sigma d^2}{n(n^2 - 1)} \quad \text{Rank correlation}$$

$$\left(\text{critical values for } n > 30: \dfrac{\pm z}{\sqrt{n - 1}}\right)$$

$$z = \dfrac{G - \mu_G}{\sigma_G} = \dfrac{G - \left(\dfrac{2n_1 n_2}{n_1 + n_2} + 1\right)}{\sqrt{\dfrac{(2n_1 n_2)(2n_1 n_2 - n_1 - n_2)}{(n_1 + n_2)^2(n_1 + n_2 - 1)}}} \quad \begin{array}{l}\text{Runs test}\\ \text{for } n > 20\end{array}$$

Ch. 14: Control Charts

R chart: Plot sample ranges

 UCL: $D_4\bar{R}$

 Centerline: $\bar{R}$

 LCL: $D_3\bar{R}$

$\bar{x}$ chart: Plot sample means

 UCL: $\bar{\bar{x}} + A_2\bar{R}$

 Centerline: $\bar{\bar{x}}$

 LCL: $\bar{\bar{x}} - A_2\bar{R}$

p chart: Plot sample proportions

 UCL: $\bar{p} + 3\sqrt{\dfrac{\bar{p}\,\bar{q}}{n}}$

 Centerline: $\bar{p}$

 LCL: $\bar{p} - 3\sqrt{\dfrac{\bar{p}\,\bar{q}}{n}}$

TABLE A-6 Critical Values of the Pearson Correlation Coefficient r

n	$\alpha = .05$	$\alpha = .01$
4	.950	.990
5	.878	.959
6	.811	.917
7	.754	.875
8	.707	.834
9	.666	.798
10	.632	.765
11	.602	.735
12	.576	.708
13	.553	.684
14	.532	.661
15	.514	.641
16	.497	.623
17	.482	.606
18	.468	.590
19	.456	.575
20	.444	.561
25	.396	.505
30	.361	.463
35	.335	.430
40	.312	.402
45	.294	.378
50	.279	.361
60	.254	.330
70	.236	.305
80	.220	.286
90	.207	.269
100	.196	.256

NOTE: To test $H_0: \rho = 0$ (no correlation) against $H_1: \rho \neq 0$ (correlation), reject H_0 if the absolute value of r is greater than or equal to the critical value in the table.

Control Chart Constants

Subgroup Size n	D_3	D_4	A_2
2	0.000	3.267	1.880
3	0.000	2.574	1.023
4	0.000	2.282	0.729
5	0.000	2.114	0.577
6	0.000	2.004	0.483
7	0.076	1.924	0.419

Inferences about μ: choosing between t and normal distributions		
t distribution:		σ not known and normally distributed population
	or	σ not known and $n > 30$
Normal distribution:		σ known and normally distributed population
	or	σ known and $n > 30$
Nonparametric method or bootstrapping: Population not normally distributed and $n \leq 30$		

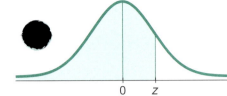

POSITIVE z Scores

TABLE A-2 (continued) Cumulative Area from the LEFT

z	.00	.01	.02	.03	.04	.05	.06	.07	.08	.09
0.0	.5000	.5040	.5080	.5120	.5160	.5199	.5239	.5279	.5319	.5359
0.1	.5398	.5438	.5478	.5517	.5557	.5596	.5636	.5675	.5714	.5753
0.2	.5793	.5832	.5871	.5910	.5948	.5987	.6026	.6064	.6103	.6141
0.3	.6179	.6217	.6255	.6293	.6331	.6368	.6406	.6443	.6480	.6517
0.4	.6554	.6591	.6628	.6664	.6700	.6736	.6772	.6808	.6844	.6879
0.5	.6915	.6950	.6985	.7019	.7054	.7088	.7123	.7157	.7190	.7224
0.6	.7257	.7291	.7324	.7357	.7389	.7422	.7454	.7486	.7517	.7549
0.7	.7580	.7611	.7642	.7673	.7704	.7734	.7764	.7794	.7823	.7852
0.8	.7881	.7910	.7939	.7967	.7995	.8023	.8051	.8078	.8106	.8133
0.9	.8159	.8186	.8212	.8238	.8264	.8289	.8315	.8340	.8365	.8389
1.0	.8413	.8438	.8461	.8485	.8508	.8531	.8554	.8577	.8599	.8621
1.1	.8643	.8665	.8686	.8708	.8729	.8749	.8770	.8790	.8810	.8830
1.2	.8849	.8869	.8888	.8907	.8925	.8944	.8962	.8980	.8997	.9015
1.3	.9032	.9049	.9066	.9082	.9099	.9115	.9131	.9147	.9162	.9177
1.4	.9192	.9207	.9222	.9236	.9251	.9265	.9279	.9292	.9306	.9319
1.5	.9332	.9345	.9357	.9370	.9382	.9394	.9406	.9418	.9429	.9441
	.9452	.9463	.9474	.9484	.9495 *	.9505	.9515	.9525	.9535	.9545
	.9554	.9564	.9573	.9582	.9591	.9599	.9608	.9616	.9625	.9633
1.8	.9641	.9649	.9656	.9664	.9671	.9678	.9686	.9693	.9699	.9706
1.9	.9713	.9719	.9726	.9732	.9738	.9744	.9750	.9756	.9761	.9767
2.0	.9772	.9778	.9783	.9788	.9793	.9798	.9803	.9808	.9812	.9817
2.1	.9821	.9826	.9830	.9834	.9838	.9842	.9846	.9850	.9854	.9857
2.2	.9861	.9864	.9868	.9871	.9875	.9878	.9881	.9884	.9887	.9890
2.3	.9893	.9896	.9898	.9901	.9904	.9906	.9909	.9911	.9913	.9916
2.4	.9918	.9920	.9922	.9925	.9927	.9929	.9931	.9932	.9934	.9936
2.5	.9938	.9940	.9941	.9943	.9945	.9946	.9948	.9949 *	.9951	.9952
2.6	.9953	.9955	.9956	.9957	.9959	.9960	.9961	.9962	.9963	.9964
2.7	.9965	.9966	.9967	.9968	.9969	.9970	.9971	.9972	.9973	.9974
2.8	.9974	.9975	.9976	.9977	.9977	.9978	.9979	.9979	.9980	.9981
2.9	.9981	.9982	.9982	.9983	.9984	.9984	.9985	.9985	.9986	.9986
3.0	.9987	.9987	.9987	.9988	.9988	.9989	.9989	.9989	.9990	.9990
3.1	.9990	.9991	.9991	.9991	.9992	.9992	.9992	.9992	.9993	.9993
3.2	.9993	.9993	.9994	.9994	.9994	.9994	.9994	.9995	.9995	.9995
3.3	.9995	.9995	.9995	.9996	.9996	.9996	.9996	.9996	.9996	.9997
3.4	.9997	.9997	.9997	.9997	.9997	.9997	.9997	.9997	.9997	.9998
3.50 and up	.9999									

NOTE: For values of z above 3.49, use 0.9999 for the area.
*Use these common values that result from interpolation:

z score	Area
	0.9500
	0.9950

Common Critical Values

Confidence Level	Critical Value
0.90	1.645
0.95	1.96
0.99	2.575

TABLE A-3 t Distribution: Critical t Values

Degrees of Freedom	Area in One Tail				
	0.005	0.01	0.025	0.05	0.10
	Area in Two Tails				
	0.01	0.02	0.05	0.10	0.20
1	63.657	31.821	12.706	6.314	3.078
2	9.925	6.965	4.303	2.920	1.886
3	5.841	4.541	3.182	2.353	1.638
4	4.604	3.747	2.776	2.132	1.533
5	4.032	3.365	2.571	2.015	1.476
6	3.707	3.143	2.447	1.943	1.440
7	3.499	2.998	2.365	1.895	1.415
8	3.355	2.896	2.306	1.860	1.397
9	3.250	2.821	2.262	1.833	1.383
10	3.169	2.764	2.228	1.812	1.372
11	3.106	2.718	2.201	1.796	1.363
12	3.055	2.681	2.179	1.782	1.356
13	3.012	2.650	2.160	1.771	1.350
14	2.977	2.624	2.145	1.761	1.345
15	2.947	2.602	2.131	1.753	1.341
16	2.921	2.583	2.120	1.746	1.337
17	2.898	2.567	2.110	1.740	1.333
18	2.878	2.552	2.101	1.734	1.330
19	2.861	2.539	2.093	1.729	1.328
20	2.845	2.528	2.086	1.725	1.325
21	2.831	2.518	2.080	1.721	1.323
22	2.819	2.508	2.074	1.717	1.321
23	2.807	2.500	2.069	1.714	1.319
24	2.797	2.492	2.064	1.711	1.318
25	2.787	2.485	2.060	1.708	1.316
26	2.779	2.479	2.056	1.706	1.315
27	2.771	2.473	2.052	1.703	1.314
28	2.763	2.467	2.048	1.701	1.313
29	2.756	2.462	2.045	1.699	1.311
30	2.750	2.457	2.042	1.697	1.310
31	2.744	2.453	2.040	1.696	1.309
32	2.738	2.449	2.037	1.694	1.309
33	2.733	2.445	2.035	1.692	1.308
34	2.728	2.441	2.032	1.691	1.307
35	2.724	2.438	2.030	1.690	1.306
36	2.719	2.434	2.028	1.688	1.306
37	2.715	2.431	2.026	1.687	1.305
38	2.712	2.429	2.024	1.686	1.304
39	2.708	2.426	2.023	1.685	1.304
40	2.704	2.423	2.021	1.684	1.303
45	2.690	2.412	2.014	1.679	1.301
50	2.678	2.403	2.009	1.676	1.299
60	2.660	2.390	2.000	1.671	1.296
70	2.648	2.381	1.994	1.667	1.294
80	2.639	2.374	1.990	1.664	1.292
90	2.632	2.368	1.987	1.662	1.291
100	2.626	2.364	1.984	1.660	1.290
200	2.601	2.345	1.972	1.653	1.286
300	2.592	2.339	1.968	1.650	1.284
400	2.588	2.336	1.966	1.649	1.284
500	2.586	2.334	1.965	1.648	1.283
1000	2.581	2.330	1.962	1.646	1.282
2000	2.578	2.328	1.961	1.646	1.282
Large	2.576	2.326	1.960	1.645	1.282

Constructing Frequency Distributions. *In Exercises 11–18, use the indicated data to construct the frequency distribution. (The data for Exercises 13–16 can be downloaded at TriolaStats.com.)*

11. Old Faithful Listed below are sorted duration times (seconds) of eruptions of the Old Faithful geyser in Yellowstone National Park. Use these times to construct a frequency distribution. Use a class width of 25 seconds and begin with a lower class limit of 125 seconds.

125 203 205 221 225 229 233 233 235 236 236 237 238 238 239 240 240
240 240 241 241 242 242 242 243 243 244 245 245 245 245 246 246 248
248 248 249 249 250 251 252 253 253 255 255 256 257 258 262 264

12. Tornadoes Listed below are the F-scale intensities of recent tornadoes in the United States. Construct a frequency distribution. Do the intensities appear to have a normal distribution?

0 4 0 0 1 1 1 0 0 0 1 2 0 1 1 0 1 0 1 1 1 1 0
0 1 0 0 1 0 0 1 1 1 3 0 0 0 2 0 3 0 0 0 0 0

13. Burger King Lunch Service Times Refer to Data Set 25 "Fast Food" and use the drive-through service times for Burger King lunches. Begin with a lower class limit of 70 seconds and use a class width of 40 seconds.

14. Burger King Dinner Service Times Refer to Data Set 25 "Fast Food" and use the drive-through service times for Burger King dinners. Begin with a lower class limit of 30 seconds and use a class width of 40 seconds.

15. Wendy's Lunch Service Times Refer to Data Set 25 "Fast Food" and use the drive-through service times for Wendy's lunches. Begin with a lower class limit of 70 seconds and use a class width of 80 seconds. Does the distribution appear to be a normal distribution?

16. Wendy's Dinner Service Times Refer to Data Set 25 "Fast Food" and use the drive-through service times for Wendy's dinners. Begin with a lower class limit of 30 seconds and use a class width of 40 seconds. Using a loose interpretation of a normal distribution, does this distribution appear to be a normal distribution?

17. Analysis of Last Digits Heights of statistics students were obtained by the author as part of an experiment conducted for class. The last digits of those heights are listed below. Construct a frequency distribution with 10 classes. Based on the distribution, do the heights appear to be reported or actually measured? What do you know about the accuracy of the results?

0 0 0 0 0 0 0 0 0 1 1 2 3 3 3 4 5 5 5
5 5 5 5 5 5 5 5 5 5 5 5 6 6 8 8 8 9

18. Analysis of Last Digits Weights of respondents were recorded as part of the California Health Interview Survey. The last digits of weights from 50 randomly selected respondents are listed below. Construct a frequency distribution with 10 classes. Based on the distribution, do the weights appear to be reported or actually measured? What do you know about the accuracy of the results?

5 0 1 0 2 0 5 0 5 0 3 8 5 0 5 0 5 6 0 0 0 0 0 0 8
5 5 0 4 5 0 0 4 0 0 0 0 8 0 9 5 3 0 5 0 0 0 5 8

Relative Frequencies for Comparisons. *In Exercises 19 and 20, construct the relative frequency distributions and answer the given questions.*

19. Oscar Winners Construct one table (similar to Table 2-9 on page 47) that includes relative frequencies based on the frequency distributions from Exercises 5 and 6, and then compare the ages of Oscar-winning actresses and actors. Are there notable differences?

20. Blood Platelet Counts Construct one table (similar to Table 2-9 on page 47) that includes relative frequencies based on the frequency distributions from Exercises 7 and 8, and then compare them. Are there notable differences?

Cumulative Frequency Distributions.   *In Exercises 21 and 22, construct the cumulative frequency distribution that corresponds to the frequency distribution in the exercise indicated.*

21. Exercise 5 (Age of Best Actress When Oscar Was Won)

22. Exercise 6 (Age of Best Actor When Oscar Was Won)

Categorical Data.   *In Exercises 23 and 24, use the given categorical data to construct the relative frequency distribution.*

23. Clinical Trial When XELJANZ (tofacitinib) was administered as part of a clinical trial for this rheumatoid arthritis treatment, 1336 subjects were given 5 mg doses of the drug, and here are the numbers of adverse reactions: 57 had headaches, 21 had hypertension, 60 had upper respiratory tract infections, 51 had nasopharyngitis, and 53 had diarrhea. Does any one of these adverse reactions appear to be much more common than the others? (*Hint:* Find the relative frequencies using only the adverse reactions, not the total number of treated subjects.)

24. Births Natural births randomly selected from four hospitals in New York State occurred on the days of the week (in the order of Monday through Sunday) with these frequencies: 52, 66, 72, 57, 57, 43, 53. Does it appear that such births occur on the days of the week with equal frequency?

Large Data Sets.   *Exercises 25–28 involve large sets of data, so technology should be used. Complete lists of the data are not included in Appendix B, but they can be downloaded from the website TriolaStats.com. Use the indicated data and construct the frequency distribution.*

25. Systolic Blood Pressure Use the systolic blood pressures of the 300 subjects included in Data Set 1 "Body Data." Use a class width of 20 mm Hg and begin with a lower class limit of 80 mm Hg. Does the frequency distribution appear to be a normal distribution?

26. Diastolic Blood Pressure Use the diastolic blood pressures of the 300 subjects included in Data Set 1 "Body Data." Use a class width of 15 mm Hg and begin with a lower class limit of 40 mm Hg. Does the frequency distribution appear to be a normal distribution?

27. Earthquake Magnitudes Use the magnitudes of the 600 earthquakes included in Data Set 21 "Earthquakes." Use a class width of 0.5 and begin with a lower class limit of 1.00. Does the frequency distribution appear to be a normal distribution?

28. Earthquake Depths Use the depths (km) of the 600 earthquakes included in Data Set 21 "Earthquakes." Use a class width of 10.0 km and begin with a lower class limit of 0.0 km. Does the frequency distribution appear to be a normal distribution?

2-1   Beyond the Basics

29. Interpreting Effects of Outliers Refer to Data Set 30 "Aluminum Cans" in Appendix B for the axial loads of aluminum cans that are 0.0111 in. thick. An axial load is the force at which the top of a can collapses. The load of 504 lb is an *outlier* because it is very far away from all of the other values. Construct a frequency distribution that includes the value of 504 lb, and then construct another frequency distribution with the value of 504 lb excluded. In both cases, start the first class at 200 lb and use a class width of 20 lb. State a generalization about the effect of an outlier on a frequency distribution.

2-2 Histograms

PART 1 Basic Concepts of Histograms

Key Concept While a frequency distribution is a useful tool for summarizing data and investigating the distribution of data, an even better tool is a *histogram*, which is a graph that is easier to interpret than a table of numbers.

DEFINITION

A **histogram** is a graph consisting of bars of equal width drawn adjacent to each other (unless there are gaps in the data). The horizontal scale represents classes of quantitative data values, and the vertical scale represents frequencies. The heights of the bars correspond to frequency values.

Important Uses of a Histogram

- Visually displays the shape of the *distribution* of the data
- Shows the location of the *center* of the data
- Shows the *spread* of the data
- Identifies *outliers*

A histogram is basically a graph of a frequency distribution. For example, Figure 2-2 shows the Minitab-generated histogram corresponding to the frequency distribution given in Table 2-2 on page 42.

Class frequencies should be used for the vertical scale and that scale should be labeled as in Figure 2-2. There is no universal agreement on the procedure for selecting which values are used for the bar locations along the horizontal scale, but it is common to use class boundaries (as shown in Figure 2-2) or class midpoints or class limits or something else. It is often easier for us mere mortals to use class midpoints for the horizontal scale. Histograms can usually be generated using technology.

Relative Frequency Histogram

A **relative frequency histogram** has the same shape and horizontal scale as a histogram, but the vertical scale uses relative frequencies (as percentages or proportions) instead of actual frequencies. Figure 2-3 is the relative frequency histogram corresponding to Figure 2-2.

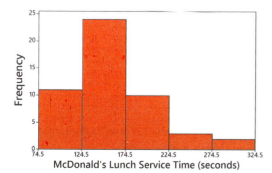

FIGURE 2-2 Histogram of McDonald's Drive-Through Lunch Service Times (seconds)

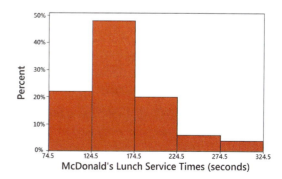

FIGURE 2-3 Relative Frequency Histogram of McDonald's Drive-Through Lunch Service Times (seconds)

Critical Thinking: Interpreting Histograms

Even though creating histograms is more fun than human beings should be allowed to have, the ultimate objective is to *understand* characteristics of the data. Explore the data by analyzing the histogram to see what can be learned about "CVDOT": the center of the data, the variation (which will be discussed at length in Section 3-2), the shape of the distribution, whether there are any outliers (values far away from the other values), and time (whether there is any change in the characteristics of the data over time). Examining Figure 2-2, we see that the histogram is centered around 160 sec or 170 sec, the values vary from around 75 sec to 325 sec, and the distribution is *very roughly* bell-shaped. There aren't any outliers, and any changes in time are irrelevant for these data.

Common Distribution Shapes

The histograms shown in Figure 2-4 depict four common distribution shapes.

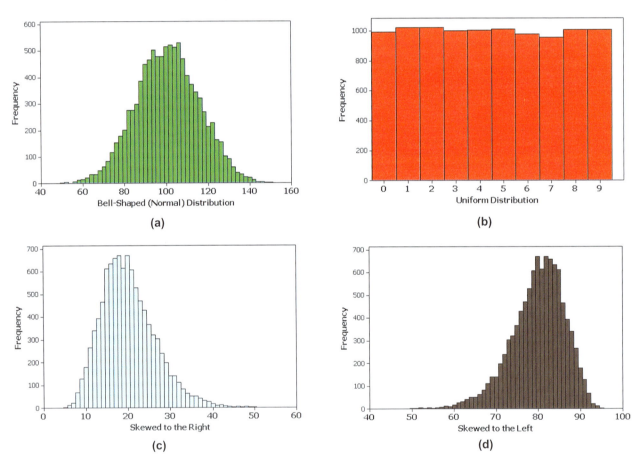

FIGURE 2-4   Common Distributions

Normal Distribution

When graphed as a histogram, a normal distribution has a "bell" shape similar to the one superimposed in Figure 2-5. Many statistical methods require that sample data come from a population having a distribution that is approximately a normal distribution, and we can often use a histogram to judge whether this requirement is satisfied. There are more advanced and less subjective methods for determining whether the distribution is a normal distribution. Normal quantile plots are very helpful for assessing normality: see Part 2 of this section.

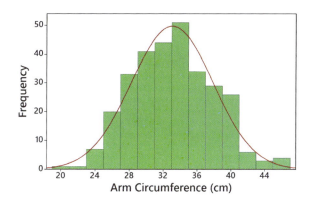

FIGURE 2-5   Bell-Shaped Distribution of Arm Circumferences
Because this histogram is roughly bell-shaped, we say that the data have a *normal distribution*. (A more rigorous definition will be given in Chapter 6.)

Uniform Distribution

The different possible values occur with approximately the same frequency, so the heights of the bars in the histogram are approximately uniform, as in Figure 2-4(b). Figure 2-4(b) depicts outcomes of digits from state lotteries.

Skewness

A distribution of data is **skewed** if it is not symmetric and extends more to one side than to the other. Data **skewed to the right** (also called *positively skewed*) have a longer right tail, as in Figure 2-4(c). Annual incomes of adult Americans are positively skewed. Data **skewed to the left** (also called *negatively skewed*) have a longer left tail, as in Figure 2-4(d). Life span data in humans are skewed to the left. (Here's a mnemonic for remembering skewness: A distribution skewed to the right resembles the toes on your right foot, and one skewed to the left resembles the toes on your left foot.) Distributions skewed to the right are more common than those skewed to the left because it's often easier to get exceptionally large values than values that are exceptionally small. With annual incomes, for example, it's impossible to get values below zero, but there are a few people who earn millions or billions of dollars in a year. Annual incomes therefore tend to be skewed to the right.

PART 2 **Assessing Normality with Normal Quantile Plots**

Some really important methods presented in later chapters have a requirement that sample data must be from a population having a normal distribution. Histograms can be helpful in determining whether the normality requirement is satisfied, but they are not very helpful with small data sets. Section 6-5 discusses methods for *assessing normality*—that is, determining whether the sample data are from a normally distributed population. Section 6-5 includes a procedure for constructing *normal quantile plots*, which are easy to generate using technology such as Statdisk, Minitab, XLSTAT, StatCrunch, or a TI-83/84 Plus calculator. Interpretation of a normal quantile plot is based on the following criteria:

Criteria for Assessing Normality with a Normal Quantile Plot

Normal Distribution: The population distribution is normal if the pattern of the points in the normal quantile plot is reasonably close to a straight line, and the points do not show some systematic pattern that is not a straight-line pattern.

Remembering Skewness:
Skewed Left: Resembles toes on left foot
Skewed Right: Resembles toes on right foot

Not a Normal Distribution: The population distribution is *not* normal if the normal quantile plot has either or both of these two conditions:

- The points do not lie reasonably close to a straight-line pattern.
- The points show some *systematic pattern* that is not a straight-line pattern.

The following are examples of normal quantile plots. Procedures for creating such plots are described in Section 6-5.

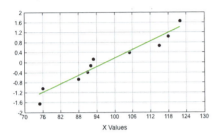 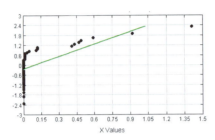 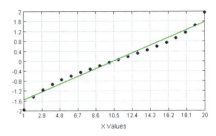

Normal Distribution: The points are reasonably close to a straight-line pattern, and there is no other systematic pattern that is not a straight-line pattern.

Not a Normal Distribution: The points do not lie reasonably close to a straight line.

Not a Normal Distribution: The points show a systematic pattern that is not a straight-line pattern.

 ## Histograms

Access tech supplements, videos, and data sets at **www.TriolaStats.com**

Statdisk

1. Click **Data** in top menu.
2. Select **Histogram** from the dropdown menu.
3. Select the desired data column.
4. Click **Plot.**
5. Check **Bar Labels** under *Plot Options* to see the frequency for each class.
6. Check **User Defined** under *Plot Options* to use your own class width and starting point.

Tip: This procedure is also an easy way to identify frequencies in a frequency distribution.

Minitab

1. Click **Graph** in top menu.
2. Select **Histogram** from the dropdown menu.
3. Select **Simple** histogram and click **OK.**
4. Click on the desired data column, then click **Select** and click **OK.**
5. Change default class width and starting point as needed by right-clicking the horizontal axis and selecting **Edit X Scale.**
   - Select the **Scale** tab to enter the location of the tick marks.
   - Select the **Binning** tab to enter the class midpoints.

StatCrunch

1. Click **Graph** in the top menu.
2. Select **Histogram** from the dropdown menu.
3. Select the desired data column.
4. To customize the histogram enter desired starting point and class width under **Bins.**
5. Click **Compute!.**

TI-83/84 Plus Calculator

1. Open the **STAT PLOTS** menu by pressing **2ND**, **Y=** .
2. Press **ENTER** to access the *Plot 1* settings screen as shown:
   a. Select **ON** and press **ENTER** .
   b. Select the bar chart option, then press **ENTER** .
   c. Enter the name of list containing data.
3. Press **ZOOM**, then **9** (ZoomStat) to generate the default histogram.
4. Press **TRACE** and use **◁ ▷** to view the class boundaries and frequencies for each class.
5. Press **WINDOW** to customize class width and boundaries. Press **GRAPH** to view histogram.

TECH CENTER *continued*

Histograms

Access tech supplements, videos, and data sets at **www.TriolaStats.com**

Excel

It is extremely difficult to generate histograms in Excel; the XLSTAT add-in should be used:

1. Select the **XLSTAT** tab in the Ribbon.
2. Click the **Visualizing Data** button.
3. Select **Histograms** from the dropdown menu.

4. Enter the range of cells containing the desired data. Click **Sample labels** if the first cell contains a data name.
5. Click **OK** to generate a default histogram.

Tip: To customize, enter the desired class boundaries in a column, select the **Options** tab, click **User Defined**, and enter the range of cells containing the boundaries in the box.

2-2   Basic Skills and Concepts

Statistical Literacy and Critical Thinking

1. Heights Heights of adult males are normally distributed. If a large sample of heights of adult males is randomly selected and the heights are illustrated in a histogram, what is the shape of that histogram?

2. More Heights The population of heights of adult males is normally distributed. If we obtain a voluntary response sample of 5000 of those heights, will a histogram of the sample heights be bell-shaped?

3. Blood Platelet Counts Listed below are blood platelet counts (1000 cells/μL) randomly selected from adults in the United States. Why does it *not* make sense to construct a histogram for this data set?

191   286   263   193   193   215   162   646   250   386

4. Blood Platelet Counts If we collect a sample of blood platelet counts much larger than the sample included with Exercise 3, and if our sample includes a single outlier, how will that outlier appear in a histogram?

Interpreting a Histogram.   *In Exercises 5–8, answer the questions by referring to the following Minitab-generated histogram, which depicts the weights (grams) of all quarters listed in Data Set 29 "Coin Weights" in Appendix B. (Grams are actually units of mass and the values shown on the horizontal scale are rounded.)*

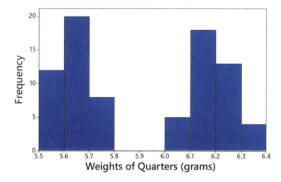

5. Sample Size What is the approximate number of quarters depicted in the three bars farthest to the left?

6. Class Width and Class Limits Give the approximate values of the class width, and the lower and upper class limits of the class depicted in the bar farthest to the left.

7. Relative Frequency Histogram How would the shape of the histogram change if the vertical scale uses relative frequencies expressed in percentages instead of the actual frequency counts as shown here?

8. Gap What is a reasonable explanation for the gap between the quarters with weights between 5.5 grams and 5.8 grams and the group of quarters with weights between 6.0 grams and 6.4 grams? (*Hint:* Refer to the columns of quarters in Data Set 29 "Coin Weights" in Appendix B.)

Constructing Histograms. *In Exercises 9–16, construct the histograms and answer the given questions.*

9. Old Faithful Use the frequency distribution from Exercise 11 in Section 2-1 on page 49 to construct a histogram. Does it appear to be the graph of data from a population with a normal distribution?

10. Tornadoes Use the frequency distribution from Exercise 12 in Section 2-1 on page 49 to construct a histogram. Does the histogram appear to be skewed? If so, identify the type of skewness.

11. Burger King Lunch Service Times Use the frequency distribution from Exercise 13 in Section 2-1 on page 49 to construct a histogram. Does the histogram appear to be skewed? If so, identify the type of skewness.

12. Burger King Dinner Service Times Use the frequency distribution from Exercise 14 in Section 2-1 on page 49 to construct a histogram. Using a strict interpretation of the criteria for being a normal distribution, does the histogram appear to depict data from a population with a normal distribution?

13. Wendy's Lunch Service Times Use the frequency distribution from Exercise 15 in Section 2-1 on page 49 to construct a histogram. Does the histogram appear to be skewed? If so, identify the type of skewness.

14. Wendy's Dinner Service Times Use the frequency distribution from Exercise 16 in Section 2-1 on page 49 to construct a histogram. In using a strict interpretation of the criteria for being a normal distribution, does the histogram appear to depict data from a population with a normal distribution?

15. Analysis of Last Digits Use the frequency distribution from Exercise 17 in Section 2-1 on page 49 to construct a histogram. What can be concluded from the distribution of the digits? Specifically, do the heights appear to be reported or actually measured?

16. Analysis of Last Digits Use the frequency distribution from Exercise 18 in Section 2-1 on page 49 to construct a histogram. What can be concluded from the distribution of the digits? Specifically, do the heights appear to be reported or actually measured?

2-2 Beyond the Basics

17. Back-to-Back Relative Frequency Histograms When using histograms to compare two data sets, it is sometimes difficult to make comparisons by looking back and forth between the two histograms. A *back-to-back relative frequency histogram* has a format that makes the comparison much easier. Instead of frequencies, we should use relative frequencies (percentages or proportions) so that the comparisons are not difficult when there are different sample sizes. Use the relative frequency distributions of the ages of Oscar-winning actresses and actors from Exercise 19 in Section 2-1 on page 49, and complete the back-to-back relative frequency histograms shown below. Then use the result to compare the two data sets.

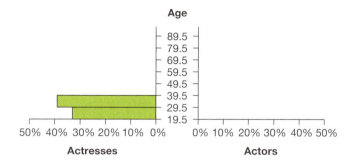

Age

Actresses   Actors

18. Interpreting Normal Quantile Plots Which of the following normal quantile plots appear to represent data from a population having a normal distribution? Explain.

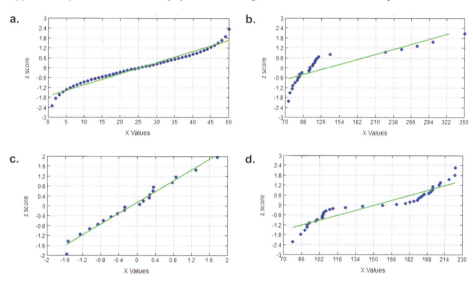

a.

b.

c.

d.

2-3   Graphs That Enlighten and Graphs That Deceive

Key Concept Section 2-2 introduced the histogram, and this section introduces other common graphs that foster understanding of data. We also discuss some graphs that are deceptive because they create impressions about data that are somehow misleading or wrong.

The era of charming and primitive hand-drawn graphs has passed, and technology now provides us with powerful tools for generating a wide variety of graphs. Here we go.

Graphs That Enlighten

Dotplots

A **dotplot** consists of a graph of *quantitative* data in which each data value is plotted as a point (or dot) above a horizontal scale of values. Dots representing equal values are stacked.

Features of a Dotplot

- Displays the shape of the distribution of data.
- It is usually possible to recreate the original list of data values.

The Power of a Graph

With annual sales around $13 billion and with roughly 50 million people using it, Pfizer's prescription drug Lipitor (generic name, atorvastatin) has become the most profitable and most widely used prescription drug ever marketed. In the early stages of its development, Lipitor was compared to other drugs (Zocor [simvastatin], Mevacor [lovastatin], Lescol [fluvastatin], and Pravachol [pravastatin]) in a process that involved controlled trials. The summary report included a graph showing a Lipitor curve that had a steeper rise than the curves for the other drugs, visually showing that Lipitor was more effective in reducing cholesterol than the other drugs. Pat Kelly, who was then a senior marketing executive for Pfizer, said, "I will never forget seeing that chart . . . . It was like 'Aha!' Now I know what this is about. We can communicate this!" The Food and Drug Administration approved Lipitor and allowed Pfizer to include the graph with each prescription. Pfizer sales personnel also distributed the graph to physicians.

EXAMPLE 1 Dotplot of Pulse Rates of Males

Figure 2-6 shows a dotplot of the pulse rates (beats per minute) of males from Data Set 1 "Body Data" in Appendix B. The two stacked dots above the position at 50 indicate that two of the pulse rates are 50. (In this dotplot, the horizontal scale allows even numbers only, but the original pulse rates are all even numbers.)

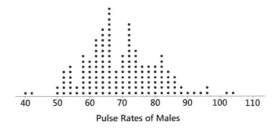

FIGURE 2-6 **Dotplot of Pulse Rates of Males**

YOUR TURN Do Exercise 5 "Pulse Rates."

Stemplots

A **stemplot** (or **stem-and-leaf plot**) represents *quantitative* data by separating each value into two parts: the stem (such as the leftmost digit) and the leaf (such as the rightmost digit). Better stemplots are often obtained by first rounding the original data values. Also, stemplots can be *expanded* to include more rows and can be *condensed* to include fewer rows, as in Exercise 21 "Expanded Stemplots".

Features of a Stemplot

- Shows the shape of the distribution of the data.
- Retains the original data values.
- The sample data are sorted (arranged in order).

EXAMPLE 2 Stemplot of Male Pulse Rates

The following stemplot displays the pulse rates of the males in Data Set 1 "Body Data" in Appendix B. The lowest pulse rate of 40 is separated into the stem of 4 and the leaf of 0. The stems and leaves are arranged in increasing order, not the order in which they occur in the original list. If you turn the stemplot on its side, you can see distribution of the IQ scores in the same way you would see it in a histogram or dotplot.

```
 4 | 02    ←——————— Pulse rates are 40 and 42
 5 | 0022222444444466888888888
 6 | 00000002222222222244444444444444446666666666666666688888
 7 | 000000002222222222222222444444444466666668888888
 8 | 00000022222222244444666688
 9 | 02466  ←——————— Pulse rates are 90, 92, 94, 96, 96
10 | 24
```

YOUR TURN Do Exercise 7 "Pulse Rates."

Time-Series Graph

A **time-series graph** is a graph of *time-series data,* which are quantitative data that have been collected at different points in time, such as monthly or yearly.

Feature of a Time-Series Graph

- Reveals information about trends over time

EXAMPLE 3 Time-Series Graph of Fatalities of Law Enforcement Officers

The time-series graph shown in Figure 2-7 depicts the yearly number of fatalities of law enforcement officers in the United States. See that a spike occurred in 2001, the year of the September 11, 2001, terrorist attacks. Except for the data from 2001, there appears to be a slight downward trend.

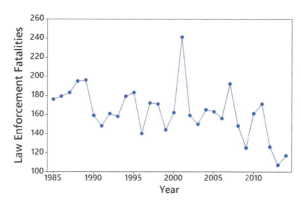

FIGURE 2-7   Time-Series Graph of Law Enforcement Fatalities

YOUR TURN Do Exercise 9 "Gender Pay Gap."

Florence Nightingale

Florence Nightingale (1820–1910) is known to many as the founder of the nursing profession, but she also saved thousands of lives by using statistics. When she encountered an unsanitary and undersupplied hospital, she improved those conditions and then used statistics to convince others of the need for more widespread medical reform. She developed original graphs to illustrate that during the Crimean War, more soldiers died as a result of unsanitary conditions than were killed in combat. Florence Nightingale pioneered the use of social statistics as well as graphics techniques.

Bar Graphs

A **bar graph** uses bars of equal width to show frequencies of categories of *categorical* (or qualitative) data. The bars may or may not be separated by small gaps.

Feature of a Bar Graph

- Shows the relative distribution of categorical data so that it is easier to compare the different categories

Pareto Charts

A **Pareto chart** is a bar graph for categorical data, with the added stipulation that the *bars are arranged in descending order* according to frequencies, so the bars decrease in height from left to right.

Features of a Pareto Chart

- Shows the relative distribution of categorical data so that it is easier to compare the different categories
- Draws attention to the more important categories

EXAMPLE 4 Pareto Chart of Boat Thefts

For the boats stolen in a recent year, Figure 2-8 shows the types most often stolen. We can see that for boat thefts, jet skis are the most serious problem.

continued

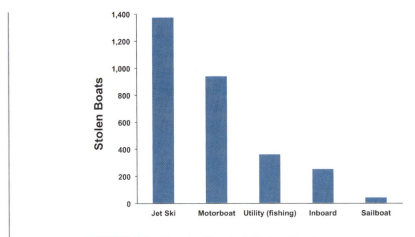

FIGURE 2-8   **Pareto Chart of Stolen Boats**

> **YOUR TURN** Do Exercise 11 "Journal Retractions."

Pie Charts

A **pie chart** is a very common graph that depicts categorical data as slices of a circle, in which the size of each slice is proportional to the frequency count for the category. Although pie charts are very common, they are not as effective as Pareto charts.

Feature of a Pie Chart

- Shows the distribution of categorical data in a commonly used format.

EXAMPLE 5   **Pie Chart of Stolen Boats**

Figure 2-9 is a pie chart of the same boat-theft data from Example 4. Construction of a pie chart involves slicing up the circle into the proper proportions that represent relative frequencies. For example, the category of jet skis accounts for 46% of the total, so the slice representing jet skis should be 46% of the total (with a central angle of $0.46 \times 360° = 166°$).

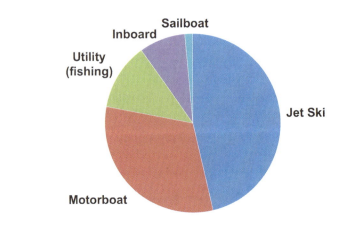

FIGURE 2-9   **Pie Chart of Stolen Boats**

> **YOUR TURN** Do Exercise 13 "Journal Retractions."

The Pareto chart in Figure 2-8 and the pie chart in Figure 2-9 depict the same data in different ways, but the Pareto chart does a better job of showing the relative sizes of the different components. Graphics expert Edwin Tufte makes the following suggestion:

> **Never use pie charts because they waste ink on components that are not data, and they lack an appropriate scale.**

Frequency Polygon

A **frequency polygon** uses line segments connected to points located directly above class midpoint values. A frequency polygon is very similar to a histogram, but a frequency polygon uses line segments instead of bars.

A variation of the basic frequency polygon is the **relative frequency polygon**, which uses relative frequencies (proportions or percentages) for the vertical scale. An advantage of relative frequency polygons is that two or more of them can be combined on a single graph for easy comparison, as in Figure 2-11 (page 62).

EXAMPLE 6 **Frequency Polygon of McDonald's Lunch Service Times**

See Figure 2-10 for the frequency polygon corresponding to the McDonald's lunch service times summarized in the frequency distribution of Table 2-2 on page 42. The heights of the points correspond to the class frequencies, and the line segments are extended to the right and left so that the graph begins and ends on the horizontal axis.

FIGURE 2-10 **Frequency Polygon of McDonald's Lunch Service Times**

YOUR TURN Do Exercise 15 "Old Faithful."

EXAMPLE 7 **Relative Frequency Polygon: McDonald's Lunch Service Times**

Figure 2-11 shows the relative frequency polygons for the drive-through lunch service times from McDonald's and Dunkin' Donuts. Figure 2-11 shows that the Dunkin' Donuts service times are generally lower (farther to the left in the graph) than those from McDonald's. This is expected, given the nature of their different menus.

continued

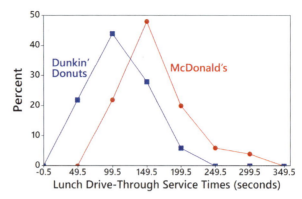

FIGURE 2-11 Relative Frequency Polygons for McDonald's and Dunkin' Donuts

Graphs That Deceive

Deceptive graphs are commonly used to mislead people, and we really don't want statistics students to be among those susceptible to such deceptions. Graphs should be constructed in a way that is fair and objective. The readers should be allowed to make their own judgments, instead of being manipulated by misleading graphs. We present two of the ways in which graphs are commonly used to misrepresent data.

Nonzero Vertical Axis

A common deceptive graph involves using a vertical scale that starts at some value greater than zero to exaggerate differences between groups.

NONZERO AXIS: Always examine a graph carefully to see whether a vertical axis begins at some point other than zero so that differences are exaggerated.

 EXAMPLE 8 Nonzero Axis

Figure 2-12(a) and Figure 2-12(b) are based on the same data from a clinical trial of OxyContin (oxycodone), a drug used to treat moderate to severe pain. The results of that clinical trial included the percentage of subjects who experienced nausea in an OxyContin treatment group and the percentage in a group given a placebo.

By using a vertical scale that starts at 10% instead of 0%, Figure 2-12(a) grossly exaggerates the difference between the two groups. Figure 2-12(a) makes it appear that those using OxyContin experience nausea at a rate that is about 12 times higher than the rate for those using a placebo, but Figure 2-12(b) shows that the true ratio is about 2:1, not 12:1. Perhaps someone wants to discourage recreational use of OxyContin by misleading people into thinking that the problem with nausea is much greater than it really is. The objective might be sincere, but the use of a misleading graph is not the way to achieve that objective.

YOUR TURN Do Exercise 17 "Self-Driving Vehicles."

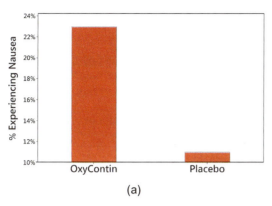
(a)

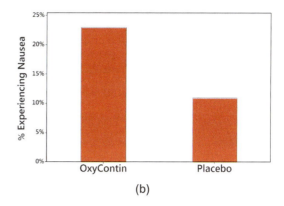
(b)

FIGURE 2-12 **Nausea in a Clinical Trial**

Pictographs

Drawings of objects, called *pictographs,* are often misleading. Data that are one-dimensional in nature (such as budget amounts) are often depicted with two-dimensional objects (such as dollar bills) or three-dimensional objects (such as stacks of coins, homes, or barrels). By using pictographs, artists can create false impressions that grossly distort differences by using these simple principles of basic geometry: (1) When you double each side of a square, its area doesn't merely double; it increases by a factor of *four*. (2) When you double each side of a cube, its volume doesn't merely double; it increases by a factor of *eight*.

> **PICTOGRAPHS:** When examining data depicted with a pictograph, determine whether the graph is misleading because objects of area or volume are used to depict amounts that are actually one-dimensional. (Histograms and bar charts represent one-dimensional data with two-dimensional bars, but they use bars with the same width so that the graph is not misleading.)

EXAMPLE 9 **Pictograph of Cigarette Smokers**

Refer to Figure 2-13 and see that the larger cigarette is about twice as long, twice as tall, and twice as deep as the smaller cigarette, so the volume of the larger cigarette is about *eight times* the volume of the smaller cigarette. (The data are from the Centers for Disease Control and Prevention.) The larger cigarette *appears* to be eight times as large as the smaller cigarette, but the actual percentages show that the 37% smoking rate in 1970 is about *twice* that of the 18% rate in 2013.

1970: 37% of U.S. adults smoked.   **2013: 18% of U.S. adults smoked.**

FIGURE 2-13 **Smoking by U.S. Adults**

YOUR TURN Do Exercise 19 "Cost of Giving Birth."

Concluding Thoughts

In addition to the graphs we have discussed in this section, there are many other useful graphs—some of which have not yet been created. The world desperately needs more people who can create original graphs that enlighten us about the nature of data. In *The Visual Display of Quantitative Information,* Edward Tufte offers these principles:

- For small data sets of 20 values or fewer, use a table instead of a graph.

- A graph of data should make us focus on the true nature of the data, not on other elements, such as eye-catching but distracting design features.

- Do not distort data; construct a graph to reveal the true nature of the data.

- Almost all of the ink in a graph should be used for the data, not for other design elements.

TECH CENTER

 ## Graphing Capabilities

Access tech supplements, videos, and data sets at **www.TriolaStats.com**

Instead of listing instructions for each type of graph, the following lists identify the graphs that can be generated with the different technologies.

Statdisk	Minitab	StatCrunch
• Histograms	• Histograms	• Histograms
• Pie Charts	• Dotplots	• Dotplots
• Scatterplots	• Stemplots	• Stemplots
	• Time-Series Graphs	• Bar Graphs
	• Bar Graphs	• Pie Charts
	• Pareto Charts	• Scatterplots
	• Pie Charts	
	• Frequency Polygons	
	• Scatterplots	

TI-83/84 Plus Calculator	Excel
• Histograms	• Histograms
• Time-Series Graphs	• Time-Series Graphs
• Frequency Polygons	• Bar Graphs
• Scatterplots	• Pareto Charts
	• Pie Charts
	• Scatterplots

2-3   Basic Skills and Concepts

Statistical Literacy and Critical Thinking

1. Body Temperatures Listed below are body temperatures (°F) of healthy adults. Why is it that a graph of these data would not be very effective in helping us understand the data?

98.6   98.6   98.0   98.0   99.0   98.4   98.4   98.4   98.4   98.6

2. Voluntary Response Data If we have a large voluntary response sample consisting of weights of subjects who chose to respond to a survey posted on the Internet, can a graph help to overcome the deficiency of having a voluntary response sample?

3. Ethics There are data showing that smoking is detrimental to good health. Given that people could be helped and lives could be saved by reducing smoking, is it ethical to graph the data in a way that is misleading by exaggerating the health risks of smoking?

4. CVDOT Section 2-1 introduced important characteristics of data summarized by the acronym CVDOT. What characteristics do those letters represent, and which graph does the best job of giving us insight into the last of those characteristics?

Dotplots. *In Exercises 5 and 6, construct the dotplot.*

5. Pulse Rates Listed below are pulse rates (beats per minute) of females selected from Data Set 1 "Body Data" in Appendix B. All of those pulse rates are even numbers. Is there a pulse rate that appears to be an outlier? What is its value?

    80 94 58 66 56 82 78 86 88 56 36 66 84 76 78 64 66 78 60 64

6. Diastolic Blood Pressure Listed below are diastolic blood pressure measurements (mm Hg) of females selected from Data Set 1 "Body Data" in Appendix B. All of the values are even numbers. Are there any outliers? If so, identify their values.

    62 70 72 88 70 66 68 70 82 74 90 62
    70 76 90 86 60 78 82 78 84 76 60 64

Stemplots. *In Exercises 7 and 8, construct the stemplot.*

7. Pulse Rates Refer to the data listed in Exercise 5. How are the data sorted in the stemplot?

8. Diastolic Blood Pressure Refer to the data listed in Exercise 6. Identify the two values that are closest to the middle when the data are sorted in order from lowest to highest. (These values are often used to find the *median*, which is defined in Section 3-1.)

Time-Series Graphs. *In Exercises 9 and 10, construct the time-series graph.*

9. Gender Pay Gap Listed below are women's median earnings as a percentage of men's median earnings for recent years beginning with 1990. Is there a trend? How does it appear to affect women?

    71.6 69.9 70.8 71.5 72.0 71.4 73.8 74.2 73.2 72.3 73.7
    76.3 76.6 75.5 76.6 77.0 76.9 77.8 77.1 77.0 77.4 77.0

10. Home Runs Listed below are the numbers of home runs in Major League Baseball for each year beginning with 1990 (listed in order by row). Is there a trend?

    3317 3383 3038 4030 3306 4081 4962 4640 5064 5528 5693 5458
    5059 5207 5451 5017 5386 4957 4878 5042 4613 4552 4934 4661

Pareto Charts. *In Exercises 11 and 12 construct the Pareto chart.*

11. Journal Retractions In a study of retractions in biomedical journals, 436 were due to error, 201 were due to plagiarism, 888 were due to fraud, 291 were duplications of publications, and 287 had other causes (based on data from "Misconduct Accounts for the Majority of Retracted Scientific Publications," by Fang, Steen, Casadevall, *Proceedings of the National Academy of Sciences of the United States of America,* Vol. 110, No. 3). Among such retractions, does misconduct (fraud, duplication, plagiarism) appear to be a major factor?

12. Getting a Job In a survey, subjects seeking a job were asked to whom they should send a thank-you note after having a job interview. Results were as follows: 40 said only the person they spent the most time with, 396 said everyone they met, 40 said only the most senior-level person, 15 said the person that they had the best conversation with, and 10 said that they don't send thank-you notes (based on data from TheLadders.com). Comment on the results.

Pie Charts. *In Exercises 13 and 14, construct the pie chart.*

13. Journal Retractions Use the data from Exercise 11 "Journal Retractions."

14. Getting a Job Use the data from Exercise 12 "Getting a Job."

Frequency Polygon. *In Exercises 15 and 16, construct the frequency polygons.*

15. Old Faithful Use the frequency distribution from Exercise 11 in Section 2-1 on page 49 to construct a frequency polygon. Does the graph suggest that the distribution is skewed? If so, how?

16. Tornadoes Use the frequency distribution from Exercise 12 in Section 2-1 on page 49 to construct a frequency polygon. Does the graph suggest that the distribution is skewed? If so, how?

Deceptive Graphs. *In Exercises 17–20, identify how the graph is deceptive.*

17. Self-Driving Vehicles In a survey of adults, subjects were asked if they felt comfortable being in a self-driving vehicle. The accompanying graph depicts the results (based on data from TE Connectivity).

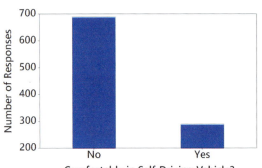

18. Subway Fare In 1986, the New York City subway fare cost $1, and as of this writing, the current cost is $2.50, so the 1986 price was multiplied by 2.5. In the accompanying graph, the large bill is 2.5 times as tall and 2.5 times as wide as the smaller bill.

1986 Subway Fare   **Current Subway Fare**

19. Cost of Giving Birth According to the Agency for Healthcare Research and Quality Healthcare Cost and Utilization Project, the typical cost of a C-section baby delivery is $4500, and the typical cost of a vaginal delivery is $2600. See the following illustration.

Cost of C-Section Delivery: $4500　　　**Cost of Vaginal Delivery: $2600**

20. Incomes and Academic Degrees The accompanying graph depicts workers with various academic degrees along with their income levels.

| $24,544 | $33,852 | $57,616 | $80,877 |
| No High School Diploma | High School Diploma | Bachelor's Degree | Advanced Degree |

2-3　Beyond the Basics

21. Expanded Stemplots A stemplot can be *condensed* by combining adjacent rows. We could use a stem of "6–7" instead of separate stems of 6 and 7. Every row in the condensed stemplot should include an asterisk to separate digits associated with the different stem values. A stemplot can be *expanded* by subdividing rows into those with leaves having digits 0 through 4 and those with leaves having digits 5 through 9. Using the body temperatures from 12 AM on Day 2 listed in Data Set 3 "Body Temperatures" in Appendix B, we see that the first three rows of an expanded stemplot have stems of 96 (for leaves between 5 and 9 inclusive), 97 (for leaves between 0 and 4 inclusive), and 97 (for leaves between 5 and 9 inclusive). Construct the complete expanded stemplot for the body temperatures from 12 AM on Day 2 listed in Data Set 3 "Body Temperatures" in Appendix B.

2-4　Scatterplots, Correlation, and Regression

Key Concept This section introduces the analysis of *paired* sample data. In Part 1 of this section we discuss *correlation* and the role of a graph called a *scatterplot*. In Part 2 we provide an introduction to the use of the *linear correlation coefficient*. In Part 3 we provide a very brief discussion of *linear regression*, which involves the equation and graph of the straight line that best fits the sample paired data.

All of the principles discussed in this section are discussed more fully in Chapter 10, but this section serves as a quick introduction to some important concepts of correlation and regression. This section does not include details for executing manual calculations, and those calculations are rarely done. Instructions for using technology to obtain results are included in Chapter 10.

PART 1 Scatterplot and Correlation

Our objective in this section is to explore whether there is a *correlation,* or association, between two variables. We begin with basic definitions.

> **DEFINITIONS**
>
> A **correlation** exists between two variables when the values of one variable are somehow associated with the values of the other variable.
>
> A **linear correlation** exists between two variables when there is a correlation and the plotted points of paired data result in a pattern that can be approximated by a straight line.
>
> A **scatterplot** (or **scatter diagram**) is a plot of paired (*x, y*) quantitative data with a horizontal *x*-axis and a vertical *y*-axis. The horizontal axis is used for the first variable (*x*), and the vertical axis is used for the second variable (*y*).

CAUTION The presence of a correlation between two variables is not evidence that one of the variables *causes* the other. We might find a correlation between beer consumption and weight, but we cannot conclude from the statistical evidence that drinking beer has a direct effect on weight.

Correlation does not imply causality!

A scatterplot can be very helpful in determining whether there is a correlation (or relationship) between the two variables. (This issue is discussed at length when the topic of correlation is considered in Section 10-1.)

EXAMPLE 1 Correlation: Waist and Arm Circumference

Data Set 1 "Body Data" in Appendix B includes waist circumferences (cm) and arm circumferences (cm) of randomly selected adult subjects. Figure 2-14 is a scatterplot of the paired waist/arm measurements. The points show a pattern of increasing values from left to right. This pattern suggests that there is a correlation or relationship between waist circumferences and arm circumferences.

YOUR TURN Do Exercise 7 "Car Weight and Fuel Consumption."

EXAMPLE 2 No Correlation: Weight and Pulse Rate

Data Set 1 "Body Data" in Appendix B includes weights (kg) and pulse rates (beats per minute) of randomly selected adult subjects. Figure 2-15 is a scatterplot of the paired weight/pulse rate measurements. The points in Figure 2-15 do not show any obvious pattern, and this lack of a pattern suggests that there is no correlation or relationship between weights and pulse rates.

YOUR TURN Do Exercise 8 "Heights of Fathers and Sons."

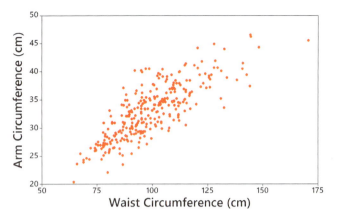

FIGURE 2-14   Waist and Arm Circumferences
Correlation: The distinct pattern of the plotted points suggests that there is a correlation between waist circumferences and arm circumferences.

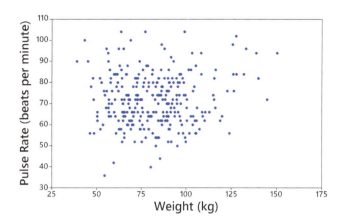

FIGURE 2-15   Weights and Pulse Rates
No Correlation: The plotted points do not show a distinct pattern, so it appears that there is no correlation between weights and pulse rates.

EXAMPLE 3   **Clusters and a Gap**

Consider the scatterplot in Figure 2-16. It depicts paired data consisting of the weight (grams) and year of manufacture for each of 72 pennies. This scatterplot shows two very distinct clusters separated by a gap, which can be explained by the inclusion of two different populations: Pre-1983 pennies are 97% copper and 3% zinc, but post-1983 pennies are 2.5% copper and 97.5% zinc. If we ignored the characteristic of the clusters, we might incorrectly think that there is a relationship between the weight of a penny and the year it was made. If we examine the two groups separately, we see that there does *not* appear to be a relationship between the weights of pennies and the years they were produced.

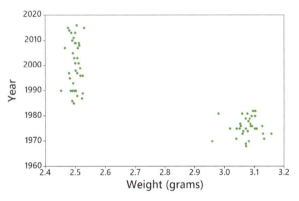

FIGURE 2-16   Weights of Pennies and Years of Production

The preceding three examples involve making decisions about a correlation based on subjective judgments of scatterplots, but Part 2 introduces the *linear correlation coefficient* as a measure that can help us make such decisions more objectively. Using paired data, we can calculate the value of the *linear correlation coefficient r.*

PART 2 Linear Correlation Coefficient r

> **DEFINITION**
>
> The **linear correlation coefficient** is denoted by r, and it measures the strength of the linear association between two variables.

The value of a linear correlation coefficient r can be manually computed by applying Formula 10-1 or Formula 10-2 found in Section 10-1 on page 473, but in practice, r is almost always found by using statistics software or a suitable calculator.

Using r for Determining Correlation

The computed value of the linear correlation coefficient is always between -1 and 1. If r is close to -1 or close to 1, there appears to be a correlation, but if r is close to 0, there does not appear to be a linear correlation. For the data depicted in the scatterplot of Figure 2-14, $r = 0.802$ (somewhat close to 1), and the data in the scatterplot of Figure 2-15 result in $r = 0.082$ (pretty close to 0). These descriptions of "close to" -1 or 1 or 0 are vague, but there are other objective criteria. For now we will use a table of special values (Table 2-11) for deciding whether there is a linear correlation. See the following example illustrating the interpretation of the linear correlation coefficient r.

○—**EXAMPLE 4** Correlation Between Shoe Print Lengths and Heights?

Consider the data in Table 2-10 (using data from Data Set 2 "Foot and Height" in Appendix B). From the accompanying scatterplot of the paired data in Table 2-10, it isn't very clear whether there is a linear correlation. The Statdisk display of the results shows that the linear correlation coefficient has the value of $r = 0.591$ (rounded).

TABLE 2-10 Shoe Print Lengths and Heights of Males

Shoe Print Length (cm)	29.7	29.7	31.4	31.8	27.6
Height (cm)	175.3	177.8	185.4	175.3	172.7

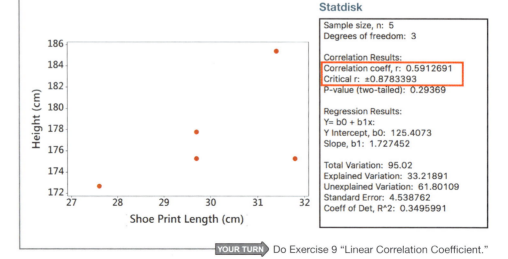

Statdisk

Sample size, n: 5
Degrees of freedom: 3

Correlation Results:
Correlation coeff, r: 0.5912691
Critical r: ±0.8783393
P-value (two-tailed): 0.29369

Regression Results:
Y= b0 + b1x:
Y Intercept, b0: 125.4073
Slope, b1: 1.727452

Total Variation: 95.02
Explained Variation: 33.21891
Unexplained Variation: 61.80109
Standard Error: 4.538762
Coeff of Det, R^2: 0.3495991

YOUR TURN Do Exercise 9 "Linear Correlation Coefficient."

In Example 4, we know from the Statdisk display that in using the five pairs of data from Table 2-10, the linear correlation coefficient is computed to be $r = 0.591$. Use the following criteria for interpreting such values.

Using Table 2-11 to Interpret r: Consider critical values from Table 2-11 as being both positive and negative, and draw a graph similar to Figure 2-17. Use the values in the table for determining whether a value of a linear correlation coefficient r is "close to" 0 or "close to" -1 or "close to" 1 by applying the following criteria:

Correlation   If the computed linear correlation coefficient r lies in the left or right tail region beyond the table value for that tail, conclude that there is sufficient evidence to support the claim of a linear correlation.

No Correlation   If the computed linear correlation coefficient r lies between the two critical values, conclude that there is not sufficient evidence to support the claim of a linear correlation.

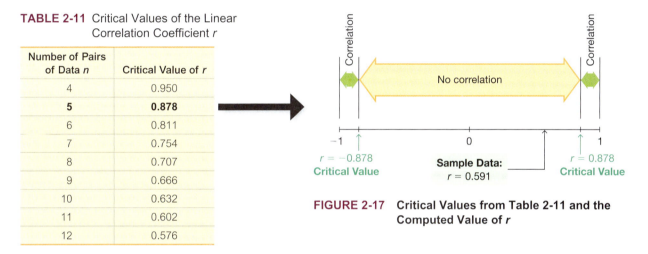

TABLE 2-11 Critical Values of the Linear Correlation Coefficient r

Number of Pairs of Data n	Critical Value of r
4	0.950
5	0.878
6	0.811
7	0.754
8	0.707
9	0.666
10	0.632
11	0.602
12	0.576

FIGURE 2-17   Critical Values from Table 2-11 and the Computed Value of r

Figure 2-17 shows that the linear correlation coefficient of $r = 0.591$ computed from the paired sample data is a value that lies between the critical values of $r = -0.878$ and $r = 0.878$ (found from Table 2-11). Figure 2-17 shows that we can consider the value of $r = 0.591$ to be close to 0 instead of being close to -1 or close to 1. Therefore, we do not have sufficient evidence to conclude that there is a linear correlation between shoe print lengths and heights of males.

P-Values for Determining Linear Correlation

In Example 4, we used the computed value of the linear correlation coefficient $r = 0.591$ and compared it to the critical r values of ± 0.878 found from Table 2-11. (See Figure 2-17.) In the real world of statistics applications, the use of such tables is almost obsolete. Section 10-1 describes a more common approach that is based on "P-values" instead of tables. The Statdisk display accompanying Example 4 shows that the P-value is 0.29369, or 0.294 when rounded. P-values are first introduced in Chapter 8, but here is a preliminary definition suitable for the context of this section:

DEFINITION

If there really is no linear correlation between two variables, the **P-value** is the probability of getting paired sample data with a linear correlation coefficient r that is at least as extreme as the one obtained from the paired sample data.

Based on Example 4 and the Statdisk displayed results showing a *P*-value of 0.294, we know that there is a 0.294 probability (or a 29.4% chance) of getting a linear correlation coefficient of $r = 0.591$ or more extreme, assuming that there is no linear correlation between shoe print length and height. (The values of r that are "more extreme" than 0.591 are the values greater than 0.591 and the values less than -0.591.)

Interpreting a *P*-Value   The *P*-value of 0.294 from Example 4 is high. It shows that there is a high chance of getting a linear correlation coefficient of $r = 0.591$ (or more extreme) by chance when there is no linear correlation between the two variables. Because the likelihood of getting $r = 0.591$ or a more extreme value is so high (29.4% chance), we conclude that there is not sufficient evidence to conclude that there is a linear correlation between shoe print lengths and heights of males.

> Only a *small P*-value, such as 0.05 or less (or a 5% chance or less), suggests that the sample results are *not* likely to occur by chance when there is no linear correlation, so a small *P*-value supports a conclusion that there is a linear correlation between the two variables.

EXAMPLE 5   Correlation Between Shoe Print Lengths and Heights?

Example 4 used only five pairs of data from Data Set 2 "Foot and Height" in Appendix B. If we use the shoe print lengths and heights from all of the 40 subjects listed in Data Set 2 in Appendix B, we get the scatterplot shown in Figure 2-18 and we get the Minitab results shown in the accompanying display. The scatterplot does show a distinct pattern instead of having points scattered about willy-nilly. Also, we see that the value of the linear correlation coefficient is $r = 0.813$, and the *P*-value is 0.000 when rounded to three decimal places. Because the *P*-value of 0.000 is *small*, we have sufficient evidence to conclude that there is a linear correlation between shoe print lengths and heights.

In Example 4 with only five pairs of data, we did not have enough evidence to conclude that there is a linear correlation, but in this example with 40 pairs of data, we have sufficient evidence to conclude that there is a linear correlation between shoe print lengths and heights.

Minitab

```
Pearson correlation of Shoe Print Length and Height = 0.813
P-Value = 0.000
```

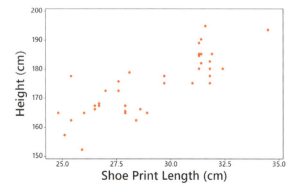

FIGURE 2-18   Scatterplot of 40 Pairs of Data

PART 3 Regression

When we do conclude that there appears to be a linear correlation between two variables (as in Example 5), we can find the equation of the straight line that best fits the sample data, and that equation can be used to predict the value of one variable when given a specific value of the other variable. Based on the results from Example 5, we can predict someone's height given the length of their shoe print (which may have been found at a crime scene).

Instead of using the straight-line equation format of $y = mx + b$ that we have all learned in prior math courses, we use the format that follows.

> **DEFINITION**
>
> Given a collection of paired sample data, the **regression line** (or *line of best fit*, or *least-squares line*) is the straight line that "best" fits the scatterplot of the data. (The specific criterion for the "best"-fitting straight line is the "least squares" property described in Section 10-2.)

The **regression equation**

$$\hat{y} = b_0 + b_1 x$$

algebraically describes the regression line.

Section 10-2 gives a good reason for using the format of $\hat{y} = b_0 + b_1 x$ instead of the format of $y = mx + b$. Section 10-2 also provides formulas that could be used to identify the values of the y-intercept b_0 and the slope b_1, but those values are usually found by using statistics software or a suitable calculator.

EXAMPLE 6 Regression Line

Example 5 included a scatterplot of the 40 pairs of shoe print lengths and heights from Data Set 2 "Foot and Height" in Appendix B. Figure 2-19 shown on the next page is that same scatterplot with the graph of the regression line included. Also shown is the Statdisk display from the 40 pairs of data.

From the Statdisk display, we see that the general form of the regression equation has a y-intercept of $b_0 = 80.9$ (rounded) and slope $b_1 = 3.22$ (rounded), so the equation of the regression line shown in Figure 2-19 is $\hat{y} = 80.9 + 3.22x$. It might be helpful to express that equation more clearly by using the names of the variables:

$$\text{Height} = 80.9 + 3.22\,(\text{Shoe Print Length})$$

Note that the *equation* shows the y-intercept of 80.9 that does not appear on the vertical scale in the *graph*. The leftmost vertical scale in Figure 2-19 is not the actual y-axis that passes through 0 on the x-axis. If the graph were extended to the left, the regression line would intercept the actual y-axis at the height of $y = 80.9$ cm.

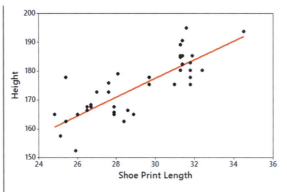

Statdisk

Correlation Results:
Correlation coeff, r: 0.812948
Critical r: ±0.3120061
P-value (two-tailed): 0.000

Regression Results:
Y= b0 + b1x:
Y Intercept, b0: 80.93041
Slope, b1: 3.218561

FIGURE 2-19   Regression Line

2-4   Basic Skills and Concepts

Statistical Literacy and Critical Thinking

1. Linear Correlation In this section we use r to denote the value of the linear correlation coefficient. Why do we refer to this correlation coefficient as being *linear*?

2. Causation A study has shown that there is a correlation between body weight and blood pressure. Higher body weights are associated with higher blood pressure levels. Can we conclude that gaining weight is a cause of increased blood pressure?

3. Scatterplot What is a scatterplot and how does it help us?

4. Estimating r For each of the following, estimate the value of the linear correlation coefficient r for the given paired data obtained from 50 randomly selected adults.

a. Their heights are measured in inches (x) and those same heights are recorded in centimeters (y).

b. Their IQ scores (x) are measured and their heights (y) are measured in centimeters.

c. Their pulse rates (x) are measured and their IQ scores are measured (y).

d. Their heights (x) are measured in centimeters and those same heights are listed again, but with negative signs (y) preceding each of these second listings.

Scatterplot.   *In Exercises 5–8, use the sample data to construct a scatterplot. Use the first variable for the x-axis. Based on the scatterplot, what do you conclude about a linear correlation?*

5. Brain Volume and IQ The table lists brain volumes (cm³) and IQ scores of five males (from Data Set 8 "IQ and Brain Size" in Appendix B).

Brain volume (cm³)	1173	1067	1347	1029	1204
IQ	101	93	94	97	113

6. Bear Measurements The table lists chest sizes (distance around chest in inches) and weights (pounds) of anesthetized bears that were measured (from Data Set 9 "Bear Measurements" in Appendix B).

Chest (in.)	26	45	54	49	35	41	41
Weight (lb)	80	344	416	348	166	220	262

7. Car Weight and Fuel Consumption The table lists weights (pounds) and highway fuel consumption amounts (mpg) for a Hyundai Elantra, Nissan Altima, VW Passat, Buick Lucerne, Mercury Grand Marquis, Honda Civic, and Honda Accord.

Weight (lb)	2895	3215	3465	4095	4180	2740	3270
Highway (mpg)	33	31	29	25	24	36	30

8. Heights of Fathers and Sons The table lists heights (in.) of fathers and the heights (in.) of their first sons (from Francis Galton).

Height of father (in.)	73.0	75.5	75.0	75.0	75.0	74.0	74.0	73.0	73.0	78.5
Height of first son (in.)	74.0	73.5	71.0	70.5	72.0	76.5	74.0	71.0	72.0	73.2

Linear Correlation Coefficient *In Exercises 9–12, the linear correlation coefficient r is provided. Use Table 2-11 on page 71 to find the critical values of r. Based on a comparison of the linear correlation coefficient r and the critical values, what do you conclude about a linear correlation?*

9. Using the data from Exercise 5 "Brain Volume and IQ," the linear correlation coefficient is $r = 0.127$.

10. Using the data from Exercise 6 "Bear Measurements," the linear correlation coefficient is $r = 0.980$.

11. Using the data from Exercise 7 "Car Weight and Fuel Consumption," the linear correlation coefficient is $r = -0.987$.

12. Using the data from Exercise 8 "Heights of Fathers and Sons," the linear correlation coefficient is $r = -0.017$.

2-4   Beyond the Basics

P-Values *In Exercises 13–16, write a statement that interprets the P-value and includes a conclusion about linear correlation.*

13. Using the data from Exercise 5 "Brain Volume and IQ," the P-value is 0.839.

14. Using the data from Exercise 6 "Bear Measurements," the P-value is 0.000.

15. Using the data from Exercise 7 "Car Weight and Fuel Consumption," the P-value is 0.000.

16. Using the data from Exercise 8 "Heights of Fathers and Sons," the P-value is 0.963.

Chapter Quick Quiz

1. Cookies Refer to the accompanying frequency distribution that summarizes the numbers of chocolate chips found in each cookie in a sample of Chips Ahoy regular chocolate chip cookies (from Data Set 28 "Chocolate Chip Cookies" in Appendix B). What is the class width? Is it possible to identify the original data values?

Chocolate Chips	Frequency
18–20	6
21–23	11
24–26	18
27–29	4
30–32	1

2. Cookies Using the same frequency distribution from Exercise 1, identify the class boundaries of the first class and identify the class limits of the first class.

3. Cookies Using the same frequency distribution from Exercise 1, how many cookies are included?

4. Cookies A stemplot of the same cookies summarized in Exercise 1 is created, and the first row of that stemplot is 1 | 99. Identify the values represented by that row of the stemplot.

5. Computers As a quality control manager at Texas Instruments, you find that defective calculators have various causes, including worn machinery, human error, bad supplies, and packaging mistreatment. Which of the following graphs would be best for describing the causes of defects: histogram; scatterplot; Pareto chart; dotplot; pie chart?

6. Distribution of Wealth In recent years, there has been much discussion about the distribution of wealth of adults in the United States. If you plan to conduct original research by somehow obtaining the amounts of wealth of 3000 randomly selected adults, what graph would be best for illustrating the *distribution* of wealth?

7. Health Test In an investigation of a relationship between systolic blood pressure and diastolic blood pressure of adult females, which of the following graphs is most helpful: histogram; pie chart; scatterplot; stemplot; dotplot?

8. Lottery In Florida's Play 4 lottery game, four digits between 0 and 9 inclusive are randomly selected each day. We normally expect that each of the 10 different digits will occur about 1/10 of the time, and an analysis of last year's results shows that this did happen. Because the results are what we normally expect, is it correct to say that the distribution of selected digits is a normal distribution?

9. Seatbelts The Beams Seatbelts company manufactures—well, you know. When a sample of seatbelts is tested for breaking point (measured in kilograms), the sample data are explored. Identify the important characteristic of data that is missing from this list: center, distribution, outliers, changing characteristics over time.

10. Seatbelts A histogram is to be constructed from the measured breaking points (in pounds) of tested car seatbelts. Identify two key features of a histogram of those values that would suggest that the data have a *normal distribution.*

Review Exercises

1. Frequency Distribution of Body Temperatures Construct a frequency distribution of the 20 body temperatures (°F) listed below. (These data are from Data Set 3 "Body Temperatures" in Appendix B.) Use a class width of 0.5°F and a starting value of 97.0°F.

97.1	97.2	97.5	97.6	97.6	97.8	98.0	98.0	98.2	98.2
98.2	98.3	98.4	98.6	98.6	98.7	98.7	98.9	99.1	99.4

2. Histogram of Body Temperatures Construct the histogram that corresponds to the frequency distribution from Exercise 1. Use class midpoint values for the horizontal scale. Does the histogram suggest that the data are from a population having a normal distribution? Why or why not?

3. Dotplot of Body Temperatures Construct a dotplot of the body temperatures listed in Exercise 1. Which does a better job of illustrating the distribution of the data: the histogram from Exercise 2 or the dotplot?

4. Stemplot of Body Temperatures Construct a stemplot of the body temperatures listed in Exercise 1. Are there any outliers?

5. Body Temperatures Listed below are the temperatures from nine males measured at 8 AM and again at 12 AM (from Data Set 3 "Body Temperatures" in Appendix B). Construct a scatterplot. Based on the graph, does there appear to be a relationship between 8 AM temperatures and 12 AM temperatures?

8 AM	98.0	97.0	98.6	97.4	97.4	98.2	98.2	96.6	97.4
12 AM	98.0	97.6	98.8	98.0	98.8	98.8	97.6	98.6	98.6

6. Environment

a. After collecting the average (mean) global temperatures for each of the most recent 100 years, we want to construct the graph that is most appropriate for these data. Which graph is best?

b. After collecting the average (mean) global temperature and the amount of carbon monoxide emissions for the most recent 100 years, we want to construct a graph to investigate the association between those two variables. Which graph is best?

c. An investigation of carbon monoxide sources includes motor vehicles, furnaces, fires, coal-burning power plants, and tobacco smoke. If we want to construct a graph that illustrates the relative importance of these sources, which graph is best?

7. It's Like Time to Do This Exercise In a Marist survey of adults, these are the words or phrases that subjects find most annoying in conversation (along with their frequencies of response): like (127); just sayin' (81); you know (104); whatever (219); obviously (35). Construct a pie chart. Identify one disadvantage of a pie chart.

8. Whatever Use the same data from Exercise 7 to construct a Pareto chart. Which graph does a better job of illustrating the data: Pareto chart or pie chart?

Cumulative Review Exercises

In Exercises 1–6, refer to the data below, which are total home game playing times (hours) for all Major League Baseball teams in a recent year (based on data from Baseball Prospectus).

236 237 238 239 241 241 242 245 245 245 246 247 247 248 248
249 250 250 250 251 252 252 253 253 258 258 258 260 262 264

1. Frequency Distribution Construct a frequency distribution. Use a class width of 5 hours and use a starting time of 235 hours.

2. Frequency Distribution For the frequency distribution from Exercise 1, find the following.

a. Class limits of the first class

b. Class boundaries of the first class

c. Class midpoint of the first class

3. Histogram Construct the histogram corresponding to the frequency distribution from Exercise 1. For the values on the horizontal axis, use the class midpoint values. Which of the following comes closest to describing the distribution: uniform, normal, skewed left, skewed right?

4. Deceptive Graph Assume that you want to create the histogram for Exercise 3 in a way that exaggerates the differences among the times. Describe how the histogram from Exercise 3 can be modified to accomplish that exaggeration.

5. Stemplot Use the total game playing times to create a stemplot. What does the stemplot reveal about the distribution of the data?

6. Data Type

a. The listed playing times are all rounded to the nearest whole number. Before rounding, are the exact playing times discrete data or continuous data?

b. For the listed times, are the data categorical or quantitative?

c. Identify the level of measurement of the listed times: nominal, ordinal, interval, or ratio.

continued

d. Which of the following best describes the sample data: voluntary response sample, random sample, convenience sample, simple sample?

e. The listed total game times are from one recent year, and the data are available for all years back to 1950. Given that the listed times are part of a larger collection of times, do the data constitute a sample or a population?

Technology Project

It was stated in this chapter that the days of charming and primitive hand-drawn graphs are well behind us, and technology now provides us with powerful tools for generating a wide variety of different graphs. This project therefore serves as a good preparation for professional presentations that will be inevitably made in the future.

The complete data sets in Appendix B can be downloaded from www.TriolaStats.com. Statdisk already includes the data sets. They can be opened by statistical software packages, such as Statdisk, Minitab, Excel, SPSS, and JMP. Use a statistical software package to open Data Set 4 "Births." Use this software with the methods of this chapter to explore and compare the birth weights of the girls and the birth weights of the boys. (Because the units are grams, they are actually measures of mass instead of weight.)

• Obtain a printed copy of the two histograms.

• Describe the natures of the two distributions (uniform, normal, skewed left, skewed right), and identify possible outliers.

• Write a brief description of your results.

Hint: The genders are coded as 1 for male and 0 for female, so *sort* (arrange in order) all of the rows using the gender column as the basis for sorting. Then the rows for boys can be separated from the rows for girls by using the copy/paste and cut functions of the software.

FROM DATA TO DECISION

Fast Food Restaurant Drive-Through Service Times: Who's Best?

Data Set 25 "Fast Food" in Appendix B includes lunch and dinner drive-through service times at McDonald's, Burger King, Wendy's, and Dunkin' Donuts restaurants. Several examples and exercises in this chapter use some of those service times. For this project, use all of the service times.

Critical Thinking

Use the methods from this chapter to address the following questions.

1. Which of the four restaurants appears to have the fastest lunch drive-through service times?

2. Which of the four restaurants appears to have the fastest dinner drive-through service times?

3. Do the lunch drive-through service times appear to be different from the dinner drive-through service times? Explain.

4. Based on the available menu items at the different restaurants, does any one of the restaurants have an inherent advantage relative to service times? Explain.

5. Considering differences in menu items, is there a restaurant that appears to be more efficient than the others? Explain.

Cooperative Group Activities

1. Out-of-class activity The Chapter Problem uses measured service times in Data Set 25 "Fast Food" in Appendix B. Go to one or more fast food restaurants and collect your own service times. Compare the results to those found in Data Set 25 in Appendix B.

2. In-class activity Using a package of purchased chocolate chip cookies, each student should be given two or three cookies. Proceed to count the number of chocolate chips in each cookie. Not all of the chocolate chips are visible, so "destructive testing" must be used through a process involving consumption. Record the numbers of chocolate chips for each cookie and combine all results. Construct a frequency distribution, histogram, dotplot, and stemplot of the results. Given that the cookies were made through a process of mass production, we might expect that the numbers of chips per cookie would not vary much. Is that indicated by the results? Explain.

3. In-class activity In class, each student should record two pulse rates by counting the number of her or his heartbeats in 1 minute. The first pulse rate should be measured while the student is seated, and the second pulse rate should be measured while the student is standing. Using the pulse rates measured while seated, construct a frequency distribution and histogram for the pulse rates of males, and then construct another frequency distribution and histogram for the pulse rates of females. Using the pulse rates measured while standing, construct a frequency distribution and histogram for the pulse rates of males, and then construct another frequency distribution and histogram for the pulse rates of females. Compare the results. Do males and females appear to have different pulse rates? Do pulse rates measured while seated appear to be different from pulse rates measured while standing? Use an appropriate graph to determine whether there is a relationship between sitting pulse rate and standing pulse rate.

4. Out-of-class activity Search newspapers and magazines to find an example of a graph that is misleading. Describe how the graph is misleading. Redraw the graph so that it depicts the information correctly. If possible, please submit your graph to www.TriolaStats.com.

5. Out-of-class activity Find Charles Joseph Minard's graph describing Napoleon's march to Moscow and back, and explain why Edward Tufte says that "it may well be the best graphic ever drawn." (See *The Visual Display of Quantitative Information* by Edward Tufte, Graphics Press). Minard's graph can be seen at www.TriolaStats.com under "Textbook Supplements."

6. Out-of-class activity In *The Visual Display of Quantitative Information* by Edward Tufte (Graphics Press), find the graph that appeared in *American Education*, and explain why Tufte says that "this may well be the worst graphic ever to find its way into print." The graph can be seen at www.TriolaStats.com under "Textbook Supplements." Construct a graph that is effective in depicting the same data.

3-1 Measures of Center

3-2 Measures of Variation

3-3 Measures of Relative
     Standing and Boxplots

3

DESCRIBING, EXPLORING, AND COMPARING DATA

CHAPTER PROBLEM   **Which carrier has the fastest smartphone data speed at airports?**

Data Set 32 "Airport Data Speeds" in Appendix B lists data speeds measured by RootMetrics at 50 different U.S. airports using the four major smartphone carriers of Verizon, Sprint, AT&T, and T-Mobile. The speeds are all in units of megabits (or 1 million bits) per second, denoted as Mbps. Because the original data speeds listed in Data Set 32 include decimal numbers such as 38.5 Mbps, unmodified dotplots of the data would be somewhat messy and not so helpful, but

if we round all of the original data sets, we get the dotplot shown in Figure 3-1. (Examination of the horizontal scale in Figure 3-1 reveals that the original data speeds have been rounded to the nearest even integer by the software used to create the dotplots.) By using the same horizontal scale and stacking the four dotplots, comparisons become much easier.

Examination of Figure 3-1 suggests that Verizon is the overall best performer with data speeds that tend to be higher

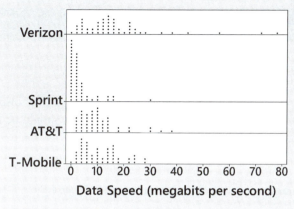

FIGURE 3-1   Dotplot of Smartphone Data Speeds

than the data speeds of the other three carriers. But instead of relying solely on subjective interpretations of a graph like Figure 3-1, this chapter introduces measures that are essential to any study of statistics. This chapter introduces the mean, median, standard deviation, and variance, which are among the most important statistics presented in this book, and they are among the most important statistics in the study of statistics. We will use these statistics for describing, exploring, and comparing the measured data speeds from Verizon, Sprint, AT&T, and T-Mobile as listed in Data Set 32.

CHAPTER OBJECTIVES

Critical Thinking and Interpretation: Going Beyond Formulas and Arithmetic

In this modern statistics course, it isn't so important to memorize formulas or manually do messy arithmetic. We can get results with a calculator or software so that we can focus on making practical sense of results through critical thinking. Although this chapter includes detailed steps for important procedures, it isn't always necessary to master those steps. It is, however, generally helpful to perform a few manual calculations before using technology, so that understanding is enhanced.

The methods and tools presented in this chapter are often called methods of **descriptive statistics**, because they summarize or *describe* relevant characteristics of data. In later chapters we use **inferential statistics** to make *inferences*, or generalizations, about populations. Here are the chapter objectives:

3-1 Measures of Center

- Develop the ability to measure the center of data by finding the mean, median, mode, and midrange.

- Determine whether an outlier has a substantial effect on the mean and median.

3-2 Measures of Variation

- Develop the ability to measure variation in a set of sample data by finding values of the range, variance, and standard deviation.

- Develop the ability to interpret values of the standard deviation by applying the *range rule of thumb* to determine whether a particular value is *significantly low* or *significantly high*.

3-3 Measures of Relative Standing and Boxplots

- Develop the ability to compute a *z* score and use the result to determine whether a given value *x* is *significantly low* or *significantly high*.

- Identify *percentile* values and *quartile* values from a set of data.

- Develop the ability to construct a boxplot from a set of data.

Key Concept The focus of this section is to obtain a value that measures the *center* of a data set. In particular, we present measures of center, including *mean* and *median*. Our objective here is not only to find the value of each measure of center, but also to interpret those values. Part 1 of this section includes core concepts that should be understood before considering Part 2.

PART 1   Basic Concepts of Measures of Center

In Part 1 of this section, we introduce the mean, median, mode, and midrange as different measures of center. Measures of center are widely used to provide representative values that "summarize" data sets.

Go Figure

$3.19: Mean amount left by the tooth fairy, based on a survey by Visa. An unlucky 10% of kids get nothing.

DEFINITION

A **measure of center** is a value at the center or middle of a data set.

There are different approaches for measuring the center, so we have different definitions for those different approaches. We begin with the mean.

Mean

The mean (or arithmetic mean) is generally the most important of all numerical measurements used to describe data, and it is what most people call an *average*.

DEFINITION

The **mean** (or **arithmetic mean**) of a set of data is the measure of center found by adding all of the data values and dividing the total by the number of data values.

Important Properties of the Mean

- Sample means drawn from the same population tend to vary less than other measures of center.
- The mean of a data set uses every data value.
- A disadvantage of the mean is that just one extreme value (outlier) can change the value of the mean substantially. (Using the following definition, we say that the mean is not *resistant*.)

DEFINITION

A statistic is **resistant** if the presence of extreme values (outliers) does not cause it to change very much.

Calculation and Notation of the Mean

The definition of the mean can be expressed as Formula 3-1, in which the Greek letter Σ (uppercase sigma) indicates that the data values should be added, so Σx represents the sum of all data values. The symbol n denotes the **sample size**, which is the number of data values.

FORMULA 3-1

$$\text{Mean} = \frac{\Sigma x}{n} \quad \begin{array}{l} \leftarrow \text{sum of all data values} \\ \leftarrow \text{number of data values} \end{array}$$

If the data are a *sample* from a population, the mean is denoted by $\bar{x}$ (pronounced "x-bar"); if the data are the entire population, the mean is denoted by μ (lowercase Greek mu).

NOTATION *Hint:* Sample statistics are usually represented by English letters, such as $\bar{x}$, and population parameters are usually represented by Greek letters, such as μ.

Σ	denotes the *sum* of a set of data values.
x	is the *variable* usually used to represent the individual data values.
n	represents the number of data values in a *sample*.
N	represents the number of data values in a *population*.
$\bar{x} = \dfrac{\Sigma x}{n}$	is the mean of a set of *sample* values.
$\mu = \dfrac{\Sigma x}{N}$	is the mean of all values in a *population*.

Class Size Paradox

There are at least two ways to obtain the mean class size, and they can have very different results. At one college, if we take the numbers of students in 737 classes, we get a mean of 40 students. But if we were to compile a list of the class sizes for each student and use this list, we would get a mean class size of 147. This large discrepancy is because there are many students in large classes, while there are few students in small classes. Without changing the number of classes or faculty, we could reduce the mean class size experienced by students by making all classes about the same size. This would also improve attendance, which is better in smaller classes.

EXAMPLE 1   Mean

Data Set 32 "Airport Data Speeds" in Appendix B includes measures of data speeds of smartphones from four different carriers. Find the mean of the first five data speeds for Verizon: 38.5, 55.6, 22.4, 14.1, and 23.1 (all in megabits per second, or Mbps).

SOLUTION

The mean is computed by using Formula 3-1. First add the data values, then divide by the number of data values:

$$\bar{x} = \frac{\Sigma x}{n} = \frac{38.5 + 55.6 + 22.4 + 14.1 + 23.1}{5} = \frac{153.7}{5}$$

$$= 30.74 \text{ Mbps}$$

The mean of the first five Verizon data speeds is 30.74 Mbps.

YOUR TURN Find the mean in Exercise 5 "Football Player Numbers."

CAUTION Never use the term *average* when referring to a measure of center. The word *average* is often used for the mean, but it is sometimes used for other measures of center. The term *average* is not used by statisticians and it will not be used throughout the remainder of this book when referring to a specific measure of center. The term *average* is not used by the statistics community or professional journals.

What the Median Is Not

Harvard biologist Stephen Jay Gould wrote, "The Median Isn't the Message." In it, he describes how he learned that he had abdominal mesothelioma, a form of cancer. He went to the library to learn more, and he was shocked to find that mesothelioma was incurable, with a median survival time of only *eight months* after it was discovered. Gould wrote this: "I suspect that most people, without training in statistics, would read such a statement as 'I will probably be dead in eight months' the very conclusion that must be avoided, since it isn't so, and since attitude (in fighting the cancer) matters so much." Gould went on to carefully interpret the value of the median. He knew that his chance of living longer than the median was good because he was young, his cancer was diagnosed early, and he would get the best medical treatment. He also reasoned that some could live much longer than eight months, and he saw no reason why he could not be in that group. Armed with this thoughtful interpretation of the median and a strong positive attitude, Gould lived for *20 years* after his diagnosis. He died of another cancer not related to the mesothelioma.

Median

The median can be thought of loosely as a "middle value" in the sense that about half of the values in a data set are less than the median and half are greater than the median. The following definition is more precise.

> **DEFINITION**
>
> The **median** of a data set is the measure of center that is the *middle value* when the original data values are arranged in order of increasing (or decreasing) magnitude.

Important Properties of the Median

- The median does not change by large amounts when we include just a few extreme values, so the median is a *resistant* measure of center.
- The median does not directly use every data value. (For example, if the largest value is changed to a much larger value, the median does not change.)

Calculation and Notation of the Median

The median of a sample is sometimes denoted by $\tilde{x}$ (pronounced "x-tilde") or M or Med; there isn't a commonly accepted notation and there isn't a special symbol for the median of a population. To find the median, first *sort* the values (arrange them in order) and then follow one of these two procedures:

1. If the number of data values is *odd*, the median is the number located in the exact middle of the sorted list.

2. If the number of data values is *even*, the median is found by computing the mean of the two middle numbers in the sorted list.

EXAMPLE 2 Median with an *Odd* Number of Data Values

Find the median of the first five data speeds for Verizon: 38.5, 55.6, 22.4, 14.1, and 23.1 (all in megabits per second, or Mbps).

SOLUTION

First sort the data values by arranging them in ascending order, as shown below:

$$14.1 \quad 22.4 \quad 23.1 \quad 38.5 \quad 55.6$$

Because there are 5 data values, the number of data values is an odd number (5), so the median is the number located in the exact middle of the sorted list, which is 23.1 Mbps. The median is therefore 23.10 Mbps. Note that the *median* of 23.10 Mbps is different from the *mean* of 30.74 Mbps found in Example 1. Note also that the result of 23.10 Mbps follows the round-off rule provided later in this section.

YOUR TURN Find the median in Exercise 5 "Football Player Numbers."

EXAMPLE 3 Median with an *Even* Number of Data Values

Repeat Example 2 after including the sixth data speed of 24.5 Mbps. That is, find the median of these data speeds: 38.5, 55.6, 22.4, 14.1, 23.1, 24.5 (all in Mbps).

SOLUTION

First arrange the values in ascending order:

14.1   22.4   23.1   24.5   38.5   55.6

Because the number of data values is an even number (6), the median is found by computing the mean of the two middle numbers, which are 23.1 and 24.5.

$$\text{Median} = \frac{23.1 + 24.5}{2} = \frac{47.6}{2} = 23.80 \text{ Mbps}$$

The median is 23.80 Mbps.

> **YOUR TURN** Find the median in Exercise 7 "Celebrity Net Worth."

Mode

The mode isn't used much with quantitative data, but it's the only measure of center that can be used with qualitative data (consisting of names, labels, or categories only).

DEFINITION

The **mode** of a data set is the value(s) that occur(s) with the greatest frequency.

Go Figure

Mohammed: The most common name in the world.

Important Properties of the Mode

- The mode can be found with qualitative data.
- A data set can have no mode or one mode or multiple modes.

Finding the Mode: A data set can have one mode, more than one mode, or no mode.

- When two data values occur with the same greatest frequency, each one is a mode and the data set is said to be **bimodal**.
- When more than two data values occur with the same greatest frequency, each is a mode and the data set is said to be **multimodal**.
- When no data value is repeated, we say that there is **no mode**.
- When you have ice cream with your pie, it is "à la mode."

EXAMPLE 4   Mode

Find the mode of these Sprint data speeds (in Mbps):

0.2   0.3   0.3   0.3   0.6   0.6   1.2

SOLUTION

The mode is 0.3 Mbps, because it is the data speed occurring most often (three times).

> **YOUR TURN** Find the mode in Exercise 7 "Celebrity Net Worth."

In Example 4, the mode is a single value. Here are other possible circumstances:

Two modes: The data speeds (Mbps) of 0.3, 0.3, 0.6, 4.0, and 4.0 have two modes: 0.3 Mbps and 4.0 Mbps.

No mode: The data speeds (Mbps) of 0.3, 1.1, 2.4, 4.0, and 5.0 have no mode because no value is repeated.

Midrange

Another measure of center is the midrange.

DEFINITION

The **midrange** of a data set is the measure of center that is the value midway between the maximum and minimum values in the original data set. It is found by adding the maximum data value to the minimum data value and then dividing the sum by 2, as in the following formula:

$$\text{Midrange} = \frac{\text{maximum data value} + \text{minimum data value}}{2}$$

Important Properties of the Midrange

- Because the midrange uses only the maximum and minimum values, it is very sensitive to those extremes so the midrange is not *resistant*.

- In practice, the midrange is rarely used, but it has three redeeming features:

  1. The midrange is very easy to compute.

  2. The midrange helps reinforce the very important point that there are several different ways to define the center of a data set.

  3. The value of the midrange is sometimes used incorrectly for the median, so confusion can be reduced by clearly defining the midrange along with the median.

EXAMPLE 5 **Midrange**

Find the midrange of these Verizon data speeds from Example 1: 38.5, 55.6, 22.4, 14.1, and 23.1 (all in Mbps)

SOLUTION

The midrange is found as follows:

$$\text{Midrange} = \frac{\text{maximum data value} + \text{minimum data value}}{2}$$

$$= \frac{55.6 + 14.1}{2} = 34.85 \text{ Mbps}$$

The midrange is 34.85 Mbps.

YOUR TURN Find the midrange in Exercise 5 "Football Player Numbers."

Rounding Measures of Center

When calculating measures of center, we often need to round the result. We use the following rule.

Round-Off Rules for Measures of Center:

- **For the mean, median, and midrange, carry one more decimal place than is present in the original set of values.**

- **For the mode, leave the value as is without rounding** (because values of the mode are the same as some of the original data values).

When applying any rounding rules, round only the final answer, *not intermediate values that occur during calculations*. For example, the mean of 2, 3, and 5 is 3.333333 . . . , which is rounded to 3.3, which has one more decimal place than the original values of 2, 3, and 5. As another example, the mean of 80.4 and 80.6 is 80.50 (one more decimal place than was used for the original values). Because the mode is one or more of the original data values, we do not round values of the mode; we simply use the same original values that are modes.

Critical Thinking

We can always calculate measures of center from a sample of numbers, but we should always think about whether it makes sense to do that. In Section 1-2 we noted that it makes no sense to do numerical calculations with data at the nominal level of measurement, because those data consist of names, labels, or categories only, so statistics such as the mean and median are meaningless. We should also think about the sampling method used to collect the data. If the sampling method is not sound, the statistics we obtain may be very misleading.

EXAMPLE 6 Critical Thinking and Measures of Center

See each of the following illustrating situations in which the mean and median are *not* meaningful statistics.

a. Zip codes of the Gateway Arch in St. Louis, White House, Air Force division of the Pentagon, Empire State Building, and Statue of Liberty: 63102, 20500, 20330, 10118, 10004. (The zip codes don't measure or count anything. The numbers are just labels for geographic locations.)

b. Ranks of selected national universities of Harvard, Yale, Duke, Dartmouth, and Brown (from *U.S. News & World Report*): 2, 3, 7, 10, 14. (The ranks reflect an ordering, but they don't measure or count anything.)

c. Numbers on the jerseys of the starting defense for the Seattle Seahawks when they won Super Bowl XLVIII: 31, 28, 41, 56, 25, 54, 69, 50, 91, 72, 29. (The numbers on the football jerseys don't measure or count anything; they are just substitutes for names.)

d. Top 5 incomes of chief executive officers (in millions of dollars): 131.2, 66.7, 64.4, 53.3, 51.5. (Such "top 5" or "top 10" lists include data that are not at all representative of the larger population.)

e. The 50 mean ages computed from the means in each of the 50 states. (If you calculate the mean of those 50 values, the result is not the mean age of people in the entire United States. The population sizes of the 50 different states must be taken into account, as described in the *weighted mean* introduced in Part 2 of this section.)

> **YOUR TURN** For Exercise 5 "Football Player Numbers," determine why the mean and median are not meaningful.

In the spirit of describing, exploring, and comparing data, we provide Table 3-1 on the next page, which summarizes the different measures of center for the smartphone data speeds referenced in the Chapter Problem. The data are listed in Data Set 32 "Airport Data Speeds" in Appendix B. Figure 3-1 on page 81 suggests that Verizon has the fastest speeds, and comparison of the means and medians in Table 3-1 also suggests that Verizon has the fastest speeds. The following chapters will describe other tools that can be used for an effective comparison.

Rounding Error Changes World Record

Rounding errors can often have disastrous results. Justin Gatlin was elated when he set the world record as the person to run 100 meters in the fastest time of 9.76 seconds. His record time lasted only 5 days, when it was revised to 9.77 seconds, so Gatlin then tied the world record instead of breaking it. His actual time was 9.766 seconds, and it should have been rounded up to 9.77 seconds, but the person doing the timing didn't know that a button had to be pressed for proper rounding. Gatlin's agent said that he (Gatlin) was very distraught and that the incident is "a total embarrassment to the IAAF (International Association of Athletics Federations) and our sport."

TABLE 3-1 Comparison of smartphone data speeds (Mbps) at airports

	Verizon	Sprint	AT&T	T-Mobile
Mean	17.60	3.71	10.70	10.99
Median	13.90	1.60	8.65	9.70
Mode	4.5, 11.1	0.3	2.7	3.2, 4.4, 5.1, 13.3, 15.0, 16.7, 27.3
Midrange	39.30	15.30	19.80	14.00

PART 2   Beyond the Basics of Measures of Center

Calculating the Mean from a Frequency Distribution

Formula 3-2 is the same calculation for the mean that was presented in Part 1, but it incorporates this approach: When working with data summarized in a frequency distribution, we make calculations possible by pretending that all sample values in each class are equal to the class midpoint. Formula 3-2 is not really a new concept; it is simply a variation of Formula 3-1 (mean).

FORMULA 3-2   MEAN FROM A FREQUENCY DISTRIBUTION

First multiply each frequency and
class midpoint; then add the products.
↓

$$\bar{x} = \frac{\Sigma(f \cdot x)}{\Sigma f} \quad \text{(Result is an approximation)}$$

↑
Sum of frequencies
(equal to n)

Example 7 illustrates the procedure for finding the mean from a frequency distribution.

EXAMPLE 7   Computing the Mean from a Frequency Distribution

The first two columns of Table 3-2 shown here are the same as the frequency distribution of Table 2-2 from Chapter 2. Use the frequency distribution in the first two columns of Table 3-2 to find the mean.

TABLE 3-2 McDonald's Lunch Service Times

Time (seconds)	Frequency f	Class Midpoint x	$f \cdot x$
75–124	11	99.5	1094.5
125–174	24	149.5	3588.0
175–224	10	199.5	1995.0
225–274	3	249.5	748.5
275–324	2	299.5	599.0
Totals:	$\Sigma f = 50$		$\Sigma(f \cdot x) = 8025.0$

SOLUTION

Remember, when working with data summarized in a frequency distribution, we make calculations possible by pretending that all sample values in each class are equal to the class midpoint. For example, consider the first class interval of 75–124 with a frequency of 11. We pretend that each of the 11 service times is 99.5 sec (the class midpoint). With the service time of 99.5 repeated 11 times, we have a total of $99.5 \cdot 11 = 1094.5$, as shown in the last column of Table 3-2. We can then add those results to find the sum of all sample values.

The bottom row of Table 3-2 shows the two components we need for the calculation of the mean (as in Formula 3-2): $\Sigma f = 50$ and $\Sigma (f \cdot x) = 8025.0$. We calculate the mean using Formula 3-2 as follows:

$$\bar{x} = \frac{\Sigma (f \cdot x)}{\Sigma f} = \frac{8025.0}{50} = 160.5 \text{ seconds}$$

The result of $\bar{x} = 160.5$ seconds is an *approximation* because it is based on the use of class midpoint values instead of the original list of service times. The mean of 160.2 seconds found by using all of the original service times is a more accurate result.

> YOUR TURN Do Exercise 29 "Age of Best Actresses."

Calculating a Weighted Mean

When different x data values are assigned different weights w, we can compute a **weighted mean**. Formula 3-3 can be used to compute the weighted mean.

FORMULA 3-3

$$\text{Weighted mean: } \bar{x} = \frac{\Sigma (w \cdot x)}{\Sigma w}$$

Formula 3-3 tells us to first multiply each weight w by the corresponding value x, then to add the products, and then finally to divide that total by the sum of the weights, Σw.

EXAMPLE 8 Computing Grade-Point Average

In her first semester of college, a student of the author took five courses. Her final grades, along with the number of credits for each course, were A (3 credits), A (4 credits), B (3 credits), C (3 credits), and F (1 credit). The grading system assigns quality points to letter grades as follows: A = 4; B = 3; C = 2; D = 1; F = 0. Compute her grade-point average.

SOLUTION

Use the numbers of credits as weights: $w = 3, 4, 3, 3, 1$. Replace the letter grades of A, A, B, C, and F with the corresponding quality points: $x = 4, 4, 3, 2, 0$. We now use Formula 3-3 as shown below. The result is a first-semester grade-point average of 3.07. (In using the preceding round-off rule, the result should be rounded to 3.1, but it is common to round grade-point averages to two decimal places.)

$$\begin{aligned} \bar{x} &= \frac{\Sigma (w \cdot x)}{\Sigma w} \\ &= \frac{(3 \times 4) + (4 \times 4) + (3 \times 3) + (3 \times 2) + (1 \times 0)}{3 + 4 + 3 + 3 + 1} \\ &= \frac{43}{14} = 3.07 \end{aligned}$$

> YOUR TURN Do Exercise 33 "Weighted Mean."

TECH CENTER

Descriptive Statistics Display Examples

Access tech supplements, videos, and data sets at **www.TriolaStats.com**

The following displays are based on Verizon data speeds from Data Set 32 "Airport Data Speeds."

Statdisk

Explore Data - Column 2

Sample Size, n: 50
Mean: 17.598
Median: 13.9
Midrange: 39.3
RMS: 23.6877
Variance, s^2: 256.5484
Standard Deviation, s: 16.01713
Mean Absolute Deviation: 10.66528
Range: 77
Coefficient of Variance: 91.02%

Minimum: 0.8
1st Quartile: 7.9
2nd Quartile: 13.9
3rd Quartile: 21.5
Maximum: 77.8

Sum: 879.9
Sum of Squares: 28055.35

95% CI for the Mean:
13.046 < mean <22.15

95% CI for the Standard Deviation:
13.3796 < SD < 19.9595

95% CI for the Variance:
179.0149 < VAR < 398.3804

TI-83/84 Plus

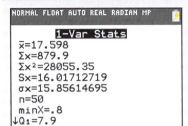

```
NORMAL FLOAT AUTO REAL RADIAN MP
          1-Var Stats
x̄=17.598
Σx=879.9
Σx²=28055.35
Sx=16.01712719
σx=15.85614695
n=50
minX=.8
↓Q₁=7.9
```

```
NORMAL FLOAT AUTO REAL RADIAN MP
          1-Var Stats
↑Sx=16.01712719
σx=15.85614695
n=50
minX=.8
Q₁=7.9
Med=13.9
Q₃=21.5
maxX=77.8
```

Minitab

Descriptive Statistics: VERIZON

Variable	N	N*	Mean	SE Mean	StDev	Minimum	Q1	Median	Q3	Maximum
VERIZON	50	0	17.60	2.27	16.02	0.80	7.85	13.90	21.68	77.80

StatCrunch

Summary statistics:

Column	n	Mean	Variance	Std. dev.	Std. err.	Median	Range	Min	Max	Q1	Q3
VERIZON	50	17.598	256.54836	16.017127	2.2651638	13.9	77	0.8	77.8	7.9	21.5

Excel - Data Analysis Toolpak

VERIZON	
Mean	17.598
Standard Error	2.26516385
Median	13.9
Mode	11.1
Standard Deviation	16.01712719
Sample Variance	256.5483633
Kurtosis	5.485898877
Skewness	2.193633453
Range	77
Minimum	0.8
Maximum	77.8
Sum	879.9
Count	50

Excel - XLSTAT Add-In

Statistic	VERIZON
Nbr. of observations	50
Minimum	0.800
Maximum	77.800
Freq. of minimum	1
Freq. of maximum	1
Range	77.000
1st Quartile	8.400
Median	13.900
3rd Quartile	21.400
Sum	879.900
Mean	17.598
Variance (n)	251.417
Variance (n-1)	256.548
Standard deviation (n)	15.856
Standard deviation (n-1)	16.017

TECH CENTER

 Descriptive Statistics

Access tech supplements, videos, and data sets at **www.TriolaStats.com**

Statdisk

1. Click **Data** in the top menu.
2. Select **Explore Data-Descriptive Statistics** from the dropdown menu.
3. Select the desired data column.
4. Click **Evaluate** to view descriptive statistics.

Minitab

1. Click **Stat** in the top menu.
2. Select **Basic Statistics** from the drop-down menu and then select **Display Descriptive Statistics**.
3. Double click on the desired data column so that it appears in the *Variables* window.
4. Click **OK** to view descriptive statistics.

TIP: Click the **Statistics** button above **OK** to select individual statistics you want displayed.

StatCrunch

1. Click on **Stat** in the top menu.
2. Select **Summary Stats** from the dropdown menu and then select **Columns**.
3. Select the desired data column.
4. Click **Compute!** to view descriptive statistics.

TIP: Customize descriptive statistics by selecting items under **Statistics**.

TI-83/84 Plus Calculator

1. Press **STAT**, then select **CALC** from the top menu.
2. Select **1-Var Stats** and press **ENTER**.
3. Enter the name of list that includes the desired data (e.g., L1).
4. Select **Calculate** and press **ENTER** to view descriptive statistics.

TIP: Press ⏷ to view additional statistics that don't fit on the initial screen.

Excel

XLSTAT Add-In

1. Click on the **XLSTAT** tab in the Ribbon and then click **Describing Data**.
2. Select **Descriptive statistics** from the dropdown menu.
3. Check the **Quantitative Data** box and enter the desired data range. If the first row of data contains a label, also check the **Sample labels** box.
4. Click **OK** to view descriptive statistics.

Excel Data Analysis Add-In

1. Click on the **Data** tab in the Ribbon and then select **Data Analysis** in the top menu.
2. Select **Descriptive Statistics** under *Analysis Tools*.
3. Enter the desired data range for **Input Range**. If the first row of data contains a label, also check the **Labels in First Row** box.
4. Check the **Summary Statistics** box and click **OK** to view descriptive statistics.

3-1 Basic Skills and Concepts

Statistical Literacy and Critical Thinking

1. Average The defunct website IncomeTaxList.com listed the "average" annual income for Florida as $35,031. What is the role of the term *average* in statistics? Should another term be used in place of *average*?

2. What's Wrong? *USA Today* published a list consisting of the state tax on each gallon of gas. If we add the 50 state tax amounts and then divide by 50, we get 27.3 cents. Is the value of 27.3 cents the mean amount of state sales tax paid by all U.S. drivers? Why or why not?

3. Measures of Center In what sense are the mean, median, mode, and midrange measures of "center"?

4. Resistant Measures Here are four of the Verizon data speeds (Mbps) from Figure 3-1: 13.5, 10.2, 21.1, 15.1. Find the mean and median of these four values. Then find the mean and median after including a fifth value of 142, which is an outlier. (One of the Verizon data speeds is 14.2 Mbps, but 142 is used here as an error resulting from an entry with a missing decimal point.) Compare the two sets of results. How much was the mean affected by the inclusion of the outlier? How much is the median affected by the inclusion of the outlier?

Critical Thinking. *For Exercises 5–20, watch out for these little buggers. Each of these exercises involves some feature that is somewhat tricky. Find the (a)* **mean,** *(b)* **median,** *(c)* **mode,** *(d)* **midrange,** *and then answer the given question.*

5. Football Player Numbers Listed below are the jersey numbers of 11 players randomly selected from the roster of the Seattle Seahawks when they won Super Bowl XLVIII. What do the results tell us?

$$89 \quad 91 \quad 55 \quad 7 \quad 20 \quad 99 \quad 25 \quad 81 \quad 19 \quad 82 \quad 60$$

6. Football Player Weights Listed below are the weights in pounds of 11 players randomly selected from the roster of the Seattle Seahawks when they won Super Bowl XLVIII (the same players from the preceding exercise). Are the results likely to be representative of all National Football League (NFL) players?

$$189 \quad 254 \quad 235 \quad 225 \quad 190 \quad 305 \quad 195 \quad 202 \quad 190 \quad 252 \quad 305$$

7. Celebrity Net Worth Listed below are the highest amounts of net worth (in millions of dollars) of celebrities. The celebrities are Tom Cruise, Will Smith, Robert De Niro, Drew Carey, George Clooney, John Travolta, Samuel L. Jackson, Larry King, Demi Moore, and Bruce Willis. What do the results tell us about the population of all celebrities? Based on the nature of the amounts, what can be inferred about their precision?

$$250 \quad 200 \quad 185 \quad 165 \quad 160 \quad 160 \quad 150 \quad 150 \quad 150 \quad 150$$

8. What Happens in Vegas . . . Listed below are prices in dollars for one night at different hotels located on Las Vegas Boulevard (the "Strip"). If you decide to stay at one of these hotels, what statistic is most relevant, other than the measures of center? Apart from price, identify one other important factor that would affect your choice.

$$212 \quad 77 \quad 121 \quad 104 \quad 153 \quad 264 \quad 195 \quad 244$$

9. Hurricanes Listed below are the numbers of Atlantic hurricanes that occurred in each year. The data are listed in order by year, starting with the year 2000. What important feature of the data is not revealed by any of the measures of center?

$$8 \quad 9 \quad 8 \quad 7 \quad 9 \quad 15 \quad 5 \quad 6 \quad 8 \quad 4 \quad 12 \quad 7 \quad 8 \quad 2$$

10. Peas in a Pod Biologists conducted experiments to determine whether a deficiency of carbon dioxide in the soil affects the phenotypes of peas. Listed below are the phenotype codes, where $1 =$ smooth-yellow, $2 =$ smooth-green, $3 =$ wrinkled-yellow, and $4 =$ wrinkled-green. Can the measures of center be obtained for these values? Do the results make sense?

$$2 \quad 1 \quad 1 \quad 1 \quad 1 \quad 1 \quad 1 \quad 4 \quad 1 \quad 2 \quad 2 \quad 1 \quad 2 \quad 3 \quad 3 \quad 2 \quad 3 \quad 1 \quad 3 \quad 1 \quad 3 \quad 1 \quad 3 \quad 2 \quad 2$$

11. TV Prices Listed below are selling prices (dollars) of TVs that are 60 inches or larger and rated as a "best buy" by *Consumer Reports* magazine. Are the resulting statistics representative of the population of all TVs that are 60 inches and larger? If you decide to buy one of these TVs, what statistic is most relevant, other than the measures of central tendency?

$$1800 \quad 1500 \quad 1200 \quad 1500 \quad 1400 \quad 1600 \quad 1500 \quad 950 \quad 1600 \quad 1150 \quad 1500 \quad 1750$$

12. Cell Phone Radiation Listed below are the measured radiation absorption rates (in W/kg) corresponding to these cell phones: iPhone 5S, BlackBerry Z30, Sanyo Vero, Optimus V, Droid Razr, Nokia N97, Samsung Vibrant, Sony Z750a, Kyocera Kona, LG G2, and Virgin Mobile Supreme. The data are from the Federal Communications Commission (FCC). The media often report about the dangers of cell phone radiation as a cause of cancer. The FCC has a standard that a cell phone absorption rate must be 1.6 W/kg or less. If you are planning to purchase a cell phone, are any of the measures of center the most important statistic? Is there another statistic that is most relevant? If so, which one?

$$1.18 \quad 1.41 \quad 1.49 \quad 1.04 \quad 1.45 \quad 0.74 \quad 0.89 \quad 1.42 \quad 1.45 \quad 0.51 \quad 1.38$$

13. Caffeine in Soft Drinks Listed below are measured amounts of caffeine (mg per 12 oz of drink) obtained in one can from each of 20 brands (7-UP, A&W Root Beer, Cherry Coke, . . . , Tab). Are the statistics representative of the population of all cans of the same 20 brands consumed by Americans?

0 0 34 34 34 45 41 51 55 36 47 41 0 0 53 54 38 0 41 47

14. Firefighter Fatalities Listed below are the numbers of heroic firefighters who lost their lives in the United States each year while fighting forest fires. The numbers are listed in order by year, starting with the year 2000. What important feature of the data is not revealed by any of the measures of center?

20 18 23 30 20 12 24 9 25 15 8 11 15 34

15. Foot Lengths Listed below are foot lengths in inches of randomly selected Army women measured in the 1988 Anthropometric Survey (ANSUR). Are the statistics representative of the current population of all Army women?

10.4 9.3 9.1 9.3 10.0 9.4 8.6 9.8 9.9 9.1 9.1

16. Most Expensive Colleges Listed below are the annual costs (dollars) of tuition and fees at the 10 most expensive colleges in the United States for a recent year (based on data from *U.S. News & World Report*). The colleges listed in order are Columbia, Vassar, Trinity, George Washington, Carnegie Mellon, Wesleyan, Tulane, Bucknell, Oberlin, and Union. What does this "top 10" list tell us about those costs for the population of all U.S. college tuitions?

49,138 47,890 47,510 47,343 46,962 46,944 46,930 46,902 46,870 46,785

17. Diamond Ring Listed below in dollars are the amounts it costs for marriage proposal packages at the different Major League Baseball stadiums. Five of the teams don't allow proposals. Are there any outliers?

39   50   50   50   55   55   75   85   100   115   175   175   200
209   250   250   350   400   450   500   500   500   500   1500   2500

18. Sales of LP Vinyl Record Albums Listed below are annual U.S. sales of vinyl record albums (millions of units). The numbers of albums sold are listed in chronological order, and the last entry represents the most recent year. Do the measures of center give us any information about a changing trend over time?

0.3   0.6   0.8   1.1   1.1   1.4   1.4   1.5   1.2   1.3   1.4   1.2   0.9   0.9
1   1.9   2.5   2.8   3.9   4.6   6.1

19. California Smokers In the California Health Interview Survey, randomly selected adults are interviewed. One of the questions asks how many cigarettes are smoked per day, and results are listed below for 50 randomly selected respondents. How well do the results reflect the smoking behavior of California adults?

9 10 10 20 40 50 0 0 0 0 0 0 0 0 0
0 0 0 0 0 0 0 0 0 0 0 0 0 0
0 0 0 0 0 0 0 0 0 0 0 0 0 0
0 0 0 0 0

20. Speed Dating In a study of speed dating conducted at Columbia University, female subjects were asked to rate the attractiveness of their male dates, and a sample of the results is listed below (1 = not attractive; 10 = extremely attractive). Can the results be used to describe the attractiveness of the population of adult males?

5 8 3 8 6 10 3 7 9 8 5 5 6 8 8 7 3 5 5 6 8 7 8 8 8 7

In Exercises 21–24, find the **mean** *and* **median** *for each of the two samples, then compare the two sets of results.*

21. Blood Pressure A sample of blood pressure measurements is taken from Data Set 1 "Body Data" in Appendix B, and those values (mm Hg) are listed below. The values are matched so that 10 subjects each have systolic and diastolic measurements. (Systolic is a measure of the force of blood being pushed through arteries, but diastolic is a measure of blood pressure when the heart is at rest between beats.) Are the measures of center the best statistics to use with these data? What else might be better?

Systolic:	118	128	158	96	156	122	116	136	126	120
Diastolic:	80	76	74	52	90	88	58	64	72	82

22. Parking Meter Theft Listed below are amounts (in millions of dollars) collected from parking meters by Brinks and others in New York City during similar time periods. A larger data set was used to convict five Brinks employees of grand larceny. The data were provided by the attorney for New York City, and they are listed on the Data and Story Library (DASL) website. Do the limited data listed here show evidence of stealing by Brinks employees?

Collection Contractor Was Brinks	1.3	1.5	1.3	1.5	1.4	1.7	1.8	1.7	1.7	1.6
Collection Contractor Was Not Brinks	2.2	1.9	1.5	1.6	1.5	1.7	1.9	1.6	1.6	1.8

23. Pulse Rates Listed below are pulse rates (beats per minute) from samples of adult males and females (from Data Set 1 "Body Data" in Appendix B). Does there appear to be a difference?

Male:	86	72	64	72	72	54	66	56	80	72	64	64	96	58	66
Female:	64	84	82	70	74	86	90	88	90	90	94	68	90	82	80

24. Bank Queues Waiting times (in seconds) of customers at the Madison Savings Bank are recorded with two configurations: single customer line; individual customer lines. Carefully examine the data to determine whether there is a difference between the two data sets that is not apparent from a comparison of the measures of center. If so, what is it?

Single Line	390	396	402	408	426	438	444	462	462	462
Individual Lines	252	324	348	372	402	462	462	510	558	600

Large Data Sets from Appendix B. *In Exercises 25–28, refer to the indicated data set in Appendix B. Use software or a calculator to find the* **means** *and* **medians***.*

25. Tornadoes Use the F-scale measurements from the tornadoes listed in Data Set 22 "Tornadoes" in Appendix B. Among the 500 tornadoes, how many have missing F-scale measurements? (*Caution:* In some technologies, missing data are represented by a constant such as −9 or 9999.)

26. Earthquakes Use the magnitudes (Richter scale) of the 600 earthquakes listed in Data Set 21 "Earthquakes" in Appendix B. In 1989, the San Francisco Bay Area was struck with an earthquake that measured 7.0 on the Richter scale. That earthquake occurred during the warm-up period for the third game of the baseball World Series. Is the magnitude of that World Series earthquake an *outlier* when considered in the context of the sample data given in Data Set 21? Explain.

27. Body Temperatures Refer to Data Set 3 "Body Temperatures" in Appendix B and use the body temperatures for 12:00 AM on day 2. Do the results support or contradict the common belief that the mean body temperature is 98.6°F?

28. Births Use the birth weights (grams) of the 400 babies listed in Data Set 4 "Births" in Appendix B. Examine the list of birth weights to make an observation about those numbers. How does that observation affect the way that the results should be rounded?

*In Exercises 29–32, find the **mean** of the data summarized in the frequency distribution. Also, compare the **computed means** to the **actual means** obtained by using the original list of data values, which are as follows: (Exercise 29) 36.2 years; (Exercise 30) 44.1 years; (Exercise 31) 224.3; (Exercise 32) 255.1.*

29.

Age (yr) of Best Actress When Oscar Was Won	Frequency
20–29	29
30–39	34
40–49	14
50–59	3
60–69	5
70–79	1
80–89	1

30.

Age (yr) of Best Actor When Oscar Was Won	Frequency
20–29	1
30–39	28
40–49	36
50–59	15
60–69	6
70–79	1

31.

Blood Platelet Count of Males (1000 cells / μL)	Frequency
0–99	1
100–199	51
200–299	90
300–399	10
400–499	0
500–599	0
600–699	1

32.

Blood Platelet Count of Females (1000 cells / μL)	Frequency
100–199	25
200–299	92
300–399	28
400–499	0
500–599	2

33. Weighted Mean A student of the author earned grades of A, C, B, A, and D. Those courses had these corresponding numbers of credit hours: 3, 3, 3, 4, and 1. The grading system assigns quality points to letter grades as follows: A = 4; B = 3; C = 2; D = 1; F = 0. Compute the grade-point average (GPA) and round the result with two decimal places. If the dean's list requires a GPA of 3.00 or greater, did this student make the dean's list?

34. Weighted Mean A student of the author earned grades of 63, 91, 88, 84, and 79 on her five regular statistics tests. She earned grades of 86 on the final exam and 90 on her class projects. Her combined homework grade was 70. The five regular tests count for 60% of the final grade, the final exam counts for 10%, the project counts for 15%, and homework counts for 15%. What is her weighted mean grade? What letter grade did she earn (A, B, C, D, or F)? Assume that a mean of 90 or above is an A, a mean of 80 to 89 is a B, and so on.

3-1 Beyond the Basics

35. Degrees of Freedom Five pulse rates randomly selected from Data Set 1 "Body Data" in Appendix B have a mean of 78.0 beats per minute. Four of the pulse rates are 82, 78, 56, and 84.

a. Find the missing value.

b. We need to create a list of n values that have a specific known mean. We are free to select any values we desire for some of the n values. How many of the n values can be freely assigned before the remaining values are determined? (The result is referred to as the *number of degrees of freedom.*)

36. Censored Data Data Set 15 "Presidents" in Appendix B lists the numbers of years that U.S. presidents lived after their first inauguration. As of this writing, five of the presidents are still alive and after their first inauguration they have lived 37 years, 25 years, 21 years, 13 years, and 5 years so far. We might use the values of 37+, 25+, 21+, 13+, and 5+, where the positive signs indicate that the actual value is equal to or greater than the current value. (These values are said to be *censored* at the current time that this list was compiled.) If you use the values in Data Set 15 and ignore the presidents who are still alive, what is the mean? If you use the values given in Data Set 15 along with the additional values of 37+, 25+, 21+, 13+, and 5+, what do we know about the mean? Do the two results differ by much?

37. Trimmed Mean Because the mean is very sensitive to extreme values, we say that it is not a *resistant* measure of center. By deleting some low values and high values, the **trimmed mean** is more resistant. To find the 10% trimmed mean for a data set, first arrange the data in order, then delete the bottom 10% of the values and delete the top 10% of the values, then calculate the mean of the remaining values. Use the axial loads (pounds) of aluminum cans listed below (from Data Set 30 "Aluminum Cans" in Appendix B) for cans that are 0.0111 in. thick. An axial load is the force at which the top of a can collapses. Identify any outliers, then compare the median, mean, 10% trimmed mean, and 20% trimmed mean.

$$247 \quad 260 \quad 268 \quad 273 \quad 276 \quad 279 \quad 281 \quad 283 \quad 284 \quad 285 \quad 286 \quad 288$$
$$289 \quad 291 \quad 293 \quad 295 \quad 296 \quad 299 \quad 310 \quad 504$$

38. Harmonic Mean The **harmonic mean** is often used as a measure of center for data sets consisting of rates of change, such as speeds. It is found by dividing the number of values n by the sum of the *reciprocals* of all values, expressed as

$$\frac{n}{\sum \frac{1}{x}}$$

(No value can be zero.) The author drove 1163 miles to a conference in Orlando, Florida. For the trip to the conference, the author stopped overnight, and the mean speed from start to finish was 38 mi/h. For the return trip, the author stopped only for food and fuel, and the mean speed from start to finish was 56 mi/h. Find the harmonic mean of 38 mi/h and 56 mi/h to find the true "average" speed for the round trip.

39. Geometric Mean The **geometric mean** is often used in business and economics for finding average rates of change, average rates of growth, or average ratios. To find the geometric mean of n values (all of which are positive), first multiply the values, then find the nth root of the product. For a 6-year period, money deposited in annual certificates of deposit had annual interest rates of 5.154%, 2.730%, 0.488%, 0.319%, 0.313%, and 0.268%. Identify the single percentage growth rate that is the same as the five consecutive growth rates by computing the geometric mean of 1.05154, 1.02730, 1.00488, 1.00319, 1.00313, and 1.00268.

40. Quadratic Mean The **quadratic mean** (or **root mean square**, or **R.M.S.**) is used in physical applications, such as power distribution systems. The quadratic mean of a set of values is obtained by squaring each value, adding those squares, dividing the sum by the number of values n, and then taking the square root of that result, as indicated below:

$$\text{Quadratic mean} = \sqrt{\frac{\sum x^2}{n}}$$

Find the R.M.S. of these voltages measured from household current: $0, 60, 110, -110, -60, 0$. How does the result compare to the mean?

41. Median When data are summarized in a frequency distribution, the median can be found by first identifying the *median class,* which is the class that contains the median. We then assume that the values in that class are evenly distributed and we interpolate. Letting n denote the sum of all class frequencies, and letting m denote the sum of the class frequencies that *precede* the median class, the median can be estimated as shown below.

$$(\text{lower limit of median class}) + (\text{class width}) \left(\frac{\left(\frac{n+1}{2} \right) - (m+1)}{\text{frequency of median class}} \right)$$

Use this procedure to find the median of the frequency distribution given in Table 3-2 on page 88. How far is that result from the median found from the original list of McDonald's lunch service times listed in Data Set 25 "Fast Food" in Appendix B?

3-2   Measures of Variation

Key Concept Variation is the single most important topic in statistics, so this is the single most important section in this book. This section presents three important measures of variation: *range, standard deviation,* and *variance.* These statistics are numbers, but our focus is not just computing those numbers but developing the ability to *interpret* and *understand* them. This section is not a study of arithmetic; it is about understanding and interpreting measures of variation, especially the standard deviation.

> **STUDY HINT** Part 1 of this section presents basic concepts of variation, and Part 2 presents additional concepts related to the standard deviation. Part 1 and Part 2 both include formulas for computation, but do not spend too much time memorizing formulas or doing arithmetic calculations. Instead, focus on *understanding* and *interpreting* values of standard deviation.

PART 1   Basic Concepts of Variation

To visualize the property of variation, see Figure 3-2, which illustrates waiting times (seconds) of customers at a bank under two different conditions: (1) All customers enter a *single* line that feeds different teller stations; (2) all customers choose to join the line at one of *several* different teller stations. Verify this important observation: The waiting times with the single line (top dotplot) have less *variation* than the waiting times with multiple lines (bottom dotplot). *Both sets of waiting times have the same mean of 100.0 seconds, they have the same median of 100.0 seconds, and they have the same mode of 100 seconds.* Those measures of center do not "see" the difference in variation.

To keep our round-off rules as consistent and as simple as possible, we will round the measures of variation using this rule:

> **ROUND-OFF RULE FOR MEASURES OF VARIATION** When rounding the value of a measure of variation, carry one more decimal place than is present in the original set of data.

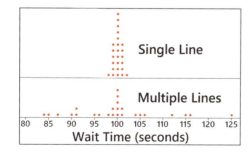

FIGURE 3-2   Dotplots of Waiting Times (seconds) with Single/Multiple Lines

Improving Quality This single line/multiple line illustration is wonderful because banks actually did change from multiple lines to a single line not because it made them more efficient, not because customer waiting times were reduced, but because customers are happier with waiting times with *less variation.* The change did not affect the measures of center, but banks instituted the change to reduce variation. An important goal of business and industry is this: ***Improve quality by reducing variation.***

Got a Second?

The time unit of 1 second is defined to be "the duration of 9,192,631,770 periods of the radiation corresponding to the transition between the two hyperfine levels of the ground state of the cesium-133 atom." That definition redefines time to be based on the behavior of atoms instead of the earth's motion. It results in accuracy of ± 1 second in 10,000,000 years, which is the most accurate measurement we use. Because it is so accurate, the second is being used to define other quantities, such as the meter. The meter was once defined as 1/10,000,000 of the distance along the surface of the earth between the North Pole and the equator (passing through Paris). The meter is now defined as the length of the distance traveled by light in a vacuum during a time interval of 1/299,792,458 sec.

When dealing with time measurement devices, the traditional standard deviation has been found to be poor because of a trend in which the mean changes over time. Instead, other special measures of variation are used, such as Allan variance, total variance, and TheoH.

Unrelated to statistics but nonetheless interesting is the fact that ads for watches usually show a watch with a time close to 10:10. That time allows the brand name to be visible, and it creates a subliminal image of a happy face. The time of 10:10 has been the industry standard since the 1940s.

Range

Let's begin with the range because it is quick and easy to compute, but it is not as important as other measures of variation.

> **DEFINITION**
>
> The **range** of a set of data values is the difference between the maximum data value and the minimum data value.
>
> $$\text{Range} = \textbf{(maximum data value)} - \textbf{(minimum data value)}$$

Important Property of the Range

- The range uses only the maximum and the minimum data values, so it is very sensitive to extreme values. The range is not *resistant*.

- Because the range uses only the maximum and minimum values, it does not take every value into account and therefore does not truly reflect the variation among all of the data values.

EXAMPLE 1   Range

Find the range of these Verizon data speeds (Mbps): 38.5, 55.6, 22.4, 14.1, 23.1. (These are the first five Verizon data speeds listed in Data Set 32 "Airport Data Speeds" in Appendix B.)

SOLUTION

The range is found by subtracting the lowest value from the largest value, so we get

$$\text{Range} = \text{(maximum value)} - \text{(minimum value)} = 55.6 - 14.1 = 41.50 \text{ Mbps}$$

The range of 41.50 Mbps is shown with one more decimal place than is present in the original data values.

YOUR TURN Find the range in Exercise 5 "Football Player Numbers."

Standard Deviation of a Sample

The *standard deviation* is the measure of variation most commonly used in statistics.

> **DEFINITION**
>
> The **standard deviation** of a set of sample values, denoted by *s,* is a measure of how much data values deviate away from the mean. It is calculated by using Formula 3-4 or 3-5. Formula 3-5 is just a different version of Formula 3-4; both formulas are algebraically the same.

The standard deviation found from sample data is a statistic denoted by *s,* but the standard deviation found from population data is a parameter denoted by σ. The formula for σ is slightly different with division by the population size N used instead of division by $n - 1$. The population standard deviation σ will be discussed later.

Notation

$s = $ *sample* standard deviation

$\sigma = $ *population* standard deviation

FORMULA 3-4

$$s = \sqrt{\frac{\Sigma (x - \bar{x})^2}{n - 1}} \quad \text{sample standard deviation}$$

FORMULA 3-5

$$s = \sqrt{\frac{n(\Sigma x^2) - (\Sigma x)^2}{n(n - 1)}} \quad \text{shortcut formula for sample standard deviation (used by calculators and software)}$$

Later we give the reasoning behind these formulas, but for now we recommend that you use Formula 3-4 for an example or two, and then learn how to find standard deviation values using a calculator or software.

Important Properties of Standard Deviation

- The standard deviation is a measure of how much data values deviate away from the *mean*.

- The value of the standard deviation s is never negative. It is zero only when all of the data values are exactly the same.

- Larger values of s indicate greater amounts of variation.

- The standard deviation s can increase dramatically with one or more outliers.

- The units of the standard deviation s (such as minutes, feet, pounds) are the same as the units of the original data values.

- The sample standard deviation s is a **biased estimator** of the population standard deviation σ, which means that values of the sample standard deviation s do not center around the value of σ. (This is explained in Part 2.)

Example 2 illustrates a calculation using Formula 3-4 because that formula better illustrates that the standard deviation is based on deviations of sample values away from the mean.

EXAMPLE 2   Calculating Standard Deviation with Formula 3-4

Use Formula 3-4 to find the standard deviation of these Verizon data speed times (in Mbps): 38.5, 55.6, 22.4, 14.1, 23.1.

SOLUTION

The left column of Table 3-3 summarizes the general procedure for finding the standard deviation using Formula 3-4, and the right column illustrates that procedure for the sample values 38.5, 55.6, 22.4, 14.1, 23.1. The result shown in Table 3-3 is 16.45 Mbps, which is rounded to one more decimal place than is present in the original list of sample values (38.5, 55.6, 22.4, 14.1, 23.1). Also, the units for the standard deviation are the same as the units of the original data. Because the original data all have units of Mbps, the standard deviation is 16.45 Mbps.

YOUR TURN Find the standard deviation in Exercise 5 "Football Player Numbers."

More Stocks, Less Risk

In their book *Investments*, authors Zvi Bodie, Alex Kane, and Alan Marcus state that "the average standard deviation for returns of portfolios composed of only one stock was 0.554. The average portfolio risk fell rapidly as the number of stocks included in the portfolio increased." They note that with 32 stocks, the standard deviation is 0.325, indicating much less variation and risk. They make the point that with only a few stocks, a portfolio has a high degree of "firm-specific" risk, meaning that the risk is attributable to the few stocks involved. With more than 30 stocks, there is very little firm-specific risk; instead, almost all of the risk is "market risk," attributable to the stock market as a whole. They note that these principles are "just an application of the well-known law of averages."

TABLE 3-3

General Procedure for Finding Standard Deviation with Formula 3-4	Specific Example Using These Sample Values: 38.5, 55.6, 22.4, 14.1, 23.1
Step 1: Compute the mean $\bar{x}$.	The sum of 38.5, 55.6, 22.4, 14.1, 23.1 is 153.7; therefore: $$\bar{x} = \frac{\Sigma x}{n} = \frac{38.5 + 55.6 + 22.4 + 14.1 + 23.1}{5}$$ $$= \frac{153.7}{5} = 30.74$$
Step 2: Subtract the mean from each individual sample value. [The result is a list of deviations of the form $(x - \bar{x})$.]	Subtract the mean of 30.74 from each sample value to get these deviations away from the mean: 7.76, 24.86, −8.34, −16.64, −7.64.
Step 3: Square each of the deviations obtained from Step 2. [This produces numbers of the form $(x - \bar{x})^2$.]	The squares of the deviations from Step 2 are: 60.2176, 618.0196, 69.5556, 276.8896, 58.3696.
Step 4: Add all of the squares obtained from Step 3. The result is $\Sigma(x - \bar{x})^2$.	The sum of the squares from Step 3 is 1083.0520.
Step 5: Divide the total from Step 4 by the number $n - 1$, which is 1 less than the total number of sample values present.	With $n = 5$ data values, $n - 1 = 4$, so we divide 1083.0520 by 4 to get this result: $$\frac{1083.0520}{4} = 270.7630.$$
Step 6: Find the square root of the result of Step 5. The result is the standard deviation, denoted by s.	The standard deviation is $\sqrt{270.7630} = 16.4548777$. Rounding the result, we get $s = 16.45$ Mbps.

EXAMPLE 3   Calculating Standard Deviation with Formula 3-5

Use Formula 3-5 to find the standard deviation of the Verizon data speeds (Mbps) of 38.5, 55.6, 22.4, 14.1, 23.1 from Example 1.

SOLUTION

Here are the components needed in Formula 3-5.

$n = 5$ (because there are 5 values in the sample)
$\Sigma x = 153.7$ (found by adding the original sample values)
$\Sigma x^2 = 5807.79$ (found by adding the squares of the sample values, as in
$\quad 38.5^2 + 55.6^2 + 22.4^2 + 14.1^2 + 23.1^2 = 5807.79$)

Using Formula 3-5, we get

$$s = \sqrt{\frac{n(\Sigma x^2) - (\Sigma x)^2}{n(n - 1)}} = \sqrt{\frac{5(5807.79) - (153.7)^2}{5(5 - 1)}} = \sqrt{\frac{5415.26}{20}} = 16.45 \text{ Mbps}$$

The result of $s = 16.45$ Mbps is the same as the result in Example 2.

> **YOUR TURN** Find the standard deviation in Exercise 5 "Football Player Numbers."

Range Rule of Thumb for Understanding Standard Deviation

The *range rule of thumb* is a crude but simple tool for understanding and interpreting standard deviation. It is based on the principle that for many data sets, the vast majority (such as 95%) of sample values lie within 2 standard deviations of the mean. We could improve the accuracy of this rule by taking into account such factors as the size of the sample and the distribution, but here we sacrifice accuracy for the sake of simplicity. The concept of *significance* that follows will be enhanced in later chapters, especially those that include the topic of hypothesis tests, which are also called tests of significance. The following range rule of thumb is based on the population mean μ and the population standard deviation σ, but for large and representative samples, we could use $\bar{x}$ and s instead.

Range Rule of Thumb for Identifying Significant Values

Significantly low values are $\mu - 2\sigma$ or lower.

Significantly high values are $\mu + 2\sigma$ or higher.

Values not significant: Between $(\mu - 2\sigma)$ and $(\mu + 2\sigma)$

See Figure 3-3, which illustrates the above criteria.

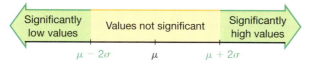

FIGURE 3-3   Range Rule of Thumb for Identifying Significant Values

Range Rule of Thumb for Estimating a Value of the Standard Deviation s
To roughly estimate the standard deviation from a collection of known sample data, use

$$s \approx \frac{\text{range}}{4}$$

EXAMPLE 4 Range Rule of Thumb for Interpreting s

Using the 50 Verizon data speeds listed in Data Set 32 "Airport Data Speeds" in Appendix B, the mean is 17.60 Mbps and the standard deviation is 16.02 Mbps. Use the range rule of thumb to find the limits separating values that are significantly low or significantly high, then determine whether the data speed of 77.8 Mbps is significantly high.

SOLUTION

With a mean of 17.60 and a standard deviation of 16.02, we use the range rule of thumb to find the limits separating values that are significantly low or significantly high, as follows:

Significantly low values are $(17.60 - 2 \times 16.02)$ or lower,
     so significantly low values are -14.44 Mbps or lower.
Significantly high values are $(17.60 + 2 \times 16.02)$ or higher,
     so significantly high values are 49.64 Mbps or higher.
Values not significant: Between -14.44 Mbps and 49.64 Mbps

INTERPRETATION

Based on these results, we expect that typical airport Verizon data speeds are between -14.44 Mbps and 49.64 Mbps. Because the given value of 77.8 Mbps falls above 49.64 Mbps, we can consider it to be significantly high.

YOUR TURN Do Exercise 33 "Pulse Rates of Females."

Variation in Faces

Researchers commented that "if everyone looked more or less the same, there would be total chaos." They studied human body measurements and found that facial traits *varied* more than other body traits, and the greatest variation occurred within the triangle formed by the eyes and mouth. They learned that facial traits vary independently of each other. For example, there is no relationship between the distance between your eyes and how big your mouth is. The researchers stated that our facial variation played an important role in human evolution. (See "Morphological and Population Genomic Evidence That Human Faces Have Evolved to Signal Individual Identity," by Sheehan and Nachman, *Nature Communications,* Vol. 5, No. 4800.)

EXAMPLE 5 **Range Rule of Thumb for Estimating *s***

Use the range rule of thumb to estimate the standard deviation of the sample of 50 Verizon data speeds listed in Data Set 32 "Airport Data Speeds" in Appendix B. Those 50 values have a minimum of 0.8 Mbps and a maximum of 77.8 Mbps.

SOLUTION

The range rule of thumb indicates that we can estimate the standard deviation by finding the range and dividing it by 4. With a minimum of 0.8 and a maximum of 77.8, the range rule of thumb can be used to estimate the standard deviation *s*, as follows:

$$s \approx \frac{\text{range}}{4} = \frac{77.8 - 0.8}{4} = 19.25 \text{ Mbps}$$

INTERPRETATION

The actual value of the standard deviation is $s = 16.02$ Mbps, so the estimate of 19.25 Mbps is in the general neighborhood of the exact result. Because this estimate is based on only the minimum and maximum values, it is generally a rough estimate that could be off by a considerable amount.

YOUR TURN Do Exercise 29 "Estimating Standard Deviation."

Standard Deviation of a Population

The definition of standard deviation and Formulas 3-4 and 3-5 apply to the standard deviation of *sample* data. A slightly different formula is used to calculate the standard deviation σ (lowercase sigma) of a *population*: Instead of dividing by $n - 1$, we divide by the population size N, as shown here:

$$\text{Population standard deviation } \sigma = \sqrt{\frac{\Sigma(x - \mu)^2}{N}}$$

Because we generally deal with sample data, we will usually use Formula 3-4, in which we divide by $n - 1$. Many calculators give both the sample standard deviation and the population standard deviation, but they use a variety of different notations.

CAUTION When using a calculator to find standard deviation, identify the notation used by your particular calculator so that you get the *sample* standard deviation, not the *population* standard deviation.

Variance of a Sample and a Population

So far, we have used the term *variation* as a general description of the amount that values vary among themselves. (The terms *dispersion* and *spread* are sometimes used instead of *variation*.) The term *variance* has a specific meaning.

DEFINITION

The **variance** of a set of values is a measure of variation equal to the square of the standard deviation.

- Sample variance: s^2 = square of the standard deviation *s*.
- Population variance: σ^2 = square of the population standard deviation σ.

Notation Here is a summary of notation for the standard deviation and variance:

s = *sample* standard deviation

s^2 = *sample* variance

σ = *population* standard deviation

σ^2 = *population* variance

Note: Articles in professional journals and reports often use SD for standard deviation and VAR for variance.

Important Properties of Variance

- The units of the variance are the *squares* of the units of the original data values. (If the original data values are in feet, the variance will have units of ft^2; if the original data values are in seconds, the variance will have units of sec^2.)

- The value of the variance can increase dramatically with the inclusion of outliers. (The variance is not *resistant*.)

- The value of the variance is never negative. It is zero only when all of the data values are the same number.

- The sample variance s^2 is an **unbiased estimator** of the population variance σ^2, as described in Part 2 of this section.

The variance is a statistic used in some statistical methods, but for our present purposes, the variance has the serious disadvantage of using units that are *different than the units of the original data set.* This makes it difficult to understand variance as it relates to the original data set. Because of this property, it is better to first focus on the standard deviation when trying to develop an understanding of variation.

PART 2 Beyond the Basics of Variation

In Part 2, we focus on making sense of the standard deviation so that it is not some mysterious number devoid of any practical significance. We begin by addressing common questions that relate to the standard deviation.

Why Is Standard Deviation Defined as in Formula 3-4?

In measuring variation in a set of sample data, it makes sense to begin with the individual amounts by which values deviate from the mean. For a particular data value x, the amount of **deviation** is $x - \bar{x}$. It makes sense to somehow combine those deviations into one number that can serve as a measure of the variation. Adding the deviations isn't good, because the sum will always be zero. To get a statistic that measures variation, it's necessary to avoid the canceling out of negative and positive numbers. One approach is to add absolute values, as in $\Sigma|x - \bar{x}|$. If we find the mean of that sum, we get the **mean absolute deviation** (or **MAD**), which is the mean distance of the data from the mean:

$$\text{Mean absolute deviation} = \frac{\Sigma|x - \bar{x}|}{n}$$

Why Not Use the Mean Absolute Deviation Instead of the Standard Deviation?

Computation of the mean absolute deviation uses absolute values, so it uses an operation that is not "algebraic." (The algebraic operations include addition, multiplication, extracting roots, and raising to powers that are integers or fractions.) The use of absolute values would be simple, but it would create algebraic difficulties in inferential methods of statistics discussed in later chapters. The standard deviation has the advantage of using only algebraic operations. Because it is based on the square root of a sum of squares, the standard deviation closely parallels distance formulas found in algebra. There are many instances where a statistical procedure is based on a similar sum of squares. Consequently, instead of using absolute values, we *square* all deviations $(x - \bar{x})$ so that they are nonnegative, and those squares are used to calculate the standard deviation.

Why Divide by $n - 1$? After finding all of the individual values of $(x - \bar{x})^2$ we combine them by finding their sum. We then divide by $n - 1$ because there are only $n - 1$ values that can assigned without constraint. With a given mean, we can use any numbers for the first $n - 1$ values, but the last value will then be automatically determined. With division by $n - 1$, sample variances s^2 tend to center around the value of the population variance σ^2; with division by n, sample variances s^2 tend to *underestimate* the value of the population variance σ^2.

How Do We Make Sense of a Value of Standard Deviation? Part 1 of this section included the range rule of thumb for interpreting a known value of a standard deviation or estimating a value of a standard deviation. (See Examples 4 and 5.) Two other approaches for interpreting standard deviation are the empirical rule and Chebyshev's theorem.

Empirical (or 68-95-99.7) Rule for Data with a Bell-Shaped Distribution

A concept helpful in interpreting the value of a standard deviation is the **empirical rule**. This rule states that *for data sets having a distribution that is approximately bell-shaped,* the following properties apply. (See Figure 3-4.)

- About 68% of all values fall within 1 standard deviation of the mean.
- About 95% of all values fall within 2 standard deviations of the mean.
- About 99.7% of all values fall within 3 standard deviations of the mean.

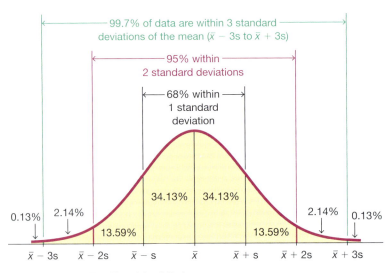

FIGURE 3-4   The Empirical Rule

EXAMPLE 6   The Empirical Rule

IQ scores have a bell-shaped distribution with a mean of 100 and a standard deviation of 15. What percentage of IQ scores are between 70 and 130?

SOLUTION

The key to solving this problem is to recognize that 70 and 130 are each exactly 2 standard deviations away from the mean of 100, as shown below:

$$2 \text{ standard deviations} = 2s = 2\,(15) = 30$$

Therefore, 2 standard deviations from the mean is

$$100 - 30 = 70$$
$$\text{or} \quad 100 + 30 = 130$$

The empirical rule tells us that about 95% of all values are within 2 standard deviations of the mean, so about 95% of all IQ scores are between 70 and 130.

> **YOUR TURN** Do Exercise 41 "The Empirical Rule."

Another concept helpful in understanding or interpreting a value of a standard deviation is **Chebyshev's theorem**. The empirical rule applies only to data sets with bell-shaped distributions, but Chebyshev's theorem applies to *any* data set. Unfortunately, results from Chebyshev's theorem are only approximate. Because the results are lower limits ("at least"), Chebyshev's theorem has limited usefulness.

Chebyshev's Theorem

The proportion of any set of data lying within K standard deviations of the mean is always *at least* $1 - 1/K^2$, where K is any positive number greater than 1. For $K = 2$ and $K = 3$, we get the following statements:

- At least 3/4 (or 75%) of all values lie within 2 standard deviations of the mean.
- At least 8/9 (or 89%) of all values lie within 3 standard deviations of the mean.

EXAMPLE 7 **Chebyshev's Theorem**

IQ scores have a mean of 100 and a standard deviation of 15. What can we conclude from Chebyshev's theorem?

SOLUTION

Applying Chebyshev's theorem with a mean of 100 and a standard deviation of 15, we can reach the following conclusions:

- At least 3/4 (or 75%) of IQ scores are within 2 standard deviations of the mean (between 70 and 130).
- At least 8/9 (or 89%) of all IQ scores are within 3 standard deviations of the mean (between 55 and 145).

> **YOUR TURN** Do Exercise 43 "Chebyshev's Theorem."

Comparing Variation in Different Samples or Populations

It's a good practice to compare two sample standard deviations only when the sample means are approximately the same. When comparing variation in samples or populations with very different means, it is better to use the *coefficient of variation*. Also use the coefficient of variation to compare variation from two samples or populations with different scales or units of values, such as the comparison of variation of *heights* of men and *weights* of men. (See Example 8.)

> **DEFINITION**
>
> The **coefficient of variation** (or **CV**) for a set of nonnegative sample or population data, expressed as a percent, describes the standard deviation relative to the mean, and is given by the following:
>
> | Sample | Population |
> |:---:|:---:|
> | $CV = \dfrac{s}{\bar{x}} \cdot 100$ | $CV = \dfrac{\sigma}{\mu} \cdot 100$ |

ROUND-OFF RULE FOR THE COEFFICIENT OF VARIATION Round the coefficient of variation to one decimal place (such as 25.3%).

EXAMPLE 8 **Verizon Data Speeds and Earthquake Magnitudes**

Compare the variation of the 50 Verizon data speeds listed in Data Set 32 "Airport Data Speeds" in Appendix B and the magnitudes of the 600 earthquakes in Data Set 21 "Earthquakes" in Appendix B. For the Verizon data speeds, $\bar{x} = 17.60$ Mbps and $s = 16.02$ Mbps; for the earthquake magnitudes, $\bar{x} = 2.572$ and $s = 0.651$. Note that we want to compare variation among *data speeds* to variation among *earthquake magnitudes*.

SOLUTION

We can compare the standard deviations if the same scales and units are used and the two means are approximately equal, but here we have different scales and different units of measurement, so we use the coefficients of variation:

$$\text{Verizon Data Speeds:} \quad CV = \frac{s}{\bar{x}} \cdot 100\% = \frac{16.02 \text{ Mbps}}{17.60 \text{ Mbps}} \cdot 100\% = 91.0\%$$

$$\text{Earthquake Magnitudes:} \quad CV = \frac{s}{\bar{x}} \cdot 100\% = \frac{0.651}{2.572} \cdot 100\% = 25.3\%$$

We can now see that the Verizon data speeds (with $CV = 91.0\%$) vary considerably more than earthquake magnitudes (with $CV = 25.3\%$).

Biased and Unbiased Estimators

The sample standard deviation s is a **biased estimator** of the population standard deviation σ, which means that values of the sample standard deviation s do *not* tend to center around the value of the population standard deviation σ. While individual values of s could equal or exceed σ, values of s generally tend to *underestimate* the value of σ. For example, consider an IQ test designed so that the population standard deviation is 15. If you repeat the process of randomly selecting 100 subjects, giving them IQ tests, and calculating the sample standard deviation s in each case, the sample standard deviations that you get will tend to be less than 15, which is the population standard deviation. There is no correction that allows us to fix the bias for all distributions of data. There is a correction that allows us to fix the bias for normally distributed populations, but it is rarely used because it is too complex and makes relatively minor corrections.

The sample variance s^2 is an **unbiased estimator** of the population variance σ^2, which means that values of s^2 tend to center around the value of σ^2 instead of systematically tending to overestimate or underestimate σ^2. Consider an IQ test designed so that the population variance is 225. If you repeat the process of randomly selecting 100 subjects, giving them IQ tests, and calculating the sample variance s^2 in each case, the sample variances that you obtain will tend to center around 225, which is the population variance.

Biased estimators and unbiased estimators will be discussed more in Section 6-3.

TECH CENTER

Measures of Variation
Access tech supplements, videos, and data sets at **www.TriolaStats.com**

Statdisk, Minitab, StatCrunch, Excel, and the TI-83/84 Plus Calculator can be used for the important calculations of this section. Use the same **Descriptive Statistics** procedures given at the end of Section 3-1 on page 91.

3-2   Basic Skills and Concepts

Statistical Literacy and Critical Thinking

1. Range Rule of Thumb for Estimating s The 20 brain volumes (cm³) from Data Set 8 "IQ and Brain Size" in Appendix B vary from a low of 963 cm³ to a high of 1439 cm³. Use the range rule of thumb to estimate the standard deviation s and compare the result to the exact standard deviation of 124.9 cm³.

2. Range Rule of Thumb for Interpreting s The 20 brain volumes (cm³) from Data Set 8 "IQ and Brain Size" in Appendix B have a mean of 1126.0 cm³ and a standard deviation of 124.9 cm³. Use the range rule of thumb to identify the limits separating values that are significantly low or significantly high. For such data, would a brain volume of 1440 cm³ be significantly high?

3. Variance The 20 subjects used in Data Set 8 "IQ and Brain Size" in Appendix B have weights with a standard deviation of 20.0414 kg. What is the variance of their weights? Be sure to include the appropriate units with the result.

4. Symbols Identify the symbols used for each of the following: (a) sample standard deviation; (b) population standard deviation; (c) sample variance; (d) population variance.

In Exercises 5–20, find the **range, variance,** *and* **standard deviation** *for the given sample data. Include appropriate units (such as "minutes") in your results. (The same data were used in Section 3-1, where we found measures of center. Here we find measures of variation.) Then answer the given questions.*

5. Football Player Numbers Listed below are the jersey numbers of 11 players randomly selected from the roster of the Seattle Seahawks when they won Super Bowl XLVIII. What do the results tell us?

89   91   55   7   20   99   25   81   19   82   60

6. Football Player Weights Listed below are the weights in pounds of 11 players randomly selected from the roster of the Seattle Seahawks when they won Super Bowl XLVIII (the same players from the preceding exercise). Are the measures of variation likely to be typical of all NFL players?

189   254   235   225   190   305   195   202   190   252   305

7. Celebrity Net Worth Listed below are the highest amounts of net worth (in millions of dollars) of celebrities. The celebrities are Tom Cruise, Will Smith, Robert De Niro, Drew Carey, George Clooney, John Travolta, Samuel L. Jackson, Larry King, Demi Moore, and Bruce Willis. Are the measures of variation typical for all celebrities?

250   200   185   165   160   160   150   150   150   150

8. What Happens in Vegas . . . Listed below are prices in dollars for one night at different hotels located on Las Vegas Boulevard (the "Strip"). How useful are the measures of variation for someone searching for a room?

212   77   121   104   153   264   195   244

9. Hurricanes Listed below are the numbers of Atlantic hurricanes that occurred in each year. The data are listed in order by year, starting with the year 2000. What important feature of the data is not revealed by any of the measures of variation?

8   9   8   7   9   15   5   6   8   4   12   7   8   2

10. Peas in a Pod Biologists conducted experiments to determine whether a deficiency of carbon dioxide in the soil affects the phenotypes of peas. Listed below are the phenotype codes, where 1 = smooth-yellow, 2 = smooth-green, 3 = wrinkled yellow, and 4 = wrinkled-green. Can the measures of variation be obtained for these values? Do the results make sense?

2   1   1   1   1   1   1   4   1   2   2   1   2   3   3   2   3   1   3   1   3   1   3   2   2

11. TV Prices Listed below are selling prices in dollars of TVs that are 60 inches or larger and rated as a "best buy" by *Consumer Reports* magazine. Are the measures of variation likely to be typical for all TVs that are 60 inches or larger?

1800   1500   1200   1500   1400   1600   1500   950   1600   1150   1500   1750

12. Cell Phone Radiation Listed below are the measured radiation absorption rates (in W/kg) corresponding to these cell phones: iPhone 5S, BlackBerry Z30, Sanyo Vero, Optimus V, Droid Razr, Nokia N97, Samsung Vibrant, Sony Z750a, Kyocera Kona, LG G2, and Virgin Mobile Supreme. The data are from the Federal Communications Commission. If one of each model of cell phone is measured for radiation and the results are used to find the measures of variation, are the results typical of the population of cell phones that are in use?

1.18   1.41   1.49   1.04   1.45   0.74   0.89   1.42   1.45   0.51   1.38

13. Caffeine in Soft Drinks Listed below are measured amounts of caffeine (mg per 12oz of drink) obtained in one can from each of 20 brands (7-UP, A&W Root Beer, Cherry Coke, . . . , Tab). Are the statistics representative of the population of all cans of the same 20 brands consumed by Americans?

0   0   34   34   34   45   41   51   55   36   47   41   0   0   53   54   38   0   41   47

14. Firefighter Fatalities Listed below are the numbers of heroic firefighters who lost their lives in the United States each year while fighting forest fires. The numbers are listed in order by year, starting with the year 2000. What important feature of the data is not revealed by any of the measures of variation?

20   18   23   30   20   12   24   9   25   15   8   11   15   34

15. Foot Lengths Listed below are foot lengths in inches of randomly selected Army women measured in the 1988 Anthropometric Survey (ANSUR). Are the statistics representative of the current population of all Army women?

10.4   9.3   9.1   9.3   10.0   9.4   8.6   9.8   9.9   9.1   9.1

16. Most Expensive Colleges Listed below in dollars are the annual costs of tuition and fees at the 10 most expensive colleges in the United States for a recent year (based on data from *U.S. News & World Report*). The colleges listed in order are Columbia, Vassar, Trinity, George Washington, Carnegie Mellon, Wesleyan, Tulane, Bucknell, Oberlin, and Union. What does this "top 10" list tell us about the variation among costs for the population of all U.S. college tuitions?

49,138 47,890 47,510 47,343 46,962 46,944 46,930 46,902 46,870 46,785

17. Diamond Ring Listed below are the amounts (dollars) it costs for marriage proposal packages at the different Major League Baseball stadiums. Five of the teams don't allow proposals. Are there any outliers, and are they likely to have much of an effect on the measures of variation?

39 50 50 50 55 55 75 85 100 115 175 175 200
209 250 250 350 400 450 500 500 500 500 1500 2500

18. Sales of LP Vinyl Record Albums Listed below are annual U.S. sales of vinyl record albums (millions of units). The numbers of albums sold are listed in chronological order, and the last entry represents the most recent year. Do the measures of variation give us any information about a changing trend over time?

0.3 0.6 0.8 1.1 1.1 1.4 1.4 1.5 1.2 1.3 1.4 1.2 0.9 0.9 1.0 1.9 2.5 2.8 3.9 4.6 6.1

19. California Smokers In the California Health Interview Survey, randomly selected adults are interviewed. One of the questions asks how many cigarettes are smoked per day, and results are listed below for 50 randomly selected respondents. How well do the results reflect the smoking behavior of California adults?

9 10 10 20 40 50 0 0 0 0 0 0 0 0 0
0 0 0 0 0 0 0 0 0 0 0 0 0 0 0
0 0 0 0 0 0 0 0 0 0 0 0 0 0 0
0 0 0 0 0

20. Speed Dating In a study of speed dating conducted at Columbia University, female subjects were asked to rate the attractiveness of their male dates, and a sample of the results is listed below (1 = not attractive; 10 = extremely attractive). Can the results be used to describe the variation among attractiveness ratings for the population of adult males?

5 8 3 8 6 10 3 7 9 8 5 5 6 8 8 7 3 5 5 6 8 7 8 8 8 7

In Exercises 21–24, find the **coefficient of variation** *for each of the two samples; then compare the variation. (The same data were used in Section 3-1.)*

21. Blood Pressure A sample of blood pressure measurements is taken from Data Set 1 "Body Data" in Appendix B, and those values (mm Hg) are listed below. The values are matched so that 10 subjects each have a systolic and diastolic measurement.

Systolic:	118	128	158	96	156	122	116	136	126	120
Diastolic:	80	76	74	52	90	88	58	64	72	82

22. Parking Meter Theft Listed below are amounts (in millions of dollars) collected from parking meters by Brinks and others in New York City during similar time periods. A larger data set was used to convict five Brinks employees of grand larceny. The data were provided by the attorney for New York City, and they are listed on the DASL Website. Do the two samples appear to have different amounts of variation?

Collection Contractor Was Brinks   1.3 1.5 1.3 1.5 1.4 1.7 1.8 1.7 1.7 1.6
Collection Contractor Was Not Brinks 2.2 1.9 1.5 1.6 1.5 1.7 1.9 1.6 1.6 1.8

23. Pulse Rates Listed below are pulse rates (beats per minute) from samples of adult males and females (from Data Set 1 "Body Data" in Appendix B). Does there appear to be a difference?

Male: 86 72 64 72 72 54 66 56 80 72 64 64 96 58 66

Female: 64 84 82 70 74 86 90 88 90 90 94 68 90 82 80

24. Bank Queues Waiting times (in seconds) of customers at the Madison Savings Bank are recorded with two configurations: single customer line; individual customer lines.

Single Line	390	396	402	408	426	438	444	462	462	462
Individual Lines	252	324	348	372	402	462	462	510	558	600

Large Data Sets from Appendix B. *In Exercises 25–28, refer to the indicated data set in Appendix B. Use software or a calculator to find the* **range, variance,** *and* **standard deviation.** *Express answers using appropriate units, such as "minutes."*

25. Tornadoes Use the F-scale measurements from the tornadoes listed in Data Set 22 "Tornadoes" in Appendix B. Be careful to account for missing data.

26. Earthquakes Use the magnitudes (Richter scale) of the 600 earthquakes listed in Data Set 21 "Earthquakes" in Appendix B. In 1989, the San Francisco Bay Area was struck with an earthquake that measured 7.0 on the Richter scale. If we add that value of 7.0 to those listed in the data set, do the measures of variation change much?

27. Body Temperatures Refer to Data Set 3 "Body Temperature" in Appendix B and use the body temperatures for 12:00 AM on day 2.

28. Births Use the birth weights (grams) of the 400 babies listed in Data Set 4 "Births" in Appendix B. Examine the list of birth weights to make an observation about those numbers. How does that observation affect the way that the results should be rounded?

Estimating Standard Deviation with the Range Rule of Thumb. *In Exercises 29–32, refer to the data in the indicated exercise. After finding the* **range** *of the data, use the range rule of thumb to estimate the value of the* **standard deviation.** *Compare the result to the standard deviation computed with all of the data.*

29. Exercise 25 "Tornadoes"   **30.** Exercise 26 "Earthquakes"

31. Exercise 27 "Body Temperatures"   **32.** Exercise 28 "Births"

Identifying Significant Values with the Range Rule of Thumb. *In Exercises 33–36, use the range rule of thumb to identify the limits separating values that are* **significantly low** *or* **significantly high.**

33. Pulse Rates of Females Based on Data Set 1 "Body Data" in Appendix B, females have pulse rates with a mean of 74.0 beats per minute and a standard deviation of 12.5 beats per minute. Is a pulse rate of 44 beats per minute significantly low or significantly high? (All of these pulse rates are measured at rest.)

34. Pulse Rates of Males Based on Data Set 1 "Body Data" in Appendix B, males have pulse rates with a mean of 69.6 beats per minute and a standard deviation of 11.3 beats per minute. Is a pulse rate of 50 beats per minute significantly low or significantly high? (All of these pulse rates are measured at rest.) Explain.

35. Foot Lengths Based on Data Set 2 "Foot and Height" in Appendix B, adult males have foot lengths with a mean of 27.32 cm and a standard deviation of 1.29 cm. Is the adult male foot length of 30 cm significantly low or significantly high? Explain.

36. Body Temperatures Based on Data Set 3 "Body Temperatures" in Appendix B, body temperatures of adults have a mean of 98.20°F and a standard deviation of 0.62°F. (The data from 12 AM on day 2 are used.) Is an adult body temperature of 100°F significantly low or significantly high?

Finding Standard Deviation from a Frequency Distribution.   *In Exercises 37–40, refer to the frequency distribution in the given exercise and find the* **standard deviation** *by using the formula below, where x represents the class midpoint, f represents the class frequency, and n represents the total number of sample values. Also, compare the computed standard deviations to these standard deviations obtained by using Formula 3-4 with the original list of data values: (Exercise 37) 11.5 years; (Exercise 38) 8.9 years; (Exercise 39) 59.5; (Exercise 40) 65.4.*

Standard deviation for frequency distribution

$$s = \sqrt{\frac{n[\Sigma(f \cdot x^2)] - [\Sigma(f \cdot x)]^2}{n(n-1)}}$$

37.

Age (yr) of Best Actress When Oscar Was Won	Frequency
20–29	29
30–39	34
40–49	14
50–59	3
60–69	5
70–79	1
80–89	1

38.

Age (yr) of Best Actor When Oscar Was Won	Frequency
20–29	1
30–39	28
40–49	36
50–59	15
60–69	6
70–79	1

39.

Blood Platelet Count of Males	Frequency
0–99	1
100–199	51
200–299	90
300–399	10
400–499	0
500–599	0
600–699	1

40.

Blood Platelet Count of Females	Frequency
100–199	25
200–299	92
300–399	28
400–499	0
500–599	2

41. The Empirical Rule Based on Data Set 1 "Body Data" in Appendix B, blood platelet counts of women have a bell-shaped distribution with a mean of 255.1 and a standard deviation of 65.4. (All units are 1000 cells/μL.) Using the empirical rule, what is the approximate percentage of women with platelet counts

a. within 2 standard deviations of the mean, or between 124.3 and 385.9?

b. between 189.7 and 320.5?

42. The Empirical Rule Based on Data Set 3 "Body Temperatures" in Appendix B, body temperatures of healthy adults have a bell-shaped distribution with a mean of 98.20°F and a standard deviation of 0.62°F. Using the empirical rule, what is the approximate percentage of healthy adults with body temperatures

a. within 1 standard deviation of the mean, or between 97.58°F and 98.82°F?

b. between 96.34°F and 100.06°F?

43. Chebyshev's Theorem Based on Data Set 1 "Body Data" in Appendix B, blood platelet counts of women have a bell-shaped distribution with a mean of 255.1 and a standard deviation of 65.4. (All units are 1000 cells/μL.) Using Chebyshev's theorem, what do we know about the percentage of women with platelet counts that are within 3 standard deviations of the mean? What are the minimum and maximum platelet counts that are within 3 standard deviations of the mean?

44. Chebyshev's Theorem Based on Data Set 3 "Body Temperatures" in Appendix B, body temperatures of healthy adults have a bell-shaped distribution with a mean of 98.20°F and a standard deviation of 0.62°F (using the data from 12 AM on day 2). Using Chebyshev's theorem, what do we know about the percentage of healthy adults with body temperatures that are within 2 standard deviations of the mean? What are the minimum and maximum body temperatures that are within 2 standard deviations of the mean?

3-2 Beyond the Basics

45. Why Divide by $n - 1$? Let a *population* consist of the values 9 cigarettes, 10 cigarettes, and 20 cigarettes smoked in a day (based on data from the California Health Interview Survey). Assume that samples of two values are randomly selected *with replacement* from this population. (That is, a selected value is replaced before the second selection is made.)

a. Find the variance σ^2 of the population {9 cigarettes, 10 cigarettes, 20 cigarettes}.

b. After listing the nine different possible samples of two values selected with replacement, find the sample variance s^2 (which includes division by $n - 1$) for each of them; then find the mean of the nine sample variances s^2.

c. For each of the nine different possible samples of two values selected with replacement, find the variance by treating each sample as if it is a population (using the formula for population variance, which includes division by n); then find the mean of those nine population variances.

d. Which approach results in values that are better estimates of σ^2: part (b) or part (c)? Why? When computing variances of samples, should you use division by n or $n - 1$?

e. The preceding parts show that s^2 is an unbiased estimator of σ^2. Is s an unbiased estimator of σ? Explain.

46. Mean Absolute Deviation Use the same population of {9 cigarettes, 10 cigarettes, 20 cigarettes} from Exercise 45. Show that when samples of size 2 are randomly selected with replacement, the samples have mean absolute deviations that do not center about the value of the mean absolute deviation of the population. What does this indicate about a sample mean absolute deviation being used as an estimator of the mean absolute deviation of a population?

3-3 Measures of Relative Standing and Boxplots

Key Concept This section introduces measures of relative standing, which are numbers showing the location of data values relative to the other values within the same data set. The most important concept in this section is the *z score*, which will be used often in following chapters. We also discuss percentiles and quartiles, which are common statistics, as well as another statistical graph called a boxplot.

PART 1 Basics of *z* Scores, Percentiles, Quartiles, and Boxplots

z Scores

A *z* score is found by converting a value to a standardized scale, as given in the following definition. This definition shows that a *z* score is the number of standard deviations that a data value is away from the mean. The *z* score is used often in Chapter 6 and later chapters.

DEFINITION

A **z score** (or **standard score** or **standardized value**) is the number of standard deviations that a given value x is above or below the mean. The z score is calculated by using one of the following:

<table>
<tr><th>Sample</th><th>Population</th></tr>
</table>

$$z = \frac{x - \bar{x}}{s} \quad \text{or} \quad z = \frac{x - \mu}{\sigma}$$

ROUND-OFF RULE FOR z SCORES Round z scores to two decimal places (such as 2.31).

This round-off rule is motivated by the format of standard tables in which z scores are expressed with two decimal places, as in Table A-2 in Appendix A. Example 1 illustrates how z scores can be used to compare values, even if they come from different populations.

Important Properties of z Scores

1. A z score is the number of standard deviations that a given value x is above or below the mean.

2. z scores are expressed as numbers with no units of measurement.

3. A data value is *significantly low* if its z score is less than or equal to -2 or the value is *significantly high* if its z score is greater than or equal to $+2$.

4. If an individual data value is less than the mean, its corresponding z score is a negative number.

EXAMPLE 1 **Comparing a Baby's Weight and Adult Body Temperature**

Which of the following two data values is more extreme relative to the data set from which it came?

- The 4000 g weight of a newborn baby (among 400 weights with sample mean $\bar{x} = 3152.0$ g and sample standard deviation $s = 693.4$ g)

- The 99°F temperature of an adult (among 106 adults with sample mean $\bar{x} = 98.20$°F and sample standard deviation $s = 0.62$°F)

SOLUTION

The 4000 g weight and the 99°F body temperature can be standardized by converting each of them to z scores as shown below.

4000 g birth weight:

$$z = \frac{x - \bar{x}}{s} = \frac{4000 \text{ g} - 3152.0 \text{ g}}{693.4 \text{ g}} = 1.22$$

99°F body temperature:

$$z = \frac{x - \bar{x}}{s} = \frac{99°F - 98.20°F}{0.62°F} = 1.29$$

The z scores show that the 4000 g birth weight is 1.22 standard deviations above the mean, and the 99°F body temperature is 1.29 standard deviations above the mean. Because the body temperature is farther above the mean, it is the more extreme value. A 99°F body temperature is slightly more extreme than a birth weight of 4000 g.

> **YOUR TURN** Do Exercise 13 "Tallest and Shortest Men."

Using z Scores to Identify Significant Values In Section 3-2 we used the range rule of thumb to conclude that a value is significantly low or significantly high if it is at least 2 standard deviations away from the mean. It follows that significantly low values have z scores less than or equal to -2 and significantly high values have z scores greater than or equal to $+2$, as illustrated in Figure 3-5. Using this criterion with the two individual values used in Example 1, we see that neither value is significant because both z scores are between -2 and $+2$.

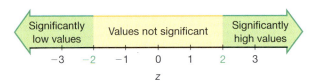

FIGURE 3-5 **Interpreting z Scores**
Significant values are those with z scores ≤ -2.00 or ≥ 2.00.

EXAMPLE 2 **Is a Platelet Count of 75 Significantly Low?**

The lowest platelet count in Data Set 1 "Body Data" in Appendix B is 75. (The platelet counts are measured in 1000 cells/μL). Is that value significantly low? Based on the platelet counts from Data Set 1 in Appendix B, assume that platelet counts have a mean of $\bar{x} = 239.4$ and a standard deviation of $s = 64.2$.

SOLUTION

The platelet count of 75 is converted to a z score as shown below:

$$z = \frac{x - \bar{x}}{s} = \frac{75 - 239.4}{64.2} = -2.56$$

INTERPRETATION

The platelet count of 75 converts to the z score of -2.56. Refer to Figure 3-5 to see that $z = -2.56$ is less than -2, so the platelet count of 75 is significantly low. (Low platelet counts are called thrombocytopenia, not for the lack of a better term.)

> **YOUR TURN** Do Exercise 9 "ACT."

A z score is a measure of position, in the sense that it describes the location of a value (in terms of standard deviations) relative to the mean. Percentiles and quartiles are other measures of position useful for comparing values within the same data set or between different sets of data.

Percentiles

Percentiles are one type of *quantiles*—or *fractiles*—which partition data into groups with roughly the same number of values in each group.

The 50th percentile, denoted P_{50}, has about 50% of the data values below it and about 50% of the data values above it, so the 50th percentile is the same as the median. There is not universal agreement on a single procedure for calculating percentiles, but we will describe relatively simple procedures for (1) finding the percentile of a data value and (2) converting a percentile to its corresponding data value. We begin with the first procedure.

Finding the Percentile of a Data Value

The process of finding the percentile that corresponds to a particular data value x is given by the following (round the result to the nearest whole number):

$$\text{Percentile of value } x = \frac{\text{number of values less than } x}{\text{total number of values}} \cdot 100$$

EXAMPLE 3   Finding a Percentile

Table 3-4 lists the same airport Verizon cell phone data speeds listed in Data Set 32 "Airport Data Speeds", but in Table 3-4 those data speeds are arranged in increasing order. Find the percentile for the data speed of 11.8 Mbps.

TABLE 3-4 *Sorted* Verizon Airport Data Speeds (Mbps)

0.8	1.4	1.8	1.9	3.2	3.6	4.5	4.5	4.6	6.2
6.5	7.7	7.9	9.9	10.2	10.3	10.9	11.1	11.1	11.6
11.8	12.0	13.1	13.5	13.7	14.1	14.2	14.7	15.0	15.1
15.5	15.8	16.0	17.5	18.2	20.2	21.1	21.5	22.2	22.4
23.1	24.5	25.7	28.5	34.6	38.5	43.0	55.6	71.3	77.8

SOLUTION

From the sorted list of airport data speeds in Table 3-4, we see that there are 20 data speeds less than 11.8 Mbps, so

$$\text{Percentile of } 11.8 = \frac{20}{50} \cdot 100 = 40$$

INTERPRETATION

A data speed of 11.8 Mbps is in the 40th percentile. This can be interpreted loosely as this: A data speed of 11.8 Mbps separates the lowest 40% of values from the highest 60% of values. We have $P_{40} = 11.8$ Mbps.

YOUR TURN   Do Exercise 17 "Percentiles."

Example 3 shows how to convert from a given sample value to the corresponding percentile. There are several different methods for the reverse procedure of converting a given percentile to the corresponding value in the data set. The procedure we will use is summarized in Figure 3-6, which uses the following notation.

Notation

n total number of values in the data set

k percentile being used (Example: For the 25th percentile, $k = 25$.)

L locator that gives the *position* of a value (Example: For the 12th value in the sorted list, $L = 12$.)

P_k kth percentile (Example: P_{25} is the 25th percentile.)

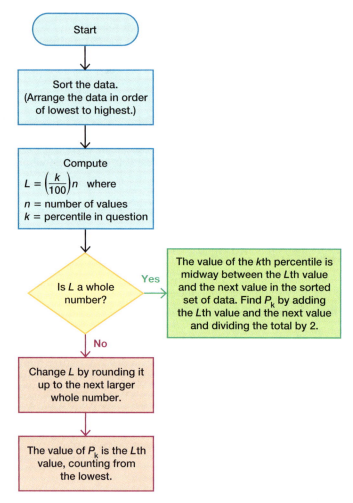

FIGURE 3-6   **Converting from the kth percentile to the corresponding data value**

EXAMPLE 4   **Converting a Percentile to a Data Value**

Refer to the sorted data speeds in Table 3-4 and use the procedure in Figure 3-6 to find the value of the 25th percentile, P_{25}.

SOLUTION

From Figure 3-6, we see that the sample data are already sorted, so we can proceed to find the value of the locator L. In this computation we use $k = 25$ because we are trying to find the value of the 25th percentile. We use $n = 50$ because there are 50 data values.

$$L = \frac{k}{100} \cdot n = \frac{25}{100} \cdot 50 = 12.5$$

Since $L = 12.5$ is not a whole number, we proceed to the next lower box in Figure 3-6, where we change L by rounding it up from 12.5 to the next larger whole number: 13. (In this book we typically round off the usual way, but this is one of two cases where we round *up* instead of rounding *off*.) From the bottom box we see that the value of P_{25} is the 13th value, counting from the lowest. In Table 3-4, the 13th value is 7.9. That is, $P_{25} = 7.9$ Mbps. Roughly speaking, about 25% of the data speeds are less than 7.9 Mbps and 75% of them are more than 7.9 Mbps.

> **YOUR TURN** Do Exercise 23 "Quartile."

 EXAMPLE 5 **Converting a Percentile to a Data Value**

Refer to the sorted data speeds in Table 3-4. Use Figure 3-6 to find the 40th percentile, denoted by P_{40}.

SOLUTION

Referring to Figure 3-6, we see that the sample data are already sorted, so we can proceed to compute the value of the locator L. In this computation, we use $k = 40$ because we are attempting to find the value of the 40th percentile, and we use $n = 50$ because there are 50 data values.

$$L = \frac{k}{100} \cdot n = \frac{40}{100} \cdot 50 = 20$$

Since $L = 20$ is a whole number, we proceed to the box in Figure 3-6 located at the right. We now see that the value of the 40th percentile is midway between the Lth (20th) value and the next value in the original set of data. That is, the value of the 40th percentile is midway between the 20th value and the 21st value. The 20th value in Table 3-4 is 11.6 and the 21st value is 11.8, so the value midway between them is 11.7 Mbps. We conclude that the 40th percentile is $P_{40} = 11.7$ Mbps.

> **YOUR TURN** Do Exercise 21 "Percentile."

Quartiles

Just as there are 99 percentiles that divide the data into 100 groups, there are three quartiles that divide the data into four groups.

DEFINITION

Quartiles are measures of location, denoted Q_1, Q_2, and Q_3, which divide a set of data into four groups with about 25% of the values in each group.

Here are descriptions of quartiles that are more accurate than those given in the preceding definition:

Q_1 **(First quartile):** Same value as P_{25}. It separates the bottom 25% of the sorted values from the top 75%. (To be more precise, at least 25% of the sorted values are less than or equal to Q_1, and at least 75% of the values are greater than or equal to Q_1.)

Q_2 **(Second quartile):** Same as P_{50} and same as the median. It separates the bottom 50% of the sorted values from the top 50%.

Q_3 **(Third quartile):** Same as P_{75}. It separates the bottom 75% of the sorted values from the top 25%. (To be more precise, at least 75% of the sorted values are less than or equal to Q_3, and at least 25% of the values are greater than or equal to Q_3.)

Nielsen Ratings for College Students

The Nielsen ratings are one of the most important measures of television viewing, and they affect billions of dollars in television advertising. In the past, the television viewing habits of college students were ignored, with the result that a large segment of the important young viewing audience was ignored. Nielsen Media Research is now including college students who do not live at home.

Some television shows have large appeal to viewers in the 18–24 age bracket, and the ratings of such shows have increased substantially with the inclusion of college students. For males, NBC's *Sunday Night Football* broadcast had an increase of 20% after male college students were included. Higher ratings ultimately translate into greater profits from charges to commercial sponsors. These ratings also give college students recognition that affects the programming they receive.

$Q_1 = P_{25}$

$Q_2 = P_{50}$

$Q_3 = P_{75}$

Finding values of quartiles can be accomplished with the same procedure used for finding percentiles. Simply use the relationships shown in the margin. In Example 4 we found that $P_{25} = 7.9$ Mbps, so it follows that $Q_1 = 7.9$ Mbps.

> **CAUTION** Just as there is not universal agreement on a procedure for finding percentiles, there is not universal agreement on a single procedure for calculating quartiles, and different technologies often yield different results. If you use a calculator or software for exercises involving quartiles, you may get results that differ somewhat from the answers obtained by using the procedures described here.

In earlier sections of this chapter we described several statistics, including the mean, median, mode, range, and standard deviation. Some other statistics are defined using quartiles and percentiles, as in the following:

Interquartile range (or IQR) $= Q_3 - Q_1$

Semi-interquartile range $= \dfrac{Q_3 - Q_1}{2}$

Midquartile $= \dfrac{Q_3 + Q_1}{2}$

10–90 percentile range $= P_{90} - P_{10}$

5-Number Summary and Boxplot

The values of the minimum, maximum and three quartiles (Q_1, Q_2, Q_3) are used for the 5-number summary and the construction of boxplot graphs.

> **DEFINITION**
>
> For a set of data, the **5-number summary** consists of these five values:
>
> **1.** Minimum
>
> **2.** First quartile, Q_1
>
> **3.** Second quartile, Q_2 (same as the median)
>
> **4.** Third quartile, Q_3
>
> **5.** Maximum

 EXAMPLE 6 Finding a 5-Number Summary

Use the Verizon airport data speeds in Table 3-4 to find the 5-number summary.

SOLUTION

Because the Verizon airport data speeds in Table 3-4 are sorted, it is easy to see that the minimum is 0.8 Mbps and the maximum is 77.8 Mbps. The value of the first quartile is $Q_1 = 7.9$ Mbps (from Example 4). The median is equal to Q_2, and it is 13.9 Mbps. Also, we can find that $Q_3 = 21.5$ Mbps by using the same procedure for finding P_{75} (as summarized in Figure 3-6). The 5-number summary is therefore 0.8, 7.9, 13.9, 21.5, and 77.8 (all in units of Mbps).

> **YOUR TURN** Find the 5-number summary in Exercise 29 "Speed Dating."

The values from the 5-number summary are used for the construction of a box-plot, defined as follows.

DEFINITION

A **boxplot** (or **box-and-whisker diagram**) is a graph of a data set that consists of a line extending from the minimum value to the maximum value, and a box with lines drawn at the first quartile Q_1, the median, and the third quartile Q_3. (See Figure 3-7.)

Procedure for Constructing a Boxplot

1. Find the 5-number summary (minimum value, Q_1, Q_2, Q_3, maximum value).

2. Construct a line segment extending from the minimum data value to the maximum data value.

3. Construct a box (rectangle) extending from Q_1 to Q_3, and draw a line in the box at the value of Q_2 (median).

CAUTION Because there is not universal agreement on procedures for finding quartiles, and because boxplots are based on quartiles, different technologies may yield different boxplots.

EXAMPLE 7 Constructing a Boxplot

Use the Verizon airport data speeds listed in Table 3-4 to construct a boxplot.

SOLUTION

The boxplot uses the 5-number summary found in Example 6: 0.8, 7.9, 13.9, 21.5, and 77.8 (all in units of Mbps). Figure 3-7 is the boxplot representing the Verizon airport data speeds listed in Table 3-4.

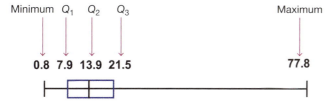

FIGURE 3-7 **Boxplot of Verizon Airport Data Speeds (Mbps)**

YOUR TURN Construct the boxplot in Exercise 29 "Speed Dating."

Skewness A boxplot can often be used to identify skewness. Recall that in Section 2-2 we stated that a distribution of data is **skewed** if it is not symmetric and extends more to one side than to the other. In a histogram of data skewed to the right (also called *positively skewed*), there is a longer right tail showing that relatively few data values are high data values; most of the data values are located at the left. The boxplot in Figure 3-7 shows that the data are skewed to the right, and most of the data values are located at the left.

Because the shape of a boxplot is determined by the five values from the 5-number summary, a boxplot is not a graph of the distribution of the data, and it doesn't show as much detailed information as a histogram or stemplot. However, boxplots are often great for comparing two or more data sets. When using two or more boxplots for comparing different data sets, graph the boxplots on the same scale so that comparisons can be easily made. Methods discussed later in this book allow us to analyze comparisons of data sets more formally than subjective conclusions based on a graph. It is always wise to construct suitable graphs, such as histograms, dotplots, and boxplots, but we should not rely solely on subjective judgments based on graphs.

EXAMPLE 8   **Comparing the Data Speeds of Verizon, Sprint, AT&T, and T-Mobile**

The Chapter Problem refers to smartphone data speeds at 50 airports, and the speeds are measured for the carriers of Verizon, Sprint, AT&T, and T-Mobile. Use the same scale to construct the four corresponding boxplots; then compare the results.

SOLUTION

The Statdisk-generated boxplots shown in Figure 3-8 suggest that the Verizon data speeds are generally faster.

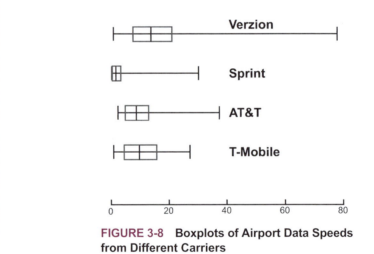

FIGURE 3-8   **Boxplots of Airport Data Speeds from Different Carriers**

YOUR TURN   Do Exercise 33 "Pulse Rates."

Outliers

When analyzing data, it is important to identify and consider outliers because they can strongly affect values of some important statistics (such as the mean and standard deviation), and they can also strongly affect important methods discussed later in this book. In Chapter 2 we described outliers as sample values that lie very far away from the vast majority of the other values in a set of data, but that description is vague and it does not provide specific objective criteria. Part 2 of this section includes a description of *modified boxplots* along with a more precise definition of outliers used in the context of creating modified boxplots.

CAUTION When analyzing data, always identify outliers and consider their effects, which can be substantial.

PART 2 Outliers and Modified Boxplots

We noted that the description of outliers is somewhat vague, but for the purposes of constructing *modified boxplots*, we can consider outliers to be data values meeting specific criteria based on quartiles and the interquartile range. (The interquartile range is often denoted by IQR, and IQR $= Q_3 - Q_1$.)

Identifying Outliers for Modified Boxplots

1. Find the quartiles Q_1, Q_2, and Q_3.

2. Find the interquartile range (IQR), where IQR $= Q_3 - Q_1$.

3. Evaluate $1.5 \times$ IQR.

4. **In a modified boxplot, a data value is an *outlier* if it is**
   above Q_3, by an amount greater than $1.5 \times$ IQR
   or   **below Q_1, by an amount greater than $1.5 \times$ IQR**

Modified Boxplots

The boxplots described earlier are called **skeletal** (or **regular**) **boxplots**, but some statistical software packages provide modified boxplots, which represent outliers as special points. A **modified boxplot** is a regular boxplot constructed with these modifications: (1) A special symbol (such as an asterisk or point) is used to identify outliers as defined above, and (2) the solid horizontal line extends only as far as the minimum data value that is not an outlier and the maximum data value that is not an outlier. (*Note: Exercises involving modified boxplots are found in the "Beyond the Basics" exercises only.*)

EXAMPLE 9 Constructing a Modified Boxplot

Use the Verizon airport data speeds in Data Set 32 "Airport Data Speeds" in Appendix B to construct a modified boxplot.

SOLUTION

Let's begin with the above four steps for identifying outliers in a modified boxplot.

1. Using the Verizon data speeds, the three quartiles are $Q_1 = 7.9$, the median is $Q_2 = 13.9$, and $Q_3 = 21.5$. (All values are in Mbps and these quartiles were found in Example 6.)

2. The interquartile range is IQR $= Q_3 - Q_1 = 21.5 - 7.9 = 13.6$.

3. $1.5 \times$ IQR $= 1.5 \times 13.6 = 20.4$.

4. Any outliers are above $Q_3 = 21.5$ by more than 20.4, or below $Q_1 = 7.9$ by more than 20.4. This means that any outliers are greater than 41.9, or less than -12.5 (which is impossible, so here there can be no outliers at the low end).

We can now examine the original Verizon airport data speeds to identify any speeds greater than 41.9, and we find these values: 43.0, 55.6, 71.3, and 77.8. The only outliers are 43.0, 55.6, 71.3, and 77.8.

We can now construct the modified boxplot shown in Figure 3-9 on the next page. In Figure 3-9, the four outliers are identified as special points, the three quartiles are shown as in a regular boxplot, and the horizontal line extends from the lowest data value that is not an outlier (0.8) to the highest data value that is not an outlier (38.5).

continued

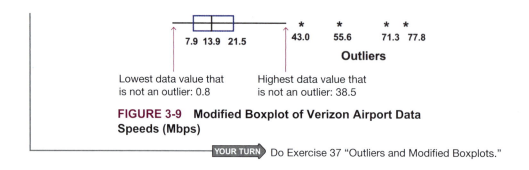

FIGURE 3-9   Modified Boxplot of Verizon Airport Data Speeds (Mbps)

YOUR TURN Do Exercise 37 "Outliers and Modified Boxplots."

CAUTION Because there is not universal agreement on procedures for finding quartiles, and because modified boxplots are based on quartiles, different technologies may yield different modified boxplots.

TECH CENTER

Boxplots, 5-Number Summary, Outliers

Access tech supplements, videos, and data sets at **www.TriolaStats.com**

Statdisk

5 Number Summary
Use **Descriptive Statistics** procedure given at the end of Section 3-1 on page 91.

Boxplots
1. Click **Data** in the top menu.
2. Select **Boxplot** from the dropdown menu.
3. Select the desired data columns.
4. Click **Boxplot** or **Modified Boxplot**.

Outliers
Create a modified boxplot using the above procedure or sort as follows:
1. Click **Data** in the top menu.
2. Select **Sort Data** from the dropdown menu.
3. Click **Sort** after making desired choices in the sort menu.
4. Examine minimum and maximum values to determine if they are far from other values.

Minitab

5 Number Summary
Use **Descriptive Statistics** procedure given at the end of Section 3-1 on page 91.

Boxplots
1. Click **Graph** in the top menu.
2. Select **Boxplot** from the dropdown menu.
3. Select **Simple** option for one boxplot or multiple boxplots, then click **OK**.
4. Double click on the desired data column(s) so that it appears in the *Graph variables* window, then click **OK**.

Outliers
Create a modified boxplot using the above procedure or sort as follows:
1. Click **Data** in the top menu.
2. Select **Sort** from the dropdown menu.
3. Double click the desired data column so that it appears in the *Sort Column(s)* window.
4. Click the *By Column* box and select the same data column. Click **OK**.
5. Examine minimum and maximum values to determine if they are far from other values.

TECH CENTER *continued*

 Boxplots, 5-Number Summary, Outliers
Access tech supplements, videos, and data sets at **www.TriolaStats.com**

| StatCrunch | TI-83/84 Plus Calculator |

StatCrunch

5 Number Summary
Use **Descriptive Statistics** procedure given at the end of Section 3-1 on page 91.

Boxplots
1. Click on **Graph** in the top menu.
2. Select **Boxplot** from the dropdown menu.
3. Select the desired data column. For a modified boxplot, check the *Use fences* box.
4. Click **Compute!**

Outliers
Create a modified boxplot using the above procedure or sort as follows:
1. Click **Data** in top menu.
2. Select **Sort** from the dropdown menu.
3. Select the desired data column.
4. Click **Compute!**
5. Examine minimum and maximum values to determine if they are far from other values.

TI-83/84 Plus Calculator

5 Number Summary
Use **Descriptive Statistics** procedure given at the end of Section 3-1 on page 91.

Boxplots
1. Open the **STAT PLOTS** menu by pressing **2ND**, **Y=**.
2. Press **ENTER** to access the *Plot 1* settings screen as shown:
   a. Select **ON** and press **ENTER**.
   b. Select the second boxplot icon, press **ENTER**. Select the first boxplot icon for a modified boxplot.
   c. Enter name of list containing data.
3. Press **ZOOM** then **9** (ZoomStat) to display the boxplot.
4. Press **TRACE** and use ◁ ▷ to view values.

Outliers
Create a modified boxplot using the above procedure or sort as follows:
1. Press **STAT**, select **SortA** (sort ascending) from the menu and press **ENTER**.
2. Enter the name of the list to be sorted and press **ENTER**.
3. To view the sorted list, press **STAT**, select **Edit** and press **ENTER**.
4. Highlight the top cell in an empty column, enter the list name and press **ENTER**.
5. Use ▲ ▼ to examine minimum and maximum values to determine if they are far from other values.

```
NORMAL FLOAT AUTO REAL RADIAN MP
Plot1  Plot2 Plot3
2a On  Off                  2b
Type: ⊔⊔ ⊠ ⊞⊞ ⊞⊞ ⊡⊡ ⊠
Xlist:VRZN  2c
Freq:1
Color:    BLUE
```

Excel

5 Number Summary
Use **Descriptive Statistics** procedure given at the end of Section 3-1 on page 91.

XLSTAT Add-In
- After Step 3 in the Descriptive Statistics procedure, click the **Outputs** tab and select **Minimum, Maximum, 1st Quartile, Median, 3rd Quartile.** Click **OK.**

Excel
The Data Analysis add-in provides only the minimum, maximum and median. To obtain quartiles use the following procedure:
1. Click on **Insert Function f_x,** select the category **Statistical** and select the function **QUARTILE.INC.**
2. Enter the range of data values in the *Array* box.
3. In the *Quart* box enter **0** to find the minimum, **1** to find the first 1st Quartile and **2,3,4** to find the remaining values.

Boxplots
XLSTAT Add-In (Required)
1. Click on the **XLSTAT** tab in the Ribbon and then click **Describing Data.**
2. Select **Descriptive Statistics** from the dropdown menu.
3. Check the **Quantitative Data** box and enter the desired data range. Selecting two or more columns will generate multiple boxplots. If the first row of data contains a label, also check the **Sample labels** box.
4. Click the **Options** tab and confirm the **Charts** box is checked.
5. Click on the **Charts (1)** tab and check the **Box plots** box under *Quantitative Data*.
6. Click **OK.**

Outliers
Create a modified boxplot using the above XLSTAT procedure or sort as follows:
1. Click **Data** tab in the top menu and select the desired range of data values.
2. Click the **Sort Smallest to Largest** (A → Z) button in the ribbon.
3. Examine minimum and maximum values to determine if they are far from other values.

3-3 Basic Skills and Concepts

Statistical Literacy and Critical Thinking

1. z Scores LeBron James, one of the most successful basketball players of all time, has a height of 6 feet 8 inches, or 203 cm. Based on statistics from Data Set 1 "Body Data" in Appendix B, his height converts to the z score of 4.07. How many standard deviations is his height above the mean?

2. Heights The boxplot shown below results from the heights (cm) of males listed in Data Set 1 "Body Data" in Appendix B. What do the numbers in that boxplot tell us?

3. Boxplot Comparison Refer to the boxplots shown below that are drawn on the same scale. One boxplot represents weights of men, and the other boxplot represents weights of women. Which boxplot represents weights of women? Explain.

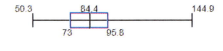

4. z Scores If your score on your next statistics test is converted to a z score, which of these z scores would you prefer: −2.00, −1.00, 0, 1.00, 2.00? Why?

z Scores. *In Exercises 5–8, express all z scores with two decimal places.*

5. ATL Data Speeds For the Verizon airport data speeds (Mbps) listed in Data Set 32 "Airport Data Speeds" in Appendix B, the highest speed of 77.8 Mbps was measured at Atlanta's (ATL) international airport. The complete list of 50 Verizon data speeds has a mean of $\bar{x} = 17.60$ Mbps and a standard deviation of $s = 16.02$ Mbps.

a. What is the difference between Verizon's data speed at Atlanta's international airport and the mean of all of Verizon's data speeds?

b. How many standard deviations is that [the difference found in part (a)]?

c. Convert Verizon's data speed at Atlanta's international airport to a z score.

d. If we consider data speeds that convert to z scores between −2 and 2 to be neither significantly low nor significantly high, is Verizon's speed at Atlanta significant?

6. PHL Data Speeds Repeat the preceding exercise using the Verizon data speed of 0.8 Mbps at Philadelphia International Airport (PHL).

7. Female Pulse Rates Pulse rates of adult females are listed in Data Set 1 "Body Data" in Appendix B. The lowest pulse rate is 36 beats per minute, the mean of the listed pulse rates is $\bar{x} = 74.0$ beats per minute, and their standard deviation is $s = 12.5$ beats per minute.

a. What is the difference between the pulse rate of 36 beats per minute and the mean pulse rate of the females?

b. How many standard deviations is that [the difference found in part (a)]?

c. Convert the pulse rate of 36 beats per minutes to a z score.

d. If we consider pulse rates that convert to z scores between −2 and 2 to be neither significantly low nor significantly high, is the pulse rate of 36 beats per minute significant?

8. Plastic Waste Data Set 31 "Garbage Weight" in Appendix B lists weights (lb) of plastic discarded by households. The highest weight is 5.28 lb, the mean of all of the weights is $\bar{x} = 1.911$ lb, and the standard deviation of the weights is $s = 1.065$ lb.

a. What is the difference between the weight of 5.28 lb and the mean of the weights?

b. How many standard deviations is that [the difference found in part (a)]?

c. Convert the weight of 5.28 lb to a z score.

d. If we consider weights that convert to z scores between -2 and 2 to be neither significantly low nor significantly high, is the weight of 5.28 lb significant?

Significant Values. *In Exercises 9–12, consider a value to be significantly low if its z score is less than or equal to -2 or consider the value to be significantly high if its z score is greater than or equal to 2.*

9. ACT The ACT test is used to assess readiness for college. In a recent year, the mean ACT score was 21.1 and the standard deviation was 5.1. Identify the ACT scores that are significantly low or significantly high.

10. MCAT In a recent year, scores on the Medical College Admission Test (MCAT) had a mean of 25.2 and a standard deviation of 6.4. Identify the MCAT scores that are significantly low or significantly high.

11. Quarters Data Set 29 "Coin Weights" lists weights (grams) of quarters manufactured after 1964. Those weights have a mean of 5.63930 g and a standard deviation of 0.06194 g. Identify the weights that are significantly low or significantly high.

12. Designing Aircraft Seats In the process of designing aircraft seats, it was found that men have hip breadths with a mean of 36.6 cm and a standard deviation of 2.5 cm (based on anthropometric survey data from Gordon, Clauser, et al.). Identify the hip breadths of men that are significantly low or significantly high.

Comparing Values. *In Exercises 13–16, use z scores to compare the given values.*

13. Tallest and Shortest Men The tallest living man at the time of this writing is Sultan Kosen, who has a height of 251 cm. The shortest living man is Chandra Bahadur Dangi, who has a height of 54.6 cm. Heights of men have a mean of 174.12 cm and a standard deviation of 7.10 cm. Which of these two men has the height that is more extreme?

14. Red Blood Cell Counts Based on Data Set 1 "Body Data" in Appendix B, males have red blood cell counts with a mean of 4.719 and a standard deviation of 0.490, while females have red blood cell counts with a mean of 4.349 and a standard deviation of 0.402. Who has the higher count relative to the sample from which it came: a male with a count of 5.58 or a female with a count of 5.23? Explain.

15. Birth Weights Based on Data Set 4 "Births" in Appendix B, newborn males have weights with a mean of 3272.8 g and a standard deviation of 660.2 g. Newborn females have weights with a mean of 3037.1 g and a standard deviation of 706.3 g. Who has the weight that is more extreme relative to the group from which they came: a male who weighs 1500 g or a female who weighs 1500 g?

16. Oscars In the 87th Academy Awards, Eddie Redmayne won for best actor at the age of 33 and Julianne Moore won for best actress at the age of 54. For all best actors, the mean age is 44.1 years and the standard deviation is 8.9 years. For all best actresses, the mean age is 36.2 years and the standard deviation is 11.5 years. (All ages are determined at the time of the awards ceremony.) Relative to their genders, who had the more extreme age when winning the Oscar: Eddie Redmayne or Julianne Moore? Explain.

Percentiles.   *In Exercises 17–20, use the following cell phone airport data speeds (Mbps) from Sprint. Find the percentile corresponding to the given data speed.*

0.2	0.3	0.3	0.3	0.3	0.3	0.3	0.4	0.4	0.4
0.5	0.5	0.5	0.5	0.5	0.6	0.6	0.7	0.8	1.0
1.1	1.1	1.2	1.2	1.6	1.6	2.1	2.1	2.3	2.4
2.5	2.7	2.7	2.7	3.2	3.4	3.6	3.8	4.0	4.0
5.0	5.6	8.2	9.6	10.6	13.0	14.1	15.1	15.2	30.4

17. 2.4 Mbps   **18.** 13.0 Mbps   **19.** 0.7 Mbps   **20.** 9.6 Mbps

In Exercises 21–28, use the same list of Sprint airport data speeds (Mbps) given for Exercises 17–20. Find the indicated percentile or quartile.

21. P_{60} **22.** Q_1 **23.** Q_3 **24.** P_{40} **25.** P_{50} **26.** P_{75} **27.** P_{25} **28.** P_{85}

Boxplots.   *In Exercises 29–32, use the given data to construct a boxplot and identify the 5-number summary.*

29. Speed Dating The following are the ratings of males by females in an experiment involving speed dating.

   2.0 3.0 4.0 5.0 6.0 6.0 7.0 7.0 7.0 7.0 7.0 7.0 8.0 8.0 8.0 8.0 9.0 9.5 10.0 10.0

30. Cell Phone Radiation Listed below are the measured radiation absorption rates (in W/kg) corresponding to these cell phones: iPhone 5S, BlackBerry Z30, Sanyo Vero, Optimus V, Droid Razr, Nokia N97, Samsung Vibrant, Sony Z750a, Kyocera Kona, LG G2, and Virgin Mobile Supreme. The data are from the Federal Communications Commission.

   1.18   1.41   1.49   1.04   1.45   0.74   0.89   1.42   1.45   0.51   1.38

31. Radiation in Baby Teeth Listed below are amounts of strontium-90 (in millibecquerels, or mBq) in a simple random sample of baby teeth obtained from Pennsylvania residents born after 1979 (based on data from "An Unexpected Rise in Strontium-90 in U.S. Deciduous Teeth in the 1990s," by Mangano et. al., *Science of the Total Environment*).

   128   130   133   137   138   142   142   144   147   149   151   151   151   155
   156   161   163   163   166   172

32. Blood Pressure Measurements Fourteen different second-year medical students at Bellevue Hospital measured the blood pressure of the same person. The systolic readings (mm Hg) are listed below.

   138   130   135   140   120   125   120   130   130   144   143   140   130   150

Boxplots from Large Data Sets in Appendix B.   *In Exercises 33–36, use the given data sets in Appendix B. Use the boxplots to compare the two data sets.*

33. Pulse Rates Use the same scale to construct boxplots for the pulse rates of males and females from Data Set 1 "Body Data" in Appendix B.

34. Ages of Oscar Winners Use the same scale to construct boxplots for the ages of the best actresses and best actors from Data Set 14 "Oscar Winner Age" in Appendix B.

35. BMI Use the body mass indexes (BMIs) for males and females listed in Data Set 1 "Body Data."

36. Lead and IQ Use the same scale to construct boxplots for the full IQ scores (IQF) for the low lead level group (group 1) and the high lead level group (group 3) in Data Set 7 "IQ and Lead" in Appendix B.

3-3   Beyond the Basics

37. Outliers and Modified Boxplots Repeat Exercise 33 "Pulse Rates" using modified boxplots. Identify any outliers as defined in Part 2 of this section.

Chapter Quick Quiz

1. Sleep Mean As part of the National Health and Nutrition Examination Survey, subjects were asked how long they slept the preceding night, and the following times (hours) were reported: 8, 7, 5, 7, 4, 7, 6, 7, 8, 8, 8, 6. Find the mean.

2. Sleep Median What is the median of the sample values listed in Exercise 1?

3. Sleep Mode What is the mode of the sample values listed in Exercise 1?

4. Sleep Variance The standard deviation of the sample values in Exercise 1 is 1.3 hours. What is the variance (including units)?

5. Sleep Outlier If the sleep time of 0 hours is included with the sample data given in Exercise 1, is it an outlier? Why or why not?

6. Sleep z Score A larger sample of 50 sleep times (hours) has a mean of 6.3 hours and a standard deviation of 1.4 hours. What is the z score for a sleep time of 5 hours?

7. Sleep Q_3 For a sample of 80 sleep times, approximately how many of those times are less than Q_3?

8. Sleep 5-Number Summary For a sample of 100 sleep times, give the *names* of the values that constitute the 5-number summary. (The actual values can't be identified; just give the *names* of those values.)

9. Estimating s A large sample of sleep times includes values ranging from a low of 4 hours to a high of 10 hours. Use the range rule of thumb to estimate the standard deviation.

10. Sleep Notation Consider a sample of sleep times taken from the population of all adults living in Alaska. Identify the symbols used for the sample mean, population mean, sample standard deviation, population standard deviation, sample variance, and the population variance.

Review Exercises

1. Old Faithful Geyser Listed below are prediction errors (minutes) that are differences between actual eruption times and predicted eruption times. Positive numbers correspond to eruptions that occurred later than predicted, and negative numbers correspond to eruptions that occurred before they were predicted. (The data are from Data Set 23 "Old Faithful" in Appendix B.) Find the (a) mean; (b) median; (c) mode; (d) midrange; (e) range; (f) standard deviation; (g) variance; (h) Q_1; (i) Q_3.

$$4 \quad -7 \quad 0 \quad 1 \quad -1 \quad 1 \quad -4 \quad -7 \quad 22 \quad 7 \quad -5 \quad 1$$

2. z Score Using the sample data from Exercise 1, find the z score corresponding to the prediction error of 0 min. Is that prediction error significantly low or high? Why or why not?

3. Boxplot Using the same prediction errors listed in Exercise 1, construct a boxplot and include the values of the 5-number summary.

4. ER Codes In an analysis of activities that resulted in brain injuries presenting at hospital emergency rooms, the following activities were identified by the codes shown in parentheses: bicycling (12); football (14); playground (22); basketball (27); swimming (40). Find the mean of 12, 14, 22, 27, and 40. What is wrong with this result?

5. Comparing Birth Weights The birth weights of a sample of males have a mean of 3272.8 g and a standard deviation of 660.2 g. The birth weights of a sample of females have a mean of 3037.1 g and a standard deviation of 706.3 g (based on Data Set 4 "Births" in Appendix B). When considered among members of the same gender, which baby has the relatively larger birth weight: a male with a birth weight of 3400 g or a female with a birth weight of 3200 g? Why?

6. Effects of an Outlier Listed below are platelet counts (1000 cells/μL) from subjects included in Data Set 1 "Body Data." Identify the outlier and then comment on the effect it has on the mean and standard deviation by finding the values of those statistics with the outlier included and then with the outlier excluded.

263   206   185   246   188   191   308   262   198   253   646

7. Interpreting a Boxplot Shown below is a boxplot of a sample of 30 maximal skull breadths (mm) measured from Egyptian skulls from around 4000 B.C. What do the numbers in the boxplot represent?

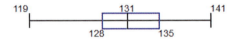

8. Estimating Standard Deviation Listed below is a sample of duration times (seconds) of eruptions of the Old Faithful geyser. Use the range rule of thumb to estimate the value of the standard deviation of all duration times, and compare the result to the standard deviation of 33.7 seconds found from a sample of 2634 duration times.

226   228   247   247   253   256   250   254   229   242   250   241   226   240   117

Cumulative Review Exercises

1. Arsenic in Rice Listed below are measured amounts (μg per serving) of arsenic in a sample of servings of brown rice [data from the Food and Drug Administration (FDA)]. Construct a frequency distribution. Use a class width of 2 μg, and use 0 μg as the lower class limit of the first class.

6.1   5.4   6.9   4.9   6.6   6.3   6.7   8.2   7.8   1.5   5.4   7.3

2. Histogram Use the frequency distribution from Exercise 1 to construct a histogram. Use class midpoint values for the horizontal scale.

3. Stemplot Use the amounts of arsenic from Exercise 1 to construct a stemplot.

4. Descriptive Statistics Use amounts of arsenic in Exercise 1 and find the following: (a) mean; (b) median; (c) standard deviation; (d) variance; (e) range. Include the appropriate units of measurement.

5. Histogram The accompanying histogram depicts outcomes of digits from the Florida Play 4 lottery. What is the major flaw in this histogram?

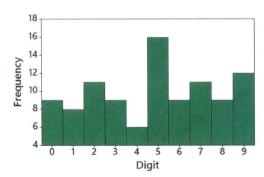

6. Normal Distribution Examine the distribution shown in the histogram from Exercise 5. Does it appear that the sample data are from a population with a normal distribution? Why or why not?

Technology Project

Words Spoken by Men and Women Refer to Data Set 24 "Word Counts" in Appendix B, which includes counts of words spoken by males and females. That data set includes 12 columns of data, but first stack all of the male word counts in one column and stack all of the female word counts in another column. Then proceed to generate histograms, any other suitable graphs, and find appropriate statistics that allow you to compare the two sets of data. Are there any outliers? Do both data sets have properties that are basically the same? Are there any significant differences? What would be a consequence of having significant differences? Write a brief report including your conclusions and supporting graphs.

FROM DATA TO DECISION

Second-Hand Smoke

Data Set 12 "Passive and Active Smoke" in Appendix B lists measures of cotinine from three groups of subjects: (1) smokers; (2) nonsmokers exposed to environmental tobacco smoke; (3) nonsmokers not exposed to environmental tobacco smoke. Cotinine is an indicator of nicotine absorption.

Critical Thinking

Use the methods from this chapter to explore and compare the cotinine measures in the three groups. Are there any notable differences? Are there any outliers? What do you conclude about the effects that smokers have on nonsmokers? Write a brief report of your conclusions, and provide supporting statistical evidence.

Cooperative Group Activities

1. In-class activity In class, each student should record two pulse rates by counting the number of heartbeats in 1 minute. The first pulse rate should be measured while the student is seated, and the second pulse rate should be measured while the student is standing. Use the methods of this chapter to compare results. Do males and females appear to have different pulse rates? Do pulse rates measured while seated appear to be different from pulse rates measured while standing?

2. Out-of-class activity In the article "Weighing Anchors" in *Omni* magazine, author John Rubin observed that when people estimate a value, their estimate is often "anchored" to (or influenced by) a preceding number, even if that preceding number is totally unrelated to the quantity being estimated. To demonstrate this, he asked people to give a quick estimate of the value of $8 \times 7 \times 6 \times 5 \times 4 \times 3 \times 2 \times 1$. The mean of the answers given was 2250, but when the order of the numbers was reversed, the mean became 512. Rubin explained that when we begin calculations with larger numbers (as in $8 \times 7 \times 6$), our estimates tend to be larger. He noted that both 2250 and 512 are far below the correct product, 40,320. The article suggests that irrelevant numbers can play a role in influencing real estate appraisals, estimates of car values, and estimates of the likelihood of nuclear war.

Conduct an experiment to test this theory. Select some subjects and ask them to quickly estimate the value of

$$8 \times 7 \times 6 \times 5 \times 4 \times 3 \times 2 \times 1$$

Then select other subjects and ask them to quickly estimate the value of

$$1 \times 2 \times 3 \times 4 \times 5 \times 6 \times 7 \times 8$$

Record the estimates along with the particular order used. Carefully design the experiment so that conditions are uniform and the two sample groups are selected in a way that minimizes any bias. Don't describe the theory to subjects until after they have provided their estimates. Compare the two sets of sample results by using the methods of this chapter. Provide a printed report that includes the data collected, the detailed methods used, the method of analysis, any relevant graphs and/or statistics, and a statement of conclusions. Include a critique of the experiment, with reasons why the results might not be correct, and describe ways in which the experiment could be improved.

3. In-class activity Complete the following table, then compare variation for smartphone data speeds. See Data Set 32 "Airport Data Speeds."

	Verizon	Sprint	AT&T	T-Mobile
Standard Deviation				
Range				

4. Out-of-class activity Record the times that cars are parked at a gas pump, and describe important characteristics of those times.

5. Out-of-class activity Several websites, such as www.gasbuddy.com, are designed to provide a list of local gas prices. Obtain a list of local gas prices and explore the data using the methods of this chapter and Chapter 2.

6. Out-of-class activity Data Set 28 "Chocolate Chip Cookies" in Appendix B includes counts of chocolate chips in five different brands of cookies. Obtain your own sample of chocolate chip cookies and proceed to count the number of chocolate chips in each cookie. Use the data to generate a histogram and any other suitable graphs. Find the descriptive statistics. Compare your chocolate chip counts to those given in Data Set 28. Are there any differences? Explain.

7. Out-of-class activity Appendix B includes many real and interesting data sets. In each group of three or four students, select a data set from Appendix B and analyze it using the methods discussed so far in this book. Write a brief report summarizing key conclusions.

8. Out-of-class activity In each group of three or four students, collect an original data set of values at the interval or ratio level of measurement. Provide the following: (1) a list of sample values; (2) software results of descriptive statistics and graphs; and (3) a written description of the nature of the data, the method of collection, and important characteristics.

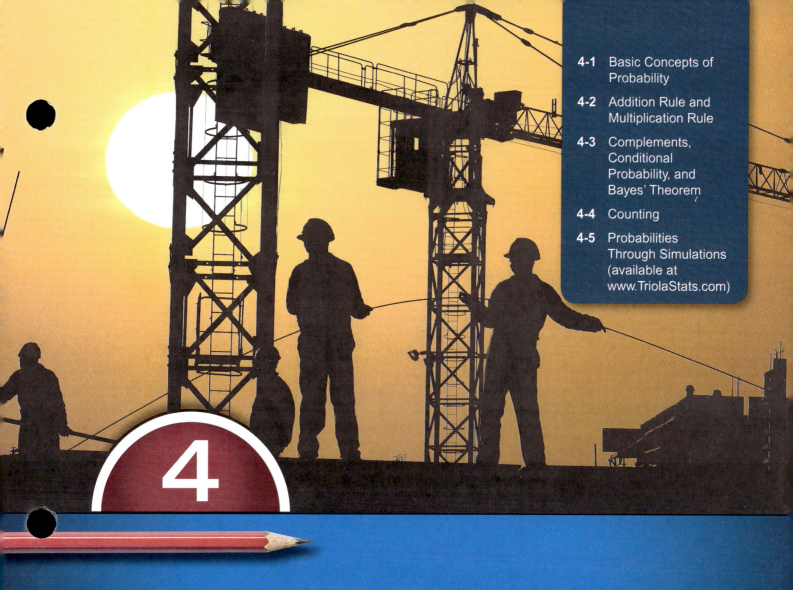

4-1 Basic Concepts of Probability

4-2 Addition Rule and Multiplication Rule

4-3 Complements, Conditional Probability, and Bayes' Theorem

4-4 Counting

4-5 Probabilities Through Simulations (available at www.TriolaStats.com)

4

PROBABILITY

CHAPTER PROBLEM **Drug testing of job applicants**

Approximately 85% of U. S. companies test employees and/or job applicants for drug use. A common and inexpensive (around $50) urine test is the EMIT (enzyme multiplied immunoassay technique) test, which tests for the presence of any of five drugs: marijuana, cocaine, amphetamines, opiates, or phencyclidine. Most companies require that positive test results be confirmed by a more reliable GC-MS (gas chromatography mass spectrometry) test.

Like nearly all medical tests, drug tests are sometimes wrong. Wrong results are of two different types: (1) false positive results and (2) false negative results. In today's society, these terms should be clearly understood. A job applicant or employee who gets a false positive result is someone who incorrectly appears to be using drugs when he or she is not actually using drugs. This type of mistake can unfairly result in job denial or termination of employment.

Analyzing the Results

Table 4-1 includes results from 555 adults in the United States. If one of the subjects from Table 4-1 is randomly selected from those who do *not* use drugs, what is the probability of a false positive result? If one of the subjects from Table 4-1 is randomly selected from those who do not use drugs, what is the probability of a true negative result? We will address such questions in this chapter.

- **Prevalence:** Proportion of the population having the condition (such as drug use or disease) being considered.

- **False positive:** *Wrong* test result that incorrectly indicates that the subject has a condition when the subject does not have that condition.

- **False negative:** *Wrong* test result that incorrectly indicates that the subject does not have a condition when the subject does have that condition.

- **True positive:** *Correct* test result that indicates that a subject has a condition when the subject does have the condition.

- **True negative:** *Correct* test result that indicates that a subject does not have a condition when the subject does not have the condition.

- **Test sensitivity:** The probability of a true positive test result, given that the subject actually has the condition being tested.

- **Test specificity:** The probability of a true negative test result, given that the subject does not have the condition being tested.

- **Positive predictive value:** Probability that a subject actually has the condition, given that the test yields a positive result (indicating that the condition is present).

- **Negative predictive value:** Probability that the subject does not actually have the condition, given that the test yields a negative result (indicating that the condition is not present).

TABLE 4-1 Results from Drug Tests of Job Applicants

	Positive Test Result (Test shows drug use.)	Negative Test Result (Test shows no drug use.)
Subject Uses Drugs	45 (True Positive)	5 (False Negative)
Subject Does Not Use Drugs	25 (False Positive)	480 (True Negative)

CHAPTER OBJECTIVES

The main objective of this chapter is to develop a sound understanding of probability values, because those values constitute the underlying foundation on which methods of inferential statistics are built. The important methods of hypothesis testing commonly use *P-values*, which are probability values expressed as numbers between 0 and 1, inclusive. Smaller probability values, such as 0.01, correspond to events that are very unlikely. Larger probability values, such as 0.99, correspond to events that are very likely. Here are the chapter objectives:

4-1 Basic Concepts of Probability

- Identify probabilities as values between 0 and 1, and interpret those values as expressions of likelihood of events.
- Develop the ability to calculate probabilities of events.
- Define the *complement* of an event and calculate the probability of that complement.

4-2 Addition Rule and Multiplication Rule

- Develop the ability to calculate the probability that in a single trial, some event *A* occurs or some event *B* occurs or they both occur. Apply the addition rule by correctly adjusting for events that are not disjoint (or are overlapping).

- Develop the ability to calculate the probability of an event *A* occurring in a first trial and an event *B* occurring in a second trial. Apply the multiplication rule by adjusting for events that are not independent.

- Distinguish between independent events and dependent events.

4-3 Complements, Conditional Probability, and Bayes' Theorem

- Compute the probability of "at least one" occurrence of an event *A*.

- Apply the multiplication rule by computing the probability of some event, given that some other event has already occurred.

4-4 Counting

- Develop the ability to apply the multiplication counting rule, factorial rule, permutations rule, and combinations rule.

- Distinguish between circumstances requiring the permutations rule and those requiring the combinations rule.

4-1   Basic Concepts of Probability

Key Concept Part 1 of this section includes basic concepts of probability. The single most important objective of this section is to learn how to *interpret* probability values, which are expressed as values between 0 and 1. A small probability, such as 0.001, corresponds to an event that rarely occurs.

Part 2 of this section includes *odds* and how they relate to probabilities. Concepts related to odds are not needed for topics in the following chapters, but odds are commonly used in situations such as lotteries and gambling.

PART 1   Basic Concepts of Probability

Role of Probability in Statistics

Probability plays a central role in the important statistical method of *hypothesis testing* introduced later in Chapter 8. Statisticians make decisions using data by rejecting explanations (such as chance) based on very low *probabilities*. See the following example illustrating the role of probability and a fundamental way that statisticians think.

EXAMPLE 1   Analyzing a Claim

Researchers have made this claim (really, they have):

> *Claim:* "We have developed a gender selection method that greatly increases the likelihood of a baby being a girl."

> *Hypothesis Used When Testing the Preceding Claim:* The method of gender selection has *no effect*, so that for couples using this method, about 50% of the births result in girls.

continued

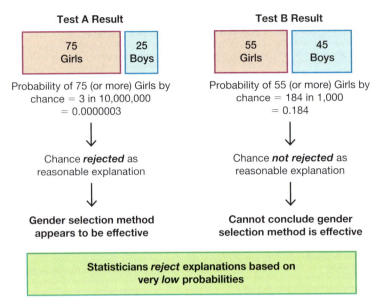

Different Gender Selection Methods
Tested with 100 Births

Test A Result	Test B Result
75 Girls **25 Boys**	**55 Girls** **45 Boys**

Probability of 75 (or more) Girls by chance = 3 in 10,000,000 = 0.0000003

Probability of 55 (or more) Girls by chance = 184 in 1,000 = 0.184

↓

↓

Chance *rejected* as reasonable explanation

Chance *not rejected* as reasonable explanation

↓

↓

Gender selection method appears to be effective

Cannot conclude gender selection method is effective

Statisticians *reject* explanations based on very *low* probabilities

FIGURE 4-1   Gender Selection Method Test Data and Conclusions

Figure 4-1 shows the sample data from two tests of 100 couples using the gender selection method and the conclusion reached for each test.

INTERPRETATION

Among the 100 babies, 75 girls and 55 girls are both greater than the 50 girls that we typically expect, but only the event of 75 girls leads us to believe that the gender selection method is effective. Even though there is a chance of getting 75 girls in 100 births with no special treatment, the probability of that happening is so small (0.0000003) that we should reject chance as a reasonable explanation. Instead, it would be generally recognized that the results provide strong support for the claim that the gender selection method is effective. This is exactly how statisticians think: They reject explanations (such as chance) based on very low *probabilities*.

YOUR TURN Do Exercise 37 "Predicting Gender."

Basics of Probability

In probability, we deal with procedures (such as generating male/female births or answering a multiple choice test question) that produce outcomes.

DEFINITIONS

An **event** is any collection of results or outcomes of a procedure.

A **simple event** is an outcome or an event that cannot be further broken down into simpler components.

The **sample space** for a procedure consists of all possible *simple* events. That is, the sample space consists of all outcomes that cannot be broken down any further.

Example 2 illustrates the concepts defined above.

> **EXAMPLE 2** **Simple Events and Sample Spaces**

In the following display, we use "b" to denote a baby boy and "g" to denote a baby girl.

Procedure	Example of Event	Sample Space: Complete List of Simple Events
Single birth	1 girl (simple event)	{b, g}
3 births	2 boys and 1 girl (bbg, bgb, and gbb are all simple events resulting in 2 boys and 1 girl)	{bbb, bbg, bgb, bgg, gbb, gbg, ggb, ggg}

Simple Events:

- With one birth, the result of 1 girl is a *simple event* and the result of 1 boy is another simple event. They are individual simple events because they cannot be broken down any further.

- With three births, the result of 2 girls followed by a boy (ggb) is a simple event.

- When rolling a single die, the outcome of 5 is a simple event, but the outcome of an even number is not a simple event.

Not a Simple Event: With three births, the event of "2 girls and 1 boy" is *not a simple event* because it can occur with these different simple events: ggb, gbg, bgg.

Sample Space: With three births, the *sample space* consists of the eight different simple events listed in the above table.

> **YOUR TURN** Do Exercise 35 "Four Children."

Three Common Approaches to Finding the Probability of an Event

We first list some basic notation, then we present three common approaches to finding the probability of an event.

Notation for Probabilities

P denotes a probability.

A, B, and C denote specific events.

$P(A)$ denotes the "probability of event A occurring."

The following three approaches for finding probabilities result in values between 0 and 1: $0 \leq P(A) \leq 1$. Figure 4-2 shows the possible values of probabilities and the more familiar and common expressions of likelihood.

1. **Relative Frequency Approximation of Probability** Conduct (or observe) a procedure and count the number of times that event A occurs. $P(A)$ is then *approximated* as follows:

$$P(A) = \frac{\text{number of times } A \text{ occurred}}{\text{number of times the procedure was repeated}}$$

When referring to relative frequency approximations of probabilities, this text will not distinguish between results that are exact probabilities and those that

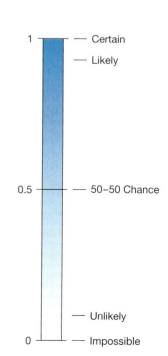

FIGURE 4-2 Possible Values for Probabilities

are approximations, so an instruction to "find the probability" could actually mean "*estimate* the probability."

2. **Classical Approach to Probability (Requires Equally Likely Outcomes)** If a procedure has *n* different simple events that are *equally likely*, and if event *A* can occur in *s* different ways, then

$$P(A) = \frac{\text{number of ways } A \text{ occurs}}{\text{number of different simple events}} = \frac{s}{n}$$

CAUTION When using the classical approach, always confirm that the outcomes are *equally likely*.

3. **Subjective Probabilities** $P(A)$, the probability of event *A*, is *estimated* by using knowledge of the relevant circumstances.

Figure 4-3 illustrates the approaches of the preceding three definitions.

1. Relative Frequency Approach: When trying to determine the probability that an individual car crashes in a year, we must examine past results to determine the number of cars in use in a year and the number of them that crashed; then we find the ratio of the number of cars that crashed to the total number of cars. For a recent year, the result is a probability of 0.0480. (See Example 3.)

2. Classical Approach: When trying to determine the probability of winning the grand prize in a lottery by selecting six different numbers between 1 and 60, each combination has an equal chance of occurring. The probability of winning is 0.0000000200, which can be found by using methods presented in Section 4-4. (See Example 4.)

3. Subjective Probability: When trying to estimate the probability of getting stuck in the next elevator that you ride, we know from personal experience that the probability is quite small. Let's estimate it to be, say, 0.001 (equivalent to 1 chance in 1000). (See Example 5.)

FIGURE 4-3 **Three Approaches to Finding a Probability**

Simulations Sometimes none of the preceding three approaches can be used. A *simulation* of a procedure is a process that behaves in the same ways as the procedure itself so that similar results are produced. Probabilities can sometimes be found by using a simulation. See the Technology Project near the end of this chapter.

ROUNDING PROBABILITIES

It is difficult to provide a universal rule for rounding probability values, but this guide will apply to most problems in this text: When expressing the value of a probability, either give the *exact* fraction or decimal or round off final decimal results to *three* significant digits. (*Suggestion:* When a probability is not a simple fraction such as 2/3 or 5/9, express it as a decimal so that the number can be better understood.) All digits in a number are significant except for the zeros that are included for proper placement of the decimal point. See the following examples.

- The probability of 0.4450323339 (from Example 6) has ten significant digits (4450323339), and it can be rounded to three significant digits as 0.445.

- The probability of 1/3 can be left as a fraction or rounded to 0.333. (Do *not* round to 0.3.)

- The probability of 2/8 can be expressed as 1/4 or 0.25. (Because 0.25 is exact, there's no need to express it with three significant digits as 0.250.)

Probabilities Expressed as Percentages? Mathematically, a probability of 0.25 is equivalent to 25%, but there are good reasons for sticking with fractions and decimals and not using percentages. Professional journals almost universally express probabilities as decimals, not as percentages. Later in this book, we will use probability values generated from statistical software, and they will always be in the form of decimals.

When finding probabilities with the relative frequency approach, we obtain an *approximation* instead of an exact value. As the total number of observations increases, the corresponding approximations tend to get closer to the actual probability. This property is commonly referred to as the *law of large numbers*.

LAW OF LARGE NUMBERS

As a procedure is repeated again and again, the relative frequency probability of an event tends to approach the actual probability.

The law of large numbers tells us that relative frequency approximations tend to get better with more observations. This law reflects a simple notion supported by common sense: A probability estimate based on only a few trials can be off by a substantial amount, but with a very large number of trials, the estimate tends to be much more accurate.

CAUTIONS

1. The law of large numbers applies to behavior over a large number of trials, and it does not apply to any one individual outcome. Gamblers sometimes foolishly lose large sums of money by incorrectly thinking that a string of losses increases the chances of a win on the next bet, or that a string of wins is likely to continue.

2. If we know nothing about the likelihood of different possible outcomes, we should not assume that they are equally likely. For example, we should not think that the probability of passing the next statistics test is 1/2, or 0.5 (because we either pass the test or do not). The actual probability depends on factors such as the amount of preparation and the difficulty of the test.

Understanding Chances of Winning the Lottery

In the New York State Lottery Mega Millions game, you must choose five different numbers from 1 to 75, and you must also select another "Mega Ball" number from 1 to 15. To win the jackpot, you must get the correct five numbers *and* the correct Mega Ball number. The chance of winning the jackpot with one ticket is 1/258,890,850. Commercials for this lottery state that "all you need is a little bit of luck," but in reality you need a ginormous amount of luck. The probability of 1/258,890,850 is not so easy to understand, so let's consider a helpful analogy suggested by Brother Donald Kelly of Marist College. A stack of 258,890,850 quarters is about 282 miles high. Commercial jets typically fly about 7 miles high, so this stack of quarters is about 40 times taller than the height of a commercial jet when it is at cruising altitude. The chance of winning the Mega Millions lottery game is equivalent to the chance of randomly selecting *one* specific quarter from that pile of quarters that is 282 miles high. Any of us who spend money on this lottery should understand that the chance of winning the jackpot is very, very, very close to zero.

How Probable?

How do we interpret such terms as *probable*, *improbable*, or *extremely improbable*? The Federal Aviation Administration (FAA) interprets these terms as follows.

- *Probable:* A probability on the order of 0.00001 or greater for each hour of flight. Such events are expected to occur several times during the operational life of each airplane.

- *Improbable:* A probability on the order of 0.00001 or less. Such events are not expected to occur during the total operational life of a single airplane of a particular type, but may occur during the total operational life of all airplanes of a particular type.

- *Extremely improbable:* A probability on the order of 0.000000001 or less. Such events are so unlikely that they need not be considered to ever occur.

EXAMPLE 3   **Relative Frequency Probability: Skydiving**

Find the probability of dying when making a skydiving jump.

SOLUTION

In a recent year, there were about 3,000,000 skydiving jumps and 21 of them resulted in deaths. We use the relative frequency approach as follows:

$$P(\text{skydiving death}) = \frac{\text{number of skydiving deaths}}{\text{total number of skydiving jumps}} = \frac{21}{3,000,0000} = 0.000007$$

Here the classical approach cannot be used because the two outcomes (dying, surviving) are not equally likely. A subjective probability can be estimated in the absence of historical data.

> **YOUR TURN** Do Exercise 25 "XSORT Gender Selection."

EXAMPLE 4   **Classical Probability: Three Children of the Same Gender**

When three children are born, the sample space of genders is as shown in Example 1: {bbb, bbg, bgb, bgg, gbb, gbg, ggb, ggg}. If boys and girls are equally likely, then those eight simple events are equally likely. Assuming that boys and girls are equally likely, find the probability of getting three children all of the same gender when three children are born. (In reality, the probability of a boy is 0.512 instead of 0.5.)

SOLUTION

The sample space {bbb, bbg, bgb, bgg, gbb, gbg, ggb, ggg} includes eight equally likely outcomes, and there are exactly two outcomes in which the three children are of the same gender: bbb and ggg. We can use the classical approach to get

$$P(\text{three children of the same gender}) = \frac{2}{8} = \frac{1}{4} \text{ or } 0.25$$

> **YOUR TURN** Do Exercise 33 "Three Children."

EXAMPLE 5   **Subjective Probability: Katy Perry's Money**

What is the probability that the next dollar bill you spend was previously spent by Katy Perry?

SOLUTION

In the absence of data on the number of dollar bills in circulation and Katy Perry's spending habits, we make a subjective estimate. Experience suggests that the probability is most likely quite small. Let's estimate it to be, say, 0.00001 (equivalent to 1 chance in 100,000). Depending on our knowledge of the relevant circumstances, that subjective estimate might be reasonably accurate or it might be grossly wrong.

> **YOUR TURN** Do Exercise 4 "Subjective Probability."

EXAMPLE 6   Texting and Driving

In a study of U.S. high school drivers, it was found that 3785 texted while driving during the previous 30 days, and 4720 did not text while driving during that same time period (based on data from "Texting While Driving . . . ," by Olsen, Shults, Eaton, *Pediatrics*, Vol. 131, No. 6). Based on these results, if a high school driver is randomly selected, find the probability that he or she texted while driving during the previous 30 days.

SOLUTION

CAUTION A common *mistake* is to blindly plug in numbers to get the wrong probability of $3785/4720 = 0.802$. We should *think* about what we are doing, as follows.

Instead of trying to determine an answer directly from the given statement, first summarize the information in a format that allows clear understanding, such as this format:

3785	texted while driving
4720	did not text while driving
8505	total number of drivers in the sample

We can now use the relative frequency approach as follows:

$$P(\text{texting while driving}) = \frac{\text{number of drivers who texted while driving}}{\text{total number of drivers in the sample}} = \frac{3785}{8505}$$

$$= 0.445$$

INTERPRETATION

There is a 0.445 probability that if a high school driver is randomly selected, he or she texted while driving during the previous 30 days.

YOUR TURN Do Exercise 27 "Mendelian Genetics."

CAUTION Don't make the common mistake of finding a probability value by mindlessly dividing a smaller number by a larger number. Instead, *think* carefully about the numbers involved and what they represent. Carefully identify the total number of items being considered, as illustrated in Example 6.

EXAMPLE 7   Thanksgiving Day

If a year is selected at random, find the probability that Thanksgiving Day in the United States will be (a) on a Wednesday or (b) on a Thursday.

SOLUTION

a. In the United States, Thanksgiving Day always falls on the fourth Thursday in November. It is therefore impossible for Thanksgiving to be on a Wednesday. When an event is impossible, its probability is 0. P(Thanksgiving on Wednesday) $= 0$.

continued

Go Figure

10^{80}: Number of particles in the observable universe. The probability of a monkey randomly hitting keys and typing Shakespeare's *Hamlet* is $10^{-216,159}$.

b. It is certain that a Thanksgiving Day in the United States will be on a Thursday. When an event is certain to occur, its probability is 1. $P(\text{Thanksgiving on Thursday}) = 1$.

Because any event imaginable is impossible, certain, or somewhere in between, it follows that the mathematical probability of any event A is 0, 1, or a number between 0 and 1 (as shown in Figure 4-2). That is, $0 \leq P(A) \leq 1$.

YOUR TURN Do Exercises 19 "Square Peg" and 20 "Death and Taxes."

Complementary Events

Sometimes we need to find the probability that an event A does *not* occur.

> **DEFINITION**
>
> The **complement** of event A, denoted by $\overline{A}$, consists of all outcomes in which event A does *not* occur.

EXAMPLE 8   Complement of Death from Skydiving

Example 3 shows that in a recent year, there were 3,000,000 skydiving jumps and 21 of them resulted in death. Find the probability of *not* dying when making a skydiving jump.

SOLUTION

Among 3,000,000 jumps there were 21 deaths, so it follows that the other 2,999,979 jumps were survived. We get

$$P(\text{not dying when making a skydiving jump}) = \frac{2,999,979}{3,000,000} = 0.999993$$

INTERPRETATION

The probability of *not* dying when making a skydiving jump is 0.999993.

YOUR TURN Do Exercise 29 "Social Networking."

Relationship Between $P(A)$ and $P(\overline{A})$ If we denote the event of dying in a skydiving jump by D, Example 3 showed that $P(D) = 0.000007$ and Example 3 showed that $P(\overline{D}) = 0.999993$. The probability of $P(\overline{D})$ could be found by just subtracting $P(D)$ from 1.

Identifying Significant Results with Probabilities: The Rare Event Rule for Inferential Statistics

> **If, under a given assumption, the probability of a particular observed event is very small and the observed event occurs *significantly less than* or *significantly greater than* what we typically expect with that assumption, we conclude that the assumption is probably not correct.**

We can use probabilities to identify values that are *significantly low* or *significantly high* as follows.

Using Probabilities to Determine When Results Are Significantly High or Significantly Low

- **Significantly *high* number of successes:** x successes among n trials is a *significantly high* number of successes if the probability of x or more successes is unlikely with a probability of 0.05 or less. That is, x is a significantly high number of successes if $P(x \text{ or more}) \le 0.05.^*$

- **Significantly *low* number of successes:** x successes among n trials is a *significantly low* number of successes if the probability of x or fewer successes is unlikely with a probability of 0.05 or less. That is, x is a significantly low number of successes if $P(x \text{ or fewer}) \le 0.05.^*$

*The value 0.05 is not absolutely rigid. Other values, such as 0.01, could be used to distinguish between results that can easily occur by chance and events that are very unlikely to occur by chance.

See Example 1 on page 133, which illustrates the following:

- Among 100 births, 75 girls is *significantly high* because the probability of 75 or more girls is 0.0000003, which is less than or equal to 0.05 (so the gender selection method appears to be effective).

- Among 100 births, 55 girls is *not significantly high* because the probability of 55 or more girls is 0.184, which is greater than 0.05 (so the gender selection does not appear to be effective).

Probability Review

- **The probability of an event is a fraction or decimal number between 0 and 1 inclusive.**
- **The probability of an impossible event is 0.**
- **The probability of an event that is certain to occur is 1.**
- **Notation: $P(A)$ = the probability of event A.**
- **Notation: $P(\overline{A})$ = the probability that event A does *not* occur.**

PART 2 Odds

Expressions of likelihood are often given as *odds*, such as 50:1 (or "50 to 1"). Here are advantages of probabilities and odds:

- Odds make it easier to deal with money transfers associated with gambling. (That is why odds are commonly used in casinos, lotteries, and racetracks.)
- Probabilities make calculations easier, so they tend to be used by statisticians, mathematicians, scientists, and researchers in all fields.

Gambling to Win

In the typical state lottery, the "house" has a 65% to 70% advantage, since only 30% to 35% of the money bet is returned as prizes. The house advantage at racetracks is usually around 15%. In casinos, the house advantage is 5.26% for roulette, 1.4% for craps, and 3% to 22% for slot machines.

The house advantage is 5.9% for blackjack, but some professional gamblers can systematically win with a 1% player advantage by using complicated card-counting techniques that require many hours of practice. If a card-counting player were to suddenly change from small bets to large bets, the dealer would recognize the card counting and the player would be ejected. Card counters try to beat this policy by working with a team. When the count is high enough, the player signals an accomplice who enters the game with large bets. A group of MIT students supposedly won millions of dollars by counting cards in blackjack.

In the three definitions that follow, the *actual odds against* and the *actual odds in favor* reflect the actual likelihood of an event, but the *payoff odds* describe the payoff amounts that are determined by casino, lottery, and racetrack operators. Racetracks and casinos are in business to make a profit, so the payoff odds will not be the same as the actual odds.

DEFINITIONS

The **actual odds against** event A occurring are the ratio $P(\overline{A})/P(A)$, usually expressed in the form of *a:b* (or "*a* to *b*"), where *a* and *b* are integers. (Reduce using the largest common factor; if $a = 16$ and $b = 4$, express the odds as 4:1 instead of 16:4.)

The **actual odds in favor** of event A occurring are the ratio $P(A)/P(\overline{A})$, which is the reciprocal of the actual odds against that event. If the odds against an event are *a:b*, then the odds in favor are *b:a*.

The **payoff odds** against event A occurring are the ratio of net profit (if you win) to the amount bet:

$$\text{Payoff odds against event A} = (\text{net profit}):(\text{amount bet})$$

EXAMPLE 9   **Actual Odds Versus Payoff Odds**

If you bet $5 on the number 13 in roulette, your probability of winning is 1/38, but the payoff odds are given by the casino as 35:1.

  a. Find the actual odds against the outcome of 13.

  b. How much net profit would you make if you win by betting $5 on 13?

  c. If the casino was not operating for profit and the payoff odds were changed to match the actual odds against 13, how much would you win if the outcome were 13?

SOLUTION

  a. With $P(13) = 1/38$ and $P(\text{not } 13) = 37/38$, we get

$$\text{Actual odds against } 13 = \frac{P(\text{not } 13)}{P(13)} = \frac{37/38}{1/38} = \frac{37}{1} \text{ or } 37{:}1$$

  b. Because the casino payoff odds against 13 are 35:1, we have

$$35{:}1 = (\text{net profit}){:}(\text{amount bet})$$

So there is a $35 profit for each $1 bet. For a $5 bet, the net profit is $175 (which is 5 × $35). The winning bettor would collect $175 plus the original $5 bet. After winning, the total amount collected would be $180, for a net profit of $175.

  c. If the casino were not operating for profit, the payoff odds would be changed to 37:1, which are the actual odds against the outcome of 13. With payoff odds of 37:1, there is a net profit of $37 for each $1 bet. For a $5 bet, the net profit would be $185. (The casino makes its profit by providing a profit of only $175 instead of the $185 that would be paid with a roulette game that is fair instead of favoring the casino.)

YOUR TURN   Do Exercise 41 "Kentucky Pick 4."

4-1 Basic Skills and Concepts

Statistical Literacy and Critical Thinking

1. New Jersey Lottery Let A denote the event of placing a $1 straight bet on the New Jersey Pick 3 lottery and winning. There are 1000 different ways that you can select the three digits (with repetition allowed) in this lottery, and only one of those three-digit numbers will be the winner. What is the value of $P(A)$? What is the value of $P(\overline{A})$?

2. Probability Rewrite the following statement so that the likelihood of rain is expressed as a value between 0 and 1: "The probability of rain today is 25%."

3. Interpreting Weather While this exercise was being created, Weather.com indicated that there was a 60% chance of rain for the author's home region. Based on that report, which of the following is the most reasonable interpretation?

a. 60% of the author's region will get rain today.

b. In the author's region, it will rain for 60% of the day.

c. There is a 0.60 probability that it will rain somewhere in the author's region at some point during the day.

4. Subjective Probability Estimate the probability that the next time you turn on a light switch, you discover that a bulb does work.

5. Identifying Probability Values Which of the following are probabilities?

$$0 \quad 3/5 \quad 5/3 \quad -0.25 \quad 250\% \quad 7:3 \quad 1 \quad 50\text{–}50 \quad 5:1 \quad 0.135 \quad 2.017$$

6. Penicillin "Who discovered penicillin: Sean Penn, William Penn, Penn Jillette, Alexander Fleming, or Louis Pasteur?" If you make a random guess for the answer to that question, what is the probability that your answer is the correct answer of Alexander Fleming?

7. Avogadro Constant If you are asked on a quiz to give the first (leftmost) nonzero digit of the Avogadro constant and, not knowing the answer, you make a random guess, what is the probability that your answer is the correct answer of 6?

8. Births Example 2 in this section includes the sample space for genders from three births. Identify the sample space for the genders from two births.

In Exercises 9–12, assume that 50 births are randomly selected. Use subjective judgment to describe the given number of girls as (a) significantly low, (b) significantly high, or (c) neither significantly low nor significantly high.

9. 47 girls.   **10.** 26 girls.   **11.** 23 girls.   **12.** 5 girls.

In Exercises 13–20, express the indicated degree of likelihood as a probability value between 0 and 1.

13. Testing If you make a random guess for the answer to a true/false test question, there is a 50–50 chance of being correct.

14. SAT Test When making a random guess for an answer to a multiple choice question on an SAT test, the possible answers are a, b, c, d, e, so there is 1 chance in 5 of being correct.

15. Luggage Based on a Harris poll, if you randomly select a traveler, there is a 43% chance that his or her luggage is black.

16. Sleepwalking Based on a report in *Neurology* magazine, 29.2% of survey respondents have sleepwalked.

17. Randomness When using a computer to randomly generate the last digit of a phone number to be called for a survey, there is 1 chance in 10 that the last digit is zero.

18. Job Applicant Mistakes Based on an Adecco survey of hiring managers who were asked to identify the biggest mistakes that job candidates make during an interview, there is a 50–50 chance that they will identify "inappropriate attire."

19. Square Peg Sydney Smith wrote in "On the Conduct of the Understanding" that it is impossible to fit a square peg in a round hole.

20. Death and Taxes Benjamin Franklin said that death is a certainty of life.

In Exercises 21–24, refer to the sample data in Table 4-1, which is included with the Chapter Problem. Assume that 1 of the 555 subjects included in Table 4-1 is randomly selected.

TABLE 4-1 Results from Drug Tests of Job Applicants

	Positive Test Result (Test shows drug use.)	Negative Test Result (Test shows no drug use.)
Subject Uses Drugs	45 (True Positive)	5 (False Negative)
Subject Does Not Use Drugs	25 (False Positive)	480 (True Negative)

21. Drug Testing Job Applicants Find the probability of selecting someone who got a result that is a false negative. Who would suffer from a false negative result? Why?

22. Drug Testing Job Applicants Find the probability of selecting someone who got a result that is a false positive. Who would suffer from a false positive result? Why?

23. Drug Testing Job Applicants Find the probability of selecting someone who uses drugs. Does the result appear to be reasonable as an estimate of the "prevalence rate" described in the Chapter Problem?

24. Drug Testing Job Applicants Find the probability of selecting someone who does not use drugs. Does the result appear to be reasonable as an estimate of the proportion of the adult population that does not use drugs?

In Exercises 25–32, find the probability and answer the questions.

25. XSORT Gender Selection MicroSort's XSORT gender selection technique was designed to increase the likelihood that a baby will be a girl. At one point before clinical trials of the XSORT gender selection technique were discontinued, 945 births consisted of 879 baby girls and 66 baby boys (based on data from the Genetics & IVF Institute). Based on these results, what is the probability of a girl born to a couple using MicroSort's XSORT method? Does it appear that the technique is effective in increasing the likelihood that a baby will be a girl?

26. YSORT Gender Selection MicroSort's YSORT gender selection technique is designed to increase the likelihood that a baby will be a boy. At one point before clinical trials of the YSORT gender selection technique were discontinued, 291 births consisted of 239 baby boys and 52 baby girls (based on data from the Genetics & IVF Institute). Based on these results, what is the probability of a boy born to a couple using MicroSort's YSORT method? Does it appear that the technique is effective in increasing the likelihood that a baby will be a boy?

27. Mendelian Genetics When Mendel conducted his famous genetics experiments with peas, one sample of offspring consisted of 428 green peas and 152 yellow peas. Based on those results, estimate the probability of getting an offspring pea that is green. Is the result reasonably close to the expected value of 3/4, as Mendel claimed?

28. Guessing Birthdays On their first date, Kelly asks Mike to guess the date of her birth, not including the year.

a. What is the probability that Mike will guess correctly? (Ignore leap years.)

b. Would it be unlikely for him to guess correctly on his first try?

c. If you were Kelly, and Mike did guess correctly on his first try, would you believe his claim that he made a lucky guess, or would you be convinced that he already knew when you were born?

d. If Kelly asks Mike to guess her age, and Mike's guess is too high by 15 years, what is the probability that Mike and Kelly will have a second date?

29. Social Networking In a Pew Research Center survey of Internet users, 3732 respondents say that they use social networking sites and 1380 respondents say that they *do not* use social networking sites. What is the probability that a randomly selected person does not use a social networking site? Does that result suggest that it is unlikely for someone to not use social networking sites?

30. Car Rollovers In a recent year in the United States, 83,600 passenger cars rolled over when they crashed, and 5,127,400 passenger cars did not roll over when they crashed. Find the probability that a randomly selected passenger car crash results in a rollover. Is it unlikely for a car to roll over in a crash?

31. Genetics: Eye Color Each of two parents has the genotype brown/blue, which consists of the pair of alleles that determine eye color, and each parent contributes one of those alleles to a child. Assume that if the child has at least one brown allele, that color will dominate and the eyes will be brown. (The actual determination of eye color is more complicated than that.)

a. List the different possible outcomes. Assume that these outcomes are equally likely.

b. What is the probability that a child of these parents will have the blue/blue genotype?

c. What is the probability that the child will have brown eyes?

32. X-Linked Genetic Disease Men have XY (or YX) chromosomes and women have XX chromosomes. X-linked recessive genetic diseases (such as juvenile retinoschisis) occur when there is a defective X chromosome that occurs *without* a paired X chromosome that is *not* defective. In the following, represent a defective X chromosome with lowercase x, so a child with the xY or Yx pair of chromosomes will have the disease and a child with XX or XY or YX or xX or Xx will not have the disease. Each parent contributes one of the chromosomes to the child.

a. If a father has the defective x chromosome and the mother has good XX chromosomes, what is the probability that a son will inherit the disease?

b. If a father has the defective x chromosome and the mother has good XX chromosomes, what is the probability that a daughter will inherit the disease?

c. If a mother has one defective x chromosome and one good X chromosome and the father has good XY chromosomes, what is the probability that a son will inherit the disease?

d. If a mother has one defective x chromosome and one good X chromosome and the father has good XY chromosomes, what is the probability that a daughter will inherit the disease?

Probability from a Sample Space. *In Exercises 33–36, use the given sample space or construct the required sample space to find the indicated probability.*

33. Three Children Use this sample space listing the eight simple events that are possible when a couple has three children (as in Example 2 on page 135): {bbb, bbg, bgb, bgg, gbb, gbg, ggb, ggg}. Assume that boys and girls are equally likely, so that the eight simple events are equally likely. Find the probability that when a couple has three children, there is exactly one girl.

34. Three Children Using the same sample space and assumption from Exercise 33, find the probability that when a couple has three children, there are exactly two girls.

35. Four Children Exercise 33 lists the sample space for a couple having three children. After identifying the sample space for a couple having four children, find the probability of getting three girls and one boy (in any order).

36. Four Children Using the same sample space and assumption from Exercise 35, find the probability that when a couple has four children, all four are of the same gender.

Using Probability to Form Conclusions. *In Exercises 37–40, use the given probability value to determine whether the sample results could easily occur by chance, then form a conclusion.*

37. Predicting Gender A study addressed the issue of whether pregnant women can correctly predict the gender of their baby. Among 104 pregnant women, 57 correctly predicted the gender of their baby (based on data from "Are Women Carrying 'Basketballs'. . . ," by Perry, DiPietro, Constigan, *Birth,* Vol. 26, No. 3). If pregnant women have no such ability, there is a 0.327 probability of getting such sample results by chance. What do you conclude?

38. Buckle Up A study of the effect of seatbelt use in head-on passenger car collisions found that drivers using a seatbelt had a 64.1% survival rate, while drivers not using a seatbelt had a 41.5% survival rate. If seatbelts have no effect on survival rate, there is less than a 0.0001 chance of getting these results (based on data from "Mortality Reduction with Air Bag and Seat Belt Use in Head-on Passenger Car Collisions," by Crandall, Olson, Sklar, *American Journal of Epidemiology,* Vol. 153, No. 3). What do you conclude?

39. Coffee Talk A study on the enhancing effect of coffee on long-term memory found that 35 participants given 200 mg of caffeine performed better on a memory test 24 hours later compared to the placebo group that received no caffeine.

a. There was a probability of 0.049 that the difference between the coffee group and the placebo group was due to chance. What do you conclude?

b. A group given a higher dose of 300 mg performed better than the 200 mg group, with a probability of 0.75 that this difference is due to chance. What do you conclude?

40. Cell Phones and Cancer A study of 420,095 Danish cell phone users resulted in 135 who developed cancer of the brain or nervous system (based on data from the *Journal of the National Cancer Institute*). When comparing this sample group to another group of people who did not use cell phones, it was found that there is a probability of 0.512 of getting such sample results by chance. What do you conclude?

4-1   Beyond the Basics

Odds. *In Exercises 41–44, answer the given questions that involve odds.*

41. Kentucky Pick 4 In the Kentucky Pick 4 lottery, you can place a "straight" bet of $1 by selecting the exact order of four digits between 0 and 9 inclusive (with repetition allowed), so the probability of winning is 1/10,000. If the same four numbers are drawn in the same order, you collect $5000, so your net profit is $4999.

a. Find the actual odds against winning.

b. Find the payoff odds.

c. The website www.kylottery.com indicates odds of 1:10,000 for this bet. Is that description accurate?

42. Finding Odds in Roulette A roulette wheel has 38 slots. One slot is 0, another is 00, and the others are numbered 1 through 36, respectively. You place a bet that the outcome is an odd number.

a. What is your probability of winning?

b. What are the actual odds against winning?

c. When you bet that the outcome is an odd number, the payoff odds are 1:1. How much profit do you make if you bet $18 and win?

d. How much profit would you make on the $18 bet if you could somehow convince the casino to change its payoff odds so that they are the same as the actual odds against winning? (*Recommendation:* Don't actually try to convince any casino of this; their sense of humor is remarkably absent when it comes to things of this sort.)

43. Kentucky Derby Odds When the horse California Chrome won the 140th Kentucky Derby, a $2 bet on a California Chrome win resulted in a winning ticket worth $7.

a. How much net profit was made from a $2 win bet on California Chrome?

b. What were the payoff odds against a California Chrome win?

c. Based on preliminary wagering before the race, bettors collectively believed that California Chrome had a 0.228 probability of winning. Assuming that 0.228 was the true probability of a California Chrome victory, what were the actual odds against his winning?

d. If the payoff odds were the actual odds found in part (c), what would be the worth of a $2 win ticket after the California Chrome win?

44. Relative Risk and Odds Ratio In a clinical trial of 2103 subjects treated with Nasonex, 26 reported headaches. In a control group of 1671 subjects given a placebo, 22 reported headaches. Denoting the proportion of headaches in the treatment group by p_t and denoting the proportion of headaches in the control (placebo) group by p_c, the *relative risk* is p_t/p_c. The relative risk is a measure of the strength of the effect of the Nasonex treatment. Another such measure is the *odds ratio,* which is the ratio of the odds in favor of a headache for the treatment group to the odds in favor of a headache for the control (placebo) group, found by evaluating the following:

$$\frac{p_t/(1 - p_t)}{p_c/(1 - p_c)}$$

The relative risk and odds ratios are commonly used in medicine and epidemiological studies. Find the relative risk and odds ratio for the headache data. What do the results suggest about the risk of a headache from the Nasonex treatment?

4-2   Addition Rule and Multiplication Rule

Key Concepts In this section we present the *addition rule* as a tool for finding $P(A \text{ or } B)$, which is the probability that either event A occurs or event B occurs (or they both occur) as the single outcome of a procedure. To find $P(A \text{ or } B)$, we begin by adding the number of ways that A can occur and the number of ways that B can occur, but add without double counting. The word "or" in the addition rule is associated with the addition of probabilities.

This section also presents the basic *multiplication rule* used for finding $P(A \text{ and } B)$, which is the probability that event A occurs and event B occurs. If the outcome of event A somehow affects the probability of event B, it is important to adjust the probability of B to reflect the occurrence of event A. The rule for finding $P(A \text{ and } B)$ is called the *multiplication rule* because it involves the multiplication of the probability of event A and the probability of event B (where, if necessary, the probability of event B is adjusted because of the outcome of event A). The word "and" in the multiplication rule is associated with the multiplication of probabilities.

In Section 4-1 we considered only *simple* events, but in this section we consider *compound events.*

Addition Rule

Notation for Addition Rule

$P(A \text{ or } B) = P(\text{in a single trial, event } A \text{ occurs or event } B \text{ occurs or they both occur})$

The word "or" used in the preceding notation is the *inclusive or*, which means either one or the other or both. The formal addition rule is often presented as a formula, but blind use of formulas is not recommended. Instead, *understand* the spirit of the rule and use that understanding, as in the intuitive addition rule that follows.

INTUITIVE ADDITION RULE

To find $P(A \text{ or } B)$, add the number of ways event A can occur and the number of ways event B can occur, but *add in such a way that every outcome is counted only once*. $P(A \text{ or } B)$ is equal to that sum, divided by the total number of outcomes in the sample space.

FORMAL ADDITION RULE

$$P(A \text{ or } B) = P(A) + P(B) - P(A \text{ and } B)$$

where $P(A \text{ and } B)$ denotes the probability that A and B both occur at the same time as an outcome in a trial of a procedure.

One way to apply the addition rule is to add the probability of event A and the probability of event B and, if there is any overlap that causes double-counting, compensate for it by subtracting the probability of outcomes that are included twice. This approach is reflected in the above formal addition rule.

EXAMPLE 1   Drug Testing of Job Applicants

Refer to Table 4-1, reproduced here for your convenience and viewing pleasure. If 1 subject is randomly selected from the 555 subjects given a drug test, find the probability of selecting a subject who had a positive test result *or* uses drugs.

TABLE 4-1 Results from Drug Tests of Job Applicants

	Positive Test Result (Test shows drug use.)	Negative Test Result (Test shows no drug use.)
Subject Uses Drugs	45 (True Positive)	5 (False Negative)
Subject Does Not Use Drugs	25 (False Positive)	480 (True Negative)

Numbers in red correspond to positive test results or subjects who use drugs, and the total of those numbers is 75.

SOLUTION

Refer to Table 4-1 and carefully count the number of subjects who tested positive (first column) or use drugs (first row), but be careful to count subjects exactly once, not twice. *When adding the frequencies from the first column and the first row, include the frequency of 45 only once.* In Table 4-1, there are $45 + 25 + 5 = 75$ subjects who had positive test results or use drugs. We get this result:

$$P(\text{positive test result or subject uses drugs}) = 75/555 = 0.135$$

YOUR TURN Do Exercise 11 "Fast Food Drive-Thru Accuracy."

Disjoint Events and the Addition Rule

The addition rule is simplified when the events are *disjoint*.

> **DEFINITION**
>
> Events A and B are **disjoint** (or **mutually exclusive**) if they cannot occur at the same time. (That is, disjoint events do not overlap.)

EXAMPLE 2   Disjoint Events

Disjoint events:

Event A—Randomly selecting someone for a clinical trial who is a male

Event B—Randomly selecting someone for a clinical trial who is a female

(The selected person *cannot* be both.)

Events that are *not* disjoint:

Event A—Randomly selecting someone taking a statistics course

Event B—Randomly selecting someone who is a female

(The selected person *can* be both.)

YOUR TURN Do Exercise 12 "Fast Food Drive-Thru Accuracy."

Whenever A and B are disjoint, $P(A \text{ and } B)$ becomes zero in the formal addition rule, so for disjoint events A and B we have $P(A \text{ or } B) = P(A) + P(B)$. But again, instead of blind use of a formula, it is better to *understand* and use the intuitive addition rule.

Here is a summary of the key points of the addition rule:

1. **To find $P(A \text{ or } B)$, first associate the word *or* with addition.**

2. **To find the value of $P(A \text{ or } B)$, add the number of ways A can occur and the number of ways B can occur, but be careful to add without double counting.**

Complementary Events and the Addition Rule

In Section 4-1 we used $\overline{A}$ to indicate that event A does not occur. Common sense dictates this principle: We are certain (with probability 1) that either an event A occurs *or* it does not occur, so it follows that $P(A \text{ or } \overline{A}) = 1$. Because events A and $\overline{A}$ must be disjoint, we can use the addition rule to express this principle as follows:

$$P(A \text{ or } \overline{A}) = P(A) + P(\overline{A}) = 1$$

This result of the addition rule leads to the following three expressions that are "equivalent" in the sense that they are just different forms of the same principle.

RULE OF COMPLEMENTARY EVENTS

$$P(A) + P(\overline{A}) = 1 \qquad P(\overline{A}) = 1 - P(A) \qquad P(A) = 1 - P(\overline{A})$$

EXAMPLE 3   Sleepwalking

Based on a journal article, the probability of randomly selecting someone who has sleepwalked is 0.292, so $P(\text{sleepwalked}) = 0.292$ (based on data from "Prevalence and Comorbidity of Nocturnal Wandering in the U.S. General Population," by Ohayon et al., *Neurology*, Vol. 78, No. 20). If a person is randomly selected, find the probability of getting someone who has *not* sleepwalked.

SOLUTION

Using the rule of complementary events, we get

$$P(\text{has } not \text{ sleepwalked}) = 1 - P(\text{sleepwalked}) = 1 - 0.292 = 0.708$$

The probability of randomly selecting someone who has not sleepwalked is 0.708.

YOUR TURN   Do Exercise 5 "LOL."

Multiplication Rule

Notation for Multiplication Rule

We begin with basic notation followed by the multiplication rule. We strongly suggest using the *intuitive* multiplication rule, because it is based on understanding instead of blind use of a formula.

Notation

$P(A \text{ and } B) = P(\text{event A occurs in a first trial and event B occurs in a second trial})$

$P(B|A)$ represents the probability of event B occurring after it is assumed that event A has already occurred. (Interpret $B|A$ as "event B occurs after event A has already occurred.")

CAUTION The notation $P(A \text{ and } B)$ has two meanings, depending on its context. For the multiplication rule, $P(A \text{ and } B)$ denotes that event A occurs in one trial and event B occurs in another trial; for the addition rule we use $P(A \text{ and } B)$ to denote that events A and B both occur in the same trial.

INTUITIVE MULTIPLICATION RULE

To find the probability that event A occurs in one trial and event B occurs in another trial, multiply the probability of event A by the probability of event B, but *be sure that the probability of event B is found by assuming that event A has already occurred.*

FORMAL MULTIPLICATION RULE

$$P(A \text{ and } B) = P(A) \cdot P(B \mid A)$$

Independence and the Multiplication Rule

When applying the multiplication rule and considering whether the probability of event *B* must be adjusted to account for the previous occurrence of event *A*, we are focusing on whether events *A* and *B* are *independent*.

DEFINITIONS

Two events *A* and *B* are **independent** if the occurrence of one does not affect the *probability* of the occurrence of the other. (Several events are independent if the occurrence of any does not affect the probabilities of the occurrence of the others.) If *A* and *B* are not independent, they are said to be **dependent.**

CAUTION Don't think that *dependence* of two events means that one is the direct *cause* of the other. Having a working light in your kitchen and having a working light in your bedroom are dependent events because they share the same power source. One of the lights may stop working for many reasons, but if one light is out, there is a higher probability that the other light will be out (because of the common power source).

 EXAMPLE 4 **Drug Screening and the Basic Multiplication Rule**

Let's use only the 50 test results from the subjects who use drugs (from Table 4-1), as shown below:

Positive Test Results:	45
Negative Test Results:	5
Total:	50

a. If 2 of these 50 subjects who use drugs are randomly selected *with replacement*, find the probability that the first selected person had a positive test result and the second selected person had a negative test result.

b. Repeat part (a) by assuming that the two subjects are selected *without* replacement.

SOLUTION

a. *With Replacement:* First selection (with 45 positive results among 50 total results):

$$P(\text{positive test result}) = \frac{45}{50}$$

Second selection (with 5 negative test results among the same 50 total results):

$$P(\text{negative test result}) = \frac{5}{50}$$

We now apply the multiplication rule as follows:

$$P(\text{1st selection is positive and 2nd is negative}) = \frac{45}{50} \cdot \frac{5}{50} = 0.0900$$

continued

Independent Jet Engines

Soon after departing from Miami, Eastern Airlines Flight 855 had one engine shut down because of a low oil pressure warning light. As the L-1011 jet turned to Miami for landing, the low pressure warning lights for the other two engines also flashed. Then an engine failed, followed by the failure of the last working engine. The jet descended without power from 13,000 ft to 4000 ft, when the crew was able to restart one engine, and the 172 people on board landed safely. With independent jet engines, the probability of all three failing is only 0.0001³, or about one chance in a trillion. The FAA found that the same mechanic who replaced the oil in all three engines failed to replace the oil plug sealing rings. The use of a single mechanic caused the operation of the engines to become dependent, a situation corrected by requiring that the engines be serviced by different mechanics.

b. *Without Replacement:* Without replacement of the first subject, the calculations are the same as in part (a), except that the second probability must be adjusted to reflect the fact that the first selection was positive and is not available for the second selection. After the first positive result is selected, we have 49 test results remaining, and 5 of them are negative. The second probability is therefore 5/49, as shown below:

$$P(\text{1st selection is positive and 2nd is negative}) = \frac{45}{50} \cdot \frac{5}{49} = 0.0918$$

> **YOUR TURN** Do Exercise 13 "Fast Food Drive-Thru Accuracy."

The key point of part (b) in Example 4 is this: *We must adjust the probability of the second event to reflect the outcome of the first event.* Because selection of the second subject is made *without* replacement of the first subject, the second probability must take into account the fact that the first selection removed a subject who tested positive, so only 49 subjects are available for the second selection, and 5 of them had a negative test result. Part (a) of Example 4 involved sampling with replacement, so the events are independent; part (b) of Example 4 involved sampling without replacement, so the events are dependent. See the following.

Sampling

In the wonderful world of statistics, sampling methods are critically important, and the following relationships hold:

- Sampling *with replacement*: Selections are *independent* events.
- Sampling *without replacement:* Selections are *dependent* events.

Exception: Treating Dependent Events as Independent

Some cumbersome calculations can be greatly simplified by using the common practice of treating events as independent when *small samples* are drawn without replacement from *large populations*. (In such cases, it is rare to select the same item twice.) Here is a common guideline routinely used with applications such as analyses of survey results:

TREATING DEPENDENT EVENTS AS INDEPENDENT: 5% GUIDELINE FOR CUMBERSOME CALCULATIONS

When sampling without replacement and the sample size is no more than 5% of the size of the population, treat the selections as being *independent* (even though they are actually dependent).

Example 5 illustrates use of the above "5% guideline for cumbersome calculations" and it also illustrates that the basic multiplication rule extends easily to three or more events.

EXAMPLE 5 Drug Screening and the 5% Guideline for Cumbersome Calculations

Assume that three adults are randomly selected *without replacement* from the 247,436,830 adults in the United States. Also assume that 10% of adults in the United States use drugs. Find the probability that the three selected adults all use drugs.

SOLUTION

Because the three adults are randomly selected without replacement, the three events are dependent, but here we can treat them as being independent by applying the 5% guideline for cumbersome calculations. The sample size of 3 is clearly no more than 5% of the population size of 247,436,830. We get

$$P(\text{all 3 adults use drugs}) = P(\text{first uses drugs } and \text{ second uses drugs } and$$
$$\text{third uses drugs})$$
$$= P(\text{first uses drugs}) \cdot P(\text{second uses drugs}) \cdot$$
$$P(\text{third uses drugs})$$
$$= (0.10)(0.10)(0.10) = 0.00100$$

There is a 0.00100 probability that all three selected adults use drugs.

> YOUR TURN Do Exercise 29 "Medical Helicopters."

In Example 5, if we treat the events as dependent without using the 5% guideline, we get the following cumbersome calculation that begins with 247,436,830 adults, with 10% of them (or 24,743,683) using drugs:

$$\left(\frac{24,743,683}{247,436,830}\right)\left(\frac{24,743,682}{247,436,829}\right)\left(\frac{24,743,681}{247,436,828}\right) = 0.0009999998909$$

$$= 0.00100 \text{ (rounded)}$$

Just imagine randomly selecting 1000 adults instead of just 3, as is commonly done in typical polls. Extending the above calculation to include 1000 factors instead of 3 factors would be what statisticians refer to as "painful."

CAUTION In any probability calculation, it is extremely important to carefully identify the event being considered. See Example 6, where parts (a) and (b) might seem quite similar but their solutions are very different.

EXAMPLE 6 Birthdays

When two different people are randomly selected from those in your class, find the indicated probability by assuming that birthdays occur on the days of the week with equal frequencies.

   a. Find the probability that the two people are born on the *same day of the week*.

   b. Find the probability that the two people are both born on *Monday*.

SOLUTION

   a. Because no particular day of the week is specified, the first person can be born on any one of the seven weekdays. The probability that the second person is born on the same day as the first person is 1/7. The probability that two people are born on the same day of the week is therefore 1/7.

   b. The probability that the first person is born on Monday is 1/7 and the probability that the second person is also born on Monday is 1/7. Because the two events are independent, the probability that both people are born on Monday is

$$\frac{1}{7} \cdot \frac{1}{7} = \frac{1}{49}$$

> YOUR TURN Do Exercise 19 "Fast Food Drive-Thru Accuracy."

To Win, Bet Boldly

The *New York Times* published an article by Andrew Pollack in which he re-ported lower than expected earn-ings for the Mirage casino in Las Vegas. He wrote that "winnings for Mirage can be particularly volatile, because it caters to high rollers, gamblers who might bet $100,000 or more on a hand of cards. The law of averages does not work as consistently for a few large bets as it does for thousands of smaller ones. . ." This reflects the most funda-mental principle of gambling: To win, place one big bet instead of many small bets! With the right game, such as craps, you have just under a 50% chance of doubling your money if you place one big bet. With many small bets, your chance of doubling your money drops substantially.

WATCH YOUR LANGUAGE! Example 6 illustrates that finding correct or relevant probability values often requires greater language skills than computational skills. In Example 6, what exactly do we mean by "same day of the week"? See how parts (a) and (b) in Example 6 are very different.

Redundancy: Important Application of Multiplication Rule

The principle of *redundancy* is used to increase the reliability of many systems. Our eyes have passive redundancy in the sense that if one of them fails, we continue to see. An im-portant finding of modern biology is that genes in an organism can often work in place of each other. Engineers often design redundant components so that the whole system will not fail because of the failure of a single component, as in the following example.

EXAMPLE 7 Airbus 310: *Redundancy* for Better Safety

Modern aircraft are now highly reliable, and one design feature contributing to that reliability is the use of *redundancy*, whereby critical components are duplicated so that if one fails, the other will work. For example, the Airbus 310 twin-engine airliner has three independent hydraulic systems, so if any one system fails, full flight control is maintained with another functioning system. For this example, we will assume that for a typical flight, the probability of a hydraulic system failure is 0.002.

   a. If the Airbus 310 were to have one hydraulic system, what is the probability that the aircraft's flight control would work for a flight?

   b. Given that the Airbus 310 actually has three independent hydraulic systems, what is the probability that on a typical flight, control can be maintained with a working hydraulic system?

SOLUTION

   a. The probability of a hydraulic system failure is 0.002, so the probability that it does *not* fail is 0.998. That is, the probability that flight control can be main-tained is as follows:

$$P(1 \text{ hydraulic system } does \ not \ fail)$$
$$= 1 - P(\text{failure}) = 1 - 0.002 = 0.998$$

   b. With three independent hydraulic systems, flight control will be maintained if the three systems do not all fail. The probability of all three hydraulic systems failing is $0.002 \cdot 0.002 \cdot 0.002 = 0.000000008$. It follows that the probabil-ity of maintaining flight control is as follows:

$$P(\text{it does } not \text{ happen that all three hydraulic systems fail})$$
$$= 1 - 0.000000008 = 0.999999992$$

INTERPRETATION

With only one hydraulic system we have a 0.002 probability of failure, but with three independent hydraulic systems, there is only a 0.000000008 probability that flight control cannot be maintained because all three systems failed. By using three hydraulic systems instead of only one, risk of failure is decreased not by a factor of 1/3, but by a factor of 1/250,000. By using three independent hydraulic systems, risk is dramatically decreased and safety is dramatically increased.

YOUR TURN Do Exercise 25 "Redundancy in Computer Hard Drives."

Rationale for the Multiplication Rule

To see the reasoning that underlies the multiplication rule, consider a pop quiz consisting of these two questions:

1. True or false: A pound of feathers is heavier than a pound of gold.

2. Who said, "By a small sample, we may judge of the whole piece"?
   (a) Judge Judy; (b) Judge Dredd; (c) Miguel de Cervantes; (d) George Gallup; (e) Gandhi

The answers are T (true) and c. (The first answer is true, because weights of feathers are in avoirdupois units where a pound is 453.59 g, but weights of gold and other precious metals are in troy units where a pound is 373.24 g. The second answer is from *Don Quixote* by Cervantes.)

Here is the sample space for the different possible answers:

$$Ta \quad Tb \quad Tc \quad Td \quad Te \quad Fa \quad Fb \quad Fc \quad Fd \quad Fe$$

If both answers are random guesses, then the above 10 possible outcomes are equally likely, so

$$P(\text{both correct}) = P(T \text{ and } c) = \frac{1}{10} = 0.1$$

With $P(T \text{ and } c) = 1/10$, $P(T) = 1/2$, and $P(c) = 1/5$, we see that

$$\frac{1}{10} = \frac{1}{2} \cdot \frac{1}{5}$$

A *tree diagram* is a graph of the possible outcomes of a procedure, as in Figure 4-4. Figure 4-4 shows that if both answers are random guesses, all 10 branches are equally likely and the probability of getting the correct pair (T,c) is $1/10$. For each response to the first question, there are 5 responses to the second. *The total number of outcomes is 5 taken 2 times, or 10.* The tree diagram in Figure 4-4 therefore provides a visual illustration for using multiplication.

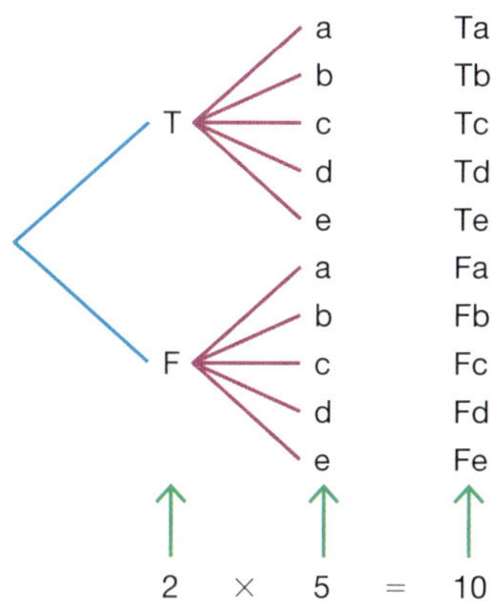

FIGURE 4-4   Tree Diagram of Test Answers

Summary of Addition Rule and Multiplication Rule

Addition Rule for $P(A \text{ or } B)$: The word *or* suggests addition, and when adding $P(A)$ and $P(B)$, we must add in such a way that every outcome is counted only once.

Multiplication Rule for $P(A \text{ and } B)$: The word *and* for two trials suggests multiplication, and when multiplying $P(A)$ and $P(B)$, we must be sure that the probability of event B takes into account the previous occurrence of event A.

4-2   Basic Skills and Concepts

Statistical Literacy and Critical Thinking

1. Notation When randomly selecting an adult, A denotes the event of selecting someone with blue eyes. What do $P(A)$ and $P(\overline{A})$ represent?

2. Notation When randomly selecting adults, let M denote the event of randomly selecting a male and let B denote the event of randomly selecting someone with blue eyes. What does $P(M \mid B)$ represent? Is $P(M \mid B)$ the same as $P(B \mid M)$?

3. Sample for a Poll There are 15,524,971 adults in Florida. If The Gallup organization randomly selects 1068 adults without replacement, are the selections independent or dependent? If the selections are dependent, can they be treated as being independent for the purposes of calculations?

4. Rule of Complements When randomly selecting an adult, let B represent the event of randomly selecting someone with type B blood. Write a sentence describing what the rule of complements is telling us: $P(B \text{ or } \overline{B}) = 1$.

Finding Complements.   *In Exercises 5–8, find the indicated complements.*

5. LOL A U.S. Cellular survey of smartphone users showed that 26% of respondents answered "yes" when asked if abbreviations (such as LOL) are annoying when texting. What is the probability of randomly selecting a smartphone user and getting a response other than "yes"?

6. Flights According to the Bureau of Transportation, 80.3% of American Airlines flights arrive on time. What is the probability of randomly selecting an American Airlines flight that does not arrive on time?

7. Flying In a Harris survey, adults were asked how often they typically travel on commercial flights, and it was found that $P(N) = 0.330$, where N denotes a response of "never." What does $P(\overline{N})$ represent, and what is its value?

8. Sobriety Checkpoint When the author observed a sobriety checkpoint conducted by the Dutchess County Sheriff Department, he saw that 676 drivers were screened and 6 were arrested for driving while intoxicated. Based on those results, we can estimate that $P(I) = 0.00888$, where I denotes the event of screening a driver and getting someone who is intoxicated. What does $P(\overline{I})$ denote, and what is its value?

In Exercises 9–20, use the data in the following table, which lists drive-thru order accuracy at popular fast food chains (data from a QSR Drive-Thru Study). Assume that orders are randomly selected from those included in the table.

	McDonald's	Burger King	Wendy's	Taco Bell
Order Accurate	329	264	249	145
Order Not Accurate	33	54	31	13

9. Fast Food Drive-Thru Accuracy If one order is selected, find the probability of getting food that is not from McDonald's.

10. Fast Food Drive-Thru Accuracy If one order is selected, find the probability of getting an order that is not accurate.

11. Fast Food Drive-Thru Accuracy If one order is selected, find the probability of getting an order from McDonald's or an order that is accurate. Are the events of selecting an order from McDonald's and selecting an accurate order disjoint events?

12. Fast Food Drive-Thru Accuracy If one order is selected, find the probability of getting an order that is not accurate or is from Wendy's. Are the events of selecting an order that is not accurate and selecting an order from Wendy's disjoint events?

13. Fast Food Drive-Thru Accuracy If two orders are selected, find the probability that they are both from Taco Bell.

a. Assume that the selections are made with replacement. Are the events independent?

b. Assume that the selections are made without replacement. Are the events independent?

14. Fast Food Drive-Thru Accuracy If two orders are selected, find the probability that both of them are not accurate.

a. Assume that the selections are made with replacement. Are the events independent?

b. Assume that the selections are made without replacement. Are the events independent?

15. Fast Food Drive-Thru Accuracy If two orders are selected, find the probability that they are both accurate.

a. Assume that the selections are made with replacement. Are the events independent?

b. Assume that the selections are made without replacement. Are the events independent?

16. Fast Food Drive-Thru Accuracy If two orders are selected, find the probability that they are both from Burger King.

a. Assume that the selections are made with replacement. Are the events independent?

b. Assume that the selections are made without replacement. Are the events independent?

17. Fast Food Drive-Thru Accuracy If one order is selected, find the probability of getting an order from McDonald's or Wendy's or an order that is not accurate.

18. Fast Food Drive-Thru Accuracy If one order is selected, find the probability of getting an order from Burger King or Taco Bell or an order that is accurate.

19. Fast Food Drive-Thru Accuracy If three *different* orders are selected, find the probability that they are all from Wendy's.

20. Fast Food Drive-Thru Accuracy If three *different* orders are selected, find the probability that they are all not accurate.

In Exercises 21–24, use these results from the "1-Panel-THC" test for marijuana use, which is provided by the company Drug Test Success: Among 143 subjects with positive test results, there are 24 false positive results; among 157 negative results, there are 3 false negative results. (Hint: Construct a table similar to Table 4-1, which is included with the Chapter Problem.)

21. Testing for Marijuana Use

a. How many subjects are included in the study?

b. How many of the subjects had a true negative result?

c. What is the probability that a randomly selected subject had a true negative result?

22. Testing for Marijuana Use If one of the test subjects is randomly selected, find the probability that the subject tested negative or used marijuana.

23. Testing for Marijuana Use If one of the test subjects is randomly selected, find the probability that the subject tested positive or did not use marijuana.

24. Testing for Marijuana Use If one of the test subjects is randomly selected, find the probability that the subject did not use marijuana. Do you think that the result reflects the general population rate of subjects who do not use marijuana?

Redundancy.   *Exercises 25 and 26 involve redundancy.*

25. Redundancy in Computer Hard Drives It is generally recognized that it is wise to back up computer data. Assume that there is a 3% rate of disk drive failure in a year (based on data from various sources, including lifehacker.com).

a. If you store all of your computer data on a single hard disk drive, what is the probability that the drive will fail during a year?

continued

b. If all of your computer data are stored on a hard disk drive with a copy stored on a second hard disk drive, what is the probability that both drives will fail during a year?

c. If copies of all of your computer data are stored on three independent hard disk drives, what is the probability that all three will fail during a year?

d. Describe the improved reliability that is gained with backup drives.

26. Redundancy in Hospital Generators Hospitals typically require backup generators to provide electricity in the event of a power outage. Assume that emergency backup generators fail 22% of the times when they are needed (based on data from Arshad Mansoor, senior vice president with the Electric Power Research Institute). A hospital has two backup generators so that power is available if one of them fails during a power outage.

a. Find the probability that both generators fail during a power outage.

b. Find the probability of having a working generator in the event of a power outage. Is that probability high enough for the hospital?

Acceptance Sampling.   *With one method of a procedure called acceptance sampling, a sample of items is randomly selected without replacement and the entire batch is accepted if every item in the sample is found to be okay. Exercises 27 and 28 involve acceptance sampling.*

27. Defective Pacemakers Among 8834 cases of heart pacemaker malfunctions, 504 were found to be caused by firmware, which is software programmed into the device (based on data from "Pacemaker and ICD Generator Malfunctions," by Maisel et al., *Journal of the American Medical Association,* Vol. 295, No. 16). If the firmware is tested in three *different* pacemakers randomly selected from this batch of 8834 and the entire batch is accepted if there are no failures, what is the probability that the firmware in the entire batch will be accepted? Is this procedure likely to result in the entire batch being accepted?

28. Something Fishy The National Oceanic and Atmospheric Administration (NOAA) inspects seafood that is to be consumed. The inspection process involves selecting seafood samples from a larger "lot." Assume a lot contains 2875 seafood containers and 288 of these containers include seafood that does not meet inspection requirements. What is the probability that 3 selected container samples all meet requirements and the entire lot is accepted based on this sample? Does this probability seem adequate?

In Exercises 29 and 30, find the probabilities and indicate when the "5% guideline for cumbersome calculations" is used.

29. Medical Helicopters In a study of helicopter usage and patient survival, results were obtained from 47,637 patients transported by helicopter and 111,874 patients transported by ground (based on data from "Association Between Helicopter vs Ground Emergency Medical Services and Survival for Adults with Major Trauma," by Galvagno et al., *Journal of the American Medical Association,* Vol. 307, No. 15).

a. If 1 of the 159,511 patients in the study is randomly selected, what is the probability that the subject was transported by helicopter?

b. If 5 of the subjects in the study are randomly selected without replacement, what is the probability that all of them were transported by helicopter?

30. Medical Helicopters In the same study cited in the preceding exercise, among the 47,637 patients transported by helicopter, 188 of them left the treatment center against medical advice, and the other 47,449 did not leave against medical advice. If 40 of the subjects transported by helicopter are randomly selected without replacement, what is the probability that none of them left the treatment center against medical advice?

4-2 Beyond the Basics

31. Surge Protectors Refer to the accompanying figure showing surge protectors p and q used to protect an expensive television. If there is a surge in the voltage, the surge protector reduces it to a safe level. Assume that each surge protector has a 0.985 probability of working correctly when a voltage surge occurs.

a. If the two surge protectors are arranged in series, what is the probability that a voltage surge will not damage the television? (Do not round the answer.)

b. If the two surge protectors are arranged in parallel, what is the probability that a voltage surge will not damage the television? (Do not round the answer.)

c. Which arrangement should be used for better protection?

Series configuration   Parallel configuration

32. Same Birthdays If 25 people are randomly selected, find the probability that no 2 of them have the same birthday. Ignore leap years.

33. Exclusive Or The *exclusive or* means either one or the other events occurs, but not both.

a. For the formal addition rule, rewrite the formula for $P(A \text{ or } B)$ assuming that the addition rule uses the *exclusive or* instead of the *inclusive or.*

b. Repeat Exercise 11 "Fast Food Drive-Thru Accuracy" using the *exclusive or* instead of the *inclusive or.*

34. Complements and the Addition Rule Refer to the table used for Exercises 9–20. Assume that one order is randomly selected. Let A represent the event of getting an order from McDonald's and let B represent the event of getting an order from Burger King. Find $P(\overline{A \text{ or } B})$, find $P(\overline{A} \text{ or } \overline{B})$, and then compare the results. In general, does $P(\overline{A \text{ or } B}) = P(\overline{A} \text{ or } \overline{B})$?

4-3 Complements, Conditional Probability, and Bayes' Theorem

Key Concept In Part 1 of this section we extend the use of the multiplication rule to include the probability that among several trials, we get *at least one* of some specified event. In Part 2 we consider *conditional probability:* the probability of an event occurring when we have additional information that some other event has already occurred. In Part 3 we provide a brief introduction to Bayes' theorem.

PART 1 Complements: The Probability of "At Least One"

When finding the probability of some event occurring "at least once," we should understand the following:

- "At least one" has the same meaning as "one or more."

- The *complement* of getting "at least one" particular event is that you get *no* occurrences of that event.

Convicted by Probability

A witness described a Los Angeles robber as a Caucasian woman with blond hair in a ponytail who escaped in a yellow car driven by an African-American male with a mustache and beard. Janet and Malcolm Collins fit this description, and they were convicted based on testimony that there is only about 1 chance in 12 million that any couple would have these characteristics. It was estimated that the probability of a yellow car is 1/10, and the other probabilities were estimated to be 1/10, 1/3, 1/10, and 1/1000. The convictions were later overturned when it was noted that no evidence was presented to support the estimated probabilities or the independence of the events. Also, because the couple was not randomly selected, a serious error was made in not considering the probability of *other* couples being in the same region with the same characteristics.

For example, not getting at least 1 girl in 10 births is the same as getting no girls, which is also the same as getting 10 boys.

> Not getting at least 1 girl in 10 births = Getting no girls = Getting 10 boys

The following steps describe the details of finding the probability of getting at least one of some event:

Finding the probability of getting *at least one* of some event:

1. **Let A = getting *at least one* of some event.**

2. **Then $\overline{A}$ = getting *none* of the event being considered.**

3. **Find $P(\overline{A})$ = probability that event A does not occur.** (This is relatively easy using the multiplication rule.)

4. **Subtract the result from 1. That is, evaluate this expression:**

$$P(\textit{at least one} \text{ occurrence of event } A)$$
$$= 1 - P(\textit{no} \text{ occurrences of event } A)$$

EXAMPLE 1   Accidental iPad Damage

A study by SquareTrade found that 6% of damaged iPads were damaged by "bags/backpacks." If 20 damaged iPads are randomly selected, find the probability of getting *at least one* that was damaged in a bag/backpack. Is the probability high enough so that we can be reasonably sure of getting at least one iPad damaged in a bag/backpack?

SOLUTION

Step 1: Let A = at least 1 of the 20 damaged iPads was damaged in a bag/backpack.

Step 2: Identify the event that is the complement of A.

$\overline{A}$ = *not* getting at least 1 iPad damaged in a bag/backpack among 20
    = all 20 iPads damaged in a way other than bag/backpack

Step 3: Find the probability of the complement by evaluating $P(\overline{A})$.

$$P(\overline{A}) = P(\text{all 20 iPads damaged in a way other than bag/backpack})$$
$$= 0.94 \cdot 0.94 \cdot \cdots \cdot 0.94$$
$$= 0.94^{20} = 0.290$$

Step 4: Find $P(A)$ by evaluating $1 - P(\overline{A})$.

$$P(A) = 1 - P(\overline{A}) = 1 - 0.290 = 0.710$$

INTERPRETATION

For a group of 20 damaged iPads, there is a 0.710 probability of getting at least 1 iPad damaged in a bag/backpack. This probability is not *very* high, so to be *reasonably sure* of getting at least 1 iPad damaged in a bag/backpack, more than 20 damaged iPads should be used.

YOUR TURN Do Exercise 7 "Births in the United States."

PART 2 Conditional Probability

We now consider the principle that the probability of an event is often affected by knowledge that some other event has occurred. For example, the probability of a golfer making a hole in one is 1/12,000 (based on past results), but if you have the additional knowledge that the selected golfer is a professional, the probability changes to 1/2375 (based on data from *USA Today*). In general, a *conditional probability* of an event is used when the probability is calculated with some additional knowledge, such as the knowledge that some other event has occurred. (Conditional probabilities were used in Section 4-2 with situations in which samples were selected without replacement.)

> **DEFINITION**
>
> A **conditional probability** of an event is a probability obtained with the additional information that some other event has already occurred.

Notation

$P(B \mid A)$ denotes the conditional probability of event B occurring, given that event A has already occurred.

INTUITIVE APPROACH FOR FINDING $P(B \mid A)$

The conditional probability of B occurring given that A has occurred can be found by *assuming that event A has occurred* and then calculating the probability that event B will occur, as illustrated in Example 2.

FORMAL APPROACH FOR FINDING $P(B \mid A)$

The probability $P(B \mid A)$ can be found by dividing the probability of events A and B both occurring by the probability of event A:

$$P(B \mid A) = \frac{P(A \text{ and } B)}{P(A)}$$

The preceding formula is a formal expression of conditional probability, but blind use of formulas is not recommended. Instead, we recommend the intuitive approach, as illustrated in Example 2.

EXAMPLE 2 Pre-Employment Drug Screening

Refer to Table 4-1 on the next page to find the following:

    a. If 1 of the 555 test subjects is randomly selected, find the probability that the subject had a positive test result, given that the subject actually uses drugs. That is, find $P(\text{positive test result} \mid \text{subject uses drugs})$.

    b. If 1 of the 555 test subjects is randomly selected, find the probability that the subject actually uses drugs, given that he or she had a positive test result. That is, find $P(\text{subject uses drugs} \mid \text{positive test result})$.

continued

Prosecutor's Fallacy

The *prosecutor's fallacy* is misunderstanding or confusion of two different conditional probabilities: (1) the probability that a defendant is innocent, given that forensic evidence shows a match; (2) the probability that forensics shows a match, given that a person is innocent. The prosecutor's fallacy has led to wrong convictions and imprisonment of some innocent people.

Lucia de Berk was a nurse who was convicted of murder and sentenced to prison in the Netherlands. Hospital administrators observed suspicious deaths that occurred in hospital wards where de Berk had been present. An expert testified that there was only 1 chance in 342 million that her presence was a coincidence. However, mathematician Richard Gill calculated the probability to be closer to 1/150, or possibly as low as 1/5. The court used the probability that the suspicious deaths could have occurred with de Berk present, given that she was innocent. The court should have considered the probability that de Berk is innocent, given that the suspicious deaths occurred when she was present. This error of the prosecutor's fallacy is subtle and can be very difficult to understand and recognize, yet it can lead to the imprisonment of innocent people.

Group Testing

During World War II, the U.S. Army tested for syphilis by giving each soldier an individual blood test that was analyzed separately. One researcher suggested mixing pairs of blood samples. After the mixed pairs were tested, those with syphilis could be identified by retesting the few blood samples that were in the pairs that tested positive. Since the total number of analyses was reduced by pairing blood specimens, why not combine them in groups of three or four or more? This technique of combining samples in groups and retesting only those groups that test positive is known as *group testing* or *pooled testing*, or *composite testing*. University of Nebraska statistician Christopher Bilder wrote an article about this topic in *Chance* magazine, and he cited some real applications. He noted that the American Red Cross uses group testing to screen for specific diseases, such as hepatitis, and group testing is used by veterinarians when cattle are tested for the bovine viral diarrhea virus.

TABLE 4-1 Results from Drug Tests of Job Applicants

	Positive Test Result (Test shows drug use.)	Negative Test Result (Test shows no drug use.)
Subject Uses Drugs	45 (True Positive)	5 (False Negative)
Subject Does Not Use Drugs	25 (False Positive)	480 (True Negative)

SOLUTION

a. **Intuitive Approach:** We want P(positive test result | subject uses drugs), the probability of getting someone with a positive test result, *given that the selected subject uses drugs*. Here is the key point: If we assume that the selected subject actually uses drugs, we are dealing only with the 50 subjects in the first row of Table 4-1. Among those 50 subjects, 45 had positive test results, so we get this result:

$$P(\text{positive test result} \mid \text{subject uses drugs}) = \frac{45}{50} = 0.900$$

Formal Approach: The same result can be found by using the formula for $P(B|A)$ given with the formal approach. We use the following notation.

$$P(B|A) = P(\text{positive test result} \mid \text{subject uses drugs})$$

where B = positive test result and A = subject uses drugs.

In the following calculation, we use P(subject uses drugs and had a positive test result) = 45/555 and P(subject uses drugs) = 50/555 to get the following results:

$$P(B|A) = \frac{P(A \text{ and } B)}{P(A)}$$

becomes

$P(\text{positive test result} \mid \text{subject uses drugs})$

$$= \frac{P(\text{subject uses drugs and had a positive test result})}{P(\text{subject uses drugs})}$$

$$= \frac{45/555}{50/555} = 0.900$$

By comparing the intuitive approach to the formal approach, it should be clear that the intuitive approach is much easier to use, and it is also less likely to result in errors. The intuitive approach is based on an *understanding* of conditional probability, instead of manipulation of a formula, and understanding is so much better.

b. Here we want P(subject uses drugs | positive test result). This is the probability that the selected subject uses drugs, *given that the subject had a positive test result*. If we assume that the subject had a positive test result, we are dealing with the 70 subjects in the first column of Table 4-1. Among those 70 subjects, 45 use drugs, so

$$P(\text{subject uses drugs} \mid \text{positive test result}) = \frac{45}{70} = 0.643$$

Again, the same result can be found by applying the formula for conditional probability, but we will leave that for those with a special fondness for manipulations with formulas.

INTERPRETATION

The first result of $P(\text{positive test result} \mid \text{subject uses drugs}) = 0.900$ indicates that a subject who uses drugs has a 0.900 probability of getting a positive test result. The second result of $P(\text{subject uses drugs} \mid \text{positive test result}) = 0.643$ indicates that for a subject who gets a positive test result, there is a 0.643 probability that this subject actually uses drugs. Note that $P(\text{positive test result} \mid \text{subject uses drugs}) \neq P(\text{subject uses drugs} \mid \text{positive test result})$. See "Confusion of the Inverse" that follows.

> **YOUR TURN** Do Exercise 13 "Denomination Effect."

Confusion of the Inverse

Note that in Example 2, $P(\text{positive test result} \mid \text{subject uses drugs}) \neq P(\text{subject uses drugs} \mid \text{positive test result})$. This example proves that in general, $P(B \mid A) \neq P(A \mid B)$. (There could be individual cases where $P(A \mid B)$ and $P(B \mid A)$ are equal, but they are generally not equal.) To incorrectly think that $P(B \mid A)$ and $P(A \mid B)$ are equal or to incorrectly use one value in place of the other is called *confusion of the inverse*.

EXAMPLE 3 **Confusion of the Inverse**

Consider these events:

$$D: \text{It is dark outdoors.}$$

$$M: \text{It is midnight.}$$

In the following, we conveniently ignore the Alaskan winter and other such anomalies.

$P(D \mid M) = 1$ (It is certain to be dark given that it is midnight.)

$P(M \mid D) = 0$ (The probability that it is exactly midnight given that it dark is almost zero.)

Here, $P(D \mid M) \neq P(M \mid D)$. Confusion of the inverse occurs when we incorrectly switch those probability values or think that they are equal.

PART 3 Bayes' Theorem

In this section we extend the discussion of conditional probability to include applications of *Bayes' theorem* (or *Bayes' rule*), which we use for revising a probability value based on additional information that is later obtained.

Let's consider a study showing that physicians often give very misleading information when they experience confusion of the inverse. They tended to confuse $P(\text{cancer} \mid \text{positive test result})$ with $P(\text{positive test result} \mid \text{cancer})$. (A positive test result indicates the patient has cancer; a negative test result indicates the patient is cancer-free.) About 95% of physicians estimated $P(\text{cancer} \mid \text{positive test result})$ to be about 10 times too high, with the result that patients were given diagnoses that were very misleading, and patients were unnecessarily distressed by the incorrect information. Let's take a closer look at this example, and let's hope that we can give physicians information in a better format that is easy to understand.

Probability of an Event That Has Never Occurred

Some events are possible but are so unlikely that they have never occurred. Here is one such problem of great interest to political scientists: Estimate the probability that your single vote will determine the winner in a U.S. presidential election. Andrew Gelman, Gary King, and John Boscardin write in the *Journal of the American Statistical Association* (Vol. 93, No. 441) that "the exact value of this probability is of only minor interest, but the number has important implications for understanding the optimal allocation of campaign resources, whether states and voter groups receive their fair share of attention from prospective presidents, and how formal 'rational choice' models of voter behavior might be able to explain why people vote at all." The authors show how the probability value of 1 in 10 million is obtained for close elections.

EXAMPLE 4 **Interpreting Medical Test Results**

Assume cancer has a 1% prevalence rate, meaning that 1% of the population has cancer. Denoting the event of having cancer by C, we have $P(C) = 0.01$ for a subject randomly selected from the population. This result is included with the following performance characteristics of the test for cancer (based on *Probabilistic Reasoning in Clinical Medicine* by David Eddy, Cambridge University Press).

- $P(C) = 0.01$ (There is a 1% prevalence rate of the cancer.)
- The false positive rate is 10%. That is, P(positive test result given that cancer is not present) $= 0.10$.
- The true positive rate is 80%. That is, P(positive test result given that cancer is present) $= 0.80$.

Find $P(C \mid$ positive test result). That is, find the probability that a subject actually has cancer given that he or she has a positive test result.

SOLUTION

Using the given information, we can construct a hypothetical population with the above characteristics. We can find the entries in Table 4-2 on the next page, as follows.

- Assume that we have 1000 subjects. With a 1% prevalence rate, 10 of the subjects are expected to have cancer. The sum of the entries in the first row of values is therefore 10.
- The other 990 subjects do not have cancer. The sum of the entries in the second row of values is therefore 990.
- Among the 990 subjects without cancer, 10% get positive test results, so 10% of the 990 cancer-free subjects in the second row get positive test results. See the entry of 99 in the second row.
- For the 990 subjects in the second row, 99 test positive, so the other 891 must test negative. See the entry of 891 in the second row.
- Among the 10 subjects with cancer in the first row, 80% of the test results are positive, so 80% of the 10 subjects in the first row test positive. See the entry of 8 in the first row.
- The other 2 subjects in the first row test negative. See the entry of 2 in the first row.

To find $P(C \mid$ positive test result), see that the first column of values includes the positive test results. In that first column, the probability of randomly selecting a subject with cancer is 8/107 or 0.0748, so $P(C \mid$ positive test result) $= 0.0748$.

INTERPRETATION

For the data given in this example, a randomly selected subject has a 1% chance of cancer, but for a randomly selected subject given a test with a positive result, the chance of cancer increases to 7.48%. Based on the data given in this example, a positive test result should not be devastating news, because there is still a good chance that the test is wrong.

TABLE 4-2 Test Results

	Positive Test Result (Test shows cancer.)	Negative Test Result (Test shows no cancer.)	Total
Cancer	8 (True Positive)	2 (False Negative)	10
No Cancer	99 (False Positive)	891 (True Negative)	990

The solution in Example 4 is not very difficult. Another approach is to compute the probability using this formula commonly given with Bayes' theorem:

$$P(A \mid B) = \frac{P(A) \cdot P(B \mid A)}{[\,P(A) \cdot P(B \mid A)\,] + [\,P(\overline{A}) \cdot P(B \mid \overline{A})\,]}$$

If we replace A with C and replace B with "positive," we get this solution for Example 4:

$$P(C \mid \text{positive}) = \frac{P(C) \cdot P(\text{positive} \mid C)}{P(C) \cdot P(\text{positive} \mid C) + P(\overline{C}) \cdot P(\text{positive} \mid \overline{C})}$$

$$= \frac{0.01 \cdot 0.80}{(0.01 \cdot 0.80) + (0.99 \cdot 0.10)} = 0.0748$$

Study Results Here is a truly fascinating fact: When 100 physicians were given the information in Example 4, 95 of them estimated $P(C \mid \text{positive})$ to be around 0.70 to 0.80, so they were wrong by a factor of 10. Physicians are extremely intelligent, but here they likely suffered from confusion of the inverse. The given rate of 80% for positive test results among those who are true positives implies that $P(\text{positive} \mid C) = 0.80$, but this is very different from $P(C \mid \text{positive})$. The physicians would have done much better if they had seen the given information in the form of a table like Table 4-2.

The importance and usefulness of Bayes' theorem is that it can be used with *sequential* events, whereby new additional information is obtained for a subsequent event, and that new information is used to revise the probability of the initial event. In this context, the terms *prior probability* and *posterior probability* are commonly used.

DEFINITIONS

A **prior probability** is an initial probability value originally obtained before any additional information is obtained.

A **posterior probability** is a probability value that has been revised by using additional information that is later obtained.

Relative to Example 4, $P(C) = 0.01$, which is the probability that a randomly selected subject has cancer. $P(C)$ is an example of a *prior probability*. Using the additional information that the subject has received a positive test result, we found that $P(C \mid \text{positive test result}) = 0.0748$, and this is a *posterior probability* because it uses that additional information of the positive test result.

Coincidences More Likely Than They Seem

Evelyn Evans won $3.9 million in the New Jersey lottery, then she won another $1.5 million only 4 months later. The *New York Times* reported that the chance of that happening was only 1 in 17 trillion. But that likelihood is misleading because it represents the chance of Evelyn Evans winning with only one ticket purchased in each of the two specific lottery drawings. A better question would be this: What is the chance of someone somewhere winning a lottery twice? Statisticians George McCabe and Steve Samuels found that over a 7-year span, there is a 53% chance of at least one past lottery winner getting lucky with another win. The chance of "1 in 17 trillion" is sensational, but the more realistic chance is 53%.

4-3   Basic Skills and Concepts

Statistical Literacy and Critical Thinking

1. Language: Complement of "At Least One" Let $A =$ the event of getting at least one defective iPhone when 3 iPhones are randomly selected with replacement from a batch. Write a statement describing event $\overline{A}$.

2. Probability of At Least One Let $A =$ the event of getting at least 1 defective iPhone when 3 iPhones are randomly selected with replacement from a batch. If 5% of the iPhones in a batch are defective and the other 95% are all good, which of the following are correct?

a. $P(\overline{A}) = (0.95)(0.95)(0.95) = 0.857$

b. $P(A) = 1 - (0.95)(0.95)(0.95) = 0.143$

c. $P(A) = (0.05)(0.05)(0.05) = 0.000125$

3. Notation When selecting one of your Facebook Friends, let event $F =$ female and let event $H =$ high school classmate. Use your own words to translate the notation $P(H|F)$ into a verbal statement.

4. Confusion of the Inverse Using the same events F and H described in Exercise 3, describe confusion of the inverse.

At Least One. *In Exercises 5–12, find the probability.*

5. Three Girls Find the probability that when a couple has three children, at least one of them is a girl. (Assume that boys and girls are equally likely.)

6. Probability of a Girl Assuming that boys and girls are equally likely, find the probability of a couple having a boy when their third child is born, given that the first two children were both girls.

7. Births in the United States In the United States, the true probability of a baby being a boy is 0.512 (based on the data available at this writing). Among the next six randomly selected births in the United States, what is the probability that at least one of them is a girl?

8. Births in China In China, where many couples were allowed to have only one child, the probability of a baby being a boy was 0.545. Among six randomly selected births in China, what is the probability that at least one of them is a girl? Could this system continue to work indefinitely? (Phasing out of this policy was begun in 2015.)

9. Phone Survey Subjects for the next presidential election poll are contacted using telephone numbers in which the last four digits are randomly selected (with replacement). Find the probability that for one such phone number, the last four digits include at least one 0.

10. At Least One Correct Answer If you make random guesses for 10 multiple choice SAT test questions (each with five possible answers), what is the probability of getting at least 1 correct? If these questions are part of a practice test and an instructor says that you must get at least one correct answer before continuing, is there a good chance you will continue?

11. At Least One Defective iPhone It has been reported that 20% of iPhones manufactured by Foxconn for a product launch did not meet Apple's quality standards. An engineer needs at least one defective iPhone so she can try to identify the problem(s). If she randomly selects 15 iPhones from a very large batch, what is the probability that she will get at least 1 that is defective? Is that probability high enough so that she can be reasonably sure of getting a defect for her work?

12. Wi-Fi Based on a poll conducted through the e-edition of *USA Today*, 67% of Internet users are more careful about personal information when using a public Wi-Fi hotspot. What is the probability that among four randomly selected Internet users, at least one is more careful about personal information when using a public Wi-Fi hotspot? How is the result affected by the additional information that the survey subjects volunteered to respond?

Denomination Effect. *In Exercises 13–16, use the data in the following table. In an experiment to study the effects of using four quarters or a $1 bill, college students were given either four quarters or a $1 bill and they could either keep the money or spend it on gum. The results are summarized in the table (based on data from "The Denomination Effect," by Priya Raghubir and Joydeep Srivastava,* **Journal of Consumer Research,** *Vol. 36).*

	Purchased Gum	Kept the Money
Students Given Four Quarters	27	16
Students Given a $1 bill	12	34

13. Denomination Effect

a. Find the probability of randomly selecting a student who spent the money, given that the student was given four quarters.

b. Find the probability of randomly selecting a student who kept the money, given that the student was given four quarters.

c. What do the preceding results suggest?

14. Denomination Effect

a. Find the probability of randomly selecting a student who spent the money, given that the student was given a $1 bill.

b. Find the probability of randomly selecting a student who kept the money, given that the student was given a $1 bill.

c. What do the preceding results suggest?

15. Denomination Effect

a. Find the probability of randomly selecting a student who spent the money, given that the student was given four quarters.

b. Find the probability of randomly selecting a student who spent the money, given that the student was given a $1 bill.

c. What do the preceding results suggest?

16. Denomination Effect

a. Find the probability of randomly selecting a student who kept the money, given that the student was given four quarters.

b. Find the probability of randomly selecting a student who kept the money, given that the student was given a $1 bill.

c. What do the preceding results suggest?

In Exercises 17–20, refer to the accompanying table showing results from a Chembio test for hepatitis C among HIV-infected patients (based on data from a variety of sources).

	Positive Test Result	Negative Test Result
Hepatitis C	335	10
No Hepatitis C	2	1153

17. False Positive Find the probability of selecting a subject with a positive test result, given that the subject does not have hepatitis C. Why is this case problematic for test subjects?

18. False Negative Find the probability of selecting a subject with a negative test result, given that the subject has hepatitis C. What would be an unfavorable consequence of this error?

19. Positive Predictive Value Find the positive predictive value for the test. That is, find the probability that a subject has hepatitis C, given that the test yields a positive result. Does the result make the test appear to be effective?

20. Negative Predictive Value Find the negative predictive value for the test. That is, find the probability that a subject does not have hepatitis C, given that the test yields a negative result. Does the result make the test appear to be effective?

21. Redundancy in Computer Hard Drives Assume that there is a 3% rate of disk drive failures in a year (based on data from various sources including lifehacker.com).

a. If all of your computer data is stored on a hard disk drive with a copy stored on a second hard disk drive, what is the probability that during a year, you can avoid catastrophe with at least one working drive? Express the result with four decimal places.

b. If copies of all of your computer data are stored on three independent hard disk drives, what is the probability that during a year, you can avoid catastrophe with at least one working drive? Express the result with six decimal places. What is wrong with using the usual round-off rule for probabilities in this case?

22. Redundancy in Stadium Generators Large stadiums rely on backup generators to provide electricity in the event of a power failure. Assume that emergency backup generators fail 22% of the times when they are needed (based on data from Arshad Mansoor, senior vice president with the Electric Power Research Institute). A stadium has three backup generators so that power is available if at least one of them works in a power failure. Find the probability of having at least one of the backup generators working given that a power failure has occurred. Does the result appear to be adequate for the stadium's needs?

23. Composite Drug Test Based on the data in Table 4-1 on page 162, assume that the probability of a randomly selected person testing positive for drug use is 0.126. If drug screening samples are collected from 5 random subjects and combined, find the probability that the combined sample will reveal a positive result. Is that probability low enough so that further testing of the individual samples is rarely necessary?

24. Composite Water Samples The Fairfield County Department of Public Health tests water for the presence of *E. coli* (*Escherichia coli*) bacteria. To reduce laboratory costs, water samples from 10 public swimming areas are combined for one test, and further testing is done only if the combined sample tests positive. Based on past results, there is a 0.005 probability of finding *E. coli* bacteria in a public swimming area. Find the probability that a combined sample from 10 public swimming areas will reveal the presence of *E. coli* bacteria. Is that probability low enough so that further testing of the individual samples is rarely necessary?

4-3   Beyond the Basics

25. Shared Birthdays Find the probability that of 25 randomly selected people, at least 2 share the same birthday.

4-4 Counting

Key Concept Probability problems typically require that we know the total number of simple events, but finding that number often requires one of the five rules presented in this section. In Section 4-2 with the addition rule, multiplication rule, and conditional probability, we encouraged intuitive rules based on understanding and we discouraged blind use of formulas, but this section requires much greater use of formulas as we consider five different methods for counting the number of possible outcomes in a variety of situations. Not all counting problems can be solved with these five methods, but they do provide a strong foundation for the most common real applications.

1. Multiplication Counting Rule

The *multiplication counting rule* is used to find the total number of possibilities from some sequence of events.

> **MULTIPLICATION COUNTING RULE**
>
> For a sequence of events in which the first event can occur n_1 ways, the second event can occur n_2 ways, the third event can occur n_3 ways, and so on, the total number of outcomes is $n_1 \cdot n_2 \cdot n_3 \ldots$.

EXAMPLE 1 Multiplication Counting Rule: Hacker Guessing a Passcode

When making random guesses for an unknown four-digit passcode, each digit can be 0, 1. . . , 9. What is the total number of different possible passcodes? Given that all of the guesses have the same chance of being correct, what is the probability of guessing the correct passcode on the first attempt?

SOLUTION

There are 10 different possibilities for each digit, so the total number of different possible passcodes is $n_1 \cdot n_2 \cdot n_3 \cdot n_4 = 10 \cdot 10 \cdot 10 \cdot 10 = 10{,}000$.

 Because all of the passcodes are equally likely, the probability of getting the correct passcode on the first attempt is $1/10{,}000$ or 0.0001.

YOUR TURN Do Exercise 5 "ATM Pin Numbers."

2. Factorial Rule

The factorial rule is used to find the total number of ways that n different items can be rearranged (order of items matters). The factorial rule uses the following notation.

NOTATION

The **factorial symbol (!)** denotes the product of decreasing positive whole numbers. For example, $4! = 4 \cdot 3 \cdot 2 \cdot 1 = 24$. By special definition, $0! = 1$.

> **FACTORIAL RULE**
>
> The number of different *arrangements* (order matters) of n different items when all n of them are selected is $\boldsymbol{n!}$.

How Many Shuffles?

After conducting extensive research, Harvard mathematician Persi Diaconis found that it takes seven shuffles of a deck of cards to get a complete mixture. The mixture is complete in the sense that all possible arrangements are equally likely. More than seven shuffles will not have a significant effect, and fewer than seven are not enough. Casino dealers rarely shuffle as often as seven times, so the decks are not completely mixed. Some expert card players have been able to take advantage of the incomplete mixtures that result from fewer than seven shuffles.

The factorial rule is based on the principle that the first item may be selected n different ways, the second item may be selected $n - 1$ ways, and so on. This rule is really the multiplication counting rule modified for the elimination of one item on each selection.

EXAMPLE 2   Factorial Rule: Travel Itinerary

A statistics researcher must personally visit the presidents of the Gallup, Nielsen, Harris, Pew, and Zogby polling companies.

a. How many different travel itineraries are possible?

b. If the itinerary is randomly selected, what is the probability that the presidents are visited in order from youngest to oldest?

SOLUTION

a. For those 5 different presidents, the number of different travel itineraries is $5! = 5 \cdot 4 \cdot 3 \cdot 2 \cdot 1 = 120$.

Note that this solution could have been done by applying the multiplication counting rule. The first person can be any one of the 5 presidents, the second person can be any one of the 4 remaining presidents, and so on. The result is again $5 \cdot 4 \cdot 3 \cdot 2 \cdot 1 = 120$. Use of the factorial rule has the advantage of including the factorial symbol, which is sure to impress.

b. There is only one itinerary with the presidents visited in order of age, so the probability is $1/120$.

> YOUR TURN Do Exercise 9 "Grading Exams."

Permutations and Combinations: Does *Order* Count?

When using different counting methods, it is essential to know whether different arrangements of the same items are counted only once or are counted separately. The terms *permutations* and *combinations* are standard in this context, and they are defined as follows:

DEFINITIONS

Permutations of items are arrangements in which different sequences of the same items are counted *separately*. (The letter arrangements of abc, acb, bac, bca, cab, and cba are all counted *separately* as six different permutations.)

Combinations of items are arrangements in which different sequences of the same items are counted as being the *same*. (The letter arrangements of abc, acb, bac, bca, cab, and cba are all considered to be the *same* combination.)

Mnemonics for Permutations and Combinations

- Remember "**P**ermutations **P**osition," where the alliteration reminds us that with permutations, the positions of the items makes a difference.

- Remember "**C**ombinations **C**ommittee," which reminds us that with members of a committee, rearrangements of the same members result in the same committee, so order does not count.

3. Permutations Rule (When All of the Items Are Different)

The permutations rule is used when there are *n* different items available for selection, we must select *r* of them without replacement, and the sequence of the items matters. The result is the total number of arrangements (or permutations) that are possible. (Remember, rearrangements of the same items are counted as different permutations.)

PERMUTATIONS RULE

When *n* different items are available and *r* of them are selected without replacement, the number of different permutations (order counts) is given by

$$_nP_r = \frac{n!}{(n-r)!}$$

EXAMPLE 3 **Permutations Rule (with Different Items): Trifecta Bet**

In a horse race, a *trifecta* bet is won by correctly selecting the horses that finish first and second and third, and you must select them in the correct order. The 140th running of the Kentucky Derby had a field of 19 horses.

a. How many different trifecta bets are possible?

b. If a bettor randomly selects three of those horses for a trifecta bet, what is the probability of winning by selecting California Chrome to win, Commanding Curve to finish second, and Danza to finish third, as they did? Do all of the different possible trifecta bets have the same chance of winning? (Ignore "dead heats" in which horses tie for a win.)

SOLUTION

a. There are *n* = 19 horses available, and we must select *r* = 3 of them without replacement. The number of different sequences of arrangements is found as shown:

$$_nP_r = \frac{n!}{(n-r)!} = \frac{19!}{(19-3)!} = 5814$$

b. There are 5814 different possible arrangements of 3 horses selected from the 19 that are available. If one of those arrangements is randomly selected, there is a probability of 1/5814 that the winning arrangement is selected.

There are 5814 different possible trifecta bets, but not all of them have the same chance of winning, because some horses tend to be faster than others. (A winning $2 trifecta bet in this race won $3424.60.)

YOUR TURN Do Exercise 11 "Scheduling Routes."

4. Permutations Rule (When Some Items Are Identical to Others)

When *n* items are all selected without replacement, but *some items are identical*, the number of possible permutations (order matters) is found by using the following rule.

Choosing Personal Security Codes

All of us use personal security codes for ATM machines, Internet accounts, and home security systems. The safety of such codes depends on the large number of different possibilities, but hackers now have sophisticated tools that can largely overcome that obstacle. Researchers found that by using variations of the user's first and last names along with 1800 other first names, they could identify 10% to 20% of the passwords on typical computer systems. When choosing a password, *do not* use a variation of any name, a word found in a dictionary, a password shorter than seven characters, telephone numbers, or social security numbers. Do include nonalphabetic characters, such as digits or punctuation marks.

PERMUTATIONS RULE (WHEN SOME ITEMS ARE IDENTICAL TO OTHERS)

The number of different permutations (order counts) when n items are available and all n of them are selected *without replacement,* but some of the items are identical to others, is found as follows:

$$\frac{n!}{n_1!n_2!\ldots n_k!}$$ where n_1 are alike, n_2 are alike, . . . , and n_k are alike.

EXAMPLE 4   Permutations Rule (with Some Identical Items): Designing Surveys

When designing surveys, pollsters sometimes repeat a question to see if a subject is thoughtlessly providing answers just to finish quickly. For one particular survey with 10 questions, 2 of the questions are identical to each other, and 3 other questions are also identical to each other. For this survey, how many different arrangements are possible? Is it practical to survey enough subjects so that every different possible arrangement is used?

SOLUTION

We have 10 questions with 2 that are identical to each other and 3 others that are also identical to each other, and we want the number of permutations. Using the rule for permutations with some items identical to others, we get

$$\frac{n!}{n_1!n_2!\ldots n_k!} = \frac{10!}{2!3!} = \frac{3,628,800}{2\cdot 6} = 302,400$$

INTERPRETATION

There are 302,400 different possible arrangements of the 10 questions. It is not practical to accommodate every possible permutation. For typical surveys, the number of respondents is somewhere around 1000.

YOUR TURN   Do Exercise 12 "Survey Reliability."

5. Combinations Rule

The combinations rule is used when there are n different items available for selection, only r of them are selected without replacement, and order does not matter. The result is the total number of combinations that are possible. (*Remember*: Rearrangements of the same items are considered to be the same combination.)

COMBINATIONS RULE:

When n different items are available, but only r of them are selected *without replacement,* the number of different combinations (order does not matter) is found as follows:

$$_nC_r = \frac{n!}{(n-r)!r!}$$

EXAMPLE 5 Combinations Rule: Lottery

In California's Fantasy 5 lottery game, winning the jackpot requires that you select 5 different numbers from 1 to 39, and the same 5 numbers must be drawn in the lottery. The winning numbers can be drawn in any order, so order does not make a difference.

a. How many different lottery tickets are possible?

b. Find the probability of winning the jackpot when one ticket is purchased.

SOLUTION

a. There are $n = 39$ different numbers available, and we must select $r = 5$ of them without replacement (because the selected numbers must be different). Because order does not count, we need to find the number of different possible *combinations*. We get

$$_nC_r = \frac{n!}{(n-r)!r!} = \frac{39!}{(39-5)!5!} = \frac{39!}{34! \cdot 5!} = 575{,}757$$

b. If you select one 5-number combination, your probability of winning is $1/575{,}757$. Typical lotteries rely on the fact that people rarely know the value of this probability and have no realistic sense for how small that probability is. This is why the lottery is often called a "tax on people who are bad at math."

> **YOUR TURN** Do Exercise 29 "Mega Millions."

Permutations or Combinations? Because choosing between permutations and combinations can often be tricky, we provide the following example that emphasizes the difference between them.

EXAMPLE 6 Permutations and Combinations: Corporate Officials and Committees

The Google company must appoint three corporate officers: chief executive officer (CEO), executive chairperson, and chief operating officer (COO). It must also appoint a Planning Committee with three different members. There are eight qualified candidates, and officers can also serve on the Planning Committee.

a. How many different ways can the officers be appointed?

b. How many different ways can the committee be appointed?

SOLUTION

Note that in part (a), order is important because the officers have very different functions. However, in part (b), the order of selection is irrelevant because the committee members all serve the same function.

a. Because order *does* count, we want the number of *permutations* of $r = 3$ people selected from the $n = 8$ available people. We get

$$_nP_r = \frac{n!}{(n-r)!} = \frac{8!}{(8-3)!} = 336$$

continued

b. Because order does *not* count, we want the number of *combinations* of $r = 3$ people selected from the $n = 8$ available people. We get

$$_nC_r = \frac{n!}{(n-r)!r!} = \frac{8!}{(8-3)!3!} = 56$$

With order taken into account, there are 336 different ways that the officers can be appointed, but without order taken into account, there are 56 different possible committees.

YOUR TURN Do Exercise 23 "Corporate Officers and Committees."

4-4 Basic Skills and Concepts

Statistical Literacy and Critical Thinking

1. Notation What does the symbol ! represent? Six different people can stand in a line 6! different ways, so what is the actual number of ways that six people can stand in a line?

2. New Jersey Pick 6 In the New Jersey Pick 6 lottery game, a bettor selects six *different* numbers, each between 1 and 49. Winning the top prize requires that the selected numbers match those that are drawn, but the order does not matter. Do calculations for winning this lottery involve permutations or combinations? Why?

3. Oregon Pick 4 In the Oregon Pick 4 lottery game, a bettor selects four numbers between 0 and 9 and any selected number can be used more than once. Winning the top prize requires that the selected numbers match those and are drawn in the same order. Do calculations for this lottery involve the combinations rule or either of the two permutation rules presented in this section? Why or why not? If not, what rule does apply?

4. Combination Lock The typical combination lock uses three numbers, each between 0 and 49. Opening the lock requires entry of the three numbers in the correct order. Is the name "combination" lock appropriate? Why or why not?

In Exercises 5–36, express all probabilities as fractions.

5. ATM Pin Numbers A thief steals an ATM card and must randomly guess the correct pin code that consists of four digits (each 0 through 9) that must be entered in the correct order. Repetition of digits is allowed. What is the probability of a correct guess on the first try?

6. Social Security Numbers A Social Security number consists of nine digits in a particular order, and repetition of digits is allowed. After seeing the last four digits printed on a receipt, if you randomly select the other digits, what is the probability of getting the correct Social Security number of the person who was given the receipt?

7. Quinela In a horse race, a quinela bet is won if you selected the two horses that finish first and second, and they can be selected in any order. The 140th running of the Kentucky Derby had a field of 19 horses. What is the probability of winning a quinela bet if random horse selections are made?

8. Soccer Shootout In soccer, a tie at the end of regulation time leads to a shootout by three members from each team. How many ways can 3 players be selected from 11 players available? For 3 selected players, how many ways can they be designated as first, second, and third?

9. Grading Exams Your professor has just collected eight different statistics exams. If these exams are graded in random order, what is the probability that they are graded in alphabetical order of the students who took the exam?

10. Radio Station Call Letters If radio station call letters must begin with either K or W and must include either two or three additional letters, how many different possibilities are there?

11. Scheduling Routes A presidential candidate plans to begin her campaign by visiting the capitals of 5 of the 50 states. If the five capitals are randomly selected without replacement, what is the probability that the route is Sacramento, Albany, Juneau, Hartford, and Bismarck, in that order?

12. Survey Reliability A survey with 12 questions is designed so that 3 of the questions are identical and 4 other questions are identical (except for minor changes in wording). How many different ways can the 12 questions be arranged?

13. Safety with Numbers The author owns a safe in which he stores all of his great ideas for the next edition of this book. The safe "combination" consists of four numbers between 0 and 99, and the safe is designed so that numbers can be repeated. If another author breaks in and tries to steal these ideas, what is the probability that he or she will get the correct combination on the first attempt? Assume that the numbers are randomly selected. Given the number of possibilities, does it seem feasible to try opening the safe by making random guesses for the combination?

14. Electricity When testing for current in a cable with five color-coded wires, the author used a meter to test two wires at a time. How many different tests are required for every possible pairing of two wires?

15. Sorting Hat At Hogwarts School of Witchcraft and Wizardry, the Sorting Hat chooses one of four houses for each first-year student. If 4 students are randomly selected from 16 available students (including Harry Potter), what is the probability that they are the four youngest students?

16. Moving Company The United Van Lines moving company has a truck filled for deliveries to five different sites. If the order of the deliveries is randomly selected, what is the probability that it is the shortest route?

17. Powerball As of this writing, the Powerball lottery is run in 44 states. Winning the jackpot requires that you select the correct five different numbers between 1 and 69 and, in a separate drawing, you must also select the correct single number between 1 and 26. Find the probability of winning the jackpot.

18. Teed Off When four golfers are about to begin a game, they often toss a tee to randomly select the order in which they tee off. What is the probability that they tee off in alphabetical order by last name?

19. ZIP Code If you randomly select five digits, each between 0 and 9, with repetition allowed, what is the probability you will get the author's ZIP code?

20. FedEx Deliveries With a short time remaining in the day, a FedEx driver has time to make deliveries at 6 locations among the 9 locations remaining. How many different routes are possible?

21. Phone Numbers Current rules for telephone *area codes* allow the use of digits 2–9 for the first digit, and 0–9 for the second and third digits. How many different area codes are possible with these rules? That same rule applies to the *exchange* numbers, which are the three digits immediately preceding the last four digits of a phone number. Given both of those rules, how many 10-digit phone numbers are possible? Given that these rules apply to the United States and Canada and a few islands, are there enough possible phone numbers? (Assume that the combined population is about 400,000,000.)

22. Classic Counting Problem A classic counting problem is to determine the number of different ways that the letters of "Mississippi" can be arranged. Find that number.

23. Corporate Officers and Committees The Digital Pet Rock Company was recently successfully funded via Kickstarter and must now appoint a president, chief executive officer (CEO), chief operating officer (COO), and chief financial officer (CFO). It must also appoint a strategic planning committee with four different members. There are 10 qualified candidates, and officers can also serve on the committee.

a. How many different ways can the four officers be appointed?

b. How many different ways can a committee of four be appointed?

c. What is the probability of randomly selecting the committee members and getting the four youngest of the qualified candidates?

24. ATM You want to obtain cash by using an ATM, but it's dark and you can't see your card when you insert it. The card must be inserted with the front side up and the printing configured so that the beginning of your name enters first.

a. What is the probability of selecting a random position and inserting the card with the result that the card is inserted correctly?

b. What is the probability of randomly selecting the card's position and finding that it is incorrectly inserted on the first attempt, but it is correctly inserted on the second attempt? (Assume that the same position used for the first attempt could also be used for the second attempt.)

c. How many random selections are required to be absolutely sure that the card works because it is inserted correctly?

25. Party Mix DJ Marty T is hosting a party tonight and has chosen 8 songs for his final set (including "Daydream Believer" by the Monkees). How many different 8-song playlists are possible (song order matters)? If the 8 songs are randomly selected, what is the probability they are in alphabetical order by song title?

26. Identity Theft with Credit Cards Credit card numbers typically have 16 digits, but not all of them are random.

a. What is the probability of randomly generating 16 digits and getting *your* MasterCard number?

b. Receipts often show the last four digits of a credit card number. If only those last four digits are known, what is the probability of randomly generating the other digits of your MasterCard number?

c. Discover cards begin with the digits 6011. If you know that the first four digits are 6011 and you also know the last four digits of a Discover card, what is the probability of randomly generating the other digits and getting all of them correct? Is this something to worry about?

27. What a Word! One of the longest words in standard statistics terminology is "homoscedasticity." How many ways can the letters in that word be arranged?

28. Phase I of a Clinical Trial A clinical test on humans of a new drug is normally done in three phases. Phase I is conducted with a relatively small number of healthy volunteers. For example, a phase I test of bexarotene involved only 14 subjects. Assume that we want to treat 14 healthy humans with this new drug and we have 16 suitable volunteers available.

a. If the subjects are selected and treated one at a time *in sequence*, how many different sequential arrangements are possible if 14 people are selected from the 16 that are available?

b. If 14 subjects are selected from the 16 that are available, and the 14 selected subjects are all treated at the same time, how many different treatment groups are possible?

c. If 14 subjects are randomly selected and treated at the same time, what is the probability of selecting the 14 youngest subjects?

29. Mega Millions As of this writing, the Mega Millions lottery is run in 44 states. Winning the jackpot requires that you select the correct five different numbers between 1 and 75 and, in a separate drawing, you must also select the correct single number between 1 and 15. Find the probability of winning the jackpot. How does the result compare to the probability of being struck by lightning in a year, which the National Weather Service estimates to be $1/960,000$?

30. Designing Experiment Clinical trials of Nasonex involved a group given placebos and another group given treatments of Nasonex. Assume that a preliminary phase I trial is to be conducted with 12 subjects, including 6 men and 6 women. If 6 of the 12 subjects are randomly selected for the treatment group, find the probability of getting 6 subjects of the same gender. Would there be a problem with having members of the treatment group all of the same gender?

31. Morse Codes The International Morse code is a way of transmitting coded text by using sequences of on/off tones. Each character is 1 or 2 or 3 or 4 or 5 segments long, and each segment is either a dot or a dash. For example, the letter *G* is transmitted as two dashes followed by a dot, as in — — • . How many different characters are possible with this scheme? Are there enough characters for the alphabet and numbers?

32. Mendel's Peas Mendel conducted some his famous experiments with peas that were either smooth yellow plants or wrinkly green plants. If four peas are randomly selected from a batch consisting of four smooth yellow plants and four wrinkly green plants, find the probability that the four selected peas are of the same type.

33. Blackjack In the game of blackjack played with one deck, a player is initially dealt 2 different cards from the 52 different cards in the deck. A winning "blackjack" hand is won by getting 1 of the 4 aces and 1 of 16 other cards worth 10 points. The two cards can be in any order. Find the probability of being dealt a blackjack hand. What approximate percentage of hands are winning blackjack hands?

34. Counting with Fingers How many different ways can you touch two or more fingers to each other on one hand?

35. Change for a Quarter How many different ways can you make change for a quarter? (Different arrangements of the same coins are not counted separately.)

36. Win $1 Billion Quicken Loans offered a prize of $1 billion to anyone who could correctly predict the winner of the NCAA basketball tournament. After the "play-in" games, there are 64 teams in the tournament.

a. How many games are required to get 1 championship team from the field of 64 teams?

b. If you make random guesses for each game of the tournament, find the probability of picking the winner in every game.

4-4   Beyond the Basics

37. Computer Variable Names A common computer programming rule is that names of variables must be between one and eight characters long. The first character can be any of the 26 letters, while successive characters can be any of the 26 letters or any of the 10 digits. For example, allowable variable names include A, BBB, and M3477K. How many different variable names are possible? (Ignore the difference between uppercase and lowercase letters.)

38. High Fives

a. Five "mathletes" celebrate after solving a particularly challenging problem during competition. If each mathlete high fives each other mathlete exactly once, what is the total number of high fives?

b. If n mathletes shake hands with each other exactly once, what is the total number of handshakes?

c. How many different ways can five mathletes be seated at a round table? (Assume that if everyone moves to the right, the seating arrangement is the same.)

d. How many different ways can n mathletes be seated at a round table?

4-5   Probabilities Through Simulations (available at www.TriolaStats.com)

The website www.TriolaStats.com includes a downloadable section that discusses the use of simulation methods for finding probabilities. Simulations are also discussed in the Technology Project near the end of this chapter.

Chapter Quick Quiz

1. **Standard Tests** Standard tests, such as the SAT or ACT or MCAT, tend to make extensive use of multiple-choice questions because they are easy to grade using software. If one such multiple choice question has possible correct answers of a, b, c, d, e, what is the probability of a wrong answer if the answer is a random guess?

2. **Rain** As the author is creating this exercise, a weather reporter stated that there is a 20% chance of rain tomorrow. What is the probability of no rain tomorrow?

3. **Months** If a month is randomly selected after mixing the pages from a calendar, what is the probability that it is a month containing the letter *y*?

4. **Social Networking** Based on data from the Pew Internet Project, 74% of adult Internet users use social networking sites. If two adult Internet users are randomly selected, what is the probability that they both use social networking sites?

5. **Subjective Probability** Estimate the probability that the next time you ride in a car, the car gets a flat tire.

In Exercises 6–10, use the following results from tests of an experiment to test the effectiveness of an experimental vaccine for children (based on data from **USA Today**)*. Express all probabilities in decimal form.*

	Developed Flu	Did Not Develop Flu
Vaccine Treatment	14	1056
Placebo	95	437

6. If 1 of the 1602 subjects is randomly selected, find the probability of getting 1 that developed flu.

7. If 1 of the 1602 subjects is randomly selected, find the probability of getting 1 who had the vaccine treatment or developed flu.

8. If 1 of the 1602 subjects is randomly selected, find the probability of getting 1 who had the vaccine treatment and developed flu.

9. Find the probability of randomly selecting 2 subjects without replacement and finding that they both developed flu.

10. Find the probability of randomly selecting 1 of the subjects and getting 1 who developed flu, given that the subject was given the vaccine treatment.

Review Exercises

In Exercises 1–10, use the data in the accompanying table and express all results in decimal form. (The data are from "Mortality Reduction with Air Bag and Seat Belt Use in Head-On Passenger Car Collisions," by Crandall, Olson, and Sklar, **American Journal of Epidemiology**, *Vol. 153, No. 3.)*

Drivers Involved in Head-On Collision of Passenger Cars

	Driver Killed	Driver Not Killed
Seatbelt used	3655	7005
Seatbelt not used	4402	3040

1. **Seatbelt Use** If one driver is randomly selected, find the probability he or she was using a seatbelt.

2. Seatbelt Use Find the probability of randomly selecting a driver and getting one who was not killed given that the driver was using a seatbelt.

3. No Seatbelt Find the probability of randomly selecting a driver and getting one who was killed given that the driver was not using a seatbelt.

4. Seatbelt Use or Driver Not Killed If one of the drivers is randomly selected, find the probability of getting a driver who used a seatbelt or was killed.

5. No Seatbelt or Driver Not Killed If one of the drivers is randomly selected, find the probability of getting someone who did not use a seatbelt or was not killed.

6. Both Using Seatbelts If 2 drivers are randomly selected *without replacement*, find the probability that they both used seatbelts.

7. Both Drivers Killed If 2 drivers are randomly selected *with replacement*, find the probability that they both were killed.

8. Complement If A represents the event of randomly selecting one driver included in the table and getting someone who was using a seatbelt, what does $\overline{A}$ represent? Find the value of $P(\overline{A})$.

9. Complement If A represents the event of randomly selecting one driver included in the table and getting someone who was not killed, what does $\overline{A}$ represent? Find the value of $P(\overline{A})$.

10. All Three Successful If 3 drivers are randomly selected without replacement, find the probability that none of them were killed.

11. Black Cars Use subjective probability to estimate the probability of randomly selecting a car and getting one that is black.

12. Vision Correction About 75% of the U.S. population uses some type of vision correction (such as glasses or contact lenses).

a. If someone is randomly selected, what is the probability that he or she does not use vision correction?

b. If four different people are randomly selected, what is the probability that they all use vision correction?

c. Would it be unlikely to randomly select four people and find that they all use vision correction? Why or why not?

13. National Statistics Day

a. If a person is randomly selected, find the probability that his or her birthday is October 18, which is National Statistics Day in Japan. Ignore leap years.

b. If a person is randomly selected, find the probability that his or her birthday is in October. Ignore leap years.

c. Estimate a subjective probability for the event of randomly selecting an adult American and getting someone who knows that October 18 is National Statistics Day in Japan.

d. Is it unlikely to randomly select an adult American and get someone who knows that October 18 is National Statistics Day in Japan?

14. Composite Sampling for Diabetes Currently, the rate for new cases of diabetes in a year is 3.4 per 1000 (based on data from the Centers for Disease Control and Prevention). When testing for the presence of diabetes, the Portland Diagnostics Laboratory saves money by combining blood samples for tests. The combined sample tests positive if at least one person has diabetes. If the combined sample tests positive, then the individual blood tests are performed. In a test for diabetes, blood samples from 10 randomly selected subjects are combined. Find the probability that the combined sample tests positive with at least 1 of the 10 people having diabetes. Is it likely that such combined samples test positive?

15. Wild Card Lottery The Wild Card lottery is run in the states of Idaho, Montana, North Dakota, and South Dakota. You must select five different numbers between 1 and 33, then you must select one of the picture cards (Ace, King, Queen, Jack), and then you must select one of the four suits (club, heart, diamond, spade). Express all answers as fractions.

a. What is the probability of selecting the correct five numbers between 1 and 33?

b. What is the probability of selecting the correct picture card?

c. What is the probability of selecting the correct suit?

d. What is the probability of selecting the correct five numbers, the correct picture card, and the correct suit?

16. Pennsylvania Cash 5 In the Pennsylvania Cash 5 lottery, winning the top prize requires that you select the correct five different numbers between 1 and 43 (in any order). What is the probability of winning the top prize? Express the answer as a fraction.

17. Redundancy Using Braun battery-powered alarm clocks, the author estimates that the probability of failure on any given day is 1/1000. (a) What is the probability that the alarm clock works for an important event? (b) When he uses two alarm clocks for important events, what is the probability that at least one of them works?

18. Exacta In a horse race, an exacta bet is won by correctly selecting the horses that finish first and second, and they must be in the order that they finished. The 140th Kentucky Derby race was run with 19 horses. If you make random selections for an exacta bet, what is the probability of winning?

Cumulative Review Exercises

1. Fatal Drunk Driving Listed below are the blood alcohol concentrations (g/dL) of drivers convicted of drunk driving in fatal car crashes (based on data from the National Highway Traffic Safety Administration).

$$0.09 \quad 0.11 \quad 0.11 \quad 0.13 \quad 0.14 \quad 0.15 \quad 0.17 \quad 0.17 \quad 0.18 \quad 0.18 \quad 0.23 \quad 0.35$$

Find the value of the following statistics and include appropriate units.

a. mean   **b.** median   **c.** midrange   **d.** range

e. standard deviation   **f.** variance

2. Fatal Drunk Driving Use the same data given in Exercise 1.

a. Identify the 5-number summary and also identify any values that appear to be outliers.

b. Construct a boxplot.

c. Construct a stemplot.

3. Organ Donors *USA Today* provided information about a survey (conducted for Donate Life America) of 5100 adult Internet users. Of the respondents, 2346 said they are willing to donate organs after death. In this survey, 100 adults were surveyed in each state and the District of Columbia, and results were weighted to account for the different state population sizes.

a. What percentage of respondents said that they are willing to donate organs after death?

b. Based on the poll results, what is the probability of randomly selecting an adult who is willing to donate organs after death?

c. What term is used to describe the sampling method of randomly selecting 100 adults from each state and the District of Columbia?

4. Sampling Eye Color Based on a study by Dr. P. Sorita Soni at Indiana University, assume that eye colors in the United States are distributed as follows: 40% brown, 35% blue, 12% green, 7% gray, 6% hazel.

a. A statistics instructor collects eye color data from her students. What is the name for this type of sample?

b. Identify one factor that might make the sample from part (a) biased and not representative of the general population of people in the United States.

c. If one person is randomly selected, what is the probability that this person will have brown or blue eyes?

d. If two people are randomly selected, what is the probability that at least one of them has brown eyes?

5. Blood Pressure and Platelets Given below are the systolic blood pressure measurements (mm Hg) and blood platelet counts (1000 cells/μL) of the first few subjects included in Data Set 1 "Body Data" in Appendix B. Construct a graph suitable for exploring an association between systolic blood pressure and blood platelet count. What does the graph suggest about that association?

Systolic	100	112	134	126	114	134	118	138	114	124
Platelet	319	187	297	170	140	192	191	286	263	193

6. New Lottery Game In the Monopoly Millionaires' Club lottery game, you pay $5 and select five different numbers between 1 and 52, and then a sixth number between 1 and 28 is randomly assigned to you. Winning requires that your five numbers match those drawn (in any order), and then your sixth assigned number must also match.

a. What is the probability of selecting the correct five numbers between 1 and 52?

b. What is the probability of getting the correct sixth number that is assigned to you?

c. What is the probability of selecting the correct five numbers and getting the correct sixth number?

Technology Project

Simulations Calculating probabilities are sometimes painfully difficult, but *simulations* provide us with a very practical alternative to calculations based on formal rules. A **simulation** of a procedure is a process that behaves the same way as the procedure so that similar results are produced. Instead of calculating the probability of getting exactly 5 boys in 10 births, you could repeatedly toss 10 coins and count the number of times that exactly 5 heads (or simulated "boys") occur. Better yet, you could do the simulation with a random number generator on a computer or calculator to randomly generate 1s (or simulated "boys") and 0s (or simulated "girls"). Let's consider this probability exercise:

> **Find the probability that among 50 randomly selected people, at least 3 have the same birthday.**

For the above problem, a simulation begins by representing birthdays by integers from 1 through 365, where 1 represents a birthday of January 1, and 2 represents January 2, and so on. We can simulate 50 birthdays by using a calculator or computer to generate 50 random numbers (with repetition allowed) between 1 and 365. Those numbers can then be sorted, so it becomes easy to examine the list to determine whether any 3 of the simulated birth dates are the same. (After sorting, equal numbers are adjacent.) We can repeat the process as many times as we wish, until we are satisfied that we have a good estimate of the probability. Use technology to simulate 20 different groups of 50 birthdays. Use the results to estimate the probability that among 50 randomly selected people, at least 3 have the same birthday.

continued

Summary of Simulation Functions:

Statdisk: Select **Data** from the top menu, select **Uniform Generator** from the dropdown menu.

Excel: Click **Insert Function** f_x, select **Math & Trig,** select **RANDBETWEEN.** Copy to additional cells.

TI-83/84 Plus: Press **MATH**, select **PROB** from the top menu, select **randInt** from the menu.

StatCrunch: Select **Data** from the top menu, select **Simulate** from the drop-down menu, select **Discrete Uniform** from the submenu.

Minitab: Select **Calc** from the top menu, select **Random Data** from the dropdown menu, select **Integer** from the submenu.

FROM DATA TO DECISION

Critical Thinking: Interpreting results from a test for smoking

It is estimated that roughly half of smokers lie when asked about their smoking involvement. Pulse CO-oximeters may be a way to get information about smoking without relying on patients' statements. Pulse CO-oximeters use light that shines through a fingernail, and it measures carbon monoxide in blood. These devices are used by firemen and emergency departments to detect carbon monoxide poisoning, but they can also be used to identify smokers. The accompanying table lists results from people aged 18–44 when the pulse CO-oximeter is set to detect a 6% or higher level of carboxyhemoglobin (based on data from "Carbon Monoxide Test Can Be Used to Identify Smoker," by Patrice Wendling, *Internal Medicine News,* Vol. 40., No. 1, and Centers for Disease Control and Prevention).

CO-Oximetry Test for Smoking

	Positive Test Result	Negative Test Result
Smoker	49	57
Nonsmoker	24	370

Analyzing the Results

1. False Positive Based on the results in the table, find the probability that a subject is not a smoker, given that the test result is positive.

2. True Positive Based on the results in the table, find the probability that a subject smokes, given that the test result is positive.

3. False Negative Based on the results in the table, find the probability that a subject smokes, given that the test result is negative.

4. True Negative Based on the results in the table, find the probability that a subject does not smoke, given that the test result is negative.

5. Sensitivity Find the *sensitivity* of the test by finding the probability of a true positive, given that the subject actually smokes.

6. Specificity Find the *specificity* of the test by finding the probability of a true negative, given that the subject does not smoke.

7. Positive Predictive Value Find the *positive predictive value* of the test by finding the probability that the subject smokes, given that the test yields a positive result.

8. Negative Predictive Value Find the *negative predictive value* of the test by finding the probability that the subject does not smoke, given that the test yields a negative result.

9. Confusion of the Inverse Find the following values, then compare them. In this case, what is confusion of the inverse?

- $P(\text{smoker} \mid \text{positive test result})$
- $P(\text{positive test result} \mid \text{smoker})$

Cooperative Group Activities

1. Out-of-class activity Divide into groups of three or four and create a new carnival game. Determine the probability of winning. Determine how much money the operator of the game can expect to gain each time the game is played.

2. In-class activity Divide into groups of three or four and use coin flipping to develop a simulation that emulates the kingdom that abides by this decree: After a mother gives birth to a son, she will not have any other children. If this decree is followed, does the proportion of girls increase?

3. In-class activity Divide into groups of three or four and use actual thumbtacks or Hershey's Kisses candies or paper cups to estimate the probability that when dropped, they will land with the point (or open side) up. How many trials are necessary to get a result that appears to be reasonably accurate when rounded to the first decimal place?

4. Out-of-class activity Marine biologists often use the *capture-recapture method* as a way to estimate the size of a population, such as the number of fish in a lake. This method involves capturing a sample from the population, tagging each member in the sample, then returning it to the population. A second sample is later captured, and the tagged members are counted along with the total size of this second sample. The results can be used to estimate the size of the population.

   Instead of capturing real fish, simulate the procedure using some uniform collection of items such as colored beads, M&Ms, or index cards. Start with a large collection of at least 200 of such items. Collect a sample of 50 and use a marker to "tag" each one. Replace the tagged items, mix the whole population, then select a second sample and proceed to estimate the population size. Compare the result to the actual population size obtained by counting all of the items.

5. Out-of-class activity Divide into groups of three or four. First, use subjective estimates for the probability of randomly selecting a car and getting each of these car colors: black, white, blue, red, silver, other. Then design a sampling plan for obtaining car colors through observation. Execute the sampling plan and obtain revised probabilities based on the observed results. Write a brief report of the results.

6. In-class activity The manufacturing process for a new computer integrated circuit has a yield of $1/6$, meaning that $1/6$ of the circuits are good and the other $5/6$ are defective. Use a die to simulate this manufacturing process, and consider an outcome of 1 to be a good integrated circuit, while outcomes of 2, 3, 4, 5, or 6 represent defective integrated circuits. Find the mean number of circuits that must be manufactured to get one that is good.

7. In-class activity The *Monty Hall problem* is based on the old television game show *Let's Make a Deal*, hosted by Monty Hall. Suppose you are a contestant who has selected one of three doors after being told that two of them conceal nothing, but that a new red Corvette is behind one of the three. Next, the host opens one of the doors you didn't select and shows that there is nothing behind it. He then offers you the choice of *sticking* with your first selection or *switching* to the other unopened door. Should you stick with your first choice or should you switch? Divide into groups of two and simulate this game to determine whether you should stick or switch. (According to *Chance* magazine, business schools at institutions such as Harvard and Stanford used this problem to help students deal with decision making.)

8. Out-of-class activity In Cumulative Review Exercise 4, it was noted that eye colors in the United States are distributed as follows: 40% brown, 35% blue, 12% green, 7% gray, 6% hazel. That distribution can form the basis for probabilities. Conduct a survey by asking fellow students to identify the color of their eyes. Does the probability of 0.4 for brown eyes appear to be consistent with your results? Why would a large sample be required to confirm that $P(\text{hazel eyes}) = 0.06$?

 5

DISCRETE PROBABILITY DISTRIBUTIONS

5-1 Probability Distributions

5-2 Binomial Probability Distributions

5-3 Poisson Probability Distributions

CHAPTER PROBLEM — **In football overtime games, does the team that wins the overtime coin toss have an advantage?**

Before 2012, National Football League (NFL) football games that were tied at the end of regulation time continued after a coin toss. The team that won the coin toss could choose to receive the ball or kick the ball. A whopping 98.1% of teams that won the coin toss chose to receive the ball. Between 1974 and 2011, there were 477 overtime games, and 17 of them ended in a tie after overtime play. Here, we consider the 460 overtime games that did not end in a tie. Among those

460 games that were decided in overtime, 252 were won by the same team that won the overtime coin toss; the teams that lost the overtime coin toss went on to win 208 games. See Figure 5-1 for a bar graph illustrating these results.

Examination of Figure 5-1 might suggest that there isn't much of a difference in wins between teams that won the overtime coin toss and teams that lost the overtime coin toss, but the graph might not be the best tool for this analysis, and this chapter will provide relevant and critical methods for addressing this issue. Starting in 2012 the NFL overtime rules were changed, but *why* were they changed? Here is the key question: *Under the old rules for overtime, does winning the coin toss become an advantage?* The result of 252 wins in 460 games is a winning rate of 54.8% for the teams that won the coin toss. Is that about the same as random chance, or is 54.8% *significantly* greater than 50%, so that teams winning the coin toss have an advantage? These questions can be answered by finding a relevant *probability* value, as shown in this chapter.

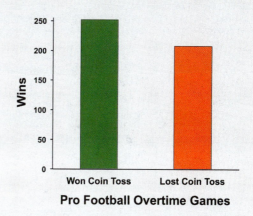

FIGURE 5-1 NFL Football Games Won and Lost in Overtime

CHAPTER OBJECTIVES

Figure 5-2 provides a visual illustration of what this chapter accomplishes. When investigating the numbers of heads in two coin tosses, we can use the following two different approaches:

- **Use real sample data to find *actual* results:** Collect numbers of heads in tosses of two coins, summarize them in a frequency distribution, and find the mean $\bar{x}$ and standard deviation s (as in Chapters 2 and 3).

- **Use probabilities to find *expected* results:** Find the probability for each possible number of heads in two tosses (as in Chapter 4), then summarize the results in a table representing a probability distribution, and then find the mean μ and standard deviation σ.

In this chapter we merge the above two approaches as we create a table describing what we *expect* to happen (instead of what did happen), then find the population mean μ and population standard deviation σ. The table at the extreme right in Figure 5-2 is a *probability distribution,* because it describes the distribution using *probabilities* instead of frequency counts. The remainder of this book and the core of inferential statistics are based on some knowledge of probability distributions. In this chapter we focus on *discrete* probability distributions.

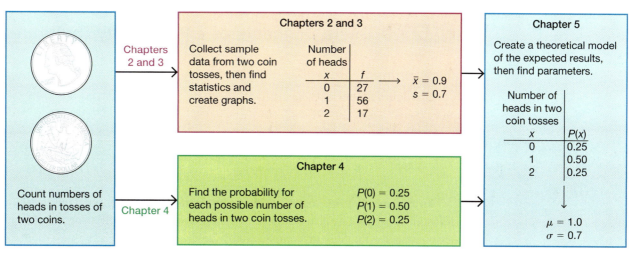

FIGURE 5-2

Here are the chapter objectives:

5-1 Probability Distributions

- Define *random variable* and *probability distribution*.

- Determine when a potential probability distribution actually satisfies the necessary requirements.

- Given a probability distribution, compute the mean and standard deviation, then use those results to determine whether results are *significantly low* or *significantly high*.

5-2 Binomial Probability Distributions

- Describe a binomial probability distribution and find probability values for a binomial distribution.

- Compute the mean and standard deviation for a binomial distribution, then use those results to determine whether results are *significantly low* or *significantly high*.

5-3 Poisson Probability Distributions

- Describe a Poisson probability distribution and find probability values for a Poisson distribution.

5-1   Probability Distributions

Key Concept This section introduces the concept of a *random variable* and the concept of a *probability distribution*. We illustrate how a *probability histogram* is a graph that visually depicts a probability distribution. We show how to find the important parameters of mean, standard deviation, and variance for a probability distribution. Most importantly, we describe how to determine whether outcomes are *significant* (significantly low or significantly high). We begin with the related concepts of *random variable* and *probability distribution*.

PART 1   Basic Concepts of a Probability Distribution

DEFINITIONS

A **random variable** is a variable (typically represented by x) that has a single numerical value, determined by chance, for each outcome of a procedure.

A **probability distribution** is a description that gives the probability for each value of the random variable. It is often expressed in the format of a table, formula, or graph.

In Section 1-2 we made a distinction between discrete and continuous data. Random variables may also be discrete or continuous, and the following two definitions are consistent with those given in Section 1-2.

DEFINITIONS

A **discrete random variable** has a collection of values that is finite or countable. (If there are infinitely many values, the number of values is countable if it is possible to count them individually, such as the number of tosses of a coin before getting heads.)

A **continuous random variable** has infinitely many values, and the collection of values is not countable. (That is, it is impossible to count the individual items because at least some of them are on a continuous scale, such as body temperatures.)

This chapter deals exclusively with discrete random variables, but the following chapters deal with continuous random variables.

Probability Distribution: Requirements

Every probability distribution must satisfy each of the following three requirements.

1. There is a *numerical* (not categorical) random variable x, and its number values are associated with corresponding probabilities.

2. $\Sigma P(x) = 1$ where x assumes all possible values. (The sum of all probabilities must be 1, but sums such as 0.999 or 1.001 are acceptable because they result from rounding errors.)

3. $0 \le P(x) \le 1$ for every individual value of the random variable x. (That is, each probability value must be between 0 and 1 inclusive.)

The second requirement comes from the simple fact that the random variable x represents all possible events in the entire sample space, so we are certain (with probability 1) that one of the events will occur. The third requirement comes from the basic principle that any probability value must be 0 or 1 or a value between 0 and 1.

EXAMPLE 1   Coin Toss

Let's consider tossing two coins, with the following random variable:

$$x = \text{number of heads when two coins are tossed}$$

The above x is a random variable because its numerical values depend on chance. With two coins tossed, the number of heads can be 0, 1, or 2, and Table 5-1 is a probability distribution because it gives the probability for each value of the random variable x and it satisfies the three requirements listed earlier:

1. The variable x is a *numerical* random variable, and its values are associated with probabilities, as in Table 5-1.

2. $\Sigma P(x) = 0.25 + 0.50 + 0.25 = 1$

3. Each value of $P(x)$ is between 0 and 1. (Specifically, 0.25 and 0.50 and 0.25 are each between 0 and 1 inclusive.)

The random variable x in Table 5-1 is a *discrete* random variable, because it has three possible values (0, 1, 2), and three is a finite number, so this satisfies the requirement of being finite or countable.

YOUR TURN Do Exercise 7 "Genetic Disorder."

TABLE 5-1 Probability Distribution for the Number of Heads in Two Coin Tosses

x: Number of Heads When Two Coins Are Tossed	P(x)
0	0.25
1	0.50
2	0.25

Notation for 0 +

In tables such as Table 5-1 or the binomial probabilities listed in Table A-1 in Appendix A, we sometimes use 0+ to represent a probability value that is positive but very small, such as 0.000000123. (When rounding a probability value for inclusion in such a table, rounding to 0 would be misleading because it would incorrectly suggest that the event is impossible.)

Probability Histogram: Graph of a Probability Distribution

There are various ways to graph a probability distribution, but for now we will consider only the **probability histogram**. Figure 5-3 is a probability histogram corresponding to Table 5-1. Notice that it is similar to a relative frequency histogram (described in Section 2-2), but the vertical scale shows *probabilities* instead of relative frequencies based on actual sample results.

In Figure 5-3, we see that the values of 0, 1, 2 along the horizontal axis are located at the centers of the rectangles. This implies that the rectangles are each 1 unit wide, so the areas of the rectangles are 0.25, 0.50, and 0.25. The *areas* of these rectangles are the same as the *probabilities* in Table 5-1. We will see in Chapter 6 and future chapters that such a correspondence between areas and probabilities is very useful.

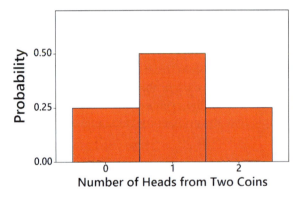

FIGURE 5-3   **Probability Histogram for Number of Heads When Two Coins Are Tossed**

Probability Formula Example 1 involves a table, but a probability distribution could also be in the form of a formula. Consider the formula

$$P(x) = \frac{1}{2(2 - x)!x!}$$ (where x can be 0, 1, or 2). Using that formula, we find that

$P(0) = 0.25, P(1) = 0.50,$ and $P(2) = 0.25.$ The probabilities found using this formula are the same as those in Table 5-1. This formula does describe a probability distribution because the three requirements are satisfied, as shown in Example 1.

EXAMPLE 2   Job Interview Mistakes

Hiring managers were asked to identify the biggest mistakes that job applicants make during an interview, and Table 5-2 is based on their responses (based on data from an Adecco survey). Does Table 5-2 describe a probability distribution?

TABLE 5-2 Job Interview Mistakes

x	P(x)
Inappropriate attire	0.50
Being late	0.44
Lack of eye contact	0.33
Checking phone or texting	0.30
Total	**1.57**

SOLUTION

Table 5-2 violates the first requirement because x is not a *numerical* random variable. Instead, the "values" of x are categorical data, not numbers. Table 5-2 also violates the second requirement because the sum of the probabilities is 1.57, but that sum should be 1. Because the three requirements are not all satisfied, we conclude that Table 5-2 does *not* describe a probability distribution.

YOUR TURN Do Exercise 9 "Pickup Line."

Parameters of a Probability Distribution

Remember that with a probability distribution, we have a description of a *population* instead of a sample, so the values of the mean, standard deviation, and variance are *parameters,* not statistics. The mean, variance, and standard deviation of a discrete probability distribution can be found with the following formulas:

FORMULA 5-1   Mean μ for a probability distribution

$$\mu = \Sigma [x \cdot P(x)]$$

FORMULA 5-2   Variance σ^2 for a probability distribution

$$\sigma^2 = \Sigma [(x - \mu)^2 \cdot P(x)] \text{ (This format is easier to understand.)}$$

FORMULA 5-3   Variance σ^2 for a probability distribution

$$\sigma^2 = \Sigma [x^2 \cdot P(x)] - \mu^2 \text{ (This format is easier for manual calculations.)}$$

FORMULA 5-4   Standard deviation σ for a probability distribution

$$\sigma = \sqrt{\Sigma [x^2 \cdot P(x)] - \mu^2}$$

When applying Formulas 5-1 through 5-4, use the following rule for rounding results.

Round-Off Rule for μ, σ, and σ^2 from a Probability Distribution

Round results by carrying *one more decimal place* than the number of decimal places used for the random variable x. If the values of x are integers, round μ, σ, and σ^2 to one decimal place.

Exceptions to Round-Off Rule In some special cases, the above round-off rule results in values that are misleading or inappropriate. For example, with four-engine jets the mean number of jet engines working successfully throughout a flight is 3.999714286, which becomes 4.0 when rounded, but that is misleading because it suggests that all jet engines always work successfully. Here we need more precision to correctly reflect the true mean, such as the precision in 3.999714.

Not at Home

Pollsters cannot simply ignore those who were not at home when they were called the first time. One solution is to make repeated callback attempts until the person can be reached. Alfred Politz and Willard Simmons describe a way to compensate for those missed calls without making repeated callbacks. They suggest weighting results based on how often people are not at home. For example, a person at home only two days out of six will have a 2/6 or 1/3 probability of being at home when called the first time. When such a person is reached the first time, his or her results are weighted to count three times as much as someone who is always home. This weighting is a compensation for the other similar people who are home two days out of six and were not at home when called the first time. This clever solution was first presented in 1949.

Expected Value

The mean of a discrete random variable x is the theoretical mean outcome for infinitely many trials. We can think of that mean as the *expected value* in the sense that it is the average value that we would expect to get if the trials could continue indefinitely.

> **DEFINITION**
>
> The **expected value** of a discrete random variable x is denoted by E, and it is the mean value of the outcomes, so $E = \mu$ and E can also be found by evaluating $\Sigma[x \cdot P(x)]$, as in Formula 5-1.

CAUTION An expected value need not be a whole number, even if the different possible values of x might all be whole numbers. The expected number of girls in five births is 2.5, even though five particular births can never result in 2.5 girls. If we were to survey many couples with five children, we *expect* that the mean number of girls will be 2.5.

EXAMPLE 3   Finding the Mean, Variance, and Standard Deviation

Table 5-1 on page 187 describes the probability distribution for the number of heads when two coins are tossed. Find the mean, variance, and standard deviation for the probability distribution described in Table 5-1 from Example 1.

SOLUTION

In Table 5-3, the two columns at the left describe the probability distribution given earlier in Table 5-1. We create the two columns at the right for the purposes of the calculations required.

Using Formulas 5-1 and 5-2 and the table results, we get

Mean: $\mu = \Sigma[x \cdot P(x)] = 1.0$

Variance: $\sigma^2 = \Sigma[(x - \mu)^2 \cdot P(x)] = 0.5$

The standard deviation is the square root of the variance, so

Standard deviation: $\sigma = \sqrt{0.5} = 0.707107 = 0.7$ (rounded)

Rounding: In Table 5-3, we use $\mu = 1.0$. If μ had been the value of 1.23456, we might round μ to 1.2, but we should use its *unrounded* value of 1.23456 in Table 5-3 calculations. Rounding in the middle of calculations can lead to results with errors that are too large.

TABLE 5-3 Calculating μ and σ for a Probability Distribution

x	$P(x)$	$x \cdot P(x)$	$(x - \mu)^2 \cdot P(x)$
0	0.25	$0 \cdot 0.25 = 0.00$	$(0 - 1.0)^2 \cdot 0.25 = 0.25$
1	0.50	$1 \cdot 0.50 = 0.50$	$(1 - 1.0)^2 \cdot 0.50 = 0.00$
2	0.25	$2 \cdot 0.25 = 0.50$	$(2 - 1.0)^2 \cdot 0.25 = 0.25$
Total		1.00	0.50
		$\uparrow$	$\uparrow$
		$\mu = \Sigma[x \cdot P(x)]$	$\sigma^2 = \Sigma[(x - \mu)^2 \cdot P(x)]$

INTERPRETATION

When tossing two coins, the mean number of heads is 1.0 head, the variance is 0.50 heads^2, and the standard deviation is 0.7 head. Also, the expected value for the number of heads when two coins are tossed is 1.0 head, which is the same value as the mean. If we were to collect data on a large number of trials with two coins tossed in each trial, we expect to get a mean of 1.0 head.

YOUR TURN Do Exercise 15 "Mean and Standard Deviation."

Making Sense of Results: Significant Values

We present the following two different approaches for determining whether a value of a random variable x is significantly low or high.

Identifying Significant Results with the Range Rule of Thumb

The range rule of thumb (introduced in Section 3-2) may be helpful in interpreting the value of a standard deviation. According to the range rule of thumb, the vast majority of values should lie within 2 standard deviations of the mean, so we can consider a value to be *significant* if it is at least 2 standard deviations away from the mean. We can therefore identify "significant" values as follows:

Range Rule of Thumb for Identifying Significant Values

Significantly low values are $(\mu - 2\sigma)$ or lower.

Significantly high values are $\mu + 2\sigma$ or higher.

Values not significant: Between $(\mu - 2\sigma)$ and $(\mu + 2\sigma)$

Figure 3-3 from Section 3-2 illustrates the above criteria:

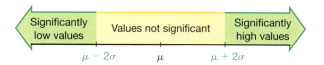

CAUTION Know that the use of the number 2 in the range rule of thumb is somewhat arbitrary, and this is a guideline, not an absolutely rigid rule.

EXAMPLE 4 **Identifying Significant Results with the Range Rule of Thumb**

In Example 3 we found that when tossing two coins, the mean number of heads is $\mu = 1.0$ head and the standard deviation is $\sigma = 0.7$ head. Use those results and the range rule of thumb to determine whether 2 heads is a significantly high number of heads.

SOLUTION

Using the range rule of thumb, the outcome of 2 heads is significantly high if it is greater than or equal to $\mu + 2\sigma$. With $\mu = 1.0$ head $\sigma = 0.7$ head, we get

$$\mu + 2\sigma = 1 + 2(0.7) = 2.4 \text{ heads}$$

Significantly high numbers of heads are 2.4 and above.

continued

INTERPRETATION

Based on these results, we conclude that 2 heads is not a significantly high number of heads (because 2 is not greater than or equal to 2.4).

> **YOUR TURN** Do Exercise 17 "Range Rule of Thumb for Significant Events."

Identifying Significant Results with Probabilities:

- **Significantly *high* number of successes:** *x* successes among *n* trials is a *significantly high* number of successes if the probability of *x* or more successes is 0.05 or less. That is, *x* is a significantly high number of successes if $P(x \text{ or more}) \leq 0.05$.*

- **Significantly *low* number of successes:** *x* successes among *n* trials is a *significantly low* number of successes if the probability of *x* or fewer successes is 0.05 or less. That is, *x* is a significantly low number of successes if $P(x \text{ or fewer}) \leq 0.05$.*

*The value 0.05 is not absolutely rigid. Other values, such as 0.01, could be used to distinguish between results that are significant and those that are not significant.

Identification of significantly low or significantly high numbers of successes is sometimes used for the purpose of rejecting assumptions, as stated in the following rare event rule.

The Rare Event Rule for Inferential Statistics

> **If, under a given assumption, the probability of a particular outcome is very small and the outcome occurs *significantly less than* or *significantly greater than* what we expect with that assumption, we conclude that the assumption is probably not correct.**

For example, if testing the assumption that boys and girls are equally likely, the outcome of 20 girls in 100 births is significantly low and would be a basis for rejecting that assumption.

EXAMPLE 5 Identifying Significant Results with Probabilities

Is 252 heads in 460 coin tosses a significantly high number of heads?

What does the result suggest about the Chapter Problem, which includes results from 460 overtime games? (Among the 460 teams that won the coin toss, 252 of them won the game. Is 252 wins in those 460 games *significantly high*?)

SOLUTION

A result of 252 heads in 460 coin tosses is greater than we expect with random chance, but we need to determine whether 252 heads is *significantly high*. Here, the relevant probability is the probability of getting *252 or more* heads in 460 coin tosses. Using methods covered later in Section 5-2, we can find that $P(252 \text{ or more heads in } 460 \text{ coin tosses}) = 0.0224$ (rounded). Because the probability of getting 252 or more heads is less than or equal to 0.05, we conclude that 252 heads in 460 coin tosses is a *significantly high* number of heads. See Figure 5-4, which is a probability histogram showing the probability for the different numbers of heads.

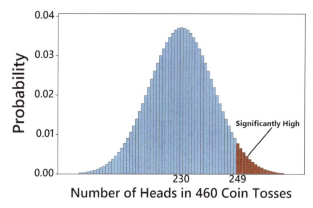

FIGURE 5-4   **Probability Histogram of Heads in 460 Coin Tosses**

INTERPRETATION

It is unlikely that we would get 252 or more heads in 460 coin tosses by chance. It follows that 252 wins by teams that won the overtime coin toss is significantly high, so winning the coin toss is an advantage. This is justification for changing the overtime rules, as was done in 2012.

> **YOUR TURN** Do Exercise 19 "Using Probabilities for Significant Events."

Not *Exactly,* but "At Least as Extreme"

It should be obvious that among 1000 tosses of a coin, 502 heads is not significantly high, whereas 900 heads is significantly high. What makes 900 heads significant while 502 heads is not significant? It is not the *exact* probabilities of 900 heads and 502 heads (they are both less than 0.026). It is the fact that the probability of 502 *or more* heads (0.462) is not low, but the probability of 900 *or more* heads (0+) is very low.

PART 2   Expected Value and Rationale for Formulas

Expected Value

In Part 1 of this section we noted that the expected value of a random variable x is equal to the mean μ. We can therefore find the expected value by computing $\Sigma [x \cdot P(x)]$, just as we do for finding the value of μ. We also noted that the concept of expected value is used in *decision theory*. In Example 6 we illustrate this use of expected value with a situation in which we must choose between two different bets. Example 6 involves a real and practical decision.

EXAMPLE 6   **Be a Better Bettor**

You have $5 to place on a bet in the Golden Nugget casino in Las Vegas. You have narrowed your choice to one of two bets:

Roulette: Bet on the number 7 in roulette.

Craps: Bet on the "pass line" in the dice game of craps.

a. If you bet $5 on the number 7 in roulette, the probability of losing $5 is $37/38$ and the probability of making a net gain of $175 is $1/38$. (The prize is $180, including your $5 bet, so the net gain is $175.) Find your expected value if you bet $5 on the number 7 in roulette.

continued

b. If you bet $5 on the pass line in the dice game of craps, the probability of losing $5 is 251/495 and the probability of making a net gain of $5 is 244/495. (If you bet $5 on the pass line and win, you are given $10 that includes your bet, so the net gain is $5.) Find your expected value if you bet $5 on the pass line.

Which of the preceding two bets is better in the sense of producing higher expected value?

SOLUTION

a. Roulette The probabilities and payoffs for betting $5 on the number 7 in roulette are summarized in Table 5-4. Table 5-4 also shows that the expected value is $\Sigma [x \cdot P(x)] = -26¢$. That is, for every $5 bet on the number 7, you can expect to *lose* an average of 26¢.

TABLE 5-4 Roulette

Event	x	P(x)	x · P(x)
Lose	−$5	37/38	−$4.868421
Win (net gain)	$175	1/38	$4.605263
Total			−$0.26 (rounded)
			(or −26¢)

b. Craps Game The probabilities and payoffs for betting $5 on the pass line in craps are summarized in Table 5-5. Table 5-5 also shows that the expected value is $\Sigma [x \cdot P(x)] = -7¢$. That is, for every $5 bet on the pass line, you can expect to lose an average of 7¢.

TABLE 5-5 Craps Game

Event	x	P(x)	x · P(x)
Lose	−$5	251/495	−$2.535353
Win (net gain)	$5	244/495	$2.464646
Total			−$0.07 (rounded)
			(or −7¢)

INTERPRETATION

The $5 bet in roulette results in an expected value of −26¢ and the $5 bet in craps results in an expected value of −7¢. Because you are better off losing 7¢ instead of losing 26¢, the craps game is better in the long run, even though the roulette game provides an opportunity for a larger payoff when playing the game once.

YOUR TURN Do Exercise 27 "Expected Value in Virginia's Pick 3 Game."

Rationale for Formulas 5-1 Through 5-4

Instead of blindly accepting and using formulas, it is much better to have some understanding of why they work. When computing the mean from a frequency distribution, f represents class frequency and N represents population size. In the expression that follows, we rewrite the formula for the mean of a frequency table so that it applies to a population. In the fraction f/N, the value of f is the frequency with which the value x occurs and N is the population size, so f/N is the probability for the value of x. When we replace f/N with $P(x)$, we make the transition from relative frequency based on a limited number of observations to probability based on infinitely many trials. This result shows why Formula 5-1 is as given earlier in this section.

$$\mu = \frac{\Sigma(f \cdot x)}{N} = \Sigma\left[\frac{f \cdot x}{N}\right] = \Sigma\left[x \cdot \frac{f}{N}\right] = \Sigma[x \cdot P(x)]$$

Similar reasoning enables us to take the variance formula from Chapter 3 and apply it to a random variable for a probability distribution; the result is Formula 5-2. Formula 5-3 is a shortcut version that will always produce the same result as Formula 5-2. Although Formula 5-3 is usually easier to work with, Formula 5-2 is easier to understand directly. Based on Formula 5-2, we can express the standard deviation as

$$\sigma = \sqrt{\Sigma[(x - \mu)^2 \cdot P(x)]}$$

or as the equivalent form given in Formula 5-4.

5-1   Basic Skills and Concepts

Statistical Literacy and Critical Thinking

1. Random Variable The accompanying table lists probabilities for the corresponding numbers of girls in four births. What is the random variable, what are its possible values, and are its values numerical?

2. Discrete or Continuous? Is the random variable given in the accompanying table discrete or continuous? Explain.

3. Probability Distribution For the accompanying table, is the sum of the values of $P(x)$ equal to 1, as required for a probability distribution? Does the table describe a probability distribution?

4. Significant For 100 births, $P(\text{exactly } 56 \text{ girls}) = 0.0390$ and $P(56 \text{ or more girls}) = 0.136$. Is 56 girls in 100 births a significantly high number of girls? Which probability is relevant to answering that question?

Number of Girls in Four Births

Number of Girls x	$P(x)$
0	0.063
1	0.250
2	0.375
3	0.250
4	0.063

Identifying Discrete and Continuous Random Variables. *In Exercises 5 and 6, refer to the given values, then identify which of the following is most appropriate:* **discrete random variable, continuous random variable,** *or* **not a random variable.**

5. a. Exact weights of the next 100 babies born in the United States

b. Responses to the survey question "Which political party do you prefer?"

c. Numbers of spins of roulette wheels required to get the number 7

d. Exact foot lengths of humans

e. Shoe sizes (such as 8 or 8½) of humans

6. a. Grades (A, B, C, D, F) earned in statistics classes

b. Heights of students in statistics classes

c. Numbers of students in statistics classes

d. Eye colors of statistics students

e. Numbers of times statistics students must toss a coin before getting heads

Identifying Probability Distributions. *In Exercises 7–14, determine whether a probability distribution is given. If a probability distribution is given, find its mean and standard deviation. If a probability distribution is not given, identify the requirements that are not satisfied.*

7. Genetic Disorder Five males with an X-linked genetic disorder have one child each. The random variable x is the number of children among the five who inherit the X-linked genetic disorder.

x	P(x)
0	0.031
1	0.156
2	0.313
3	0.313
4	0.156
5	0.031

8. Male Color Blindness When conducting research on color blindness in males, a researcher forms random groups with five males in each group. The random variable x is the number of males in the group who have a form of color blindness (based on data from the National Institutes of Health).

x	P(x)
0	0.659
1	0.287
2	0.050
3	0.004
4	0.001
5	0+

9. Pickup Line Ted is not particularly creative. He uses the pickup line "If I could rearrange the alphabet, I'd put U and I together." The random variable x is the number of women Ted approaches before encountering one who reacts positively.

x	P(x)
1	0.001
2	0.009
3	0.030
4	0.060

10. Fun Ways to Flirt In a Microsoft Instant Messaging survey, respondents were asked to choose the most fun way to flirt, and the accompanying table is based on the results.

	P(x)
E-mail	0.06
In person	0.55
Instant message	0.24
Text message	0.15

11. Fun Ways to Flirt A sociologist randomly selects single adults for different groups of three, and the random variable x is the number in the group who say that the most fun way to flirt is in person (based on a Microsoft Instant Messaging survey).

x	P(x)
0	0.091
1	0.334
2	0.408
3	0.166

12. Self-Driving Vehicle Groups of adults are randomly selected and arranged in groups of three. The random variable x is the number in the group who say that they would feel comfortable in a self-driving vehicle (based on a TE Connectivity survey).

x	P(x)
0	0.358
1	0.439
2	0.179
3	0.024

13. Cell Phone Use In a survey, cell phone users were asked which ear they use to hear their cell phone, and the table is based on their responses (based on data from "Hemispheric Dominance and Cell Phone Use," by Seidman et al., *JAMA Otolaryngology—Head & Neck Surgery,* Vol. 139, No. 5).

	P(x)
Left	0.636
Right	0.304
No preference	0.060

14. Casino Games When betting on the pass line in the dice game of craps at the Mohegan Sun casino in Connecticut, the table lists the probabilities for the number of bets that must be placed in order to have a win.

x	P(x)
1	0.493
2	0.250
3	0.127
4	0.064

Genetics. *In Exercises 15–20, refer to the accompanying table, which describes results from groups of 8 births from 8 different sets of parents. The random variable x represents the number of girls among 8 children.*

Number of Girls x	P(x)
0	0.004
1	0.031
2	0.109
3	0.219
4	0.273
5	0.219
6	0.109
7	0.031
8	0.004

15. Mean and Standard Deviation Find the mean and standard deviation for the numbers of girls in 8 births.

16. Range Rule of Thumb for Significant Events Use the range rule of thumb to determine whether 1 girl in 8 births is a significantly low number of girls.

17. Range Rule of Thumb for Significant Events Use the range rule of thumb to determine whether 6 girls in 8 births is a significantly high number of girls.

18. Using Probabilities for Significant Events

a. Find the probability of getting exactly 7 girls in 8 births.

b. Find the probability of getting 7 or more girls in 8 births.

c. Which probability is relevant for determining whether 7 is a significantly high number of girls in 10 births: the result from part (a) or part (b)?

d. Is 7 a significantly high number of girls in 8 births? Why or why not?

19. Using Probabilities for Significant Events

a. Find the probability of getting exactly 6 girls in 8 births.

b. Find the probability of getting 6 or more girls in 8 births.

c. Which probability is relevant for determining whether 6 is a significantly high number of girls in 8 births: the result from part (a) or part (b)?

d. Is 6 a significantly high number of girls in 8 births? Why or why not?

20. Using Probabilities for Significant Events

a. Find the probability of getting exactly 1 girl in 8 births.

b. Find the probability of getting 1 or fewer girls in 8 births.

c. Which probability is relevant for determining whether 1 is a significantly low number of girls in 8 births: the result from part (a) or part (b)?

d. Is 1 a significantly low number of girls in 8 births? Why or why not?

Sleepwalking. *In Exercises 21–25, refer to the accompanying table, which describes the numbers of adults in groups of five who reported sleepwalking (based on data from "Prevalence and Comorbidity of Nocturnal Wandering In the U.S. Adult General Population," by Ohayon et al.,* **Neurology,** *Vol. 78, No. 20).*

x	P(x)
0	0.172
1	0.363
2	0.306
3	0.129
4	0.027
5	0.002

21. Mean and Standard Deviation Find the mean and standard deviation for the numbers of sleepwalkers in groups of five.

22. Range Rule of Thumb for Significant Events Use the range rule of thumb to determine whether 4 is a significantly high number of sleepwalkers in a group of 5 adults.

23. Range Rule of Thumb for Significant Events Use the range rule of thumb to determine whether 3 is a significantly high number of sleepwalkers in a group of 5 adults.

24. Using Probabilities for Identifying Significant Events

a. Find the probability of getting exactly 4 sleepwalkers among 5 adults.

b. Find the probability of getting 4 or more sleepwalkers among 5 adults.

c. Which probability is relevant for determining whether 4 is a significantly high number of sleepwalkers among 5 adults: the result from part (a) or part (b)?

d. Is 4 a significantly high number of sleepwalkers among 5 adults? Why or why not?

25. Using Probabilities for Identifying Significant Events

a. Find the probability of getting exactly 1 sleepwalker among 5 adults.

b. Find the probability of getting 1 or fewer sleepwalkers among 5 adults.

c. Which probability is relevant for determining whether 1 is a significantly low number of sleepwalkers among 5 adults: the result from part (a) or part (b)?

d. Is 1 a significantly low number of sleepwalkers among 5 adults? Why or why not?

5-1 Beyond the Basics

26. Expected Value for the Ohio Pick 4 Lottery In the Ohio Pick 4 lottery, you can bet $1 by selecting four digits, each between 0 and 9 inclusive. If the same four numbers are drawn in the same order, you win and collect $5000.

a. How many different selections are possible?

b. What is the probability of winning?

c. If you win, what is your net profit?

d. Find the expected value for a $1 bet.

e. If you bet $1 on the pass line in the casino dice game of craps, the expected value is −1.4¢. Which bet is better in the sense of producing a higher expected value: a $1 bet in the Ohio Pick 4 lottery or a $1 bet on the pass line in craps?

27. Expected Value in Virginia's Pick 3 Game In Virginia's Pick 3 lottery game, you can pay $1 to select a three-digit number from 000 through 999. If you select the same sequence of three digits that are drawn, you win and collect $500.

a. How many different selections are possible?

b. What is the probability of winning?

c. If you win, what is your net profit?

d. Find the expected value.

e. If you bet $1 in Virginia's Pick 4 game, the expected value is −50¢. Which bet is better in the sense of a producing a higher expected value: A $1 bet in the Virginia Pick 3 game or a $1 bet in the Virginia Pick 4 game?

28. Expected Value in Roulette When playing roulette at the Venetian casino in Las Vegas, a gambler is trying to decide whether to bet $5 on the number 27 or to bet $5 that the outcome is any one of these five possibilities: 0, 00, 1, 2, 3. From Example 6, we know that the expected value of the $5 bet for a single number is −26¢. For the $5 bet that the outcome is 0, 00, 1, 2, or 3, there is a probability of 5/38 of making a net profit of $30 and a 33/38 probability of losing $5.

a. Find the expected value for the $5 bet that the outcome is 0, 00, 1, 2, or 3.

b. Which bet is better: a $5 bet on the number 27 or a $5 bet that the outcome is any one of the numbers 0, 00, 1, 2, or 3? Why?

29. Expected Value for Life Insurance There is a 0.9986 probability that a randomly selected 30-year-old male lives through the year (based on data from the U.S. Department of Health and Human Services). A Fidelity life insurance company charges $161 for insuring that the male will live through the year. If the male does not survive the year, the policy pays out $100,000 as a death benefit.

a. From the perspective of the 30-year-old male, what are the monetary values corresponding to the two events of surviving the year and not surviving?

b. If a 30-year-old male purchases the policy, what is his expected value?

c. Can the insurance company expect to make a profit from many such policies? Why?

30. Expected Value for Life Insurance There is a 0.9968 probability that a randomly selected 50-year-old female lives through the year (based on data from the U.S. Department of Health and Human Services). A Fidelity life insurance company charges $226 for insuring that the female will live through the year. If she does not survive the year, the policy pays out $50,000 as a death benefit.

a. From the perspective of the 50-year-old female, what are the values corresponding to the two events of surviving the year and not surviving.

b. If a 50-year-old female purchases the policy, what is her expected value?

c. Can the insurance company expect to make a profit from many such policies? Why?

5-2 Binomial Probability Distributions

Key Concept Section 5-1 introduced the important concept of a discrete probability distribution. Among the various discrete probability distributions that exist, the focus of this section is the *binomial probability distribution*. Part 1 of this section introduces the binomial probability distribution along with methods for finding probabilities. Part 2 presents easy methods for finding the mean and standard deviation of a binomial distribution. As in other sections, we stress the importance of *interpreting* probability values to determine whether events are *significantly low* or *significantly high.*

PART 1 Basics of Binomial Probability Distribution

Binomial probability distributions allow us to deal with circumstances in which the outcomes belong to *two* categories, such as heads/tails or acceptable/defective or survived/died.

Go Figure

9,000,000: Number of other people with the same birthday as you.

DEFINITION

A **binomial probability distribution** results from a procedure that meets these four requirements:

1. The procedure has a *fixed number of trials*. (A trial is a single observation.)

2. The trials must be *independent*, meaning that the outcome of any individual trial doesn't affect the probabilities in the other trials.

3. Each trial must have all outcomes classified into exactly *two categories,* commonly referred to as *success* and *failure.*

4. The probability of a success remains the same in all trials.

Notation for Binomial Probability Distributions

S and F (success and failure) denote the two possible categories of all outcomes.

$P(S) = p$ (p = probability of a success)

$P(F) = 1 - p = q$ (q = probability of a failure)

n the fixed number of trials

x a specific number of successes in n trials, so x can be any whole number between 0 and n, inclusive

p probability of *success* in *one* of the n trials

q probability of *failure* in *one* of the n trials

$P(x)$ probability of getting exactly x successes among the n trials

The word *success* as used here is arbitrary and does not necessarily represent something good. Either of the two possible categories may be called the success S as long as its probability is identified as p. (The value of q can always be found from $q = 1 - p$. If $p = 0.95$, then $q = 1 - 0.95 = 0.05$.)

> **CAUTION** When using a binomial probability distribution, always be sure that x and p are *consistent* in the sense that they both refer to the *same* category being called a success.

 EXAMPLE 1 **Twitter**

When an adult is randomly selected (with replacement), there is a 0.85 probability that this person knows what Twitter is (based on results from a Pew Research Center survey). Suppose that we want to find the probability that exactly three of five randomly selected adults know what Twitter is.

a. Does this procedure result in a binomial distribution?

b. If this procedure does result in a binomial distribution, identify the values of $n, x, p,$ and q.

SOLUTION

a. This procedure does satisfy the requirements for a binomial distribution, as shown below.

    1. The number of trials (5) is fixed.

    2. The 5 trials are independent because the probability of any adult knowing Twitter is not affected by results from other selected adults.

3. Each of the 5 trials has two categories of outcomes: The selected person knows what Twitter is or that person does not know what Twitter is.

4. For each randomly selected adult, there is a 0.85 probability that this person knows what Twitter is, and that probability remains the same for each of the five selected people.

b. Having concluded that the given procedure does result in a binomial distribution, we now proceed to identify the values of n, x, p, and q.

1. With five randomly selected adults, we have $n = 5$.

2. We want the probability of exactly three who know what Twitter is, so $x = 3$.

3. The probability of success (getting a person who knows what Twitter is) for one selection is 0.85, so $p = 0.85$.

4. The probability of failure (not getting someone who knows what Twitter is) is 0.15, so $q = 0.15$.

Again, it is very important to be sure that x and p both refer to the same concept of "success." In this example, we use x to count the number of people who know what Twitter is, so p must be the probability that the selected person knows what Twitter is. Therefore, x and p do use the same concept of success: knowing what Twitter is.

YOUR TURN Do Exercise 5 "Clinical Trial of YSORT."

Treating Dependent Events as Independent

When selecting a sample (as in a survey), we usually sample without replacement. Sampling without replacement results in dependent events, which violates a requirement of a binomial distribution. However, we can often treat the events as if they were independent by applying the following 5% guideline introduced in Section 4-2:

5% Guideline for Cumbersome Calculations

When sampling without replacement and the sample size is no more than 5% of the size of the population, treat the selections as being *independent* **(even though they are actually dependent).**

Methods for Finding Binomial Probabilities

We now proceed with three methods for finding the probabilities corresponding to the random variable x in a binomial distribution. The first method involves calculations using the *binomial probability formula* and is the basis for the other two methods. The second method involves the use of software or a calculator, and the third method involves the use of the Appendix Table A-1. (With technology so widespread, such tables are becoming obsolete.) If using technology that automatically produces binomial probabilities, we recommend that you solve one or two exercises using Method 1 to better understand the basis for the calculations.

Method 1: Using the Binomial Probability Formula In a binomial probability distribution, probabilities can be calculated by using Formula 5-5.

FORMULA 5-5 Binomial Probability Formula

$$P(x) = \frac{n!}{(n-x)!x!} \cdot p^x \cdot q^{n-x} \qquad \text{for } x = 0, 1, 2, \ldots, n$$

where

n = number of trials

x = number of successes among n trials

p = probability of success in any one trial

q = probability of failure in any one trial $(q = 1 - p)$

Formula 5-5 can also be expressed as $P(x) = {}_nC_x \, p^x \, q^{n-x}$. With x items identical to themselves, and $n - x$ other items identical to themselves, the number of permutations is ${}_nC_x = n!/[(n-x)!x!]$, so the two sides of this equation are interchangeable. The factorial symbol !, introduced in Section 4-4, denotes the product of decreasing factors. Two examples of factorials are $3! = 3 \cdot 2 \cdot 1 = 6$ and $0! = 1$ (by definition).

⊖⊢ **EXAMPLE 2** **Twitter**

Given that there is a 0.85 probability that a randomly selected adult knows what Twitter is, use the binomial probability formula to find the probability that when five adults are randomly selected, exactly three of them know what Twitter is. That is, apply Formula 5-5 to find $P(3)$ given that $n = 5, x = 3, p = 0.85$, and $q = 0.15$.

SOLUTION

Using the given values of n, x, p, and q in the binomial probability formula (Formula 5-5), we get

$$P(3) = \frac{5!}{(5-3)!3!} \cdot 0.85^3 \cdot 0.15^{5-3}$$

$$= \frac{5!}{2!3!} \cdot 0.614125 \cdot 0.0225$$

$$= (10)(0.614125)(0.0225) = 0.138178$$

$$= 0.138 \text{ (rounded to three significant digits)}$$

The probability of getting exactly three adults who know Twitter among five randomly selected adults is 0.138.

YOUR TURN Do Exercise 13 "Guessing Answers."

Calculation hint: When computing a probability with the binomial probability formula, it's helpful to get a single number for $n!/[(n-x)!x!]$ or ${}_nC_x$, a single number for p^x, and a single number for q^{n-x}, then simply multiply the three factors together as shown in the third line of the calculation in the preceding example. Don't round when you find those three factors; round only at the end, and round to three significant digits.

Method 2: Using Technology Technologies can be used to find binomial probabilities. The screen displays list binomial probabilities for $n = 5$ and $p = 0.85$, as in Example 2. Notice that in each display, the probability distribution is given as a table.

Minitab

x	P(X = x)
0	0.000076
1	0.002152
2	0.024384
3	0.138178
4	0.391505
5	0.443705

Statdisk

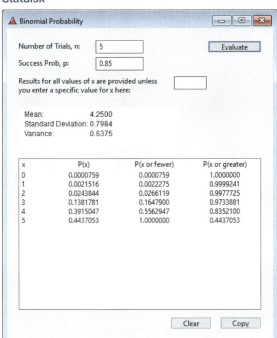

Excel

	A	B
1	x	P(x)
2	0	7.594E-05
3	1	0.0021516
4	2	0.0243844
5	3	0.1381781
6	4	0.3915047
7	5	0.4437053

TI-83/84 Plus CE

NORMAL FLOAT AUTO REAL RADIAN MP

L1	L2	L3	L4	L5
0	7.6E-5		------	------
1	.00215			
2	.02438			
3	.13818			
4	.3915			
5	.44371			
------	------			

EXAMPLE 3   Overtime Rule in Football

In the Chapter Problem, we noted that between 1974 and 2011, there were 460 NFL football games decided in overtime, and 252 of them were won by the team that won the overtime coin toss. Is the result of 252 wins in the 460 games equivalent to random chance, or is 252 wins *significantly high?* We can answer that question by finding the probability of 252 wins or more in 460 games, assuming that wins and losses are equally likely.

SOLUTION

Using the notation for binomial probabilities, we have $n = 460, p = 0.5, q = 0.5$, and we want to find the sum of all probabilities for each value of x from 252 through 460. The formula is not practical here, because we would need to apply it 209 times—we don't want to go there. Table A-1 (Binomial Probabilities) doesn't apply because $n = 460$, which is way beyond the scope of that table. Instead, we wisely choose to use technology.

The Statdisk display on the next page shows that the probability of 252 or more wins in 460 overtime games is 0.0224 (rounded), which is low (such as less than 0.05). This shows that it is unlikely that we would get 252 or more wins by chance. If we effectively rule out chance, we are left with the more reasonable explanation that the team winning the overtime coin toss has a better chance of winning the game.

continued

Statdisk

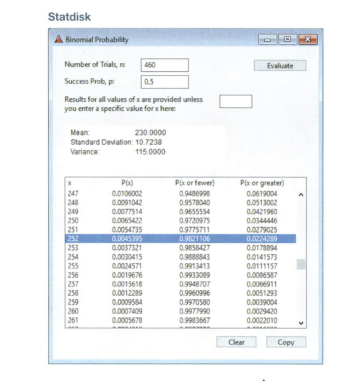

YOUR TURN Do Exercise 23 "Smartphones."

Example 3 illustrates well the power and ease of using technology. Example 3 also illustrates the rare event rule of statistical thinking: If under a given assumption (such as the assumption that winning the overtime coin toss has no effect), the probability of a particular observed event (such as 252 or more wins in 460 games) is extremely small (such as 0.05 or less), we conclude that the assumption is probably not correct.

Method 3: Using Table A-1 in Appendix A This method can be skipped if technology is available. Table A-1 in Appendix A lists binomial probabilities for select values of n and p. It cannot be used if $n > 8$ or if the probability p is not one of the 13 values included in the table.

To use the table of binomial probabilities, we must first locate n and the desired corresponding value of x. At this stage, one row of numbers should be isolated. Now align that row with the desired probability of p by using the column across the top. The isolated number represents the desired probability. A very small probability, such as 0.000064, is indicated by 0+.

EXAMPLE 4   Devil of a Problem

Based on a Harris poll, 60% of adults believe in the devil. Assuming that we randomly select five adults, use Table A-1 to find the following:

    a. The probability that exactly three of the five adults believe in the devil

    b. The probability that the number of adults who believe in the devil is at least two

SOLUTION

    a. The following excerpt from the table shows that when $n = 5$ and $p = 0.6$, the probability for $x = 3$ is given by $P(3) = 0.346$.

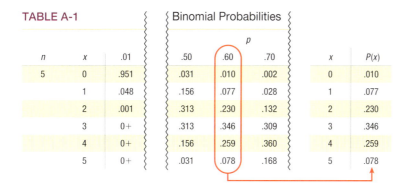

TABLE A-1 Binomial Probabilities

n	x	.01		.50	.60	.70		x	P(x)
5	0	.951		.031	.010	.002		0	.010
	1	.048		.156	.077	.028		1	.077
	2	.001		.313	.230	.132		2	.230
	3	0+		.313	.346	.309		3	.346
	4	0+		.156	.259	.360		4	.259
	5	0+		.031	.078	.168		5	.078

b. The phrase "at least two" successes means that the number of successes is 2 or 3 or 4 or 5.

$$P(\text{at least 2 believe in the devil}) = P(2 \text{ or } 3 \text{ or } 4 \text{ or } 5)$$
$$= P(2) + P(3) + P(4) + P(5)$$
$$= 0.230 + 0.346 + 0.259 + 0.078$$
$$= 0.913$$

YOUR TURN Do Exercise 15 "SAT Test" using Table A-1.

PART 2 Using Mean and Standard Deviation for Critical Thinking

Section 5-1 included formulas for finding the mean, variance, and standard deviation from *any* discrete probability distribution. A binomial distribution is a particular type of discrete probability distribution, so we could use those same formulas, but if we know the values of *n* and *p,* it is much easier to use the following:

For Binomial Distributions

Formula 5-6   Mean: $\mu = np$

Formula 5-7   Variance: $\sigma^2 = npq$

Formula 5-8   Standard Deviation: $\sigma = \sqrt{npq}$

As in earlier sections, finding values for μ and σ can be great fun, but it is especially important to *interpret* and *understand* those values, so the range rule of thumb and the rare event rule for inferential statistics can be very helpful. Here is a brief summary of the range rule of thumb: Values are significantly low or high if they differ from the mean by more than 2 standard deviations, as described by the following:

Range Rule of Thumb

Significantly low values $\leq (\mu - 2\sigma)$

Significantly high values $\geq (\mu + 2\sigma)$

Values not significant: Between $(\mu - 2\sigma)$ and $(\mu + 2\sigma)$

Proportions of Males/Females

It is well known that when a baby is born, boys and girls are not equally likely. It is currently believed that 105 boys are born for every 100 girls, so the probability of a boy is 0.512. Kristen Navara of the University of Georgia conducted a study showing that around the world, more boys are born than girls, but the difference becomes smaller as people are located closer to the equator. She used latitudes, temperatures, unemployment rates, gross and national products from 200 countries and conducted a statistical analysis showing that the proportions of boys appear to be affected only by latitude and its related weather. So far, no one has identified a reasonable explanation for this phenomenon.

EXAMPLE 5   **Using Parameters to Determine Significance**

The Chapter Problem and Example 3 involve $n = 460$ overtime wins in NFL football games. We get $p = 0.5$ and $q = 0.5$ by assuming that winning the overtime coin toss does not provide an advantage, so both teams have the same 0.5 chance of winning the game in overtime.

a. Find the mean and standard deviation for the number of wins in groups of 460 games.

b. Use the range rule of thumb to find the values separating the numbers of wins that are significantly low or significantly high.

c. Is the result of 252 overtime wins in 460 games significantly high?

SOLUTION

a. With $n = 460$, $p = 0.5$, and $q = 0.5$, Formulas 5-6 and 5-8 can be applied as follows:

$$\mu = np = (460)(0.5) = 230.0 \text{ games}$$

$$\sigma = \sqrt{npq} = \sqrt{(460)(0.5)(0.5)} = 10.7 \text{ games (rounded)}$$

For random groups of 460 overtime games, the mean number of wins is 230.0 games, and the standard deviation is 10.7 games.

b. The values separating numbers of wins that are significantly low or significantly high are the values that are two standard deviations away from the mean. With $\mu = 230.0$ games and $\sigma = 10.7$ games, we get

$$\mu - 2\sigma = 230.0 - 2(10.7) = 208.6 \text{ games}$$

$$\mu + 2\sigma = 230.0 + 2(10.7) = 251.4 \text{ games}$$

Significantly low numbers of wins are 208.6 games or fewer, significantly high numbers of wins are 251.4 games or greater, and values not significant are between 208.6 games and 251.4 games.

c. The result of 252 wins is significantly high because it is greater than the value of 251.4 games found in part (b).

YOUR TURN   Do Exercise 29 "Gender Selection."

Instead of the range rule of thumb, we could also use probabilities to determine when values are significantly high or low.

Using Probabilities to Determine When Results Are Significantly High or Low

- **Significantly *high* number of successes:** x successes among n trials is *significantly high* if the probability of x or more successes is 0.05 or less. That is, x is a significantly high number of successes if $P(x \text{ or more}) \le 0.05$.*

- **Significantly *low* number of successes:** x successes among n trials is *significantly low* if the probability of x or fewer successes is 0.05 or less. That is, x is a significantly low number of successes if $P(x \text{ or fewer}) \le 0.05$.*

*The value 0.05 is not absolutely rigid. Other values, such as 0.01, could be used to distinguish between results that are significant and those that are not significant.

Rationale for the Binomial Probability Formula

The binomial probability formula is the basis for all three methods presented in this section. Instead of accepting and using that formula blindly, let's see why it works.

In Example 2, we used the binomial probability formula to find the probability of getting exactly three adults who know Twitter when five adults are randomly selected. With $P(\text{knows Twitter}) = 0.85$, we can use the multiplication rule from Section 4-2 to find the probability that the first three adults know Twitter and the last two adults do not know Twitter. We get the following result:

$P(3 \text{ adults know Twitter followed by 2 adults who do not know Twitter})$

$= 0.85 \cdot 0.85 \cdot 0.85 \cdot 0.15 \cdot 0.15$

$= 0.85^3 \cdot 0.15^2$

$= 0.0138$

This result gives a probability of randomly selecting five adults and finding that the first three know Twitter and the last two do not. However, the probability of 0.0138 is not the probability of getting exactly three adults who know Twitter because it was found by assuming a particular sequence. Other different sequences are possible.

In Section 4-4 we saw that with three subjects identical to each other (such as adults who know Twitter) and two other subjects identical to each other (such as adults who do not know Twitter), the total number of arrangements, or permutations, is $5!/[(5-3)!\,3!]$ or 10. Each of those 10 different arrangements has a probability of $0.85^3 \cdot 0.15^2$, so the total probability is as follows:

$$P(3 \text{ adults know Twitter among 5}) = \frac{5!}{(5-3)!3!} \cdot 0.85^3 \cdot 0.15^2$$

$$= 0.138$$

This particular result can be generalized as the binomial probability formula (Formula 5-5). That is, the binomial probability formula is a combination of the multiplication rule of probability and the counting rule for the number of arrangements of n items when x of them are identical to each other and the other $n - x$ are identical to each other.

The number of outcomes with exactly x successes among n trials

The probability of x successes among n trials for any one particular order

$$P(x) = \frac{n!}{(n-x)!x!} \cdot p^x \cdot q^{n-x}$$

Go Figure

$5: Cost of the ticket on the first commercial airline flight in the United States in 1914, which flew 21 miles.

Binomial Distributions
Access tech supplements, videos, and data sets at **www.TriolaStats.com**

Statdisk	Minitab	StatCrunch
1. Click **Analysis** in the top menu. 2. Select **Probability Distributions** from the dropdown menu and select **Binomial Distribution** from the submenu. 3. Enter the values for *n, p* and click **Evaluate**. *Tip:* Enter a specific value for *x* to get a single probability.	1. Enter the values of *x* for which you want probabilities (such as 0, 1, 2, 3, 4, 5) in column C1. 2. Select **Calc** from the top menu. 3. Select **Probability Distributions** from the dropdown menu and **Binomial** from the submenu. 4. Select **Probability,** enter the number of trials, enter the event probability, and select **C1** for *Input Column*. 5. Click **OK**.	1. Click **Stat** in the top menu. 2. Select **Calculators** from the dropdown menu and **Binomial** from the submenu. 3. In the dialog box, enter the desired values for *n, p, x*. Select = or the desired inequality for *x*. 4. Click **Compute**.

TI-83/84 Plus Calculator	Excel
1. Press **2ND** then **VARS** keys to access the *DISTR* (distributions) menu. 2. Select **binompdf** and click **ENTER**. 3. Enter the values for trials *n*, probability *p,* and number of successes *x* to complete the command **binompdf(*n, p, x*)**. Press **ENTER**. *Tip:* Omitting a value for *x* provides a list for all probabilities corresponding to *x* = 0, 1, 2...*n*. Press **STO** then **2ND** then ② to save the probabilities as list *L2*. You can then manually enter the values of *x* in list *L1* for calculations. *Tip:* Select **binomcdf** in Step 2 for *cumulative* probabilities.	1. Enter the values of *x* for which you want probabilities (such as 0, 1, 2, 3, 4, 5) in column A. 2. Select cell **B1**, click **Insert Function** *f*ₓ, select the category **Statistical**, select the function **BINOM.DIST** and click **OK**. 3. Enter **A1** for *Number_s* and then enter the number of trials *n* and probability *p*. 4. Enter **0** in the *Cumulative* box. 5. Click **OK** and the probability will appear in cell B1. 6. Copy **B1** down the column to obtain the probability for each value of *x* listed in column A. *Tip:* Enter **1** in Step 4 for the *cumulative* binomial distribution.

5-2 Basic Skills and Concepts

Statistical Literacy and Critical Thinking

1. Drone Deliveries Based on a Pitney Bowes survey, assume that 42% of consumers are comfortable having drones deliver their purchases. Suppose we want to find the probability that when five consumers are randomly selected, exactly two of them are comfortable with the drones. What is wrong with using the multiplication rule to find the probability of getting two consumers comfortable with drones followed by three consumers not comfortable, as in this calculation: $(0.42)(0.42)(0.58)(0.58)(0.58) = 0.0344$?

2. Notation Assume that we want to find the probability that when five consumers are randomly selected, exactly two of them are comfortable with delivery by drones. Also assume that 42% of consumers are comfortable with the drones (based on a Pitney Bowes survey). Identify the values of *n, x, p,* and *q*.

3. Independent Events Based on a Pitney Bowes survey, when 1009 consumers were asked if they are comfortable with drones delivering their purchases, 42% said yes. Consider the probability that among 30 different consumers randomly selected from the 1009 who were surveyed, there are at least 10 who are comfortable with the drones. Given that the subjects surveyed were selected without replacement, are the 30 selections independent? Can they be treated as being independent? Can the probability be found by using the binomial probability formula? Explain.

4. Notation of 0 + Using the same survey from Exercise 3, the probability of randomly selecting 30 of the 1009 consumers and getting exactly 24 who are comfortable with the drones is represented as 0+. What does 0+ indicate? Does 0+ indicate that it is impossible to get exactly 24 consumers who are comfortable with the drones?

Identifying Binomial Distributions.   *In Exercises 5–12, determine whether the given procedure results in a binomial distribution (or a distribution that can be treated as binomial). For those that are not binomial, identify at least one requirement that is not satisfied.*

5. Clinical Trial of YSORT The YSORT method of gender selection, developed by the Genetics & IVF Institute, was designed to increase the likelihood that a baby will be a boy. When 291 couples used the YSORT method and gave birth to 291 babies, the weights of the babies were recorded.

6. Clinical Trial of YSORT The YSORT method of gender selection, developed by the Genetics & IVF Institute, was designed to increase the likelihood that a baby will be a boy. When 291 couples use the YSORT method and give birth to 291 babies, the genders of the babies are recorded.

7. LOL In a U.S. Cellular survey of 500 smartphone users, subjects are asked if they find abbreviations (such as LOL or BFF) annoying, and each response was recorded as "yes" or "other."

8. LOL In a U.S. Cellular survey of 500 smartphone users, subjects are asked if they find abbreviations (such as LOL or BFF) annoying, and each response was recorded as "yes," "no," or "not sure."

9. Surveying Senators The Senate members of the 113th Congress include 80 males and 20 females. Forty different senators are randomly selected without replacement, and the gender of each selected senator is recorded.

10. Surveying Senators Ten different senators from the 113th Congress are randomly selected without replacement, and the numbers of terms that they have served are recorded.

11. Credit Card Survey In an *AARP Bulletin* survey, 1019 different adults were randomly selected without replacement. Respondents were asked if they have one or more credit cards, and responses were recorded as "yes" and "no."

12. Investigating Dates In a survey sponsored by TGI Friday's, 1000 different adult respondents were randomly selected without replacement, and each was asked if they investigate dates on social media before meeting them. Responses consist of "yes" or "no."

Binomial Probability Formula.   *In Exercises 13 and 14, answer the questions designed to help understand the rationale for the binomial probability formula.*

13. Guessing Answers Standard tests, such as the SAT, ACT, or Medical College Admission Test (MCAT), typically use multiple choice questions, each with five possible answers (a, b, c, d, e), one of which is correct. Assume that you guess the answers to the first three questions.

a. Use the multiplication rule to find the probability that the first two guesses are wrong and the third is correct. That is, find $P(\text{WWC})$, where W denotes a wrong answer and C denotes a correct answer.

b. Beginning with WWC, make a complete list of the different possible arrangements of two wrong answers and one correct answer, then find the probability for each entry in the list.

c. Based on the preceding results, what is the probability of getting exactly one correct answer when three guesses are made?

14. News Source Based on data from a Harris Interactive survey, 40% of adults say that they prefer to get their news online. Four adults are randomly selected.

continued

a. Use the multiplication rule to find the probability that the first three prefer to get their news online and the fourth prefers a different source. That is, find $P(OOOD)$, where O denotes a preference for online news and D denotes a preference for a news source different from online.

b. Beginning with OOOD, make a complete list of the different possible arrangements of those four letters, then find the probability for each entry in the list.

c. Based on the preceding results, what is the probability of getting exactly three adults who prefer to get their news online and one adult who prefers a different news source.

SAT Test.   *In Exercises 15–20, assume that random guesses are made for eight multiple choice questions on an SAT test, so that there are n = 8 trials, each with probability of success (correct) given by p = 0.20. Find the indicated probability for the number of correct answers.*

15. Find the probability that the number x of correct answers is exactly 7.

16. Find the probability that the number x of correct answers is at least 4.

17. Find the probability that the number x of correct answers is fewer than 3.

18. Find the probability that the number x of correct answers is no more than 2.

19. Find the probability of no correct answers.

20. Find the probability that at least one answer is correct.

In Exercises 21–24, assume that when adults with smartphones are randomly selected, 54% use them in meetings or classes (based on data from an LG Smartphone survey).

21. If 8 adult smartphone users are randomly selected, find the probability that exactly 6 of them use their smartphones in meetings or classes.

22. If 20 adult smartphone users are randomly selected, find the probability that exactly 15 of them use their smartphones in meetings or classes.

23. If 10 adult smartphone users are randomly selected, find the probability that at least 8 of them use their smartphones in meetings or classes.

24. If 12 adult smartphone users are randomly selected, find the probability that fewer than 3 of them use their smartphones in meetings or classes.

In Exercises 25–28, find the probabilities and answer the questions.

25. *Whitus v. Georgia* In the classic legal case of *Whitus v. Georgia*, a jury pool of 90 people was supposed to be randomly selected from a population in which 27% were minorities. Among the 90 people selected, 7 were minorities. Find the probability of getting 7 or fewer minorities if the jury pool was randomly selected. Is the result of 7 minorities significantly low? What does the result suggest about the jury selection process?

26. Vision Correction A survey sponsored by the Vision Council showed that 79% of adults need correction (eyeglasses, contacts, surgery, etc.) for their eyesight. If 20 adults are randomly selected, find the probability that at least 19 of them need correction for their eyesight. Is 19 a significantly high number of adults requiring eyesight correction?

27. See You Later Based on a Harris Interactive poll, 20% of adults believe in reincarnation. Assume that six adults are randomly selected, and find the indicated probability.

a. What is the probability that exactly five of the selected adults believe in reincarnation?

b. What is the probability that all of the selected adults believe in reincarnation?

c. What is the probability that at least five of the selected adults believe in reincarnation?

d. If six adults are randomly selected, is five a significantly high number who believe in reincarnation?

28. Too Young to Tat Based on a Harris poll, among adults who regret getting tattoos, 20% say that they were too young when they got their tattoos. Assume that five adults who regret getting tattoos are randomly selected, and find the indicated probability.

a. Find the probability that none of the selected adults say that they were too young to get tattoos.

b. Find the probability that exactly one of the selected adults says that he or she was too young to get tattoos.

c. Find the probability that the number of selected adults saying they were too young is 0 or 1.

d. If we randomly select five adults, is 1 a significantly low number who say that they were too young to get tattoos?

Significance with Range Rule of Thumb.   *In Exercises 29 and 30, assume that different groups of couples use the XSORT method of gender selection and each couple gives birth to one baby. The XSORT method is designed to increase the likelihood that a baby will be a girl, but assume that the method has no effect, so the probability of a girl is 0.5.*

29. Gender Selection Assume that the groups consist of 36 couples.

a. Find the mean and standard deviation for the numbers of girls in groups of 36 births.

b. Use the range rule of thumb to find the values separating results that are significantly low or significantly high.

c. Is the result of 26 girls a result that is significantly high? What does it suggest about the effectiveness of the XSORT method?

30. Gender Selection Assume that the groups consist of 16 couples.

a. Find the mean and standard deviation for the numbers of girls in groups of 16 births.

b. Use the range rule of thumb to find the values separating results that are significantly low or significantly high.

c. Is the result of 11 girls a result that is significantly high? What does it suggest about the effectiveness of the XSORT method?

Significance with Range Rule of Thumb.   *In Exercises 31 and 32, assume that hybridization experiments are conducted with peas having the property that for offspring, there is a 0.75 probability that a pea has green pods (as in one of Mendel's famous experiments).*

31. Hybrids Assume that offspring peas are randomly selected in groups of 10.

a. Find the mean and standard deviation for the numbers of peas with green pods in the groups of 10.

b. Use the range rule of thumb to find the values separating results that are significantly low or significantly high.

c. Is the result of 9 peas with green pods a result that is significantly high? Why or why not?

32. Hybrids Assume that offspring peas are randomly selected in groups of 16.

a. Find the mean and standard deviation for the numbers of peas with green pods in the groups of 16.

b. Use the range rule of thumb to find the values separating results that are significantly low or significantly high.

c. Is a result of 7 peas with green pods a result that is significantly low? Why or why not?

Composite Sampling.   *Exercises 33 and 34 involve the method of composite sampling, whereby a medical testing laboratory saves time and money by combining blood samples for tests so that only one test is conducted for several people. A combined sample tests positive* *if at least one person has the disease. If a combined sample tests positive, then individual blood tests are used to identify the individual with the disease or disorder.*

33. HIV It is estimated that worldwide, 1% of those aged 15–49 are infected with the human immunodeficiency virus (HIV) (based on data from the National Institutes of Health). In tests for HIV, blood samples from 36 people are combined. What is the probability that the combined sample tests positive for HIV? Is it unlikely for such a combined sample to test positive?

34. Anemia Based on data from Bloodjournal.org, 10% of women 65 years of age and older have anemia, which is a deficiency of red blood cells. In tests for anemia, blood samples from 8 women 65 and older are combined. What is the probability that the combined sample tests positive for anemia? Is it likely for such a combined sample to test positive?

Acceptance Sampling.   *Exercises 35 and 36 involve the method of acceptance sampling, whereby a shipment of a large number of items is accepted based on test results from a sample of the items.*

35. Aspirin The MedAssist Pharmaceutical Company receives large shipments of aspirin tablets and uses this acceptance sampling plan: Randomly select and test 40 tablets, then accept the whole batch if there is only one or none that doesn't meet the required specifications. If one shipment of 5000 aspirin tablets actually has a 3% rate of defects, what is the probability that this whole shipment will be accepted? Will almost all such shipments be accepted, or will many be rejected?

36. AAA Batteries AAA batteries are made by companies including Duracell, Energizer, Eveready, and Panasonic. When purchasing bulk orders of AAA batteries, a toy manufacturer uses this acceptance sampling plan: Randomly select 50 batteries and determine whether each is within specifications. The entire shipment is accepted if at most 2 batteries do not meet specifications. A shipment contains 2000 AAA batteries, and 2% of them do not meet specifications. What is the probability that this whole shipment will be accepted? Will almost all such shipments be accepted, or will many be rejected?

Ultimate Binomial Exercises!   *Exercises 37–40 involve finding binomial probabilities, finding parameters, and determining whether values are significantly high or low by using the range rule of thumb and probabilities.*

37. M&Ms Data Set 27 "M&M Weights" in Appendix B includes data from 100 M&M candies, and 19 of them are green. Mars, Inc. claims that 16% of its plain M&M candies are green. For the following, assume that the claim of 16% is true, and assume that a sample consists of 100 M&Ms.

a. Use the range rule of thumb to identify the limits separating values that are significantly low and those that are significantly high. Based on the results, is the result of 19 green M&Ms significantly high?

b. Find the probability of exactly 19 green M&Ms.

c. Find the probability of 19 or more green M&Ms.

d. Which probability is relevant for determining whether the result of 19 green M&Ms is significantly high: the probability from part (b) or part (c)? Based on the relevant probability, is the result of 19 green M&Ms significantly high?

e. What do the results suggest about the 16% claim by Mars, Inc.?

38. Politics The County Clerk in Essex, New Jersey, was accused of cheating by not using randomness in assigning line positions on voting ballots. Among 41 different ballots, Democrats were assigned the top line 40 times. Assume that Democrats and Republicans are assigned the top line using a method of random selection so that they are equally likely to get that top line.

a. Use the range rule of thumb to identify the limits separating values that are significantly low and those that are significantly high. Based on the results, is the result of 40 top lines for Democrats significantly high?

b. Find the probability of exactly 40 top lines for Democrats.

c. Find the probability of 40 or more top lines for Democrats.

d. Which probability is relevant for determining whether 40 top lines for Democrats is significantly high: the probability from part (b) or part (c)? Based on the relevant probability, is the result of 40 top lines for Democrats significantly high?

e. What do the results suggest about how the clerk met the requirement of assigning the line positions using a random method?

39. Perception and Reality In a presidential election, 611 randomly selected voters were surveyed, and 308 of them said that they voted for the winning candidate (based on data from ICR Survey Research Group). The actual percentage of votes for the winning candidate was 43%. Assume that 43% of voters actually did vote for the winning candidate, and assume that 611 voters are randomly selected.

a. Use the range rule of thumb to identify the limits separating values that are significantly low and those that are significantly high. Based on the results, is the 308 voters who said that they voted for the winner significantly high?

b. Find the probability of exactly 308 voters who actually voted for the winner.

c. Find the probability of 308 or more voters who actually voted for the winner.

d. Which probability is relevant for determining whether the value of 308 voters is significantly high: the probability from part (b) or part (c)? Based on the relevant probability, is the result of 308 voters who said that they voted for the winner significantly high?

e. What is an important observation about the survey results?

40. Hybrids One of Mendel's famous experiments with peas resulted in 580 offspring, and 152 of them were yellow peas. Mendel claimed that under the same conditions, 25% of offspring peas would be yellow. Assume that Mendel's claim of 25% is true, and assume that a sample consists of 580 offspring peas.

a. Use the range rule of thumb to identify the limits separating values that are significantly low and those that are significantly high. Based on the results, is the result of 152 yellow peas either significantly low or significantly high?

b. Find the probability of exactly 152 yellow peas.

c. Find the probability of 152 or more yellow peas.

d. Which probability is relevant for determining whether 152 peas is significantly high: the probability from part (b) or part (c)? Based on the relevant probability, is the result of 152 yellow peas significantly high?

e. What do the results suggest about Mendel's claim of 25%?

5-2   Beyond the Basics

41. Geometric Distribution If a procedure meets all the conditions of a binomial distribution except that the number of trials is not fixed, then the **geometric distribution** can be used. The probability of getting the first success on the xth trial is given by $P(x) = p(1 - p)^{x-1}$, where p is the probability of success on any one trial. Subjects are randomly selected for the National Health and Nutrition Examination Survey conducted by the National Center for Health Statistics, Centers for Disease Control and Prevention. The probability that someone is a universal donor (with group O and type Rh negative blood) is 0.06. Find the probability that the first subject to be a universal blood donor is the fifth person selected.

42. Multinomial Distribution The binomial distribution applies only to cases involving two types of outcomes, whereas the **multinomial distribution** involves more than two categories. Suppose we have three types of mutually exclusive outcomes denoted by A, B, and C. Let $P(A) = p_1$, $P(B) = p_2$, and $P(C) = p_3$. In n independent trials, the probability of x_1 outcomes of type A, x_2 outcomes of type B, and x_3 outcomes of type C is given by

$$\frac{n!}{(x_1)!(x_2)!(x_3)!} \cdot p_1^{x_1} \cdot p_2^{x_2} \cdot p_3^{x_3}$$

A roulette wheel in the Venetian casino in Las Vegas has 18 red slots, 18 black slots, and 2 green slots. If roulette is played 15 times, find the probability of getting 7 red outcomes, 6 black outcomes, and 2 green outcomes.

43. Hypergeometric Distribution If we sample from a small finite population without replacement, the binomial distribution should not be used because the events are not independent. If sampling is done without replacement and the outcomes belong to one of two types, we can use the **hypergeometric distribution**. If a population has A objects of one type (such as lottery numbers you selected), while the remaining B objects are of the other type (such as lottery numbers you didn't select), and if n objects are sampled without replacement (such as six drawn lottery numbers), then the probability of getting x objects of type A and $n - x$ objects of type B is

$$P(x) = \frac{A!}{(A - x)!x!} \cdot \frac{B!}{(B - n + x)!(n - x)!} \div \frac{(A + B)!}{(A + B - n)!n!}$$

In New Jersey's Pick 6 lottery game, a bettor selects six numbers from 1 to 49 (without repetition), and a winning six-number combination is later randomly selected. Find the probabilities of getting exactly two winning numbers with one ticket. (*Hint:* Use $A = 6$, $B = 43$, $n = 6$, and $x = 2$.)

5-3 Poisson Probability Distributions

Key Concept In Section 5-1 we introduced general discrete probability distributions and in Section 5-2 we considered binomial probability distributions, which is one particular category of discrete probability distributions. In this section we introduce *Poisson probability distributions,* which are another category of discrete probability distributions.

The following definition states that Poisson distributions are used with occurrences of an event over a specified interval, and here are some applications:

- Number of Internet users logging onto a website in one day
- Number of patients arriving at an emergency room in one hour
- Number of Atlantic hurricanes in one year

DEFINITION

A **Poisson probability distribution** is a discrete probability distribution that applies to occurrences of some event *over a specified interval*. The random variable x is the number of occurrences of the event in an interval. The interval can be time, distance, area, volume, or some similar unit. The probability of the event occurring x times over an interval is given by Formula 5-9.

FORMULA 5-9   Poisson Probability Distribution

$$P(x) = \frac{\mu^x \cdot e^{-\mu}}{x!}$$

where $e \approx 2.71828$

μ = mean number of occurrences of the event in the intervals

Go Figure

42: Number of years before we run out of oil.

Requirements for the Poisson Probability Distribution

1. The random variable x is the number of occurrences of an event *in some interval.*
2. The occurrences must be *random.*
3. The occurrences must be *independent* of each other.
4. The occurrences must be *uniformly distributed* over the interval being used.

Parameters of the Poisson Probability Distribution

- The mean is μ.
- The standard deviation is $\sigma = \sqrt{\mu}$.

Properties of the Poisson Probability Distribution

1. A particular Poisson distribution is determined only by the mean μ.
2. A Poisson distribution has possible x values of 0, 1, 2, . . . with no upper limit.

EXAMPLE 1   **Atlantic Hurricanes**

For the 55-year period since 1960, there were 336 Atlantic hurricanes. Assume that the Poisson distribution is a suitable model.

    a. Find μ, the mean number of hurricanes per year.

    b. Find the probability that in a randomly selected year, there are exactly 8 hurricanes. That is, find $P(8)$, where $P(x)$ is the probability of x Atlantic hurricanes in a year.

    c. In this 55-year period, there were actually 5 years with 8 Atlantic hurricanes. How does this actual result compare to the probability found in part (b)? Does the Poisson distribution appear to be a good model in this case?

SOLUTION

    a. The Poisson distribution applies because we are dealing with the occurrences of an event (hurricanes) over some interval (a year). The mean number of hurricanes per year is

$$\mu = \frac{\text{number of hurricanes}}{\text{number of years}} = \frac{336}{55} = 6.1$$

continued

b. Using Formula 5-9, the probability of $x = 8$ hurricanes in a year is as follows (with $x = 8$, $\mu = 6.1$, and $e = 2.71828$):

$$P(8) = \frac{\mu^x \cdot e^{-\mu}}{x!} = \frac{6.1^8 \cdot 2.71828^{-6.1}}{8!}$$

$$= \frac{(1{,}917{,}073.13)(0.0022428769)}{40{,}320} = 0.107$$

The probability of exactly 8 hurricanes in a year is $P(8) = 0.107$.

c. The probability of $P(8) = 0.107$ from part (b) is the likelihood of getting 8 Atlantic hurricanes in 1 year. In 55 years, the expected number of years with 8 Atlantic hurricanes is $55 \times 0.107 = 5.9$ years. The expected number of years with 8 hurricanes is 5.9, which is reasonably close to the 5 years that actually had 8 hurricanes, so in this case, the Poisson model appears to work quite well.

YOUR TURN Do Exercise 5 "Hurricanes."

Poisson Distribution as Approximation to Binomial

The Poisson distribution is sometimes used to approximate the binomial distribution when n is large and p is small. One rule of thumb is to use such an approximation when the following two requirements are both satisfied.

Requirements for Using Poisson as an Approximation to Binomial

1. $n \geq 100$

2. $np \leq 10$

If both requirements are satisfied and we want to use the Poisson distribution as an approximation to the binomial distribution, we need a value for μ. That value can be calculated by using Formula 5-6 (from Section 5-2):

FORMULA 5-6   Mean for Poisson as an Approximation to Binomial

$\mu = np$

EXAMPLE 2 Maine Pick 4

In the Maine Pick 4 game, you pay 50¢ to select a sequence of four digits (0–9), such as 1377. If you play this game once every day, find the probability of winning at least once in a year with 365 days.

SOLUTION

The time interval is a day, and playing once each day results in $n = 365$ games. Because there is one winning set of numbers among the 10,000 that are possible (from 0000 to 9999), the probability of a win is $p = 1/10{,}000$. With $n = 365$ and $p = 1/10{,}000$, the conditions $n \geq 100$ and $np \leq 10$ are both satisfied, so we can use the Poisson distribution as an approximation to the binomial distribution. We first need the value of μ, which is found as follows:

$$\mu = np = 365 \cdot \frac{1}{10{,}000} = 0.0365$$

Having found the value of μ, we can proceed to find the probability for specific values of x. Because we want the probability that x is "at least 1," we will use the clever strategy of first finding $P(0)$, the probability of no wins in 365 days. The probability of at least one win can then be found by subtracting that result from 1. We find $P(0)$ by using $x = 0$, $\mu = 0.0365$, and $e = 2.71828$, as shown here:

$$P(0) = \frac{\mu^x \cdot e^{-\mu}}{x!} = \frac{0.0365^0 \cdot 2.71828^{-0.0365}}{0!} = \frac{1 \cdot 0.9642}{1} = 0.9642$$

Using the Poisson distribution as an approximation to the binomial distribution, we find that there is a 0.9642 probability of no wins, so the probability of at least one win is $1 - 0.9642 = 0.0358$. If we use the binomial distribution, we get a probability of 0.0358, so the Poisson distribution works quite well here.

YOUR TURN Do Exercise 17 "Powerball: Poisson Approximation to Binomial."

TECH CENTER

Poisson Distributions
Access tech supplements, videos, and data sets at **www.TriolaStats.com**

Statdisk
1. Click **Analysis** in the top menu.
2. Select **Probability Distributions** from the dropdown menu and select **Poisson Distribution** from the submenu.
3. Enter the value of the mean and click **Evaluate.**

Minitab
1. Enter the values of x for which you want probabilities (such as 0, 1, 2, 3, 4, 5) in column C1.
2. Select **Calc** from the top menu.
3. Select **Probability Distributions** from dropdown menu and **Poisson** from the submenu.
4. Select **Probability,** enter the mean, and select **C1** for *Input Column*.
5. Click **OK.**

StatCrunch
1. Click **Stat** in the top menu.
2. Select **Calculators** from the dropdown menu and **Poisson** from the submenu.
3. In the dialog box enter the value of the mean and the value of x. Select = or the desired inequality for x.
4. Click **Compute.**

TI-83/84 Plus Calculator
1. Press **2ND** then **VARS** keys to access the *DISTR* (distributions) menu.
2. Select **poissonpdf** and press **ENTER**.
3. Enter the values for mean (μ) and x to complete the command **poissonpdf (μ, x).** Press **ENTER**.

Tip: Select **poissoncdf** in Step 2 for *cumulative* probability.

Excel
1. Enter the values of x for which you want probabilities (such as 0, 1, 2, 3, 4, 5) in column A.
2. Select cell **B1,** click **Insert Function** f_x, select the category **Statistical,** select the function **POISSON.DIST** and click **OK.**
3. Enter **A1** for X and then enter the value of the mean.
4. Enter **0** in the *Cumulative* box.
5. Click **OK** and the probability will appear in cell B1.
5. Copy B1 down the column to obtain the probability for each value of x listed in column A.

Tip: Enter **1** in Step 4 for the *cumulative* Poisson distribution.

5-3 Basic Skills and Concepts

Statistical Literacy and Critical Thinking

1. Notation In analyzing hits by V-1 buzz bombs in World War II, South London was partitioned into 576 regions, each with an area of 0.25 km^2. A total of 535 bombs hit the combined area of 576 regions. Assume that we want to find the probability that a randomly selected region had exactly two hits. In applying Formula 5-9, identify the values of μ, x, and e. Also, briefly describe what each of those symbols represents.

2. Tornadoes During a recent 64-year period, New Mexico had a total of 153 tornadoes that measured 1 or greater on the Fujita scale. Let the random variable x represent the number of such tornadoes to hit New Mexico in one year, and assume that it has a Poisson distribution. What is the mean number of such New Mexico tornadoes in one year? What is the standard deviation? What is the variance?

3. Poisson Probability Distribution The random variable x represents the number of phone calls the author receives in a day, and it has a Poisson distribution with a mean of 7.2 calls. What are the possible values of x? Is a value of $x = 2.3$ possible? Is x a discrete random variable or a continuous random variable?

4. Probability if 0 For Formula 5-9, what does $P(0)$ represent? Simplify Formula 5-9 for the case in which $x = 0$.

Hurricanes. *In Exercises 5–8, assume that the Poisson distribution applies; assume that the mean number of Atlantic hurricanes in the United States is 6.1 per year, as in Example 1; and proceed to find the indicated probability.*

5. Hurricanes

a. Find the probability that in a year, there will be 5 hurricanes.

b. In a 55-year period, how many years are expected to have 5 hurricanes?

c. How does the result from part (b) compare to the recent period of 55 years in which 8 years had 5 hurricanes? Does the Poisson distribution work well here?

6. Hurricanes

a. Find the probability that in a year, there will be no hurricanes.

b. In a 55-year period, how many years are expected to have no hurricanes?

c. How does the result from part (b) compare to the recent period of 55 years in which there were no years without any hurricanes? Does the Poisson distribution work well here?

7. Hurricanes

a. Find the probability that in a year, there will be 7 hurricanes.

b. In a 55-year period, how many years are expected to have 7 hurricanes?

c. How does the result from part (b) compare to the recent period of 55 years in which 7 years had 7 hurricanes? Does the Poisson distribution work well here?

8. Hurricanes

a. Find the probability that in a year, there will be 4 hurricanes.

b. In a 55-year period, how many years are expected to have 4 hurricanes?

c. How does the result from part (b) compare to the recent period of 55 years in which 10 years had 4 hurricanes? Does the Poisson distribution work well here?

In Exercises 9–16, use the Poisson distribution to find the indicated probabilities.

9. Births In a recent year, NYU-Langone Medical Center had 4221 births. Find the mean number of births per day, then use that result to find the probability that in a day, there are 15 births. Does it appear likely that on any given day, there will be exactly 15 births?

10. Murders In a recent year, there were 333 murders in New York City. Find the mean number of murders per day, then use that result to find the probability that in a day, there are no murders. Does it appear that there are expected to be many days with no murders?

11. Radioactive Decay Radioactive atoms are unstable because they have too much energy. When they release their extra energy, they are said to decay. When studying cesium-137, a nuclear engineer found that over 365 days, 1,000,000 radioactive atoms decayed to 977,287 radioactive atoms; therefore 22,713 atoms decayed during 365 days.

a. Find the mean number of radioactive atoms that decayed in a day.

b. Find the probability that on a given day, exactly 50 radioactive atoms decayed.

12. Deaths from Horse Kicks A classical example of the Poisson distribution involves the number of deaths caused by horse kicks to men in the Prussian Army between 1875 and 1894. Data for 14 corps were combined for the 20-year period, and the 280 corps-years included a total of 196 deaths. After finding the mean number of deaths per corps-year, find the probability that a randomly selected corps-year has the following numbers of deaths: (**a**) 0, (**b**) 1, (**c**) 2, (**d**) 3, (**e**) 4. The actual results consisted of these frequencies: 0 deaths (in 144 corps-years); 1 death (in 91 corps-years); 2 deaths (in 32 corps-years); 3 deaths (in 11 corps-years); 4 deaths (in 2 corps-years). Compare the actual results to those expected by using the Poisson probabilities. Does the Poisson distribution serve as a good tool for predicting the actual results?

13. World War II Bombs In Exercise 1 "Notation" we noted that in analyzing hits by V-1 buzz bombs in World War II, South London was partitioned into 576 regions, each with an area of 0.25 km^2. A total of 535 bombs hit the combined area of 576 regions.

a. Find the probability that a randomly selected region had exactly 2 hits.

b. Among the 576 regions, find the expected number of regions with exactly 2 hits.

c. How does the result from part (b) compare to this actual result: There were 93 regions that had exactly 2 hits?

14. Disease Cluster Neuroblastoma, a rare form of cancer, occurs in 11 children in a million, so its probability is 0.000011. Four cases of neuroblastoma occurred in Oak Park, Illinois, which had 12,429 children.

a. Assuming that neuroblastoma occurs as usual, find the mean number of cases in groups of 12,429 children.

b. Using the unrounded mean from part (a), find the probability that the number of neuroblastoma cases in a group of 12,429 children is 0 or 1.

c. What is the probability of more than one case of neuroblastoma?

d. Does the cluster of four cases appear to be attributable to random chance? Why or why not?

15. Car Fatalities The recent rate of car fatalities was 33,561 fatalities for 2969 billion miles traveled (based on data from the National Highway Traffic Safety Administration). Find the probability that for the next billion miles traveled, there will be at least one fatality. What does the result indicate about the likelihood of at least one fatality?

16. Checks In a recent year, the author wrote 181 checks. Find the probability that on a randomly selected day, he wrote at least one check.

5-3   Beyond the Basics

17. Powerball: Poisson Approximation to Binomial There is a $1/292,201,338$ probability of winning the Powerball lottery jackpot with a single ticket. Assume that you purchase a ticket in each of the next 5200 different Powerball games that are run over the next 50 years. Find the probability of winning the jackpot with at least one of those tickets. Is there a good chance that you would win the jackpot at least once in 50 years?

Chapter Quick Quiz

1. Is a probability distribution defined if the only possible values of a random variable are 0, 1, 2, 3, and $P(0) = P(1) = P(2) = P(3) = 1/3$?

2. There are 80 questions from an SAT test, and they are all multiple choice with possible answers of a, b, c, d, e. For each question, only one answer is correct. Find the mean and standard deviation for the numbers of correct answers for those who make random guesses for all 80 questions.

3. Are the values found in Exercise 2 statistics or parameters? Why?

4. Using the same SAT questions described in Exercise 2, is 20 a significantly high number of correct answers for someone making random guesses?

5. Using the same SAT questions described in Exercise 2, is 8 a significantly low number of correct answers for someone making random guesses?

x	P(x)
0	0+
1	0.006
2	0.051
3	0.205
4	0.409
5	0.328

In Exercises 6–10, use the following: Five American Airlines flights are randomly selected, and the table in the margin lists the probabilities for the number that arrive on time (based on data from the Department of Transportation). Assume that five flights are randomly selected.

6. Does the table describe a probability distribution?

7. Find the mean of the number of flights among five that arrive on time.

8. Based on the table, the standard deviation is 0.9 flight. What is the variance? Include appropriate units.

9. What does the probability of 0+ indicate? Does it indicate that among five randomly selected flights, it is impossible for none of them to arrive on time?

10. What is the probability that fewer than three of the five flights arrive on time?

Review Exercises

In Exercises 1–5, assume that 74% of randomly selected adults have a credit card (based on results from an **AARP Bulletin** *survey). Assume that a group of five adults is randomly selected.*

1. Credit Cards Find the probability that exactly three of the five adults have credit cards.

2. Credit Cards Find the probability that at least one of the five adults has a credit card. Does the result apply to five adult friends who are vacationing together? Why or why not?

3. Credit Cards Find the mean and standard deviation for the numbers of adults in groups of five who have credit cards.

4. Credit Cards If all five of the adults have credit cards, is five significantly high? Why or why not?

5. Credit Cards If the group of five adults includes exactly 1 with a credit card, is that value of 1 significantly low?

6. Security Survey In a *USA Today* poll, subjects were asked if passwords should be replaced with biometric security, such as fingerprints. The results from that poll have been used to create the accompanying table. Does this table describe a probability distribution? Why or why not?

Response	P(x)
Yes	0.53
No	0.17
Not Sure	0.30

7. Brand Recognition In a study of brand recognition of Sony, groups of four consumers are interviewed. If x is the number of people in the group who recognize the Sony brand name, then x can be 0, 1, 2, 3, or 4, and the corresponding probabilities are 0.0016, 0.0250, 0.1432, 0.3892, and 0.4096. Does the given information describe a probability distribution? Why or why not?

8. Family/Partner Groups of people aged 15–65 are randomly selected and arranged in groups of six. The random variable x is the number in the group who say that their family and/or partner contribute most to their happiness (based on a Coca-Cola survey). The accompanying table lists the values of x along with their corresponding probabilities. Does the table describe a probability distribution? If so, find the mean and standard deviation.

x	P(x)
0	0+
1	0.003
2	0.025
3	0.111
4	0.279
5	0.373
6	0.208

9. Detecting Fraud The Brooklyn District Attorney's office analyzed the leading (leftmost) digits of check amounts in order to identify fraud. The leading digit of 1 is expected to occur 30.1% of the time, according to "Benford's law," which applies in this case. Among 784 checks issued by a suspect company, there were none with amounts that had a leading digit of 1.

a. If there is a 30.1% chance that the leading digit of the check amount is 1, what is the expected number of checks among 784 that should have a leading digit of 1?

b. Assume that groups of 784 checks are randomly selected. Find the mean and standard deviation for the numbers of checks with amounts having a leading digit of 1.

c. Use the results from part (b) and the range rule of thumb to identify the values that are significantly low.

d. Given that the 784 actual check amounts had no leading digits of 1, is there very strong evidence that the suspect checks are very different from the expected results? Why or why not?

10. Poisson: Deaths Currently, an average of 7 residents of the village of Westport (population 760) die each year (based on data from the U.S. National Center for Health Statistics).

a. Find the mean number of deaths per day.

b. Find the probability that on a given day, there are no deaths.

c. Find the probability that on a given day, there is more than one death.

d. Based on the preceding results, should Westport have a contingency plan to handle more than one death per day? Why or why not?

Cumulative Review Exercises

1. Planets The planets of the solar system have the numbers of moons listed below in order from the sun. (Pluto is not included because it was uninvited from the solar system party in 2006.) Include appropriate units whenever relevant.

$$0 \quad 0 \quad 1 \quad 2 \quad 17 \quad 28 \quad 21 \quad 8$$

a. Find the mean.

b. Find the median.

c. Find the mode.

continued

d. Find the range.

e. Find the standard deviation.

f. Find the variance.

g. Use the range rule of thumb to identify the values separating significant values from those that are not significant.

h. Based on the result from part (g), do any of the planets have a number of moons that is significantly low or significantly high? Why or why not?

i. What is the level of measurement of the data: nominal, ordinal, interval, or ratio?

j. Are the data discrete or continuous?

2. South Carolina Pick 3 In South Carolina's Pick 3 lottery game, you can pay $1 to select a sequence of three digits, such as 227. If you buy only one ticket and win, your prize is $500 and your net gain is $499.

a. If you buy one ticket, what is the probability of winning?

b. If you play this game once every day, find the mean number of wins in years with exactly 365 days.

c. If you play this game once every day, find the probability of winning exactly once in 365 days.

d. Find the expected value for the purchase of one ticket.

3. Tennis Challenge In a recent U.S. Open tennis tournament, there were 879 challenges made by singles players, and 231 of them resulted in referee calls that were overturned. The accompanying table lists the results by gender.

	Challenge Upheld with Overturned Call	Challenge Rejected with No Change
Challenges by Men	152	412
Challenges by Women	79	236

a. If 1 of the 879 challenges is randomly selected, what is the probability that it resulted in an overturned call?

b. If one of the overturned calls is randomly selected, what is the probability that the challenge was made by a woman?

c. If two different challenges are randomly selected without replacement, find the probability that they both resulted in an overturned call.

d. If 1 of the 879 challenges is randomly selected, find the probability that it was made by a man or was upheld with an overturned call.

e. If one of the challenges is randomly selected, find the probability that it was made by a man, given that the challenge was upheld with an overturned call.

4. Job Applicants The Society for Human Resource Management conducted a survey of 347 human resource professionals and found that 73% of them reported that their companies do criminal background checks of all job applicants.

a. Find the number of respondents who reported that their companies do criminal background checks of all job applicants.

b. Identify the sample and the population.

c. Is the value of 73% a statistic or a parameter? Explain.

5. Bar Graph Fox News broadcast a graph similar to the one shown here. The graph is intended to compare the number of people actually enrolled in a government health plan (left bar) and the goal for the number of enrollees (right bar). Does the graph depict the data correctly or is it somehow misleading? Explain.

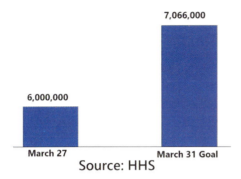

Source: HHS

6. Washing Hands Based on results from a Bradley Corporation poll, assume that 70% of adults always wash their hands after using a public restroom.

a. Find the probability that among 8 randomly selected adults, exactly 5 always wash their hands after using a public restroom.

b. Find the probability that among 8 randomly selected adults, at least 7 always wash their hands after using a public restroom.

c. For groups of 8 randomly selected adults, find the mean and standard deviation of the numbers in the groups who always wash their hands after using a public restroom.

d. If 8 adults are randomly selected and it is found that exactly 1 of them washes hands after using a public restroom, is that a significantly low number?

Technology Project

Overbooking Flights American Airlines Flight 171 from New York's JFK airport to LAX airport in Los Angeles uses an Airbus A321 aircraft with 189 seats available for passengers. American Airlines can overbook by accepting more reservations than there are seats available. If the flight is not overbooked, the airline loses revenue from empty seats, but if too many seats are sold, the airline loses money from the compensation it must pay to the bumped passengers. Assume that there is a 0.0995 probability that a passenger with a reservation will not show up for the flight (based on data from the IBM research paper "Passenger-Based Predictive Modeling of Airline No-Show Rates," by Lawrence, Hong, and Cherrier). Also assume that American Airlines accepts 205 reservations for the 189 seats that are available.

• Find the probability that when 205 reservations are accepted for American Airlines Flight 171, there are more passengers showing up than there are seats available. Is the probability of overbooking small enough so that it does not happen very often, or does it seem too high so that changes must be made to make it lower?

• Use trial and error to find the maximum number of reservations that could be accepted so that the probability of having more passengers than seats is 0.05 or less.

FROM DATA TO DECISION

Critical Thinking: Did Mendel's results from plant hybridization experiments contradict his theory?

Gregor Mendel conducted original experiments to study the genetic traits of pea plants. In 1865 he wrote "Experiments in Plant Hybridization," which was published in *Proceedings of the Natural History Society*. Mendel presented a theory that when there are two inheritable traits, one of them will be dominant and the other will be recessive. Each parent contributes one gene to an offspring and, depending on the combination of genes, that offspring could inherit the dominant trait or the recessive trait. Mendel conducted an experiment using pea plants. The pods of pea plants can be green or yellow. When one pea carrying a dominant green gene and a recessive yellow gene is crossed with another pea carrying the same green/yellow genes, the offspring can inherit any one of these four combinations of genes: (1) green/green; (2) green/yellow; (3) yellow/green; (4) yellow/yellow. Because green is dominant and yellow is recessive, the offspring pod will be green if either of the two inherited genes is green. The offspring can have a yellow pod only if it inherits the yellow gene from each of the two parents. Given these conditions, we expect that 3/4 of the offspring peas should have green pods; that is, $P(\text{green pod}) = 3/4$.

When Mendel conducted his famous hybridization experiments using parent pea plants with the green/yellow combination of genes, he obtained 580 offspring. According to Mendel's theory, 3/4 of the offspring should have green pods, but the actual number of plants with green pods was 428. So the proportion of offspring with green pods to the total number of offspring is $428/580 = 0.738$. Mendel *expected* a proportion of 3/4 or 0.75, but his *actual result* is a proportion of 0.738.

a. Assuming that $P(\text{green pod}) = 3/4$, find the probability that among 580 offspring, the number of peas with green pods is *exactly* 428.

b. Assuming that $P(\text{green pod}) = 3/4$, find the probability that among 580 offspring, the number of peas with green pods is 428 *or fewer*.

c. Which of the two preceding probabilities should be used for determining whether 428 is a significantly low number of peas with green pods?

d. Use probabilities to determine whether 428 peas with green pods is a significantly low number. (*Hint:* See "Identifying Significant Results with Probabilities" in Section 5-1.)

Cooperative Group Activities

1. In-class activity Win $1,000,000! The James Randi Educational Foundation offers a $1,000,000 prize to anyone who can show "under proper observing conditions, evidence of any paranormal, supernatural, or occult power or event." Divide into groups of three. Select one person who will be tested for extrasensory perception (ESP) by trying to correctly identify a digit (0–9) randomly selected by another member of the group. Conduct at least 20 trials. Another group member should record the randomly selected digit, the digit guessed by the subject, and whether the guess was correct or wrong. Construct the table for the probability distribution of randomly generated digits, construct the relative frequency table for the random digits that were actually obtained, and construct a relative frequency table for the guesses that were made. After comparing the three tables, what do you conclude? What proportion of guesses is correct? Does it seem that the subject has the ability to select the correct digit significantly more often than would be expected by chance?

2. In-class activity See the preceding activity and *design an experiment* that would be effective in testing someone's claim that he or she has the ability to identify the color of a card selected from a standard deck of playing cards. Describe the experiment with great detail. Because the prize of $1,000,000 is at stake, we want to be careful to avoid the serious mistake of concluding that the person has a paranormal power when that power is not actually present. There will likely be some chance that the subject could make random guesses and be correct every time, so identify a probability that is reasonable for the event of the subject passing the test with random guesses. Be sure that the test is designed so that this probability is equal to or less than the probability value selected.

3. In-class activity Suppose we want to identify the probability distribution for the number of children in families with at least one child. For each student in the class, find the number of brothers and sisters and record the total number of children (including the student) in each family. Construct the relative frequency table for the result obtained. (The values of the random variable x will be 1, 2, 3, ...) What is wrong with using this relative frequency table as an estimate of the probability distribution for the number of children in randomly selected families?

4. Out-of-class activity The analysis of the last digits of data can sometimes reveal whether the data have been collected through actual measurements or reported by the subjects. Refer to an almanac or the Internet and find a collection of data (such as lengths of rivers in the world), then analyze the distribution of last digits to determine whether the values were obtained through actual measurements.

5. Out-of-class activity In the past, leading (leftmost) digits of the amounts on checks have been analyzed for fraud. For checks not involving fraud, the leading digit of 1 is expected about 30.1% of the time. Obtain a random sample of actual check amounts and record the leading digits. Compare the actual number of checks with amounts that have a leading digit of 1 to the 30.1% rate expected. Do the actual checks conform to the expected rate, or is there a substantial discrepancy? Explain.

6. Out-of-class activity The photos shown below depict famous statisticians with first names of David and John, not necessarily in the order shown. Conduct a survey by asking this question: "Which man is named David and which man is named John?" Do the respondents appear to give results significantly different from what is expected with random guesses? (See "Who Do You Look Like? Evidence of Facial Stereotypes for Male Names," by Lea, Thomas, Lamkin, and Bell, *Psychonomic Bulletin & Review,* Vol. 14, Issue 5.)

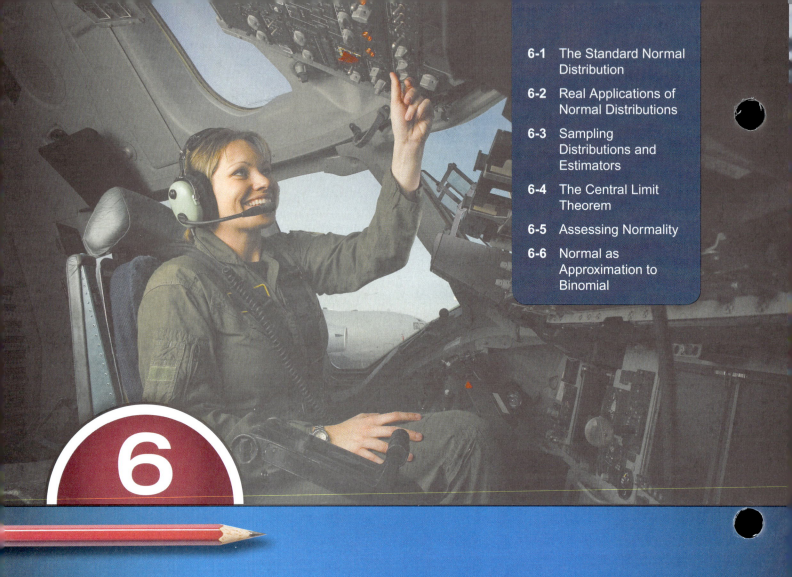

6-1 The Standard Normal Distribution

6-2 Real Applications of Normal Distributions

6-3 Sampling Distributions and Estimators

6-4 The Central Limit Theorem

6-5 Assessing Normality

6-6 Normal as Approximation to Binomial

6

NORMAL PROBABILITY DISTRIBUTIONS

Ergonomics 101

This chapter introduces statistical tools with many important real-world applications, including ergonomics. *Ergonomics* is a discipline focused on the design of tools and equipment so that they can be used safely, comfortably, and efficiently. Here are examples of ergonomic problems:

- Section 4.4.2 of the Americans with Disabilities Act relates to height clearances with this statement: "Walks, halls, corridors, passageways, aisles, or other circulation spaces

shall have 80 in. (2030 mm) minimum clear head room." What percentage of male adults are taller than 80 in.?

- British Airways and many other airline carriers have a requirement that cabin crew must have heights between 62 in. and 73 in. What percentage of adult females are shorter than 62 in.?

- The U.S. Army requires that women must be between 58 in. and 80 in. tall. What percentage of women meet this requirement?

- The elevator in the San Francisco Airport rental car facility has a placard indicating a maximum load of 4000 lb or 27 passengers. How likely is it that 27 random passengers will exceed the maximum load of 4000 lb?

Ergonomic problems often involve extremely important safety issues, and here are real cases that proved to be fatal:

- "We have an emergency for Air Midwest fifty-four eighty," said pilot Katie Leslie, just before her Beech plane crashed in Charlotte, North Carolina, resulting in the death of all 21 crew and passengers. Excessive total weight of the passengers was suspected as a factor that contributed to the crash.

- After 20 passengers perished when the *Ethan Allen* tour boat capsized on New York's Lake George, an investigation showed that although the number of passengers was below the maximum allowed, the boat should have been certified for a much smaller number of passengers.

- A water taxi sank in Baltimore's Inner Harbor, killing 5 of the 25 people on board. The boat was certified to carry 25 passengers, but their total weight exceeded the safe load of 3500 lb, so the number of passengers should have been limited to 20.

CHAPTER OBJECTIVES

Chapter 5 introduced *discrete* probability distributions, but in this chapter we introduce *continuous* probability distributions, and most of this chapter focuses on *normal distributions*. Here are the chapter objectives:

6-1 The Standard Normal Distribution

- Describe the characteristics of a standard normal distribution.
- Find the probability of some range of z values in a standard normal distribution.
- Find z scores corresponding to regions under the curve representing a standard normal distribution.

6-2 Real Applications of Normal Distributions

- Develop the ability to describe a normal distribution (not necessarily a standard normal distribution).
- Find the probability of some range of values in a normal distribution.
- Find x scores corresponding to regions under the curve representing a normal distribution.

6-3 Sampling Distributions and Estimators

- Develop the ability to describe a *sampling distribution of a statistic*.
- Determine whether a statistic serves as a good estimator of the corresponding population parameter.

6-4 The Central Limit Theorem

- Describe what the central limit theorem states.
- Apply the central limit theorem by finding the probability that a sample mean falls within some specified range of values.
- Identify conditions for which it is appropriate to use a normal distribution for the distribution of sample means.

6-5 Assessing Normality

- Develop the ability to examine histograms, outliers, and normal quantile plots to determine whether sample data appear to be from a population having a distribution that is approximately normal.

6-6 Normal as Approximation to Binomial

- Identify conditions for which it is appropriate to use a normal distribution as an approximation to a binomial probability distribution.

- Use the normal distribution for approximating probabilities for a binomial distribution.

6-1 The Standard Normal Distribution

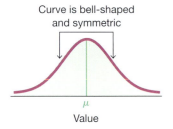

Curve is bell-shaped and symmetric

μ

Value

FIGURE 6-1   The Normal Distribution

Key Concept In this section we present the *standard normal distribution*, which is a specific normal distribution having the following three properties:

1. Bell-shaped: The graph of the standard normal distribution is bell-shaped (as in Figure 6-1).

2. $\mu = 0$: The standard normal distribution has a mean equal to 0.

3. $\sigma = 1$: The standard normal distribution has a standard deviation equal to 1.

In this section we develop the skill to find areas (or probabilities or relative frequencies) corresponding to various regions under the graph of the standard normal distribution. In addition, we find z scores that correspond to areas under the graph. These skills become important in the next section as we study nonstandard normal distributions and the real and important applications that they involve.

Normal Distributions

There are infinitely many different normal distributions, depending on the values used for the mean and standard deviation. We begin with a brief introduction to this general family of normal distributions.

> **DEFINITION**
>
> If a continuous random variable has a distribution with a graph that is symmetric and bell-shaped, as in Figure 6-1, and it can be described by the equation given as Formula 6-1, we say that it has a **normal distribution.**

FORMULA 6-1

$$y = \frac{e^{-\frac{1}{2}\left(\frac{x-\mu}{\sigma}\right)^2}}{\sigma\sqrt{2\pi}}$$

Fortunately, we won't actually use Formula 6-1, but examining the right side of the equation reveals that any particular normal distribution is determined by two parameters: the population mean, μ, and population standard deviation, σ. (In Formula 6-1, x is a variable that can change, $\pi = 3.14159\ldots$ and $e = 2.71828\ldots$.) Once specific values are selected for μ and σ, Formula 6-1 is an equation relating x and y, and we can graph that equation to get a result that will look like Figure 6-1. And that's about all we need to know about Formula 6-1!

Uniform Distributions

The major focus of this chapter is the concept of a normal probability distribution, but we begin with a *uniform distribution* so that we can see the following two very important properties:

1. The area under the graph of a continuous probability distribution is equal to 1.

2. There is a correspondence between area and probability, so *probabilities* can be found by identifying the corresponding *areas* in the graph using this formula for the area of a rectangle:

$$\text{Area} = \text{height} \times \text{width}$$

DEFINITION

A continuous random variable has a **uniform distribution** if its values are spread *evenly* over the range of possibilities. The graph of a uniform distribution results in a rectangular shape.

Density Curve The graph of any continuous probability distribution is called a **density curve,** and any density curve must satisfy the requirement that the total area under the curve is exactly 1. This requirement that the area must equal 1 simplifies probability problems, so the following statement is really important:

Because the total area under any density curve is equal to 1, there is a correspondence between *area* and *probability*.

EXAMPLE 1 Waiting Times for Airport Security

During certain time periods at JFK airport in New York City, passengers arriving at the security checkpoint have waiting times that are uniformly distributed between 0 minutes and 5 minutes, as illustrated in Figure 6-2 on the next page.

Refer to Figure 6-2 to see these properties:

■ All of the different possible waiting times are *equally likely*.

■ Waiting times can be *any* value between 0 min and 5 min, so it is possible to have a waiting time of 1.234567 min.

■ By assigning the probability of 0.2 to the height of the vertical line in Figure 6-2, the *enclosed area is exactly 1*. (In general, we should make the height of the vertical line in a uniform distribution equal to 1/range.)

continued

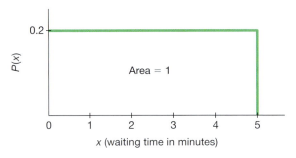

FIGURE 6-2 **Uniform Distribution of Waiting Time**

YOUR TURN Do Exercise 5 "Continuous Uniform Distribution."

Power of Small Samples

The Environmental Protection Agency (EPA) had discovered that Chrysler automobiles had malfunctioning carburetors, with the result that carbon monoxide emissions were too high. Chryslers with 360- and 400-cubic-inch displacements and two-barrel carburetors were involved. The EPA ordered Chrysler to fix the problem, but Chrysler refused, and the case of *Chrysler Corporation vs. The Environmental Protection Agency* followed. That case led to the conclusion that there was "substantial evidence" that the Chryslers produced excessive levels of carbon monoxide. The EPA won the case, and Chrysler was forced to recall and repair 208,000 vehicles. In discussing this case in an article in *AMSTAT News,* Chief Statistician for the EPA Barry Nussbaum wrote this: "Sampling is expensive, and environmental sampling is usually quite expensive. At the EPA, we have to do the best we can with small samples or develop models. . . . What was the sample size required to affect such a recall (of the 208,000 Chryslers)? The answer is a mere 10. It is both an affirmation of the power of inferential statistics and a challenge to explain how such a (small) sample could possibly suffice."

EXAMPLE 2 Waiting Times for Airport Security

Given the uniform distribution illustrated in Figure 6-2, find the probability that a randomly selected passenger has a waiting time of at least 2 minutes.

SOLUTION

The shaded area in Figure 6-3 represents waiting times of at least 2 minutes. Because the total area under the density curve is equal to 1, there is a correspondence between area and probability. We can easily find the desired *probability* by using *areas* as follows:

$$P(\text{wait time of at least 2 min}) = \text{height} \times \text{width of shaded area in Figure 6-3}$$
$$= 0.2 \times 3$$
$$= 0.6$$

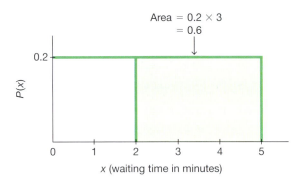

FIGURE 6-3 **Using Area to Find Probability**

INTERPRETATION

The probability of randomly selecting a passenger with a waiting time of at least 2 minutes is 0.6.

YOUR TURN Do Exercise 7 "Continuous Uniform Distribution."

Standard Normal Distribution

The density curve of a uniform distribution is a horizontal straight line, so we can find the area of any rectangular region by applying this formula:

$$\text{Area} = \text{height} \times \text{width.}$$

Because the density curve of a normal distribution has a more complicated bell shape, as shown in Figure 6-1, it is more difficult to find areas. However, the basic principle is the same: *There is a correspondence between area and probability*. In Figure 6-4 we show that for a standard normal distribution, the area under the density curve is equal to 1. In Figure 6-4, we use "*z* Score" as a label for the horizontal axis, and this is common for the standard normal distribution, defined as follows.

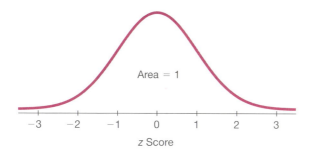

FIGURE 6-4   Standard Normal Distribution

> **DEFINITION**
>
> The **standard normal distribution** is a normal distribution with the parameters of $\mu = 0$ and $\sigma = 1$. The total area under its density curve is equal to 1 (as in Figure 6-4).

Finding Probabilities When Given *z* Scores

It is not easy to manually find areas in Figure 6-4, but we can find areas (or probabilities) for many different regions in Figure 6-4 by using technology, or we can also use Table A-2 (in Appendix A and the *Formulas and Tables* insert card). Key features of the different methods are summarized in Table 6-1, which follows. (StatCrunch provides options for a cumulative left region, a cumulative right region, or the region between two boundaries.) Because calculators and software generally give more accurate results than Table A-2, we *strongly* recommend using technology. (When there are discrepancies, answers in Appendix D will generally include results based on technology as well as answers based on Table A-2.)

If using Table A-2, it is essential to understand these points:

1. Table A-2 is designed only for the *standard* normal distribution, which is a normal distribution with a mean of 0 and a standard deviation of 1.

2. Table A-2 is on two pages, with the left page for *negative z* scores and the right page for *positive z* scores.

3. Each value in the body of the table is a *cumulative area from the left* up to a vertical boundary above a specific *z* score.

4. When working with a graph, avoid confusion between *z* scores and areas.

   ***z* score:** *Distance* **along the horizontal scale of the standard normal distribution (corresponding to the number of standard deviations above or below the mean); refer to the leftmost column and top row of Table A-2.**

   Area: *Region* **under the curve; refer to the values in the** *body* **of Table A-2.**

5. The part of the *z* score denoting hundredths is found across the top row of Table A-2.

TABLE 6-1 Formats Used for Finding Normal Distribution Areas

Cumulative Area from the Left The following provide the *cumulative area from the left* up to a vertical line above a specific value of *z*: • **Table A-2** • **Statdisk** • **Minitab** • **Excel** • **StatCrunch**	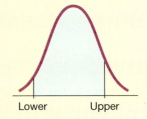 *z* **Cumulative Left Region**
Area Between Two Boundaries The following provide the area bounded on the left and bounded on the right by vertical lines above specific values. • **TI-83/84 Plus calculator** • **StatCrunch**	Lower Upper **Area Between Two Boundaries**

CAUTION When working with a normal distribution, be careful to avoid confusion between *z* scores and areas.

The following examples illustrate procedures that can be used with real and important applications introduced in the following sections.

EXAMPLE 3   Bone Density Test

A bone mineral density test can be helpful in identifying the presence or likelihood of osteoporosis, a disease causing bones to become more fragile and more likely to break. The result of a bone density test is commonly measured as a *z* score. The population of *z* scores is normally distributed with a mean of 0 and a standard deviation of 1, so these test results meet the requirements of a standard normal distribution, and the graph of the bone density test scores is as shown in Figure 6-5.

A randomly selected adult undergoes a bone density test. Find the probability that this person has a bone density test score less than 1.27.

SOLUTION

Note that the following are the *same* (because of the aforementioned correspondence between probability and area):

- *Probability* that the bone density test score is less than 1.27

- Shaded *area* shown in Figure 6-5

So we need to find the area in Figure 6-5 below $z = 1.27$. If using technology, see the Tech Center instructions included at the end of this section. If using Table A-2, begin with the *z* score of 1.27 by locating 1.2 in the left column; next find the value in the adjoining row of probabilities that is directly below 0.07, as shown in the accompanying excerpt. Table A-2 shows that there is an area of 0.8980 corresponding to $z = 1.27$. We want the area *below* 1.27, and Table A-2 gives the cumulative area from the left, so the desired area is 0.8980. Because of the correspondence between area and probability, we know that the probability of a z score below 1.27 is 0.8980.

INTERPRETATION

The *probability* that a randomly selected person has a bone density test result below 1.27 is 0.8980, shown as the shaded region in Figure 6-5. Another way to interpret this result is to conclude that 89.80% of people have bone density levels below 1.27.

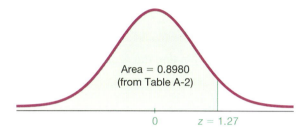

Area = 0.8980
(from Table A-2)

0   z = 1.27

FIGURE 6-5   **Finding Area to the Left of z = 1.27**

TABLE A-2 *(continued)* Cumulative Area from the LEFT

z	.00	.01	.02	.03	.04	.05	.06	.07	.08	.09
0.0	.5000	.5040	.5080	.5120	.5160	.5199	.5239	.5279	.5319	.5359
0.1	.5398	.5438	.5478	.5517	.5557	.5596	.5636	.5675	.5714	.5753
0.2	.5793	.5832	.5871	.5910	.5948	.5987	.6026	.6064	.6103	.6141
1.0	.8413	.8438	.8461	.8485	.8508	.8531	.8554	.8577	.8599	.8621
1.1	.8643	.8665	.8686	.8708	.8729	.8749	.8770	.8790	.8810	.8830
1.2	.8849	.8869	.8888	.8907	.8925	.8944	.8962	.8980	.8997	.9015
1.3	.9032	.9049	.9066	.9082	.9099	.9115	.9131	.9147	.9162	.9177
1.4	.9192	.9207	.9222	.9236	.9251	.9265	.9279	.9292	.9306	.9319

 YOUR TURN Do Exercise 9 "Standard Normal Distribution."

EXAMPLE 4 **Bone Density Test: Finding the Area to the *Right* of a Value**

Using the same bone density test from Example 3, find the probability that a randomly selected person has a result above −1.00. A value above −1.00 is considered to be in the "normal" range of bone density readings.

SOLUTION

We again find the desired *probability* by finding a corresponding *area*. We are looking for the area of the region to the right of z = −1.00 that is shaded in Figure 6-6 on the next page. The Statdisk display on the next page shows that the area to the right of z = −1.00 is 0.841345.

If we use Table A-2, we should know that it is designed to apply only to cumulative areas from the *left*. Referring to the page with *negative z* scores, we find that the cumulative area from the left up to z = −1.00 is 0.1587, as shown in Figure 6-6. Because the total area under the curve is 1, we can find the shaded area by

continued

subtracting 0.1587 from 1. The result is 0.8413. Even though Table A-2 is designed only for cumulative areas from the left, we can use it to find cumulative areas from the right, as shown in Figure 6-6.

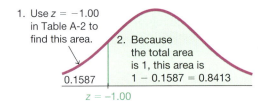

1. Use $z = -1.00$ in Table A-2 to find this area.

2. Because the total area is 1, this area is $1 - 0.1587 = 0.8413$

0.1587

$z = -1.00$

FIGURE 6-6   Finding the Area to the Right of $z = -1$

Statdisk

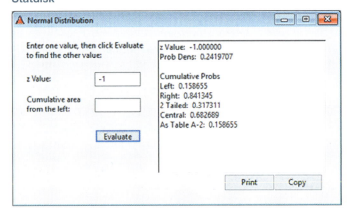

INTERPRETATION

Because of the correspondence between probability and area, we conclude that the *probability* of randomly selecting someone with a bone density reading above -1 is 0.8413 (which is the *area* to the right of $z = -1.00$). We could also say that 84.13% of people have bone density levels above -1.00.

Example 4 illustrates a way that Table A-2 can be used indirectly to find a cumulative area from the right. The following example illustrates another way that we can find an area indirectly by using Table A-2.

EXAMPLE 5   Bone Density Test: Finding the Area Between Two Values

A bone density test reading between -1.00 and -2.50 indicates that the subject has osteopenia, which is some bone loss. Find the probability that a randomly selected subject has a reading between -1.00 and -2.50.

SOLUTION

We are again dealing with normally distributed values having a mean of 0 and a standard deviation of 1. The values between -1.00 and -2.50 correspond to the shaded region in the third graph included in Figure 6-7. Table A-2 cannot be used to find that area directly, but we can use this table to find the following:

■ The area to the left of $z = -1.00$ is 0.1587.

■ The area to the left of $z = -2.50$ is 0.0062.

■ The area *between* $z = -2.50$ and $z = -1.00$ (the shaded area at the far right in Figure 6-7) is the difference between the areas found in the preceding two steps:

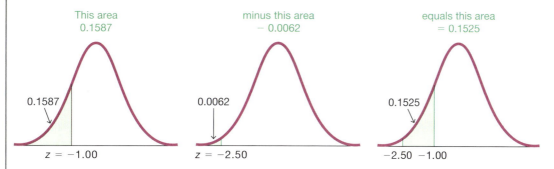

FIGURE 6-7   Finding the Area Between Two z Scores

INTERPRETATION

Using the correspondence between probability and area, we conclude that there is a probability of 0.1525 that a randomly selected subject has a bone density reading between -1.00 and -2.50. Another way to interpret this result is to state that 15.25% of people have osteopenia, with bone density readings between -1.00 and -2.50.

YOUR TURN   Do Exercise 11 "Standard Normal Distribution."

Example 5 can be generalized as the following rule:

The area corresponding to the region *between* two z scores can be found by finding the difference between the two areas found in Table A-2.

Figure 6-8 illustrates this general rule. The shaded region B can be found by calculating the *difference* between two areas found from Table A-2.

HINT Don't try to memorize a rule or formula for this case. Focus on *understanding* by using a graph. Draw a graph, shade the desired area, and then get creative to think of a way to find the desired area by working with cumulative areas from the left.

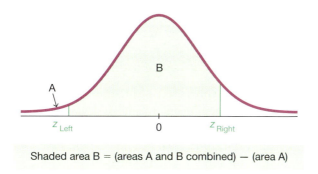

Shaded area B = (areas A and B combined) − (area A)

FIGURE 6-8   Finding the Area Between Two z Scores

Probabilities such as those in the preceding examples can also be expressed with the following notation.

Notation

$P(a < z < b)$ denotes the probability that the z score is between a and b.

$P(z > a)$ denotes the probability that the z score is greater than a.

$P(z < a)$ denotes the probability that the z score is less than a.

With this notation, $P(-2.50 < z < -1.00) = 0.1525$, states in symbols that the probability of a z score falling between -2.50 and -1.00 is 0.1525 (as in Example 5).

Finding *z* Scores from Known Areas

Examples 3, 4, and 5 all involved the standard normal distribution, and they were all examples with this same format: Given z scores, find areas (or probabilities). In many cases, we need a method for reversing the format: Given a known area (or probability), find the corresponding z score. In such cases, it is really important to avoid confusion between z scores and areas. Remember, z scores are *distances* along the horizontal scale, but areas (or probabilities) are regions under the density curve. (Table A-2 lists z-scores in the left column and across the top row, but areas are found in the *body* of the table.) We should also remember that z scores positioned in the left half of the curve are always negative. If we already know a probability and want to find the corresponding z score, we use the following procedure.

Procedure for Finding a *z* Score from a Known Area

1. Draw a bell-shaped curve and identify the region under the curve that corresponds to the given probability. If that region is not a cumulative region from the left, work instead with a known region that is a cumulative region from the left.

2. Use technology or Table A-2 to find the z score. With Table A-2, use the cumulative area from the left, locate the closest probability in the *body* of the table, and identify the corresponding z score.

Special Cases in Table A-2

z Score	Cumulative Area from the Left
1.645	0.9500
−1.645	0.0500
2.575	0.9950
−2.575	0.0050
Above 3.49	0.9999
Below −3.49	0.0001

Special Cases In the solution to Example 6 that follows, Table A-2 leads to a z score of 1.645, which is midway between 1.64 and 1.65. When using Table A-2, we can usually avoid interpolation by simply selecting the closest value. The accompanying table lists special cases that are often used in a wide variety of applications. (For one of those special cases, the value of $z = 2.576$ gives an area slightly closer to the area of 0.9950, but $z = 2.575$ has the advantage of being the value exactly midway between $z = 2.57$ and $z = 2.58$.) Except in these special cases, we can usually select the closest value in the table. (If a desired value is midway between two table values, select the larger value.) For z scores above 3.49, we can use 0.9999 as an approximation of the cumulative area from the left; for z scores below -3.49, we can use 0.0001 as an approximation of the cumulative area from the left.

EXAMPLE 6 **Bone Density Test: Finding a Test Score**

Use the same bone density test scores used in earlier examples. Those scores are normally distributed with a mean of 0 and a standard deviation of 1, so they meet the requirements of a standard normal distribution. Find the bone density score corresponding to P_{95}, the 95th percentile. That is, find the bone density score that separates the bottom 95% from the top 5%. See Figure 6-9.

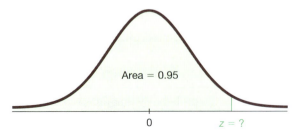

FIGURE 6-9 **Finding the 95th Percentile**

SOLUTION

Figure 6-9 shows the *z* score that is the 95th percentile, with 95% of the area (or 0.95) below it.

Technology: We could find the *z* score using technology. The accompanying Excel display shows that the *z* score with an area of 0.95 to its left is $z = 1.644853627$, or 1.645 when rounded.

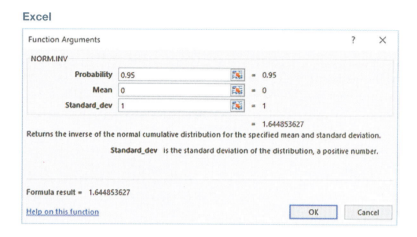

Excel

Table A-2: If using Table A-2, search for the area of 0.95 *in the body* of the table and then find the corresponding *z* score. In Table A-2 we find the areas of 0.9495 and 0.9505, but there's an asterisk with a special note indicating that 0.9500 corresponds to a *z* score of 1.645. We can now conclude that the *z* score in Figure 6-9 is 1.645, so the 95th percentile is $z = 1.645$.

INTERPRETATION

For bone density test scores, 95% of the scores are less than or equal to 1.645, and 5% of them are greater than or equal to 1.645.

YOUR TURN Do Exercise 37 "Finding Bone Density Scores."

EXAMPLE 7   **Bone Density Test**

Using the same bone density test described in Example 3, we have a standard normal distribution with a mean of 0 and a standard deviation of 1. Find the bone density test score that separates the bottom 2.5% and find the score that separates the top 2.5%.

SOLUTION

The required *z* scores are shown in Figure 6-10 on the next page. Those *z* scores can be found using technology. If using Table A-2 to find the *z* score located to the left, we search the *body of the table* for an area of 0.025. The result is $z = -1.96$. To find the *z* score located to the right, we search *the body of* Table A-2 for an area of 0.975. (Remember that Table A-2 always gives cumulative areas from the *left*.) The result is $z = 1.96$. The values of $z = -1.96$ and $z = 1.96$ separate the bottom 2.5% and the top 2.5%, as shown in Figure 6-10.

continued

Residents of New York City believed that taxi cabs became scarce around rush hour in the late afternoon. Their complaints could not be addressed, because there were no data to support that alleged shortage. However, GPS units were installed on cabs and officials could then track their locations. After analyzing the GPS data, it was found that 20% fewer cabs were in service between 4:00 PM and 5:00 PM than in the preceding hour. Subjective beliefs and anecdotal stories were now substantiated with objective data.

Two factors were found to be responsible for the late afternoon cab shortage. First, the 12-hour shifts were scheduled to change at 5:00 PM so that drivers on both shifts would get an equal share at a rush hour. Second, rising rents in Manhattan forced many cab companies to house their cabs in Queens, so drivers had to start returning around 4:00 PM so that they could make it back in time and avoid fines for being late. In recent years, the shortage of cabs has been alleviated with the growth of companies such as Uber and Lyft.

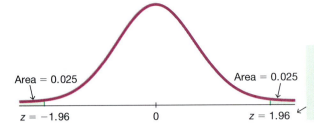

To find this z score, locate the cumulative area to the left in Table A–2. Locate 0.975 in the body of Table A–2.

FIGURE 6-10 Finding z Scores

INTERPRETATION

For the population of bone density test scores, 2.5% of the scores are equal to or less than −1.96 and 2.5% of the scores are equal to or greater than 1.96. Another interpretation is that 95% of all bone density test scores are between −1.96 and 1.96.

YOUR TURN Do Exercise 39 "Finding Bone Density Scores."

Critical Values For a normal distribution, a *critical value* is a z score on the borderline separating those z scores that are *significantly low* or *significantly high*. Common critical values are $z = -1.96$ and $z = 1.96$, and they are obtained as shown in Example 7. In Example 7, values of $z = -1.96$ or lower are significantly low because only 2.5% of the population have scores at or below −1.96, and the values at or above $z = 1.96$ are significantly high because only 2.5% of the population have scores at or above 1.96. Critical values will become extremely important in subsequent chapters. The following notation is used for critical z values found by using the standard normal distribution.

DEFINITION

For the standard normal distribution, a **critical value** is a z score on the borderline separating those z scores that are *significantly low* or *significantly high*.

Notation

The expression z_α denotes the z score with an area of α to its right. (α is the Greek letter alpha.)

 EXAMPLE 8 Finding the Critical Value z_α

Find the value of $z_{0.025}$. (Let $\alpha = 0.025$ in the expression z_α.)

SOLUTION

The notation of $z_{0.025}$ is used to represent the z score with an area of 0.025 to its *right*. Refer to Figure 6-10 and note that the value of $z = 1.96$ has an area of 0.025 to its right, so $z_{0.025} = 1.96$. Note that $z_{0.025}$ corresponds to a cumulative left area of 0.975.

YOUR TURN Do Exercise 41 "Critical Values."

CAUTION When finding a value of z_α for a particular value of α, note that α is the area to the *right* of z_α, but Table A-2 and some technologies give cumulative areas to the *left* of a given z score. To find the value of z_α, resolve that conflict by using the value of $1 - \alpha$. For example, to find $z_{0.1}$, refer to the z score with an area of 0.9 to its left.

Examples 3 through 7 in this section are based on the real application of the bone density test, with scores that are normally distributed with a mean of 0 and standard deviation of 1, so that these scores have a standard normal distribution. Apart from the bone density test scores, it is rare to find such convenient parameters, because typical normal distributions have means different from 0 and standard deviations different from 1. In the next section we present methods for working with such normal distributions.

TECH CENTER

Finding z Scores/Areas (Standard Normal)
Access tech supplements, videos, and data sets at **www.TriolaStats.com**

Statdisk
1. Click **Analysis** in the top menu.
2. Select **Probability Distributions** from the dropdown menu and select **Normal Distribution** from the submenu.
3. Enter the desired z score or cumulative area from the left of the z score and click **Evaluate**.

Minitab

1. Click **Calc** in the top menu.
2. Select **Probability Distributions** from the dropdown menu and select **Normal** from the submenu.

Finding Cumulative Area to the Left of a z Score
- Select **Cumulative probability,** enter mean of **0** and standard deviation of **1**.
- Select **Input Constant,** enter the desired z score, and click **OK**.

Finding z Score from a Known Probability
- Select **Inverse cumulative probability,** enter mean of **0** and standard deviation of **1**.
- Select **Input Constant,** enter the total area to the left of the z score, and click **OK**.

StatCrunch

1. Click **Stat** in the top menu.
2. Select **Calculators** from the dropdown menu and **Normal** from the submenu.
3. In the calculator box enter mean of **0** and standard deviation of **1**.
4. Enter the desired z score (middle box) or known probability/area (rightmost box). Select the desired inequality.
5. Click **Compute**

TI-83/84 Plus Calculator

Unlike most other technologies, the TI-83/84 Plus bases areas on the region between two z scores, rather than cumulative regions from the left.

Finding Area Between Two z Scores

1. Press **2ND** then **VARS** keys to access the *DISTR* (distributions) menu.
2. Select **normalcdf** and press **ENTER**.
3. Enter the desired lower z score and upper z score. Enter **0** for μ and **1** for σ to complete the command **normalcdf(lower z,upper z,μ,σ)**. Press **ENTER**.

TIP: If there is no lower z score, enter −999999; if there is no upper z score, enter 999999.

Finding z Score from a Known Probability

1. Press **2ND** then **VARS** keys to access the *DISTR* (distributions) menu.
2. Select **invNorm** and press **ENTER**.
3. Enter the area to the left of the z score, **0** for μ, and **1** for σ to complete the command **invNorm(area,μ,σ)**. Press **ENTER**.

Excel

Finding Cumulative Area to the Left of a z Score

1. Click **Insert Function** f_x, select the category **Statistical,** select the function **NORM.DIST,** and click **OK**.
2. For x enter the z score, enter **0** for *Mean,* enter **1** for *Standard_dev,* and enter **1** for *Cumulative*.
3. Click **OK**.

Finding z Score from a Known Probability

1. Click **Insert Function** f_x, select the category **Statistical,** and select the function **NORM.INV**.
2. Enter the probability, enter **0** for *Mean,* and enter **1** for *Standard_dev*.
3. Click **OK**.

6-1 Basic Skills and Concepts

Statistical Literacy and Critical Thinking

1. Normal Distribution What's wrong with the following statement? "Because the digits 0, 1, 2, . . . , 9 are the normal results from lottery drawings, such randomly selected numbers have a normal distribution."

2. Normal Distribution A normal distribution is informally described as a probability distribution that is "bell-shaped" when graphed. Draw a rough sketch of a curve having the bell shape that is characteristic of a normal distribution.

3. Standard Normal Distribution Identify the two requirements necessary for a normal distribution to be a *standard* normal distribution.

4. Notation What does the notation z_α indicate?

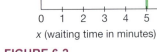

FIGURE 6-2

Continuous Uniform Distribution. *In Exercises 5–8, refer to the continuous uniform distribution depicted in Figure 6-2 and described in Example 1. Assume that a passenger is randomly selected, and find the probability that the waiting time is within the given range.*

5. Greater than 3.00 minutes

6. Less than 4.00 minutes

7. Between 2 minutes and 3 minutes

8. Between 2.5 minutes and 4.5 minutes

Standard Normal Distribution. *In Exercises 9–12, find the area of the shaded region. The graph depicts the standard normal distribution of bone density scores with mean 0 and standard deviation 1.*

9.

z = 0.44

10.

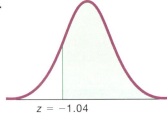

z = −1.04

11.

z = −0.84   z = 1.28

12.

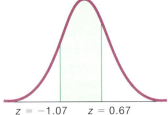

z = −1.07   z = 0.67

Standard Normal Distribution. *In Exercises 13–16, find the indicated z score. The graph depicts the standard normal distribution of bone density scores with mean 0 and standard deviation 1.*

13.

0.8907

0   z

14.

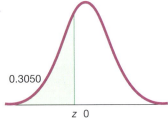

0.3050

z   0

15.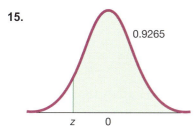

16.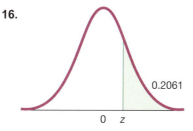

Standard Normal Distribution. *In Exercises 17–36, assume that a randomly selected subject is given a bone density test. Those test scores are normally distributed with a mean of 0 and a standard deviation of 1. In each case, draw a graph, then find the probability of the given bone density test scores. If using technology instead of Table A-2, round answers to four decimal places.*

17. Less than -1.23

18. Less than -1.96

19. Less than 1.28

20. Less than 2.56

21. Greater than 0.25

22. Greater than 0.18

23. Greater than -2.00

24. Greater than -3.05

25. Between 2.00 and 3.00

26. Between 1.50 and 2.50

27. Between and -2.55 and -2.00

28. Between -2.75 and -0.75

29. Between -2.00 and 2.00

30. Between -3.00 and 3.00

31. Between -1.00 and 5.00

32. Between -4.27 and 2.34

33. Less than 4.55

34. Greater than -3.75

35. Greater than 0

36. Less than 0

Finding Bone Density Scores. *In Exercises 37–40 assume that a randomly selected subject is given a bone density test. Bone density test scores are normally distributed with a mean of 0 and a standard deviation of 1. In each case, draw a graph, then find the bone density test score corresponding to the given information. Round results to two decimal places.*

37. Find P_{99}, the 99th percentile. This is the bone density score separating the bottom 99% from the top 1%.

38. Find P_{10}, the 10th percentile. This is the bone density score separating the bottom 10% from the top 90%.

39. If bone density scores in the bottom 2% and the top 2% are used as cutoff points for levels that are too low or too high, find the two readings that are cutoff values.

40. Find the bone density scores that can be used as cutoff values separating the lowest 3% and highest 3%.

Critical Values. *In Exercises 41–44, find the indicated critical value. Round results to two decimal places.*

41. $z_{0.10}$

42. $z_{0.02}$

43. $z_{0.04}$

44. $z_{0.15}$

Basis for the Range Rule of Thumb and the Empirical Rule. *In Exercises 45–48, find the indicated area under the curve of the standard normal distribution; then convert it to a percentage and fill in the blank. The results form the basis for the range rule of thumb and the empirical rule introduced in Section 3-2.*

45. About _____% of the area is between $z = -1$ and $z = 1$ (or within 1 standard deviation of the mean).

46. About _____% of the area is between $z = -2$ and $z = 2$ (or within 2 standard deviations of the mean).

47. About _____% of the area is between $z = -3$ and $z = 3$ (or within 3 standard deviations of the mean).

48. About _____% of the area is between $z = -3.5$ and $z = 3.5$ (or within 3.5 standard deviations of the mean).

6-1 Beyond the Basics

49. Significance For bone density scores that are normally distributed with a mean of 0 and a standard deviation of 1, find the *percentage* of scores that are

a. *significantly high* (or at least 2 standard deviations above the mean).

b. *significantly low* (or at least 2 standard deviations below the mean).

c. *not significant* (or less than 2 standard deviations away from the mean).

50. Distributions In a continuous uniform distribution,

$$\mu = \frac{\text{minimum} + \text{maximum}}{2} \quad \text{and} \quad \sigma = \frac{\text{range}}{\sqrt{12}}$$

a. Find the mean and standard deviation for the distribution of the waiting times represented in Figure 6-2, which accompanies Exercises 5–8.

b. For a continuous uniform distribution with $\mu = 0$ and $\sigma = 1$, the minimum is $-\sqrt{3}$ and the maximum is $\sqrt{3}$. For this continuous uniform distribution, find the probability of randomly selecting a value between –1 and 1, and compare it to the value that would be obtained by incorrectly treating the distribution as a standard normal distribution. Does the distribution affect the results very much?

6-2 Real Applications of Normal Distributions

Key Concept Now we really get real as we extend the methods of the previous section so that we can work with any *nonstandard normal distribution* (with a mean different from 0 and/or a standard deviation different from 1). The key is a simple conversion (Formula 6-2) that allows us to "standardize" any normal distribution so that *x* values can be transformed to *z* scores; then the methods of the preceding section can be used.

FORMULA 6-2

$$z = \frac{x - \mu}{\sigma} \quad (\text{round } z \text{ scores to 2 decimal places})$$

Figure 6-11 illustrates the conversion from a nonstandard to a standard normal distribution. The area in *any* normal distribution bounded by some score *x* (as in Figure 6-11a) is the *same* as the area bounded by the corresponding *z* score in the standard normal distribution (as in Figure 6-11b).

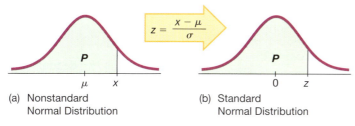

(a) Nonstandard
    Normal Distribution

(b) Standard
    Normal Distribution

FIGURE 6-11 **Converting Distributions**

Some calculators and software do not require the use of Formula 6-2 to convert to z scores because probabilities can be found directly. However, if using Table A-2, we must first convert values to standard z scores.

When finding areas with a nonstandard normal distribution, use the following procedure.

Procedure for Finding Areas with a Nonstandard Normal Distribution

1. Sketch a normal curve, label the mean and any specific x values, and then *shade* the region representing the desired probability.

2. For each relevant value x that is a boundary for the shaded region, use Formula 6-2 to convert that value to the equivalent z score. (With many technologies, this step can be skipped.)

3. Use technology (software or a calculator) or Table A-2 to find the area of the shaded region. This area is the desired probability.

The following example illustrates the above procedure.

EXAMPLE 1 **What Proportion of Men Are Taller Than the 72 in. Height Requirement for Showerheads (According to Most Building Codes)?**

Heights of men are normally distributed with a mean of 68.6 in. and a standard deviation of 2.8 in. (based on Data Set 1 "Body Data" in Appendix B). Find the percentage of men who are taller than a showerhead at 72 in.

SOLUTION

Step 1: See Figure 6-12, which incorporates this information: Men have heights that are normally distributed with a mean of 68.6 in. and a standard deviation of 2.8 in. The shaded region represents the men who are taller than the showerhead height of 72 in.

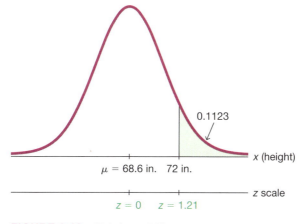

FIGURE 6-12 **Heights of Men**

continued

Step 2: We can convert the showerhead height of 72 in. to the z score of 1.21 by using Formula 6-2 as follows:

$$z = \frac{x - \mu}{\sigma} = \frac{72 - 68.6}{2.8} = 1.21 \text{ (rounded to two decimal places)}$$

Step 3: Technology: Technology can be used to find that the area to the right of 72 in. in Figure 6-12 is 0.1123 rounded. (With many technologies, Step 2 can be skipped. See technology instructions at the end of this section.) The result of 0.1123 from technology is more accurate than the result of 0.1131 found by using Table A-2.

Table A-2: Use Table A-2 to find that the cumulative area to the *left* of $z = 1.21$ is 0.8869. (Remember, Table A-2 is designed so that all areas are cumulative areas from the *left*.) Because the total area under the curve is 1, it follows that the shaded area in Figure 6-12 is $1 - 0.8869 = 0.1131$.

INTERPRETATION

The proportion of men taller than the showerhead height of 72 in. is 0.1123, or 11.23%. About 11% of men may find the design to be unsuitable. (*Note:* Some NBA teams have been known to intentionally use lower showerheads in the locker rooms of visiting basketball teams.)

YOUR TURN Do Exercise 13 "Seat Designs."

 EXAMPLE 2 **Air Force Height Requirement**

The U.S. Air Force requires that pilots have heights between 64 in. and 77 in. Heights of women are normally distributed with a mean of 63.7 in. and a standard deviation of 2.9 in. (based on Data Set 1 "Body Data" in Appendix B). What percentage of women meet that height requirement?

SOLUTION

Figure 6-13 shows the shaded region representing heights of women between 64 in. and 77 in.

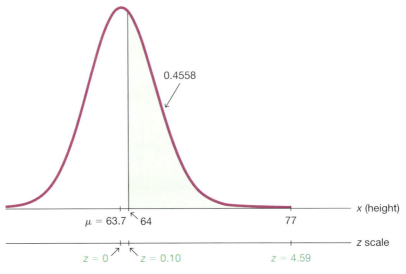

FIGURE 6-13   **Heights of Women**

Step 1: See Figure 6-13, which incorporates this information: Women have heights that are normally distributed with a mean of 63.7 in. and a standard deviation of 2.9 in. The shaded region represents the women with heights between 64 in. and 77 in.

Step 2: With some technologies, the shaded area in Figure 6-13 can be found directly and it is not necessary to convert the x scores of 64 in. and 77 in. to z scores. (See Step 3.)

   If using Table A-2, we cannot find the shaded area directly, but we can find it indirectly by using the same procedures from Section 6-1, as follows: (1) Find the cumulative area from the left up to 77 in. (or $z = 4.59$); (2) find the cumulative area from the left up to 64 in. (or $z = 0.10$); (3) find the difference between those two areas. The heights of 77 in. and 64 in. are converted to z scores by using Formula 6-2 as follows:

$$\text{For } x = 77 \text{ in.: } z = \frac{x - \mu}{\sigma} = \frac{77 - 63.7}{2.9} = 4.59$$
$$(z = 4.59 \text{ yields an area of } 0.9999.)$$

$$\text{For } x = 64 \text{ in.: } z = \frac{x - \mu}{\sigma} = \frac{64 - 63.7}{2.9} = 0.10$$
$$(z = 0.10 \text{ yields an area of } 0.5398.)$$

Step 3: Technology: To use technology, refer to the Tech Center instructions at the end of this section. Technology will show that the shaded area in Figure 6-13 is 0.4588

Table A-2: Refer to Table A-2 with $z = 4.59$ and find that the cumulative area to the *left* of $z = 4.59$ is 0.9999. (Remember, Table A-2 is designed so that all areas are cumulative areas from the *left*.) Table A-2 also shows that $z = 0.10$ corresponds to an area of 0.5398. Because the areas of 0.9999 and 0.5398 are *cumulative areas from the left*, we find the shaded area in Figure 6-13 as follows:

$$\text{Shaded area in Figure } 6-13 = 0.9999 - 0.5398 = 0.4601$$

There is a relatively small discrepancy between the area of 0.4588 found from technology and the area of 0.4601 found from Table A-2. The area obtained from technology is more accurate because it is based on unrounded z scores, whereas Table A-2 requires z scores rounded to two decimal places.

INTERPRETATION

Expressing the result as a percentage, we conclude that about 46% of women satisfy the requirement of having a height between 64 in. and 77 in. About 54% of women do not meet that requirement and they are not eligible to be pilots in the U.S. Air Force.

YOUR TURN Do Exercise 15 "Seat Designs."

Finding Values from Known Areas

Here are helpful hints for those cases in which the area (or probability or percentage) is known and we must find the relevant value(s):

1. Graphs are extremely helpful in visualizing, understanding, and successfully working with normal probability distributions, so they should always be used.

2. *Don't confuse z scores and areas.* Remember, z scores are *distances* along the horizontal scale, but areas are *regions* under the normal curve. Table A-2 lists z scores in the left columns and across the top row, but areas are found in the body of the table.

3. *Choose the correct (right/left) side of the graph.* A value separating the *top* 10% from the others will be located on the right side of the graph, but a value separating the *bottom* 10% will be located on the left side of the graph.

4. A *z* score must be *negative* whenever it is located in the *left* half of the normal distribution.

5. Areas (or probabilities) are always between 0 and 1, and they are never negative.

Procedure for Finding Values from Known Areas or Probabilities

1. Sketch a normal distribution curve, write the given probability or percentage in the appropriate region of the graph, and identify the *x* value(s) being sought.

2. If using technology, refer to the Tech Center instructions at the end of this section. If using Table A-2, refer to the *body* of Table A-2 to find the area to the left of *x*, then identify the *z* score corresponding to that area.

3. If you know *z* and must convert to the equivalent *x* value, use Formula 6-2 by entering the values for μ, σ, and the *z* score found in Step 2, then solve for *x*. Based on Formula 6-2, we can solve for *x* as follows:

$$x = \mu + (z \cdot \sigma) \quad \text{(another form of Formula 6–2)}$$

$\uparrow$

(If *z* is located to the left of the mean, be sure that it is a negative number.)

4. Refer to the sketch of the curve to verify that the solution makes sense in the context of the graph and in the context of the problem.

The following example uses this procedure for finding a value from a known area.

EXAMPLE 3 Designing an Aircraft Cockpit

When designing equipment, one common criterion is to use a design that accommodates 95% of the population. In Example 2 we saw that only 46% of women satisfy the height requirements for U.S. Air Force pilots. What would be the maximum acceptable height of a woman if the requirements were changed to allow the *shortest* 95% of women to be pilots? That is, find the 95th percentile of heights of women. Assume that heights of women are normally distributed with a mean of 63.7 in. and a standard deviation of 2.9 in. In addition to the maximum allowable height, should there also be a minimum required height? Why?

SOLUTION

Step 1: Figure 6-14 shows the normal distribution with the height *x* that we want to identify. The shaded area represents the shortest 95% of women.

Step 2: Technology: Technology will provide the value of *x* in Figure 6-14. For example, see the accompanying Excel display showing that $x = 68.47007552$ in., or 68.5 in. when rounded.

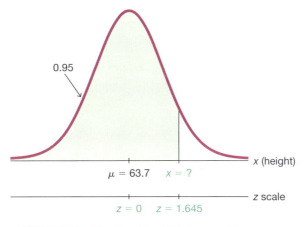

0.95

$\mu = 63.7$ $x = ?$ x (height)

z scale

$z = 0$ $z = 1.645$

FIGURE 6-14   Finding the 95th Percentile

Table A-2: If using Table A-2, search for an area of 0.9500 *in the body* of the table. (The area of 0.9500 shown in Figure 6-14 is a cumulative area from the left, and that is exactly the type of area listed in Table A-2.) The area of 0.9500 is between the Table A-2 areas of 0.9495 and 0.9505, but there is an asterisk and footnote indicating that an area of 0.9500 corresponds to $z = 1.645$.

Step 3: With $z = 1.645$, $\mu = 63.7$ in., and $\sigma = 2.9$ in., we can solve for x by using Formula 6-2:

$$z = \frac{x - \mu}{\sigma} \quad \text{becomes} \quad 1.645 = \frac{x - 63.7}{2.9}$$

The result of $x = 68.4705$ in. can be found directly or by using the following version of Formula 6-2:

$$x = \mu + (z \cdot \sigma) = 63.7 + (1.645 \cdot 2.9) = 68.4705 \text{ in.}$$

Step 4: The solution of $x = 68.5$ in. (rounded) in Figure 6-14 is reasonable because it is greater than the mean of 63.7in.

INTERPRETATION

A requirement of a height less than 68.5 in. would allow 95% of women to be eligible as U.S. Air Force pilots. There should also be a *minimum* height requirement so that the pilot can easily reach all controls.

YOUR TURN Do Exercise 17 "Seat Designs."

Significance

In Chapter 4 we saw that probabilities can be used to determine whether values are *significantly high* or *significantly low.* Chapter 4 referred to x successes among n trials, but we can adapt those criteria to apply to continuous variables as follows:

Significantly high: The value x is *significantly high* if $P(x$ or greater$) \leq 0.05.$*

Significantly low: The value x is *significantly low* if $P(x$ or less$) \leq 0.05.$*

---

*The value of 0.05 is not absolutely rigid, and other values such as 0.01 could be used instead.

EXAMPLE 4 **Significantly Low or Significantly High Female Pulse Rates**

Use the preceding criteria to identify pulse rates of women that are significantly low or significantly high. Based on Data Set 1 "Body Data" in Appendix B, assume that women have normally distributed pulse rates with a mean of 74.0 beats per minute and a standard deviation of 12.5 beats per minute.

SOLUTION

Step 1: We begin with the graph shown in Figure 6-15. We have entered the mean of 74.0, and we have identified the x values separating the lowest 5% and the highest 5%.

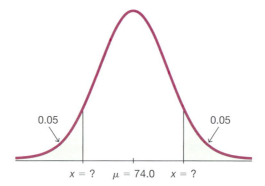

FIGURE 6-15 **Pulse Rates of Women**

Step 2: Technology: To use technology, refer to the Tech Center instructions at the end of this section. Technology will show that the values of x in Figure 6-15 are 53.4 beats per minute and 94.6 beats per minute when rounded.

Table A-2: If using Table A-2, we must work with cumulative areas from the left. For the leftmost value of x, the cumulative area from the left is 0.05, so search for an area of 0.05 *in the body* of the table to get $z = -1.645$ (identified by the asterisk between 0.0505 and 0.0495). For the rightmost value of x, the cumulative area from the left is 0.95, so search for an area of 0.9500 *in the body* of the table to get $z = 1.645$ (identified by the asterisk between 0.9495 and 0.9505). Having found the two z scores, we now proceed to convert them to pulse rates.

Step 3: We now solve for the two values of x by using Formula 6-2 directly or by using the following version of Formula 6-2:

Leftmost value of x: $x = \mu + (z \cdot \sigma) = 74.0 + (-1.645 \cdot 12.5) = 53.4$

Rightmost value of x: $x = \mu + (z \cdot \sigma) = 74.0 + (1.645 \cdot 12.5) = 94.6$

Step 4: Referring to Figure 6-15, we see that the leftmost value of $x = 53.4$ is reasonable because it is less than the mean of 74.0. Also, the rightmost value of 94.6 is reasonable because it is above the mean of 74.0.

INTERPRETATION

Here are the pulse rates of women that are significant:

• Significantly low: 53.4 beats per minute or lower

• Significantly high: 94.6 beats per minute or higher

Physicians could use these results to investigate health issues that could cause pulse rates to be significantly low or significantly high.

YOUR TURN Do Exercise 19 "Significance."

TECH CENTER

 Finding *x* Values/Areas

Access tech supplements, videos, and data sets at **www.TriolaStats.com**

Statdisk	Minitab	StatCrunch
1. Click **Analysis** in the top menu. 2. Select **Probability Distributions** from the dropdown menu and select **Normal Distribution** from the submenu. 3. Enter the desired *z* value or cumulative area from the left of the *z* score and click **Evaluate.** *TIP:* Statdisk does not work directly with nonstandard normal distributions, so use corresponding *z* scores.	1. Click **Calc** in the top menu. 2. Select **Probability Distributions** from the dropdown menu and select **Normal** from the submenu. **Finding Cumulative Area to the Left of an *x* Value** 3. Select **Cumulative probability,** enter the mean and standard deviation. 4. Select **Input Constant,** enter the desired *x* value, and click **OK.** **Finding *x* Value from a Known Probability** 3. Select **Inverse cumulative probability,** enter the mean and standard deviation. 4. Select **Input Constant,** enter the total area to the left of the *x* value, and click **OK.**	1. Click **Stat** in the top menu. 2. Select **Calculators** from the dropdown menu and **Normal** from the submenu. 3. In the calculator box enter the mean and standard deviation. 4. Enter the desired *x* value (middle box) or probability (rightmost box). 5. Click **Compute**

TI-83/84 Plus Calculator	Excel
Unlike most other technologies, the TI-83/84 Plus bases areas on the region between two z scores, rather than cumulative regions from the left. **Finding Area Between Two *x* Values** 1. Press **2ND** then **VARS** keys to access the *DISTR* (distributions) menu. 2. Select **normalcdf** and press **ENTER**. 3. Enter the desired *lower x* value and *upper x* value. Enter the mean (μ) and standard deviation (σ) to complete the command **normalcdf(lower x,upper x, μ, σ).** Press **ENTER**. *TIP:* If there is no lower *x* value, enter −999999; if there is no upper *x* value, enter 999999. **Finding *x* Value Corresponding to a Known Area** 1. Press **2ND** then **VARS** keys to access the *DISTR* (distributions) menu. 2. Select **invNorm** and press **ENTER**. 3. Enter the *area* to the left of the *x* value, enter the mean (μ), and the standard deviation (σ) to complete the command **invNorm(area, μ, σ).** Press **ENTER**.	**Finding Cumulative Area to the Left of an *x* Value** 1. Click **Insert Function** f_x, select the category **Statistical,** select the function **NORM.DIST,** and **click OK.** 2. For *x* enter the *x* value, enter *Mean,* enter *Standard_dev,* and enter **1** for *Cumulative.* 3. Click **OK.** **Finding *x* Value Corresponding to a Known Probability** 1. Click **Insert Function** f_x, select the category **Statistical,** select the function **NORM.INV,** and click **OK.** 2. Enter the probability or the area to the left of the desired *x* value, enter *Mean,* and enter *Standard_dev.* 3. Click **OK.**

6-2   Basic Skills and Concepts

Statistical Literacy and Critical Thinking

1. Birth Weights Based on Data Set 4 "Births" in Appendix B, birth weights are normally distributed with a mean of 3152.0 g and a standard deviation of 693.4 g.

a. What are the values of the mean and standard deviation after converting all birth weights to z scores using $z = (x - \mu)/\sigma$?

b. The original birth weights are in grams. What are the units of the corresponding z scores?

2. Birth Weights Based on Data Set 4 "Births" in Appendix B, birth weights are normally distributed with a mean of 3152.0 g and a standard deviation of 693.4 g.

a. For the bell-shaped graph, what is the area under the curve?

b. What is the value of the median?

c. What is the value of the mode?

d. What is the value of the variance?

3. Normal Distributions What is the difference between a standard normal distribution and a nonstandard normal distribution?

4. Random Digits Computers are commonly used to randomly generate digits of telephone numbers to be called when conducting a survey. Can the methods of this section be used to find the probability that when one digit is randomly generated, it is less than 3? Why or why not? What is the probability of getting a digit less than 3?

IQ Scores.   *In Exercises 5–8, find the area of the shaded region. The graphs depict IQ scores of adults, and those scores are normally distributed with a mean of 100 and a standard deviation of 15 (as on the Wechsler IQ test).*

5.

118

6.
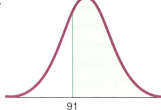
91

7.

79   133

8.

112 124

IQ Scores.   *In Exercises 9–12, find the indicated IQ score and round to the nearest whole number. The graphs depict IQ scores of adults, and those scores are normally distributed with a mean of 100 and a standard deviation of 15 (as on the Wechsler IQ test).*

9.
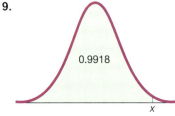
0.9918
x

10.
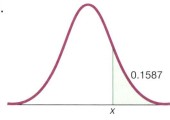
0.1587
x

11.

0.9798

x

12.

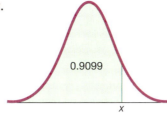

0.9099

x

Seat Designs. *In Exercises 13–20, use the data in the table below for sitting adult males and females (based on anthropometric survey data from Gordon, Churchill, et al.). These data are used often in the design of different seats, including aircraft seats, train seats, theater seats, and classroom seats. (Hint: Draw a graph in each case.)*

Sitting Back-to-Knee Length (inches)

	Mean	St. Dev.	Distribution
Males	23.5 in.	1.1 in.	Normal
Females	22.7 in.	1.0 in.	Normal

13. Find the probability that a male has a back-to-knee length less than 21 in.

14. Find the probability that a female has a back-to-knee length greater than 24.0 in.

15. Find the probability that a female has a back-to-knee length between 22.0 in. and 24.0 in.

16. Find the probability that a male has a back-to-knee length between 22.0 in. and 24.0 in.

17. For males, find P_{90}, which is the length separating the bottom 90% from the top 10%.

18. For females, find the first quartile Q_1, which is the length separating the bottom 25% from the top 75%.

19. Significance Instead of using 0.05 for identifying significant values, use the criteria that a value *x* is *significantly high* if $P(x$ or greater$) \leq 0.01$ and a value is *significantly low* if $P(x$ or less$) \leq 0.01$. Find the back-to-knee lengths for males, separating significant values from those that are not significant. Using these criteria, is a male back-to-knee length of 26 in. significantly high?

20. Significance Instead of using 0.05 for identifying significant values, use the criteria that a value *x* is *significantly high* if $P(x$ or greater$) \leq 0.025$ and a value is *significantly low* if $P(x$ or less$) \leq 0.025$. Find the female back-to-knee length, separating significant values from those that are not significant. Using these criteria, is a female back-to-knee length of 20 in. significantly low?

In Exercises 21–24, use these parameters (based on Data Set 1 "Body Data" in Appendix B):

• *Men's heights are normally distributed with mean 68.6 in. and standard deviation 2.8 in.*

• *Women's heights are normally distributed with mean 63.7 in. and standard deviation 2.9 in.*

21. Navy Pilots The U.S. Navy requires that fighter pilots have heights between 62 in. and 78 in.

a. Find the percentage of women meeting the height requirement. Are many women not qualified because they are too short or too tall?

b. If the Navy changes the height requirements so that all women are eligible except the shortest 3% and the tallest 3%, what are the new height requirements for women?

22. Air Force Pilots The U.S. Air Force requires that pilots have heights between 64 in. and 77 in.

a. Find the percentage of men meeting the height requirement.

b. If the Air Force height requirements are changed to exclude only the tallest 2.5% of men and the shortest 2.5% of men, what are the new height requirements?

23. Mickey Mouse Disney World requires that people employed as a Mickey Mouse character must have a height between 56 in. and 62 in.

a. Find the percentage of men meeting the height requirement. What does the result suggest about the genders of the people who are employed as Mickey Mouse characters?

b. If the height requirements are changed to exclude the tallest 50% of men and the shortest 5% of men, what are the new height requirements?

24. Executive Jet Doorway The Gulfstream 100 is an executive jet that seats six, and it has a doorway height of 51.6 in.

a. What percentage of adult men can fit through the door without bending?

b. Does the door design with a height of 51.6 in. appear to be adequate? Why didn't the engineers design a larger door?

c. What doorway height would allow 40% of men to fit without bending?

25. Eye Contact In a study of facial behavior, people in a control group are timed for eye contact in a 5-minute period. Their times are normally distributed with a mean of 184.0 seconds and a standard deviation of 55.0 seconds (based on data from "Ethological Study of Facial Behavior in Nonparanoid and Paranoid Schizophrenic Patients," by Pittman, Olk, Orr, and Singh, *Psychiatry*, Vol. 144, No. 1). For a randomly selected person from the control group, find the probability that the eye contact time is greater than 230.0 seconds, which is the mean for paranoid schizophrenics. Based on personal experience, does the result appear to be the proportion of people who are paranoid schizophrenics?

26. Designing a Work Station A common design requirement is that an environment must fit the range of people who fall between the 5th percentile for women and the 95th percentile for men. In designing an assembly work table, we must consider *sitting knee height*, which is the distance from the bottom of the feet to the top of the knee. Males have sitting knee heights that are normally distributed with a mean of 21.4 in. and a standard deviation of 1.2 in.; females have sitting knee heights that are normally distributed with a mean of 19.6 in. and a standard deviation of 1.1 in. (based on data from the Department of Transportation).

a. What is the minimum table clearance required to satisfy the requirement of fitting 95% of men? Why is the 95th percentile for women ignored in this case?

b. The author is writing this exercise at a table with a clearance of 23.5 in. above the floor. What percentage of men fit this table, and what percentage of women fit this table? Does the table appear to be made to fit almost everyone?

27. Jet Ejection Seats The U.S. Air Force once used ACES-II ejection seats designed for men weighing between 140 lb and 211 lb. Given that women's weights are normally distributed with a mean of 171.1 lb and a standard deviation of 46.1 lb (based on data from the National Health Survey), what percentage of women have weights that are within those limits? Were many women excluded with those past specifications?

28. Quarters After 1964, quarters were manufactured so that their weights have a mean of 5.67 g and a standard deviation of 0.06 g. Some vending machines are designed so that you can adjust the weights of quarters that are accepted. If many counterfeit coins are found, you can narrow the range of acceptable weights with the effect that most counterfeit coins are rejected along with some legitimate quarters.

a. If you adjust your vending machines to accept weights between 5.60 g and 5.74 g, what percentage of legal quarters are rejected? Is that percentage too high?

b. If you adjust vending machines to accept all legal quarters except those with weights in the top 2.5% and the bottom 2.5%, what are the limits of the weights that are accepted?

29. Low Birth Weight The University of Maryland Medical Center considers "low birth weights" to be those that are less than 5.5 lb or 2495 g. Birth weights are normally distributed

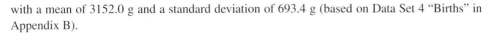

with a mean of 3152.0 g and a standard deviation of 693.4 g (based on Data Set 4 "Births" in Appendix B).

a. If a birth weight is randomly selected, what is the probability that it is a "low birth weight"?

b. Find the weights considered to be *significantly* low, using the criterion of a birth weight having a probability of 0.05 or less.

c. Compare the results from parts (a) and (b).

30. Body Temperatures Based on the sample results in Data Set 3 "Body Temperatures" in Appendix B, assume that human body temperatures are normally distributed with a mean of 98.20°F and a standard deviation of 0.62°F.

a. According to emedicinehealth.com, a body temperature of 100.4°F or above is considered to be a fever. What percentage of normal and healthy persons would be considered to have a fever? Does this percentage suggest that a cutoff of 100.4°F is appropriate?

b. Physicians want to select a minimum temperature for requiring further medical tests. What should that temperature be, if we want only 2.0% of healthy people to exceed it? (Such a result is a *false positive*, meaning that the test result is positive, but the subject is not really sick.)

31. Durations of Pregnancies The lengths of pregnancies are normally distributed with a mean of 268 days and a standard deviation of 15 days.

a. In a letter to "Dear Abby," a wife claimed to have given birth 308 days after a brief visit from her husband, who was working in another country. Find the probability of a pregnancy lasting 308 days or longer. What does the result suggest?

b. If we stipulate that a baby is *premature* if the duration of pregnancy is in the lowest 3%, find the duration that separates premature babies from those who are not premature. Premature babies often require special care, and this result could be helpful to hospital administrators in planning for that care.

32. Water Taxi Safety When a water taxi sank in Baltimore's Inner Harbor, an investigation revealed that the safe passenger load for the water taxi was 3500 lb. It was also noted that the mean weight of a passenger was assumed to be 140 lb. Assume a "worst-case" scenario in which all of the passengers are adult men. Assume that weights of men are normally distributed with a mean of 188.6 lb and a standard deviation of 38.9 lb (based on Data Set 1 "Body Data" in Appendix B).

a. If one man is randomly selected, find the probability that he weighs less than 174 lb (the new value suggested by the National Transportation and Safety Board).

b. With a load limit of 3500 lb, how many male passengers are allowed if we assume a mean weight of 140 lb?

c. With a load limit of 3500 lb, how many male passengers are allowed if we assume the updated mean weight of 188.6 lb?

d. Why is it necessary to periodically review and revise the number of passengers that are allowed to board?

Large Data Sets. *In Exercises 33 and 34, refer to the data sets in Appendix B and use software or a calculator.*

33. Pulse Rates of Males Refer to Data Set 1 "Body Data" in Appendix B and use the pulse rates of males.

a. Find the mean and standard deviation, and verify that the pulse rates have a distribution that is roughly normal.

b. Treating the unrounded values of the mean and standard deviation as parameters, and assuming that male pulse rates are normally distributed, find the pulse rate separating the lowest 2.5% and the pulse rate separating the highest 2.5%. These values could be helpful when physicians try to determine whether pulse rates are significantly low or significantly high.

34. Weights of M&Ms Refer to Data Set 27 "M&M Weights" in Appendix B and use the weights (grams) of all of the listed M&Ms.

a. Find the mean and standard deviation, and verify that the data have a distribution that is roughly normal.

b. Treating the unrounded values of the mean and standard deviation as parameters, and assuming that the weights are normally distributed, find the weight separating the lowest 0.5% and the weight separating the highest 0.5%. These values could be helpful when quality control specialists try to control the manufacturing process.

6-2 Beyond the Basics

35. Curving Test Scores A professor gives a test and the scores are normally distributed with a mean of 60 and a standard deviation of 12. She plans to curve the scores.

a. If she curves by adding 15 to each grade, what is the new mean and standard deviation?

b. Is it fair to curve by adding 15 to each grade? Why or why not?

c. If the grades are curved so that grades of B are given to scores above the bottom 70% and below the top 10%, find the numerical limits for a grade of B.

d. Which method of curving the grades is fairer: adding 15 to each original score or using a scheme like the one given in part (c)? Explain.

36. Outliers For the purposes of constructing modified boxplots as described in Section 3-3, outliers are defined as data values that are above Q_3 by an amount greater than $1.5 \times IQR$ or below Q_1 by an amount greater than $1.5 \times IQR$, where IQR is the interquartile range. Using this definition of outliers, find the probability that when a value is randomly selected from a normal distribution, it is an outlier.

6-3 Sampling Distributions and Estimators

Key Concept We now consider the concept of a *sampling distribution of a statistic*. Instead of working with values from the original population, we want to focus on the values of *statistics* (such as sample proportions or sample means) obtained from the population. Figure 6-16 shows the key points that we need to know, so try really, really hard to understand the story that Figure 6-16 tells.

A Short Story Among the population of all adults, exactly 70% do not feel comfortable in a self-driving vehicle (the author just knows this). In a TE Connectivity survey of 1000 adults, 69% said that they did not feel comfortable in a self-driving vehicle. Empowered by visions of multitudes of driverless cars, 50,000 people became so enthusiastic that they each conducted their own individual survey of 1000 randomly selected adults on the same topic. Each of these 50,000 newbie surveyors reported the percentage that they found, with results such as 68%, 72%, 70%. The author obtained each of the 50,000 sample percentages, he changed them to proportions, and then he constructed the histogram shown in Figure 6-17. Notice anything about the *shape* of the histogram? It's *normal* (unlike the 50,000 newbie surveyors). Notice anything about the mean of the sample proportions? They are centered about the value of 0.70, which happens to be the population proportion. Moral: When samples of the same size are taken from the same population, the following two properties apply:

1. Sample proportions tend to be normally distributed.

2. The mean of sample proportions is the same as the population mean.

The implications of the preceding properties will be extensive in the chapters that follow. What a happy ending!

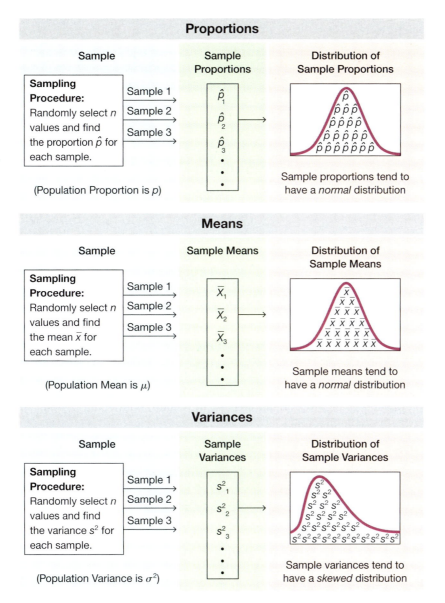

FIGURE 6-16   General Behavior of Sampling Distributions

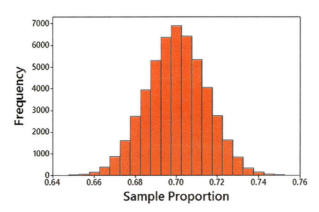

FIGURE 6-17   Histogram of 50,000 Sample Proportions

Let's formally define *sampling distribution,* the main character in the preceding short story.

DEFINITION

The **sampling distribution of a statistic** (such as a sample proportion or sample mean) is the distribution of all values of the statistic when all possible samples of the same size n are taken from the same population. (The sampling distribution of a statistic is typically represented as a probability distribution in the format of a probability histogram, formula, or table.)

Sampling Distribution of Sample Proportion

The preceding general definition of a sampling distribution of a statistic can now be restated for the specific case of a sample proportion:

DEFINITION

The **sampling distribution of the sample proportion** is the distribution of sample proportions (or the distribution of the variable $\hat{p}$), with all samples having the same sample size n taken from the same population. (The sampling distribution of the sample proportion is typically represented as a probability distribution in the format of a probability histogram, formula, or table.)

We need to distinguish between a population proportion p and a sample proportion, and the following notation is common and will be used throughout the remainder of this book, so it's very important.

Notation for Proportions

$p = population$ proportion

$\hat{p} = sample$ proportion

HINT $\hat{p}$ is pronounced "p-hat." When symbols are used above a letter, as in $\bar{x}$ and $\hat{p}$, they represent *statistics,* not parameters.

Behavior of Sample Proportions

1. The distribution of sample proportions tends to approximate a normal distribution.

2. Sample proportions *target* the value of the population proportion in the sense that the mean of all of the sample proportions $\hat{p}$ is equal to the population proportion p; the expected value of the sample proportion is equal to the population proportion.

EXAMPLE 1 **Sampling Distribution of the Sample Proportion**

Consider repeating this process: Roll a die 5 times and find the proportion of *odd* numbers (1 or 3 or 5). What do we know about the behavior of all sample proportions that are generated as this process continues indefinitely?

SOLUTION

Figure 6-18 illustrates a process of rolling a die 5 times and finding the proportion of odd numbers. (Figure 6-18 shows results from repeating this process 10,000 times, but the true sampling distribution of the sample proportion involves repeating the process indefinitely.) Figure 6-18 shows that the sample proportions are approximately normally distributed. (Because the values of 1, 2, 3, 4, 5, 6 are all equally likely, the proportion of odd numbers in the population is 0.5, and Figure 6-18 shows that the sample proportions have a mean of 0.50.)

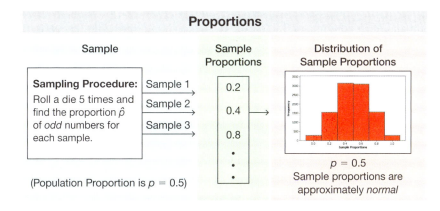

FIGURE 6-18   Sample Proportions from 10,000 Trials

> YOUR TURN Do Exercise 10 "Sampling Distribution of the Sample Proportion."

Sampling Distribution of the Sample Mean

We now consider sample means.

DEFINITION

The **sampling distribution of the sample mean** is the distribution of all possible sample means (or the distribution of the variable $\bar{x}$), with all samples having the same sample size n taken from the same population. (The sampling distribution of the sample mean is typically represented as a probability distribution in the format of a probability histogram, formula, or table.)

Behavior of Sample Means

1. The distribution of sample means tends to be a normal distribution. (This will be discussed further in the following section, but the distribution tends to become closer to a normal distribution as the sample size increases.)

2. The sample means *target* the value of the population mean. (That is, the mean of the sample means is the population mean. The expected value of the sample mean is equal to the population mean.)

CHAPTER 6 Normal Probability Distributions

EXAMPLE 2 **Sampling Distribution of the Sample Mean**

A friend of the author has three children with ages of 4, 5, and 9. Let's consider the population consisting of $\{4, 5, 9\}$. (We don't usually know all values in a population, and we don't usually work with such a small population, but it works well for the purposes of this example.) If two ages are randomly selected with replacement from the population $\{4, 5, 9\}$, identify the sampling distribution of the sample mean by creating a table representing the probability distribution of the sample mean. Do the values of the sample mean target the value of the population mean?

SOLUTION

If two values are randomly selected with replacement from the population $\{4, 5, 9\}$, the leftmost column of Table 6-2 lists the nine different possible samples. The second column lists the corresponding sample means. The nine samples are equally likely with a probability of $1/9$. We saw in Section 5-1 that a probability distribution gives the probability for each value of a random variable, as in the second and third columns of Table 6-2. The second and third columns of Table 6-2 constitute a probability distribution for the random variable representing sample means, so the second and third columns represent the sampling distribution of the sample mean. In Table 6-2, some of the sample mean values are repeated, so we combined them in Table 6-3.

TABLE 6-2 Sampling Distribution of Mean

Sample	Sample Mean $\bar{x}$	Probability
4, 4	4.0	1/9
4, 5	4.5	1/9
4, 9	6.5	1/9
5, 4	4.5	1/9
5, 5	5.0	1/9
5, 9	7.0	1/9
9, 4	6.5	1/9
9, 5	7.0	1/9
9, 9	9.0	1/9

TABLE 6-3 Sampling Distribution of Mean (Condensed)

Sample Mean $\bar{x}$	Probability
4.0	1/9
4.5	2/9
5.0	1/9
6.5	2/9
7.0	2/9
9.0	1/9

INTERPRETATION

Because Table 6-3 lists the possible values of the sample mean along with their corresponding probabilities, Table 6-3 is an example of a sampling distribution of a sample mean.

The value of the mean of the population $\{4, 5, 9\}$ is $\mu = 6.0$. Using either Table 6-2 or 6-3, we could calculate the mean of the sample values and we get 6.0. Because the mean of the sample means (6.0) is equal to the mean of the population (6.0), we conclude that the values of the sample mean do *target* the value of the population mean. It's unfortunate that this sounds so much like doublespeak, but this illustrates that *the mean of the sample means is equal to the population mean* μ.

HINT Read the last sentence of the above paragraph a few times until it makes sense.

If we were to create a probability histogram from Table 6-2, it would not have the bell shape that is characteristic of a normal distribution, but that's because we are working with such small samples. If the population of $\{4, 5, 9\}$ were much larger and if we were selecting samples much larger than $n = 2$ as in this example, we would get a probability histogram that is much closer to being bell-shaped, indicating a normal distribution, as in Example 3.

YOUR TURN Do Exercise 11 "Sampling Distribution of the Sample Mean."

EXAMPLE 3 Sampling Distribution of the Sample Mean

Consider repeating this process: Roll a die 5 times to randomly select 5 values from the population $\{1, 2, 3, 4, 5, 6\}$, then find the mean $\bar{x}$ of the results. What do we know about the behavior of all sample means that are generated as this process continues indefinitely?

SOLUTION

Figure 6-19 illustrates a process of rolling a die 5 times and finding the mean of the results. Figure 6-19 shows results from repeating this process 10,000 times, but the true sampling distribution of the mean involves repeating the process indefinitely. Because the values of 1, 2, 3, 4, 5, 6 are all equally likely, the population has a mean of $\mu = 3.5$. The 10,000 sample means included in Figure 6-19 have a mean of 3.5. If the process is continued indefinitely, the mean of the sample means will be 3.5. Also, Figure 6-19 shows that the distribution of the sample means is approximately a normal distribution.

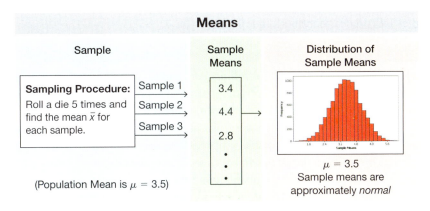

FIGURE 6-19   Sample Means from 10,000 Trials

Sampling Distribution of the Sample Variance

Let's now consider the sampling distribution of sample variances.

DEFINITION

The **sampling distribution of the sample variance** is the distribution of sample variances (the variable s^2), with all samples having the same sample size n taken from the same population. (The sampling distribution of the sample variance is typically represented as a probability distribution in the format of a table, probability histogram, or formula.)

CAUTION When working with population standard deviations or variances, be sure to evaluate them correctly. In Section 3-2 we saw that the computations for *population* standard deviations or variances involve division by the population size N instead of $n - 1$, as shown here.

Population standard deviation: $\quad \sigma = \sqrt{\dfrac{\sum (x - \mu)^2}{N}}$

Population variance: $\quad \sigma^2 = \dfrac{\sum (x - \mu)^2}{N}$

Because the calculations are typically performed with software or calculators, be careful to correctly distinguish between the variance of a sample and the variance of a population.

Behavior of Sample Variances

1. The distribution of sample variances tends to be a distribution skewed to the right.

2. The sample variances *target* the value of the population variance. (That is, the mean of the sample variances is the population variance. The expected value of the sample variance is equal to the population variance.)

EXAMPLE 4 **Sampling Distribution of the Sample Variance**

Consider repeating this process: Roll a die 5 times and find the variance s^2 of the results. What do we know about the behavior of all sample variances that are generated as this process continues indefinitely?

SOLUTION

Figure 6-20 illustrates a process of rolling a die 5 times and finding the variance of the results. Figure 6-20 shows results from repeating this process 10,000 times, but the true sampling distribution of the sample variance involves repeating the process indefinitely. Because the values of 1, 2, 3, 4, 5, 6 are all equally likely, the population has a variance of $\sigma^2 = 2.9$, and the 10,000 sample variances included in Figure 6-20 have a mean of 2.9. If the process is continued indefinitely, the mean of the sample variances will be 2.9. Also, Figure 6-20 shows that the distribution of the sample variances is a skewed distribution, not a normal distribution with its characteristic bell shape.

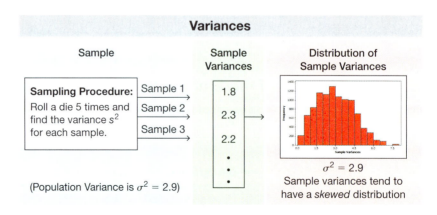

Variances

| Sample | Sample Variances | Distribution of Sample Variances |

Sampling Procedure: Roll a die 5 times and find the variance s^2 for each sample.

Sample 1 → 1.8
Sample 2 → 2.3
Sample 3 → 2.2

(Population Variance is $\sigma^2 = 2.9$)

$\sigma^2 = 2.9$
Sample variances tend to have a *skewed* distribution

FIGURE 6-20 **Sample Variances from 10,000 Trials**

YOUR TURN Do Exercise 14 "Sampling Distribution of the Variance."

Estimators: Unbiased and Biased

The preceding examples show that sample proportions, means, and variances tend to *target* the corresponding population parameters. More formally, we say that sample proportions, means, and variances are *unbiased estimators*. See the following two definitions.

> **DEFINITIONS**
>
> An **estimator** is a statistic used to infer (or estimate) the value of a population parameter.
>
> An **unbiased estimator** is a statistic that targets the value of the corresponding population parameter in the sense that the sampling distribution of the statistic has a mean that is equal to the corresponding population parameter.

Unbiased Estimators These statistics are unbiased estimators. That is, they each target the value of the corresponding population parameter (with a sampling distribution having a mean equal to the population parameter):

- Proportion $\hat{p}$

- Mean $\bar{x}$

- Variance s^2

Biased Estimators These statistics are biased estimators. That is, they do *not* target the value of the corresponding population parameter:

- Median

- Range

- Standard deviation s

    Important Note: The sample standard deviations do not target the population standard deviation σ, but the bias is relatively small in large samples, so **s is often used to estimate σ** even though s is a biased estimator of σ.

EXAMPLE 5 **Sampling Distribution of the Sample Range**

As in Example 2, consider samples of size $n = 2$ randomly selected from the population $\{4, 5, 9\}$.

   a. List the different possible samples along with the probability of each sample, then find the range for each sample.

   b. Describe the sampling distribution of the sample range in the format of a table summarizing the probability distribution.

   c. Based on the results, do the sample ranges target the population range, which is $9 - 4 = 5$?

   d. What do these results indicate about the sample range as an estimator of the population range?

SOLUTION

   a. In Table 6-4 we list the nine different possible samples of size $n = 2$ selected with replacement from the population $\{4, 5, 9\}$. The nine samples are equally likely, so each has probability $1/9$. Table 6-4 also shows the range for each of the nine samples.

continued

TABLE 6-4 Sampling Distribution of Range

Sample	Sample Range	Probability
4, 4	0	1/9
4, 5	1	1/9
4, 9	5	1/9
5, 4	1	1/9
5, 5	0	1/9
5, 9	4	1/9
9, 4	5	1/9
9, 5	4	1/9
9, 9	0	1/9

b. The last two columns of Table 6-4 list the values of the range along with the corresponding probabilities, so the last two columns constitute a table summarizing the probability distribution. Table 6-4 therefore describes the *sampling distribution* of the sample range.

c. The mean of the sample ranges in Table 6-4 is 20/9 or 2.2. The population of $\{4, 5, 9\}$ has a range of $9 - 4 = 5$. Because the mean of the sample ranges (2.2) is not equal to the population range (5), the sample ranges do *not* target the value of the population range.

d. Because the sample ranges do not target the population range, the sample range is a *biased estimator* of the population range.

INTERPRETATION

Because the sample range is a biased estimator of the population range, a sample range should generally not be used to estimate the value of the population range.

YOUR TURN Do Exercise 13 "Sampling Distribution of the Range."

Why Sample with Replacement? All of the examples in this section involved sampling *with replacement*. Sampling *without replacement* would have the very practical advantage of avoiding wasteful duplication whenever the same item is selected more than once. Many of the statistical procedures discussed in the following chapters are based on the assumption that sampling is conducted with replacement because of these two very important reasons:

1. When selecting a relatively small sample from a large population, it makes no significant difference whether we sample with replacement or without replacement.

2. Sampling with replacement results in *independent* events that are unaffected by previous outcomes, and independent events are easier to analyze and result in simpler calculations and formulas.

6-3 Basic Skills and Concepts

Statistical Literacy and Critical Thinking

1. Births There are about 11,000 births each day in the United States, and the proportion of boys born in the United States is 0.512. Assume that each day, 100 births are randomly selected and the proportion of boys is recorded.

a. What do you know about the mean of the sample proportions?

b. What do you know about the shape of the distribution of the sample proportions?

2. Sampling with Replacement The Orangetown Medical Research Center randomly selects 100 births in the United States each day, and the proportion of boys is recorded for each sample.

a. Do you think the births are randomly selected with replacement or without replacement?

b. Give two reasons why statistical methods tend to be based on the assumption that sampling is conducted *with* replacement, instead of without replacement.

3. Unbiased Estimators Data Set 4 "Births" in Appendix B includes birth weights of 400 babies. If we compute the values of sample statistics from that sample, which of the following statistics are *unbiased* estimators of the corresponding population parameters: sample mean; sample median; sample range; sample variance; sample standard deviation; sample proportion?

4. Sampling Distribution Data Set 4 "Births" in Appendix B includes a sample of birth weights. If we explore this sample of 400 birth weights by constructing a histogram and finding the mean and standard deviation, do those results describe the sampling distribution of the mean? Why or why not?

5. Good Sample? A geneticist is investigating the proportion of boys born in the world population. Because she is based in China, she obtains sample data from that country. Is the resulting sample proportion a good estimator of the population proportion of boys born worldwide? Why or why not?

6. College Presidents There are about 4200 college presidents in the United States, and they have annual incomes with a distribution that is skewed instead of being normal. Many different samples of 40 college presidents are randomly selected, and the mean annual income is computed for each sample.

a. What is the approximate shape of the distribution of the sample means (uniform, normal, skewed, other)?

b. What value do the sample means target? That is, what is the mean of all such sample means?

In Exercises 7–10, use the same population of {4, 5, 9} that was used in Examples 2 and 5. As in Examples 2 and 5, assume that samples of size n = 2 are randomly selected with replacement.

7. Sampling Distribution of the Sample Variance

a. Find the value of the population variance σ^2.

b. Table 6-2 describes the sampling distribution of the sample mean. Construct a similar table representing the sampling distribution of the sample variance s^2. Then combine values of s^2 that are the same, as in Table 6-3 (*Hint:* See Example 2 on page 258 for Tables 6-2 and 6-3, which describe the sampling distribution of the sample mean.)

c. Find the mean of the sampling distribution of the sample variance.

d. Based on the preceding results, is the sample variance an unbiased estimator of the population variance? Why or why not?

8. Sampling Distribution of the Sample Standard Deviation For the following, round results to three decimal places.

a. Find the value of the population standard deviation σ.

b. Table 6-2 describes the sampling distribution of the sample mean. Construct a similar table representing the sampling distribution of the sample standard deviation s. Then combine values of s that are the same, as in Table 6-3 (*Hint:* See Example 2 on page 258 for Tables 6-2 and 6-3, which describe the sampling distribution of the sample mean.)

c. Find the mean of the sampling distribution of the sample standard deviation.

d. Based on the preceding results, is the sample standard deviation an unbiased estimator of the population standard deviation? Why or why not?

9. Sampling Distribution of the Sample Median

a. Find the value of the population median.

b. Table 6-2 describes the sampling distribution of the sample mean. Construct a similar table representing the sampling distribution of the sample median. Then combine values of the median that are the same, as in Table 6-3. (*Hint:* See Example 2 on page 258 for Tables 6-2 and 6-3, which describe the sampling distribution of the sample mean.)

c. Find the mean of the sampling distribution of the sample median.

d. Based on the preceding results, is the sample median an unbiased estimator of the population median? Why or why not?

10. Sampling Distribution of the Sample Proportion

a. For the population, find the proportion of odd numbers.

b. Table 6-2 describes the sampling distribution of the sample mean. Construct a similar table representing the sampling distribution of the sample proportion of odd numbers. Then combine values of the sample proportion that are the same, as in Table 6-3. (*Hint:* See Example 2 on page 258 for Tables 6-2 and 6-3, which describe the sampling distribution of the sample mean.)

c. Find the mean of the sampling distribution of the sample proportion of odd numbers.

d. Based on the preceding results, is the sample proportion an unbiased estimator of the population proportion? Why or why not?

In Exercises 11–14, use the population of {34, 36, 41, 51} of the amounts of caffeine (mg / 12 oz) in Coca-Cola Zero, Diet Pepsi, Dr Pepper, and Mellow Yello Zero. Assume that random samples of size n = 2 are selected with replacement.

11. Sampling Distribution of the Sample Mean

a. After identifying the 16 different possible samples, find the mean of each sample, then construct a table representing the sampling distribution of the sample mean. In the table, combine values of the sample mean that are the same. (*Hint:* See Table 6-3 in Example 2 on page 258.)

b. Compare the mean of the population { 34, 36, 41, 51 } to the mean of the sampling distribution of the sample mean.

c. Do the sample means target the value of the population mean? In general, do sample means make good estimators of population means? Why or why not?

12. Sampling Distribution of the Median Repeat Exercise 11 using medians instead of means.

13. Sampling Distribution of the Range Repeat Exercise 11 using ranges instead of means.

14. Sampling Distribution of the Variance Repeat Exercise 11 using variances instead of means.

15. Births: Sampling Distribution of Sample Proportion When two births are randomly selected, the sample space for genders is bb, bg, gb, and gg (where b = boy and g = girl). Assume that those four outcomes are equally likely. Construct a table that describes the sampling distribution of the sample proportion of girls from two births. Does the mean of the sample proportions equal the proportion of girls in two births? Does the result suggest that a sample proportion is an unbiased estimator of a population proportion?

16. Births: Sampling Distribution of Sample Proportion For three births, assume that the genders are equally likely. Construct a table that describes the sampling distribution of the sample proportion of girls from three births. Does the mean of the sample proportions equal the proportion of girls in three births? (*Hint:* See Exercise 15 for two births.)

17. SAT and ACT Tests Because they enable efficient procedures for evaluating answers, multiple choice questions are commonly used on standardized tests, such as the SAT or ACT.

Such questions typically have five choices, one of which is correct. Assume that you must make random guesses for two such questions. Assume that both questions have correct answers of "a."

a. After listing the 25 different possible samples, find the proportion of correct answers in each sample, then construct a table that describes the sampling distribution of the sample proportions of correct responses.

b. Find the mean of the sampling distribution of the sample proportion.

c. Is the mean of the sampling distribution [from part (b)] equal to the population proportion of correct responses? Does the mean of the sampling distribution of proportions *always* equal the population proportion?

18. Hybridization A hybridization experiment begins with four peas having yellow pods and one pea having a green pod. Two of the peas are randomly selected *with replacement* from this population.

a. After identifying the 25 different possible samples, find the proportion of peas with yellow pods in each of them, then construct a table to describe the sampling distribution of the proportions of peas with yellow pods.

b. Find the mean of the sampling distribution.

c. Is the mean of the sampling distribution [from part (b)] equal to the population proportion of peas with yellow pods? Does the mean of the sampling distribution of proportions *always* equal the population proportion?

6-3 Beyond the Basics

19. Using a Formula to Describe a Sampling Distribution Exercise 15 "Births" requires the construction of a table that describes the sampling distribution of the proportions of girls from two births. Consider the formula shown here, and evaluate that formula using sample proportions (represented by x) of 0, 0.5, and 1. Based on the results, does the formula describe the sampling distribution? Why or why not?

$$P(x) = \frac{1}{2(2 - 2x)!(2x)!} \quad \text{where } x = 0, 0.5, 1$$

20. Mean Absolute Deviation Is the mean absolute deviation of a sample a good statistic for estimating the mean absolute deviation of the population? Why or why not? (*Hint:* See Example 5.)

6-4   The Central Limit Theorem

Key Concept In the preceding section we saw that the sampling distribution of sample means tends to be a normal distribution as the sample size increases. In this section we introduce and apply the *central limit theorem*. The central limit theorem allows us to use a normal distribution for some very meaningful and important applications.

Given any population with *any* distribution (uniform, skewed, whatever), the distribution of sample means $\bar{x}$ can be approximated by a normal distribution when the samples are large enough with $n > 30$. (There are some special cases of very non-normal distributions for which the requirement of $n > 30$ isn't quite enough, so the number 30 should be higher in those cases, but those cases are rare.)

CENTRAL LIMIT THEOREM

For all samples of the same size n with $n > 30$, the sampling distribution of $\bar{x}$ can be approximated by a normal distribution with mean μ and standard deviation $\sigma / \sqrt{n}$.

 EXAMPLE 1 **Earthquake Depths**

Figures 6-21 and 6-22 illustrate the central limit theorem.

- **Original data:** Figure 6-21 is a histogram of the depths (km) of 1392 earthquakes, and this histogram shows that those depths have a distribution that is clearly *not* normal.

- **Sample means:** Figure 6-22 is a histogram of 100 *sample means.* Each sample includes 50 earthquake depths (km), and this histogram shows that the sample means have a distribution that is approximately normal.

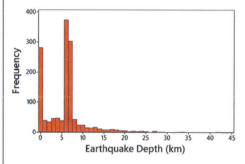

FIGURE 6-21 **Nonnormal Distribution: Depths (km) of 1392 Earthquakes**

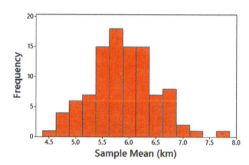

FIGURE 6-22 **Approximately Normal Distribution: Means from Samples of Size $n = 50$ of Earthquake Depths**

INTERPRETATION

The original earthquake depths depicted in Figure 6-21 have a skewed distribution, but when we collect samples and compute their means, those sample means tend to have a distribution that is *normal.*

A Universal Truth Example 1 and the central limit theorem are truly remarkable because they describe a rule of nature that works throughout the universe. If we could send a spaceship to a distant planet "in a galaxy far, far away," and if we collect samples of rocks (all of the same large sample size) and weigh them, the sample means would have a distribution that is approximately normal. Think about the significance of that! Whoa.

The following key points form the foundation for estimating population parameters and hypothesis testing—topics discussed at length in the following chapters.

KEY ELEMENTS

The Central Limit Theorem and the Sampling Distribution of $\bar{x}$

Given

1. Population (with any distribution) has mean μ and standard deviation σ.

2. Simple random samples all of the same size n are selected from the population.

Practical Rules for Real Applications Involving a Sample Mean $\bar{x}$

Requirements: Population has a normal distribution *or* $n > 30$:

$$\text{Mean of all values of } \bar{x}: \qquad \mu_{\bar{x}} = \mu$$

$$\text{Standard deviation of all values of } \bar{x}: \qquad \sigma_{\bar{x}} = \frac{\sigma}{\sqrt{n}}$$

$$z \text{ score conversion of } \bar{x}: \qquad z = \frac{\bar{x} - \mu}{\dfrac{\sigma}{\sqrt{n}}}$$

Original population is *not* normally distributed *and* $n \leq 30$: The distribution of $\bar{x}$ cannot be approximated well by a normal distribution, and the methods of this section do not apply. Use other methods, such as nonparametric methods (Chapter 13) or bootstrapping methods (Section 7-4).

Considerations for Practical Problem Solving

1. **Check Requirements:** When working with the mean from a sample, verify that the normal distribution can be used by confirming that the original population has a normal distribution or the sample size is $n > 30$.

2. **Individual Value or Mean from a Sample?** Determine whether you are using a normal distribution with a *single* value x or the mean $\bar{x}$ from a sample of n values. See the following.

   - Individual value: When working with an *individual* value from a normally distributed population, use the methods of Section 6-2 with $z = \dfrac{x - \mu}{\sigma}$.

   - Mean from a sample of values: When working with a mean for some *sample* of n values, be sure to use the value of $\sigma/\sqrt{n}$ for the standard deviation of the sample means, so use $z = \dfrac{\bar{x} - \mu}{\dfrac{\sigma}{\sqrt{n}}}$.

The following new notation is used for the mean and standard deviation of the distribution of $\bar{x}$.

NOTATION FOR THE SAMPLING DISTRIBUTION OF $\bar{x}$

If all possible simple random samples of size n are selected from a population with mean μ and standard deviation σ, the mean of all sample means is denoted by $\mu_{\bar{x}}$ and the standard deviation of all sample means is denoted by $\sigma_{\bar{x}}$.

$$\text{Mean of all values of } \bar{x}: \qquad \mu_{\bar{x}} = \mu$$

$$\text{Standard deviation of all values of } \bar{x}: \qquad \sigma_{\bar{x}} = \frac{\sigma}{\sqrt{n}}$$

Note: $\sigma_{\bar{x}}$ is called the *standard error of the mean* and is sometimes denoted as SEM.

Applying the Central Limit Theorem

Many practical problems can be solved with the central limit theorem. Example 2 is a good illustration of the central limit theorem because we can see the difference between working with an *individual* value in part (a) and working with the *mean* for a sample in part (b). Study Example 2 carefully to understand the fundamental difference between the procedures used in parts (a) and (b). In particular, note that when working with an *individual* value, we use $z = \dfrac{x - \mu}{\sigma}$, but when working with the mean $\bar{x}$ for a collection of *sample* values, we use $z = \dfrac{\bar{x} - \mu}{\sigma/\sqrt{n}}$.

EXAMPLE 2 Safe Loading of Elevators

The elevator in the car rental building at San Francisco International Airport has a placard stating that the maximum capacity is "4000 lb—27 passengers." Because $4000/27 = 148$, this converts to a mean passenger weight of 148 lb when the elevator is full. We will assume a worst-case scenario in which the elevator is filled with 27 adult males. Based on Data Set 1 "Body Data" in Appendix B, assume that adult males have weights that are normally distributed with a mean of 189 lb and a standard deviation of 39 lb.

a. Find the probability that 1 randomly selected adult male has a weight greater than 148 lb.

b. Find the probability that a sample of 27 randomly selected adult males has a mean weight greater than 148 lb.

SOLUTION

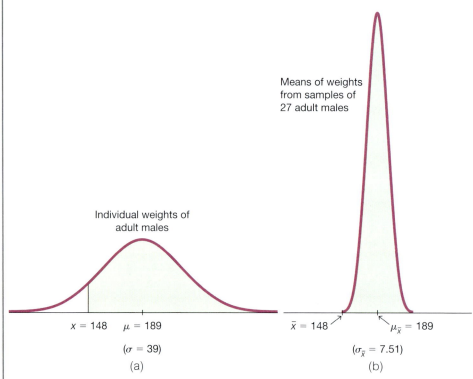

FIGURE 6-23   Elevator Weights

a. **Approach Used for an Individual Value:** Use the methods presented in Section 6-2 because we are dealing with an *individual* value from a normally distributed population. We seek the area of the green-shaded region in Figure 6-23(a).

Technology: If using technology (as described at the end of Section 6-2), we find that the green-shaded area is 0.8534.

Table A-2: If using Table A-2, we convert the weight of $x = 148$ lb to the corresponding z score of $z = -1.05$, as shown here:

$$z = \frac{x - \mu}{\sigma} = \frac{148 - 189}{39} = -1.05$$

We refer to Table A-2 to find that the cumulative area to the *left* of $z = -1.05$ is 0.1469, so the green-shaded area in Figure 6-23(a) is $1 - 0.1469 = 0.8531$. (The result of 0.8534 from technology is more accurate than the result found from Table A-2.)

b. **Approach Used for the Mean of Sample Values:** Use the central limit theorem because we are dealing with the mean of a sample of 27 males, not an individual male.

REQUIREMENT CHECK FOR PART B We can use the normal distribution if the original population is normally distributed or $n > 30$. The sample size is not greater than 30, but the original population of weights of males has a normal distribution, so samples of any size will yield means that are normally distributed. ☑

Because we are now dealing with a distribution of sample means, we must use the parameters $\mu_{\bar{x}}$ and $\sigma_{\bar{x}}$, which are evaluated as follows:

$$\mu_{\bar{x}} = \mu = 189$$

$$\sigma_{\bar{x}} = \frac{\sigma}{\sqrt{n}} = \frac{39}{\sqrt{27}} = 7.51$$

We want to find the green-shaded area shown in Figure 6-23(b).

Technology: If using technology, the green-shaded area in Figure 6-23(b) is 0.99999998.

Table A-2: If using Table A-2, we convert the value of $\bar{x} = 148$ to the corresponding z score of $z = -5.46$, as shown here:

$$z = \frac{x - \mu_{\bar{x}}}{\sigma_{\bar{x}}} = \frac{148 - 189}{\dfrac{39}{\sqrt{27}}} = \frac{-41}{7.51} = -5.46$$

From Table A-2 we find that the cumulative area to the left of $z = -5.46$ is 0.0001, so the green-shaded area of Figure 6-23(b) is $1 - 0.0001 = 0.9999$. We are quite sure that 27 randomly selected adult males have a mean weight greater than 148 lb.

INTERPRETATION

There is a 0.8534 probability that an individual male will weigh more than 148 lb, and there is a 0.99999998 probability that 27 randomly selected males will have a mean weight of more than 148 lb. Given that the safe capacity of the elevator is 4000 lb, it is almost certain that it will be overweight if it is filled with 27 randomly selected adult males.

YOUR TURN Do Exercise 5 "Using the Central Limit Theorem."

The Fuzzy Central Limit Theorem

In *The Cartoon Guide to Statistics*, by Gonick and Smith, the authors describe the Fuzzy Central Limit Theorem as follows: "Data that are influenced by many small and unrelated random effects are approximately normally distributed. This explains why the normal is everywhere: stock market fluctuations, student weights, yearly temperature averages, SAT scores: All are the result of many different effects." People's heights, for example, are the results of hereditary factors, environmental factors, nutrition, health care, geographic region, and other influences, which, when combined, produce normally distributed values.

Example 2 shows that we can use the same basic procedures from Section 6-2, but we must remember to correctly adjust the standard deviation when working with a *sample mean* instead of an individual sample value. The calculations used in Example 2 are exactly the type of calculations used by engineers when they design ski lifts, escalators, airplanes, boats, amusement park rides, and other devices that carry people.

Introduction to Hypothesis Testing

Carefully examine the conclusions that are reached in the next example illustrating the type of thinking that is the basis for the important procedure of hypothesis testing (formally introduced in Chapter 8). Example 3 uses the rare event rule for inferential statistics, first presented in Section 4-1:

Identifying Significant Results with Probabilities: The Rare Event Rule for Inferential Statistics

If, under a given assumption, the probability of a particular observed event is very small and the observed event occurs *significantly less than* or *significantly greater than* what we typically expect with that assumption, we conclude that the assumption is probably not correct.

The following example illustrates the above rare event rule and it uses the author's all-time favorite data set.

 EXAMPLE 3 Body Temperatures

Assume that the population of human body temperatures has a mean of 98.6°F, as is commonly believed. Also assume that the population standard deviation is 0.62°F (based on data from University of Maryland researchers). If a sample of size $n = 106$ is randomly selected, find the probability of getting a mean of 98.2°F or lower. (The value of 98.2°F was actually obtained from researchers; see the midnight temperatures for Day 2 in Data Set 3 "Body Temperatures" in Appendix B.)

SOLUTION

We work under the assumption that the population of human body temperatures has a mean of 98.6°F. We weren't given the distribution of the population, but because the sample size $n = 106$ exceeds 30, we use the central limit theorem and conclude that the distribution of sample means is a normal distribution with these parameters:

$$\mu_{\bar{x}} = \mu = 98.6 \text{ (by assumption)}$$

$$\sigma_{\bar{x}} = \frac{\sigma}{\sqrt{n}} = \frac{0.62}{\sqrt{106}} = 0.0602197$$

Figure 6-24 shows the shaded area (see the tiny left tail of the graph) corresponding to the probability we seek. Having already found the parameters that apply to the distribution shown in Figure 6-24, we can now find the shaded area by using the same procedures developed in Section 6-2.

Technology: If we use technology to find the shaded area in Figure 6-24, we get 0.0000000000155, which can be expressed as 0+.

Table A-2: If we use Table A-2 to find the shaded area in Figure 6-24, we must first convert the score of $x = 98.20$°F to the corresponding z score:

$$z = \frac{\bar{x} - \mu_{\bar{x}}}{\sigma_{\bar{x}}} = \frac{98.20 - 98.6}{0.0602197} = -6.64$$

Referring to Table A-2, we find that $z = -6.64$ is off the chart, but for values of z below -3.49, we use an area of 0.0001 for the cumulative left area up to $z = -3.49$. We therefore conclude that the shaded region in Figure 6-24 is 0.0001.

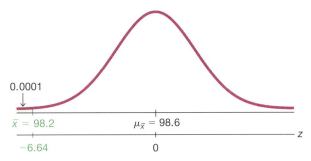

FIGURE 6-24   Means of Body Temperatures from Samples of Size $n = 106$

INTERPRETATION

The result shows that if the mean of our body temperatures is really 98.6°F, as we assumed, then there is an extremely small probability of getting a sample mean of 98.2°F or lower when 106 subjects are randomly selected. University of Maryland researchers did obtain such a sample mean, and after confirming that the sample is sound, there are two feasible explanations: (1) The population mean really is 98.6°F and their sample represents a chance event that is extremely rare; (2) the population mean is actually lower than the assumed value of 98.6°F and so their sample is typical. Because the probability is so low, it is more reasonable to conclude that the population mean is lower than 98.6°F. In reality it appears that the true mean body temperature is closer to 98.2°F!

This is the type of reasoning used in *hypothesis testing,* to be introduced in Chapter 8. For now, we should focus on the use of the central limit theorem for finding the probability of 0.0001, but we should also observe that this theorem will be used later in applying some very important concepts in statistics.

YOUR TURN Do Exercise 9 "Elevator Safety."

Correction for a Finite Population

In applying the central limit theorem, our use of $\sigma_{\bar{x}} = \sigma/\sqrt{n}$ assumes that the population has infinitely many members. When we sample with replacement, the population is effectively infinite. When sampling without replacement from a finite population, we may need to adjust $\sigma_{\bar{x}}$. Here is a common rule:

> **When sampling without replacement and the sample size n is greater than 5% of the finite population size N (that is, $n > 0.05N$), adjust the standard deviation of sample means $\sigma_{\bar{x}}$ by multiplying it by this *finite population correction factor*:**

$$\sqrt{\frac{N - n}{N - 1}}$$

Except for Exercise 21 "Correcting for a Finite Population," the examples and exercises in this section assume that the finite population correction factor does *not* apply, because we are sampling with replacement, or the population is infinite, or the sample size doesn't exceed 5% of the population size.

6-4   Basic Skills and Concepts

Statistical Literacy and Critical Thinking

1. Requirements A researcher collects a simple random sample of grade-point averages of statistics students, and she calculates the mean of this sample. Under what conditions can that sample mean be treated as a value from a population having a normal distribution?

2. Small Sample Weights of golden retriever dogs are normally distributed. Samples of weights of golden retriever dogs, each of size $n = 15$, are randomly collected and the sample means are found. Is it correct to conclude that the sample means cannot be treated as being from a normal distribution because the sample size is too small? Explain.

3. Notation In general, what do the symbols $\mu_{\bar{x}}$ and $\sigma_{\bar{x}}$ represent? What are the values of $\mu_{\bar{x}}$ and $\sigma_{\bar{x}}$ for samples of size 64 randomly selected from the population of IQ scores with population mean of 100 and standard deviation of 15?

4. Annual Incomes Annual incomes are known to have a distribution that is skewed to the right instead of being normally distributed. Assume that we collect a large $(n > 30)$ random sample of annual incomes. Can the distribution of incomes in that sample be approximated by a normal distribution because the sample is large? Why or why not?

Using the Central Limit Theorem.   *In Exercises 5–8, assume that females have pulse rates that are normally distributed with a mean of 74.0 beats per minute and a standard deviation of 12.5 beats per minute (based on Data Set 1 "Body Data" in Appendix B).*

5. a. If 1 adult female is randomly selected, find the probability that her pulse rate is less than 80 beats per minute.

b. If 16 adult females are randomly selected, find the probability that they have pulse rates with a mean less than 80 beats per minute.

c. Why can the normal distribution be used in part (b), even though the sample size does not exceed 30?

6. a. If 1 adult female is randomly selected, find the probability that her pulse rate is greater than 70 beats per minute.

b. If 25 adult females are randomly selected, find the probability that they have pulse rates with a mean greater than 70 beats per minute.

c. Why can the normal distribution be used in part (b), even though the sample size does not exceed 30?

7. a. If 1 adult female is randomly selected, find the probability that her pulse rate is between 72 beats per minute and 76 beats per minute.

b. If 4 adult females are randomly selected, find the probability that they have pulse rates with a mean between 72 beats per minute and 76 beats per minute.

c. Why can the normal distribution be used in part (b), even though the sample size does not exceed 30?

8. a. If 1 adult female is randomly selected, find the probability that her pulse rate is between 78 beats per minute and 90 beats per minute.

b. If 16 adult females are randomly selected, find the probability that they have pulse rates with a mean between 78 beats per minute and 90 beats per minute.

c. Why can the normal distribution be used in part (b), even though the sample size does not exceed 30?

9. Elevator Safety Example 2 referred to an elevator with a maximum capacity of 4000 lb. When rating elevators, it is common to use a 25% safety factor, so the elevator should actually be able to carry a load that is 25% greater than the stated limit. The maximum capacity of 4000 lb becomes 5000 lb after it is increased by 25%, so 27 adult male passengers can have a mean weight of up to 185 lb. If the elevator is loaded with 27 adult male passengers, find the probability that it is overloaded because they have a mean weight greater than 185 lb. (As in Example 2, assume that weights of males are normally distributed with a mean of 189 lb and a standard deviation of 39 lb.) Does this elevator appear to be safe?

10. Elevator Safety Exercise 9 uses $\mu = 189$ lb, which is based on Data Set 1 "Body Data" in Appendix B. Repeat Exercise 9 using $\mu = 174$ lb (instead of 189 lb), which is the assumed mean weight that was commonly used just a few years ago. What do you conclude about the effect of using an outdated mean that is substantially lower than it should be?

11. Mensa Membership in Mensa requires a score in the top 2% on a standard intelligence test. The Wechsler IQ test is designed for a mean of 100 and a standard deviation of 15, and scores are normally distributed.

a. Find the minimum Wechsler IQ test score that satisfies the Mensa requirement.

b. If 4 randomly selected adults take the Wechsler IQ test, find the probability that their mean score is at least 131.

c. If 4 subjects take the Wechsler test and they have a mean of 132 but the individual scores are lost, can we conclude that all 4 of them are eligible for Mensa?

12. Designing Manholes According to the website www.torchmate.com, "manhole covers must be a minimum of 22 in. in diameter, but can be as much as 60 in. in diameter." Assume that a manhole is constructed to have a circular opening with a diameter of 22 in. Men have shoulder breadths that are normally distributed with a mean of 18.2 in. and a standard deviation of 1.0 in. (based on data from the National Health and Nutrition Examination Survey).

a. What percentage of men will fit into the manhole?

b. Assume that the Connecticut's Eversource company employs 36 men who work in manholes. If 36 men are randomly selected, what is the probability that their mean shoulder breadth is less than 18.5 in.? Does this result suggest that money can be saved by making smaller manholes with a diameter of 18.5 in.? Why or why not?

13. Water Taxi Safety Passengers died when a water taxi sank in Baltimore's Inner Harbor. Men are typically heavier than women and children, so when loading a water taxi, assume a worst-case scenario in which all passengers are men. Assume that weights of men are normally distributed with a mean of 189 lb and a standard deviation of 39 lb (based on Data Set 1 "Body Data" in Appendix B). The water taxi that sank had a stated capacity of 25 passengers, and the boat was rated for a load limit of 3500 lb.

a. Given that the water taxi that sank was rated for a load limit of 3500 lb, what is the maximum mean weight of the passengers if the boat is filled to the stated capacity of 25 passengers?

b. If the water taxi is filled with 25 randomly selected men, what is the probability that their mean weight exceeds the value from part (a)?

c. After the water taxi sank, the weight assumptions were revised so that the new capacity became 20 passengers. If the water taxi is filled with 20 randomly selected men, what is the probability that their mean weight exceeds 175 lb, which is the maximum mean weight that does not cause the total load to exceed 3500 lb?

d. Is the new capacity of 20 passengers safe?

14. Vending Machines Quarters are now manufactured so that they have a mean weight of 5.670 g and a standard deviation of 0.062 g, and their weights are normally distributed. A vending machine is configured to accept only those quarters that weigh between 5.550 g and 5.790 g.

a. If 1 randomly selected quarter is inserted into the vending machine, what is the probability that it will be accepted?

continued

b. If 4 randomly selected quarters are inserted into the vending machine, what is the probability that their mean weight is between 5.550 g and 5.790 g?

c. If you own the vending machine, which result is more important: the result from part (a) or part (b)?

15. Southwest Airlines Seats Southwest Airlines currently has a seat width of 17 in. Men have hip breadths that are normally distributed with a mean of 14.4 in. and a standard deviation of 1.0 in. (based on anthropometric survey data from Gordon, Churchill, et al.).

a. Find the probability that if an individual man is randomly selected, his hip breadth will be greater than 17 in.

b. Southwest Airlines uses a Boeing 737 for some of its flights, and that aircraft seats 122 passengers. If the plane is full with 122 randomly selected men, find the probability that these men have a mean hip breadth greater than 17 in.

c. Which result should be considered for any changes in seat design: the result from part (a) or part (b)?

16. Coke Cans Assume that cans of Coke are filled so that the actual amounts are normally distributed with a mean of 12.00 oz and a standard deviation of 0.11 oz.

a. Find the probability that a single can of Coke has at least 12.19 oz.

b. The 36 cans of Coke in Data Set 26 "Cola Weights and Volumes" in Appendix B have a mean of 12.19 oz. Find the probability that 36 random cans of Coke have a mean of at least 12.19 oz.

c. Given the result from part (b), is it reasonable to believe that the cans are actually filled with a mean equal to 12.00 oz? If the mean is not equal to 12.00 oz, are consumers being cheated?

17. Redesign of Ejection Seats When women were finally allowed to become pilots of fighter jets, engineers needed to redesign the ejection seats because they had been originally designed for men only. The ACES-II ejection seats were designed for men weighing between 140 lb and 211 lb. Weights of women are now normally distributed with a mean of 171 lb and a standard deviation of 46 lb (based on Data Set 1 "Body Data" in Appendix B).

a. If 1 woman is randomly selected, find the probability that her weight is between 140 lb and 211 lb.

b. If 25 different women are randomly selected, find the probability that their mean weight is between 140 lb and 211 lb.

c. When redesigning the fighter jet ejection seats to better accommodate women, which probability is more relevant: the result from part (a) or the result from part (b)? Why?

18. Loading a Tour Boat The Ethan Allen tour boat capsized and sank in Lake George, New York, and 20 of the 47 passengers drowned. Based on a 1960 assumption of a mean weight of 140 lb for passengers, the boat was rated to carry 50 passengers. After the boat sank, New York State changed the assumed mean weight from 140 lb to 174 lb.

a. Given that the boat was rated for 50 passengers with an assumed mean of 140 lb, the boat had a passenger load limit of 7000 lb. Assume that the boat is loaded with 50 male passengers, and assume that weights of men are normally distributed with a mean of 189 lb and a standard deviation of 39 lb (based on Data Set 1 "Body Data" in Appendix B). Find the probability that the boat is overloaded because the 50 male passengers have a mean weight greater than 140 lb.

b. The boat was later rated to carry only 14 passengers, and the load limit was changed to 2436 lb. If 14 passengers are all males, find the probability that the boat is overloaded because their mean weight is greater than 174 lb (so that their total weight is greater than the maximum capacity of 2436 lb). Do the new ratings appear to be safe when the boat is loaded with 14 male passengers?

19. Doorway Height The Boeing 757-200 ER airliner carries 200 passengers and has doors with a height of 72 in. Heights of men are normally distributed with a mean of 68.6 in. and a standard deviation of 2.8 in. (based on Data Set 1 "Body Data" in Appendix B).

a. If a male passenger is randomly selected, find the probability that he can fit through the doorway without bending.

b. If half of the 200 passengers are men, find the probability that the mean height of the 100 men is less than 72 in.

c. When considering the comfort and safety of passengers, which result is more relevant: the probability from part (a) or the probability from part (b)? Why?

d. When considering the comfort and safety of passengers, why are women ignored in this case?

20. Loading Aircraft Before every flight, the pilot must verify that the total weight of the load is less than the maximum allowable load for the aircraft. The Bombardier Dash 8 aircraft can carry 37 passengers, and a flight has fuel and baggage that allows for a total passenger load of 6200 lb. The pilot sees that the plane is full and all passengers are men. The aircraft will be overloaded if the mean weight of the passengers is greater than $6200 \text{ lb}/37 = 167.6$ lb. What is the probability that the aircraft is overloaded? Should the pilot take any action to correct for an overloaded aircraft? Assume that weights of men are normally distributed with a mean of 189 lb and a standard deviation of 39 lb (based on Data Set 1 "Body Data" in Appendix B).

6-4 Beyond the Basics

21. Correcting for a Finite Population In a study of babies born with very low birth weights, 275 children were given IQ tests at age 8, and their scores approximated a normal distribution with $\mu = 95.5$ and $\sigma = 16.0$ (based on data from "Neurobehavioral Outcomes of School-age Children Born Extremely Low Birth Weight or Very Preterm," by Anderson et al., *Journal of the American Medical Association*, Vol. 289, No. 24). Fifty of those children are to be randomly selected without replacement for a follow-up study.

a. When considering the distribution of the mean IQ scores for samples of 50 children, should $\sigma_{\bar{x}}$ be corrected by using the finite population correction factor? Why or why not? What is the value of $\sigma_{\bar{x}}$?

b. Find the probability that the mean IQ score of the follow-up sample is between 95 and 105.

6-5 Assessing Normality

Key Concept The following chapters include important statistical methods requiring that sample data are from a population having a *normal* distribution. In this section we present criteria for determining whether the requirement of a normal distribution is satisfied. The criteria involve (1) visual inspection of a histogram to see if it is roughly bell-shaped; (2) identifying any outliers; and (3) constructing a *normal quantile plot*.

 PART 1 **Basic Concepts of Assessing Normality**

When trying to determine whether a collection of data has a distribution that is approximately normal, we can visually inspect a histogram to see if it is approximately bell-shaped (as discussed in Section 2-2), we can identify outliers, and we can also use a *normal quantile plot* (discussed briefly in Section 2-2).

> **DEFINITION**
>
> A **normal quantile plot** (or **normal probability plot**) is a graph of points (x, y) where each x value is from the original set of sample data, and each y value is the corresponding z score that is expected from the standard normal distribution.

Procedure for Determining Whether It Is Reasonable to Assume That Sample Data Are from a Population Having a Normal Distribution

1. *Histogram:* Construct a histogram. If the histogram departs dramatically from a bell shape, conclude that the data do not have a normal distribution.

2. *Outliers:* Identify outliers. If there is more than one outlier present, conclude that the data might not have a normal distribution. (Just one outlier could be an error or the result of chance variation, but be careful, because even a single outlier can have a dramatic effect on results.)

3. *Normal quantile plot:* If the histogram is basically symmetric and the number of outliers is 0 or 1, use technology to generate a *normal quantile plot*. Apply the following criteria to determine whether the distribution is normal. (These criteria can be used loosely for small samples, but they should be used more strictly for large samples.)

   Normal Distribution: The population distribution is normal if the pattern of the points is reasonably close to a straight line and the points do not show some systematic pattern that is not a straight-line pattern.

   Not a Normal Distribution: The population distribution is *not* normal if either or both of these two conditions applies:

   - The points do not lie reasonably close to a straight line.

   - The points show some *systematic pattern* that is not a straight-line pattern.

Histograms and Normal Quantile Plots

In Part 2 of this section we describe the process of constructing a normal quantile plot, but for now we focus on interpreting a normal quantile plot. The following displays show histograms of data along with the corresponding normal quantile plots.

   Normal: The first case shows a histogram of IQ scores that is close to being bell-shaped, so the histogram suggests that the IQ scores are from a normal distribution. The corresponding normal quantile plot shows points that are reasonably close to a straight-line pattern, and the points do not show any other systematic pattern that is not a straight line. It is safe to assume that these IQ scores are from a population that has a normal distribution.

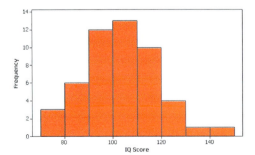
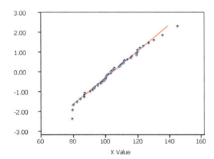

Uniform: The second case shows a histogram of data having a uniform (flat) distribution. The corresponding normal quantile plot suggests that the points are not normally distributed. Although the pattern of points is reasonably close to a straight-line pattern, *there is another systematic pattern that is not a straight-line pattern.* We conclude that these sample values are from a population having a distribution that is not normal.

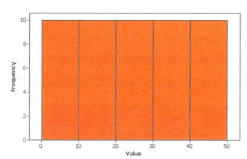 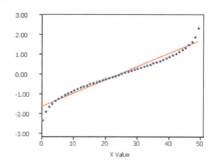

Skewed: The third case shows a histogram of the amounts of rainfall (in inches) in Boston for every Monday during one year. The shape of the histogram is skewed to the right, not bell-shaped. The corresponding normal quantile plot shows points that are not at all close to a straight-line pattern. These rainfall amounts are from a population having a distribution that is not normal.

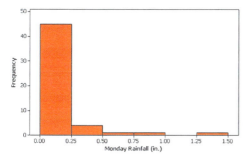 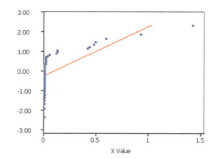

Tools for Determining Normality

- **Histogram/Outliers:** If the requirement of a normal distribution is not too strict, simply look at a histogram and find the number of outliers. If the histogram is roughly bell-shaped and the number of outliers is 0 or 1, treat the population as if it has a normal distribution.

- **Normal Quantile Plot:** Normal quantile plots can be difficult to construct on your own, but they can be generated with a TI-83/84 Plus calculator or suitable software, such as Statdisk, Minitab, Excel, or StatCrunch.

- **Advanced Methods:** In addition to the procedures discussed in this section, there are other more advanced procedures for assessing normality, such as the chi-square goodness-of-fit test, the Kolmogorov-Smirnov test, the Lilliefors test, the Anderson-Darling test, the Jarque-Bera test, and the Ryan-Joiner test (discussed briefly in Part 2).

PART 2 Manual Construction of Normal Quantile Plots

The following is a relatively simple procedure for manually constructing a normal quantile plot, and it is the same procedure used by Statdisk and the TI-83/84 Plus calculator. Some statistical packages use various other approaches, but the interpretation of the graph is essentially the same.

Manual Construction of a Normal Quantile Plot

Step 1: First sort the data by arranging the values in order from lowest to highest.

Step 2: With a sample of size n, each value represents a proportion of $1/n$ of the sample. Using the known sample size n, find the values of $\frac{1}{2n}, \frac{3}{2n}, \frac{5}{2n}$, and so on, until you get n values. These values are the cumulative areas to the left of the corresponding sample values.

Step 3: Use the standard normal distribution (software or a calculator or Table A-2) to find the z scores corresponding to the cumulative left areas found in Step 2. (These are the z scores that are expected from a normally distributed sample.)

Step 4: Match the original sorted data values with their corresponding z scores found in Step 3, then plot the points (x, y), where each x is an original sample value and y is the corresponding z score.

Step 5: Examine the normal quantile plot and use the criteria given in Part 1. Conclude that the population has a normal distribution if the pattern of the points is reasonably close to a straight line and the points do not show some systematic pattern that is not a straight-line pattern.

EXAMPLE 1   Old Faithful Eruption Times

Data Set 23 "Old Faithful" in Appendix B includes duration times (seconds) of eruptions of the Old Faithful Geyser. Let's consider this sample of five eruption times: 125, 229, 236, 257, 234. With only five values, a histogram will not be very helpful in revealing the distribution of the data. Instead, construct a normal quantile plot for these five values and determine whether they appear to come from a population that is normally distributed.

SOLUTION

The following steps correspond to those listed in the procedure above for constructing a normal quantile plot.

Step 1: First, sort the data by arranging them in order. We get 125, 229, 234, 236, 257.

Step 2: With a sample of size $n = 5$, each value represents a proportion of $1/5$ of the sample, so we proceed to identify the cumulative areas to the left of the corresponding sample values. The cumulative left areas, which are expressed in general as $\frac{1}{2n}, \frac{3}{2n}, \frac{5}{2n}$, and so on, become these specific areas for this example with $n = 5$: $\frac{1}{10}, \frac{3}{10}, \frac{5}{10}, \frac{7}{10}, \frac{9}{10}$. These cumulative left areas expressed in decimal form are 0.1, 0.3, 0.5, 0.7, and 0.9.

Step 3: We now use technology (or Table A-2) with the cumulative left areas of 0.1000, 0.3000, 0.5000, 0.7000, and 0.9000 to find these corresponding z scores: $-1.28, -0.52, 0, 0.52$, and 1.28. (For example, the z score of -1.28 has an area of 0.1000 to its left.)

Step 4: We now pair the original sorted eruption duration times with their corresponding z scores. We get these (x, y) coordinates, which are plotted in the accompanying Statdisk display:

$$(125, -1.28), (229, -0.52), (234, 0), (236, 0.52), (257, 1.28)$$

Statdisk

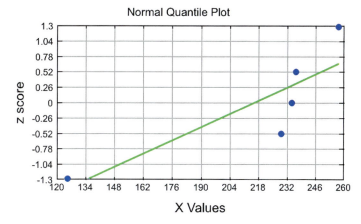

INTERPRETATION

We examine the normal quantile plot in the Statdisk display. The points do not appear to lie reasonably close to the straight line, so we conclude that the sample of
five eruption times does *not* appear to be from a normally distributed population.

YOUR TURN Do Exercise 17 "Female Arm Circumference."

Ryan-Joiner Test The Ryan-Joiner test is one of several formal tests of normality,
each having its own advantages and disadvantages. Statdisk has a feature of **Normality Assessment** that displays a histogram, normal quantile plot, the number of potential outliers, and results from the Ryan-Joiner test.

EXAMPLE 2 **Old Faithful Eruption Times**

Example 1 used only five of the eruption times listed in Data Set 23 "Old Faithful"
in Appendix B. We can use the **Normality Assessment** feature of Statdisk with all
250 eruption times to get the accompanying display.

Statdisk

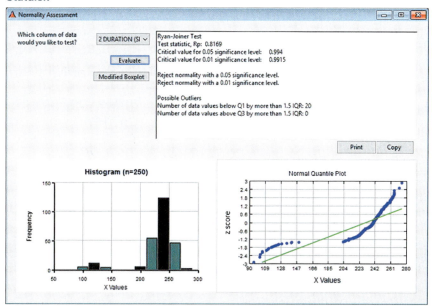

continued

Let's use the display with the three criteria for assessing normality.

1. *Histogram:* We can see that the histogram is *skewed* to the left and is far from being bell-shaped.

2. *Outliers:* The display shows that there are 20 possible outliers. If we examine a sorted list of the 250 eruption times, the 20 lowest times do appear to be outliers.

3. *Normal quantile plot:* Whoa! The points in the normal quantile plot are very far from a straight-line pattern. We conclude that the 250 eruption times do *not* appear to be from a population with a normal distribution.

> **YOUR TURN** Do Exercise 19 "Brain Volumes."

Data Transformations Many data sets have a distribution that is not normal, but we can *transform* the data so that the modified values have a normal distribution. One common transformation is to transform each value of x by taking its logarithm. (You can use natural logarithms or logarithms with base 10. If any original values are 0, take logarithms of values of $x + 1$). If the distribution of the logarithms of the values is a normal distribution, the distribution of the original values is called a **lognormal distribution**. (See Exercises 21 "Transformations" and 22 "Lognormal Distribution.") In addition to transformations with logarithms, there are other transformations, such as replacing each x value with $\sqrt{x}$, or $1/x$, or x^2. In addition to getting a required normal distribution when the original data values are not normally distributed, such transformations can be used to correct deficiencies, such as a requirement (found in later chapters) that different data sets have the same variance.

TECH CENTER

 Normal Quantile Plots
Access tech supplements, videos, and data sets at **www.TriolaStats.com**

Statdisk	Minitab	StatCrunch

Statdisk

1. Click **Data** in the top menu.
2. Select **Normal Quantile Plot** from the dropdown menu.
3. Select the desired data column and click **Plot**.

TIP: Select **Normality Assessment** in the dropdown menu under **Data** to obtain the normal quantile plot along with other results helpful in assessing normality.

Minitab

Minitab generates a probability plot that is similar to the normal quantile plot and can be interpreted using the same criteria given in this section.

Probability Plot

1. Click **Stat** in the top menu.
2. Select **Basic Statistics** from the dropdown menu and select **Normality Test** from the sub-menu.
3. Select the desired column in the *Variable* box and click **OK**.

Probability Plot with Boundaries

1. Click **Graph** in the top menu.
2. Select **Probability Plot** from the dropdown menu, select **Single**, and click **OK**.
3. Select the desired column in the *Graph Variables* box and click **OK**.
4. If the points all lie within the boundaries, conclude that the data are normally distributed. If points are outside the boundaries, conclude that the data are not normally distributed.

StatCrunch

StatCrunch generates a QQ Plot that is similar to the normal quantile plot and can be interpreted using the same criteria given in this section.

1. Click **Graph** in the top menu.
2. Select **QQ Plot** in the dropdown menu.
3. Select the desired data column.
4. Click **Compute!**

TECH CENTER *continued*

Normal Quantile Plots
Access tech supplements, videos, and data sets at **www.TriolaStats.com**

TI-83/84 Plus Calculator

1. Open the **STAT PLOTS** menu by pressing **2ND**, **Y=**.
2. Press **ENTER** to access the Plot 1 settings screen as shown:
   a. Select **ON** and press **ENTER**.
   b. Select last graph type, press **ENTER**.
   c. Enter name of list containing data.
   d. For *Data Axis* select **X**.
3. Press **ZOOM** then **9** (ZoomStat) to generate the normal quantile plot.
4. Press **WINDOW** to customize graph and then press **GRAPH** to view the normal quantile plot.

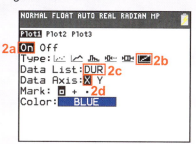

Excel

XLSTAT Add-In (Required)

1. Click on the **XLSTAT** tab in the Ribbon and then click **Describing Data**.
2. Select **Normality tests** from the dropdown menu.
3. Enter the desired data range in the *Data* box. If the first row of data contains a label, check the **Sample labels** box.
4. Click the **Charts** tab and confirm that the **Normal Q-Q plots** box is checked.
5. Click **OK** and scroll down the results to view the Normal Q-Q plot.

6-5 Basic Skills and Concepts

Statistical Literacy and Critical Thinking

1. Normal Quantile Plot Data Set 1 "Body Data" in Appendix B includes the heights of 147 randomly selected women, and heights of women are normally distributed. If you were to construct a histogram of the 147 heights of women in Data Set 1, what shape do you expect the histogram to have? If you were to construct a normal quantile plot of those same heights, what pattern would you expect to see in the graph?

2. Normal Quantile Plot After constructing a histogram of the ages of the 147 women included in Data Set 1 "Body Data" in Appendix B, you see that the histogram is far from being bell-shaped. What do you now know about the pattern of points in the normal quantile plot?

3. Small Sample Data set 29 "Coin Weights" in Appendix B includes weights of 20 one-dollar coins. Given that the sample size is less than 30, what requirement must be met in order to treat the sample mean as a value from a normally distributed population? Identify three tools for verifying that requirement.

4. Assessing Normality The accompanying histogram is constructed from the diastolic blood pressure measurements of the 147 women included in Data Set 1 "Body Data" in Appendix B. If you plan to conduct further statistical tests and there is a loose requirement of a normally distributed population, what do you conclude about the population distribution based on this histogram?

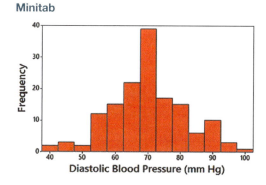

Interpreting Normal Quantile Plots. *In Exercises 5–8, examine the normal quantile plot and determine whether the sample data appear to be from a population with a normal distribution.*

5. Ages of Presidents The normal quantile plot represents the ages of presidents of the United States at the times of their inaugurations. The data are from Data Set 15 "Presidents" in Appendix B.

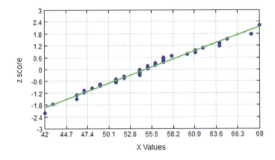

6. Diet Pepsi The normal quantile plot represents weights (pounds) of the contents of cans of Diet Pepsi from Data Set 26 "Cola Weights and Volumes" in Appendix B.

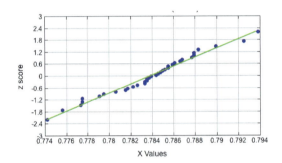

7. Dunkin' Donuts Service Times The normal quantile plot represents service times during the dinner hours at Dunkin' Donuts (from Data Set 25 "Fast Food" in Appendix B).

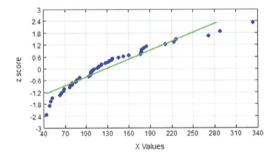

8. Tornadoes The normal quantile plot represents the distances (miles) that tornadoes traveled (from Data Set 22 "Tornadoes" in Appendix B).

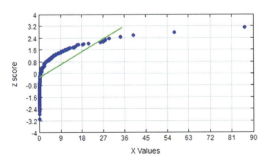

Determining Normality. *In Exercises 9–12, refer to the indicated sample data and determine whether they appear to be from a population with a normal distribution. Assume that this requirement is loose in the sense that the population distribution need not be exactly normal, but it must be a distribution that is roughly bell-shaped.*

9. Cookies The numbers of chocolate chips in Chips Ahoy (reduced fat) cookies, as listed in Data Set 28 "Chocolate Chip Cookies" in Appendix B.

10. Ages of Best Actresses The ages of actresses at the times that they won Oscars, as listed in Data Set 14 "Oscar Winner Age" in Appendix B.

11. Garbage The weights (pounds) of discarded yard waste, as listed in Data Set 31 "Garbage Weight" in Appendix B.

12. Diet Coke The weights (lb) of the contents in cans of Diet Coke, as listed in Data Set 26 "Cola Weights and Volumes" in Appendix B.

Using Technology to Generate Normal Quantile Plots. *In Exercises 13–16, use the data from the indicated exercise in this section. Use software (such as Statdisk, Minitab, Excel, or StatCrunch) or a TI-83/84 Plus calculator to generate a normal quantile plot. Then determine whether the data come from a normally distributed population.*

13. Exercise 9 "Cookies"   **14.** Exercise 10 "Ages of Best Actresses"

15. Exercise 11 "Garbage"   **16.** Exercise 12 "Diet Coke"

Constructing Normal Quantile Plots. *In Exercises 17–20, use the given data values to identify the corresponding z scores that are used for a normal quantile plot, then identify the coordinates of each point in the normal quantile plot. Construct the normal quantile plot, then determine whether the data appear to be from a population with a normal distribution.*

17. Female Arm Circumferences A sample of arm circumferences (cm) of females from Data Set 1 "Body Data" in Appendix B: 40.7, 44.3, 34.2, 32.5, 38.5.

18. Earthquake Depths A sample of depths (km) of earthquakes is obtained from Data Set 21 "Earthquakes" in Appendix B: 17.3, 7.0, 7.0, 7.0, 8.1, 6.8.

19. Brain Volumes A sample of human brain volumes (cm^3) is obtained from those listed in Data Set 8 "IQ and Brain Size" in Appendix B: 1027, 1029, 1034, 1070, 1079, 1079, 963, 1439.

20. McDonald's Dinner Service Times A sample of drive-through service times (seconds) at McDonald's during dinner hours, as listed in Data Set 25 "Fast Food" in Appendix B: 84, 121, 119, 146, 266, 181, 123, 152, 162.

6-5   Beyond the Basics

21. Transformations The heights (in inches) of men listed in Data Set 1 "Body Data" in Appendix B have a distribution that is approximately normal, so it appears that those heights are from a normally distributed population.

a. If 2 inches is added to each height, are the new heights also normally distributed?

b. If each height is converted from inches to centimeters, are the heights in centimeters also normally distributed?

c. Are the logarithms of normally distributed heights also normally distributed?

22. Lognormal Distribution The following are the values of net worth (in thousands of dollars) of recent members of the executive branch of the U.S. government. Test these values for normality, then take the logarithm of each value and test for normality. What do you conclude?

237,592   16,068   15,350   11,712   7304   6037   4483   4367   2658   1361   311

6-6   Normal as Approximation to Binomial

Key Concept Section 5-2 introduced binomial probability distributions, and this section presents a method for using a normal distribution as an approximation to a binomial probability distribution, so that some problems involving proportions can be solved by using a normal distribution. Here are the two main points of this section:

- Given probabilities p and q (where $q = 1 - p$) and sample size n, if the conditions $np \geq 5$ and $nq \geq 5$ are both satisfied, then probabilities from a binomial probability distribution can be approximated reasonably well by using a normal distribution having these parameters:

$$\mu = np$$
$$\sigma = \sqrt{npq}$$

- The binomial probability distribution is *discrete* (with whole numbers for the random variable x), but the normal approximation is *continuous*. To compensate, we use a "continuity correction" with a whole number x represented by the interval from $x - 0.5$ to $x + 0.5$.

Brief Review of Binomial Probability Distribution In Section 5-2 we saw that a *binomial probability distribution* has (1) a fixed number of trials; (2) trials that are independent; (3) trials that are each classified into two categories commonly referred to as *success* and *failure*; and (4) trials with the property that the probability of success remains constant. Section 5-2 also introduced the following notation.

Notation

$n =$ the fixed number of trials

$x =$ the specific number of successes in n trials

$p =$ probability of *success* in *one* of the n trials

$q =$ probability of *failure* in *one* of the n trials (so $q = 1 - p$)

Rationale for Using a Normal Approximation We saw in Section 6-3 that the sampling distribution of a sample proportion tends to approximate a normal distribution. Also, see the following probability histogram for the binomial distribution with $n = 580$ and $p = 0.25$. (In one of Mendel's famous hybridization experiments, he expected 25% of his 580 peas to be yellow, but he got 152 yellow peas, for a rate of 26.2%.) The bell-shape of this graph suggests that we can use a normal distribution to approximate the binomial distribution.

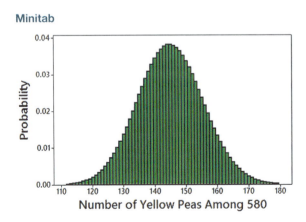

Minitab

Number of Yellow Peas Among 580

Normal Distribution as an Approximation to the Binomial Distribution

Requirements

1. The sample is a simple random sample of size n from a population in which the proportion of successes is p, or the sample is the result of conducting n independent trials of a binomial experiment in which the probability of success is p.

2. $np \geq 5$ and $nq \geq 5$.

(The requirements of $np \geq 5$ and $nq \geq 5$ are common, but some recommend using 10 instead of 5.)

Normal Approximation

If the above requirements are satisfied, then the probability distribution of the random variable x can be approximated by a normal distribution with these parameters:

- $\mu = np$
- $\sigma = \sqrt{npq}$

Continuity Correction

When using the normal approximation, adjust the discrete whole number x by using a *continuity correction* so that any individual value x is represented in the normal distribution by the interval from $x - 0.5$ to $x + 0.5$.

Procedure for Using a Normal Distribution to Approximate a Binomial Distribution

1. Check the requirements that $np \geq 5$ and $nq \geq 5$.

2. Find $\mu = np$ and $\sigma = \sqrt{npq}$ to be used for the normal distribution.

3. Identify the discrete whole number x that is relevant to the binomial probability problem being considered, and represent that value by the region bounded by $x - 0.5$ and $x + 0.5$.

4. Graph the normal distribution and shade the desired area bounded by $x - 0.5$ or $x + 0.5$ as appropriate.

EXAMPLE 1   Was Mendel Wrong?

In one of Mendel's famous hybridization experiments, he expected that among 580 offspring peas, 145 of them (or 25%) would be yellow, but he actually got 152 yellow peas. Assuming that Mendel's rate of 25% is correct, find the probability of getting 152 or more yellow peas by random chance. That is, given $n = 580$ and $p = 0.25$, find $P(\text{at least } 152 \text{ yellow peas})$. Is 152 yellow peas *significantly high?*

SOLUTION

Step 1: Requirement check: With $n = 580$ and $p = 0.25$, we get $np = (580)(0.25) = 145$ and $nq = (580)(0.75) = 435$, so the requirements that $np \geq 5$ and $nq \geq 5$ are both satisfied.

Step 2: We now find μ and σ needed for the normal distribution:

$$\mu = np = 580 \cdot 0.25 = 145$$

$$\sigma = \sqrt{npq} = \sqrt{580 \cdot 0.25 \cdot 0.75} = 10.4283$$

continued

Step 3: We want the probability of at least 152 yellow peas, so the discrete whole number relevant to this example is $x = 152$. We use the continuity correction as we represent the discrete value of 152 in the graph of the normal distribution by the interval between 151.5 and 152.5 (as shown in the top portion of Figure 6-25).

Step 4: See the bottom portion of Figure 6-25, which shows the normal distribution and the area to the right of 151.5 (representing "152 or more" yellow peas).

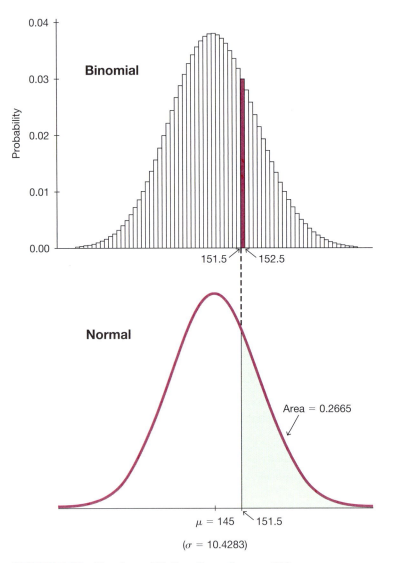

FIGURE 6-25   Number of Yellow Peas Among 580

We want the area to the right of 151.5 in the bottom portion of Figure 6-25.

Technology: If using technology, we find that the shaded area is 0.2665

Table A-2: If using Table A-2, we must first find the z score using $x = 151.5$, $\mu = 145$, and $\sigma = 10.4283$ as follows:

$$z = \frac{x - \mu}{\sigma} = \frac{151.5 - 145}{10.4283} = 0.62$$

Using Table A-2, we find that $z = 0.62$ corresponds to a cumulative left area of 0.7324, so the shaded region in the bottom portion of Figure 6-25 is $1 - 0.7324 = 0.2676$. (The result of 0.2665 from technology is more accurate.)

INTERPRETATION

Mendel's result of 152 yellow peas is greater than the 145 yellow peas he expected with his theory of hybrids, but with $P(152$ or more yellow peas$) = 0.2665$, we see that 152 yellow peas is *not significantly high*. That is a result that could easily occur with a true rate of 25% for yellow peas. This experiment does not contradict Mendel's theory.

YOUR TURN Do Exercise 9 "White Cars."

Continuity Correction

DEFINITION

When we use the normal distribution (which is a *continuous* probability distribution) as an approximation to the binomial distribution (which is *discrete*), a **continuity correction** is made to a discrete whole number x in the binomial distribution by representing the discrete whole number x by the *interval* from $x - 0.5$ to $x + 0.5$ (that is, adding and subtracting 0.5).

Example 1 used a continuity correction when the discrete value of 152 was represented in the normal distribution by the area between 151.5 and 152.5. Because we wanted the probability of "152 or more" yellow peas, we used the area to the right of 151.5. Here are other uses of the continuity correction:

Statement About the *Discrete* Value	Area of the *Continuous* Normal Distribution
At least 152 (includes 152 and above)	To the *right* of 151.5
More than 152 (doesn't include 152)	To the *right* of 152.5
At most 152 (includes 152 and below)	To the *left* of 152.5
Fewer than 152 (doesn't include 152)	To the *left* of 151.5
Exactly 152	Between 151.5 and 152.5

EXAMPLE 2   Exactly 252 Yellow Peas

Using the same information from Example 1, find the probability of *exactly* 152 yellow peas among the 580 offspring peas. That is, given $n = 580$ and assuming that $p = 0.25$, find $P(\text{exactly 152 yellow peas})$. Is this result useful for determining whether 152 yellow peas is *significantly high*?

SOLUTION

See Figure 6-26 on the next page, which shows the normal distribution with $\mu = 145$ and $\sigma = 10.4283$. The shaded area approximates the probability of *exactly* 152 yellow peas. That region is the vertical strip between 151.5 and 152.5, as shown. We can find that area by using the same methods introduced in Section 6-2.

Technology: Using technology, the shaded area is 0.0305.

Table A-2: Using Table A-2, we convert 151.5 and 152.5 to $z = 0.62$ and $z = 0.72$, which yield cumulative left areas of 0.7324 and 0.7642. Because they are both cumulative left areas, the shaded region in Figure 6-26 is $0.7642 - 0.7324 = 0.0318$. The probability of exactly 152 yellow peas is 0.0318.

continued

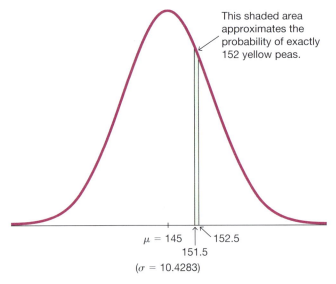

This shaded area approximates the probability of exactly 152 yellow peas.

$\mu = 145$ 151.5   152.5

$(\sigma = 10.4283)$

FIGURE 6-26   **Probability of Exactly 152 Yellow Peas**

INTERPRETATION

In Section 4-1 we saw that *x* successes among *n* trials is significantly high if the probability of *x or more* successes is unlikely with a probability of 0.05 or less. In determining whether Mendel's result of 152 yellow peas contradicts his theory that 25% of the offspring should be yellow peas, we should consider the probability of *152 or more* yellow peas, not the probability of exactly 152 peas. The result of 0.0305 is not the relevant probability; the relevant probability is 0.2665 found in Example 1. In general, the relevant result is the probability of getting a result *at least as extreme* as the one obtained.

YOUR TURN Do Exercise 11 "Red Cars."

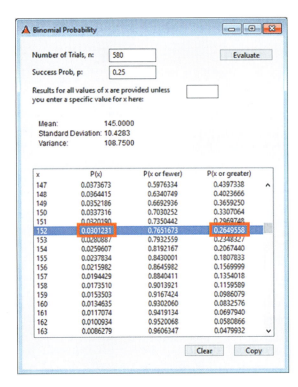

Technology for Binomial Probabilities

This topic of using a normal distribution to approximate a binomial distribution was once quite important, but we can now use technology to find binomial probabilities that were once beyond our capabilities. For example, see the accompanying Statdisk display showing that for Example 1, the probability of 152 or more yellow peas is 0.2650, and for Example 2, the probability of exactly 152 yellow peas is 0.0301, so there is no real need to use a normal approximation. However, there are cases where we need to use a normal approximation, and Section 8-2 uses a normal approximation to a binomial distribution for an important statistical method introduced in that section.

6-6 Basic Skills and Concepts

Statistical Literacy and Critical Thinking

1. Continuity Correction In testing the assumption that the probability of a baby boy is 0.512, a geneticist obtains a random sample of 1000 births and finds that 502 of them are boys. Using the continuity correction, describe the area under the graph of a normal distribution corresponding to the following. (For example, the area corresponding to "the probability of at least 502 boys" is this: the area to the right of 501.5.)

a. The probability of 502 or fewer boys

b. The probability of exactly 502 boys

c. The probability of more than 502 boys

2. Checking Requirements Common tests such as the SAT, ACT, Law School Admission test (LSAT), and Medical College Admission Test (MCAT) use multiple choice test questions, each with possible answers of a, b, c, d, e, and each question has only one correct answer. We want to find the probability of getting at least 25 correct answers for someone who makes random guesses for answers to a block of 100 questions. If we plan to use the methods of this section with a normal distribution used to approximate a binomial distribution, are the necessary requirements satisfied? Explain.

3. Notation Common tests such as the SAT, ACT, LSAT, and MCAT tests use multiple choice test questions, each with possible answers of a, b, c, d, e, and each question has only one correct answer. For people who make random guesses for answers to a block of 100 questions, identify the values of p, q, μ, and σ. What do μ and σ measure?

4. Distribution of Proportions Each week, Nielsen Media Research conducts a survey of 5000 households and records the proportion of households tuned to *60 Minutes*. If we obtain a large collection of those proportions and construct a histogram of them, what is the approximate shape of the histogram?

Using Normal Approximation. *In Exercises 5–8, do the following: If the requirements of np ≥ 5 and nq ≥ 5 are both satisfied, estimate the indicated probability by using the normal distribution as an approximation to the binomial distribution; if np < 5 or nq < 5, then state that the normal approximation should not be used.*

5. Births of Boys With $n = 20$ births and $p = 0.512$ for a boy, find P(fewer than 8 boys).

6. Births of Boys With $n = 8$ births and $p = 0.512$ for a boy, find P(exactly 5 boys).

7. Guessing on Standard Tests With $n = 20$ guesses and $p = 0.2$ for a correct answer, find P(at least 6 correct answers).

8. Guessing on Standard Tests With $n = 50$ guesses and $p = 0.2$ for a correct answer, find P(exactly 12 correct answers).

Car Colors. *In Exercises 9–12, assume that 100 cars are randomly selected. Refer to the accompanying graph, which shows the top car colors and the percentages of cars with those colors (based on PPG Industries).*

9. White Cars Find the probability that fewer than 20 cars are white. Is 20 a significantly low number of white cars?

10. Black Cars Find the probability that at least 25 cars are black. Is 25 a significantly high number of black cars?

11. Red Cars Find the probability of exactly 14 red cars. Why can't the result be used to determine whether 14 is a significantly high number of red cars?

12. Gray Cars Find the probability of exactly 10 gray cars. Why can't the result be used to determine whether 10 is a significantly low number of gray cars?

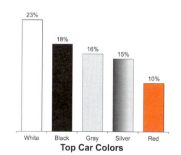

13. Tennis Replay In the year that this exercise was written, there were 879 challenges made to referee calls in professional tennis singles play. Among those challenges, 231 challenges were upheld with the call overturned. Assume that in general, 25% of the challenges are successfully upheld with the call overturned.

a. If the 25% rate is correct, find the probability that among the 879 challenges, the number of overturned calls is exactly 231.

b. If the 25% rate is correct, find the probability that among the 879 challenges, the number of overturned calls is 231 or more. If the 25% rate is correct, is 231 overturned calls among 879 challenges a result that is significantly high?

14. Tennis Replay Repeat the preceding exercise after changing the assumed rate of overturned calls from 25% to 22%.

15. Smartphones Based on an LG smartphone survey, assume that 51% of adults with smartphones use them in theaters. In a separate survey of 250 adults with smartphones, it is found that 109 use them in theaters.

a. If the 51% rate is correct, find the probability of getting 109 or fewer smartphone owners who use them in theaters.

b. Is the result of 109 significantly low?

16. Eye Color Based on a study by Dr. P. Sorita at Indiana University, assume that 12% of us have green eyes. In a study of 650 people, it is found that 86 of them have green eyes.

a. Find the probability of at least 86 people with green eyes among 650 randomly selected people.

b. Is 86 people with green eyes significantly high?

17. Mendelian Genetics When Mendel conducted his famous genetics experiments with peas, one sample of offspring consisted of 929 peas, with 705 of them having red flowers. If we assume, as Mendel did, that under these circumstances, there is a 3/4 probability that a pea will have a red flower, we would expect that 696.75 (or about 697) of the peas would have red flowers, so the result of 705 peas with red flowers is more than expected.

a. If Mendel's assumed probability is correct, find the probability of getting 705 or more peas with red flowers.

b. Is 705 peas with red flowers significantly high?

c. What do these results suggest about Mendel's assumption that 3/4 of peas will have red flowers?

18. Sleepwalking Assume that 29.2% of people have sleepwalked (based on "Prevalence and Comorbidity of Nocturnal Wandering in the U.S. Adult General Population," by Ohayon et al., *Neurology*, Vol. 78, No. 20). Assume that in a random sample of 1480 adults, 455 have sleepwalked.

a. Assuming that the rate of 29.2% is correct, find the probability that 455 or more of the 1480 adults have sleepwalked.

b. Is that result of 455 or more significantly high?

c. What does the result suggest about the rate of 29.2%?

19. Voters Lying? In a survey of 1002 people, 701 said that they voted in a recent presidential election (based on data from ICR Research Group). Voting records showed that 61% of eligible voters actually did vote.

a. Given that 61% of eligible voters actually did vote, find the probability that among 1002 randomly selected eligible voters, at least 701 actually did vote.

b. What does the result suggest?

20. Cell Phones and Brain Cancer In a study of 420,095 cell phone users in Denmark, it was found that 135 developed cancer of the brain or nervous system. For those not using cell phones, there is a 0.000340 probability of a person developing cancer of the brain or nervous system. We therefore expect about 143 cases of such cancers in a group of 420,095 randomly selected people.

a. Find the probability of 135 or fewer cases of such cancers in a group of 420,095 people.

b. What do these results suggest about media reports that suggest cell phones cause cancer of the brain or nervous system?

6-6   Beyond the Basics

21. Births The probability of a baby being born a boy is 0.512. Consider the problem of finding the probability of exactly 7 boys in 11 births. Solve that problem using (1) normal approximation to the binomial using Table A-2; (2) normal approximation to the binomial using technology instead of Table A-2; (3) using technology with the binomial distribution instead of using a normal approximation. Compare the results. Given that the requirements for using the normal approximation are just barely met, are the approximations off by very much?

22. Overbooking a Boeing 767-300 A Boeing 767-300 aircraft has 213 seats. When someone buys a ticket for a flight, there is a 0.0995 probability that the person will not show up for the flight (based on data from an IBM research paper by Lawrence, Hong, and Cherrier). How many reservations could be accepted for a Boeing 767-300 for there to be at least a 0.95 probability that all reservation holders who show will be accommodated?

Chapter Quick Quiz

Bone Density Test.   *In Exercises 1–4, assume that scores on a bone mineral density test are normally distributed with a mean of 0 and a standard deviation of 1.*

1. Bone Density Sketch a graph showing the shape of the distribution of bone density test scores.

2. Bone Density Find the score separating the lowest 9% of scores from the highest 91%.

3. Bone Density For a randomly selected subject, find the probability of a score greater than -2.93.

4. Bone Density For a randomly selected subject, find the probability of a score between 0.87 and 1.78.

5. Notation

a. Identify the values of μ and σ for the standard normal distribution.

b. What do the symbols $\mu_{\bar{x}}$ and $\sigma_{\bar{x}}$ represent?

In Exercises 6–10, assume that women have diastolic blood pressure measures that are normally distributed with a mean of 70.2 mm Hg and a standard deviation of 11.2 mm Hg (based on Data Set 1 "Body Data" in Appendix B).

6. Diastolic Blood Pressure Find the probability that a randomly selected woman has a normal diastolic blood pressure level, which is below 80 mm Hg.

7. Diastolic Blood Pressure Find the probability that a randomly selected woman has a diastolic blood pressure level between 60 mm Hg and 80 mm Hg.

8. Diastolic Blood Pressure Find P_{90}, the 90th percentile for the diastolic blood pressure levels of women.

9. Diastolic Blood Pressure If 16 women are randomly selected, find the probability that the mean of their diastolic blood pressure levels is less than 75 mm Hg.

10. Diastolic Blood Pressure The accompanying normal quantile plot was constructed from the diastolic blood pressure levels of a sample of women. What does this graph suggest about diastolic blood pressure levels of women?

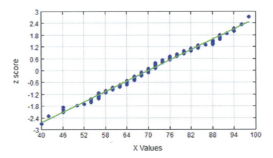

Review Exercises

1. Bone Density Test A bone mineral density test is used to identify a bone disease. The result of a bone density test is commonly measured as a z score, and the population of z scores is normally distributed with a mean of 0 and a standard deviation of 1.

a. For a randomly selected subject, find the probability of a bone density test score less than 1.54.

b. For a randomly selected subject, find the probability of a bone density test score greater than -1.54.

c. For a randomly selected subject, find the probability of a bone density test score between -1.33 and 2.33.

d. Find Q_1, the bone density test score separating the bottom 25% from the top 75%.

e. If the mean bone density test score is found for 9 randomly selected subjects, find the probability that the mean is greater than 0.50.

2. Biometric Security In designing a security system based on eye (iris) recognition, we must consider the standing eye heights of women, which are normally distributed with a mean of 59.7 in. and a standard deviation of 2.5 in. (based on anthropometric survey data from Gordon, Churchill, et al.).

a. If an eye recognition security system is positioned at a height that is uncomfortable for women with standing eye heights less than 54 in., what percentage of women will find that height uncomfortable?

b. In positioning the eye recognition security system, we want it to be suitable for the lowest 95% of standing eye heights of women. What standing eye height of women separates the lowest 95% of standing eye heights from the highest 5%?

3. Biometric Security Standing eye heights of men are normally distributed with a mean of 64.3 in. and a standard deviation of 2.6 in. (based on anthropometric survey data from Gordon, Churchill, et al.).

a. If an eye recognition security system is positioned at a height that is uncomfortable for men with standing eye heights greater than 70 in., what percentage of men will find that height uncomfortable?

b. In positioning the eye recognition security system, we want it to be suitable for the tallest 98% of standing eye heights of men. What standing eye height of men separates the tallest 98% of standing eye heights from the lowest 2%?

4. Sampling Distributions Scores on the Gilliam Autism Rating Scale (GARS) are normally distributed with a mean of 100 and a standard deviation of 15. A sample of 64 GARS scores is randomly selected and the sample mean is computed.

a. Describe the distribution of such sample means.

b. What is the mean of all such sample means?

c. What is the standard deviation of all such sample means?

5. Unbiased Estimators

a. What is an unbiased estimator?

b. For the following statistics, identify those that are unbiased estimators: mean, median, range, variance, proportion.

c. Determine whether the following statement is true or false: "The sample standard deviation is a biased estimator, but the bias is relatively small in large samples, so s is often used to estimate σ."

6. Disney Monorail The Mark VI monorail used at Disney World has doors with a height of 72 in. Heights of men are normally distributed with a mean of 68.6 in. and a standard deviation of 2.8 in. (based on Data Set 1 "Body Data" in Appendix B).

a. What percentage of adult men can fit through the doors without bending? Does the door design with a height of 72 in. appear to be adequate? Explain.

b. What doorway height would allow 99% of adult men to fit without bending?

7. Disney Monorail Consider the same Mark VI monorail described in the preceding exercise. Again assume that heights of men are normally distributed with a mean of 68.6 in. and a standard deviation of 2.8 in.

a. In determining the suitability of the monorail door height, why does it make sense to consider men while women are ignored?

b. Mark VI monorail cars have a capacity of 60 passengers. If a car is loaded with 60 randomly selected men, what is the probability that their mean height is less than 72 in.?

c. Why can't the result from part (b) be used to determine how well the doorway height accommodates men?

8. Assessing Normality Listed below are the recent salaries (in millions of dollars) of players on the San Antonio Spurs professional basketball team.

a. Do these salaries appear to come from a population that has a normal distribution? Why or why not?

b. Can the mean of this sample be treated as a value from a population having a normal distribution? Why or why not?

12.5   10.4   9.3   7.0   4.0   3.1   2.9   2.1   1.8   1.4   1.1   1.1   0.9

9. Hybridization Experiment In one of Mendel's experiments with plants, 1064 offspring consisted of 787 plants with long stems. According to Mendel's theory, 3/4 of the offspring plants should have long stems. Assuming that Mendel's proportion of 3/4 is correct, find the probability of getting 787 or fewer plants with long stems among 1064 offspring plants. Based on the result, is 787 offspring plants with long stems significantly low? What does the result imply about Mendel's claimed proportion of 3/4?

10. Tall Clubs The social organization Tall Clubs International has a requirement that women must be at least 70 in. tall. Assume that women have normally distributed heights with a mean of 63.7 in. and a standard deviation of 2.9 in. (based on Data Set 1 "Body Data" in Appendix B).

a. Find the percentage of women who satisfy the height requirement.

b. If the height requirement is to be changed so that the tallest 2.5% of women are eligible, what is the new height requirement?

Cumulative Review Exercises

In Exercises 1–3, use the following recent annual salaries (in millions of dollars) for players on the N.Y. Knicks professional basketball team.

> 23.4   22.5   11.5   7.1   6.0   4.1   3.3   2.8   2.6   1.7   1.6   1.3   0.9   0.9   0.6

1. NY Knicks Salaries

a. Find the mean $\bar{x}$.

b. Find the median.

c. Find the standard deviation s.

d. Find the variance.

e. Convert the highest salary to a z score.

f. What level of measurement (nominal, ordinal, interval, ratio) describes this data set?

g. Are the salaries discrete data or continuous data?

2. NY Knicks Salaries

a. Find Q_1, Q_2, and Q_3.

b. Construct a boxplot.

c. Based on the accompanying normal quantile plot of the salaries, what do you conclude about the sample of salaries?

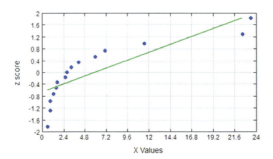

3. NY Knicks Salaries Round each of the salaries to the nearest million dollars, then construct a dotplot. Do the values appear to be from a population having a normal distribution?

4. Blue Eyes Assume that 35% of us have blue eyes (based on a study by Dr. P. Soria at Indiana University).

a. Let B denote the event of selecting someone who has blue eyes. What does the event $\overline{B}$ denote?

b. Find the value of $P(\overline{B})$.

c. Find the probability of randomly selecting three different people and finding that all of them have blue eyes.

d. Find the probability that among 100 randomly selected people, at least 40 have blue eyes.

e. If 35% of us really do have blue eyes, is a result of 40 people with blue eyes among 100 randomly selected people a result that is significantly high?

5. Foot Lengths of Women Assume that foot lengths of women are normally distributed with a mean of 9.6 in. and a standard deviation of 0.5 in., based on data from the U.S. Army Anthropometry Survey (ANSUR).

a. Find the probability that a randomly selected woman has a foot length less than 10.0 in.

b. Find the probability that a randomly selected woman has a foot length between 8.0 in. and 11.0 in.

c. Find P_{95}.

d. Find the probability that 25 women have foot lengths with a mean greater than 9.8 in.

Technology Projects

Some methods in this chapter are easy with technology but very difficult without it. The two projects that follow illustrate how easy it is to use technology for assessing normality and finding binomial probabilities.

1. Assessing Normality It is often necessary to determine whether sample data appear to be from a normally distributed population, and that determination is helped with the construction of a histogram and normal quantile plot. Refer to Data Set 1 "Body Data" in Appendix B. For each of the 13 columns of data (not including age or gender), determine whether the data appear to be from a normally distributed population. Use Statdisk or any other technology. (Download a free copy of Statdisk from www.statdisk.org.)

2. Binomial Probabilities Section 6-6 described a method for using a normal distribution to approximate a binomial distribution. Many technologies are capable of generating probabilities for a binomial distribution. Instructions for these different technologies are found in the Tech Center at the end of Section 5-2 on page 208. Instead of using a normal approximation to a binomial distribution, use technology to find the exact binomial probabilities in Exercises 9–12 of Section 6-6.

FROM DATA TO DECISION

Critical Thinking: Designing a campus dormitory elevator

An Ohio college student died when he tried to escape from a dormitory elevator that was overloaded with 24 passengers. The elevator was rated for a maximum weight of 2500 pounds.

Let's consider this elevator with an allowable weight of 2500 pounds. Let's also consider parameters for weights of adults, as shown in the accompanying table (based on Data Set 1 "Body Data" in Appendix B).

continued

Weights of Adults

	Males	Females
μ	189 lb	171 lb
σ	39 lb	46 lb
Distribution	Normal	Normal

We could consider design features such as the type of music that could be played on the elevator. We could select songs such as "Imagine," or "Daydream Believer." Instead, we will focus on the critical design feature of weight.

a. First, elevators commonly have a 25% margin of error, so they can safely carry a load that is 25% greater than the stated load. What amount is 25% greater than 2500 pounds? Let's refer to this amount as "the maximum safe load" while the 2500 pound limit is the "placard maximum load."

b. Now we need to determine the maximum number of passengers that should be allowed. Should we base our calculations on the maximum safe load or the 2500 pound placard maximum load?

c. The weights given in the accompanying table are weights of adults not including clothing or textbooks. Add another 10 pounds for each student's clothing and textbooks. What is the maximum number of elevator passengers that should be allowed?

d. Do you think that weights of college students are different from weights of adults from the general population? If so, how? How would that affect the elevator design?

Cooperative Group Activities

1. Out-of-class activity Use the Internet to find "Pick 4" lottery results for 50 different drawings. Find the 50 different means. Graph a histogram of the original 200 digits that were selected, and graph a histogram of the 50 sample means. What important principle do you observe?

2. In-class activity Divide into groups of three or four students and address these issues affecting the design of manhole covers.

• Which of the following is most relevant for determining whether a manhole cover diameter of 24 in. is large enough: weights of men, weights of women, heights of men, heights of women, hip breadths of men, hip breadths of women, shoulder breadths of men, shoulder breadths of women?

• Why are manhole covers usually round? (This was once a popular interview question asked of applicants at IBM, and there are at least three good answers. One good answer is sufficient here.)

3. Out-of-class activity Divide into groups of three or four students. In each group, develop an original procedure to illustrate the central limit theorem. The main objective is to show that when you randomly select samples from a population, the means of those samples tend to be *normally* distributed, regardless of the nature of the population distribution. For this illustration, begin with some population of values that does *not* have a normal distribution.

4. In-class activity Divide into groups of three or four students. Using a coin to simulate births, each individual group member should simulate 25 births and record the number of simulated girls. Combine all results from the group and record n = total number of births and x = number of girls. Given batches of n births, compute the mean and standard deviation for the number of girls. Is the simulated result unusual? Why or why not?

5. In-class activity Divide into groups of three or four students. Select a set of data from one of these data sets in Appendix B: 2, 3, 4, 5, 6, 7, 13, 16, 17, 18, 19, 20, 24, 27, 30, 32. (These are the data sets that were not used in examples or exercises in Section 6-5). Use the methods of Section 6-5 to construct a histogram and normal quantile plot, then determine whether the data set appears to come from a normally distributed population.

6. Out-of-class activity Divide into groups of three or four students. Each student should obtain a sample of cars and record the number of cars that are white. Combine the results and use methods of this chapter to compare the results to those expected from the graph of top car colors that accompanies Exercises 9–12 in Section 6-6.

7-1 Estimating a
Population Proportion

7-2 Estimating a
Population Mean

7-3 Estimating a
Population Standard
Deviation or Variance

7-4 Bootstrapping: Using
Technology for
Estimates

7

ESTIMATING PARAMETERS AND DETERMINING SAMPLE SIZES

CHAPTER PROBLEM ── **Surveys: The Window to Evolving Technologies**

Importance of Surveys In our modern data-driven world, it is essential that we have the ability to analyze and understand the polls and surveys that play a central and increasing role in guiding entertainment, politics, product development, and just about every other facet of our lives. This chapter presents the tools for developing that ability. Here are four recent surveys that focus on evolving technologies:

- **Self-Driving Vehicles:** In a TE Connectivity survey of 1000 adults, 29% said that they would feel comfortable in a self-driving vehicle.

- **Cell Phone Ownership:** In a Pew Research Center poll of 2076 adults, 91% said that they own a cell phone.

- **Facebook:** In a Gallup poll of 1487 adults, 43% said that they have a Facebook page.

continued

- **Biometric Security:** In a *USA Today* survey of 510 people, 53% said that we should replace passwords with biometric security, such as fingerprints.

Because surveys are now so pervasive and extensive, and because they are often accepted without question, we should analyze them by considering issues such as the following:

- What method was used to select the survey subjects?

- How do we use sample results to estimate values of population parameters?

- How accurate are survey sample results likely to be?

- Typical media reports about surveys are missing an extremely important element of relevant information. What is usually missing?

- How do we correctly interpret survey results?

For example, the "biometric security" poll cited above is based on a *voluntary response sample* (described in Section 1-1), so its fundamental validity is very questionable. The other three surveys all involve sound sampling methods, so with these surveys we can proceed to consider the other issues listed above, and that is the focus of this chapter.

CHAPTER OBJECTIVES

In this chapter we begin the study of methods of *inferential statistics.* The following are the major activities of inferential statistics, and this chapter introduces methods for the first activity of using sample data to estimate population parameters. Chapter 8 will introduce the basic methods for testing claims (or hypotheses) about population parameters.

Major Activities of Inferential Statistics

1. Use sample data to *estimate values of population parameters* (such as a population proportion or population mean).

2. Use sample data to *test hypotheses* (or claims) made about population parameters.

Here are the chapter objectives:

7-1 Estimating a Population Proportion

- Construct a confidence interval estimate of a population proportion and interpret such confidence interval estimates.

- Identify the requirements necessary for the procedure that is used, and determine whether those requirements are satisfied.

- Develop the ability to determine the sample size necessary to estimate a population proportion.

7-2 Estimating a Population Mean

- Construct a confidence interval estimate of a population mean, and be able to interpret such confidence interval estimates.

- Determine the sample size necessary to estimate a population mean.

7-3 Estimating a Population Standard Deviation or Variance

- Develop the ability to construct a confidence interval estimate of a population standard deviation or variance, and be able to interpret such confidence interval estimates.

7-4 Bootstrapping: Using Technology for Estimates

- Develop the ability to use technology along with the bootstrapping method to construct a confidence interval estimate of a population proportion, population mean, and population standard deviation and population variance.

7-1   Estimating a Population Proportion

Key Concept This section presents methods for using a sample proportion to make an inference about the value of the corresponding population proportion. This section focuses on the population proportion p, but we can also work with probabilities or percentages. When working with percentages, we will perform calculations with the equivalent proportion value. Here are the three main concepts included in this section:

- **Point Estimate:** The sample proportion (denoted by $\hat{p}$) is the best *point estimate* (or single value estimate) of the population proportion p.

- **Confidence Interval:** We can use a sample proportion to construct a *confidence interval* estimate of the true value of a population proportion, and we should know how to construct and interpret such confidence intervals.

- **Sample Size:** We should know how to find the sample size necessary to estimate a population proportion.

The concepts presented in this section are used in the following sections and chapters, so it is important to understand this section quite well.

PART 1 Point Estimate, Confidence Interval, and Sample Size

Point Estimate

If we want to estimate a population proportion with a single value, the best estimate is the sample proportion $\hat{p}$. Because $\hat{p}$ consists of a single value that is equivalent to a point on a line, it is called a *point estimate.*

> **DEFINITION**
>
> A **point estimate** is a single value used to estimate a population parameter.

> The sample proportion $\hat{p}$ is the best *point estimate* of the population proportion p.

Unbiased Estimator We use $\hat{p}$ as the point estimate of p because it is unbiased and it is the most consistent of the estimators that could be used. (An unbiased estimator is a statistic that targets the value of the corresponding population parameter in the sense that the sampling distribution of the statistic has a mean that is equal to the corresponding population parameter. The statistic $\hat{p}$ targets the population proportion p.) The sample proportion $\hat{p}$ is the most consistent estimator of p in the sense that the standard deviation of sample proportions tends to be smaller than the standard deviation of other unbiased estimators of p.

Push Polling

Push polling is the practice of political campaigning under the guise of a poll. Its name is derived from its objective of pushing voters away from opposition candidates by asking loaded questions designed to discredit them. This survey question was used in one campaign: "Please tell me if you would be more likely or less likely to vote for Roy Romer if you knew that Governor Romer appoints a parole board which has granted early release to an average of four convicted felons per day every day since Romer took office." The National Council on Public Polls says that push polls are unethical. Reputable pollsters do not approve of push polling.

EXAMPLE 1 Facebook

The Chapter Problem included reference to a Gallup poll in which 1487 adults were surveyed and 43% of them said that they have a Facebook page. Based on that result, find the best point estimate of the proportion of *all* adults who have a Facebook page.

SOLUTION

Because the sample proportion is the best point estimate of the population proportion, we conclude that the best point estimate of p is 0.43. (If using the sample results to estimate the *percentage* of all adults who have a Facebook page, the best point estimate is 43%.)

> **YOUR TURN** Find the point estimate in Exercise 13 "Mickey D's."

Confidence Interval

Why Do We Need Confidence Intervals? In Example 1 we saw that 0.43 is our *best* point estimate of the population proportion p, but we have no indication of how *good* that best estimate is. A confidence interval gives us a much better sense of how good an estimate is.

> **DEFINITION**
>
> A **confidence interval** (or **interval estimate**) is a range (or an interval) of values used to estimate the true value of a population parameter. A confidence interval is sometimes abbreviated as CI.

> **DEFINITION**
>
> The **confidence level** is the probability $1 - \alpha$ (such as 0.95, or 95%) that the confidence interval actually does contain the population parameter, assuming that the estimation process is repeated a large number of times. (The confidence level is also called the **degree of confidence**, or the **confidence coefficient.**)

The following table shows the relationship between the confidence level and the corresponding value of α. The confidence level of 95% is the value used most often.

Most Common Confidence Levels	Corresponding Values of α
90% (or 0.90) confidence level:	$\alpha = 0.10$
95% (or 0.95) confidence level:	$\alpha = 0.05$
99% (or 0.99) confidence level:	$\alpha = 0.01$

Here's an example of a confidence interval found later in Example 3:

The 0.95 (or 95%) confidence interval estimate of the population proportion p is $0.405 < p < 0.455$.

Interpreting a Confidence Interval

We must be careful to interpret confidence intervals correctly. There is a correct interpretation and many different and creative incorrect interpretations of the confidence interval $0.405 < p < 0.455$.

Correct: **"We are 95% confident that the interval from 0.405 to 0.455 actually does contain the true value of the population proportion p."**
This is a short and acceptable way of saying that if we were to select many different random samples of size 1487 (from Example 3) and construct the corresponding confidence intervals, 95% of them would contain the population proportion p. In this correct interpretation, the confidence level of 95% refers to the *success rate of the process* used to estimate the population proportion.

Wrong: **"There is a 95% chance that the true value of p will fall between 0.405 and 0.455."**
This is wrong because p is a population parameter with a fixed value; it is not a random variable with values that vary.

Wrong: **"95% of sample proportions will fall between 0.405 and 0.455."**
This is wrong because the values of 0.405 and 0.455 result from one sample; they are not parameters describing the behavior of all samples.

Confidence Level: The Process Success Rate A confidence level of 95% tells us that the *process* we are using should, in the long run, result in confidence interval limits that contain the true population proportion 95% of the time. Suppose that the true proportion of adults with Facebook pages is $p = 0.50$. See Figure 7-1, which shows that 19 out of 20 (or 95%) different confidence intervals contain the assumed value of $p = 0.50$. Figure 7-1 is trying to tell this story: With a 95% confidence level, we expect about 19 out of 20 confidence intervals (or 95%) to contain the true value of p.

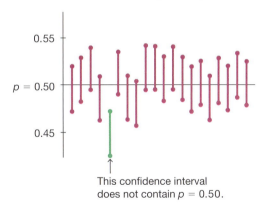

This confidence interval does not contain $p = 0.50$.

FIGURE 7-1   Confidence Intervals from 20 Different Samples

Critical Values

Critical values are formally defined on the next page and they are based on the following observations:

1. When certain requirements are met, the sampling distribution of sample proportions can be approximated by a normal distribution, as shown in Figure 7-2.

2. A z score associated with a sample proportion has a probability of $\alpha/2$ of falling in the right tail portion of Figure 7-2.

3. The z score at the boundary of the right-tail region is commonly denoted by $z_{\alpha/2}$ and is referred to as a *critical value* because it is on the borderline separating z scores that are significantly high.

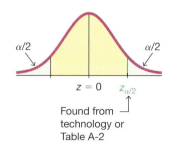

Found from technology or Table A-2

FIGURE 7-2

Critical Value $z_{\alpha/2}$ in the Standard Normal Distribution

Shakespeare's Vocabulary

According to Bradley Efron and Ronald Thisted, Shakespeare's writings included 31,534 different words. They used probability theory to conclude that Shakespeare probably knew at least another 35,000 words that he didn't use in his writings. The problem of estimating the size of a population is an important problem often encountered in ecology studies, but the result given here is another interesting application. (See "Estimating the Number of Unseen Species: How Many Words Did Shakespeare Know?"; in *Biometrika,* Vol. 63, No. 3.)

DEFINITION

A **critical value** is the number on the borderline separating sample statistics that are significantly high or low from those that are not significant. The number $z_{\alpha/2}$ is a critical value that is a *z* score with the property that it is at the border that separates an area of $\alpha/2$ in the right tail of the standard normal distribution (as in Figure 7-2).

EXAMPLE 2   Finding a Critical Value

Find the critical value $z_{\alpha/2}$ corresponding to a 95% confidence level.

SOLUTION

A 95% confidence level corresponds to $\alpha = 0.05$, so $\alpha/2 = 0.025$. Figure 7-3 shows that the area in each of the green-shaded tails is $\alpha/2 = 0.025$. We find $z_{\alpha/2} = 1.96$ by noting that the cumulative area to its left must be $1 - 0.025$, or 0.975. We can use technology or refer to Table A-2 to find that the cumulative left area of 0.9750 corresponds to $z = 1.96$. For a 95% confidence level, the critical value is therefore $z_{\alpha/2} = 1.96$.

Note that when finding the critical z score for a 95% confidence level, we use a cumulative left area of 0.9750 (*not* 0.95). Think of it this way:

This is our confidence level:	The area in *both* tails is:	The area in the *right* tail is:	The cumulative area from the left, excluding the right tail, is:
95% $\rightarrow$	$\alpha = 0.05$ $\rightarrow$	$\alpha/2 = 0.025$ $\rightarrow$	$1 - 0.025 = 0.975$

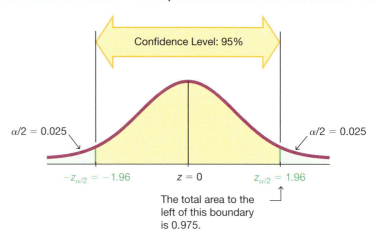

FIGURE 7-3   Finding the Critical Value $z_{\alpha/2}$ for a 95% Confidence Level

YOUR TURN Do Exercise 5 "Finding Critical Values."

Example 2 showed that a 95% confidence level results in a critical value of $z_{\alpha/2} = 1.96$. This is the most common critical value, and it is listed with two other common values in the table that follows.

Confidence Level	α	Critical Value, $z_{\alpha/2}$
90%	0.10	1.645
95%	0.05	1.96
99%	0.01	2.575

Margin of Error

We now formally define the *margin of error E* that we have all heard about so often in media reports.

Go Figure

$1,000,000: Estimated difference in lifetime earnings for someone with a college degree and someone without a college degree.

DEFINITION

When data from a simple random sample are used to estimate a population proportion p, the difference between the sample proportion $\hat{p}$ and the population proportion p is an error. The maximum likely amount of that error is the **margin of error,** denoted by **E.** There is a probability of $1 - \alpha$ (such as 0.95) that the difference between $\hat{p}$ and p is E or less. The margin of error E is also called the *maximum error of the estimate* and can be found by multiplying the critical value and the estimated standard deviation of sample proportions, as shown in Formula 7-1.

FORMULA 7-1

$$E = z_{\alpha/2}\sqrt{\frac{\hat{p}\hat{q}}{n}} \quad \text{margin of error for proportions}$$

↑   ↑

Critical value   Estimated standard deviation of sample proportions

KEY ELEMENTS

Confidence Interval for Estimating a Population Proportion p

Objective

Construct a confidence interval used to estimate a population proportion p.

Notation

$p = $ *population* proportion
$\hat{p} = $ *sample* proportion
$n = $ number of sample values

$E = $ margin of error
$z_{\alpha/2} = $ critical value: the z score separating an area of $\alpha/2$ in the right tail of the standard normal distribution

Requirements

1. The sample is a simple random sample.

2. The conditions for the binomial distribution are satisfied: There is a fixed number of trials, the trials are independent, there are two categories of outcomes, and the probabilities remain constant for each trial (as in Section 5-2).

3. There are at least 5 successes and at least 5 failures. (This requirement is a way to verify that $np \geq 5$ and $nq \geq 5$, so the normal distribution is a suitable approximation to the binomial distribution.)

Confidence Interval Estimate of p

$$\hat{p} - E < p < \hat{p} + E \quad \text{where} \quad E = z_{\alpha/2}\sqrt{\frac{\hat{p}\hat{q}}{n}}$$

The confidence interval is often expressed in the following two equivalent formats:

$$\hat{p} \pm E \quad \text{or} \quad (\hat{p} - E, \hat{p} + E)$$

Round-Off Rule for Confidence Interval Estimates of p

Round the confidence interval limits for p to three significant digits.

Bias in Internet Surveys?

Capitalizing on the widespread use of technology and social media, there is a growing trend to conduct surveys using only the Internet instead of using inperson interviews or phone calls to randomly selected subjects. Internet surveys are faster and much less expensive, and they provide important advantages in survey design and administration. But are Internet surveys biased because they use only subjects randomly selected from the 90% of the U.S. population that uses the Internet? The Pew Research Center studied this issue by comparing results from online polls to polls that included the offline population. It was found that the differences were generally quite small, but topics related to the Internet and technology resulted in much larger differences. We should be careful to consider consequences of bias with Internet surveys.

Procedure for Constructing a Confidence Interval for p

1. Verify that the requirements in the preceding Key Elements box are satisfied.

2. Use technology or Table A-2 to find the critical value $z_{\alpha/2}$ that corresponds to the desired confidence level.

3. Evaluate the margin of error $E = z_{\alpha/2}\sqrt{\hat{p}\hat{q}/n}$.

4. Using the value of the calculated margin of error E and the value of the sample proportion $\hat{p}$, find the values of the *confidence interval limits* $\hat{p} - E$ and $\hat{p} + E$. Substitute those values in the general format for the confidence interval.

5. Round the resulting confidence interval limits to three significant digits.

EXAMPLE 3   **Constructing a Confidence Interval: Poll Results**

In the Chapter Problem we noted that a Gallup poll of 1487 adults showed that 43% of the respondents have Facebook pages. The sample results are $n = 1487$ and $\hat{p} = 0.43$.

a. Find the margin of error E that corresponds to a 95% confidence level.

b. Find the 95% confidence interval estimate of the population proportion p.

c. Based on the results, can we safely conclude that fewer than 50% of adults have Facebook pages? Assuming that you are a newspaper reporter, write a brief statement that accurately describes the results and includes all of the relevant information.

SOLUTION

REQUIREMENT CHECK   (1) The polling methods used by the Gallup organization result in samples that can be considered to be simple random samples. (2) The conditions for a binomial experiment are satisfied because there is a fixed number of trials (1487), the trials are independent (because the response from one person doesn't affect the probability of the response from another person), there are two categories of outcome (subject has a Facebook page or does not), and the probability remains constant, because P(having a Facebook page) is fixed for a given point in time. (3) With 43% of the respondents having Facebook pages, the number with Facebook pages is 639 (or 43% of 1487). If 639 of the 1487 subjects have Facebook pages, the other 848 do not, so the number of successes (639) and the number of failures (848) are both at least 5. The check of requirements has been successfully completed. ✓

Technology   The confidence interval and margin of error can be easily found using technology. From the Statdisk display on the next page, we can see the required entries on the left and the results displayed on the right. Like most technologies, Statdisk requires a value for the number of successes, so we simply find 43% of 1487 and round the result of 639.41 to the whole number 639. The results show that the margin of error is $E = 0.025$(rounded) and the confidence interval is $0.405 < p < 0.455$ (rounded). (The Wilson Score confidence interval included in the display will be discussed later in Part 2 of this section.)

Statdisk

Confidence Interval: Proportion One Sample				

Confidence Level: `0.95`   Margin of error, E = 0.0251611

Sample Size, n: `1487`   95% Confidence Interval (using normal approx):
0.4045632 < p < 0.4548854

Number of Successes, x: `639`

[Evaluate]   Wilson Score Confidence Interval:
0.404776 < p < 0.4550347

[Print] [Copy]

Manual Calculation Here is how to find the confidence interval with manual calculations:

a. The margin of error is found by using Formula 7-1 with $z_{\alpha/2} = 1.96$ (as found in Example 2), $\hat{p} = 0.43$, $\hat{q} = 0.57$, and $n = 1487$.

$$E = z_{\alpha/2}\sqrt{\frac{\hat{p}\hat{q}}{n}} = 1.96\sqrt{\frac{(0.43)(0.57)}{1487}} = 0.0251636$$

b. Constructing the confidence interval is really easy now that we know that $\hat{p} = 0.43$ and $E = 0.0251636$. Simply substitute those values to obtain this result:

$$\hat{p} - E < p < \hat{p} + E$$
$$0.43 - 0.0251636 < p < 0.43 + 0.0251636$$
$$0.405 < p < 0.455 \quad \text{(rounded to three significant digits)}$$

This same result could be expressed in the format of 0.43 ± 0.025 or (0.405, 0.455). If we want the 95% confidence interval for the true population *percentage,* we could express the result as $40.5\% < p < 45.5\%$.

c. Based on the confidence interval obtained in part (b), it does appear that fewer than 50% of adults have a Facebook page because the interval of values from 0.405 to 0.455 is an interval that is completely below 0.50.

    Here is one statement that summarizes the results: 43% of adults have Facebook pages. That percentage is based on a Gallup poll of 1487 randomly selected adults in the United States. In theory, in 95% of such polls, the percentage should differ by no more than 2.5 percentage points in either direction from the percentage that would be found by interviewing all adults.

> **YOUR TURN** Find the confidence interval in Exercise 13 "Mickey D's."

Analyzing Polls Example 3 deals with a typical poll. When analyzing results from polls, consider the following:

1. The sample should be a simple random sample, not an inappropriate sample (such as a voluntary response sample).

2. The confidence level should be provided. (It is often 95%, but media reports usually fail to identify the confidence level.)

3. The sample size should be provided. (It is often provided by the media, but not always.)

4. Except for relatively rare cases, the quality of the poll results depends on the sampling method and the size of the sample, but the size of the *population* is usually not a factor.

CAUTION Never think that poll results are unreliable if the *sample size* is a small percentage of the *population size*. The population size is usually not a factor in determining the reliability of a poll.

Finding the Point Estimate and *E* from a Confidence Interval

Sometimes we want to better understand a confidence interval that might have been obtained from a journal article or technology. If we already know the confidence interval limits, the sample proportion (or the best point estimate) $\hat{p}$ and the margin of error E can be found as follows:

Point estimate of p:

$$\hat{p} = \frac{\text{(upper confidence interval limit)} + \text{(lower confidence interval limit)}}{2}$$

Margin of error:

$$E = \frac{\text{(upper confidence interval limit)} - \text{(lower confidence interval limit)}}{2}$$

> **EXAMPLE 4**   **Finding the Sample Proportion and Margin of Error**
>
> The article "High-Dose Nicotine Patch Therapy," by Dale, Hurt, et al. (*Journal of the American Medical Association,* Vol. 274, No. 17) includes this statement: "Of the 71 subjects, 70% were abstinent from smoking at 8 weeks (95% confidence interval [CI], 58% to 81%)." Use that statement to find the point estimate $\hat{p}$ and the margin of error E.
>
> **SOLUTION**
>
> We get the 95% confidence interval of $0.58 < p < 0.81$ from the given statement of "58% to 81%." The point estimate $\hat{p}$ is the value midway between the upper and lower confidence interval limits, so we get
>
> $$\hat{p} = \frac{\text{(upper confidence limit)} + \text{(lower confidence limit)}}{2}$$
>
> $$= \frac{0.81 + 0.58}{2} = 0.695$$
>
> The margin of error can be found as follows:
>
> $$E = \frac{\text{(upper confidence limit)} - \text{(lower confidence limit)}}{2}$$
>
> $$= \frac{0.81 - 0.58}{2} = 0.115$$

Using Confidence Intervals for Hypothesis Tests

A confidence interval can be used to *informally* address some claim made about a population proportion. For example, if sample results consist of 70 heads in 100 tosses of a coin, the resulting 95% confidence interval of $0.610 < p < 0.790$ can be used to *informally* support a claim that the proportion of heads is *different from* 50% (because 0.50 is not contained within the confidence interval).

Determining Sample Size

If we plan to collect sample data in order to estimate some population proportion, how do we know *how many* sample units we must collect? If we solve the formula for the margin of error E (Formula 7-1) for the sample size n, we get Formula 7-2 below. Formula 7-2 requires $\hat{p}$ as an estimate of the population proportion p, but if no such estimate is known (as is often the case), we replace $\hat{p}$ by 0.5 and replace $\hat{q}$ by 0.5, with the result given in Formula 7-3. Replacing $\hat{p}$ and $\hat{q}$ with 0.5 results in the largest possible sample size, so we are sure that the sample size is adequate for estimating p.

KEY ELEMENTS

Finding the Sample Size Required to Estimate a Population Proportion

Objective

Determine how large the sample size n should be in order to estimate the population proportion p.

Notation

p = population proportion $\hat{p}$ = sample proportion

n = number of sample values E = desired margin of error

$z_{\alpha/2}$ = z score separating an area of $\alpha/2$ in the right tail of the standard normal distribution

Requirements

The sample must be a simple random sample of independent sample units.

When an estimate $\hat{p}$ is known: **Formula 7-2** $n = \dfrac{[z_{\alpha/2}]^2 \hat{p}\hat{q}}{E^2}$

When no estimate $\hat{p}$ is known: **Formula 7-3** $n = \dfrac{[z_{\alpha/2}]^2 0.25}{E^2}$

If a reasonable estimate of $\hat{p}$ can be made by using previous samples, a pilot study, or someone's expert knowledge, use Formula 7-2. If nothing is known about the value of $\hat{p}$, use Formula 7-3.

Round-Off Rule for Determining Sample Size

If the computed sample size n is not a whole number, round the value of n up to the next *larger* whole number, so the sample size is sufficient instead of being slightly insufficient. For example, round 1067.11 to 1068.

EXAMPLE 5   **What Percentage of Adults Make Online Purchases?**

When the author was conducting research for this chapter, he could find no information about the percentage of adults who make online purchases, yet that information is extremely important to online stores as well as brick and mortar stores. If the author were to conduct his own survey, how many adults must be surveyed in order to be 95% confident that the sample percentage is in error by no more than three percentage points?

   a. Assume that a recent poll showed that 80% of adults make online purchases.

   b. Assume that we have no prior information suggesting a possible value of the population proportion.

continued

Curbstoning

The glossary for the Census defines *curbstoning* as "the practice by which a census enumerator fabricates a questionnaire for a residence without actually visiting it." Curbstoning occurs when a census enumerator sits on a curbstone (or anywhere else) and fills out survey forms by making up responses. Because data from curbstoning are not real, they can affect the validity of the Census. The extent of curbstoning has been investigated in several studies, and one study showed that about 4% of Census enumerators practiced curbstoning at least some of the time. The methods of Section 7-1 assume that the sample data have been collected in an appropriate way, so if much of the sample data have been obtained through curbstoning, then the resulting confidence interval estimates might be very flawed.

SOLUTION

a. With a 95% confidence level, we have $\alpha = 0.05$, so $z_{\alpha/2} = 1.96$. Also, the margin of error is $E = 0.03$, which is the decimal equivalent of "three percentage points." The prior survey suggests that $\hat{p} = 0.80$, so $\hat{q} = 0.20$ (found from $\hat{q} = 1 - 0.80$). Because we have an estimated value of $\hat{p}$, we use Formula 7-2 as follows:

$$n = \frac{[z_{\alpha/2}]^2\, \hat{p}\hat{q}}{E^2} = \frac{[1.96]^2\,(0.80)(0.20)}{0.03^2}$$
$$= 682.951 = 683 \text{ (rounded up)}$$

We must obtain a simple random sample that includes at least 683 adults.

b. With no prior knowledge of $\hat{p}$ (or $\hat{q}$), we use Formula 7-3 as follows:

$$n = \frac{[z_{\alpha/2}]^2 \cdot 0.25}{E^2} = \frac{[1.96]^2 \cdot 0.25}{0.03^2}$$
$$= 1067.11 = 1068 \text{ (rounded up)}$$

We must obtain a simple random sample that includes at least 1068 adults.

INTERPRETATION

To be 95% confident that our sample percentage is within three percentage points of the true percentage for all adults, we should obtain a simple random sample of 1068 adults, assuming no prior knowledge. By comparing this result to the sample size of 683 found in part (a), we can see that if we have no knowledge of a prior study, a larger sample is required to achieve the same results as when the value of $\hat{p}$ can be estimated.

> **YOUR TURN** Do Exercise 31 "Lefties."

CAUTION Try to avoid these three common errors when calculating sample size:

1. Don't make the mistake of using $E = 3$ as the margin of error corresponding to "three percentage points." If the margin of error is three percentage points, use $E = 0.03$.

2. Be sure to substitute the critical z score for $z_{\alpha/2}$. For example, when working with 95% confidence, be sure to replace $z_{\alpha/2}$ with 1.96. Don't make the mistake of replacing $z_{\alpha/2}$ with 0.95 or 0.05.

3. Be sure to round up to the next higher integer; don't round off using the usual round-off rules. Round 1067.11 to 1068.

Role of the Population Size N Formulas 7-2 and 7-3 are remarkable because they show that the sample size does not depend on the size (N) of the population; the sample size depends on the desired confidence level, the desired margin of error, and sometimes the known estimate of $\hat{p}$. (See Exercise 39 "Finite Population Correction Factor" for dealing with cases in which a relatively large sample is selected without replacement from a finite population, so the sample size n does depend on the population size N.)

PART 2 Better-Performing Confidence Intervals

Disadvantage of Wald Confidence Interval

Coverage Probability The **coverage probability** of a confidence interval is the actual proportion of such confidence intervals that contain the true population proportion. If we select a specific confidence level, such as 0.95 (or 95%), we would like to get the *actual* coverage probability equal to our *desired* confidence level. However, for the confidence interval described in Part 1 (called a "Wald confidence interval"), the actual coverage probability is usually less than or equal to the confidence level that we select, and it could be substantially less. For example, if we select a 95% confidence level, we usually get 95% or *fewer* of confidence intervals containing the population proportion *p*. (This is sometimes referred to as being "too liberal.") For this reason, the Wald confidence interval is rarely used in professional applications and professional journals.

Better-Performing Confidence Intervals

Important note about exercises: Except for some Beyond the Basics exercises, the exercises for this Section 7-1 are based on the method for constructing a Wald confidence interval as described in Part 1, not the confidence intervals described here. It is recommended that students learn the methods presented earlier, but recognize that there are better methods available, and they can be used with suitable technology.

Plus Four Method The *plus four confidence interval* performs better than the Wald confidence interval in the sense that its coverage probability is closer to the confidence level that is used. The plus four confidence interval uses this very simple procedure: Add 2 to the number of successes *x*, add 2 to the number of failures (so that the number of trials *n* is increased by 4), and then find the Wald confidence interval as described in Part 1 of this section. The plus four confidence interval is very easy to calculate and it has coverage probabilities similar to those for the Wilson score confidence interval that follows.

Wilson Score Another confidence interval that performs better than the Wald CI is the *Wilson score confidence interval:*

$$\frac{\hat{p} + \frac{z_{\alpha/2}^2}{2n} \pm z_{\alpha/2}\sqrt{\frac{\hat{p}\hat{q} + \frac{z_{\alpha/2}^2}{4n}}{n}}}{1 + \frac{z_{\alpha/2}^2}{n}}$$

The Wilson score confidence interval performs better than the Wald CI in the sense that the coverage probability is closer to the confidence level. With a confidence level of 95%, the Wilson score confidence interval would get us closer to a 0.95 probability of containing the parameter *p*. However, given its complexity, it is easy to see why this superior Wilson score confidence interval is not used much in introductory statistics courses. The complexity of the above expression can be circumvented by using some technologies, such as Statdisk, that provide Wilson score confidence interval results.

Clopper-Pearson Method The Clopper-Pearson method is an "exact" method in the sense that it is based on the exact binomial distribution instead of an approximation of a distribution. It is criticized for being *too conservative* in this sense: When we select a specific confidence level, the coverage probability is usually greater than or equal to the selected confidence level. Select a confidence level of 0.95, and the actual coverage probability is usually 0.95 or greater, so that 95% or more of such confidence intervals will contain *p*. Calculations with this method are too messy to consider here.

Which Method Is Best? There are other methods for constructing confidence intervals that are not discussed here. There isn't universal agreement on which method is best for constructing a confidence interval estimate of *p*.

- The Wald confidence interval is best as a teaching tool for introducing students to confidence intervals.

- The plus four confidence interval is almost as easy as Wald and it performs better than Wald by having a coverage probability closer to the selected confidence level.

Again, note that except for some Beyond the Basic exercises, the exercises that follow are based on the Wald confidence interval given earlier, not the better-performing confidence intervals discussed here.

TECH CENTER

 ## Proportions: Confidence Intervals and Sample Size Determination
Access tech supplements, videos, and data sets at **www.TriolaStats.com**

Statdisk

Confidence Interval
1. Click **Analysis** in the top menu.
2. Select **Confidence Intervals** from the dropdown menu and select **Proportion One Sample** from the submenu.
3. Enter the confidence level, sample size, and number of successes.
4. Click **Evaluate**.

Sample Size Determination
1. Click **Analysis** in the top menu.
2. Select **Sample Size Determination** from the dropdown menu and select **Estimate Proportion** from the submenu.
3. Enter the confidence level, margin of error *E*, estimate of *p* if known, and population size if known.
4. Click **Evaluate.**

StatCrunch

Confidence Interval
1. Click **Stat** in the top menu.
2. Select **Proportion Stats** from the dropdown menu and select **One Sample—With Summary** from the submenu.
3. Enter the number of successes and number of observations.
4. Select **Confidence interval for p** and enter the confidence level. Select the **Standard-Wald** method.
5. Click **Compute!**

Sample Size Determination
Not available.

Minitab

Confidence Interval
1. Click **Stat** in the top menu.
2. Select **Basic Statistics** from the dropdown menu and select **1 Proportion** from the submenu.
3. Select **Summarized data** in the dropdown window and enter the number of events (successes) and number of trials. Confirm *Perform hypothesis test* is not checked.
4. Click the **Options** button and enter the desired confidence level. For *Alternative Hypothesis* select ≠ and for *Method* select **Normal approximation** for the methods of this section.
5. Click **OK** twice.

Sample Size Determination

Minitab determines sample size using the binomial distribution (not normal distribution), so the results will differ from those found using the methods of this section.
1. Click **Stat** in the top menu.
2. Select **Power and Sample Size** from the dropdown menu and select **Sample Size for Estimation** from the submenu.
3. For *Parameter* select **Proportion (Binomial)** and enter an estimate of the proportion if known or enter **0.5** if not known.
4. Select **Estimate sample sizes** from the dropdown menu and enter the margin of error for confidence intervals.
5. Click the **Options** button to enter the confidence level and select a **two-sided** confidence interval.
6. Click **OK** twice.

TECH CENTER *continued*

Proportions: Confidence Intervals and Sample Size Determination
Access tech supplements, videos, and data sets at **www.TriolaStats.com**

TI-83/84 Plus Calculator

Confidence Interval

1. Press **STAT**, then select **TESTS** in the top menu.
2. Select **1-PropZInt** in the menu and press **ENTER**.
3. Enter the number of successes *x*, number of observations *n*, and confidence level (*C-Level*).
4. Select **Calculate** and press **ENTER**.

Sample Size Determination
Not available.

Excel

XLSTAT Add-In (Required)

1. Click on the **XLSTAT** tab in the Ribbon and then click **Parametric Tests.**
2. Select **Tests for one proportion** from the dropdown menu.
3. Under *Data Format* select **Frequency** if you know the number of successes *x* or select **Proportion** if you know the sample proportion $\hat{p}$.
4. Enter the frequency or sample proportion, sample size, and **0.5** for *Test proportion*.
5. Check **z test** and uncheck **Continuity correction.**
6. Click the **Options** tab.
7. Under *Alternative hypothesis* select ≠ **D**. Enter **0** for *Hypothesized difference* and enter the desired significance level (enter **5** for 95% confidence interval). Under *Variance (confidence interval)* select **Sample** and under *Confidence interval* select **Wald.**
8. Click **OK** to display the result under "confidence interval on the proportion (Wald)."

Sample Size Determination
Not available.

7-1 Basic Skills and Concepts

Statistical Literacy and Critical Thinking

1. Poll Results in the Media *USA Today* provided results from a poll of 1000 adults who were asked to identify their favorite pie. Among the 1000 respondents, 14% chose chocolate pie, and the margin of error was given as ±4 percentage points. What important feature of the poll was omitted?

2. Margin of Error For the poll described in Exercise 1, describe what is meant by the statement that "the margin of error was given as ±4 percentage points."

3. Notation For the poll described in Exercise 1, what values do $\hat{p}$, $\hat{q}$, *n*, *E*, and *p* represent? If the confidence level is 95%, what is the value of α?

4. Confidence Levels Given specific sample data, such as the data given in Exercise 1, which confidence interval is wider: the 95% confidence interval or the 80% confidence interval? Why is it wider?

Finding Critical Values. *In Exercises 5–8, find the critical value $z_{\alpha/2}$ that corresponds to the given confidence level.*

5. 90%   **6.** 99%   **7.** 99.5%   **8.** 98%

Formats of Confidence Intervals. *In Exercises 9–12, express the confidence interval using the indicated format. (The confidence intervals are based on the proportions of red, orange, yellow, and blue M&Ms in Data Set 27 "M&M Weights" in Appendix B.)*

9. Red M&Ms Express $0.0434 < p < 0.217$ in the form of $\hat{p} \pm E$.

10. Orange M&Ms Express $0.179 < p < 0.321$ in the form of $\hat{p} \pm E$.

11. Yellow M&Ms Express the confidence interval (0.0169, 0.143) in the form of $\hat{p} - E < p < \hat{p} + E$.

12. Blue M&Ms Express the confidence interval 0.270 ± 0.073 in the form of $\hat{p} - E < p < \hat{p} + E$.

Constructing and Interpreting Confidence Intervals. *In Exercises 13–16, use the given sample data and confidence level. In each case, (a) find the best point estimate of the population proportion p; (b) identify the value of the margin of error E; (c) construct the confidence interval; (d) write a statement that correctly interprets the confidence interval.*

13. Mickey D's In a study of the accuracy of fast food drive-through orders, McDonald's had 33 orders that were not accurate among 362 orders observed (based on data from *QSR* magazine). Construct a 95% confidence interval for the proportion of orders that are not accurate.

14. Eliquis The drug Eliquis (apixaban) is used to help prevent blood clots in certain patients. In clinical trials, among 5924 patients treated with Eliquis, 153 developed the adverse reaction of nausea (based on data from Bristol-Myers Squibb Co.). Construct a 99% confidence interval for the proportion of adverse reactions.

15. Survey Return Rate In a study of cell phone use and brain hemispheric dominance, an Internet survey was e-mailed to 5000 subjects randomly selected from an online group involved with ears. 717 surveys were returned. Construct a 90% confidence interval for the proportion of returned surveys.

16. Medical Malpractice In a study of 1228 randomly selected medical malpractice lawsuits, it was found that 856 of them were dropped or dismissed (based on data from the Physicians Insurers Association of America). Construct a 95% confidence interval for the proportion of medical malpractice lawsuits that are dropped or dismissed.

Critical Thinking. *In Exercises 17–28, use the data and confidence level to construct a confidence interval estimate of p, then address the given question.*

17. Births A random sample of 860 births in New York State included 426 boys. Construct a 95% confidence interval estimate of the proportion of boys in all births. It is believed that among all births, the proportion of boys is 0.512. Do these sample results provide strong evidence against that belief?

18. Mendelian Genetics One of Mendel's famous genetics experiments yielded 580 peas, with 428 of them green and 152 yellow.

a. Find a 99% confidence interval estimate of the *percentage* of green peas.

b. Based on his theory of genetics, Mendel expected that 75% of the offspring peas would be green. Given that the percentage of offspring green peas is not 75%, do the results contradict Mendel's theory? Why or why not?

19. Fast Food Accuracy In a study of the accuracy of fast food drive-through orders, Burger King had 264 accurate orders and 54 that were not accurate (based on data from *QSR* magazine).

a. Construct a 99% confidence interval estimate of the *percentage* of orders that are not accurate.

b. Compare the result from part (a) to this 99% confidence interval for the percentage of orders that are not accurate at Wendy's: $6.2\% < p < 15.9\%$. What do you conclude?

20. OxyContin The drug OxyContin (oxycodone) is used to treat pain, but it is dangerous because it is addictive and can be lethal. In clinical trials, 227 subjects were treated with OxyContin and 52 of them developed nausea (based on data from Purdue Pharma L.P.).

a. Construct a 95% confidence interval estimate of the *percentage* of OxyContin users who develop nausea.

b. Compare the result from part (a) to this 95% confidence interval for 5 subjects who developed nausea among the 45 subjects given a placebo instead of OxyContin: $1.93\% < p < 20.3\%$. What do you conclude?

21. Touch Therapy When she was 9 years of age, Emily Rosa did a science fair experiment in which she tested professional touch therapists to see if they could sense her energy field. She flipped a coin to select either her right hand or her left hand, and then she asked the therapists to identify the selected hand by placing their hand just under Emily's hand without seeing it and without touching it. Among 280 trials, the touch therapists were correct 123 times (based on data in "A Close Look at Therapeutic Touch," *Journal of the American Medical Association,* Vol. 279, No. 13).

a. Given that Emily used a coin toss to select either her right hand or her left hand, what proportion of correct responses would be expected if the touch therapists made random guesses?

b. Using Emily's sample results, what is the best point estimate of the therapists success rate?

c. Using Emily's sample results, construct a 99% confidence interval estimate of the proportion of correct responses made by touch therapists.

d. What do the results suggest about the ability of touch therapists to select the correct hand by sensing an energy field?

22. Medication Usage In a survey of 3005 adults aged 57 through 85 years, it was found that 81.7% of them used at least one prescription medication (based on data from "Use of Prescription and Over-the-Counter Medications and Dietary Supplements Among Older Adults in the United States," by Qato et al., *Journal of the American Medical Association,* Vol. 300, No. 24).

a. How many of the 3005 subjects used at least one prescription medication?

b. Construct a 90% confidence interval estimate of the *percentage* of adults aged 57 through 85 years who use at least one prescription medication.

c. What do the results tell us about the proportion of college students who use at least one prescription medication?

23. Cell Phones and Cancer A study of 420,095 Danish cell phone users found that 0.0321% of them developed cancer of the brain or nervous system. Prior to this study of cell phone use, the rate of such cancer was found to be 0.0340% for those not using cell phones. The data are from the *Journal of the National Cancer Institute.*

a. Use the sample data to construct a 90% confidence interval estimate of the *percentage* of cell phone users who develop cancer of the brain or nervous system.

b. Do cell phone users appear to have a rate of cancer of the brain or nervous system that is different from the rate of such cancer among those not using cell phones? Why or why not?

24. Nonvoters Who Say They Voted In a survey of 1002 people, 70% said that they voted in a recent presidential election (based on data from ICR Research Group). Voting records show that 61% of eligible voters actually did vote.

a. Find a 98% confidence interval estimate of the proportion of people who say that they voted.

b. Are the survey results consistent with the actual voter turnout of 61%? Why or why not?

25. Lipitor In clinical trials of the drug Lipitor (atorvastatin), 270 subjects were given a placebo and 7 of them had allergic reactions. Among 863 subjects treated with 10 mg of the drug, 8 experienced allergic reactions. Construct the two 95% confidence interval estimates of the percentages of allergic reactions. Compare the results. What do you conclude?

26. Gender Selection Before its clinical trials were discontinued, the Genetics & IVF Institute conducted a clinical trial of the XSORT method designed to increase the probability of conceiving a girl and, among the 945 babies born to parents using the XSORT method, there

continued

were 879 girls. The YSORT method was designed to increase the probability of conceiving a boy and, among the 291 babies born to parents using the YSORT method, there were 239 boys. Construct the two 95% confidence interval estimates of the percentages of success. Compare the results. What do you conclude?

27. Smoking Stopped In a program designed to help patients stop smoking, 198 patients were given *sustained* care, and 82.8% of them were no longer smoking after one month. Among 199 patients given *standard* care, 62.8% were no longer smoking after one month (based on data from "Sustained Care Intervention and Postdischarge Smoking Cessation Among Hospitalized Adults," by Rigotti et al., *Journal of the American Medical Association,* Vol. 312, No. 7). Construct the two 95% confidence interval estimates of the percentages of success. Compare the results. What do you conclude?

28. Measured Results vs. Reported Results The same study cited in the preceding exercise produced these results after six months for the 198 patients given sustained care: 25.8% were no longer smoking, and these results were biochemically confirmed, but 40.9% of these patients *reported* that they were no longer smoking. Construct the two 95% confidence intervals. Compare the results. What do you conclude?

Using Appendix B Data Sets. *In Exercises 29 and 30, use the indicated data set in Appendix B.*

29. Heights of Presidents Refer to Data Set 15 "Presidents" in Appendix B. Treat the data as a sample and find the proportion of presidents who were taller than their opponents. Use that result to construct a 95% confidence interval estimate of the population percentage. Based on the result, does it appear that greater height is an advantage for presidential candidates? Why or why not?

30. Green M&Ms Data Set 27 "M&M Weights" in Appendix B includes data from 100 M&M plain candies, and 19 of them are green. The Mars candy company claims that 16% of its M&M plain candies are green. Use the sample data to construct a 95% confidence interval estimate of the percentage of green M&Ms. What do you conclude about the claim of 16%?

Determining Sample Size. *In Exercises 31–38, use the given data to find the minimum sample size required to estimate a population proportion or percentage.*

31. Lefties Find the sample size needed to estimate the percentage of California residents who are left-handed. Use a margin of error of three percentage points, and use a confidence level of 99%.

a. Assume that $\hat{p}$ and $\hat{q}$ are unknown.

b. Assume that based on prior studies, about 10% of Californians are left-handed.

c. How do the results from parts (a) and (b) change if the entire United States is used instead of California?

32. Chickenpox You plan to conduct a survey to estimate the percentage of adults who have had chickenpox. Find the number of people who must be surveyed if you want to be 90% confident that the sample percentage is within two percentage points of the true percentage for the population of all adults.

a. Assume that nothing is known about the prevalence of chickenpox.

b. Assume that about 95% of adults have had chickenpox.

c. Does the added knowledge in part (b) have much of an effect on the sample size?

33. Bachelor's Degree in Four Years In a study of government financial aid for college students, it becomes necessary to estimate the percentage of full-time college students who earn a bachelor's degree in four years or less. Find the sample size needed to estimate that percentage. Use a 0.05 margin of error, and use a confidence level of 95%.

continued

a. Assume that nothing is known about the percentage to be estimated.

b. Assume that prior studies have shown that about 40% of full-time students earn bachelor's degrees in four years or less.

c. Does the added knowledge in part (b) have much of an effect on the sample size?

34. Astrology A sociologist plans to conduct a survey to estimate the percentage of adults who believe in astrology. How many people must be surveyed if we want a confidence level of 99% and a margin of error of four percentage points?

a. Assume that nothing is known about the percentage to be estimated.

b. Use the information from a previous Harris survey in which 26% of respondents said that they believed in astrology.

35. Airline Seating You are the operations manager for American Airlines and you are considering a higher fare level for passengers in aisle seats. You want to estimate the percentage of passengers who now prefer aisle seats. How many randomly selected air passengers must you survey? Assume that you want to be 95% confident that the sample percentage is within 2.5 percentage points of the true population percentage.

a. Assume that nothing is known about the percentage of passengers who prefer aisle seats.

b. Assume that a prior survey suggests that about 38% of air passengers prefer an aisle seat (based on a 3M Privacy Filters survey).

36. iOS Marketshare You plan to develop a new iOS social gaming app that you believe will surpass the success of Angry Birds and Facebook combined. In forecasting revenue, you need to estimate the percentage of all smartphone and tablet devices that use the iOS operating system versus Android and other operating systems. How many smartphones and tablets must be surveyed in order to be 99% confident that your estimate is in error by no more than two percentage points?

a. Assume that nothing is known about the percentage of portable devices using the iOS operating system.

b. Assume that a recent survey suggests that about 43% of smartphone and tablets are using the iOS operating system (based on data from NetMarketShare).

c. Does the additional survey information from part (b) have much of an effect on the sample size that is required?

37. Video Games An investor is considering funding of a new video game. She wants to know the worldwide percentage of people who play video games, so a survey is being planned. How many people must be surveyed in order to be 90% confident that the estimated percentage is within three percentage points of the true population percentage?

a. Assume that nothing is known about the worldwide percentage of people who play video games.

b. Assume that about 16% of people play video games (based on a report by Spil Games).

c. Given that the required sample size is relatively small, could you simply survey the people that you know?

38. Women Who Give Birth An epidemiologist plans to conduct a survey to estimate the percentage of women who give birth. How many women must be surveyed in order to be 99% confident that the estimated percentage is in error by no more than two percentage points?

a. Assume that nothing is known about the percentage to be estimated.

b. Assume that a prior study conducted by the U.S. Census Bureau showed that 82% of women give birth.

c. What is wrong with surveying randomly selected adult women?

7-1 Beyond the Basics

39. Finite Population Correction Factor For Formulas 7-2 and 7-3 we assume that the population is infinite or very large and that we are sampling with replacement. When we sample without replacement from a relatively small population with size N, we modify E to include the *finite population correction factor* shown here, and we can solve for n to obtain the result given here. Use this result to repeat part (b) of Exercise 38, assuming that we limit our population to a county with 2500 women who have completed the time during which they can give birth.

$$E = z_{\alpha/2}\sqrt{\frac{\hat{p}\hat{q}}{n}}\sqrt{\frac{N-n}{N-1}} \qquad n = \frac{N\hat{p}\hat{q}[z_{\alpha/2}]^2}{\hat{p}\hat{q}[z_{\alpha/2}]^2 + (N-1)E^2}$$

40. One-Sided Confidence Interval A one-sided claim about a population proportion is a claim that the proportion is less than (or greater than) some specific value. Such a claim can be formally addressed using a *one-sided confidence interval* for p, which can be expressed as $p < \hat{p} + E$ or $p > \hat{p} - E$, where the margin of error E is modified by replacing $z_{\alpha/2}$ with z_{α}. (Instead of dividing α between two tails of the standard normal distribution, put all of it in one tail.) The Chapter Problem refers to a Gallup poll of 1487 adults showing that 43% of the respondents have Facebook pages. Use that data to construct a one-sided 95% confidence interval that would be suitable for helping to determine whether the proportion of all adults having Facebook pages is less than 50%.

41. Coping with No Success According to the *Rule of Three,* when we have a sample size n with $x = 0$ successes, we have 95% confidence that the true population proportion has an upper bound of $3/n$. (See "A Look at the Rule of Three," by Jovanovic and Levy, *American Statistician,* Vol. 51, No. 2.)

a. If n independent trials result in no successes, why can't we find confidence interval limits by using the methods described in this section?

b. If 40 couples use a method of gender selection and each couple has a baby girl, what is the 95% upper bound for p, the proportion of all babies who are boys?

7-2 Estimating a Population Mean

Key Concept The main goal of this section is to present methods for using a sample mean $\bar{x}$ to make an inference about the value of the corresponding population mean μ. There are three main concepts included in this section:

- **Point Estimate:** The sample mean $\bar{x}$ is the best *point estimate* (or single value estimate) of the population mean μ.

- **Confidence Interval:** Use sample data to construct and interpret a *confidence interval* estimate of the true value of a population mean μ.

- **Sample Size:** Find the sample size necessary to estimate a population mean.

Part 1 of this section deals with the very realistic and commonly used case in which we want to estimate μ and the population standard deviation σ is not known. Part 2 includes a brief discussion of the procedure used when σ is known, which is very rare.

PART 1 **Estimating a Population Mean When σ Is Not Known**

It's rare that we want to estimate the unknown value of a population mean μ but we somehow know the value of the population standard deviation σ, so Part 1 focuses on the realistic situation in which σ is not known.

Point Estimate As discussed in Section 6-3, the sample mean $\bar{x}$ is an *unbiased estimator* of the population mean μ. Also, for many populations, sample means tend to vary less than other measures of center. For these reasons, the sample mean $\bar{x}$ is usually the best point estimate of the population mean μ.

> **The sample mean $\bar{x}$ is the best *point estimate* of the population mean μ.**

Because even the best point estimate gives us no indication of how accurate it is, we use a *confidence interval* (or *interval estimate*), which consists of a range (or an interval) of values instead of just a single value.

Confidence Interval The accompanying box includes the key elements for constructing a confidence interval estimate of a population mean μ in the common situation where σ is not known.

KEY ELEMENTS

Confidence Interval for Estimating a Population Mean with σ Not Known

Objective

Construct a confidence interval used to estimate a population mean.

Notation

μ = population mean
$\bar{x}$ = sample mean
s = sample standard deviation

n = number of sample values
E = margin of error

Requirements

1. The sample is a simple random sample.

2. Either or both of these conditions are satisfied: The population is normally distributed or $n > 30$.

Confidence Interval

Formats: $\bar{x} - E < \mu < \bar{x} + E$ or $\bar{x} \pm E$ or $(\bar{x} - E, \bar{x} + E)$

- **Margin of Error:** $E = t_{\alpha/2} \cdot \dfrac{s}{\sqrt{n}}$ (Use df $= n - 1$.)

- **Confidence Level:** The confidence interval is associated with a confidence level, such as 0.95 (or 95%), and α is the complement of the confidence level. For a 0.95 (or 95%) confidence level, $\alpha = 0.05$.

- **Critical Value:** $t_{\alpha/2}$ is the critical t value separating an area of $\alpha/2$ in the right tail of the Student t distribution.

- **Degrees of Freedom:** df $= n - 1$ is the number of degrees of freedom used when finding the critical value.

Round-Off Rule

1. *Original Data:* When using an *original set of data* values, round the confidence interval limits to one more decimal place than is used for the original set of data.

2. *Summary Statistics:* When using the *summary statistics* of n, $\bar{x}$, and s, round the confidence interval limits to the same number of decimal places used for the sample mean.

Requirement of "Normality or $n > 30$"

Normality The method for finding a confidence interval estimate of μ is *robust* against a departure from normality, which means that the normality requirement is loose. The distribution need not be perfectly bell-shaped, but it should appear to be somewhat symmetric with one mode and no outliers.

Sample Size $n > 30$ This is a common guideline, but sample sizes of 15 to 30 are adequate if the population appears to have a distribution that is not far from being normal and there are no outliers. For some population distributions that are extremely far from normal, the sample size might need to be larger than 30. This text uses the simplified criterion of $n > 30$ as justification for treating the distribution of sample means as a normal distribution.

Student *t* Distribution

In this section we use a *Student t distribution*, which is commonly referred to as a "*t* distribution." It was developed by William Gosset (1876–1937), who was a Guinness Brewery employee who needed a distribution that could be used with small samples. The brewery prohibited publication of research results, but Gosset got around this by publishing under the pseudonym "Student." (Strictly in the interest of better serving his readers, the author visited the Guinness Brewery and felt obligated to sample some of the product.) Here are some key points about the Student *t* distribution:

- **Student *t* Distribution** If a population has a normal distribution, then the distribution of

$$t = \frac{\bar{x} - \mu}{\frac{s}{\sqrt{n}}}$$

  is a **Student *t* distribution** for all samples of size *n*. A Student *t* distribution is commonly referred to as a ***t* distribution.**

- **Degrees of Freedom** Finding a critical value $t_{\alpha/2}$ requires a value for the **degrees of freedom** (or **df**). In general, the number of degrees of freedom for a collection of sample data is the number of sample values that can vary after certain restrictions have been imposed on all data values. (*Example:* If 10 test scores have the restriction that their mean is 80, then their sum must be 800, and we can freely assign values to the first 9 scores, but the 10th score would then be determined, so in this case there are 9 degrees of freedom.) For the methods of this section, the number of degrees of freedom is the sample size minus 1.

Degrees of freedom $= n - 1$

- **Finding Critical Value $t_{\alpha/2}$** A critical value $t_{\alpha/2}$ can be found using technology or Table A-3. Technology can be used with any number of degrees of freedom, but Table A-3 can be used for select numbers of degrees of freedom only.
  If using Table A-3 to find a critical value of $t_{\alpha/2}$, but the table does not include the exact number of degrees of freedom, you could use the closest value, or you could be conservative by using the next lower number of degrees of freedom found in the table, or you could interpolate.

- The Student *t* distribution is different for different sample sizes. (See Figure 7-4 for the cases $n = 3$ and $n = 12$.)

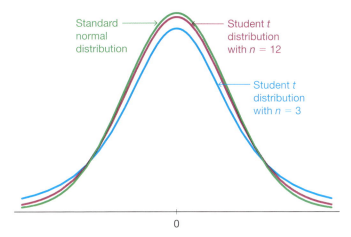

FIGURE 7-4   Student *t* Distributions for *n* = 3 and *n* = 12
The Student *t* distribution has the same general shape and symmetry as the standard normal distribution, but it has the greater variability that is expected with small samples.

- The Student *t* distribution has the same general symmetric bell shape as the standard normal distribution, but has more variability (with wider distributions), as we expect with small samples.

- The Student *t* distribution has a mean of *t* = 0 (just as the standard normal distribution has a mean of *z* = 0).

- The standard deviation of the Student *t* distribution varies with the sample size, but it is greater than 1 (unlike the standard normal distribution, which has $\sigma = 1$).

- As the sample size *n* gets larger, the Student *t* distribution gets closer to the standard normal distribution.

Procedure for Constructing a Confidence Interval for μ

Confidence intervals can be easily constructed with technology or they can be manually constructed by using the following procedure.

1. Verify that the two requirements are satisfied: The sample is a simple random sample and the population is normally distributed or *n* > 30.

2. With σ unknown (as is usually the case), use *n* − 1 degrees of freedom and use technology or a *t* distribution table (such as Table A-3) to find the critical value $t_{\alpha/2}$ that corresponds to the desired confidence level.

3. Evaluate the margin of error using $E = t_{\alpha/2} \cdot s / \sqrt{n}$.

4. Using the value of the calculated margin of error *E* and the value of the sample mean $\bar{x}$, substitute those values in one of the formats for the confidence interval: $\bar{x} - E < \mu < \bar{x} + E$ or $\bar{x} \pm E$ or $(\bar{x} - E, \bar{x} + E)$.

5. Round the resulting confidence interval limits as follows: With an *original set of data* values, round the confidence interval limits to one more decimal place than is used for the original set of data, but when using the *summary statistics* of *n*, $\bar{x}$, and *s*, round the confidence interval limits to the same number of decimal places used for the sample mean.

Estimating Sugar in Oranges

In Florida, members of the citrus industry make extensive use of statistical methods. One particular application involves the way in which growers are paid for oranges used to make orange juice. An arriving truckload of oranges is first weighed at the receiving plant, and then a sample of about a dozen oranges is randomly selected. The sample is weighed and then squeezed, and the amount of sugar in the juice is measured. Based on the sample results, an estimate is made of the total amount of sugar in the entire truckload. Payment for the load of oranges is based on the estimate of the amount of sugar because sweeter oranges are more valuable than those less sweet, even though the amounts of juice may be the same.

EXAMPLE 1 Finding a Critical Value $t_{\alpha/2}$

Find the critical value $t_{\alpha/2}$ corresponding to a 95% confidence level, given that the sample has size $n = 15$.

SOLUTION

Because $n = 15$, the number of degrees of freedom is $n - 1 = 14$. The 95% confidence level corresponds to $\alpha = 0.05$, so there is an area of 0.025 in each of the two tails of the t distribution, as shown in Figure 7-5.

Using Technology Technology can be used to find that for 14 degrees of freedom and an area of 0.025 in each tail, the critical value is $t_{\alpha/2} = t_{0.025} = 2.145$.

Using Table A-3 To find the critical value using Table A-3, use the column with 0.05 for the "Area in Two Tails" (or use the same column with 0.025 for the "Area in One Tail"). The number of degrees of freedom is df $= n - 1 = 14$. We get $t_{\alpha/2} = t_{0.025} = 2.145$.

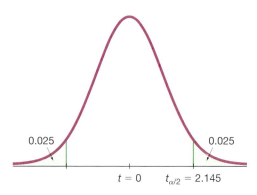

FIGURE 7-5 Critical Value $t_{\alpha/2}$

YOUR TURN Find the critical value for Exercise 2 "Degrees of Freedom."

EXAMPLE 2 Confidence Interval Using Birth Weights

Listed below are weights (hectograms or hg) of randomly selected girls at birth, based on data from the National Center for Health Statistics. Here are the summary statistics: $n = 15$, $\bar{x} = 30.9$ hg, $s = 2.9$ hg. Use the sample data to construct a 95% confidence interval for the mean birth weight of girls.

    33 28 33 37 31 32 31 28 34 28 33 26 30 31 28

SOLUTION

REQUIREMENT CHECK We must first verify that the requirements are satisfied. (1) The sample is a simple random sample. (2) Because the sample size is $n = 15$, the requirement that "the population is normally distributed or the sample size is greater than 30" can be satisfied only if the sample data appear to be from a normally distributed population, so we need to investigate normality. The accompanying normal quantile plot shows that the sample data appear to be from a normally distributed population, so this second requirement is satisfied. ✓

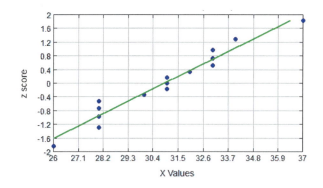

Using Technology Technology can be used to automatically construct the confidence interval. (See instructions near the end of this section.) Shown here is the StatCrunch display resulting from the 15 birth weights. The display shows the lower confidence interval limit (29.247163) and the upper confidence interval limit (32.486171). After rounding to one decimal place (as in the sample mean), we can express the 95% confidence interval as 29.2 hg $< \mu <$ 32.5 hg.

StatCrunch

95% confidence interval results:					
μ : Mean of variable					
Variable	Sample Mean	Std. Err.	DF	L. Limit	U. Limit
BirthWt	30.866667	0.75508856	14	29.247163	32.486171

Using t Distribution Table Using Table A-3, the critical value is $t_{0.025} = 2.145$, as shown in Example 1. We now find the margin of error E as shown here:

$$E = t_{\alpha/2}\frac{s}{\sqrt{n}} = 2.145 \cdot \frac{2.9}{\sqrt{15}} = 1.606126$$

With $\bar{x} = 30.9$ hg and $E = 1.606126$ hg, we construct the confidence interval as follows:

$$\bar{x} - E < \mu < \bar{x} + E$$

$$30.9 - 1.606126 < \mu < 30.9 + 1.606126$$

$$29.3 \text{ hg} < \mu < 32.5 \text{ hg} \quad \text{(rounded to one decimal place)}$$

The lower confidence interval limit of 29.3 hg is actually 29.2 hg if we use technology or if we use summary statistics with more decimal places than the one decimal place used in the preceding calculation.

> **INTERPRETATION**

We are 95% confident that the limits of 29.2 hg and 32.5 hg actually do contain the value of the population mean μ. If we were to collect many different random samples of 15 newborn girls and find the mean weight in each sample, about 95% of the resulting confidence intervals should contain the value of the mean weight of all newborn girls.

> **YOUR TURN** Do Exercise 15 "Genes."

Interpreting the Confidence Interval

The confidence interval is associated with a **confidence level,** such as 0.95 (or 95%). When interpreting a confidence interval estimate of μ, know that the confidence level gives us the *success rate of the procedure* used to construct the confidence interval.

Captured Tank Serial Numbers Reveal Population Size

During World War II, Allied intelligence specialists wanted to determine the number of tanks Germany was producing. Traditional spy techniques provided unreliable results, but statisticians obtained accurate estimates by analyzing serial numbers on captured tanks. As one example, records show that Germany actually produced 271 tanks in June 1941. The estimate based on serial numbers was 244, but traditional intelligence methods resulted in the extreme estimate of 1550. (See "An Empirical Approach to Economic Intelligence in World War II," by Ruggles and Brodie, *Journal of the American Statistical Association,* Vol. 42.)

Estimating Crowd Size

There are sophisticated methods of analyzing the size of a crowd. Aerial photographs and measures of people density can be used with reasonably good accuracy. However, reported crowd size estimates are often simple guesses. After the Boston Red Sox won the World Series for the first time in 86 years, Boston city officials estimated that the celebration parade was attended by 3.2 million fans. Boston police provided an estimate of around 1 million, but it was admittedly based on guesses by police commanders. A photo analysis led to an estimate of around 150,000. Boston University Professor Farouk El-Baz used images from the U.S. Geological Survey to develop an estimate of at most 400,000. MIT physicist Bill Donnelly said that "it's a serious thing if people are just putting out any number. It means other things aren't being vetted that carefully."

For example, the 95% confidence interval estimate of 29.2 hg $< \mu <$ 32.5 hg can be interpreted as follows:

> **"We are 95% confident that the interval from 29.2 hg to 32.5 hg actually does contain the true value of μ."**

By "95% confident" we mean that if we were to select many different samples of the same size and construct the corresponding confidence intervals, in the long run, 95% of the confidence intervals should actually contain the value of μ.

Finding a Point Estimate and Margin of Error *E* from a Confidence Interval

Technology and journal articles often express a confidence interval in a format such as (10.0, 30.0). The sample mean $\bar{x}$ is the value midway between those limits, and the margin of error E is one-half the difference between those limits (because the upper limit is $\bar{x} + E$ and the lower limit is $\bar{x} - E$, the distance separating them is 2E).

$$\text{Point estimate of } \mu: \bar{x} = \frac{(\text{upper confidence limit}) + (\text{lower confidence limit})}{2}$$

$$\text{Margin of error: } E = \frac{(\text{upper confidence limit}) - (\text{lower confidence limit})}{2}$$

For example, the confidence interval (10.0, 30.0) yields $\bar{x} = 20.0$ and $E = 10.0$.

Using Confidence Intervals to Describe, Explore, or Compare Data

In some cases, confidence intervals might be among the different tools used to describe, explore, or compare data sets, as in the following example.

EXAMPLE 3   Second-Hand Smoke

Figure 7-6 shows graphs of confidence interval estimates of the mean cotinine level in each of three samples: (1) people who smoke; (2) people who don't smoke but are exposed to tobacco smoke at home or work; (3) people who don't smoke and are not exposed to smoke. (The sample data are listed in Data Set 12 "Passive and Active Smoke" in Appendix B.) Because cotinine is produced by the body when nicotine is absorbed, cotinine is a good indication of nicotine intake. Figure 7-6 helps us see the effects of second-hand smoke. In Figure 7-6, we see that the confidence interval for smokers does not overlap the other confidence intervals, so it appears that the mean cotinine level of smokers is different from that of the other two groups. The two nonsmoking groups have confidence intervals that do overlap, so it is possible that they have the same mean cotinine level. It is helpful to compare confidence intervals or their graphs, but such comparisons should not be used for making formal and final conclusions about equality of means. Chapters 9 and 12 introduce better methods for formal comparisons of means.

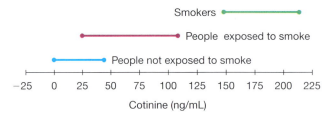

FIGURE 7-6   Comparing Confidence Intervals

YOUR TURN Do Exercise 25 "Pulse Rates."

CAUTION Confidence intervals can be used *informally* to compare different data sets, but *the overlapping of confidence intervals should not be used for making formal and final conclusions about equality of means.*

Determining Sample Size

If we want to collect a sample to be used for estimating a population mean μ, *how many* sample values do we need? When determining the sample size needed to estimate a population mean, we must have an estimated or known value of the population standard deviation σ, so that we can use Formula 7-4 shown in the accompanying Key Elements box.

KEY ELEMENTS

Finding the Sample Size Required to Estimate a Population Mean

Objective

Determine the sample size n required to estimate the value of a population mean μ.

Notation

μ = population mean

σ = population standard deviation

$\bar{x}$ = sample mean

E = desired margin of error

$z_{\alpha/2}$ = z score separating an area of $\alpha/2$ in the right tail of the standard normal distribution

Requirement

The sample must be a simple random sample.

Sample Size

The required sample size is found by using Formula 7-4.

$$\text{Formula 7-4} \qquad n = \left[\frac{z_{\alpha/2}\sigma}{E} \right]^2$$

Round-Off Rule

If the computed sample size n is not a whole number, round the value of n up to the next *larger* whole number.

Population Size Formula 7-4 does not depend on the size (N) of the population (except for cases in which a relatively large sample is selected without replacement from a finite population).

Rounding The sample size must be a whole number because it is the number of sample values that must be found, but Formula 7-4 usually gives a result that is not a whole number. The round-off rule is based on the principle that when rounding is necessary, the required sample size should be rounded *upward* so that it is at least adequately large instead of being slightly too small.

Dealing with Unknown σ When Finding Sample Size Formula 7-4 requires that we substitute a known value for the population standard deviation σ, but in reality, it is usually unknown. When determining a required sample size (not constructing a confidence interval), here are some ways that we can work around the problem of not knowing the value of σ:

1. Use the range rule of thumb (see Section 3-2) to estimate the standard deviation as follows: $\sigma \approx$ range/4, where the range is determined from sample data. (With a sample of 87 or more values randomly selected from a normally distributed population, range/4 will yield a value that is greater than or equal to σ at least 95% of the time.)

2. Start the sample collection process without knowing σ and, using the first several values, calculate the sample standard deviation s and use it in place of σ. The estimated value of σ can then be improved as more sample data are obtained, and the required sample size can be adjusted as you collect more sample data.

3. Estimate the value of σ by using the results of some other earlier study. In addition, we can sometimes be creative in our use of other known results. For example, Wechsler IQ tests are designed so that the standard deviation is 15. Statistics students have IQ scores with a standard deviation less than 15, because they are a more homogeneous group than people randomly selected from the general population. We do not know the specific value of σ for statistics students, but we can be safe by using $\sigma = 15$. Using a value for σ that is larger than the true value will make the sample size larger than necessary, but using a value for σ that is too small would result in a sample size that is inadequate. *When determining the sample size n, any errors should always be conservative in the sense that they make the sample size too large instead of too small.*

EXAMPLE 4 **IQ Scores of Statistics Students**

Assume that we want to estimate the mean IQ score for the population of statistics students. How many statistics students must be randomly selected for IQ tests if we want 95% confidence that the sample mean is within 3 IQ points of the population mean?

SOLUTION

For a 95% confidence interval, we have $\alpha = 0.05$, so $z_{\alpha/2} = 1.96$. Because we want the sample mean to be within 3 IQ points of μ, the margin of error is $E = 3$. Also, we can assume that $\sigma = 15$ (see the discussion that immediately precedes this example). Using Formula 7-4, we get

$$n = \left[\frac{z_{\alpha/2}\sigma}{E}\right]^2 = \left[\frac{1.96 \cdot 15}{3}\right]^2 = 96.04 = 97 \quad (\text{rounded } up)$$

INTERPRETATION

Among the thousands of statistics students, we need to obtain a simple random sample of at least 97 of their IQ scores. With a simple random sample of only 97 statistics students, we will be 95% confident that the sample mean $\bar{x}$ is within 3 IQ points of the true population mean μ.

YOUR TURN Do Exercise 29 "Mean IQ of College Professors."

PART 2 Estimating a Population Mean When σ Is Known

In the real world of professional statisticians and professional journals and reports, it is extremely rare that we want to estimate an unknown value of a population mean μ but we somehow know the value of the population standard deviation σ. If we somehow do know the value of σ, the confidence interval is constructed using the standard normal distribution instead of the Student t distribution, so the same procedure from Part 1 can be used with this margin of error:

$$\text{Margin of error: } E = z_{\alpha/2} \cdot \frac{\sigma}{\sqrt{n}} \quad \text{(used with known } \sigma \text{)}$$

EXAMPLE 5 Confidence Interval Estimate of μ with Known σ

Use the same 15 birth weights of girls given in Example 2, for which $n = 15$ and $\bar{x} = 30.9$ hg. Construct a 95% confidence interval estimate of the mean birth weight of all girls by assuming that σ is known to be 2.9 hg.

SOLUTION

REQUIREMENT CHECK The requirements were checked in Example 2. The requirements are satisfied. ☑

With a 95% confidence level, we have $\alpha = 0.05$, and we get $z_{\alpha/2} = 1.96$ (as in Example 2 from Section 7-1). Using $z_{\alpha/2} = 1.96$, $\sigma = 2.9$ hg, and $n = 15$, we find the value of the margin of error E:

$$E = z_{\alpha/2} \cdot \frac{\sigma}{\sqrt{n}}$$

$$= 1.96 \cdot \frac{2.9}{\sqrt{15}} = 1.46760$$

With $\bar{x} = 30.9$ and $E = 1.46760$, we find the 95% confidence interval as follows:

$$\bar{x} - E < \mu < \bar{x} + E$$

$$30.9 - 1.46760 < \mu < 30.9 + 1.46760$$

$$29.4 \text{ hg} < \mu < 32.4 \text{ hg} \quad \text{(rounded to one decimal place)}$$

The confidence interval found here using the normal distribution is slightly narrower than the confidence interval found using the t distribution in Example 2. Because $z_{\alpha/2} = 1.96$ is smaller than $t_{\alpha/2} = 2.145$, the margin of error E is smaller and the confidence interval is narrower. The critical value $t_{\alpha/2}$ is larger because the t distribution incorporates the greater amount of variation that we get with smaller samples.

Remember, this example illustrates the situation in which the population standard deviation σ is known, which is rare. The more realistic situation with σ unknown is considered in Part 1 of this section.

YOUR TURN Do Exercise 37 "Birth Weights of Girls."

Choosing the Appropriate Distribution

When constructing a confidence interval estimate of the population mean μ, it is important to use the correct distribution. Table 7-1 on the next page summarizes the key points to consider.

TABLE 7-1 Choosing between Student *t* and *z* (Normal) Distributions

Conditions	Method
σ not known and normally distributed population *or* σ not known and $n > 30$	Use Student *t* distribution.
σ known and normally distributed population *or* σ known and $n > 30$ (In reality, σ is rarely known.)	Use normal (*z*) distribution.
Population is not normally distributed and $n \leq 30$.	Use the bootstrapping method (Section 7-4) or a nonparametric method.

TECH CENTER

Means: Confidence Intervals and Sample Size Determination

Access tech supplements, videos, and data sets at **www.TriolaStats.com**

Statdisk

Confidence Interval

1. Click **Analysis** in the top menu.
2. Select **Confidence Intervals** from the dropdown menu and select **Mean One-Sample** from the submenu.
3. *Using Summary Statistics:* Select the **Use Summary Statistics** tab and enter the desired confidence level, sample size, sample mean, and sample standard deviation.

   Using Sample Data: Select the **Use Data** tab and select the desired data column.
4. Click **Evaluate**.

TIP: Statdisk will automatically choose between the normal and *t* distributions, depending on whether a value for the population standard deviation is entered.

Sample Size Determination

1. Click **Analysis** in the top menu.
2. Select **Sample Size Determination** from the dropdown menu and select **Estimate Mean** from the submenu.
3. Enter the confidence level, margin of error *E*, and population standard deviation. Also enter the known population size if you are sampling without replacement from a finite population.
4. Click **Evaluate**.

Minitab

Confidence Interval

1. Click **Stat** in the top menu.
2. Select **Basic Statistics** from the dropdown menu and select **1-Sample t** from the submenu.
3. *Using Summary Statistics:* Select **Summarized data** from the dropdown menu and enter the sample size, sample mean, and standard deviation.

   Using Sample Data: Select **One or more samples, each in a column** from the dropdown menu and select the desired data column(s).
4. Confirm *Perform hypothesis test* is not checked.
5. Click the **Options** button and enter the confidence level. For *Alternative Hypothesis* select ≠.
6. Click **OK** twice.

Sample Size Determination

1. Click **Stat** in the top menu.
2. Select **Power and Sample Size** from the dropdown menu and select **Sample Size for Estimation** from the submenu.
3. For *Parameter* select **Mean (Normal)** and enter the standard deviation.
4. Select **Estimate sample sizes** from the dropdown menu and enter the desired margin of error for confidence intervals.
5. Click the **Options** button to enter the confidence level and select a **two-sided** confidence interval.
6. Click **OK** twice.

StatCrunch

Confidence Interval

1. Click **Stat** in the top menu.
2. Select **T Stats** in the dropdown menu, then select **One Sample** from the submenu.
3. *Using Summary Statistics:* Select *With Summary* from the submenu and enter the sample mean, sample standard deviation, and sample size.

   Using Sample Data: Select *With Data* from the submenu and select the desired data column(s).
4. Select **Confidence interval for *μ*** and enter the desired confidence level.
5. Click **Compute!**

Sample Size Determination
Not available.

TECH CENTER *continued*

 Means: Confidence Intervals and Sample Size Determination
Access tech supplements, videos, and data sets at **www.TriolaStats.com**

TI-83/84 Plus Calculator

Confidence Interval

1. Press **STAT**, then select **TESTS** in the top menu.
2. Select **TInterval** from the menu if σ is *not* known. (Choose **ZInterval** if σ is known.)
3. *Using Summary Statistics:* Select **Stats**, press **ENTER**, and enter the values for sample mean $\bar{x}$, sample standard deviation *Sx*, and sample size *n*.

   Using Sample Data: Select **Data**, press **ENTER**, and enter the name of the list containing the sample data. *Freq* should be set to **1**.
4. Enter the desired confidence level, *C-Level*.
5. Select **Calculate** and press **ENTER**.

Sample Size Determination
Not available.

Excel

Confidence Interval

XLSTAT Add-In (Required)
Requires original sample data, does not work with summary data.

1. Click on the **XLSTAT** tab in the Ribbon and then click **Parametric Tests.**
2. Select **One-Sample t-test and z-test** from the dropdown menu.
3. Under *Data* enter the range of cells containing the sample data. For *Data format* select **One Sample.** If the first row of data contains a label, also check the **Column labels** box.
4. Select **Student's t test.**
5. Click the **Options** tab.
6. Under *Alternative hypothesis* select ≠ **Theoretical Mean.** Enter **0** for *Theoretical mean* and enter the desired significance level (enter **5** for 95% confidence interval).
7. Click **OK** to display the result under "confidence interval on the mean."

Sample Size Determination
Not available.

7-2   Basic Skills and Concepts

Statistical Literacy and Critical Thinking

In Exercises 1–3, refer to the accompanying screen display that results from the Verizon airport data speeds (Mbps) from Data Set 32 "Airport Data Speeds" in Appendix B. The confidence level of 95% was used.

1. Airport Data Speeds Refer to the accompanying screen display.

a. Express the confidence interval in the format that uses the "less than" symbol. Given that the original listed data use one decimal place, round the confidence interval limits accordingly.

b. Identify the best point estimate of μ and the margin of error.

c. In constructing the confidence interval estimate of μ, why is it not necessary to confirm that the sample data appear to be from a population with a normal distribution?

TI-83/84 Plus

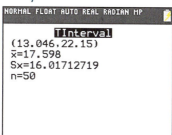

2. Degrees of Freedom

a. What is the number of degrees of freedom that should be used for finding the critical value $t_{\alpha/2}$?

b. Find the critical value $t_{\alpha/2}$ corresponding to a 95% confidence level.

c. Give a brief general description of the number of degrees of freedom.

3. Interpreting a Confidence Interval The results in the screen display are based on a 95% confidence level. Write a statement that correctly interprets the confidence interval.

4. Normality Requirement What does it mean when we say that the confidence interval methods of this section are *robust* against departures from normality?

Using Correct Distribution.   *In Exercises 5–8, assume that we want to construct a confidence interval. Do one of the following, as appropriate: (a) Find the critical value $t_{\alpha/2}$, (b) find the critical value $z_{\alpha/2}$, or (c) state that neither the normal distribution nor the t distribution applies.*

5. Miami Heat Salaries Confidence level is 95%, σ is not known, and the normal quantile plot of the 17 salaries (thousands of dollars) of Miami Heat basketball players is as shown.

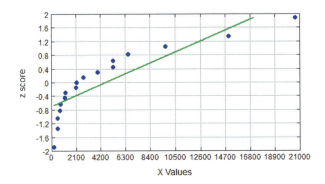

6. Denver Bronco Salaries Confidence level is 90%, σ is not known, and the histogram of 61 player salaries (thousands of dollars) is as shown.

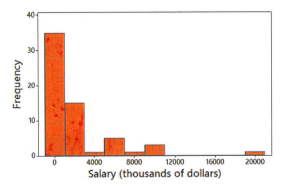

7. Denver Bronco Salaries Confidence level is 99%, $\sigma = 3342$ thousand dollars, and the histogram of 61 player salaries (thousands of dollars) is shown in Exercise 6.

8. Birth Weights Here are summary statistics for randomly selected weights of newborn girls: $n = 205$, $\bar{x} = 30.4$ hg, $s = 7.1$ hg (based on Data Set 4 "Births" in Appendix B). The confidence level is 95%.

Confidence Intervals.   *In Exercises 9–24, construct the confidence interval estimate of the mean.*

9. Birth Weights of Girls Use these summary statistics given in Exercise 8: $n = 205$, $\bar{x} = 30.4$ hg, $s = 7.1$ hg. Use a 95% confidence level. Are the results very different from those found in Example 2 with only 15 sample values?

10. Birth Weights of Boys Use these summary statistics for birth weights of 195 boys: $\bar{x} = 32.7$ hg, $s = 6.6$ hg (based on Data Set 4 "Births" in Appendix B). Use a 95% confidence level. Are the results very different from those found in Exercise 9? Does it appear that boys and girls have very different birth weights?

11. Mean Body Temperature Data Set 3 "Body Temperatures" in Appendix B includes a sample of 106 body temperatures having a mean of 98.20°F and a standard deviation of 0.62°F. Construct a 95% confidence interval estimate of the mean body temperature for the entire population. What does the result suggest about the common belief that 98.6°F is the mean body temperature?

12. Atkins Weight Loss Program In a test of weight loss programs, 40 adults used the Atkins weight loss program. After 12 months, their mean weight *loss* was found to be 2.1 lb, with a standard deviation of 4.8 lb. Construct a 90% confidence interval estimate of the mean weight loss for all such subjects. Does the Atkins program appear to be effective? Does it appear to be practical?

13. Insomnia Treatment A clinical trial was conducted to test the effectiveness of the drug zopiclone for treating insomnia in older subjects. Before treatment with zopiclone, 16 subjects had a mean wake time of 102.8 min. After treatment with zopiclone, the 16 subjects had a mean wake time of 98.9 min and a standard deviation of 42.3 min (based on data from "Cognitive Behavioral Therapy vs Zopiclone for Treatment of Chronic Primary Insomnia in Older Adults," by Sivertsen et al., *Journal of the American Medical Association,* Vol. 295, No. 24). Assume that the 16 sample values appear to be from a normally distributed population and construct a 98% confidence interval estimate of the mean wake time for a population with zopiclone treatments. What does the result suggest about the mean wake time of 102.8 min before the treatment? Does zopiclone appear to be effective?

14. Garlic for Reducing Cholesterol In a test of the effectiveness of garlic for lowering cholesterol, 49 subjects were treated with raw garlic. Cholesterol levels were measured before and after the treatment. The changes (before minus after) in their levels of LDL cholesterol (in mg/dL) had a mean of 0.4 and a standard deviation of 21.0 (based on data from "Effect of Raw Garlic vs Commercial Garlic Supplements on Plasma Lipid Concentrations in Adults with Moderate Hypercholesterolemia," by Gardner et al., *Archives of Internal Medicine,* Vol. 167). Construct a 98% confidence interval estimate of the mean net change in LDL cholesterol after the garlic treatment. What does the confidence interval suggest about the effectiveness of garlic in reducing LDL cholesterol?

15. Genes Samples of DNA are collected, and the four DNA bases of A, G, C, and T are coded as 1, 2, 3, and 4, respectively. The results are listed below. Construct a 95% confidence interval estimate of the mean. What is the practical use of the confidence interval?

2   2   1   4   3   3   3   3   4   1

16. Arsenic in Rice Listed below are amounts of arsenic (μg, or micrograms, per serving) in samples of brown rice from California (based on data from the Food and Drug Administration). Use a 90% confidence level. The Food and Drug Administration also measured amounts of arsenic in samples of brown rice from Arkansas. Can the confidence interval be used to describe arsenic levels in Arkansas?

5.4   5.6   8.4   7.3   4.5   7.5   1.5   5.5   9.1   8.7

17. Speed Dating In a study of speed dating conducted at Columbia University, male subjects were asked to rate the attractiveness of their female dates, and a sample of the results is listed below (1 = not attractive; 10 = extremely attractive). Use a 99% confidence level. What do the results tell us about the mean attractiveness ratings made by the population of all adult females?

7   8   2   10   6   5   7   8   8   9   5   9

18. Speed Dating In a study of speed dating conducted at Columbia University, female subjects were asked to rate the attractiveness of their male dates, and a sample of the results is listed below (1 = not attractive; 10 = extremely attractive). Use a 99% confidence level. Can the result be used to estimate the mean amount of attractiveness of the population of all adult males?

5   8   3   8   6   10   3   7   9   8   5   5   6   8   8   7   3   5   5   6   8   7   8   8   87

19. Mercury in Sushi An FDA guideline is that the mercury in fish should be below 1 part per million (ppm). Listed below are the amounts of mercury (ppm) found in tuna sushi sampled at different stores in New York City. The study was sponsored by the *New York Times,* and the stores (in order) are D'Agostino, Eli's Manhattan, Fairway, Food Emporium, Gourmet Garage, Grace's Marketplace, and Whole Foods. Construct a 98% confidence interval estimate of the mean amount of mercury in the population. Does it appear that there is too much mercury in tuna sushi?

0.56   0.75   0.10   0.95   1.25   0.54   0.88

20. Years in College Listed below are the numbers of years it took for a random sample of college students to earn bachelor's degrees (based on data from the National Center for Education Statistics). Construct a 95% confidence interval estimate of the mean time required for all college students to earn bachelor's degrees. Does it appear that college students typically earn bachelor's degrees in four years? Is there anything about the data that would suggest that the confidence interval might not be a good result?

4   4   4   4   4   4   4.5   4.5   4.5   4.5   4.5   4.5
6   6   8   9   9   13   13   15

21. Celebrity Net Worth Listed below are the amounts of net worth (in millions of dollars) of these ten wealthiest celebrities: Tom Cruise, Will Smith, Robert De Niro, Drew Carey, George Clooney, John Travolta, Samuel L. Jackson, Larry King, Demi Moore, and Bruce Willis. Construct a 98% confidence interval. What does the result tell us about the population of all celebrities? Do the data appear to be from a normally distributed population as required?

250   200   185   165   160   160   150   150   150   150

22. Caffeine in Soft Drinks Listed below are measured amounts of caffeine (mg per 12 oz of drink) obtained in one can from each of 20 brands (7UP, A&W Root Beer, Cherry Coke, . . . , TaB). Use a confidence level of 99%. Does the confidence interval give us good information about the population of all cans of the same 20 brands that are consumed? Does the sample appear to be from a normally distributed population? If not, how are the results affected?

0   0   34   34   34   45   41   51   55   36   47   41   0   0   53   54   38   0   41   47

23. Student Evaluations Listed below are student evaluation ratings of courses, where a rating of 5 is for "excellent." The ratings were obtained at the University of Texas at Austin. (See Data Set 17 "Course Evaluations" in Appendix B.) Use a 90% confidence level. What does the confidence interval tell us about the population of college students in Texas?

3.8   3.0   4.0   4.8   3.0   4.2   3.5   4.7   4.4   4.2   4.3   3.8   3.3   4.0   3.8

24. Flight Arrivals Listed below are arrival delays (minutes) of randomly selected American Airlines flights from New York (JFK) to Los Angeles (LAX). Negative numbers correspond to flights that arrived before the scheduled arrival time. Use a 95% confidence interval. How good is the on-time performance?

−5   −32   −13   −9   −19   49   −30   −23   14   −21   −32   11

Appendix B Data Sets.   In Exercises 25–28, use the Appendix B data sets to construct the confidence interval estimates of the mean.

25. Pulse Rates Refer to Data Set 1 "Body Data" and construct a 95% confidence interval estimate of the mean pulse rate of adult females; then do the same for adult males. Compare the results.

26. Airport Data Speeds Refer to Data Set 32 "Airport Data Speeds" and construct a 95% confidence interval estimate of the mean speed for Sprint, then do the same for T-Mobile. Compare the results.

27. Fast Food Service Times Refer to Data Set 25 "Fast Food" and construct a 95% confidence interval estimate of the mean drive-through service time for McDonald's at dinner; then do the same for Burger King at dinner. Compare the results.

28. Chocolate Chip Cookies Refer to Data Set 28 "Chocolate Chip Cookies" and construct a 95% confidence interval estimate of the mean number of chocolate chips in Chips Ahoy regular cookies; then do the same for Keebler cookies. Compare the results.

Sample Size. *In Exercises 29–36, find the sample size required to estimate the population mean.*

29. Mean IQ of College Professors The Wechsler IQ test is designed so that the mean is 100 and the standard deviation is 15 for the population of normal adults. Find the sample size necessary to estimate the mean IQ score of college professors. We want to be 99% confident that our sample mean is within 4 IQ points of the true mean. The mean for this population is clearly greater than 100. The standard deviation for this population is less than 15 because it is a group with less variation than a group randomly selected from the general population; therefore, if we use $\sigma = 15$ we are being conservative by using a value that will make the sample size at least as large as necessary. Assume then that $\sigma = 15$ and determine the required sample size. Does the sample size appear to be practical?

30. Mean IQ of Attorneys See the preceding exercise, in which we can assume that $\sigma = 15$ for the IQ scores. Attorneys are a group with IQ scores that vary less than the IQ scores of the general population. Find the sample size needed to estimate the mean IQ of attorneys, given that we want 98% confidence that the sample mean is within 3 IQ points of the population mean. Does the sample size appear to be practical?

31. Mean Grade-Point Average Assume that all grade-point averages are to be standardized on a scale between 0 and 4. How many grade-point averages must be obtained so that the sample mean is within 0.01 of the population mean? Assume that a 95% confidence level is desired. If we use the range rule of thumb, we can estimate σ to be range$/4 = (4 - 0)/4 = 1$. Does the sample size seem practical?

32. Mean Weight of Male Statistics Students Data Set 1 "Body Data" in Appendix B includes weights of 153 randomly selected adult males, and those weights have a standard deviation of 17.65 kg. Because it is reasonable to assume that weights of male statistics students have less variation than weights of the population of adult males, let $\sigma = 17.65$ kg. How many male statistics students must be weighed in order to estimate the mean weight of all male statistics students? Assume that we want 90% confidence that the sample mean is within 1.5 kg of the population mean. Does it seem reasonable to assume that weights of male statistics students have less variation than weights of the population of adult males?

33. Mean Age of Female Statistics Students Data Set 1 "Body Data" in Appendix B includes ages of 147 randomly selected adult females, and those ages have a standard deviation of 17.7 years. Assume that ages of female statistics students have less variation than ages of females in the general population, so let $\sigma = 17.7$ years for the sample size calculation. How many female statistics student ages must be obtained in order to estimate the mean age of all female statistics students? Assume that we want 95% confidence that the sample mean is within one-half year of the population mean. Does it seem reasonable to assume that ages of female statistics students have less variation than ages of females in the general population?

34. Mean Pulse Rate of Females Data Set 1 "Body Data" in Appendix B includes pulse rates of 147 randomly selected adult females, and those pulse rates vary from a low of 36 bpm to a high of 104 bpm. Find the minimum sample size required to estimate the mean pulse rate of adult females. Assume that we want 99% confidence that the sample mean is within 2 bpm of the population mean.

a. Find the sample size using the range rule of thumb to estimate σ.

b. Assume that $\sigma = 12.5$ bpm, based on the value of $s = 12.5$ bpm for the sample of 147 female pulse rates.

c. Compare the results from parts (a) and (b). Which result is likely to be better?

35. Mean Pulse Rate of Males Data Set 1 "Body Data" in Appendix B includes pulse rates of 153 randomly selected adult males, and those pulse rates vary from a low of 40 bpm to a high of 104 bpm. Find the minimum sample size required to estimate the mean pulse rate of adult males. Assume that we want 99% confidence that the sample mean is within 2 bpm of the population mean.

continued

a. Find the sample size using the range rule of thumb to estimate σ.

b. Assume that $\sigma = 11.3$ bpm, based on the value of $s = 11.3$ bpm for the sample of 153 male pulse rates.

c. Compare the results from parts (a) and (b). Which result is likely to be better?

36. Mean Body Temperature Data Set 3 "Body Temperatures" in Appendix B includes 106 body temperatures of adults for Day 2 at 12 AM, and they vary from a low of 96.5°F to a high of 99.6°F. Find the minimum sample size required to estimate the mean body temperature of all adults. Assume that we want 98% confidence that the sample mean is within 0.1°F of the population mean.

a. Find the sample size using the range rule of thumb to estimate σ.

b. Assume that $\sigma = 0.62$°F, based on the value of $s = 0.62$°F for the sample of 106 body temperatures.

c. Compare the results from parts (a) and (b). Which result is likely to be better?

7-2 Beyond the Basics

Confidence Interval with Known σ. *In Exercises 37 and 38, find the confidence interval using the known value of σ.*

37. Birth Weights of Girls Construct the confidence interval for Exercise 9 "Birth Weights of Girls," assuming that σ is known to be 7.1 hg.

38. Birth Weights of Boys Construct the confidence interval for Exercise 10 "Birth Weights of Boys," assuming that σ is known to be 6.6 hg.

39. Finite Population Correction Factor If a simple random sample of size n is selected without replacement from a finite population of size N, and the sample size is more than 5% of the population size ($n > 0.05N$), better results can be obtained by using the finite population correction factor, which involves multiplying the margin of error E by $\sqrt{(N-n)/(N-1)}$. For the sample of 100 weights of M&M candies in Data Set 27 "M&M Weights" in Appendix B, we get $\bar{x} = 0.8565$ g and $s = 0.0518$ g. First construct a 95% confidence interval estimate of μ, assuming that the population is large; then construct a 95% confidence interval estimate of the mean weight of M&Ms in the full bag from which the sample was taken. The full bag has 465 M&Ms. Compare the results.

7-3 Estimating a Population Standard Deviation or Variance

Key Concept This section presents methods for using a sample standard deviation s (or a sample variance s^2) to estimate the value of the corresponding population standard deviation σ (or population variance σ^2). Here are the main concepts included in this section:

- **Point Estimate:** The sample variance s^2 is the best *point estimate* (or single value estimate) of the population variance σ^2. The sample standard deviation s is commonly used as a point estimate of σ, even though it is a biased estimator, as described in Section 6-3.

- **Confidence Interval:** When constructing a *confidence interval* estimate of a population standard deviation (or population variance), we construct the confidence interval using the χ^2 *distribution*. (The Greek letter χ is pronounced "kigh.")

Chi-Square Distribution

Here are key points about the χ^2 (chi-square or chi-squared) distribution:

- In a normally distributed population with variance σ^2, if we randomly select independent samples of size n and, for each sample, compute the sample variance s^2, the sample statistic $\chi^2 = (n - 1)s^2/\sigma^2$ has a sampling distribution called the **chi-square distribution,** as shown in Formula 7-5.

    FORMULA 7-5

    $$\chi^2 = \frac{(n - 1)s^2}{\sigma^2}$$

- **Critical Values of** χ^2 We denote a right-tailed critical value by χ^2_R and we denote a left-tailed critical value by χ^2_L. Those critical values can be found by using technology or Table A-4, and they require that we first determine a value for the number of *degrees of freedom.*

- **Degrees of Freedom** For the methods of this section, the number of degrees of freedom is the sample size minus 1.

    Degrees of freedom: df $= n - 1$

- The chi-square distribution is skewed to the right, unlike the normal and Student t distributions (see Figure 7-7).

- The values of chi-square can be zero or positive, but they cannot be negative, as shown in Figure 7-7.

- The chi-square distribution is different for each number of degrees of freedom, as illustrated in Figure 7-8. As the number of degrees of freedom increases, the chi-square distribution approaches a normal distribution.

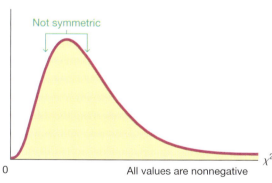

FIGURE 7-7   Chi-Square Distribution

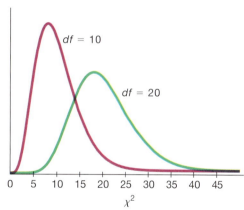

FIGURE 7-8   Chi-Square Distribution for df $= 10$ and df $= 20$

Because the chi-square distribution is not symmetric, a confidence interval estimate of σ^2 does not fit a format of $s^2 - E < \sigma^2 < s^2 + E$, so we must do separate calculations for the upper and lower confidence interval limits. If using Table A-4 for finding critical values, note the following design feature of that table:

> **In Table A-4, each critical value of χ^2 in the body of the table corresponds to an area given in the top row of the table, and each area in that top row is a *cumulative area to the right* of the critical value.**

CAUTION Table A-2 for the standard normal distribution provides cumulative areas from the *left,* but Table A-4 for the chi-square distribution uses cumulative areas from the *right.*

 EXAMPLE 1 Finding Critical Values of X^2

A simple random sample of 22 IQ scores is obtained (as in Example 2, which follows). Construction of a confidence interval for the population standard deviation σ requires the left and right critical values of X^2 corresponding to a confidence level of 95% and a sample size of $n = 22$. Find X_L^2 (the critical value of X^2 separating an area of 0.025 in the left tail), and find X_R^2 (the critical value of X^2 separating an area of 0.025 in the right tail).

SOLUTION

With a sample size of $n = 22$, the number of degrees of freedom is df $= n - 1 = 21$. See Figure 7-9.

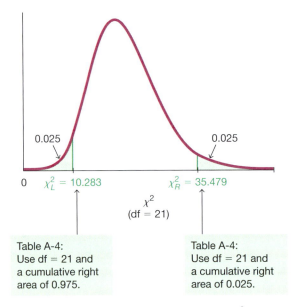

0.025

0.025

$0 \quad X_L^2 = 10.283 \qquad\qquad X_R^2 = 35.479$

X^2
(df = 21)

Table A-4:
Use df = 21 and
a cumulative right
area of 0.975.

Table A-4:
Use df = 21 and
a cumulative right
area of 0.025.

FIGURE 7-9   Finding Critical Values of X^2

The critical value to the right ($X_R^2 = 35.479$) is obtained from Table A-4 in a straightforward manner by locating 21 in the degrees-of-freedom column at the left and 0.025 across the top row. The leftmost critical value of $X_L^2 = 10.283$ also corresponds to 21 in the degrees-of-freedom column, but we must locate 0.975 (or $1 - 0.025$) across the top row because the values in the top row are always *areas to the right* of the critical value. Refer to Figure 7-9 and see that the total area to the right of $X_L^2 = 10.283$ is 0.975.

YOUR TURN Find the critical values in Exercise 5 "Nicotine in Menthol Cigarettes."

When obtaining critical values of X^2 from Table A-4, if a number of degrees of freedom is not found in the table, you can be conservative by using the next lower number of degrees of freedom, or you can use the closest critical value in the table, or you can get an approximate result with interpolation. For numbers of degrees of freedom greater than 100, use the equation given in Exercise 23 "Finding Critical Values" on page 342, or use a more extensive table, or use technology.

Although s^2 is the best point estimate of σ^2, there is no indication of how good it is, so we use a confidence interval that gives us a range of values associated with a confidence level.

KEY ELEMENTS

Confidence Interval for Estimating a Population Standard Deviation or Variance

Objective

Construct a confidence interval estimate of a population standard deviation or variance.

Notation

σ = population standard deviation

s = sample standard deviation

n = number of sample values

χ_L^2 = left-tailed critical value of χ^2

σ^2 = population variance

s^2 = sample variance

E = margin of error

χ_R^2 = right-tailed critical value of χ^2

Requirements

1. The sample is a simple random sample.

2. The population must have normally distributed values (even if the sample is large). The requirement of a normal distribution is much stricter here than in earlier sections, so large departures from normal distributions can result in large errors. (If the normality requirement is not satisfied, use the bootstrap method described in Section 7-4.)

Confidence Interval for the Population Variance σ^2

$$\frac{(n-1)s^2}{\chi_R^2} < \sigma^2 < \frac{(n-1)s^2}{\chi_L^2}$$

Confidence Interval for the Population Standard Deviation σ

$$\sqrt{\frac{(n-1)s^2}{\chi_R^2}} < \sigma < \sqrt{\frac{(n-1)s^2}{\chi_L^2}}$$

Round-Off Rule

1. *Original Data:* When using the *original set of data* values, round the confidence interval limits to one more decimal place than is used for the original data.

2. *Summary Statistics:* When using the *summary statistics (n, s)*, round the confidence interval limits to the same number of decimal places used for the sample standard deviation.

CAUTION A confidence interval can be expressed in a format such as $11.0 < \sigma < 20.4$ or a format of $(11.0, 20.4)$, *but it cannot be expressed in a format of $s \pm E$.*

Procedure for Constructing a Confidence Interval for σ or σ^2

Confidence intervals can be easily constructed with technology or they can be constructed by using Table A-4 with the following procedure.

1. Verify that the two requirements are satisfied: The sample is a random sample from a normally distributed population.

continued

2. Using $n - 1$ degrees of freedom, find the critical values χ_R^2 and χ_L^2 that correspond to the desired confidence level (as in Example 1).

3. To get a confidence interval estimate of σ^2, use the following:

$$\frac{(n-1)s^2}{\chi_R^2} < \sigma^2 < \frac{(n-1)s^2}{\chi_L^2}$$

4. To get a confidence interval estimate of σ, take the square root of each component of the above confidence interval.

5. Round the confidence interval limits using the round-off rule given in the preceding Key Elements box.

Using Confidence Intervals for Comparisons or Hypothesis Tests

Comparisons Confidence intervals can be used *informally* to compare the variation in different data sets, but *the overlapping of confidence intervals should not be used for making formal and final conclusions about equality of variances or standard deviations.*

 EXAMPLE 2 Confidence Interval for Estimating σ of IQ Scores

Data Set 7 "IQ and Lead" in Appendix B lists IQ scores for subjects in three different lead exposure groups. The 22 full IQ scores for the group with medium exposure to lead (Group 2) have a standard deviation of 14.29263. Consider the sample to be a simple random sample and construct a 95% confidence interval estimate of σ, the standard deviation of the population from which the sample was obtained.

SOLUTION

REQUIREMENT CHECK

Step 1: Check requirements. (1) The sample can be treated as a simple random sample. (2) The accompanying histogram has a shape very close to the bell shape of a normal distribution, so the requirement of normality is satisfied. ✅

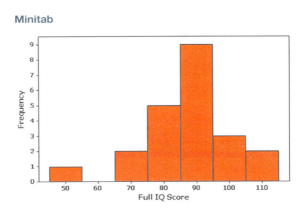

Minitab

Step 2: Using Technology The confidence interval can be found using technology. The StatCrunch display shows the lower and upper confidence interval limits for the 95% confidence interval estimate of σ^2, so we get $120.9 < \sigma^2 < 417.2$. Taking square roots, we get $11.0 < \sigma < 20.4$.

StatCrunch

95% confidence interval results:
σ^2 : Variance of variable

Variable	Sample Var.	DF	L. Limit	U. Limit
IQ	204.27922	21	120.91318	417.18431

Using Table A-4 If using Table A-4, we first use the sample size of $n = 22$ to find degrees of freedom: df $= n - 1 = 21$. In Table A-4, refer to the row corresponding to 21 degrees of freedom, and refer to the columns with areas of 0.975 and 0.025. (For a 95% confidence level, we divide $\alpha = 0.05$ equally between the two tails of the chi-square distribution, and we refer to the values of 0.975 and 0.025 across the top row of Table A-4.) The critical values are $\chi_L^2 = 10.283$ and $\chi_R^2 = 35.479$ (as shown in Example 1).

Step 3: Using the critical values of 10.283 and 35.479, the sample standard deviation of $s = 14.29263$ and the sample size of $n = 22$, we construct the 95% confidence interval by evaluating the following:

$$\frac{(n-1)s^2}{\chi_R^2} < \sigma^2 < \frac{(n-1)s^2}{\chi_L^2}$$

$$\frac{(22-1)(14.29263)^2}{35.479} < \sigma^2 < \frac{(22-1)(14.29263)^2}{10.283}$$

Step 4: Evaluating the expression above results in $120.9 < \sigma^2 < 417.2$. Finding the square root of each part (before rounding), then rounding to one decimal place, yields this 95% confidence interval estimate of the population standard deviation: $11.0 < \sigma < 20.4$.

INTERPRETATION

Based on this result, we have 95% confidence that the limits of 11.0 and 20.4 contain the true value of σ. The confidence interval can also be expressed as (11.0, 20.4), *but it cannot be expressed in a format of $s \pm E$.*

YOUR TURN Find the confidence interval in Exercise 5 "Nicotine in Menthol Cigarettes."

Rationale for the Confidence Interval See Figure 7-9 on page 334 to make sense of this statement: If we select random samples of size n from a normally distributed population with variance σ^2, there is a probability of $1 - \alpha$ that the statistic $(n-1)s^2/\sigma^2$ will fall between the critical values of χ_L^2 and χ_R^2. It follows that there is a $1 - \alpha$ probability that both of the following are true:

$$\frac{(n-1)s^2}{\sigma^2} < \chi_R^2 \quad \text{and} \quad \frac{(n-1)s^2}{\sigma^2} > \chi_L^2$$

Multiply both of the preceding inequalities by σ^2, then divide each inequality by the appropriate critical value of χ^2, so the two preceding inequalities can be expressed in these equivalent forms:

$$\frac{(n-1)s^2}{\chi_R^2} < \sigma^2 \quad \text{and} \quad \frac{(n-1)s^2}{\chi_L^2} > \sigma^2$$

continued

The two preceding inequalities can be combined into one inequality to get the format of the confidence interval used in this section:

$$\frac{(n-1)s^2}{\chi_R^2} < \sigma^2 < \frac{(n-1)s^2}{\chi_L^2}$$

TABLE 7-2 Finding Sample Size

σ	
To be 95% confident that s is within . . .	of the value of σ, the sample size n should be at least
1%	19,205
5%	768
10%	192
20%	48
30%	21
40%	12
50%	8
To be 99% confident that s is within . . .	of the value of σ, the sample size n should be at least
1%	33,218
5%	1,336
10%	336
20%	85
30%	38
40%	22
50%	14

Determining Sample Size

The procedures for finding the sample size necessary to estimate σ are much more complex than the procedures given earlier for means and proportions. For normally distributed populations, Table 7-2 or the formula given in Exercise 24 "Finding Sample Size" on page 342 can be used.

Statdisk also provides sample sizes. With Statdisk, select **Analysis, Sample Size Determination,** and then **Estimate Standard Deviation.** Minitab, Excel, StatCrunch, and the TI-83/84 Plus calculator do not provide such sample sizes.

EXAMPLE 3 **Finding Sample Size for Estimating σ**

We want to estimate the standard deviation σ of all IQ scores of people with exposure to lead. We want to be 99% confident that our estimate is within 5% of the true value of σ. How large should the sample be? Assume that the population is normally distributed.

SOLUTION

From Table 7-2, we can see that 99% confidence and an error of 5% for σ correspond to a sample of size 1336. We should obtain a simple random sample of 1336 IQ scores from the population of subjects exposed to lead.

YOUR TURN Do Exercise 19 "IQ of Statistics Professors."

TECH CENTER

Confidence Interval Estimate for Standard Deviation or Variance
Access tech supplements, videos, and data sets at **www.TriolaStats.com**

Statdisk	**Minitab**	**Excel**
Confidence Interval 1. Verify the distribution is normal using **Data—Normality Assessment.** 2. Click **Analysis** in the top menu. 3. Select **Confidence Intervals** from the dropdown menu and select **Standard Deviation One Sample** from the submenu. 4. Enter the desired confidence level. 5. *Using Summary Statistics:* Select the **Use Summary Statistics** tab and enter the sample size and sample standard deviation. *Using Sample Data:* Select the **Use Data** tab and select the desired data column. 6. Click **Evaluate.**	**Confidence Interval** 1. Click **Stat** in the top menu. 2. Select **Basic Statistics** from the dropdown menu and select **1 Variance** from the submenu. 3. *Using Summary Statistics:* Select **Sample standard deviation** or **Sample Variance** from the dropdown menu and enter sample size and sample standard deviation or sample variance. *Using Sample Data:* Select **One or more samples, each in a column** from the dropdown menu and select the desired data column(s). 4. Confirm *Perform hypothesis test* is not checked. 5. Click the **Options** button, enter the desired confidence level, and for *Alternative Hypothesis* select $\neq$. 6. Click **OK** twice. The results include a confidence interval for the standard deviation and a confidence interval for the variance.	Neither Excel nor XLSTAT has a function for generating a confidence interval estimate of standard deviation or variance.

TECH CENTER *continued*

Confidence Interval Estimate for Standard Deviation or Variance
Access tech supplements, videos, and data sets at **www.TriolaStats.com**

StatCrunch

Confidence Interval
1. Click **Stat** in the top menu.
2. Select **Variance Stats** in the dropdown menu, then select **One Sample** from the submenu.
3. *Using Summary Statistics:* Select *With Summary* from the submenu and enter the sample variance and sample size.

   Using Sample Data: Select *With Data* from the submenu and select the desired data column(s).
4. Select **Confidence interval for σ^2** and enter the desired confidence level.
5. Click **Compute!**

TI-83/84 Plus Calculator

1. Download and install the Michael Lloyd programs *S2INT* and *ZZINEWT* (available at www.TriolaStats.com) on your TI-83/84 Plus Calculator.
2. Press **PRGM**, select **S2INT** from the menu, and press **ENTER** twice.
3. Enter the sample variance Sx^2, sample size n, and confidence level *C-Level*. Press **ENTER** after each entry.
4. Wait for the σx^2 confidence interval to be displayed. (Be patient; it may take a while!).
5. Press **ENTER** to view the σx confidence interval.

7-3 Basic Skills and Concepts

Statistical Literacy and Critical Thinking

1. Brain Volume Using all of the brain volumes listed in Data Set 8 "IQ and Brain Size", we get this 95% confidence interval estimate: $9027.8 < \sigma^2 < 33{,}299.8$, and the units of measurement are $(\text{cm}^3)^2$. Identify the corresponding confidence interval estimate of σ and include the appropriate units. Given that the original values are whole numbers, round the limits using the round-off rule given in this section. Write a statement that correctly interprets the confidence interval estimate of σ.

2. Expressing Confidence Intervals Example 2 showed how the statistics of $n = 22$ and $s = 14.3$ result in this 95% confidence interval estimate of σ: $11.0 < \sigma < 20.4$. That confidence interval can also be expressed as (11.0, 20.4), but it cannot be expressed as 15.7 ± 4.7. Given that 15.7 ± 4.7 results in values of 11.0 and 20.4, why is it wrong to express the confidence interval as 15.7 ± 4.7?

3. Last Digit Analysis The dotplot below depicts the last digits of the weights of 153 males in Data Set 1 "Body Data." Do those digits appear to be from a normally distributed population? If not, does the large sample size of $n = 153$ justify treating the values as if they were from a normal distribution? Can the sample be used to construct a 95% confidence interval estimate of σ for the population of all such digits?

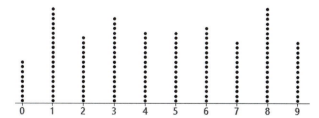

4. Normality Requirement What is different about the normality requirement for a confidence interval estimate of σ and the normality requirement for a confidence interval estimate of μ?

Finding Critical Values and Confidence Intervals.   *In Exercises 5–8, use the given information to find the number of degrees of freedom, the critical values X_L^2 and X_R^2, and the confidence interval estimate of σ. The samples are from Appendix B and it is reasonable to assume that a simple random sample has been selected from a population with a normal distribution.*

5. Nicotine in Menthol Cigarettes 95% confidence; $n = 25$, $s = 0.24$ mg.

6. Weights of Pennies 95% confidence; $n = 37$, $s = 0.01648$ g.

7. Platelet Counts of Women 99% confidence; $n = 147$, $s = 65.4$.

8. Heights of Men 99% confidence; $n = 153$, $s = 7.10$ cm.

Finding Confidence Intervals.   *In Exercises 9–16, assume that each sample is a simple random sample obtained from a population with a normal distribution.*

9. Body Temperature Data Set 3 "Body Temperatures" in Appendix B includes a sample of 106 body temperatures having a mean of 98.20°F and a standard deviation of 0.62°F (for day 2 at 12 AM). Construct a 95% confidence interval estimate of the standard deviation of the body temperatures for the entire population.

10. Atkins Weight Loss Program In a test of weight loss programs, 40 adults used the Atkins weight loss program. After 12 months, their mean weight *loss* was found to be 2.1 lb, with a standard deviation of 4.8 lb. Construct a 90% confidence interval estimate of the standard deviation of the weight loss for all such subjects. Does the confidence interval give us information about the effectiveness of the diet?

11. Insomnia Treatment A clinical trial was conducted to test the effectiveness of the drug zopiclone for treating insomnia in older subjects. After treatment with zopiclone, 16 subjects had a mean wake time of 98.9 min and a standard deviation of 42.3 min (based on data from "Cognitive Behavioral Therapy vs Zopiclone for Treatment of Chronic Primary Insomnia in Older Adults," by Sivertsen et al., *Journal of the American Medical Association,* Vol. 295, No. 24). Assume that the 16 sample values appear to be from a normally distributed population and construct a 98% confidence interval estimate of the standard deviation of the wake times for a population with zopiclone treatments. Does the result indicate whether the treatment is effective?

12. Garlic for Reducing Cholesterol In a test of the effectiveness of garlic for lowering cholesterol, 49 subjects were treated with raw garlic. Cholesterol levels were measured before and after the treatment. The changes (before minus after) in their levels of LDL cholesterol (in mg/dL) had a mean of 0.4 and a standard deviation of 21.0 (based on data from "Effect of Raw Garlic vs Commercial Garlic Supplements on Plasma Lipid Concentrations in Adults with Moderate Hypercholesterolemia," by Gardner et al., *Archives of Internal Medicine,* Vol. 167). Construct a 98% confidence interval estimate of the standard deviation of the changes in LDL cholesterol after the garlic treatment. Does the result indicate whether the treatment is effective?

13. Speed Dating In a study of speed dating conducted at Columbia University, male subjects were asked to rate the attractiveness of their female dates, and a sample of the results is listed below (1 = not attractive; 10 = extremely attractive). Construct a 95% confidence interval estimate of the standard deviation of the population from which the sample was obtained.

$$7 \quad 8 \quad 2 \quad 10 \quad 6 \quad 5 \quad 7 \quad 8 \quad 8 \quad 9 \quad 5 \quad 9$$

14. Speed Dating In a study of speed dating conducted at Columbia University, female subjects were asked to rate the attractiveness of their male dates, and a sample of the results is listed below (1 = not attractive; 10 = extremely attractive). Construct a 95% confidence interval estimate of the standard deviation of the population from which the sample was obtained.

$$5 \quad 8 \quad 3 \quad 8 \quad 6 \quad 10 \quad 3 \quad 7 \quad 9 \quad 8 \quad 5 \quad 5 \quad 6 \quad 8 \quad 8 \quad 7 \quad 3 \quad 5 \quad 5 \quad 6 \quad 8 \quad 7 \quad 8 \quad 8 \quad 8 \quad 7$$

15. Highway Speeds Listed below are speeds (mi/h) measured from southbound traffic on I-280 near Cupertino, California (based on data from SigAlert). This simple random sample was obtained at 3:30 PM on a weekday. Use the sample data to construct a 95% confidence interval estimate of the population standard deviation. Does the confidence interval describe the standard deviation for all times during the week?

62 61 61 57 61 54 59 58 59 69 60 67

16. Comparing Waiting Lines

a. The values listed below are waiting times (in minutes) of customers at the Jefferson Valley Bank, where customers enter a single waiting line that feeds three teller windows. Construct a 95% confidence interval for the population standard deviation σ.

6.5 6.6 6.7 6.8 7.1 7.3 7.4 7.7 7.7 7.7

b. The values listed below are waiting times (in minutes) of customers at the Bank of Providence, where customers may enter any one of three different lines that have formed at three teller windows. Construct a 95% confidence interval for the population standard deviation σ.

4.2 5.4 5.8 6.2 6.7 7.7 7.7 8.5 9.3 10.0

c. Interpret the results found in parts (a) and (b). Do the confidence intervals suggest a difference in the variation among waiting times? Which arrangement seems better: the single-line system or the multiple-line system?

Large Data Sets from Appendix B. *In Exercises 17 and 18, use the data set in Appendix B. Assume that each sample is a simple random sample obtained from a population with a normal distribution.*

17. Student Evaluations Refer to Data Set 17 "Course Evaluations" in Appendix B.

a. Use the 93 course evaluations to construct a 95% confidence interval estimate of the standard deviation of the population from which the sample was obtained.

b. Repeat part (a) using the 93 professor evaluations.

c. Compare the results from part (a) and part (b).

18. Birth Weights Refer to Data Set 4 "Births" in Appendix B.

a. Use the 205 birth weights of girls to construct a 95% confidence interval estimate of the standard deviation of the population from which the sample was obtained.

b. Repeat part (a) using the 195 birth weights of boys.

c. Compare the results from part (a) and part (b).

Determining Sample Size. *In Exercises 19–22, assume that each sample is a simple random sample obtained from a normally distributed population. Use Table 7-2 on page 338 to find the indicated sample size.*

19. IQ of Statistics Professors You want to estimate σ for the population of IQ scores of statistics professors. Find the minimum sample size needed to be 95% confident that the sample standard deviation s is within 1% of σ. Is this sample size practical?

20. Space Mountain You want to estimate σ for the population of waiting times for the Space Mountain ride in Walt Disney World. You want to be 99% confident that the sample standard deviation is within 1% of σ. Find the minimum sample size. Is this sample size practical?

21. Statistics Student Incomes You want to estimate the standard deviation of the annual incomes of all current statistics students. Find the minimum sample size needed to be 95% confident that the sample standard deviation is within 20% of the population standard deviation. Are those incomes likely to satisfy the requirement of a normal distribution?

22. Quarters When setting specifications of quarters to be accepted in a vending machine, you must estimate the standard deviation of the population of quarters in use. Find the minimum sample size needed to be 99% confident that the sample standard deviation is within 10% of the population standard deviation.

7-3 Beyond the Basics

23. Finding Critical Values In constructing confidence intervals for σ or σ^2, Table A-4 can be used to find the critical values χ_L^2 and χ_R^2 only for select values of n up to 101, so the number of degrees of freedom is 100 or smaller. For larger numbers of degrees of freedom, we can approximate χ_L^2 and χ_R^2 by using

$$\chi^2 = \frac{1}{2}\left[\pm z_{\alpha/2} + \sqrt{2k - 1}\right]^2$$

where k is the number of degrees of freedom and $z_{\alpha/2}$ is the critical z score described in Section 7-1. Use this approximation to find the critical values χ_L^2 and χ_R^2 for Exercise 8 "Heights of Men," where the sample size is 153 and the confidence level is 99%. How do the results compare to the actual critical values of $\chi_L^2 = 110.846$ and $\chi_R^2 = 200.657$?

24. Finding Sample Size Instead of using Table 7-2 for determining the sample size required to estimate a population standard deviation σ, the following formula can be used

$$n = \frac{1}{2}\left(\frac{z_{\alpha/2}}{d}\right)^2$$

where $z_{\alpha/2}$ corresponds to the confidence level and d is the decimal form of the percentage error. For example, to be 95% confident that s is within 15% of the value of σ, use $z_{\alpha/2} = 1.96$ and $d = 0.15$ to get a sample size of $n = 86$. Find the sample size required to estimate σ, assuming that we want 98% confidence that s is within 15% of σ.

7-4 Bootstrapping: Using Technology for Estimates

Key Concept The preceding sections presented methods for estimating population proportions, means, and standard deviations (or variances). All of those methods have certain requirements that limit the situations in which they can be used. When some of the requirements are not satisfied, we can often use the bootstrap method to estimate a parameter with a confidence interval. The bootstrap method typically requires the use of software.

Sampling Requirement The preceding methods of this chapter all have a requirement that the sample must be a simple random sample. If the sample is not collected in an appropriate way, there's a good chance that *nothing* can be done to get a usable confidence interval estimate of a parameter. *Bootstrap methods do not correct for poor sampling methods.*

Requirements Listed below are important requirements from the preceding sections of this chapter:

- **CI for Proportion (Section 7-1):** There are at least 5 successes and at least 5 failures, or $np \geq 5$ and $nq \geq 5$.

- **CI for Mean (Section 7-2):** The population is normally distributed or $n > 30$.

- **CI for σ or σ^2 (Section 7-3):** The population must have normally distributed values, even if the sample is large.

When the above requirements are not satisfied, we should not use the methods presented in the preceding sections of this chapter, but we can use the bootstrap method instead. The bootstrap method does not require large samples. This method does not require the sample to be collected from a normal or any other particular distribution, and so it is called a **nonparametric** or **distribution-free method**; other nonparametric methods are included in Chapter 13.

DEFINITION

Given a simple random sample of size *n*, a **bootstrap sample** is another random sample of *n* values obtained *with replacement* from the original sample.

Without replacement, every sample would be identical to the original sample, so the proportions or means or standard deviations or variances would all be the same, and there would be no confidence "interval."

CAUTION Note that a bootstrap sample involves sampling *with replacement,* so that when a sample value is selected, it is replaced before the next selection is made.

EXAMPLE 1 Bootstrap Sample of Incomes

When the author collected annual incomes of current statistics students, he obtained these results (in thousands of dollars): 0, 2, 3, 7.

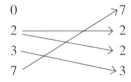

Original Sample   Bootstrap Sample

The sample of { 7, 2, 2, 3 } is one bootstrap sample obtained from the original sample. Other bootstrap samples may be different.

    Incomes tend to have distributions that are skewed instead of being normal, so we should not use the methods of Section 7-2 with a small sample of incomes. This is a situation in which the bootstrap method comes to the rescue.

YOUR TURN Do Exercise 3 "Bootstrap Sample."

Why Is It Called "Bootstrap"? The term "bootstrap" is used because the data "pull themselves up by their own bootstraps" to generate new data sets. In days of yore, "pulling oneself up by one's bootstraps" meant that an impossible task was somehow accomplished, and the bootstrap method described in this section might seem impossible, but it works!

How Many? In the interest of providing manageable examples that don't occupy multiple pages each, the examples in this section involve very small data sets and no more than 20 bootstrap samples, but we should use at least 1000 bootstrap samples when we use bootstrap methods in serious applications. Professional statisticians commonly use 10,000 or more bootstrap samples.

Bootstrap Procedure for a Confidence Interval Estimate of a Parameter

1. Given a simple random sample of size *n*, obtain many (such as 1000 or more) bootstrap samples of the same size *n*.

2. For the parameter to be estimated, find the corresponding statistic for each of the bootstrap samples. (Example: For a confidence estimate of μ, find the *sample mean* $\bar{x}$ from each bootstrap sample.)

3. Sort the list of sample statistics from low to high.

continued

How Many People Do You Know?

It's difficult for anyone to count the number of people he or she knows, but statistical methods can be used to estimate the mean number of people that we all know. The simple approach of just asking someone how many people are known has worked poorly in the past. A much better approach is to select a representative sample of people and ask each person how many people he or she knows who are named Mario, Ginny, Rachel, or Todd. (Uncommon names are more effective because people with more common names are more difficult to accurately recall.) Responses are then used to project the total number of people that are known. (If sample subjects know a mean of 1.76 people with those names, and we know that 0.288% of the population has those names, then the mean number of people known is 1.76/0.00288 = 611.) According to one estimate, the mean number of people known is 611, and the median is 472. (See "How Many People Do You Know? Efficiently Estimating Personal Network Size," by McCormick, Salganik, and Zheng, *Journal of the American Statistical Association,* Vol. 105, No. 4.)

4. Using the sorted list of the statistics, create the confidence interval by finding corresponding percentile values. Procedures for finding percentiles are given in Section 3-3. (Example: Using a list of sorted sample means, the 90% confidence interval limits are P_5 and P_{95}. The 90% confidence interval estimate of μ is $P_5 < \mu < P_{95}$.)

Usefulness of Results For the purpose of illustrating the bootstrap procedure, Examples 2, 3, and 4 all involve very small samples with only 20 bootstrap samples. Consequently, the resulting confidence intervals include almost the entire range of sample values, and those confidence intervals are not very useful. Larger samples with 1000 or more bootstrap samples will provide much better results than those from Examples 2, 3, and 4.

Proportions

When working with proportions, it is very helpful to represent the data from the two categories by using 0's and 1's, as in the following example.

─┤ **EXAMPLE 2** **Eye Color Survey: Bootstrap CI for Proportion**

In a survey, four randomly selected subjects were asked if they have brown eyes, and here are the results: 0, 0, 1, 0 (where $0 =$ no and $1 =$ yes). Use the bootstrap resampling procedure to construct a 90% confidence interval estimate of the population proportion p, the proportion of people with brown eyes in the population.

SOLUTION

REQUIREMENT CHECK The sample is a simple random sample. (There is no requirement of at least 5 successes and at least 5 failures or $np \geq 5$ and $nq \geq 5$. There is no requirement that the sample must be from a normally distributed population.) ✓

Step 1: In Table 7-3, we created 20 bootstrap samples from the original sample of 0, 0, 1, 0.

Step 2: Because we want a confidence interval estimate of the population proportion p, we want the sample proportion $\hat{p}$ for each of the 20 bootstrap samples, and those sample proportions are shown in the column to the right of the bootstrap samples.

Step 3: The column of data shown farthest to the right is a list of the 20 sample proportions arranged in order ("sorted") from lowest to highest.

Step 4: Because we want a confidence level of 90%, we want to find the percentiles P_5 and P_{95}. Recall that P_5 separates the lowest 5% of values, and P_{95} separates the top 5% of values. Using the methods from Section 3-3 for finding percentiles, we use the *sorted* list of bootstrap sample proportions to find that $P_5 = 0.00$ and $P_{95} = 0.75$. The 90% confidence interval estimate of the population proportion is $0.00 < p < 0.75$.

TABLE 7–3 Bootstrap Samples for p

Bootstrap Sample				$\hat{p}$	Sorted $\hat{p}$	
1	0	0	1	0.50	0.00	$\rightarrow P_5 = 0.00$
1	0	1	0	0.50	0.00	
0	1	1	1	0.75	0.00	
0	0	0	0	0.00	0.00	
0	1	0	0	0.25	0.25	
1	0	0	0	0.25	0.25	
0	1	0	1	0.50	0.25	
1	0	0	0	0.25	0.25	
0	0	0	0	0.00	0.25	
0	0	1	1	0.50	0.25	**90% Confidence Interval:**
0	0	0	1	0.25	0.25	$0.00 < p < 0.75$
0	0	1	0	0.25	0.25	
1	1	1	0	0.75	0.50	
0	0	0	0	0.00	0.50	
0	0	0	0	0.00	0.50	
0	1	1	0	0.50	0.50	
0	0	1	0	0.25	0.50	
1	0	0	0	0.25	0.75	
1	1	1	0	0.75	0.75	$\rightarrow P_{95} = 0.75$
0	0	0	1	0.25	0.75	

INTERPRETATION

The confidence interval of $0.00 < p < 0.75$ is quite wide. After all, every confidence interval for every proportion must fall between 0 and 1, so the 90% confidence interval of $0.00 < p < 0.75$ doesn't seem to be helpful, but it is based on only four sample values.

YOUR TURN Do Exercise 5 "Online Buying."

HINT: Example 2 uses only 20 bootstrap samples, but effective use of the bootstrap method typically requires the use of software to generate 1000 or more bootstrap samples.

Means

In Section 7-2 we noted that when constructing a confidence interval estimate of a population mean, there is a requirement that the sample is from a normally distributed population or the sample size is greater than 30. The bootstrap method can be used when this requirement is not satisfied.

EXAMPLE 3 **Incomes: Bootstrap CI for Mean**

When the author collected a simple random sample of annual incomes of his statistics students, he obtained these results (in thousands of dollars): 0, 2, 3, 7. Use the bootstrap resampling procedure to construct a 90% confidence interval estimate of the mean annual income of the population of all of the author's statistics students.

SOLUTION

REQUIREMENT CHECK The sample is a simple random sample and there is no requirement that the sample must be from a normally distributed population. Because distributions of incomes are typically skewed instead of normal, we should not use the methods of Section 7-2 for finding the confidence interval, but the bootstrap method can be used. ✓

Step 1: In Table 7-4, we created 20 bootstrap samples (with replacement!) from the original sample of 0, 2, 3, 7. (Here we use only 20 bootstrap samples so we have a manageable example that doesn't occupy many pages of text, but we usually want at least 1000 bootstrap samples.)

Step 2: Because we want a confidence interval estimate of the population mean μ, we want the sample mean $\bar{x}$ for each of the 20 bootstrap samples, and those sample means are shown in the column to the right of the bootstrap samples.

Step 3: The column of data shown farthest to the right is a list of the 20 sample means arranged in order ("sorted") from lowest to highest.

Step 4: Because we want a confidence level of 90%, we want to find the percentiles P_5 and P_{95}. Recall that P_5 separates the lowest 5% of values, and P_{95} separates the top 5% of values. Using the methods from Section 3-3 for finding percentiles, we use the *sorted* list of bootstrap sample means to find that $P_5 = 1.75$ and $P_{95} = 4.875$. The 90% confidence interval estimate of the population mean is $1.75 < \mu < 4.875$, where the values are in thousands of dollars.

TABLE 7-4 Bootstrap Samples for μ

Bootstrap Sample				$\bar{x}$	Sorted $\bar{x}$	
3	3	0	2	2.00	1.75	
0	3	2	2	1.75	1.75	$\rightarrow P_5 = 1.75$
7	0	2	7	4.00	1.75	
3	2	7	3	3.75	2.00	
0	0	7	2	2.25	2.00	
7	0	0	3	2.50	2.25	
3	0	3	2	2.00	2.50	
3	7	3	7	5.00	2.50	
0	3	2	2	1.75	2.50	
0	3	7	0	2.50	2.75	**90% Confidence Interval:**
0	7	2	2	2.75	3.00	$1.75 < \mu < 4.875$
7	2	2	3	3.50	3.25	
7	2	3	7	4.75	3.25	
2	7	2	7	4.50	3.50	
0	7	2	3	3.00	3.75	
7	3	7	2	4.75	4.00	
3	7	0	3	3.25	4.50	
0	0	3	7	2.50	4.75	
3	3	7	0	3.25	4.75	
2	0	2	3	1.75	5.00	$\rightarrow P_{95} = 4.875$

YOUR TURN Do Exercise 7 "Freshman 15."

Standard Deviations

In Section 7-3 we noted that when constructing confidence interval estimates of population standard deviations or variances, there is a requirement that the sample must
be from a population with normally distributed values. Even if the sample is large,
this normality requirement is much stricter than the normality requirement used for
estimating population means. Consequently, the bootstrap method becomes more important for confidence interval estimates of σ or σ^2.

EXAMPLE 4   Incomes: Bootstrap CI for Standard Deviation

Use these same incomes (thousands of dollars) from Example 3: 0, 2, 3, 7. Use the
bootstrap resampling procedure to construct a 90% confidence interval estimate of
the population standard deviation σ, the standard deviation of the annual incomes of
the population of the author's statistics students.

SOLUTION

REQUIREMENT CHECK   The same requirement check used in Example 3 applies here. ✅
     The same basic procedure used in Example 3 is used here. Example 3 already
includes 20 bootstrap samples, so here we find the *standard deviation* of each
bootstrap sample, and then we sort them to get this sorted list of sample standard
deviations:

1.26	1.26	1.26	1.41	1.41	2.22	2.31	2.38	2.63	2.63
2.87	2.87	2.89	2.94	2.99	3.30	3.32	3.32	3.32	3.56

The 90% confidence interval limits are found from this sorted list of standard deviations by finding P_5 and P_{95}. Using the methods from Section 3-3, we get $P_5 = 1.26$
and $P_{95} = 3.44$. The 90% confidence interval estimate of the population standard
deviation σ is $1.26 < \sigma < 3.44$, where the values are in thousands of dollars.

YOUR TURN   Do part (b) of Exercise 8 "Cell Phone Radiation."

Again, know that for practical reasons, the examples of this section involved very
small data sets and no more than 20 bootstrap samples, but use at least 1000 bootstrap
samples. The use of 10,000 or more bootstrap samples is common.

TECH CENTER

Bootstrap Resampling
Access tech supplements, videos, and data sets at **www.TriolaStats.com**

Statdisk	Minitab	StatCrunch
1. Enter the listed sample values in Column 1 of the Sample Editor.	Minitab does not yet have a bootstrap function, but bootstrap samples can be generated through a series of existing Minitab commands.	1. Enter the listed sample values in column *var1*.
2. Click **Analysis** in the top menu and select **Bootstrap Resampling** from the dropdown menu.		2. Click **Stat** in the top menu.
3. Enter the number of desired bootstrap samples, then click **Resample.**	Visit www.TriolaStats.com for more information.	3. Select **Resample-Statistic** in the dropdown menu.
4. The sorted sample means are listed in Column 2; the sorted sample standard deviations are listed in Column 3.		4. Under *Columns to resample* select **var1**.
		5. Under *Statistic* click **Build** and create the desired expression, such as **mean(var1)** or **std(var1)**. Click **Okay** when done.
		6. Complete the dialog box and click **Compute!**

TECH CENTER *continued*

 Bootstrap Resampling

Access tech supplements, videos, and data sets at **www.TriolaStats.com**

TI-83/84 Plus Calculator	Excel
Not available.	**XLSTAT Add-In (Required)** 1. Click on the **XLSTAT** tab in the Ribbon and then click **Describing data.** 2. Select **Resampled statistics** from the dropdown menu. 3. Under *Quantitative Data* enter the range of cells containing the sample values. If the first row contains a label, check the **Sample Labels** box. 4. Under *Method* select **Bootstrap.** 5. Enter the desired number of samples. 6. Click the **Outputs** tab and enter the confidence level (%) of **95**; confirm **Standard bootstrap interval** is selected. Also select **Mean** and **Standard deviation ($n-1$).** 7. Click **OK.**

7-4 Basic Skills and Concepts

Statistical Literacy and Critical Thinking

1. Replacement Why does the bootstrap method require sampling with replacement? What would happen if we used the methods of this section but sampled without replacement?

2. Bootstrap Sample Here is a random sample of taxi-out times (min) for American Airlines flights leaving JFK airport: 12, 19, 13, 43, 15. For this sample, what is a bootstrap sample?

3. Bootstrap Sample Given the sample data from Exercise 2, which of the following are *not* possible bootstrap samples?

a. 12, 19, 13, 43, 15   **b.** 12, 19, 15   **c.** 12, 12, 12, 43, 43

d. 14, 20, 12, 19, 15   **e.** 12, 13, 13, 12, 43, 15, 19

4. How Many? The examples in this section all involved no more than 20 bootstrap samples. How many should be used in real applications?

In Exercises 5–8, use the relatively small number of given bootstrap samples to construct the confidence interval.

5. Online Buying In a *Consumer Reports* Research Center survey, women were asked if they purchase books online, and responses included these: no, yes, no, no. Letting "yes" = 1 and letting "no" = 0, here are ten bootstrap samples for those responses: {0, 0, 0, 0}, {1, 0, 1, 0}, {1, 0, 1, 0}, {0, 0, 0, 0}, {0, 0, 0, 0}, {0, 1, 0, 0}, {0, 0, 0, 0}, {0, 0, 0, 0}, {0, 1, 0, 0}, {1, 1, 0, 0}. Using only the ten given bootstrap samples, construct a 90% confidence interval estimate of the proportion of women who said that they purchase books online.

6. Seating Choice In a 3M Privacy Filters poll, respondents were asked to identify their favorite seat when they fly, and the results include these responses: window, window, other, other. Letting "window" = 1 and letting "other" = 0, here are ten bootstrap samples for those responses: {0, 0, 0, 0}, {0, 1, 0, 0}, {0, 1, 0, 1}, {0, 0, 1, 0}, {1, 1, 1, 0}, {0, 1, 1, 0}, {1, 0, 0, 1}, {0, 1, 1, 1}, {1, 0, 1, 0}, {1, 0, 0, 1}. Using only the ten given bootstrap samples, construct an 80% confidence interval estimate of the proportion of respondents who indicated their favorite seat is "window."

7. Freshman 15 Here is a sample of amounts of weight change (kg) of college students in their freshman year (from Data Set 6 "Freshman 15" in Appendix B): 11, 3, 0, −2, where −2 represents a *loss* of 2 kg and positive values represent weight gained. Here are ten bootstrap

continued

samples: $\{11, 11, 11, 0\}$, $\{11, -2, 0, 11\}$, $\{11, -2, 3, 0\}$, $\{3, -2, 0, 11\}$, $\{0, 0, 0, 3\}$, $\{3, -2, 3, -2\}$, $\{11, 3, -2, 0\}$, $\{-2, 3, -2, 3\}$, $\{-2, 0, -2, 3\}$, $\{3, 11, 11, 11\}$.

a. Using only the ten given bootstrap samples, construct an 80% confidence interval estimate of the mean weight change for the population.

b. Using only the ten given bootstrap samples, construct an 80% confidence interval estimate of the standard deviation of the weight changes for the population.

8. Cell Phone Radiation Here is a sample of measured radiation emissions (cW/kg) for cell phones (based on data from the Environmental Working Group): 38, 55, 86, 145. Here are ten bootstrap samples: $\{38, 145, 55, 86\}$, $\{86, 38, 145, 145\}$, $\{145, 86, 55, 55\}$, $\{55, 55, 55, 145\}$, $\{86, 86, 55, 55\}$, $\{38, 38, 86, 86\}$, $\{145, 38, 86, 55\}$, $\{55, 86, 86, 86\}$, $\{145, 86, 55, 86\}$, $\{38, 145, 86, 55\}$.

a. Using only the ten given bootstrap samples, construct an 80% confidence interval estimate of the population mean.

b. Using only the ten given bootstrap samples, construct an 80% confidence interval estimate of the population standard deviation.

In Exercises 9–22, use technology to create the large number of bootstrap samples.

9. Freshman 15 Repeat Exercise 7 "Freshman 15" using a confidence level of 90% for parts (a) and (b) and using 1000 bootstrap samples instead of the 10 that were given in Exercise 7.

10. Cell Phone Radiation Repeat Exercise 8 "Cell Phone Radiation" using a confidence level of 90% for parts (a) and (b) and using 1000 bootstrap samples instead of the 10 that were given in Exercise 8.

11. Speed Dating Use these male measures of female attractiveness given in Exercise 17 "Speed Dating" in Section 7-2 on page 329: 7, 8, 2, 10, 6, 5, 7, 8, 8, 9, 5, 9. Use the bootstrap method with 1000 bootstrap samples.

a. Construct a 99% confidence interval estimate of the population mean. Is the result dramatically different from the 99% confidence interval found in Exercise 17 in Section 7-2?

b. Construct a 95% confidence interval estimate of the population standard deviation. Is the result dramatically different from the 95% confidence interval found in Exercise 13 "Speed Dating" in Section 7-3 on page 340?

12. Speed Dating Use these female measures of male attractiveness given in Exercise 18 "Speed Dating" in Section 7-2 on page 329: 5, 8, 3, 8, 6, 10, 3, 7, 9, 8, 5, 5, 6, 8, 8, 7, 3, 5, 5, 6, 8, 7, 8, 8, 8, 7. Use the bootstrap method with 1000 bootstrap samples.

a. Construct a 99% confidence interval estimate of the population mean. Is the result dramatically different from the 99% confidence interval found in Exercise 18 in Section 7-2?

b. Construct a 95% confidence interval estimate of the population standard deviation. Is the result dramatically different from the 95% confidence interval found in Exercise 14 "Speed Dating" in Section 7-3 on page 340?

13. Mickey D's In a study of the accuracy of fast food drive-through orders, McDonald's had 33 orders that were not accurate among 362 orders observed (based on data from *QSR* magazine). Use the bootstrap method to construct a 95% confidence interval estimate of the proportion of orders that are not accurate. Use 1000 bootstrap samples. How does the result compare to the confidence interval found in Exercise 13 "Mickey D's" from Section 7-1?

14. Eliquis The drug Eliquis (apixaban) is used to help prevent blood clots in certain patients. In clinical trials, among 5924 patients treated with Eliquis, 153 developed the adverse reaction of nausea (based on data from Bristol-Myers Squibb Co.). Use the bootstrap method to construct a 99% confidence interval estimate of the proportion patients who experience nausea. Use 1000 bootstrap samples. How does the result compare to the confidence interval found in Exercise 14 "Eliquis" from Section 7-1 on page 312?

15. Survey Return Rate In a study of cell phone use and brain hemispheric dominance, an Internet survey was e-mailed to 5000 subjects randomly selected from an online otological group (focused on ears), and 717 surveys were returned. Use the bootstrap method to construct a 90% confidence interval estimate of the proportion of returned surveys. Use 1000 bootstrap samples. How does the result compare to the confidence interval found in Exercise 15 "Survey Return Rate" from Section 7-1 on page 312?

16. Medical Malpractice In a study of 1228 randomly selected medical malpractice lawsuits, it was found that 856 of them were dropped or dismissed (based on data from the Physicians Insurers Association of America). Use the bootstrap method to construct a 95% confidence interval estimate of the proportion of lawsuits that are dropped or dismissed. Use 1000 bootstrap samples. How does the result compare to the confidence interval found in Exercise 16 "Medical Malpractice" from Section 7-1 on page 312?

17. Student Evaluations Listed below are student evaluation ratings of courses, where a rating of 5 is for "excellent." The ratings were obtained at the University of Texas at Austin. (See Data Set 17 "Course Evaluations" in Appendix B.) Using the bootstrap method with 1000 bootstrap samples, construct a 90% confidence interval estimate of μ. How does the result compare to the confidence interval found in Exercise 23 "Student Evaluations" in Section 7-2 on page 330?

   3.8   3.0   4.0   4.8   3.0   4.2   3.5   4.7   4.4   4.2   4.3   3.8   3.3   4.0   3.8

18. Caffeine in Soft Drinks Listed below are measured amounts of caffeine (mg per 12 oz of drink) obtained in one can from each of 20 brands. Using the bootstrap method with 1000 bootstrap samples, construct a 99% confidence interval estimate of μ. How does the result compare to the confidence interval found in Exercise 22 "Caffeine in Soft Drinks" in Section 7-2 on page 330?

   0   0   34   34   34   45   41   51   55   36   47   41   0   0   53   54   38   0   41   47

19. Old Faithful Use these Old Faithful eruption duration times (seconds):

125 203 205 221 225 229 233 233 235 236 236 237 238 238 239 240 240 240 240 241 241 242 242 242 243 243 244 245 245 245 245 246 246 248 248 248 249 249 250 251 252 253 253 255 255 256 257 258 262 264

a. Use the bootstrap method with 1000 bootstrap samples to find a 95% confidence interval estimate of μ.

b. Find the 95% confidence interval estimate of μ found by using the methods of Section 7-2.

c. Compare the results.

20. Old Faithful Repeat Exercise 19 "Old Faithful" using the standard deviation instead of the mean. Compare the confidence interval to the one that would be found using the methods of Section 7-3. If the two confidence intervals are very different, which one is better? Why?

21. Analysis of Last Digits Weights of respondents were recorded as part of the California Health Interview Survey. The last digits of weights from 50 randomly selected respondents are listed below.

   5   0   1   0   2   0   5   0   5   0   3   8   5   0   5   0   5   6   0   0   0   0   0   0   8
   5   5   0   4   5   0   0   4   0   0   0   0   8   0   9   5   3   0   5   0   0   0   5   8

a. Use the bootstrap method with 1000 bootstrap samples to find a 95% confidence interval estimate of σ.

b. Find the 95% confidence interval estimate of σ found by using the methods of Section 7-3.

c. Compare the results. If the two confidence intervals are different, which one is better? Why?

22. Analysis of Last Digits Repeat Exercise 21 "Analysis of Last Digits" using the mean instead of the standard deviation. Compare the confidence interval to the one that would be found using the methods of Section 7-2.

7-4   Beyond the Basics

23. Effect of the Number of Bootstrap Samples Repeat Exercise 21 "Analysis of Last Digits" using 10,000 bootstrap samples instead of 1000. What happens?

24. Distribution Shapes Use the sample data given in Exercise 21 "Analysis of Last Digits."

a. Do the original sample values appear to be from a normally distributed population? Explain.

b. Do the 1000 bootstrap samples appear to have means that are from a normally distributed population? Explain.

c. Do the 1000 bootstrap samples appear to have standard deviations that are from a normally distributed population? Explain.

Chapter Quick Quiz

1. Celebrities and the Law Here is a 95% confidence interval estimate of the proportion of adults who say that the law goes easy on celebrities: $0.692 < p < 0.748$ (based on data from a Rasmussen Reports survey). What is the best point estimate of the proportion of adults in the population who say that the law goes easy on celebrities?

2. Interpreting CI Write a brief statement that correctly interprets the confidence interval given in Exercise 1 "Celebrities and the Law."

3. Critical Value For the survey described in Exercise 1 "Celebrities and the Law," find the critical value that would be used for constructing a 99% confidence interval estimate of the population proportion.

4. Loose Change *USA Today* reported that 40% of people surveyed planned to use accumulated loose change for paying bills. The margin of error was given as ± 3.1 percentage points. Identify the confidence interval that corresponds to that information.

5. Sample Size for Proportion Find the sample size required to estimate the percentage of college students who take a statistics course. Assume that we want 95% confidence that the proportion from the sample is within four percentage points of the true population percentage.

6. Sample Size for Mean Find the sample size required to estimate the mean IQ of professional musicians. Assume that we want 98% confidence that the mean from the sample is within three IQ points of the true population mean. Also assume that $\sigma = 15$.

7. Requirements A quality control analyst has collected a random sample of 12 smartphone batteries and she plans to test their voltage level and construct a 95% confidence interval estimate of the mean voltage level for the population of batteries. What requirements must be satisfied in order to construct the confidence interval using the method with the t distribution?

8. Degrees of Freedom In general, what does "degrees of freedom" refer to? For the sample data described in Exercise 7 "Requirements," find the number of degrees of freedom, assuming that you want to construct a confidence interval estimate of μ using the t distribution.

9. Critical Value Refer to Exercise 7 "Requirements" and assume that the requirements are satisfied. Find the critical value that would be used for constructing a 95% confidence interval estimate of μ using the t distribution.

10. Which Method? Refer to Exercise 7 "Requirements" and assume that sample of 12 voltage levels appears to be from a population with a distribution that is substantially far from being normal. Should a 95% confidence interval estimate of σ be constructed using the χ^2 distribution? If not, what other method could be used to find a 95% confidence interval estimate of σ?

Review Exercises

1. Online News In a Harris poll of 2036 adults, 40% said that they prefer to get their news online. Construct a 95% confidence interval estimate of the *percentage* of all adults who say that they prefer to get their news online. Can we safely say that fewer than 50% of adults prefer to get their news online?

2. Computers In order to better plan for student resources, the chairperson of the mathematics department at Broward College wants to estimate the percentage of students who own a computer. If we want to estimate that percentage based on survey results, how many students must we survey in order to be 90% confident that we are within four percentage points of the population percentage? Assume that the number of students is very large.

3. Earthquake Magnitudes Listed below are Richter scale magnitudes of randomly selected earthquakes.

a. Identify the best point estimate of the population mean μ.

b. Construct a 95% confidence interval estimate of the mean magnitude of the population of earthquakes.

c. Write a statement that interprets the confidence interval.

        3.09   2.76   2.65   3.44   3.01   2.94   3.45   2.72   2.69   2.89   2.71   2.76

4. Lefties There have been several studies conducted in an attempt to identify ways in which left-handed people are different from those who are right handed. Assume that you want to estimate the mean IQ of all left-handed adults. How many random left-handed adults must be tested in order to be 99% confident that the mean IQ of the sample group is within four IQ points of the mean IQ of all left-handed adults? Assume that σ is known to be 15.

5. Distributions Identify the distribution (normal, Student t, chi-square) that should be used in each of the following situations. If none of the three distributions can be used, what other method could be used?

a. In constructing a confidence interval of μ, you have 75 sample values and they appear to be from a population with a skewed distribution. The population standard deviation is not known.

b. In constructing a confidence interval estimate of μ, you have 75 sample values and they appear to be from a population with a skewed distribution. The population standard deviation is known to be 18.2 cm.

c. In constructing a confidence interval estimate of σ, you have 75 sample values and they appear to be from a population with a skewed distribution.

d. In constructing a confidence interval estimate of σ, you have 75 sample values and they appear to be from a population with a normal distribution.

e. In constructing a confidence interval estimate of p, you have 1200 survey respondents and 5% of them answered "yes" to the first question.

6. Sample Size You have been hired by your new employer to survey adults about printed newspaper subscriptions.

a. If you want to estimate the percentage of adults who have a paid subscription to a printed newspaper, how many adults must you survey if you want 95% confidence that your percentage has a margin of error of three percentage points?

b. If you want to estimate the mean amount that adults have spent on printed newspapers within the past year, how many adults must you survey if you want 95% confidence that your sample mean is in error by no more than $5? (Based on results from a pilot study, assume that the standard deviation of amounts spent on printed newspapers in the last year is $47.)

c. If you plan to obtain the estimates described in parts (a) and (b) with a single survey having several questions, how many adults must be surveyed?

7. Wristwatch Accuracy Students of the author collected data measuring the accuracy of wristwatches. The times (sec) below show the discrepancy between the real time and the time indicated on the wristwatch. Negative values correspond to watches that are running ahead of the actual time. The data appear to be from a normally distributed population. Construct a 95% confidence interval estimate of the mean discrepancy for the population of wristwatches.

$$-85 \quad 325 \quad 20 \quad 305 \quad -93 \quad 15 \quad 282 \quad 27 \quad 555 \quad 570 \quad -241 \quad 36$$

8. Wristwatch Accuracy Use the sample data from Exercise 7 "Wristwatch Accuracy" and construct a 95% confidence interval estimate of σ.

9. Bootstrap for Wristwatch Accuracy Repeat Exercise 7 "Wristwatch Accuracy" using 1000 bootstrap samples.

10. CI for Proportion In a TE Connectivity survey of 1000 randomly selected adults, 2% said that they "did not know" when asked if they felt comfortable being in a self-driving vehicle. There is a need to construct a 95% confidence interval estimate of the proportion of all adults in the population who don't know.

a. Find the confidence interval using the normal distribution as an approximation to the binomial distribution.

b. Find the confidence interval using 1000 bootstrap samples.

c. Compare the results.

Cumulative Review Exercises

Flight Arrivals. *Listed below are the arrival delay times (min) of randomly selected American Airlines flights that departed from JFK in New York bound for LAX in Los Angeles. Negative values correspond to flights that arrived early and ahead of the scheduled arrival time. Use these values for Exercises 1–4.*

$$-30 \quad -23 \quad 14 \quad -21 \quad -32 \quad 11 \quad -23 \quad 28 \quad 103 \quad -19 \quad -5 \quad -46$$

1. Statistics Find the mean, median, standard deviation, and range. Are the results statistics or parameters?

2. Range Rule of Thumb Use the results from Exercise 1 "Statistics" with the range rule of thumb to find arrival times separating those that are significantly low and those that are significantly high. Is the arrival delay time of 103 min significantly high?

3. Level of Measurement What is the level of measurement of these data (nominal, ordinal, interval, ratio)? Are the original unrounded arrival times continuous data or discrete data?

4. Confidence Interval Construct a 95% confidence interval estimate of the mean arrival delay time for the population of all American Airlines flights from JFK to LAX.

5. Normal Distribution Using a larger data set than the one given for Exercises 1-4, assume that airline arrival delays are normally distributed with a mean of -5.0 min and a standard deviation of 30.4 min.

a. Find the probability that a randomly selected flight has an arrival delay time of more than 15 min.

b. Find the value of Q_3, the arrival delay time that is the third quartile.

6. Sample Size Find the sample size necessary to estimate the mean arrival delay time for all American Airlines flights from JFK to LAX. Assume that we want 95% confidence that the sample mean is in error by no more than 5 min. Based on a larger sample than the one given for Exercises 1–4, assume that all arrival delay times have a standard deviation of 30.4 min.

7. On-Time Arrivals According to the Bureau of Transportation, American Airlines had an on-time arrival rate of 80.3% for a given year. Assume that this statistic is based on a sample of 1000 randomly selected American Airlines flights. Find the 99% confidence interval estimate of the *percentage* of all American Airlines flights that arrive on time.

8. Normality Assessment A random sample consists of 48 times (min) required for American Airlines flights to taxi out for takeoff. All of the flights are American Airlines flights from New York (JFK) to Los Angeles and they all occurred in January of a recent year. The 48 taxi-out times are depicted in the histogram and normal quantile plot shown below. Based on those graphs, does it appear that the taxi-out times are from a population having a normal distribution? Give an explanation for the distribution shown. Do the taxi-out times appear to satisfy the requirements necessary for construction of a confidence interval estimate of the standard deviation of the population of all such times?

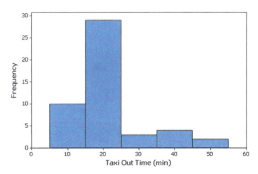 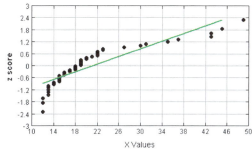

Technology Project

Earthquake Depths Data Set 21 "Earthquakes" in Appendix B includes the depths (km) of the sources of 600 earthquakes. Use technology for the following.
a. Find the mean and standard deviation of the 600 depths.
b. Generate a histogram and normal quantile plot of the 600 depths. Does it appear that the depths are from a population having a normal distribution? Explain.
c. In obtaining a 95% confidence interval estimate of the depth of all earthquakes, are the requirements for using a *t* distribution satisfied? Explain.
d. Find a 95% confidence interval estimate of the depth of all earthquakes using the *t* distribution.
e. Find a 95% confidence interval estimate of the depth of all earthquakes using 1000 bootstrap samples.
f. Compare the results from parts (d) and (e). Which confidence interval is likely to be better? Why?

FROM DATA TO DECISION

Critical Thinking: What does the survey tell us?

Surveys have become an integral part of our lives. Because it is so important that every citizen has the ability to interpret survey results, surveys are the focus of this project.

The Pew Research Center recently conducted a survey of 1007 U.S. adults and found that 85% of those surveyed know what Twitter is.

Analyzing the Data

1. Use the survey results to construct a 95% confidence interval estimate of the *percentage* of all adults who know what Twitter is.

2. Identify the margin of error for this survey.

3. Explain why it would or would not be okay for a newspaper to make this statement: "Based on results from a recent survey, more than 3 out of 4 adults know what Twitter is."

4. Assume that you are a newspaper reporter. Write a description of the survey results for your newspaper.

5. A common criticism of surveys is that they poll only a very small percentage of the population and therefore cannot be accurate. Is a sample of only 1007 adults taken from a population of all adults a sample size that is too small? Write a brief explanation of why the sample size of 1007 is or is not too small.

6. In reference to another survey, the president of a company wrote to the Associated Press about a nationwide survey of 1223 subjects. Here is what he wrote:

> When you or anyone else attempts to tell me and my associates that 1223 persons account for our opinions and tastes here in America, I get mad as hell! How dare you! When you or anyone else tells me that 1223 people represent America, it is astounding and unfair and should be outlawed.

The writer of that letter then proceeds to claim that because the sample size of 1223 people represents 120 million people, his single letter represents 98,000 (120 million divided by 1223) who share the same views. Do you agree or disagree with this claim? Write a response that either supports or refutes this claim.

Cooperative Group Activities

1. Out-of-class activity Collect sample data, and use the methods of this chapter to construct confidence interval estimates of population parameters. Here are some suggestions for parameters:

- Proportion of students at your college who can raise one eyebrow without raising the other eyebrow

- Mean age of cars driven by statistics students and/or the mean age of cars driven by faculty

- Mean length of words in *New York Times* editorials and mean length of words in editorials found in your local newspaper

- Proportion of students at your college who can correctly identify the president, vice president, and secretary of state

- Proportion of students at your college who are over the age of 18 and are registered to vote

- Mean age of full-time students at your college

- Mean number of hours that students at your college study each week

- Proportion of student cars that are painted white

- Proportion of cars that are red

2. In-class activity Without using any measuring device, each student should draw a line believed to be 3 in. long and another line believed to be 3 cm long. Then use rulers to measure and record the lengths of the lines drawn. Find the means and standard deviations of the two sets of lengths. Use the sample data to construct a confidence interval for the length of the line estimated to be 3 in., and then do the same for the length of the line estimated to be 3 cm. Do the confidence interval limits actually contain the correct length? Compare the results. Do the estimates of the 3-in. line appear to be more accurate than those for the 3-cm line?

3. In-class activity Assume that a method of gender selection can affect the probability of a baby being a girl, so that the probability becomes 1/4. Each student should simulate 20 births by drawing 20 cards from a shuffled deck. Replace each card after it has been drawn, then reshuffle. Consider the hearts to be girls and consider all other cards to be boys. After making 20 selections and recording the "genders" of the babies, construct a confidence interval estimate of the proportion of girls. Does the result appear to be effective in identifying the true value of the population proportion? (If decks of cards are not available, use some other way to simulate the births, such as using the random number generator on a calculator or using digits from phone numbers or Social Security numbers.)

4. Out-of-class activity Groups of three or four students should go to the library and collect a sample consisting of the ages of books (based on copyright dates). Plan and describe the sampling procedure, execute the sampling procedure, then use the results to construct a confidence interval estimate of the mean age of all books in the library.

5. In-class activity Each student should write an estimate of the age of the current president of the United States. All estimates should be collected, and the sample mean and standard deviation should be calculated. Then use the sample results to construct a confidence interval. Do the confidence interval limits contain the correct age of the president?

6. In-class activity A class project should be designed to conduct a test in which each student is given a taste of Coke and a taste of Pepsi. The student is then asked to identify which sample is Coke. After all of the results are collected, analyze the claim that the success rate is better than the rate that would be expected with random guesses.

7. In-class activity Each student should estimate the length of the classroom. The values should be based on visual estimates, with no actual measurements being taken. After the estimates have been collected, construct a confidence interval, then measure the length of the room. Does the confidence interval contain the actual length of the classroom? Is there a "collective wisdom," whereby the class mean is approximately equal to the actual room length?

8. In-class activity Divide into groups of three or four. Examine a sample of different issues of a current magazine and find the proportion of pages that include advertising. Based on the results, construct a 95% confidence interval estimate of the percentage of all such pages that have advertising. Compare results with other groups.

9. In-class activity Divide into groups of two. First find the sample size required to estimate the proportion of times that a coin turns up heads when tossed, assuming that you want 80% confidence that the sample proportion is within 0.08 of the true population proportion. Then toss a coin the required number of times and record your results. What percentage of such confidence intervals should actually contain the true value of the population proportion, which we know is $p = 0.5$? Verify this last result by comparing your confidence interval with the confidence intervals found in other groups.

10. Out-of-class activity Identify a topic of general interest and coordinate with all members of the class to conduct a survey. Instead of conducting a "scientific" survey using sound principles of random selection, use a convenience sample consisting of respondents who are readily available, such as friends, relatives, and other students. Analyze and interpret the results. Identify the population. Identify the shortcomings of using a convenience sample, and try to identify how a sample of subjects randomly selected from the population might be different.

11. Out-of-class activity Each student should find an article in a professional journal that includes a confidence interval of the type discussed in this chapter. Write a brief report describing the confidence interval and its role in the context of the article.

8-1 Basics of Hypothesis Testing

8-2 Testing a Claim About a Proportion

8-3 Testing a Claim About a Mean

8-4 Testing a Claim About a Standard Deviation or Variance

8

HYPOTHESIS TESTING

CHAPTER PROBLEM

Drones: Are Most Consumers Uncomfortable with Drones Delivering Their Purchases?

Recent decades have brought us incredible advances in technology, including drones, GPS devices, smartphones, HDTVs, biometric security devices, and the clapper sound-activated light switch. To leverage the latest technologies, Amazon has been pursuing Amazon Prime Air, which uses drones to deliver packages to customers in 30 minutes or less. The use of drones in this manner has generated much discussion and debate.

We can use methods of statistics to analyze surveys so that we can better evaluate and implement new technologies. In a Pitney Bowes survey, 1009 consumers were asked if they are comfortable with having drones deliver their purchases, and 54% (or 545) of them responded with "no." Based on this survey result, it is reasonable to claim that the *majority* of consumers are not comfortable with drones delivering their purchases, but do the survey results really justify that claim?

After all, the surveyors obtained responses from only 1009 consumers among the 247,696,327 consumers in the United States.

The claim that the majority of consumers are not comfortable with drone deliveries can be addressed by using the method of *hypothesis testing* that is presented in this chapter. We have the claim that $p > 0.5$, which is the symbolic form of the verbal claim that the majority (more than half or 0.5) of consumers are not comfortable with drone deliveries. This chapter will present the standard methods for testing such claims. With this knowledge, we will be better prepared to answer this question: Are we ready to accept drones as delivery vehicles?

CHAPTER OBJECTIVES

Here are the chapter objectives:

8-1 Basics of Hypothesis Testing

- Develop the ability to identify the null and alternative hypotheses when given some claim about a population parameter (such as a proportion, mean, standard deviation, or variance).

- Develop the ability to calculate a test statistic, find critical values, calculate *P*-values, and state a final conclusion that addresses the original claim. Here are the components that should be included in the hypothesis test:

  - Statements of the null and alternative hypotheses expressed in symbolic form
  - Value of the test statistic
  - Selection of the sampling distribution to be used for the hypothesis test
  - Identification of a *P*-value and/or critical value(s)
  - Statement of a conclusion rejecting the null hypothesis or failing to reject the null hypothesis
  - Statement of a final conclusion that uses simple and nontechnical terms to address the original claim

8-2 Testing a Claim About a Proportion

- Develop the ability to use sample data to conduct a formal hypothesis test of a claim about a population proportion. The procedure should include the components listed above with the objectives for Section 8-1.

8-3 Testing a Claim About a Mean

- Develop the ability to use sample data to conduct a formal hypothesis test of a claim made about a population mean. The procedure should include the same components listed above with the objectives for Section 8-1.

8-4 Testing a Claim About a Standard Deviation or Variance

- Develop the ability to use sample data to conduct a formal hypothesis test of a claim made about a population standard deviation or variance. The procedure should include the same components listed above with the objectives for Section 8-1.

8-1 Basics of Hypothesis Testing

Key Concept In this section we present key components of a formal hypothesis test. The concepts in this section are general and apply to hypothesis tests involving proportions, means, or standard deviations or variances. In Part 1, we begin with the "big picture" to understand the basic underlying approach to hypothesis tests. Then we describe null and alternative hypotheses, significance level, types of tests (two-tailed, left-tailed, right-tailed), test statistic, *P*-value, critical values, and statements of conclusions. In Part 2 we describe types of errors (type I and type II). In Part 3 we describe the *power* of a hypothesis test.

PART 1 Basic Concepts of Hypothesis Testing

We begin with two very basic definitions.

> **DEFINITIONS**
>
> In statistics, a **hypothesis** is a claim or statement about a property of a population.
>
> A **hypothesis test** (or **test of significance**) is a procedure for testing a claim about a property of a population.

The "property of a population" referred to in the preceding definitions is often the value of a population parameter, so here are some examples of typical hypotheses (or claims):

- $\mu < 98.6°\text{F}$ "The mean body temperature of humans is less than 98.6°F."

- $p > 0.5$ "The proportion of consumers not comfortable with drone deliveries is greater than 0.5."

- $\sigma = 15$ "The population of college students has IQ scores with a standard deviation equal to 15."

EXAMPLE 1 The Majority of Consumers Are Not Comfortable with Drone Deliveries

Consider the claim from the Chapter Problem that "the majority of consumers are not comfortable with drone deliveries." Using p to denote the proportion of consumers not comfortable with drone deliveries, the "majority" claim is equivalent to the claim that the proportion is greater than half, or $p > 0.5$. The expression $p > 0.5$ is the symbolic form of the original claim. (In the Chapter Problem, a survey of 1009 consumers included 54% who said that they are not comfortable with drone delivery.)

The Big Picture In Example 1, we have the claim that the population proportion p is such that $p > 0.5$. Among 1009 consumers, how many do we need to get a *significantly high* number who are not comfortable with drone delivery? A result of 506 (or 50.1%) is just barely more than half, so 506 is clearly *not significantly high*. A result of 1006 (or 99.7%) is clearly *significantly high*. But what about the result of 545 (or 54.0%) that was actually obtained in the Pitney Bowes survey? Is 545 (or 54.0%) *significantly high*? The method of hypothesis testing allows us to answer that key question.

Using Technology It is easy to obtain hypothesis-testing results using technology. The accompanying screen displays show results from four different technologies, so *we* *can use computers or calculators to do all of the computational heavy lifting.* Examining the four screen displays, we see some common elements. They all display a "test statistic" of $z = 2.55$ (rounded), and they all include a "*P*-value" of 0.005 (rounded). These two results are important, but *understanding* the hypothesis-testing procedure is critically important. Focus on *understanding* how the hypothesis-testing procedure works and learn the associated terminology. Only then will results from technology make sense.

Statdisk

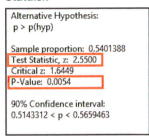

```
Alternative Hypothesis:
p > p(hyp)

Sample proportion:  0.5401388
Test Statistic, z:  2.5500
Critical z:  1.6449
P-Value:  0.0054

90% Confidence interval:
0.5143312 < p < 0.5659463
```

Minitab

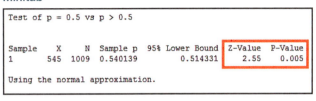

```
Test of p = 0.5 vs p > 0.5

Sample    X     N   Sample p  95% Lower Bound   Z-Value  P-Value
1        545  1009  0.540139        0.514331      2.55    0.005

Using the normal approximation.
```

TI-83/84 Plus

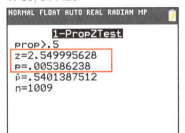

```
NORMAL FLOAT AUTO REAL RADIAN MP

        1-PropZTest
prop>.5
z=2.549995628
p=.005386238
p̂=.5401387512
n=1009
```

StatCrunch

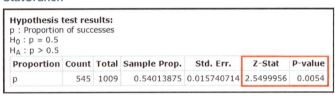

Hypothesis test results:
p : Proportion of successes
$H_0 : p = 0.5$
$H_A : p > 0.5$

Proportion	Count	Total	Sample Prop.	Std. Err.	Z-Stat	P-value
p	545	1009	0.54013875	0.015740714	2.5499956	0.0054

Significance Hypothesis tests are also called *tests of significance.* In Section 4-1 we used probabilities to determine when sample results are *significantly low* or *significantly high.* This chapter formalizes those concepts in a unified procedure that is used often throughout many different fields of application. Figure 8-1 on the next page summarizes the procedures used in two slightly different methods for conducting a formal hypothesis test. We will proceed to conduct a formal test of the claim from Example 1 that $p > 0.5$. In testing that claim, we will use the sample data from the survey cited in the Chapter Problem, with $x = 545$ consumers not comfortable with drone delivery among $n = 1009$ consumers surveyed.

Steps 1, 2, 3: Use the Original Claim to Create a Null Hypothesis H_0 and an Alternative Hypothesis H_1

The objective of Steps 1, 2, 3 is to identify the *null hypothesis* and *alternative hypothesis* so that the formal hypothesis test includes these standard components that are often used in many different disciplines. The null hypothesis includes the working assumption for the purposes of conducting the test.

> **DEFINITIONS**
>
> The **null hypothesis** (denoted by H_0) is a statement that the value of a population parameter (such as proportion, mean, or standard deviation) is *equal to* some claimed value.
>
> The **alternative hypothesis** (denoted by H_1 or H_a or H_A) is a statement that the parameter has a value that somehow differs from the null hypothesis. For the methods of this chapter, the symbolic form of the alternative hypothesis must use one of these symbols: $<, >, \neq$.

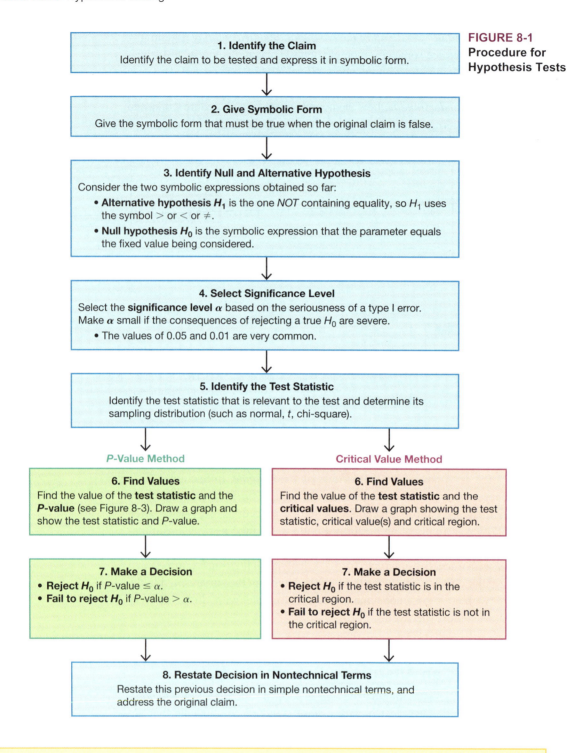

FIGURE 8-1
Procedure for Hypothesis Tests

1. Identify the Claim
Identify the claim to be tested and express it in symbolic form.

2. Give Symbolic Form
Give the symbolic form that must be true when the original claim is false.

3. Identify Null and Alternative Hypothesis
Consider the two symbolic expressions obtained so far:
- **Alternative hypothesis H_1** is the one *NOT* containing equality, so H_1 uses the symbol $>$ or $<$ or $\neq$.
- **Null hypothesis H_0** is the symbolic expression that the parameter equals the fixed value being considered.

4. Select Significance Level
Select the **significance level α** based on the seriousness of a type I error. Make α small if the consequences of rejecting a true H_0 are severe.
- The values of 0.05 and 0.01 are very common.

5. Identify the Test Statistic
Identify the test statistic that is relevant to the test and determine its sampling distribution (such as normal, t, chi-square).

P-Value Method

6. Find Values
Find the value of the **test statistic** and the **P-value** (see Figure 8-3). Draw a graph and show the test statistic and *P*-value.

7. Make a Decision
- **Reject H_0** if *P*-value $\leq \alpha$.
- **Fail to reject H_0** if *P*-value $> \alpha$.

Critical Value Method

6. Find Values
Find the value of the **test statistic** and the **critical values**. Draw a graph showing the test statistic, critical value(s) and critical region.

7. Make a Decision
- **Reject H_0** if the test statistic is in the critical region.
- **Fail to reject H_0** if the test statistic is not in the critical region.

8. Restate Decision in Nontechnical Terms
Restate this previous decision in simple nontechnical terms, and address the original claim.

Confidence Interval Method

Construct a confidence interval with a confidence level selected as in Table 8-1.

Because a confidence interval estimate of a population parameter contains the likely values of that parameter, reject a claim that the population parameter has a value that is not included in the confidence interval.

Table 8-1 Confidence Level for Confidence Interval

		Two-Tailed Test	One-Tailed Test
Significance	0.01	99%	98%
Level for	0.05	95%	90%
Hypothesis	0.10	90%	80%
Test			

The term *null* is used to indicate *no* change or no effect or no difference. We conduct the hypothesis test by assuming that the parameter is *equal to* some specified value so that we can work with a single distribution having a specific value.

Example: Here is an example of a null hypothesis involving a proportion:

$$H_0: p = 0.5$$

Example: Here are different examples of alternative hypotheses involving proportions:

$$H_1: p > 0.5 \quad H_1: p < 0.5 \quad H_1: p \neq 0.5$$

Given the claim from Example 1 that "the majority of consumers are uncomfortable with drone delivery," we can apply Steps 1, 2, and 3 in Figure 8-1 as follows.

Step 1: Identify the claim to be tested and express it in symbolic form. Using p to denote the probability of selecting a consumer uncomfortable with drone delivery, the claim that "the majority is uncomfortable with drone delivery" can be expressed in symbolic form as $p > 0.5$.

Step 2: Give the symbolic form that must be true when the original claim is false. If the original claim of $p > 0.5$ is false, then $p \leq 0.5$ must be true.

Step 3: This step is in two parts: Identify the alternative hypothesis H_1 and identify the null hypothesis H_0.

- Identify H_1: Using the two symbolic expressions $p > 0.5$ and $p \leq 0.5$, the alternative hypothesis H_1 is the one that does not contain equality. Of those two expressions, $p > 0.5$ does not contain equality, so we get

$$H_1: p > 0.5$$

- Identify H_0: The null hypothesis H_0 is the symbolic expression that the parameter *equals* the fixed value being considered, so we get

$$H_0: p = 0.5$$

The first three steps yield the null and alternative hypotheses:

$$H_0: p = 0.5 \text{ (null hypothesis)}$$

$$H_1: p > 0.5 \text{ (alternative hypothesis)}$$

Note About Forming Your Own Claims (Hypotheses) If you are conducting a study and want to use a hypothesis test to *support* your claim, your claim must be worded so that it becomes the alternative hypothesis (and can be expressed using only the symbols $<$, $>$, or $\neq$). You can never support a claim that a parameter is *equal to* a specified value.

Step 4: Select the Significance Level α

DEFINITION

The **significance level** α for a hypothesis test is the probability value used as the cutoff for determining when the sample evidence constitutes *significant* evidence against the null hypothesis. By its nature, the significance level α is the probability of mistakenly rejecting the null hypothesis when it is true:

Significance level $\alpha = P$ (rejecting H_0 when H_0 is true)

The significance level α is the same α introduced in Section 7-1, where we defined "critical value." Common choices for α are 0.05, 0.01, and 0.10; 0.05 is most common.

Step 5: Identify the Statistic Relevant to the Test and Determine Its Sampling Distribution (such as normal, t, or χ^2)

Table 8-2 lists parameters along with the corresponding sampling distributions.

Example: The claim $p > 0.5$ is a claim about the population proportion p, so use the normal distribution, provided that the requirements are satisfied. (With $n = 1009$, $p = 0.5$, and $q = 0.5$ from Example 1, $np \geq 5$ and $nq \geq 5$ are both true.)

TABLE 8-2

Parameter	Sampling Distribution	Requirements	Test Statistic
Proportion p	Normal (z)	$np \geq 5$ and $nq \geq 5$	$z = \dfrac{\hat{p} - p}{\sqrt{\dfrac{pq}{n}}}$
Mean μ	t	σ not known and normally distributed population or σ not known and $n > 30$	$t = \dfrac{\bar{x} - \mu}{\dfrac{s}{\sqrt{n}}}$
Mean μ	Normal (z)	σ known and normally distributed population or σ known and $n > 30$	$z = \dfrac{\bar{x} - \mu}{\dfrac{\sigma}{\sqrt{n}}}$
St. dev. σ or variance σ^2	χ^2	Strict requirement: normally distributed population	$\chi^2 = \dfrac{(n-1)s^2}{\sigma^2}$

Step 6: Find the Value of the Test Statistic, Then Find Either the *P*-Value or the Critical Value(s)

DEFINITION

The **test statistic** is a value used in making a decision about the null hypothesis. It is found by converting the sample statistic (such as $\hat{p}$, $\bar{x}$, or s) to a score (such as z, t, or χ^2) with the assumption that the null hypothesis is true.

In this chapter we use the test statistics listed in the last column of Table 8-2.

Example: From Example 1 we have a claim made about the population proportion p, we have $n = 1009$ and $x = 545$, so $\hat{p} = x/n = 0.540$. With the null hypothesis of H_0: $p = 0.5$, we are working with the assumption that $p = 0.5$, and it follows that $q = 1 - p = 0.5$. We can evaluate the test statistic as shown below (or technology can find the test statistic for us). The test statistic of $z = 2.55$ from each of the previous technology displays is more accurate than the result of $z = 2.54$ shown below. (If we replace 0.540 with $545/1009 = 0.54013875$, we get $z = 2.55$.)

$$z = \frac{\hat{p} - p}{\sqrt{\dfrac{pq}{n}}} = \frac{0.540 - 0.5}{\sqrt{\dfrac{(0.5)(0.5)}{1009}}} = 2.54$$

Finding the *P*-value and/or critical value(s) requires that we first consider whether the hypothesis test is two-tailed, left-tailed, or right-tailed, which are described as follows.

Two-Tailed, Left-Tailed, Right-Tailed

> **DEFINITION**
>
> The **critical region** (or **rejection region**) is the area corresponding to all values of the test statistic that cause us to reject the null hypothesis.

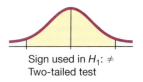

Sign used in H_1: $\neq$
Two-tailed test

Depending on the claim being tested, the critical region could be in the two extreme tails, it could be in the left tail, or it could be in the right tail.

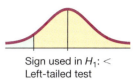

Sign used in H_1: $<$
Left-tailed test

- **Two-tailed test:** The critical region is in the two extreme regions (tails) under the curve (as in the top graph in Figure 8-2).

- **Left-tailed test:** The critical region is in the extreme left region (tail) under the curve (as in the middle graph in Figure 8-2).

- **Right-tailed test:** The critical region is in the extreme right region (tail) under the curve (as in the bottom graph in Figure 8-2).

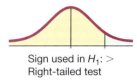

Sign used in H_1: $>$
Right-tailed test

FIGURE 8-2   Critical Region in Two-Tailed, Left-Tailed, and Right-Tailed Tests

HINT Look at the symbol used in the alternative hypothesis H_1.

- The symbol $>$ points to the right and the test is right-tailed.
- The symbol $<$ points to the left and the test is left-tailed.
- The symbol $\neq$ is used for a two-tailed test.

Example: With H_0: $p = 0.5$ and H_1: $p > 0.5$, we reject the null hypothesis and support the alternative hypothesis only if the sample proportion is greater than 0.5 by a significant amount, so the hypothesis test in this case is *right-tailed.*

P-Value Method

With the **_P_-value method** of testing hypotheses, we make a decision by comparing the *P*-value to the significance level.

> **DEFINITION**
>
> In a hypothesis test, the **_P_-value** is the probability of getting a value of the test statistic that is *at least as extreme* as the test statistic obtained from the sample data, assuming that the null hypothesis is true.

To find the *P*-value, first find the area beyond the test statistic, then use the procedure given in Figure 8-3 on the next page. That procedure can be summarized as follows:

- Critical region in left tail: *P*-value = area to the *left* of the test statistic
- Critical region in right tail: *P*-value = area to the *right* of the test statistic
- Critical region in two tails: *P*-value = *twice* the area in the tail beyond the test statistic

Example: Using the data from the Chapter Problem, the test statistic is $z = 2.55$, and it has a normal distribution area of 0.0054 to its right, so a right-tailed test with test statistic $z = 2.55$ has a *P*-value of 0.0054. See the different technology displays given earlier, and note that each of them provides the same *P*-value of 0.005 after rounding.

Journal Bans P-Values!

The *P*-value method of testing hypotheses has received widespread acceptance in the research community, but the editors of the journal *Basic and Applied Social Psychology* took a dramatic stance when they said that they would no longer publish articles that included *P*-values. In an editorial, David Trafimow and Michael Marks stated their belief that "the *P*-value bar is too easy to pass and sometimes serves as an excuse for lower quality research." David Trafimow stated that he did not know which statistical method should replace the use of *P*-values.

Many reactions to the *P*-value ban acknowledged that although *P*-values can be misused and misinterpreted, their use as a valuable research tool remains.

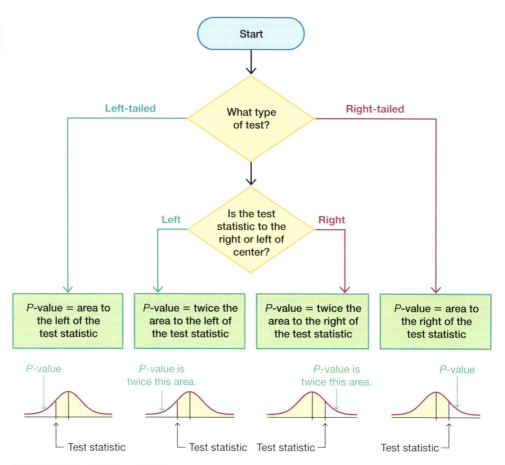

FIGURE 8-3   Finding *P*-Values

CAUTION Don't confuse a *P*-value with the parameter p or the statistic $\hat{p}$. Know the following notation:

P-value = probability of a test statistic at least as extreme as the one obtained

p = population proportion

$\hat{p}$ = sample proportion

P-Value and Hypothesis Testing Controversy

The standard method of testing hypotheses and the use of *P*-values have very widespread acceptance and use, but not everyone is convinced that these methods are sound. Editors of the *Journal of Basic and Applied Social Psychology* took a strong stand when they said that they would no longer publish articles that included *P*-values. They said that *P*-values are an excuse for lower-quality research and the *P*-value criterion is too easy to pass. In the past, *P*-values have been misinterpreted and misused, so a serious and important statistical analysis should not rely solely on *P*-value results. Instead, it would be wise to consider other aspects, such as the following.

- *Sample Size:* Very large samples could result in small *P*-values, suggesting that results are significant when the results don't really make much of a practical difference.

- *Power:* Part 3 of this section discusses the concept of *power,* and it is often helpful to analyze power as part of an analysis.

■ *Other Factors:* Instead of relying on just one outcome such as the *P*-value, it is generally better to also consider other results, such as a confidence interval, results from simulations, practical significance, design of the study, quality of the sample, consequences of type I and type II errors (discussed in Part 2 of this section), and replication of results.

This chapter presents the same methods of hypothesis testing and the same use of *P*-values that are currently being used, but again, it should be stressed that important applications should also consider other factors, such as those listed above.

Critical Value Method

With the **critical value method** (or traditional method) of testing hypotheses, we make a decision by comparing the test statistic to the critical value(s).

DEFINITION

In a hypothesis test, the **critical value(s)** separates the critical region (where we reject the null hypothesis) from the values of the test statistic that do not lead to rejection of the null hypothesis.

Critical values depend on the null hypothesis, the sampling distribution, and the significance level α.

Example: The critical region in Figure 8-4 is shaded in green. Figure 8-4 shows that with a significance level of $\alpha = 0.05$, the critical value is $z = 1.645$.

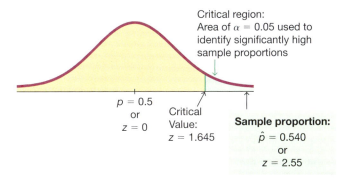

FIGURE 8-4   Critical Region, Critical Value, and Test Statistic

Step 7: Make a Decision to Either Reject H_0 or Fail to Reject H_0

Decision Criteria for the *P*-Value Method:

■ If *P*-value $\leq \alpha$, reject H_0. ("If the *P* is low, the null must go.")

■ If *P*-value $> \alpha$, fail to reject H_0.

Example: With significance level $\alpha = 0.05$ and *P*-value $= 0.005$, we have the condition that *P*-value $\leq \alpha$, so reject H_0. Remember, the *P*-value is the probability of getting a sample result at least as extreme as the one obtained, so if the *P*-value is low (less than or equal to α), the sample statistic is significantly low or significantly high.

Decision Criteria for the Critical Value Method:

- If the test statistic is in the critical region, reject H_0.

- If the test statistic is not in the critical region, fail to reject H_0.

Example: With test statistic $z = 2.55$ and the critical region from $z = 1.645$ to infinity, the test statistic falls within the critical region, so reject H_0.

Step 8: Restate the Decision Using Simple and Nontechnical Terms

Without using technical terms not understood by most people, state a final conclusion that addresses the original claim with wording that can be understood by those without knowledge of statistical procedures.

Example: There is sufficient evidence to support the claim that the majority of consumers are uncomfortable with drone deliveries.

Wording the Final Conclusion For help in wording the final conclusion, refer to Table 8-3, which lists the four possible circumstances and their corresponding conclusions. Note that only the first case leads to wording indicating *support* for the original conclusion. If you want to support some claim, state it in such a way that it becomes the alternative hypothesis, and then hope that the null hypothesis gets rejected.

TABLE 8-3 Wording of the Final Conclusion

Condition	Conclusion
Original claim does not include equality, and you reject H_0.	"There is sufficient evidence to *support* the claim that . . . (original claim)."
Original claim does not include equality, and you fail to reject H_0.	"There is not sufficient evidence to support the claim that . . . (original claim)."
Original claim includes equality, and you reject H_0.	"There is sufficient evidence to warrant *rejection* of the claim that . . . (original claim)."
Original claim includes equality, and you fail to reject H_0.	"There is not sufficient evidence to warrant rejection of the claim that . . . (original claim)."

Accept or Fail to Reject? We should say that we "fail to reject the null hypothesis" instead of saying that we "accept the null hypothesis." The term *accept* is misleading, because it implies incorrectly that the null hypothesis has been proved, but we can never prove a null hypothesis. The phrase *fail to reject* says more correctly that the available evidence isn't strong enough to warrant rejection of the null hypothesis.

Multiple Negatives Final conclusions can include as many as three negative terms. (*Example:* "There is *not* sufficient evidence to warrant *rejection* of the claim of *no* difference between 0.5 and the population proportion.") For such confusing conclusions, it is better to restate them to be understandable. Instead of saying that "there is not sufficient evidence to warrant rejection of the claim of no difference between 0.5 and the population proportion," a better statement would be this: "Until stronger evidence is obtained, continue to assume that the population proportion is equal to 0.5."

CAUTION Never conclude a hypothesis test with a statement of "reject the null hypothesis" or "fail to reject the null hypothesis." Always make sense of the conclusion with a statement that uses simple nontechnical wording that addresses the original claim.

Confidence Intervals for Hypothesis Tests

In this section we have described the individual components used in a hypothesis test, but the following sections will combine those components in comprehensive procedures. We can test claims about population parameters by using the *P*-value method or the critical value method summarized in Figure 8-1, or we can use confidence intervals.

A confidence interval estimate of a population parameter contains the likely values of that parameter. If a confidence interval does not include a claimed value of a population parameter, reject that claim. For two-tailed hypothesis tests, construct a confidence interval with a confidence level of $1 - \alpha$, but for a one-tailed hypothesis test with significance level α, construct a confidence interval with a confidence level of $1 - 2\alpha$. (See Table 8-1 on page 360 for common cases.) (For a left-tailed test or a right-tailed test, we could also use a one-sided confidence interval; see Exercise 40 in Section 7-1.) After constructing the confidence interval, use this criterion:

> **A confidence interval estimate of a population parameter contains the likely values of that parameter. We should therefore reject a claim that the population parameter has a value that is not included in the confidence interval.**

Equivalent Methods

In some cases, a conclusion based on a confidence interval may be different from a conclusion based on a hypothesis test. The *P*-value method and critical value method are equivalent in the sense that they always lead to the same conclusion. The following table shows that for the methods included in this chapter, a confidence interval estimate of a proportion might lead to a conclusion different from that of a hypothesis test.

Parameter	Is a confidence interval *equivalent* to a hypothesis test in the sense that they always lead to the same conclusion?
Proportion	No
Mean	Yes
Standard deviation or variance	Yes

PART 2 Type I and Type II Errors

When testing a null hypothesis, we arrive at a conclusion of rejecting it or failing to reject it. Our conclusions are sometimes correct and sometimes wrong (even if we apply all procedures correctly). Table 8-4 on the next page includes two different types of errors and we distinguish between them by calling them type I and type II errors, as described here:

- **Type I error:** The mistake of rejecting the null hypothesis when it is actually true. The symbol α (alpha) is used to represent the probability of a type I error.

$$\alpha = P(\text{type I error}) = P(\text{rejecting } H_0 \text{ when } H_0 \text{ is true})$$

- **Type II error:** The mistake of failing to reject the null hypothesis when it is actually false. The symbol β (beta) is used to represent the probability of a type II error.

$$\beta = P(\text{type II error}) = P(\text{failing to reject } H_0 \text{ when } H_0 \text{ is false})$$

MEMORY HINT FOR TYPE I AND TYPE II ERRORS Remember "routine for fun," and use the consonants from those words (**R**ou**T**i**N**e **F**o**R** **F**u**N**) to remember that a type I error is RTN: Reject True Null (hypothesis), and a type II error is FRFN: Fail to Reject a False Null (hypothesis).

Aspirin Not Helpful for Geminis and Libras

Physician Richard Peto submitted an article to *Lancet,* a British medical journal. The article showed that patients had a better chance of surviving a heart attack if they were treated with aspirin within a few hours of their heart attacks. *Lancet* editors asked Peto to break down his results into subgroups to see if recovery worked better or worse for different groups, such as males or females. Peto believed that he was being asked to use too many subgroups, but the editors insisted. Peto then agreed, but he supported his objections by showing that when his patients were categorized by signs of the zodiac, aspirin was useless for Gemini and Libra heart attack patients, but aspirin is a lifesaver for those born under any other sign. This shows that when conducting multiple hypothesis tests with many different subgroups, there is a very large chance of getting some wrong results.

TABLE 8-4 Type I and Type II Errors

		True State of Nature	
		Null hypothesis is true	Null hypothesis is false
Preliminary Conclusion	Reject H_0	Type I error: Reject a true H_0. P (type I error) $= \alpha$	Correct decision
	Fail to reject H_0	Correct decision	Type II error: Fail to reject a false H_0. P (type II error) $= \beta$

HINT FOR DESCRIBING TYPE I AND TYPE II ERRORS Descriptions of a type I error and a type II error refer to the *null hypothesis* being true or false, but when wording a statement representing a type I error or a type II error, *be sure that the conclusion addresses the original claim* (which may or may not be the null hypothesis). See Example 2.

EXAMPLE 2 **Describing Type I and Type II Errors**

Consider the claim that a medical procedure designed to increase the likelihood of a baby girl is effective, so that the probability of a baby girl is $p > 0.5$. Given the following null and alternative hypotheses, write statements describing (a) a type I error, and (b) a type II error.

H_0: $p = 0.5$

H_1: $p > 0.5$ (original claim that will be addressed in the final conclusion)

SOLUTION

a. **Type I Error:** A type I error is the mistake of rejecting a true null hypothesis, so the following is a type I error: In reality $p = 0.5$, but sample evidence leads us to conclude that $p > 0.5$. (In this case, a type I error is to conclude that the medical procedure is effective when in reality it has no effect.)

b. **Type II Error:** A type II error is the mistake of failing to reject the null hypothesis when it is false, so the following is a type II error: In reality $p > 0.5$, but we fail to support that conclusion. (In this case, a type II error is to conclude that the medical procedure has no effect, when it really is effective in increasing the likelihood of a baby girl.)

YOUR TURN Do Exercise 31 "Type I and Type II Errors."

Controlling Type I and Type II Errors Step 4 in our standard procedure for testing hypotheses is to select a significance level α (such as 0.05), which is the probability of a type I error. The values of α, β, and the sample size n are all related, so if you choose any two of them, the third is automatically determined (although β can't be determined until an alternative value of the population parameter has been specified along with α and n). One common practice is to select the significance level α, then select a sample size that is practical, so the value of β is determined. Generally, try to use the largest α that you can tolerate, but for type I errors with more serious consequences, select smaller values of α. Then choose a sample size n as large as is reasonable, based on considerations of time, cost, and other relevant factors. Another common practice is to select α and β so the required sample size n is automatically determined. (See Example 4 "Finding the Sample Size Required to Achieve 80% Power" in Part 3 of this section.)

PART 3 Power of a Hypothesis Test

We use β to denote the probability of failing to reject a false null hypothesis, so $P(\text{type II error}) = \beta$. It follows that $1 - \beta$ is the probability of rejecting a false null hypothesis, so $1 - \beta$ is a probability that is one measure of the effectiveness of a hypothesis test.

DEFINITION

The **power** of a hypothesis test is the probability $1 - \beta$ of rejecting a false null hypothesis. The value of the power is computed by using a particular significance level α and a *particular* value of the population parameter that is an alternative to the value assumed true in the null hypothesis.

Because determination of power requires a particular value that is an alternative to the value assumed in the null hypothesis, a hypothesis test can have many different values of power, depending on the particular values of the population parameter chosen as alternatives to the null hypothesis.

EXAMPLE 3   Power of a Hypothesis Test

Consider these preliminary results from the XSORT method of gender selection: There were 13 girls among the 14 babies born to couples using the XSORT method. If we want to test the claim that girls are more likely ($p > 0.5$) with the XSORT method, we have the following null and alternative hypotheses:

$$H_0: p = 0.5 \quad H_1: p > 0.5$$

Let's use a significance level of $\alpha = 0.05$. In addition to all of the given test components, finding power requires that we select a particular value of p that is an alternative to the value assumed in the null hypothesis $H_0: p = 0.5$. Find the values of power corresponding to these alternative values of p: 0.6, 0.7, 0.8, and 0.9.

SOLUTION

The values of power in the following table were found by using Minitab, and exact calculations are used instead of a normal approximation to the binomial distribution.

Specific Alternative Value of p	β	Power of Test = $1 - \beta$
0.6	0.820	0.180
0.7	0.564	0.436
0.8	0.227	0.773
0.9	0.012	0.988

INTERPRETATION

On the basis of the power values listed above, we see that this hypothesis test has power of 0.180 (or 18.0%) of rejecting $H_0: p = 0.5$ when the population proportion p is actually 0.6. That is, if the true population proportion is actually equal to 0.6, there is an 18.0% chance of making the correct conclusion of rejecting the false null hypothesis that $p = 0.5$. That low power of 18.0% is not so good.

continued

Process of Drug Approval

Gaining Food and Drug Administration (FDA) approval for a new drug is expensive and time-consuming. Here are the different stages of getting approval for a new drug:

- **Phase I study:** The safety of the drug is tested with a small (20–100) group of volunteers.

- **Phase II:** The drug is tested for effectiveness in randomized trials involving a larger (100–300) group of subjects. This phase often has subjects randomly assigned to either a treatment group or a placebo group.

- **Phase III:** The goal is to better understand the effectiveness of the drug as well as its adverse reactions. This phase typically involves 1,000–3,000 subjects, and it might require several years of testing.

Lisa Gibbs wrote in *Money* magazine that "the (drug) industry points out that for every 5,000 treatments tested, only 5 make it to clinical trials and only 1 ends up in drugstores." Total cost estimates vary from a low of $40 million to as much as $1.5 billion.

There is a 0.436 probability of rejecting $p = 0.5$ when the true value of p is actually 0.7. It makes sense that this test is more effective in rejecting the claim of $p = 0.5$ when the population proportion is actually 0.7 than when the population proportion is actually 0.6. (When identifying animals assumed to be horses, there's a better chance of rejecting an elephant as a horse—because of the greater difference—than rejecting a mule as a horse.) In general, increasing the difference between the assumed parameter value and the actual parameter value results in an increase in power, as shown in the table on the previous page.

> **YOUR TURN** Do Exercise 34 "Calculating Power."

Because the calculations of power are quite complicated, the use of technology is strongly recommended. (In this section, only Exercises 33, 34, and 35 involve power.)

Power and the Design of Experiments

Just as 0.05 is a common choice for a significance level, a power of at least 0.80 is a common requirement for determining that a hypothesis test is effective. (Some statisticians argue that the power should be higher, such as 0.85 or 0.90.) When designing an experiment, we might consider how much of a difference between the claimed value of a parameter and its true value is an important amount of difference. If testing the effectiveness of the XSORT gender selection method, a change in the proportion of girls from 0.5 to 0.501 is not very important, whereas a change in the proportion of girls from 0.5 to 0.9 would be very important. Such magnitudes of differences affect power. When designing an experiment, a goal of having a power value of at least 0.80 can often be used to determine the minimum required sample size, as in the following example.

EXAMPLE 4   **Finding the Sample Size Required to Achieve 80% Power**

Here is a statement similar to one in an article from the *Journal of the American Medical Association:* "The trial design assumed that with a 0.05 significance level, 153 randomly selected subjects would be needed to achieve 80% power to detect a reduction in the coronary heart disease rate from 0.5 to 0.4." From that statement, we know the following:

- Before conducting the experiment, the researchers selected a significance level of 0.05 and a power of at least 0.80.

- The researchers decided that a reduction in the proportion of coronary heart disease from 0.5 to 0.4 is an important difference that they wanted to detect (by correctly rejecting the false null hypothesis).

- Using a significance level of 0.05, power of 0.80, and the alternative proportion of 0.4, technology such as Minitab is used to find that the required minimum sample size is 153.

The researchers can then proceed by obtaining a sample of at least 153 randomly selected subjects. Because of factors such as dropout rates, the researchers are likely to need somewhat more than 153 subjects. (See Exercise 35.)

> **YOUR TURN** Do Exercise 35 "Finding Sample Size to Achieve Power."

8-1   Basic Skills and Concepts

Statistical Literacy and Critical Thinking

1. Vitamin C and Aspirin A bottle contains a label stating that it contains Spring Valley pills with 500 mg of vitamin C, and another bottle contains a label stating that it contains Bayer pills with 325 mg of aspirin. When testing claims about the mean contents of the pills, which would have more serious implications: rejection of the Spring Valley vitamin C claim or rejection of the Bayer aspirin claim? Is it wise to use the same significance level for hypothesis tests about the mean amount of vitamin C and the mean amount of aspirin?

2. Estimates and Hypothesis Tests Data Set 3 "Body Temperatures" in Appendix B includes sample body temperatures. We could use methods of Chapter 7 for making an estimate, or we could use those values to test the common belief that the mean body temperature is 98.6°F. What is the difference between estimating and hypothesis testing?

3. Mean Height of Men A formal hypothesis test is to be conducted using the claim that the mean height of men is equal to 174.1 cm.

a. What is the null hypothesis, and how is it denoted?

b. What is the alternative hypothesis, and how is it denoted?

c. What are the possible conclusions that can be made about the null hypothesis?

d. Is it possible to conclude that "there is sufficient evidence to support the claim that the mean height of men is equal to 174.1 cm"?

4. Interpreting *P*-value The Ericsson method is one of several methods claimed to increase the likelihood of a baby girl. In a clinical trial, results could be analyzed with a formal hypothesis test with the alternative hypothesis of $p > 0.5$, which corresponds to the claim that the method increases the likelihood of having a girl, so that the proportion of girls is greater than 0.5. If you have an interest in establishing the success of the method, which of the following *P*-values would you prefer: 0.999, 0.5, 0.95, 0.05, 0.01, 0.001? Why?

Identifying H_0 and H_1.   *In Exercises 5–8, do the following:*

a. Express the original claim in symbolic form.

b. Identify the null and alternative hypotheses.

5. Online Data Claim: Most adults would erase all of their personal information online if they could. A GFI Software survey of 565 randomly selected adults showed that 59% of them would erase all of their personal information online if they could.

6. Cell Phone Claim: Fewer than 95% of adults have a cell phone. In a Marist poll of 1128 adults, 87% said that they have a cell phone.

7. Pulse Rates Claim: The mean pulse rate (in beats per minute, or bpm) of adult males is equal to 69 bpm. For the random sample of 153 adult males in Data Set 1 "Body Data" in Appendix B, the mean pulse rate is 69.6 bpm and the standard deviation is 11.3 bpm.

8. Pulse Rates Claim: The standard deviation of pulse rates of adult males is more than 11 bpm. For the random sample of 153 adult males in Data Set 1 "Body Data" in Appendix B, the pulse rates have a standard deviation of 11.3 bpm.

Conclusions. *In Exercises 9–12, refer to the exercise identified. Make* **subjective** *estimates to decide whether results are significantly low or significantly high, then state a conclusion about the original claim. For example, if the claim is that a coin favors heads and sample results consist of 11 heads in 20 flips, conclude that there is not sufficient evidence to support the claim that the coin favors heads (because it is easy to get 11 heads in 20 flips by chance with a fair coin).*

9. Exercise 5 "Online Data"   **10.** Exercise 6 "Cell Phone"

11. Exercise 7 "Pulse Rates"   **12.** Exercise 8 "Pulse Rates"

Test Statistics. *In Exercises 13–16, refer to the exercise identified and find the value of the test statistic. (Refer to Table 8-2 on page 362 to select the correct expression for evaluating the test statistic.)*

13. Exercise 5 "Online Data"   **14.** Exercise 6 "Cell Phone"

15. Exercise 7 "Pulse Rates"   **16.** Exercise 8 "Pulse Rates"

P-Values. *In Exercises 17–20, do the following:*

a. Identify the hypothesis test as being two-tailed, left-tailed, or right-tailed.

b. Find the *P*-value. (See Figure 8-3 on page 364.)

c. Using a significance level of $\alpha = 0.05$, should we reject H_0 or should we fail to reject H_0?

17. The test statistic of $z = 1.00$ is obtained when testing the claim that $p > 0.3$.

18. The test statistic of $z = -2.50$ is obtained when testing the claim that $p < 0.75$.

19. The test statistic of $z = 2.01$ is obtained when testing the claim that $p \neq 0.345$.

20. The test statistic of $z = -1.94$ is obtained when testing the claim that $p = 3/8$.

Critical Values. *In Exercises 21–24, refer to the information in the given exercise and do the following.*

a. Find the critical value(s).

b. Using a significance level of $\alpha = 0.05$, should we reject H_0 or should we fail to reject H_0?

21. Exercise 17   **22.** Exercise 18

23. Exercise 19   **24.** Exercise 20

Final Conclusions. *In Exercises 25–28, use a significance level of $\alpha = 0.05$ and use the given information for the following:*

a. State a conclusion about the null hypothesis. (Reject H_0 or fail to reject H_0.)

b. Without using technical terms or symbols, state a final conclusion that addresses the original claim.

25. Original claim: More than 58% of adults would erase all of their personal information online if they could. The hypothesis test results in a *P*-value of 0.3257.

26. Original claim: Fewer than 90% of adults have a cell phone. The hypothesis test results in a *P*-value of 0.0003.

27. Original claim: The mean pulse rate (in beats per minute) of adult males is 72 bpm. The hypothesis test results in a *P*-value of 0.0095.

28. Original claim: The standard deviation of pulse rates of adult males is more than 11 bpm. The hypothesis test results in a *P*-value of 0.3045.

Type I and Type II Errors. *In Exercises 29–32, provide statements that identify the type I error and the type II error that correspond to the given claim. (Although conclusions are usually expressed in verbal form, the answers here can be expressed with statements that include symbolic expressions such as $p = 0.1$.)*

29. The proportion of people who write with their left hand is equal to 0.1.

30. The proportion of people with blue eyes is equal to 0.35.

31. The proportion of adults who use the Internet is greater than 0.87.

32. The proportion of people who require no vision correction is less than 0.25.

8-1 Beyond the Basics

33. Interpreting Power Chantix (varenicline) tablets are used as an aid to help people stop smoking. In a clinical trial, 129 subjects were treated with Chantix twice a day for 12 weeks, and 16 subjects experienced abdominal pain (based on data from Pfizer, Inc.). If someone claims that more than 8% of Chantix users experience abdominal pain, that claim is supported with a hypothesis test conducted with a 0.05 significance level. Using 0.18 as an alternative value of p, the power of the test is 0.96. Interpret this value of the power of the test.

34. Calculating Power Consider a hypothesis test of the claim that the Ericsson method of gender selection is effective in increasing the likelihood of having a baby girl, so that the claim is $p > 0.5$. Assume that a significance level of $\alpha = 0.05$ is used, and the sample is a simple random sample of size $n = 64$.

a. Assuming that the true population proportion is 0.65, find the power of the test, which is the probability of rejecting the null hypothesis when it is false. (*Hint:* With a 0.05 significance level, the critical value is $z = 1.645$, so any test statistic in the right tail of the accompanying top graph is in the rejection region where the claim is supported. Find the sample proportion $\hat{p}$ in the top graph, and use it to find the power shown in the bottom graph.)

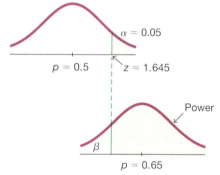

b. Explain why the green-shaded region of the bottom graph represents the power of the test.

35. Finding Sample Size to Achieve Power Researchers plan to conduct a test of a gender selection method. They plan to use the alternative hypothesis of $H_1: p > 0.5$ and a significance level of $\alpha = 0.05$. Find the sample size required to achieve at least 80% power in detecting an increase in p from 0.50 to 0.55. (This is a very difficult exercise. *Hint:* See Exercise 34.)

8-2 Testing a Claim About a Proportion

Key Concept This section describes a complete procedure for testing a claim made about a population proportion p. We illustrate hypothesis testing with the *P*-value method, the critical value method, and the use of confidence intervals. The methods of this section can be used with claims about population proportions, probabilities, or the decimal equivalents of percentages.

There are different methods for testing a claim about a population proportion. Part 1 of this section is based on the use of a normal approximation to a binomial distribution, and this method serves well as an introduction to basic concepts, but it is not a method used by professional statisticians. Part 2 discusses other methods that might require the use of technology.

PART 1 Normal Approximation Method

The following box includes the key elements used for testing a claim about a population proportion by using a normal distribution as an approximation to a binomial distribution.

The test statistic above does not include a correction for continuity (as described in Section 6-6) because its effect tends to be very small with large samples.

KEY ELEMENTS

Testing a Claim About a Population Proportion (Normal Approximation Method)

Objective

Conduct a formal hypothesis test of a claim about a population proportion p.

Notation

n = sample size or number of trials

$\hat{p} = \dfrac{x}{n}$ (*sample* proportion)

p = population proportion (p is the value used in the statement of the null hypothesis)

$q = 1 - p$

Requirements

1. The sample observations are a simple random sample.

2. The conditions for a *binomial distribution* are satisfied:

- There is a fixed number of trials.

- The trials are independent.

- Each trial has two categories of "success" and "failure."

- The probability of a success remains the same in all trials.

3. The conditions $np \geq 5$ and $nq \geq 5$ are both satisfied, so **the binomial distribution of sample proportions can be approximated by a normal distribution with** $\mu = np$ and $\sigma = \sqrt{npq}$ (as described in Section 6-6). Note that p used here is the *assumed* proportion used in the claim, not the sample proportion $\hat{p}$.

Test Statistic for Testing a Claim About a Proportion

$$z = \frac{\hat{p} - p}{\sqrt{\dfrac{pq}{n}}}$$

P-values: P-values are automatically provided by technology. If technology is not available, use the standard normal distribution (Table A-2) and refer to Figure 8-3 on page 364.

Critical values: Use the standard normal distribution (Table A-2).

Equivalent Methods

When testing claims about proportions, the confidence interval method is not equivalent to the *P*-value and critical value methods, so the confidence interval method could result in a different conclusion. (Both the *P*-value method and the critical value method use the same standard deviation based on the *claimed proportion p*, so they are equivalent to each other, but the confidence interval method uses an estimated standard deviation based on the *sample proportion.*) *Recommendation:* Use a confidence interval to *estimate* a population proportion, but use the *P*-value method or critical value method for *testing a claim* about a proportion. See Exercise 34.

Claim: Most Consumers Uncomfortable with Drone Deliveries

The Chapter Problem cited a Pitney Bowes survey in which 1009 consumers were asked if they are comfortable with having drones deliver their purchases, and 54% (or 545) of them responded with "no." Use these results to test the claim that most consumers are uncomfortable with drone deliveries. We interpret "most" to mean "more than half" or "greater than 0.5."

REQUIREMENT CHECK We first check the three requirements.

1. The 1009 consumers are randomly selected.

2. There is a fixed number (1009) of independent trials with two categories (the subject is uncomfortable with drone deliveries or is not).

3. The requirements $np \geq 5$ and $nq \geq 5$ are both satisfied with $n = 1009$, $p = 0.5$, and $q = 0.5$. [The value of $p = 0.5$ comes from the claim. We get $np = (1009)(0.5) = 504.5$, which is greater than or equal to 5, and we get $nq = (1009)(0.5) = 504.5$, which is also greater than or equal to 5.]

The three requirements are satisfied. ☑

Solution: *P*-Value Method

Technology: Computer programs and calculators usually provide a *P*-value, so the *P*-value method is used. See the accompanying TI-83/84 Plus calculator results showing the alternative hypothesis of "prop > 0.5," the test statistic of $z = 2.55$ (rounded), and the *P*-value of 0.0054 (rounded).

Table A-2: If technology is not available, Figure 8-1 on page 360 in the preceding section lists the steps for using the *P*-value method. Using those steps from Figure 8-1, we can test the claim as follows.

Step 1: The original claim is that most consumers are uncomfortable with drone deliveries, and that claim can be expressed in symbolic form as $p > 0.5$.

Step 2: The opposite of the original claim is $p \leq 0.5$.

Step 3: Of the preceding two symbolic expressions, the expression $p > 0.5$ does not contain equality, so it becomes the alternative hypothesis. The null hypothesis is the statement that p equals the fixed value of 0.5. We can therefore express H_0 and H_1 as follows:

$$H_0: p = 0.5$$

$$H_1: p > 0.5 \text{ (original claim)}$$

TI-83/84 Plus

continued

Is 0.05 a Bad Choice?

The value of 0.05 is a very common choice for serving as the cutoff separating results considered to be significant from those that are not. Science writer John Timmer wrote in *Ars Technica* that some problems with conclusions in science are attributable to the fact that statistics is sometimes weak because of the common use of 0.05 for a significance level. He gives examples of particle physics and genetics experiments in which *P*-values must be much lower than 0.05. He cites a study by statistician Valen Johnson, who suggested that we should raise standards by requiring that experiments use a *P*-value of 0.005 or lower. We do know that the choice of 0.05 is largely arbitrary, and lowering the significance level will result in fewer conclusions of significance, along with fewer wrong conclusions.

Step 4: For the significance level, we select $\alpha = 0.05$, which is a very common choice.

Step 5: Because we are testing a claim about a population proportion p, the sample statistic $\hat{p}$ is relevant to this test. The sampling distribution of sample proportions $\hat{p}$ can be approximated by a normal distribution in this case (as described in Section 6-3).

Step 6: The test statistic $z = 2.55$ can be found by using technology or it can be calculated by using $\hat{p} = 545/1009$ (sample proportion), $n = 1009$ (sample size), $p = 0.5$ (assumed in the null hypothesis), and $q = 1 - 0.5 = 0.5$.

$$z = \frac{\hat{p} - p}{\sqrt{\dfrac{pq}{n}}} = \frac{\dfrac{545}{1009} - 0.5}{\sqrt{\dfrac{(0.5)(0.5)}{1009}}} = 2.55$$

The *P*-value can be found from technology or it can be found by using the following procedure, which is shown in Figure 8-3 on page 364:

Left-tailed test: *P*-value = area to left of test statistic z

Right-tailed test: *P*-value = area to right of test statistic z

Two-tailed test: *P*-value = *twice* the area of the extreme region bounded by the test statistic z

Because this hypothesis test is right-tailed with a test statistic of $z = 2.55$, the *P*-value is the area to the right of $z = 2.55$. Referring to Table A-2, we see that the cumulative area to the *left* of $z = 2.55$ is 0.9946, so the area to the right of that test statistic is $1 - 0.9946 = 0.0054$. We get *P*-value $= 0.0054$. Figure 8-5 shows the test statistic and *P*-value for this example.

Step 7: Because the *P*-value of 0.0054 is less than or equal to the significance level of $\alpha = 0.05$, we reject the null hypothesis.

Step 8: Because we reject H_0: $p = 0.5$, we support the alternative hypothesis of $p > 0.5$. We conclude that there is sufficient sample evidence to support the claim that more than half of consumers are uncomfortable with drone deliveries. (See Table 8-3 on page 366 for help with wording this final conclusion.)

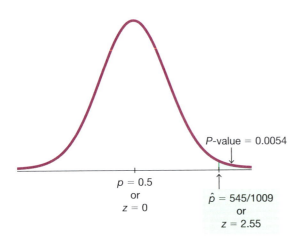

FIGURE 8-5   *P*-Value Method

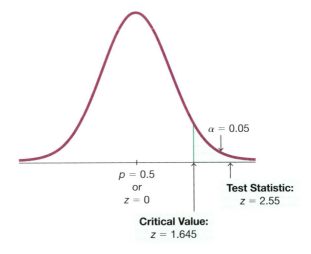

FIGURE 8-6   Critical Value Method

CAUTION Don't confuse the following notation.

- *P*-value = probability of getting a test statistic at least as extreme as the one representing the sample data, assuming that the null hypothesis H_0 is true
- p = population proportion
- $\hat{p}$ = sample proportion

Solution: Critical Value Method

The critical value method of testing hypotheses is summarized in Figure 8-1 on page 360. When using the critical value method with the claim given in Section 8-1 Example 1, "The Majority of Consumers Are Not Comfortable with Drone Deliveries," Steps 1 through 5 are the same as in Steps 1 through 5 for the *P*-value method, as shown on the previous pages. We continue with Step 6 of the critical value method.

Step 6: The test statistic is computed to be $z = 2.55$, as shown for the preceding *P*-value method. With the critical value method, we now find the critical values (instead of the *P*-value). This is a right-tailed test, so the area of the critical region is an area of $\alpha = 0.05$ in the right tail. Referring to Table A-2 and applying the methods of Section 6-1, we find that the critical value is $z = 1.645$, which is at the boundary of the critical region, as shown in Figure 8-6.

Step 7: Because the test statistic does fall within the critical region, we reject the null hypothesis.

Step 8: Because we reject H_0: $p = 0.5$, we conclude that there is sufficient sample evidence to support the claim that most (more than half) consumers are uncomfortable with drone deliveries. (See Table 8-3 on page 366 for help with wording this final conclusion.)

Solution: Confidence Interval Method

The claim given in Section 8-1 Example 1, "The Majority of Consumers Are Not Comfortable with Drone Deliveries," can be tested with a 0.05 significance level by constructing a 90% confidence interval. (See Table 8-1 on page 360 to see why the 0.05 significance level corresponds to a 90% confidence interval.)

The 90% confidence interval estimate of the population proportion p is found using the sample data consisting of $n = 1009$ and $\hat{p} = 545/1009$. Using the methods of Section 7-1 we get: $0.514 < p < 0.566$. The entire range of values in this confidence interval is greater than 0.5. Because we are 90% confident that the limits of 0.514 and 0.566 contain the true value of p, the sample data appear to support the claim that most (more than 0.5) consumers are uncomfortable with drone deliveries. In this case, the conclusion is the same as with the *P*-value method and the critical value method, but that is not always the case. It is possible that a conclusion based on the confidence interval can be different from the conclusion based on the *P*-value method or critical value method.

Finding the Number of Successes *x*

When using technology for hypothesis tests of proportions, we must usually enter the sample size n and the number of successes x, but in real applications the sample proportion $\hat{p}$ is often given instead of x. The number of successes x can be found by evaluating $x = n\hat{p}$, as illustrated in Example 1. Note that in Example 1, the result of 5587.712 adults must be rounded to the nearest whole number of 5588.

EXAMPLE 1 Finding the Number of Successes *x*

A study of sleepwalking or "nocturnal wandering" was described in *Neurology* magazine, and it included information that 29.2% of 19,136 American adults have sleepwalked. What is the actual number of adults who have sleepwalked?

SOLUTION

The number of adults who have sleepwalked is 29.2% of 19,136, or $0.292 \times 19,136 = 5587.712$, but the result must be a whole number, so we round the product to the nearest whole number of 5588.

> **YOUR TURN** Do part (a) of Exercise 1 "Number and Proportion."

CAUTION When conducting hypothesis tests of claims about proportions, slightly different results can be obtained when calculating the test statistic using a given sample proportion instead of using a rounded value of *x* found by using $x = n\hat{p}$.

 EXAMPLE 2 Fewer Than 30% of Adults Have Sleepwalked?

Using the same sleepwalking data from Example 1 ($n = 19,136$ and $\hat{p} = 29.2\%$), would a reporter be justified in stating that "fewer than 30% of adults have sleepwalked"? Let's use a 0.05 significance level to test the claim that for the adult population, the proportion of those who have sleepwalked is less than 0.30.

SOLUTION

REQUIREMENT CHECK (1) The sample is a simple random sample. (2) There is a fixed number (19,136) of independent trials with two categories (a subject has sleepwalked or has not). (3) The requirements $np \geq 5$ and $nq \geq 5$ are both satisfied with $n = 19,136$ and $p = 0.30$. [We get $np = (19,136)(0.30) = 5740.8$, which is greater than or equal to 5, and we also get $nq = (19,136)(0.70) = 13,395.2$, which is greater than or equal to 5.] The three requirements are all satisfied. ✓

Step 1: The original claim is expressed in symbolic form as $p < 0.30$.

Step 2: The opposite of the original claim is $p \geq 0.30$.

Step 3: Because $p < 0.30$ does not contain equality, it becomes H_1. We get

$$H_0: p = 0.30 \text{ (null hypothesis)}$$

$$H_1: p < 0.30 \text{ (alternative hypothesis and original claim)}$$

Step 4: The significance level is $\alpha = 0.05$.

Step 5: Because the claim involves the proportion p, the statistic relevant to this test is the sample proportion $\hat{p}$ and the sampling distribution of sample proportions can be approximated by the normal distribution.

Step 6: Technology If using technology, the test statistic and the *P*-value will be provided. See the accompanying results from StatCrunch showing that the test statistic is $z = -2.41$ (rounded) and the *P*-value $= 0.008$.

StatCrunch

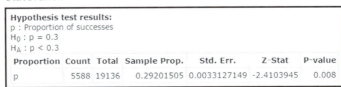

Hypothesis test results:
p : Proportion of successes
H_0 : p = 0.3
H_A : p < 0.3

Proportion	Count	Total	Sample Prop.	Std. Err.	Z-Stat	P-value
p	5588	19136	0.29201505	0.0033127149	-2.4103945	0.008

Table A-2 If technology is not available, proceed as follows to conduct the hypothesis test using the *P*-value method summarized in Figure 8-1 on page 360.

The test statistic $z = -2.41$ is calculated as follows:

$$z = \frac{\hat{p} - p}{\sqrt{\dfrac{pq}{n}}} = \frac{\dfrac{5588}{19{,}136} - 0.30}{\sqrt{\dfrac{(0.30)(0.70)}{19{,}136}}} = -2.41$$

Refer to Figure 8-3 on page 364 for the procedure for finding the *P*-value. For this left-tailed test, the *P*-value is the area to the left of the test statistic. Using Table A-2, we see that the area to the left of $z = -2.41$ is 0.0080, so the *P*-value is 0.0080.

Step 7: Because the *P*-value of 0.0080 is less than or equal to the significance level of 0.05, we reject the null hypothesis.

> INTERPRETATION
>
> Because we reject the null hypothesis, we support the alternative hypothesis. We therefore conclude that there is sufficient evidence to support the claim that fewer than 30% of adults have sleepwalked.

YOUR TURN Do Exercise 9 "Mickey D's."

Lie Detectors and the Law

Why not simply require all criminal suspects to take polygraph (lie detector) tests and eliminate trials by jury? According to the Council of Scientific Affairs of the American Medical Association, when lie detectors are used to determine guilt, accuracy can range from 75% to 97%. However, a high accuracy rate of 97% can still result in a high percentage of false positives, so it is possible that 50% of innocent subjects incorrectly appear to be guilty. Such a high chance of false positives rules out the use of polygraph tests as the single criterion for determining guilt.

Critical Value Method If we were to repeat Example 2 using the critical value method of testing hypotheses, we would see that in Step 6 the critical value is $z = -1.645$, which can be found from technology or Table A-2. In Step 7 we would reject the null hypothesis because the test statistic of $z = -2.41$ would fall within the critical region bounded by $z = -1.645$. We would then reach the same conclusion given in Example 2.

Confidence Interval Method If we were to repeat Example 2 using the confidence interval method, we would use a 90% confidence level because we have a left-tailed test. (See Table 8-1.) We get this 90% confidence interval: $0.287 < p < 0.297$. Because the entire range of the confidence interval falls below 0.30, there is sufficient evidence to support the claim that fewer than 30% of adults have sleepwalked.

PART 2 Exact Methods for Testing Claims About a Population Proportion *p*

Instead of using the normal distribution as an *approximation* to the binomial distribution, we can get *exact* results by using the binomial probability distribution itself. Binomial probabilities are a real nuisance to calculate manually, but technology makes this approach quite simple. Also, this exact approach does not require that $np \geq 5$ and $nq \geq 5$, so we have a method that applies when that requirement is not satisfied. To test hypotheses using the exact method, find *P*-values as follows:

Exact Method Identify the sample size *n*, the number of successes *x*, and the claimed value of the population proportion *p* (used in the null hypothesis); then find the *P*-value by using technology for finding binomial probabilities as follows:

Left-tailed test: *P*-value = $P(x$ or fewer successes among n trials$)$
Right-tailed test: *P*-value = $P(x$ or more successes among n trials$)$
Two-tailed test: *P*-value = twice the smaller of the preceding left-tailed and right-tailed values

Note: There is no universally accepted method for the two-tailed exact case, so this case can be treated with other different approaches, some of which are quite complex. For example, Minitab uses a "likelihood ratio test" that is different from the above approach that is commonly used.

EXAMPLE 3 Using the Exact Method

In testing a method of gender selection, 10 randomly selected couples are treated with the method, they each have a baby, and 9 of the babies are girls. Use a 0.05 significance level to test the claim that with this method, the probability of a baby being a girl is greater than 0.75.

SOLUTION

REQUIREMENT CHECK   The normal approximation method described in Part 1 of this section requires that $np \geq 5$ and $nq \geq 5$, but $nq = (10)(0.25) = 2.5$, so the requirement is violated. The exact method has only the requirements of being a simple random sample and satisfying the conditions for binomial distribution and those two requirements are satisfied. ☑

Here are the null and alternative hypotheses:

$$H_0\text{: } p = 0.75 \quad (\text{null hypothesis})$$

$$H_1\text{: } p > 0.75 \quad (\text{alternative hypothesis and original claim})$$

Instead of using the normal distribution, we use technology to find probabilities in a binomial distribution with $p = 0.75$. Because this is a right-tailed test, the *P*-value is the probability of 9 or more successes among 10 trials, assuming that $p = 0.75$. See the accompanying Statdisk display of exact probabilities from the binomial distribution. This Statdisk display shows that the probability of 9 or more successes is 0.2440252 when rounded to seven decimal places, so the *P*-value is 0.2440252. The *P*-value is high (greater than 0.05), so we fail to reject the null hypothesis. There is not sufficient evidence to support the claim that with the gender selection method, the probability of a girl is greater than 0.75.

> **YOUR TURN**   Do Exercise 33 "Exact Method."

Statdisk

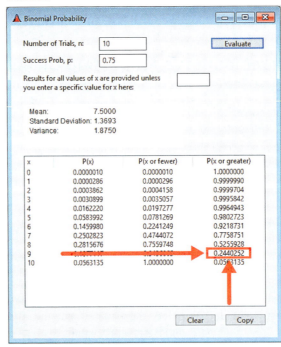

Improving the Exact Method A criticism of the exact method is that it is *too conservative* in the sense that the actual probability of a type I error is always less than or equal to α, and it could be much lower than α.

> **With the exact method, the *actual* probability of a type I error is less than or equal to α, which is the *desired* probability of a type I error.**

A **simple continuity correction** improves the conservative behavior of the exact method with an adjustment to the P-value that is obtained by subtracting from it the value that is one-half the binomial probability at the boundary, as shown below. (See Exercise 33 "Exact Method.") This method is easy to apply if technology is available for finding binomial probabilities.

Simple Continuity Correction to the Exact Method

Left-tailed test: $P\text{-value} = P(x \text{ or fewer}) - \dfrac{1}{2}P(\text{exactly } x)$

Right-tailed test: $P\text{-value} = P(x \text{ or more}) - \dfrac{1}{2}P(\text{exactly } x)$

Two-tailed test: $P\text{-value} =$ twice the smaller of the preceding left-tailed and right-tailed values

The above "simple continuity correction" is described in "Modifying the Exact Test for a Binomial Proportion and Comparisons with Other Approaches," by Alan Huston, *Journal of Applied Statistics,* Vol. 33, No. 7. For another improvement that uses weighted tail areas based on a measure of skewness, see the preceding article by Alan Huston.

Win $1,000,000 for ESP

Magician James Randi instituted an educational foundation that offers a prize of $1 million to anyone who can demonstrate paranormal, supernatural, or occult powers. Anyone possessing power such as fortune telling, ESP (extrasensory perception), or the ability to contact the dead can win the prize by passing testing procedures. A preliminary test is followed by a formal test, but so far, no one has passed the preliminary test. The formal test would be designed with sound statistical methods, and it would likely involve analysis with a formal hypothesis test. According to the foundation, "We consult competent statisticians when an evaluation of the results, or experiment design, is required."

TECH CENTER

Hypothesis Test: Proportion
Access tech supplements, videos, and data sets at **www.TriolaStats.com**

Statdisk	Minitab	StatCrunch
1. Click **Analysis** in the top menu. 2. Select **Hypothesis Testing** from the dropdown menu and select **Proportion One Sample** from the submenu. 3. Under *Alternative Hypothesis* select the format used for the alternative hypothesis, enter significance level, claimed proportion (from null hypothesis), sample size, and number of successes. 4. Click **Evaluate.**	1. Click **Stat** in the top menu. 2. Select **Basic Statistics** from the dropdown menu and select **1 Proportion** from the submenu. 3. Select **Summarized data** from the dropdown menu and enter number of events and number of trials (n). 4. Check the **Perform Hypothesis Test** box and enter the proportion used in the null hypothesis. 5. Click the **Options** button and enter the confidence level. (Enter 95.0 for a significance level of 0.05.) For *Alternative Hypothesis* select the format used for the alternative hypothesis. 6. For *Method* select **Normal approximation** to use the same method in this Section and click **OK** twice.	1. Click **Stat** in the top menu. 2. Select **Proportion Stats** from the dropdown menu, then select **One Sample—With Summary** from the submenu. 3. Enter the number of successes and number of observations (n). 4. Select **Hypothesis test for p** and for H_0 enter the claimed value of the population proportion (from the null hypothesis). For H_A select the format used for the alternative hypothesis. 5. Click **Compute!**

continued

TECH CENTER *continued*

 Hypothesis Test: Proportion

Access tech supplements, videos, and data sets at **www.TriolaStats.com**

TI-83/84 Plus Calculator	Excel
1. Press **STAT**, then select **TESTS** in the top menu.	**XLSTAT Add-In (Required)**
2. Select **1-PropZTest** in the menu and press **ENTER**.	1. Click on the **XLSTAT** tab in the Ribbon and then click **Parametric Tests**.
3. Enter the claimed population proportion p_0, number of successes x, and sample size n. For *prop* select the format used for the alternative hypothesis.	2. Select **Tests for one proportion** from the dropdown menu.
4. Select **Calculate** and press **ENTER**.	3. Under *Data format* select **Frequency** if you know the number of successes x or select **Proportion** if you know the sample proportion $\hat{p}$.

1. Press **STAT**, then select **TESTS** in the top menu.
2. Select **1-PropZTest** in the menu and press **ENTER**.
3. Enter the claimed population proportion p_0, number of successes x, and sample size n. For *prop* select the format used for the alternative hypothesis.
4. Select **Calculate** and press **ENTER**.

XLSTAT Add-In (Required)
1. Click on the **XLSTAT** tab in the Ribbon and then click **Parametric Tests.**
2. Select **Tests for one proportion** from the dropdown menu.
3. Under *Data format* select **Frequency** if you know the number of successes x or select **Proportion** if you know the sample proportion $\hat{p}$.
4. Enter the frequency or sample proportion, sample size, and claimed value for the population proportion (*Test proportion*).
5. Check **z test** and uncheck **Continuity correction** for the methods of this section.
6. Click the **Options** tab.
7. Under *Alternative hypothesis* select the format used for the alternative hypothesis. For *Hypothesized Difference* enter **0** and enter the desired significance level (enter **5** for a 0.05 significance level). Under *Variance (confidence interval)* select **Test Proportion** and under *Confidence Interval* select **Wald.**
8. Click **OK** to display the result. The test statistic is labeled z *(Observed value)* and the *P*-value is below that. Critical values will also be displayed.

8-2   Basic Skills and Concepts

Statistical Literacy and Critical Thinking

In Exercises 1–4, use these results from a **USA Today** *survey in which 510 people chose to respond to this question that was posted on the* **USA Today** *website: "Should Americans replace passwords with biometric security (fingerprints, etc)?" Among the respondents, 53% said "yes." We want to test the claim that more than half of the population believes that passwords should be replaced with biometric security.*

1. Number and Proportion

a. Identify the actual number of respondents who answered "yes."

b. Identify the sample proportion and the symbol used to represent it.

2. Null and Alternative Hypotheses Identify the null hypothesis and alternative hypothesis.

3. Equivalence of Methods If we use the same significance level to conduct the hypothesis test using the *P*-value method, the critical value method, and a confidence interval, which method is not equivalent to the other two?

4. Requirements and Conclusions

a. Are any of the three requirements violated? Can the methods of this section be used to test the claim?

b. It was stated that we can easily remember how to interpret *P*-values with this: "If the *P* is low, the null must go." What does this mean?

c. Another memory trick commonly used is this: "If the *P* is high, the null will fly." Given that a hypothesis test never results in a conclusion of proving or supporting a null hypothesis, how is this memory trick misleading?

d. Common significance levels are 0.01 and 0.05. Why would it be unwise to use a significance level with a number like 0.0483?

Using Technology. *In Exercises 5–8, identify the indicated values or interpret the given display. Use the normal distribution as an approximation to the binomial distribution, as described in Part 1 of this section. Use a 0.05 significance level and answer the following:*

a. Is the test two-tailed, left-tailed, or right-tailed?

b. What is the test statistic?

c. What is the *P*-value?

d. What is the null hypothesis, and what do you conclude about it?

e. What is the final conclusion?

5. Adverse Reactions to Drug The drug Lipitor (atorvastatin) is used to treat high cholesterol. In a clinical trial of Lipitor, 47 of 863 treated subjects experienced headaches (based on data from Pfizer). The accompanying TI-83/84 Plus calculator display shows results from a test of the claim that fewer than 10% of treated subjects experience headaches.

6. Self-Driving Vehicles In a TE Connectivity survey of 1000 adults, 29% said that they would feel comfortable in a self-driving vehicle. The accompanying StatCrunch display results from testing the claim that more than 1/4 of adults feel comfortable in a self-driving vehicle.

TI-83/84 Plus

StatCrunch

Hypothesis test results:
p : Proportion of successes
$H_0 : p = 0.25$
$H_A : p > 0.25$

Proportion	Count	Total	Sample Prop.	Std. Err.	Z-Stat	P-value
p	290	1000	0.29	0.013693064	2.921187	0.0017

7. Cell Phone Ownership A Pew Research Center poll of 2076 randomly selected adults showed that 91% of them own cell phones. The following Minitab display results from a test of the claim that 92% of adults own cell phones.

Minitab

Test of p = 0.92 vs p ≠ 0.92

Sample	X	N	Sample p	95% CI	Z-Value	P-Value
1	1889	2076	0.909923	(0.897608, 0.922238)	−1.69	0.091

8. Biometric Security In a *USA Today* survey of 510 people, 53% said that we should replace passwords with biometric security, such as fingerprints. The accompanying Statdisk display results from a test of the claim that half of us say that we should replace passwords with biometric security.

Statdisk

Sample proportion: 0.5294118
Test Statistic, z: 1.3284
Critical z: ±1.9600
P-Value: 0.1840

Testing Claims About Proportions. *In Exercises 9–32, test the given claim. Identify the null hypothesis, alternative hypothesis, test statistic, P-value, or critical value(s), then state the conclusion about the null hypothesis, as well as the final conclusion that addresses the original claim. Use the P-value method unless your instructor specifies otherwise. Use the normal distribution as an approximation to the binomial distribution, as described in Part 1 of this section.*

9. Mickey D's In a study of the accuracy of fast food drive-through orders, McDonald's had 33 orders that were not accurate among 362 orders observed (based on data from *QSR* magazine). Use a 0.05 significance level to test the claim that the rate of inaccurate orders is equal to 10%. Does the accuracy rate appear to be acceptable?

10. Eliquis The drug Eliquis (apixaban) is used to help prevent blood clots in certain patients. In clinical trials, among 5924 patients treated with Eliquis, 153 developed the adverse reaction of nausea (based on data from Bristol-Myers Squibb Co.). Use a 0.05 significance level to test the claim that 3% of Eliquis users develop nausea. Does nausea appear to be a problematic adverse reaction?

11. Stem Cell Survey Adults were randomly selected for a *Newsweek* poll. They were asked if they "favor or oppose using federal tax dollars to fund medical research using stem cells obtained from human embryos." Of those polled, 481 were in favor, 401 were opposed, and 120 were unsure. A politician claims that people don't really understand the stem cell issue and their responses to such questions are random responses equivalent to a coin toss. Exclude the 120 subjects who said that they were unsure, and use a 0.01 significance level to test the claim that the proportion of subjects who respond in favor is equal to 0.5. What does the result suggest about the politician's claim?

12. M&Ms Data Set 27 "M&M Weights" in Appendix B lists data from 100 M&Ms, and 27% of them are blue. The Mars candy company claims that the percentage of blue M&Ms is equal to 24%. Use a 0.05 significance level to test that claim. Should the Mars company take corrective action?

13. OxyContin The drug OxyContin (oxycodone) is used to treat pain, but it is dangerous because it is addictive and can be lethal. In clinical trials, 227 subjects were treated with OxyContin and 52 of them developed nausea (based on data from Purdue Pharma L.P.). Use a 0.05 significance level to test the claim that more than 20% of OxyContin users develop nausea. Does the rate of nausea appear to be too high?

14. Medical Malpractice In a study of 1228 randomly selected medical malpractice lawsuits, it was found that 856 of them were dropped or dismissed (based on data from the Physicians Insurers Association of America). Use a 0.01 significance level to test the claim that most medical malpractice lawsuits are dropped or dismissed. Should this be comforting to physicians?

15. Survey Return Rate In a study of cell phone use and brain hemispheric dominance, an Internet survey was e-mailed to 5000 subjects randomly selected from an online group involved with ears. 717 surveys were returned. Use a 0.01 significance level to test the claim that the return rate is less than 15%.

16. Drug Screening The company Drug Test Success provides a "1-Panel-THC" test for marijuana usage. Among 300 tested subjects, results from 27 subjects were wrong (either a false positive or a false negative). Use a 0.05 significance level to test the claim that less than 10% of the test results are wrong. Does the test appear to be good for most purposes?

17. Births A random sample of 860 births in New York State included 426 boys. Use a 0.05 significance level to test the claim that 51.2% of newborn babies are boys. Do the results support the belief that 51.2% of newborn babies are boys?

18. Mendelian Genetics When Mendel conducted his famous genetics experiments with peas, one sample of offspring consisted of 428 green peas and 152 yellow peas. Use a 0.01 significance level to test Mendel's claim that under the same circumstances, 25% of offspring peas will be yellow. What can we conclude about Mendel's claim?

19. Lie Detectors Trials in an experiment with a polygraph yield 98 results that include 24 cases of wrong results and 74 cases of correct results (based on data from experiments conducted by researchers Charles R. Honts of Boise State University and Gordon H. Barland of the Department of Defense Polygraph Institute). Use a 0.05 significance level to test the claim that such polygraph results are correct less than 80% of the time. Based on the results, should polygraph test results be prohibited as evidence in trials?

20. Tennis Instant Replay The Hawk-Eye electronic system is used in tennis for displaying an instant replay that shows whether a ball is in bounds or out of bounds so players can challenge calls made by referees. In a recent U.S. Open, singles players made 879 challenges and 231 of them were successful, with the call overturned. Use a 0.01 significance level to test the claim that fewer than $1/3$ of the challenges are successful. What do the results suggest about the ability of players to see calls better than referees?

21. Touch Therapy When she was 9 years of age, Emily Rosa did a science fair experiment in which she tested professional touch therapists to see if they could sense her energy field. She flipped a coin to select either her right hand or her left hand, and then she asked the therapists to identify the selected hand by placing their hand just under Emily's hand without seeing it and without touching it. Among 280 trials, the touch therapists were correct 123 times (based on data in "A Close Look at Therapeutic Touch," *Journal of the American Medical Association,* Vol. 279, No. 13). Use a 0.10 significance level to test the claim that touch therapists use a method equivalent to random guesses. Do the results suggest that touch therapists are effective?

22. Touch Therapy Repeat the preceding exercise using a 0.01 significance level. Does the conclusion change?

23. Cell Phones and Cancer In a study of 420,095 Danish cell phone users, 135 subjects developed cancer of the brain or nervous system (based on data from the *Journal of the National Cancer Institute* as reported in *USA Today*). Test the claim of a somewhat common belief that such cancers are affected by cell phone use. That is, test the claim that cell phone users develop cancer of the brain or nervous system at a rate that is different from the rate of 0.0340% for people who do not use cell phones. Because this issue has such great importance, use a 0.005 significance level. Based on these results, should cell phone users be concerned about cancer of the brain or nervous system?

24. Store Checkout-Scanner Accuracy In a study of store checkout-scanners, 1234 items were checked for pricing accuracy; 20 checked items were found to be overcharges, and 1214 checked items were not overcharges (based on data from "UPC Scanner Pricing Systems: Are They Accurate?" by Goodstein, *Journal of Marketing,* Vol. 58). Use a 0.05 significance level to test the claim that with scanners, 1% of sales are overcharges. (Before scanners were used, the overcharge rate was estimated to be about 1%.) Based on these results, do scanners appear to help consumers avoid overcharges?

25. Super Bowl Wins Through the sample of the first 49 Super Bowls, 28 of them were won by teams in the National Football Conference (NFC). Use a 0.05 significance level to test the claim that the probability of an NFC team Super Bowl win is greater than one-half.

26. Testing Effectiveness of Nicotine Patches In one study of smokers who tried to quit smoking with nicotine patch therapy, 39 were smoking one year after the treatment and 32 were not smoking one year after the treatment (based on data from "High-Dose Nicotine Patch Therapy," by Dale et al., *Journal of the American Medical Association,* Vol. 274, No. 17). Use a 0.05 significance level to test the claim that among smokers who try to quit with nicotine patch therapy, the majority are smoking one year after the treatment. Do these results suggest that the nicotine patch therapy is not effective?

27. Overtime Rule in Football Before the overtime rule in the National Football League was changed in 2011, among 460 overtime games, 252 were won by the team that won the coin toss at the beginning of overtime. Using a 0.05 significance level, test the claim that the coin toss is fair in the sense that neither team has an advantage by winning it. Does the coin toss appear to be fair?

28. Postponing Death An interesting and popular hypothesis is that individuals can temporarily postpone death to survive a major holiday or important event such as a birthday. In a study, it was found that there were 6062 deaths in the week before Thanksgiving, and 5938 deaths the week after Thanksgiving (based on data from "Holidays, Birthdays, and Postponement of Cancer Death," by Young and Hade, *Journal of the American Medical Association,* Vol. 292, No. 24). If people can postpone death until after Thanksgiving, then the proportion of deaths in the week before should be less than 0.5. Use a 0.05 significance level to test the claim that the proportion of deaths in the week before Thanksgiving is less than 0.5. Based on the result, does there appear to be any indication that people can temporarily postpone death to survive the Thanksgiving holiday?

29. Is Nessie Real? This question was posted on the America Online website: Do you believe the Loch Ness monster exists? Among 21,346 responses, 64% were "yes." Use a 0.01 significance level to test the claim that most people believe that the Loch Ness monster exists. How is the conclusion affected by the fact that Internet users who saw the question could decide whether to respond?

30. Smoking Stopped In a program designed to help patients stop smoking, 198 patients were given *sustained* care, and 82.8% of them were no longer smoking after one month (based on data from "Sustained Care Intervention and Postdischarge Smoking Cessation Among

continued

Hospitalized Adults," by Rigotti et al., *Journal of the American Medical Association,* Vol. 312, No. 7). Use a 0.01 significance level to test the claim that 80% of patients stop smoking when given sustained care. Does sustained care appear to be effective?

31. Bias in Jury Selection In the case of *Casteneda v. Partida,* it was found that during a period of 11 years in Hidalgo County, Texas, 870 people were selected for grand jury duty and 39% of them were Americans of Mexican ancestry. Among the people eligible for grand jury duty, 79.1% were Americans of Mexican ancestry. Use a 0.01 significance level to test the claim that the selection process is biased against Americans of Mexican ancestry. Does the jury selection system appear to be biased?

32. Medication Usage In a survey of 3005 adults aged 57 through 85 years, it was found that 81.7% of them used at least one prescription medication (based on data from "Use of Prescription and Over-the-Counter Medications and Dietary Supplements Among Older Adults in the United States," by Qato et al., *Journal of the American Medical Association,* Vol. 300, No. 24). Use a 0.01 significance level to test the claim that more than $3/4$ of adults use at least one prescription medication. Does the rate of prescription use among adults appear to be high?

8-2 Beyond the Basics

33. Exact Method For each of the three different methods of hypothesis testing (identified in the left column), enter the *P*-values corresponding to the given alternative hypothesis and sample data. Note that the entries in the last column correspond to the Chapter Problem. How do the results agree with the large sample size?

	$H_1: p \neq 0.5$ $n = 10, x = 9$	$H_1: p \neq 0.4$ $n = 10, x = 9$	$H_1: p > 0.5$ $n = 1009, x = 545$
Normal approximation			
Exact			
Exact with simple continuity correction			

34. Using Confidence Intervals to Test Hypotheses When analyzing the last digits of telephone numbers in Port Jefferson, it is found that among 1000 randomly selected digits, 119 are zeros. If the digits are randomly selected, the proportion of zeros should be 0.1.

a. Use the critical value method with a 0.05 significance level to test the claim that the proportion of zeros equals 0.1.

b. Use the *P*-value method with a 0.05 significance level to test the claim that the proportion of zeros equals 0.1.

c. Use the sample data to construct a 95% confidence interval estimate of the proportion of zeros. What does the confidence interval suggest about the claim that the proportion of zeros equals 0.1?

d. Compare the results from the critical value method, the *P*-value method, and the confidence interval method. Do they all lead to the same conclusion?

35. Power For a hypothesis test with a specified significance level α, the probability of a type I error is α, whereas the probability β of a type II error depends on the particular value of p that is used as an alternative to the null hypothesis.

a. Using an alternative hypothesis of $p < 0.4$, using a sample size of $n = 50$, and assuming that the true value of p is 0.25, find the power of the test. See Exercise 34 "Calculating Power" in Section 8-1. [*Hint:* Use the values $p = 0.25$ and $pq/n = (0.25)(0.75)/50$.]

b. Find the value of β, the probability of making a type II error.

c. Given the conditions cited in part (a), find the power of the test. What does the power tell us about the effectiveness of the test?

8-3 Testing a Claim About a Mean

Key Concept Testing a claim about a population mean is one of the most important methods presented in this book. Part 1 of this section deals with the very realistic and commonly used case in which the population standard deviation σ is not known. Part 2 includes a brief discussion of the procedure used when σ is known, which is very rare.

PART 1 Testing a Claim About μ with σ Not Known

In reality, it is very rare that we test a claim about an unknown value of a population mean μ but we somehow know the value of the population standard deviation σ. The realistic situation is that we test a claim about a population mean and the value of the population standard deviation σ is not known. When σ is not known, we estimate it with the sample standard deviation s. From the central limit theorem (Section 6-4), we know that the distribution of sample means $\bar{x}$ is approximately a normal distribution with mean $\mu_{\bar{x}} = \mu$ and standard deviation $\sigma_{\bar{x}} = \sigma/\sqrt{n}$, but if σ is unknown, we estimate it with $s/\sqrt{n}$, which is used in the test statistic for a "t test." This test statistic has a distribution called the Student t distribution. The requirements, test statistic, P-value, and critical values are summarized in the Key Elements box that follows.

Equivalent Methods

For the t test described in this section, the P-value method, the critical value method, and the confidence interval method are all equivalent in the sense that they all lead to the same conclusions.

KEY ELEMENTS

Testing Claims About a Population Mean with σ Not Known

Objective

Use a formal hypothesis test to test a claim about a population mean μ.

Notation

n = sample size $\bar{x}$ = *sample* mean
s = *sample* standard deviation $\mu_{\bar{x}}$ = *population* mean (this value is taken from the claim and is used in the statement of the null hypothesis H_0)

Requirements

1. The sample is a simple random sample.

2. Either or both of these conditions are satisfied: The population is normally distributed or $n > 30$.

Test Statistic for Testing a Claim About a Mean

$t = \dfrac{\bar{x} - \mu_{\bar{x}}}{\dfrac{s}{\sqrt{n}}}$ (Round t to three decimal places, as in Table A-3.)

P-values: Use technology or use the Student t distribution (Table A-3) with degrees of freedom given by df $= n - 1$. (Figure 8-3 on page 364 summarizes the procedure for finding P-values.)

Critical values: Use the Student t distribution (Table A-3) with degrees of freedom given by df $= n - 1$. (When Table A-3 doesn't include the number of degrees of freedom, you could be conservative by using the next lower number of degrees of freedom found in the table, you could use the closest number of degrees of freedom in the table, or you could interpolate.)

Go Figure

There are now 2.7 zettabytes (10^{21}) of data in our digital universe.

Requirement of Normality or $n > 30$ This t test is *robust* against a departure from normality, meaning that the test works reasonably well if the departure from normality is not too extreme. Verify that there are no outliers and that the histogram or dotplot has a shape that is not very far from a normal distribution.

If the original population is not itself normally distributed, we use the condition $n > 30$ for justifying use of the normal distribution, but there is no exact specific minimum sample size that works for all cases. Sample sizes of 15 to 30 are sufficient if the population has a distribution that is not far from normal, but some populations have distributions that are extremely far from normal, and sample sizes greater than 30 might be necessary. In this text we use the simplified criterion of $n > 30$ as justification for treating the distribution of sample means as a normal distribution, regardless of how far the distribution departs from a normal distribution.

Important Properties of the Student t Distribution

Here is a brief review of important properties of the Student t distribution first presented in Section 7-2:

1. The Student t distribution is different for different sample sizes (see Figure 7-4 in Section 7-2).

2. The Student t distribution has the same general bell shape as the standard normal distribution; its wider shape reflects the greater variability that is expected when s is used to estimate σ.

3. The Student t distribution has a mean of $t = 0$ (just as the standard normal distribution has a mean of $z = 0$).

4. The standard deviation of the Student t distribution varies with the sample size and is greater than 1 (unlike the standard normal distribution, which has $\sigma = 1$).

5. As the sample size n gets larger, the Student t distribution gets closer to the standard normal distribution.

P-Value Method with Technology

If suitable technology is available, the P-value method of testing hypotheses is the way to go.

EXAMPLE 1 Adult Sleep: *P*-Value Method with Technology

The author obtained times of sleep for randomly selected adult subjects included in the National Health and Nutrition Examination Study, and those times (hours) are listed below. Here are the unrounded statistics for this sample: $n = 12$, $\bar{x} = 6.83333333$ hours, $s = 1.99240984$ hours. A common recommendation is that adults should sleep between 7 hours and 9 hours each night. Use the P-value method with a 0.05 significance level to test the claim that the mean amount of sleep for adults is less than 7 hours.

$$4 \quad 8 \quad 4 \quad 4 \quad 8 \quad 6 \quad 9 \quad 7 \quad 7 \quad 10 \quad 7 \quad 8$$

SOLUTION

REQUIREMENT CHECK   (1) The sample is a simple random sample. (2) The second requirement is that "the population is normally distributed or $n > 30$." The sample size is $n = 12$, which does not exceed 30, so we must determine whether the sample data appear to be from a normally distributed population. The accompanying histogram and normal quantile plot, along with the apparent absence of outliers, indicate that the sample appears to be from a population with a distribution that is approximately normal. Both requirements are satisfied. ✓

Statdisk

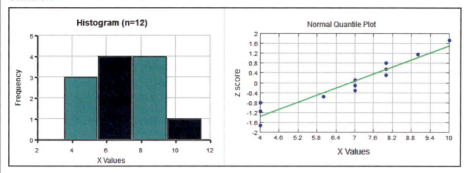

Here are the steps that follow the procedure summarized in Figure 8-1 on page 360.

Step 1: The claim that "the mean amount of adult sleep is less than 7 hours" becomes $\mu < 7$ hours when expressed in symbolic form.

Step 2: The alternative (in symbolic form) to the original claim is $\mu \geq 7$ hours.

Step 3: Because the statement $\mu < 7$ hours does not contain the condition of equality, it becomes the alternative hypothesis H_1. The null hypothesis H_0 is the statement that $\mu = 7$ hours.

$$H_0: \mu = 7 \text{ hours (null hypothesis)}$$

$$H_1: \mu < 7 \text{ hours (alternative hypothesis and original claim)}$$

Step 4: As specified in the statement of the problem, the significance level is $\alpha = 0.05$.

Step 5: Because the claim is made about the *population mean* μ, the sample statistic most relevant to this test is the *sample mean* $\bar{x}$, and we use the t distribution.

Step 6: The sample statistics of $n = 12$, $\bar{x} = 6.83333333$ hours, $s = 1.99240984$ hours are used to calculate the test statistic as follows, but technologies provide the test statistic of $t = -0.290$. In calculations such as the following, it is good to carry extra decimal places and not round.

$$t = \frac{\bar{x} - \mu_{\bar{x}}}{\frac{s}{\sqrt{n}}} = \frac{6.83333333 - 7}{\frac{1.99240984}{\sqrt{12}}} = -0.290$$

P-Value with Technology We could use technology to obtain the P-value. Shown on the next page are results from several technologies, and we can see that the P-value is 0.3887 (rounded). (SPSS shows a two-tailed P-value of 0.777, so it must be halved for this one-tailed test.)

Step 7: Because the P-value of 0.3887 is greater than the significance level of $\alpha = 0.05$, we fail to reject the null hypothesis.

INTERPRETATION

Step 8: Because we fail to reject the null hypothesis, we conclude that there is not sufficient evidence to support the claim that the mean amount of adult sleep is less than 7 hours.

YOUR TURN   Do Exercise 13 "Course Evaluations."

TI-83/84 Plus

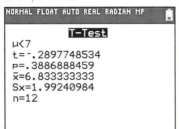

```
NORMAL FLOAT AUTO REAL RADIAN MP

            T-Test
μ<7
t=-.2897748534
p=.3886888459
x̄=6.833333333
Sx=1.99240984
n=12
```

Statdisk

t Test
Test Statistic, t: -0.2898
Critical t: -1.7959
P-Value: 0.3887

90% Confidence interval:
5.800414 < μ < 7.866252

Excel (XLSTAT)

Difference	-0.1667
t (Observed value)	-0.2898
t (Critical value)	-1.7959
DF	11
p-value (one-tailed)	0.3887
alpha	0.05

Minitab

Test of μ = 7 vs < 7

Variable	N	Mean	StDev	SE Mean	95% Upper Bound	T	P
Sleep	12	6.833	1.992	0.575	7.866	-0.29	0.389

JMP

Hypothesized Value	7
Actual Estimate	6.83333
DF	11
Std Dev	1.99241

t Test	
Test Statistic	-0.2898
Prob > |t|	0.7774
Prob > t	0.6113
Prob < t	0.3887

StatCrunch

Hypothesis test results:
μ : Mean of variable
$H_0 : \mu = 7$
$H_A : \mu < 7$

Variable	Sample Mean	Std. Err.	DF	T-Stat	P-value
Sleep	6.8333333	0.57515918	11	-0.28977485	0.3887

SPSS

	Test Value = 7					
				Mean	95% Confidence Interval of the Difference	
	t	df	Sig. (2-tailed)	Difference	Lower	Upper
SLEEP	-.290	11	.777	-.16667	-1.4326	1.0993

Examine the technology displays to see that only two of them include critical values, but they all include *P*-values. This is a major reason why the *P*-value method of testing hypotheses has become so widely used in recent years.

P-Value Method Without Technology

If suitable technology is not available, we can use Table A-3 to identify a *range of values* containing the *P*-value. In using Table A-3, keep in mind that it is designed for positive values of *t* and right-tail areas only, but left-tail areas correspond to the same *t* values with negative signs.

EXAMPLE 2   Adult Sleep: *P*-Value Method Without Technology

Example 1 is a left-tailed test with a test statistic of $t = -0.290$ (rounded) and a sample size of $n = 12$, so the number of degrees of freedom is df $= n - 1 = 11$. Using the test statistic of $t = -0.290$ with Table A-3, examine the values of *t* in the row for df $= 11$ to see that 0.290 is less than all of the listed *t* values in the row, which indicates that the area in the left tail below the test statistic of $t = -0.290$ is greater than 0.10. In this case, Table A-3 allows us to conclude that the *P*-value > 0.10, but technology provided the *P*-value of 0.3887. With the *P*-value > 0.10, the conclusions are the same as in Example 1.

YOUR TURN Do Exercise 21 "Lead in Medicine."

HINT Because using Table A-3 to find a range of values containing the *P*-value can be a bit tricky, the critical value method (see Example 3) might be easier than the *P*-value method if suitable technology is not available.

Critical Value Method

EXAMPLE 3 Adult Sleep: Critical Value Method

Example 1 is a left-tailed test with test statistic $t = -0.290$ (rounded). The sample size is $n = 12$, so the number of degrees of freedom is df $= n - 1 = 11$. Given the significance level of $\alpha = 0.05$, refer to the row of Table A-3 corresponding to 11 degrees of freedom, and refer to the column identifying an "area in one tail" of 0.05 (the significance level). The intersection of the row and column yields the critical value of $t = 1.796$, but this test is left-tailed, so the actual critical value is $t = -1.796$. Figure 8-7 shows that the test statistic of $t = -0.290$ does not fall within the critical region bounded by the critical value $t = -1.796$, so we fail to reject the null hypothesis. The conclusions are the same as those given in Example 1.

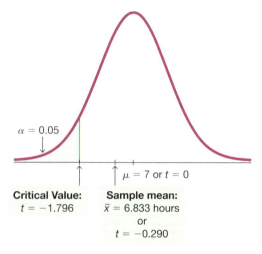

$\alpha = 0.05$

$\mu = 7$ or $t = 0$

Critical Value: **Sample mean:**
$t = -1.796$ $\bar{x} = 6.833$ hours
                              or
                          $t = -0.290$

FIGURE 8-7 *t* Test: Critical Value Method

Confidence Interval Method

EXAMPLE 4 Adult Sleep: Confidence Interval Method

Example 1 is a left-tailed test with significance level $\alpha = 0.05$, so we should use 90% as the confidence level (as indicated by Table 8-1 on page 360). For the sample data given in Example 1, here is the 90% confidence interval estimate of μ: 5.80 hours $< \mu < 7.87$ hours. In testing the claim that $\mu < 7$ hours, we use H_0: $\mu = 7$ hours, but the assumed value of $\mu = 7$ hours is contained within the confidence interval limits, so the confidence interval is telling us that 7 hours could be the value of μ. We don't have sufficient evidence to reject H_0: $\mu = 7$ hours, so we fail to reject this null hypothesis and we get the same conclusions given in Example 1.

EXAMPLE 5 Is the Mean Body Temperature Really 98.6°F?

Data Set 3 "Body Temperatures" in Appendix B includes measured body temperatures with these statistics for 12 AM on day 2: $n = 106$, $\bar{x} = 98.20°F$, $s = 0.62°F$. (This is the author's favorite data set.) Use a 0.05 significance level to test the common belief that the population mean is 98.6°F.

continued

Human Lie Detectors

Researchers tested 13,000 people for their ability to determine when someone is lying. They found 31 people with exceptional skills at identifying lies. These human lie detectors had accuracy rates around 90%. They also found that federal officers and sheriffs were quite good at detecting lies, with accuracy rates around 80%. Psychology Professor Maureen O'Sullivan questioned those who were adept at identifying lies, and she said that "all of them pay attention to nonverbal cues and the nuances of word usages and apply them differently to different people. They could tell you eight things about someone after watching a two-second tape. It's scary, the things these people notice." Methods of statistics can be used to distinguish between people unable to detect lying and those with that ability.

SOLUTION

REQUIREMENT CHECK   (1) With the study design used, we can treat the sample as a simple random sample. (2) The second requirement is that "the population is normally distributed or $n > 30$." The sample size is $n = 106$, so the second requirement is satisfied and there is no need to investigate the normality of the data. Both requirements are satisfied. ☑

Here are the steps that follow the procedure summarized in Figure 8-1.

Step 1: The claim that "the population mean is 98.6°F" becomes $\mu = 98.6°F$ when expressed in symbolic form.

Step 2: The alternative (in symbolic form) to the original claim is $\mu \neq 98.6°F$.

Step 3: Because the statement $\mu \neq 98.6°F$ does not contain the condition of equality, it becomes the alternative hypothesis H_1. The null hypothesis H_0 is the statement that $\mu = 98.6°F$.

$$H_0: \mu = 98.6°F \text{ (null hypothesis and original claim)}$$

$$H_1: \mu \neq 98.6°F \text{ (alternative hypothesis)}$$

Step 4: As specified in the statement of the problem, the significance level is $\alpha = 0.05$.

Step 5: Because the claim is made about the *population mean* μ, the sample statistic most relevant to this test is the *sample mean* $\bar{x}$. We use the t distribution because the relevant sample statistic is $\bar{x}$ and the requirements for using the t distribution are satisfied.

Step 6: The sample statistics are used to calculate the test statistic as follows, but technologies use unrounded values to provide the test statistic of $t = -6.61$.

$$t = \frac{\bar{x} - \mu_{\bar{x}}}{\frac{s}{\sqrt{n}}} = \frac{98.20 - 98.6}{\frac{0.62}{\sqrt{106}}} = -6.64$$

P-Value The *P*-value is 0.0000 or 0+ (or "less than 0.01" if using Table A-3).

Critical Values: The critical values are ± 1.983 (or ± 1.984 if using Table A-3).

Confidence Interval: The 95% confidence interval is $98.08°F < \mu < 98.32°F$.

Step 7: All three approaches lead to the same conclusion: Reject H_0.

- **P-Value:** The *P*-value of 0.0000 is less than the significance level of $\alpha = 0.05$.

- **Critical Values:** The test statistic $t = -6.64$ falls in the critical region bounded by ± 1.983.

- **Confidence Interval:** The claimed mean of 98.6°F does not fall within the confidence interval of $98.08°F < \mu < 98.32°F$.

INTERPRETATION

Step 8: There is sufficient evidence to warrant *rejection* of the common belief that the population mean is 98.6°F.

Alternative Methods Used When Population Is Not Normal and $n \leq 30$

The methods of this section include two requirements: (1) The sample is a simple random sample; (2) either the population is normally distributed or $n > 30$. If we have sample data that are not collected in an appropriate way, such as a voluntary response sample, it is likely that there is nothing that can be done to salvage the data, and the methods of this section should not be used. If the data are a simple random sample but the second condition is violated, there are alternative methods that could be used, including these three alternative methods:

- **Bootstrap Resampling** Use the confidence interval method of testing hypotheses, but obtain the confidence interval using bootstrap resampling, as described in Section 7-4. Be careful to use the appropriate confidence level, as indicated by Table 8-1 on page 360. Reject the null hypothesis if the confidence interval limits do not contain the value of the mean claimed in the null hypothesis. See Example 6.

- **Sign Test** See Section 13-2.

- **Wilcoxon Signed-Ranks Test** See Section 13-3.

EXAMPLE 6 **Bootstrap Resampling**

Listed below is a random sample of times (seconds) of tobacco use in animated children's movies (from Data Set 11 "Alcohol and Tobacco in Movies" in Appendix B). Use a 0.05 significance level to test the claim that the sample is from a population with a mean greater than 1 minute, or 60 seconds.

   0 223 0 176 0 548 0 37 158 51 0 0 299 37 0 11 0 0 0 0

SOLUTION

REQUIREMENT CHECK   The t test described in Part 1 of this section requires that the population is normally distributed or $n > 30$, but we have $n = 20$ and the accompanying normal quantile plot shows that the sample does not appear to be from a normally distributed population. The t test should *not* be used. ✓

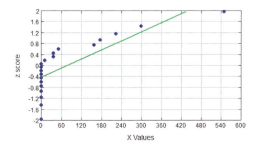

Instead of incorrectly using the t test, we use the bootstrap resampling method described in Section 7-4. After obtaining 1000 bootstrap samples and finding the mean of each sample, we sort the means. Because the test is right-tailed with a 0.05 significance level, we use the 1000 sorted sample means to find the 90% confidence interval limits of $P_5 = 29.9$ seconds and $P_{95} = 132.9$ seconds. (These values can vary somewhat.) The 90% confidence interval is 29.9 seconds $< \mu <$ 132.9 seconds. Because the assumed mean of 60 seconds is contained within those confidence interval limits, we fail to reject H_0: $\mu = 60$ seconds. There is not sufficient evidence to support H_1: $\mu > 60$ seconds.

INTERPRETATION

There is not sufficient evidence to support the claim that the given sample is from a population with a mean greater than 60 seconds.

PART 2 Testing a Claim About μ When σ Is Known

In reality, it is very rare to test a claim about an unknown population mean while the population standard deviation is somehow known. For this case, the procedure is essentially the same as in Part 1 of this section, the requirements are the same, but the test statistic, P-value, and critical values are as follows.

KEY ELEMENTS

Testing a Claim About a Mean (When σ Is Known)

Test Statistic

$$z = \frac{\bar{x} - \mu_{\bar{x}}}{\frac{\sigma}{\sqrt{n}}}$$

P-value: Provided by technology, or use the standard normal distribution (Table A-2) with the procedure in Figure 8-3 on page 364.

Critical values: Use the standard normal distribution (Table A-2).

If we repeat Example 1 "Adult Sleep" with the assumption that the value of $\sigma = 1.99240984$ hours is known, the test statistic is

$$z = \frac{\bar{x} - \mu_{\bar{x}}}{\frac{\sigma}{\sqrt{n}}} = \frac{6.83333333 - 7}{\frac{1.99240984}{\sqrt{12}}} = -0.29$$

Using this test statistic of $z = -0.29$, we can proceed to find the *P*-value. (See Figure 8-3 on page 364 for the flowchart summarizing the procedure for finding *P*-values.) Example 1 refers to a left-tailed test, so the *P*-value is the area to the *left* of $z = -0.29$, which is 0.3859 (found from Table A-2). Because the *P*-value of 0.3859 is greater than the significance level of $\alpha = 0.05$, we fail to reject the null hypothesis, as we did in Example 1. The conclusions from Example 1 are the same here.

With known σ, we get the *P*-value = 0.3859, but with unknown σ, the *P*-value = 0.3887. With known σ, the normal distribution (z) provides a smaller *P*-value, so the sample data are given greater significance.

TECH CENTER

 Hypothesis Test: Mean
Access tech supplements, videos, and data sets at **www.TriolaStats.com**

Statdisk

1. Click **Analysis** in the top menu.
2. Select **Hypothesis Testing** from the dropdown menu and select **Mean-One Sample** from the submenu.
3. Select the desired format for the *Alternative Hypothesis,* enter the desired significance level and claimed mean (from the null hypothesis).
4. *Using Summary Statistics:* Select the **Use Summary Statistics** tab and enter the sample size, sample mean, and sample standard deviation.

    Using Sample Data: Select the **Use Data** tab and select the desired data column.
5. Click **Evaluate**.

TECH CENTER *continued*

Hypothesis Test: Mean
Access tech supplements, videos, and data sets at **www.TriolaStats.com**

Minitab
1. Click **Stat** in the top menu.
2. Select **Basic Statistics** from the dropdown menu and select **1-Sample t** from the submenu.
3. *Using Summary Statistics:* Select **Summarized Data** from the dropdown menu and enter the sample size, sample mean, and sample standard deviation.
Using Sample Data: Select **One or more samples, each in a column** from the dropdown menu and select the desired data column.
4. Check the **Perform Hypothesis Test** box and enter the mean used in the null hypothesis.
5. Click the **Options** button and enter the confidence level. (Enter 95.0 for a significance level of 0.05.) For *Alternative Hypothesis* select the format used for the alternative hypothesis.
6. Click **OK** twice.

TIP: Another procedure is to click on **Assistant** in the top menu, then select **Hypothesis Tests** and **1-Sample t.** Complete the dialog box to get results, including *P*-value and other helpful information.

StatCrunch
1. Click **Stat** in the top menu.
2. Select **T Stats** from the dropdown menu, then select **One Sample** from the submenu.
3. *Using Summary Statistics:* Select **With Summary** from the submenu and enter the sample mean, sample standard deviation, and sample size.
Using Sample Data: Select **With Data** from the submenu and select the desired data column.
4. Select **Hypothesis test for μ** and for H_0 enter the claimed value of the population mean (from the null hypothesis). For H_A select the format used for the alternative hypothesis.
5. Click **Compute!**

TI-83/84 Plus Calculator
1. Press **STAT**, then select **TESTS** in the top menu.
2. Select **T-Test** in the menu and press **ENTER**.
3. Enter the claimed value of the population mean μ_0.
4. *Using Summary Statistics:* Select **Stats,** press **ENTER** and enter the sample mean $\overline{x}$, sample standard deviation Sx, and sample size n.
Using Sample Data: Select **Data,** press **ENTER**, and enter the name of the list containing the sample data. *Freq* should be set to **1**.
5. For μ select the format used for the alternative hypothesis.
6. Select **Calculate** and press **ENTER**.

Excel
XLSTAT Add-In (Required)
Requires original sample data; does not work with summary data.
1. Click on the **XLSTAT** tab in the Ribbon and then click **Parametric Tests.**
2. Select **One-Sample t-test and z-test** from the dropdown menu.
3. Under *Data* enter the range of cells containing the sample data. For *Data format* select **One Sample.** If the first row of data contains a label, also check the **Column labels** box.
4. Check the **Student's t test** box.
5. Click the **Options** tab.
6. Under *Alternative hypothesis* select the desired format ($\neq$ for two-tailed test, $<$ for left-tailed test, $>$ for right-tailed test).
7. For *Theoretical Mean*, enter the claimed value of the population mean (from the null hypothesis). Enter the desired significance level (enter **5** for a significance level of 0.05).
8. Click **OK** to display the results. The test statistic is identified as *t(Observed value)* or *z(Observed value)*. The *P*-value and critical value(s) are also displayed.

8-3 Basic Skills and Concepts

Statistical Literacy and Critical Thinking

1. Video Games: Checking Requirements Twelve different video games showing alcohol use were observed. The duration times of alcohol use were recorded, with the times (seconds) listed below (based on data from "Content and Ratings of Teen-Rated Video Games," by Haninger and Thompson, *Journal of the American Medical Association,* Vol. 291, No. 7). What requirements must be satisfied to test the claim that the sample is from a population with a mean greater than 90 sec? Are the requirements all satisfied?

$$84 \quad 14 \quad 583 \quad 50 \quad 0 \quad 57 \quad 207 \quad 43 \quad 178 \quad 0 \quad 2 \quad 57$$

2. df If we are using the sample data from Exercise 1 for a *t* test of the claim that the population mean is greater than 90 sec, what does df denote, and what is its value?

3. *t* Test Exercise 2 refers to a *t* test. What is a *t* test? Why is the letter *t* used? What is unrealistic about the *z* test methods in Part 2 of this section?

4. Confidence Interval Assume that we will use the sample data from Exercise 1 "Video Games" with a 0.05 significance level in a test of the claim that the population mean is greater than 90 sec. If we want to construct a confidence interval to be used for testing that claim, what confidence level should be used for the confidence interval? If the confidence interval is found to be 21.1 sec $< \mu <$ 191.4 sec, what should we conclude about the claim?

Finding *P*-values. *In Exercises 5–8, either use technology to find the P-value or use Table A-3 to find a range of values for the P-value.*

5. Airport Data Speeds The claim is that for Verizon data speeds at airports, the mean is $\mu = 14.00$ mbps. The sample size is $n = 13$ and the test statistic is $t = -1.625$

6. Body Temperatures The claim is that for 12 AM body temperatures, the mean is $\mu < 98.6°F$. The sample size is $n = 4$ and the test statistic is $t = -2.503$.

7. Old Faithful The claim is that for the duration times (sec) of eruptions of the Old Faithful geyser, the mean is $\mu = 240$ sec. The sample size is $n = 6$ and the test statistic is $t = 1.340$.

8. Tornadoes The claim is that for the widths (yd) of tornadoes, the mean is $\mu < 140$ yd. The sample size is $n = 21$ and the test statistic is $t = -0.024$.

Technology. *In Exercises 9–12, test the given claim by using the display provided from technology. Use a 0.05 significance level. Identify the null and alternative hypotheses, test statistic, P-value (or range of P-values), or critical value(s), and state the final conclusion that addresses the original claim.*

9. Airport Data Speeds Data Set 32 "Airport Data Speeds" in Appendix B includes Sprint data speeds (mbps). The accompanying TI-83/84 Plus display results from using those data to test the claim that they are from a population having a mean less than 4.00 Mbps. Conduct the hypothesis test using these results.

10. Body Temperatures Data Set 3 "Body Temperatures" in Appendix B includes 93 body temperatures measured at 12 AM on day 1 of a study, and the accompanying XLSTAT display results from using those data to test the claim that the mean body temperature is equal to 98.6°F. Conduct the hypothesis test using these results.

11. Old Faithful Data Set 23 "Old Faithful" in Appendix B includes data from 250 random eruptions of the Old Faithful geyser. The National Park Service makes predictions of times to the next eruption, and the data set includes the errors (minutes) in those predictions. The accompanying Statdisk display results from using the prediction errors (minutes) to test the claim that the mean prediction error is equal to zero. Comment on the accuracy of the predictions.

For Exercise 9

```
NORMAL FLOAT AUTO REAL RADIAN MP

            T-Test
μ<4
t=-.3662917532
p=.3578621222
x̄=3.71
Sx=5.598296024
n=50
```

For Exercise 10

Difference	-0.476
t (Observed value)	-7.102
|t| (Critical value)	1.986
DF	92
p-value (Two-tailed)	< 0.0001
alpha	0.05

For Exercise 11

t Test	
Test Statistic, t:	-8.7201
Critical t:	±1.9695
P-Value:	0.0000

12. Tornadoes Data Set 22 "Tornadoes" in Appendix B includes data from 500 random tornadoes. The accompanying StatCrunch display results from using the tornado lengths (miles) to test the claim that the mean tornado length is greater than 2.5 miles.

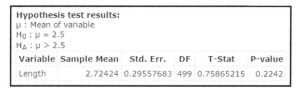

Hypothesis test results:					
μ : Mean of variable					
$H_0 : \mu = 2.5$					
$H_A : \mu > 2.5$					
Variable	Sample Mean	Std. Err.	DF	T-Stat	P-value
Length	2.72424	0.29557683	499	0.75865215	0.2242

Testing Hypotheses. *In Exercises 13–24, assume that a simple random sample has been selected and test the given claim. Unless specified by your instructor, use either the P-value method or the critical value method for testing hypotheses. Identify the null and alternative hypotheses, test statistic, P-value (or range of P-values), or critical value(s), and state the final conclusion that addresses the original claim.*

13. Course Evaluations Data Set 17 "Course Evaluations" in Appendix B includes data from student evaluations of courses. The summary statistics are $n = 93, \bar{x} = 3.91, s = 0.53$. Use a 0.05 significance level to test the claim that the population of student course evaluations has a mean equal to 4.00.

14. Speed Dating Data Set 18 "Speed Dating" in Appendix B includes "attractive" ratings of male dates made by the female dates. The summary statistics are $n = 199, \bar{x} = 6.19, s = 1.99$. Use a 0.01 significance level to test the claim that the population mean of such ratings is less than 7.00.

15. Garlic for Reducing Cholesterol In a test of the effectiveness of garlic for lowering cholesterol, 49 subjects were treated with raw garlic. Cholesterol levels were measured before and after the treatment. The changes (before minus after) in their levels of LDL cholesterol (in mg/dL) have a mean of 0.4 and a standard deviation of 21.0 (based on data from "Effect of Raw Garlic vs Commercial Garlic Supplements on Plasma Lipid Concentrations in Adults with Moderate Hypercholesterolemia," by Gardner et al., *Archives of Internal Medicine,* Vol. 167, No. 4). Use a 0.05 significance level to test the claim that with garlic treatment, the mean change in LDL cholesterol is greater than 0. What do the results suggest about the effectiveness of the garlic treatment?

16. Earthquake Depths Data Set 21 "Earthquakes" in Appendix B lists earthquake depths, and the summary statistics are $n = 600, \bar{x} = 5.82$ km, $s = 4.93$ km. Use a 0.01 significance level to test the claim of a seismologist that these earthquakes are from a population with a mean equal to 5.00 km.

17. Is the Diet Practical? When 40 people used the Weight Watchers diet for one year, their mean weight loss was 3.0 lb and the standard deviation was 4.9 lb (based on data from "Comparison of the Atkins, Ornish, Weight Watchers, and Zone Diets for Weight Loss and Heart Disease Reduction," by Dansinger et al., *Journal of the American Medical Association,* Vol. 293, No. 1). Use a 0.01 significance level to test the claim that the mean weight loss is greater than 0. Based on these results, does the diet appear to have statistical significance? Does the diet appear to have practical significance?

18. How Many English Words? A simple random sample of 10 pages from *Merriam-Webster's Collegiate Dictionary* is obtained. The numbers of words defined on those pages are found, with these results: $n = 10, \bar{x} = 53.3$ words, $s = 15.7$ words. Given that this dictionary has 1459 pages with defined words, the claim that there are more than 70,000 defined words is equivalent to the claim that the mean number of words per page is greater than 48.0 words. Assume a normally distributed population. Use a 0.01 significance level to test the claim that the mean number of words per page is greater than 48.0 words. What does the result suggest about the claim that there are more than 70,000 defined words?

19. Cans of Coke Data Set 26 "Cola Weights and Volumes" in Appendix B includes volumes (ounces) of a sample of cans of regular Coke. The summary statistics are $n = 36, \bar{x} = 12.19$ oz, $s = 0.11$ oz. Use a 0.05 significance level to test the claim that cans of Coke have a mean volume of 12.00 ounces. Does it appear that consumers are being cheated?

20. Insomnia Treatment A clinical trial was conducted to test the effectiveness of the drug zopiclone for treating insomnia in older subjects. Before treatment with zopiclone, 16 subjects had a mean wake time of 102.8 min. After treatment with zopiclone, the 16 subjects had a mean wake time of 98.9 min and a standard deviation of 42.3 min (based on data from "Cognitive Behavioral Therapy vs Zopiclone for Treatment of Chronic Primary Insomnia in Older Adults," by Sivertsen et al., *Journal of the American Medical Association,* Vol. 295, No. 24). Assume that the 16 sample values appear to be from a normally distributed population, and test the claim that after treatment with zopiclone, subjects have a mean wake time of less than 102.8 min. Does zopiclone appear to be effective?

21. Lead in Medicine Listed below are the lead concentrations (in $\mu g/g$) measured in different Ayurveda medicines. Ayurveda is a traditional medical system commonly used in India. The lead concentrations listed here are from medicines manufactured in the United States (based on data from "Lead, Mercury, and Arsenic in US and Indian Manufactured Ayurvedic Medicines Sold via the Internet," by Saper et al., *Journal of the American Medical Association,* Vol. 300, No. 8). Use a 0.05 significance level to test the claim that the mean lead concentration for all such medicines is less than 14 $\mu g/g$.

> 3.0   6.5   6.0   5.5   20.5   7.5   12.0   20.5   11.5   17.5

22. Got a Minute? Students of the author estimated the length of one minute without reference to a watch or clock, and the times (seconds) are listed below. Use a 0.05 significance level to test the claim that these times are from a population with a mean equal to 60 seconds. Does it appear that students are reasonably good at estimating one minute?

> 69   81   39   65   42   21   60   63   66   48   64   70   96   91   65

23. Car Booster Seats The National Highway Traffic Safety Administration conducted crash tests of child booster seats for cars. Listed below are results from those tests, with the measurements given in hic (standard head injury condition units). The safety requirement is that the hic measurement should be less than 1000 hic. Use a 0.01 significance level to test the claim that the sample is from a population with a mean less than 1000 hic. Do the results suggest that all of the child booster seats meet the specified requirement?

> 774   649   1210   546   431   612

24. Heights of Supermodels Listed below are the heights (cm) for the simple random sample of female supermodels Lima, Bundchen, Ambrosio, Ebanks, Iman, Rubik, Kurkova, Kerr, Kroes, Swanepoel, Prinsloo, Hosk, Kloss, Robinson, Heatherton, and Refaeli. Use a 0.01 significance level to test the claim that supermodels have heights with a mean that is greater than the mean height of 162 cm for women in the general population. Given that there are only 16 heights represented, can we really conclude that supermodels are taller than the typical woman?

> 178   177   176   174   175   178   175   178   178   177   180   176   180   178   180   176

Large Data Sets from Appendix B.   *In Exercises 25–28, use the data set from Appendix B to test the given claim. Use the P-value method unless your instructor specifies otherwise.*

25. Pulse Rates Use the pulse rates of adult females listed in Data Set 1 "Body Data" to test the claim that the mean is less than 75 bpm. Use a 0.05 significance level.

26. Earthquake Magnitudes Use the magnitudes of the earthquakes listed in Data Set 21 "Earthquakes" and test the claim that the sample is from a population with a mean greater than 2.50. Use a 0.01 significance level.

27. Diastolic Blood Pressure for Women Use the diastolic blood pressure measurements for adult females listed in Data Set 1 "Body Data" and test the claim that the adult female population has a mean diastolic blood pressure level less than 90 mm Hg. A diastolic blood pressure above 90 is considered to be hypertension. Use a 0.05 significance level. Based on the result, can we conclude that none of the adult females in the sample have hypertension?

28. Diastolic Blood Pressure for Men Repeat the preceding exercise for adult males instead of adult females.

8-3 Beyond the Basics

29. Hypothesis Test with Known σ How do the results from Exercise 13 "Course Evaluations" change if σ is known to be 0.53? Does the knowledge of σ have much of an effect?

30. Hypothesis Test with Known σ How do the results from Exercise 14 "Speed Dating" change if σ is known to be 1.99? Does the knowledge of σ have much of an effect?

31. Finding Critical t Values When finding critical values, we often need significance levels other than those available in Table A-3. Some computer programs approximate critical t values by calculating $t = \sqrt{df \cdot (e^{A^2/df} - 1)}$ where $df = n - 1$, $e = 2.718$, $A = z(8 \cdot df + 3)/(8 \cdot df + 1)$, and z is the critical z score. Use this approximation to find the critical t score for Exercise 12 "Tornadoes," using a significance level of 0.05. Compare the results to the critical t score of 1.648 found from technology. Does this approximation appear to work reasonably well?

32. Interpreting Power For the sample data in Example 1 "Adult Sleep" from this section, Minitab and StatCrunch show that the hypothesis test has power of 0.4943 of supporting the claim that $\mu < 7$ hours of sleep when the actual population mean is 6.0 hours of sleep. Interpret this value of the power, then identify the value of β and interpret that value. (For the t test in this section, a "noncentrality parameter" makes calculations of power much more complicated than the process described in Section 8-1, so software is recommended for power calculations.)

8-4 Testing a Claim About a Standard Deviation or Variance

Key Concept This section presents methods for conducting a formal hypothesis test of a claim made about a population standard deviation σ or population variance σ^2. The methods of this section use the chi-square distribution that was first introduced in Section 7-3. The assumptions, test statistic, P-value, and critical values are summarized as follows.

KEY ELEMENTS

Testing Claims About σ or σ^2

Objective

Conduct a hypothesis test of a claim made about a population standard deviation σ or population variance σ^2.

Notation

n = sample size

s = *sample* standard deviation s^2 = *sample* variance

σ = *population* standard deviation σ^2 = *population* variance

Requirements

1. The sample is a simple random sample.
2. The population has a normal distribution. (This is a fairly strict requirement.)

Test Statistic

$$\chi^2 = \frac{(n-1)s^2}{\sigma^2}$$ (round to three decimal places, as in Table A-4)

P-values: Use technology or Table A-4 with degrees of freedom: $df = n - 1$.

Critical values: Use Table A-4 with degrees of freedom $df = n - 1$.

CAUTION The X^2 (chi-square) test of this section is not *robust* against a departure from normality, meaning that the test does not work well if the population has a distribution that is far from normal. The condition of a normally distributed population is therefore a much stricter requirement when testing claims about σ or σ^2 than when testing claims about a population mean μ.

Equivalent Methods

When testing claims about σ or σ^2, the *P*-value method, the critical value method, and the confidence interval method are all equivalent in the sense that they will always lead to the same conclusion.

Properties of the Chi-Square Distribution

The chi-square distribution was introduced in Section 7-3, where we noted the following important properties.

1. All values of X^2 are nonnegative, and the distribution is not symmetric (see Figure 8-8).

2. There is a different X^2 distribution for each number of degrees of freedom (see Figure 8-9).

3. The critical values are found in Table A-4 using

$$\text{degrees of freedom} = n - 1$$

Here is an important note if using Table A-4 for finding critical values:

In Table A-4, each critical value of X^2 in the body of the table corresponds to an area given in the top row of the table, and each area in that top row is a *cumulative area to the right* of the critical value.

CAUTION Table A-4 for the chi-square distribution uses cumulative areas from the *right* (unlike Table A-2 for the standard normal distribution, which provides cumulative areas from the *left*.) See Example 1 in Section 7-3.

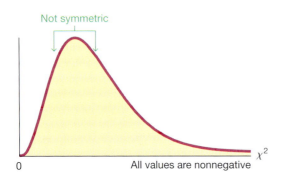

FIGURE 8-8   Properties of the Chi-Square Distribution

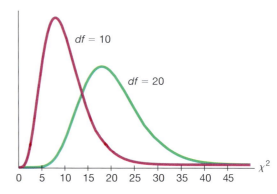

FIGURE 8-9   Chi-Square Distribution for df = 10 and df = 20

EXAMPLE 1 *P*-Value Method: Do Supermodel Heights Vary Less?

Listed below are the heights (cm) for the simple random sample of female supermodels Lima, Bundchen, Ambrosio, Ebanks, Iman, Rubik, Kurkova, Kerr, Kroes, Swanepoel, Prinsloo, Hosk, Kloss, Robinson, Heatherton, and Refaeli. Use a 0.01 significance level to test the claim that supermodels have heights with a standard deviation that is less than $\sigma = 7.5$ cm for the population of women. Does it appear that heights of supermodels vary less than heights of women from the population?

    178   177   176   174   175   178   175   178   178   177   180   176   180   178   180   176

SOLUTION

REQUIREMENT CHECK   (1) The sample is a simple random sample. (2) In checking for normality, we see that the sample has no outliers, the accompanying normal quantile plot shows points that are reasonably close to a straight-line pattern, and there is no other pattern that is not a straight line. Both requirements are satisfied. ✅

Technology Technology capable of conducting this test will typically display the *P*-value. StatCrunch can be used as described at the end of this section, and the result will be as shown in the accompanying display. (Instead of using the assumed value of σ for H_0 and H_1, StatCrunch uses σ^2. For the null hypothesis, $\sigma = 7.5$ is equivalent to $\sigma^2 = 7.5^2 = 56.25$.) The display shows that the test statistic is $X^2 = 0.907$ (rounded) and the *P*-value is less than 0.0001.

Statdisk

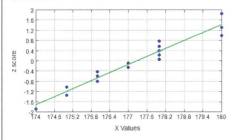

StatCrunch

Hypothesis test results:
σ^2 : Variance of population
$H_0 : \sigma^2 = 56.25$
$H_A : \sigma^2 < 56.25$

Variance	Sample Var.	DF	Chi-Square Stat	P-value
σ^2	3.4000004	15	0.90666677	<0.0001

Step 1: The claim that "the standard deviation is less than 7.5 cm" is expressed in symbolic form as $\sigma < 7.5$ cm.

Step 2: If the original claim is false, then $\sigma \geq 7.5$ cm.

Step 3: The expression $\sigma < 7.5$ cm does not contain equality, so it becomes the alternative hypothesis. The null hypothesis is the statement that $\sigma = 7.5$ cm.

$$H_0: \sigma = 7.5 \text{ cm}$$

$$H_1: \sigma < 7.5 \text{ cm} \quad \text{(original claim)}$$

Step 4: The significance level is $\alpha = 0.01$.

Step 5: Because the claim is made about σ, we use the X^2 (chi-square) distribution.

Step 6: The StatCrunch display shows the test statistic of $X^2 = 0.907$ (rounded) and it shows that the *P*-value is less than 0.0001.

Step 7: Because the *P*-value is less than the significance level of $\alpha = 0.01$, we reject H_0.

continued

INTERPRETATION

Step 8: There is sufficient evidence to support the claim that female supermodels have heights with a standard deviation that is less than 7.5 cm for the population of women. It appears that heights of supermodels do vary less than heights of women in the general population.

YOUR TURN Do Exercise 7 "Body Temperature."

Critical Value Method

Technology typically provides a *P*-value, so the *P*-value method is used. If technology is not available, the *P*-value method of testing hypotheses is a bit challenging because Table A-4 allows us to find only a range of values for the *P*-value. Instead, we could use the critical value method. Steps 1 through 5 in Example 1 would be the same. In Step 6, the test statistic is calculated by using $\sigma = 7.5$ cm (as assumed in the null hypothesis in Example 1), $n = 16$, and $s = 1.843909$ cm, which is the unrounded standard deviation computed from the original list of 16 heights. We get this test statistic:

$$\chi^2 = \frac{(n-1)s^2}{\sigma^2} = \frac{(16-1)(1.843909)^2}{7.5^2} = 0.907$$

The critical value of $\chi^2 = 5.229$ is found from Table A-4, and it corresponds to 15 degrees of freedom and an "area to the right" of 0.99 (based on the significance level of 0.01 for a left-tailed test). See Figure 8-10. In Step 7 we reject the null hypothesis because the test statistic of $\chi^2 = 0.907$ falls in the critical region, as shown in Figure 8-10. We conclude that there is sufficient evidence to support the claim that supermodels have heights with a standard deviation that is less than 7.5 cm for the population of women.

Confidence Interval Method

As stated earlier, when testing claims about σ or σ^2, the *P*-value method, the critical value method, and the confidence interval method are all equivalent in the sense that they will always lead to the same conclusion. See Example 2.

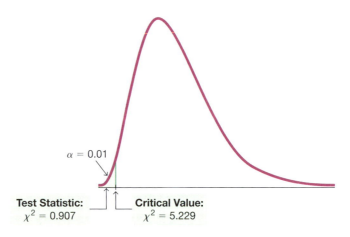

$\alpha = 0.01$

Test Statistic: $\chi^2 = 0.907$ Critical Value: $\chi^2 = 5.229$

FIGURE 8-10   Testing the Claim That $\sigma < 7.5$ cm

EXAMPLE 2   Supermodel Heights: Confidence Interval Method

Repeat the hypothesis test in Example 1 by constructing a suitable confidence interval.

SOLUTION

First, we should be careful to select the correct confidence level. Because the hypothesis test is left-tailed and the significance level is 0.01, we should use a confidence level of 98%, or 0.98. (See Table 8-1 on page 360 for help in selecting the correct confidence level.)

Using the methods described in Section 7-3, we can use the sample data listed in Example 1 to construct a 98% confidence interval estimate of σ. We use $n = 16$, $s = 1.843909$ cm, $\chi_L^2 = 5.229$, and $\chi_R^2 = 30.578$. (The critical values χ_L^2 and χ_R^2 are found in Table A-4. Use the row with df $= n - 1 = 15$. The 0.98 confidence level corresponds to $\alpha = 0.02$, and we divide that area of 0.02 equally between the two tails so that the areas to the *right* of the critical values are 0.99 and 0.01. Refer to Table A-4 and use the columns with areas of 0.99 and 0.01 and use the 15th row.)

$$\sqrt{\frac{(n-1)s^2}{\chi_R^2}} < \sigma < \sqrt{\frac{(n-1)s^2}{\chi_L^2}}$$

$$\sqrt{\frac{(16-1)(1.843909^2)}{30.578}} < \sigma < \sqrt{\frac{(16-1)(1.843909^2)}{5.229}}$$

$$1.3 \text{ cm} < \sigma < 3.1 \text{ cm}$$

With this confidence interval, we can support the claim that $\sigma < 7.5$ cm because all values of the confidence interval are less than 7.5 cm. We reach the same conclusion found with the *P*-value method and the critical value method.

Alternative Method Used When Population Is Not Normal

The methods of this section include two requirements: (1) The sample is a simple random sample; (2) the population is normally distributed. If sample data are not collected in a random manner, the methods of this section do not apply. If the sample appears to be from a population not having a normal distribution, we could use the confidence interval method of testing hypotheses, but obtain the confidence interval using bootstrap resampling, as described in Section 7-4. Be careful to use the appropriate confidence level, as indicated by Table 8-1 on page 360. Reject the null hypothesis if the confidence interval limits do not contain the value of the mean claimed in the null hypothesis. See the Technology Project near the end of this chapter.

TECH CENTER

Hypothesis Test: Standard Deviation or Variance
Access tech supplements, videos, and data sets at **www.TriolaStats.com**

Statdisk

1. Click **Analysis** in the top menu.
2. Select **Hypothesis Testing** from the dropdown menu and select **Standard Deviation One Sample** from the submenu.
3. Select the desired format for the *Alternative Hypothesis,* then enter the desired significance level and claimed standard deviation (from the null hypothesis).
4. *Using Summary Statistics:* Select the **Use Summary Statistics** tab and enter the sample size and sample standard deviation.

   Using sample data: Select the **Use Data** tab and select the desired data column.
5. Click **Evaluate.**

continued

Hypothesis Test: Standard Deviation or Variance

Access tech supplements, videos, and data sets at **www.TriolaStats.com**

Minitab

1. Click **Stat** in the top menu.
2. Select **Basic Statistics** from the dropdown menu and select **1 Variance** from the submenu.
3. *Using Summary Statistics:* Select **Sample standard deviation** (or **variance**) from the dropdown menu and enter the sample size and sample standard deviation (or variance).

   Using Sample Data: Select **One or more samples, each in a column** from the dropdown menu and select the desired data column.
4. Check the **Perform Hypothesis Test** box and enter the standard deviation (or variance) used in the null hypothesis.
5. Click the **Options** button and enter the confidence level. (Enter 95.0 for a significance level of 0.05.) For *Alternative Hypothesis* select the format used for the alternative hypothesis.
6. Click **OK** twice.

TIP: Another procedure is to click on **Assistant** in the top menu, then select **Hypothesis Tests** and **1-Sample Standard Deviation.** Complete the dialog box to get results, including *P*-value and other helpful information.

StatCrunch

1. Click **Stat** in the top menu.
2. Select **Variance Stats** from the dropdown menu, then select **One Sample** from the submenu.
3. *Using Summary Statistics:* Select **With Summary** from the submenu and enter the sample variance and sample size.

   Using Sample Data: Select **With Data** from the submenu and select the desired data column.
4. Select **Hypothesis test for** σ^2 and for H_0 enter the claimed value of the population variance. For H_A select the format used for the alternative hypothesis.
5. Click **Compute!**

TI-83/84 Plus Calculator

1. Download and install the Michael Lloyd programs *S2TEST* and *ZZINEWT* (available at www.TriolaStats.com) on your TI-83/84 Plus Calculator.
2. Press the **PRGM** button, then select **S2TEST** from the menu and press **ENTER** twice.
3. Enter the population variance σx^2, sample variance Sx^2, and sample size n. Press **ENTER** after each entry.
4. For SIGMA^2 select the desired format for the alternative hypothesis ($\neq$ for two-tailed test, $<$ for left-tailed test, $>$ for right-tailed test).
5. Press **ENTER** and the test statistic and the *P*-value will be provided.

Excel

XLSTAT Add-In (Required)

Requires original sample data; does not work with summary data.

1. Click on the **XLSTAT** tab in the Ribbon and then click **Parametric Tests.**
2. Select **One-sample variance test** from the dropdown menu.
3. Under *Data* enter the range of cells containing the sample data. For *Data format* select **One column per sample.** If the first row of data contains a label, also check the **Column labels** box.
4. For *Range,* enter a cell location such as D5 where the results will be displayed.
5. Click the **Options** tab.
6. Under *Alternative hypothesis* select the desired format ($\neq$ for two-tailed test, $<$ for left-tailed test, $>$ for right-tailed test).
7. For *Theoretical Variance,* enter the claimed value of the population variance. Enter the desired significance level (enter **5** for a significance level of 0.05).
8. Click **OK** to display the results. The test statistic is identified as Chi-square. The *P*-value is also displayed.

TIP: The above procedure is based on testing a claim about a population *variance;* to test a claim about a population standard deviation, use the same procedure and enter σ^2 for the *Theoretical Variance.*

8-4   Basic Skills and Concepts

Statistical Literacy and Critical Thinking

1. Cans of Coke Data Set 26 "Cola Weights and Volumes" in Appendix B includes volumes (oz) of regular Coke. Based on that data set, assume that the cans are produced so that the volumes have a standard deviation of 0.115 oz. A new filling process is being tested for filling cans of cola, and a random sample of volumes is listed below. The sample has these summary statistics: $n = 10$, $\bar{x} = 12.0004$ oz, $s = 0.2684$ oz. If we want to use the sample data to test the claim that the sample is from a population with a standard deviation equal to 0.115 oz, what requirements must be satisfied? How does the normality requirement for a hypothesis test of a claim about a standard deviation differ from the normality requirement for a hypothesis test of a claim about a mean?

   12.078   11.851   12.108   11.760   12.142   11.779   12.397   11.504   12.147   12.238

2. Cans of Coke Use the data and the claim given in Exercise 1 to identify the null and alternative hypotheses and the test statistic. What is the sampling distribution of the test statistic?

3. Cans of Coke For the sample data from Exercise 1, we get "P-value < 0.01" when testing the claim that the new filling process results in volumes with the same standard deviation of 0.115 oz.

a. What should we conclude about the null hypothesis?

b. What should we conclude about the original claim?

c. What do these results suggest about the new filling process?

4. Cans of Coke: Confidence Interval If we use the data given in Exercise 1, we get this 95% confidence interval estimate of the standard deviation of volumes with the new filling process: 0.1846 oz $< \sigma <$ 0.4900 oz. What does this confidence interval tell us about the new filling process?

Testing Claims About Variation.   *In Exercises 5–16, test the given claim. Identify the null hypothesis, alternative hypothesis, test statistic, P-value, or critical value(s), then state the conclusion about the null hypothesis, as well as the final conclusion that addresses the original claim. Assume that a simple random sample is selected from a normally distributed population.*

5. Pulse Rates of Men A simple random sample of 153 men results in a standard deviation of 11.3 beats per minute (based on Data Set 1 "Body Data" in Appendix B). The normal range of pulse rates of adults is typically given as 60 to 100 beats per minute. If the range rule of thumb is applied to that normal range, the result is a standard deviation of 10 beats per minute. Use the sample results with a 0.05 significance level to test the claim that pulse rates of men have a standard deviation equal to 10 beats per minute; see the accompanying StatCrunch display for this test. What do the results indicate about the effectiveness of using the range rule of thumb with the "normal range" from 60 to 100 beats per minute for estimating σ in this case?

> **One sample variance hypothesis test:**
> σ^2 : Variance of variable
> $H_0 : \sigma^2 = 100$
> $H_A : \sigma^2 \neq 100$
>
> **Hypothesis test results:**
>
> | Variable | Sample Var. | DF | Chi-Square Stat | P-value |
> |----------|-------------|-----|-----------------|---------|
> | PULSE | 128.40282 | 152 | 195.17229 | 0.0208 |

6. Pulse Rates of Women Repeat the preceding exercise using the pulse rates of women listed in Data Set 1 "Body Data" in Appendix B. For the sample of pulse rates of women, $n = 147$ and $s = 12.5$. See the accompanying JMP display that results from using the original list of pulse rates instead of the summary statistics. (*Hint:* The bottom three rows of the display provide P-values for a two-tailed test, a left-tailed test, and a right-tailed test, respectively.) What do the results indicate about the effectiveness of using the range rule of thumb with the "normal range" from 60 to 100 beats per minute for estimating σ in this case?

> ⊿ **Test Standard Deviation**
>
> | | |
> |---|---|
> | Hypothesized Value | 10 |
> | Actual Estimate | 12.5436 |
> | DF | 146 |
>
> | Test | ChiSquare |
> |---|---|
> | Test Statistic | 229.7176 |
> | Min PValue | <.0001* |
> | Prob < ChiSq | 1.0000 |
> | Prob > ChiSq | <.0001* |

7. Body Temperature Example 5 in Section 8-3 involved a test of the claim that humans have body temperatures with a mean equal to 98.6°F. The sample of 106 body temperatures has a standard deviation of 0.62°F. The conclusion in that example would change if the sample standard deviation s were 2.08°F or greater. Use a 0.01 significance level to test the claim that the sample of 106 body temperatures is from a population with a standard deviation less than 2.08°F. What does the result tell us about the validity of the hypothesis test in Example 5 in Section 8-3?

8. Birth Weights A simple random sample of birth weights of 30 girls has a standard deviation of 829.5 hg. Use a 0.01 significance level to test the claim that birth weights of girls have the same standard deviation as birth weights of boys, which is 660.2 hg (based on Data Set 4 "Births" in Appendix B).

9. Cola Cans A random sample of 20 aluminum cola cans with thickness 0.0109 in. is selected and the axial loads are measured and the standard deviation is 18.6 lb. The axial load is the pressure applied to the top that causes the can to crush. Use a 0.05 significance level to test the claim that cans with thickness 0.0109 in. have axial loads with the same standard deviation as the axial loads of cans that are 0.0111 in. thick. The thicker cans have axial loads with a standard deviation of 27.8 lb (based on Data Set 30 "Aluminum Cans" in Appendix B). Does the thickness of the cans appear to affect the variation of the axial loads?

10. Statistics Test Scores Tests in the author's statistics classes have scores with a standard deviation equal to 14.1. One of his last classes had 27 test scores with a standard deviation of 9.3. Use a 0.01 significance level to test the claim that this class has less variation than other past classes. Does a lower standard deviation suggest that this last class is doing better?

11. Coffee Vending Machines The Brazil vending machine dispenses coffee, and a random sample of 27 filled cups have contents with a mean of 7.14 oz and a standard deviation of 0.17 oz. Use a 0.05 significance level to test the claim that the machine dispenses amounts with a standard deviation greater than the standard deviation of 0.15 oz specified in the machine design.

12. Spoken Words Couples were recruited for a study of how many words people speak in a day. A random sample of 56 males resulted in a mean of 16,576 words and a standard deviation of 7871 words. Use a 0.01 significance level to test the claim that males have a standard deviation that is greater than the standard deviation of 7460 words for females (based on Data Set 24 "Word Counts").

13. Aircraft Altimeters The Skytek Avionics company uses a new production method to manufacture aircraft altimeters. A simple random sample of new altimeters resulted in the errors listed below. Use a 0.05 level of significance to test the claim that the new production method has errors with a standard deviation greater than 32.2 ft, which was the standard deviation for the old production method. If it appears that the standard deviation is greater, does the new production method appear to be better or worse than the old method? Should the company take any action?

$$-42 \quad 78 \quad -22 \quad -72 \quad -45 \quad 15 \quad 17 \quad 51 \quad -5 \quad -53 \quad -9 \quad -109$$

14. Bank Lines The Jefferson Valley Bank once had a separate customer waiting line at each teller window, but it now has a single waiting line that feeds the teller windows as vacancies occur. The standard deviation of customer waiting times with the old multiple-line configuration was 1.8 min. Listed below is a simple random sample of waiting times (minutes) with the single waiting line. Use a 0.05 significance level to test the claim that with a single waiting line, the waiting times have a standard deviation less than 1.8 min. What improvement occurred when banks changed from multiple waiting lines to a single waiting line?

$$6.5 \quad 6.6 \quad 6.7 \quad 6.8 \quad 7.1 \quad 7.3 \quad 7.4 \quad 7.7 \quad 7.7 \quad 7.7$$

15. Fast Food Drive-Through Service Times Listed below are drive-through service times (seconds) recorded at McDonald's during dinner times (from Data Set 25 "Fast Food" in Appendix B). Assuming that dinner service times at Wendy's have standard deviation $\sigma = 55.93$ sec, use a 0.01 significance level to test the claim that service times at McDonald's have the same variation as service times at Wendy's. Should McDonald's take any action?

$$121 \quad 119 \quad 146 \quad 266 \quad 333 \quad 308 \quad 333 \quad 308$$

16. Mint Specs Listed below are weights (grams) from a simple random sample of "wheat" pennies (from Data Set 29 "Coin Weights" in Appendix B). U.S. Mint specifications now require a standard deviation of 0.0230 g for weights of pennies. Use a 0.01 significance level to test the claim that wheat pennies are manufactured so that their weights have a standard deviation equal to 0.0230 g. Does the Mint specification appear to be met?

       2.5024   2.5298   2.4998   2.4823   2.5163   2.5222   2.4900   2.4907   2.5017

Large Data Sets from Appendix B. *In Exercises 17 and 18, use the data set from Appendix B to test the given claim. Identify the null hypothesis, alternative hypothesis, test statistic, P-value, or critical value(s), then state the conclusion about the null hypothesis, as well as the final conclusion that addresses the original claim.*

17. Fast Food Drive-Through Service Times Repeat Exercise 15 using the 50 service times for McDonald's during dinner times in Data Set 25 "Fast Food."

18. Mint Specs Repeat Exercise 16 using the weights of the 37 post-1983 pennies included in Data Set 29 "Coin Weights" in Appendix B.

8-4   Beyond the Basics

19. Finding Critical Values of χ^2 For large numbers of degrees of freedom, we can approximate critical values of χ^2 as follows:

$$\chi^2 = \frac{1}{2}(z + \sqrt{2k - 1})^2$$

Here k is the number of degrees of freedom and z is the critical value(s) found from technology or Table A-2. In Exercise 12 "Spoken Words" we have df $= 55$, so Table A-4 does not list an exact critical value. If we want to approximate a critical value of χ^2 in the right-tailed hypothesis test with $\alpha = 0.01$ and a sample size of 56, we let $k = 55$ with $z = 2.33$ (or the more accurate value of $z = 2.326348$ found from technology). Use this approximation to estimate the critical value of χ^2 for Exercise 12. How close is it to the critical value of $\chi^2 = 82.292$ obtained by using Statdisk and Minitab?

20. Finding Critical Values of χ^2 Repeat Exercise 19 using this approximation (with k and z as described in Exercise 19):

$$\chi^2 = k\left(1 - \frac{2}{9k} + z\sqrt{\frac{2}{9k}}\right)^3$$

Chapter Quick Quiz

1. Distributions Using the methods of this chapter, identify the distribution that should be used for testing a claim about the given population parameter.

a. Mean

b. Proportion

c. Standard deviation

2. Tails Determine whether the given claim involves a hypothesis test that is left-tailed, two-tailed, or right-tailed.

a. $p \neq 0.5$

b. $\mu < 98.6°F$

c. $\sigma > 15$ cm

3. Instagram Poll In a Pew Research Center poll of Internet users aged 18–29, 53% said that they use Instagram. We want to use a 0.05 significance level to test the claim that the majority of Internet users aged 18–29 use Instagram.

a. Identify the null and alternative hypotheses.

b. Using a sample size of 532, find the value of the test statistic.

c. Technology is used to find that the P-value for the test is 0.0827. What should we conclude about the null hypothesis?

d. What should we conclude about the original claim?

4. *P*-Value Find the P-value in a test of the claim that the mean annual income of a CIA agent is greater than $81,623 (based on data from payscale.com) given that the test statistic is $t = 1.304$ for a sample of 40 CIA agents.

5. Conclusions True or false: In hypothesis testing, it is *never* valid to form a conclusion of supporting the null hypothesis.

6. Conclusions True or false: The conclusion of "fail to reject the null hypothesis" has exactly the same meaning as "accept the null hypothesis."

7. Uncertainty True or false: If correct methods of hypothesis testing are used with a large simple random sample that satisfies the test requirements, the conclusion will always be true.

8. Chi-Square Test In a test of the claim that $\sigma = 15$ for the population of IQ scores of professional athletes, we find that the rightmost critical value is $\chi_R^2 = 40.646$. Is the leftmost critical χ_L^2 value equal to -40.646?

9. Robust Explain what is meant by the statements that the t test for a claim about μ is robust, but the χ^2 test for a claim about σ^2 is not robust.

10. Equivalent Methods Which of the following statements are true?

a. When testing a claim about a population mean μ, the P-value method, critical value method, and confidence interval method are all equivalent in the sense that they always yield the same conclusions.

b. When testing a claim about a population proportion p, the P-value method, critical value method, and confidence interval method are all equivalent in the sense that they always yield the same conclusions.

c. When testing a claim about any population parameter, the P-value method, critical value method, and confidence interval method are all equivalent in the sense that they always yield the same conclusions.

Review Exercises

1. True/False Characterize each of the following statements as being true or false.

a. In a hypothesis test, a very high P-value indicates strong support of the alternative hypothesis.

b. The Student t distribution can be used to test a claim about a population mean whenever the sample data are randomly selected from a normally distributed population.

c. When using a χ^2 distribution to test a claim about a population standard deviation, there is a very loose requirement that the sample data are from a population having a normal distribution.

d. When conducting a hypothesis test about the claimed proportion of adults who have current passports, the problems with a convenience sample can be overcome by using a larger sample size.

e. When repeating the same hypothesis test with different random samples of the same size, the conclusions will all be the same.

2. Politics A county clerk in Essex County, New Jersey, selected candidates for positions on election ballots. Democrats were selected first in 40 of 41 ballots. Because he was supposed to use a method of random selection, Republicans claimed that instead of using randomness, he used a method that favored Democrats. Use a 0.01 significance level to test the claim that the ballot selection method favors Democrats.

3. Oscar-Winning Actresses Data Set 14 "Oscar Winner Age" in Appendix B lists ages of actresses when they won Oscars, and the summary statistics are $n = 87$, $\bar{x} = 36.2$ years, and $s = 11.5$ years. Use a 0.01 significance level to test the claim that the mean age of actresses when they win Oscars is greater than 30 years.

4. Red Blood Cell Count A simple random sample of 40 adult males is obtained, and the red blood cell count (in cells per microliter) is measured for each of them, with these results: $n = 40$, $\bar{x} = 4.932$ million cells per microliter, $s = 0.504$ million cells per microliter (from Data Set 1 "Body Data" in Appendix B). Use a 0.01 significance level to test the claim that the sample is from a population with a mean less than 5.4 million cells per microliter, which is often used as the upper limit of the range of normal values. Does the result suggest that each of the 40 males has a red blood cell count below 5.4 million cells per microliter?

5. Perception and Reality In a presidential election, 308 out of 611 voters surveyed said that they voted for the candidate who won (based on data from ICR Survey Research Group). Use a 0.05 significance level to test the claim that among all voters, the percentage who believe that they voted for the winning candidate is equal to 43%, which is the actual percentage of votes for the winning candidate. What does the result suggest about voter perceptions?

6. BMI for Miss America A claimed trend of thinner Miss America winners has generated charges that the contest encourages unhealthy diet habits among young women. Listed below are body mass indexes (BMI) for recent Miss America winners. Use a 0.01 significance level to test the claim that recent winners are from a population with a mean BMI less than 20.16, which was the BMI for winners from the 1920s and 1930s. Given that BMI is a measure of the relative amounts of body fat and height, do recent winners appear to be significantly smaller than those from the 1920s and 1930s?

$$19.5 \quad 20.3 \quad 19.6 \quad 20.2 \quad 17.8 \quad 17.9 \quad 19.1 \quad 18.8 \quad 17.6 \quad 16.8$$

7. BMI for Miss America Use the same BMI indexes given in Exercise 6. Use a 0.01 significance level to test the claim that recent Miss America winners are from a population with a standard deviation equal to 1.34, which was the standard deviation of BMI for winners from the 1920s and 1930s. Do recent winners appear to have variation that is different from that of the 1920s and 1930s?

8. Type I Error and Type II Error

a. In general, what is a type I error? In general, what is a type II error?

b. For the hypothesis test in Exercise 6 "BMI for Miss America," write a statement that would be a type I error, and write another statement that would be a type II error.

Cumulative Review Exercises

1. Lightning Deaths Listed below are the numbers of deaths from lightning strikes in the United States each year for a sequence of 14 recent and consecutive years. Find the values of the indicated statistics.

$$51 \quad 44 \quad 51 \quad 43 \quad 32 \quad 38 \quad 48 \quad 45 \quad 27 \quad 34 \quad 29 \quad 26 \quad 28 \quad 23$$

a. Mean   **b.** Median   **c.** Standard deviation   **d.** Variance   **e.** Range

f. What important feature of the data is not revealed from an examination of the statistics, and what tool would be helpful in revealing it?

2. Lightning Deaths Refer to the sample data in Cumulative Review Exercise 1.

a. What is the level of measurement of the data (nominal, ordinal, interval, ratio)?

b. Are the values discrete or continuous?

c. Are the data categorical or quantitative?

d. Is the sample a simple random sample?

3. Confidence Interval for Lightning Deaths Use the sample values given in Cumulative Review Exercise 1 to construct a 99% confidence interval estimate of the population mean. Assume that the population has a normal distribution. Write a brief statement that interprets the confidence interval.

4. Hypothesis Test for Lightning Deaths Refer to the sample data given in Cumulative Review Exercise 1 and consider those data to be a random sample of annual lightning deaths from recent years. Use those data with a 0.01 significance level to test the claim that the mean number of annual lightning deaths is less than the mean of 72.6 deaths from the 1980s. If the mean is now lower than in the past, identify one of the several factors that could explain the decline.

5. Lightning Deaths The accompanying bar chart shows the numbers of lightning strike deaths broken down by gender for a recent period of nine years. What is wrong with the graph?

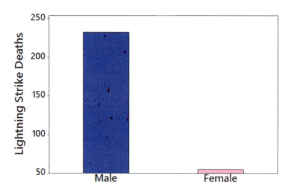

6. Lightning Deaths The graph in Cumulative Review Exercise 5 was created by using data consisting of 232 male deaths from lightning strikes and 55 female deaths from lightning strikes. Assume that these data are randomly selected lightning deaths and proceed to test the claim that the proportion of male deaths is greater than $1/2$. Use a 0.01 significance level. Any explanation for the result?

7. Lightning Deaths The graph in Cumulative Review Exercise 5 was created by using data consisting of 232 male deaths from lightning strikes and 55 female deaths from lightning strikes. Assume that these data are randomly selected lightning deaths and proceed to construct a 95% confidence interval estimate of the proportion of males among all lightning deaths. Based on the result, does it seem feasible that males and females have equal chances of being killed by lightning?

8. Lightning Deaths Based on the results given in Cumulative Review Exercise 6, assume that for a randomly selected lightning death, there is a 0.8 probability that the victim is a male.

a. Find the probability that three random people killed by lightning strikes are all males.

b. Find the probability that three random people killed by lightning strikes are all females.

c. Find the probability that among three people killed by lightning strikes, at least one is a male.

d. If five people killed by lightning strikes are randomly selected, find the probability that exactly three of them are males.

e. A study involves random selection of different groups of 50 people killed by lightning strikes. For those groups, find the mean and standard deviation for the numbers of male victims.

f. For the same groups described in part (e), would 46 be a significantly high number of males in a group? Explain.

Technology Project

Bootstrapping and Robustness Consider the probability distribution defined by the formula $P(x) = \dfrac{3x^2}{1000}$ where x can be any value between 0 and 10 inclusive (not just integers). The accompanying graph of this probability distribution shows that its shape is very far from the bell shape of a normal distribution. This probability distribution has parameters $\mu = 7.5$ and $\sigma = 1.93649$. Listed below is a simple random sample of values from this distribution, and the normal quantile plot for this sample is shown. Given the very non-normal shape of the distribution, it is not surprising to see the normal quantile plot with points that are far from a straight-line pattern, confirming that the sample does not appear to be from a normally distributed population.

$$8.69 \quad 2.03 \quad 9.09 \quad 7.15 \quad 9.05 \quad 9.40 \quad 6.30 \quad 7.89 \quad 7.98 \quad 7.67$$

$$7.77 \quad 7.17 \quad 8.86 \quad 8.29 \quad 9.21 \quad 7.80 \quad 7.70 \quad 8.12 \quad 9.11 \quad 7.64$$

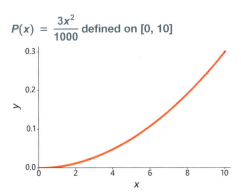

$P(x) = \dfrac{3x^2}{1000}$ defined on [0, 10]

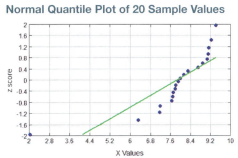

Normal Quantile Plot of 20 Sample Values

a. Mean Test the claim that the 20 given sample values are from a population having a mean equal to 7.5, which is the known population mean. Because the sample is not from a normally distributed population and because $n = 20$ does not satisfy the requirement of $n > 30$, we should not use the methods of Section 8-3. Instead, test the claim by using the confidence interval method based on a bootstrap sample of size 1000. (See Section 7-4.) Use a 0.05 significance level. Does the bootstrap confidence interval contain the known population mean of 7.5? Is the bootstrap method effective for this test? What happens if we conduct this test by throwing all caution to the wind and constructing the 95% confidence interval by using the t distribution as described in Section 7-2?

b. Standard Deviation Test the claim that the 20 sample values are from a population with a standard deviation equal to 1.93649, which is the known population standard deviation. Use the confidence interval method based on a bootstrap sample of size 1000. (See Section 7-4.) Use a 0.05 significance level. Does the bootstrap confidence interval contain the known population standard deviation of 1.93649? Is the bootstrap method effective for this test? What happens if we conduct the test by throwing all caution to the wind and constructing the 95% confidence interval by using the χ^2 distribution as described in Section 7-3?

FROM DATA TO DECISION

Critical Thinking: Testing the Salk Vaccine

The largest health experiment ever conducted involved a test of the Salk vaccine designed to protect children from the devastating effects of polio. The test included 201,229 children who were given the Salk vaccine, and 33 of them developed polio. The claim that the Salk vaccine is effective is equivalent to the claim that the proportion of vaccinated children who develop polio is less than 0.0000573, which was the rate of polio among children not given the Salk vaccine. (*Note:* The actual Salk vaccine experiment involved another group of 200,745 children who were injected with an ineffective salt solution instead of the Salk vaccine. This study design with a treatment group and placebo group is very common and very effective. Methods for comparing two proportions are presented in Chapter 9.)

continued

Analyzing the Results

a. Test the given claim using a 0.05 significance level. Does the Salk vaccine appear to be effective?

b. For the hypothesis test from part (a), consider the following two errors:

- Concluding that the Salk vaccine is effective when it is not effective.

- Concluding that the Salk vaccine is not effective when it is effective.

Determine which of the above two errors is a type I error and determine which is a type II error. Which error would have worse consequences? How could the hypothesis test be conducted in order to reduce the chance of making the more serious error?

Cooperative Group Activities

1. Out-of-class activity Here is the breakdown of the most common car colors from PPG Industries: 23% are white, 18% are black, 16% are gray, 15% are silver, and 10% are red. After selecting one of the given colors, groups of three or four students should go to the college parking lot and randomly select cars to test the claim that the percentage for the chosen color is as claimed.

2. Out-of-class activity In the United States, 40% of us have brown eyes, according to Dr. P. Sorita Soni at Indiana University. Groups of three or four students should randomly select people and identify the color of their eyes. The claim that 40% of us have brown eyes can then be tested.

3. In-class activity Without using any measuring device, each student should draw a line believed to be 3 in. long and another line believed to be 3 cm long. Then use rulers to measure and record the lengths of the lines drawn. Find the means and standard deviations of the two sets of lengths. Test the claim that the lines estimated to be 3 in. have a mean length that is equal to 3 in. Test the claim that the lines estimated to be 3 cm have a mean length that is equal to 3 cm. Compare the results. Do the estimates of the 3-in. line appear to be more accurate than those for the 3-cm line? What do these results suggest?

4. In-class activity Assume that a method of gender selection can affect the probability of a baby being a girl so that the probability becomes 1/4. Each student should simulate 20 births by drawing 20 cards from a shuffled deck. Replace each card after it has been drawn, then reshuffle. Consider the hearts to be girls and consider all other cards to be boys. After making 20 selections and recording the "genders" of the babies, use a 0.10 significance level to test the claim that the proportion of girls is equal to 1/4. How many students are expected to get results leading to the wrong conclusion that the proportion is not 1/4? How does that relate to the probability of a type I error? Does this procedure appear to be effective in identifying the effectiveness of the gender selection method? (If decks of cards are not available, use some other way to simulate the births, such as using the random number generator on a calculator or using digits from phone numbers or Social Security numbers.)

5. Out-of-class activity Groups of three or four students should go to the library and collect a sample consisting of the ages of books (based on copyright dates). Plan and describe the sampling plan, execute the sampling procedure, and then use the results to test the claim that the mean age of books in the library is greater than 20 years.

6. In-class activity Each student should write an estimate of the age of the current president of the United States. All estimates should be collected, and the sample mean and standard deviation should be calculated. Then test the hypothesis that the mean of all such estimates is equal to the actual current age of the president.

7. In-class activity A class project should be designed to conduct a test in which each student is given a taste of Coke and a taste of Pepsi. The student is then asked to identify which sample is Coke. After all of the results are collected, test the claim that the success rate is better than the rate that would be expected with random guesses.

8. In-class activity Each student should estimate the length of the classroom. The values should be based on visual estimates, with no actual measurements being taken. After the estimates have been collected, measure the length of the room, then test the claim that the sample mean is equal to the actual length of the classroom. Is there a "collective wisdom," whereby the class mean is approximately equal to the actual room length?

9. Out-of-class activity Using one wristwatch that is reasonably accurate, set the time to be exact. Visit www.time.gov to set the exact time. If you cannot set the time to the nearest second, record the error for the watch you are using. Now compare the time on this watch to the time on other watches that have not been set to the exact time. Record the errors with negative signs for watches that are ahead of the actual time and positive signs for those watches that are behind the actual time. Use the data to test the claim that the mean error of all wristwatches is equal to 0. Do we collectively run on time, or are we early or late? Also test the claim that the standard deviation of errors is less than 1 min. What are the practical implications of a standard deviation that is excessively large?

10. In-class activity In a group of three or four people, conduct an extrasensory perception (ESP) experiment by selecting one of the group members as the subject. Draw a circle on one small piece of paper and draw a square on another sheet of the same size. Repeat this experiment 20 times: Randomly select the circle or the square and place it in the subject's hand behind his or her back so that it cannot be seen, then ask the subject to identify the shape (without seeing it); record whether the response is correct. Test the claim that the subject has ESP because the proportion of correct responses is significantly greater than 0.5.

11. In-class activity After dividing into groups of between 10 and 20 people, each group member should record the number of heartbeats in a minute. After calculating the sample mean and standard deviation, each group should proceed to test the claim that the mean is greater than 48 beats per minute, which is the author's result. (When people exercise, they tend to have lower pulse rates, and the author runs 5 miles a few times each week. What a guy!)

12. Out-of-class activity In groups of three or four, collect data to determine whether subjects have a Facebook page, then combine the results and test the claim that more than 3/4 of students have a Facebook page.

13. Out-of-class activity Each student should find an article in a professional journal that includes a hypothesis test of the type discussed in this chapter. Write a brief report describing the hypothesis test and its role in the context of the article.

9-1 Two Proportions

9-2 Two Means: Independent Samples

9-3 Two Dependent Samples (Matched Pairs)

9-4 Two Variances or Standard Deviations

INFERENCES FROM TWO SAMPLES

CHAPTER PROBLEM

Car License Plate Laws: Are They Enforced Equally in Different States?

Many drivers have learned that speed limit laws are not strictly enforced and enforcement seems to vary from state to state. Nobody is ticketed for going three miles over the speed limit. Some police seem to operate with the rule of "9 you're fine, 10 you're mine." In practice, speed limit laws are often treated more as guidelines.

When the author recently purchased a new car, he was surprised when the dealer told him that he would be given two license plates, but only the rear plate would be installed. The salesman said that state law requires license plates on the front and rear of the car, but police don't bother to ticket drivers with only a rear license plate. Are the license plate laws enforced as laws or are they treated as guidelines? Is adherence to the laws different in different states? To help answer these questions, the author collected the sample data shown in Table 9-1. The cars included in Table 9-1 are non-commercial

TABLE 9-1 Numbers of Cars with Rear and Front License Plates

	Connecticut	New York
Cars with rear license plate only	239	9
Cars with front and rear license plates	1810	541
Total	**2049**	**550**

passenger cars. Connecticut and New Year are contiguous states, both having laws that require front and rear license plates.

Analysis The proportion of Connecticut "illegal" cars with rear license plates only is 239/2049, or 11.7%. The proportion of New York "illegal" cars with rear license plates only is 9/550, or 1.6%. The sample percentages of 11.7% and 1.6% are obviously different, but are they *significantly* different? That question will be addressed in Section 9-1, which includes methods for comparing two sample proportions.

Because there are so many studies with so many different applications involving a comparison of *two* samples, the methods of this chapter are really, really important.

CHAPTER OBJECTIVES

Inferential statistics involves forming conclusions (or inferences) about a population parameter. Two major activities of inferential statistics are estimating values of population parameters using confidence intervals (as in Chapter 7) and testing claims made about population parameters (as in Chapter 8). Chapters 7 and 8 both involved methods for dealing with a sample from *one* population, and this chapter extends those methods to situations involving *two* populations. Here are the chapter objectives:

9-1 Two Proportions

- Conduct a formal hypothesis test of a claim made about two population proportions.
- Construct a confidence interval estimate of the difference between two population proportions.

9-2 Two Means: Independent Samples

- Distinguish between a situation involving two independent samples and a situation involving two samples that are not independent.
- Conduct a formal hypothesis test of a claim made about two means from independent populations.
- Construct a confidence interval estimate of the difference between two population means.

9-3 Two Dependent Samples (Matched Pairs)

- Identify sample data consisting of matched pairs.
- Conduct a formal hypothesis test of a claim made about the mean of the differences between matched pairs.
- Construct a confidence interval estimate of the mean difference between matched pairs.

9-4 Two Variances or Standard Deviations

- Develop the ability to conduct a formal hypothesis test of a claim made about two population standard deviations or variances.

9-1   Two Proportions

Key Concept In this section we present methods for (1) testing a claim made about two population proportions and (2) constructing a confidence interval estimate of the difference between two population proportions. The methods of this chapter can also be used with probabilities or the decimal equivalents of percentages.

KEY ELEMENTS

Inferences About Two Proportions

Objectives

1. **Hypothesis Test:** Conduct a hypothesis test of a claim about two population proportions.

2. **Confidence Interval:** Construct a confidence interval estimate of the difference between two population proportions.

Notation for Two Proportions

For population 1 we let

$p_1 = population$ proportion

$n_1 = $ size of the first sample

$x_1 = $ number of successes in the first sample

$\hat{p}_1 = \dfrac{x_1}{n_1}$ (*sample* proportion)

$\hat{q}_1 = 1 - \hat{p}_1$ (complement of $\hat{p}_1$)

The corresponding notations $p_2, n_2, x_2, \hat{p}_2,$ and $\hat{q}_2$ apply to population 2.

Pooled Sample Proportion

The **pooled sample proportion** is denoted by $\bar{p}$ and it combines the two sample proportions into one proportion, as shown here:

$$\bar{p} = \frac{x_1 + x_2}{n_1 + n_2}$$

$$\bar{q} = 1 - \bar{p}$$

Requirements

1. The sample proportions are from two simple random samples.

2. The two samples are *independent*. (Samples are *independent* if the sample values selected from one population are not related to or somehow naturally

paired or matched with the sample values from the other population.)

3. For each of the two samples, there are at least 5 successes and at least 5 failures. (That is, $n\hat{p} \geq 5$ and $n\hat{q} \geq 5$ for each of the two samples).

Test Statistic for Two Proportions (with H_0: $p_1 = p_2$)

$$z = \frac{(\hat{p}_1 - \hat{p}_2) - (p_1 - p_2)}{\sqrt{\dfrac{\bar{p}\bar{q}}{n_1} + \dfrac{\bar{p}\bar{q}}{n_2}}}$$ where $p_1 - p_2 = 0$ (assumed in the null hypothesis)

Where $\bar{p} = \dfrac{x_1 + x_2}{n_1 + n_2}$ (*pooled* sample proportion)   and $\bar{q} = 1 - \bar{p}$

P-Value:	*P*-values are automatically provided by technology. If technology is not available, use Table A-2 (standard normal distribution) and find the *P*-value using the procedure given in Figure 8-3 on page 364.
Critical Values:	Use Table A-2. (Based on the significance level α, find critical values by using the same procedures introduced in Section 8-1.)

Confidence Interval Estimate of $p_1 - p_2$

The confidence interval estimate of the difference $p_1 - p_2$ is

$$(\hat{p}_1 - \hat{p}_2) - E < (p_1 - p_2) < (\hat{p}_1 - \hat{p}_2) + E$$

where the margin of error E is given by $E = z_{\alpha/2}\sqrt{\dfrac{\hat{p}_1\hat{q}_1}{n_1} + \dfrac{\hat{p}_2\hat{q}_2}{n_2}}$.

Rounding: Round the confidence interval limits to three significant digits.

Equivalent Methods

When testing a claim about two population proportions:

- The *P*-value method and the critical value method are equivalent.

- The confidence interval method is *not* equivalent to the *P*-value method or the critical value method.

Recommendation: If you want to *test a claim* about two population proportions, use the *P*-value method or critical value method; if you want to *estimate* the difference between two population proportions, use the confidence interval method.

Hypothesis Tests

For tests of hypotheses made about two population proportions, we consider only tests having a null hypothesis of $p_1 = p_2$ (so the null hypothesis is H_0: $p_1 = p_2$). With the assumption that $p_1 = p_2$, the estimates of $\hat{p}_1$ and $\hat{p}_2$ are combined to provide the best estimate of the common value of $\hat{p}_1$ and $\hat{p}_2$, and that combined value is the pooled sample proportion $\bar{p}$ given in the preceding Key Elements box. The following example will help clarify the roles of $x_1, n_1, \hat{p}_1, \bar{p}$, and so on. Note that with the assumption of equal population proportions, the best estimate of the common population proportion is obtained by pooling both samples into one big sample, so that $\bar{p}$ is the estimator of the common population proportion.

P-Value Method

 **EXAMPLE 1** **Proportions of Cars with Rear License Plates Only: Are the Proportions the Same in Connecticut and New York?**

Table 9-1 in the Chapter Problem includes two sample proportions of cars with rear license plates only:

$$\text{Connecticut:}\quad \hat{p}_1 = 239/2049 = 0.117$$

$$\text{New York:}\quad \hat{p}_2 = 9/550 = 0.016$$

Use a 0.05 significance level and the *P*-value method to test the claim that Connecticut and New York have the same proportion of cars with rear license plates only.

continued

The Lead Margin of Error

Authors Stephen Ansolabehere and Thomas Belin wrote in their article "Poll Faulting" (*Chance* magazine) that "our greatest criticism of the reporting of poll results is with the margin of error of a single proportion (usually ±3%) when media attention is clearly drawn to the *lead* of one candidate." They point out that the lead is really the *difference* between two proportions ($p_1 - p_2$) and go on to explain how they developed the following rule of thumb: The lead is approximately $\sqrt{3}$ times larger than the margin of error for any one proportion. For a typical pre-election poll, a reported ±3% margin of error translates to about ±5% for the lead of one candidate over the other. They write that the margin of error for the lead should be reported.

SOLUTION

REQUIREMENT CHECK   We first verify that the three necessary requirements are satisfied. (1) The two samples are simple random samples (trust the author!). (2) The two samples are independent because cars in the samples are not matched or paired in any way. (3) Let's consider a "success" to be a car with a rear license plate only. For Connecticut, the number of successes is 239 and the number of failures (cars with front and rear license plates) is 1810, so they are both at least 5. For New York, there are 9 successes and 541 failures, and they are both at least 5. The requirements are satisfied. ☑

The following steps are from the *P*-value method of testing hypotheses, which is summarized in Figure 8-1 on page 360.

Step 1: The claim that "Connecticut and New York have the same proportion of cars with rear license plates only" can be expressed as $p_1 = p_2$.

Step 2: If $p_1 = p_2$ is false, then $p_1 \neq p_2$.

Step 3: Because the claim of $p_1 \neq p_2$ does not contain equality, it becomes the alternative hypothesis. The null hypothesis is the statement of equality, so we have

$$H_0: p_1 = p_2 \quad H_1: p_1 \neq p_2$$

Step 4: The significance level was specified as $\alpha = 0.05$, so we use $\alpha = 0.05$. (Who are we to argue?)

Step 5: This step and the following step can be circumvented by using technology; see the display that follows this example. If not using technology, we use the normal distribution (with the test statistic given in the Key Elements box) as an approximation to the binomial distribution. We estimate the common value of p_1 and p_2 with the pooled sample estimate $\bar{p}$ calculated as shown below, with extra decimal places used to minimize rounding errors in later calculations.

$$\bar{p} = \frac{x_1 + x_2}{n_1 + n_2} = \frac{239 + 9}{2049 + 550} = 0.09542132$$

With $\bar{p} = 0.09542132$, it follows that $\bar{q} = 1 - 0.09542132 = 0.90457868$.

Step 6: Because we assume in the null hypothesis that $p_1 = p_2$, the value of $p_1 - p_2$ is 0 in the following calculation of the test statistic:

$$z = \frac{(\hat{p}_1 - \hat{p}_2) - (p_1 - p_2)}{\sqrt{\dfrac{\bar{p}\bar{q}}{n_1} + \dfrac{\bar{p}\bar{q}}{n_2}}}$$

$$= \frac{\left(\dfrac{239}{2049} - \dfrac{9}{550}\right) - 0}{\sqrt{\dfrac{(0.09542132)(0.90457868)}{2049} + \dfrac{(0.09542132)(0.90457868)}{550}}} = 7.11$$

This is a two-tailed test, so the *P*-value is twice the area to the right of the test statistic $z = 7.11$ (as indicated by Figure 8-3 on page 364). Refer to Table A-2 and find that the area to the right of the test statistic $z = 7.11$ is 0.0001, so the *P*-value is 0.0002. Technology provides a more accurate *P*-value of 0.00000000000119, which is often expressed as 0.0000 or "*P*-value < 0.0001." The test statistic and *P*-value are shown in Figure 9-1(a).

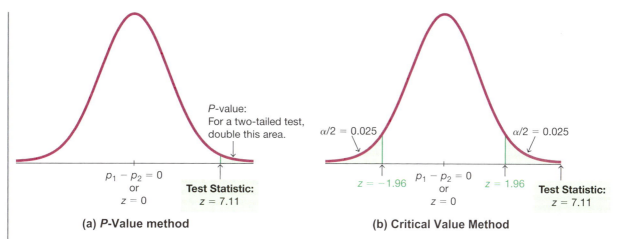

FIGURE 9-1 Hypothesis Test with Two Proportions

Step 7: Because the P-value of 0.0000 is less than the significance level of $\alpha = 0.05$, we reject the null hypothesis of $p_1 = p_2$. ("If the P is *low*, the null must go.")

INTERPRETATION

We must address the original claim that "Connecticut and New York have the same proportion of cars with rear license plates only." Because we reject the null hypothesis, we conclude that there is sufficient evidence to warrant rejection of the claim that $p_1 = p_2$. That is, there is sufficient evidence to conclude that Connecticut and New York have *different* proportions of cars with rear license plates only. It's reasonable to speculate that enforcement of the license plate laws is much stricter in New York than in Connecticut, and that is why Connecticut car owners are less likely to install the front license plate.

YOUR TURN Do Exercise 7 "License Plate Laws."

Technology Software and calculators usually provide a P-value, so the P-value method is typically used for testing a claim about two proportions. See the accompanying Statdisk results from Example 1 showing the test statistic of $z = 7.11$ (rounded) and the P-value of 0.0000.

Statdisk

Pooled proportion: 0.0954213
Test Statistic, z: 7.1074
Critical z: ±1.9600
P-Value: 0.0000

95% Confidence interval:
0.0827974 < p1-p2 < 0.1177599

Critical Value Method

The critical value method of testing hypotheses (see Figure 8-1 on page 360) can also be used for Example 1. In Step 6, instead of finding the P-value, find the critical values. With a significance level of $\alpha = 0.05$ in a two-tailed test based on the normal distribution, we refer to Table A-2 and find that an area of $\alpha = 0.05$ divided equally between the two tails corresponds to the critical values of $z = \pm 1.96$. In Figure 9-1(b) we can see that the test statistic of $z = 7.11$ falls within the critical region beyond the critical value of 1.96. We again reject the null hypothesis. The conclusions are the same as in Example 1.

Confidence Intervals

Using the format given in the preceding Key Elements box, we can construct a confidence interval estimate of the difference between population proportions ($p_1 - p_2$). If a confidence interval estimate of $p_1 - p_2$ does not include 0, we have evidence suggesting that p_1 and p_2 have different values. The confidence interval uses a standard deviation based on *estimated* values of the population proportions, whereas a hypothesis test uses a standard deviation based on the *assumption* that the two population proportions are equal. Consequently, a conclusion based on a confidence interval might be different from a conclusion based on a hypothesis test. See the caution that follows Example 2.

EXAMPLE 2   Confidence Interval for Claim About Two Proportions

Use the sample data given in Example 1 to construct a 95% confidence interval estimate of the difference between the two population proportions. What does the result suggest about the claim that "Connecticut and New York have the same proportion of cars with rear license plates only"?

SOLUTION

REQUIREMENT CHECK   We are using the same data from Example 1, and the same requirement check applies here, so the requirements are satisfied. ☑

The confidence interval can be found using technology; see the preceding Statdisk display. If not using technology, proceed as follows.

With a 95% confidence level, $z_{\alpha/2} = 1.96$ (from Table A-2). We first calculate the value of the margin of error E as shown here.

$$E = z_{\alpha/2}\sqrt{\frac{\hat{p}_1\hat{q}_1}{n_1} + \frac{\hat{p}_2\hat{q}_2}{n_2}} = 1.96\sqrt{\frac{\left(\frac{239}{2049}\right)\left(\frac{1810}{2049}\right)}{2049} + \frac{\left(\frac{9}{550}\right)\left(\frac{541}{550}\right)}{550}} = 0.017482$$

With $\hat{p}_1 = 239/2049 = 0.116642$ and $\hat{p}_2 = 9/550 = 0.016364$, we get $\hat{p}_1 - \hat{p}_2 = 0.100278$. With $\hat{p}_1 - \hat{p}_2 = 0.100278$ and $E = 0.017482$, the confidence interval is evaluated as follows, with the confidence interval limits rounded to three significant digits:

$$(\hat{p}_1 - \hat{p}_2) - E < (p_1 - p_2) < (\hat{p}_1 - \hat{p}_2) + E$$

$$0.100278 - 0.017482 < (p_1 - p_2) < 0.100278 + 0.017482$$

$$0.0828 < (p_1 - p_2) < 0.118$$

See the preceding Statdisk display showing the same confidence interval obtained here.

INTERPRETATION

The confidence interval limits do not contain 0, suggesting that there is a significant difference between the two proportions. The confidence interval suggests that the value of p_1 is greater than the value of p_2, so there does appear to be sufficient evidence to warrant rejection of the claim that "Connecticut and New York have the same proportion of cars with rear license plates only." This "suggestion" can be supported with the formal hypothesis test in Example 1.

CAUTION **Use of One Confidence Interval** Don't test for equality of two population proportions by determining whether there is an overlap between two individual
confidence interval estimates of the two individual population proportions. When
compared to the confidence interval estimate of $p_1 - p_2$, the analysis of overlap
between two individual confidence intervals is more conservative (by rejecting
equality less often), and it has less power (because it is less likely to reject $p_1 - p_2$
when in reality $p_1 \neq p_2$). See Exercise 25 "Overlap of Confidence Intervals."

What Can We Do When the Requirements Are Not Satisfied?

Bad Samples If we violate the requirement that we have two simple random samples,
we could be in big trouble. For example, if we have two convenience samples, there is
probably nothing that can be done to salvage them.

Fewer Than 5 Successes or Fewer Than 5 Failures in a Hypothesis Test If we violate the requirement that each of the two samples has at least 5 successes and at least
5 failures, we can use *Fisher's exact test,* which provides an *exact P*-value instead of
using the method based on a normal distribution approximation. Fisher's exact test
involves very complicated calculations, so the use of technology is strongly recommended. Statdisk, Minitab, XLSTAT, and StatCrunch all have the ability to perform
Fisher's exact test. (See Section 11-2.)

Fewer Than 5 Successes or Fewer Than 5 Failures in a Confidence Interval If we
violate the requirement that each of the two samples has at least 5 successes and at
least 5 failures, we can use bootstrap resampling methods to construct a confidence
interval. See Section 7-4.

Rationale: Why Do the Procedures of This Section Work?

Hypothesis Tests With $n_1\hat{p}_1 \geq 5$ and $n_1\hat{q}_1 \geq 5$, the distribution of $\hat{p}_1$ can be approximated by a normal distribution with mean p_1, standard deviation $\sqrt{p_1q_1/n_1}$, and variance p_1q_1/n_1 (based on Sections 6-6 and 7-1). This also applies to the second sample.
The distributions of $\hat{p}_1$ and $\hat{p}_2$ are each approximated by a normal distribution, so
the difference $\hat{p}_1 - \hat{p}_2$ will also be approximated by a normal distribution with mean
$p_1 - p_2$ and variance

$$\sigma^2_{(\hat{p}_1-\hat{p}_2)} = \sigma^2_{\hat{p}_1} + \sigma^2_{\hat{p}_2} = \frac{p_1q_1}{n_1} + \frac{p_2q_2}{n_2}$$

(The result above is based on this property: The variance of the *differences* between
two independent random variables is the *sum* of their individual variances.)

The pooled estimate of the common value of p_1 and p_2 is $\bar{p} = (x_1 + x_2)/(n_1 + n_2)$. If we replace p_1 and p_2 by $\bar{p}$ and replace q_1 and q_2 by $\bar{q} = 1 - \bar{p}$, the variance above leads to the following standard deviation:

$$\sigma_{(\hat{p}_1-\hat{p}_2)} = \sqrt{\frac{\bar{p}\,\bar{q}}{n_1} + \frac{\bar{p}\,\bar{q}}{n_2}}$$

We now know that the distribution of $\hat{p}_1 - \hat{p}_2$ is approximately normal, with mean
$p_1 - p_2$ and standard deviation as shown above, so the z test statistic has the form
given in the Key Elements box near the beginning of this section.

Confidence Interval The form of the confidence interval requires an expression for
the variance different from the one given above. When constructing a confidence

continued on page 423

Inferences with Two Proportions

Access tech supplements, videos, and data sets at **www.TriolaStats.com**

Statdisk	Minitab	StatCrunch

Statdisk

Hypothesis Testing

1. Click **Analysis** in the top menu.
2. Select **Hypothesis Testing** from the dropdown menu and **Proportion Two Samples** from the submenu.
3. Select the desired format for *Alternative Hypothesis* and enter the significance level. For both samples, enter sample size and number of successes.
4. Click **Evaluate.**

Confidence Intervals

1. Click **Analysis** in the top menu.
2. Select **Confidence Intervals** from the dropdown menu and **Proportion Two Samples** from the submenu.
3. Enter the desired confidence level. For both samples, enter sample size and number of successes.
4. Click **Evaluate.**

Minitab

1. Click **Stat** in the top menu.
2. Select **Basic Statistics** from the dropdown menu and select **2 Proportions** from the submenu.
3. Select **Summarized data** from the dropdown menu and enter the number of events and number of trials for both samples.
4. Click the **Options** button and enter the confidence level. If testing a hypothesis, enter **0** for *Hypothesized Difference* and select the desired format for *the Alternative Hypothesis*.
5. For *Test Method* select **Use the pooled estimate of the proportion.**
6. Click **OK** twice.

TIP: Another procedure is to click on **Assistant** in the top menu, then select **Hypothesis Tests** and **2-Sample % Defective.** Complete the dialog box to get results, including *P*-value and other helpful information.

StatCrunch

1. Click **Stat** in the top menu.
2. Select **Proportion Stats** from the dropdown menu, then select **Two Sample—With Summary** from the submenu.
3. Enter the number of successes and number of observations for both samples.
4. *Hypothesis Testing*: Select **Hypothesis test for p_1-p_2.** Enter **0** for the hypothesized difference (H_0) and select the desired format for the alternative hypothesis (H_A).

   Confidence Intervals: Select **Confidence interval for p_1-p_2** and enter the confidence level.
5. Click **Compute!**

TI-83/84 Plus Calculator	Excel

TI-83/84 Plus Calculator

Hypothesis Testing:

1. Press **STAT**, then select **TESTS** in the top menu.
2. Select **2-PropZTest** in the menu and press **ENTER**.
3. Enter the number of successes (*x*) and number of observations (*n*) for both samples. Select the desired format for the alternative hypothesis (*p1*).
4. Select **Calculate** and press **ENTER**.

Confidence Intervals:

1. Press **STAT**, then select **TESTS** in the top menu.
2. Select **2-PropZInt** in the menu and press **ENTER**.
3. Enter the number of successes (*x*) and number of observations (*n*) for both samples. Enter the desired confidence level (*C-Level*).
4. Select **Calculate** and press **ENTER**.

Excel

XLSTAT Add-In (Required)

1. Click on the **XLSTAT** tab in the Ribbon and then click **Parametric Tests.**
2. Select **Tests for two proportions** from the dropdown menu.
3. Under *Data Format* select **Frequency** if you know the number of successes *x* or select **Proportion** if you know the sample proportion $\hat{p}$.
4. Enter the frequency or sample proportion and sample size for both samples.
5. Check **z test,** uncheck **Continuity Correction,** and uncheck **Monte Carlo method.**
6. Click the **Options** tab.
7. *Hypothesis Testing:*

   Under *Alternative hypothesis* select the desired format ($\neq$ for two-tailed test, $<$ for left-tailed test, $>$ for right-tailed test). *For Hypothesized difference* enter **0.** Enter the desired significance level (enter **5** for 0.05 significance level). Under *Variance* select **pq(1/n1 + 1/n2).**

   Confidence Intervals:

   Under *Alternative hypothesis* select $\neq$ **D** for a two-tailed test. *For Hypothesized difference* enter **0.** Enter the desired significance level (enter **5** for a 95% confidence level). Under *Variance* select **p1q1/n1 + p2q2/n2.**
8. Click **OK** to display the results that include the test statistic labeled *z(Observed value)*, *P*-value, and confidence interval.

interval estimate of the difference between two proportions, we don't assume that the two proportions are equal, and we estimate the standard deviation as

$$\sqrt{\frac{\hat{p}_1\hat{q}_1}{n_1} + \frac{\hat{p}_2\hat{q}_2}{n_2}}$$

In the test statistic

$$z = \frac{(\hat{p}_1 - \hat{p}_2) - (p_1 - p_2)}{\sqrt{\frac{\hat{p}_1\hat{q}_1}{n_1} + \frac{\hat{p}_2\hat{q}_2}{n_2}}}$$

use the positive and negative values of z (for two tails) and solve for $p_1 - p_2$. The results are the limits of the confidence interval given in the Key Elements box near the beginning of this section.

9-1 Basic Skills and Concepts

Statistical Literacy and Critical Thinking

1. Verifying Requirements In the largest clinical trial ever conducted, 401,974 children were randomly assigned to two groups. The treatment group consisted of 201,229 children given the Salk vaccine for polio, and 33 of those children developed polio. The other 200,745 children were given a placebo, and 115 of those children developed polio. If we want to use the methods of this section to test the claim that the rate of polio is less for children given the Salk vaccine, are the requirements for a hypothesis test satisfied? Explain.

2. Notation For the sample data given in Exercise 1, consider the Salk vaccine treatment group to be the first sample. Identify the values of n_1, $\hat{p}_1$, $\hat{q}_1$, n_2, $\hat{p}_2$, $\hat{q}_2$, $\bar{p}$, and $\bar{q}$. Round all values so that they have six significant digits.

3. Hypotheses and Conclusions Refer to the hypothesis test described in Exercise 1.

a. Identify the null hypothesis and the alternative hypothesis.

b. If the P-value for the test is reported as "less than 0.001," what should we conclude about the original claim?

4. Using Confidence Intervals

a. Assume that we want to use a 0.05 significance level to test the claim that $p_1 < p_2$. Which is better: A hypothesis test or a confidence interval?

b. In general, when dealing with inferences for two population proportions, which two of the following are equivalent: confidence interval method; P-value method; critical value method?

c. If we want to use a 0.05 significance level to test the claim that $p_1 < p_2$, what confidence level should we use?

d. If we test the claim in part (c) using the sample data in Exercise 1, we get this confidence interval: $-0.000508 < p_1 - p_2 < -0.000309$. What does this confidence interval suggest about the claim?

Interpreting Displays. *In Exercises 5 and 6, use the results from the given displays.*

5. Testing Laboratory Gloves The *New York Times* published an article about a study in which Professor Denise Korniewicz and other Johns Hopkins researchers subjected laboratory gloves to stress. Among 240 vinyl gloves, 63% leaked viruses. Among 240 latex gloves, 7% leaked viruses. See the accompanying display of the Statdisk results. Using a 0.01 significance level, test the claim that vinyl gloves have a greater virus leak rate than latex gloves.

Statdisk

> Pooled proportion: 0.35
> Test Statistic, z: 12.8231
> Critical z: 2.3264
> P-Value: 0.0000
>
> 98% Confidence interval:
> 0.4762035 < p1-p2 < 0.6404632

continued

6. Treating Carpal Tunnel Syndrome Carpal tunnel syndrome is a common wrist complaint resulting from a compressed nerve, and it is often the result of extended use of repetitive wrist movements, such as those associated with the use of a keyboard. In a randomized controlled trial, 73 patients were treated with surgery and 67 were found to have successful treatments. Among 83 patients treated with splints, 60 were found to have successful treatments (based on data from "Splinting vs Surgery in the Treatment of Carpal Tunnel Syndrome," by Gerritsen et al., *Journal of the American Medical Association,* Vol. 288, No. 10). Use the accompanying StatCrunch display with a 0.01 significance level to test the claim that the success rate is better with surgery.

StatCrunch

Difference	0.1949
z (Observed value)	3.1226
z (Critical value)	2.3263
p-value (one-tailed)	0.0009
alpha	0.01

Testing Claims About Proportions.   *In Exercises 7–22, test the given claim. Identify the null hypothesis, alternative hypothesis, test statistic, P-value or critical value(s), then state the conclusion about the null hypothesis, as well as the final conclusion that addresses the original claim.*

7. License Plate Laws The Chapter Problem involved passenger cars in Connecticut and passenger cars in New York, but here we consider passenger cars and commercial trucks. Among 2049 Connecticut passenger cars, 239 had only rear license plates. Among 334 Connecticut trucks, 45 had only rear license plates (based on samples collected by the author). A reasonable hypothesis is that passenger car owners violate license plate laws at a higher rate than owners of commercial trucks. Use a 0.05 significance level to test that hypothesis.

a. Test the claim using a hypothesis test.

b. Test the claim by constructing an appropriate confidence interval.

8. Accuracy of Fast Food Drive-Through Orders In a study of Burger King drive-through orders, it was found that 264 orders were accurate and 54 were not accurate. For McDonald's, 329 orders were found to be accurate while 33 orders were not accurate (based on data from *QSR* magazine). Use a 0.05 significance level to test the claim that Burger King and McDonald's have the same accuracy rates.

a. Test the claim using a hypothesis test.

b. Test the claim by constructing an appropriate confidence interval.

c. Relative to accuracy of orders, does either restaurant chain appear to be better?

9. Smoking Cessation Programs Among 198 smokers who underwent a "sustained care" program, 51 were no longer smoking after six months. Among 199 smokers who underwent a "standard care" program, 30 were no longer smoking after six months (based on data from "Sustained Care Intervention and Postdischarge Smoking Cessation Among Hospitalized Adults," by Rigotti et al., *Journal of the American Medical Association,* Vol. 312, No. 7). We want to use a 0.01 significance level to test the claim that the rate of success for smoking cessation is greater with the sustained care program.

a. Test the claim using a hypothesis test.

b. Test the claim by constructing an appropriate confidence interval.

c. Does the difference between the two programs have practical significance?

10. Tennis Challenges Since the Hawk-Eye instant replay system for tennis was introduced at the U.S. Open in 2006, men challenged 2441 referee calls, with the result that 1027 of the calls were overturned. Women challenged 1273 referee calls, and 509 of the calls were overturned. We want to use a 0.05 significance level to test the claim that men and women have equal success in challenging calls.

a. Test the claim using a hypothesis test.

b. Test the claim by constructing an appropriate confidence interval.

c. Based on the results, does it appear that men and women have equal success in challenging calls?

11. Dreaming in Black and White A study was conducted to determine the proportion of people who dream in black and white instead of color. Among 306 people over the age of 55, 68 dream in black and white, and among 298 people under the age of 25, 13 dream in black and white (based on data from "Do We Dream in Color?" by Eva Murzyn, *Consciousness and Cognition,* Vol. 17, No. 4). We want to use a 0.01 significance level to test the claim that the proportion of people over 55 who dream in black and white is greater than the proportion of those under 25.

a. Test the claim using a hypothesis test.

b. Test the claim by constructing an appropriate confidence interval.

c. An explanation given for the results is that those over the age of 55 grew up exposed to media that was mostly displayed in black and white. Can the results from parts (a) and (b) be used to verify that explanation?

12. Clinical Trials of OxyContin OxyContin (oxycodone) is a drug used to treat pain, but it is well known for its addictiveness and danger. In a clinical trial, among subjects treated with OxyContin, 52 developed nausea and 175 did not develop nausea. Among other subjects given placebos, 5 developed nausea and 40 did not develop nausea (based on data from Purdue Pharma L.P.). Use a 0.05 significance level to test for a difference between the rates of nausea for those treated with OxyContin and those given a placebo.

a. Use a hypothesis test.

b. Use an appropriate confidence interval.

c. Does nausea appear to be an adverse reaction resulting from OxyContin?

13. Are Seat Belts Effective? A simple random sample of front-seat occupants involved in car crashes is obtained. Among 2823 occupants not wearing seat belts, 31 were killed. Among 7765 occupants wearing seat belts, 16 were killed (based on data from "Who Wants Airbags?" by Meyer and Finney, *Chance,* Vol. 18, No. 2). We want to use a 0.05 significance level to test the claim that seat belts are effective in reducing fatalities.

a. Test the claim using a hypothesis test.

b. Test the claim by constructing an appropriate confidence interval.

c. What does the result suggest about the effectiveness of seat belts?

14. Cardiac Arrest at Day and Night A study investigated survival rates for in-hospital patients who suffered cardiac arrest. Among 58,593 patients who had cardiac arrest during the day, 11,604 survived and were discharged. Among 28,155 patients who suffered cardiac arrest at night, 4139 survived and were discharged (based on data from "Survival from In-Hospital Cardiac Arrest During Nights and Weekends," by Peberdy et al., *Journal of the American Medical Association,* Vol. 299, No. 7). We want to use a 0.01 significance level to test the claim that the survival rates are the same for day and night.

a. Test the claim using a hypothesis test.

b. Test the claim by constructing an appropriate confidence interval.

c. Based on the results, does it appear that for in-hospital patients who suffer cardiac arrest, the survival rate is the same for day and night?

15. Is Echinacea Effective for Colds? Rhinoviruses typically cause common colds. In a test of the effectiveness of echinacea, 40 of the 45 subjects treated with echinacea developed rhinovirus infections. In a placebo group, 88 of the 103 subjects developed rhinovirus infections (based on data from "An Evaluation of Echinacea Angustifolia in Experimental Rhinovirus Infections," by Turner et al., *New England Journal of Medicine,* Vol. 353, No. 4). We want to use a 0.05 significance level to test the claim that echinacea has an effect on rhinovirus infections.

a. Test the claim using a hypothesis test.

b. Test the claim by constructing an appropriate confidence interval.

c. Based on the results, does echinacea appear to have any effect on the infection rate?

16. Bednets to Reduce Malaria In a randomized controlled trial in Kenya, insecticide-treated bednets were tested as a way to reduce malaria. Among 343 infants using bednets, 15 developed malaria. Among 294 infants not using bednets, 27 developed malaria (based on data from "Sustainability of Reductions in Malaria Transmission and Infant Mortality in Western Kenya with Use of Insecticide-Treated Bednets," by Lindblade et al., *Journal of the American Medical Association,* Vol. 291, No. 21). We want to use a 0.01 significance level to test the claim that the incidence of malaria is lower for infants using bednets.

a. Test the claim using a hypothesis test.

b. Test the claim by constructing an appropriate confidence interval.

c. Based on the results, do the bednets appear to be effective?

17. Cell Phones and Handedness A study was conducted to investigate the association between cell phone use and hemispheric brain dominance. Among 216 subjects who prefer to use their left ear for cell phones, 166 were right-handed. Among 452 subjects who prefer to use their right ear for cell phones, 436 were right-handed (based on data from "Hemispheric Dominance and Cell Phone Use," by Seidman et al., *JAMA Otolaryngology—Head & Neck Surgery,* Vol. 139, No. 5). We want to use a 0.01 significance level to test the claim that the rate of right-handedness for those who prefer to use their left ear for cell phones is less than the rate of right-handedness for those who prefer to use their right ear for cell phones. (Try not to get too confused here.)

a. Test the claim using a hypothesis test.

b. Test the claim by constructing an appropriate confidence interval.

18. Denomination Effect A trial was conducted with 75 women in China given a 100-yuan bill, while another 75 women in China were given 100 yuan in the form of smaller bills (a 50-yuan bill plus two 20-yuan bills plus two 5-yuan bills). Among those given the single bill, 60 spent some or all of the money. Among those given the smaller bills, 68 spent some or all of the money (based on data from "The Denomination Effect," by Raghubir and Srivastava, *Journal of Consumer Research,* Vol. 36). We want to use a 0.05 significance level to test the claim that when given a single large bill, a smaller proportion of women in China spend some or all of the money when compared to the proportion of women in China given the same amount in smaller bills.

a. Test the claim using a hypothesis test.

b. Test the claim by constructing an appropriate confidence interval.

c. If the significance level is changed to 0.01, does the conclusion change?

19. Headache Treatment In a study of treatments for very painful "cluster" headaches, 150 patients were treated with oxygen and 148 other patients were given a placebo consisting of ordinary air. Among the 150 patients in the oxygen treatment group, 116 were free from headaches 15 minutes after treatment. Among the 148 patients given the placebo, 29 were free from headaches 15 minutes after treatment (based on data from "High-Flow Oxygen for Treatment of Cluster Headache," by Cohen, Burns, and Goadsby, *Journal of the American Medical Association,* Vol. 302, No. 22). We want to use a 0.01 significance level to test the claim that the oxygen treatment is effective.

continued

a. Test the claim using a hypothesis test.

b. Test the claim by constructing an appropriate confidence interval.

c. Based on the results, is the oxygen treatment effective?

20. Does Aspirin Prevent Heart Disease? In a trial designed to test the effectiveness of aspirin in preventing heart disease, 11,037 male physicians were treated with aspirin and 11,034 male physicians were given placebos. Among the subjects in the aspirin treatment group, 139 experienced myocardial infarctions (heart attacks). Among the subjects given placebos, 239 experienced myocardial infarctions (based on data from "Final Report on the Aspirin Component of the Ongoing Physicians' Health Study," *New England Journal of Medicine,* Vol. 321: 129–135). Use a 0.05 significance level to test the claim that aspirin has no effect on myocardial infarctions.

a. Test the claim using a hypothesis test.

b. Test the claim by constructing an appropriate confidence interval.

c. Based on the results, does aspirin appear to be effective?

21. Lefties In a random sample of males, it was found that 23 write with their left hands and 217 do not. In a random sample of females, it was found that 65 write with their left hands and 455 do not (based on data from "The Left-Handed: Their Sinister History," by Elaine Fowler Costas, Education Resources Information Center, Paper 399519). We want to use a 0.01 significance level to test the claim that the rate of left-handedness among males is less than that among females.

a. Test the claim using a hypothesis test.

b. Test the claim by constructing an appropriate confidence interval.

c. Based on the results, is the rate of left-handedness among males less than the rate of left-handedness among females?

22. Ground vs. Helicopter for Serious Injuries A study investigated rates of fatalities among patients with serious traumatic injuries. Among 61,909 patients transported by helicopter, 7813 died. Among 161,566 patients transported by ground services, 17,775 died (based on data from "Association Between Helicopter vs Ground Emergency Medical Services and Survival for Adults With Major Trauma," by Galvagno et al., *Journal of the American Medical Association,* Vol. 307, No. 15). Use a 0.01 significance level to test the claim that the rate of fatalities is higher for patients transported by helicopter.

a. Test the claim using a hypothesis test.

b. Test the claim by constructing an appropriate confidence interval.

c. Considering the test results and the actual sample rates, is one mode of transportation better than the other? Are there other important factors to consider?

9-1 Beyond the Basics

23. Determining Sample Size The sample size needed to estimate the difference between two population proportions to within a margin of error E with a confidence level of $1 - \alpha$ can be found by using the following expression:

$$E = z_{\alpha/2}\sqrt{\frac{p_1 q_1}{n_1} + \frac{p_2 q_2}{n_2}}$$

continued

Replace n_1 and n_2 by n in the preceding formula (assuming that both samples have the same size) and replace each of p_1, q_1, p_2, and q_2 by 0.5 (because their values are not known). Solving for n results in this expression:

$$n = \frac{z_{\alpha/2}^2}{2E^2}$$

Use this expression to find the size of each sample if you want to estimate the difference between the proportions of men and women who own smartphones. Assume that you want 95% confidence that your error is no more than 0.03.

24. Yawning and Fisher's Exact Test In one segment of the TV series *MythBusters,* an experiment was conducted to test the common belief that people are more likely to yawn when they see others yawning. In one group, 34 subjects were exposed to yawning, and 10 of them yawned. In another group, 16 subjects were not exposed to yawning, and 4 of them yawned. We want to test the belief that people are more likely to yawn when they are exposed to yawning.

a. Why can't we test the claim using the methods of this section?

b. If we ignore the requirements and use the methods of this section, what is the *P*-value? How does it compare to the *P*-value of 0.5128 that would be obtained by using Fisher's exact test?

c. Comment on the conclusion of the *Mythbusters* segment that yawning is contagious.

25. Overlap of Confidence Intervals In the article "On Judging the Significance of Differences by Examining the Overlap Between Confidence Intervals," by Schenker and Gentleman (*American Statistician,* Vol. 55, No. 3), the authors consider sample data in this statement: "Independent simple random samples, each of size 200, have been drawn, and 112 people in the first sample have the attribute, whereas 88 people in the second sample have the attribute."

a. Use the methods of this section to construct a 95% confidence interval estimate of the difference $p_1 - p_2$. What does the result suggest about the equality of p_1 and p_2?

b. Use the methods of Section 7-1 to construct individual 95% confidence interval estimates for each of the two population proportions. After comparing the overlap between the two confidence intervals, what do you conclude about the equality of p_1 and p_2?

c. Use a 0.05 significance level to test the claim that the two population proportions are equal. What do you conclude?

d. Based on the preceding results, what should you conclude about the equality of p_1 and p_2? Which of the three preceding methods is least effective in testing for the equality of p_1 and p_2?

26. Equivalence of Hypothesis Test and Confidence Interval Two different simple random samples are drawn from two different populations. The first sample consists of 20 people with 10 having a common attribute. The second sample consists of 2000 people with 1404 of them having the same common attribute. Compare the results from a hypothesis test of $p_1 = p_2$ (with a 0.05 significance level) and a 95% confidence interval estimate of $p_1 - p_2$.

9-2 Two Means: Independent Samples

Key Concept This section presents methods for using sample data from two independent samples to test hypotheses made about two population means or to construct confidence interval estimates of the difference between two population means. In Part 1 we discuss situations in which the standard deviations of the two populations are unknown and are not assumed to be equal. In Part 2 we briefly discuss two other situations: (1) The two population standard deviations are unknown but are assumed to be equal; (2) the unrealistic case in which two population standard deviations are both known.

PART 1 Independent Samples: σ_1 and σ_2 Unknown and Not Assumed Equal

This section involves two *independent* samples, and the following section deals with samples that are *dependent*. It is important to know the difference between independent samples and dependent samples.

DEFINITIONS

Two samples are **independent** if the sample values from one population are not related to or somehow naturally paired or matched with the sample values from the other population.

Two samples are **dependent** (or consist of **matched pairs**) if the sample values are somehow matched, where the matching is based on some inherent relationship. (That is, each pair of sample values consists of two measurements from the same subject—such as before/after data—or each pair of sample values consists of matched pairs—such as husband/wife data—where the matching is based on some meaningful relationship.) *Caution*: "Dependence" does not require a direct cause/effect relationship.

HINT: If the two samples have different sample sizes with no missing data, they must be independent. If the two samples have the same sample size, the samples may or may not be independent.

Here is an example of independent samples and another example of dependent samples:

- *Independent Samples:* **Heights of Men and Women** Data Set 1 "Body Data" in Appendix B includes the following heights (cm) of samples of men and women, and the two samples are not matched according to some inherent relationship. They are actually two independent samples that just happen to be listed in a way that might cause us to incorrectly think that they are matched.

Heights (cm) of Men	172	154	156	158	169
Heights (cm) of Women	186	161	179	167	179

- *Dependent Samples:* **Heights of Husbands and Wives** Students of the author collected data consisting of the heights (cm) of husbands and the heights (cm) of their wives. Five of those pairs of heights are listed below. These two samples are dependent, because the height of each husband is *matched* with the height of his wife.

Height (cm) of Husband	175	180	173	176	178
Height (cm) of Wife	160	165	163	162	166

For inferences about means from two independent populations, the following box summarizes key elements of a hypothesis test and a confidence interval estimate of the difference between the population means.

Do Real Estate Agents Get You the Best Prices?

When a real estate agent sells a home, does he or she get the best price for the seller? This question was addressed by Steven Levitt and Stephen Dubner in *Freakonomics*. They collected data from thousands of homes near Chicago, including homes owned by the agents themselves. Here is what they write: "There's one way to find out: measure the difference between the sales data for houses that belong to real-estate agents themselves and the houses they sold on behalf of clients. Using the data from the sales of those 100,000 Chicago homes, and controlling for any number of variables—location, age and quality of the house, aesthetics, and so on—it turns out that a real-estate agent keeps her own home on the market an average of ten days longer and sells it for an extra 3-plus percent, or $10,000 on a $300,000 house." A conclusion such as this can be obtained by using the methods of this section.

Inferences About Two Means: Independent Samples

Objectives

1. **Hypothesis Test:** Conduct a hypothesis test of a claim about two independent population means.

2. **Confidence Interval:** Construct a confidence interval estimate of the difference between two independent population means.

Notation

For population 1 we let

$\mu_1 = population$ mean $\bar{x}_1 = sample$ mean

$\sigma_1 = population$ standard deviation $s_1 = sample$ standard deviation

$n_1 = $ size of the first sample

The corresponding notations μ_2, σ_2, $\bar{x}_2$, s_2, and n_2, apply to population 2.

Requirements

1. The values of σ_1 and σ_2 are unknown and we do not assume that they are equal.

2. The two samples are *independent*.

3. Both samples are *simple random samples*.

4. Either or both of these conditions are satisfied: The two sample sizes are both *large* (with $n_1 > 30$ and $n_2 > 30$)

or both samples come from populations having normal distributions. (The methods used here are *robust* against departures from normality, so for small samples, the normality requirement is loose in the sense that the procedures perform well as long as there are no outliers and departures from normality are not too extreme.)

Hypothesis Test Statistic for Two Means: Independent Samples (with $H_0: \mu_1 = \mu_2$)

$$t = \frac{(\bar{x}_1 - \bar{x}_2) - (\mu_1 - \mu_2)}{\sqrt{\dfrac{s_1^2}{n_1} + \dfrac{s_2^2}{n_2}}} \quad \text{(where } \mu_1 - \mu_2 \text{ is often assumed to be } 0)$$

Degrees of Freedom When finding critical values or *P*-values, use the following for determining the number of degrees of freedom, denoted by df. (Although these two methods typically result in different numbers of degrees of freedom, the conclusion of a hypothesis test is rarely affected by the choice.)

1. Use this simple and conservative estimate:

$$\mathbf{df} = \text{smaller of } n_1 - 1 \text{ and } n_2 - 1$$

2. Technologies typically use the more accurate but more difficult estimate given in Formula 9-1.

FORMULA 9-1

$$df = \frac{(A + B)^2}{\dfrac{A^2}{n_1 - 1} + \dfrac{B^2}{n_2 - 1}}$$

where $A = \dfrac{s_1^2}{n_1}$ and $B = \dfrac{s_2^2}{n_2}$

Note: Answers in Appendix D include technology answers based on Formula 9-1 along with "Table" answers based on using Table A-3 with the simple estimate of df given in option 1 above.

P-Values: *P*-values are automatically provided by technology. If technology is not available, refer to the *t* distribution in Table A-3. Use the procedure summarized in Figure 8-3 on page 364.

Critical Values: Refer to the *t* distribution in Table A-3.

Confidence Interval Estimate of $\mu_1 - \mu_2$: Independent Samples

The confidence interval estimate of the difference $\mu_1 - \mu_2$ is

$$(\bar{x}_1 - \bar{x}_2) - E < (\mu_1 - \mu_2) < (\bar{x}_1 - \bar{x}_2) + E$$

where

$$E = t_{\alpha/2}\sqrt{\frac{s_1^2}{n_1} + \frac{s_2^2}{n_2}}$$

and the number of degrees of freedom df is as described for hypothesis tests.
(In this book, we use df = smaller of $n_1 - 1$ and $n_2 - 1$.)

Equivalent Methods

The *P*-value method of hypothesis testing, the critical value method of hypothesis testing, and confidence intervals all use the same distribution and standard error, so they are all equivalent in the sense that they result in the same conclusions.

P-Value Method

 EXAMPLE 1 **Are Male Professors and Female Professors Rated Differently by Students?**

Listed below are student course evaluation scores for courses taught by female professors and male professors (from Data Set 17 "Course Evaluations" in Appendix B). Use a 0.05 significance level to test the claim that the two samples are from populations with the same mean. Does there appear to be a difference in evaluation scores of courses taught by female professors and male professors?

Female	4.3	4.3	4.4	4.0	3.4	4.7	2.9	4.0	4.3	3.4	3.4	3.3			
Male	4.5	3.7	4.2	3.9	3.1	4.0	3.8	3.4	4.5	3.8	4.3	4.4	4.1	4.2	4.0

SOLUTION

REQUIREMENT CHECK (1) The values of the two population standard deviations are not known and we are not making an assumption that they are equal. (2) The two samples are independent because the female professors and male professors are not matched or paired in any way. (3) The samples are simple random samples. (4) Both samples are small (30 or fewer), so we need to determine whether both samples come from populations having normal distributions. Normal quantile plots of the two samples suggest that the samples are from populations having distributions that are not far from normal. The requirements are all satisfied. ☑

Using the *P*-value method summarized in Figure 8-1 on page 360, we can test the claim as follows.

Step 1: The claim that "the two samples are from populations with the same mean" can be expressed as $\mu_1 = \mu_2$.

Step 2: If the original claim is false, then $\mu_1 \neq \mu_2$.

Step 3: The alternative hypothesis is the expression not containing equality, and the null hypothesis is an expression of equality, so we have

$$H_0: \mu_1 = \mu_2 \qquad H_1: \mu_1 \neq \mu_2$$

continued

We now proceed with the assumption that $\mu_1 = \mu_2$, or $\mu_1 - \mu_2 = 0$.

Step 4: The significance level is $\alpha = 0.05$.

Step 5: Because we have two independent samples and we are testing a claim about the two population means, we use a t distribution with the test statistic given earlier in this section.

Step 6: The test statistic is calculated using these statistics (with extra decimal places) obtained from the listed sample data: Females: $n = 12$, $\bar{x} = 3.866667$, $s = 0.563001$; males: $n = 15$, $\bar{x} = 3.993333$, $s = 0.395450$.

$$t = \frac{(\bar{x}_1 - \bar{x}_2) - (\mu_1 - \mu_2)}{\sqrt{\frac{s_1^2}{n_1} + \frac{s_2^2}{n_2}}} = \frac{(3.866667 - 3.993333) - 0}{\sqrt{\frac{0.563001^2}{12} + \frac{0.395450^2}{15}}} = -0.660$$

P-Value With test statistic $t = -0.660$, we refer to Table A-3 (t Distribution). The number of degrees of freedom is the smaller of $n_1 - 1$ and $n_2 - 1$, or the smaller of $(12 - 1)$ and $(15 - 1)$, which is 11. With df $= 11$ and a two-tailed test, Table A-3 indicates that the P-value is greater than 0.20. Technology will provide the P-value of 0.5172 when using the original data or unrounded sample statistics.

Step 7: Because the P-value is greater than the significance level of 0.05, we fail to reject the null hypothesis. ("If the P is *low*, the null must go.")

> **INTERPRETATION**

Step 8: There is not sufficient evidence to warrant rejection of the claim that female professors and male professors have the same mean course evaluation score.

> **YOUR TURN** Do Exercise 7 "Color and Creativity."

Technology The tricky part about the preceding P-value approach is that Table A-3 can give only a range for the P-value, and determining that range is often somewhat difficult. Technology automatically provides the P-value, so technology makes the P-value method quite easy. See the accompanying XLSTAT display showing the test statistic of $t = -0.660$ (rounded) and the P-value of 0.5172.

XLSTAT

Difference	-0.1267
t (Observed value)	-0.6599
|t| (Critical value)	2.0926
DF	19
p-value (Two-tailed)	0.5172
alpha	0.05

Critical Value Method

If technology is not available, the critical value method of testing a claim about two means is generally easier than the P-value method. Example 1 can be solved using the critical value method. When finding critical values in Table A-3, we use df $=$ smaller of $n_1 - 1$ and $n_2 - 1$ as a relatively easy way to avoid using the really messy calculation required with Formula 9-1. In Example 1 with sample sizes of $n_1 = 12$ and $n_2 = 15$, the number of degrees of freedom is 11, which is the smaller of 11 and 14. In Table A-3 with df $= 11$ and $\alpha = 0.05$ in two tails, we get critical values of $t = \pm 2.201$. Technology can be used to find the more accurate critical values of $t = \pm 2.093$. Figure 9-2 shows the more accurate test statistic and critical values found from technology. The test statistic of $t = -0.660$ falls between the critical values, so the test statistic is not in the critical region and we fail to reject the null hypothesis, as we did in Example 1.

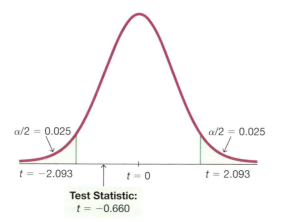

$\alpha/2 = 0.025$

$\alpha/2 = 0.025$

$t = -2.093$ $t = 0$ $t = 2.093$

Test Statistic:
$t = -0.660$

FIGURE 9-2 **Hypothesis Test of Means from Two Independent Populations**

Confidence Intervals

 EXAMPLE 2 **Confidence Interval for Female and Male Course Evaluation Scores**

Using the sample data given in Example 1, construct a 95% confidence interval estimate of the difference between the mean course evaluation score for female professors and the mean course evaluation score for male professors.

SOLUTION

REQUIREMENT CHECK Because we are using the same data from Example 1, the same requirement check applies here, so the requirements are satisfied. ☑

We first find the value of the margin of error E. In Table A-3 with df $= 11$ and $\alpha = 0.05$ in two tails, we get critical values of $t = \pm 2.201$. (Technology can be used to find the more accurate critical values of $t = \pm 2.093$.)

$$E = t_{\alpha/2}\sqrt{\frac{s_1^2}{n_1} + \frac{s_2^2}{n_2}} = 2.201\sqrt{\frac{0.563001^2}{12} + \frac{0.395450^2}{15}} = 0.422452$$

Using $E = 0.422452$, $\bar{x}_1 = 3.866667$, and $\bar{x}_2 = 3.993333$, we can now find the confidence interval as follows:

$$(\bar{x}_1 - \bar{x}_2) - E < (\mu_1 - \mu_2) < (\bar{x}_1 - \bar{x}_2) + E$$

$$-0.55 < (\mu_1 - \mu_2) < 0.30$$

If we use technology to obtain more accurate results, we get the confidence interval of $-0.53 < (\mu_1 - \mu_2) < 0.27$, so we can see that the confidence interval above is quite good, even though we used a simplified method for finding the number of degrees of freedom (instead of getting more accurate results by using Formula 9-1 to compute the number of degrees of freedom).

INTERPRETATION

We are 95% confident that the limits of -0.53 and 0.27 actually do contain the difference between the two population means. Because those limits contain 0, this confidence interval suggests that there is not a significant difference between the mean course evaluation score for female professors and the mean course evaluation score for male professors.

Super Bowls

Students were invited to a Super Bowl game, and half of them were given large 4-liter snack bowls while the other half were given smaller 2-liter bowls. Those using the large bowls consumed 56% more than those using the smaller bowls. (See "Super Bowls: Serving Bowl Size and Food Consumption," by Wansink and Cheney, *Journal of the American Medical Association,* Vol. 293, No. 14.)

A separate study showed that there is "a significant increase in fatal motor vehicle crashes during the hours following the Super Bowl telecast in the United States." Researchers analyzed 20,377 deaths on 27 Super Bowl Sundays and 54 other Sundays used as controls. They found a 41% increase in fatalities after Super Bowl games. (See "Do Fatal Crashes Increase Following a Super Bowl Telecast?" by Redelmeier and Stewart, *Chance,* Vol. 18, No. 1.)

PART 2 Alternative Methods

Part 1 of this section dealt with situations in which the two population standard deviations are unknown and are not assumed to be equal. In Part 2 we address two other situations:

1. The two population standard deviations are unknown but are assumed to be equal.

2. The two population standard deviations are both known.

Alternative Method: Assume That $\sigma_1 = \sigma_2$ and *Pool* the Sample Variances

Even when the specific values of σ_1 and σ_2 are not known, if it can be assumed that they have the *same* value, the sample variances s_1^2 and s_2^2 can be *pooled* to obtain an estimate of the common population variance σ^2. The **pooled estimate of σ^2** is denoted by s_p^2 and is a weighted average of s_1^2 and s_2^2, which is used in the test statistic for this case:

$$\textbf{Test Statistic} \quad t = \frac{(\bar{x}_1 - \bar{x}_2) - (\mu_1 - \mu_2)}{\sqrt{\dfrac{s_p^2}{n_1} + \dfrac{s_p^2}{n_2}}}$$

$$\text{where} \quad s_p^2 = \frac{(n_1 - 1)s_1^2 + (n_2 - 1)s_2^2}{(n_1 - 1) + (n_2 - 1)} \quad \text{(pooled sample variance)}$$

and the number of degrees of freedom is df $= n_1 + n_2 - 2$.

The requirements for this case are the same as in Part 1, except the first requirement is that σ_1 and σ_2 are not known but they are assumed to be equal. Confidence intervals are found by evaluating $(\bar{x}_1 - \bar{x}_2) - E < (\mu_1 - \mu_2) < (\bar{x}_1 - \bar{x}_2) + E$ with the following margin of error E.

$$\textbf{Margin of Error for Confidence Interval} \quad E = t_{\alpha/2}\sqrt{\frac{s_p^2}{n_1} + \frac{s_p^2}{n_2}}$$

where s_p^2 is as given in the test statistic above, and df $= n_1 + n_2 - 2$.

When Should We Assume That $\sigma_1 = \sigma_2$? If we use randomness to assign subjects to treatment and placebo groups, we know that the samples are drawn from the same population. So if we conduct a hypothesis test assuming that two population means are equal, it is not unreasonable to assume that the samples are from populations with the same standard deviations (but we should still check that assumption).

Advantage of Pooling The advantage of this alternative method of pooling sample variances is that the number of degrees of freedom is a little higher, so hypothesis tests have more power and confidence intervals are a little narrower.

Alternative Method Used When σ_1 and σ_2 Are Known

In reality, the population standard deviations σ_1 and σ_2 are almost never known, but if they are somehow known, the test statistic and confidence interval are based on the normal distribution instead of the t distribution. The requirements are the same as those given in Part 1, except for the first requirement that σ_1 and σ_2 are known. Critical values and P-values are found using technology or Table A-2, and the test statistic for this case is as follows:

$$\textbf{Test Statistic} \quad z = \frac{(\bar{x}_1 - \bar{x}_2) - (\mu_1 - \mu_2)}{\sqrt{\dfrac{\sigma_1^2}{n_1} + \dfrac{\sigma_2^2}{n_2}}}$$

Confidence intervals are found by evaluating

$$(\bar{x}_1 - \bar{x}_2) - E < (\mu_1 - \mu_2) < (\bar{x}_1 - \bar{x}_2) + E, \text{ where:}$$

Margin of Error for Confidence Interval $E = z_{\alpha/2}\sqrt{\dfrac{\sigma_1^2}{n_1} + \dfrac{\sigma_2^2}{n_2}}$

What if One Standard Deviation Is Known and the Other Is Unknown? If σ_1 is known but σ_2 is unknown, use the procedures in Part 1 of this section with these changes: Replace s_1 with the known value of σ_1 and use the number of degrees of freedom found from the expression below. (See "The Two-Sample t Test with One Variance Unknown," by Maity and Sherman, *The American Statistician*, Vol. 60, No. 2.)

$$df = \frac{\left(\dfrac{\sigma_1^2}{n_1} + \dfrac{s_2^2}{n_2}\right)^2}{\dfrac{(s_2^2/n_2)^2}{n_2 - 1}}$$

Recommended Strategy for Two Independent Means

Here is the recommended strategy for the methods of this section:

> **Assume that σ_1 and σ_2 are unknown, do *not* assume that $\sigma_1 = \sigma_2$, and use the test statistic and confidence interval given in Part 1 of this section.**

Using Statistics to Identify Thieves

Methods of statistics can be used to determine that an employee is stealing, and they can also be used to estimate the amount stolen. For comparable time periods, samples of sales have means that are significantly different. The mean sale amount decreases significantly. There is a significant increase in "no sale" register openings. There is a significant decrease in the ratio of cash receipts to checks. (See "How to Catch a Thief," by Manly and Thomson, *Chance*, Vol. 11, No. 4.)

TECH CENTER

Inferences with Two Means: Independent Samples

Access tech supplements, videos, and data sets at **www.TriolaStats.com**

Statdisk	Minitab

Statdisk

Hypothesis Testing

1. Click **Analysis** in the top menu.
2. Select **Hypothesis Testing** from the dropdown menu and **Mean-Two Independent Samples** from the submenu.
3. Select the desired format for *Alternative Hypothesis* and enter the significance level. For both samples enter the sample statistics, or click the **Use Data** tab to use columns of data.
4. Under *Method of Analysis* select **Unequal variances: No Pool**.
5. Click **Evaluate**.

Confidence Intervals

1. Click **Analysis** in the top menu.
2. Select **Confidence Intervals** from the dropdown menu and **Mean-Two Independent Samples** from the submenu.
3. Enter the desired confidence level. For both samples enter the sample statistics, or click the **Use Data** tab to use columns of data.
4. Under *Method of Analysis* select **Unequal variances: No Pool**.
5. Click **Evaluate**.

Minitab

1. Click **Stat** in the top menu.
2. Select **Basic Statistics** from the dropdown menu and select **2-sample t** from the submenu.
3. *Using Summary Statistics:* Select **Summarized data** from the dropdown menu and enter the sample size, sample mean, and sample standard deviation for each sample.

   Using Sample Data: Select **Each sample is in its own column** from the dropdown menu and select the desired data columns.
4. Click the **Options** button and enter the confidence level. Enter **0** for *Hypothesized Difference* and select the desired format for *the Alternative Hypothesis*.
5. Leave **Assume equal variances** unchecked. Check this box only if you want to assume the populations have equal variances—this is not recommended.
6. Click **OK** twice.

Tip: Another procedure is to click on **Assistant** in the top menu, select **Hypothesis Tests** and **2-Sample t**. Complete the dialog box to get results, including *P*-value and other helpful information.

continued

TECH CENTER *continued*

Inferences with Two Means: Independent Samples
Access tech supplements, videos, and data sets at **www.TriolaStats.com**

StatCrunch

1. Click **Stat** in the top menu.
2. Select **T Stats** from the dropdown menu, then select **Two Sample** from the submenu.
3. *Using Summary Statistics:* Select **With Summary** from the submenu and enter the sample mean, sample standard deviation, and sample size for each sample.

   Using Sample Data: Select **With Data** from the submenu and select the desired data column for each sample.
4. Leave **Pool variances** unchecked. Check this box only if you want to assume the populations have equal variances—this is not recommended.
5. *Hypothesis Testing:* Select **Hypothesis test for $\mu_1 - \mu_2$**. For hypothesized difference (H_0) enter **0** and select the desired format for the alternative hypothesis (H_A).

   Confidence Intervals: Select **Confidence interval for $\mu_1 - \mu_2$** and enter the confidence level.
6. Click **Compute!**

TI-83/84 Plus Calculator

Hypothesis Testing:
1. Press **STAT**, then select **TESTS** in the top menu.
2. Select **2-SampTTest** in the menu and press **ENTER**.
3. Select **Data** if you have sample data in lists or **Stats** if you have summary statistics. Press **ENTER** and enter the list names (leave *Freq* = **1**) or summary statistics.
4. For μ_1 select the desired format for the alternative hypothesis.
5. For pooled select **No**. Select *Yes* only if the population variances are believed to be equal.
6. Select **Calculate** and press **ENTER**.

Confidence Intervals:
1. Press **STAT**, then select **TESTS** in the top menu.
2. Select **2-SampTInt** in the menu and press **ENTER**.
3. Select **Data** if you have sample data in lists or **Stats** if you have summary statistics. Press **ENTER** and enter the list names (leave *Freq* = **1**) or summary statistics.
4. For *C-Level* enter the desired confidence level.
5. For pooled select **No**. Select *Yes* only if the population variances are believed to be equal.
6. Select **Calculate** and press **ENTER**.

Excel

Hypothesis Test

XLSTAT Add-In

Requires original sample data; does not work with summary data.
1. Click on the **XLSTAT** tab in the Ribbon and then click **Parametric Tests.**
2. Select **Two-sample t-test and z-test** from the dropdown menu.
3. Under *Sample 1 & 2* enter the range of cells containing the sample data. For *Data format* select **One column per sample.**
4. Select **Student's t test.**
5. If the first row of data contains a label, check the **Column labels** box.
6. Click the **Options** tab.
7. Under *Alternative hypothesis* select the desired format ($\neq$ for two-tailed test, $<$ for left-tailed test, $>$ for right-tailed test). Enter **0** for *Hypothesized difference (D)* and enter the desired significance level (enter **5** for 0.05 significance level).
8. Uncheck the **Assume equality** box and uncheck the **Cochran-Cox** box. Uncheck the **Use an *F*-test** box.
9. Click **OK** to display the test statistic (labeled *t Observed value*) and *P*-value.

TECH CENTER *continued*

 Inferences with Two Means: Independent Samples

Access tech supplements, videos, and data sets at **www.TriolaStats.com**

Excel

Excel (Data Analysis Add-In)

1. Click on **Data** in the ribbon, then click on the **Data Analysis** tab.
2. Select **t-Test: Two-Sample Assuming Unequal Variances** and click **OK.**
3. Enter the data range for each variable in the *Variable Range* boxes. If the first row contains a label, check the **Labels** box.
4. Enter **0** for *Hypothesized Mean Difference.*
5. Enter the desired significance level in the *Alpha* box and click **OK.** The test statistic is labeled *t Stat* and *P*-value is labeled *P.*

Confidence Interval

XLSTAT Add-In (Required)

Requires original sample data; does not work with summary data.

1-5. Follow above Steps 1–5 for **Hypothesis Test** using XLSTAT Add-In.
6. Click the **Options** tab.
7. Under *Alternative hypothesis* select the two-tailed option $\neq$. Enter **0** for *Hypothesized difference (D)* and enter the desired significance level (enter **5** for 95% confidence level).
8. Uncheck the **Assume equality** box and uncheck the **Cochran-Cox box**. Uncheck the **Use an *F*-test** box.
9. Click **OK** to display the confidence interval.

9-2 Basic Skills and Concepts

Statistical Literacy and Critical Thinking

1. Independent and Dependent Samples Which of the following involve independent samples?

a. Data Set 14 "Oscar Winner Age" in Appendix B includes pairs of ages of actresses and actors at the times that they won Oscars for Best Actress and Best Actor categories. The pair of ages of the winners is listed for each year, and each pair consists of ages matched according to the year that the Oscars were won.

b. Data Set 15 "Presidents" in Appendix B includes heights of elected presidents along with the heights of their main opponents. The pair of heights is listed for each election.

c. Data Set 26 "Cola Weights and Volumes" in Appendix B includes the volumes of the contents in 36 cans of regular Coke and the volumes of the contents in 36 cans of regular Pepsi.

2. Confidence Interval for Hemoglobin Large samples of women and men are obtained, and the hemoglobin level is measured in each subject. Here is the 95% confidence interval for the difference between the two population means, where the measures from women correspond to population 1 and the measures from men correspond to population 2: $-1.76 \text{ g/dL} < \mu_1 - \mu_2 < -1.62 \text{ g/dL}.$

a. What does the confidence interval suggest about equality of the mean hemoglobin level in women and the mean hemoglobin level in men?

b. Write a brief statement that interprets that confidence interval.

c. Express the confidence interval with measures from men being population 1 and measures from women being population 2.

3. Hypothesis Tests and Confidence Intervals for Hemoglobin

a. Exercise 2 includes a confidence interval. If you use the *P*-value method or the critical value method from Part 1 of this section to test the claim that women and men have the same mean hemoglobin levels, will the hypothesis tests and the confidence interval result in the same conclusion?

b. In general, if you conduct a hypothesis test using the methods of Part 1 of this section, will the *P*-value method, the critical value method, and the confidence interval method result in the same conclusion?

c. Assume that you want to use a 0.01 significance level to test the claim that the mean hemoglobin level in women is *less* than the mean hemoglobin level in men. What *confidence level* should be used if you want to test that claim using a confidence interval?

4. Degrees of Freedom
For Example 1 on page 431, we used df = smaller of $n_1 - 1$ and $n_2 - 1$, we got df = 11, and the corresponding critical values are $t = \pm 2.201$. If we calculate df using Formula 9-1, we get df = 19.063, and the corresponding critical values are ± 2.093. How is using the critical values of $t = \pm 2.201$ more "conservative" than using the critical values of ± 2.093?

In Exercises 5–20, assume that the two samples are independent simple random samples selected from normally distributed populations, and do not assume that the population standard deviations are equal. (Note: Answers in Appendix D include technology answers based on Formula 9-1 along with "Table" answers based on Table A-3 with df equal to the smaller of $n_1 - 1$ and $n_2 - 1$.)

5. Regular Coke and Diet Coke
Data Set 26 "Cola Weights and Volumes" in Appendix B includes weights (lb) of the contents of cans of Diet Coke ($n = 36$, $\bar{x} = 0.78479$ lb, $s = 0.00439$ lb) and of the contents of cans of regular Coke ($n = 36$, $\bar{x} = 0.81682$ lb, $s = 0.00751$ lb).

a. Use a 0.05 significance level to test the claim that the contents of cans of Diet Coke have weights with a mean that is less than the mean for regular Coke.

b. Construct the confidence interval appropriate for the hypothesis test in part (a).

c. Can you explain why cans of Diet Coke would weigh less than cans of regular Coke?

6. Coke and Pepsi
Data Set 26 "Cola Weights and Volumes" in Appendix B includes volumes of the contents of cans of regular Coke ($n = 36$, $\bar{x} = 12.19$ oz, $s = 0.11$ oz) and volumes of the contents of cans of regular Pepsi ($n = 36$, $\bar{x} = 12.29$ oz, $s = 0.09$ oz).

a. Use a 0.05 significance level to test the claim that cans of regular Coke and regular Pepsi have the same mean volume.

b. Construct the confidence interval appropriate for the hypothesis test in part (a).

c. What do you conclude? Does there appear to be a difference? Is there practical significance?

7. Color and Creativity
Researchers from the University of British Columbia conducted trials to investigate the effects of color on creativity. Subjects with a red background were asked to think of creative uses for a brick; other subjects with a blue background were given the same task. Responses were scored by a panel of judges and results from scores of creativity are given below. Higher scores correspond to more creativity. The researchers make the claim that "blue enhances performance on a creative task."

a. Use a 0.01 significance level to test the claim that blue enhances performance on a creative task.

b. Construct the confidence interval appropriate for the hypothesis test in part (a). What is it about the confidence interval that causes us to reach the same conclusion from part (a)?

Red Background:	$n = 35$, $\bar{x} = 3.39$, $s = 0.97$
Blue Background:	$n = 36$, $\bar{x} = 3.97$, $s = 0.63$

8. Color and Cognition Researchers from the University of British Columbia conducted a study to investigate the effects of color on cognitive tasks. Words were displayed on a computer screen with background colors of red and blue. Results from scores on a test of word recall are given below. Higher scores correspond to greater word recall.

a. Use a 0.05 significance level to test the claim that the samples are from populations with the same mean.

b. Construct a confidence interval appropriate for the hypothesis test in part (a). What is it about the confidence interval that causes us to reach the same conclusion from part (a)?

c. Does the background color appear to have an effect on word recall scores? If so, which color appears to be associated with higher word memory recall scores?

Red Background	$n = 35, \bar{x} = 15.89, s = 5.90$
Blue Background	$n = 36, \bar{x} = 12.31, s = 5.48$

9. Magnet Treatment of Pain People spend around $5 billion annually for the purchase of magnets used to treat a wide variety of pains. Researchers conducted a study to determine whether magnets are effective in treating back pain. Pain was measured using the visual analog scale, and the results given below are among the results obtained in the study (based on data from "Bipolar Permanent Magnets for the Treatment of Chronic Lower Back Pain: A Pilot Study," by Collacott, Zimmerman, White, and Rindone, *Journal of the American Medical Association*, Vol. 283, No. 10). Higher scores correspond to greater pain levels.

a. Use a 0.05 significance level to test the claim that those treated with magnets have a greater mean reduction in pain than those given a sham treatment (similar to a placebo).

b. Construct the confidence interval appropriate for the hypothesis test in part (a).

c. Does it appear that magnets are effective in treating back pain? Is it valid to argue that magnets might appear to be effective if the sample sizes are larger?

Reduction in Pain Level After Magnet Treatment: $n = 20, \bar{x} = 0.49, s = 0.96$

Reduction in Pain Level After Sham Treatment: $n = 20, \bar{x} = 0.44, s = 1.4$

10. Second-Hand Smoke Data Set 12 "Passive and Active Smoke" in Appendix B includes cotinine levels measured in a group of nonsmokers exposed to tobacco smoke ($n = 40$, $\bar{x} = 60.58$ ng/mL, $s = 138.08$ ng/mL) and a group of nonsmokers not exposed to tobacco smoke ($n = 40$, $\bar{x} = 16.35$ ng/mL, $s = 62.53$ ng/mL). Cotinine is a metabolite of nicotine, meaning that when nicotine is absorbed by the body, cotinine is produced.

a. Use a 0.05 significance level to test the claim that nonsmokers exposed to tobacco smoke have a higher mean cotinine level than nonsmokers not exposed to tobacco smoke.

b. Construct the confidence interval appropriate for the hypothesis test in part (a).

c. What do you conclude about the effects of second-hand smoke?

11. BMI We know that the mean weight of men is greater than the mean weight of women, and the mean height of men is greater than the mean height of women. A person's body mass index (BMI) is computed by dividing weight (kg) by the square of height (m). Given below are the BMI statistics for random samples of females and males taken from Data Set 1 "Body Data" in Appendix B.

a. Use a 0.05 significance level to test the claim that females and males have the same mean BMI.

b. Construct the confidence interval that is appropriate for testing the claim in part (a).

c. Do females and males appear to have the same mean BMI?

Female BMI: $n = 70, \bar{x} = 29.10, s = 7.39$

Male BMI: $n = 80, \bar{x} = 28.38, s = 5.37$

12. IQ and Lead Exposure Data Set 7 "IQ and Lead" in Appendix B lists full IQ scores for a random sample of subjects with low lead levels in their blood and another random sample of subjects with high lead levels in their blood. The statistics are summarized below.

a. Use a 0.05 significance level to test the claim that the mean IQ score of people with low blood lead levels is higher than the mean IQ score of people with high blood lead levels.

b. Construct a confidence interval appropriate for the hypothesis test in part (a).

c. Does exposure to lead appear to have an effect on IQ scores?

$$\text{Low Blood Lead Level:} \quad n = 78, \bar{x} = 92.88462, s = 15.34451$$

$$\text{High Blood Lead Level:} \quad n = 21, \bar{x} = 86.90476, s = 8.988352$$

13. Are Male Professors and Female Professors Rated Differently?

a. Use a 0.05 significance level to test the claim that two samples of course evaluation scores are from populations with the same mean. Use these summary statistics: Female professors: $n = 40, \bar{x} = 3.79, s = 0.51$; male professors: $n = 53, \bar{x} = 4.01, s = 0.53$. (Using the raw data in Data Set 17 "Course Evaluations" will yield different results.)

b. Using the summary statistics given in part (a), construct a 95% confidence interval estimate of the difference between the mean course evaluation score for female professors and male professors.

c. Example 1 used similar sample data with samples of size 12 and 15, and Example 1 led to the conclusion that there is not sufficient evidence to warrant rejection of the null hypothesis. Do the larger samples in this exercise affect the results much?

14. Seat Belts A study of seat belt use involved children who were hospitalized after motor vehicle crashes. For a group of 123 children who were wearing seat belts, the number of days in intensive care units (ICU) has a mean of 0.83 and a standard deviation of 1.77. For a group of 290 children who were not wearing seat belts, the number of days spent in ICUs has a mean of 1.39 and a standard deviation of 3.06 (based on data from "Morbidity Among Pediatric Motor Vehicle Crash Victims: The Effectiveness of Seat Belts," by Osberg and Di Scala, *American Journal of Public Health,* Vol. 82, No. 3).

a. Use a 0.05 significance level to test the claim that children wearing seat belts have a lower mean length of time in an ICU than the mean for children not wearing seat belts.

b. Construct a confidence interval appropriate for the hypothesis test in part (a).

c. What important conclusion do the results suggest?

15. Are Quarters Now Lighter? Weights of quarters are carefully considered in the design of the vending machines that we have all come to know and love. Data Set 29 "Coin Weights" in Appendix B includes weights of a sample of pre-1964 quarters ($n = 40, \bar{x} = 6.19267$ g, $s = 0.08700$ g) and weights of a sample of post-1964 quarters ($n = 40, \bar{x} = 5.63930$ g, $s = 0.06194$ g).

a. Use a 0.05 significance level to test the claim that pre-1964 quarters have a mean weight that is greater than the mean weight of post-1964 quarters.

b. Construct a confidence interval appropriate for the hypothesis test in part (a).

c. Do post-1964 quarters appear to weigh less than before 1964? If so, why aren't vending machines affected very much by the difference?

16. Bad Stuff in Children's Movies Data Set 11 "Alcohol and Tobacco in Movies" in Appendix B includes lengths of times (seconds) of tobacco use shown in animated children's movies. For the Disney movies, $n = 33, \bar{x} = 61.6$ sec, $s = 118.8$ sec. For the other movies, $n = 17$, $\bar{x} = 49.3$ sec, $s = 69.3$ sec. The sorted times for the non-Disney movies are listed below.

continued

a. Use a 0.05 significance level to test the claim that Disney animated children's movies and other animated children's movies have the same mean time showing tobacco use.

b. Construct a confidence interval appropriate for the hypothesis test in part (a).

c. Conduct a quick visual inspection of the listed times for the non-Disney movies and comment on the normality requirement. How does the normality of the 17 non-Disney times affect the results?

0 0 0 0 0 0 1 5 6 17 24 55 91 117 155 162 205

17. Are Male Professors and Female Professors Rated Differently? Listed below are student evaluation scores of female professors and male professors from Data Set 17 "Course Evaluations" in Appendix B. Test the claim that female professors and male professors have the same mean evaluation ratings. Does there appear to be a difference?

Females	4.4	3.4	4.8	2.9	4.4	4.9	3.5	3.7	3.4	4.8
Males	4.0	3.6	4.1	4.1	3.5	4.6	4.0	4.3	4.5	4.3

18. Car and Taxi Ages When the author visited Dublin, Ireland (home of Guinness Brewery employee William Gosset, who first developed the *t* distribution), he recorded the ages of randomly selected passenger cars and randomly selected taxis. The ages can be found from the license plates. (There is no end to the fun of traveling with the author.) The ages (in years) are listed below. We might expect that taxis would be newer, so test the claim that the mean age of cars is greater than the mean age of taxis.

Car Ages	4	0	8	11	14	3	4	4	3	5	8	3	3	7	4	6	6	1	8	2	15	11	4	1	6	1	8
Taxi Ages	8	8	0	3	8	4	3	3	6	11	7	7	6	9	5	10	8	4	3	4							

19. Is Old Faithful Not Quite So Faithful? Listed below are time intervals (min) between eruptions of the Old Faithful geyser. The "recent" times are within the past few years, and the "past" times are from 1995. Does it appear that the mean time interval has changed? Is the conclusion affected by whether the significance level is 0.05 or 0.01?

Recent	78	91	89	79	57	100	62	87	70	88	82	83	56	81	74	102	61
Past	89	88	97	98	64	85	85	96	87	95	90	95					

20. Blanking Out on Tests Many students have had the unpleasant experience of panicking on a test because the first question was exceptionally difficult. The arrangement of test items was studied for its effect on anxiety. The following scores are measures of "debilitating test anxiety," which most of us call panic or blanking out (based on data from "Item Arrangement, Cognitive Entry Characteristics, Sex and Test Anxiety as Predictors of Achievement in Examination Performance," by Klimko, *Journal of Experimental Education*, Vol. 52, No. 4.) Is there sufficient evidence to support the claim that the two populations of scores have different means? Is there sufficient evidence to support the claim that the arrangement of the test items has an effect on the score? Is the conclusion affected by whether the significance level is 0.05 or 0.01?

Questions Arranged from Easy to Difficult					Questions Arranged from Difficult to Easy			
24.64	39.29	16.32	32.83	28.02	33.62	34.02	26.63	30.26
33.31	20.60	21.13	26.69	28.90	35.91	26.68	29.49	35.32
26.43	24.23	7.10	32.86	21.06	27.24	32.34	29.34	33.53
28.89	28.71	31.73	30.02	21.96	27.62	42.91	30.20	32.54
25.49	38.81	27.85	30.29	30.72				

Larger Data Sets. *In Exercises 21–24, use the indicated Data Sets in Appendix B. The complete data sets can be found at www.TriolaStats.com. Assume that the two samples are independent simple random samples selected from normally distributed populations. Do not assume that the population standard deviations are equal.*

21. Do Men Talk Less Than Women? Refer to Data Set 24 "Word Counts" and use the measured word counts from men in the third column and the measured word counts from women in the fourth column. Use a 0.05 significance level to test the claim that men talk less than women.

22. Do Men and Women Have the Same Mean Diastolic Blood Pressure? Refer to Data Set 1 "Body Data" and use a 0.05 significance level to test the claim that women and men have the same mean diastolic blood pressure.

23. Birth Weights Refer to Data Set 4 "Births" and use the birth weights of boys and girls. Test the claim that at birth, girls have a lower mean weight than boys.

24. Birth Length of Stay Refer to Data Set 4 "Births" and use the "lengths of stay" for boys and girls. A length of stay is the number of days the child remained in the hospital. Test the claim boys and girls have the same mean length of stay.

9-2 Beyond the Basics

25. Pooling Repeat Exercise 12 "IQ and Lead" by assuming that the two population standard deviations are equal, so $\sigma_1 = \sigma_2$. Use the appropriate method from Part 2 of this section. Does pooling the standard deviations yield results showing greater significance?

26. Degrees of Freedom In Exercise 20 "Blanking Out on Tests," using the "smaller of $n_1 - 1$ and $n_2 - 1$" for the number of degrees of freedom results in df $= 15$. Find the number of degrees of freedom using Formula 9-1. In general, how are hypothesis tests and confidence intervals affected by using Formula 9-1 instead of the "smaller of $n_1 - 1$ and $n_2 - 1$"?

27. No Variation in a Sample An experiment was conducted to test the effects of alcohol. Researchers measured the breath alcohol levels for a treatment group of people who drank ethanol and another group given a placebo. The results are given below (based on data from "Effects of Alcohol Intoxication on Risk Taking, Strategy, and Error Rate in Visuomotor Performance," by Streufert et al., *Journal of Applied Psychology,* Vol. 77, No. 4). Use a 0.05 significance level to test the claim that the two sample groups come from populations with the same mean.

$$\text{Treatment Group:} \quad n_1 = 22, \bar{x}_1 = 0.049, s_1 = 0.015$$

$$\text{Placebo Group:} \quad n_2 = 22, \bar{x}_2 = 0.000, s_2 = 0.000$$

9-3 Two Dependent Samples (Matched Pairs)

Key Concept This section presents methods for testing hypotheses and constructing confidence intervals involving the mean of the differences of the values from two populations that are dependent in the sense that the data consist of matched pairs. The pairs must be matched according to some relationship, such as before/after measurements from the same subjects or IQ scores of husbands and wives.

Good Experimental Design

Suppose we want to test the effectiveness of the Kaplan SAT preparatory course. It would be better to use before/after scores from a single group of students who took the course than to use scores from one group of students who did not take the Kaplan course and another group who took the course. The advantage of using matched pairs (before/after scores) is that we reduce extraneous variation, which could occur with the two different independent samples. This strategy for designing an experiment can be generalized by the following design principle:

> When designing an experiment or planning an observational study, using dependent samples with matched pairs is generally better than using two independent samples.

Déjà Vu All Over Again The methods of hypothesis testing in this section are the *same methods* for testing a claim about a population mean (Part 1 of Section 8-3), except that here we use the *differences* from the matched pairs of sample data.

There are no exact procedures for dealing with dependent samples, but the following approximation methods are commonly used.

KEY ELEMENTS

Inferences About Differences from Matched Pairs

Objectives

1. **Hypothesis Test:** Use the differences from two dependent samples (matched pairs) to test a claim about the mean of the population of all such differences.

2. **Confidence Interval:** Use the differences from two dependent samples (matched pairs) to construct a confidence interval estimate of the mean of the population of all such differences.

Notation for Dependent Samples

d = individual difference between the two values in a single matched pair

μ_d = mean value of the differences d for the *population* of all matched pairs of data

$\bar{d}$ = mean value of the differences d for the paired *sample* data

s_d = standard deviation of the differences d for the paired *sample* data

n = number of *pairs* of sample data

Requirements

1. The sample data are dependent (matched pairs).

2. The matched pairs are a simple random sample.

3. Either or both of these conditions are satisfied: The number of pairs of sample data is large ($n > 30$) or the

pairs of values have differences that are from a population having a distribution that is approximately normal. These methods are *robust* against departures for normality, so the normality requirement is loose.

Test Statistic for Dependent Samples (with H_0: $\mu_d = 0$)

$$t = \frac{\bar{d} - \mu_d}{\dfrac{s_d}{\sqrt{n}}}$$

P-Values: P-values are automatically provided by technology or the t distribution in Table A-3 can be used. Use the procedure given in Figure 8-3 on page 364.

Critical Values: Use Table A-3 (t distribution). For degrees of freedom, use df $= n - 1$.

Confidence Intervals for Dependent Samples

$$\bar{d} - E < \mu_d < \bar{d} + E$$

where $E = t_{\alpha/2}\dfrac{s_d}{\sqrt{n}}$ (Degrees of freedom: df $= n - 1$.)

Crest and Dependent Samples

In the late 1950s, Procter & Gamble introduced Crest toothpaste as the first such product with fluoride. To test the effectiveness of Crest in reducing cavities, researchers conducted experiments with several sets of twins. One of the twins in each set was given Crest with fluoride, while the other twin continued to use ordinary toothpaste without fluoride. It was believed that each pair of twins would have similar eating, brushing, and genetic characteristics. Results showed that the twins who used Crest had significantly fewer cavities than those who did not. This use of twins as dependent samples allowed the researchers to control many of the different variables affecting cavities.

Procedures for Inferences with Dependent Samples

1. Verify that the sample data consist of dependent samples (or matched pairs), and verify that the requirements in the preceding Key Elements box are satisfied.

2. Find the difference d for each pair of sample values. (*Caution:* Be sure to subtract in a consistent manner, such as "before – after.")

3. Find the value of $\bar{d}$ (mean of the differences) and s_d (standard deviation of the differences).

4. For hypothesis tests and confidence intervals, use the same t test procedures used for a single population mean (described in Part 1 of Section 8-3).

Equivalent Methods

Because the hypothesis test and confidence interval in this section use the same distribution and standard error, they are *equivalent* in the sense that they result in the same conclusions. Consequently, a null hypothesis that the mean difference equals 0 can be tested by determining whether the confidence interval includes 0.

 EXAMPLE 1 **Are Best Actresses Generally Younger Than Best Actors?**

Here we consider one aspect of how we treat women and men differently based on their ages. Data Set 14 "Oscar Winner Age" in Appendix B lists ages of actresses when they won Oscars in the category of Best Actress, along with the ages of actors when they won Oscars in the category of Best Actor. The ages are matched according to the year that the awards were presented. Table 9-2 includes a small random selection of the available data so that we can better illustrate the procedures of this section. Use the sample data in Table 9-2 with a 0.05 significance level to test the claim that for the population of ages of Best Actresses and Best Actors, the differences have a mean less than 0 (indicating that Best Actresses are generally younger than Best Actors).

TABLE 9-2   Ages of Best Actresses and Best Actors

Actress (years)	28	28	31	29	35
Actor (years)	62	37	36	38	29
Difference d	−34	−9	−5	−9	6

SOLUTION

REQUIREMENT CHECK   We address the three requirements listed earlier in the Key Elements box. (1) The samples are dependent because the values are matched by the year in which the awards were given. (2) The pairs of data are randomly selected. We will consider the data to be a simple random sample. (3) Because the number of pairs of data is $n = 5$, which is not large, we should check for normality of the differences and we should check for outliers. There are no outliers, and a normal quantile plot would show that the points approximate a straight-line pattern with no other pattern, so the differences satisfy the loose requirement of being from a normally distributed population. All requirements are satisfied. ✓

We will follow the same method of hypothesis testing that we used for testing a claim about a mean (see Figure 8-1 on page 360), but we use *differences* instead of raw sample data.

Step 1: The claim that the differences have a mean less than 0 can be expressed as $\mu_d < 0$ year.

Step 2: If the original claim is not true, we have $\mu_d \geq 0$ year.

Step 3: The null hypothesis must express equality and the alternative hypothesis cannot include equality, so we have

$$H_0: \mu_d = 0 \text{ year} \quad H_1: \mu_d < 0 \text{ year (original claim)}$$

Step 4: The significance level is $\alpha = 0.05$.

Step 5: We use the Student t distribution.

Step 6: Before finding the value of the test statistic, we must first find the values of $\bar{d}$ and s_d. We use the differences from Table 9-2 (-34, -9, -5, -9, 6) to find these sample statistics: $\bar{d} = -10.2$ years and $s_d = 14.7$ years. Using these sample statistics and the assumption from the null hypothesis that $\mu_d = 0$ year, we can now find the value of the test statistic. (The value of $t = -1.557$ is obtained if unrounded values of $\bar{d}$ and s_d are used; technology will provide a test statistic of $t = -1.557$.)

$$t = \frac{\bar{d} - \mu_d}{\dfrac{s_d}{\sqrt{n}}} = \frac{-10.2 - 0}{\dfrac{14.7}{\sqrt{5}}} = -1.552$$

P-Value Method Because we are using a t distribution, we refer to Table A-3 for the row with df = 4 and we see that the test statistic $t = -1.552$ corresponds to an "Area in One Tail" that is greater than 0.05, so P-value > 0.05. Technology would provide P-value = 0.0973. See Figure 9-3 (a).

Critical Value Method Refer to Table A-3 to find the critical value of $t = -2.132$ as follows: Use the column for 0.05 (Area in One Tail), and use the row with degrees of freedom of $n - 1 = 4$. The critical value $t = -2.132$ is negative because this test is left-tailed where all values of t are negative. See Figure 9-3(b).

Step 7: If we use the P-value method, we fail to reject H_0 because the P-value is greater than the significance level of 0.05. If we use the critical value method, we fail to reject H_0 because the test statistic does not fall in the critical region.

Twins in Twinsburg

During the first weekend in August of each year, Twinsburg, Ohio, celebrates its annual "Twins Days in Twinsburg" festival. Thousands of twins from around the world have attended this festival in the past. Scientists saw the festival as an opportunity to study identical twins. Because they have the same basic genetic structure, identical twins are ideal for studying the different effects of heredity and environment on a variety of traits, such as male baldness, heart disease, and deafness—traits that were recently studied at one Twinsburg festival. A study of twins showed that myopia (near-sightedness) is strongly affected by hereditary factors, not by environmental factors such as watching television, surfing the Internet, or playing computer or video games.

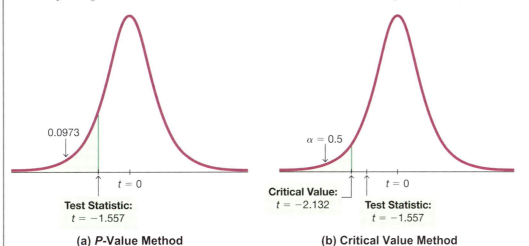

0.0973

Test Statistic:
$t = -1.557$

(a) *P-Value Method*

$\alpha = 0.5$

Critical Value:
$t = -2.132$

Test Statistic:
$t = -1.557$

(b) **Critical Value Method**

FIGURE 9-3   Hypothesis Test with Dependent Samples

INTERPRETATION

We conclude that there is not sufficient evidence to support $\mu_d < 0$. There is not sufficient evidence to support the claim that for the population of ages of Best Actresses and Best Actors, the differences have a mean less than 0. There is not sufficient evidence to conclude that Best Actresses are generally younger than Best Actors.

YOUR TURN Do Exercise 5 "Oscar Hypothesis Test."

Technology Software and calculators typically provide a *P*-value, so the *P*-value method of testing hypotheses is usually used. See the accompanying Statdisk results showing the test statistic of $t = -1.5566$ and the *P*-value of 0.0973. Because the *P*-value of 0.0973 is greater than the significance level of 0.05, we fail to reject the null hypothesis and we conclude that there is not sufficient evidence to support the claim that for the population of ages of Best Actresses and Best Actors, the differences have a mean less than 0.

Statdisk

Sample size, n: 5
Difference Mean, d: -10.2
Difference Standard Deviation, sd: 14.65264
Test Statistic, t: -1.5566
Critical t: -2.1318
P-Value: 0.0973

90% Confidence interval:
-24.16971 < μd < 3.769712

EXAMPLE 2 Confidence Interval for Estimating the Mean of the Age Differences

Using the same sample data in Table 9-2, construct a 90% confidence interval estimate of μ_d, which is the mean of the age differences. By using a confidence level of 90%, we get a result that could be used for the hypothesis test in Example 1. (Because the hypothesis test is one-tailed with a significance level of $\alpha = 0.05$, the confidence level should be 90%. See Table 8-1 on page 360.)

SOLUTION

REQUIREMENT CHECK The solution for Example 1 includes verification that the requirements are satisfied. ✓

The preceding Statdisk display shows the 90% confidence interval. It is found using the values of $\bar{d} = -10.2$ years, $s_d = 14.7$ years, and $t_{\alpha/2} = 2.132$ (found from Table A-3 with $n - 1 = 4$ degrees of freedom and an area of 0.10 divided equally between the two tails). We first find the value of the margin of error E.

$$E = t_{\alpha/2}\frac{s_d}{\sqrt{n}} = 2.132 \cdot \frac{14.7}{\sqrt{5}} = 14.015853$$

We now find the confidence interval.

$$\bar{d} - E < \mu_d < \bar{d} + E$$
$$-10.2 - 14.015853 < \mu_d < -10.2 + 14.015853$$
$$-24.2 \text{ years} < \mu_d < 3.8 \text{ years}$$

INTERPRETATION

We have 90% confidence that the limits of −24.2 years and 3.8 years contain the true value of the mean of the age differences. In the long run, 90% of such samples will lead to confidence interval limits that actually do contain the true population mean of the differences. See that the confidence interval includes the value of 0 year, so it is very possible that the mean of the differences is equal to 0 year, indicating that there is no significant difference between ages of Best Actresses and Best Actors. Remember, this conclusion is based on the very small sample included in Table 9-2.

Alternative Method Used When Population Is Not Normal and $n \leq 30$

Bootstrap The Key Elements box near the beginning of this section included the following requirement: The number of pairs of sample data is large ($n > 30$) or the pairs of values have differences that are from a population having a distribution that is approximately normal. If that condition is violated, we can use the "Bootstrap Procedure for a Confidence Interval Estimate of a Parameter" included in Section 7-4. Use that procedure to find the *mean* of each bootstrap sample, then sort those means, then use percentiles to find the confidence interval that can be used for hypothesis tests. See Exercise 26 "Bootstrap."

TECH CENTER

Inferences with Two Means: Dependent Samples
Access tech supplements, videos, and data sets at **www.TriolaStats.com**

Statdisk	Minitab
Requires *paired sample data* entered in columns.	Requires *paired sample data* entered in columns.
Hypothesis Testing	1. Click **Stat** in the top menu.
1. Click **Analysis** in the top menu.	2. Select **Basic Statistics** from the dropdown menu and select **Paired t** from the submenu.
2. Select **Hypothesis Testing** from the dropdown menu and **Mean-Matched Pairs** from the submenu.	3. Select **Each sample is in a column** from the dropdown menu and select the desired data columns.
3. Select the desired format for *Alternative Hypothesis*, enter the significance level, and select the data columns to compare.	4. Click the **Options** button and enter the confidence level. Enter **0** for the *Hypothesized Difference* and select the desired format for the *Alternative Hypothesis*.
4. Click **Evaluate**.	5. Click **OK** twice.
Confidence Intervals	
1. Click **Analysis** in the top menu.	
2. Select **Confidence Intervals** from the dropdown menu and **Mean-Matched Pairs** from the submenu.	
3. Enter the desired confidence level and select the data columns to compare.	
4. Click **Evaluate**.	

StatCrunch

Requires *paired sample data* entered in columns.
1. Click **Stat** in the top menu.
2. Select **T Stats** from the dropdown menu, then select **Paired** from the submenu.
3. Select the columns containing paired sample data.
4. *Hypothesis Testing*: Select **Hypothesis test for $\mu_D = \mu_1 - \mu_2$**. For hypothesized difference (H_0) enter **0** and select the desired format for the alternative hypothesis (H_A).

   Confidence Intervals: Select **Confidence interval for $\mu_D = \mu_1 - \mu_2$** and enter the confidence level.
5. Click **Compute!**

continued

TECH CENTER *continued*

Inferences with Two Means: Dependent Samples

Access tech supplements, videos, and data sets at **www.TriolaStats.com**

TI-83/84 Plus Calculator

Caution: Do not use the menu item **2-SampTTest** because it applies only to *independent samples*.

1. Enter the data for the first variable in list *L1* and the data for the second variable in list *L2*.
2. Create a list of differences and store the list in *L3* by entering **L1** ⊖ **L2** **STO▸** **L3.**
3. Press **STAT**, then select **TESTS** in the top menu.

Hypothesis Testing

4. Choose **T-Test** and press **ENTER**.
5. Select **Data** and press **ENTER**.
6. For μ_0 enter **0**.
7. For *List* enter **L3** (leave *Freq* = **1**).
8. For μ select the desired format for the alternative hypothesis.
9. Select **Calculate** and press **ENTER**.

Confidence Interval

4. Choose **TInterval** and press **ENTER**.
5. Select **Data**, press **ENTER**, and enter **L3** for list name (leave *Freq* = **1**).
6. Enter the desired confidence level *C-Level*.
7. Select **Calculate** and press **ENTER**.

Excel

Requires *paired sample data* entered in columns.

Hypothesis Test

XLSTAT Add-In

1. Click on the **XLSTAT** tab in the Ribbon and then click **Parametric Tests.**
2. Select **Two-sample t-test and z-test** from the dropdown menu.
3. Under *Sample 1 & 2*, enter the range of cells containing the paired sample data. For *Data format* select **Paired samples.**
4. Select **Student's t test.**
5. If the first row of data contains a label, check the **Column labels** box.
6. Click the **Options** tab.
7. Under *Alternative hypothesis* select the desired format ($\neq$ for two-tailed test, $<$ for left-tailed test, $>$ for right-tailed test). Enter **0** for *Hypothesized difference (D)* and enter the desired significance level (enter **5** for 0.05 significance level).
8. Click **OK** to display the test statistic (labeled *t Observed value*) and *P*-value.

Excel (Data Analysis Add-In)

1. Click on **Data** in the ribbon, then click on the **Data Analysis** tab.
2. Select **t-Test: Paired Two Sample for Means** and click **OK**.
3. Enter the data range for each variable in the *Variable Range* boxes. If the first row contains a label, check the **Labels** box.
4. Enter **0** for *Hypothesized Mean Difference*.
5. Enter the desired significance level in the *Alpha* box.
6. Click **OK**. The results include the test statistic (labeled *t Stat*), *P*-values for a one-tail and two-tail test, and critical values for one-tail and two-tail test.

Confidence Interval

XLSTAT Add-In (Required)

1–5. Follow above Steps 1–5 for **Hypothesis Test** using the XLSTAT Add-In.
6. Click the **Options** tab.
7. Under *Alternative hypothesis* select the two-tailed option $\neq$. Enter **0** for *Hypothesized difference (D) and* enter the desired significance level (enter **5** for 95% confidence level).
8. Click **OK** to display the confidence interval.

9-3 Basic Skills and Concepts

Statistical Literacy and Critical Thinking

1. True? For the methods of this section, which of the following statements are true?

a. When testing a claim with ten matched pairs of heights, hypothesis tests using the *P*-value method, critical value method, and confidence interval method will all result in the same conclusion.

b. The methods of this section are *robust* against departures from normality, which means that the distribution of sample differences must be very close to a normal distribution.

c. If we want to use a confidence interval to test the claim that $\mu_d < 0$ with a 0.01 significance level, the confidence interval should have a confidence level of 98%.

d. The methods of this section can be used with annual incomes of 50 randomly selected attorneys in North Carolina and 50 randomly selected attorneys in South Carolina.

e. With ten matched pairs of heights, the methods of this section require that we use $n = 20$.

2. Notation Listed below are body temperatures from five different subjects measured at 8 AM and again at 12 AM (from Data Set 3 "Body Temperatures" in Appendix B). Find the values of $\bar{d}$ and s_d. In general, what does μ_d represent?

Temperature (°F) at 8 AM	97.8	99.0	97.4	97.4	97.5
Temperature (°F) at 12 AM	98.6	99.5	97.5	97.3	97.6

3. Units of Measure If the values listed in Exercise 2 are changed so that they are expressed in Celsius degrees instead of Fahrenheit degrees, how are hypothesis test results affected?

4. Degrees of Freedom If we use the sample data in Exercise 2 for constructing a 99% confidence interval, what is the number of degrees of freedom that should be used for finding the critical value of $t_{\alpha/2}$? What is the critical value $t_{\alpha/2}$?

In Exercises 5–16, use the listed paired sample data, and assume that the samples are simple random samples and that the differences have a distribution that is approximately normal.

5. Oscar Hypothesis Test

a. Example 1 on page 444 in this section used only five pairs of data from Data Set 14 "Oscar Winner Age" in Appendix B. Repeat the hypothesis test of Example 1 using the data given below. Use a 0.05 significance level as in Example 1.

b. Construct the confidence interval that could be used for the hypothesis test described in part (a). What feature of the confidence interval leads to the same conclusion reached in part (a)?

Actress (years)	28	28	31	29	35	26	26	41	30	34
Actor (years)	62	37	36	38	29	34	51	39	37	42

6. Heights of Presidents A popular theory is that presidential candidates have an advantage if they are taller than their main opponents. Listed are heights (cm) of presidents along with the heights of their main opponents (from Data Set 15 "Presidents").

a. Use the sample data with a 0.05 significance level to test the claim that for the population of heights of presidents and their main opponents, the differences have a mean greater than 0 cm.

b. Construct the confidence interval that could be used for the hypothesis test described in part (a). What feature of the confidence interval leads to the same conclusion reached in part (a)?

Height (cm) of President	185	178	175	183	193	173
Height (cm) of Main Opponent	171	180	173	175	188	178

7. Body Temperatures Listed below are body temperatures from seven different subjects measured at two different times in a day (from Data Set 3 "Body Temperatures" in Appendix B).

a. Use a 0.05 significance level to test the claim that there is no difference between body temperatures measured at 8 AM and at 12 AM.

b. Construct the confidence interval that could be used for the hypothesis test described in part (a). What feature of the confidence interval leads to the same conclusion reached in part (a)?

Body Temperature (°F) at 8 AM	96.6	97.0	97.0	97.8	97.0	97.4	96.6
Body Temperature (°F) at 12 AM	99.0	98.4	98.0	98.6	98.5	98.9	98.4

8. The Spoken Word Listed below are the numbers of words spoken in a day by each member of six different couples. The data are randomly selected from the first two columns in Data Set 24 "Word Counts" in Appendix B.

a. Use a 0.05 significance level to test the claim that among couples, males speak fewer words in a day than females.

b. Construct the confidence interval that could be used for the hypothesis test described in part (a). What feature of the confidence interval leads to the same conclusion reached in part (a)?

Male	15,684	26,429	1,411	7,771	18,876	15,477	14,069	25,835
Female	24,625	13,397	18,338	17,791	12,964	16,937	16,255	18,667

9. Heights of Mothers and Daughters Listed below are heights (in.) of mothers and their first daughters. The data are from a journal kept by Francis Galton. (See Data Set 5 "Family Heights" in Appendix B.) Use a 0.05 significance level to test the claim that there is no difference in heights between mothers and their first daughters.

Height of Mother	68.0	60.0	61.0	63.5	69.0	64.0	69.0	64.0	63.5	66.0
Height of Daughter	68.5	60.0	63.5	67.5	68.0	65.5	69.0	68.0	64.5	63.0

10. Heights of Fathers and Sons Listed below are heights (in.) of fathers and their first sons. The data are from a journal kept by Francis Galton. (See Data Set 5 "Family Heights" in Appendix B.) Use a 0.05 significance level to test the claim that there is no difference in heights between fathers and their first sons.

Height of Father	72.0	66.0	69.0	70.0	70.0	70.0	70.0	75.0	68.2	65.0
Height of Son	73.0	68.0	68.0	71.0	70.0	70.0	71.0	71.0	70.0	63.0

11. Speed Dating: Attributes Listed below are "attribute" ratings made by participants in a speed dating session. Each attribute rating is the sum of the ratings of five attributes (sincerity, intelligence, fun, ambition, shared interests). The listed ratings are from Data Set 18 "Speed Dating" in Appendix B. Use a 0.05 significance level to test the claim that there is a difference between female attribute ratings and male attribute ratings.

Rating of Male by Female	29	38	36	37	30	34	35	23	43
Rating of Female by Male	36	34	34	33	31	17	31	30	42

12. Speed Dating: Attractiveness Listed below are "attractiveness" ratings made by participants in a speed dating session. Each attribute rating is the sum of the ratings of five attributes (sincerity, intelligence, fun, ambition, shared interests). The listed ratings are from Data Set 18 "Speed Dating." Use a 0.05 significance level to test the claim that there is a difference between female attractiveness ratings and male attractiveness ratings.

Rating of Male by Female	4.0	8.0	7.0	7.0	6.0	8.0	6.0	4.0	2.0	5.0	9.5	7.0
Rating of Female by Male	6.0	8.0	7.0	9.0	5.0	7.0	5.0	4.0	6.0	8.0	6.0	5.0

13. Friday the 13th Researchers collected data on the numbers of hospital admissions resulting from motor vehicle crashes, and results are given below for Fridays on the 6th of a month and Fridays on the following 13th of the same month (based on data from "Is Friday the 13th Bad for Your Health?" by Scanlon et al., *British Medical Journal,* Vol. 307, as listed in the *Data and Story Line* online resource of data sets). Construct a 95% confidence interval estimate of the mean of the population of differences between hospital admissions on days that are Friday the 6th of a month and days that are Friday the 13th of a month. Use the confidence interval to test the claim that when the 13th day of a month falls on a Friday, the numbers of hospital admissions from motor vehicle crashes are not affected.

Friday the 6th	9	6	11	11	3	5
Friday the 13th	13	12	14	10	4	12

14. Two Heads Are Better Than One Listed below are brain volumes (cm³) of twins from Data Set 8 "IQ and Brain Size" in Appendix B. Construct a 99% confidence interval estimate of the mean of the differences between brain volumes for the first-born and the second-born twins. What does the confidence interval suggest?

First Born	1005	1035	1281	1051	1034	1079	1104	1439	1029	1160
Second Born	963	1027	1272	1079	1070	1173	1067	1347	1100	1204

15. Hypnotism for Reducing Pain A study was conducted to investigate the effectiveness of hypnotism in reducing pain. Results for randomly selected subjects are given in the accompanying table (based on "An Analysis of Factors That Contribute to the Efficacy of Hypnotic Analgesia," by Price and Barber, *Journal of Abnormal Psychology,* Vol. 96, No. 1). The values are before and after hypnosis; the measurements are in centimeters on a pain scale. Higher values correspond to greater levels of pain. Construct a 95% confidence interval for the mean of the "before/after" differences. Does hypnotism appear to be effective in reducing pain?

Subject	A	B	C	D	E	F	G	H
Before	6.6	6.5	9.0	10.3	11.3	8.1	6.3	11.6
After	6.8	2.4	7.4	8.5	8.1	6.1	3.4	2.0

16. Self-Reported and Measured Male Heights As part of the National Health and Nutrition Examination Survey, the Department of Health and Human Services obtained self-reported heights (in.) and measured heights (in.) for males aged 12–16. Listed below are sample results. Construct a 99% confidence interval estimate of the mean difference between reported heights and measured heights. Interpret the resulting confidence interval, and comment on the implications of whether the confidence interval limits contain 0.

Reported	68	71	63	70	71	60	65	64	54	63	66	72
Measured	67.9	69.9	64.9	68.3	70.3	60.6	64.5	67.0	55.6	74.2	65.0	70.8

Larger Data Sets. *In Exercises 17–24, use the indicated Data Sets in Appendix B. The complete data sets can be found at www.TriolaStats.com. Assume that the paired sample data are simple random samples and the differences have a distribution that is approximately normal.*

17. Oscars Repeat Exercise 5 "Oscar Hypothesis Test" using all of the sample data from Data Set 14 "Oscar Winner Age" in Appendix B. Note that the pairs of data consist of ages that are matched according to the year in which the Oscars were won. Again use a significance level of 0.05.

18. Heights of Presidents Repeat Exercise 6 "Heights of Presidents" using all of the sample data from Data Set 15 "Presidents" in Appendix B.

19. Body Temperatures Repeat Exercise 7 "Body Temperatures" using all of the 8 AM and 12 AM body temperatures on Day 2 as listed in Data Set 3 "Body Temperatures" in Appendix B.

20. The Spoken Word Repeat Exercise 8 "The Spoken Word" using all of the data in the first two columns of Data Set 24 "Word Counts" in Appendix B.

21. Heights of Mothers and Daughters Repeat Exercise 9 "Heights of Mothers and Daughters" using all of the heights of mothers and daughters listed in Data Set 5 "Family Heights" in Appendix B.

22. Heights of Fathers and Sons Repeat Exercise 10 "Heights of Fathers and Sons" using all of the heights of fathers and sons listed in Data Set 5 "Family Heights" in Appendix B.

23. Speed Dating: Attributes Repeat Exercise 11 "Speed Dating: Attributes" using all of the attribute ratings by females and males. The ratings are listed in Data Set 18 "Speed Dating" in Appendix B.

24. Speed Dating: Attractiveness Repeat Exercise 12 "Speed Dating: Attractiveness" using all of the attractiveness ratings by females and males. The ratings are listed in Data Set 18 "Speed Dating" in Appendix B.

9-3 Beyond the Basics

25. Body Temperatures Refer to Data Set 3 "Body Temperatures" in Appendix B and use all of the matched pairs of body temperatures at 8 AM and 12 AM on Day 1. When using a 0.05 significance level for testing a claim of a difference between the temperatures at 8 AM and at 12 AM on Day 1, how are the hypothesis test results and confidence interval results affected if the temperatures are converted from degrees Fahrenheit to degrees Celsius? What is the relationship between the confidence interval limits for the body temperatures in degrees Fahrenheit and the confidence interval limits for the body temperatures in degrees Celsius? *Hint:* $C = \dfrac{5}{9}(F - 32)$.

26. Bootstrap

a. If paired sample data (x, y) are such that the values of x do not appear to be from a population with a normal distribution, and the values of y do not appear to be from a population with a normal distribution, does it follow that the values of $\overline{d}$ will not appear to be from a population with a normal distribution?

b. For the hypothesis test described in Exercise 25 "Body Temperatures," use the temperatures in degrees Fahrenheit and find the 95% confidence interval estimate of μ_d based on 1000 bootstrap samples. Generate the bootstrap samples using the values of $\overline{d}$.

9-4 Two Variances or Standard Deviations

Key Concept In this section we present the F test for testing claims made about two population variances (or standard deviations). The F test (named for statistician Sir Ronald Fisher) uses the F distribution introduced in this section. The F test requires that both populations have normal distributions. Instead of being robust, this test is *very* sensitive to departures from normal distributions, so the normality requirement is quite strict. Part 1 describes the F test procedure for conducting a hypothesis test, and Part 2 gives a brief description of two alternative methods for comparing variation in two samples.

PART 1 *F* Test with Two Variances or Standard Deviations

The following Key Elements box includes elements of a hypothesis test of a claim about two population variances or two population standard deviations. The procedure is based on using the two sample variances, but the *same procedure* is used for claims made about two population standard deviations.

The actual *F* test could be two-tailed, left-tailed, or right-tailed, but we can make computations much easier by stipulating that the larger of the two sample variances is denoted by s_1^2. It follows that the smaller sample variance is denoted as s_2^2. This stipulation of denoting the larger sample variance by s_1^2 allows us to avoid the somewhat messy problem of finding a critical value of *F* for the left tail.

KEY ELEMENTS

Hypothesis Test with Two Variances or Standard Deviations

Objective

Conduct a hypothesis test of a claim about two population variances or standard deviations. (Any claim made about two population standard deviations can be restated with an equivalent claim about two population variances, so the same procedure is used for two population standard deviations or two population variances.)

Notation

s_1^2 = *larger* of the two sample variances

n_1 = size of the sample with the *larger* variance

σ_1^2 = variance of the population from which the sample with the *larger* variance was drawn

The symbols s_2^2, n_2, and σ_2^2 are used for the other sample and population.

Requirements

1. The two populations are *independent*.

2. The two samples are simple random samples.

3. Each of the two populations must be *normally distributed*, regardless of their sample sizes. This

F test is *not robust* against departures from normality, so it performs poorly if one or both of the populations have a distribution that is not normal. The requirement of normal distributions is quite strict for this *F* test.

Test Statistic for Hypothesis Tests with Two Variances (with H_0: $\sigma_1^2 = \sigma_2^2$)

$F = \dfrac{s_1^2}{s_2^2}$ (where s_1^2 is the *larger* of the two sample variances)

P-Values: *P*-values are automatically provided by technology. If technology is not available, use the computed value of the *F* test statistic with Table A-5 to find a range for the *P*-value.

Critical Values: Use Table A-5 to find critical *F* values that are determined by the following:

1. The significance level α (Table A-5 includes critical values for $\alpha = 0.025$ and $\alpha = 0.05$.)

2. **Numerator degrees of freedom** = $n_1 - 1$ (determines *column* of Table A-5)

3. **Denominator degrees of freedom** = $n_2 - 1$ (determines *row* of Table A-5) For significance level $\alpha = 0.05$, refer to Table A-5 and use the right-tail

area of 0.025 or 0.05, depending on the type of test, as shown below:

- *Two-tailed test:* Use Table A-5 with 0.025 in the right tail. (The significance level of 0.05 is divided between the two tails, so the area in the right tail is 0.025.)

- *One-tailed test:* Use Table A-5 with $\alpha = 0.05$ in the right tail.

Find the critical F value for the right tail: Because we are stipulating that the larger sample variance is s_1^2, all one-tailed tests will be right-tailed and all two-tailed tests will require that we find only the critical value located to the right. (We have no need to find the critical value at the left tail, which is not very difficult. See Exercise 19 "Finding Lower Critical *F* Values.")

Explore the Data! Because the F test requirement of normal distributions is quite strict, be sure to examine the distributions of the two samples using histograms and normal quantile plots, and confirm that there are no outliers. (See "Assessing Normality" in Section 6-5.)

F Distribution

For two normally distributed populations with equal variances ($\sigma_1^2 = \sigma_2^2$), the sampling distribution of the test statistic $F = s_1^2/s_2^2$ is the **F distribution** shown in Figure 9-4 (provided that we have not yet imposed the stipulation that the larger sample variance is s_1^2). If you repeat the process of selecting samples from two normally distributed populations with equal variances, the distribution of the ratio s_1^2/s_2^2 is the F distribution.

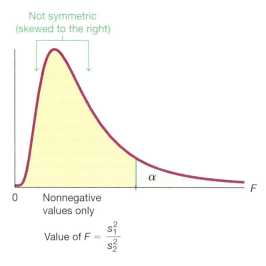

FIGURE 9-4 F Distribution

There is a different F distribution for each different pair of degrees of freedom for the numerator and denominator.

See Figure 9-4 and note these properties of the F distribution:

- The F distribution is not symmetric.
- Values of the F distribution cannot be negative.
- The exact shape of the F distribution depends on the two different degrees of freedom.

Interpreting the Value of the F Test Statistic

If the two populations have equal variances, then the ratio s_1^2/s_2^2 will tend to be close to 1. Because we are stipulating that s_1^2 is the larger sample variance, the ratio s_1^2/s_2^2 will be a *large* number whenever s_1^2 and s_2^2 are far apart in value. Consequently, a value of F near 1 will be evidence in favor of $\sigma_1^2 = \sigma_2^2$, but a large value of F will be evidence against $\sigma_1^2 = \sigma_2^2$.

Large **values of F are evidence *against* $\sigma_1^2 = \sigma_2^2$.**

EXAMPLE 1   Course Evaluation Scores

Listed below are the same student course evaluation scores used in Example 1 in Section 9-2, where we tested the claim that the two samples are from populations with the same *mean*. Use the same data with a 0.05 significance level to test the claim that course evaluation scores of female professors and male professors have the same *variation*.

Female	4.3	4.3	4.4	4.0	3.4	4.7	2.9	4.0	4.3	3.4	3.4	3.3			
Male	4.5	3.7	4.2	3.9	3.1	4.0	3.8	3.4	4.5	3.8	4.3	4.4	4.1	4.2	4.0

SOLUTION

REQUIREMENT CHECK   (1) The two populations are independent of each other. The two samples are not matched in any way. (2) Given the design for the study, we assume that the two samples can be treated as simple random samples. (3) A normal quantile plot of each set of sample course evaluation scores shows that both samples appear to be from populations with a normal distribution. The requirements are satisfied. ☑

For females, we get $s = 0.5630006$ and for males we get $s = 0.3954503$. We can conduct the test using either variances or standard deviations. Because we stipulate in this section that the larger variance is denoted by s_1^2, we let $s_1^2 = 0.5630006^2$ and $s_2^2 = 0.3954503^2$.

Step 1: The claim that male and female professors have the same variation can be expressed symbolically as $\sigma_1^2 = \sigma_2^2$ or as $\sigma_1 = \sigma_2$. We will use $\sigma_1 = \sigma_2$.

Step 2: If the original claim is false, then $\sigma_1 \neq \sigma_2$.

Step 3: Because the null hypothesis is the statement of equality and because the alternative hypothesis cannot contain equality, we have

$$H_0: \sigma_1 = \sigma_2 \text{ (original claim)} \quad H_1: \sigma_1 \neq \sigma_2$$

Step 4: The significance level is $\alpha = 0.05$.

Step 5: Because this test involves two population variances, we use the F distribution.

Step 6: The test statistic is

$$F = \frac{s_1^2}{s_2^2} = \frac{0.5630006^2}{0.3954503^2} = 2.0269$$

***P*-Value Method** Due to the format of Table A-5, the *P*-value method is a bit tricky without technology, but here we go. For a two-tailed test with significance level 0.05, there is an area of 0.025 in the right tail, so we use the two pages for the F distribution (Table A-5) with "0.025 in the right tail." With numerator degrees of freedom $= n_1 - 1 = 11$ and denominator degrees of freedom $= n_2 - 1 = 14$, we find that the critical value of F is between 3.1469 and 3.0502. The test statistic of $F = 2.0269$ is less than the critical value, so we know that the area to the right of the test statistic is greater than 0.025, and it follows that for this two-tailed test, *P*-value > 0.05.

Critical Value Method As with the *P*-value method, we find that the critical value is between 3.1469 and 3.0502. The test statistic $F = 2.0269$ is less than the critical value (which is between 3.1469 and 3.0502), so the test statistic does not fall in the critical region. See Figure 9-5.

continued

Step 7: Figure 9-5 shows that the test statistic $F = 2.0269$ does not fall within the critical region, so we fail to reject the null hypothesis of equal variances. There is not sufficient evidence to warrant rejection of the claim of equal standard deviations.

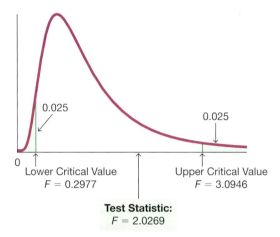

FIGURE 9-5   *F* **Test of Equal Standard Deviations**

INTERPRETATION

Step 8: There is not sufficient evidence to warrant rejection of the claim that the two standard deviations are equal. Course evaluation scores of female professors and male professors appear to have the same amount of variation.

YOUR TURN   Do Exercise 5 "Color and Creativity."

XLSTAT

Ratio	2.0269
F (Observed value)	2.0269
F (Critical value)	3.0946
DF1	11
DF2	14
p-value (Two-tailed)	0.2139
alpha	0.05

Technology Software and calculators usually provide a *P*-value, so the *P*-value method is used. See the accompanying XLSTAT results showing the test statistic of $F = 2.0269$ and the *P*-value of 0.2139. Because the *P*-value of 0.2139 is greater than the significance level of $\alpha = 0.05$, we fail to reject the null hypothesis, and we get the same conclusions in Example 1.

Caution: Part 2 of Section 9-2 includes methods for testing claims about two population means, and one of those methods has a requirement that $\sigma_1 = \sigma_2$. Using the *F* test is *not* recommended as a way to decide whether this requirement is met. For Section 9-2, using the *F* test runs the risk of using differences that are too small to have an effect on the *t* test for two independent samples. That approach is often described as being analogous to sending someone out to sea in a rowboat (the preliminary *F* test) to determine whether the sea is safe for an ocean liner (the *t* test).

PART 2   **Alternative Methods**

Part 1 of this section presents the *F* test for testing claims made about the standard deviations (or variances) of two independent populations. Because that test is so sensitive to departures from normality, we now briefly describe two alternative methods that are not so sensitive to departures from normality.

Count Five

The *count five* method is a relatively simple alternative to the *F* test, and it does not require normally distributed populations. (See "A Quick, Compact, Two-Sample Dispersion Test: Count Five," by McGrath and Yeh, *American Statistician*, Vol. 59, No. 1.) If the two sample sizes are equal, and if one sample has at least five of the largest mean absolute deviations (MAD), then we conclude that its population has a larger variance. See Exercise 17 "Count Five Test" for the specific procedure.

Levene-Brown-Forsythe Test

The *Levene-Brown-Forsythe test* (or modified Levene's test) is another alternative to the *F* test, and it is much more robust against departures from normality. This test begins with a transformation of each set of sample values. Within the first sample, replace each *x* value with $|x - \text{median}|$, and apply the same transformation to the second sample. Using the transformed values, conduct a *t* test of equality of means for independent samples, as described in Part 1 of Section 9-2. Because the transformed values are now deviations, the *t* test for equality of means is actually a test comparing variation in the two samples. See Exercise 18 "Levene-Brown-Forsythe Test."

There are other alternatives to the *F* test, as well as adjustments that improve the performance of the *F* test. See "Fixing the *F* Test for Equal Variances," by Shoemaker, *American Statistician,* Vol. 57, No. 2.

Go Figure

Every minute there are more than 100,000 tweets, 170 million e-mails, and 12 million instant messages.

TECH CENTER

 Inferences from Two Standard Deviations

Access tech supplements, videos, and data sets at **www.TriolaStats.com**

Statdisk

1. Click **Analysis** in the top menu.
2. Select **Hypothesis Testing** from the dropdown menu and **Standard Deviation Two Samples** from the submenu.
3. Select the desired format for *Alternative Hypothesis* and enter the significance level.
4. *Using Summary Statistics:* Select the **Use Summary Statistics** tab and enter the sample size and sample standard deviation for each sample.

   Using Sample Data: Select the **Use Data** tab and select the desired data columns.
5. Click **Evaluate**.

Minitab

1. Click **Stat** in the top menu.
2. Select **Basic Statistics** from the dropdown menu and select **2 Variances** from the submenu.
3. *Using Summary Statistics:* Select **Sample variances** or **Sample standard deviations** from the dropdown menu and enter the sample sizes and sample variances or standard deviations.

   Using sample data: Select **Each sample is in its own column** from the dropdown menu and select the desired data columns.
4. Click the **Options** button and select the desired ratio. Enter the desired confidence level and enter **1** for *Hypothesized Ratio*. Select the desired format for *the Alternative Hypothesis*.
5. Check the box labeled **Use test and confidence intervals based on normal distribution**.
6. Click **OK** twice.

TIP: Another procedure is to click on **Assistant** in the top menu, then select **Hypothesis Tests** and **2-Sample Standard Deviation**. Complete the dialog box to get results, including *P*-value and other helpful information.

StatCrunch

1. Click **Stat** in the top menu.
2. Select **Variance Stats** from the dropdown menu, then select **Two Sample** from the submenu.
3. *Using Summary Statistics:* Select **With Summary** from the submenu and enter the sample variances and sample sizes.

   Using Sample Data: Select **With Data** from the submenu and select the desired data column for each sample.
4. Select **Hypothesis test for σ_1^2 / σ_2^2**. Enter the hypothesized ratio (H_0) and select the desired format for the alternative hypothesis (H_A).
5. Click **Compute!**

TI-83/84 Plus Calculator

1. Press **STAT**, then select **TESTS** in the top menu.
2. Select **2-SampFTest** in the menu and press **ENTER**.
3. Select **Data** if you have sample data in lists or **Stats** if you have summary statistics. Press **ENTER** and enter the list names (leave *Freq* = **1**) or summary statistics.
4. For σ_1 select the desired format for the alternative hypothesis.
5. Select **Calculate** and press **ENTER**.

continued

TECH CENTER *continued*

Inferences from Two Standard Deviations
Access tech supplements, videos, and data sets at **www.TriolaStats.com**

Excel

XLSTAT Add-In

Requires original sample data, does not work with summary data.

1. Click on the **XLSTAT** tab in the Ribbon and then click **Parametric Tests.**
2. Select **Two-sample comparison of variances** from the dropdown menu.
3. Under *Sample 1 & 2* enter the range of cells containing the sample data. For *Data format* select **One column per sample.** If the first row of data contains a label, also check the **Column labels** box.
4. Select **Fisher's F-test.**
5. Click the **Options** tab.
6. Under *Alternative hypothesis* select the desired format ($\neq$ for two-tailed test, $<$ for left-tailed test, $>$ for right-tailed test). Enter the *Hypothesized ratio* (usually **1**) and enter the desired significance level (enter **5** for 0.05 significance level).
7. Click **OK** to display the test statistic (labeled *F Observed value*) and *P*-value.

Excel (Data Analysis Add-In)

1. Click on **Data** in the ribbon, then click on the **Data Analysis** tab.
2. Select **F-Test Two-Sample for Variances** and click **OK.**
3. Enter the data range for each variable in the *Variable Range* boxes. If the first row contains a label, check the **Labels** box.
4. Enter the desired significance level in the *Alpha* box and click **OK.** The results include the *F* test statistic, *P*-value for a one-tail test, and critical value for a one-tail test.

TIP: For a two-tailed test make two adjustments: (1) Enter the value that is half the significance level (Step 4) and (2) double the *P*-value given in the results.

9-4   Basic Skills and Concepts

Statistical Literacy and Critical Thinking

1. *F* Test Statistic

a. If s_1^2 represents the larger of two sample variances, can the *F* test statistic ever be less than 1?

b. Can the *F* test statistic ever be a negative number?

c. If testing the claim that $\sigma_1^2 \neq \sigma_2^2$, what do we know about the two samples if the test statistic *F* is very close to 1?

d. Is the *F* distribution symmetric, skewed left, or skewed right?

2. *F* Test If using the sample data in Data Set 1 "Body Data" in Appendix B for a test of the claim that heights of men and heights of women have different variances, we find that $s = 7.48296$ cm for women and $s = 7.10098$ cm for men.

a. Find the values of s_1^2 and s_2^2 and express them with appropriate units of measure.

b. Identify the null and alternative hypotheses.

c. Find the value of the *F* test statistic and round it to four decimal places.

d. The *P*-value for this test is 0.5225. What do you conclude about the stated claim?

3. Testing Normality For the hypothesis test described in Exercise 2, the sample sizes are $n_1 = 147$ and $n_2 = 153$. When using the *F* test with these data, is it correct to reason that there is no need to check for normality because $n_1 > 30$ and $n_2 > 30$?

4. Robust What does it mean when we say that the F test described in this section is *not robust* against departures from normality?

In Exercises 5–16, test the given claim.

5. Color and Creativity Researchers from the University of British Columbia conducted trials to investigate the effects of color on creativity. Subjects with a red background were asked to think of creative uses for a brick; other subjects with a blue background were given the same task. Responses were scored by a panel of judges and results from scores of creativity are given below. Use a 0.05 significance level to test the claim that creative task scores have the same variation with a red background and a blue background.

Red Background:	$n = 35, \bar{x} = 3.39, s = 0.97$
Blue Background:	$n = 36, \bar{x} = 3.97, s = 0.63$

6. Color and Recall Researchers from the University of British Columbia conducted trials to investigate the effects of color on the accuracy of recall. Subjects were given tasks consisting of words displayed on a computer screen with background colors of red and blue. The subjects studied 36 words for 2 minutes, and then they were asked to recall as many of the words as they could after waiting 20 minutes. Results from scores on the word recall test are given below. Use a 0.05 significance level to test the claim that variation of scores is the same with the red background and blue background.

Accuracy Scores

Red Background:	$n = 35, \bar{x} = 15.89, s = 5.90$
Blue Background:	$n = 36, \bar{x} = 12.31, s = 5.48$

7. Testing Effects of Alcohol Researchers conducted an experiment to test the effects of alcohol. Errors were recorded in a test of visual and motor skills for a treatment group of 22 people who drank ethanol and another group of 22 people given a placebo. The errors for the treatment group have a standard deviation of 2.20, and the errors for the placebo group have a standard deviation of 0.72 (based on data from "Effects of Alcohol Intoxication on Risk Taking, Strategy, and Error Rate in Visuomotor Performance," by Streufert et al., *Journal of Applied Psychology,* Vol. 77, No. 4). Use a 0.05 significance level to test the claim that the treatment group has errors that vary significantly more than the errors of the placebo group.

8. Second-Hand Smoke Data Set 12 "Passive and Active Smoke" includes cotinine levels measured in a group of smokers ($n = 40, \bar{x} = 172.48$ ng/mL, $s = 119.50$ ng/mL) and a group of nonsmokers not exposed to tobacco smoke ($n = 40, \bar{x} = 16.35$ ng/mL, $s = 62.53$ ng/mL). Cotinine is a metabolite of nicotine, meaning that when nicotine is absorbed by the body, cotinine is produced.

a. Use a 0.05 significance level to test the claim that the variation of cotinine in smokers is greater than the variation of cotinine in nonsmokers not exposed to tobacco smoke.

b. The 40 cotinine measurements from the nonsmoking group consist of these values (all in ng/mL): 1, 1, 90, 244, 309, and 35 other values that are all 0. Does this sample appear to be from a normally distributed population? If not, how are the results from part (a) affected?

9. Coke and Diet Coke Data Set 26 "Cola Weights and Volumes" in Appendix B includes the weights (in pounds) of cola for a sample of cans of regular Coke ($n = 36, \bar{x} = 0.81682$ lb, $s = 0.00751$ lb) and the weights of cola for a sample of cans of Diet Coke ($n = 36, \bar{x} = 0.78479$ lb, $s = 0.00439$ lb). Use a 0.05 significance level to test the claim that variation is the same for both types of Coke.

10. IQ and Lead Exposure Data Set 7 "IQ and Lead" in Appendix B lists full IQ scores for a random sample of subjects with low lead levels in their blood and another random sample of subjects with high lead levels in their blood. The statistics are summarized on the top of the next page. Use a 0.05 significance level to test the claim that IQ scores of people with low lead levels vary more than IQ scores of people with high lead levels.

continued

Low Lead Level: $n = 78, \bar{x} = 92.88462, s = 15.34451$

High Lead Level: $n = 21, \bar{x} = 86.90476, s = 8.988352$

11. Magnet Treatment of Pain Researchers conducted a study to determine whether magnets are effective in treating back pain, with results given below (based on data from "Bipolar Permanent Magnets for the Treatment of Chronic Lower Back Pain: A Pilot Study," by Collacott, Zimmerman, White, and Rindone, *Journal of the American Medical Association,* Vol. 283, No. 10). The values represent measurements of pain using the visual analog scale. Use a 0.05 significance level to test the claim that those given a sham treatment (similar to a placebo) have pain reductions that vary more than the pain reductions for those treated with magnets.

Reduction in Pain Level After Sham Treatment: $n = 20, \bar{x} = 0.44, s = 1.4$

Reduction in Pain Level After Magnet Treatment: $n = 20, \bar{x} = 0.49, s = 0.96$

12. Car and Taxi Ages When the author visited Dublin, Ireland (home of Guinness Brewery employee William Gosset, who first developed the *t* distribution), he recorded the ages of randomly selected passenger cars and randomly selected taxis. The ages (in years) are listed below. Use a 0.05 significance level to test the claim that in Dublin, car ages and taxi ages have the same variation.

Car Ages	4	0	8	11	14	3	4	4	3	5	8	3	3	7	4	6	6	1	8	2	15	11	4	1	6	1	8
Taxi Ages	8	8	0	3	8	4	3	3	6	11	7	7	6	9	5	10	8	4	3	4							

13. Professor Evaluation Scores Listed below are student evaluation scores of female professors and male professors from Data Set 17 "Course Evaluations" in Appendix B. Use a 0.05 significance level to test the claim that female professors and male professors have evaluation scores with the same variation.

Female	4.4	3.4	4.8	2.9	4.4	4.9	3.5	3.7	3.4	4.8
Males	4.0	3.6	4.1	4.1	3.5	4.6	4.0	4.3	4.5	4.3

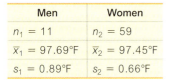

Men	Women
$n_1 = 11$	$n_2 = 59$
$\bar{x}_1 = 97.69°F$	$\bar{x}_2 = 97.45°F$
$s_1 = 0.89°F$	$s_2 = 0.66°F$

14. Body Temperatures of Men and Women If we use the body temperatures from 8 AM on Day 2 as listed in Data Set 3 "Body Temperatures" in Appendix B, we get the statistics given in the accompanying table. Use these data with a 0.05 significance level to test the claim that men have body temperatures that vary more than the body temperatures of women.

15. Old Faithful Listed below are time intervals (min) between eruptions of the Old Faithful geyser. The "recent" times are within the past few years, and the "past" times are from 1995. Does it appear that the variation of the times between eruptions has changed?

Recent	78	91	89	79	57	100	62	87	70	88	82	83	56	81	74	102	61
Past	89	88	97	98	64	85	85	96	87	95	90	95					

16. Blanking Out on Tests Many students have had the unpleasant experience of panicking on a test because the first question was exceptionally difficult. The arrangement of test items was studied for its effect on anxiety. The following scores are measures of "debilitating test anxiety," which most of us call panic or blanking out (based on data from "Item Arrangement, Cognitive Entry Characteristics, Sex and Test Anxiety as Predictors of Achievement in Examination Performance," by Klimko, *Journal of Experimental Education,* Vol. 52, No. 4.) Using a 0.05 significance level, test the claim that the two populations of scores have different amounts of variation.

Questions Arranged from Easy to Difficult				
24.64	39.29	16.32	32.83	28.02
33.31	20.60	21.13	26.69	28.90
26.43	24.23	7.10	32.86	21.06
28.89	28.71	31.73	30.02	21.96
25.49	38.81	27.85	30.29	30.72

Questions Arranged from Difficult to Easy			
33.62	34.02	26.63	30.26
35.91	26.68	29.49	35.32
27.24	32.34	29.34	33.53
27.62	42.91	30.20	32.54

9-4 Beyond the Basics

17. Count Five Test for Comparing Variation in Two Populations Repeat Exercise 16 "Blanking Out on Tests", but instead of using the F test, use the following procedure for the "count five" test of equal variations (which is not as complicated as it might appear).

a. For each value x in the first sample, find the absolute deviation $|x - \bar{x}|$, then sort the absolute deviation values. Do the same for the second sample.

b. Let c_1 be the count of the number of absolute deviation values in the first sample that are greater than the largest absolute deviation value in the second sample. Also, let c_2 be the count of the number of absolute deviation values in the second sample that are greater than the largest absolute deviation value in the first sample. (One of these counts will always be zero.)

c. If the sample sizes are equal ($n_1 = n_2$), use a critical value of 5. If $n_1 \neq n_2$, calculate the critical value shown below.

$$\frac{\log(\alpha/2)}{\log\left(\dfrac{n_1}{n_1 + n_2}\right)}$$

d. If $c_1 \geq$ critical value, then conclude that $\sigma_1^2 > \sigma_2^2$. If $c_2 \geq$ critical value, then conclude that $\sigma_2^2 > \sigma_1^2$. Otherwise, fail to reject the null hypothesis of $\sigma_1^2 = \sigma_2^2$.

18. Levene-Brown-Forsythe Test Repeat Exercise 16 "Blanking Out on Tests" using the Levene-Brown-Forsythe test.

19. Finding Lower Critical F Values For hypothesis tests that are two-tailed, the methods of Part 1 require that we need to find only the upper critical value. Let's denote the upper critical value by F_R, where the subscript indicates the critical value for the right tail. The lower critical value F_L (for the left tail) can be found as follows: (1) Interchange the degrees of freedom used for finding F_R, then (2) using the degrees of freedom found in Step 1, find the F value from Table A-5; (3) take the reciprocal of the F value found in Step 2, and the result is F_L. Find the critical values F_L and F_R for Exercise 16 "Blanking Out on Tests."

Chapter Quick Quiz

In Exercises 1–5, use the following survey results: Randomly selected subjects were asked if they were aware that the Earth has lost half of its wildlife population during the past 50 years. Among 1121 women, 23% said that they were aware. Among 1084 men, 26% said that they were aware (based on data from a Harris poll).

1. Biodiversity Identify the null and alternative hypotheses resulting from the claim that for the people who were aware of the statement, the proportion of women is equal to the proportion of men.

2. Biodiversity Find the values of x_1 (the number of women who were aware of the statement), x_2 (the number of men who were aware of the statement), $\hat{p}_1$, $\hat{p}_2$, and the pooled proportion $\bar{p}$ obtained when testing the claim given in Exercise 1.

3. Biodiversity When testing the claim that $p_1 = p_2$, a test statistic of $z = -1.64$ is obtained. Find the P-value for the hypothesis test.

4. Biodiversity When using the given sample data to construct a 95% confidence interval estimate of the difference between the two population proportions, the result of $(-0.0659, 0.00591)$ is obtained from technology.

a. Express that confidence interval in a format that uses the symbol $<$.

b. What feature of the confidence interval is a basis for deciding whether there is a significant difference between the proportion of women aware of the statement and the proportion of men who are aware?

5. Biodiversity Assume that a P-value of 0.1 is obtained when testing the claim given in Exercise 1 "Biodiversity." What should be concluded about the null hypothesis? What should be the final conclusion?

6. True? Determine whether the following statement is true: When random samples of 50 men and 50 women are obtained and we want to test the claim that men and women have different mean annual incomes, there is no need to confirm that the samples are from populations with normal distributions.

7. True? When we collect random samples to test the claim that the proportion of female CIA agents stationed in the United States is equal to the proportion of female CIA agents stationed outside the United States, there is a requirement that $np \geq 30$ and $nq \geq 30$.

8. Dependent or Independent? Listed below are systolic blood pressure measurements (mm Hg) taken from the right and left arms of the same woman at different times (based on data from "Consistency of Blood Pressure Differences Between the Left and Right Arms," by Eguchi et al., *Archives of Internal Medicine,* Vol. 167). Are the data dependent or independent?

Right arm	102	101	94	79	79
Left arm	175	169	182	146	144

9. Hypotheses Identify the null and alternative hypotheses for using the sample data from Exercise 8 in testing the claim that for differences between right-arm systolic blood pressure amounts and left-arm systolic blood pressure amounts, those differences are from a population with a mean equal to 0.

10. Test Statistics Identify the test statistic that should be used for testing the following given claims.

a. The mean of the differences between IQ scores of husbands and IQ scores of their wives is equal to 0.

b. The mean age of female CIA agents is equal to the mean age of male CIA agents.

c. The proportion of left-handed men is equal to the proportion of left-handed women.

d. The variation among pulse rates of women is equal to the variation among pulse rates of men.

Review Exercises

1. Denomination Effect In the article "The Denomination Effect" by Priya Raghubir and Joydeep Srivastava, *Journal of Consumer Research,* Vol. 36, researchers reported results from studies conducted to determine whether people have different spending characteristics when they have larger bills, such as a $20 bill, instead of smaller bills, such as twenty $1 bills. In one trial, 89 undergraduate business students from two different colleges were randomly assigned to two different groups. In the "dollar bill" group, 46 subjects were given dollar bills; the "quarter" group consisted of 43 subjects given quarters. All subjects from both groups were given a choice of keeping the money or buying gum or mints. The article includes the claim that "money in a large denomination is less likely to be spent relative to an equivalent amount in smaller denominations." Test that claim using a 0.05 significance level with the following sample data from the study.

	Group 1	Group 2
	Subjects Given $1 Bill	Subjects Given 4 Quarters
Spent the money	$x_1 = 12$	$x_2 = 27$
Subjects in group	$n_1 = 46$	$n_2 = 43$

2. Denomination Effect Construct the confidence interval that could be used to test the claim in Exercise 1. What feature of the confidence interval leads to the same conclusion from Exercise 1?

3. Heights Listed below are heights (cm) randomly selected from the sample of women and heights (cm) randomly selected from the sample of men (from Data Set 1 "Body Data" in Appendix B). Use a 95% confidence level to estimate the magnitude of the difference between the mean height of women and the mean height of men.

Women:	160.3	167.7	166.9	153.3	160.0	177.3	169.1	134.5	163.3	171.1
Men:	190.3	169.8	179.8	179.8	177.0	178.5	173.5	178.7	179.0	181.3

4. Heights Use a 0.01 significance level with the sample data from Exercise 3 to test the claim that women have heights with a mean that is less than the mean height of men.

5. Before/After Treatment Results Captopril is a drug designed to lower systolic blood pressure. When subjects were treated with this drug, their systolic blood pressure readings (in mm Hg) were measured before and after the drug was taken. Results are given in the accompanying table (based on data from "Essential Hypertension: Effect of an Oral Inhibitor of Angiotensin-Converting Enzyme," by MacGregor et al., *British Medical Journal*, Vol. 2). Using a 0.01 significance level, is there sufficient evidence to support the claim that captopril is effective in lowering systolic blood pressure?

Subject	A	B	C	D	E	F	G	H	I	J	K	L
Before	200	174	198	170	179	182	193	209	185	155	169	210
After	191	170	177	167	159	151	176	183	159	145	146	177

6. Eyewitness Accuracy of Police Does stress affect the recall ability of police eyewitnesses? This issue was studied in an experiment that tested eyewitness memory a week after a nonstressful interrogation of a cooperative suspect and a stressful interrogation of an uncooperative and belligerent suspect. The numbers of details recalled a week after the incident were recorded, and the summary statistics are given below (based on data from "Eyewitness Memory of Police Trainees for Realistic Role Plays," by Yuille et al., *Journal of Applied Psychology*, Vol. 79, No. 6). Use a 0.01 significance level to test the claim in the article that "stress decreases the amount recalled."

$$\text{Nonstress:} \quad n = 40, \bar{x} = 53.3, s = 11.6$$

$$\text{Stress:} \quad n = 40, \bar{x} = 45.3, s = 13.2$$

7. Are Flights Cheaper When Scheduled Earlier? Listed below are the costs (in dollars) of flights from New York (JFK) to Los Angeles (LAX). Use a 0.01 significance level to test the claim that flights scheduled one day in advance cost more than flights scheduled 30 days in advance. What strategy appears to be effective in saving money when flying?

	Delta	Jet Blue	American	Virgin	Alaska	United
1 day in advance	501	634	633	646	633	642
30 days in advance	148	149	156	156	252	313

8. Variation of Heights Use the sample data given in Exercise 3 "Heights" and test the claim that women and men have heights with the same variation. Use a 0.05 significance level.

Cumulative Review Exercises

Family Heights.   *In Exercises 1–5, use the following heights (in.) The data are matched so that each column consists of heights from the same family.*

Father	68.0	68.0	65.5	66.0	67.5	70.0	68.0	71.0
Mother	64.0	60.0	63.0	59.0	62.0	69.0	65.5	66.0
Son	71.0	64.0	71.0	68.0	70.0	71.0	71.7	71.0

1. a. Are the three samples independent or dependent? Why?

b. Find the mean, median, range, standard deviation, and variance of the heights of the sons.

c. What is the level of measurement of the sample data (nominal, ordinal, interval, ratio)?

d. Are the original unrounded heights discrete data or continuous data?

2. Scatterplot Construct a scatterplot of the father/son heights, then interpret it.

3. Confidence Interval Construct a 95% confidence interval estimate of the mean height of sons. Write a brief statement that interprets the confidence interval.

4. Hypothesis Test Use a 0.05 significance level to test the claim that differences between heights of fathers and their sons have a mean of 0 in.

5. Assessing Normality Interpret the normal quantile plot of heights of fathers.

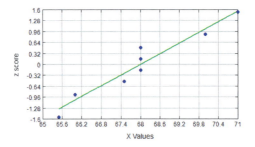

6. Braking Reaction Times: Histogram Listed below are sorted braking reaction times (in 1/10,000 sec) for male and female subjects (based on data from the RT-2S Brake Reaction Time Tester). Construct a histogram for the reaction times of males. Use a class width of 8 and use 28 as the lower limit of the first class. For the horizontal axis, use class midpoint values. Does it appear that the data are from a population with a normal distribution?

Male	28	30	31	34	34	36	36	36	36	38	39	40	40	40	40	41	41	41
	42	42	44	46	47	48	48	49	51	53	54	54	56	57	60	61	61	63
Female	22	24	34	36	36	37	39	41	41	43	43	45	45	47	53	54	54	55
	56	57	57	57	58	61	62	63	66	67	68	71	72	76	77	78	79	80

7. Braking Reaction Times: Normal? The accompanying normal quantile plot is obtained by using the braking reaction times of females listed in Exercise 6. Interpret this graph.

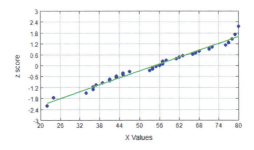

8. Braking Reaction Times: Boxplots Use the same data from Exercise 6 and use the same scale to construct a boxplot of the braking reaction times of males and another boxplot for the braking reaction times of females. What do the boxplots suggest?

9. Braking Reaction Times: Hypothesis Test Use the sample data from Exercise 6 with a 0.01 significance level to test the claim that males and females have the same mean.

10. Braking Reaction Times: Confidence Intervals

a. Construct a 99% confidence interval estimate of the mean braking reaction time of males, construct a 99% confidence interval estimate of the mean braking reaction time of females, then compare the results.

b. Construct a 99% confidence interval estimate of the difference between the mean braking reaction time of males and the mean braking reaction time of females.

c. Which is better for comparing the mean reaction times of males and females: the results from part (a) or the results from part (b)?

Technology Project

Statdisk, Minitab, Excel, StatCrunch, the TI-83/84 Plus calculator, and many other technologies are all capable of generating normally distributed data drawn from a population with a specified mean and standard deviation. Bone density test scores are measured as z scores having a normal distribution with a mean of 0 and a standard deviation of 1. Generate two sets of sample data that represent simulated bone density scores, as shown below.

• **Treatment Group:** Generate 10 sample values from a normally distributed population of bone density scores with mean 0 and standard deviation 1.

• **Placebo Group:** Generate 15 sample values from a normally distributed population of bone density scores with mean 0 and standard deviation 1.

Statdisk:	Select **Data,** then **Normal Generator.**
Minitab:	Select **Calc, Random Data, Normal.**
Excel:	Select **Data Analysis, Random Number Generation.**
TI-83/84 Plus:	Press **MATH,** select **PROB,** then use **randNorm** function with the format of ($\bar{x}$, s, n).
StatCrunch:	Click on **Data,** select **Simulate,** select **Normal.**

Because each of the two samples consists of random selections from a normally distributed population with a mean of 0 and a standard deviation of 1, the data are generated so that both data sets really come from the same population, so there should be no difference between the two sample means.

a. After generating the two data sets, use a 0.10 significance level to test the claim that the two samples come from populations with the same mean.

b. If this experiment is repeated many times, what is the expected percentage of trials leading to the conclusion that the two population means are different? How does this relate to a type I error?

c. If your generated data leads to the conclusion that the two population means are different, would this conclusion be correct or incorrect in reality? How do you know?

d. If part (a) is repeated 20 times, what is the probability that none of the hypothesis tests leads to rejection of the null hypothesis?

e. Repeat part (a) 20 times. How often was the null hypothesis of equal means rejected? Is this the result you expected?

Critical Thinking: Did the NFL Rule Change Have the Desired Effect?

Among 460 overtime National Football League (NFL) games between 1974 and 2011, 252 of the teams that won the over-time coin toss went on to win the game. During those years, a team could win the coin toss and march down the field to win the game with a field goal, and the other team would never get possession of the ball. That just didn't seem fair. Starting in 2012, the overtime rules were changed. In the first three years with the new overtime rules, 47 games were

decided in overtime and the team that won the coin toss won 24 of those games.

Analyzing the Results

1. First *explore* the two proportions of overtime wins. Does there appear to be a difference? If so, how?

2. Create a claim to be tested, then test it. Use a hypothesis test as well as a confidence interval.

3. What do you conclude about the effectiveness of the over-time rule change?

Cooperative Group Activities

1. Out-of-class activity The Chapter Problem is based on observations of cars with rear li-cense plates only in states with laws that require both front and rear license plates. Work to-gether in groups of three or four and collect data in your state. Use a hypothesis test to test the claim that in your state, the proportion of cars with only rear license plates is the same as the proportion of 239/2049 from Connecticut. (Connecticut students can compare the proportion they get to the proportion of 239/2049 obtained by the author.)

2. Out-of-class activity Survey couples and record the number of credit cards each person has. Analyze the paired data to determine whether the males in couple relationships have more credit cards than the females. Try to identify reasons for any discrepancy.

3. Out-of-class activity Measure and record the height of the male and the height of the female from each of several different couples. Estimate the mean of the differences. Compare the result to the difference between the mean height of men and the mean height of women in-cluded in Data Set 1 "Body Data" in Appendix B. Do the results suggest that height is a factor when people select couple partners?

4. Out-of-class activity Are estimates influenced by anchoring numbers? Refer to the related Chapter 3 Cooperative Group Activity on page 129. In Chapter 3 we noted that, according to author John Rubin, when people must estimate a value, their estimate is often "anchored" to (or influenced by) a preceding number. In that Chapter 3 activity, some subjects were asked to quickly estimate the value of $8 \times 7 \times 6 \times 5 \times 4 \times 3 \times 2 \times 1$, and others were asked to quickly estimate the value of $1 \times 2 \times 3 \times 4 \times 5 \times 6 \times 7 \times 8$. In Chapter 3, we could compare the two sets of results by using statistics (such as the mean) and graphs (such as box-plots). The methods of this chapter now allow us to compare the results with a formal hypoth-esis test. Specifically, collect your own sample data and test the claim that when we begin with larger numbers (as in $8 \times 7 \times 6$), our estimates tend to be larger.

5. In-class activity Divide into groups according to gender, with about 10 or 12 students in each group. Each group member should record his or her pulse rate by counting the number of heartbeats in 1 minute, then the group statistics $(n, \bar{x}, s)$ should be calculated. The groups should test the null hypothesis of no difference between their mean pulse rate and the mean of the pulse rates for the population from which subjects of the same gender were selected for Data Set 1 "Body Data" in Appendix B.

6. Out-of-class activity Randomly select a sample of male students and a sample of female students and ask each selected person a yes/no question, such as whether they support a death penalty for people convicted of murder, or whether they believe that the federal government should fund stem cell research. Record the response, the gender of the respondent, and the gender of the person asking the question. Use a formal hypothesis test to determine whether there is a difference between the proportions of *yes* responses from males and females. Also, determine whether the responses appear to be influenced by the gender of the interviewer.

7. Out-of-class activity Construct a short survey of just a few questions, including a question asking the subject to report his or her height. After the subject has completed the survey, measure the subject's height (without shoes) using an accurate measuring system. Record the gender, reported height, and measured height of each subject. Do male subjects appear to exaggerate their heights? Do female subjects appear to exaggerate their heights? Do the errors for males appear to have the same mean as the errors for females?

8. In-class activity Without using any measuring device, ask each student to draw a line believed to be 3 in. long and another line believed to be 3 cm long. Then use rulers to measure and record the lengths of the lines drawn. Record the errors along with the genders of the students making the estimates. Test the claim that when estimating the length of a 3-in. line, the mean error from males is equal to the mean error from females. Also, do the results show that we have a better understanding of the British system of measurement (inches) than the SI system (centimeters)?

9. Out-of-class activity Obtain simple random samples of cars in the student and faculty parking lots, and test the claim that students and faculty have the same proportions of foreign cars.

10. Out-of-class activity Obtain sample data to test the claim that in the college library, science books have a mean age that is less than the mean age of fiction novels.

11. Out-of-class activity Conduct experiments and collect data to test the claim that there are no differences in taste between ordinary tap water and different brands of bottled water.

12. Out-of-class activity Collect sample data and test the claim that people who exercise tend to have pulse rates that are lower than those who do not exercise.

13. Out-of-class activity Collect sample data and test the claim that the proportion of female students who smoke is equal to the proportion of male students who smoke.

14. Out-of-class activity Collect sample data to test the claim that women carry more pocket change than men.

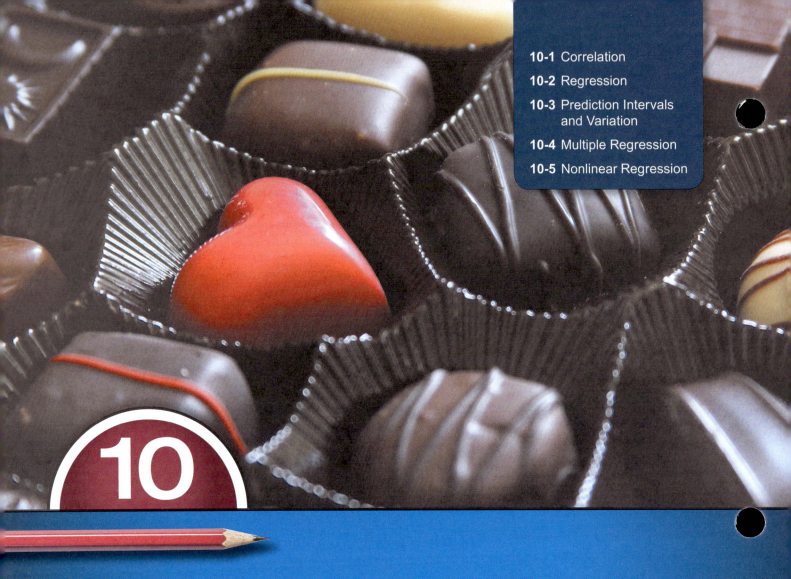

10-1 Correlation

10-2 Regression

10-3 Prediction Intervals and Variation

10-4 Multiple Regression

10-5 Nonlinear Regression

10

CORRELATION AND REGRESSION

CHAPTER PROBLEM

Eat More Chocolate and Win a Nobel Prize?

If you want to win a Nobel Prize, should you study subjects like physics, chemistry, and economics, or should you just eat more chocolate? Table 10-1 lists chocolate consumption (kg per capita) and the numbers of Nobel Laureates (per 10 million people) for each of 23 different countries. See Data Set 16 "Nobel Laureates and Chocolate" in Appendix B, where the countries are identified. Section 2-4 presented methods for constructing scatterplots, and Figure 10-1 is the scatterplot of

the chocolate/Nobel paired data. Using the methods of this chapter, we can address questions such as these:

- Is there a *correlation* between chocolate consumption and the rate of Nobel Laureates?

- If there is a correlation between chocolate consumption and the rate of Nobel Laureates, can we describe it with an equation so that we can predict the rate of Nobel Laureates if given the rate of chocolate consumption?

TABLE 10-1 Chocolate Consumption and Nobel Laureates

Chocolate	Nobel
4.5	5.5
10.2	24.3
4.4	8.6
2.9	0.1
3.9	6.1
0.7	0.1
8.5	25.3
7.3	7.6
6.3	9.0
11.6	12.7
2.5	1.9
8.8	12.7
3.7	3.3
1.8	1.5
4.5	11.4
9.4	25.5
3.6	3.1
2.0	1.9
3.6	1.7
6.4	31.9
11.9	31.5
9.7	18.9
5.3	10.8

- If a country increases chocolate consumption, is it likely to experience an increase in the rate of Nobel Laureates?

The last of the above questions involves at least as much common sense as statistical knowledge. Like every topic in statistics, common sense or critical thinking proves to be an indispensable tool.

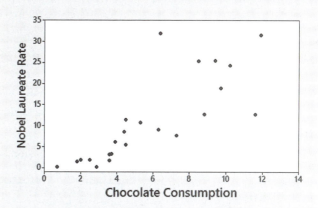

FIGURE 10-1   Scatterplot of Chocolate Consumption and Nobel Laureate Rate

CHAPTER OBJECTIVES

A major focus of this chapter is to analyze *paired* sample data. In Section 9-3 we considered sample data consisting of matched pairs, but the goal in Section 9-3 was to make inferences about the *mean of the differences* from the matched pairs. In this chapter we again consider paired sample data, but the objective is fundamentally different from that of Section 9-3. In this chapter we present methods for determining whether there is a *correlation,* or association, between two variables. For linear correlations, we can identify an equation of a straight line that best fits the data, and we can use that equation to predict the value of one variable given the value of the other variable. Here are the chapter objectives:

10-1 Correlation

- Use paired data to find the value of the linear correlation coefficient *r*.

- Determine whether there is sufficient evidence to support a conclusion that there is a linear correlation between two variables.

10-2 Regression

- Use paired sample data to find the equation of the regression line.

- Find the best predicted value of a variable given some value of the other variable.

10-3 Prediction Intervals and Variation

- Use paired sample data to determine the value of the coefficient of determination r^2, and to interpret that value.

- Use a given value of one variable to find a prediction interval for the other variable.

10-4 Multiple Regression

- Interpret results from technology to determine whether a multiple regression equation is suitable for making predictions.

- Compare results from different combinations of predictor variables and identify the combination that results in the best multiple regression equation.

10-5 Nonlinear Regression

- Use paired data to identify the linear, quadratic, logarithmic, exponential, and power models.

- Determine which model best fits the paired data.

10-1 Correlation

Key Concept In Part 1 we introduce the *linear correlation coefficient r*, which is a number that measures how well paired sample data fit a straight-line pattern when graphed. We use the sample of paired data (sometimes called **bivariate data**) to find the value of *r* (usually using technology), and then we use that value to decide whether there is a linear correlation between the two variables. In this section we consider only *linear* relationships, which means that when graphed in a scatterplot, the points approximate a *straight-line* pattern. In Part 2, we discuss methods for conducting a formal hypothesis test that can be used to decide whether there is a linear correlation between all population values for the two variables.

PART 1 Basic Concepts of Correlation

We begin with the basic definition of *correlation*, a term commonly used in the context of an association between two variables.

> **DEFINITIONS**
>
> A **correlation** exists between two variables when the values of one variable are somehow associated with the values of the other variable.
>
> A **linear correlation** exists between two variables when there is a correlation and the plotted points of paired data result in a pattern that can be approximated by a straight line.

Table 10-1, for example, includes paired sample data consisting of data for chocolate consumption and Nobel Laureates. We will determine whether there is a linear correlation between the variable *x* (chocolate consumption) and the variable *y* (Nobel Laureates). Instead of blindly jumping into the calculation of the linear correlation coefficient *r*, it is wise to first *explore* the data.

Explore!

Because it is always wise to explore sample data before applying a formal statistical procedure, we should use a scatterplot to explore the paired data visually. Figure 10-1 in the Chapter Problem shows a scatterplot of the data. We observe that there are no outliers, which are data points that are far away from all the other points. The points in Figure 10-1 do appear to show a pattern of increasing values of the Nobel rate corresponding to increasing values of the chocolate consumption rate.

Interpreting Scatterplots

Figure 10-2 shows four scatterplots with different characteristics.

- Figure 10-2(a): Distinct straight-line, or linear, pattern. We say that there is a *positive* linear correlation between x and y, since as the x values increase, the corresponding y values also increase.

- Figure 10-2(b): Distinct straight-line, or linear pattern. We say that there is a *negative* linear correlation between x and y, since as the x values increase, the corresponding y values decrease.

- Figure 10-2(c): No distinct pattern, which suggests that there is no correlation between x and y.

- Figure 10-2(d): Distinct pattern suggesting a correlation between x and y, but the pattern is not that of a straight line.

Measure the Strength of the Linear Correlation with *r*

Because conclusions based on visual examinations of scatterplots are largely subjective, we need more objective measures. We use the linear correlation coefficient r, which is a number that measures the strength of the linear association between the two variables.

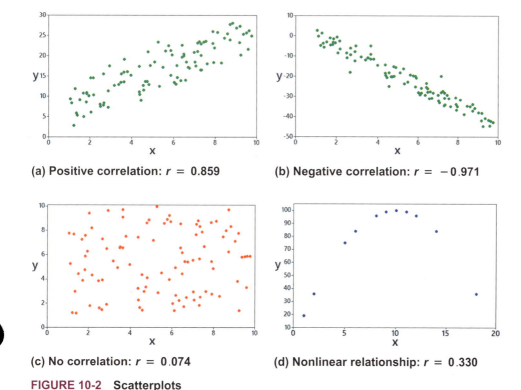

(a) **Positive correlation:** $r = 0.859$

(b) **Negative correlation:** $r = -0.971$

(c) **No correlation:** $r = 0.074$

(d) **Nonlinear relationship:** $r = 0.330$

FIGURE 10-2 Scatterplots

DEFINITION

The **linear correlation coefficient r** measures the strength of the linear correlation between the paired quantitative x values and y values in a *sample*. The linear correlation coefficient r is computed by using Formula 10-1 or Formula 10-2, included in the following Key Elements box. [The linear correlation coefficient is sometimes referred to as the **Pearson product moment correlation coefficient** in honor of Karl Pearson (1857–1936), who originally developed it.]

Because the linear correlation coefficient r is calculated using sample data, it is a sample statistic used to measure the strength of the linear correlation between x and y. If we had every pair of x and y values from an entire population, the result of Formula 10-1 or Formula 10-2 would be a population parameter, represented by ρ (Greek letter rho).

KEY ELEMENTS

Calculating and Interpreting the Linear Correlation Coefficient r

Objective

Determine whether there is a linear correlation between two variables.

Notation for the Linear Correlation Coefficient

n	number of *pairs* of sample data.
Σ	denotes addition of the items indicated.
Σx	sum of all x values.
Σx^2	indicates that each x value should be squared and then those squares added.
$(\Sigma x)^2$	indicates that the x values should be added and the total then squared. Avoid confusing Σx^2 and $(\Sigma x)^2$.
Σxy	indicates that each x value should first be multiplied by its corresponding y value. After obtaining all such products, find their sum.
r	linear correlation coefficient for *sample* data.
ρ	linear correlation coefficient for a *population* of paired data.

Requirements

Given any collection of sample paired quantitative data, the linear correlation coefficient r can always be computed, but the following requirements should be satisfied when using the sample paired data to make a conclusion about linear correlation in the corresponding population of paired data.

1. The sample of paired (x, y) data is a simple random sample of quantitative data. (It is important that the sample data have not been collected using some inappropriate method, such as using a voluntary response sample.)

2. Visual examination of the scatterplot must confirm that the points approximate a straight-line pattern.*

3. Because results can be strongly affected by the presence of outliers, any outliers must be removed if they are known to be errors. The effects of any other outliers should be considered by calculating r with and without the outliers included.*

Note: Requirements 2 and 3 above are simplified attempts at checking this formal requirement: The pairs of (x, y) data must have a **bivariate normal distribution.** Normal distributions are discussed in Chapter 6, but this assumption basically requires that for any fixed value of x, the corresponding values of y have a distribution that is approximately normal, and for any fixed value of y, the values of x have a distribution that is approximately normal. This requirement is usually difficult to check, so for now, we will use Requirements 2 and 3 as listed above.

continued

Formulas for Calculating *r*

FORMULA 10-1

$$r = \frac{n(\Sigma xy) - (\Sigma x)(\Sigma y)}{\sqrt{n(\Sigma x^2) - (\Sigma x)^2}\sqrt{n(\Sigma y^2) - (\Sigma y)^2}} \quad \text{(Good format for calculations)}$$

FORMULA 10-2

$$r = \frac{\Sigma(z_x z_y)}{n-1} \quad \text{(Good format for understanding)}$$

where z_x denotes the *z* score for an individual sample value *x* and z_y is the *z* score for the corresponding sample value *y*.

Rounding the Linear Correlation Coefficient *r*

Round the linear correlation coefficient *r* to three decimal places so that its value can be directly compared to critical values in Table A-6.

Interpreting the Linear Correlation Coefficient *r*

- *Using P-Value from Technology to Interpret r:* Use the *P*-value and significance level α as follows:

  P-value $\leq \alpha$: Supports the claim of a linear correlation.

  P-value $> \alpha$: Does not support the claim of a linear correlation.

- *Using Table A-6 to Interpret r:* Consider critical values from Table A-6 or technology as being both positive and negative, draw a graph similar to Figure 10-3 that accompanies Example 4 on page 477, and then use the following decision criteria:

  Correlation If the computed linear correlation coefficient *r* lies in the left tail beyond the leftmost critical value or if it lies in the right tail beyond the rightmost critical value (that is, $|r| \geq$ critical value), conclude that there is sufficient evidence to support the claim of a linear correlation.

  No Correlation If the computed linear correlation coefficient lies *between* the two critical values (that is, $|r| <$ critical value), conclude that there is not sufficient evidence to support the claim of a linear correlation.

CAUTION Remember, the methods of this section apply to a *linear* correlation. If you conclude that there does not appear to be a linear correlation, it is possible that there might be some other association that is not linear, as in Figure 10-2(d). Always generate a scatterplot to see relationships that might not be linear.

Properties of the Linear Correlation Coefficient *r*

1. The value of *r* is always between -1 and 1 inclusive. That is, $-1 \leq r \leq 1$.

2. If all values of either variable are converted to a different scale, the value of *r* does not change.

3. The value of *r* is not affected by the choice of *x* or *y*. Interchange all *x* values and *y* values, and the value of *r* will not change.

4. *r* measures the strength of a *linear* relationship. It is not designed to measure the strength of a relationship that is not linear [as in Figure 10-2(d)].

5. *r* is very sensitive to outliers in the sense that a single outlier could dramatically affect its value.

Calculating the Linear Correlation Coefficient *r*

The following three examples illustrate three different methods for finding the value of the linear correlation coefficient *r*, but you need to use only one method. *The use of technology (as in Example 1) is strongly recommended.* If manual calculations are absolutely necessary, Formula 10-1 is recommended (as in Example 2). If a better understanding of *r* is desired, Formula 10-2 is recommended (as in Example 3).

EXAMPLE 1 Finding *r* Using Technology

To better illustrate the calculation of *r*, we use Table 10-2, which lists five of the paired data values from Table 10-1 after they have been rounded. Use technology to find the value of the correlation coefficient *r* for the data in Table 10-2.

TABLE 10-2 Chocolate Consumption and Nobel Laureates

Chocolate	5	6	4	4	5
Nobel	6	9	3	2	11

SOLUTION

The value of *r* will be automatically calculated with software or a calculator. See the accompanying technology displays showing that $r = 0.795$ (rounded).

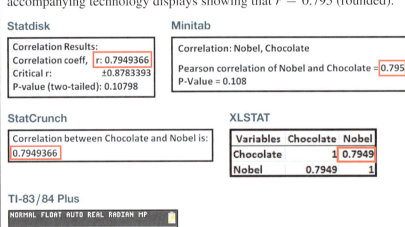

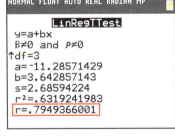

  YOUR TURN Do Exercise 13 "Internet and Nobel Laureates."

EXAMPLE 2 Finding *r* Using Formula 10-1

Use Formula 10-1 to find the value of the linear correlation coefficient *r* for the five pairs of chocolate/Nobel data listed in Table 10-2.

SOLUTION

Using Formula 10-1, the value of r is calculated as shown below. Here, the variable x is used for the chocolate values, and the variable y is used for the Nobel values. Because there are five pairs of data, $n = 5$. Other required values are computed in Table 10-3.

TABLE 10-3 Calculating r with Formula 10-1

x (Chocolate)	y (Nobel)	x^2	y^2	xy
5	6	25	36	30
6	9	36	81	54
4	3	16	9	12
4	2	16	4	8
5	11	25	121	55
$\Sigma x = 24$	$\Sigma y = 31$	$\Sigma x^2 = 118$	$\Sigma y^2 = 251$	$\Sigma xy = 159$

Using Formula 10-1 with the paired data from Table 10-3, r is calculated as follows:

$$r = \frac{n\Sigma xy - (\Sigma x)(\Sigma y)}{\sqrt{n(\Sigma x^2) - (\Sigma x)^2}\sqrt{n(\Sigma y^2) - (\Sigma y)^2}}$$

$$= \frac{5(159) - (24)(31)}{\sqrt{5(118) - (24)^2}\sqrt{5(251) - (31)^2}}$$

$$= \frac{51}{\sqrt{14}\sqrt{294}} = 0.795$$

YOUR TURN Do Exercise 13 "Internet and Nobel Laureates."

EXAMPLE 3 Finding r Using Formula 10-2

Use Formula 10-2 to find the value of the linear correlation coefficient r for the five pairs of chocolate/Nobel data listed in Table 10-2.

SOLUTION

If manual calculations are absolutely necessary, Formula 10-1 is much easier than Formula 10-2, but Formula 10-2 has the advantage of making it easier to *understand* how r works. (See the *rationale* for r discussed later in this section.) As in Example 2, the variable x is used for the chocolate values, and the variable y is used for the Nobel values. In Formula 10-2, each sample value is replaced by its corresponding z score. For example, using unrounded numbers, the chocolate values have a mean of $\bar{x} = 4.8$ and a standard deviation of $s_x = 0.836660$, so the first chocolate value of 5 is converted to a z score of 0.239046 as shown here:

$$z_x = \frac{x - \bar{x}}{s_x} = \frac{5 - 4.8}{0.836660} = 0.239046$$

Table 10-4 lists the z scores for all of the chocolate values (see the third column) and the z scores for all of the Nobel values (see the fourth column). The last column of Table 10-4 lists the products $z_x \cdot z_y$.

continued

Speeding Out-of-Towners Ticketed More?

Are police more likely to issue a ticket to a speeding driver who is out-of-town than to a local driver? George Mason University researchers Michael Makowsky and Thomas Stratmann addressed this question by examining more than 60,000 warnings and tickets issued by Massachusetts police in one year. They found that out-of-town drivers from Massachusetts were 10% more likely to be ticketed than local drivers, and the 10% figure rose to 20% for out-of-state drivers. They also found a statistical association between a town's finances and speeding tickets. When compared to local drivers, out-of-town drivers had a 37% greater chance of being ticketed when speeding in a town in which voters had rejected a proposition to raise taxes more than the 2.5% amount allowed by the state levy limit.

TABLE 10-4 Calculating r with Formula 10-2

x (Chocolate)	y (Nobel)	z_x	z_y	$z_x \cdot z_y$
5	6	0.239046	−0.052164	−0.012470
6	9	1.434274	0.730297	1.047446
4	3	−0.956183	−0.834625	0.798054
4	2	−0.956183	−1.095445	1.047446
5	11	0.239046	1.251937	0.299270
				$\Sigma(z_x \cdot z_y) = 3.179746$

Using $\Sigma(z_x \cdot z_y) = 3.179746$ from Table 10-4, the value of r is calculated by using Formula 10-2, as shown below.

$$r = \frac{\Sigma(z_x \cdot z_y)}{n-1} = \frac{3.179746}{4} = 0.795$$

YOUR TURN Do Exercise 13 "Internet and Nobel Laureates."

Is There a Linear Correlation?

We know from the preceding three examples that the value of the linear correlation coefficient is $r = 0.795$ for the five pairs of sample data in Table 10-2, but if we use the 23 pairs of sample data in Table 10-1 from the Chapter Problem, we get the following Statdisk results, including $r = 0.801$ (rounded).

Statdisk

```
Correlation Results:
Correlation coeff, r:    0.8006078
Critical r:              ±0.4132467
P-value (two-tailed):    0.000
```

We now proceed to interpret the meaning of $r = 0.801$ from the 23 pairs of chocolate/Nobel data, and our goal in this section is to decide whether there appears to be a linear correlation between chocolate consumption and numbers of Nobel Laureates. Using the criteria given in the preceding Key Elements box, we can base our interpretation on a P-value or a critical value from Table A-6. See the criteria for "Interpreting the Linear Correlation Coefficient r" given in the preceding Key Elements box.

EXAMPLE 4   Is There a Linear Correlation?

Using the value of $r = 0.801$ for the 23 pairs of data in Table 10-1 and using a significance level of 0.05, is there sufficient evidence to support a claim that there is a linear correlation between chocolate consumption and numbers of Nobel Laureates?

SOLUTION

REQUIREMENT CHECK The first requirement of a simple random sample of quantitative data is questionable. The data are quantitative, but examining the original data, it does not appear that they are randomly selected pairs. Because this requirement is not satisfied, the results obtained may not be valid. The second requirement of a scatterplot showing a straight-line pattern is satisfied. See the scatterplot in Figure 10-1 on page 469. The scatterplot of Figure 10-1 also shows that the third requirement of no outliers is satisfied. ✓

We can base our conclusion about correlation on either the P-value obtained from technology or the critical value found in Table A-6. (See the criteria for "Interpreting the Linear Correlation Coefficient r" given in the preceding Key Elements box.)

- *Using P-Value from Technology to Interpret r:* Use the *P*-value and significance level α as follows:

  P-value $\leq \alpha$: Supports the claim of a linear correlation.

  P-value $> \alpha$: Does not support the claim of a linear correlation.

The preceding Statdisk display shows that the *P*-value is 0.000 when rounded. Because that *P*-value is less than or equal to the significance level of 0.05, we conclude that *there is sufficient evidence to support the conclusion that for countries, there is a linear correlation between chocolate consumption and numbers of Nobel Laureates.*

- *Using Table A-6 to Interpret r:* Consider critical values from Table A-6 as being both positive and negative, and draw a graph similar to Figure 10-3. For the 23 pairs of data in Table 10-1, Table A-6 yields a critical value that is between $r = 0.396$ and $r = 0.444$; technology yields a critical value of $r = 0.413$. We can now compare the computed value of $r = 0.801$ to the critical values of $r = \pm 0.413$, as shown in Figure 10-3.

  Correlation If the computed linear correlation coefficient r lies in the left or right tail region beyond the critical value for that tail, conclude that there is sufficient evidence to support the claim of a linear correlation.

  No Correlation If the computed linear correlation coefficient lies between the two critical values, conclude that there is not sufficient evidence to support the claim of a linear correlation.

Because Figure 10-3 shows that the computed value of $r = 0.801$ lies beyond the upper critical value, we conclude that *there is sufficient evidence to support the claim of a linear correlation between chocolate consumption and number of Nobel Laureates* for different countries.

INTERPRETATION

Although we have found a linear correlation, it would be absurd to think that eating more chocolate would help win a Nobel Prize. See the following discussion under "*Interpreting r with Causation: Don't Go There!*" Also, the requirement of a simple random sample is not satisfied, so the conclusion of a linear correlation is questionable.

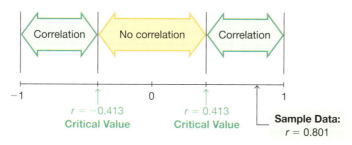

FIGURE 10-3   **Critical *r* Values and the Computed *r* Value**

YOUR TURN   Do Exercise 15 "Pizza and the Subway."

Interpreting *r*: Explained Variation

If we conclude that there is a linear correlation between *x* and *y*, we can find a linear equation that expresses *y* in terms of *x*, and that equation can be used to predict values of *y* for given values of *x*. In Section 10-2 we will describe a procedure for finding such equations and show how to predict values of *y* when given values of *x*. But a

Teacher Evaluations Correlate with Grades

Student evaluations of faculty are often used to measure teaching effectiveness. Many studies reveal a correlation, with higher student grades being associated with higher faculty evaluations. One study at Duke University involved student evaluations collected before and after final grades were assigned. The study showed that "grade expectations or received grades caused a change in the way students perceived their teacher and the quality of instruction." It was noted that with student evaluations, "the incentives for faculty to manipulate their grading policies in order to enhance their evaluations increase." It was concluded that "the ultimate consequence of such manipulations is the degradation of the quality of education in the United States." (See "Teacher Course Evaluations and Student Grades: An Academic Tango," by Valen Johnson, *Chance*, Vol. 15, No. 3.)

predicted value of *y* will not necessarily be the exact result that occurs because in addition to *x*, there are other factors affecting *y*, such as random variation and other characteristics not included in the study. In Section 10-3 we will present a rationale and more details about this principle:

The value of r^2 is the proportion of the variation in *y* that is explained by the linear relationship between *x* and *y*.

EXAMPLE 5   Explained Variation

Using the 23 pairs of chocolate/Nobel data from Table 10-1 in the Chapter Problem, we get $r = 0.801$. What proportion of the variation in numbers of Nobel Laureates can be explained by the variation in the consumption of chocolate?

SOLUTION

With $r = 0.801$ we get $r^2 = 0.642$.

INTERPRETATION

We conclude that 0.642 (or about 64%) of the variation in numbers of Nobel Laureates can be explained by the linear relationship between chocolate consumption and numbers of Nobel Laureates. This implies that about 36% of the variation in numbers of Nobel Laureates cannot be explained by rates of chocolate consumption.

Interpreting *r* with Causation: Don't Go There!

In Example 4 we concluded that there is sufficient evidence to support the claim of a linear correlation between chocolate consumption and number of Nobel Laureates for different countries. We should *not* make any conclusion that includes a statement about a cause-effect relationship between the two variables. We should not conclude that winning Nobel Prizes causes people to consume more chocolate; we should not conclude that consuming more chocolate causes an increase in Nobel Laureates. See the first of the following common errors, and know this:

Correlation does not imply causality!

We noted in the Chapter Problem that we should use common sense when interpreting results. Clearly, it would be absurd to think that eating more chocolate would help win a Nobel Prize.

Common Errors Involving Correlation

Here are three of the most common errors made in interpreting results involving correlation:

1. *Assuming that correlation implies causality.* One classic example involves paired data consisting of the stork population in Copenhagen and the number of human births. For several years, the data suggested a linear correlation. *Bulletin:* Storks do not actually cause births, and births do not cause storks. Both variables were affected by another variable lurking in the background. (A **lurking variable** is one that affects the variables being studied but is not included in the study.) Here, an increasing human population resulted in more births and increased construction of thatched roofs that attracted storks!

2. *Using data based on averages.* Averages suppress individual variation and may inflate the correlation coefficient. One study produced a 0.4 linear correlation

coefficient for paired data relating income and education among individuals, but the linear correlation coefficient became 0.7 when regional averages were used.

3. *Ignoring the possibility of a nonlinear relationship.* If there is no linear correlation, there might be some other correlation that is not linear, as in Figure 10-2(d).

PART 2 **Formal Hypothesis Test**

Hypotheses If conducting a formal hypothesis test to determine whether there is a significant linear correlation between two variables, use the following null and alternative hypotheses that use ρ to represent the linear correlation coefficient of the population:

$$\textbf{Null Hypothesis} \qquad H_0\colon \rho = 0 \ (\text{No correlation})$$

$$\textbf{Alternative Hypothesis} \quad H_1\colon \rho \neq 0 \ (\text{Correlation})$$

Test Statistic The same methods of Part 1 can be used with the test statistic r, or the t test statistic can be found using the following:

$$\textbf{Test Statistic } t = \frac{r}{\sqrt{\dfrac{1 - r^2}{n - 2}}} \quad (\text{with } n - 2 \text{ degrees of freedom})$$

If the above t test statistic is used, P-values and critical values can be found using technology or Table A-3 as described in earlier chapters. See the following example.

EXAMPLE 6 **Hypothesis Test Using the *P*-Value from the *t* Test**

Use the paired chocolate/Nobel data from Table 10-1 on page 469 to conduct a formal hypothesis test of the claim that there is a linear correlation between the two variables. Use a 0.05 significance level with the P-value method of testing hypotheses.

SOLUTION

REQUIREMENT CHECK The requirements were addressed in Example 4. ✅

To claim that there is a linear correlation is to claim that the population linear correlation coefficient ρ is different from 0. We therefore have the following hypotheses:

$$H_0\colon \rho = 0 \quad (\text{There is no linear correlation.})$$

$$H_1\colon \rho \neq 0 \quad (\text{There is a linear correlation.})$$

The linear correlation coefficient is $r = 0.801$ (from technology) and $n = 23$ (because there are 23 pairs of sample data), so the test statistic is

$$t = \frac{r}{\sqrt{\dfrac{1 - r^2}{n - 2}}} = \frac{0.801}{\sqrt{\dfrac{1 - 0.801^2}{23 - 2}}} = 6.131$$

Technologies use more precision to obtain the more accurate test statistic of $t = 6.123$. With $n - 2 = 21$ degrees of freedom, Table A-3 shows that the test statistic of $t = 6.123$ yields a P-value that is less than 0.01. Technologies show that the P-value is 0.000 when rounded. Because the P-value of 0.000 is less than the significance level of 0.05, we reject H_0. ("If the P is low, the null must go." The P-value of 0.000 is low.)

INTERPRETATION

We conclude that for countries, there is sufficient evidence to support the claim of a linear correlation between chocolate consumption and Nobel Laureates.

YOUR TURN Do Exercise 17 "CSI Statistics."

Palm Reading

Some people believe that the length of their palm's lifeline can be used to predict longevity. In a letter published in the *Journal of the American Medical Association,* authors M. E. Wilson and L. E. Mather refuted that belief with a study of cadavers. Ages at death were recorded, along with the lengths of palm lifelines. The authors concluded that there is no correlation between age at death and length of lifeline. Palmistry lost, hands down.

One-Tailed Tests The examples and exercises in this section generally involve two-tailed tests, but one-tailed tests can occur with a claim of a positive linear correlation or a claim of a negative linear correlation. In such cases, the hypotheses will be as shown here.

Claim of Negative Correlation (Left-Tailed Test)	Claim of Positive Correlation (Right-Tailed Test)
$H_0: \rho = 0$	$H_0: \rho = 0$
$H_1: \rho < 0$	$H_1: \rho > 0$

For these one-tailed tests, the *P*-value method can be used as in earlier chapters.

Rationale for Methods of This Section We have presented Formulas 10-1 and 10-2 for calculating *r* and have illustrated their use. Those formulas are given below along with some other formulas that are "equivalent," in the sense that they all produce the same values.

FORMULA 10-1
$$r = \frac{n\Sigma xy - (\Sigma x)(\Sigma y)}{\sqrt{n(\Sigma x^2) - (\Sigma x)^2}\sqrt{n(\Sigma y^2) - (\Sigma y)^2}}$$

FORMULA 10-2
$$r = \frac{\Sigma(z_x z_y)}{n - 1}$$

$$r = \frac{\Sigma(x - \bar{x})(y - \bar{y})}{(n - 1)s_x s_y} \qquad r = \frac{\Sigma\left[\frac{(x - \bar{x})(y - \bar{y})}{s_x \quad s_y}\right]}{n - 1}$$

$$r = \frac{s_{xy}}{\sqrt{s_{xx}}\sqrt{s_{yy}}}$$

We will use Formula 10-2 to help us understand the reasoning that underlies the development of the linear correlation coefficient. Because Formula 10-2 uses *z* scores, the value of $\Sigma(z_x z_y)$ does not depend on the scale that is used for the *x* and *y* values. Figure 10-1 on page 469 shows the scatterplot of the chocolate/Nobel data from Table 10-1, and Figure 10-4 on the next page shows the scatterplot of the *z* scores from the same sample data. Compare Figure 10-1 to Figure 10-4 and see that they are essentially the same scatterplots with different scales. The red lines in Figure 10-4 form the same coordinate axes that we have all come to know and love from earlier mathematics courses. Those red lines partition Figure 10-4 into four quadrants.

If the points of the scatterplot approximate an uphill line (as in the figure), individual values of the product $z_x \cdot z_y$ tend to be positive (because most of the points are found in the first and third quadrants, where the values of z_x and z_y are either both positive or both negative), so $\Sigma(z_x z_y)$ tends to be positive. If the points of the scatterplot approximate a downhill line, most of the points are in the second and fourth quadrants, where z_x and z_y are opposite in sign, so $\Sigma(z_x z_y)$ tends to be negative. Points that follow no linear pattern tend to be scattered among the four quadrants, so the value of $\Sigma(z_x z_y)$ tends to be close to 0.

We can therefore use $\Sigma(z_x z_y)$ as a measure of how the points are configured among the four quadrants. A large positive sum suggests that the points are predominantly in the first and third quadrants (corresponding to a positive linear correlation), a

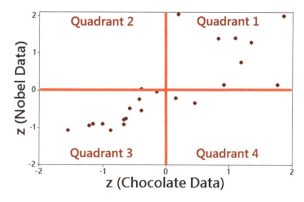

FIGURE 10-4 Scatterplot of *z* Scores from Chocolate/Nobel
Data in Table 10-1

large negative sum suggests that the points are predominantly in the second and fourth quadrants (corresponding to a negative linear correlation), and a sum near 0 suggests that the points are scattered among the four quadrants (with no linear correlation). We divide $\Sigma (z_x z_y)$ by $n-1$ to get an average instead of a statistic that becomes larger simply because there are more data values. (The reasons for dividing by $n-1$ instead of n are essentially the same reasons that relate to the standard deviation.) The end result is Formula 10-2, which can be algebraically manipulated into any of the other expressions for *r*.

TECH CENTER

Correlation
Access tech supplements, videos, and data sets at **www.TriolaStats.com**

Statdisk	Minitab	StatCrunch
1. Click **Analysis** in the top menu.	1. Click **Stat** in the top menu.	1. Click **Stat** in the top menu.
2. Select **Correlation and Regression** from the dropdown menu.	2. Select **Basic Statistics** from the dropdown menu and select **Correlation** from the submenu.	2. Select **Regression** from the dropdown menu, then select **Simple Linear** from the submenu.
3. Enter the desired significance level and select the columns to be evaluated.	3. Select the columns to be evaluated under *Variables*.	3. Select the columns to be used for the *x* variable and *y* variable.
4. Click **Evaluate**.	4. Select **Pearson correlation** for *Method* and check the **Display p-values** box.	4. Click **Compute!**
5. Click **Scatterplot** to obtain a scatterplot with the regression line included.	5. Click **OK**.	5. Click the arrow at the bottom of the results window to view the scatterplot.

Minitab Scatterplot

1. Click **Stat** in the top menu.
2. Select **Regression—Fitted Line Plot** from the dropdown menu.
3. Select the desired columns for the *y* variable and *x* variable.
4. Select **Linear** under *Type of Regression Model* and click **OK**.

TIP: Another procedure is to click on **Assistant** in the top menu, then select **Regression** and **Simple Regression.** Complete the dialog box to get results.

continued

TECH CENTER *continued*

Correlation

Access tech supplements, videos, and data sets at **www.TriolaStats.com**

TI-83/84 Plus Calculator

1. Press **STAT**, then select **TESTS** in the top menu.
2. Select **LinRegTTest** in the menu and press **ENTER**.
3. Enter the list names for the x and y variables. Enter **1** for *Freq* and for β & ρ select $\neq$ **0** to test the null hypothesis of no correlation.
4. Select **Calculate** and press **ENTER**.

Scatterplot

1. Open the **STAT PLOTS** menu by pressing **2ND**, **Y=**.
2. Press **ENTER** to access the Plot 1 settings screen as shown:
   a. Select **ON** and press **ENTER**.
   b. Select first chart option (scatterplot), then press **ENTER**.
   c. Enter name of lists containing data for the x and y variables.
3. Press **ZOOM** then **9** (ZoomStat) to generate the scatterplot.

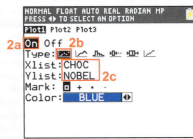

Excel

XLSTAT Add-In

1. Click on the **XLSTAT** tab in the Ribbon and then click **Modeling Data.**
2. Select **Linear Regression** from the dropdown menu.
3. Enter the range of cells containing the *Y/Dependent variable* data and *X/Explanatory variable* data. Check the **Quantitative** box under *X/Explanatory variable*. If the first data row includes a label, check the **Variable labels** box.
4. Click the **Outputs** tab and ensure **Correlations** and **Analysis of Variance** are both checked.
5. Click **OK,** and the linear correlation coefficient, *P*-value, a scatterplot, and hypothesis test results will be displayed. The linear coefficient *r* is found in the *Correlation matrix* and the *P*-value is found in the *Analysis of Variance* table under *Pr > F.*

Excel

1. Click **Insert Function** f_x, select the category **Statistical,** and select the function **CORREL.** Click **OK.**
2. For *Array1* enter the data range for the independent *x* variable. For *Array2* enter the data range for the dependent *y* variable.
3. Click **OK** for the linear correlation coefficient *r*.

Scatterplot (Excel)

1. Select the data range.
2. Click the **Insert** tab in the Ribbon.
3. In the *Charts* section of the top menu select the **Scatter** chart type.
4. Right click on the chart to customize.

10-1 Basic Skills and Concepts

Statistical Literacy and Critical Thinking

1. Notation Twenty different statistics students are randomly selected. For each of them, their body temperature (°C) is measured and their head circumference (cm) is measured.

a. For this sample of paired data, what does *r* represent, and what does ρ represent?

b. Without doing any research or calculations, estimate the value of *r*.

c. Does *r* change if the body temperatures are converted to Fahrenheit degrees?

2. Interpreting r For the same two variables described in Exercise 1, if we find that $r = 0$, does that indicate that there is no association between those two variables?

3. Global Warming If we find that there is a linear correlation between the concentration of carbon dioxide (CO_2) in our atmosphere and the global mean temperature, does that indicate that changes in CO_2 cause changes in the global mean temperature? Why or why not?

4. Scatterplots Match these values of r with the five scatterplots shown below: 0.268, 0.992, -1, 0.746, and 1.

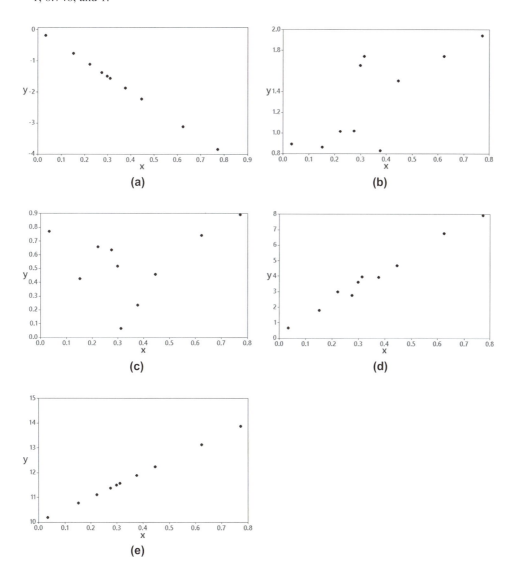

(a)   (b)

(c)   (d)

(e)

Interpreting r. *In Exercises 5–8, use a significance level of $\alpha = 0.05$ and refer to the accompanying displays.*

5. Bear Weight and Chest Size Fifty-four wild bears were anesthetized, and then their weights and chest sizes were measured and listed in Data Set 9 "Bear Measurements" in Appendix B; results are shown in the accompanying Statdisk display. Is there sufficient evidence to support the claim that there is a linear correlation between the weights of bears and their chest sizes? When measuring an anesthetized bear, is it easier to measure chest size than weight? If so, does it appear that a measured chest size can be used to predict the weight?

```
Correlation Results:
Correlation coeff,  r: 0.963141
Critical r:              ±0.2680855
P-value (two-tailed): 0.000
```

6. Casino Size and Revenue The *New York Times* published the sizes (square feet) and revenues (dollars) of seven different casinos in Atlantic City. Is there sufficient evidence to support the claim that there is a linear correlation between size and revenue? Do the results suggest that a casino can increase its revenue by expanding its size?

> Correlation between Size and Revenue is:
> 0.44456896

7. Garbage Data Set 31 "Garbage Weight" in Appendix B includes weights of garbage discarded in one week from 62 different households. The paired weights of paper and glass were used to obtain the XLSTAT results shown here. Is there sufficient evidence to support the claim that there is a linear correlation between weights of discarded paper and glass?

XLSTAT

Correlation matrix (Pearson):

Variables	Paper	Glass
Paper	1	0.1174
Glass	0.1174	1

8. Cereal Killers The amounts of sugar (grams of sugar per gram of cereal) and calories (per gram of cereal) were recorded for a sample of 16 different cereals. TI-83/84 Plus calculator results are shown here. Is there sufficient evidence to support the claim that there is a linear correlation between sugar and calories in a gram of cereal? Explain.

TI-83/84 Plus

```
NORMAL FLOAT AUTO REAL RADIAN MP
        LinRegTTest
 y=a+bx
 β≠0 and ρ≠0
↑df=14
 a=-188.3423729
 b=5.789830508
 s=11.17312583
 r²=.5858430396
 r=.7654038409
```

Explore!   *Exercises 9 and 10 provide two data sets from "Graphs in Statistical Analysis," by F. J. Anscombe, the* **American Statistician,** *Vol. 27. For each exercise,*

a. Construct a scatterplot.

b. Find the value of the linear correlation coefficient r, then determine whether there is sufficient evidence to support the claim of a linear correlation between the two variables.

c. Identify the feature of the data that would be missed if part (b) was completed without constructing the scatterplot.

9.

x	10	8	13	9	11	14	6	4	12	7	5
y	9.14	8.14	8.74	8.77	9.26	8.10	6.13	3.10	9.13	7.26	4.74

10.

x	10	8	13	9	11	14	6	4	12	7	5
y	7.46	6.77	12.74	7.11	7.81	8.84	6.08	5.39	8.15	6.42	5.73

11. Outlier Refer to the accompanying Minitab-generated scatterplot.

a. Examine the pattern of all 10 points and subjectively determine whether there appears to be a correlation between x and y.

b. After identifying the 10 pairs of coordinates corresponding to the 10 points, find the value of the correlation coefficient r and determine whether there is a linear correlation.

continued

c. Now remove the point with coordinates (10, 10) and repeat parts (a) and (b).

d. What do you conclude about the possible effect from a single pair of values?

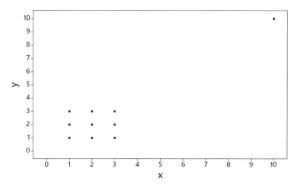

12. Clusters Refer to the following Minitab-generated scatterplot. The four points in the lower left corner are measurements from women, and the four points in the upper right corner are from men.

a. Examine the pattern of the four points in the lower left corner (from women) only, and subjectively determine whether there appears to be a correlation between x and y for women.

b. Examine the pattern of the four points in the upper right corner (from men) only, and subjectively determine whether there appears to be a correlation between x and y for men.

c. Find the linear correlation coefficient using only the four points in the lower left corner (for women). Will the four points in the upper left corner (for men) have the same linear correlation coefficient?

d. Find the value of the linear correlation coefficient using all eight points. What does that value suggest about the relationship between x and y?

e. Based on the preceding results, what do you conclude? Should the data from women and the data from men be considered together, or do they appear to represent two different and distinct populations that should be analyzed separately?

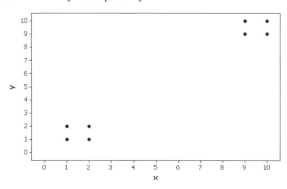

Testing for a Linear Correlation. *In Exercises 13–28, construct a scatterplot, and find the value of the linear correlation coefficient r. Also find the P-value or the critical values of r from Table A-6. Use a significance level of $\alpha = 0.05$. Determine whether there is sufficient evidence to support a claim of a linear correlation between the two variables. (Save your work because the same data sets will be used in Section 10-2 exercises.)*

13. Internet and Nobel Laureates Listed below are numbers of Internet users per 100 people and numbers of Nobel Laureates per 10 million people (from Data Set 16 "Nobel Laureates and Chocolate" in Appendix B) for different countries. Is there sufficient evidence to conclude that there is a linear correlation between Internet users and Nobel Laureates?

Internet Users	79.5	79.6	56.8	67.6	77.9	38.3
Nobel Laureates	5.5	9.0	3.3	1.7	10.8	0.1

continued

14. Old Faithful Listed below are duration times (seconds) and time intervals (min) to the next eruption for randomly selected eruptions of the Old Faithful geyser in Yellowstone National Park. Is there sufficient evidence to conclude that there is a linear correlation between duration times and interval after times?

Duration	242	255	227	251	262	207	140
Interval After	91	81	91	92	102	94	91

15. Pizza and the Subway The "pizza connection" is the principle that the price of a slice of pizza in New York City is always about the same as the subway fare. Use the data listed below to determine whether there is a significant linear correlation between the cost of a slice of pizza and the subway fare.

Year	1960	1973	1986	1995	2002	2003	2009	2013	2015
Pizza Cost	0.15	0.35	1.00	1.25	1.75	2.00	2.25	2.30	2.75
Subway Fare	0.15	0.35	1.00	1.35	1.50	2.00	2.25	2.50	2.75
CPI	30.2	48.3	112.3	162.2	191.9	197.8	214.5	233.0	237.2

16. CPI and the Subway Use CPI/subway data from the preceding exercise to determine whether there is a significant linear correlation between the CPI (Consumer Price Index) and the subway fare.

17. CSI Statistics Police sometimes measure shoe prints at crime scenes so that they can learn something about criminals. Listed below are shoe print lengths, foot lengths, and heights of males (from Data Set 2 "Foot and Height" in Appendix B). Is there sufficient evidence to conclude that there is a linear correlation between shoe print lengths and heights of males? Based on these results, does it appear that police can use a shoe print length to estimate the height of a male?

Shoe Print (cm)	29.7	29.7	31.4	31.8	27.6
Foot Length (cm)	25.7	25.4	27.9	26.7	25.1
Height (cm)	175.3	177.8	185.4	175.3	172.7

18. CSI Statistics Use the paired foot length and height data from the preceding exercise. Is there sufficient evidence to conclude that there is a linear correlation between foot lengths and heights of males? Based on these results, does it appear that police can use foot length to estimate the height of a male?

19. Lemons and Car Crashes Listed below are annual data for various years. The data are weights (metric tons) of lemons imported from Mexico and U.S. car crash fatality rates per 100,000 population [based on data from "The Trouble with QSAR (or How I Learned to Stop Worrying and Embrace Fallacy)," by Stephen Johnson, *Journal of Chemical Information and Modeling*, Vol. 48, No. 1]. Is there sufficient evidence to conclude that there is a linear correlation between weights of lemon imports from Mexico and U.S. car fatality rates? Do the results suggest that imported lemons cause car fatalities?

Lemon Imports	230	265	358	480	530
Crash Fatality Rate	15.9	15.7	15.4	15.3	14.9

20. Revised mpg Ratings Listed below are combined city-highway fuel economy ratings (in mi/gal) for different cars. The old ratings are based on tests used before 2008 and the new ratings are based on tests that went into effect in 2008. Is there sufficient evidence to conclude that there is a linear correlation between the old ratings and the new ratings? What do the data suggest about the old ratings?

Old	16	27	17	33	28	24	18	22	20	29	21
New	15	24	15	29	25	22	16	20	18	26	19

21. Oscars Listed below are ages of Oscar winners matched by the years in which the awards were won (from Data Set 14 "Oscar Winner Age" in Appendix B). Is there sufficient evidence to conclude that there is a linear correlation between the ages of Best Actresses and Best Actors? Should we expect that there would be a correlation?

Best Actress	28	30	29	61	32	33	45	29	62	22	44	54
Best Actor	43	37	38	45	50	48	60	50	39	55	44	33

22. Crickets and Temperature A classic application of correlation involves the association between the temperature and the number of times a cricket chirps in a minute. Listed below are the numbers of chirps in 1 min and the corresponding temperatures in °F (based on data from *The Song of Insects,* by George W. Pierce, Harvard University Press). Is there sufficient evidence to conclude that there is a linear correlation between the number of chirps in 1 min and the temperature?

Chirps in 1 min	882	1188	1104	864	1200	1032	960	900
Temperature (°F)	69.7	93.3	84.3	76.3	88.6	82.6	71.6	79.6

23. Weighing Seals with a Camera Listed below are the overhead widths (cm) of seals measured from photographs and the weights (kg) of the seals (based on "Mass Estimation of Weddell Seals Using Techniques of Photogrammetry," by R. Garrott of Montana State University). The purpose of the study was to determine if weights of seals could be determined from overhead photographs. Is there sufficient evidence to conclude that there is a linear correlation between overhead widths of seals from photographs and the weights of the seals?

Overhead Width	7.2	7.4	9.8	9.4	8.8	8.4
Weight	116	154	245	202	200	191

24. Manatees Listed below are numbers of registered pleasure boats in Florida (tens of thousands) and the numbers of manatee fatalities from encounters with boats in Florida for each of several recent years. The values are from Data Set 10 "Manatee Deaths" in Appendix B. Is there sufficient evidence to conclude that there is a linear correlation between numbers of registered pleasure boats and numbers of manatee boat fatalities?

Pleasure Boats	99	99	97	95	90	90	87	90	90
Manatee Fatalities	92	73	90	97	83	88	81	73	68

25. Tips Listed below are amounts of bills for dinner and the amounts of the tips that were left. The data were collected by students of the author. Is there sufficient evidence to conclude that there is a linear correlation between the bill amounts and the tip amounts? If everyone were to tip with the same percentage, what should be the value of r?

Bill (dollars)	33.46	50.68	87.92	98.84	63.60	107.34
Tip (dollars)	5.50	5.00	8.08	17.00	12.00	16.00

26. POTUS Media periodically discuss the issue of heights of winning presidential candidates and heights of their main opponents. Listed below are those heights (cm) from several recent presidential elections (from Data Set 15 "Presidents" in Appendix B). Is there sufficient evidence to conclude that there is a linear correlation between heights of winning presidential candidates and heights of their main opponents? Should there be such a correlation?

President	178	182	188	175	179	183	192	182	177	185	188	188	183	188
Opponent	180	180	182	173	178	182	180	180	183	177	173	188	185	175

27. Sports Diameters (cm), circumferences (cm), and volumes (cm³) from balls used in different sports are listed in the table below. Is there sufficient evidence to conclude that there is a linear correlation between diameters and circumferences? Does the scatterplot confirm a *linear* association?

	Baseball	Basketball	Golf	Soccer	Tennis	Ping-Pong	Volleyball	Softball
Diameter	7.4	23.9	4.3	21.8	7.0	4.0	20.9	9.7
Circumference	23.2	75.1	13.5	68.5	22.0	12.6	65.7	30.5
Volume	212.2	7148.1	41.6	5424.6	179.6	33.5	4780.1	477.9

28. Sports Repeat the preceding exercise using diameters and volumes.

Appendix B Data Sets. *In Exercises 29–34, use the data from Appendix B to construct a scatterplot, find the value of the linear correlation coefficient r, and find either the P-value or the critical values of r from Table A-6 using a significance level of α = 0.05. Determine whether there is sufficient evidence to support the claim of a linear correlation between the two variables. (Save your work because the same data sets will be used in Section 10-2 exercises.)*

29. Internet and Nobel Laureates Use all of the paired Internet/Nobel data listed in Data Set 16 "Nobel Laureates and Chocolate" in Appendix B.

30. Old Faithful Use all of the paired duration/interval after times listed in Data Set 23.

31. CSI Statistics Use all of the shoe print lengths and heights of the 19 males from Data Set 2 "Foot and Height" in Appendix B.

32. CSI Statistics Use all of the foot lengths and heights of the 19 males from Data Set 2 "Foot and Height" in Appendix B.

33. Word Counts of Men and Women Refer to Data Set 24 "Word Counts" in Appendix B and use the word counts measured from men and women in couple relationships listed in the first two columns of Data Set 24.

34. Earthquakes Refer to Data Set 21 "Earthquakes" in Appendix B and use the depths and magnitudes from the earthquakes. Does it appear that depths of earthquakes are associated with their magnitudes?

10-1 Beyond the Basics

35. Transformed Data In addition to testing for a linear correlation between x and y, we can often use *transformations* of data to explore other relationships. For example, we might replace each x value by x^2 and use the methods of this section to determine whether there is a linear correlation between y and x^2. Given the paired data in the accompanying table, construct the scatterplot and then test for a linear correlation between y and each of the following. Which case results in the largest value of r?

a. x **b.** x^2 **c.** $\log x$ **d.** $\sqrt{x}$ **e.** $1/x$

x	2	3	20	50	95
y	0.3	0.5	1.3	1.7	2.0

36. Finding Critical r Values Table A-6 lists critical values of r for selected values of n and α. More generally, critical r values can be found by using the formula

$$r = \frac{t}{\sqrt{t^2 + n - 2}}$$

where the t value is found from the table of critical t values (Table A-3) assuming a two-tailed case with $n - 2$ degrees of freedom. Use the formula for r given here and in Table A-3 (with $n - 2$ degrees of freedom) to find the critical r values corresponding to $H_1: \rho \neq 0$, $\alpha = 0.02$, and $n = 27$.

10-2   Regression

Key Concept This section presents methods for finding the equation of the straight line that best fits the points in a scatterplot of paired sample data. That best-fitting straight line is called the *regression line,* and its equation is called the *regression equation*. We can use the regression equation to make predictions for the value of one of the variables, given some specific value of the other variable. In Part 2 of this section we discuss marginal change, influential points, and residual plots as tools for analyzing correlation and regression results.

PART 1   Basic Concepts of Regression

In some cases, two variables are related in a *deterministic* way, meaning that given a value for one variable, the value of the other variable is exactly determined without any error, as in the equation $y = 2.54x$ for converting a distance x from inches to centimeters. Such equations are considered in algebra courses, but statistics courses focus on *probabilistic* models, which are equations with a variable that is not determined completely by the other variable. For example, the height of a child cannot be determined completely by the height of the father and/or mother. Sir Francis Galton (1822–1911) studied the phenomenon of heredity and showed that when tall or short couples have children, the heights of those children tend to *regress,* or revert to the more typical mean height for people of the same gender. We continue to use Galton's "regression" terminology, even though our data do not involve the same height phenomena studied by Galton.

DEFINITIONS

Given a collection of paired sample data, the **regression line** (or *line of best fit,* or *least-squares line*) is the straight line that "best" fits the scatterplot of the data. (The specific criterion for the "best-fitting" straight line is the "least-squares" property described later.)

The **regression equation**

$$\hat{y} = b_0 + b_1 x$$

algebraically describes the regression line. The regression equation expresses a relationship between x (called the **explanatory variable,** or **predictor variable,** or **independent variable**) and $\hat{y}$ (called the **response variable** or **dependent variable**).

The preceding definition shows that in statistics, the typical equation of a straight line $y = mx + b$ is expressed in the form $\hat{y} = b_0 + b_1 x$, where b_0 is the y-intercept and b_1 is the slope. The values of the slope b_1 and y-intercept b_0 can be easily found by using any one of the many computer programs and calculators designed to provide those values, as illustrated in Example 1. The values of b_1 and b_0 can also be found with manual calculations, as shown in Example 2.

KEY ELEMENTS

Finding the Equation of the Regression Line

Objective

Find the equation of a regression line.

Notation for the Equation of a Regression Line

	Sample Statistic	Population Parameter
y-intercept of regression equation	b_0	β_0
Slope of regression equation	b_1	β_1
Equation of the regression line	$\hat{y} = b_0 + b_1x$	$y = \beta_0 + \beta_1x$

Requirements

1. The sample of paired (x, y) data is a *random* sample of quantitative data.

2. Visual examination of the scatterplot shows that the points approximate a straight-line pattern.*

3. Outliers can have a strong effect on the regression equation, so remove any outliers if they are known to be errors. Consider the effects of any outliers that are not known errors.*

Note: Requirements 2 and 3 above are simplified attempts at checking these formal requirements for regression analysis:

- For each fixed value of x, the corresponding values of y have a normal distribution.

- For the different fixed values of x, the distributions of the corresponding y-values all have the same standard deviation. (This is violated if part of the scatterplot shows points very close to the regression line while another portion of the scatterplot shows points that are much farther away from the regression line. See the discussion of residual plots in Part 2 of this section.)

- For the different fixed values of x, the distributions of the corresponding y values have means that lie along the same straight line.

The methods of this section are not seriously affected if departures from normal distributions and equal standard deviations are not too extreme.

Formulas for Finding the Slope b_1 and y-Intercept b_0 in the Regression Equation $\hat{y} = b_0 + b_1x$

FORMULA 10-3   **Slope:** $b_1 = r\dfrac{s_y}{s_x}$ where r is the linear correlation coefficient, s_y is the standard deviation of the y values, and s_x is the standard deviation of the x values.

FORMULA 10-4   **y-intercept:** $b_0 = \bar{y} - b_1\bar{x}$

The slope b_1 and y-intercept b_0 can also be found using the following formulas that are useful for manual calculations or writing computer programs:

$$b_1 = \frac{n(\Sigma xy) - (\Sigma x)(\Sigma y)}{n(\Sigma x^2) - (\Sigma x)^2} \qquad b_0 = \frac{(\Sigma y)(\Sigma x^2) - (\Sigma x)(\Sigma xy)}{n(\Sigma x^2) - (\Sigma x)^2}$$

Rounding the Slope b_1 and the y-Intercept b_0

Round b_1 and b_0 to three significant digits. It's difficult to provide a simple universal rule for rounding values of b_1 and b_0, but this rule will work for most situations in this book. (Depending on how you round, this book's answers to examples and exercises may be slightly different from your answers.)

EXAMPLE 1 **Using Technology to Find the Regression Equation**

Refer to the sample data given in Table 10-1 in the Chapter Problem on page 469. Use technology to find the equation of the regression line in which the explanatory variable (or x variable) is chocolate consumption and the response variable (or y variable) is the corresponding Nobel Laureate rate.

SOLUTION

REQUIREMENT CHECK (1) The data are assumed to be a simple random sample. (2) Figure 10-1 is a scatterplot showing a pattern of points. This pattern is very roughly a straight-line pattern. (3) There are no outliers. The requirements are satisfied. ✓

Technology The use of technology is recommended for finding the equation of a regression line. Shown below are the results from different technologies. Minitab and XLSTAT provide the actual equation; the other technologies list the values of the y-intercept and the slope. All of these technologies show that the regression equation can be expressed as $\hat{y} = -3.37 + 2.49x$, where $\hat{y}$ is the predicted Nobel Laureate rate and x is the amount of chocolate consumption.

Statdisk

Regression Results:
Y= b0 + b1x:
Y Intercept, b0: −3.366668
Slope, b1: 2.493134

Excel (XLSTAT)

Equation of the model:

Nobel = -3.36667+2.49313*Choc

Minitab

Regression Equation

Nobel = −3.37 + 2.493 Choc

TI-83/84 Plus

SPSS

Model		Unstandardized Coefficients		Standardized Coefficients		
		B	Std. Error	Beta	t	Sig.
1	(Constant)	-3.367	2.700		-1.247	.226
	Choc	2.493	.407	.801	6.123	.000

JMP

△ Parameter Estimates

Term	Estimate	Std Error	t Ratio	Prob>|t|
Intercept	-3.366668	2.700151	-1.25	0.2262
Choc	2.4931337	0.407174	6.12	<.0001*

StatCrunch

Simple linear regression results:
Dependent Variable: Nobel
Independent Variable: Choc
Nobel = -3.3666676 + 2.4931337 Choc

We should know that the regression equation is an *estimate* of the true regression equation for the population of paired data. This estimate is based on one particular set of sample data, but another sample drawn from the same population would probably lead to a slightly different equation.

YOUR TURN ▷ Do Exercise 13 "Internet and Nobel Laureates."

 EXAMPLE 2 **Using Manual Calculations to Find the Regression Equation**

Refer to the sample data given in Table 10-1 in the Chapter Problem on page 469. Use Formulas 10-3 and 10-4 to find the equation of the regression line in which the explanatory variable (or x variable) is chocolate consumption and the response variable (or y variable) is the corresponding number of Nobel Laureates.

continued

SOLUTION

REQUIREMENT CHECK The requirements are verified in Example 1. ☑

We begin by finding the slope b_1 using Formula 10-3 as follows (with extra digits included for greater accuracy). Remember, r is the linear correlation coefficient, s_y is the standard deviation of the sample y values, and s_x is the standard deviation of the sample x values.

$$b_1 = r\frac{s_y}{s_x} = 0.800608 \cdot \frac{10.211601}{3.279201} = 2.493135$$

After finding the slope b_1, we can now use Formula 10-4 to find the y-intercept as follows:

$$b_0 = \bar{y} - b_1\bar{x} = 11.104348 - (2.493135)(5.804348) = -3.366675$$

After rounding, the slope is $b_1 = 2.49$ and the y-intercept is $b_0 = -3.37$. We can now express the regression equation as $\hat{y} = -3.37 + 2.49x$, where $\hat{y}$ is the predicted Nobel Laureate rate and x is the amount of chocolate consumption.

YOUR TURN Do Exercise 13 "Internet and Nobel Laureates."

EXAMPLE 3 **Graphing the Regression Line**

Graph the regression equation $\hat{y} = -3.37 + 2.49x$ (found in Examples 1 and 2) on the scatterplot of the chocolate/Nobel data from Table 10-1 and examine the graph to subjectively determine how well the regression line fits the data.

SOLUTION

Shown below is the Minitab display of the scatterplot with the graph of the regression line included. We can see that the regression line fits the points well, but the points are not very close to the line.

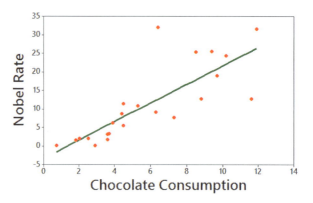

Making Predictions

Regression equations are often useful for *predicting* the value of one variable, given some specific value of the other variable. When making predictions, we should consider the following:

1. **Bad Model:** If the regression equation does not appear to be useful for making predictions, *don't* use the regression equation for making predictions. For bad models, the best predicted value of a variable is simply its sample mean.

2. **Good Model:** Use the regression equation for predictions only if the graph of the regression line on the scatterplot confirms that the regression line fits the points reasonably well.

3. **Correlation:** Use the regression equation for predictions only if the linear correlation coefficient *r* indicates that there is a linear correlation between the two variables (as described in Section 10-1).

4. **Scope:** Use the regression line for predictions only if the data do not go much beyond the scope of the available sample data. (Predicting too far beyond the scope of the available sample data is called *extrapolation,* and it could result in bad predictions.)

Figure 10-5 summarizes a strategy for predicting values of a variable *y* when given some value of *x*. Figure 10-5 shows that if the regression equation is a good model, then we substitute the value of *x* into the regression equation to find the predicted value of *y*. However, if the regression equation is not a good model, the best predicted value of *y* is simply $\bar{y}$, the mean of the *y* values. Remember, this strategy applies to *linear* patterns of points in a scatterplot. If the scatterplot shows a pattern that is nonlinear (not a straight-line) pattern, other methods apply.

Strategy for Predicting Values of *y*

Is the regression equation a good model?
- The regression line graphed in the scatterplot shows that the line fits the points well.
- *r* indicates that there is a linear correlation.
- The prediction is not much beyond the scope of the available sample data.

Yes.
The regression equation is a good model.

No.
The regression equation *is not* a good model.

Substitute the given value of *x* into the regression equation $\hat{y} = b_0 + b_1 x$

Regardless of the value of *x*, the best predicted value of *y* is the value of $\bar{y}$ (the mean of the *y* values).

FIGURE 10-5 Recommended Strategy for Predicting Values of *y*

EXAMPLE 4 **Making Predictions**

a. Use the chocolate/Nobel data from Table 10-1 on page 469 to predict the Nobel rate for a country with chocolate consumption of 10 kg per capita.

b. Predict the IQ score of an adult who is exactly 175 cm tall.

SOLUTION

a. **Good Model: Use the Regression Equation for Predictions.** The regression line fits the points well, as shown in Example 3. Also, there is a linear correlation between chocolate consumption and the Nobel Laureate rate, as shown in Section 10-1. Because the regression equation $\hat{y} = -3.37 + 2.49x$ is a good model, substitute $x = 10$ into the regression equation to get a predicted Nobel Laureate rate of 21.5 Nobel Laureates per 10 million people.

b. **Bad Model: Use $\bar{y}$ for predictions.** Knowing that there is no correlation between height and IQ score, we know that a regression equation is not a good model, so the best predicted value of IQ score is the mean, which is 100.

continued

Postponing Death

Several studies addressed the ability of people to postpone their death until after an important event. For example, sociologist David Phillips analyzed death rates of Jewish men who died near Passover, and he found that the death rate dropped dramatically in the week before Passover, but rose the week after. Other researchers of cancer patients concluded that there is "no pattern to support the concept that 'death takes a holiday.'" (See "Holidays, Birthdays, and Postponement of Cancer Death," by Young and Hade, *Journal of the American Medical Association,* Vol. 292, No. 24.) Based on records of 1.3 million deaths, this more recent study found no relationship between the time of death and Christmas, Thanksgiving, or the person's birthday. The findings were disputed by David Phillips, who said that the study focused on cancer patients, but they are least likely to have psychosomatic effects.

Note that in part (a), the paired data result in a *good* regression model, so the predicted Nobel rate is found by substituting the value of $x = 10$ into the regression equation. However, in part (b) there is no correlation between height and IQ, so the best predicted IQ score is the mean of $\bar{y} = 100$.

Key point: Use the regression equation for predictions only if it is a good model. If the regression equation is not a good model, use the predicted value of $\bar{y}$.

YOUR TURN Do Exercise 5 "Speed Dating."

PART 2 Beyond the Basics of Regression

In Part 2 we consider the concept of marginal change, which is helpful in interpreting a regression equation; then we consider the effects of outliers and special points called *influential points*. We also consider residual plots.

Interpreting the Regression Equation: Marginal Change

We can use the regression equation to see the effect on one variable when the other variable changes by some specific amount.

> **DEFINITION**
>
> In working with two variables related by a regression equation, the **marginal change** in a variable is the amount that it changes when the other variable changes by exactly one unit. The slope b_1 in the regression equation represents the marginal change in y that occurs when x changes by one unit.

Let's consider the 23 pairs of chocolate/Nobel data included in Table 10-1. Those 23 pairs of data result in this regression equation: $\hat{y} = -3.37 + 2.49x$. The slope of 2.49 tells us that if we increase x (chocolate consumption) by 1 (kg per capita), the predicted Nobel Laureate rate will increase by 2.49 (per 10 million people). That is, for every additional 1 kg per capita increase in chocolate consumption, we expect the Nobel Laureate rate to increase by 2.49 per 10 million people.

Outliers and Influential Points

A correlation/regression analysis of bivariate (paired) data should include an investigation of *outliers* and *influential points*, defined as follows.

> **DEFINITIONS**
>
> In a scatterplot, an **outlier** is a point lying far away from the other data points.
>
> Paired sample data may include one or more **influential points,** which are points that strongly affect the graph of the regression line.

To determine whether a point is an outlier, examine the scatterplot to see if the point is far away from the others. Here's how to determine whether a point is an influential point: First graph the regression line resulting from the data with the point included, then graph the regression line resulting from the data with the point excluded. If the regression line changes by a considerable amount, the point is influential.

EXAMPLE 5   Influential Point

Consider the 23 pairs of chocolate/Nobel data from Table 10-1 in the Chapter Problem. The scatterplot located to the left below shows the regression line. If we include an additional pair of data, $x = 50$ and $y = 0$, we get the regression line shown to the right below. The additional point $(50, 0)$ is an influential point because the graph of the regression line did change considerably, as shown by the regression line located to the right below. Compare the two graphs to see clearly that the addition of this one pair of values has a very dramatic effect on the regression line, so that additional point is an influential point. The additional point is also an outlier because it is far from the other points.

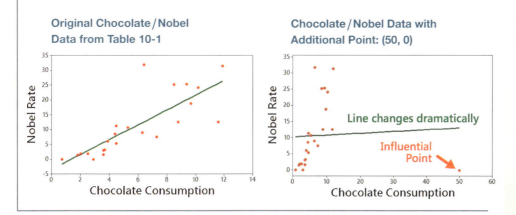

Original Chocolate/Nobel Data from Table 10-1

Chocolate/Nobel Data with Additional Point: (50, 0)

Residuals and the Least-Squares Property

We stated that the regression equation represents the straight line that "best" fits the data. The criterion to determine the line that is better than all others is based on the vertical distances between the original data points and the regression line. Such distances are called *residuals*.

DEFINITION

For a pair of sample x and y values, the **residual** is the difference between the *observed* sample value of y and the y value that is *predicted* by using the regression equation. That is,

$$\text{Residual} = \text{observed } y - \text{predicted } y = y - \hat{y}$$

So far, this definition hasn't yet won any prizes for simplicity, but you can easily understand residuals by referring to Figure 10-6 on the next page, which corresponds to the paired sample data shown in the margin. In Figure 10-6, the residuals are represented by the dashed lines. The paired data are plotted as red points in Figure 10-6.

Consider the sample point with coordinates of $(8, 4)$. If we substitute $x = 8$ into the regression equation $\hat{y} = 1 + x$, we get a predicted value of $\hat{y} = 9$. But for $x = 8$, the actual observed sample value is $y = 4$. The difference $y - \hat{y} = 4 - 9 = -5$ is a residual.

x	8	12	20	24
y	4	24	8	32

x	8	12	20	24
y	4	24	8	32

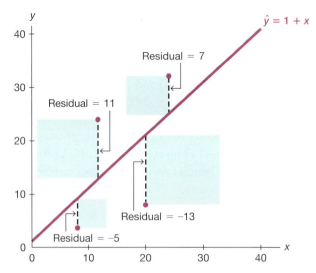

FIGURE 10-6   Residuals and Squares of Residuals

The regression equation represents the line that "best" fits the points according to the following least-squares property.

> **DEFINITION**
>
> A straight line satisfies the **least-squares property** if the sum of the squares of the residuals is the smallest sum possible.

From Figure 10-6, we see that the residuals are $-5, 11, -13$, and 7, so the sum of their squares is

$$(-5)^2 + 11^2 + (-13)^2 + 7^2 = 364$$

We can visualize the least-squares property by referring to Figure 10-6, where the squares of the residuals are represented by the shaded square areas. The sum of the shaded square areas is 364, which is the smallest sum possible. Use any other straight line, and the shaded squares will combine to produce an area larger than the combined shaded area of 364.

Fortunately, we need not deal directly with the least-squares property when we want to find the equation of the regression line. Calculus has been used to build the least-squares property into Formulas 10-3 and 10-4. Because the derivations of these formulas require calculus, we don't include the derivations in this text.

Residual Plots

In this section and the preceding section we listed simplified requirements for the effective analyses of correlation and regression results. We noted that we should always begin with a scatterplot, and we should verify that the pattern of points is approximately a straight-line pattern. We should also consider outliers. A *residual plot* can be another helpful tool for analyzing correlation and regression results and for checking the requirements necessary for making inferences about correlation and regression.

> **DEFINITION**
>
> A **residual plot** is a scatterplot of the (x, y) values after each of the y-coordinate values has been replaced by the residual value $y - \hat{y}$ (where $\hat{y}$ denotes the predicted value of y). That is, a residual plot is a graph of the points $(x, y - \hat{y})$.

To construct a residual plot, draw a horizontal reference line through the residual value of 0, then plot the paired values of $(x, y - \hat{y})$. Because the manual construction of residual plots can be tedious, the use of technology is strongly recommended. When analyzing a residual plot, look for a pattern in the way the points are configured, and use these criteria:

- The residual plot should not have any obvious pattern (not even a straight-line pattern). (This lack of a pattern confirms that a scatterplot of the sample data is a straight-line pattern instead of some other pattern.)

- The residual plot should not become much wider (or thinner) when viewed from left to right. (This confirms the requirement that for the different fixed values of x, the distributions of the corresponding y values all have the same standard deviation.)

EXAMPLE 6 Residual Plot

The chocolate/Nobel data from Table 10-1 are used to obtain the accompanying Statdisk-generated residual plot. When the first sample x value of 4.5 is substituted into the regression equation of $\hat{y} = -3.37 + 2.49x$ (found in Examples 1 and 2), we get the predicted value of $\hat{y} = 7.84$. For the first x value of 4.5, the actual corresponding y value is 5.5, so the value of the residual is

$$\text{Observed } y - \text{predicted } y = y - \hat{y} = 5.5 - 7.84 = -2.34$$

Using the x value of 4.5 and the residual of -2.34, we get the coordinates of the point $(4.5, -2.34)$, which is one of the points in the residual plot shown here.

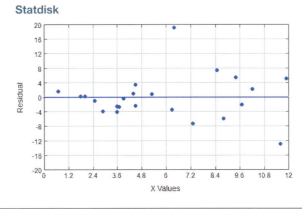

See the three residual plots below. The leftmost residual plot suggests that the regression equation is a good model. The middle residual plot shows a distinct pattern, suggesting that the sample data do not follow a straight-line pattern as required. The rightmost residual plot becomes thicker, which suggests that the requirement of equal standard deviations is violated.

Residual Plot Suggesting That the Regression Equation Is a Good Model

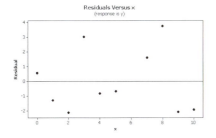

Residual Plot with an Obvious Pattern, Suggesting That the Regression Equation Is Not a Good Model

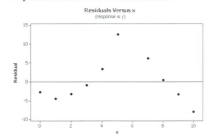

Residual Plot That Becomes Wider, Suggesting That the Regression Equation Is Not a Good Model

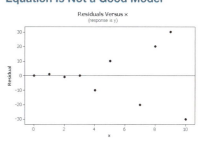

Regression
Access tech supplements, videos, and data sets at www.TriolaStats.com

Statdisk

1. Click **Analysis** in the top menu.
2. Select **Correlation and Regression** from the dropdown menu.
3. Enter the desired significance level and select the columns to be evaluated.
4. Click **Evaluate.**
5. Click **Scatterplot** to obtain a scatterplot with the regression line.

Minitab

1. Click **Stat** in the top menu.
2. Select **Regression** from the dropdown menu and select **Regression—Fit Regression Model** from the submenu.
3. Under *Responses* select the column that contains the dependent *y* values. Under *Continuous predictors* select the column that contains the independent *x* values.
4. Click **OK.** The regression equation is included in the results.

Scatterplot

1. Click **Stat** in the top menu.
2. Select **Regression—Fitted Line Plot** from the dropdown menu.
3. Select the desired columns for the *y* variable and *x* variable.
4. Select **Linear** under *Type of Regression Model* and click **OK.**

TIP: Another procedure is to click on **Assistant** in the top menu, then select **Regression,** and **Simple Regression.** Complete the dialog box to get results, including the regression equation.

StatCrunch

1. Click **Stat** in the top menu.
2. Select **Regression** from the dropdown menu, then select **Simple Linear** from the submenu.
3. Select the columns to be used for the *x* variable and *y* variable.
4. Click **Compute!**
5. Click the arrow at the bottom of the results window to view the scatterplot with regression line.

TI-83/84 Plus Calculator

1. Press **STAT**, then select **TESTS** in the top menu.
2. Select **LinRegTTest** in the menu and press **ENTER**.
3. Enter the list names for the *x* and *y* variables. Enter **1** for *Freq* and for β & ρ select $\neq$ **0** to test the null hypothesis of no correlation.
4. Select **Calculate** and press **ENTER** to view results, which include the *y*-intercept (a) and slope (b) of the regression equation.

Excel

XLSTAT Add-In

1. Click on the **XLSTAT** tab in the Ribbon and then click **Modeling Data.**
2. Select **Linear Regression** from the dropdown menu.
3. Enter the range of cells containing the *Y/Dependent variable* data and *X/Explanatory variable* data. Check the **Quantitative** box under *X/Explanatory variable.* If the first data row includes a label, check the **Variable labels** box.
4. Click the **Outputs** tab and ensure **Correlations** and **Analysis of Variance** and **Prediction and residuals** are checked.
5. Click **OK,** and the equation of the regression line will be displayed in the results.

Excel (Data Analysis Add-In)

1. Click on the **Data** tab in the Ribbon and then click the **Data Analysis** tab.
2. Select **Regression** under *Analysis Tools* and click **OK.**
3. For *Input Y Range* enter the data range for the dependent *y* variable. For *Input X Range* enter the data range for the independent *x* variable.
4. Check the **Labels** box if the first row contains a label.
5. Check the **Line Fit Plots** box and **Residuals Plots** box and click **OK** to display the results. In the *Coefficients* table, the slope is labeled *X Variable 1* and the *y*-intercept is labeled *Intercept.*

TIP: The displayed graph will include a scatterplot of the original sample points along with the points that would be predicted by the regression equation. You can obtain the regression line by connecting the "predicted *y*" points.

10-2 Basic Skills and Concepts

Statistical Literacy and Critical Thinking

1. Notation Different hotels on Las Vegas Boulevard ("the strip") in Las Vegas are randomly selected, and their ratings and prices were obtained from Travelocity. Using technology, with x representing the ratings and y representing price, we find that the regression equation has a slope of 130 and a y-intercept of -368.

a. What is the equation of the regression line?

b. What does the symbol $\hat{y}$ represent?

2. Notation What is the difference between the regression equation $\hat{y} = b_0 + b_1 x$ and the regression equation $y = \beta_0 + \beta_1 x$?

3. Best-Fit Line

a. What is a residual?

b. In what sense is the regression line the straight line that "best" fits the points in a scatterplot?

4. Correlation and Slope What is the relationship between the linear correlation coefficient r and the slope b_1 of a regression line?

Making Predictions. *In Exercises 5–8, let the predictor variable x be the first variable given. Use the given data to find the regression equation and the best predicted value of the response variable. Be sure to follow the prediction procedure summarized in Figure 10-5 on page 493. Use a 0.05 significance level.*

5. Speed Dating For 50 randomly selected speed dates, attractiveness ratings by males of their female date partners (x) are recorded along with the attractiveness ratings by females of their male date partners (y); the ratings are from Data Set 18 "Speed Dating" in Appendix B. The 50 paired ratings yield $\bar{x} = 6.5$, $\bar{y} = 5.9$, $r = -0.277$, P-value $= 0.051$, and $\hat{y} = 8.18 - 0.345x$. Find the best predicted value of $\hat{y}$ (attractiveness rating by female of male) for a date in which the attractiveness rating by the male of the female is $x = 8$.

6. Bear Measurements Head widths (in.) and weights (lb) were measured for 20 randomly selected bears (from Data Set 9 "Bear Measurements" in Appendix B). The 20 pairs of measurements yield $\bar{x} = 6.9$ in., $\bar{y} = 214.3$ lb, $r = 0.879$, P-value $= 0.000$, and $\hat{y} = -212 + 61.9x$. Find the best predicted value of $\hat{y}$ (weight) given a bear with a head width of 6.5 in.

7. Height and Weight Heights (cm) and weights (kg) are measured for 100 randomly selected adult males (from Data Set 1 "Body Data" in Appendix B). The 100 paired measurements yield $\bar{x} = 173.79$ cm, $\bar{y} = 85.93$ kg, $r = 0.418$, P-value $= 0.000$, and $\hat{y} = -106 + 1.10x$. Find the best predicted value of $\hat{y}$ (weight) given an adult male who is 180 cm tall.

8. Best Supporting Actors and Actresses For 30 recent Academy Award ceremonies, ages of Best Supporting Actors (x) and ages of Best Supporting Actresses (y) are recorded. The 30 paired ages yield $\bar{x} = 52.1$ years, $\bar{y} = 37.3$ years, $r = 0.076$, P-value $= 0.691$, and $\hat{y} = 34.4 + 0.0547x$. Find the best predicted value of $\hat{y}$ (age of Best Supporting Actress) in 1982, when the age of the Best Supporting Actor (x) was 46 years.

Finding the Equation of the Regression Line. *In Exercises 9 and 10, use the given data to find the equation of the regression line. Examine the scatterplot and identify a characteristic of the data that is ignored by the regression line.*

9.

x	10	8	13	9	11	14	6	4	12	7	5
y	9.14	8.14	8.74	8.77	9.26	8.10	6.13	3.10	9.13	7.26	4.74

10.

x	10	8	13	9	11	14	6	4	12	7	5
y	7.46	6.77	12.74	7.11	7.81	8.84	6.08	5.39	8.15	6.42	5.73

11. Effects of an Outlier Refer to the Minitab-generated scatterplot given in Exercise 11 of Section 10-1 on page 485.

a. Using the pairs of values for all 10 points, find the equation of the regression line.

b. After removing the point with coordinates (10, 10), use the pairs of values for the remaining 9 points and find the equation of the regression line.

c. Compare the results from parts (a) and (b).

12. Effects of Clusters Refer to the Minitab-generated scatterplot given in Exercise 12 of Section 10-1 on page 485.

a. Using the pairs of values for all 8 points, find the equation of the regression line.

b. Using only the pairs of values for the 4 points in the lower left corner, find the equation of the regression line.

c. Using only the pairs of values for the 4 points in the upper right corner, find the equation of the regression line.

d. Compare the results from parts (a), (b), and (c).

Regression and Predictions.   *Exercises 13–28 use the same data sets as Exercises 13–28 in Section 10-1. In each case, find the regression equation, letting the first variable be the predictor (x) variable. Find the indicated predicted value by following the prediction procedure summarized in Figure 10-5 on page 493.*

13. Internet and Nobel Laureates Find the best predicted Nobel Laureate rate for Japan, which has 79.1 Internet users per 100 people. How does it compare to Japan's Nobel Laureate rate of 1.5 per 10 million people?

Internet Users	79.5	79.6	56.8	67.6	77.9	38.3
Nobel Laureates	5.5	9.0	3.3	1.7	10.8	0.1

14. Old Faithful Using the listed duration and interval after times, find the best predicted "interval after" time for an eruption with a duration of 253 seconds. How does it compare to an actual eruption with a duration of 253 seconds and an interval after time of 83 minutes?

Duration	242	255	227	251	262	207	140
Interval After	91	81	91	92	102	94	91

15. Pizza and the Subway Use the pizza costs and subway fares to find the best predicted subway fare, given that the cost of a slice of pizza is $3.00. Is the best predicted subway fare likely to be implemented?

Year	1960	1973	1986	1995	2002	2003	2009	2013	2015
Pizza Cost	0.15	0.35	1.00	1.25	1.75	2.00	2.25	2.30	2.75
Subway Fare	0.15	0.35	1.00	1.35	1.50	2.00	2.25	2.50	2.75
CPI	30.2	48.3	112.3	162.2	191.9	197.8	214.5	233.0	237.2

16. CPI and the Subway Use the CPI/subway fare data from the preceding exercise and find the best predicted subway fare for a time when the CPI reaches 500. What is wrong with this prediction?

17. CSI Statistics Use the shoe print lengths and heights to find the best predicted height of a male who has a shoe print length of 31.3 cm. Would the result be helpful to police crime scene investigators in trying to describe the male?

Shoe Print (cm)	29.7	29.7	31.4	31.8	27.6
Foot Length (cm)	25.7	25.4	27.9	26.7	25.1
Height (cm)	175.3	177.8	185.4	175.3	172.7

18. CSI Statistics Use the foot lengths and heights to find the best predicted height of a male who has a foot length of 28 cm. Would the result be helpful to police crime scene investigators in trying to describe the male?

19. Lemons and Car Crashes Using the listed lemon/crash data, find the best predicted crash fatality rate for a year in which there are 500 metric tons of lemon imports. Is the prediction worthwhile?

Lemon Imports	230	265	358	480	530
Crash Fatality Rate	15.9	15.7	15.4	15.3	14.9

20. Revised mpg Ratings Using the listed old/new mpg ratings, find the best predicted new mpg rating for a car with an old rating of 30 mpg. Is there anything to suggest that the prediction is likely to be quite good?

Old	16	27	17	33	28	24	18	22	20	29	21
New	15	24	15	29	25	22	16	20	18	26	19

21. Oscars Using the listed actress/actor ages, find the best predicted age of the Best Actor given that the age of the Best Actress is 54 years. Is the result reasonably close to the Best Actor's (Eddie Redmayne) actual age of 33 years, which happened in 2015, when the Best Actress was Julianne Moore, who was 54 years of age?

Best Actress	28	30	29	61	32	33	45	29	62	22	44	54
Best Actor	43	37	38	45	50	48	60	50	39	55	44	33

22. Crickets and Temperature Find the best predicted temperature at a time when a cricket chirps 3000 times in 1 minute. What is wrong with this predicted temperature?

Chirps in 1 min	882	1188	1104	864	1200	1032	960	900
Temperature (°F)	69.7	93.3	84.3	76.3	88.6	82.6	71.6	79.6

23. Weighing Seals with a Camera Using the listed width/weight data, find the best predicted weight of a seal if the overhead width measured from a photograph is 2 cm. Can the prediction be correct? If not, what is wrong?

Overhead Width	7.2	7.4	9.8	9.4	8.8	8.4
Weight	116	154	245	202	200	191

24. Manatees Use the listed boat/manatee data. In a year not included in the data below, there were 970,000 registered pleasure boats in Florida. Find the best predicted number of manatee fatalities resulting from encounters with boats. Is the result reasonably close to 79, which was the actual number of manatee fatalities?

Pleasure Boats	99	99	97	95	90	90	87	90	90
Manatee Fatalities	92	73	90	97	83	88	81	73	68

25. Tips Using the bill/tip data, find the best predicted tip amount for a dinner bill of $100. What tipping rule does the regression equation suggest?

Bill (dollars)	33.46	50.68	87.92	98.84	63.60	107.34
Tip (dollars)	5.50	5.00	8.08	17.00	12.00	16.00

26. POTUS Using the president/opponent heights, find the best predicted height of an opponent of a president who is 190 cm tall. Does it appear that heights of opponents can be predicted from the heights of the presidents?

President	178	182	188	175	179	183	192	182	177	185	188	188	183	188
Opponent	180	180	182	173	178	182	180	180	183	177	173	188	185	175

27. Sports Using the diameter/circumference data, find the best predicted circumference of a marble with a diameter of 1.50 cm. How does the result compare to the actual circumference of 4.7 cm?

	Baseball	Basketball	Golf	Soccer	Tennis	Ping-Pong	Volleyball	Softball
Diameter	7.4	23.9	4.3	21.8	7.0	4.0	20.9	9.7
Circumference	23.2	75.1	13.5	68.5	22.0	12.6	65.7	30.5
Volume	212.2	7148.1	41.6	5424.6	179.6	33.5	4780.1	477.9

28. Sports Using the diameter/volume data from the preceding exercise, find the best predicted volume of a marble with a diameter of 1.50 cm. How does the result compare to the actual volume of 1.8 cm^3?

Large Data Sets.   *Exercises 29–32 use the same Appendix B data sets as Exercises 29–32 in Section 10-1. In each case, find the regression equation, letting the first variable be the predictor (x) variable. Find the indicated predicted values following the prediction procedure summarized in Figure 10-5 on page 493.*

29. Internet and Nobel Laureates Repeat Exercise 13 using all of the paired Internet/Nobel data listed in Data Set 16 "Nobel Laureates and Chocolate" in Appendix B.

30. Old Faithful Repeat Exercise 14 using all of the paired duration/interval after times listed in Data Set 23.

31. CSI Statistics Repeat Exercise 17 using the shoe print lengths and heights of the 19 males from Data Set 2 "Foot and Height."

32. CSI Statistics Repeat Exercise 18 using the foot lengths and heights of the 19 males from Data Set 2 "Foot and Height."

33. Word Counts of Men and Women Refer to Data Set 24 "Word Counts" in Appendix B and use the word counts measured from men (*x*) and women (*y*) in couple relationships listed in the first two columns of Data Set 24. Find the best prediction for the number of words spoken by a female, given that her male partner speaks 16,000 words in a day.

34. Earthquakes Refer to Data Set 21 "Earthquakes" in Appendix B and use the depths (*x*) and magnitudes (*y*) from the earthquakes. Find the best predicted magnitude of an earthquake with a depth of 5 km.

10-2   Beyond the Basics

35. Least-Squares Property According to the least-squares property, the regression line minimizes the sum of the squares of the residuals. Refer to the data in Table 10-1 on page 469.

a. Find the sum of squares of the residuals.

b. Show that the regression equation $\hat{y} = -3 + 2.5x$ results in a larger sum of squares of residuals.

10-3   Prediction Intervals and Variation

Key Concept In Section 10-2 we presented a method for using a regression equation to find a predicted value of y, but it would be great to have a way of determining the *accuracy* of such predictions. In this section we introduce the *prediction interval,* which is an interval estimate of a predicted value of y. See the following definitions for the distinction between *confidence interval* and *prediction interval.*

> **DEFINITIONS**
>
> A **prediction interval** is a range of values used to estimate a *variable* (such as a predicted value of y in a regression equation).
>
> A **confidence interval** is a range of values used to estimate a population *parameter* (such as p or μ or σ).

   In Example 4(a) from the preceding section, we showed that when using the 23 pairs of chocolate consumption and Nobel Laureate rates, the regression equation is $\hat{y} = -3.37 + 2.49x$. Given a country with chocolate consumption of $x = 10$ kg per capita, the best predicted Nobel Laureate rate is 21.5 Nobel Laureates per 10 million people (which is found by substituting $x = 10$ in the regression equation). For $x = 10$, the "best" predicted Nobel Laureate rate is 21.5, but we have no sense of the accuracy of that estimate, so we need an interval estimate. A prediction interval estimate of a predicted value $\hat{y}$ can be found using the components in the following Key Elements box. Given the nature of the calculations, the use of technology is strongly recommended.

KEY ELEMENTS

Prediction Intervals

Objective

Find a prediction interval, which is an interval estimate of a predicted value of y.

Requirement

For each fixed value of x, the corresponding sample values of y are normally distributed about the regression line, and those normal distributions have the same variance.

continued

Formulas for Creating a Prediction Interval

Given a fixed and known value x_0, the prediction interval for an individual y value is

$$\hat{y} - E < y < \hat{y} + E$$

where the margin of error is

$$E = t_{\alpha/2} s_e \sqrt{1 + \frac{1}{n} + \frac{n(x_0 - \bar{x})^2}{n(\Sigma x^2) - (\Sigma x)^2}}$$

and x_0 is a given value of x, $t_{\alpha/2}$ has $n - 2$ degrees of freedom, and s_e is the **standard error of estimate** found from Formula 10-5 or Formula 10-6. (The standard error of estimate s_e is a measure of variation of the residuals, which are the differences between the observed sample y values and the predicted values $\hat{y}$ that are found from the regression equation.)

FORMULA 10-5 $s_e = \sqrt{\dfrac{\Sigma(y - \hat{y})^2}{n - 2}}$

FORMULA 10-6 $s_e = \sqrt{\dfrac{\Sigma y^2 - b_0 \Sigma y - b_1 \Sigma xy}{n - 2}}$ (This is an equivalent form of Formula 10-5 that is good for manual calculations or writing computer programs.)

 EXAMPLE 1   **Chocolate and Nobel Laureates: Finding a Prediction Interval**

For the paired chocolate/Nobel data in Table 10-1 from the Chapter Problem, we found that there is sufficient evidence to support the claim of a linear correlation between those two variables, and the regression equation is $\hat{y} = -3.37 + 2.49x$.

    a. If a country has a chocolate consumption amount given by $x = 10$ kg per capita, find the best predicted value of the Nobel Laureate rate.

    b. Use a chocolate consumption amount of $x = 10$ kg per capita to construct a 95% prediction interval for the Nobel Laureate rate.

SOLUTION

    a. Substitute $x = 10$ into the regression equation $\hat{y} = -3.37 + 2.49x$ to get a predicted value of $\hat{y} = 21.5$ Nobel Laureates per 10 million people.

    b. The accompanying StatCrunch and Minitab displays provide the 95% prediction interval, which is $7.8 < y < 35.3$ when rounded.

StatCrunch

```
Predicted values:
X value  Pred. Y  s.e.(Pred. y)   95% C.I. for mean       95% P.I. for new
10       21.56467 2.1502909 (17.092895, 26.036445)(7.7944348, 35.334905)
```

Minitab

```
Prediction for Nobel
Regression Equation
Nobel = -3.37 + 2.493 Chocolate

Variable    Setting
Chocolate        10

   Fit    SE Fit        95% CI              95% PI
21.5647  2.15029  (17.0929, 26.0364)  (7.79443, 35.3349)
```

The same 95% prediction interval could be manually calculated using these components:

$x_0 = 10$ (given)

$s_e = 6.262665$ (provided by many technologies, including Statdisk, Minitab, Excel, StatCrunch, and TI-83/84 Plus calculator)

$\hat{y} = 21.5$ (predicted value of y found by substituting $x = 10$ into the regression equation)

$t_{\alpha/2} = 2.080$ (from Table A-3 with df $= 21$ and an area of 0.05 in two tails)

$n = 23, \bar{x} = 5.804348, \Sigma x = 133.5, \Sigma x^2 = 1011.45$

INTERPRETATION

The 95% prediction interval is $7.8 < y < 35.3$. This means that if we select some country with a chocolate consumption rate of 10 kg per capita ($x = 10$), we have 95% confidence that the limits of 7.8 and 35.3 contain the Nobel Laureate rate. That is a wide range of values. The prediction interval would be much narrower and our estimated Nobel rate would be much better if we were using a much larger set of sample data instead of using only the 23 pairs of values listed in Table 10-1.

YOUR TURN   Do Exercise 13 "Boats."

Explained and Unexplained Variation

Assume that we have a sample of paired data having the following properties shown in Figure 10-7:

- There is sufficient evidence to support the claim of a linear correlation between x and y.

- The equation of the regression line is $\hat{y} = 3 + 2x$.

- The mean of the y values is given by $\bar{y} = 9$.

- One of the pairs of sample data is $x = 5$ and $y = 19$.

- The point $(5, 13)$ is one of the points on the regression line, because substituting $x = 5$ into the regression equation of $\hat{y} = 3 + 2x$ yields $\hat{y} = 13$.

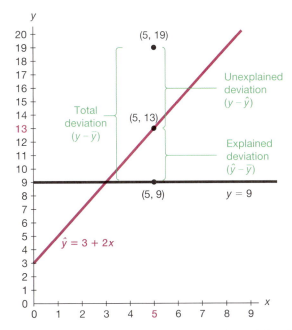

FIGURE 10-7 **Total, Explained, and Unexplained Deviation**

Figure 10-7 shows that the point (5, 13) lies on the regression line, but the point (5, 19) from the original data set does not lie on the regression line. If we completely ignore correlation and regression concepts and want to predict a value of y given a value of x and a collection of paired (x, y) data, our best guess would be the mean $\bar{y} = 9$. But in this case there is a linear correlation between x and y, so a better way to predict the value of y when $x = 5$ is to substitute $x = 5$ into the regression equation to get $\hat{y} = 13$. We can *explain* the discrepancy between $\bar{y} = 9$ and $\hat{y} = 13$ by noting that there is a linear relationship best described by the regression line. Consequently, when $x = 5$, the predicted value of y is 13, not the mean value of 9. For $x = 5$, the predicted value of y is 13, but the observed sample value of y is actually 19. The discrepancy between $\hat{y} = 13$ and $y = 19$ cannot be explained by the regression line, and it is called a *residual* or *unexplained deviation,* which can be expressed in the general format of $y - \hat{y}$.

As in Section 3-2, where we defined the standard deviation, we again consider a *deviation* to be a difference between a value and the mean. (In this case, the mean is $\bar{y} = 9$.) Examine Figure 10-7 carefully and note these specific deviations from $\bar{y} = 9$:

$$\textit{Total deviation} \ (\text{from } \bar{y} = 9) \ \text{of the point} \ (5, 19) = y - \bar{y} = 19 - 9 = 10$$

$$\textit{Explained deviation} \ (\text{from } \bar{y} = 9) \ \text{of the point} \ (5, 19) = \hat{y} - \bar{y} = 13 - 9 = 4$$

$$\textit{Unexplained deviation} \ (\text{from } \bar{y} = 9) \ \text{of the point} \ (5, 19) = y - \hat{y} = 19 - 13 = 6$$

These deviations from the mean are generalized and formally defined as follows.

DEFINITIONS

Assume that we have a collection of paired data containing the sample point (x, y), that $\hat{y}$ is the predicted value of y (obtained by using the regression equation), and that the mean of the sample y values is $\bar{y}$.

The **total deviation** of (x, y) is the vertical distance $y - \bar{y}$, which is the distance between the point (x, y) and the horizontal line passing through the sample mean $\bar{y}$.

The **explained deviation** is the vertical distance $\hat{y} - \bar{y}$, which is the distance between the predicted y value and the horizontal line passing through the sample mean $\bar{y}$.

The **unexplained deviation** is the vertical distance $y - \hat{y}$, which is the vertical distance between the point (x, y) and the regression line. (The distance $y - \hat{y}$ is also called a *residual,* as defined in Section 10-2.)

In Figure 10-7 we can see the following relationship for an individual point (x, y):

$$(\text{total deviation}) = (\text{explained deviation}) + (\text{unexplained deviation})$$

$$(y - \bar{y}) \quad = \quad (\hat{y} - \bar{y}) \quad + \quad (y - \hat{y})$$

The expression above involves deviations away from the mean, and it applies to any one particular point (x, y). If we sum the squares of deviations using all points (x, y), we get amounts of *variation.* The same relationship applies to the sums of squares shown in Formula 10-7, even though the expression above is not algebraically equivalent to Formula 10-7. In Formula 10-7, the **total variation** is the sum of the squares of the total deviation values, the **explained variation** is the sum of the squares of the explained deviation values, and the **unexplained variation** is the sum of the squares of the unexplained deviation values.

FORMULA 10-7

$$(\text{total variation}) = (\text{explained variation}) + (\text{unexplained variation})$$
$$\Sigma(y - \bar{y})^2 \;\; = \;\; \Sigma(\hat{y} - \bar{y})^2 \;\; + \;\; \Sigma(y - \hat{y})^2$$

Coefficient of Determination

In Section 10-1 we saw that the linear correlation coefficient r can be used to find the proportion of the total variation in y that can be explained by the linear correlation. This statement was made in Section 10-1:

> **The value of r^2 is the proportion of the variation in y that is explained by the linear relationship between x and y.**

This statement about the explained variation is formalized with the following definition.

DEFINITION

The **coefficient of determination** is the proportion of the variation in y that is explained by the regression line. It is computed as

$$r^2 = \frac{\text{explained variation}}{\text{total variation}}$$

We can compute r^2 by using the definition just given with Formula 10-7, or we can simply square the linear correlation coefficient r. Go with squaring r.

EXAMPLE 2 **Chocolate/Nobel Data: Finding a Coefficient of Determination**

If we use the 23 pairs of chocolate/Nobel data from Table 10-1 in the Chapter Problem, we find that the linear correlation coefficient is $r = 0.801$. Find the coefficient of determination. Also, find the percentage of the total variation in y (Nobel rate) that can be explained by the linear correlation between chocolate consumption and Nobel rate.

SOLUTION

With $r = 0.801$ the coefficient of determination is $r^2 = 0.642$.

INTERPRETATION

Because r^2 is the proportion of total variation that can be explained, we conclude that 64.2% of the total variation in the Nobel rate can be explained by chocolate consumption, and the other 35.8% cannot be explained by chocolate consumption. The other 35.8% might be explained by some other factors and/or random variation. But common sense suggests that it is somewhat silly to seriously think that a country's rate of Nobel Laureates is affected by the amount of chocolate consumed.

YOUR TURN Do Exercise 5 "Crickets and Temperature."

Prediction Intervals

Access tech supplements, videos, and data sets at **www.TriolaStats.com**

Statdisk	Minitab	StatCrunch

Statdisk provides the intercept and slope of the regression equation, the standard error of estimate (labeled "Standard Error"), and the coefficient of determination. These results are helpful in finding a prediction interval, but the actual prediction interval is not provided.

1. Click **Analysis** in the top menu.
2. Select **Correlation and Regression** from the dropdown menu.
3. Enter the desired significance level and select the two columns to be evaluated.
4. Click **Evaluate.**

1. Complete the Minitab Regression procedure from Section 10-2 to get the regression equation. Minitab will automatically use this equation in this procedure.
2. Click **Stat** in the top menu.
3. Select **Regression** from the dropdown menu and select **Regression—Predict** from the submenu.
4. Select **Enter individual values** from the dropdown menu.
5. Enter the desired value(s) for the *x* variable.
6. Click the **Options** button and change the confidence level to the desired value.
7. Click **OK** twice.

1. Click **Stat** in the top menu.
2. Select **Regression** from the dropdown menu, then select **Simple Linear** from the submenu.
3. Select the columns to be used for the *x* variable and *y* variable.
4. For *Prediction of Y* enter the desired *x* value(s) and significance level.
5. Click **Compute!**

TI-83/84 Plus Calculator

TI-83/84 Plus results include the intercept (a) and slope of the regression equation (b), the standard error of estimate (s), and the coefficient of determination (r^2). These results are helpful in finding a prediction interval, but the actual prediction interval is not provided.

1. Press **STAT**, then select **TESTS** in the top menu.
2. Select **LinRegTTest** in the menu and press **ENTER**.
3. Enter the list names for the *x* and *y* variables. Enter **1** for *Freq* and for β & ρ select $\neq$ **0** to test the null hypothesis of no correlation.
4. Select **Calculate** and press **ENTER** to view results.

Excel

XLSTAT Add-In

1. Enter the sample data in columns of the worksheet.
2. Enter the desired value(s) for *x* to be used for the prediction interval in a column.
3. Click on the **XLSTAT** tab in the Ribbon and then click **Modeling Data.**
4. Select **Linear Regression** from the dropdown menu.
5. Enter the range of cells containing the *Y/Dependent variable* data and *X/Explanatory variable* data. Check the **Quantitative** box under *X/Explanatory variable.* If the first data row includes a label, check the **Variable labels** box.
6. Click the **Options** tab and enter the desired confidence interval, such as **95.**
7. Click the **Prediction** tab.
8. Check the **Prediction** box and in the *Quantitative* box enter the cell range containing the desired value(s) of *x* from Step 2. The first cell in the range must contain a value, not a label.
9. Click **OK.** The prediction interval(s) are in the *Predictions for the new observations* table.

Excel (Data Analysis Add-In)

Excel provides the intercept and slope of the regression equation, the standard error of estimate s_e (labeled "Standard Error"), and the coefficient of determination (labeled "R Square"). These results are helpful in finding a prediction interval, but the actual prediction interval is not provided.

1. Click on the **Data** tab in the Ribbon and then click the **Data Analysis** tab.
2. Select **Regression** under *Analysis Tools* and click **OK.**
3. For *Input Y Range* enter the data range for the dependent *y* variable. For *Input X Range* enter the data range for the independent *x* variable.
4. Check the **Labels** box if the first row contains a label.
5. Click **OK** to display the results.

10-3   Basic Skills and Concepts

Statistical Literacy and Critical Thinking

1. s_e Notation Using Data Set 1 "Body Data" in Appendix B, if we let the predictor variable x represent heights of males and let the response variable y represent weights of males, the sample of 153 heights and weights results in $s_e = 16.27555$ cm. In your own words, describe what that value of s_e represents.

2. Prediction Interval Using the heights and weights described in Exercise 1, a height of 180 cm is used to find that the predicted weight is 91.3 kg, and the 95% prediction interval is (59.0 kg, 123.6 kg). Write a statement that interprets that prediction interval. What is the major advantage of using a prediction interval instead of simply using the predicted weight of 91.3 kg? Why is the terminology of *prediction interval* used instead of *confidence interval*?

3. Coefficient of Determination Using the heights and weights described in Exercise 1, the linear correlation coefficient r is 0.394. Find the value of the coefficient of determination. What practical information does the coefficient of determination provide?

4. Standard Error of Estimate A random sample of 118 different female statistics students is obtained and their weights are measured in kilograms and in pounds. Using the 118 paired weights (weight in kg, weight in lb), what is the value of s_e? For a female statistics student who weighs 100 lb, the predicted weight in kilograms is 45.4 kg. What is the 95% prediction interval?

Interpreting the Coefficient of Determination. *In Exercises 5–8, use the value of the linear correlation coefficient r to find the coefficient of determination and the percentage of the total variation that can be explained by the linear relationship between the two variables.*

5. Crickets and Temperature $r = 0.874$ (x = number of cricket chirps in 1 minute, y = temperature in °F)

6. Pizza and Subways $r = 0.992$ (x = cost of a slice of pizza, y = subway fare in New York City)

7. Weight/Waist $r = 0.885$ (x = weight of male, y = waist size of male)

8. Bears $r = 0.783$ (x = head width of a bear, y = weight of a bear)

Interpreting a Computer Display. *In Exercises 9–12, refer to the display obtained by using the paired data consisting of Florida registered boats (tens of thousands) and numbers of manatee deaths from encounters with boats in Florida for different recent years (from Data Set 10 in Appendix B). Along with the paired boat/manatee sample data, Stat-Crunch was also given the value of 85 (tens of thousands) boats to be used for predicting manatee fatalities.*

StatCrunch

```
Manatees = -49.048987 + 1.4062442 Boats
Sample size: 24
R (correlation coefficient) = 0.85014394
Estimate of error standard deviation: 9.6605284

Predicted values:
X value  Pred. Y  s.e.(Pred. y)  95% C.I. for mean       95% P.I. for new
85       70.481772  1.9724935(66.391071, 74.572473)(50.033706, 90.929839)
```

9. Testing for Correlation Use the information provided in the display to determine the value of the linear correlation coefficient. Is there sufficient evidence to support a claim of a linear correlation between numbers of registered boats and numbers of manatee deaths from encounters with boats?

10. Identifying Total Variation What percentage of the total variation in manatee fatalities can be explained by the linear correlation between registered boats and manatee fatalities?

11. Predicting Manatee Fatalities Using $x = 85$ (for 850,000 registered boats), what is the single value that is the best predicted number of manatee fatalities resulting from encounters with boats?

12. Finding a Prediction Interval For a year with 850,000 ($x = 85$) registered boats in Florida, identify the 95% prediction interval estimate of the number of manatee fatalities resulting from encounters with boats. Write a statement interpreting that interval.

Finding a Prediction Interval. *In Exercises 13–16, use the paired data consisting of registered Florida boats (tens of thousands) and manatee fatalities from boat encounters listed in Data Set 10 "Manatee Deaths" in Appendix B. Let x represent number of registered boats and let y represent the corresponding number of manatee deaths. Use the given number of registered boats and the given confidence level to construct a prediction interval estimate of manatee deaths.*

13. Boats Use $x = 85$ (for 850,000 registered boats) with a 99% confidence level.

14. Boats Use $x = 98$ (for 980,000 registered boats) with a 95% confidence level.

15. Boats Use $x = 96$ (for 960,000 registered boats) with a 95% confidence level.

16. Boats Use $x = 87$ (for 870,000 registered boats) with a 99% confidence level.

Variation and Prediction Intervals. *In Exercises 17–20, find the (a) explained variation, (b) unexplained variation, and (c) indicated prediction interval. In each case, there is sufficient evidence to support a claim of a linear correlation, so it is reasonable to use the regression equation when making predictions.*

17. Altitude and Temperature Listed below are altitudes (thousands of feet) and outside air temperatures (°F) recorded by the author during Delta Flight 1053 from New Orleans to Atlanta. For the prediction interval, use a 95% confidence level with the altitude of 6327 ft (or 6.327 thousand feet).

Altitude (thousands of feet)	3	10	14	22	28	31	33
Temperature (°F)	57	37	24	−5	−30	−41	−54

18. Town Courts Listed below are amounts of court income and salaries paid to the town justices (based on data from the *Poughkeepsie Journal*). All amounts are in thousands of dollars, and all of the towns are in Dutchess County, New York. For the prediction interval, use a 99% confidence level with a court income of $800,000.

Court Income	65	404	1567	1131	272	252	111	154	32
Justice Salary	30	44	92	56	46	61	25	26	18

19. Crickets and Temperature The table below lists numbers of cricket chirps in 1 minute and the temperature in °F. For the prediction interval, use 1000 chirps in 1 minute and use a 90% confidence level.

Chirps in 1 min	882	1188	1104	864	1200	1032	960	900
Temperature (°F)	69.7	93.3	84.3	76.3	88.6	82.6	71.6	79.6

20. Weighing Seals with a Camera The table below lists overhead widths (cm) of seals measured from photographs and the weights (kg) of the seals (based on "Mass Estimation of Weddell Seals Using Techniques of Photogrammetry," by R. Garrott of Montana State University). For the prediction interval, use a 99% confidence level with an overhead width of 9.0 cm.

Overhead Width	7.2	7.4	9.8	9.4	8.8	8.4
Weight	116	154	245	202	200	191

10-3   Beyond the Basics

21. Confidence Interval for Mean Predicted Value Example 1 in this section illustrated the procedure for finding a prediction interval for an *individual* value of y. When using a specific value x_0 for predicting the *mean* of all values of y, the confidence interval is as follows:

$$\hat{y} - E < \bar{y} < \hat{y} + E$$

where

$$E = t_{\alpha/2} \cdot s_e \sqrt{\frac{1}{n} + \frac{n(x_0 - \bar{x})^2}{n(\Sigma x^2) - (\Sigma x)^2}}$$

The critical value $t_{\alpha/2}$ is found with $n - 2$ degrees of freedom. Using the 23 pairs of chocolate/Nobel data from Table 10-1 on page 469 in the Chapter Problem, find a 95% confidence interval estimate of the mean Nobel rate given that the chocolate consumption is 10 kg per capita.

10-4   Multiple Regression

Key Concept So far in this chapter we have discussed the linear correlation between *two* variables, but this section presents methods for analyzing a linear relationship with *more than two* variables. We focus on these two key elements: (1) finding the multiple regression equation, and (2) using the value of adjusted R^2 and the P-value as measures of how well the multiple regression equation fits the sample data. Because the required calculations are so difficult, manual calculations are impractical and a threat to mental health, so this section emphasizes the use and interpretation of results from technology.

PART 1 **Basic Concepts of a Multiple Regression Equation**

As in the preceding sections of this chapter, we will consider *linear* relationships only. The following *multiple regression equation* describes linear relationships involving more than two variables.

> **DEFINITION**
>
> A **multiple regression equation** expresses a linear relationship between a response variable y and two or more predictor variables $(x_1, x_2, \ldots, x_k)$. The general form of a multiple regression equation obtained from sample data is
>
> $$\hat{y} = b_0 + b_1 x_1 + b_2 x_2 + \cdots + b_k x_k$$

The following Key Elements box includes the key components of this section. For notation, see that the coefficients $b_0, b_1, b_2, \ldots, b_k$ are sample *statistics* used to estimate the corresponding population parameters $\beta_0, \beta_1, \beta_2, \ldots, \beta_k$. Also, note that the multiple regression equation is a natural extension of the format $\hat{y} = b_0 + b_1 x_1$ used in Section 10-2 for regression equations with a single independent variable x_1. In Section 10-2, it would have been reasonable to question why we didn't use the more common and familiar format of $y = mx + b$, and we can now see that using $\hat{y} = b_0 + b_1 x_1$ allows us to easily extend that format to include additional predictor variables.

Finding a Multiple Regression Equation

Objective

Use sample matched data from three or more variables to find a multiple regression equation that is useful for predicting values of the response variable y.

Notation

$\hat{y} = b_0 + b_1 x_1 + b_2 x_2 + \cdots + b_k x_k$ (multiple regression equation found from *sample* data)

$y = \beta_0 + \beta_1 x_1 + \beta_2 x_2 + \cdots + \beta_k x_k$ (multiple regression equation for the *population* of data)

$\hat{y} =$ predicted value of y (computed using the multiple regression equation)

$k =$ number of *predictor* variables (also called *independent variables* or x variables)

$n =$ sample size (number of values for any one of the variables)

Requirements

For any specific set of x values, the regression equation is associated with a random error often denoted by ε. We assume that such errors are normally distributed with a mean of 0 and a standard deviation of σ and that the random errors are independent.

Procedure for Finding a Multiple Regression Equation

Manual calculations are not practical, so technology must be used. (See the "Tech Center" instructions at the end of this section.)

In 1886, Francis Galton was among the first to study genetics using the methods of regression we are now considering. He wrote the article "Regression Towards Mediocrity in Hereditary Stature," claiming that heights of offspring regress or revert back toward a mean. Although we continue to use the term "regression," current applications extend far beyond those involving heights.

 EXAMPLE 1 **Predicting Weight**

Data Set 1 "Body Data" in Appendix B includes heights (cm), waist circumferences (cm), and weights (kg) from a sample of 153 males. Find the multiple regression equation in which the response variable (y) is the weight of a male and the predictor variables are height (x_1) and waist circumference (x_2).

SOLUTION

Using Statdisk with the sample data in Data Set 1, we obtain the results shown in the display on the top of the next page. The coefficients b_0, b_1, and b_2 are used in the multiple regression equation:

$$\hat{y} = -149 + 0.769x_1 + 1.01x_2$$

or

$$\text{Weight} = -149 + 0.769\,\text{Height} + 1.01\,\text{Waist}$$

The obvious advantage of the second format above is that it is easier to keep track of the roles that the variables play.

Statdisk

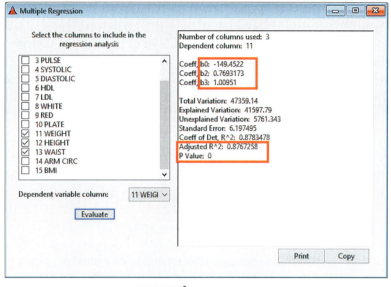

YOUR TURN Do Exercise 13 "Predicting Nicotine in Cigarettes"

If a multiple regression equation fits the sample data well, it can be used for predictions. For example, if we determine that the multiple regression equation in Example 1 is suitable for predictions, we can use the height and waist circumference of a male to predict his weight. But how do we determine whether the multiple regression equation fits the sample data well? Two very helpful tools are the values of adjusted R^2 and the P-value.

R^2 and Adjusted R^2

R^2 denotes the **multiple coefficient of determination,** which is a measure of how well the multiple regression equation fits the sample data. A perfect fit would result in $R^2 = 1$, and a very good fit results in a value near 1. A very poor fit results in a value of R^2 close to 0. The value of $R^2 = 0.878$ ("Coeff of Det, R^2") in the Statdisk display for Example 1 indicates that 87.8% of the variation in weights of males can be explained by their heights and waist circumferences. However, the multiple coefficient of determination R^2 has a serious flaw: As more variables are included, R^2 increases. (R^2 could remain the same, but it usually increases.) The largest R^2 is obtained by simply including *all* of the available variables, but the best multiple regression equation does not necessarily use all of the available variables. Because of that flaw, it is better to use the *adjusted coefficient of determination,* which is R^2 adjusted for the number of variables and the sample size.

> **DEFINITION**
>
> The **adjusted coefficient of determination** is the multiple coefficient of determination R^2 modified to account for the number of variables and the sample size. It is calculated by using Formula 10-8.

FORMULA 10-8

$$\text{Adjusted } R^2 = 1 - \frac{(n-1)}{[n-(k+1)]}(1 - R^2)$$

where

n = sample size

k = number of predictor (x) variables

The preceding Statdisk display shows the adjusted coefficient of determination as "Adjusted R^2" = 0.877 (rounded). If we use Formula 10-8 with $R^2 = 0.8783478$, $n = 153$, and $k = 2$, we get adjusted $R^2 = 0.877$ (rounded). When comparing this multiple regression equation to others, it is better to use the adjusted R^2 of 0.877. When considering the adjusted R^2 of 0.877 by itself, we see that it is fairly high (close to 1), suggesting that the regression equation is a good fit with the sample data.

P-Value

The *P*-value is a measure of the overall significance of the multiple regression equation. The displayed *P*-value of 0 (rounded) is small, indicating that the multiple regression equation has good overall significance and is usable for predictions. We can predict weights of males based on their heights and waist circumferences. Like the adjusted R^2, this *P*-value is a good measure of how well the equation fits the sample data. The *P*-value results from a test of the null hypothesis that $\beta_1 = \beta_2 = 0$. Rejection of $\beta_1 = \beta_2 = 0$ implies that at least one of β_1 and β_2 is not 0, indicating that this regression equation is effective in predicting weights of males. A complete analysis of results might include other important elements, such as the significance of the individual coefficients, but we are keeping things simple (!) by limiting our discussion to the three key components—multiple regression equation, adjusted R^2, and *P*-value.

Finding the Best Multiple Regression Equation

When trying to find the best multiple regression equation, we should not necessarily include all of the available predictor variables. Finding the best multiple regression equation requires abundant use of judgment and common sense, and there is no exact and automatic procedure that can be used to find the best multiple regression equation. *Determination of the best multiple regression equation is often quite difficult and is beyond the scope of this section,* but the following guidelines are helpful.

Guidelines for Finding the Best Multiple Regression Equation

1. *Use common sense and practical considerations to include or exclude variables.* For example, when trying to find a good multiple regression equation for predicting the height of a daughter, we should exclude the height of the physician who delivered the daughter, because that height is obviously irrelevant.

2. *Consider the P-value.* Select an equation having overall significance, as determined by a low *P*-value found in the technology results display.

3. *Consider equations with high values of adjusted R^2, and try to include only a few variables.* Instead of including almost every available variable, try to include relatively few predictor (*x*) variables. Use these guidelines:

   - Select an equation having a value of adjusted R^2 with this property: If an additional predictor variable is included, the value of adjusted R^2 does not increase very much.

   - For a particular number of predictor (*x*) variables, select the equation with the largest value of adjusted R^2.

   - In excluding predictor (*x*) variables that don't have much of an effect on the response (*y*) variable, it might be helpful to find the linear correlation coefficient *r* for each pair of variables being considered. If two predictor values have a very high linear correlation coefficient (called *multicollinearity*), there is no need to include them both, and we should exclude the variable with the lower value of adjusted R^2.

The following example illustrates that common sense and *critical thinking* are essential tools for effective use of methods of statistics.

EXAMPLE 2   Predicting Height from Footprint Evidence

Data Set 2 "Foot and Height" in Appendix B includes the age, foot length, shoe print length, shoe size, and height for each of 40 different subjects. Using those sample data, find the regression equation that is best for predicting height. Is the "best" regression equation a *good* equation for predicting height?

SOLUTION

Using the response variable of height and possible predictor variables of age, foot length, shoe print length, and shoe size, there are 15 different possible combinations of predictor variables. Table 10-5 includes key results from five of those combinations. Blind and thoughtless application of regression methods would suggest that the best regression equation uses all four of the predictor variables, because that combination yields the highest adjusted R^2 value of 0.7585. However, given the objective of using evidence to estimate the height of a suspect, we use *critical thinking* as follows.

1. Delete the variable of age, because criminals rarely leave evidence identifying their ages.

2. Delete the variable of shoe size, because it is really a rounded form of foot length.

3. For the remaining variables of foot length and shoe print length, use only foot length because its adjusted R^2 value of 0.7014 is greater than 0.6520 for shoe print length, and it is not very much less than the adjusted R^2 value of 0.7484 for both foot length and shoe print length. In this case, it is better to use one predictor variable instead of two.

4. Although it appears that the use of the single variable of foot length is best, we also note that criminals usually wear shoes, so shoe print lengths are more likely to be found than foot lengths.

TABLE 10-5 Select Key Results from Data Set 2 "Foot and Height" in Appendix B

Predictor Variables	Adjusted R^2	P-Value	
Age	0.1772	0.004	← **Not best:** Adjusted R^2 is far less than 0.7014 for Foot Length.
Foot Length	**0.7014**	**0.000**	← **Best:** High adjusted R^2 and lowest P-value.
Shoe Print Length	0.6520	0.000	← **Not best:** Adjusted R^2 is less than 0.7014 for Foot Length.
Foot Length/Shoe Print Length	0.7484	0.000	← **Not best:** The adjusted R^2 value is not very much higher than 0.7014 for the single variable of Foot Length.
Age/Foot Length/ Shoe Print Length/ Shoe Size	0.7585	0.000	← **Not best:** There are other cases using fewer variables with adjusted R^2 that are not too much smaller.

INTERPRETATION

Blind use of regression methods suggests that when estimating the height of a subject, we should use all of the available data by including all four predictor variables of age, foot length, shoe print length, and shoe size, but other practical considerations suggest that it is best to use the single predictor variable of foot length. So the best regression equation appears to be this: Height $= 64.1 + 4.29$ (Foot Length). However, given that criminals usually wear shoes, it is best to use the

continued

single predictor variable of shoe print length, so the best practical regression equation appears to be this: Height = 80.9 + 3.22 (Shoe Print Length). The *P*-value of 0.000 suggests that the regression equation yields a good model for estimating height.

Because the results of this example are based on sample data from only 40 subjects, estimates of heights will not be very accurate. As is usually the case, better results could be obtained by using larger samples.

YOUR TURN Do Exercise 13 "Predicting Nicotine in Cigarettes."

Tests of Regression Coefficients The preceding guidelines for finding the best multiple regression equation are based on the adjusted R^2 and the *P*-value, but we could also conduct individual hypothesis tests based on values of the regression coefficients. Consider the regression coefficient of β_1. A test of the null hypothesis $\beta_1 = 0$ can tell us whether the corresponding predictor variable should be included in the regression equation. Rejection of $\beta_1 = 0$ suggests that β_1 has a nonzero value and is therefore helpful for predicting the value of the response variable. Procedures for such tests are described in Exercise 17.

Predictions With Multiple Regression

When we discussed regression in Section 10-2, we listed (on page 492) four points to consider when using regression equations to make predictions. These same points should be considered when using multiple regression equations.

PART 2 Dummy Variables and Logistic Regression

So far in this chapter, all variables have represented continuous data, but many situations involve a variable with only *two* possible qualitative values (such as male/female or dead/alive or cured/not cured). To obtain regression equations that include such variables, we must somehow assign numbers to the two different categories. A common procedure is to represent the two possible values by 0 and 1, where 0 represents a "failure" and 1 represents a "success." For disease outcomes, 1 is often used to represent the event of the disease or death, and 0 is used to represent the nonevent.

> **DEFINITION**
>
> A **dummy variable** is a variable having only the values of 0 and 1 that are used to represent the two different categories of a qualitative variable.

A dummy variable is sometimes called a *dichotomous variable.* The word "dummy" is used because the variable does not actually have any quantitative value, but we use it as a substitute to represent the different categories of the qualitative variable.

Dummy Variable as a Predictor Variable

Procedures of regression analysis differ dramatically, depending on whether the dummy variable is a predictor (*x*) variable or the response (*y*) variable. If we include a dummy variable as another *predictor* (*x*) variable, we can use the same methods of Part 1 in this section, as illustrated in Example 3.

EXAMPLE 3 **Using a Dummy Variable as a Predictor Variable**

Table 10-6 is adapted from Data Set 5 "Family Heights" in Appendix B and it is in a more convenient format for this example. Use the dummy variable of sex (coded as $0 =$ female, $1 =$ male). Given that a father is 69 in. tall and a mother is 63 in. tall, find the multiple regression equation and use it to predict the height of (a) a daughter and (b) a son.

TABLE 10-6 Heights (inches) of Fathers, Mothers, and Their Children

Height of Father	Height of Mother	Height of Child	Sex of Child (1 = Male)
66.5	62.5	70.0	1
70.0	64.0	68.0	1
67.0	65.0	69.7	1
68.7	70.5	71.0	1
69.5	66.0	71.0	1
70.0	65.0	73.0	1
69.0	66.0	70.0	1
68.5	67.0	73.0	1
65.5	60.0	68.0	1
69.5	66.5	70.5	1
70.5	63.0	64.5	0
71.0	65.0	62.0	0
70.5	62.0	60.0	0
66.0	66.0	67.0	0
68.0	61.0	63.5	0
68.0	63.0	63.0	0
71.0	62.0	64.5	0
65.5	63.0	63.5	0
64.0	60.0	60.0	0
71.0	63.0	63.5	0

SOLUTION

Using the methods of multiple regression from Part 1 of this section and computer software, we get this regression equation:

$$\text{Height of child} = 36.5 - 0.0336 \,(\text{Height of father})$$
$$+ 0.461 \,(\text{Height of mother}) + 6.14 \,(\text{Sex})$$

where the value of the dummy variable of sex is either 0 for a daughter or 1 for a son.

a. To find the predicted height of a *daughter*, we substitute 0 for the sex variable, and we also substitute 69 in. for the father's height and 63 in. for the mother's height. The result is a predicted height of 63.2 in. for a daughter.

b. To find the predicted height of a *son*, we substitute 1 for the sex variable, and we also substitute 69 in. for the father's height and 63 in. for the mother's height. The result is a predicted height of 69.4 in. for a son.

The coefficient of 6.14 in the regression equation shows that when given the height of a father and the height of a mother, a son will have a predicted height that is 6.14 in. more than the height of a daughter.

Icing the Kicker

Just as a kicker in football is about to attempt a field goal, it is a common strategy for the opposing coach to call a time-out to "ice" the kicker. The theory is that the kicker has time to think and become nervous and less confident, but does the practice actually work? In "The Cold-Foot Effect" by Scott M. Berry in *Chance* magazine, the author wrote about his statistical analysis of results from two National Football League (NFL) seasons. He uses a logistic regression model with variables such as wind, clouds, precipitation, temperature, the pressure of making the kick, and whether a time-out was called prior to the kick. He writes that "the conclusion from the model is that icing the kicker works—it is likely icing the kicker reduces the probability of a successful kick."

Logistic Regression In Example 3, we could use the same methods of Part 1 in this section because the dummy variable of sex is a *predictor* variable. However, if the dummy variable is the response (*y*) variable, we cannot use the methods in Part 1 of this section, and we should use a different method known as **logistic regression.** This section does not include detailed procedures for using logistic regression, but many books are devoted to this topic. Example 4 briefly illustrates the method of logistic regression.

EXAMPLE 4   **Logistic Regression**

Let a sample data set consist of the heights (cm) and arm circumferences (cm) of women and men as listed in Data Set 1 "Body Data" in Appendix B. Let the *response y* variable represent gender (0 = female, 1 = male). Using the gender values of *y* and the combined list of corresponding heights and arm circumferences, logistic regression could be used to obtain this model:

$$\ln\left(\frac{p}{1-p}\right) = -40.6 + 0.242(\text{HT}) + 0.000129(\text{ArmCirc})$$

In the expression above, *p* is the probability of a male, so *p* = 1 indicates that the subject is definitely a male, and *p* = 0 indicates that the subject is definitely not a male (so the subject is a female). [To solve for *p*, substitute values for height and arm circumference to get a value *v*, then $p = e^v/(1 + e^v)$.] See the following two sets of results.

- If we use the model above and substitute a height of 183 cm (or 72.0 in.) and an arm circumference of 33 cm (or 13.0 in.), we can solve for *p* to get *p* = 0.976, indicating that such a person has a 97.6% chance of being a male.

- In contrast, a smaller person with a height of 150 cm (or 59.1 in.) and an arm circumference of 20 cm (or 7.9 in.) results in a probability of *p* = 0.0134, indicating that such a small person is very unlikely to be a male.

TECH CENTER

Multiple Regression
Access tech supplements, videos, and data sets at **www.TriolaStats.com**

Statdisk	Minitab	StatCrunch
1. Click **Analysis** in the top menu.	1. Click **Stat** in the top menu.	1. Click **Stat** in the top menu.
2. Select **Multiple Regression** from the dropdown menu.	2. Select **Regression** from the dropdown menu and select **Regression—Fit Regression Model** from the submenu.	2. Select **Regression** from the dropdown menu, then select **Multiple Linear** from the submenu.
3. Select the columns to be included in the regression analysis. For *Dependent variable column,* select the column to be used for the dependent *y* variable.	3. Under *Responses* select the column that contains the dependent *y* values. Under *Continuous predictors* select the columns that contain the variables you want included as predictor *x* variables.	3. Select the columns to be used for the *x* variable and the column to be used for the *y* variable.
4. Click **Evaluate.**	4. Click **OK.** The regression equation is included in the results.	4. Click **Compute!**

TECH CENTER *continued*

Multiple Regression

Access tech supplements, videos, and data sets at **www.TriolaStats.com**

TI-83/84 Plus Calculator

Requires program *A2MULREG* (available at TriolaStats.com)

1. Data must be entered as columns in *Matrix D*, with the first column containing values of the dependent *y* variable:
Manually enter data: Press **2ND** then **x⁻¹** to get to the *MATRIX* menu, select **EDIT** from the top menu, select **[D]**, and press **ENTER**. Enter the number of rows and columns needed, press **ENTER**, and proceed to enter the sample values.

   Using existing lists: Lists can be combined and stored in *Matrix D*. Press **2ND** then **x⁻¹** to get to the *MATRIX* menu, select **MATH** from the top menu, and select the item **List → matr.** Enter the list names (the first list must contain values for the dependent *y* variable), followed by the matrix name **[D]**, all separated by **,**. *Important:* The matrix name must be entered by pressing **2ND** then **x⁻¹**, selecting **[D]**, and pressing **ENTER**. The following is a summary of the commands used to create a matrix from three lists (L1, L2, L3): **List → matr(L1, L2, L3, [D]).**

2. Press **PRGM**, select *A2MULREG*, press **ENTER** three times, select **MULT REGRESSION,** and press **ENTER**.

3. Enter the number of independent *x* variables, then enter the column number of each independent *x* variable. Press **ENTER** after each entry.

4. The results will be displayed, including *P*-value and adjusted R^2. Press **ENTER** to view additional results, including values used in the multiple regression equation.

5. Press **ENTER** to select the **QUIT** option.

Excel

XLSTAT Add-In

1. Click on the **XLSTAT** tab in the Ribbon and then click **Modeling Data.**

2. Select **Linear Regression** from the drop-down menu.

3. Enter the range of cells containing the *Y/Dependent variable* data and *X/Explanatory variable* data (multiple columns). Check the **Quantitative** box under *X/Explanatory variable*. If the first data row includes a label, check the **Variable labels** box.

4. Click the **Outputs** tab and ensure **Correlations** and **Analysis of Variance** are both checked.

5. Click **OK,** and the equation of the multiple regression line will be displayed in the results.

Excel (Data Analysis Add-In)

1. Click on the **Data** tab in the Ribbon and then click the **Data Analysis** tab. Select **Regression** under *Analysis Tools.*

2. For *Input Y Range* enter the data range for the dependent *y* variable. For *Input X Range* enter the data range for the independent *x* variables. *The x variable data must be located in adjacent columns.*

3. Check the **Labels** box if the first row contains a label.

4. Click **OK** to display the results.

10-4   Basic Skills and Concepts

Statistical Literacy and Critical Thinking

1. Terminology Using the lengths (in.), chest sizes (in.), and weights (lb) of bears from Data Set 9 "Bear Measurements" in Appendix B, we get this regression equation: Weight = −274 + 0.426 Length +12.1 Chest Size. Identify the response and predictor variables.

2. Best Multiple Regression Equation For the regression equation given in Exercise 1, the *P*-value is 0.000 and the adjusted R^2 value is 0.925. If we were to include an additional predictor variable of neck size (in.), the *P*-value becomes 0.000 and the adjusted R^2 becomes 0.933. Given that the adjusted R^2 value of 0.933 is larger than 0.925, is it better to use the regression equation with the three predictor variables of length, chest size, and neck size? Explain.

3. Adjusted Coefficient of Determination For Exercise 2, why is it better to use values of adjusted R^2 instead of simply using values of R^2?

4. Interpreting R^2 For the multiple regression equation given in Exercise 1, we get $R^2 = 0.928$. What does that value tell us?

Interpreting a Computer Display.   *In Exercises 5–8, we want to consider the correlation between heights of fathers and mothers and the heights of their sons. Refer to the StatCrunch display and answer the given questions or identify the indicated items. The display is based on Data Set 5 "Family Heights" in Appendix B.*

5. Height of Son Identify the multiple regression equation that expresses the height of a son in terms of the height of his father and mother.

6. Height of Son Identify the following:

a. The *P*-value corresponding to the overall significance of the multiple regression equation

b. The value of the multiple coefficient of determination R^2

c. The adjusted value of R^2

7. Height of Son Should the multiple regression equation be used for predicting the height of a son based on the height of his father and mother? Why or why not?

8. Height of Son A son will be born to a father who is 70 in. tall and a mother who is 60 in. tall. Use the multiple regression equation to predict the height of the son. Is the result likely to be a good predicted value? Why or why not?

City Fuel Consumption: Finding the Best Multiple Regression Equation.   *In Exercises 9–12, refer to the accompanying table, which was obtained using the data from 21 cars listed in Data Set 20 "Car Measurements" in Appendix B. The response (y) variable is CITY (fuel consumption in mi/gal). The predictor (x) variables are WT (weight in pounds), DISP (engine displacement in liters), and HWY (highway fuel consumption in mi/gal).*

Predictor (x) Variables	P-Value	R^2	Adjusted R^2	Regression Equation
WT/DISP/HWY	0.000	0.943	0.933	CITY = 6.86 − 0.00128 WT − 0.257 DISP + 0.652 HWY
WT/DISP	0.000	0.748	0.720	CITY = 38.0 − 0.00395 WT − 1.29 DISP
WT/HWY	0.000	0.942	0.935	CITY = 6.69 − 0.00159 WT + 0.670 HWY
DISP/HWY	0.000	0.935	0.928	CITY = 1.87 − 0.625 DISP + 0.706 HWY
WT	0.000	0.712	0.697	CITY = 41.8 − 0.00607 WT
DISP	0.000	0.659	0.641	CITY = 29.0 − 2.98 DISP
HWY	0.000	0.924	0.920	CITY = −3.15 + 0.819 HWY

9. If only one predictor (*x*) variable is used to predict the city fuel consumption, which single variable is best? Why?

10. If exactly two predictor (*x*) variables are to be used to predict the city fuel consumption, which two variables should be chosen? Why?

11. Which regression equation is best for predicting city fuel consumption? Why?

12. A Honda Civic weighs 2740 lb, it has an engine displacement of 1.8 L, and its highway fuel consumption is 36 mi/gal. What is the best predicted value of the city fuel consumption? Is that predicted value likely to be a good estimate? Is that predicted value likely to be very accurate?

Appendix B Data Sets. *In Exercises 13–16, refer to the indicated data set in Appendix B and use technology to obtain results.*

13. Predicting Nicotine in Cigarettes Refer to Data Set 13 "Cigarette Contents" in Appendix B and use the tar, nicotine, and CO amounts for the cigarettes that are 100 mm long, filtered, nonmenthol, and nonlight (the last set of measurements). Find the best regression equation for predicting the amount of nicotine in a cigarette. Why is it best? Is the best regression equation a good regression equation for predicting the nicotine content? Why or why not?

14. Predicting Nicotine in Cigarettes Repeat the preceding exercise using the sample data from the Menthol cigarettes listed in Data Set 13 "Cigarette Contents" in Appendix B.

15. Predicting IQ Score Refer to Data Set 8 "IQ and Brain Size" in Appendix B and find the best regression equation with IQ score as the response (y) variable. Use predictor variables of brain volume and/or body weight. Why is this equation best? Based on these results, can we predict someone's IQ score if we know their brain volume and body weight? Based on these results, does it appear that people with larger brains have higher IQ scores?

16. Full IQ Score Refer to Data Set 7 "IQ and Lead" in Appendix B and find the best regression equation with IQ FULL (full IQ score) as the response (y) variable. Use predictor variables of IQ VERB (verbal IQ score) and IQ PERF (performance IQ score). Why is this equation best? Based on these results, can we predict someone's full IQ score if we know their verbal IQ score and their performance IQ score? Is such a prediction likely to be very accurate?

10-4 Beyond the Basics

17. Testing Hypotheses About Regression Coefficients If the coefficient β_1 has a nonzero value, then it is helpful in predicting the value of the response variable. If $\beta_1 = 0$, it is not helpful in predicting the value of the response variable and can be eliminated from the regression equation. To test the claim that $\beta_1 = 0$ use the test statistic $t = (b_1 - 0)/s_{b_1}$. Critical values or P-values can be found using the t distribution with $n - (k + 1)$ degrees of freedom, where k is the number of predictor (x) variables and n is the number of observations in the sample. The standard error s_{b_1} is often provided by software. For example, see the accompanying StatCrunch display for Example 1, which shows that $s_{b_1} = 0.071141412$ (found in the column with the heading of "Std. Err." and the row corresponding to the first predictor variable of height). Use the sample data in Data Set 1 "Body Data" and the StatCrunch display to test the claim that $\beta_1 = 0$. Also test the claim that $\beta_2 = 0$. What do the results imply about the regression equation?

Parameter estimates:

Parameter	Estimate	Std. Err.	Alternative	DF	T-Stat	P-value
Intercept	-149.45217	12.523494	≠ 0	150	-11.933743	<0.0001
Height	0.76931731	0.071141412	≠ 0	150	10.813917	<0.0001
Waist	1.0095102	0.033812346	≠ 0	150	29.856261	<0.0001

18. Confidence Intervals for a Regression Coefficients A confidence interval for the regression coefficient β_1 is expressed as

$$b_1 - E < \beta_1 < b_1 + E$$

where

$$E = t_{\alpha/2}s_{b_1}$$

continued

The critical t score is found using $n - (k + 1)$ degrees of freedom, where k, n, and s_{b_1} are described in Exercise 17. Using the sample data from Example 1, $n = 153$ and $k = 2$, so df $= 150$ and the critical t scores are ± 1.976 for a 95% confidence level. Use the sample data for Example 1, the Statdisk display in Example 1 on page 513, and the StatCrunch display in Exercise 17 to construct 95% confidence interval estimates of β_1 (the coefficient for the variable representing height) and β_2 (the coefficient for the variable representing waist circumference). Does either confidence interval include 0, suggesting that the variable be eliminated from the regression equation?

 19. Dummy Variable Refer to Data Set 9 "Bear Measurements" in Appendix B and use the sex, age, and weight of the bears. For sex, let 0 represent female and let 1 represent male. Letting the response (y) variable represent weight, use the variable of age and the dummy variable of sex to find the multiple regression equation. Use the equation to find the predicted weight of a bear with the characteristics given below. Does sex appear to have much of an effect on the weight of a bear?

a. Female bear that is 20 years of age

b. Male bear that is 20 years of age

10-5 Nonlinear Regression

Key Concept The preceding sections of this chapter deal with *linear* relationships only, but not all in the world is linear. This section is a brief introduction to methods for finding some *nonlinear* functions that fit sample data. We focus on the use of technology because the required calculations are quite complex.

Shown below are five basic generic models considered in this section. Each of the five models is given with a generic formula along with an example of a specific function and its graph.

Linear: $y = a + bx$
Example: $y = 1 + 2x$

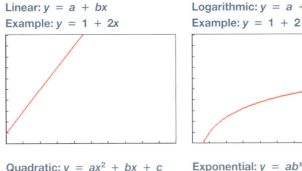

Logarithmic: $y = a + b \ln x$
Example: $y = 1 + 2 \ln x$

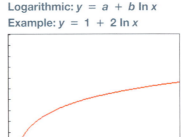

Power: $y = ax^b$
Example: $y = 3x^{2.5}$

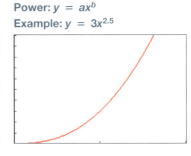

Quadratic: $y = ax^2 + bx + c$
Example: $y = x^2 - 8x + 18$

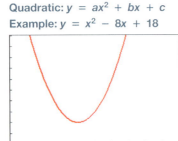

Exponential: $y = ab^x$
Example: $y = 2^x$

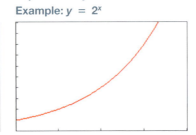

Here are three basic rules for identifying a good mathematical model:

1. ***Look for a pattern in the graph.*** Construct a graph, compare it to those shown here, and identify the model that appears to be most similar.

2. ***Compare values of R^2.*** For each model being considered, use technology to find the value of the coefficient of determination R^2. Choose functions that result in larger values of R^2, because such larger values correspond to functions that better fit the observed sample data.

- Don't place much importance on small differences, such as the difference between $R^2 = 0.984$ and $R^2 = 0.989$.

- Unlike in Section 10-4, we don't need to use values of adjusted R^2. Because the examples of this section all involve a single predictor variable, it makes sense to compare values of R^2.

- In addition to R^2, another measure used to assess the quality of a model is the sum of squares of the residuals. See Exercise 18 "Sum of Squares Criterion".

3. **Think.** Use common sense. Don't use a model that leads to predicted values that are unrealistic. Use the model to calculate future values, past values, and values for missing data, and then determine whether the results are realistic and make sense. Don't go too far beyond the scope of the available sample data.

EXAMPLE 1   Finding the Best Population Model

Table 10-7 lists the population of the United States for different 20-year intervals. Find a mathematical model for the population size, then predict the size of the U.S. population in the year 2040.

TABLE 10-7 Population (in millions) of the United States

Year	1800	1820	1840	1860	1880	1900	1920	1940	1960	1980	2000
Coded Year	1	2	3	4	5	6	7	8	9	10	11
Population	5	10	17	31	50	76	106	132	179	227	281

SOLUTION

First, we "code" the year values by using 1, 2, 3,…, instead of 1800, 1820, 1840,…. The reason for this coding is to use values of x that are much smaller and much less likely to cause computational difficulties.

1. *Look for a pattern in the graph.* Examine the pattern of the data values in the TI-83/84 Plus display (shown in the margin), and compare that pattern to the generic models shown earlier in this section. The pattern of those points is clearly not a straight line, so we rule out a linear model. Good candidates for the model appear to be the quadratic, exponential, and power functions.

2. *Find and compare values of R^2.* The TI-83/84 display for the quadratic model is shown in the margin. For the quadratic model, $R^2 = 0.9992$ (rounded), which is quite high. Table 10-8 includes this result with results from two other potential models. In comparing the values of the coefficient R^2, it appears that the quadratic model is best because it has the highest value of 0.9992. If we select the quadratic function as the best model, we conclude that the equation $y = 2.77x^2 - 6.00x + 10.01$ best describes the relationship between the year x (coded with $x = 1$ representing 1800, $x = 2$ representing 1820, and so on) and the population y (in millions).

    Based on its R^2 value of 0.9992, the quadratic model appears to be best, but the other values of R^2 are also quite high. Our general knowledge of population growth might suggest that the exponential model is most appropriate. (With a constant birth rate and no limiting factors, population will grow exponentially.)

TABLE 10-8 Models for the Population Data

Model	R^2	Equation
Quadratic	**0.9992**	$y = 2.77x^2 - 6.00x + 10.01$
Exponential	0.9631	$y = 5.24(1.48^x)$
Power	0.9764	$y = 3.35x^{1.77}$

continued

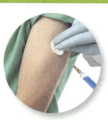

Clinical Trial Cut Short

What do you do when you're testing a new treatment and, before your study ends, you find that it is clearly effective? You should cut the study short and inform all participants of the treatment's effectiveness. This happened when hydroxyurea was tested as a treatment for sickle cell anemia. The study was scheduled to last about 40 months, but the effectiveness of the treatment became obvious and the study was stopped after 36 months. (See "Trial Halted as Sickle Cell Treatment Proves Itself," by Charles Marwick, *Journal of the American Medical Association,* Vol. 273, No. 8.)

TI-83/84 Plus

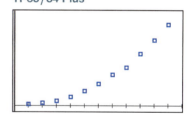

To predict the U.S. population for the year 2040, first note that the year 2040 is coded as $x = 13$ (see Table 10-7). Substituting $x = 13$ into the quadratic model of $y = 2.77x^2 - 6.00x + 10.01$ results in $y = 400$, which indicates that the U.S. population is estimated to be 400 million in the year 2040.

3. *Think.* The forecast result of 400 million in 2040 seems reasonable. (As of this writing, the latest figures from the U.S. Bureau of the Census use much more sophisticated methods to project that the U.S. population in 2040 will be 380 million.) However, there is considerable danger in making estimates for times that are beyond the scope of the available data. For example, the quadratic model suggests that in 1492, the U.S. population was 671 million, which is a result statisticians refer to as *ridiculous.* The quadratic model appears to be good for the available data (1800–2000), but other models might be better if it is necessary to make future population estimates.

> **YOUR TURN** Do Exercise 5 "Dropping the Ball."

 EXAMPLE 2 Interpreting R^2

In Example 1, we obtained the value of $R^2 = 0.9992$ for the quadratic model. Interpret that value as it relates to the predictor variable of year and the response variable of population size.

SOLUTION

In the context of the year/population data from Table 10-7, the value of $R^2 = 0.9992$ can be interpreted as follows: 99.92% of the variation in the population size can be explained by the quadratic regression equation (given in Example 1) that relates year and population size.

> **YOUR TURN** Do Exercise 3 "Interpreting R^2."

In "Modeling the U.S. Population" (*AMATYC Review,* Vol. 20, No. 2), Sheldon Gordon makes this important point that applies to all uses of statistical methods:

"The best choice (of a model) depends on the set of data being analyzed and requires an exercise in judgment, not just computation."

TECH CENTER

 Nonlinear Regression

Access tech supplements, videos, and data sets at **www.TriolaStats.com**

Statdisk	Minitab	StatCrunch
Statdisk can find the quadratic model using the Multiple Regression *function. The following procedure features data from Table 10-7.* 1. Enter the population data from Table 10-7 in column 1 of the Sample Editor. 2. Enter the corresponding coded year values (1, 2, 3…, 11) in column 2. 3. Enter the squares of the coded year values (1, 4, 9,…121) in column 3. 4. Click **Analysis** in the top menu. 5. Select **Multiple Regression** from the dropdown menu. 6. Select columns 1, 2, 3 and select column 1 as the dependent variable. 7. Click **Evaluate.** Statdisk provides the coefficients for the regression equation and value for R^2.	1. Click **Stat** in the top menu. 2. Select **Regression** from the dropdown menu and select **Fitted Line Plot** from the submenu. 3. Select the column to be used for the *Response y* variable and the column to be used for the *Predictor x* variable. 4. Choose the desired type of regression model: *linear, quadratic* or *cubic.* 5. Click **OK.**	*StatCrunch can find the model for a quadratic function (polynomial of order 2).* 1. Click **Stat** in the top menu. 2. Select **Regression** from the dropdown menu, then select **Polynomial** from the submenu. 3. Select the column to be used for the x variable and y variable. 4. For a quadratic function select **2** under *Poly. order.* 5. Click **Compute!**

TECH CENTER *continued*

Nonlinear Regression

Access tech supplements, videos, and data sets at **www.TriolaStats.com**

TI-83/84 Plus Calculator

1. Turn on the *Stat Diagnostics* feature by pressing the **MODE** button, scrolling down to **Stat Diagnostics**, highlighting **ON**, and pressing **ENTER**.
2. Press **STAT**, then select **CALC** in the top menu.
3. Select the desired model from the list of available options, then press **ENTER**.
4. Enter the desired data list names for *x* and *y* variables (for TI-83 calculators, enter the list names separated by **,**).
5. Select **Calculate** and press **ENTER**.

*TIP: For TI-83 Plus calculators, turn Stat Diagnostics ON by pressing **2ND** **0** for the Catalog menu. Scroll down to **DiagnosticON** and press **ENTER** twice.*

Excel

The XLSTAT add-in cannot be used to create nonlinear regression models, so Excel itself must be used.

1. Select the range of cells containing paired data.
2. Click the **Insert** tab in the Ribbon and select **Scatter** in the *Charts* section.
3. Right-click on any data point on the scatterplot and select **Add Trendline...**
4. Selected desired model and check **Display Equation on chart** and **Display R-squared value on chart.**

10-5   Basic Skills and Concepts

Statistical Literacy and Critical Thinking

1. Identifying a Model and R^2 Different samples are collected, and each sample consists of IQ scores of 25 statistics students. Let *x* represent the standard deviation of the 25 IQ scores in a sample, and let *y* represent the variance of the 25 IQ scores in a sample. What formula best describes the relationship between *x* and *y*? Which of the five models describes this relationship? What should be the value of R^2?

2. Super Bowl and R^2 Let *x* represent years coded as 1, 2, 3, . . . for years starting in 1980, and let *y* represent the numbers of points scored in each Super Bowl from 1980. Using the data from 1980 to the last Super Bowl at the time of this writing, we obtain the following values of R^2 for the different models: linear: 0.147; quadratic: 0.255; logarithmic: 0.176; exponential: 0.175; power: 0.203. Based on these results, which model is best? Is the best model a good model? What do the results suggest about predicting the number of points scored in a future Super Bowl game?

3. Interpreting R^2 In Exercise 2, the quadratic model results in $R^2 = 0.255$. Identify the percentage of the variation in Super Bowl points that can be explained by the quadratic model relating the variable of year and the variable of points scored. (*Hint:* See Example 2.) What does the result suggest about the usefulness of the quadratic model?

4. Interpreting a Graph The accompanying graph plots the numbers of points scored in each Super Bowl to the last Super Bowl at the time of this writing. The graph of the quadratic equation that best fits the data is also shown in red. What feature of the graph justifies the value of $R^2 = 0.255$ for the quadratic model?

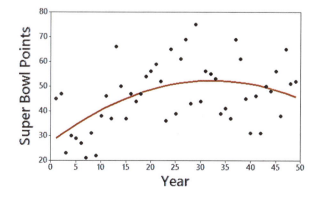

Finding the Best Model.   *In Exercises 5–16, construct a scatterplot and identify the mathematical model that best fits the given data. Assume that the model is to be used only for the scope of the given data, and consider only linear, quadratic, logarithmic, exponential, and power models.*

5. Dropping the Ball The table lists the distance *d* (meters) above the ground for an object dropped in a low-gravity vacuum from a height of 300 m. The time *t* (sec) is the time after the object has been released.

t (seconds)	1	2	3	4	5
d (meters)	295.1	280.5	256.1	222.0	178.1

6. Topsoil The Dirt Guy Topsoil company in Durham, CT, sells (you guessed it) topsoil. It is sold by the "yard," which is actually a cubic yard, and the variable *x* is the length (yd) of each side of a cube of topsoil.

x (yd)	1	2	3	4	6
Cost (dollars)	25	200	675	1600	5400

7. CD Yields The table lists the value *y* (in dollars) of $1000 deposited in a certificate of deposit at Bank of New York (based on rates currently in effect).

Year	1	2	3	4	5
Value	1012.20	1024.55	1037.05	1049.70	1062.51

8. Sound Intensity The table lists intensities of sounds as multiples of a basic reference sound. A scale similar to the decibel scale is used to measure the sound intensity.

Sound Intensity	316	500	750	2000	5000
Scale Value	25.0	27.0	28.75	33.0	37.0

9. Bacterial Growth In a carefully controlled experiment, bacteria are allowed to grow for a week. The number of bacteria are recorded at the end of each day with these results: 20, 40, 80, 160, 320, 640, 1280.

10. Deaths from Motor Vehicle Crashes Listed below are the numbers of deaths in the United States resulting from motor vehicle crashes. Use the best model to find the projected number of such deaths for the year 2025.

Year	1975	1980	1985	1990	1995	2000	2005	2010
Deaths	44,525	51,091	43,825	44,599	41,817	41,945	43,443	32,708

11. Richter Scale The table lists different amounts (metric tons) of the explosive TNT and the corresponding value measured on the Richter scale resulting from explosions of the TNT.

TNT	2	10	15	50	100	500
Richter Scale	3.4	3.9	4.0	4.4	4.6	5.0

12. Benford's Law According to Benford's law, a variety of different data sets include numbers with leading (first) digits that occur with the proportions listed in the following table.

Leading Digit	1	2	3	4	5	6	7	8	9
Proportion	0.301	0.176	0.125	0.097	0.079	0.067	0.058	0.051	0.046

13. Stock Market Listed on the top of the next page in order by row are the annual high values of the Dow Jones Industrial Average for each year beginning with 1990. Find the best model and then predict the value for the year 2014 (the last year listed). Is the predicted value close to the actual value of 18,054?

3000	3169	3413	3794	3978	5216	6561	8259	9374	11,568
11,401	11,350	10,635	10,454	10,855	10,941	12,464	14,198	13,338	10,606
11,625	12,929	13,589	16,577	18,054					

14. Sunspot Numbers Listed below in order by row are annual sunspot numbers beginning with 1980. Is the best model a good model? Carefully examine the scatterplot and identify the pattern of the points. Which of the models fits that pattern?

154.6	140.5	115.9	66.6	45.9	17.9	13.4	29.2	100.2	157.6
142.6	145.7	94.3	54.6	29.9	17.5	8.6	21.5	64.3	93.3
119.6	123.3	123.3	65.9	40.4	29.8	15.2	7.5	2.9	3.1
16.5	55.7	57.6	64.7	79.3					

15. Carbon Dioxide Listed below are mean amounts of carbon dioxide concentrations (parts per million) in our atmosphere for each decade, beginning with the 1880s. Find the best model and then predict the value for 2090–2099. Comment on the result.

292   294   297   300   304   307   309   314   320   331   345   360   377

16. Global Warming Listed below are mean annual temperatures (°C) of the earth for each decade, beginning with the 1880s. Find the best model and then predict the value for 2090–2099. Comment on the result.

13.819	13.692	13.741	13.788	13.906	14.016	14.052
13.983	13.938	14.014	14.264	14.396	14.636	

10-5   Beyond the Basics

17. Moore's Law In 1965, Intel cofounder Gordon Moore initiated what has since become known as *Moore's law:* The number of transistors per square inch on integrated circuits will double approximately every 18 months. In the table below, the first row lists different years and the second row lists the number of transistors (in thousands) for different years.

1971	1974	1978	1982	1985	1989	1993	1997	2000	2002	2003	2007	2011
2.3	5	29	120	275	1180	3100	7500	42,000	220,000	410,000	789,000	2,600,000

a. Assuming that Moore's law is correct and transistors per square inch double every 18 months, which mathematical model best describes this law: linear, quadratic, logarithmic, exponential, power? What specific function describes Moore's law?

b. Which mathematical model best fits the listed sample data?

c. Compare the results from parts (a) and (b). Does Moore's law appear to be working reasonably well?

18. Sum of Squares Criterion In addition to the value of R^2, another measurement used to assess the quality of a model is the *sum of squares of the residuals*. Recall from Section 10-2 that a residual is the difference between an observed y value and the value of y predicted from the model, which is denoted as $\hat{y}$. Better models have smaller sums of squares. Refer to the data in Table 10-7 on page 523.

a. Find $\Sigma (y - \hat{y})^2$, the sum of squares of the residuals resulting from the linear model.

b. Find the sum of squares of residuals resulting from the quadratic model.

c. Verify that according to the sum of squares criterion, the quadratic model is better than the linear model.

Chapter Quick Quiz

The following exercises are based on the following sample data consisting of numbers of enrolled students (in thousands) and numbers of burglaries for randomly selected large colleges in a recent year (based on data from the **New York Times***).*

Enrollment (thousands)	53	28	27	36	42
Burglaries	86	57	32	131	157

1. Conclusion The linear correlation coefficient r is found to be 0.499, the P-value is 0.393, and the critical values for a 0.05 significance level are ± 0.878. What should you conclude?

2. Switched Variables Which of the following change if the two variables of enrollment and burglaries are switched: the value of $r = 0.499$, the P-value of 0.393, the critical values of ± 0.878?

3. Change in Scale Exercise 1 stated that r is found to be 0.499. Does that value change if the actual enrollment values of 53,000, 28,000, 27,000, 36,000, and 42,000 are used instead of 53, 28, 27, 36, and 42?

4. Values of r If you had computed the value of the linear correlation coefficient to be 1.500, what should you conclude?

5. Predictions The sample data result in a linear correlation coefficient of $r = 0.499$ and the regression equation $\hat{y} = 3.83 + 2.39x$. What is the best predicted number of burglaries, given an enrollment of 50 (thousand), and how was it found?

6. Predictions Repeat the preceding exercise, assuming that the linear correlation coefficient is $r = 0.997$.

7. Explained Variation Given that the linear correlation coefficient r is found to be 0.499, what is the proportion of the variation in numbers of burglaries that is explained by the linear relationship between enrollment and number of burglaries?

8. Linear Correlation and Relationships True or false: If there is no linear correlation between enrollment and number of burglaries, then those two variables are not related in any way.

9. Causality True or false: If the sample data lead us to the conclusion that there is sufficient evidence to support the claim of a linear correlation between enrollment and number of burglaries, then we could also conclude that higher enrollments cause increases in numbers of burglaries.

10. Interpreting Scatterplot If the sample data were to result in the scatterplot shown here, what is the value of the linear correlation coefficient r?

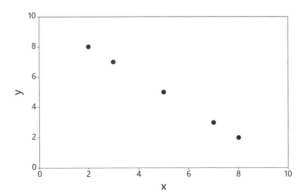

Review Exercises

1. Cigarette Tar and Nicotine The table below lists measured amounts (mg) of tar, carbon monoxide (CO), and nicotine in king size cigarettes of different brands (from Data Set 13 "Cigarette Contents" in Appendix B).

a. Is there is sufficient evidence to support a claim of a linear correlation between tar and nicotine?

b. What percentage of the variation in nicotine can be explained by the linear correlation between nicotine and tar?

c. Letting y represent the amount of nicotine and letting x represent the amount of tar, identify the regression equation.

d. The Raleigh brand king size cigarette is not included in the table, and it has 23 mg of tar. What is the best predicted amount of nicotine? How does the predicted amount compare to the actual amount of 1.3 mg of nicotine?

Tar	25	27	20	24	20	20	21	24
CO	18	16	16	16	16	16	14	17
Nicotine	1.5	1.7	1.1	1.6	1.1	1.0	1.2	1.4

2. Cigarette Nicotine and Carbon Monoxide Refer to the table of data given in Exercise 1 and use the amounts of nicotine and carbon monoxide (CO).

a. Construct a scatterplot using nicotine for the x scale, or horizontal axis. What does the scatterplot suggest about a linear correlation between amounts of nicotine and carbon monoxide?

b. Find the value of the linear correlation coefficient and determine whether there is sufficient evidence to support a claim of a linear correlation between amounts of nicotine and carbon monoxide.

c. Letting y represent the amount of carbon monoxide and letting x represent the amount of nicotine, find the regression equation.

d. The Raleigh brand king size cigarette is not included in the table, and it has 1.3 mg of nicotine. What is the best predicted amount of carbon monoxide? How does the predicted amount compare to the actual amount of 15 mg of carbon monoxide?

3. Time and Motion In a physics experiment at Doane College, a soccer ball was thrown upward from the bed of a moving truck. The table below lists the time (sec) that has lapsed from the throw and the height (m) of the soccer ball. What do you conclude about the relationship between time and height? What horrible mistake would be easy to make if the analysis is conducted without a scatterplot?

Time (sec)	0.0	0.2	0.4	0.6	0.8	1.0	1.2	1.4	1.6	1.8
Height (m)	0.0	1.7	3.1	3.9	4.5	4.7	4.6	4.1	3.3	2.1

4. Multiple Regression with Cigarettes Usc the sample data given in Review Exercise 1 "Cigarette Tar and Nicotine."

a. Find the multiple regression equation with the response (y) variable of amount of nicotine and predictor (x) variables of amounts of tar and carbon monoxide.

b. Identify the value of the multiple coefficient of determination R^2, the adjusted R^2, and the P-value representing the overall significance of the multiple regression equation.

c. Use a 0.05 significance level and determine whether the regression equation can be used to predict the amount of nicotine given the amounts of tar and carbon monoxide.

d. The Raleigh brand king size cigarette is not included in the table, and it has 23 mg of tar and 15 mg of carbon monoxide. What is the best predicted amount of nicotine? How does the predicted amount compare to the actual amount of 1.3 mg of nicotine?

Cumulative Review Exercises

Stocks and Sunspots. *Listed below are annual high values of the Dow Jones Industrial Average (DJIA) and annual mean sunspot numbers for eight recent years. Use the data for Exercises 1–5. A sunspot number is a measure of sunspots or groups of sunspots on the surface of the sun. The DJIA is a commonly used index that is a weighted mean calculated from different stock values.*

DJIA	14,198	13,338	10,606	11,625	12,929	13,589	16,577	18,054
Sunspot Number	7.5	2.9	3.1	16.5	55.7	57.6	64.7	79.3

1. Data Analysis Use only the sunspot numbers for the following.

a. Find the mean, median, range, standard deviation, and variance.

b. Are the sunspot numbers categorical data or quantitative data?

c. What is the level of measurement of the data? (nominal, ordinal, interval, ratio)

2. Correlation Use a 0.05 significance level to test for a linear correlation between the DJIA values and the sunspot numbers. Is the result as you expected? Should anyone consider investing in stocks based on sunspot numbers?

3. z Scores Using only the sunspot numbers, identify the highest number and convert it to a z score. In the context of these sample data, is that highest value "significantly high"? Why or why not?

4. Hypothesis Test The mean sunspot number for the past three centuries is 49.7. Use a 0.05 significance level to test the claim that the eight listed sunspot numbers are from a population with a mean equal to 49.7.

5. Confidence Interval Construct a 95% confidence interval estimate of the mean sunspot number. Write a brief statement interpreting the confidence interval.

6. Cell Phones and Driving In the author's home town of Madison, CT, there were 2733 police traffic stops in a recent year, and 7% of them were attributable to improper use of cell phones. Use a 0.05 significance level to test the claim that the sample is from a population in which fewer than 10% of police traffic stops are attributable to improper cell phone use.

7. Ages of Moviegoers The table below shows the distribution of the ages of moviegoers (based on data from the Motion Picture Association of America). Use the data to estimate the mean, standard deviation, and variance of ages of moviegoers. *Hint:* For the open-ended category of "60 and older," assume that the category is actually 60–80.

Age	2–11	12–17	18–24	25–39	40–49	50–59	60 and older
Percent	7	15	19	19	15	11	14

8. Ages of Moviegoers Based on the data from Cumulative Review Exercise 7, assume that ages of moviegoers are normally distributed with a mean of 35 years and a standard deviation of 20 years.

a. What is the percentage of moviegoers who are younger than 30 years of age?

b. Find P_{25}, which is the 25th percentile.

c. Find the probability that a simple random sample of 25 moviegoers has a mean age that is less than 30 years.

d. Find the probability that for a simple random sample of 25 moviegoers, each of the moviegoers is younger than 30 years of age. For a particular movie and showtime, why might it not be unusual to have 25 moviegoers all under the age of 30?

Technology Project

Speed Dating Data Set 18 "Speed Dating" in Appendix B includes data from 199 dates. Due to the large size of this data set, the data are available at www.TriolaStats.com. Download the data set and proceed to investigate correlations between pairs of variables using the data in the 5th, 7th, and 9th columns, which are all based on responses by *females*. Use the "like" measures by females as the *y* variable in each case.

1. Is there a correlation between "like" measures and attractiveness measures?

2. Is there a correlation between "like" measures and attribute measures?

3. Is there a correlation between attractiveness measures and attribute measures?

4. If Section 10-5 (Multiple Regression) was covered, investigate correlation and regression using the "like" measures as the *y* variable and use the attractiveness measures and attribute measures as the other two *x* variables.

5. Repeat the above using the 6th, 8th, and 10th columns, which are all based on responses by *males*.

6. Based on the results, what do you conclude? Write a brief report and include appropriate computer results.

FROM DATA TO DECISION

Critical Thinking: Is the pain medicine Duragesic effective in reducing pain?

Listed below are measures of pain intensity before and after using the drug Duragesic (fentanyl) (based on data from Janssen Pharmaceutical Products, L.P.). The data are listed in order by row, and corresponding measures are from the same subject before and after treatment. For example, the first subject had a measure of 1.2 before treatment and a measure of 0.4 after treatment. Each pair of measurements is from one subject, and the intensity of pain was measured using the standard visual analog score. A higher score corresponds to higher pain intensity.

Pain Intensity Before Duragesic Treatment

1.2	1.3	1.5	1.6	8.0	3.4	3.5	2.8	2.6	2.2
3.0	7.1	2.3	2.1	3.4	6.4	5.0	4.2	2.8	3.9
5.2	6.9	6.9	5.0	5.5	6.0	5.5	8.6	9.4	10.0
7.6									

Pain Intensity After Duragesic Treatment

0.4	1.4	1.8	2.9	6.0	1.4	0.7	3.9	0.9	1.8
0.9	9.3	8.0	6.8	2.3	0.4	0.7	1.2	4.5	2.0
1.6	2.0	2.0	6.8	6.6	4.1	4.6	2.9	5.4	4.8
4.1									

Analyzing the Results

1. Correlation Use the given data to construct a scatterplot, then use the methods of Section 10-1 to test for a linear correlation between the pain intensity before and after treatment. If there does appear to be a linear correlation, can we conclude that the drug treatment is effective?

2. Regression Use the given data to find the equation of the regression line. Let the response (*y*) variable be the pain intensity after treatment. What would be the equation of the regression line for a treatment having absolutely no effect?

3. Two Independent Samples The methods of Section 9-2 can be used to test the claim that two populations have the same mean. Identify the specific claim that the treatment is effective, then use the methods of Section 9-2 to test that claim. The methods of Section 9-2 are based on the requirement that the samples are independent. Are they independent in this case?

4. Matched Pairs The methods of Section 9-3 can be used to test a claim about matched data. Identify the specific claim that the treatment is effective, then use the methods of Section 9-3 to test that claim.

5. Best Method? Which of the preceding results is best for determining whether the drug treatment is effective in reducing pain? Based on the preceding results, does the drug appear to be effective?

Cooperative Group Activities

1. In-class activity For each student in the class, measure shoe print length and height. Test for a linear correlation and identify the equation of the regression line. Measure the shoe print length of the professor and use it to estimate his or her height. How close is the estimated height to the actual height?

2. Out-of-class activity Each student should estimate the number of footsteps that he or she would walk between the door of the classroom and the door used to exit the building. After recording all of the estimates, each student should then count the number of footsteps while walking from the classroom door to the door used to exit the building. After all of the estimates and actual counts have been compiled, explore correlation and regression using the tools presented in this chapter.

3. In-class activity Divide into groups of 8 to 12 people. For each group member, measure the person's height and also measure his or her navel height, which is the height from the floor to the navel. Is there a correlation between height and navel height? If so, find the regression equation with height expressed in terms of navel height. According to one theory, the average person's ratio of height to navel height is the golden ratio: $(1 + \sqrt{5})/2 \approx 1.6$. Does this theory appear to be reasonably accurate?

4. In-class activity Divide into groups of 8 to 12 people. For each group member, measure height and arm span. For the arm span, the subject should stand with arms extended, like the wings on an airplane. Using the paired sample data, is there a correlation between height and arm span? If so, find the regression equation with height expressed in terms of arm span. Can arm span be used as a reasonably good predictor of height?

5. In-class activity Divide into groups of 8 to 12 people. For each group member, use a string and ruler to measure head circumference and forearm length. Is there a relationship between these two variables? If so, what is it?

6. In-class activity Use a ruler as a device for measuring reaction time. One person should suspend the ruler by holding it at the top while the subject holds his or her thumb and forefinger at the bottom edge, ready to catch the ruler when it is released. Record the distance that the ruler falls before it is caught. Convert that distance to the time (in seconds) that it took the subject to react and catch the ruler. (If the distance is measured in inches, use $t = \sqrt{d/192}$. If the distance is measured in centimeters, use $t = \sqrt{d/487.68}$.) Test each subject once with the right hand and once with the left hand, and record the paired data. Test for a correlation. Find the equation of the regression line. Does the equation of the regression line suggest that the dominant hand has a faster reaction time?

7. In-class activity Divide into groups of 8 to 12 people. Record the pulse rate of each group member while he or she is seated. Then record the pulse rate of each group member while he or she is standing. Is there a relationship between sitting and standing pulse rate? If so, what is it?

8. In-class activity Divide into groups of three or four people. Appendix B includes many data sets not yet included in examples or exercises in this chapter. Search Appendix B for a pair of variables of interest, then investigate correlation and regression. State your conclusions and try to identify practical applications.

9. Out-of-class activity Divide into groups of three or four people. Investigate the relationship between two variables by collecting your own paired sample data and using the methods of this chapter to determine whether there is a significant linear correlation. Also identify the regression equation and describe a procedure for predicting values of one of the variables when given values of the other variable. Suggested topics:

• Is there a relationship between taste and cost of different brands of chocolate chip cookies (or colas)? Taste can be measured on some number scale, such as 1 to 10.

• Is there a relationship between salaries of professional baseball (or basketball or football) players and their season achievements?

• Is there a relationship between student grade-point averages and the amount of television watched? If so, what is it?

11-1 Goodness-of-Fit

11-2 Contingency Tables

11

GOODNESS-OF-FIT AND CONTINGENCY TABLES

CHAPTER PROBLEM
Cybersecurity: Detecting Intrusions into Computer Systems

According to Benford's law, many data sets have the property that the leading (leftmost) digits of numbers have a distribution described by the top two rows of Table 11-1 on the next page. Data sets with values having leading digits that conform to Benford's law include numbers of Twitter followers, population sizes, amounts on tax returns, lengths of rivers, and check amounts. In the *New York Times* article "Following Benford's Law, or Looking Out for No. 1," Malcolm Browne writes that

"the income tax agencies of several nations and several states, including California, are using detection software (to identify computer system intrusions) based on Benford's Law, as are a score of large companies and accounting businesses."

It now appears that Benford's law may be helpful in detecting attacks on computer systems by analyzing inter-arrival times, which are times between consecutive arrivals of Internet traffic. The basic idea is to detect anomalies in inter-arrival times of Internet traffic flow by analyzing leading digits of those times and determining whether the distribution of those leading digits is a significant departure from the distribution that follows Benford's law. (See "Benford's Law Behavior of Internet Traffic," by Arshadi and Jahangir, *Journal of Network and Computer Applications,* Vol. 40, No. 2014.) Major advantages of this approach are that it is relatively simple, it doesn't require difficult

computations, it can be done in real time, and hackers would not be able to configure their malware to avoid detection.

In the bottom two rows of Table 11-1, we list leading digits of inter-arrival times of Internet traffic flow. One of the bottom two rows represents normal Internet traffic, and the other row is from Internet traffic with an intrusion by a hacker. Section 11-1 will present methods for identifying which of the two bottom rows is signaling that an intrusion has occurred.

TABLE 11-1 Benford's Law: Distribution of Leading Digits

Leading Digit	1	2	3	4	5	6	7	8	9
Benford's Law	30.1%	17.6%	12.5%	9.7%	7.9%	6.7%	5.8%	5.1%	4.6%
Sample 1	76	62	29	33	19	27	28	21	22
Sample 2	69	40	42	26	25	16	16	17	20

CHAPTER OBJECTIVES

Chapters 7 and 8 introduced important methods of inferential statistics, including confidence intervals for estimating population parameters (Chapter 7) and methods for testing hypotheses or claims (Chapter 8). In Chapters 9 and 10, we considered inferences involving two populations and correlation/regression with paired data. In this chapter we use statistical methods for analyzing categorical (or qualitative, or attribute) data that can be separated into different cells. The methods of this chapter use the same χ^2 (chi-square) distribution that was introduced in Section 7-3 and again in Section 8-4. See Section 7-3 or Section 8-4 for a quick review of properties of the χ^2 distribution. Here are the chapter objectives:

11-1 Goodness-of-Fit

- Use frequency counts of categorical data partitioned into different categories and determine whether the data fit some claimed distribution.

11-2 Contingency Tables

- Use categorical data summarized as frequencies in a two-way table with at least two rows and at least two columns to conduct a formal test of independence between the row variable and column variable.

- Be able to conduct a formal test of a claim that different populations have the same proportions of some characteristics.

Goodness-of-Fit

Key Concept By "goodness-of-fit" we mean that sample data consisting of observed frequency counts arranged in a single row or column (called a *one-way frequency table*) agree with some particular distribution (such as normal or uniform) being considered. We will use a hypothesis test for the claim that the observed frequency counts agree with the claimed distribution.

> **DEFINITION**
>
> A **goodness-of-fit test** is used to test the hypothesis that an observed frequency distribution fits (or conforms to) some claimed distribution.

KEY ELEMENTS

Testing for Goodness-of-Fit

Objective

Conduct a goodness-of-fit test, which is a hypothesis test to determine whether a single row (or column) of frequency counts agrees with some specific distribution (such as uniform or normal).

Notation

O represents the *observed frequency* of an outcome, found from the sample data.
E represents the *expected frequency* of an outcome, found by assuming that the distribution is as claimed.
k represents the *number of different categories* or cells.
n represents the total *number of trials* (or the total of observed sample values).
p represents the *probability* that a sample value falls within a particular category.

Requirements

1. The data have been randomly selected.
2. The sample data consist of frequency counts for each of the different categories.
3. For each category, the *expected* frequency is at least 5. (The expected frequency for a category is the frequency that would occur if the data actually have the distribution that is being claimed. There is no requirement that the *observed* frequency for each category must be at least 5.)

Null and Alternative Hypotheses

H_0: The frequency counts agree with the claimed distribution.
H_1: The frequency counts do not agree with the claimed distribution.

Test Statistic for Goodness-of-Fit Tests

$$\chi^2 = \sum \frac{(O - E)^2}{E}$$

***P*-values:** *P*-values are typically provided by technology, or a range of *P*-values can be found from Table A-4.

Critical values:

1. Critical values are found in Table A-4 by using $k - 1$ degrees of freedom, where k is the number of categories.
2. Goodness-of-fit hypothesis tests are always *right-tailed*.

Finding Expected Frequencies

Conducting a goodness-of-fit test requires that we identify the *observed* frequencies denoted by O, then find the frequencies *expected* (denoted by E) with the claimed distribution. There are two different approaches for finding expected frequencies E:

- **If the expected frequencies are all equal: Calculate $E = n/k$.**

- **If the expected frequencies are not all equal: Calculate $E = np$ for each individual category.**

As good as these two preceding formulas for E might be, it is better to use an informal approach by simply asking, "How can the observed frequencies be split up among the different categories so that there is perfect agreement with the claimed distribution?" Also, note that the *observed* frequencies are all whole numbers because they represent actual counts, but the *expected* frequencies need not be whole numbers.

Examples:

a. **Equally Likely** A single die is rolled 45 times with the following results. Assuming that the die is fair and all outcomes are equally likely, find the expected frequency E for each empty cell.

Outcome	1	2	3	4	5	6
Observed Frequency O	13	6	12	9	3	2
Expected Frequency E						

With $n = 45$ outcomes and $k = 6$ categories, the expected frequency for each cell is the same: $E = n/k = 45/6 = 7.5$. If the die is fair and the outcomes are all equally likely, we expect that each outcome should occur about 7.5 times.

b. **Not Equally Likely** Using the same results from part (a), suppose that we claim that instead of being fair, the die is loaded so that the outcome of 1 occurs 50% of the time and the other five outcomes occur 10% of the time. The probabilities are listed in the second row below. Using $n = 45$ and the probabilities listed below, we find that for the first cell, $E = np = (45)(0.5) = 22.5$. Each of the other five cells will have the expected value of $E = np = (45)(0.1) = 4.5$.

Outcome	1	2	3	4	5	6
Probability	0.5	0.1	0.1	0.1	0.1	0.1
Observed Frequency O	13	6	12	9	3	2
Expected Frequency E	22.5	4.5	4.5	4.5	4.5	4.5

Measuring Disagreement with the Claimed Distribution

We know that sample frequencies typically differ somewhat from the values we theoretically expect, so we consider the key question:

> **Are the differences between the actual *observed* frequencies O and the theoretically *expected* frequencies E significant?**

To measure the discrepancy between the O and E values, we use the test statistic given in the preceding Key Elements box. (Later we will explain how this test statistic was developed, but it has differences of $O - E$ as a key component.)

$$\chi^2 = \sum \frac{(O - E)^2}{E}$$

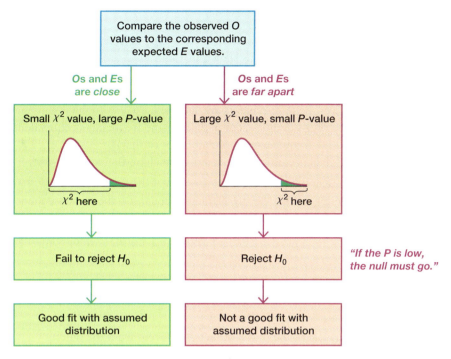

FIGURE 11-1 Relationships Among the χ^2 Test Statistic, *P*-Value, and Goodness-of-Fit

The χ^2 test statistic is based on differences between the observed and expected values. If the observed and expected values are *close,* the χ^2 test statistic will be small and the *P*-value will be large. If the observed and expected frequencies are *far apart,* the χ^2 test statistic will be large and the *P*-value will be small. Figure 11-1 summarizes this relationship. The hypothesis tests of this section are always right-tailed, because the critical value and critical region are located at the extreme right of the distribution. If you are confused, just remember this mnemonic:

"If the *P* is low, the null must go."

(If the *P*-value is small, reject the null hypothesis that the distribution is as claimed.)

 **EXAMPLE 1** **Last Digits of Weights**

A random sample of 100 weights of Californians is obtained, and the last digits of those weights are summarized in Table 11-2 (based on data from the California Department of Public Health). When obtaining weights of subjects, it is extremely important to actually measure their weights instead of asking them to report their weights. By analyzing the *last digits* of weights, researchers can verify that they were obtained through actual measurements instead of being reported. When people report weights, they tend to round down and they often round *way* down, so a weight of 197 lb might be rounded and reported as a more desirable 170 lb. Reported weights tend to have many last digits consisting of 0 or 5. In contrast, if people are actually weighed, the weights tend to have last digits that are uniformly distributed, with 0, 1, 2, . . . , 9 all occurring with roughly the same frequencies. We could subjectively examine the frequencies in Table 11-2 to see that the digits of 0 and 5 do seem to occur much more often than the other digits, but we will proceed with a formal hypothesis test to reinforce that subjective conclusion.

Test the claim that the sample is from a population of weights in which the last digits do *not* occur with the same frequency. Based on the results, what can we conclude about the procedure used to obtain the weights?

TABLE 11-2
Last Digits of Weights

Last Digit	Frequency
0	46
1	1
2	2
3	3
4	3
5	30
6	4
7	0
8	8
9	3

continued

SOLUTION

REQUIREMENT CHECK   (1) The data come from randomly selected subjects. (2) The data do consist of frequency counts, as shown in Table 11-2. (3) With 100 sample values and 10 categories that are claimed to be equally likely, each expected frequency is 10, so each expected frequency does satisfy the requirement of being a value of at least 5. All of the requirements are satisfied. ✅

The claim that the digits do not occur with the same frequency is equivalent to the claim that the relative frequencies or probabilities of the 10 cells ($p_0, p_1, \ldots, p_9$) are not all equal. (This is equivalent to testing the claim that the distribution of digits is not a uniform distribution.)

Step 1: The original claim is that the digits do not occur with the same frequency. That is, at least one of the probabilities, $p_0, p_1, \ldots, p_9$, is different from the others.

Step 2: If the original claim is false, then all of the probabilities are the same. That is, $p_0 = p_1 = p_2 = p_3 = p_4 = p_5 = p_6 = p_7 = p_8 = p_9$.

Step 3: The null hypothesis must contain the condition of equality, so we have

$$H_0: p_0 = p_1 = p_2 = p_3 = p_4 = p_5 = p_6 = p_7 = p_8 = p_9$$

H_1: At least one of the probabilities is different from the others.

Step 4: No significance level was specified, so we select the common choice of $\alpha = 0.05$.

Step 5: Because we are testing a claim about the distribution of the last digits being a uniform distribution (with all of the digits having the same probability), we use the goodness-of-fit test described in this section. The χ^2 distribution is used with the test statistic given in the preceding Key Elements box.

XLSTAT

Chi-square (Observed value)	212.8000
Chi-square (Critical value)	16.9190
DF	9
p-value	< 0.0001
alpha	0.05

Step 6: The observed frequencies O are listed in Table 11-2. Each corresponding expected frequency E is equal to 10 (because the 100 digits would be uniformly distributed among the 10 categories). The Excel add-in XLSTAT is used to obtain the results shown in the accompanying screen display, and Table 11-3 shows the manual computation of the χ^2 test statistic. The test statistic is $\chi^2 = 212.800$. The critical value is $\chi^2 = 16.919$ (found in Table A-4 with $\alpha = 0.05$ in the right tail and degrees of freedom equal to $k - 1 = 9$). The P-value is less than 0.0001. The test statistic and critical value are shown in Figure 11-2.

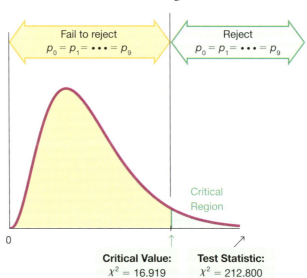

FIGURE 11-2   Test of $p_0 = p_1 = p_2 = p_3 = p_4 = p_5 = p_6 = p_7 = p_8 = p_9$

Step 7: If we use the P-value method of testing hypotheses, we see that the P-value is small (less than 0.0001), so we reject the null hypothesis. If we use the critical value method of testing hypotheses, Figure 11-2 shows that the test statistic falls in the critical region, so there is sufficient evidence to reject the null hypothesis.

Step 8: There is sufficient evidence to support the claim that the last digits do not occur with the same relative frequency.

TABLE 11-3 Calculating the χ^2 Test Statistic for the Last Digits of Weights

Last Digit	Observed Frequency O	Expected Frequency E	$O - E$	$(O - E)^2$	$\dfrac{(O - E)^2}{E}$
0	46	10	36	1296	129.6
1	1	10	−9	81	8.1
2	2	10	−8	64	6.4
3	3	10	−7	49	4.9
4	3	10	−7	49	4.9
5	30	10	20	400	40.0
6	4	10	−6	36	3.6
7	0	10	−10	100	10.0
8	8	10	−2	4	0.4
9	3	10	−7	49	4.9

$$\chi^2 = \sum \frac{(O - E)^2}{E} = 212.8$$

INTERPRETATION

This goodness-of-fit test suggests that the last digits do not provide a good fit with the claimed uniform distribution of equally likely frequencies. Instead of actually weighing the subjects, it appears that the subjects reported their weights. In fact, the weights are from the California Health Interview Survey (CHIS), and the title of that survey indicates that subjects were interviewed, not measured. Because those weights are reported, the reliability of the data is very questionable.

> **YOUR TURN** Do Exercise 20 "Last Digits of Weights."

Example 1 involves a situation in which the expected frequencies E for the different categories are all equal. The methods of this section can also be used when the expected frequencies are different, as in Example 2.

EXAMPLE 2 **Benford's Law: Detecting Computer Intrusions**

The Chapter Problem introduced *Benford's law,* whereby a variety of different data sets include numbers with leading (first) digits that follow the distribution shown in the first two rows of Table 11-4. The bottom row lists the frequencies of leading digits of Internet traffic inter-arrival times (from the Chapter Problem). Do the frequencies in the bottom row fit the distribution according to Benford's Law?

TABLE 11-4 Leading Digits of Internet Traffic Inter-Arrival Times

Leading Digit	1	2	3	4	5	6	7	8	9
Benford's Law: Distribution of Leading Digits	30.1%	17.6%	12.5%	9.7%	7.9%	6.7%	5.8%	5.1%	4.6%
Sample 2 of Leading Digits	69	40	42	26	25	16	16	17	20

continued

continued

Which Car Seats Are Safest?

Many people believe that the back seat of a car is the safest place to sit, but is it? University of Buffalo researchers analyzed more than 60,000 fatal car crashes and found that the middle back seat is the safest place to sit in a car. They found that sitting in that seat makes a passenger 86% more likely to survive than those who sit in the front seats, and they are 25% more likely to survive than those sitting in either of the back seats nearest the windows. An analysis of seat belt use showed that when not wearing a seat belt in the back seat, passengers are three times more likely to die in a crash than those wearing seat belts in that same seat. Passengers concerned with safety should sit in the middle back seat and wear a seat belt.

SOLUTION

REQUIREMENT CHECK (1) The sample data are randomly selected from a larger population. (2) The sample data do consist of frequency counts. (3) Each expected frequency is at least 5. The lowest expected frequency is $271 \cdot 0.46 = 12.466$. All of the requirements are satisfied. ✓

Step 1: The original claim is that the leading digits fit the distribution given as Benford's law. Using subscripts corresponding to the leading digits, we can express this claim as $p_1 = 0.301$ and $p_2 = 0.176$ and $p_3 = 0.125$ and ... and $p_9 = 0.046$.

Step 2: If the original claim is false, then at least one of the proportions does not have the value as claimed.

Step 3: The null hypothesis must contain the condition of equality, so we have

H_0: $p_1 = 0.301$ and $p_2 = 0.176$ and $p_3 = 0.125$ and ... and $p_9 = 0.046$.

H_1: At least one of the proportions is not equal to the given claimed value.

Step 4: The significance level is not specified, so we use the common choice of $\alpha = 0.05$.

Step 5: Because we are testing a claim that the distribution of leading digits fits the distribution given by Benford's law, we use the goodness-of-fit test described in this section. The X^2 distribution is used with the test statistic given in the preceding Key Elements box.

TI-84 Plus C

```
NORMAL FLOAT AUTO REAL RADIAN MP     
        χ²GOF-Test
χ²=11.27917666
p=.1863768489
df=8
CNTRB={1.937331172 1.241…
```

Step 6: Table 11-5 shows the calculations of the components of the X^2 test statistic for the leading digits of 1 and 2. If we include all nine leading digits, we get the test statistic of $X^2 = 11.2792$, as shown in the accompanying TI-84 Plus C calculator display. The critical value is $X^2 = 15.507$ (found in Table A-4, with $\alpha = 0.05$ in the right tail and degrees of freedom equal to $k - 1 = 8$). The TI-84 Plus C calculator display shows the value of the test statistic as well as the P-value of 0.186. (The entire bottom row of the display can be viewed by scrolling to the right. CNTRB is an abbreviated form of "contribution," and the values are the individual contributions to the total value of the X^2 test statistic.)

TABLE 11-5 Calculating the X^2 Test Statistic for Leading Digits in Table 11-4

Leading Digit	Observed Frequency O	Expected Frequency $E = np$	$O - E$	$(O - E)^2$	$\dfrac{(O - E)^2}{E}$
1	69	$271 \cdot 0.301 = 81.5710$	-12.5710	158.0300	1.9373
2	40	$271 \cdot 0.176 = 47.6960$	-7.6960	59.2284	1.2418

Step 7: The P-value of 0.186 is greater than the significance level of 0.05, so there is not sufficient evidence to reject the null hypothesis. (Also, the test statistic of $X^2 = 11.2792$ does not fall in the critical region bounded by the critical value of 15.507, so there is not sufficient evidence to reject the null hypothesis.)

Step 8: There is not sufficient evidence to warrant rejection of the claim that the 271 leading digits fit the distribution given by Benford's law.

INTERPRETATION

The sample of leading digits does not provide enough evidence to conclude that the Benford's law distribution is not being followed. There is not sufficient evidence to support a conclusion that the leading digits are from inter-arrival times that are not from normal traffic, so there is not sufficient evidence to conclude that an Internet intrusion has occurred.

YOUR TURN Do Exercise 21 "Detecting Fraud."

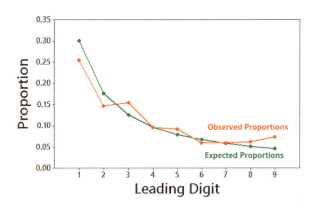

FIGURE 11-3   **Inter-Arrival Times: Observed Proportions and Proportions Expected with Benford's Law**

In Figure 11-3 we use a green line to graph the expected proportions given by Benford's law (as in Table 11-4) along with a red line for the observed proportions from Table 11-4. Figure 11-3 allows us to visualize the "goodness-of-fit" between the distribution given by Benford's law and the frequencies that were observed. In Figure 11-3, the green and red lines agree reasonably well, so it appears that the observed data fit the expected values reasonably well.

Rationale for the Test Statistic Examples 1 and 2 show that the χ^2 test statistic is a measure of the discrepancy between observed and expected frequencies. Simply summing the differences $O - E$ between observed and expected values tells us nothing, because that sum is always 0. Squaring the $O - E$ values gives us a better statistic. (The reasons for squaring the $O - E$ values are essentially the same as the reasons for squaring the $x - \bar{x}$ values in the formula for standard deviation.) The value of $\Sigma(O - E)^2$ measures only the magnitude of the differences, but we need to find the magnitude of the differences relative to what was expected. We need a type of average instead of a cumulative total. This relative magnitude is found through division by the expected frequencies, as in the test statistic $\Sigma(O - E)^2/E$.

The theoretical distribution of $\Sigma(O - E)^2/E$ is a discrete distribution because the number of possible values is finite. The distribution can be approximated by a chi-square distribution, which is continuous. This approximation is generally considered acceptable, provided that all expected values E are at least 5. (There are ways of circumventing the problem of an expected frequency that is less than 5, such as combining some categories so that all expected frequencies are at least 5. Also, there are different procedures that can be used when not all expected frequencies are at least 5.)

The number of degrees of freedom reflects the fact that we can freely assign frequencies to $k - 1$ categories before the frequency for every category is determined. (Although we say that we can "freely" assign frequencies to $k - 1$ categories, we cannot have negative frequencies, nor can we have frequencies so large that their sum exceeds the total of the observed frequencies for all categories combined.)

TECH CENTER

Goodness-of-Fit Test

Access tech supplements, videos, and data sets at **www.TriolaStats.com**

Statdisk	Minitab	StatCrunch
1. Click **Analysis** in the top menu. 2. Select **Goodness-of-Fit** from the dropdown menu. 3. Select **Equal Expected Frequencies** or **Unequal Expected Frequencies.** 4. Enter the desired significance level and select the column containing the observed frequencies. For *Unequal Expected Frequencies* also indicate if data are in the format of counts or proportions and select the column containing expected data. 5. Click **Evaluate.**	1. Click **Stat** in the top menu. 2. Select **Tables** from the dropdown menu and select **Chi-Square Goodness-of-Fit Test** from the submenu. 3. Click **Observed Counts** and select the column containing the observed frequencies. 4. Under *Test* select **Equal Proportions** if expected frequencies are all equal. For unequal expected frequencies or proportions, select **Proportions specified by historical counts** and select the column containing the expected frequencies or proportions). 5. Click **OK.**	1. Click **Stat** in the top menu. 2. Select **Goodness-of-fit** from the dropdown menu, then select **Chi-Square Test** from the submenu. 3. Select the column with the observed frequencies. 4. Select the column containing the expected frequencies if expected frequencies are not all equal. Otherwise, click **All cells in equal proportion.** 5. Click **Compute!**

TI-83/84 Plus Calculator	Excel
TI-83/84 calculators require expected frequencies. Expected proportions cannot be used. 1. Enter the observed values in a list (*L1*) and expected frequencies in a separate list (*L2*). 2. Press **STAT**, then select **TESTS** in the top menu. 3. Select χ^2 **GOF-Test** in the menu and press **ENTER**. 4. Enter the list names for the observed and expected frequencies. For *df* enter the degrees of freedom, which is 1 less than the number of categories. 5. Select **Calculate** and press **ENTER**. *TIP:* TI-83 calculators require the program **X2GOF**, which is available at TriolaStats.com.	**XLSTAT Add-In** 1. Click on the **XLSTAT** tab in the Ribbon and then click **Parametric Tests.** 2. Select **Multinomial goodness of fit test** from the dropdown menu. 3. In the *Frequencies* box enter the range of cells containing the observed frequencies. In the *Expected frequencies* box enter the range of cells containing the expected frequencies. If using expected proportions, check the **Proportions** box under *Data format*. 4. Check the **Chi-square test** box. If the data range includes a data label, also check the **Column labels** box. 5. Enter a significance level and click **OK.** The test statistic is labeled *Chi-Square (Observed Value)*. **Excel** 1. Click **Insert Function** f_x, select the category **Statistical,** and select the function **CHISQ.TEST.** 2. For *Actual_range* enter the cell range for observed frequencies. For *Expected_range* enter the expected frequencies. 3. Click **OK** for the *P*-value.

11-1 Basic Skills and Concepts

Statistical Literacy and Critical Thinking

1. Cybersecurity The table below lists leading digits of 317 inter-arrival Internet traffic times for a computer, along with the frequencies of leading digits expected with Benford's law (from Table 11-1 in the Chapter Problem).

a. Identify the notation used for observed and expected values.

b. Identify the observed and expected values for the leading digit of 2.

c. Use the results from part (b) to find the contribution to the χ^2 test statistic from the category representing the leading digit of 2.

Leading Digit	1	2	3	4	5	6	7	8	9
Benford's Law	30.1%	17.6%	12.5%	9.7%	7.9%	6.7%	5.8%	5.1%	4.6%
Leading Digits of Inter-Arrival Traffic Times	76	62	29	33	19	27	28	21	22

2. Cybersecurity When using the data from Exercise 1 to test for goodness-of-fit with the distribution described by Benford's law, identify the null and alternative hypotheses.

3. Cybersecurity The accompanying Statdisk results shown in the margin are obtained from the data given in Exercise 1. What should be concluded when testing the claim that the leading digits have a distribution that fits well with Benford's law?

Test Statistic, X^2:	20.9222
Critical X^2:	15.5073
P-Value:	0.0074

4. Cybersecurity What do the results from the preceding exercises suggest about the possibility that the computer has been hacked? Is there any corrective action that should be taken?

In Exercises 5–20, conduct the hypothesis test and provide the test statistic and the P-value and/or critical value, and state the conclusion.

5. Testing a Slot Machine The author purchased a slot machine (Bally Model 809) and tested it by playing it 1197 times. There are 10 different categories of outcomes, including no win, win jackpot, win with three bells, and so on. When testing the claim that the observed outcomes agree with the expected frequencies, the author obtained a test statistic of $\chi^2 = 8.185$. Use a 0.05 significance level to test the claim that the actual outcomes agree with the expected frequencies. Does the slot machine appear to be functioning as expected?

6. Flat Tire and Missed Class A classic story involves four carpooling students who missed a test and gave as an excuse a flat tire. On the makeup test, the instructor asked the students to identify the particular tire that went flat. If they really didn't have a flat tire, would they be able to identify the same tire? The author asked 41 other students to identify the tire they would select. The results are listed in the following table (except for one student who selected the spare). Use a 0.05 significance level to test the author's claim that the results fit a uniform distribution. What does the result suggest about the likelihood of four students identifying the same tire when they really didn't have a flat?

Tire	Left Front	Right Front	Left Rear	Right Rear
Number Selected	11	15	8	6

7. Loaded Die The author drilled a hole in a die and filled it with a lead weight, then proceeded to roll it 200 times. Here are the observed frequencies for the outcomes of 1, 2, 3, 4, 5, and 6, respectively: 27, 31, 42, 40, 28, and 32. Use a 0.05 significance level to test the claim that the outcomes are not equally likely. Does it appear that the loaded die behaves differently than a fair die?

8. Bias in Clinical Trials? Researchers investigated the issue of race and equality of access to clinical trials. The following table shows the population distribution and the numbers of participants in clinical trials involving lung cancer (based on data from "Participation in Cancer

continued

Clinical Trials," by Murthy, Krumholz, and Gross, *Journal of the American Medical Association*, Vol. 291, No. 22). Use a 0.01 significance level to test the claim that the distribution of clinical trial participants fits well with the population distribution. Is there a race/ethnic group that appears to be very underrepresented?

Race/ethnicity	White non-Hispanic	Hispanic	Black	Asian/Pacific Islander	American Indian/ Alaskan Native
Distribution of Population	75.6%	9.1%	10.8%	3.8%	0.7%
Number in Lung Cancer Clinical Trials	3855	60	316	54	12

9. Mendelian Genetics Experiments are conducted with hybrids of two types of peas. If the offspring follow Mendel's theory of inheritance, the seeds that are produced are yellow smooth, green smooth, yellow wrinkled, and green wrinkled, and they should occur in the ratio of 9:3:3:1, respectively. An experiment is designed to test Mendel's theory, with the result that the offspring seeds consist of 307 that are yellow smooth, 77 that are green smooth, 98 that are yellow wrinkled, and 18 that are green wrinkled. Use a 0.05 significance level to test the claim that the results contradict Mendel's theory.

10. Do World War II Bomb Hits Fit a Poisson Distribution? In analyzing hits by V-1 buzz bombs in World War II, South London was subdivided into regions, each with an area of 0.25 km^2. Shown below is a table of actual frequencies of hits and the frequencies expected with the Poisson distribution. (The Poisson distribution is described in Section 5-3.) Use the values listed and a 0.05 significance level to test the claim that the actual frequencies fit a Poisson distribution. Does the result prove that the data conform to the Poisson distribution?

Number of Bomb Hits	0	1	2	3	4 or more
Actual Number of Regions	229	211	93	35	8
Expected Number of Regions (from Poisson Distribution)	227.5	211.4	97.9	30.5	8.7

11. Police Calls The police department in Madison, Connecticut, released the following numbers of calls for the different days of the week during a February that had 28 days: Monday (114); Tuesday (152); Wednesday (160); Thursday (164); Friday (179); Saturday (196); Sunday (130). Use a 0.01 significance level to test the claim that the different days of the week have the same frequencies of police calls. Is there anything notable about the observed frequencies?

12. Police Calls Repeat Exercise 11 using these observed frequencies for police calls received during the month of March: Monday (208); Tuesday (224); Wednesday (246); Thursday (173); Friday (210); Saturday (236); Sunday (154). What is a fundamental error with this analysis?

13. Kentucky Derby The table below lists the frequency of wins for different post positions through the 141st running of the Kentucky Derby horse race. A post position of 1 is closest to the inside rail, so the horse in that position has the shortest distance to run. (Because the number of horses varies from year to year, only the first 10 post positions are included.) Use a 0.05 significance level to test the claim that the likelihood of winning is the same for the different post positions. Based on the result, should bettors consider the post position of a horse racing in the Kentucky Derby?

Post Position	1	2	3	4	5	6	7	8	9	10
Wins	19	14	11	15	15	7	8	12	5	11

14. California Daily 4 Lottery The author recorded all digits selected in California's Daily 4 Lottery for the 60 days preceding the time that this exercise was created. The frequencies of the digits from 0 through 9 are 21, 30, 31, 33, 19, 23, 21, 16, 24, and 22. Use a 0.05 significance level to test the claim of lottery officials that the digits are selected in a way that they are equally likely.

15. World Series Games The table below lists the numbers of games played in 105 Major League Baseball (MLB) World Series. This table also includes the expected proportions for the numbers of games in a World Series, assuming that in each series, both teams have about the same chance of winning. Use a 0.05 significance level to test the claim that the actual numbers of games fit the distribution indicated by the expected proportions.

Games Played	4	5	6	7
World Series Contests	21	23	23	38
Expected Proportion	2/16	4/16	5/16	5/16

16. Baseball Player Births In his book *Outliers,* author Malcolm Gladwell argues that more baseball players have birth dates in the months immediately following July 31, because that was the age cutoff date for nonschool baseball leagues. Here is a sample of frequency counts of months of birth dates of American-born Major League Baseball players starting with January: 387, 329, 366, 344, 336, 313, 313, 503, 421, 434, 398, 371. Using a 0.05 significance level, is there sufficient evidence to warrant rejection of the claim that American-born Major League Baseball players are born in different months with the same frequency? Do the sample values appear to support Gladwell's claim?

Exercises 17–20 are based on data sets included in Appendix B. The complete data sets can be found at www.TriolaStats.com.

17. Admissions for Birth Data Set 4 "Births" includes the days of the weeks that prospective mothers were admitted to a hospital to give birth. A physician claims that because many births are induced or involve cesarean section, they are scheduled for days other than Saturday or Sunday, so births do not occur on the seven different days of the week with equal frequency. Use a 0.01 significance level to test that claim.

18. Discharges After Birth Data Set 4 "Births" includes the days of the weeks that newborn babies were discharged from the hospital. A hospital administrator claims that such discharges occur on the seven different days of the week with equal frequency. Use a 0.01 significance level to test that claim.

19. M&M Candies Mars, Inc. claims that its M&M plain candies are distributed with the following color percentages: 16% green, 20% orange, 14% yellow, 24% blue, 13% red, and 13% brown. Refer to Data Set 27 "M&M Weights" in Appendix B and use the sample data to test the claim that the color distribution is as claimed by Mars, Inc. Use a 0.05 significance level.

20. Last Digits of Weights Data Set 1 "Body Data" in Appendix B includes weights (kg) of 300 subjects. Use a 0.05 significance level to test the claim that the sample is from a population of weights in which the last digits do *not* occur with the same frequency. When people *report* their weights instead of being measured, they tend to round so that the last digits do not occur with the same frequency. Do the results suggest that the weights were reported?

Benford's Law. *According to Benford's law, a variety of different data sets include numbers with leading (first) digits that follow the distribution shown in the table below. In Exercises 21–24, test for goodness-of-fit with the distribution described by Benford's law.*

Leading Digit	1	2	3	4	5	6	7	8	9
Benford's Law: Distribution of Leading Digits	30.1%	17.6%	12.5%	9.7%	7.9%	6.7%	5.8%	5.1%	4.6%

21. Detecting Fraud When working for the Brooklyn district attorney, investigator Robert Burton analyzed the leading digits of the amounts from 784 checks issued by seven suspect companies. The frequencies were found to be 0, 15, 0, 76, 479, 183, 8, 23, and 0, and those digits correspond to the leading digits of 1, 2, 3, 4, 5, 6, 7, 8, and 9, respectively. If the observed frequencies are substantially different from the frequencies expected with Benford's law, the check amounts appear to result from fraud. Use a 0.01 significance level to test for goodness-of-fit with Benford's law. Does it appear that the checks are the result of fraud?

22. Author's Check Amounts Exercise 21 lists the observed frequencies of leading digits from amounts on checks from seven suspect companies. Here are the observed frequencies of the leading digits from the amounts on the most recent checks written by the author at the time this exercise was created: 83, 58, 27, 21, 21, 21, 6, 4, 9. (Those observed frequencies correspond to the leading digits of 1, 2, 3, 4, 5, 6, 7, 8, and 9, respectively.) Using a 0.01 significance level, test the claim that these leading digits are from a population of leading digits that conform to Benford's law. Does the conclusion change if the significance level is 0.05?

23. Tax Cheating? Frequencies of leading digits from IRS tax files are 152, 89, 63, 48, 39, 40, 28, 25, and 27 (corresponding to the leading digits of 1, 2, 3, 4, 5, 6, 7, 8, and 9, respectively, based on data from Mark Nigrini, who provides software for Benford data analysis). Using a 0.05 significance level, test for goodness-of-fit with Benford's law. Does it appear that the tax entries are legitimate?

24. Author's Computer Files The author recorded the leading digits of the sizes of the electronic document files for the current edition of this book. The leading digits have frequencies of 55, 25, 17, 24, 18, 12, 12, 3, and 4 (corresponding to the leading digits of 1, 2, 3, 4, 5, 6, 7, 8, and 9, respectively). Using a 0.05 significance level, test for goodness-of-fit with Benford's law.

11-1   Beyond the Basics

 25. Testing Goodness-of-Fit with a Normal Distribution Refer to Data Set 1 "Body Data" in Appendix B for the heights of females.

Height (cm)	Less than 155.45	155.45 − 162.05	162.05 − 168.65	Greater than 168.65
Frequency				

a. Enter the observed frequencies in the table above.

b. Assuming a normal distribution with mean and standard deviation given by the sample mean and standard deviation, use the methods of Chapter 6 to find the probability of a randomly selected height belonging to each class.

c. Using the probabilities found in part (b), find the expected frequency for each category.

d. Use a 0.01 significance level to test the claim that the heights were randomly selected from a normally distributed population. Does the goodness-of-fit test suggest that the data are from a normally distributed population?

11-2   Contingency Tables

Key Concept We now consider methods for analyzing *contingency tables* (or two-way frequency tables), which include frequency counts for categorical data arranged in a table with at least two rows and at least two columns. In Part 1 of this section, we present a method for conducting a hypothesis test of the null hypothesis that the row and column variables are independent of each other. This test of independence is widely used in real-world applications. In Part 2, we will consider three variations of the basic method presented in Part 1: (1) test of homogeneity, (2) Fisher's exact test, and (3) McNemar's test for matched pairs.

PART 1   Basic Concepts of Testing for Independence

In this section we use standard statistical methods to analyze frequency counts in a contingency table (or two-way frequency table).

DEFINITION

A **contingency table** (or **two-way frequency table**) is a table consisting of frequency counts of categorical data corresponding to two different variables. (One variable is used to categorize rows, and a second variable is used to categorize columns.)

The word *contingent* has a few different meanings, one of which refers to a *dependence* on some other factor. We use the term *contingency table* because we test for *independence* between the row and column variables. We first define a *test of independence* and we provide key elements of the test in the Key Elements box that follows.

DEFINITION

In a **test of independence,** we test the null hypothesis that in a contingency table, the row and column variables are independent. (That is, there is no dependency between the row variable and the column variable.)

KEY ELEMENTS

Contingency Table

Objective

Conduct a hypothesis test of independence between the row variable and column variable in a contingency table.

Notation

O represents the *observed frequency* in a cell of a contingency table.
E represents the *expected frequency* in a cell, found by assuming that the row and column variables are independent.
r represents the number of rows in a contingency table (not including labels or row totals).
c represents the number of columns in a contingency table (not including labels or column totals).

Requirements

1. The sample data are randomly selected.
2. The sample data are represented as frequency counts in a two-way table.
3. For every cell in the contingency table, the expected frequency E is at least 5. (There is no requirement that every *observed* frequency must be at least 5.)

Null and Alternative Hypotheses

The null and alternative hypotheses are as follows:

H_0: The row and column variables are independent.
H_1: The row and column variables are dependent.

Test Statistic for a Test of Independence

$$\chi^2 = \sum \frac{(O - E)^2}{E}$$

where O is the observed frequency in a cell and E is the expected frequency in a cell that is found by evaluating

$$E = \frac{(\text{row total})\,(\text{column total})}{(\text{grand total})}$$

P-values

P-values are typically provided by technology, or a range of P-values can be found from Table A-4.

continued

Critical values:

1. The critical values are found in Table A-4 using

$$\text{Degrees of freedom} = (r - 1)(c - 1)$$

where r is the number of rows and c is the number of columns.

2. Tests of independence with a contingency table are always *right-tailed*.

The distribution of the test statistic X^2 can be approximated by the chi-square distribution, provided that all cells have expected frequencies that are at least 5. The number of degrees of freedom $(r - 1)(c - 1)$ reflects the fact that because we know the total of all frequencies in a contingency table, we can freely assign frequencies to only $r - 1$ rows and $c - 1$ columns before the frequency for every cell is determined. However, we cannot have negative frequencies or frequencies so large that any row (or column) sum exceeds the total of the observed frequencies for that row (or column).

Observed and Expected Frequencies The test statistic allows us to measure the amount of disagreement between the frequencies actually observed and those that we would theoretically expect when the two variables are independent. Large values of the X^2 test statistic are in the rightmost region of the chi-square distribution, and they reflect significant differences between observed and expected frequencies. As in Section 11-1, if observed and expected frequencies are close, the X^2 test statistic will be small and the *P*-value will be large. If observed and expected frequencies are far apart, the X^2 test statistic will be large and the *P*-value will be small. These relationships are summarized and illustrated in Figure 11-4.

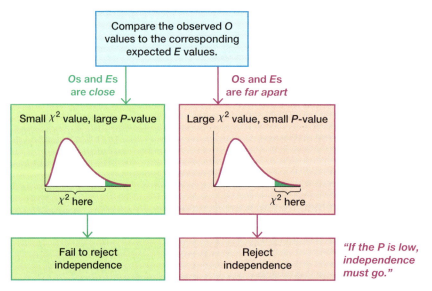

FIGURE 11-4   Relationships Among Key Components in a Test of Independence

Finding Expected Values *E*

An individual expected frequency *E* for a cell can be found by simply multiplying the total of the row frequencies by the total of the column frequencies, then dividing by the grand total of all frequencies, as shown in Example 1.

$$E = \frac{(\text{row total}) \, (\text{column total})}{(\text{grand total})}$$

 EXAMPLE 1 Finding Expected Frequency

Table 11-6 is a contingency table with four rows and two columns. The cells of the table contain frequency counts. The frequency counts are the observed values, and the expected values are shown in parentheses. The row variable identifies the treatment used for a stress fracture in a foot bone, and the column variable identifies the outcome as a success or failure (based on data from "Surgery Unfounded for Tarsal Navicular Stress Fracture," by Bruce Jancin, *Internal Medicine News*, Vol. 42, No. 14). Refer to Table 11-6 and find the expected frequency for the cell in the first row and first column, where the observed frequency is 54.

SOLUTION

TABLE 11-6 Treatments for Stress Fracture in a Foot Bone

	Success	Failure
Surgery	54 ($E = 47.478$)	12 ($E = 18.522$)
Weight-Bearing Cast	41 ($E = 66.182$)	51 ($E = 25.818$)
Non–Weight-Bearing Cast for 6 Weeks	70 ($E = 52.514$)	3 ($E = 20.486$)
Non–Weight-Bearing Cast for Less Than 6 Weeks	17 ($E = 15.826$)	5 ($E = 6.174$)

The first cell lies in the first row (with a total frequency of 66) and the first column (with total frequency of 182). The "grand total" is the sum of all frequencies in the table, which is 253. The expected frequency of the first cell is

$$E = \frac{(\text{row total}) \, (\text{column total})}{(\text{grand total})} = \frac{(66) \, (182)}{253} = 47.478$$

INTERPRETATION

We know that the first cell has an *observed* frequency of $O = 54$ and an *expected* frequency of $E = 47.478$. We can interpret the expected value by stating that if we assume that success is independent of the treatment, then we expect to find that 47.478 of the subjects would be treated with surgery and that treatment would be successful. There is a discrepancy between $O = 54$ and $E = 47.478$, and such discrepancies are key components of the test statistic that is a collective measure of the overall disagreement between the observed frequencies and the frequencies expected with independence between the row and column variables.

YOUR TURN Do part (a) of Exercise 1 "Handedness and Cell Phone Use."

Example 2 illustrates the procedure for conducting a hypothesis test of independence between the row and column variables in a contingency table.

Polls and Psychologists

Poll results can be dramatically affected by the wording of questions. A phrase such as "over the last few years" is interpreted differently by different people. Over the last few years (actually, since 1980), survey researchers and psychologists have been working together to improve surveys by decreasing bias and increasing accuracy. In one case, psychologists studied the finding that 10 to 15 percent of those surveyed say they voted in the last election when they did not. They experimented with theories of faulty memory, a desire to be viewed as responsible, and a tendency of those who usually vote to say that they voted in the most recent election, even if they did not. Only the last theory was actually found to be part of the problem.

EXAMPLE 2 **Does the Choice of Treatment for a Fracture Affect Success?**

Use the same sample data from Example 1 with a 0.05 significance level to test the claim that success of the treatment is independent of the type of treatment. What does the result indicate about the increasing trend to use surgery?

SOLUTION

REQUIREMENT CHECK (1) On the basis of the study description, we will treat the subjects as being randomly selected and randomly assigned to the different treatment groups. (2) The results are expressed as frequency counts in Table 11-6. (3) The expected frequencies are all at least 5. (The lowest expected frequency is 6.174.) The requirements are satisfied. ☑

The null hypothesis and alternative hypothesis are as follows:

H_0: Success is independent of the treatment.

H_1: Success and the treatment are dependent.

The significance level is $\alpha = 0.05$.

Because the data in Table 11-6 are in the form of a contingency table, we use the χ^2 distribution with this test statistic:

$$\chi^2 = \sum \frac{(O-E)^2}{E} = \frac{(54-47.478)^2}{47.478} + \cdots + \frac{(5-6.174)^2}{6.174}$$

$$= 58.393$$

XLSTAT

Chi-square (Observed value)	58.3933
Chi-square (Critical value)	7.8147
DF	3
p-value	< 0.0001
alpha	0.05

P-Value from Technology If using technology, results typically include the χ^2 test statistic and the P-value. For example, see the accompanying XLSTAT display showing the test statistic is $\chi^2 = 58.393$ and the P-value is less than 0.0001.

P-Value from Table A-4 If using Table A-4 instead of technology, first find the number of degrees of freedom: $(r-1)(c-1) = (4-1)(2-1) = 3$ degrees of freedom. Because the test statistic of $\chi^2 = 58.393$ exceeds the highest value (12.838) in Table A-4 for the row corresponding to 3 degrees of freedom, we know that P-value < 0.005.

Because the P-value is less than the significance level of 0.05, we reject the null hypothesis of independence between success and treatment.

Critical Value If using the critical value method of hypothesis testing, the critical value of $\chi^2 = 7.815$ is found from Table A-4, with $\alpha = 0.05$ in the right tail and the number of degrees of freedom given by $(r-1)(c-1) = (4-1)(2-1) = 3$. The test statistic and critical value are shown in Figure 11-5. Because the test statistic does fall within the critical region, we reject the null hypothesis of independence between success and treatment.

INTERPRETATION

It appears that success is dependent on the treatment. Although the results of this test do not tell us which treatment is best, we can see from Table 11-6 that the success rates of 81.8%, 44.6%, 95.9%, and 77.3% suggest that the best treatment is to use a non–weight-bearing cast for 6 weeks. These results suggest that the increasing use of surgery is a treatment strategy that is not supported by the evidence.

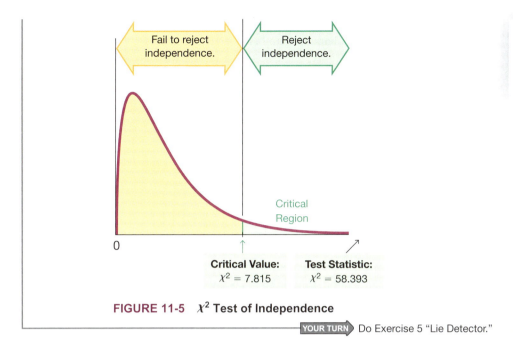

FIGURE 11-5 x^2 **Test of Independence**

> **YOUR TURN** Do Exercise 5 "Lie Detector."

Rationale for Expected Frequencies E To better understand expected frequencies, pretend that we know only the row and column totals in Table 11-6. Let's assume that the row and column variables are independent and that 1 of the 253 study subjects is randomly selected. The probability of getting someone counted in the first cell of Table 11-6 is found as follows:

$$P(\text{surgery}) = 66/253 \quad \text{and} \quad P(\text{success}) = 182/253$$

If the row and column variables are independent, as we are assuming, we can use the multiplication rule for independent events (see Section 4-2) as follows:

$$P(\text{surgery treatment } and \text{ success}) = \frac{66}{253} \cdot \frac{182}{253} = 0.187661$$

With a probability of 0.187661 for the first cell, we expect that among 253 subjects, there are $253 \cdot 0.187661 = 47.478$ subjects in the first cell. If we generalize these calculations, we get the following:

$$\text{Expected frequency } E = (\text{grand total}) \cdot \frac{(\text{row total})}{(\text{grand total})} \cdot \frac{(\text{column total})}{(\text{grand total})}$$

This expression can be simplified to

$$E = \frac{(\text{row total}) \, (\text{column total})}{(\text{grand total})}$$

PART 2 Test of Homogeneity, Fisher's Exact Test, and McNemar's Test for Matched Pairs

Test of Homogeneity

In Part 1 of this section, we focused on the test of *independence* between the row and column variables in a contingency table. In Part 1, the sample data are from one population, and individual sample results are categorized with the row and column variables. In a *chi-square test of homogeneity,* we have samples randomly selected from

Safest Seats in a Commercial Jet

A study by aviation writer and researcher David Noland showed that sitting farther back in a commercial jet will increase your chances of surviving in the event of a crash. The study suggests that the chance of surviving is not the same for each seat, so a goodness-of-fit test would lead to rejection of the null hypothesis that every seat has the same probability of a passenger surviving. Records from the 20 commercial jet crashes that occurred since 1971 were analyzed. It was found that if you sit in business or first class, you have a 49% chance of surviving a crash; if you sit in coach over the wing or ahead of the wing, you have a 56% chance of surviving; and if you sit in the back behind the wing, you have a 69% chance of surviving.

In commenting on this study, David Noland stated that he does not seek a rear seat when he flies. He says that because the chance of a crash is so small, he doesn't worry about where he sits, but he prefers a window seat.

different populations, and we want to determine whether those populations have the same proportions of some characteristic being considered. (The word *homogeneous* means "having the same quality," and in this context, we are testing to determine whether the proportions are the same.) Section 9-1 presented a procedure for testing a claim about *two* populations with categorical data having two possible outcomes, but a chi-square test of homogeneity allows us to use two or more populations with outcomes from several categories.

> **DEFINITION**
>
> A **chi-square test of homogeneity** is a test of the claim that *different populations* have the same proportions of some characteristics.

Sampling from Different Populations In a typical test of independence, as described in Part 1 of this section, sample subjects are randomly selected from one population (such as people treated for stress fractures in a foot bone) and values of two different variables are observed (such as success/failure for people receiving different treatments). In a typical chi-square test of homogeneity, subjects are randomly selected from the different populations separately.

Procedure In conducting a test of homogeneity, we can use the same notation, requirements, test statistic, critical value, and procedures given in the Key Elements box from Part 1 on page 547 of this section, with this exception: Instead of testing the null hypothesis of independence between the row and column variables, we test the null hypothesis that *the different populations have the same proportion of some characteristic.*

EXAMPLE 3   The Lost Wallet Experiment

Table 11-7 lists results from a *Reader's Digest* experiment in which 12 wallets were intentionally lost in each of 16 different cities, including New York City, London, Amsterdam, and so on. Use a 0.05 significance level with the data from Table 11-7 to test the null hypothesis that the cities have the same proportion of returned wallets. The *Reader's Digest* headline "Most Honest Cities: The *Reader's Digest* Lost Wallet Test" implies that whether a wallet is returned is dependent on the city in which it was lost. Test the claim that the proportion of returned wallets is not the same in the 16 different cities.

TABLE 11-7 Lost Wallet Experiment

City	A	B	C	D	E	F	G	H	I	J	K	L	M	N	O	P
Wallet Returned	8	5	7	11	5	8	6	7	3	1	4	2	4	6	4	9
Wallet Not Returned	4	7	5	1	7	4	6	5	9	11	8	10	8	6	8	3

SOLUTION

REQUIREMENT CHECK (1) Based on the description of the study, we will treat the subjects as being randomly selected and randomly assigned to the different cities. (2) The results are expressed as frequency counts in Table 11-7. (3) The expected frequencies are all at least 5. (All expected values are either 5.625 or 6.375.) The requirements are satisfied. ✓

The null hypothesis and alternative hypothesis are as follows:

H_0: Whether a lost wallet is returned is independent of the city in which it was lost.

H_1: A lost wallet being returned depends on the city in which it was lost.

The accompanying StatCrunch display shows the test statistic of $X^2 = 35.388$ (rounded) and the P-value of 0.002 (rounded). Because the P-value of 0.002 is less than the significance level of 0.05, we reject the null hypothesis of independence between the two variables. ("If the P is low, the null must go.")

StatCrunch

Chi-Square test:

Statistic	DF	Value	P-value
Chi-square	15	35.388235	0.0022

INTERPRETATION

We reject the null hypothesis of independence, so it appears that the proportion of returned wallets depends on the city in which they were lost. There is sufficient evidence to conclude that the proportion of returned wallets is not the same in the 16 different cities.

Fisher's Exact Test

The procedures for testing hypotheses with contingency tables have the requirement that every cell must have an expected frequency of at least 5. This requirement is necessary for the X^2 distribution to be a suitable approximation to the exact distribution of the X^2 test statistic. *Fisher's exact test* is often used for a 2 × 2 contingency table with one or more expected frequencies that are below 5. Fisher's exact test provides an *exact* P-value and does not require an approximation technique. Because the calculations are quite complex, it's a good idea to use technology when using Fisher's exact test. Statdisk, Minitab, XLSTAT, and StatCrunch all have the ability to perform Fisher's exact test.

EXAMPLE 4 Does Yawning Cause Others to Yawn?

The MythBusters show on the Discovery Channel tested the theory that when someone yawns, others are more likely to yawn. The results are summarized in Table 11-8. The methods of Part 1 in this Section should not be used because one of the cells has an expected frequency of 4.480, which violates the requirement that every cell must have an expected frequency E of at least 5. Using Fisher's exact test results in a P-value of 0.513, so there is not sufficient evidence to support the myth that people exposed to yawning actually yawn more than those not exposed to yawning. (For testing the claim of no difference, the P-value is 1.000, indicating that there is not a significant difference between the two groups.)

TABLE 11-8 Yawning Theory Experiment

		Subject Exposed to Yawning?	
		Yes	No
Did Subject Yawn?	Yes	10	4
	No	24	12

McNemar's Test for Matched Pairs

The methods in Part 1 of this section are based on independent data. For 2 × 2 tables consisting of frequency counts that result from matched pairs, the frequency counts within each matched pair are not independent and, for such cases, we can use McNemar's test of the null hypothesis that the frequencies from the discordant (different) categories occur in the same proportion.

Table 11-9 shows a general format for summarizing results from data consisting of frequency counts from matched pairs. Table 11-9 refers to two different treatments (such as two different eyedrop solutions) applied to two different parts of each subject (such as left eye and right eye). We should be careful when reading a table such as Table 11-9. If $a = 100$, then 100 subjects were cured with both treatments. If $b = 50$ in Table 11-9, then each of 50 subjects had no cure with treatment X but they were each cured with treatment Y. The total number of subjects is $a + b + c + d$, and each of those subjects yields results from each of two parts of a matched pair. Remember, the entries in Table 11-9 are frequency counts of subjects, not the total number of individual components in the matched pairs. If 500 people have each eye treated with two different ointments, the value of $a + b + c + d$ is 500 (the number of subjects), not 1000 (the number of treated eyes).

TABLE 11-9 2×2 Table with Frequency Counts from Matched Pairs

		Treatment X	
		Cured	Not Cured
Treatment Y	Cured	a	b
	Not Cured	c	d

McNemar's test requires that for a table such as Table 11-9, the frequencies are such that $b + c \geq 10$. The test is a right-tailed chi-square test with the following test statistic:

$$\chi^2 = \frac{(|b - c| - 1)^2}{b + c}$$

P-values are typically provided by software, and critical values can be found in Table A-4 using 1 degree of freedom. *Caution:* When applying McNemar's test, be careful to use only the two frequency counts from *discordant* (different) pairs, such as the frequency b in Table 11-9 (with different pairs of cured/not cured) and frequency c in Table 11-9 (with different pairs of not cured/cured).

EXAMPLE 5   **Are Hip Protectors Effective?**

A randomized controlled trial was designed to test the effectiveness of hip protectors in preventing hip fractures in the elderly. Nursing home residents each wore protection on one hip, but not the other. Results are summarized in Table 11-10 (based on data from "Efficacy of Hip Protector to Prevent Hip Fracture in Nursing Home Residents," by Kiel et al., *Journal of the American Medical Association,* Vol. 298, No. 4). McNemar's test can be used to test the null hypothesis that the following two proportions are the same:

- The proportion of subjects with no hip fracture on the protected hip and a hip fracture on the unprotected hip.

- The proportion of subjects with a hip fracture on the protected hip and no hip fracture on the unprotected hip.

TABLE 11-10 Randomized Controlled Trial of Hip Protectors

		No Hip Protector Worn	
		No Hip Fracture	Hip Fracture
Hip Protector Worn	No Hip Fracture	309	10
	Hip Fracture	15	2

Using the discordant (different) pairs with the general format from Table 11-9 we have $b = 10$ and $c = 15$, so the test statistic is calculated as follows:

$$\chi^2 = \frac{(|b - c| - 1)^2}{b + c} = \frac{(|10 - 15| - 1)^2}{10 + 15} = 0.640$$

With a 0.05 significance level and degrees of freedom given by df $= 1$, we refer to Table A-4 to find the critical value of $\chi^2 = 3.841$ for this right-tailed test. The test statistic of $\chi^2 = 0.640$ does not exceed the critical value of $\chi^2 = 3.841$, so we fail to reject the null hypothesis. (Also, the P-value is 0.424, which is greater than 0.05, indicating that we fail to reject the null hypothesis.) The proportion of hip fractures with the protectors worn is not significantly different from the proportion of hip fractures without the protectors worn. The hip protectors do not appear to be effective in preventing hip fractures.

TECH CENTER

Contingency Tables
Access tech supplements, videos, and data sets at **www.TriolaStats.com**

Statdisk

1. Click **Analysis** in the top menu.
2. Select **Contingency Tables** from the dropdown menu.
3. Enter the desired significance level and select the columns to be included in the analysis.
4. Click **Evaluate.**

Minitab

1. Click **Stat** in the top menu.
2. Select **Tables** from the dropdown menu and select **Chi-Square Test for Association.**
3. Select **Summarized data in a two-way table** from the dropdown box.
4. Select the columns containing the observed frequencies.
5. Click **OK.**

TIP: Observed frequencies must be entered in columns just as they appear in the contingency table.

StatCrunch

1. Click **Stat** in the top menu.
2. Select **Tables** from the dropdown menu, then select **Contingency— With Summary** from the submenu.
3. Select the columns of data to be included in the analysis.
4. For *Row labels* select the column containing the row names.
5. Click **Compute!** The test statistic and *P*-value are displayed at the bottom of the results.

TIP: You must enter row names in the first column.

TI-83/84 Plus Calculator

1. Enter the contingency data as a matrix:

   Manually enter data: Press **2ND** then **x⁻¹** to get to the *MATRIX* menu, select **EDIT** from the top menu, select a matrix letter, and press **ENTER**. Enter the number of rows and columns needed, press **ENTER**, and proceed to enter the sample values.

   Using existing lists: Lists can be combined and stored in a matrix. Press **2ND** then **x⁻¹** to get to the *MATRIX* menu, select **MATH** from the top menu, and select the item **List → matr.** Enter the list names (the first list must contain values for the dependent *y* variable), followed by the matrix name, all separated by commas. *Important:* The matrix name must be entered by pressing **2ND** then **x⁻¹**, selecting the matrix letter, and pressing **ENTER**. The following is a summary of the commands used to create a matrix from three lists (L1, L2, L3): **List → matr(L1, L2, L3,[D]).**
2. Press **STAT**, then select **TESTS** in the top menu.
3. Select χ^2**-Test** in the menu and press **ENTER**.
4. For *Observed* enter the matrix created in Step 1 by pressing **2ND** then **x⁻¹** and selecting the matrix letter. *Expected* shows the matrix that will be used to automatically store the expected frequencies that are calculated.
5. Select **Calculate** and press **ENTER**.

Excel

XLSTAT Add-In

1. Click on the **XLSTAT** tab in the Ribbon and then click **Correlation/ Association tests**.
2. Select **Tests on contingency tables** from the dropdown menu.
3. In the *Contingency table* box enter the range of cells containing the frequency counts of the contingency table. If the range includes data labels, check the **Labels included** box.
4. Under *Data format* select **Contingency table**.
5. Click the **Options** tab.
6. Check the **Chi-square test** box and enter a significance level.
7. Click **OK** to display the results.

11-2   Basic Skills and Concepts

Statistical Literacy and Critical Thinking

1. Handedness and Cell Phone Use The accompanying table is from a study conducted with the stated objective of addressing cell phone safety by understanding why we use a particular ear for cell phone use. (See "Hemispheric Dominance and Cell Phone Use," by Seidman, Siegel, Shah, and Bowyer, *JAMA Otolaryngology—Head & Neck Surgery,* Vol. 139, No. 5.) The goal was to determine whether the ear choice is associated with auditory or language brain hemispheric dominance. Assume that we want to test the claim that handedness and cell phone ear preference are independent of each other.

a. Use the data in the table to find the expected value for the cell that has an observed frequency of 3. Round the result to three decimal places.

b. What does the expected value indicate about the requirements for the hypothesis test?

Ear Preference for Cell Phone Use

	Right Ear	Left Ear	No Preference
Right-Handed	436	166	40
Left-Handed	16	50	3

2. Hypotheses Refer to the data given in Exercise 1 and assume that the requirements are all satisfied and we want to conduct a hypothesis test of independence using the methods of this section. Identify the null and alternative hypotheses.

3. Hypothesis Test The accompanying TI-83/84 Plus calculator display results from the hypothesis test described in Exercise 1. Assume that the hypothesis test requirements are all satisfied. Identify the test statistic and the P-value (expressed in standard form and rounded to three decimal places), and then state the conclusion about the null hypothesis.

4. Right-Tailed, Left-Tailed, Two-Tailed Is the hypothesis test described in Exercise 1 right-tailed, left-tailed, or two-tailed? Explain your choice.

In Exercises 5–18, test the given claim.

5. Lie Detector The table below includes results from polygraph (lie detector) experiments conducted by researchers Charles R. Honts (Boise State University) and Gordon H. Barland (Department of Defense Polygraph Institute). In each case, it was known if the subject lied or did not lie, so the table indicates when the polygraph test was correct. Use a 0.05 significance level to test the claim that whether a subject lies is independent of the polygraph test indication. Do the results suggest that polygraphs are effective in distinguishing between truths and lies?

	Did the Subject Actually Lie?	
	No (Did Not Lie)	Yes (Lied)
Polygraph test indicated that the subject lied.	15	42
Polygraph test indicated that the subject did not lie.	32	9

6. Splint or Surgery? A randomized controlled trial was designed to compare the effectiveness of splinting versus surgery in the treatment of carpal tunnel syndrome. Results are given in the table below (based on data from "Splinting vs. Surgery in the Treatment of Carpal Tunnel Syndrome," by Gerritsen et al., *Journal of the American Medical Association,* Vol. 288, No. 10). The results are based on evaluations made one year after the treatment. Using a 0.01 significance level, test the claim that success is independent of the type of treatment. What do the results suggest about treating carpal tunnel syndrome?

	Successful Treatment	Unsuccessful Treatment
Splint Treatment	60	23
Surgery Treatment	67	6

7. Texting and Drinking In a study of high school students at least 16 years of age, researchers obtained survey results summarized in the accompanying table (based on data from "Texting While Driving and Other Risky Motor Vehicle Behaviors Among U.S. High School Students," by O'Malley, Shults, and Eaton, *Pediatrics,* Vol. 131, No. 6). Use a 0.05 significance level to test the claim of independence between texting while driving and driving when drinking alcohol. Are those two risky behaviors independent of each other?

	Drove When Drinking Alcohol?	
	Yes	No
Texted While Driving	731	3054
No Texting While Driving	156	4564

8. Texting and Seat Belt Use In a study of high school students at least 16 years of age, researchers obtained survey results summarized in the accompanying table (based on data from "Texting While Driving and Other Risky Motor Vehicle Behaviors Among U.S. High School Students," by O'Malley, Shults, and Eaton, *Pediatrics,* Vol. 131, No. 6). Use a 0.05 significance level to test the claim of independence between texting while driving and irregular seat belt use. Are those two risky behaviors independent of each other?

	Irregular Seat Belt Use?	
	Yes	No
Texted While Driving	1737	2048
No Texting While Driving	1945	2775

9. Four Quarters the Same as $1? In a study of the "denomination effect," 43 college students were each given one dollar in the form of four quarters, while 46 other college students were each given one dollar in the form of a dollar bill. All of the students were then given two choices: (1) keep the money; (2) spend the money on gum. The results are given in the accompanying table (based on "The Denomination Effect," by Priya Raghubir and Joydeep Srivastava, *Journal of Consumer Research,* Vol. 36.) Use a 0.05 significance level to test the claim that whether students purchased gum or kept the money is independent of whether they were given four quarters or a $1 bill. Is there a "denomination effect"?

	Purchased Gum	Kept the Money
Students Given Four Quarters	27	16
Students Given a $1 Bill	12	34

10. Overtime Rule in Football The accompanying table lists results of overtime football games before and after the overtime rule was changed in the National Football League in 2011. Use a 0.05 significance level to test the claim of independence between winning an overtime game and whether playing under the old rule or the new rule. What do the results suggest about the effectiveness of the rule change?

	Before Rule Change	After Rule Change
Overtime Coin Toss Winner Won the Game	252	24
Overtime Coin Toss Winner Lost the Game	208	23

11. Tennis Challenges The table below shows results since 2006 of challenged referee calls in the U.S. Open. Use a 0.05 significance level to test the claim that the gender of the tennis player is independent of whether the call is overturned. Do players of either gender appear to be better at challenging calls?

	Was the Challenge to the Call Successful?	
	Yes	No
Men	161	376
Women	68	152

12. Nurse a Serial Killer? Alert nurses at the Veteran's Affairs Medical Center in Northampton, Massachusetts, noticed an unusually high number of deaths at times when another nurse, Kristen Gilbert, was working. Those same nurses later noticed missing supplies of the drug epinephrine, which is a synthetic adrenaline that stimulates the heart. Kristen Gilbert was arrested and charged with four counts of murder and two counts of attempted murder. When seeking a grand jury indictment, prosecutors provided a key piece of evidence consisting of the table below. Use a 0.01 significance level to test the defense claim that deaths on shifts are independent of whether Gilbert was working. What does the result suggest about the guilt or innocence of Gilbert?

	Shifts With a Death	Shifts Without a Death
Gilbert Was Working	40	217
Gilbert Was Not Working	34	1350

13. Soccer Strategy In soccer, serious fouls in the penalty box result in a penalty kick with one kicker and one defending goalkeeper. The table below summarizes results from 286 kicks during games among top teams (based on data from "Action Bias Among Elite Soccer Goalkeepers: The Case of Penalty Kicks," by Bar-Eli et al., *Journal of Economic Psychology,* Vol. 28, No. 5). In the table, jump direction indicates which way the goalkeeper jumped, where the kick direction is from the perspective of the goalkeeper. Use a 0.05 significance level to test the claim that the direction of the kick is independent of the direction of the goalkeeper jump. Do the results support the theory that because the kicks are so fast, goalkeepers have no time to react, so the directions of their jumps are independent of the directions of the kicks?

	Goalkeeper Jump		
	Left	Center	Right
Kick to Left	54	1	37
Kick to Center	41	10	31
Kick to Right	46	7	59

14. Is Seat Belt Use Independent of Cigarette Smoking? A study of seat belt users and nonusers yielded the randomly selected sample data summarized in the given table (based on data from "What Kinds of People Do Not Use Seat Belts?" by Helsing and Comstock, *American Journal of Public Health,* Vol. 67, No. 11). Test the claim that the amount of smoking is independent of seat belt use. A plausible theory is that people who smoke more are less concerned about their health and safety and are therefore less inclined to wear seat belts. Is this theory supported by the sample data?

	Number of Cigarettes Smoked per Day			
	0	1–14	15–34	35 and over
Wear Seat Belts	175	20	42	6
Don't Wear Seat Belts	149	17	41	9

15. Clinical Trial of Echinacea In a clinical trial of the effectiveness of echinacea for preventing colds, the results in the table below were obtained (based on data from "An Evaluation of *Echinacea Angustifolia* in Experimental Rhinovirus Infections," by Turner et al., *New England Journal of Medicine,* Vol. 353, No. 4). Use a 0.05 significance level to test the claim that getting a cold is independent of the treatment group. What do the results suggest about the effectiveness of echinacea as a prevention against colds?

	Treatment Group		
	Placebo	Echinacea: 20% Extract	Echinacea: 60% Extract
Got a Cold	88	48	42
Did Not Get a Cold	15	4	10

16. Injuries and Motorcycle Helmet Color A case-control (or retrospective) study was conducted to investigate a relationship between the colors of helmets worn by motorcycle drivers and whether they are injured or killed in a crash. Results are given in the table below (based on data from "Motorcycle Rider Conspicuity and Crash Related Injury: Case-Control Study," by Wells et al., *BMJ USA*, Vol. 4). Test the claim that injuries are independent of helmet color. Should motorcycle drivers choose helmets with a particular color? If so, which color appears best?

	Color of Helmet				
	Black	White	Yellow/Orange	Red	Blue
Controls (not injured)	491	377	31	170	55
Cases (injured or killed)	213	112	8	70	26

17. Survey Refusals A study of people who refused to answer survey questions provided the randomly selected sample data shown in the table below (based on data from "I Hear You Knocking But You Can't Come In," by Fitzgerald and Fuller, *Sociological Methods and Research*, Vol. 11, No. 1). At the 0.01 significance level, test the claim that the cooperation of the subject (response or refusal) is independent of the age category. Does any particular age group appear to be particularly uncooperative?

	Age					
	18–21	22–29	30–39	40–49	50–59	60 and over
Responded	73	255	245	136	138	202
Refused	11	20	33	16	27	49

18. Baseball Player Births In his book *Outliers*, author Malcolm Gladwell argues that more American-born baseball players have birth dates in the months immediately following July 31 because that was the age cutoff date for nonschool baseball leagues. The table below lists months of births for a sample of American-born baseball players and foreign-born baseball players. Using a 0.05 significance level, is there sufficient evidence to warrant rejection of the claim that months of births of baseball players are independent of whether they are born in America? Do the data appear to support Gladwell's claim?

	Jan.	Feb.	March	April	May	June	July	Aug.	Sept.	Oct.	Nov.	Dec.
Born in America	387	329	366	344	336	313	313	503	421	434	398	371
Foreign Born	101	82	85	82	94	83	59	91	70	100	103	82

19. Car License Plates California, Connecticut, and New York are states with laws requiring that cars have license plates on the front and rear. The author randomly selected cars in those states and the results are given in the accompanying table. Use a 0.05 significance level to test the claim of independence between the state and whether a car has front and rear license plates. Does it appear that the license plate laws are followed at the same rates in the three states?

	California	Connecticut	New York
Car with Rear Plate Only	35	45	9
Car with Front and Rear Plates	528	289	541

20. Is the Home Field Advantage Independent of the Sport? Winning team data were collected for teams in different sports, with the results given in the table on the top of the next page (based on data from "Predicting Professional Sports Game Outcomes from Intermediate Game Scores," by Copper, DeNeve, and Mosteller, *Chance*, Vol. 5, No. 3–4). Use a 0.10 significance level to test the claim that home/visitor wins are independent of the sport. Given that among the four sports included here, baseball is the only sport in which the home team can modify field dimensions to favor its own players, does it appear that baseball teams are effective in using this advantage?

continued

	Basketball	Baseball	Hockey	Football
Home Team Wins	127	53	50	57
Visiting Team Wins	71	47	43	42

11-2   Beyond the Basics

21. Equivalent Tests A X^2 test involving a 2×2 table is equivalent to the test for the difference between two proportions, as described in Section 9-1. Using the claim and table in Exercise 9 "Four Quarters the Same as $1?" verify that the X^2 test statistic and the z test statistic (found from the test of equality of two proportions) are related as follows: $z^2 = X^2$. Also show that the critical values have that same relationship.

22. Using Yates's Correction for Continuity The chi-square distribution is continuous, whereas the test statistic used in this section is discrete. Some statisticians use *Yates's correction for continuity* in cells with an expected frequency of less than 10 or in all cells of a contingency table with two rows and two columns. With Yates's correction, we replace

$$\sum \frac{(O - E)^2}{E} \quad \text{with} \quad \sum \frac{(|O - E| - 0.5)^2}{E}$$

Given the contingency table in Exercise 9 "Four Quarters the Same as $1?" find the value of the X^2 test statistic using Yates's correction in all cells. What effect does Yates's correction have?

Chapter Quick Quiz

Exercises 1–5 refer to the sample data in the following table, which summarizes the last digits of the heights (cm) of 300 randomly selected subjects (from Data Set 1 "Body Data" in Appendix B). Assume that we want to use a 0.05 significance level to test the claim that the data are from a population having the property that the last digits are all equally likely.

Last Digit	0	1	2	3	4	5	6	7	8	9
Frequency	30	35	24	25	35	36	37	27	27	24

1. What are the null and alternative hypotheses corresponding to the stated claim?

2. When testing the claim in Exercise 1, what are the observed and expected frequencies for the last digit of 7?

3. Is the hypothesis test left-tailed, right-tailed, or two-tailed?

4. If using a 0.05 significance level to test the stated claim, find the number of degrees of freedom.

5. Given that the P-value for the hypothesis test is 0.501, what do you conclude? Does it appear that the heights were obtained through measurement or that the subjects reported their heights?

Questions 6–10 refer to the sample data in the following table, which describes the fate of the passengers and crew aboard the **Titanic** *when it sank on April 15, 1912. Assume that the data are a sample from a large population and we want to use a 0.05 significance level to test the claim that surviving is independent of whether the person is a man, woman, boy, or girl.*

	Men	Women	Boys	Girls
Survived	332	318	29	27
Died	1360	104	35	18

6. Identify the null and alternative hypotheses corresponding to the stated claim.

7. What distribution is used to test the stated claim (normal, *t*, *F*, chi-square, uniform)?

8. Is the hypothesis test left-tailed, right-tailed, or two-tailed?

9. Find the number of degrees of freedom.

10. Given that the *P*-value for the hypothesis test is 0.000 when rounded to three decimal places, what do you conclude? What do the results indicate about the rule that women and children should be the first to be saved?

Review Exercises

1. Motor Vehicle Fatalities The table below lists motor vehicle fatalities by day of the week for a recent year (based on data from the Insurance Institute for Highway Safety). Use a 0.01 significance level to test the claim that auto fatalities occur on the different days of the week with the same frequency. Provide an explanation for the results.

Day	Sun.	Mon.	Tues.	Wed.	Thurs.	Fri.	Sat.
Frequency	5304	4002	4082	4010	4268	5068	5985

2. Tooth Fillings The table below shows results from a study in which some patients were treated with fillings that contain mercury and others were treated with fillings that do not contain mercury (based on data from "Neuropsychological and Renal Effects of Dental Amalgam in Children," by Bellinger et al., *Journal of the American Medical Association*, Vol. 295, No. 15). Use a 0.05 significance level to test for independence between the type of filling and the presence of any adverse health conditions. Do fillings that contain mercury appear to affect health conditions?

	Fillings With Mercury	Fillings Without Mercury
Adverse Health Condition Reported	135	145
No Adverse Health Condition Reported	132	122

3. American Idol Contestants on the TV show *American Idol* competed to win a singing contest. At one point, the website WhatNotToSing.com listed the actual numbers of eliminations for different orders of singing, and the expected number of eliminations was also listed. The results are in the table below. Use a 0.05 significance level to test the claim that the actual eliminations agree with the expected numbers. Does there appear to be support for the claim that the leadoff singers appear to be at a disadvantage?

Singing Order	1	2	3	4	5	6	7–12
Actual Eliminations	20	12	9	8	6	5	9
Expected Eliminations	12.9	12.9	9.9	7.9	6.4	5.5	13.5

4. Clinical Trial of Lipitor Lipitor is the trade name of the drug atorvastatin, which is used to reduce cholesterol in patients. (Until its patent expired in 2011, this was the largest-selling drug in the world, with annual sales of $13 billion.) Adverse reactions have been studied in clinical trials, and the table below summarizes results for infections in patients from different treatment groups (based on data from Parke-Davis). Use a 0.01 significance level to test the claim that getting an infection is independent of the treatment. Does the atorvastatin (Lipitor) treatment appear to have an effect on infections?

	Placebo	Atorvastatin 10 mg	Atorvastatin 40 mg	Atorvastatin 80 mg
Infection	27	89	8	7
No Infection	243	774	71	87

5. Weather-Related Deaths For a recent year, the numbers of weather-related U.S. deaths for each month were 28, 17, 12, 24, 88, 61, 104, 32, 20, 13, 26, 25 (listed in order beginning with January). Use a 0.01 significance level to test the claim that weather-related deaths occur in the different months with the same frequency. Provide an explanation for the result.

Cumulative Review Exercises

1. Weather-Related Deaths Review Exercise 5 involved weather-related U.S. deaths. Among the 450 deaths included in that exercise, 320 are males. Use a 0.05 significance level to test the claim that among those who die in weather-related deaths, the percentage of males is equal to 50%. Provide an explanation for the results.

2. Chocolate and Happiness In a survey sponsored by the Lindt chocolate company, 1708 women were surveyed and 85% of them said that chocolate made them happier.

a. Is there anything potentially wrong with this survey?

b. Of the 1708 women surveyed, what is the number of them who said that chocolate made them happier?

3. Chocolate and Happiness Use the results from part (b) of Cumulative Review Exercise 2 to construct a 99% confidence interval estimate of the *percentage* of women who say that chocolate makes them happier. Write a brief statement interpreting the result.

4. Chocolate and Happiness Use the results from part (b) of Cumulative Review Exercise 2 to test the claim that when asked, more than 80% of women say that chocolate makes them happier. Use a 0.01 significance level.

5. One Big Bill or Many Smaller Bills In a study of the "denomination effect," 150 women in China were given either a single 100 yuan bill or a total of 100 yuan in smaller bills. The value of 100 yuan is about $15. The women were given the choice of spending the money on specific items or keeping the money. The results are summarized in the table below (based on "The Denomination Effect," by Priya Raghubir and Joydeep Srivastava, *Journal of Consumer Research,* Vol. 36). Use a 0.05 significance level to test the claim that the form of the 100 yuan is independent of whether the money was spent. What does the result suggest about a denomination effect?

	Spent the Money	Kept the Money
Women Given a Single 100-Yuan Bill	60	15
Women Given 100 Yuan in Smaller Bills	68	7

6. Probability Refer to the results from the 150 subjects in Cumulative Review Exercise 5.

a. Find the probability that if 1 of the 150 subjects is randomly selected, the result is a woman who spent the money.

b. Find the probability that if 1 of the 150 subjects is randomly selected, the result is a woman who spent the money or was given a single 100-yuan bill.

c. If two different women are randomly selected, find the probability that they both spent the money.

7. Car Repair Costs Listed below are repair costs (in dollars) for cars crashed at 6 mi/h in full-front crash tests and the same cars crashed at 6 mi/h in full-rear crash tests (based on data from the Insurance Institute for Highway Safety). The cars are the Toyota Camry, Mazda 6, Volvo S40, Saturn Aura, Subaru Legacy, Hyundai Sonata, and Honda Accord. Is there sufficient evidence to conclude that there is a linear correlation between the repair costs from full-front crashes and full-rear crashes?

Front	936	978	2252	1032	3911	4312	3469
Rear	1480	1202	802	3191	1122	739	2767

8. Forward Grip Reach and Ergonomics When designing cars and aircraft, we must consider the forward grip reach of women. Women have normally distributed forward grip reaches with a mean of 686 mm and a standard deviation of 34 mm (based on anthropometric survey data from Gordon, Churchill, et al.).

a. If a car dashboard is positioned so that it can be reached by 95% of women, what is the shortest forward grip reach that can access the dashboard?

b. If a car dashboard is positioned so that it can be reached by women with a grip reach greater than 650 mm, what percentage of women cannot reach the dashboard? Is that percentage too high?

c. Find the probability that 16 randomly selected women have forward grip reaches with a mean greater than 680 mm. Does this result have any effect on the design?

Technology Project

Use Statdisk, Minitab, Excel, StatCrunch, a TI-83/84 Plus calculator, or any other software package or calculator capable of generating equally likely random digits between 0 and 9 inclusive. Generate 5000 digits and record the results in the accompanying table. Use a 0.05 significance level to test the claim that the sample digits come from a population with a uniform distribution (so that all digits are equally likely). Does the random number generator appear to be working as it should?

Digit	0	1	2	3	4	5	6	7	8	9
Frequency										

Statdisk: Select **Data,** then **Uniform Generator.**

Minitab: Select **Calc, Random Data, Integer.**

Excel: Click **Insert function** f_x, then select category **Math & Trig** and function **RANDBETWEEN**. Click and drag cell down the column to generate additional random numbers.

TI-83/84 Plus: Press **MATH,** select **PROB,** then use **randInt** function with the format of **randInt(*lower, upper, n*).**

StatCrunch: Select **Data, Simulate, Discrete Uniform.**

Critical Thinking: Was Allstate wrong?

The Allstate insurance company once issued a press release listing zodiac signs along with the corresponding numbers of automobile crashes, as shown in the first and last columns in the table below.

In the original press release, Allstate included comments such as one stating that Virgos are worried and shy,

and they were involved in 211,650 accidents, making them the worst offenders. Allstate quickly issued an apology and retraction. In a press release, Allstate included this: "Astrological signs have absolutely no role in how we base coverage and set rates. Rating by astrology would not be actuarially sound."

Zodiac Sign	Dates	Length (days)	Crashes
Capricorn	Jan. 18–Feb. 15	29	128,005
Aquarius	Feb. 16–March 11	24	106,878
Pisces	March 12–April 16	36	172,030
Aries	April 17–May 13	27	112,402
Taurus	May 14–June 19	37	177,503
Gemini	June 20–July 20	31	136,904
Cancer	July 21–Aug. 9	20	101,539
Leo	Aug. 10–Sept. 15	37	179,657
Virgo	Sept. 16–Oct. 30	45	211,650
Libra	Oct. 31–Nov. 22	23	110,592
Scorpio	Nov. 23–Nov. 28	6	26,833
Ophiuchus	Nov. 29–Dec. 17	19	83,234
Sagittarius	Dec. 18–Jan. 17	31	154,477

Analyzing the Results

The original Allstate press release did not include the lengths (days) of the different zodiac signs. The preceding table lists those lengths in the third column. A reasonable explanation for the different numbers of crashes is that they should be proportional to the lengths of the zodiac signs. For example,

people are born under the Capricorn sign on 29 days out of the 365 days in the year, so they are expected to have $29/365$ of the total number of crashes. Use the methods of this chapter to determine whether this appears to explain the results in the table. Write a brief report of your findings.

Cooperative Group Activities

1. Out-of-class activity Divide into groups of four or five students. The Chapter Problem noted that according to Benford's law, a variety of different data sets include numbers with leading (first) digits that follow the distribution shown in the table below. Collect original data and use the methods of Section 11-1 to support or refute the claim that the data conform reasonably well to Benford's law. Here are some suggestions: (1) leading digits of smartphone passcodes; (2) leading digits of the prices of stocks; (3) leading digits of the numbers of Facebook friends; (5) leading digits of the lengths of rivers in the world.

Leading Digit	1	2	3	4	5	6	7	8	9
Benford's Law	30.1%	17.6%	12.5%	9.7%	7.9%	6.7%	5.8%	5.1%	4.6%

2. Out-of-class activity Divide into groups of four or five students and collect past results from a state lottery. Such results are often available on websites for individual state lotteries. Use the methods of Section 11-1 to test the claim that the numbers are selected in such a way that all possible outcomes are equally likely.

3. Out-of-class activity Divide into groups of four or five students. Each group member should survey at least 15 male students and 15 female students at the same college by asking two questions: (1) Which political party does the subject favor most? (2) If the subject were to make up an absence excuse of a flat tire, which tire would he or she say went flat if the instructor asked? (See Exercise 6 in Section 11-1.) Ask the subject to write the two responses on an index card, and also record the gender of the subject and whether the subject wrote with the right or left hand. Use the methods of this chapter to analyze the data collected. Include these claims:

• The four possible choices for a flat tire are selected with equal frequency.

• The tire identified as being flat is independent of the gender of the subject.

• Political party choice is independent of the gender of the subject.

• Political party choice is independent of whether the subject is right- or left-handed.

• The tire identified as being flat is independent of whether the subject is right- or left-handed.

• Gender is independent of whether the subject is right- or left-handed.

• Political party choice is independent of the tire identified as being flat.

4. Out-of-class activity Divide into groups of four or five students. Each group member should select about 15 other students and first ask them to "randomly" select four digits each. After the four digits have been recorded, ask each subject to write the last four digits of his or her Social Security number (for security, write these digits in any order). Take the "random" sample results of individual digits and mix them into one big sample, then mix the individual Social Security digits into a second big sample. Using the "random" sample set, test the claim that students select digits randomly. Then use the Social Security digits to test the claim that they come from a population of random digits. Compare the results. Does it appear that students can randomly select digits? Are they likely to select any digits more often than others? Are they likely to select any digits less often than others? Do the last digits of Social Security numbers appear to be randomly selected?

5. In-class activity Divide into groups of three or four students. Each group should be given a die along with the instruction that it should be tested for "fairness." Is the die fair or is it biased? Describe the analysis and results.

6. Out-of-class activity Divide into groups of two or three students. The analysis of last digits of data can sometimes reveal whether values are the results of actual measurements or whether they are reported estimates. Find the lengths of rivers in the world, then analyze the last digits to determine whether those lengths appear to be actual measurements or whether they appear to be reported estimates. Instead of lengths of rivers, you could use other variables, such as the following:

• Heights of mountains

• Heights of tallest buildings

• Lengths of bridges

• Heights of roller coasters

12-1 One-Way ANOVA
12-2 Two-Way ANOVA

12

ANALYSIS OF VARIANCE

CHAPTER PROBLEM — Does exposure to lead affect IQ scores of children?

An important environment/health study involved children who lived within 7 km (about 4 miles) of a large ore smelter in El Paso, Texas. A smelter is used to melt the ore in order to separate the metals in it. Because the smelter emitted lead pollution, there was concern that these children would somehow suffer. The focus of this Chapter Problem is to investigate the possible effect of lead exposure on "performance" IQ scores as measured by the Wechsler intelligence scale. (A full IQ score is a combination of a performance IQ score and a verbal IQ score. The performance test includes components such as picture analysis, picture arrangement, and matching patterns.)

Data from the study are included in Data Set 7 "IQ and Lead" in Appendix B. Based on measured blood lead levels, the children were partitioned into a low lead level group, a medium lead level group, or a high lead level group. (See Data Set 7 for the specific blood lead-level cutoff values.)

TABLE 12-1 Performance IQ Scores of Children

Low Blood Lead Level															
85	90	107	85	100	97	101	64	111	100	76	136	100	90	135	104
149	99	107	99	113	104	101	111	118	99	122	87	118	113	128	121
111	104	51	100	113	82	146	107	83	108	93	114	113	94	106	92
79	129	114	99	110	90	85	94	127	101	99	113	80	115	85	112
112	92	97	97	91	105	84	95	108	118	118	86	89	100		

Medium Blood Lead Level															
78	97	107	80	90	83	101	121	108	100	110	111	97	51	94	80
101	92	100	77	108	85										

High Blood Lead Level															
93	100	97	79	97	71	111	99	85	99	97	111	104	93	90	107
108	78	95	78	86											

The performance IQ scores are included in Table 12-1 (based on data from "Neuropsychological Dysfunction in Children with Chronic Low-Level Lead Absorption," by P. J. Landrigan, R. H. Whitworth, R. W. Baloh, N. W. Staehling, W. F. Barthel, and B. F. Rosenblum, *Lancet,* Vol. 1, Issue 7909).

Before jumping into the application of a particular statistical method, we should first explore the data. Sample statistics are included in the table below. See also the following boxplots of the three sets of performance IQ scores. Informal and subjective comparisons show that the low group has a mean that is somewhat higher than the means of the medium and high groups. The boxplots all overlap, so differences do not

appear to be dramatic. But we need more formal methods that allow us to recognize any significant differences. We could use the methods of Section 9-2 to compare means from samples collected from *two* different populations, but here we need to compare means from samples collected from *three* different populations. When we have samples from three or more populations, we can test for equality of the population means by using the method of *analysis of variance,* to be introduced in Section 12-1. In Section 12-1, we will use analysis of variance to test the claim that the three samples are from populations with the same mean.

	Low Blood Lead Level	Medium Blood Lead Level	High Blood Lead Level
Sample Size n	78	22	21
$\bar{x}$	102.7	94.1	94.2
s	16.8	15.5	11.4
Distribution	Approximately normal	Approximately normal	Approximately normal
Outliers	Potential low outlier of 51 and high outliers of 146 and 149, but they are not very far from the other data values.	None	None

Minitab Boxplots of Performance IQ Scores

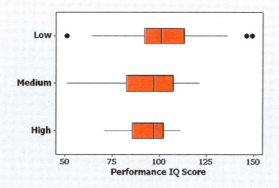

CHAPTER OBJECTIVES

Section 9-2 includes methods for testing equality of means from *two* independent populations, but this chapter presents a method for testing equality of *three or more* population means. Here are the chapter objectives:

12-1 One-Way ANOVA

- Apply the method of one-way analysis of variance to conduct a hypothesis test of equality of three or more population means. The focus of this section is the interpretation of results from technology.

12-2 Two-Way ANOVA

- Analyze sample data from populations separated into categories using two characteristics (or factors), such as gender and eye color.

- Apply the method of two-way analysis of variance to the following: (1) test for an *interaction* between two factors, (2) test for an effect from the *row* factor, and (3) test for an effect from the *column* factor. The focus of this section is the interpretation of results from technology.

12-1   One-Way ANOVA

Key Concept In this section we introduce the method of *one-way analysis of variance*, which is used for tests of hypotheses that three or more populations have means that are all equal, as in H_0: $\mu_1 = \mu_2 = \mu_3$. Because the calculations are very complicated, we emphasize the interpretation of results obtained by using technology.

F Distribution

The analysis of variance (ANOVA) methods of this chapter require the F distribution, which was first introduced in Section 9-4. In Section 9-4 we noted that the F distribution has the following properties (see Figure 12-1):

There is a different F distribution for each different pair of degrees of freedom for numerator and denominator.

1. The F distribution is not symmetric. It is skewed right.
2. Values of the F distribution cannot be negative.
3. The exact shape of the F distribution depends on the two different degrees of freedom.

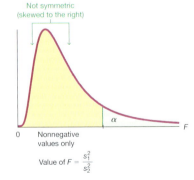

FIGURE 12-1   ***F* Distribution**

PART 1 Basics of One-Way Analysis of Variance

When testing for equality of three or more population means, use the method of one-way analysis of variance.

DEFINITION

One-way analysis of variance (ANOVA) is a method of testing the equality of three or more population means by analyzing sample variances. One-way analysis of variance is used with data categorized with *one* **factor** (or **treatment**), so there is one characteristic used to separate the sample data into the different categories.

The term *treatment* is used because early applications of analysis of variance involved agricultural experiments in which different plots of farmland were treated with different fertilizers, seed types, insecticides, and so on. Table 12-1 uses the one "treatment" (or factor) of blood lead level. That factor has three different categories: low, medium, and high blood lead levels (as defined in Data Set 7 in Appendix B).

KEY ELEMENTS

One-Way Analysis of Variance for Testing Equality of Three or More Population Means

Objective

Use samples from three or more different populations to test a claim that the populations all have the same mean.

Requirements

1. The populations have distributions that are approximately normal. (This is a loose requirement, because the method works well unless a population has a distribution that is very far from normal. If a population does have a distribution that is far from normal, use the Kruskal-Wallis test described in Section 13-5.)

2. The populations have the same variance σ^2 (or standard deviation σ). This is a loose requirement, because the method works well unless the population variances differ by large amounts. Statistician George E. P. Box showed that as long as the sample sizes are equal (or nearly equal), the largest variance can be up to nine times the smallest variance and the results of ANOVA will continue to be essentially reliable.

3. The samples are simple random samples of quantitative data.

4. The samples are independent of each other. (The samples are not matched or paired in any way.)

5. The different samples are from populations that are categorized in only one way.

Procedure for Testing $H_0: \mu_1 = \mu_2 = \mu_3 = \cdots = \mu_k$

1. Use technology to obtain results that include the test statistic and *P*-value.

2. Identify the *P*-value from the display. (The ANOVA test is right-tailed because only large values of the test statistic cause us to reject equality of the population means.)

3. Form a conclusion based on these criteria that use the significance level α:

   • **Reject:** If the *P*-value $\leq \alpha$, reject the null hypothesis of equal means and conclude that at least one of the population means is different from the others.

   • **Fail to Reject:** If the *P*-value $> \alpha$, fail to reject the null hypothesis of equal means.

Because the calculations required for one-way analysis of variance are messy, we recommend using technology with this study strategy:

1. Understand that a small *P*-value (such as 0.05 or less) leads to rejection of the null hypothesis of equal means. ("If the *P* is low, the null must go.") With a large *P*-value (such as greater than 0.05), fail to reject the null hypothesis of equal means.

2. Develop an understanding of the underlying rationale by studying the examples in this section.

EXAMPLE 1 Lead and Performance IQ Scores

Use the performance IQ scores listed in Table 12-1 and a significance level of $\alpha = 0.05$ to test the claim that the three samples come from populations with means that are all equal.

SOLUTION

REQUIREMENT CHECK (1) Based on the three samples listed in Table 12-1, the three populations appear to have distributions that are approximately normal, as indicated by normal quantile plots. (2) The three samples in Table 12-1 have standard deviations that are not dramatically different, so the three population variances appear to be about the same. (3) On the basis of the study design, we can treat the samples as simple random samples. (4) The samples are independent of each other; the performance IQ scores are not matched in any way. (5) The three samples are from populations categorized according to the single factor of lead level (low, medium, high). The requirements are satisfied. ☑

The null hypothesis and the alternative hypothesis are as follows:

$$H_0: \mu_1 = \mu_2 = \mu_3$$

H_1: At least one of the means is different from the others

The significance level is $\alpha = 0.05$.

Step 1: Use technology to obtain ANOVA results, such as one of those shown in the accompanying displays.

Statdisk

Source:	DF:	SS:	MS:	Test Stat, F:	Critical F:	P-Value:
Treatment:	2	2022.729906	1011.364953	4.071122	3.073087	0.01951
Error:	118	29314.046953	248.424127			
Total:	120	31336.77686				

Minitab

One-way ANOVA: Low, Medium, High

Source	DF	SS	MS	F	P
Factor	2	2023	1011	4.07	0.020
Error	118	29314	248		
Total	120	31337			

S = 15.76   R-Sq = 6.45%   R-Sq(adj) = 4.87%

StatCrunch

ANOVA table

Source	df	SS	MS	F-Stat	P-value
Treatments	2	2022.7299	1011.3649	4.071122	0.0195
Error	118	29314.047	248.42413		
Total	120	31336.777			

Excel

ANOVA						
Source of Variation	SS	df	MS	F	P-value	F crit
Between Groups	2022.729906	2	1011.364953	4.071122103	0.019510383	3.073090341
Within Groups	29314.04695	118	248.4241267			
Total	31336.77686	120				

TI-83/84 Plus

```
NORMAL FLOAT AUTO REAL RADIAN MP

        One-way ANOVA
 F=4.071122103
 p=.0195103826
 Factor
   df=2
   SS=2022.72991
   MS=1011.36495
 Error
↓ df=118
```

SPSS

	Sum of Squares	df	Mean Square	F	Sig.
Between Groups	2022.730	2	1011.365	4.071	.020
Within Groups	29314.047	118	248.424		
Total	31336.777	120			

JMP

Source	DF	Sum of Squares	Mean Square	F Ratio
Model	2	2022.730	1011.36	4.0711
Error	118	29314.047	248.42	Prob > F
C. Total	120	31336.777		0.0195*

Step 2: In addition to the test statistic of $F = 4.0711$, the displays all show that the P-value is 0.020 when rounded.

Step 3: Because the P-value of 0.020 is less than the significance level of $\alpha = 0.05$, we reject the null hypothesis of equal means. (If the P is low, the null must go.)

INTERPRETATION

There is sufficient evidence to warrant rejection of the claim that the three samples come from populations with means that are all equal. Using the samples of measurements listed in Table 12-1, we conclude that those values come from populations having means that are not all the same. On the basis of this ANOVA test, we cannot conclude that any particular mean is different from the others, but we can informally note that the sample mean for the low blood lead group is higher than the means for the medium and high blood lead groups. It appears that greater blood lead levels are associated with lower performance IQ scores.

YOUR TURN Do Exercise 5 "Lead and Verbal IQ Scores."

CAUTION When we conclude that there is sufficient evidence to reject the claim of equal population means, we cannot conclude from ANOVA that any particular mean is different from the others. (There are several other methods that can be used to identify the specific means that are different, and some of them are discussed in Part 2 of this section.)

How is the *P*-Value Related to the Test Statistic? *Larger* values of the test statistic result in *smaller P*-values, so the ANOVA test is right-tailed. Figure 12-2 on the next page shows the relationship between the F test statistic and the P-value. Assuming that the populations have the same variance σ^2 (as required for the test), the F test statistic is the ratio of these two estimates of σ^2: (1) variation *between* samples (based on variation among sample means); and (2) variation *within* samples (based on the sample variances).

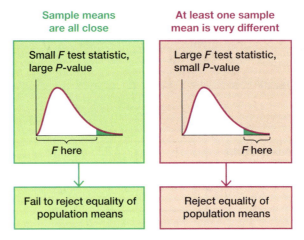

FIGURE 12-2   **Relationship Between the *F* Test Statistic and the *P*-Value**

Test Statistic for One-Way ANOVA: $F = \dfrac{\text{variance between samples}}{\text{variance within samples}}$

The numerator of the *F* test statistic measures variation between sample means. The estimate of variance in the denominator depends only on the sample variances and is not affected by differences among the sample means. Consequently, sample means that are close in value to each other result in a small *F* test statistic and a large *P*-value, so we conclude that there are no significant differences among the sample means. Sample means that are very far apart in value result in a large *F* test statistic and a small *P*-value, so we reject the claim of equal means.

Why Not Just Test Two Samples at a Time?   If we want to test for equality among three or more population means, why do we need a new procedure when we can test for equality of two means using the methods presented in Section 9-2? For example, if we want to use the sample data from Table 12-1 to test the claim that the three populations have the same mean, why not simply pair them off and test two at a time by testing $H_0: \mu_1 = \mu_2$, $H_0: \mu_2 = \mu_3$, and $H_0: \mu_1 = \mu_3$? For the data in Table 12-1, the approach of testing equality of two means at a time requires three different hypothesis tests. If we use a 0.05 significance level for each of those three hypothesis tests, the actual overall confidence level could be as low as 0.95^3 (or 0.857). In general, as we increase the number of individual tests of significance, we increase the risk of finding a difference by chance alone (instead of a real difference in the means). The risk of a type I error—finding a difference in one of the pairs when no such difference actually exists—is far too high. The method of analysis of variance helps us avoid that particular pitfall (rejecting a true null hypothesis) by using *one test* for equality of several means, instead of several tests that each compare two means at a time.

> **CAUTION** When testing for equality of three or more populations, use analysis of variance. (Using multiple hypothesis tests with two samples at a time could adversely affect the significance level.)

PART 2 ## Calculations and Identifying Means That Are Different

Calculating the Test Statistic *F* with Equal Sample Sizes *n*

Table 12-2 can be very helpful in understanding the methods of ANOVA. In Table 12-2, compare Data Set A to Data Set B to see that Data Set A is the same as Data Set B with this notable exception: The Sample 1 values each differ by 10. If the data sets all have the same sample size (as in $n = 4$ for Table 12-2), the following calculations aren't too difficult, as shown here.

TABLE 12-2 Effect of a Mean on the F Test Statistic

Add 10 to data in Sample 1

Data Set A			Data Set B		
Sample 1	Sample 2	Sample 3	Sample 1	Sample 2	Sample 3
7	6	4	17	6	4
3	5	7	13	5	7
6	5	6	16	5	6
6	8	7	16	8	7
$n_1 = 4$	$n_2 = 4$	$n_3 = 4$	$n_1 = 4$	$n_2 = 4$	$n_3 = 4$
$\bar{x}_1 = 5.5$	$\bar{x}_2 = 6.0$	$\bar{x}_3 = 6.0$	$\bar{x}_1 = 15.5$	$\bar{x}_2 = 6.0$	$\bar{x}_3 = 6.0$
$s_1^2 = 3.0$	$s_2^2 = 2.0$	$s_3^2 = 2.0$	$s_1^2 = 3.0$	$s_2^2 = 2.0$	$s_3^2 = 2.0$
Data Set A			**Data Set B**		

Step 1: Variance *between* samples	$ns_{\bar{x}}^2 = 4(0.0833) = 0.3332$	$ns_{\bar{x}}^2 = 4(30.0833) = 120.3332$
Step 2: Variance *within* samples	$s_p^2 = \dfrac{3.0 + 2.0 + 2.0}{3} = 2.3333$	$s_p^2 = \dfrac{3.0 + 2.0 + 2.0}{3} = 2.3333$
Step 3: F test statistic	$F = \dfrac{ns_{\bar{x}}^2}{s_p^2} = \dfrac{0.3332}{2.3333} = 0.1428$	$F = \dfrac{ns_{\bar{x}}^2}{s_p^2} = \dfrac{120.3332}{2.3333} = 51.5721$
P-value	$P\text{-value} = 0.8688$	$P\text{-value} = 0.0000118$

Step 1: Find the Variance Between Samples

Calculate the variance *between* samples by evaluating $ns_{\bar{x}}^2$ where $s_{\bar{x}}^2$ is the variance of the sample means and n is the size of each of the samples. That is, consider the sample means to be an ordinary set of values and calculate the variance. (From the central limit theorem, $\sigma_{\bar{x}} = \sigma / \sqrt{n}$ can be solved for σ to get $\sigma = \sqrt{n} \cdot \sigma_{\bar{x}}$, so that we can estimate σ^2 with $ns_{\bar{x}}^2$.) For example, the sample means for Data Set A in Table 12-2 are 5.5, 6.0, and 6.0, and these three values have a variance of $s_{\bar{x}}^2 = 0.0833$, so that

$$\text{variance between samples} = ns_{\bar{x}}^2 = 4(0.0833) = 0.3332$$

Step 2: Find the Variance Within Samples

Estimate the variance *within* samples by calculating s_p^2, which is the pooled variance obtained by finding the mean of the sample variances. The sample variances in Table 12-2 are 3.0, 2.0, and 2.0, so that

$$\text{variance within samples} = s_p^2 = \frac{3.0 + 2.0 + 2.0}{3} = 2.3333$$

Step 3: Calculate the Test Statistic

Evaluate the F test statistic as follows:

$$F = \frac{\text{variance between samples}}{\text{variance within samples}} = \frac{ns_{\bar{x}}^2}{s_p^2} = \frac{0.3332}{2.3333} = 0.1428$$

Finding the Critical Value

The critical value of F is found by assuming a right-tailed test because large values of F correspond to significant differences among means. With k samples each having n values, the numbers of degrees of freedom are as follows.

Degrees of Freedom (using k = number of samples and n = sample size)

$$\text{Numerator degrees of freedom} = k - 1$$

$$\text{Denominator degrees of freedom} = k(n - 1)$$

For Data Set A in Table 12-2, $k = 3$ and $n = 4$, so the degrees of freedom are 2 for the numerator and $3(4 - 1) = 9$ for the denominator. With $\alpha = 0.05$, 2 degrees of freedom for the numerator, and 9 degrees of freedom for the denominator, the critical F value from Table A-5 is 4.2565. If we were to use the critical value method of hypothesis testing with Data Set A in Table 12-2, we would see that this right-tailed test has a test statistic of $F = 0.1428$ and a critical value of $F = 4.2565$, so the test statistic is not in the critical region. We therefore fail to reject the null hypothesis of equal means.

Understanding the Effect of a Mean on the F Test Statistic To really understand how the method of analysis of variance works, consider Data Set A and Data Set B in Table 12-2 and note the following.

- The three samples in Data Set A are identical to the three samples in Data Set B, except for this: Each value in Sample 1 of Data Set B is 10 more than the corresponding value in Data Set A.

- Adding 10 to each data value in the first sample of Data Set A has a significant effect on the test statistic, with F changing from 0.1428 to 51.5721.

- Adding 10 to each data value in the first sample of Data Set A has a dramatic effect on the P-value, which changes from 0.8688 (not significant) to 0.0000118 (significant).

- The three sample *means* in Data Set A (5.5, 6.0, 6.0) are very close, but the sample means in Data Set B (15.5, 6.0, 6.0) are not close.

- The three sample variances in Data Set A are identical to those in Data Set B.

- The *variance between samples* in Data Set A is 0.3332, but for Data Set B it is 120.3332 (indicating that the sample means in B are farther apart).

- The *variance within samples* is 2.3333 in both Data Set A and Data Set B, because the variance *within* a sample isn't affected when we add a constant to every sample value. *The change in the F test statistic and the P-value is attributable only to the change in* $\bar{x}_1$. This illustrates the key point underlying the method of one-way analysis of variance:

> **The F test statistic is very sensitive to sample *means*, even though it is obtained through two different estimates of the common population variance.**

Calculations with Unequal Sample Sizes

While the calculations for cases with equal sample sizes are somewhat reasonable, they become much more complicated when the sample sizes are not all the same, but the same basic reasoning applies. Instead of providing the relevant messy formulas required for cases with unequal sample sizes, we wisely and conveniently assume that technology should be used to obtain the P-value for the analysis of variance. We become unencumbered by complex computations and we can focus on checking requirements and interpreting results.

We calculate an F test statistic that is the ratio of two different estimates of the common population variance σ^2. With unequal sample sizes, we must use *weighted* measures that take the sample sizes into account. The test statistic is essentially the same as the one given earlier, and its interpretation is also the same as described earlier.

Designing Experiments

With one-way (or single-factor) analysis of variance, we use one factor as the basis for partitioning the data into different categories. If we conclude that the differences among the means are significant, we can't be absolutely sure that the differences can be explained by the factor being used. It is possible that the variation of some other unknown factor is responsible. One way to reduce the effect of the extraneous factors is to design the experiment so that it has a **completely randomized design,** in which each sample value is given the same chance of belonging to the different factor groups. For example, you might assign subjects to two different treatment groups and a third placebo group through a process of random selection equivalent to picking slips of paper from a bowl. Another way to reduce the effect of extraneous factors is to use a **rigorously controlled design,** in which sample values are carefully chosen so that all other factors have no variability. In general, good results require that the experiment be carefully designed and executed.

Identifying Which Means Are Different

After conducting an analysis of variance test, we might conclude that there is sufficient evidence to reject a claim of equal population means, but we cannot conclude from ANOVA that any *particular* means are different from the others. There are several formal and informal procedures that can be used to identify the specific means that are different. Here are two *informal* methods for comparing means:

- Construct boxplots of the different samples and examine any overlap to see if one or more of the boxplots is very different from the others.

- Construct confidence interval estimates of the means for each of the different samples, then compare those confidence intervals to see if one or more of them does not overlap with the others.

There are several formal procedures for identifying which means are different. Some of the tests, called **range tests,** allow us to identify subsets of means that are not significantly different from each other. Other tests, called **multiple comparison tests,** use pairs of means, but they make adjustments to overcome the problem of having a significance level that increases as the number of individual tests increases. There is no consensus on which test is best, but some of the more common tests are the Duncan test, Student-Newman-Keuls test (or SNK test), Tukey test (or Tukey honestly significant difference test), Scheffé test, Dunnett test, least significant difference test, and the Bonferroni test. Let's consider the Bonferroni test to see one example of a multiple comparison test. Here is the procedure.

Bonferroni Multiple Comparison Test

Step 1: Do a separate t test for each pair of samples, but make the adjustments described in the following steps.

continued

Go Figure

$5816: The extra cost incurred by a private employer each year attributable to an employee who smokes. That total includes the costs of smoking breaks and health care costs due to the larger number of health problems suffered by smokers.

Step 2: For an estimate of the variance σ^2 that is common to all of the involved populations, use the value of MS(error), which uses all of the available sample data. The value of MS(error) is typically provided with the results when conducting the analysis of variance test. Using the value of MS(error), calculate the value of the test statistic t, as shown below. The particular test statistic calculated below is based on the choice of Sample 1 and Sample 2; change the subscripts and use another pair of samples until all of the different possible pairs of samples have been tested.

$$t = \frac{\bar{x}_1 - \bar{x}_2}{\sqrt{MS(error) \cdot \left(\dfrac{1}{n_1} + \dfrac{1}{n_2}\right)}}$$

Step 3: After calculating the value of the test statistic t for a particular pair of samples, find either the critical t value or the P-value, but make the following adjustment so that the overall significance level does not increase.

P-Value Use the test statistic t with df $= N - k$, where N is the total number of sample values and k is the number of samples, and find the P-value using technology or Table A-3, but adjust the P-value by multiplying it by the number of different possible pairings of two samples. (For example, with three samples, there are three different possible pairings, so adjust the P-value by multiplying it by 3.)

Critical Value When finding the critical value, adjust the significance level α by dividing it by the number of different possible pairings of two samples. (For example, with three samples, there are three different possible pairings, so adjust the significance level by dividing it by 3.)

Note that in Step 3 of the preceding Bonferroni procedure, either an individual test is conducted with a much lower significance level or the P-value is greatly increased. Rejection of equality of means therefore requires differences that are much farther apart. This adjustment in Step 3 compensates for the fact that we are doing several tests instead of only one test.

EXAMPLE 2 Bonferroni Test

Example 1 in this section used analysis of variance with the sample data in Table 12-1. We concluded that there is sufficient evidence to warrant rejection of the claim of equal means. Use the Bonferroni test with a 0.05 significance level to identify which mean is different from the others.

SOLUTION

The Bonferroni test requires a separate t test for each of three different possible pair of samples. Here are the null hypotheses to be tested:

$$H_0: \mu_1 = \mu_2 \qquad H_0: \mu_1 = \mu_3 \qquad H_0: \mu_2 = \mu_3$$

We begin with $H_0: \mu_1 = \mu_2$. Using the sample data given in Table 12-1 and carrying some extra decimal places for greater accuracy in the calculations, we have $n_1 = 78$ and $\bar{x}_1 = 102.705128$. Also, $n_2 = 22$ and $\bar{x}_2 = 94.136364$. From the technology results shown in Example 1 we also know that MS(error) $= 248.424127$.

We now evaluate the test statistic using the unrounded sample means:

$$t = \frac{\bar{x}_1 - \bar{x}_2}{\sqrt{\text{MS(error)} \cdot \left(\frac{1}{n_1} + \frac{1}{n_2}\right)}}$$

$$= \frac{102.705128 - 94.136364}{\sqrt{248.424127 \cdot \left(\frac{1}{78} + \frac{1}{22}\right)}} = 2.252$$

The number of degrees of freedom is df $= N - k = 121 - 3 = 118$. ($N = 121$ because there are 121 different sample values in all three samples combined, and $k = 3$ because there are three different samples.) With a test statistic of $t = 2.252$ and with df $= 118$, the two-tailed P-value is 0.026172, but we adjust this P-value by multiplying it by 3 (the number of different possible pairs of samples) to get a final P-value of 0.078516, or 0.079 when rounded. Because this P-value is not small (less than 0.05), we fail to reject the null hypothesis. It appears that Samples 1 and 2 do not have significantly different means.

Instead of continuing with separate hypothesis tests for the other two pairings, see the SPSS display showing all of the Bonferroni test results. In these results, low lead levels are represented by 1, medium levels are represented by 2, and high levels are represented by 3. (The first row of numerical results corresponds to the results found here; see the value of 0.079, which was previously calculated.) The display shows that the pairing of low/high yields a P-value of 0.090, so there is not a significant difference between the means from the low and high blood lead levels. Also, the SPSS display shows that the pairing of medium/high yields a P-value of 1.000, so there is not a significant difference between the means from the medium and high blood lead levels.

SPSS Bonferroni Results

(I) Level	(J) Level	Mean Difference (I-J)	Std. Error	Sig.	95% Confidence Interval	
					Lower Bound	Upper Bound
1.00	2.00	8.56876	3.80486	.079	-.6717	17.8092
	3.00	8.51465	3.87487	.090	-.8958	17.9251
2.00	1.00	-8.56876	3.80486	.079	-17.8092	.6717
	3.00	-.05411	4.80851	1.000	-11.7320	11.6238
3.00	1.00	-8.51465	3.87487	.090	-17.9251	.8958
	2.00	.05411	4.80851	1.000	-11.6238	11.7320

INTERPRETATION

Although the analysis of variance test tells us that at least one of the means is different from the others, the Bonferroni test results do not identify any one particular sample mean that is significantly different from the others. In the original article discussing these results, the authors state that "our findings indicate that a chronic absorption of particulate lead . . . may result in subtle but statistically significant impairment in the non-verbal cognitive and perceptual motor skills measured by the performance scale of the Wechsler intelligence tests." That statement confirms these results: From analysis of variance we know that at least one mean is different from the others, but the Bonferroni test failed to identify any one particular mean as being significantly different [although the sample means of 102.7 (low blood lead level), 94.1 (medium blood lead level), and 94.2 (high blood lead level) suggest that medium and high blood lead levels seem to be associated with lower mean performance IQ scores than the low blood level group].

YOUR TURN Do Exercise 18 "Bonferroni Test."

Go Figure

Since the beginning of humankind, about 100 billion humans have been born. There are roughly 7 billion humans alive now, so about 7% of all humans are still alive.

TECH CENTER

One-Way Analysis of Variance

Access tech supplements, videos, and data sets at **www.TriolaStats.com**

Statdisk	Minitab	StatCrunch
1. Click **Analysis** in the top menu. 2. Select **One-Way Analysis of Variance** from the dropdown menu. 3. Enter the desired significance level and select at least 3 columns to be included in the analysis. 4. Click **Evaluate**.	1. Click **Stat** in the top menu. 2. Select **ANOVA** from the dropdown menu and select **One-Way** from the submenu. 3. Select **Response data are in a separate column for each factor level.** 4. In the *Response* box select the columns to be included in the analysis. 5. Click the **Options** button and check the **Assume equal variances** box. 6. Click **OK** twice.	1. Click **Stat** in the top menu. 2. Select **ANOVA** from the dropdown menu, then select **One Way** from the submenu. 3. Select the columns to be included in the analysis. 4. Click **Compute!**

TI-83/84 Plus Calculator

1. Press **STAT**, then select **TESTS** in the top menu.
2. Select **ANOVA** in the menu and press **ENTER**.
3. Enter list names that include the data to be included in the analysis. Separate list names with **,** so the command appears in the format **ANOVA(L1, L2, L3).**
4. Press **ENTER** and use the arrow buttons to scroll through the results.

Excel

XLSTAT Add-In

Requires all data to be stacked in a single column with the corresponding category name for each data value in a separate column.

1. Click on the **XLSTAT** tab in the Ribbon and then click **Modeling Data.**
2. Select **ANOVA** from the dropdown menu.
3. Enter the range of cells containing the *Y/Dependent variable* data values.
4. Select **Qualitative** box and enter the range of cells containing the qualitative values (category names) for the *X/Explanatory variable*.
5. If the first data row includes a label, check the **Variable labels** box.
6. Click **OK.** The Analysis of Variance table includes the *F* test statistic and *P*-value.

Excel Data Analysis Add-In

1. Click on the **Data** tab in the Ribbon and then select **Data Analysis** in the top menu.
2. Select **Anova: Single Factor** under *Analysis Tools* and click **OK.**
3. Enter the desired data range for **Input Range.**
4. For *Grouped By* select **Columns** if data for each category are contained in separate columns; select **Rows** if data are organized by rows.
5. Check the **Labels in First Row** box if the first cell contains a category label.
6. Click **OK** for the results, including the *F* test statistic and *P*-value.

12-1   Basic Skills and Concepts

Statistical Literacy and Critical Thinking

In Exercises 1–4, use the following listed arrival delay times (minutes) for American Airline flights from New York to Los Angeles. Negative values correspond to flights that arrived early. Also shown are the SPSS results for analysis of variance. Assume that we plan to use a 0.05 significance level to test the claim that the different flights have the same mean arrival delay time.

Flight 1	−32	−25	−26	−6	5	−15	−17	−36
Flight 19	−5	−32	−13	−9	−19	49	−30	−23
Flight 21	−23	28	103	−19	−5	−46	13	−3

SPSS

	Sum of Squares	df	Mean Square	F	Sig.
Between Groups	2575.000	2	1287.500	1.334	.285
Within Groups	20271.500	21	965.310		
Total	22846.500	23			

1. ANOVA

a. What characteristic of the data above indicates that we should use *one-way* analysis of variance?

b. If the objective is to test the claim that the three flights have the same *mean* arrival delay time, why is the method referred to as analysis of *variance?*

2. Why Not Test Two at a Time? Refer to the sample data given in Exercise 1. If we want to test for equality of the three means, why don't we use three separate hypothesis tests for $\mu_1 = \mu_2$, $\mu_1 = \mu_3$, and $\mu_2 = \mu_3$?

3. Test Statistic What is the value of the test statistic? What distribution is used with the test statistic?

4. *P*-Value If we use a 0.05 significance level in analysis of variance with the sample data given in Exercise 1, what is the *P*-value? What should we conclude? If a passenger abhors late flight arrivals, can that passenger be helped by selecting one of the flights?

In Exercises 5–16, use analysis of variance for the indicated test.

5. Lead and Verbal IQ Scores Example 1 used measured *performance* IQ scores for three different blood lead levels. If we use the same three categories of blood lead levels with measured *verbal* IQ scores, we get the accompanying Minitab display. (The data are listed in Data Set 7 "IQ and Lead" in Appendix B.) Using a 0.05 significance level, test the claim that the three categories of blood lead level have the same mean verbal IQ score. Does exposure to lead appear to have an effect on verbal IQ scores?

Minitab

Source	DF	SS	MS	F	P
LEAD	2	142	71	0.39	0.677
Error	118	21441	182		
Total	120	21584			

6. Lead and Full IQ Scores Example 1 used measured *performance* IQ scores for three different blood lead levels. If we use the same three categories of blood lead levels with the *full* IQ scores, we get the accompanying Excel display. (The data are listed in Data Set 7 "IQ and Lead" in Appendix B.) Using a 0.05 significance level, test the claim that the three categories of blood lead level have the same mean full IQ score. Does it appear that exposure to lead has an effect on full IQ scores?

Excel

ANOVA						
Source of Variation	SS	df	MS	F	P-value	F crit
Between Groups	938.3653	2	469.1827	2.303395	0.104395	3.07309
Within Groups	24035.63	118	203.6918			
Total	24974	120				

7. Fast Food Dinner Service Times Data Set 25 "Fast Food" in Appendix B lists drive-through service times (seconds) for dinners at McDonald's, Burger King, and Wendy's. Using those times with a TI-83/84 Plus calculator yields the following display. Using a 0.05

continued

significance level, test the claim that the three samples are from populations with the same mean. What do you conclude?

TI-83/84 Plus

8. Birth Weights Data Set 4 "Births" in Appendix B lists birth weights from babies born at Albany Medical Center, Bellevue Hospital in New York City, Olean General Hospital, and Strong Memorial Hospital in Rochester, New York. After partitioning the birth weights according to the hospital, we get the StatCrunch display shown here. Use a 0.05 significance level to test the claim that the different hospitals have different mean birth weights. Do birth weights appear to be different in urban and rural areas?

StatCrunch

ANOVA table

Source	DF	SS	MS	F-Stat	P-value
Columns	3	1701400	567133.33	1.1810493	0.3167
Error	396	1.90157e8	480194.44		
Total	399	1.918584e8			

9. Female Pulse Rates and Age Using the pulse rates of females from Data Set 1 "Body Data" in Appendix B after they are partitioned into the three age brackets of 18–25, 26–40, and 41–80, we get the following Statdisk display. Using a 0.05 significance level, test the claim that females from the three age brackets have the same mean pulse rate. What do you conclude?

Statdisk

Source:	DF:	SS:	MS:	Test Stat, F:	Critical F:	P-Value:
Treatment:	2	2280.049935	1140.024967	7.933788	3.058925	0.000539
Error:	144	20691.705167	143.692397			
Total:	146	22971.755102				

10. Male Pulse Rates and Age Using the pulse rates of males from Data Set 1 "Body Data" in Appendix B after they are partitioned into the three age brackets of 18–25, 26–40, and 41–80, we get the following SPSS display. Using a 0.05 significance level, test the claim that males from the three age brackets have the same mean pulse rate. What do you conclude?

SPSS

	Sum of Squares	df	Mean Square	F	Sig.
Between Groups	333.464	2	166.732	1.304	.275
Within Groups	19183.765	150	127.892		
Total	19517.229	152			

11. Triathlon Times Jeff Parent is a statistics instructor who participates in triathlons. Listed below are times (in minutes and seconds) he recorded while riding a bicycle for five stages through each mile of a 3-mile loop. Use a 0.05 significance level to test the claim that it takes the same time to ride each of the miles. Does one of the miles appear to have a hill?

Mile 1	3:15	3:24	3:23	3:22	3:21
Mile 2	3:19	3:22	3:21	3:17	3:19
Mile 3	3:34	3:31	3:29	3:31	3:29

12. Arsenic in Rice Listed below are amounts of arsenic in samples of brown rice from three different states. The amounts are in micrograms of arsenic and all samples have the same serving size. The data are from the Food and Drug Administration. Use a 0.05 significance level to test the claim that the three samples are from populations with the same mean. Do the amounts of arsenic appear to be different in the different states? Given that the amounts of arsenic in the samples from Texas have the highest mean, can we conclude that brown rice from Texas poses the greatest health problem?

Arkansas	4.8	4.9	5.0	5.4	5.4	5.4	5.6	5.6	5.6	5.9	6.0	6.1
California	1.5	3.7	4.0	4.5	4.9	5.1	5.3	5.4	5.4	5.5	5.6	5.6
Texas	5.6	5.8	6.6	6.9	6.9	6.9	7.1	7.3	7.5	7.6	7.7	7.7

13. Flight Departure Delays Listed below are departure delay times (minutes) for American Airline flights from New York to Los Angeles. Negative values correspond to flights that departed early. Use a 0.05 significance level to test the claim that the different flights have the same mean departure delay time. What notable feature of the data can be identified by visually examining the data?

Flight 1	−2	−1	−2	2	−2	0	−2	−3
Flight 19	19	−4	−5	−1	−4	73	0	1
Flight 21	18	60	142	−1	−11	−1	47	13

14. Speed Dating Listed below are attribute ratings of males by females who participated in speed dating events (from Data Set 18 "Speed Dating" in Appendix B). Use a 0.05 significance level to test the claim that females in the different age brackets give attribute ratings with the same mean. Does age appear to be a factor in the female attribute ratings?

Age 20–22	38	42	30.0	39	47	43	33	31	32	28
Age 23–26	39	31	36.0	35	41	45	36	23	36	20
Age 27–29	36	42	35.5	27	37	34	22	47	36	32

In Exercises 15 and 16, use the data set in Appendix B.

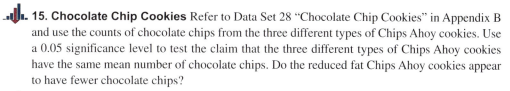 **15. Chocolate Chip Cookies** Refer to Data Set 28 "Chocolate Chip Cookies" in Appendix B and use the counts of chocolate chips from the three different types of Chips Ahoy cookies. Use a 0.05 significance level to test the claim that the three different types of Chips Ahoy cookies have the same mean number of chocolate chips. Do the reduced fat Chips Ahoy cookies appear to have fewer chocolate chips?

16. Secondhand Smoke Refer to Data Set 12 "Passive and Active Smoke" in Appendix B and use the measured serum cotinine levels (in mg/mL) from the three groups of subjects (smokers, nonsmokers exposed to tobacco smoke, and nonsmokers not exposed to tobacco smoke). When nicotine is absorbed by the body, cotinine is produced. Use a 0.05 significance level to test the claim that the three samples are from populations with the same mean. What do the results suggest about the effects of secondhand smoke?

12-1 Beyond the Basics

17. Tukey Test A display of the Bonferroni test results from Table 12-1 (which is part of the Chapter Problem) is provided on page 577. Shown on the top of the next page is the SPSS-generated display of results from the Tukey test using the same data. Compare the Tukey test results to those from the Bonferroni test.

continued

SPSS

(I) Level	(J) Level	Mean Difference (I-J)	Std. Error	Sig.	95% Confidence Interval	
					Lower Bound	Upper Bound
1.00	2.00	8.56876	3.80486	.067	-.4626	17.6002
	3.00	8.51465	3.87487	.076	-.6830	17.7123
2.00	1.00	-8.56876	3.80486	.067	-17.6002	.4626
	3.00	-.05411	4.80851	1.000	-11.4678	11.3596
3.00	1.00	-8.51465	3.87487	.076	-17.7123	.6830
	2.00	.05411	4.80851	1.000	-11.3596	11.4678

18. Bonferroni Test Shown below are weights (kg) of poplar trees obtained from trees planted in a rich and moist region. The trees were given different treatments identified in the table below. The data are from a study conducted by researchers at Pennsylvania State University and were provided by Minitab, Inc. Also shown are partial results from using the Bonferroni test with the sample data.

No Treatment	Fertilizer	Irrigation	Fertilizer and Irrigation
1.21	0.94	0.07	0.85
0.57	0.87	0.66	1.78
0.56	0.46	0.10	1.47
0.13	0.58	0.82	2.25
1.30	1.03	0.94	1.64

a. Use a 0.05 significance level to test the claim that the different treatments result in the same mean weight.

b. What do the displayed Bonferroni SPSS results tell us?

c. Use the Bonferroni test procedure with a 0.05 significance level to test for a significant difference between the mean amount of the irrigation treatment group and the group treated with both fertilizer and irrigation. Identify the test statistic and either the *P*-value or critical values. What do the results indicate?

Bonferroni Results from SPSS

(I) TREATMENT	(J) TREATMENT	Mean Difference (I-J)	Std. Error	Sig.	95% Confidence Interval	
					Lower Bound	Upper Bound
1.00	2.00	-.02200	.26955	1.000	-.8329	.7889
	3.00	.23600	.26955	1.000	-.5749	1.0469
	4.00	-.84400*	.26955	.039	-1.6549	-.0331

12-2 Two-Way ANOVA

Key Concept Section 12-1 considered data partitioned using *one* factor, but this section describes the method of *two-way analysis of variance,* which is used with data partitioned into categories according to *two* factors. The method of this section requires that we first test for an *interaction* between the two factors; then we test for an effect from the row factor, and we test for an effect from the column factor.

Table 12-3 is an example of pulse rates (beats per minute) categorized with *two* factors:

TABLE 12-3 Pulse Rates with Two Factors: Age Bracket and Gender

	Female	Male
18–29	104 82 80 78 80 84 82 66 70 78	72 64 72 64 64 70 72 64 54 52
30–49	66 74 96 86 98 88 82 72 80 80	80 90 58 74 96 72 58 66 80 92
50-80	94 72 82 86 72 90 64 72 72 100	54 102 52 52 62 82 82 60 52 74

1. Age Bracket (years): One factor is age bracket (18–29, 30–49, 50–80).

2. Gender: The second factor is gender (female, male).

The subcategories in Table 12-3 are called *cells,* so Table 12-3 has six cells containing ten values each.

   In analyzing the sample data in Table 12-3, we have already discussed one-way analysis of variance for a single factor, so it might seem reasonable to simply proceed with one-way ANOVA for the factor of age bracket and another one-way ANOVA for the factor of gender, but that approach wastes information and totally ignores a very important feature: the possible effect of an *interaction* between the two factors.

> **DEFINITION**
>
> There is an **interaction** between two factors if the effect of one of the factors changes for different categories of the other factor.

   As an example of an *interaction* between two factors, consider food pairings. Peanut butter and jelly interact well, but ketchup and ice cream interact in a way that results in a bad taste, so we rarely see someone eating ice cream topped with ketchup. In general, consider an interaction effect to be an effect due to the combination of the two factors.

Explore Data with Means and an Interaction Graph

Let's explore the data in Table 12-3 by calculating the mean for each cell and by constructing a graph. The individual cell means are shown in Table 12-4. Those means vary from a low of 64.8 to a high of 82.2, so they vary considerably. Figure 12-3 is an *interaction graph,* which shows graphs of those means. We can interpret an interaction graph as follows:

■ **Interaction Effect:** An interaction effect is suggested when line segments are far from being parallel.

■ **No Interaction Effect:** If the line segments are approximately *parallel,* as in Figure 12-3, it appears that the different categories of a variable have the same effect for the different categories of the other variable, so there does not appear to be an interaction effect.

TABLE 12-4 Means of Cells from Table 12-3

	Female	Male
18–29	80.4	64.8
30–49	82.2	76.6
50-80	80.4	67.2

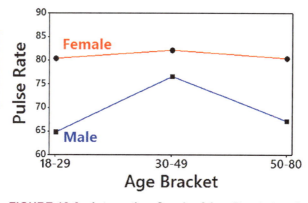

FIGURE 12-3 **Interaction Graph of Age Bracket and Gender: Means from Table 12-4**

Instead of relying only on subjective judgments made by examining the means in Table 12-4 and the interaction graph in Figure 12-3, we will proceed with the more objective procedure of two-way analysis of variance. Here are the requirements and basic procedure for two-way analysis of variance (ANOVA). The procedure is also summarized in Figure 12-4, which follows the Key Elements box on the next page.

Two-Way Analysis of Variance

Objective

With sample data categorized with two factors (a row variable and a column variable), use two-way analysis of variance to conduct the following three tests:

1. Test for an effect from an interaction between the row factor and the column factor.

2. Test for an effect from the row factor.

3. Test for an effect from the column factor.

Requirements

1. Normality For each cell, the sample values come from a population with a distribution that is approximately normal. (This procedure is robust against reasonable departures from normal distributions.)

2. Variation The populations have the same variance σ^2 (or standard deviation σ). (This procedure is robust against reasonable departures from the requirement of equal variances.)

3. Sampling The samples are simple random samples of quantitative data.

4. Independence The samples are independent of each other. (This procedure does not apply to samples lacking independence.)

5. Two-Way The sample values are categorized two ways. (This is the basis for the name of the method: *two-way* analysis of variance.)

6. Balanced Design All of the cells have the same number of sample values. (This is called a *balanced* design. This section does not include methods for a design that is not balanced.)

Procedure for Two-Way ANOVA (See Figure 12-4)

Step 1: *Interaction Effect:* In two-way analysis of variance, begin by testing the null hypothesis that there is no interaction between the two factors. Use technology to find the *P*-value corresponding to the following test statistic:

$$F = \frac{\text{MS(interaction)}}{\text{MS(error)}}$$

Conclusion:

- **Reject:** If the *P*-value corresponding to the above test statistic is small (such as less than or equal to 0.05), reject the null hypothesis of no interaction. Conclude that there is an interaction effect.

- **Fail to Reject:** If the *P*-value is large (such as greater than 0.05), fail to reject the null hypothesis of no interaction between the two factors. Conclude that there is no interaction effect.

Step 2: *Row/Column Effects:* If we conclude that there is an interaction effect, then we should stop now; we should not proceed with the two additional tests. (If there is an interaction between factors, we shouldn't consider the effects of either factor without considering those of the other.)

If we conclude that there is no interaction effect, then we should proceed with the following two hypothesis tests.

Row Factor

For the row factor, test the null hypothesis H_0: There are no effects from the row factor (that is, the row values are from populations with the same mean). Find the *P*-value corresponding to the test statistic $F = \text{MS(row)}/\text{MS(error)}$.

Conclusion:

- **Reject:** If the *P*-value corresponding to the test statistic is small (such as less than or equal to 0.05), reject the null hypothesis of no effect from the row factor. Conclude that there is an effect from the row factor.

- **Fail to Reject:** If the *P*-value is large (such as greater than 0.05), fail to reject the null hypothesis of no effect from the row factor. Conclude that there is no effect from the row factor.

Column Factor

For the column factor, test the null hypothesis H_0: There are no effects from the column factor (that is, the column values are from populations with the same mean). Find the *P*-value corresponding to the test statistic $F = MS(\text{column})/MS(error)$.

Conclusion:

- **Reject:** If the *P*-value corresponding to the test statistic is small (such as less than or equal to 0.05), reject the null hypothesis of no effect from the column factor. Conclude that there is an effect from the column factor.

- **Fail to Reject:** If the *P*-value is large (such as greater than 0.05), fail to reject the null hypothesis of no effect from the column factor. Conclude that there is no effect from the column factor.

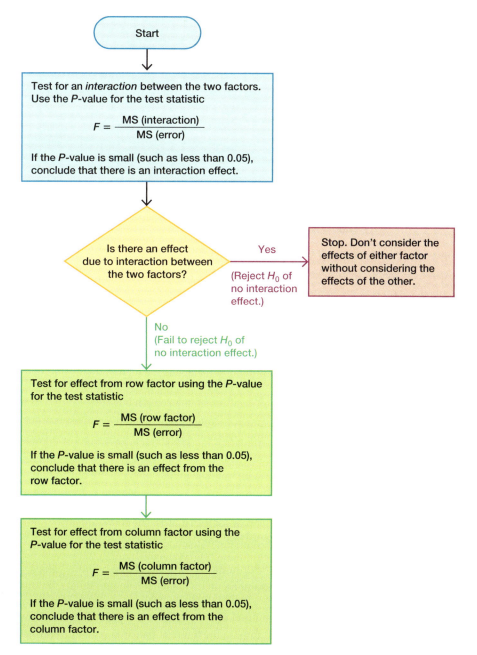

FIGURE 12-4 **Procedure for Two-Way Analysis of Variance**

EXAMPLE 1 Pulse Rates

Given the pulse rates in Table 12-3 on page 582 (from Data Set 1 "Body Data" in Appendix B), use two-way analysis of variance to test for an interaction effect, an effect from the row factor of age bracket, and an effect from the column factor of gender. Use a 0.05 significance level.

SOLUTION

REQUIREMENT CHECK (1) For each cell, the sample values appear to be from a population with a distribution that is approximately normal, as indicated by normal quantile plots. (2) The variances of the cells (100.3, 51.7, 103.5, 183.2, 138.5, 293.5) differ considerably, but the test is robust against departures from equal variances. (3) The samples are simple random samples of subjects. (4) The samples are independent of each other; the subjects are not matched in any way. (5) The sample values are categorized in two ways (age bracket and gender). (6) All of the cells have the same number (ten) of sample values. The requirements are satisfied. ☑

The calculations are quite involved, so we use technology. The StatCrunch two-way analysis of variance display for the data in Table 12-3 is shown here.

StatCrunch

ANOVA table

Source	DF	SS	MS	F-Stat	P-value
Age	2	526.93333	263.46667	1.8156202	0.1725
Gender	1	1972.2667	1972.2667	13.591424	0.0005
Interaction	2	272.53333	136.26667	0.93905054	0.3973
Error	54	7836	145.11111		
Total	59	10607.733			

Step 1: Interaction Effect: We begin by testing the null hypothesis that there is no interaction between the two factors. Using StatCrunch for the data in Table 12-3, we get the results shown in the preceding StatCrunch display and we can see that the test statistic for the interaction is $F = 0.9391$. This test statistic can be calculated as follows:

$$F = \frac{MS(\text{interaction})}{MS(\text{error})} = \frac{136.26667}{145.11111} = 0.9391$$

Interpretation: The corresponding P-value is shown in the StatCrunch display as 0.3973, so we fail to reject the null hypothesis of no interaction between the two factors. It does not appear that pulse rates are affected by an interaction between age bracket (18–29, 30–49, 50–80) and gender. There does not appear to be an interaction effect.

Step 2: Row/Column Effects: Because there does not appear to be an interaction effect, we proceed to test for effects from the row and column factors. The two hypothesis tests use these null hypotheses:

H_0: There are no effects from the row factor (that is, the row values are from populations with equal means).

H_0: There are no effects from the column factor (that is, the column values are from populations with equal means).

Row Factor: For the row factor (age bracket), we refer to the preceding StatCrunch display of results to see that the test statistic for the row factor is $F = 1.8156$ (rounded). This test statistic can be calculated as follows:

$$F = \frac{\text{MS(age bracket)}}{\text{MS(error)}} = \frac{263.46667}{145.11111} = 1.8156$$

Conclusion: The corresponding *P*-value is shown in the StatCrunch display as 0.1725. Because that *P*-value is greater than the significance level of 0.05, we fail to reject the null hypothesis of no effects from age bracket. That is, pulse rates do not appear to be affected by the age bracket.

Column Factor: For the column factor (gender), we refer to the preceding StatCrunch display of results to see that the test statistic for the column factor is $F = 13.5914$ (rounded). This test statistic can be calculated as follows:

$$F = \frac{\text{MS(gender)}}{\text{MS(error)}} = \frac{1972.2667}{145.11111} = 13.5914$$

Conclusion: The corresponding *P*-value is shown in the StatCrunch display as 0.0005. Because that *P*-value is less than the significance level of 0.05, we reject the null hypothesis of no effects from gender. Pulse rates do appear to be affected by gender.

> **INTERPRETATION**

On the basis of the sample data in Table 12-3, we conclude that pulse rates appear to be affected by gender, but not by age bracket and not by an interaction between age bracket and gender.

YOUR TURN Do Exercise 5 "Pulse Rates."

CAUTION Two-way analysis of variance is not one-way analysis of variance done twice. When conducting a two-way analysis of variance, be sure to test for an *interaction* between the two factors.

TECH CENTER

Two-Way Analysis of Variance
Access tech supplements, videos, and data sets at **www.TriolaStats.com**

Statdisk	Minitab	StatCrunch
1. Click **Analysis** in the top menu. 2. Select **Two-Way Analysis of Variance** from the dropdown menu. 3. Enter the number of categories for row variables and column variables. 4. Enter the number of values in each cell and click **Continue**. 5. In the table, enter or paste the data in the *Value* column. 6. Click **Evaluate**.	1. Enter all of the sample values in column C1. 2. Enter the corresponding row numbers (or names) in column C2. 3. Enter the corresponding column numbers (or names) in column C3. 4. Click **Stat** in the top menu. 5. Select **ANOVA** from the dropdown menu and select **General Linear Model—Fit General Linear Model**. 6. For *Responses* select **C1** and select **C2** and **C3** as *Factors*. 7. Click the **Model** button. 8. Under *Factors and covariates* select **C2** and **C3** and click the **Add** button. 9. Click **OK** twice. See *Analysis of Variance* in the results. *TIP:* Use descriptive labels rather than C1, C2, and C3 to avoid confusion.	1. Enter all sample values in one column named "Responses." 2. Enter corresponding *row* numbers (or names) in a second column named "Row Factor." 3. Enter the corresponding *column* numbers (or names) in a third column named "Column Factor." 4. Click **Stat** in the top menu. 5. Select **ANOVA** from the dropdown menu, then select **Two Way** from the submenu. 6. Select the columns to be used for responses, row factor and column factor. 6. Click **Compute!**

continued

TECH CENTER *continued*

Two-Way Analysis of Variance

Access tech supplements, videos, and data sets at **www.TriolaStats.com**

TI-83/84 Plus Calculator

Requires program *A1ANOVA* (available at www.TriolaStats.com).

1. The program A1ANOVA requires that we first create a Matrix [D] containing the sample data:
   Manually enter data: Press **2ND** then **x⁻¹** to get to the *MATRIX* menu, select **EDIT** from the top menu, select **[D]**, and press **ENTER**.
   Enter the number of rows and columns needed, press **ENTER**, and proceed to enter the sample values.

   Using existing lists: Lists can be combined and stored in a matrix. Press **2ND** then **x⁻¹** to get to the *MATRIX* menu, select **MATH** from the top menu, and select the item **List → matr**. Enter the list names, followed by the matrix name **[D]**, all separated by commas. *Important:* The matrix name must be entered by pressing **2ND** then **x⁻¹**, selecting **[D]**, and pressing **ENTER**. The following is a summary of the commands used to create a matrix from three lists (L1, L2, L3):
   List → matr(L1, L2, L3,[D]).

2. Press **PRGM**, then select **A1ANOVA** and press **ENTER** twice.

3. Select **RAN BLOCK DESIGN** and press **ENTER** twice. Select **Continue** and press **ENTER**.

4. The program will work with data in Matrix **[D]** and display the results. The results do not fit on a single screen, so press **ENTER** to see the remaining results.

TIP: In the results, F(A) is the *F* test statistic for the row factor, F(B) is the *F* test statistic for the column factor, and F(AB) is the *F* test statistic for the interaction effect.

Excel

XLSTAT Add-In

Requires all data to be stacked in a single column with the corresponding category names for each data value in two separate and adjacent columns. The row names should be in one of those columns and the column names should be in the other column.

1. Click on the **XLSTAT** tab in the Ribbon and then click **Modeling Data.**

2. Select **ANOVA** from the dropdown menu.

3. Enter the range of cells containing sample values in the *Y/Dependent variables* box.

4. Select **Qualitative** box and enter the range of cells containing the row and column names in the *X/Explanatory variables* box, such as B1:C30.

5. If a variable label is included in the data range, check the **Variable labels** box.

6. Click the **Options** tab and confirm the **Interactions/Level** box is checked and set to **2.**

7. Click the **Output** tab and check the box labeled **Type I/II/III SS.**

8. Click **OK.** Click **All** in the *Factors and interactions* window and click **OK** for the results. Look for key results under the heading of "Type I Sum of Squares analysis." *P*-values are labeled *Pr > F.*

Excel Data Analysis Add-In
More than one entry per cell

For two-way tables with more than one entry per cell, entries from the same cell must be listed down a column, not across a row. Enter the labels corresponding to the data set in column A and row 1, as shown in this example:

	A	B	C	D
1		Low	Medium	High
2	Male	85	78	93
3	Male	90	107	97
⋮	⋮	⋮	⋮	⋮

1. Click on the **Data** tab in the Ribbon and then select **Data Analysis** in the top menu.

2. Select **Anova: Two-Factor With Replication** under *Analysis Tools* and click **OK.**

3. Enter the desired data range for **Input Range.**

4. In *Rows per sample* enter the number of values in each cell.

5. Click **OK.**

12-2   Basic Skills and Concepts

Statistical Literacy and Critical Thinking

1. Two-Way ANOVA The pulse rates in Table 12-3 from Example 1 are reproduced below with fabricated data (in **red**) used for the pulse rates of females aged 30–49. What characteristic of the data suggests that the appropriate method of analysis is *two-way* analysis of variance? That is, what is "two-way" about the data entered in this table?

	Female										Male									
18–29	104	82	80	78	80	84	82	66	70	78	72	64	72	64	64	70	72	64	54	52
30–49	46	54	76	66	78	68	62	52	60	60	80	90	58	74	96	72	58	66	80	92
50–80	94	72	82	86	72	90	64	72	72	100	54	102	52	52	62	82	82	60	52	74

2. Two-Way ANOVA If we have a goal of using the data described in Exercise 1 to (1) determine whether age bracket has an effect on pulse rates and (2) to determine whether gender has an effect on pulse rates, should we use one-way analysis of variance for the two individual tests? Why or why not?

3. Interaction

a. What is an interaction between two factors?

b. In general, when using two-way analysis of variance, if we find that there is an interaction effect, how does that affect the procedure?

c. Shown below is an interaction graph constructed from the data in Exercise 1. What does the graph suggest?

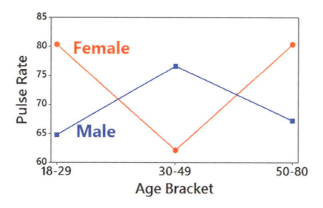

4. Balanced Design Does the table given in Exercise 1 constitute a *balanced design?* Why or why not?

5. Pulse Rates If we use the data given in Exercise 1 with two-way analysis of variance, we get the accompanying display. What do you conclude?

Statdisk

Source:	DF:	SS:	MS:	Test Stat, F:	Critical F:	P-Value:
Interaction:	2	2779.2	1389.6	9.58	3.1682	0.0003
Row Variable:	2	206.9333	103.4667	0.7130	3.1682	0.4947
Column Variable:	1	345.6	345.6	2.3816	4.0195	0.1286

6. Weights The weights (kg) in the following table are from Data Set 1 "Body Data" in Appendix B. Results from two-way analysis of variance are also shown. Use the displayed results and use a 0.05 significance level. What do you conclude?

	Female						Male				
18–29	63.4	57.8	52.6	46.9	61.7	61.5	71.6	64.9	144.9	96.4	80.7	84.4
	77.2	50.4	97.0	76.1			63.9	79.0	99.4	64.1		
30–49	110.5	84.6	133.3	90.2	125.7	105.3	96.2	56.4	107.4	99.5	64.8	94.7
	115.5	75.3	92.8	57.7			74.2	112.8	72.6	91.4		
50–80	103.2	48.3	87.8	101.3	67.8	45.2	84.8	127.5	89.9	75.3	110.2	72.3
	79.8	60.1	68.5	43.3			77.2	86.5	71.3	73.1		

StatCrunch

ANOVA table

Source	DF	SS	MS	F-Stat	P-value
Age	2	3727.7583	1863.8792	4.3546612	0.0176
Gender	1	1013.526	1013.526	2.3679444	0.1297
Interaction	2	3137.611	1568.8055	3.6652678	0.0322
Error	54	23113.044	428.01933		
Total	59	30991.939			

7. Heights The heights (cm) in the following table are from Data Set 1 "Body Data" in Appendix B. Results from two-way analysis of variance are also shown. Use the displayed results and use a 0.05 significance level. What do you conclude?

	Female					Male				
18–29	161.2	170.2	162.9	155.5	168.0	172.8	178.7	183.1	175.9	161.8
	153.3	152.0	154.9	157.4	159.5	177.5	170.5	180.1	178.6	178.5
30–49	169.1	170.6	171.1	159.6	169.8	170.1	165.4	178.5	168.5	180.3
	169.5	156.5	164.0	164.8	155.6	178.2	174.4	174.6	162.8	174.4
50–80	146.7	160.9	163.3	176.1	163.1	181.9	166.6	171.7	170.0	169.1
	151.6	164.7	153.3	160.3	134.5	182.9	176.3	166.7	166.3	160.5

XLSTAT

Source	DF	Sum of squares	Mean squares	F	Pr > F
Age	2	222.0610	111.0305	2.0403	0.1399
Gender (1=M)	1	2365.0482	2365.0482	43.4607	<0.0001
Age*Gender (1=M)	2	195.5803	97.7902	1.7970	0.1756

8. Pancake Experiment Listed below are ratings of pancakes made by experts (based on data from Minitab). Different pancakes were made with and without a supplement and with different amounts of whey. The results from two-way analysis of variance are shown. Use the displayed results and a 0.05 significance level. What do you conclude?

	Whey											
	0%			10%			20%			30%		
No Supplement	4.4	4.5	4.3	4.6	4.5	4.8	4.5	4.8	4.8	4.6	4.7	5.1
Supplement	3.3	3.2	3.1	3.8	3.7	3.6	5.0	5.3	4.8	5.4	5.6	5.3

Minitab

Two-way ANOVA: Quality versus Supplement, Whey

Source	DF	SS	MS	F	P
Supplement	1	0.5104	0.51042	17.01	0.001
Whey	3	6.6912	2.23042	74.35	0.000
Interaction	3	3.7246	1.24153	41.38	0.000
Error	16	0.4800	0.03000		
Total	23	11.4062			

9. Marathon Times Listed below are New York City Marathon running times (in seconds) for randomly selected runners who completed the marathon. Are the running times affected by an interaction between gender and age bracket? Are running times affected by gender? Are running times affected by age bracket? Use a 0.05 significance level.

Times (in seconds) for New York City Marathon Runners

	Age		
	21–29	**30–39**	**40 and over**
Male	13,615	14,677	14,528
	18,784	16,090	17,034
	14,256	14,086	14,935
	10,905	16,461	14,996
	12,077	20,808	22,146
Female	16,401	15,357	17,260
	14,216	16,771	25,399
	15,402	15,036	18,647
	15,326	16,297	15,077
	12,047	17,636	25,898

10. Smoking, Gender, and Body Temperature The table below lists body temperatures obtained from randomly selected subjects (based on Data Set 3 "Body Temperatures" in Appendix B). Using a 0.05 significance level, test for an interaction between gender and smoking, test for an effect from gender, and test for an effect from smoking. What do you conclude?

	Smokes				Does not smoke			
Male	98.8	97.6	98.0	98.5	98.4	97.8	98.0	97.0
Female	98.0	98.5	98.3	98.7	97.7	98.0	98.2	99.1

12-2 Beyond the Basics

11. Transformations of Data Example 1 illustrated the use of two-way ANOVA to analyze the sample data in Table 12-3 on page 582. How are the results affected in each of the following cases?

a. The same constant is added to each sample value.

b. Each sample value is multiplied by the same nonzero constant.

c. The format of the table is transposed so that the row and column factors are interchanged.

d. The first sample value in the first cell is changed so that it becomes an outlier.

Chapter Quick Quiz

1. Cola Weights Data Set 26 "Cola Weights and Volumes" in Appendix B lists the weights (lb) of the contents of cans of cola from four different samples: (1) regular Coke, (2) Diet Coke, (3) regular Pepsi, and (4) Diet Pepsi. The results from analysis of variance are shown on the top of the next page. What is the null hypothesis for this analysis of variance test? Based on the displayed results, what should you conclude about H_0? What do you conclude about equality of the mean weights from the four samples?

continued

Minitab

Source	DF	Adj SS	Adj MS	F-Value	P-Value
Factor	3	0.047979	0.015993	503.06	0.000
Error	140	0.004451	0.000032		
Total	143	0.052430			

2. Cola Weights For the four samples described in Exercise 1, the sample of regular Coke has a mean weight of 0.81682 lb, the sample of Diet Coke has a mean weight of 0.78479 lb, the sample of regular Pepsi has a mean weight of 0.82410 lb, and the sample of Diet Pepsi has a mean weight of 0.78386 lb. If we use analysis of variance and reach a conclusion to reject equality of the four sample means, can we then conclude that any of the specific samples have means that are significantly different from the others?

3. Cola Weights For the analysis of variance test described in Exercise 1, is that test left-tailed, right-tailed, or two-tailed?

4. Cola Weights Identify the value of the test statistic in the display included with Exercise 1. In general, do larger test statistics result in larger P-values, smaller P-values, or P-values that are unrelated to the value of the test statistic?

5. Cola Weights The displayed results from Exercise 1 are from one-way analysis of variance. What is it about this test that characterizes it as one-way analysis of variance instead of two-way analysis of variance?

6. One-Way ANOVA In general, what is one-way analysis of variance used for?

7. One vs. Two What is the fundamental difference between one-way analysis of variance and two-way analysis of variance?

8. Estimating Length Given below is a Minitab display resulting from two-way analysis of variance with sample data consisting of 18 different student visual estimates of the length of a classroom. The values are categorized according to sex and major (math, business, liberal arts). What do you conclude about an interaction between sex and major?

Minitab

Source	DF	SS	MS	F	P
Sex	1	29.389	29.3889	0.78	0.395
Major	2	10.111	5.0556	0.13	0.876
Interaction	2	14.111	7.0556	0.19	0.832
Error	12	453.333	37.7778		
Total	17	506.944			

9. Estimating Length Using the same results displayed in Exercise 8, does it appear that the length estimates are affected by the sex of the subject?

10. Estimating Length Using the same results displayed in Exercise 8, does it appear that the length estimates are affected by the subject's major?

Review Exercises

1. Speed Dating Data Set 18 "Speed Dating" in Appendix B lists attribute ratings of females by males who participated in speed dating events, and some of those values are included in the table on the top of the next page. Analysis of variance is used with the values in that table, and the StatCrunch results are shown on the next page following the data.. Use a 0.05 significance level to test the claim that males in the different age brackets give attribute ratings with the same mean. Does age appear to be a factor in the male attribute ratings?

Age 20–22	32	34	37	40.5	33	28	31	50	39	41
Age 23–26	40	21	14	32	26	34	31	34	34	34
Age 27–29	31	39	27	34	43	31	30	38	37	34

StatCrunch

ANOVA table

Source	DF	SS	MS	F-Stat	P-value
Columns	2	222.95	111.475	2.7346508	0.0829
Error	27	1100.625	40.763889		
Total	29	1323.575			

2. Author Readability Pages were randomly selected by the author from *The Bear and the Dragon* by Tom Clancy, *Harry Potter and the Sorcerer's Stone* by J. K. Rowling, and *War and Peace* by Leo Tolstoy. The Flesch Reading Ease scores for those pages are listed below. Do the authors appear to have the same level of readability?

Clancy	58.2	73.4	73.1	64.4	72.7	89.2	43.9	76.3	76.4	78.9	69.4	72.9
Rowling	85.3	84.3	79.5	82.5	80.2	84.6	79.2	70.9	78.6	86.2	74.0	83.7
Tolstoy	69.4	64.2	71.4	71.6	68.5	51.9	72.2	74.4	52.8	58.4	65.4	73.6

3. Car Crash Tests Data Set 19 "Car Crash Tests" in Appendix B lists results from car crash tests. The data set includes crash test loads (pounds) on the left femur and right femur. When those loads are partitioned into the three car size categories of small, midsize, and large, the two-way analysis of results from XLSTAT are as shown below. (The row factor of femur has the two values of left femur and right femur, and the column factor of size has the three values of small, midsize, and large.) Use a 0.05 significance level to apply the methods of two-way analysis of variance. What do you conclude?

XLSTAT

Source	DF	Sum of squares	Mean squares	F	Pr > F
Femur	1	166068.5952	166068.5952	1.3896	0.2462
Size	2	532911.8571	266455.9286	2.2296	0.1222
Femur*Size	2	410435.7619	205217.8810	1.7171	0.1940

4. Speed Dating Listed below are attribute ratings of males by females who participated in speed dating events (from Data Set 18 "Speed Dating" in Appendix B). Use a 0.05 significance level to apply the methods of two-way analysis of variance. What do you conclude?

		Males														
		20–23					24–26					27–30				
Females	20–23	42	24	40	32	30	37	47	33	32	21	22	30	32	43	28
	24–26	34	31	25	36	30	36	41	36	33	48	34	32	27	43	35
	27–30	35	31	40	31	32	36	25	42	28	42	37	40	21	34	23

Cumulative Review Exercises

In Exercises 1–5, refer to the following list of departure delay times (min) of American Airline flights from JFK airport in New York to LAX airport in Los Angeles. Assume that the data are samples randomly selected from larger populations.

Flight 3	22	−11	7	0	−5	3	−8	8
Flight 19	19	−4	−5	−1	−4	73	0	1
Flight 21	18	60	142	−1	−11	−1	47	13

1. Exploring the Data Include appropriate units in all answers.

a. Find the mean for each of the three flights.

b. Find the standard deviation for each of the three flights.

c. Find the variance for each of the three flights.

d. Are there any obvious outliers?

e. What is the level of measurement of the data (nominal, ordinal, interval, ratio)?

2. Comparing Two Means Treating the data as samples from larger populations, test the claim that there is a difference between the mean departure delay time for Flight 3 and Flight 21.

3. Normal Quantile Plot The accompanying normal quantile plot was obtained from the Flight 19 departure delay times. What does this graph tell us?

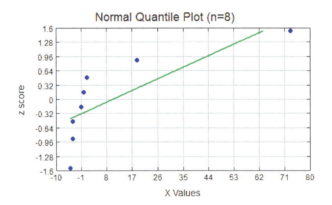

4. Confidence Interval Use the departure delay times for Flight 3 and construct a 95% confidence interval estimate of the population mean. Write a brief statement that interprets the confidence interval.

5. ANOVA The XLSTAT display below results from using the one-way analysis of variance test with the three samples.

a. What is the null hypothesis?

b. Assuming a 0.05 significance level, what conclusion is indicated by the displayed results?

XLSTAT

Source	DF	Sum of squares	Mean squares	F	Pr > F
Q1	2	4263.0833	2131.5417	1.9104	0.1729

6. Quarters Assume that weights of quarters minted after 1964 are normally distributed with a mean of 5.670 g and a standard deviation of 0.062 g (based on U.S. Mint specifications).

a. Find the probability that a randomly selected quarter weighs between 5.600 g and 5.700 g.

b. If 25 quarters are randomly selected, find the probability that their mean weight is greater than 5.675 g.

c. Find the probability that when eight quarters are randomly selected, they all weigh less than 5.670 g.

d. If a vending machine is designed to accept quarters with weights above the 10th percentile P_{10}, find the weight separating acceptable quarters from those that are not acceptable.

7. Job Priority Survey *USA Today* reported on an Adecco Staffing survey of 1000 randomly selected adults. Among those respondents, 20% chose health benefits as being most important to their job.

a. What is the number of respondents who chose health benefits as being most important to their job?

b. Construct a 95% interval estimate of the proportion of all adults who choose health benefits as being most important to their job.

c. Based on the result from part (b), can we safely conclude that the true proportion is different from 1/4? Why?

8. Win 4 Lottery Shown below is a histogram of digits selected in California's Win 4 lottery. Each drawing involves the random selection (with replacement) of four digits between 0 and 9 inclusive.

a. What is fundamentally wrong with the graph?

b. Does the display depict a normal distribution? Why or why not? What should be the shape of the histogram?

c. Identify the frequencies, then test the claim that the digits are selected from a population in which the digits are all equally likely. Is there a problem with the lottery?

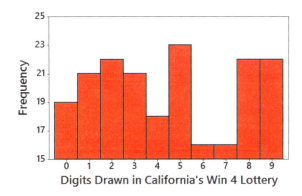

Technology Project

Does Weight Change with Age? Refer to Data Set 1 "Body Data" in Appendix B and use the weights of males partitioned into the three different age brackets of 18–25, 26–40, and 41–80. Test the claim that men in those three age brackets have the same mean weight.

   Sorting One challenge in this project is identifying the weights of men in the three age brackets. First, use the *sort* feature of your technology to sort all of the columns using *Gender* as the basis for sorting. You can then delete all of the rows representing females. Next, sort all of the columns using *Age* as the basis for sorting. It will then be much easier to identify the weights in the different age brackets.

FROM DATA TO DECISION

Critical Thinking: The Age/Gender Gap
Data Set 18 "Speed Dating" in Appendix B includes data from a study of speed dating, and Review Exercise 4 "Speed Dating" includes some of those data in a two-way table with different age brackets for men and women. Use the methods of this chapter to analyze the attractiveness ratings in the data set to determine whether there is a gender gap related to age brackets. Do older women appear to be more attracted to younger men? Do older men appear to be more attracted to younger women? Or is there no difference?

Cooperative Group Activities

1. Out-of-class activity Flesch Reading Ease scores and Flesch-Kincaid Grade Level scores measure readability of text. Some programs, such as Microsoft Word, include features that allow you to automatically obtain readability scores. Divide into groups of three or four students. Using at least three different writing samples, such as the *New York Times, USA Today,* and the *Onion,* obtain readability scores for ten samples of text from each source. Use the methods of this chapter to determine whether there are any differences.

2. In-class activity Divide the class into three groups. One group should record the pulse rate of each member while he or she remains seated. The second group should record the pulse rate of each member while he or she is standing. The third group should record the pulse rate of each member immediately after he or she stands and sits 10 times. Analyze the results. What do the results indicate?

3. In-class activity Ask each student in the class to estimate the length of the classroom. Specify that the length is the distance between the whiteboard and the opposite wall. On the same sheet of paper, each student should also write his or her gender (male/female) and major. Then divide into groups of three or four, and use the data from the entire class to address these questions:

• Is there a significant difference between the mean estimate for males and the mean estimate for females?

• Is there sufficient evidence to reject equality of the mean estimates for different majors? Describe how the majors were categorized.

• Does an interaction between gender and major have an effect on the estimated length?

• Does gender appear to have an effect on estimated length?

• Does major appear to have an effect on estimated length?

4. Out-of-class activity Biographyonline.net includes information on the lives of notable artists, politicians, scientists, actors, and others. Design and conduct an observational study that begins with choosing samples from select groups, followed by a comparison of life spans of people from the different groups. Do any particular groups appear to have life spans that are different from those of the other groups? Can you explain such differences?

5. Out-of-class activity Divide into groups of three or four students. Each group should survey other students at the same college by asking them to identify their major and gender. You might include other factors, such as employment (none, part-time, full-time) and age (under 21, 21–30, over 30). For each surveyed subject, determine the number of Twitter followers or Facebook friends.

• Does gender appear to have an effect on the number of followers/friends?

• Does major have an effect on the number of followers/friends?

• Does an interaction between gender and major have an effect on the number of followers/friends?

13-1 Basics of Nonparametric Tests

13-2 Sign Test

13-3 Wilcoxon Signed-Ranks Test for Matched Pairs

13-4 Wilcoxon Rank-Sum Test for Two Independent Samples

13-5 Kruskal-Wallis Test for Three or More Samples

13-6 Rank Correlation

13-7 Runs Test for Randomness

13

NONPARAMETRIC TESTS

CHAPTER PROBLEM — Do better televisions cost more?

Table 13-1 on the next page lists ranks and costs (hundreds of dollars) of LCD (liquid crystal display) televisions with screen sizes of at least 60 inches (based on data from *Consumer Reports*). The ranks are based on "overall scores" determined by *Consumer Reports,* and lower rank numbers correspond to "better" televisions with higher overall scores. Among the televisions included in Table 13-1, the best television has a rank of 1 and it costs $2300. Is there a correlation between ranks and

cost? If so, does it appear that better televisions cost more? Do you get what you pay for?

It would be wise to begin the analysis with a basic exploration of the data. Because we want to address the issue of correlation, we create the scatterplot shown on the next page. Clearly, there is no distinct straight-line pattern, so there does not appear to be a linear correlation. We could move beyond this subjective judgment and proceed to compute a linear

continued

correlation coefficient *r*, but let's consider the nature of the data. Specifically, the ranks simply identify an order and they do not really measure or count anything. Instead of using the linear correlation method from Section 10-1, we can use the rank correlation method described in Section 13-6. We can then provide objective results that are better than a subjective judgment.

TABLE 13-1 Ranks and Costs of LCD Televisions

Quality Rank	1	2	3	4	5	6	7	8	9	10
Cost (hundreds of dollars)	23	50	23	20	32	25	14	16	40	22

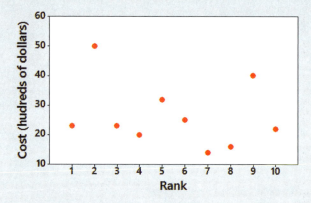

CHAPTER OBJECTIVES

Here are the chapter objectives:

13-1 Basics of Nonparametric Tests

- Develop the ability to describe the difference between parametric tests and nonparametric tests.
- Identify advantages and disadvantages of nonparametric tests.
- Know how nonparametric tests are generally less *efficient* than the corresponding parametric tests.
- Develop the ability to convert data into *ranks*.

13-2 Sign Test

- Develop the ability to conduct a sign test for claims involving matched pairs of sample data, or claims involving nominal data, or claims made about the median of a population.

13-3 Wilcoxon Signed-Ranks Test for Matched Pairs

- Develop the ability to apply the Wilcoxon signed-ranks test for sample data consisting of matched pairs.

13-4 Wilcoxon Rank-Sum Test for Two Independent Samples

- Develop the ability to apply the Wilcoxon rank-sum test for sample data from two independent populations.

13-5 Kruskal-Wallis Test for Three or More Samples

- Develop the ability to apply the Kruskal-Wallis test for sample data from three or more independent populations.

13-6 Rank Correlation

- Develop the ability to compute the value of the rank correlation coefficient r_s, and use it to determine whether there is a correlation between two variables.

13-7 Runs Test for Randomness

- Develop the ability to use the runs test for randomness to determine whether sample data occur in a random sequence.

13-1 Basics of Nonparametric Tests

This chapter introduces methods of *nonparametric* tests, which do not have the stricter requirements of corresponding parametric tests, which are based on samples from populations with specific parameters such as μ or σ.

> **DEFINITIONS**
>
> **Parametric tests** have requirements about the distribution of the populations involved; **nonparametric (**or **distribution-free) tests** do not require that samples come from populations with normal distributions or any other particular distributions.

Misleading Terminology The term *distribution-free test* correctly indicates that a test does not require a particular distribution. The term *nonparametric tests* is misleading in the sense that it suggests that the tests are not based on a parameter, but there are some nonparametric tests that are based on a parameter such as the median. Due to the widespread use of the term *nonparametric test,* we use that terminology, but we define it to be a test that does not require a particular distribution. (The author likes the term *nondistribution test,* but he wasn't first in line when definitions were given out.)

Advantages and Disadvantages

Advantages of Nonparametric Tests

1. Because nonparametric tests have less rigid requirements than parametric tests, they can be applied to a wider variety of situations.

2. Nonparametric tests can be applied to more data types than parametric tests. For example, nonparametric tests can be used with data consisting of ranks, and they can be used with categorical data, such as genders of survey respondents.

Disadvantages of Nonparametric Tests

1. Nonparametric tests tend to waste information because exact numerical data are often reduced to a qualitative form. For example, with the nonparametric sign test (Section 13-2), weight losses by dieters are recorded simply as negative signs, and the actual magnitudes of the weight losses are ignored.

2. Nonparametric tests are not as *efficient* as parametric tests, so a nonparametric test generally needs stronger evidence (such as a larger sample or greater differences) in order to reject a null hypothesis.

Efficiency of Nonparametric Tests When the requirements of population distributions are satisfied, nonparametric tests are generally less efficient than their corresponding parametric tests. For example, Section 13-6 presents the concept of *rank correlation,* which has an efficiency rating of 0.91 when compared to linear correlation in Section 10-1. This means that with all other things being equal, the nonparametric rank correlation method in Section 13-6 requires 100 sample observations to achieve the same results as 91 sample observations analyzed through the parametric linear correlation in Section 10-1, assuming the stricter requirements for using the parametric test are met. Table 13-2 lists nonparametric tests along with the corresponding parametric test and **efficiency** rating. Table 13-2 shows that several nonparametric tests have efficiency ratings above 0.90, so the lower efficiency might not be an important factor in choosing between parametric and nonparametric tests. However, because parametric tests do have higher efficiency ratings than their nonparametric counterparts, it's generally better to use the parametric tests when their required assumptions are satisfied.

TABLE 13-2 Efficiency: Comparison of Parametric and Nonparametric Tests

Application	Parametric Test	Nonparametric Test	Efficiency Rating of Nonparametric Test with Normal Populations
Matched pairs of sample data	*t* test	Sign test or	0.63
		Wilcoxon signed-ranks test	0.95
Two independent samples	*t* test	Wilcoxon rank-sum test	0.95
Three or more independent samples	Analysis of variance (*F* test)	Kruskal-Wallis test	0.95
Correlation	Linear correlation	Rank correlation test	0.91
Randomness	No parametric test	Runs test	No basis for comparison

Ranks

Sections 13-2 through 13-5 use methods based on ranks, defined as follows.

DEFINITION

Data are *sorted* when they are arranged according to some criterion, such as smallest to largest or best to worst. A **rank** is a number assigned to an individual sample item according to its order in the sorted list. The first item is assigned a rank of 1, the second item is assigned a rank of 2, and so on.

Handling Ties Among Ranks If a tie in ranks occurs, one very common procedure is to find the mean of the ranks involved in the tie and then assign this mean rank to each of the tied items, as in the following example.

| EXAMPLE 1 | Handling Ties Among Ranks |

The numbers 4, 5, 5, 5, 10, 11, 12, and 12 are given ranks of 1, 3, 3, 3, 5, 6, 7.5, and 7.5, respectively. The table below illustrates the procedure for handling ties.

Sorted Data	Preliminary Ranking	Rank
4	1	1
5	2	3
5	3 Mean is 3.	3
5	4	3
10	5	5
11	6	6
12	7 Mean is 7.5.	7.5
12	8	7.5

13-2   Sign Test

Key Concept This section introduces the *sign test,* which involves converting data values to positive and negative signs, then testing to determine whether either sign occurs significantly more often than the other sign.

DEFINITION

The **sign test** is a nonparametric (distribution-free) test that uses positive and negative signs to test different claims, including these:

1. Claims involving matched pairs of sample data
2. Claims involving nominal data with two categories
3. Claims about the median of a single population

Basic Concept of the Sign Test The basic idea underlying the sign test is to analyze the frequencies of positive and negative signs to determine whether they are significantly different. For example, consider the results of clinical trials of the XSORT method of gender selection. Among 726 couples who used the XSORT method in trying to have a baby girl, 668 couples did have baby girls. Is 668 girls in 726 births *significant?* Common sense should suggest that 668 girls in 726 births is significant, but what about 365 girls in 726 births? Or 400 girls in 726 births? The sign test allows us to determine when such results are significant. Figure 13-1 summarizes the sign test procedure.

For consistency and simplicity, we will use a test statistic based on the number of times that the *less frequent* sign occurs.

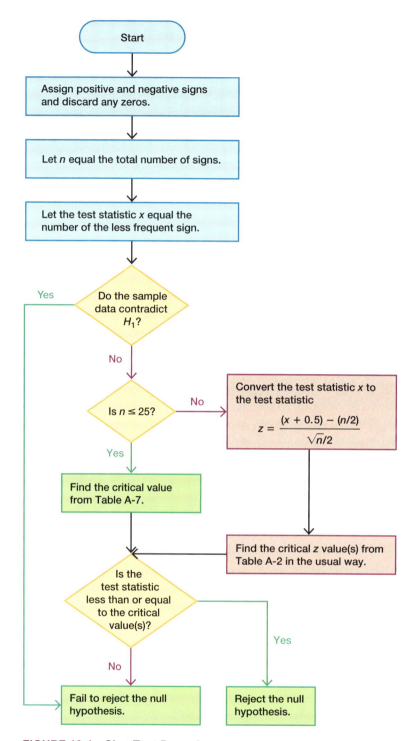

FIGURE 13-1   Sign Test Procedure

KEY ELEMENTS

Sign Test

Objective

Use positive and negative signs to test a claim falling into one of the following three categories:

1. Matched Pairs

- Subtract the second value in each pair from the first, record the sign of the difference, and ignore any 0s.

2. Nominal Data with Two Categories

- Represent each member of one category by a positive sign and represent each member of the other category by a negative sign.

3. Median of a Single Population

- Subtract the median from each sample value, record the sign of the difference, and ignore any 0s.

Notation

x = the number of times the *less frequent* sign occurs
n = the total number of positive and negative signs combined

Requirements

The sample data are a simple random sample.

Note: There is *no* requirement that the sample data come from a population with a particular distribution, such as a normal distribution.

Test Statistic

If $n \leq 25$: Test statistic is x = the number of times the less frequent sign occurs.
If $n > 25$: Test statistic is

$$z = \frac{(x + 0.5) - \left(\frac{n}{2}\right)}{\frac{\sqrt{n}}{2}}$$

P-Values

P-values are often provided by technology, or *P*-values can often be found using the z test statistic.

Critical Values

1. If $n \leq 25$, critical x values are found in Table A-7.   **2.** If $n > 25$, critical z values are found in Table A-2.

Hint: Because x or z is based on the *less* frequent sign, all one-sided tests are treated as if they were left-tailed tests.

CAUTION When using the sign test in a one-tailed test, be very careful to avoid making the wrong conclusion when one sign occurs significantly more often or less often than the other sign, but the sample data *contradict* the alternative hypothesis. A sample of 7% boys can never be used to support the claim that boys occur *more than* 50% of the time, as in Example 1.

EXAMPLE 1 **Data Contradicting the Alternative Hypothesis**

Among 945 couples who used the XSORT method of gender selection, 66 had boys, so the sample proportion of boys is 66/945, or 0.0698 (based on data from the Genetics & IVF Institute). Consider the claim that the XSORT method of gender selection *increases* the likelihood of baby *boys* so that the probability of a boy is $p > 0.5$. This claim of $p > 0.5$ becomes the alternative hypothesis.

Using common sense, we see that with a sample proportion of boys of 0.0698, we can never support a claim that $p > 0.5$. (We would need a sample proportion of boys *greater* than 0.5 by a significant amount.) Here, the sample proportion of 66/945, or 0.0698, *contradicts* the alternative hypothesis because it is not greater than 0.5.

INTERPRETATION

An alternative hypothesis can never be supported with data that contradict it. The sign test will show that 66 boys in 945 births is significant, but it is significant in the wrong direction. We can never support a claim that $p > 0.5$ with a sample proportion of 66/945, or 0.0698, which is *less than* 0.5.

YOUR TURN   Exercise 3 "Contradicting H_1."

Claims About Matched Pairs

When using the sign test with data that are matched pairs, we convert the raw data to positive and negative signs as follows:

1. Subtract each value of the second variable from the corresponding value of the first variable.

2. Record only the *sign* of the difference found in Step 1. Exclude *ties* by deleting any matched pairs in which both values are equal.

The main concept underlying this use of the sign test is as follows:

> **If the two sets of data have equal medians, the number of positive signs should be approximately equal to the number of negative signs.**

EXAMPLE 2 **Is There a Gender Difference in Ages of Best Actresses and Best Actors?**

Table 13-3 (from Data Set 14 "Oscar Winner Age" in Appendix B) lists ages of Oscar-winning Best Actresses and ages of Oscar-winning Best Actors. The ages are paired according to the year that the awards were presented. Use the sign test with the sample data in Table 13-3 to test the claim that there is no difference between ages of Best Actresses and Best Actors.

TABLE 13-3 Ages of Best Actresses and Best Actors

Best Actress	28	63	29	41	30	41	28	26	29	29
Best Actor	62	52	41	39	49	41	44	51	54	50
Sign of Difference	−	+	−	+	−	0	−	−	−	−

SOLUTION

REQUIREMENT CHECK The only requirement of the sign test is that the sample data are a simple random sample, and that requirement is satisfied.

If there is no difference between ages of Best Actresses and Best Actors, the numbers of positive and negative signs should be approximately equal. In Table 13-3 we have 2 positive signs, 7 negative signs, and 1 difference of 0. We discard the difference of 0 and proceed with the 2 positive signs and 7 negative signs. The sign test tells us whether or not the numbers of positive and negative signs are approximately equal.

The null hypothesis is the claim of no difference between ages of Best Actresses and Best Actors, and the alternative hypothesis is the claim that there is a difference.

H_0: There is no difference. (The median of the differences is equal to 0.)

H_1: There is a difference. (The median of the differences is not equal to 0.)

Following the sign test procedure summarized in Figure 13-1, we let $n = 9$ (the total number of positive and negative signs) and we let $x = 2$ (the number of the less frequent sign, or the smaller of 2 and 7).

The sample data do not contradict H_1, because there is a difference between the 2 positive signs and the 7 negative signs. The sample data show a difference, and we need to continue with the test to determine whether that difference is significant.

Figure 13-1 shows that with $n = 9$, we should proceed to find the critical value from Table A-7. We refer to Table A-7, where the critical value of 1 is found for $n = 9$ and $\alpha = 0.05$ in two tails.

Since $n \leq 25$, the test statistic is $x = 2$ (and we do not convert x to a z score). With a test statistic of $x = 2$ and a critical x value of 1, we fail to reject the null hypothesis of no difference. (See Note 2 included with Table A-7: "Reject the null hypothesis if the number of the less frequent sign (x) is less than or equal to the value in the table." Because $x = 2$ is *not* less than or equal to the critical value of 1, we fail to reject the null hypothesis.) There is not sufficient evidence to warrant rejection of the claim that the median of the differences is equal to 0.

INTERPRETATION

We conclude that there is not sufficient evidence to reject the claim of no difference between ages of Best Actresses and Best Actors.

YOUR TURN Do Exercise 5 "Speed Dating: Attributes."

Claims Involving Nominal Data with Two Categories

In Chapter 1 we defined nominal data to be data that consist of names, labels, or categories only. The nature of nominal data limits the calculations that are possible, but we can identify the *proportion* of the sample data that belong to a particular category, and we can test claims about the corresponding population proportion p. The following example uses nominal data consisting of genders (girls/boys). The sign test is used by representing girls with positive $(+)$ signs and boys with negative $(-)$ signs. (Those signs are chosen arbitrarily—honest.)

EXAMPLE 3 **Gender Selection**

The Genetics & IVF Institute conducted a clinical trial of its methods for gender selection for babies. Before the clinical trials were concluded, 879 of 945 babies born to parents using the XSORT method of gender selection were girls. Use the sign test and a 0.05 significance level to test the claim that this method of gender selection is effective in increasing the likelihood of a baby girl.

SOLUTION

REQUIREMENT CHECK The only requirement is that the sample is a simple random sample. Based on the design of this experiment, we can assume that the sample data are a simple random sample. ✓

Let p denote the population proportion of baby girls. The claim that girls are more likely with the XSORT method can be expressed as $p > 0.5$, so the null and alternative hypotheses are as follows:

continued

$$H_0\text{: } p = 0.5 \text{ (the proportion of girls is equal to 0.5)}$$

$$H_1\text{: } p > 0.5 \text{ (girls are more likely)}$$

Denoting girls by positive signs $(+)$ and boys by negative signs $(-)$, we have 879 positive signs and 66 negative signs. Using the sign test procedure summarized in Figure 13-1, we let the test statistic x be the smaller of 879 and 66, so $x = 66$ boys. *Instead of trying to determine whether 879 girls is high enough to be significantly high, we proceed with the equivalent goal of trying to determine whether 66 boys is low enough to be significantly low, so we treat the test as a* left-tailed *test.*

The sample data do not contradict the alternative hypothesis because the sample proportion of girls is $879/945$, or 0.930, which is greater than 0.5, as in the above alternative hypothesis. Continuing with the procedure in Figure 13-1, we note that the value of $n = 945$ is greater than 25, so the test statistic $x = 66$ is converted (using a correction for continuity) to the test statistic z as follows:

$$z = \frac{(x + 0.5) - \left(\dfrac{n}{2}\right)}{\dfrac{\sqrt{n}}{2}}$$

$$= \frac{(66 + 0.5) - \left(\dfrac{945}{2}\right)}{\dfrac{\sqrt{945}}{2}} = -26.41$$

***P*-Value** We could use the test statistic of $z = -26.41$ to find the left-tailed *P*-value of 0.0000 (Table: 0.0001). That low *P*-value causes us to reject the null hypothesis.

Critical Value With $\alpha = 0.05$ in a left-tailed test, the critical value is $z = -1.645$. Figure 13-2 shows that the test statistic $z = -26.41$ is in the critical region bounded by $z = -1.645$, so we reject the null hypothesis that the proportion of girls is equal to 0.5.

There is sufficient sample evidence to support the claim that girls are more likely with the XSORT method.

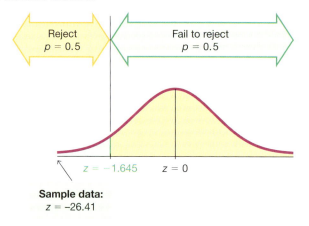

FIGURE 13-2   **Testing Effectiveness of the XSORT Gender Selection Method**

INTERPRETATION

The XSORT method of gender selection does appear to be associated with an increase in the likelihood of a girl, so this method appears to be effective (but this hypothesis test does not prove that the XSORT method is the *cause* of the increase).

YOUR TURN Do Exercise 9 "Stem Cell Survey."

Claims About the Median of a Single Population

The next example illustrates the procedure for using the sign test in testing a claim about the median of a single population. See how the negative and positive signs are based on the claimed value of the median.

EXAMPLE 4   Body Temperatures

Data Set 3 "Body Temperatures" in Appendix B includes measured body temperatures of adults. Use the 106 temperatures listed for 12 AM on Day 2 with the sign test to test the claim that the median is less than 98.6°F. Of the 106 subjects, 68 had temperatures below 98.6°F, 23 had temperatures above 98.6°F, and 15 had temperatures equal to 98.6°F.

SOLUTION

REQUIREMENT CHECK   The only requirement is that the sample is a simple random sample. Based on the design of this experiment, we assume that the sample data are a simple random sample. ✓

The claim that the median is less than 98.6°F is the alternative hypothesis, while the null hypothesis is the claim that the median is equal to 98.6°F.

$$H_0: \text{Median is equal to } 98.6°F. \ (\text{median} = 98.6°F)$$

$$H_1: \text{Median is less than } 98.6°F. \ (\text{median} < 98.6°F)$$

Following the procedure outlined in Figure 13-1, we use a negative sign to represent each temperature below 98.6°F, and we use a positive sign for each temperature above 98.6°F. We discard the 15 data values of 98.6, since they result in differences of zero. We have 68 negative signs and 23 positive signs, so $n = 91$ and $x = 23$ (the number of the less frequent sign). The sample data do not contradict the alternative hypothesis, because most of the 91 temperatures are below 98.6 °F. The value of n exceeds 25, so we convert the test statistic x to the test statistic z:

$$z = \frac{(x + 0.5) - \left(\dfrac{n}{2}\right)}{\dfrac{\sqrt{n}}{2}}$$

$$= \frac{(23 + 0.5) - \left(\dfrac{91}{2}\right)}{\dfrac{\sqrt{91}}{2}} = -4.61$$

P-Value In this left-tailed test, the test statistic of $z = -4.61$ yields a P-value of 0.0000 (Table: 0.0001). Because that P-value is so small, we reject the null hypothesis.

Critical Value In this left-tailed test with $\alpha = 0.05$, use Table A-2 to get the critical z value of -1.645. From Figure 13-3 on the next page we see that the test statistic of $z = -4.61$ is within the critical region, so reject the null hypothesis.

continued

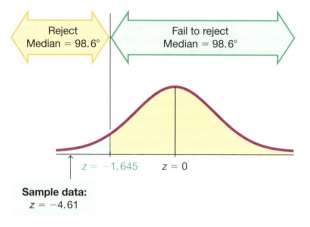

FIGURE 13-3   **Testing the Claim That the Median Is Less Than 98.6°F**

INTERPRETATION

There is sufficient sample evidence to support the claim that the median body temperature of healthy adults is less than 98.6°F. It is not equal to 98.6°F, as is commonly believed.

YOUR TURN   Do Exercise 13 "Earthquake Magnitudes."

In Example 4, the sign test of the claim that the median is below 98.6°F results in a test statistic of $z = -4.61$ and a P-value of 0.00000202. However, a parametric test of the claim that $\mu < 98.6°F$ results in a test statistic of $t = -6.611$ with a P-value of 0.000000000813. Because the P-value from the sign test is not as low as the P-value from the parametric test, we see that the sign test isn't as sensitive as the parametric test. Both tests lead to rejection of the null hypothesis, but the sign test doesn't consider the sample data to be as extreme, partly because the sign test uses only information about the *direction* of the data, ignoring the *magnitudes* of the data values. The next section introduces the Wilcoxon signed-ranks test, which largely overcomes that disadvantage.

Rationale for the Test Statistic Used When $n > 25$ When finding critical values for the sign test, we use Table A-7 only for n up to 25. When $n > 25$, the test statistic z is based on a normal approximation to the binomial probability distribution with $p = q = 1/2$. In Section 6-6 we saw that the normal approximation to the binomial distribution is acceptable when both $np \geq 5$ and $nq \geq 5$. In Section 5-2 we saw that $\mu = np$ and $\sigma = \sqrt{npq}$ for binomial probability distributions. Because this sign test assumes that $p = q = 1/2$, we meet the $np \geq 5$ and $nq \geq 5$ prerequisites whenever $n \geq 10$. Also, with the assumption that $p = q = 1/2$, we get $\mu = np = n/2$ and $\sigma = \sqrt{npq} = \sqrt{n/4} = \sqrt{n}/2$, so the standard z score

$$z = \frac{x - \mu}{\sigma}$$

becomes

$$z = \frac{x - \left(\dfrac{n}{2}\right)}{\dfrac{\sqrt{n}}{2}}$$

We replace x by $x + 0.5$ as a correction for continuity. That is, the values of x are discrete, but since we are using a continuous probability distribution, a discrete value such as 10 is actually represented by the interval from 9.5 to 10.5. Because x represents the less frequent sign, we act conservatively by concerning ourselves only with $x + 0.5$; we get the test statistic z shown below and in the Key Elements box.

$$z = \frac{(x + 0.5) - \left(\frac{n}{2}\right)}{\frac{\sqrt{n}}{2}}$$

TECH CENTER

Sign Test
Access tech supplements, videos, and data sets at **www.TriolaStats.com**

Statdisk	Minitab	StatCrunch

Statdisk

1. Click **Analysis** in the top menu.
2. Select **Sign Test** from the dropdown menu.
3. *Known number of signs*
   Select **Given Number of Signs,** choose the format for the claim, enter the significance level and numbers of positive and negative signs.

   Pairs of values
   Select **Given Pairs of Values,** choose the format for the claim, enter the significance level, and select the two data columns to include.
4. Click **Evaluate.**

Minitab

Minitab requires a single column of values.
Matched pairs: *Enter a column consisting of the differences.*
Nominal data in two categories: *Enter a 1 for each value of one category and −1 for each value of the other category. Enter 0 for the claimed value of the median.*
Individual values to be tested with claimed median: *Enter sample values in a single column.*

1. Click **Stat** in the top menu.
2. Select **Nonparametrics** from the dropdown menu and select **1-Sample Sign** from the submenu.
3. Under *Variables* select the column containing the data to be analyzed.
4. Select **Test Median** and enter the claimed median value.
5. Choose the format of the alternative hypothesis.
6. Click **OK.**

StatCrunch

StatCrunch is able to test the claim that a single list of values is from a population with a median equal to some specified value.

1. Click **Stat** in the top menu.
2. Select **Nonparametrics** from the dropdown menu and **Sign Test** from the submenu.
3. Select the column containing the data to be analyzed.
4. Select **Hypothesis test for median** and for H_0 enter the claimed median value. For H_A select the desired format.
5. Click **Compute!**

TI-83/84 Plus Calculator

The TI-83/84 Plus calculator does not have a function dedicated to the sign test, but the calculator's binomcdf *function can be used to find the P-value for a sign test.*

1. Press **2ND** then **VARS** keys to access the *DISTR* (distributions) menu.
2. Select **binomcdf** and click **ENTER**.
3. Enter the values for trials *n, p,* and *x* to complete the command **binomcdf(n,p,x).** For *trials* enter the total number of positive and negative signs. For *p* enter **0.5.** For *x* enter the number of the less frequent sign.
4. Press **ENTER**. The result is the probability of getting *x* or fewer trials. *Double this value for two-tailed tests.*

TIP: The final result is the P-value, so reject the null hypothesis if the P-value is less than or equal to the significance level. Otherwise, fail to reject the null hypothesis.

continued

TECH CENTER *continued*

Sign Test
Access tech supplements, videos, and data sets at **www.TriolaStats.com**

Excel

XLSTAT Add-In

1. Click on the **XLSTAT** tab in the Ribbon and then click **Nonparametric tests.**
2. Select **Comparison of two samples** from the dropdown menu.
3. Enter the data range for each sample in the *Sample 1 & 2 boxes.* Check the **Column labels** box if the data range includes labels.
4. Select **Paired samples** under *Data format.*
5. Check the **sign test** option only.
6. Click the **Options** tab.
7. Under Alternative Hypothesis select **Sample 1 – Sample 2 ≠ D.** Confirm *Hypothesized difference (D)* is **0.**
8. Enter a significance level and check the **Exact p-value** box.
9. Click **OK.**

Excel

Excel does not have a function dedicated to the sign test, but can be used to find the P-value for a sign test.

1. Click on the **Insert Function f_x** button, select the category **Statistical,** and select the function **BINOM.DIST** and click **OK.**
2. For *Number_s* enter the number of times the *less frequent* sign occurs. For *Trials* enter the total number of positive and negative signs. For *probability_s* enter **0.5.** For *Cumulative* enter **1** for "True."
3. Click **OK.** The single-tail P-value will be displayed. *Double this value for two-tailed tests.*

TIP: The final result is the *P*-value, so reject the null hypothesis if the *P*-value is less than or equal to the significance level. Otherwise, fail to reject the null hypothesis.

13-2   Basic Skills and Concepts

Statistical Literacy and Critical Thinking

1. Sign Test for Freshman 15 The table below lists some of the weights (kg) from Data Set 6 "Freshman 15" in Appendix B. Those weights were measured from college students in September and later in April of their freshman year. Assume that we plan to use the sign test to test the claim of no difference between September weights and April weights. What requirements must be satisfied for this test? Is there any requirement that the populations must have a normal distribution or any other specific distribution? In what sense is this sign test a "distribution-free test"?

September weight (kg)	67	53	64	74	67	70	55	74	62	57
April weight (kg)	66	52	68	77	67	71	60	82	65	58

2. Identifying Signs For the sign test described in Exercise 1, identify the number of positive signs, the number of negative signs, the number of ties, the sample size *n* that is used for the sign test, and the value of the test statistic.

3. Contradicting H_1 An important step in conducting the sign test is to determine whether the sample data contradict the alternative hypothesis H_1. For the sign test described in Exercise 1, identify the null hypothesis and the alternative hypothesis, and explain how the sample data contradict or do not contradict the alternative hypothesis.

4. Efficiency of the Sign Test Refer to Table 13-2 on page 600 and identify the efficiency of the sign test. What does that value tell us about the sign test?

Matched Pairs.   *In Exercises 5–8, use the sign test for the data consisting of matched pairs.*

5. Speed Dating: Attributes Listed below are "attribute" ratings made by couples participating in a speed dating session. Each attribute rating is the sum of the ratings of five attributes (sincerity, intelligence, fun, ambition, shared interests). The listed ratings are from Data Set 18 "Speed Dating" in Appendix B. Use a 0.05 significance level to test the claim that there is a difference between female attribute ratings and male attribute ratings.

Rating of Male by Female	29	38	36	37	30	34	35	23	43
Rating of Female by Male	36	34	34	33	31	17	31	30	42

6. Speed Dating: Attractiveness Listed below are "attractiveness" ratings (1 = not attractive; 10 = extremely attractive) made by couples participating in a speed dating session. The listed ratings are from Data Set 18 "Speed Dating". Use a 0.05 significance level to test the claim that there is a difference between female attractiveness ratings and male attractiveness ratings.

Rating of Male by Female	4	8	7	7	6	8	6	4	2	5	9.5	7
Rating of Female by Male	6	8	7	9	5	7	5	4	6	8	6	5

7. Speed Dating: Attributes Repeat Exercise 5 using all of the attribute ratings in Data Set 18 "Speed Dating" in Appendix B.

8. Speed Dating: Attractiveness Repeat Exercise 6 using all of the attractiveness ratings in Data Set 18 "Speed Dating" in Appendix B.

Nominal Data.   *In Exercises 9–12, use the sign test for the claim involving nominal data.*

9. Stem Cell Survey *Newsweek* conducted a poll in which respondents were asked if they "favor or oppose using federal tax dollars to fund medical research using stem cells obtained from human embryos." Of those polled, 481 were in favor, 401 were opposed, and 120 were unsure. Use a 0.01 significance level to test the claim that there is no difference between the proportions of those opposed and those in favor.

10. Medical Malpractice In a study of 1228 randomly selected medical malpractice lawsuits, it was found that 856 of them were dropped or dismissed (based on data from the Physicians Insurers Association of America). Use a 0.01 significance level to test the claim that there is a difference between the rate of medical malpractice lawsuits that go to trial and the rate of such lawsuits that are dropped or dismissed.

11. Births A random sample of 860 births in New York State included 426 boys and 434 girls. Use a 0.05 significance level to test the claim that when babies are born, boys and girls are equally likely.

12. Overtime Rule in Football Before the overtime rule in the National Football League was changed in 2011, among 460 overtime games, 252 were won by the team that won the coin toss at the beginning of overtime. Using a 0.05 significance level, test the claim that the coin toss is fair in the sense that neither team has an advantage by winning it. Does the coin toss appear to be fair?

Appendix B Data Sets.   *In Exercises 13–16, refer to the indicated data set in Appendix B and use the sign test for the claim about the median of a population.*

13. Earthquake Magnitudes Refer to Data Set 21 "Earthquakes" in Appendix B for the earthquake magnitudes. Use a 0.01 significance level to test the claim that the median is equal to 2.00.

14. Earthquake Depths Refer to Data Set 21 "Earthquakes" in Appendix B for the earthquake depths (km). Use a 0.01 significance level to test the claim that the median is equal to 5.0 km.

15. Testing for Median Weight of Quarters Refer to Data Set 29 "Coin Weights" in Appendix B for the weights (g) of randomly selected quarters that were minted after 1964. The

continued

quarters are supposed to have a median weight of 5.670 g. Use a 0.01 significance level to test the claim that the median is equal to 5.670 g. Do quarters appear to be minted according to specifications?

 16. Old Faithful Refer to Data Set 23 "Old Faithful" in Appendix B for the time intervals before eruptions of the Old Faithful geyser. Use a 0.05 significance level to test the claim that those times are from a population with a median of 90 minutes.

13-2 Beyond the Basics

17. Procedures for Handling Ties In the sign test procedure described in this section, we exclude ties (represented by 0 instead of a sign of + or −). A second approach is to treat half of the 0s as positive signs and half as negative signs. (If the number of 0s is odd, exclude one so that they can be divided equally.) With a third approach, in two-tailed tests make half of the 0s positive and half negative; in one-tailed tests make all 0s either positive or negative, whichever supports the null hypothesis. Repeat Example 4 "Body Temperatures" using the second and third approaches to handling ties. Do the different approaches lead to very different test statistics, P-values, and conclusions?

18. Finding Critical Values Table A-7 lists critical values for limited choices of α. Use Table A-1 to add a new column in Table A-7 (from $n = 1$ to $n = 8$) that represents a significance level of 0.03 in one tail or 0.06 in two tails. For any particular n, use $p = 0.5$, because the sign test requires the assumption that P(positive sign) $= P$(negative sign) $= 0.5$. The probability of x or fewer like signs is the sum of the probabilities for values up to and including x.

13-3 Wilcoxon Signed-Ranks Test for Matched Pairs

Key Concept This section introduces the *Wilcoxon signed-ranks test,* which begins with the conversion of the sample data into ranks. This test can be used for the two different applications described in the following definition.

> **DEFINITION**
>
> The **Wilcoxon signed-ranks test** is a nonparametric test that uses ranks for these applications:
> 1. Testing a claim that a population of matched pairs has the property that the matched pairs have differences with a median equal to zero
> 2. Testing a claim that a single population of individual values has a median equal to some claimed value

When testing a claimed value of a median for a population of individual values, we create matched pairs by pairing each sample value with the claimed median, so the same procedure is used for both of the applications above.

Claims Involving Matched Pairs

The sign test (Section 13-2) can be used with matched pairs, but the sign test uses only the *signs* of the differences. By using ranks instead of signs, the Wilcoxon signed-ranks test takes the magnitudes of the differences into account, so it includes and uses more information than the sign test and therefore tends to yield conclusions that better reflect the true nature of the data.

KEY ELEMENTS

Wilcoxon Signed-Ranks Test

Objective: Use the Wilcoxon signed-ranks test for the following tests:

- **Matched Pairs:** Test the claim that a population of matched pairs has the property that the matched pairs have differences with a median equal to zero.

- **One Population of Individual Values:** Test the claim that a population has a median equal to some claimed value. (By pairing each sample value with the claimed median, we again work with matched pairs.)

Notation

T = the smaller of the following two sums:

1. The sum of the positive ranks of the nonzero differences d

2. The absolute value of the sum of the negative ranks of the nonzero differences d

(Details for evaluating T are given in the procedure following this Key Elements box.)

Requirements

1. The data are a simple random sample.

2. The population of differences has a distribution that is approximately *symmetric,* meaning that the left half of its histogram is roughly a mirror image of its right half. (For a sample of matched pairs, obtain differences by subtracting the second value from the first value in each

pair; for a sample of individual values, obtain differences by subtracting the value of the claimed median from each sample value.)

Note: There is *no* requirement that the data have a normal distribution.

Test Statistic

If $n \leq 30$, the test statistic is T.

If $n > 30$, the test statistic is $z = \dfrac{T - \dfrac{n(n+1)}{4}}{\sqrt{\dfrac{n(n+1)(2n+1)}{24}}}$

P-Values

P-values are often provided by technology, or *P*-values can be found using the z test statistic and Table A-2.

Critical Values

1. If $n \leq 30$, the critical T value is found in Table A-8.

2. If $n > 30$, the critical z values are found in Table A-2.

The following procedure requires that you sort data, then assign ranks. When working with larger data sets, sorting and ranking become tedious, but technology can be used to automate that process. Stemplots can also be very helpful in sorting data.

Wilcoxon Signed-Ranks Procedure To see how the following steps are applied, refer to the sample of matched pairs listed in the first two rows of Table 13-4 on the next page. Assume that we want to test the null hypothesis that the matched pairs are from a population of matched pairs with differences having a median equal to zero.

TABLE 13-4 Ages of Best Actresses and Best Actors

Best Actress	28	63	29	41	30	41	28	26	29	29
Best Actor	62	52	41	39	49	41	44	51	54	50
d (difference)	−34	+11	−12	+2	−19	0	−16	−25	−25	−21
Rank of $\lvert d \rvert$	9	2	3	1	5	×	4	7.5	7.5	6
Signed rank	−9	+2	−3	+1	−5	×	−4	−7.5	−7.5	−6

Wilcoxon Signed-Ranks Procedure

Step 1: For each pair of data, find the difference d by subtracting the second value from the first value. Discard any pairs that have a difference of 0.

EXAMPLE: The third row of Table 13-4 lists the differences found by subtracting the ages of the Best Actors from the ages of the Best Actresses, and the difference of 0 will be ignored in the following steps.

Step 2: *Ignore the signs of the differences,* then sort the differences from lowest to highest and replace the differences by the corresponding rank value (as described in Section 13-1). When differences have the same numerical value, assign to them the mean of the ranks involved in the tie.

EXAMPLE: The fourth row of Table 13-4 shows the ranks of the values of $\lvert d \rvert$. Ignoring the difference of 0, the smallest value of $\lvert d \rvert$ is 2, so it is assigned the rank of 1. The next smallest value of $\lvert d \rvert$ is 11, so it is assigned a rank of 2, and so on. (Tie: There are two $\lvert d \rvert$ values of 25, so we find the mean of the ranks of 7 and 8, which is 7.5. Each of those $\lvert d \rvert$ values is assigned a rank of 7.5.)

Step 3: Attach to each rank the sign of the difference from which it came. That is, insert the signs that were ignored in Step 2.

EXAMPLE: The bottom row of Table 13-4 lists the same ranks found in the fourth row, but the signs of the differences shown in the third row are inserted.

Step 4: Find the sum of the ranks that are positive. Also find the absolute value of the sum of the negative ranks.

EXAMPLE: The bottom row of Table 13-4 lists the signed ranks. The sum of the positive ranks is $2 + 1 = 3$. The sum of the negative ranks is $(-9) + (-3) + (-5) + (-4) + (-7.5) + (-7.5) + (-6) = -42$ and the absolute value of this sum is 42. The two rank sums are 3 and 42.

Step 5: Let T be the *smaller* of the two sums found in Step 4. Either sum could be used, but for a simplified procedure we arbitrarily select the smaller of the two sums.

EXAMPLE: The data in Table 13-4 result in the rank sums of 3 and 42, so the smaller of those two sums is 3.

Step 6: Let n be the number of pairs of data for which the difference d is not 0.

EXAMPLE: The data in Table 13-4 have 9 differences that are not 0, so $n = 9$.

Step 7: Determine the test statistic and critical values based on the sample size, as shown in the preceding Key Elements box.

EXAMPLE: For the data in Table 13-4 the test statistic is $T = 3$. The sample size is $n = 9$, so the critical value is found in Table A-8. Using a 0.05 significance level with a two-tailed test, the critical value from Table A-8 is 6.

Step 8: When forming the conclusion, reject the null hypothesis if the sample data lead to a test statistic that is in the critical region—that is, the test statistic is less than or equal to the critical value(s). Otherwise, fail to reject the null hypothesis.

EXAMPLE: If the test statistic is T (instead of z), reject the null hypothesis if T is less than or equal to the critical value. Fail to reject the null hypothesis if T is greater than the critical value. For the sample of matched pairs in the first two rows of Table 13-4, $T = 3$ and the critical value is 6, so we reject the null hypothesis that the matched pairs are from a population of matched pairs with differences having a median equal to zero

EXAMPLE 1 Ages of Best Actresses and Best Actors

The first two rows of Table 13-4 include ages of Best Actresses and Best Actors (from Data Set 14 "Oscar Winner Age" in Appendix B). The data are matched by the year in which the awards were won. Use the sample data in the first two rows of Table 13-4 to test the claim that there is no difference between ages of Best Actresses and Best Actors. Use the Wilcoxon signed-ranks test with a 0.05 significance level.

SOLUTION

REQUIREMENT CHECK (1) The data are a simple random sample. (2) The histogram of the differences in the third row of Table 13-4 is shown in the accompanying display. The left side of the graph should be roughly a mirror image of the right side, which does not appear to be the case. But with only 10 differences, the difference between the left and right sides is not too extreme, so we will consider this requirement to be satisfied. ✓

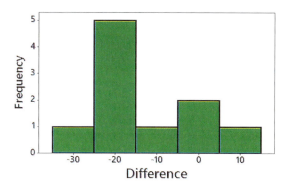

The null hypothesis is the claim of no difference between the ages of Best Actresses and Best Actors, and the alternative hypothesis is the claim that there is a difference.

H_0: There is no difference. (The median of the differences is equal to 0.)

H_1: There is a difference. (The median of the differences is not equal to 0.)

Test Statistic Because we are using the Wilcoxon signed-ranks test, the test statistic is calculated by using the eight-step procedure presented earlier in this section. Those steps include examples illustrating the calculation of the test statistic with the sample data in Table 13-4, and the result is the test statistic of $T = 3$.

Critical Value The sample size is $n = 9$, so the critical value is found in Table A-8. Using a 0.05 significance level with a two-tailed test, the critical value from Table A-8 is found to be 6.

continued

Conclusion Table A-8 includes a note stating that we should reject the null hypothesis if the test statistic T is less than or equal to the critical value. Because the test statistic of $T = 3$ is less than or equal to the critical value of 6, we reject the null hypothesis.

INTERPRETATION

We conclude that there appears to be a difference between ages of Best Actresses and Best Actors.

YOUR TURN Do Exercise 5 "Speed Dating: Attributes."

Claims About the Median of a Single Population

The Wilcoxon signed-ranks test can also be used to test a claim that a single population has some claimed value of the median. The preceding procedures can be used with one simple adjustment:

> **When testing a claim about the median of a single population, create matched pairs by pairing each sample value with the claimed value of the median. The preceding procedure can then be used.**

EXAMPLE 2 Body Temperatures

Data Set 3 "Body Temperatures" in Appendix B includes measured body temperatures of adults. Use the 106 temperatures listed for 12 AM on Day 2 with the Wilcoxon signed-ranks test to test the claim that the median is less than 98.6°F. Use a 0.05 significance level.

Statdisk

Num Unequal pairs: 91

Using Approximation
Test Statistic, T: 661.0000
Mean, μ: 2093
Standard Deviation: 252.6589
Test Statistic, z: -5.6677
Critical z: ±1.959962

SOLUTION

REQUIREMENT CHECK (1) The design of the experiment that led to the data in Data Set 3 justifies treating the sample as a simple random sample. (2) The requirement of an approximately symmetric distribution of differences is satisfied, because a histogram of those differences is approximately symmetric.

By pairing each individual sample value with the median of 98.6°F, we are working with matched pairs. Shown in the margin is the Statdisk display showing the test statistic of $T = 661$, which converts to the test statistic $z = -5.67$. (The display is from a two-tailed test; for this left-tailed test, the critical value is -1.645.) The test statistic of $z = -5.67$ yields a P-value of 0.000, so we reject the null hypothesis that the population of differences between body temperatures and the claimed median of 98.6°F is zero. There is sufficient evidence to support the claim that the median body temperature is less than 98.6°F. This is the same conclusion that results from the sign test in Example 4 in Section 13-2.

YOUR TURN Do Exercise 9 "Earthquake Magnitudes."

Rationale: In Example 1, the unsigned ranks of 1 through 9 have a total of 45, so if there are no significant differences, each of the two signed-rank totals should be around $45 \div 2$, or 22.5. That is, the negative ranks and positive ranks should split up as 22.5–22.5 or something close, such as 24–21. Table A-8, the table of critical values, shows that at the 0.05 significance level with 9 pairs of data, the critical value is 6, so a split of 6–39 represents a significant departure from the null hypothesis, and any split that is farther apart will also represent a significant departure from the null hypothesis. Conversely, splits like 7–38 do not represent significant departures from a 22.5–22.5 split, and they would not justify rejecting the null hypothesis. The Wilcoxon signed-ranks test is based on the lower rank total, so instead of analyzing both numbers constituting the split, we consider only the lower number.

The sum of all the ranks $1 + 2 + 3 + \cdots + n$ is equal to $n(n + 1)/2$. If this rank sum is to be divided equally between two categories (positive and negative), each of the two totals should be near $n(n + 1)/4$, which is half of $n(n + 1)/2$. Recognition of this principle helps us understand the test statistic used when $n > 30$.

TECH CENTER

Wilcoxon Signed-Ranks Test
Access tech supplements, videos, and data sets at **www.TriolaStats.com**

Statdisk	Minitab	StatCrunch
1. Click **Analysis** in the top menu.	1. Create a column consisting of the differences between the matched pairs. To do this, enter paired data in columns *C1* and *C2*, select **Edit— Command Line Editor,** and enter the command **Let C3 = C1 − C2.**	1. Click **Stat** in the top menu.
2. Select **Wilcoxon Tests** from the dropdown menu and **Wilcoxon (Matched Pairs)** from the submenu.		2. Select **Nonparametrics** from the dropdown menu and **Wilcoxon Signed Ranks** from the submenu.
3. Enter a significance level and select the two data columns to include.	2. Click **Stat** in the top menu.	3. Choose **Paired** and select the columns containing the paired data to be analyzed.
4. Click **Evaluate.**	3. Select **Nonparametrics** from the dropdown menu and select **1-Sample Wilcoxon** from the submenu.	4. Select **Hypothesis test for median** and for H_0 enter **0** for the claimed median value. For H_A select the desired format.
	4. Under *Variables* select the column containing the differences between matched pairs (**C3**).	5. Click **Compute!**
	5. Select **Test Median** and enter the median value **0.**	
	6. Choose the format of the alternative hypothesis.	
	7. Click **OK.**	

TI-83/84 Plus Calculator

Requires programs *SRTEST* and *ZZRANK* (available at TriolaStats.com)

1. Create a list of differences between values in the matched pairs. To do this, enter paired data in lists *L1* and *L2* and store the differences in *L3* by entering *L1* ⊖ *L2* STO▸ *L3*.
2. Press PRGM, select **SRTEST,** and press ENTER twice.
3. For **DATA =** enter the name of the list containing the differences (*L3*) and press ENTER.
4. The sample size (*N*), sum of positive ranks (*T+*), and sum of negative ranks (*T−*) will be displayed. Press ENTER to see the mean and standard deviation. Press ENTER again to see the *z* score.
5. If $n \leq 30$, get the critical *T* value from Table A-8. If $n > 30$, get the critical *z* values from Table A-2.

TIP: The list name *L1* (and *L2 . . . L6)* can be quickly entered by pressing 2ND ①.

Excel

XLSTAT Add-In (Required)

1. Click on the **XLSTAT** tab in the Ribbon and then click **Nonparametric tests.**
2. Select **Comparison of two samples** from the dropdown menu.
3. Enter the data range for each sample in the *Sample 1 & 2 boxes*. Check the **Column labels** box if the data range includes labels.
4. Select **Paired samples** under *Data format.*
5. Check the **Wilcoxon signed-rank test** option only.
6. Click the **Options** tab.
7. Under *Alternative Hypothesis* select the desired format. Confirm *Hypothesized difference (D)* is **0.**
8. Enter a significance level and check the **Exact p-value** box.
9. Click **OK.**

13-3   Basic Skills and Concepts

Statistical Literacy and Critical Thinking

1. Wilcoxon Signed-Ranks Test for Body Temperatures The table below lists body temperatures of seven subjects at 8 AM and at 12 AM (from Data Set 3 "Body Temperatures in Appendix B). The data are matched pairs because each pair of temperatures is measured from the same person. Assume that we plan to use the Wilcoxon signed-ranks test to test the claim of no difference between body temperatures at 8 AM and 12 AM.

a. What requirements must be satisfied for this test?

b. Is there any requirement that the samples must be from populations having a normal distribution or any other specific distribution?

c. In what sense is this sign test a "distribution-free test"?

Temperature (°F) at 8 AM	98.0	97.6	97.2	98.0	97.0	98.0	98.2
Temperature (°F) at 12 AM	97.0	98.8	97.6	98.0	97.7	98.8	97.6

2. Body Temperatures For the matched pairs listed in Exercise 1, identify the following components used in the Wilcoxon signed-ranks test:

a. Differences d

b. The ranks corresponding to the nonzero values of $|d|$

c. The signed ranks

d. The sum of the positive ranks and the sum of the absolute values of the negative ranks

e. The value of the test statistic T

f. The critical value of T (assuming a 0.05 significance level in a test of no difference between body temperatures at 8 AM and 12 AM)

3. Sign Test vs. Wilcoxon Signed-Ranks Test Using the data in Exercise 1, we can test for no difference between body temperatures at 8 AM and 12 AM by using the sign test or the Wilcoxon signed-ranks test. In what sense does the Wilcoxon signed-ranks test incorporate and use more information than the sign test?

4. Efficiency of the Wilcoxon Signed-Ranks Test Refer to Table 13-2 on page 600 and identify the efficiency of the Wilcoxon signed-ranks test. What does that value tell us about the test?

Using the Wilcoxon Signed-Ranks Test. *In Exercises 5–8, refer to the sample data for the given exercises in Section 13-2 on page 611. Use the Wilcoxon signed-ranks test to test the claim that the matched pairs have differences that come from a population with a median equal to zero. Use a 0.05 significance level.*

5. Exercise 5 "Speed Dating: Attributes"

6. Exercise 6 "Speed Dating: Attractiveness"

7. Exercise 7 "Speed Dating: Attributes"

8. Exercise 8 "Speed Dating: Attractiveness"

In Exercises 9–12, refer to the sample data from the given exercises in Section 13-2 on page 611. Use the Wilcoxon signed-ranks test for the claim about the median of a population.

9. Exercise 13 "Earthquake Magnitudes"

10. Exercise 14 "Earthquake Depths"

11. Exercise 15 "Testing for Median Weight of Quarters"

12. Exercise 16 "Old Faithful"

13-3 Beyond the Basics

13. Rank Sums Exercise 12 uses Data Set 23 "Old Faithful" in Appendix B, and the sample size is 250.

a. If we have sample paired data with 250 nonzero differences, what are the smallest and largest possible values of T?

b. If we have sample paired data with 250 nonzero differences, what is the expected value of T if the population consists of matched pairs with differences having a median of 0?

c. If we have sample paired data with 250 nonzero differences and the sum of the positive ranks is 1234, find the absolute value of the sum of the negative ranks.

d. If we have sample paired data with n nonzero differences and one of the two rank sums is k, find an expression for the other rank sum.

Wilcoxon Rank-Sum Test for Two Independent Samples

Key Concept This section describes the *Wilcoxon rank-sum test,* which uses ranks of values from two *independent* samples to test the null hypothesis that the samples are from populations having equal medians. The Wilcoxon rank-sum test is equivalent to the **Mann-Whitney U test** (see Exercise 13), which is included in some textbooks and technologies (such as Minitab, StatCrunch, and XLSTAT). Here is the basic idea underlying the Wilcoxon rank-sum test: If two samples are drawn from identical populations and the individual values are all *ranked* as one combined collection of values, then the high and low ranks should fall evenly between the two samples. If the low ranks are found predominantly in one sample and the high ranks are found predominantly in the other sample, we have an indication that the two populations have different medians.

Unlike the parametric t tests for two independent samples in Section 9-2, the Wilcoxon rank-sum test does *not* require normally distributed populations and it can be used with data at the ordinal level of measurement, such as data consisting of ranks. In Table 13-2 we noted that the Wilcoxon rank-sum test has a 0.95 efficiency rating when compared to the parametric test. Because this test has such a high efficiency rating and involves easier calculations, it is often preferred over the parametric t test, even when the requirement of normality is satisfied.

> **CAUTION** Don't confuse the Wilcoxon rank-sum test for two *independent* samples with the Wilcoxon signed-ranks test for matched pairs. Use "Internal Revenue Service" as the mnemonic for IRS to remind yourself of "**I**ndependent: **R**ank **S**um."

> **DEFINITION**
>
> The **Wilcoxon rank-sum test** is a nonparametric test that uses ranks of sample data from two independent populations to test this null hypothesis:
>
> H_0: Two independent samples come from populations with equal medians.
>
> (The alternative hypothesis H_1 can be any one of the following three possibilities: The two populations have *different* medians, or the first population has a median *greater than* the median of the second population, or the first population has a median *less than* the median of the second population.)

Wilcoxon Rank-Sum Test

Objective

Use the Wilcoxon rank-sum test with samples from two independent populations for the following null and alternative hypotheses:

H_0: The two samples come from populations with equal medians.

H_1: The median of the first population is different from (or greater than, or less than)
the median from the second population.

Notation

n_1 = size of Sample 1
n_2 = size of Sample 2
R_1 = sum of ranks for Sample 1
R_2 = sum of ranks for Sample 2
R = same as R_1 (sum of ranks for Sample 1)

μ_R = mean of the sample R values that is expected when
the two populations have equal medians
σ_R = standard deviation of the sample R values that is expected with two populations having equal medians

Requirements

1. There are two independent simple random samples.

2. Each of the two samples has more than 10 values. (For
samples with 10 or fewer values, special tables are
available in reference books, such as *CRC Standard*

Probability and Statistics Tables and Formulae,
published by CRC Press.)

Note: There is *no* requirement that the two populations have
a normal distribution or any other particular distribution.

Test Statistic

$$z = \frac{R - \mu_R}{\sigma_R}$$

where $\mu_R = \dfrac{n_1(n_1 + n_2 + 1)}{2}$ and $\sigma_R = \sqrt{\dfrac{n_1 n_2(n_1 + n_2 + 1)}{12}}$

n_1 = size of the sample from which the rank sum R is found
n_2 = size of the other sample
R = sum of ranks of the sample with size n_1

P-Values

P-values can be found from technology or by using the z test statistic and Table A-2.

Critical Values

Critical values can be found in Table A-2 (because the test statistic is based on the normal distribution).

Procedure for Finding the Value of the Test Statistic

To see how the following steps are applied, refer to the sample data listed in Table 13-5.
The data are from Data Set 17 "Course Evaluations" in Appendix B.

Step 1: Temporarily combine the two samples into one big sample, then replace
each sample value with its rank. (The lowest value gets a rank of 1, the next
lowest value gets a rank of 2, and so on. If values are tied, assign to them the

mean of the ranks involved in the tie. See Section 13-1 for a description of ranks and the procedure for handling ties.)

EXAMPLE: In Table 13-5, the ranks of the 27 student course evaluations are shown in parentheses. The rank of 1 is assigned to the lowest sample value of 2.9, the rank of 2 is assigned to the next lowest value of 3.1, and the rank of 3 is assigned to the next lowest value of 3.3. The next lowest value is 3.4, which occurs four times, so we find the mean of the ranks of 4, 5, 6, 7, which is 5.5, and we assign the rank of 5.5 to each of those four tied values.

Step 2: Find the sum of the ranks for either one of the two samples.

EXAMPLE: In Table 13-5, the sum of the ranks from the first sample is 159.5. (That is, $R_1 = 20.5 + 20.5 + 23.5 + \cdots + 3 = 159.5$.)

Step 3: Calculate the value of the z test statistic as shown in the preceding Key Elements box, where either sample can be used as "Sample 1." (If both sample sizes are greater than 10, then the sampling distribution of R is approximately normal with mean μ_R and standard deviation σ_R, and the test statistic is as shown in the preceding Key Elements box.)

EXAMPLE: Calculations of μ_R and σ_R and z are shown in Example 1, which follows.

TABLE 13-5
Student Course Evaluations

Female Professor	Male Professor
4.3 (20.5)	4.5 (25.5)
4.3 (20.5)	3.7 (8)
4.4 (23.5)	4.2 (17.5)
4.0 (13.5)	3.9 (11)
3.4 (5.5)	3.1 (2)
4.7 (27)	4.0 (13.5)
2.9 (1)	3.8 (9.5)
4.0 (13.5)	3.4 (5.5)
4.3 (20.5)	4.5 (25.5)
3.4 (5.5)	3.8 (9.5)
3.4 (5.5)	4.3 (20.5)
3.3 (3)	4.4 (23.5)
	4.1 (16)
	4.2 (17.5)
	4.0 (13.5)
$n_1 = 12$	$n_2 = 15$
$R_1 = 159.5$	$R_2 = 218.5$

EXAMPLE 1 Course Evaluation Ratings for Female and Male Professors

Table 13-5 lists course evaluation ratings for courses taught by female and male professors (from Data Set 17 "Course Evaluations" in Appendix B). Use a 0.05 significance level to test the claim that female and male professors have the same median course evaluation rating.

SOLUTION

REQUIREMENT CHECK (1) The sample data are two independent simple random samples. (2) The sample sizes are 12 and 15, so both sample sizes are greater than 10. The requirements are satisfied. ✔

The null and alternative hypotheses are as follows:

H_0: The median of the course evaluation ratings for courses taught by female professors is equal to the median of the course evaluation ratings for courses taught by male professors.

H_1: The median course evaluation rating for courses taught by female professors is different from the median course evaluation rating for courses taught by male professors.

Rank the combined list of all 27 course ratings, beginning with a rank of 1 (assigned to the lowest value of 2.9). The ranks corresponding to the individual sample values are shown in parentheses in Table 13-5. R denotes the sum of the ranks for the sample we choose as Sample 1. If we choose the course evaluations of female professors as Sample 1, we get

$$R = 20.5 + 20.5 + 23.5 + \cdots + 3 = 159.5$$

Because there are course evaluation ratings for 12 female professors, we have $n_1 = 12$. Also, $n_2 = 15$ because there are course evaluation ratings of 15 male

continued

professors. The values of μ_R and σ_R and the test statistic z can now be found as follows.

$$\mu_R = \frac{n_1(n_1 + n_2 + 1)}{2} = \frac{12(12 + 15 + 1)}{2} = 168$$

$$\sigma_R = \sqrt{\frac{n_1 n_2(n_1 + n_2 + 1)}{12}} = \sqrt{\frac{(12)(15)(12 + 15 + 1)}{12}} = 20.4939$$

$$z = \frac{R - \mu_R}{\sigma_R} = \frac{159.5 - 168}{20.4939} = -0.41$$

The test is two-tailed because a large positive value of z would indicate that disproportionately more higher ranks are found in Sample 1, and a large negative value of z would indicate that disproportionately more lower ranks are found in Sample 1. In either case, we would have strong evidence against the claim that the two samples come from populations with equal medians.

The significance of the test statistic z can be treated as in previous chapters. We are testing (with $\alpha = 0.05$) the hypothesis that the two populations have equal medians, so we have a two-tailed test.

P-Value Using the unrounded z score, the P-value is 0.678, so we fail to reject the null hypothesis that the populations of female professors and male professors have the same median course evaluation rating.

Critical Values If we use the critical values of $z = \pm 1.96$, we see that the test statistic of $z = -0.41$ does *not* fall within the critical region, so we fail to reject the null hypothesis that the populations of female professors and male professors have the same median course evaluation rating.

INTERPRETATION

There is not sufficient evidence to warrant rejection of the claim that female professors and male professors have the same median course evaluation rating. Based on the available sample data, it appears that female professors and male professors teach courses that are rated about the same.

YOUR TURN   Do Exercise 5 "Student Evaluations of Professors."

In Example 1, if we interchange the two sets of sample values and consider the course evaluations of male professors to be the first sample, then $R = 218.5$, $\mu_R = 210$, $\sigma_R = 20.4939$, and $z = 0.41$, so the conclusion is exactly the same.

Statdisk

Total Number of Values: 93
Rank Sum 1: 1619.5000
Rank Sum 2: 2751.5000

Mean, μ: 1880
Standard Deviation: 128.8669
Test Statistic, z: -2.0215
Critical z: ±1.959962

EXAMPLE 2   Course Evaluation Ratings for Female and Male Professors

Example 1 uses 27 student course evaluation ratings from Data Set 17 "Course Evaluations" in Appendix B. Repeating Example 1 using the course evaluation ratings for all 93 professors in Data Set 17 would not be much fun. The larger sample sizes encourage the use of technology. If we use Statdisk to repeat Example 1 using the course evaluation ratings for all 93 professors in Data Set 17, we get the accompanying display. We can see that the test statistic is $z = -2.02$ (rounded). The unrounded test statistic of $z = -2.0215$ can be used to find that the P-value in this two-tailed test is 0.043. Also, the test statistic falls in the critical region bounded by the critical values of -1.96 and 1.96. We reject the null hypothesis of equal medians. Based on the larger sample of 93 professors, it does appear that female professors and male professors have different median course evaluation ratings.

YOUR TURN   Do Exercise 9 "Student Evaluations of Professors."

Wilcoxon Rank-Sum Test

Access tech supplements, videos, and data sets at **www.TriolaStats.com**

Statdisk	Minitab	StatCrunch
1. Click **Analysis** in the top menu. 2. Select **Wilcoxon Tests** from the dropdown menu and **Wilcoxon (Independent Samples)** from the submenu. 3. Enter a significance level and select the two data columns to include. 4. Click **Evaluate.**	1. Enter the two sets of sample data into columns *C1* and *C2*. 2. Click **Stat** in the top menu. 3. Select **Nonparametrics** from the drop-down menu and select **Mann-Whitney** from the submenu. 4. For *First Sample* select **C1** and for *Second Sample* select **C2**. 5. Enter the confidence level (95.0 corresponds to a significance level of $\alpha = 0.05$). 6. For *Alternative*, choose the format of the alternative hypothesis (*not equal* corresponds to a two-tailed hypothesis test). 7. Click **OK.**	1. Click **Stat** in the top menu. 2. Select **Nonparametrics** from the dropdown menu and **Mann-Whitney** from the submenu. 3. Select the columns to be used for the two samples. 4. Select **Hypothesis test for m1-m2** and for H_0 enter the value of the claimed difference. For H_A select the desired format. 5. Click **Compute!**

TI-83/84 Plus Calculator	Excel
Requires programs *RSTEST* and *ZZRANK* (available at TriolaStats.com) 1. Enter the two sets of sample data in list *L1* and *L2*. 2. Press **PRGM**, select **RSTEST,** and press **ENTER** twice. 3. For **GROUP A** = enter **L1** and press **ENTER**. For **GROUP B** = enter **L2** and press **ENTER**. 4. The rank sum *R*, mean, standard deviation and test statistic *z* will be calculated based on the sample with the fewer number of values. Press **ENTER** again to get the test statistic *z*. Find the critical value by referring to Table A-2 or using the *normalcdf* function as described in the Tech Center in Section 6-1. *TIP: The list name L1 (and L2 . . . L6) can be quickly entered by pressing* **2ND** **①**.	**XLSTAT Add-In (Required)** 1. Click on the **XLSTAT** tab in the Ribbon and then click **Nonparametric tests.** 2. Select **Comparison of two samples** from the dropdown menu. 3. Enter the data range for each sample in the *Sample 1 & 2* boxes. Check the **Column labels** box if the data range includes labels. 4. Select **One column per sample** under *Data format*. 5. Check the **Mann-Whitney test** option only. 6. Click the **Options** tab. 7. Under Alternative Hypothesis select **Sample 1 − Sample 2 ≠ D.** Confirm *Hypothesized difference (D)* is **0.** 8. Enter a significance level and check the **Exact p-value** box. 9. Click **OK.** *TIP: Because XLSTAT uses a different procedure than the one described in this section, results may be somewhat different, especially for small samples.*

13-4 Basic Skills and Concepts

Statistical Literacy and Critical Thinking

1. Student Evaluations of Professors Example 1 in this section used samples of course evaluations, and the table below lists student evaluations of female professors and male professors (from Data Set 17 "Course Evaluations" in Appendix B). Are the requirements for using the Wilcoxon rank-sum test satisfied? Why or why not?

Female	3.9	3.4	3.7	4.1	3.7	3.5	4.4	3.4	4.8	4.1	2.3	4.2	3.6	4.4
Male	3.8	3.4	4.9	4.1	3.2	4.2	3.9	4.9	4.7	4.4	4.3	4.1		

2. Rank Sum After ranking the combined list of professor evaluations given in Exercise 1, find the sum of the ranks for the female professors.

3. What Are We Testing? Refer to the sample data in Exercise 1. Assuming that we use the Wilcoxon rank-sum test with those data, identify the null hypothesis and all possible alternative hypotheses.

4. Efficiency Refer to Table 13-2 on page 600 and identify the efficiency of the Wilcoxon rank-sum test. What does that value tell us about the test?

Wilcoxon Rank-Sum Test. *In Exercises 5–8, use the Wilcoxon rank-sum test.*

5. Student Evaluations of Professors Use the sample data given in Exercise 1 and test the claim that evaluation ratings of female professors have the same median as evaluation ratings of male professors. Use a 0.05 significance level.

6. Radiation in Baby Teeth Listed below are amounts of strontium-90 (in millibecquerels, or mBq, per gram of calcium) in a simple random sample of baby teeth obtained from Pennsylvania residents and New York residents born after 1979 (based on data from "An Unexpected Rise in Strontium-90 in U.S. Deciduous Teeth in the 1990s," by Mangano et al., *Science of the Total Environment*). Use a 0.05 significance level to test the claim that the median amount of strontium-90 from Pennsylvania residents is the same as the median from New York residents.

Pennsylvania	155	142	149	130	151	163	151	142	156	133	138	161
New York	133	140	142	131	134	129	128	140	140	140	137	143

7. Clinical Trials of Lipitor The sample data below are changes in LDL cholesterol levels in clinical trials of Lipitor (atorvastatin). It was claimed that Lipitor had an effect on LDL cholesterol. (The data are based on results given in a Parke-Davis memo from David G. Orloff, M.D., the medical team leader for clinical trials of Lipitor. Pfizer declined to provide the author with the original data values.) Negative values represent decreases in LDL cholesterol. Use a 0.05 significance level to test the claim that for those treated with 20 mg of Lipitor and those treated with 80 mg of Lipitor, changes in LDL cholesterol have the same median. What do the results suggest?

Group Treated with 20 mg of Lipitor:												
−28	−32	−29	−39	−31	−35	−25	−36	−35	−26	−29	−34	−30

Group Treated with 80 mg of Lipitor:												
−42	−41	−38	−42	−41	−41	−40	−44	−32	−37	−41	−37	−34	−31

8. Blanking Out on Tests In a study of students blanking out on tests, the arrangement of test items was studied for its effect on anxiety. The following scores are measures of "debilitating test anxiety" (based on data from "Item Arrangement, Cognitive Entry Characteristics, Sex and Test Anxiety as Predictors of Achievement in Examination Performance," by Klimko, *Journal of Experimental Education*, Vol. 52, No. 4.) Is there sufficient evidence to support the claim that the two samples are from populations with different medians? Is there sufficient evidence to support the claim that the arrangement of the test items has an effect on the score? Use a 0.01 significance level.

Questions Arranged from Easy to Difficult				
24.64	39.29	16.32	32.83	28.02
33.31	20.60	21.13	26.69	28.90
26.43	24.23	7.10	32.86	21.06
28.89	28.71	31.73	30.02	21.96
25.49	38.81	27.85	30.29	30.72

Questions Arranged from Difficult to Easy			
33.62	34.02	26.63	30.26
35.91	26.68	29.49	35.32
27.24	32.34	29.34	33.53
27.62	42.91	30.20	32.54

Appendix B Data Sets. *In Exercises 9–12, refer to the indicated data set in Appendix B and use the Wilcoxon rank-sum test.*

9. Student Evaluations of Professors Repeat Exercise 5 "Student Evaluations of Professors" using all of the student evaluations of professors given in Data Set 17 "Course Evaluations" in Appendix B.

10. Do Men Talk the Same Amount as Women Refer to Data Set 24 "Word Counts" in Appendix B and use the measured word counts from men in the third column and the measured word counts from women in the fourth column. Use a 0.01 significance level to test the claim that contrary to a popular myth, the median of the numbers of words spoken by men in a day is the same as the median of the numbers of words spoken by women in a day.

11. IQ and Lead Exposure Data Set 7 "IQ and Lead" in Appendix B lists *full* IQ scores for a random sample of subjects with "medium" lead levels in their blood and another random sample of subjects with "high" lead levels in their blood. Use a 0.05 significance level to test the claim that subjects with medium lead levels have a higher median of the full IQ scores than subjects with high lead levels. Does lead level appear to affect full IQ scores?

12. IQ and Lead Exposure Data Set 7 "IQ and Lead" in Appendix B lists *performance* IQ scores for a random sample of subjects with low lead levels in their blood and another random sample of subjects with high lead levels in their blood. Use a 0.05 significance level to test the claim that subjects with low lead levels have a higher median of the performance IQ score than those with high lead levels. Does lead exposure appear to have an adverse effect?

13-4 Beyond the Basics

13. Using the Mann-Whitney U Test The Mann-Whitney U test is equivalent to the Wilcoxon rank-sum test for independent samples in the sense that they both apply to the same situations and always lead to the same conclusions. In the Mann-Whitney U test we calculate

$$z = \frac{U - \frac{n_1 n_2}{2}}{\sqrt{\frac{n_1 n_2 (n_1 + n_2 + 1)}{12}}}$$

where

$$U = n_1 n_2 + \frac{n_1 (n_1 + 1)}{2} - R$$

and R is the sum of the ranks for Sample 1. Use the student course evaluation ratings in Table 13-5 on page 621 to find the z test statistic for the Mann-Whitney U test. Compare this value to the z test statistic found using the Wilcoxon rank-sum test.

14. Finding Critical Values Assume that we have two treatments (A and B) that produce quantitative results, and we have only two observations for treatment A and two observations for treatment B. We cannot use the Wilcoxon signed ranks test given in this section because both sample sizes do not exceed 10.

	Rank			Rank Sum for Treatment A
1	2	3	4	
A	A	B	B	3

a. Complete the accompanying table by listing the five rows corresponding to the other five possible outcomes, and enter the corresponding rank sums for treatment A.

b. List the possible values of R and their corresponding probabilities. (Assume that the rows of the table from part (a) are equally likely.)

c. Is it possible, at the 0.10 significance level, to reject the null hypothesis that there is no difference between treatments A and B? Explain.

13-5   Kruskal-Wallis Test for Three or More Samples

Key Concept This section describes the *Kruskal-Wallis test,* which uses *ranks* of data from three or more independent simple random samples to test the null hypothesis that the samples come from populations with the same median.

Section 12-1 described one-way analysis of variance (ANOVA) as a method for testing the null hypothesis that three or more populations have the same *mean,* but ANOVA requires that all of the involved populations have normal distributions. The Kruskal-Wallis test for equal *medians* does not require normal distributions, so it is a distribution-free or nonparametric test.

> **DEFINITION**
>
> The **Kruskal-Wallis test** (also called the **H test**) is a nonparametric test that uses ranks of combined simple random samples from three or more independent populations to test the null hypothesis that the populations have the same median. (The alternative hypothesis is the claim that the populations have medians that are not all equal.)

In applying the Kruskal-Wallis test, we compute the test statistic H, which has a distribution that can be approximated by the chi-square distribution provided that each sample has at least five observations. (For a quick review of the key features of the chi-square distribution, see Section 7-3.)

The H test statistic measures the variance of the rank sums $R_1, R_2, \ldots, R_k$ from the different samples. If the ranks are distributed evenly among the sample groups, then H should be a relatively small number. If the samples are very different, then the ranks will be excessively low in some groups and high in others, with the net effect that H will be large. Consequently, only large values of H lead to rejection of the null hypothesis that the samples come from identical populations. *The Kruskal-Wallis test is therefore a right-tailed test.*

KEY ELEMENTS

Kruskal-Wallis Test

Objective

Use the Kruskal-Wallis test with simple random samples from three or more independent populations for the following null and alternative hypotheses:

H_0: The samples come from populations with the same median.

H_1: The samples come from populations with medians that are not all equal.

Notation

N = total number of observations in all samples combined
k = number of different samples
R_1 = sum of ranks for Sample 1
n_1 = number of observations in Sample 1

For Sample 2, the sum of ranks is R_2 and the number of observations is n_2, and similar notation is used for the other samples.

Requirements

1. We have at least three independent simple random samples.

2. Each sample has at least five observations. (If samples have fewer than five observations, refer to special tables of critical values, such as *CRC Standard Probability* *and Statistics Tables and Formulae,* published by CRC Press.)

Note: There is *no* requirement that the populations have a normal distribution or any other particular distribution.

Test Statistic

$$H = \frac{12}{N(N+1)} \left(\frac{R_1^2}{n_1} + \frac{R_2^2}{n_2} + \cdots + \frac{R_k^2}{n_k} \right) - 3(N+1)$$

P-Values

P-values are often provided by technology. By using the test statistic H and the number of degrees of freedom $(k-1)$, Table A-4 can be used to find a range of values for the *P*-value.

Critical Values

1. The test is *right-tailed* and critical values can be found from technology or from the chi-square distribution in Table A-4.

2. df $= k - 1$ (where df is the number of degrees of freedom and k is the number of different samples)

Procedure for Finding the Value of the *H* Test Statistic To see how the following steps are applied, refer to the sample data in Table 13-6 on the next page. Table 13-6 includes only some of the data in Data Set 7 "IQ and Lead" in Appendix B. This shortened data set is more suitable for illustrating the method of the Kruskal-Wallis test.

Step 1: Temporarily combine all samples into one big sample and assign a rank to each sample value. (Sort the values from lowest to highest, and in cases of ties, assign to each observation the mean of the ranks involved.)

EXAMPLE: In Table 13-6, the numbers in parentheses are the ranks of the combined data set. The rank of 1 is assigned to the lowest value of 64, the rank of 2 is assigned to the next lowest value of 78, and so on. In the case of ties, each of the tied values is assigned the mean of the ranks involved in the tie.

Step 2: For each sample, find the sum of the ranks and find the sample size.

EXAMPLE: In Table 13-6, the sum of the ranks from the first sample is 86, the sum of the ranks for the second sample is 50.5, and the sum of the ranks for the third sample is 53.5.

Step 3: Calculate H using the results of Step 2 and the notation and test statistic given in the preceding Key Elements box.

EXAMPLE: The test statistic is computed in Example 1.

TABLE 13-6 Performance IQ Scores (Ranks in parentheses)

Low Blood Lead Level	Medium Blood Lead Level	High Blood Lead Level
85 (**6.5**)	78 (**2**)	93 (**10**)
90 (**8.5**)	97 (**12.5**)	100 (**15.5**)
107 (**18.5**)	107 (**18.5**)	97 (**12.5**)
85 (**6.5**)	80 (**4**)	79 (**3**)
100 (**15.5**)	90 (**8.5**)	97 (**12.5**)
97 (**12.5**)	83 (**5**)	
101 (**17**)		
64 (**1**)		
$n_1 = 8$	$n_2 = 6$	$n_3 = 5$
$R_1 = 86$	$R_2 = 50.5$	$R_3 = 53.5$

EXAMPLE 1   Effect of Lead on IQ Score

Table 13-6 lists performance (non-verbal) IQ scores from samples of subjects with low blood lead level, medium blood lead level, and high blood lead level (from Data Set 7 "IQ and Lead" in Appendix B). Use a 0.05 significance level to test the claim that the three samples of performance IQ scores come from populations with medians that are all equal.

SOLUTION

REQUIREMENT CHECK   (1) Each of the three samples is a simple random independent sample. (2) Each sample size is at least 5. The requirements are satisfied. ☑
    The null and alternative hypotheses are as follows:

H_0: The population of subjects with low lead exposure, the population with medium lead exposure, and the population with high lead exposure all have performance IQ scores with the same median.

H_1: The three populations of performance IQ scores have three medians that are not all the same.

Test Statistic First combine all of the sample data and rank them, then find the sum of the ranks for each category. In Table 13-6, ranks are shown in parentheses next to the original sample values. Next, find the sample size (n) and sum of ranks (R) for each sample. Those values are shown at the bottom of Table 13-6. Because the total number of observations is 19, we have $N = 19$. We can now evaluate the test statistic as follows:

$$H = \frac{12}{N(N+1)}\left(\frac{R_1^2}{n_1} + \frac{R_2^2}{n_2} + \cdots + \frac{R_k^2}{n_k}\right) - 3(N+1)$$

$$= \frac{12}{19(19+1)}\left(\frac{86^2}{8} + \frac{50.5^2}{6} + \frac{53.5^2}{5}\right) - 3(19+1)$$

$$= 0.694$$

Because each sample has at least five observations, the distribution of H is approximately a chi-square distribution with $k - 1$ degrees of freedom. The number of samples is $k = 3$, so we have $3 - 1 = 2$ degrees of freedom.

P-Value With $H = 0.694$ and df $= 2$, Table A-4 shows that the P-value is greater than 0.10. Using technology, we get P-value $= 0.707$. Because the P-value is greater than the significance level of 0.05, we fail to reject the null hypothesis of equal population medians.

Critical Value Refer to Table A-4 to find the critical value of 5.991, which corresponds to 2 degrees of freedom and a 0.05 significance level (with an area of 0.05 in the right tail). Figure 13-4 shows that the test statistic $H = 0.694$ does not fall within the critical region bounded by 5.991, so we fail to reject the null hypothesis of equal population medians.

Figure 13-4 shows the test statistic of $H = 0.694$ and the critical value of $\chi^2 = 5.991$. (The chi-square distribution has the general shape shown in Figure 13-4 whenever the number of degrees of freedom is 1 or 2.) The test statistic does not fall in the critical region, so we fail to reject the null hypothesis of equal medians.

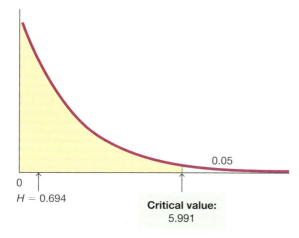

FIGURE 13-4 Chi-Square Distribution for Example 1

INTERPRETATION

There is not sufficient evidence to reject the claim that the three samples of performance IQ scores come from populations with medians that are all equal. The population medians do not appear to be significantly different.

YOUR TURN   Do Exercise 5 "Speed Dating."

Rationale: The Kruskal-Wallis H test statistic is the rank version of the F test statistic used in analysis of variance discussed in Chapter 12. When we deal with ranks R instead of original values x, many components are predetermined. For example, the sum of all ranks can be expressed as $N(N + 1)/2$, where N is the total number of values in all samples combined. The expression

$$H = \frac{12}{N(N + 1)} \Sigma n_i (\overline{R}_i - \overline{\overline{R}})^2$$

where

$$\overline{R}_i = \frac{R_i}{n_i} \quad and \quad \overline{\overline{R}} = \frac{\Sigma R_i}{\Sigma n_i}$$

combines weighted variances of ranks to produce the H test statistic given here, and this expression for H is algebraically equivalent to the expression for H given earlier as the test statistic.

TECH CENTER

Kruskal-Wallis Test

Access tech supplements, videos, and data sets at **www.TriolaStats.com**

Statdisk	Minitab	StatCrunch
1. Click **Analysis** in the top menu. 2. Select **Kruskal-Wallis Test** from the dropdown menu. 3. Enter a significance level and select the columns to be included in the analysis. 4. Click **Evaluate.**	1. List all of the sample data in column *C1* and identify the sample (using names or numbers) for the corresponding value in a second column *C2*. • For the data of Table 13-6 in this section, enter the 19 sample values in *C1* and in *C2* enter eight *1*s followed by six *2*s followed by five *3*s. 2. Click **Stat** in the top menu. 3. Select **Nonparametrics** from the dropdown menu and select **Kruskal-Wallis** from the submenu. 4. For *Response* select column **C1** and for *Factor* select column **C2**. 5. Click **OK.**	1. Click **Stat** in the top menu. 2. Select **Nonparametrics** in the dropdown menu and **Kruskal-Wallis** in the submenu. 3. Select the columns to be used in the analysis. 4. Check **Adjust for ties** under *Options*. 5. Click **Compute!** *TIP: Checking Adjust for ties provides more accurate results that may be different from those in this section of the textbook.*

TI-83/84 Plus Calculator	Excel
Requires programs *KWTEST* and *ZZRANK* (available at TriolaStats.com) 1. Data must be entered as columns in *Matrix A*: *Manually enter data:* Press **2ND** then **x⁻¹** to get to the *MATRIX* menu, select **EDIT** from the top menu, select **[A]**, and press **ENTER**. Enter the number of rows and columns needed, press **ENTER**, and proceed to enter the sample values. *Using existing lists:* Lists can be combined and stored in *Matrix A*. Press **2ND** then **x⁻¹** to get to the *MATRIX* menu, select **MATH** from the top menu, and select the item **List → matr.** Enter the list names followed by the matrix name **[A]**, all separated by commas. *Important:* The matrix name must be entered by pressing **2ND** then **x⁻¹**, selecting **[A]**, and pressing **ENTER**. The following is a summary of the commands used to create a matrix from three lists (*L1, L2, L3*): **List → matr(L1, L2, L3, [A]).** 2. Press **PRGM**, select **KWTEST**, and press **ENTER** twice. The value of the *H* test statistic and the number of degrees of freedom will be provided. Refer to Table A-4 to find the critical value. *TIP: If the samples have different sizes, some of the matrix entries will be zeros. If any of the original data values are zero, add some convenient constant to all of the sample values so that no zeros are present among the original data values.*	**XLSTAT Add-In (Required)** 1. Click on the **XLSTAT** tab in the Ribbon and then click **Nonparametric tests.** 2. Select **Comparison of k samples** from the dropdown menu. 3. In the *Samples* box enter the data range for the sample values. If the range includes labels, check the **Column labels** box. 4. Select **One column per sample** under *Data format*. 5. Check the **Kruskal-Wallis test** option only. 6. Click the **Options** tab. 7. Enter a significance level and check the **Asymptotic p-value** box. 8. Click **OK.**

13-5 Basic Skills and Concepts

Statistical Literacy and Critical Thinking

1. Speed Dating Listed on the top of the next page are attribute ratings of males by females who participated in speed dating events (from Data Set 18 "Speed Dating" in Appendix B). In using the Kruskal-Wallis test, we must rank all of the data combined, and then we must find the sum of the ranks for each sample. Find the sum of the ranks for each of the three samples.

Age 20–22	38	42	30.0	39	47	43	33	31	32	28
Age 23–26	39	31	36.0	35	41	45	36	23	36	20
Age 27–29	36	42	35.5	27	37	34	22	47	36	32

2. Requirements Assume that we want to use the data from Exercise 1 with the Kruskal-Wallis test. Are the requirements satisfied? Explain.

3. Notation For the data given in Exercise 1, identify the values of n_1, n_2, n_3, and N.

4. Efficiency Refer to Table 13-2 on page 600 and identify the efficiency of the Kruskal-Wallis test. What does that value tell us about the test?

Using the Kruskal-Wallis Test. *In Exercises 5–8, use the Kruskal-Wallis test.*

5. Speed Dating Use the sample data from Exercise 1 to test the claim that females from the different age brackets give attribute ratings with the same median. Use a 0.05 significance level.

6. Arsenic in Rice Listed below are amounts of arsenic in samples of brown rice from three different states. The amounts are in micrograms of arsenic and all samples have the same serving size. The data are from the Food and Drug Administration. Use a 0.01 significance level to test the claim that the three samples are from populations with the same median.

Arkansas	4.8	4.9	5.0	5.4	5.4	5.4	5.6	5.6	5.6	5.9	6.0	6.1
California	1.5	3.7	4.0	4.5	4.9	5.1	5.3	5.4	5.4	5.5	5.6	5.6
Texas	5.6	5.8	6.6	6.9	6.9	6.9	7.1	7.3	7.5	7.6	7.7	7.7

7. Car Crash Measurements Use the following listed chest deceleration measurements (in g, where g is the force of gravity) from samples of small, midsize, and large cars. (These values are from Data Set 19 "Car Crash Tests" in Appendix B.) Use a 0.05 significance level to test the claim that the different size categories have the same median chest deceleration in the standard crash test. Do the data suggest that larger cars are safer?

Small	44	39	37	54	39	44	42
Midsize	36	53	43	42	52	49	41
Large	32	45	41	38	37	38	33

8. Highway Fuel Consumption Listed below are highway fuel consumption amounts (mi/gal) for cars categorized by the sizes of small, midsize, and large (from Data Set 20 "Car Measurements" in Appendix B). Using a 0.05 significance level, test the claim that the three size categories have the same median highway fuel consumption. Does the size of a car appear to affect highway fuel consumption?

Small	28	26	23	24	26	24	25
Midsize	28	31	26	30	28	29	31
Large	34	36	28	40	33	35	26

Appendix B Data Sets. *In Exercises 9–12, use the Kruskal-Wallis test with the data set in Appendix B.*

9. Fast Food Dinner Service Times Data Set 25 "Fast Food" in Appendix B lists drive-through service times (seconds) for dinners at McDonald's, Burger King, and Wendy's. Using a 0.05 significance level, test the claim that the three samples are from populations with the same median.

10. Passive and Active Smoke Data Set 12 "Passive and Active Smoke" in Appendix B lists measured cotinine levels from a sample of subjects who smoke, another sample of subjects who do not smoke but are exposed to environmental tobacco smoke, and a third sample of subjects who do not smoke and are not exposed to environmental tobacco smoke. Cotinine is

continued

produced when the body absorbs nicotine. Use a 0.01 significance level to test the claim that the three samples are from populations with the same median. What do the results suggest about a smoker who argues that he absorbs as much nicotine as people who don't smoke?

 11. Birth Weights Data Set 4 "Births" in Appendix B lists birth weights from babies born at Albany Medical Center, Bellevue Hospital in New York City, Olean General Hospital, and Strong Memorial Hospital in Rochester, New York. Use a 0.05 significance level to test the claim that the four different hospitals have different birth weights with different medians.

12. Chocolate Chip Cookies Refer to Data Set 28 "Chocolate Chip Cookies" in Appendix B and use the counts of chocolate chips from the three different types of Chips Ahoy cookies. Use a 0.01 significance level to test the claim that the three different types of Chips Ahoy cookies have the same median number of chocolate chips.

13-5 Beyond the Basics

13. Correcting the _H_ Test Statistic for Ties In using the Kruskal-Wallis test, there is a correction factor that should be applied whenever there are many ties: Divide _H_ by

$$1 - \frac{\Sigma T}{N^3 - N}$$

First combine all of the sample data into one list, and then, in that combined list, identify the different groups of sample values that are tied. For each individual group of tied observations, identify the _number_ of sample values that are tied and designate that number as _t_, then calculate $T = t^3 - t$. Next, add the _T_ values to get ΣT. The value of _N_ is the total number of observations in all samples combined. Use this procedure to find the corrected value of _H_ for Example 1 in this section on page 628. Does the corrected value of _H_ differ substantially from the value found in Example 1?

13-6 Rank Correlation

Key Concept This section describes the nonparametric method of the _rank correlation test,_ which uses _ranks_ of paired data to test for an association between two variables. In Section 10-1, paired sample data were used to compute values for the linear correlation coefficient _r_, but in this section we use _ranks_ as the basis for computing the rank correlation coefficient r_s. As in Chapter 10, we should begin an analysis of paired data by exploring with a scatterplot so that we can identify any patterns in the data as well as outliers.

> **DEFINITION**
>
> The **rank correlation test** (or **Spearman's rank correlation test**) is a nonparametric test that uses ranks of sample data consisting of matched pairs. It is used to test for an association between two variables.

We use the notation r_s for the rank correlation coefficient so that we don't confuse it with the linear correlation coefficient _r_. The subscript _s_ does _not_ refer to a standard deviation; it is used in honor of Charles Spearman (1863–1945), who originated the rank correlation approach. In fact, r_s is often called **Spearman's rank correlation coefficient.** Key components of the rank correlation test are given in the following Key Elements box, and the procedure is summarized in Figure 13-5 on page 634.

Rank Correlation

Objective

Compute the rank correlation coefficient r_s and use it to test for an association between two variables. The null and alternative hypotheses are as follows:

H_0: $\rho_s = 0$ (There is no correlation.)
H_1: $\rho_s \neq 0$ (There is a correlation.)

Notation

r_s = rank correlation coefficient for sample paired data (r_s is a sample statistic)

ρ_s = rank correlation coefficient for all the population data (ρ_s is a population parameter)

n = number of pairs of sample data

d = difference between ranks for the two values within an individual pair

Requirements

1. The paired data are a simple random sample.

2. The data are ranks or can be converted to ranks.

Note: Unlike the parametric methods of Section 10-1, there is *no* requirement that the sample pairs of data have a bivariate normal distribution (as described in Section 10-1). There is *no* requirement of a normal distribution for any population.

Test Statistic

Within each sample, first convert the data to *ranks,* then find the exact value of the rank correlation coefficient r_s by using Formula 10-1:

FORMULA 10-1

$$r_s = \frac{n(\Sigma xy) - (\Sigma x)(\Sigma y)}{\sqrt{n(\Sigma x^2) - (\Sigma x)^2}\sqrt{n(\Sigma y^2) - (\Sigma y)^2}}$$

Simpler Test Statistic if There Are No Ties: After converting the data in each sample to ranks, if there are no ties among ranks for the first variable and there are no ties among ranks for the second variable, the exact value of the test statistic can be calculated using Formula 10-1 or with the following relatively simple formula, but it is probably easier to use Formula 10-1 with technology:

$$r_s = 1 - \frac{6\Sigma d^2}{n(n^2 - 1)}$$

P-Values

P-values are sometimes provided by technology, but use them only if they result from Spearman's rank correlation. (Do not use *P*-values resulting from tests of *linear* correlation; see the "caution" on the top of page 635.)

Critical Values

1. If $n \leq 30$, critical values are found in Table A-9.

2. If $n > 30$, critical values of r_s are found using Formula 13-1.

FORMULA 13-1

$$r_s = \frac{\pm z}{\sqrt{n - 1}} \text{ (critical values for } n > 30)$$

where the value of z corresponds to the significance level. (For example, if $\alpha = 0.05$, $z = 1.96$.)

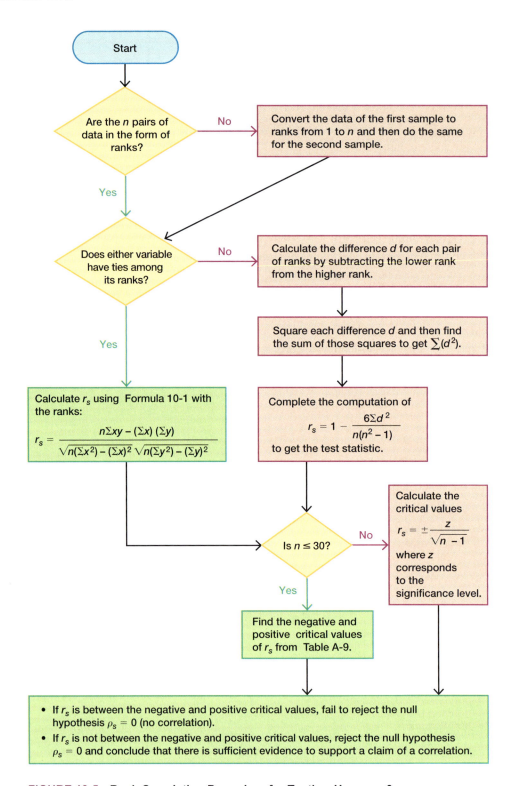

FIGURE 13-5   **Rank Correlation Procedure for Testing H_0: $\rho_s = 0$**

CAUTION *Do not use P-values from linear correlation for methods of rank correlation.* When working with data having ties among ranks, the rank correlation coefficient r_s can be calculated using Formula 10-1. Technology can be used instead of manual calculations with Formula 10-1, but the displayed *P*-values from *linear* correlation do not apply to the methods of *rank* correlation.

Advantages of Rank Correlation: Rank correlation has these advantages over the parametric methods discussed in Chapter 10 :

1. Rank correlation can be used with paired data that are ranks or can be converted to ranks. Unlike the parametric methods of Chapter 10, the method of rank correlation does *not* require a normal distribution for any population.

2. Rank correlation can be used to detect some (not all) relationships that are not linear.

Disadvantage of Rank Correlation: Efficiency A minor disadvantage of rank correlation is its efficiency rating of 0.91, as described in Section 13-1. This efficiency rating shows that with all other circumstances being equal, the nonparametric approach of rank correlation requires 100 pairs of sample data to achieve the same results as only 91 pairs of sample observations analyzed through the parametric approach, assuming that the stricter requirements of the parametric approach are met.

EXAMPLE 1 **Do Better Televisions Cost More?**

Table 13-1 from the Chapter Problem lists ranks and costs (hundreds of dollars) of LCD televisions with screens of at least 60 inches (based on data from *Consumer Reports*). Find the value of the rank correlation coefficient and use it to determine whether there is sufficient evidence to support the claim of a correlation between quality and price. Use a 0.05 significance level. Based on the result, does it appear that you can get better quality by spending more?

TABLE 13-1 Ranks and Costs of LCD Televisions

Quality Rank		1	2	3	4	5	6	7	8	9	10
Cost (hundreds of dollars)		23	50	23	20	32	25	14	16	40	22

SOLUTION

REQUIREMENT CHECK The sample data are a simple random sample from the televisions that were tested. The data are ranks or can be converted to ranks. ✓

The quality ranks are consecutive integers and are not from a population that is normally distributed, so we use the rank correlation coefficient to test for a relationship between quality and price. The null and alternative hypotheses are as follows:

$H_0: \rho_s = 0$ (There is *no* correlation between quality and price.)

$H_1: \rho_s \neq 0$ (There is a correlation between quality and price.)

Following the procedure of Figure 13-5, we begin by converting the costs in Table 13-1 into their corresponding ranks shown in Table 13-7 on the next page. The lowest cost of $1400 in Table 13-1 is assigned a rank of 1. Because the fifth and sixth costs are tied at $2300, we assign the rank of 5.5 to each of them (where 5.5 is the mean of ranks 5 and 6). The ranks corresponding to the costs from Table 13-1 are shown in the second row of Table 13-7.

continued

Direct Link Between Smoking and Cancer

When we find a statistical correlation between two variables, we must be extremely careful to avoid the mistake of concluding that there is a cause-effect link. The tobacco industry has consistently emphasized that correlation does not imply causality as they denied that tobacco products cause cancer. However, Dr. David Sidransky of Johns Hopkins University and other researchers found a direct physical link that involves mutations of a specific gene among smokers. Molecular analysis of genetic changes allows researchers to determine whether cigarette smoking is the cause of a cancer. (See "Association Between Cigarette Smoking and Mutation of the p53 Gene in Squamous-Cell Carcinoma of the Head and Neck," by Brennan, Boyle, et al., *New England Journal of Medicine*, Vol 332, No. 11.) Although statistical methods cannot prove that smoking *causes* cancer, statistical methods can be used to identify an association, and physical proof of causation can then be sought by researchers.

TABLE 13-7 Ranks of Data from Table 13-1

Quality Rank	1	2	3	4	5	6	7	8	9	10
Cost Rank	5.5	10	5.5	3	8	7	1	2	9	4

Because there are ties among ranks, we must use Formula 10-1 to find that the rank correlation coefficient r_s is equal to -0.274.

Formula 10-1

$$r_s = \frac{n\Sigma xy - (\Sigma x)(\Sigma y)}{\sqrt{n(\Sigma x^2) - (\Sigma x)^2}\sqrt{n(\Sigma y^2) - (\Sigma y)^2}}$$

$$= \frac{10(280) - (55)(55)}{\sqrt{10(385) - (55)^2}\sqrt{10(384.5) - (55)^2}} = -0.274$$

Now we refer to Table A-9 to find the critical values of ± 0.648 (based on $\alpha = 0.05$ and $n = 10$). Because the test statistic $r_s = -0.274$ is between the critical values of -0.648 and 0.648, we fail to reject the null hypothesis. There is not sufficient evidence to support a claim of a correlation between quality and cost. Based on the given sample data, it appears that you don't necessarily get better quality by paying more.

> YOUR TURN Do Exercise 7 "Chocolate and Nobel Prizes."

EXAMPLE 2 Large Sample Case

Refer to the measured systolic and diastolic blood pressure measurements of 147 randomly selected females in Data Set 1 "Body Data" in Appendix B and use a 0.05 significance level to test the claim that among women, there is a correlation between systolic blood pressure and diastolic blood pressure.

SOLUTION

REQUIREMENT CHECK The data are a simple random sample and can be converted to ranks. ☑

Test Statistic The value of the rank correlation coefficient is $r_s = 0.354$, which can be found by using technology.

Critical Values Because there are 147 pairs of data, we have $n = 147$. Because n exceeds 30, we find the critical values from Formula 13-1 instead of Table A-9. With $\alpha = 0.05$ in two tails, we let $z = 1.96$ to get the critical values of -0.162 and 0.162, as shown below.

$$r_s = \frac{\pm z}{\sqrt{n-1}} = \frac{\pm 1.96}{\sqrt{147-1}} = \pm 0.162$$

The test statistic of $r_s = 0.354$ is not between the critical values of -0.162 and 0.162, so we reject the null hypothesis of $r_s = 0$. There is sufficient evidence to support the claim that among women, there is a correlation between systolic blood pressure and diastolic blood pressure.

> YOUR TURN Do Exercise 13 "Chocolate and Nobel Prizes."

Detecting Nonlinear Patterns *Rank correlation* methods sometimes allow us to detect relationships that we cannot detect with the *linear correlation* methods of Chapter 10. See the following scatterplot that shows an S-shaped pattern of points suggesting that there is a correlation between x and y. The methods of Chapter 10 result in the linear correlation coefficient of $r = 0.590$ and critical values of ± 0.632, suggesting that there is not sufficient evidence to support the claim of a linear correlation between x and y. But if we use the methods of this section, we get $r_s = 1$ and critical values of ± 0.648, suggesting that there is sufficient evidence to support the claim of a correlation between x and y.

With rank correlation, we can sometimes detect relationships that are not linear.

Nonlinear Pattern

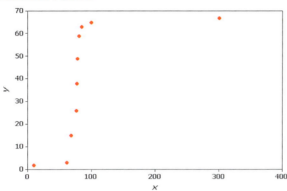

TECH CENTER

 Rank Correlation

Access tech supplements, videos, and data sets at **www.TriolaStats.com**

Statdisk	Minitab	StatCrunch
1. Click **Analysis** from the top menu. 2. Select **Rank Correlation** from the dropdown menu. 3. Enter a significance level and select the two data columns to be included. 4. Click **Evaluate.**	1. Enter the paired data in columns *C1* and *C2*. 2. Click **Stat** in the top menu. 3. Select **Basic Statistics** from the drop-down menu and select **Correlation** from the submenu. 4. Select the columns to be included under *Variables*. 5. Select **Spearman rho** for *Method* and check the **Display p-values** box. 6. Click **OK.**	1. Click **Stat** in the top menu. 2. Select **Nonparametrics** from the dropdown menu and **Spearman's Correlation** from the submenu. 3. Select the columns to be used in the analysis. 4. Under *Display* check **Two-sided P-value.** 5. Click **Compute!**

TI-83/84 Plus Calculator

The TI-83/84 Plus calculator is not designed to calculate rank correlation, but we can replace each value with its corresponding rank and calculate the value of the linear correlation coefficient r.

1. Replace each sample value with its corresponding rank and enter the paired ranks in lists *L1* and *L2*.
2. Press **STAT**, then select **TESTS** in the top menu.
3. Select **LinRegTTest** in the menu and press **ENTER**.
4. Enter the list names for the *x* and *y* variables. Enter **1** for *Freq* and for β & ρ select $\neq$ **0** to test the null hypothesis of no correlation.
5. Select **Calculate** and press **ENTER**. Because the calculation of *r* is done using ranks, the value shown as *r* is actually the rank correlation coefficient r_s. Ignore the *P*-value because it is using the methods of Chapter 10, not the methods of this section.

continued

Rank Correlation

Access tech supplements, videos, and data sets at **www.TriolaStats.com**

Excel

XLSTAT Add-In

1. Click on the **XLSTAT** tab in the Ribbon and then click **Correlation/Association tests.**
2. Select **Correlation tests** from the dropdown menu.
3. For **Observations/variables table** enter the cell range for the data values. If the data range includes a data label, check the **Variable labels** box.
4. For *Type of correlation* select **Spearman.**
5. Enter the desired significance level.
6. Click **OK.** The rank correlation coefficient is displayed in the *Correlation Matrix*. If the value displayed is in **bold font,** we can reject the claim of no correlation.

Excel

Excel does not have a function that calculates the rank correlation coefficient from the original sample values, but the following procedure can be used.

1. Replace each of the original sample values with its corresponding rank.
2. Click **Insert Function f_x,** select the category **Statistical,** select the function **CORREL,** and click **OK.**
3. For *Array1* enter the data range for the first variable. For *Array2* enter the data range for the second variable.
4. Click **OK** for the rank correlation coefficient r_s.

13-6   Basic Skills and Concepts

Statistical Literacy and Critical Thinking

1. Regression If the methods of this section are used with paired sample data, and the conclusion is that there is sufficient evidence to support the claim of a correlation between the two variables, can we use the methods of Section 10-2 to find the regression equation that can be used for predictions? Why or why not?

2. Level of Measurement Which of the levels of measurement (nominal, ordinal, interval, ratio) describe data that *cannot* be used with the methods of rank correlation? Explain.

3. Notation What do r, r_s, ρ, and ρ_s denote? Why is the subscript s used? Does the subscript s represent the same standard deviation s introduced in Section 3-2?

4. Efficiency Refer to Table 13-2 on page 600 and identify the efficiency of the rank correlation test. What does that value tell us about the test?

In Exercises 5 and 6, use the scatterplot to find the value of the rank correlation coefficient r_s and the critical values corresponding to a 0.05 significance level used to test the null hypothesis of $\rho_s = 0$. Determine whether there is a correlation.

5. Distance/Time Data for a Dropped Object

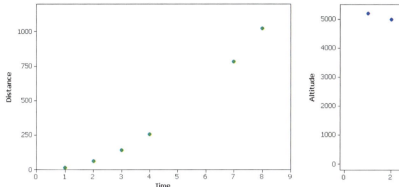

6. Altitude/Time Data for a Descending Aircraft

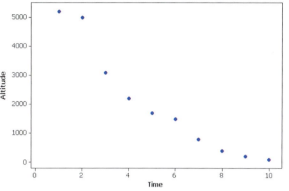

Testing for Rank Correlation. *In Exercises 7–12, use the rank correlation coefficient to test for a correlation between the two variables. Use a significance level of $\alpha = 0.05$.*

7. Chocolate and Nobel Prizes The table below lists chocolate consumption (kg per capita) and the numbers of Nobel Laureates (per 10 million people) for several different countries (from Data Set 16 in Appendix B). Is there a correlation between chocolate consumption and the rate of Nobel Laureates? How could such a correlation be explained?

Chocolate	11.6	2.5	8.8	3.7	1.8	4.5	9.4	3.6	2.0	3.6	6.4
Nobel	12.7	1.9	12.7	3.3	1.5	11.4	25.5	3.1	1.9	1.7	31.9

8. Ages of Best Actresses and Best Actors Listed below are ages of Best Actresses and Best Actors at the times they won Oscars (from Data Set 14 "Oscar Winner Age" in Appendix B). Do these data suggest that there is a correlation between ages of Best Actresses and Best Actors?

Actress	61	32	33	45	29	62	22	44	54
Actor	45	50	48	60	50	39	55	44	33

9. Pizza and the Subway The "pizza connection" is the principle that the price of a slice of pizza in New York City is always about the same as the subway fare. Use the data listed below to determine whether there is a correlation between the cost of a slice of pizza and the subway fare.

Year	1960	1973	1986	1995	2002	2003	2009	2013	2015
Pizza Cost	0.15	0.35	1.00	1.25	1.75	2.00	2.25	2.30	2.75
Subway Fare	0.15	0.35	1.00	1.35	1.50	2.00	2.25	2.50	2.75
CPI	30.2	48.3	112.3	162.2	191.9	197.8	214.5	233.0	237.2

10. CPI and the Subway Use CPI/subway data from the preceding exercise to test for a correlation between the CPI (Consumer Price Index) and the subway fare.

11. Measuring Seals from Photos Listed below are the overhead widths (in cm) of seals measured from photographs and the weights of the seals (in kg). The data are based on "Mass Estimation of Weddell Seals Using Techniques of Photogrammetry," by R. Garrott of Montana State University. The purpose of the study was to determine if weights of seals could be determined from overhead photographs. Is there sufficient evidence to conclude that there is a correlation between overhead widths and the weights of the seals?

Overhead width (cm)	7.2	7.4	9.8	9.4	8.8	8.4
Weight (kg)	116	154	245	202	200	191

12. Crickets and Temperature The association between the temperature and the number of times a cricket chirps in 1 min was studied. Listed below are the numbers of chirps in 1 min and the corresponding temperatures in degrees Fahrenheit (based on data from *The Song of Insects* by George W. Pierce, Harvard University Press). Is there sufficient evidence to conclude that there is a relationship between the number of chirps in 1 min and the temperature?

Chirps in 1 min	882	1188	1104	864	1200	1032	960	900
Temperature (°F)	69.7	93.3	84.3	76.3	88.6	82.6	71.6	79.6

Appendix B Data Sets. *In Exercises 13–16, use the data in Appendix B to test for rank correlation with a 0.05 significance level.*

13. Chocolate and Nobel Prizes Repeat Exercise 7 using all of the paired chocolate/Nobel data in Data Set 16 "Nobel Laureates and Chocolate" in Appendix B.

14. Ages of Best Actresses and Best Actors Repeat Exercise 8 using all of the paired ages from Data Set 14 "Oscar Winner Age" in Appendix B.

15. Blood Pressure Refer to the measured systolic and diastolic blood pressure measurements of 153 randomly selected males in Data Set 1 "Body Data" in Appendix B and use a 0.05 significance level to test the claim that among men, there is a correlation between systolic blood pressure and diastolic blood pressure.

16. IQ and Brain Volume Refer to Data Set 8 "IQ and Brain Size" in Appendix B and test for a correlation between brain volume (cm^3) and IQ score.

13-6 Beyond the Basics

17. Finding Critical Values An alternative to using Table A–9 to find critical values for rank correlation is to compute them using this approximation:

$$r_s = \pm\sqrt{\frac{t^2}{t^2 + n - 2}}$$

Here, t is the critical t value from Table A-3 corresponding to the desired significance level and $n - 2$ degrees of freedom. Use this approximation to find critical values of r_s for Exercise 15 "Blood Pressure." How do the resulting critical values compare to the critical values that would be found by using Formula 13-1 on page 633?

13-7 Runs Test for Randomness

Key Concept This section describes the *runs test for randomness,* which is used to determine whether a sequence of sample data has a random order. This test requires a criterion for categorizing each data value into one of two separate categories, and it analyzes *runs* of those two categories to determine whether the runs appear to result from a random process, or whether the runs suggest that the order of the data is not random.

> **DEFINITIONS**
>
> After characterizing each data value as one of two separate categories, a **run** is a sequence of data having the same characteristic; the sequence is preceded and followed by data with a different characteristic or by no data at all.
>
> The **runs test** uses the number of runs in a sequence of sample data to test for randomness in the order of the data.

Runs Test for Randomness

Objective

Apply the runs test for randomness to a *sequence* of sample data to test for randomness in the *order* of the data. Use the following null and alternative hypotheses:

H_0: The data are in a random order. H_1: The data are in an order that is not random.

Notation

n_1 = number of elements in the sequence that have one particular characteristic. (The characteristic chosen for n_1 is arbitrary.)

n_2 = number of elements in the sequence that have the other characteristic

G = number of runs

Requirements

1. The sample data are arranged according to some ordering scheme, such as the order in which the sample values were obtained.

2. Each data value can be categorized into one of *two* separate categories (such as male/female).

Test Statistic and Critical Values

For Small Samples and $\alpha = 0.05$: If $n_1 \leq 20$ and $n_2 \leq 20$ and the significance level is $\alpha = 0.05$, the test statistic, critical values, and decision criteria are as follows:

- **Test statistic:** number of runs G

- **Critical values of G:** Use Table A-10.

For Large Samples or $\alpha \neq 0.05$: If $n_1 > 20$ or $n_2 > 20$ or $\alpha \neq 0.05$, the test statistic, critical values, and decision criteria are as follows:

- **Test statistic:** $z = \dfrac{G - \mu_G}{\sigma_G}$

  where $\mu_G = \dfrac{2n_1 n_2}{n_1 + n_2} + 1$

  and $\sigma_G = \sqrt{\dfrac{(2n_1 n_2)\,(2n_1 n_2 - n_1 - n_2)}{(n_1 + n_2)^2 (n_1 + n_2 - 1)}}$

- **Decision criteria:** Reject randomness if the number of runs G is such that

  - $G \leq$ smaller critical value found in Table A-10.

  - or $G \geq$ larger critical value found in Table A-10.

- **Critical values of z:** Use Table A-2.

- **Decision criteria:** Reject randomness if the test statistic z is such that

  - $z \leq$ negative critical z score (such as -1.96)

  - or $z \geq$ positive critical z score (such as 1.96).

CAUTION The runs test for randomness is based on the *order* in which the data occur; it is *not* based on the *frequency* of the data. For example, a sequence of 3 men and 20 women might appear to be random, but the issue of whether 3 men and 20 women constitute a *biased* sample (with disproportionately more women) is *not* addressed by the runs test.

Sports Hot Streaks

It is a common belief that athletes often have "hot streaks"— that is, brief periods of extraordinary success. Stanford University psychologist Amos Tversky and other researchers used statistics to analyze the thousands of shots taken by the Philadelphia 76ers for one full season and half of another. They found that the number of "hot streaks" was no different than you would expect from random trials with the outcome of each trial independent of any preceding results. That is, the probability of making a basket doesn't depend on the preceding make or miss.

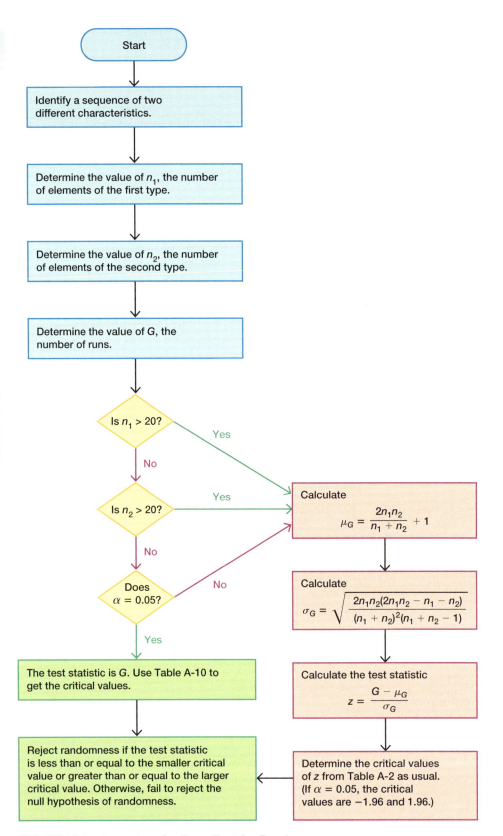

FIGURE 13-6 **Procedure for Runs Test for Randomness**

Fundamental Principle of the Runs Test

Here is the key idea underlying the runs test:

Reject randomness if the number of runs is very *low* or very *high*.

- Example: The sequence of genders FFFFFMMMMM is not random because it has only 2 runs, so the number of runs is very *low*.
- Example: The sequence of genders FMFMFMFMFM is not random because there are 10 runs, which is very *high*.

The exact criteria for determining whether a number of runs is very high or low are found in the Key Elements box. The procedure for the runs test for randomness is also summarized in Figure 13-6.

EXAMPLE 1   Small Sample: Political Parties of Presidents

Listed below are the recent political parties of presidents of the United States. The letter R represents a Republican president and the letter D represents a Democratic president. Use a 0.05 significance level to test for randomness in the sequence.

R R D D R D D R R D R R D R D

SOLUTION

REQUIREMENT CHECK   (1) The data are arranged in order. (2) Each data value is categorized into one of two separate categories (Republican/Democrat). The requirements are satisfied. ✓

We will follow the procedure summarized in Figure 13-6. The sequence of two characteristics (Republican/Democrat) has been identified. We must now find the values of n_1, n_2, and the number of runs G. The sequence is shown below with spacing adjusted to better identify the different runs.

RR	DD	R	DD	RR	D	RR	D	R	D
1st run	2nd run	3rd run	4th run	5th run	6th run	7th run	8th run	9th run	10th run

The above display shows that there are 8 Republican presidents and 7 Democratic presidents, and the number of runs is 10. We represent those results with the following notation.

n_1 = number of Republican presidents = 8
n_2 = number of Democratic presidents = 7
G = number of runs = 10

Because $n_1 \leq 20$ and $n_2 \leq 20$ and the significance level is $\alpha = 0.05$, the test statistic is $G = 10$ (the number of runs), and we refer to Table A-10 to find the critical values of 4 and 13. Because $G = 10$ is neither less than or equal to the critical value of 4, nor is it greater than or equal to the critical value of 13, *we do not reject randomness.* There is not sufficient evidence to reject randomness in the sequence of political parties of recent presidents.

EXAMPLE 2 Large Sample: Randomness of Births

Data Set 4 "Births" in Appendix B lists genders from 100 births at Bellevue Hospital Center. Let's consider the sequence of the listed genders indicated below, where $1 =$ boy and $0 =$ girl. (The complete list of 100 genders can be seen in Data Set 4.) Use a 0.05 significance level to test the claim that the sequence is random.

$$1 \quad 0 \quad 1 \quad 0 \quad 0 \quad 0 \quad 0 \quad 1 \ldots 1 \quad 0 \quad 1$$

SOLUTION

REQUIREMENT CHECK (1) The data are arranged in order. (2) Each data value is categorized into one of two separate categories (boy/girl). The requirements are satisfied. ✓

The null and alternative hypotheses are as follows:

H_0: The sequence is random.

H_1: The sequence is not random.

Examination of the sequence of 100 genders results in these values:

$$n_1 = \text{number of boys} = 51$$
$$n_2 = \text{number of girls} = 49$$
$$G = \text{number of runs} = 59$$

Since $n_1 > 20$, we need to calculate the test statistic z, so we must first evaluate μ_G and σ_G as follows:

$$\mu_G = \frac{2n_1n_2}{n_1 + n_2} + 1 = \frac{2(51)(49)}{51 + 49} + 1 = 50.98$$

$$\sigma_G = \sqrt{\frac{(2n_1n_2)(2n_1n_2 - n_1 - n_2)}{(n_1 + n_2)^2(n_1 + n_2 - 1)}}$$

$$= \sqrt{\frac{(2)(51)(49)[2(51)(49) - 51 - 49]}{(51 + 49)^2(51 + 49 - 1)}} = 4.972673$$

We now find the test statistic:

$$z = \frac{G - \mu_G}{\sigma_G} = \frac{59 - 50.98}{4.972673} = 1.61$$

We can use the test statistic $z = 1.61$ to find the *P*-value of 0.1074, which is greater than the significance level of $\alpha = 0.05$, so we fail to reject the null hypothesis of randomness.

Also, because the significance level is $\alpha = 0.05$ and we have a two-tailed test, the critical values are $z = -1.96$ and $z = 1.96$. The test statistic of $z = 1.61$ does not fall within the critical region, so we again fail to reject the null hypothesis of randomness.

INTERPRETATION

We do not have sufficient evidence to reject randomness of the sequence of 100 births at Bellevue Hospital Center (from Data Set 4 "Births" in Appendix B).

YOUR TURN Do Exercise 9 "Testing for Randomness of Super Bowl Victories."

Testing for Randomness Above and Below the Mean or Median

We can test for randomness in the way numerical data fluctuate above or below a mean or median. To test for randomness above and below the median, for example, use the sample data to find the value of the median, then replace each individual value with the letter A if it is *above* the median and replace it with B if it is *below* the median. Delete any values that are equal to the median. (It is helpful to write the A's and B's directly above or below the numbers they represent because this makes checking easier and also reduces the chance of having the wrong number of letters.) After finding the sequence of A and B letters, we can proceed to apply the runs test as described earlier.

Economists use the runs test for randomness above and below the median to identify trends or cycles. An upward economic trend would contain a predominance of B's at the beginning and A's at the end, so the number of runs would be very small. A downward trend would have A's dominating at the beginning and B's at the end, with a small number of runs. A cyclical pattern would yield a sequence that systematically changes, so the number of runs would tend to be large.

TECH CENTER

Runs Test for Randomness
Access tech supplements, videos, and data sets at **www.TriolaStats.com**

Statdisk	Minitab	StatCrunch
1. Determine the number of elements in the first category, the number of elements in the second category, and count the number of runs. 2. Click **Analysis** in the top menu. 3. Select **Runs Test** from the dropdown menu. 4. Enter a significance level and the values for number of runs, number of elements in the first category *(Element 1)*, and number of elements in the second category *(Element 2).* 5. Click **Evaluate.**	*Minitab will do a runs test with a sequence of numerical data only. The Minitab Student Laboratory Manual and Workbook provides additional information on how to circumvent this constraint.* 1. Enter numerical data in column *C1.* 2. Click **Stat** in the top menu. 3. Select **Nonparametrics** from the dropdown menu and select **Runs Test** from the submenu. 4. Select column **C1** under *Variables.* 5. Select to test above and below the mean or enter a desired value to be used. 6. Click **OK.**	Not available.

TI-83/84 Plus Calculator	Excel
Not available.	Not available.

13-7   Basic Skills and Concepts

Statistical Literacy and Critical Thinking

In Exercises 1–4, use the following sequence of political party affiliations of recent presidents of the United States, where **R** *represents Republican and* **D** *represents Democrat.*

R R R R D R D R R D R R R D D R D D R R D R R D R D

1. Testing for Bias Can the runs test be used to show the proportion of Republicans is significantly greater than the proportion of Democrats?

2. Notation Identify the values of n_1, n_2, and G that would be used in the runs test for randomness.

3. Runs Test If we use a 0.05 significance level to test for randomness, what are the critical values from Table A-10? Based on those values and the number of runs from Exercise 2, what should be concluded about randomness?

4. Good Sample? Given the sequence of data, if we fail to reject randomness, does it follow that the sampling method is suitable for statistical methods? Explain.

Using the Runs Test for Randomness. *In Exercises 5–10, use the runs test with a significance level of $\alpha = 0.05$. (All data are listed in order by row.)*

5. Law Enforcement Fatalities Listed below are numbers of law enforcement fatalities for 20 recent and consecutive years. First find the mean, identify each value as being above the mean (A) or below the mean (B), then test for randomness above and below the mean. Is there a trend?

183	140	172	171	144	162	241	159	150	165
163	156	192	148	125	161	171	126	107	117

6. Odd and Even Digits in Pi A *New York Times* article about the calculation of decimal places of π noted that "mathematicians are pretty sure that the digits of π are indistinguishable from any random sequence." Given below are the first 25 decimal places of π. Test for randomness in the way that odd (O) and even (E) digits occur in the sequence. Based on the result, does the statement from the *New York Times* appear to be accurate?

1 4 1 5 9 2 6 5 3 5 8 9 7 9 3 2 3 8 4 6 2 6 4 3 3

7. Draft Lottery In 1970, a lottery was used to determine who would be drafted into the U.S. Army. The 366 dates in the year were placed in individual capsules, they were mixed, and then capsules were selected to identify birth dates of men to be drafted first. The first 30 results are listed below. Test for randomness before and after the middle of the year, which is July 1.

Sept. 14	Apr. 24	Dec. 30	Feb. 14	Oct. 18	Sept. 6	Oct. 26	Sept. 7	Nov. 22
Dec. 6	Aug 31	Dec. 7	July 8	Apr. 11	July 12	Dec. 29	Jan. 15	Sept. 26
Nov. 1	June 4	Aug. 10	June 26	July 24	Oct. 5	Feb. 19	Dec. 14	July 21
June 5	Mar. 2	Mar. 31						

8. Newspapers Media experts claim that daily print newspapers are declining because of Internet access. Listed below are the numbers of daily print newspapers in the United States for a recent sequence of years. First find the median, then test for randomness of the numbers above and below the median. What do the results suggest?

1611	1586	1570	1556	1548	1533	1520	1509	1489	1483	1480
1468	1457	1456	1457	1452	1437	1422	1408	1387	1382	

Runs Test with Large Samples. *In Exercises 9–12, use the runs test with a significance level of $\alpha = 0.05$. (All data are listed in order by row.)*

9. Testing for Randomness of Super Bowl Victories Listed below are the conference designations of teams that won the Super Bowl, where N denotes a team from the NFC and A denotes a team from the AFC. Do the results suggest that either conference is superior?

N	N	A	A	A	N	A	A	A	A	A	N	A	A	A	N	N	A	N	N	N	N	N	N	N
N	N	N	N	N	N	A	A	N	A	A	N	A	A	A	A	N	A	N	N	N	A	N	A	

10. Baseball World Series Victories Test the claim that the sequence of World Series wins by American League and National League teams is random. Given on the next page are recent results, with A = American League and N = National League.

A	N	A	N	N	N	A	A	A	A	N	A	A	A	A	N	A	N	N	A	A	N	N	A
A	A	A	N	A	N	N	A	A	A	A	A	N	A	N	A	N	A	N	A	A	A	A	A
A	A	N	N	A	N	A	N	N	A	A	N	N	N	A	N	A	N	A	N	A	A	A	N
N	A	A	N	N	N	N	A	A	A	N	A	N	A	N	A	A	A	N	A	N	A	A	A
N	A	N	A	A	N	A	N	A	N	N	N	A	N										

11. Stock Market: Testing for Randomness Above and Below the Median Listed below are the annual high values of the Dow Jones Industrial Average for a recent sequence of years. Find the median, then test for randomness below and above the median. What does the result suggest about the stock market as an investment consideration?

969	995	943	985	969	842	951	1036	1052	892	882	1015
1000	908	898	1000	1024	1071	1287	1287	1553	1956	2722	2184
2791	3000	3169	3413	3794	3978	5216	6561	8259	9374	11568	11401
11350	10635	10454	10855	10941	12464	14198	13279	10580	11625	12929	13589
16577	18054										

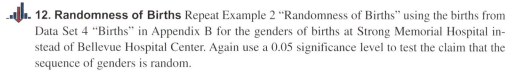 **12. Randomness of Births** Repeat Example 2 "Randomness of Births" using the births from Data Set 4 "Births" in Appendix B for the genders of births at Strong Memorial Hospital instead of Bellevue Hospital Center. Again use a 0.05 significance level to test the claim that the sequence of genders is random.

13-7 Beyond the Basics

13. Finding Critical Values

a. Using all of the elements A, A, A, B, B, B, B, B, B, list the 84 different possible sequences.

b. Find the number of runs for each of the 84 sequences.

c. Use the results from parts (a) and (b) to find your own critical values for G.

d. Compare your results to those given in Table A-10.

Chapter Quick Quiz

1. Speed Dating Some of the nonparametric methods in this chapter use ranks of data. Find the ranks corresponding to these attractiveness ratings (1 = not attractive; 10 = extremely attractive) of males by females who participated in a speed dating event (from Data Set 18 "Speed Dating"): 5, 7, 7, 8, 7.

2. Efficiency What does it mean when we say that the rank correlation test has an efficiency rating of 0.91 when compared to the parametric test for linear correlation?

3. Nonparametric Tests

a. Which of the following terms is sometimes used instead of "nonparametric test": *normality test; abnormality test; distribution-free test; last testament; test of patience*?

b. Why is the term that is the answer to part (a) better than "nonparametric test"?

4. Foot Length/Height Listed below are foot lengths (cm) and heights (cm) of males from Data Set 2 "Foot and Height" in Appendix B. Which method of nonparametric statistics should be used? What characteristic of the data is investigated with this test?

| Foot Length | 27.8 | 25.7 | 26.7 | 25.9 | 26.4 | 29.2 | 26.8 | 28.1 | 25.4 | 27.9 |
| Height | 180.3 | 175.3 | 184.8 | 177.8 | 182.3 | 185.4 | 180.3 | 175.3 | 177.8 | 185.4 |

5. Foot Length/Height When analyzing the paired data in Exercise 4, are the *P*-values and conclusions from the nonparametric test and the parametric test always the same?

6. Foot Length/Height For the sample data given in Exercise 4, identify at least one advantage of using the appropriate nonparametric test over the parametric test.

7. Runs Test Assume that we use the runs test of randomness above and below the mean for the monetary value of Google at the end of each year for 20 years and we find that $G = 2$. What does that value tell us about the monetary value of Google?

8. Runs Test Consider sample data consisting of genders of criminals charged with hacking computer systems of corporations. Determine whether the following are true or false.

a. If the runs test suggests that sample data occur in a random order, then it follows that the data have been randomly selected.

b. If the runs test suggests that sample data occur in a random order, then there is not a significant difference between the proportions of males and females.

9. Sign Test and Wilcoxon Signed-Ranks Test What is a major advantage of the Wilcoxon signed-ranks test over the sign test when analyzing data consisting of matched pairs?

10. Which Test? Three different judges give the same singers ratings on a scale of 0 to 10. What method of this chapter can be used to determine whether one of the judges is tougher or more lenient than the others, as indicated by a median rating that is significantly different from the others?

Review Exercises

Using Nonparametric Tests. *In Exercises 1–10, use a 0.05 significance level with the indicated test. If no particular test is specified, use the appropriate nonparametric test from this chapter.*

1. Job Stress and Income Listed below are job stress scores and median annual salaries (thousands of dollars) for various jobs, including firefighters, airline pilots, police officers, and university professors (based on data from "Job Rated Stress Score" from CareerCast.com). Do these data suggest that there is a correlation between job stress and annual income? Does it appear that jobs with more stress have higher salaries?

Stress	71.59	60.46	50.82	6.94	8.10	50.33	49.2	48.8	11.4
Median salary	45.6	98.4	57.0	69.0	35.4	46.1	42.5	37.1	31.2

2. Presidents, Popes, Monarchs Listed below are numbers of years that U.S. presidents, popes, and British monarchs lived after their inauguration, election, or coronation, respectively. Assume that the data are samples randomly selected from larger populations. Test the claim that the three samples are from populations with the same median.

	10	29	26	28	15	23	17	25	0	20	4	1	24	16	12
Presidents	4	10	17	16	0	7	24	12	4	18	21	11	2	9	36
	12	28	3	16	9	25	23	32							
	2	9	21	3	6	10	18	11	6	25	23	6	2	15	32
Popes	25	11	8	17	19	5	15	0	26						
Monarchs	17	6	13	12	13	33	59	10	7	63	9	25	36	15	

3. World Series The last 110 baseball World Series ended with 63 wins by American League teams and 47 wins by National League teams. Use the sign test to test the claim that in each World Series, the American League team has a 0.5 probability of winning.

4. California Lottery Listed below are consecutive first-digits drawn in the California Daily 4 lottery. Test for randomness of even and odd integers. Does the lottery appear to be working as it should?

| 5 | 2 | 2 | 8 | 4 | 8 | 8 | 7 | 1 | 0 | 6 | 4 | 1 | 5 | 1 | 5 | 5 | 3 | 1 | 4 | 1 | 5 | 0 | 0 | 3 | 9 | 6 | 6 | 3 | 7 |

5. Speed Dating In a study of speed dating conducted at Columbia University, female subjects were asked to rate the attractiveness of their male dates, and a sample of the results is listed below (1 = not attractive; 10 = extremely attractive). Use the sign test to test the claim that the sample is from a population with a median equal to 5.

| 5 | 8 | 3 | 8 | 6 | 10 | 3 | 7 | 9 | 8 | 5 | 5 | 6 | 8 | 8 | 7 | 3 | 5 | 5 | 6 | 8 | 7 | 8 | 8 | 8 | 7 |

6. Speed Dating Repeat the preceding exercise using the Wilcoxon signed-ranks test.

7. Old Faithful Listed below are time intervals (min) between eruptions of the Old Faithful geyser. The "recent" times are within the past few years, and the "past" times are from 1995. Test the claim that the two samples are from populations with the same median. Does the conclusion change with a 0.01 significance level?

Recent	78	91	89	79	57	100	62	87	70	88	82	83	56	81	74	102	61
Past (1995)	89	88	97	98	64	85	85	96	87	95	90	95					

8. Airline Fares Listed below are the costs (in dollars) of eight different flights from New York (JFK) to San Francisco for Virgin America, US Airways, United Airlines, JetBlue, Delta, American Airlines, Alaska Airlines, and Sun Country Airlines. (Each pair of costs is for the same flight.) Use the sign test to test the claim that there is no difference in cost between flights scheduled 1 day in advance and those scheduled 30 days in advance. What appears to be a wise scheduling strategy?

Flights scheduled one day in advance	584	490	584	584	584	606	628	717
Flight scheduled 30 days in advance	254	308	244	229	284	509	394	258

9. Airline Fares Refer to the same data from the preceding exercise. Use the Wilcoxon signed-ranks test to test the claim that differences between fares for flights scheduled 1 day in advance and those scheduled 30 days in advance have a median equal to 0. What do the results suggest?

10. Student and *U.S. News & World Report* Rankings of Colleges Each year, *U.S. News & World Report* publishes rankings of colleges based on statistics such as admission rates, graduation rates, class size, faculty–student ratio, faculty salaries, and peer ratings of administrators. Economists Christopher Avery, Mark Glickman, Caroline Minter Hoxby, and Andrew Metrick took an alternative approach of analyzing the college choices of 3240 high-achieving school seniors. They examined the colleges that offered admission along with the colleges that the students chose to attend. The table below lists rankings for a small sample of colleges. Find the value of the rank correlation coefficient and use it to determine whether there is a correlation between the student rankings and the rankings of the magazine.

Student ranks	1	2	3	4	5	6	7	8
U.S. News & World Report ranks	1	2	5	4	7	6	3	8

Cumulative Review Exercises

In Exercises 1–3, use the data listed below. The values are departure delay times (minutes) for American Airlines flights from New York to Los Angeles. Negative values correspond to flights that departed early.

Flight 1 (min)	−2	−1	−2	2	−2	0	−2	−3
Flight 19 (min)	19	−4	−5	−1	−4	73	0	1
Flight 21 (min)	18	60	142	−1	−11	−1	47	13

1. Flight Departure Delays Compare the three samples using means, medians, and standard deviations.

2. Test for Normality Use the departure delay times for Flight 19 and test for normality using a normal quantile plot.

3. Departure Delay Times Use a nonparametric test to test the claim that the three samples are from populations with the same median departure delay time.

4. Drug Tests There is a 3.9% rate of positive drug test results among workers in the United States (based on data from Quest Diagnostics). Assuming that this statistic is based on a sample of size 2000, construct a 95% confidence interval estimate of the *percentage* of positive drug test results. Write a brief statement that interprets the confidence interval.

5. Drug Tests Use the data from the preceding exercise and test the claim that the rate of positive drug test results among workers in the United States is greater than 3.0%. Use a 0.05 significance level.

6. Randomness Refer to the following ages at inauguration of the elected presidents of the United States (from Data Set 15 "Presidents" in Appendix B). Test for randomness above and below the mean. Do the results suggest an upward trend or a downward trend?

57	61	57	57	58	57	61	54	68	49	64	48	65	52	46	54	49	47
55	54	42	51	56	55	51	54	51	60	62	43	55	56	52	69	64	46
54	47																

7. Sample Size Advances in technology are dramatically affecting different aspects of our lives. For example, the number of daily print newspapers is decreasing because of easy access to Internet and television news. To help address such issues, we want to estimate the percentage of adults in the United States who use a computer at least once each day. Find the sample size needed to estimate that percentage. Assume that we want 95% confidence that the sample percentage is within two percentage points of the true population percentage.

8. Mean and Median In a recent year, the players on the New York Yankees baseball team had salaries with a mean of $7,052,129 and a median of $2,500,000. Explain how the mean and median can be so far apart.

9. Fear of Heights Among readers of a *USA Today* website, 285 chose to respond to this posted question: "Are you afraid of heights in tall buildings?" Among those who chose to respond, 46% answered "yes" and 54% answered "no." Use a 0.05 significance level to test the claim that the majority of the population is not afraid of heights in tall buildings. What is wrong with this hypothesis test?

10. Cell Phones and Crashes: Analyzing Newspaper Report In an article from the Associated Press, it was reported that researchers "randomly selected 100 New York motorists who had been in an accident and 100 who had not been in an accident. Of those in accidents, 13.7 percent owned a cellular phone, while just 10.6 percent of the accident-free drivers had a phone in the car." What is wrong with these results?

Technology Project

Past attempts to identify or contact extraterrestrial intelligent life have involved efforts to send radio messages carrying information about us earthlings. Dr. Frank Drake of Cornell University developed such a radio message that could be transmitted as a series of pulses and gaps. The pulses and gaps can be considered to be 1s and 0s. Listed below is a message consisting of 77 entries of 0s and 1s. If we factor 77 into the prime numbers of 7 and 11 and then make an 11 × 7 grid and put a dot at those positions corresponding to a pulse of 1, we can get a simple picture of something. Assume that the sequence of 77 entries of 1s and 0s is sent as a radio message that is intercepted by extraterrestrial life with enough intelligence to have studied this book. If the radio message is tested using the methods of this chapter, will the sequence appear to be "random noise" or will it be identified as a pattern that is not random? Also, construct the image represented by the digits and identify it.

0	0	1	1	1	0	0	0	0	1	1	1	0	0	0	0	0	1	0	0	0
1	1	1	1	1	1	1	0	0	1	1	1	0	0	0	0	1	1	1	0	0
0	0	1	1	1	0	0	0	1	0	0	0	1	0	1	0	0	0	0	1	0
1	0	0	0	0	1	0	1	0	0	0	0	1	0							

FROM DATA TO DECISION

Critical Thinking: Was the draft lottery random?

On December 1, 1969, during the Vietnam War, a lottery was used to determine who would be drafted into the U.S. Army, but the lottery generated considerable controversy. The different dates in a year were placed in 366 individual capsules. First, the 31 January capsules were placed in a box; then the 29 February capsules were added and the two months were mixed. Next, the 31 March capsules were added and the three months were mixed. This process continued until all months were included. The first capsule selected was September 14, so men born on that date were drafted first. The accompanying list shows the 366 priority dates in the order of selection. This list is available for download at www.TriolaStats.com.

Analyzing the Results

a. Use the runs test to test the sequence for randomness above and below the median of 183.5.

b. Use the Kruskal-Wallis test to test the claim that the 12 months had priority numbers drawn from the same population.

c. Calculate the 12 monthly means. Then plot those 12 means on a graph. (The horizontal scale lists the 12 months, and the vertical scale ranges from 100 to 260.) Note any pattern suggesting that the original priority numbers were not randomly selected.

d. Based on the results from parts (a), (b), and (c), decide whether this particular draft lottery was fair. Write a statement explaining why you believe that it was or was not fair. If you decided that this lottery was unfair, describe a process for selecting lottery numbers that would have been fair.

Day	Jan.	Feb.	Mar.	Apr.	May	Jun.	Jul.	Aug.	Sep.	Oct.	Nov.	Dec.
1	305	86	108	32	330	249	93	111	225	359	19	129
2	159	144	29	271	298	228	350	45	161	125	34	328
3	251	297	267	83	40	301	115	261	49	244	348	157
4	215	210	275	81	276	20	279	145	232	202	266	165
5	101	214	293	269	364	28	188	54	82	24	310	56
6	224	347	139	253	155	110	327	114	6	87	76	10
7	306	91	122	147	35	85	50	168	8	234	51	12
8	199	181	213	312	321	366	13	48	184	283	97	105
9	194	338	317	219	197	335	277	106	263	342	80	43
10	325	216	323	218	65	206	284	21	71	220	282	41
11	329	150	136	14	37	134	248	324	158	237	46	39
12	221	68	300	346	133	272	15	142	242	72	66	314

continued

Day	Jan.	Feb.	Mar.	Apr.	May	Jun.	Jul.	Aug.	Sep.	Oct.	Nov.	Dec.
13	318	152	259	124	295	69	42	307	175	138	126	163
14	238	4	354	231	178	356	331	198	1	294	127	26
15	17	89	169	273	130	180	322	102	113	171	131	320
16	121	212	166	148	55	274	120	44	207	254	107	96
17	235	189	33	260	112	73	98	154	255	288	143	304
18	140	292	332	90	278	341	190	141	246	5	146	128
19	58	25	200	336	75	104	227	311	177	241	203	240
20	280	302	239	345	183	360	187	344	63	192	185	135
21	186	363	334	62	250	60	27	291	204	243	156	70
22	337	290	265	316	326	247	153	339	160	117	9	53
23	118	57	256	252	319	109	172	116	119	201	182	162
24	59	236	258	2	31	358	23	36	195	196	230	95
25	52	179	343	351	361	137	67	286	149	176	132	84
26	92	365	170	340	357	22	303	245	18	7	309	173
27	355	205	268	74	296	64	289	352	233	264	47	78
28	77	299	223	262	308	222	88	167	257	94	281	123
29	349	285	362	191	226	353	270	61	151	229	99	16
30	164		217	208	103	209	287	333	315	38	174	3
31	211		30		313		193	11		79		100

Cooperative Group Activities

1. Out-of-class activity Half of the students should make up results for 200 coin flips and the other half should collect results from 200 actual tosses of a coin. Then use the runs test to determine whether the results appear to be random.

2. In-class activity Use the existing seating arrangement in your class and apply the runs test to determine whether the students are arranged randomly according to gender. After recording the seating arrangement, analysis can be done in subgroups of three or four students.

3. In-class activity Divide into groups of 8 to 12 people. For each group member, *measure* his or her height and *measure* his or her arm span. For the arm span, the subject should stand with arms extended, like the wings on an airplane. Divide the following tasks among subgroups of three or four people.

a. Use rank correlation with the paired sample data to determine whether there is a correlation between height and arm span.

b. Use the sign test to test for a difference between the two variables.

c. Use the Wilcoxon signed-ranks test to test for a difference between the two variables.

4. In-class activity Do Activity 3 using pulse rate instead of arm span. Measure pulse rates by counting the number of heartbeats in 1 min.

5. Out-of-class activity Divide into groups of three or four students. Investigate the relationship between two variables by collecting your own paired sample data and using the methods of Section 13-6 to determine whether there is a correlation. Suggested topics:

• Is there a correlation between taste and cost of different brands of chocolate chip cookies (or colas)? (Taste can be measured on some number scale, such as 1 to 10.)

• Is there a correlation between salaries of professional baseball (or basketball or football) players and their season achievements (such as batting average or points scored)?

• Is there a correlation between car fuel consumption rates and car weights?

• Is there a correlation between the lengths of men's (or women's) feet and their heights?

• Is there a correlation between student grade-point averages and the amount of television watched?

• Is there a correlation between heights of fathers (or mothers) and heights of their first sons (or daughters)?

6. Out-of-class activity See this chapter's "From Data to Decision" project, which involves analysis of the 1970 lottery used for drafting men into the U.S. Army. Because the 1970 results raised concerns about the randomness of selecting draft priority numbers, design a new procedure for generating the 366 priority numbers. Use your procedure to generate the 366 numbers and test your results using the techniques suggested in parts (a), (b), and (c) of the "From Data to Decision" project. How do your results compare to those obtained in 1970? Does your random selection process appear to be better than the one used in 1970? Write a report that clearly describes the process you designed. Also include your analyses and conclusions.

7. Out-of-class activity Divide into groups of three or four. Survey other students by asking them to identify their major and gender. For each surveyed subject, determine the number of Twitter followers or Facebook friends. Use the sample data to address these questions:

• Do the numbers of Twitter followers or Facebook friends appear to be the same for both genders?

• Do the numbers of Twitter followers or Facebook friends appear to be the same for the different majors?

8. In-class activity Divide into groups of 8 to 12 people. For each group member, measure the person's height and also measure his or her navel height, which is the height from the floor to the navel. Use the rank correlation coefficient to determine whether there is a correlation between height and navel height.

9. In-class activity Divide into groups of three or four people. Appendix B includes many data sets not yet addressed by the methods of this chapter. Search Appendix B for variables of interest, then investigate using appropriate methods of nonparametric statistics. State your conclusions and try to identify practical applications.

14-1 Control Charts for Variation and Mean

14-2 Control Charts for Attributes

14

STATISTICAL PROCESS CONTROL

CHAPTER PROBLEM

Aviation Safety: Is the production of aircraft altimeters out of statistical control?

The Orange Avionics Company manufactures instruments and navigation devices used in aircraft, including altimeters that provide pilots with readings of their heights above sea level. The accuracy of altimeters is important because pilots rely on them to maintain altitudes with safe heights above mountains, towers, and buildings, as well as vertical separation from other aircraft. The accuracy of altimeters is especially important when pilots fly approaches to landing while not being able

to see the ground. Pilots and passengers have been killed in crashes caused by wrong altimeter readings that led pilots to believe that they were safely above the ground when they were actually flying dangerously low.

Because aircraft altimeters are so critically important to aviation safety, their accuracy is carefully controlled by government regulations. Federal Aviation Administration (FAA) Regulation 91.411 requires periodic testing of aircraft altimeters, and those altimeters must comply with specifications included in Appendix E to Part 43 of the FAA regulations. One of those FAA specifications is that an altimeter must give a reading with an error of no more than 30 ft when tested for an altitude of 2000 ft.

At the Orange Avionics Company, five altimeters are randomly selected from production on each of 20 consecutive production days, and Table 14-1 lists the errors (in feet) when they are tested in a pressure chamber that simulates an altitude of 2000 ft. On day 1, for example, the actual altitude

readings for the four selected altimeters are 1999 ft, 2007 ft, 2007 ft, and 1999 ft, and 2002 ft, so the corresponding errors (in feet) are –1, 7, 7, –1, and 2.

In this chapter we evaluate this altimeter manufacturing process by analyzing the behavior of the errors over time. We will see how methods of statistics can be used to monitor a manufacturing process with the goal of identifying and correcting serious problems. In addition to helping companies stay in business, methods of statistics can positively affect our safety in very significant ways.

TABLE 14-1 Aircraft Altimeter Errors (in feet)

Available for download at www.TriolaStats.com

Day	Error (ft)					Mean $\bar{x}$	Median	Range R	Standard Deviation s
1	−1	7	7	−1	2	2.8	2	8	4.02
2	9	2	3	−1	−5	1.6	2	14	5.18
3	5	0	4	1	3	2.6	3	5	2.07
4	0	10	1	4	3	3.6	3	10	3.91
5	10	8	5	16	3	8.4	8	13	5.03
6	6	9	3	−6	−16	−0.8	3	25	10.18
7	14	4	2	14	8	8.4	8	12	5.55
8	20	9	4	−4	28	11.4	9	32	12.72
9	−12	−4	21	−5	−5	−1.0	−5	33	12.71
10	−21	−12	7	2	19	−1.0	2	40	15.76
11	2	20	12	−3	−13	3.6	2	33	12.86
12	−1	9	5	12	−4	4.2	5	16	6.69
13	−16	−21	−14	6	7	−7.6	−14	28	13.13
14	−21	−27	−3	−1	−9	−12.2	−9	26	11.37
15	−27	0	−37	−15	4	−15.0	−15	41	17.42
16	0	29	18	−27	25	9.0	18	56	22.99
17	29	−24	9	20	48	16.4	20	72	26.73
18	1	30	8	−15	13	7.4	8	45	16.47
19	2	−5	−23	−25	−4	−11.0	−5	27	12.19
20	3	−46	31	−20	−3	−7.0	−3	77	28.50

CHAPTER OBJECTIVES

This chapter presents methods for constructing and interpreting *control charts* that are commonly used to monitor changing characteristics of data over time. A control chart is a graph with a centerline, an upper control limit, and a lower control limit. A control chart can be used to determine whether a process is statistically stable (or within statistical control) with only natural variation and no patterns, cycles, or unusual points.

Here are the chapter objectives:

14-1 Control Charts for Variation and Mean

- Develop the ability to construct a run chart.

- Develop the ability to construct a control chart for R (range).

- Develop the ability to construct a control chart for $\bar{x}$.

- Identify out-of-control criteria and apply them to determine whether process data are within statistical control.

14-2 Control Charts for Attributes

- Develop the ability to construct a control chart for p, the proportion corresponding to some attribute, such as being a defect.

- Identify out-of-control criteria and apply them to determine whether attribute data are within statistical control with only natural variation and no patterns, cycles, or unusual points.

14-1 Control Charts for Variation and Mean

Key Concept This section presents run charts, R charts, and $\bar{x}$ charts as tools that enable us to monitor characteristics of data over time. We can use such charts to determine whether a process is statistically stable (or within statistical control).

Process Data

The following definition formally describes the type of data that will be considered in this chapter.

> **DEFINITION**
>
> **Process data** are data arranged according to some time sequence. They are measurements of a characteristic of goods or services that result from some combination of equipment, people, materials, methods, and conditions.

 EXAMPLE 1 Aircraft Altimeter Errors as Process Data

Table 14-1 includes process data consisting of the measured errors (feet) of aircraft altimeters. Because the values in Table 14-1 are arranged according to the time at which they were selected, they are process data.

It is important to know that companies have gone bankrupt because they allowed manufacturing processes to deteriorate without constant monitoring. This section introduces three tools commonly used to monitor process data: run charts, R charts, and $\bar{x}$ charts. We begin with run charts.

Run Chart

A run chart is one of several different tools commonly used to monitor a process to ensure that desired characteristics don't change. A run chart is basically the same as a time-series graph, which was introduced in Section 2-3.

DEFINITION

A **run chart** is a sequential plot of *individual* data values over time. One axis (usually the vertical axis) is used for the data values, and the other axis (usually the horizontal axis) is used for the time sequence.

⊙⟩ **EXAMPLE 2** **Run Chart of Aircraft Altimeter Errors**

Treating the 100 aircraft altimeter errors from Table 14-1 as a string of 100 consecutive measurements, construct a run chart using a vertical axis for the errors and a horizontal axis to identify the chronological order of the errors.

SOLUTION

Figure 14-1 is the Minitab-generated run chart for the data in Table 14-1. In Figure 14-1, the horizontal scale identifies the sample number, so the number 1 corresponds to the first altimeter error, the number 2 corresponds to the second error, and so on. The vertical scale represents the measured errors.

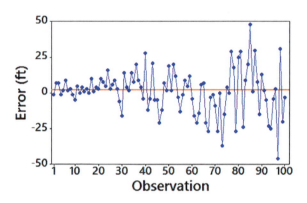

FIGURE 14-1   Run Chart of Altimeter Errors

INTERPRETATION

Examine Figure 14-1 and try to identify any *patterns*. From Figure 14-1 we see that as time progresses from left to right, the points appear to exhibit *greater variation*. If this pattern continues, some errors will become unacceptably large. This pattern of increasing variation is a classic issue in quality control, and failure to recognize it has caused companies to go out of business.

YOUR TURN Do Exercise 6 "Pepsi Cans: Run Chart".

DEFINITION

A process is **statistically stable** (or **within statistical control**) if it has only natural variation, with no patterns, cycles, or unusual points.

Interpreting Run Charts

A run chart with no obvious pattern suggests that the data are from a process that is *statistically stable,* and the data can be treated as if they came from a population with a constant mean, standard deviation, distribution, and other characteristics. Figure 14-1 shows a pattern of increasing variation, and that is one of several criteria for determining that a process is *not statistically stable* (or out of statistical control). Violating *one or more* of the following criteria indicates that a process is not statistically stable or out of statistical control.

Flynn Effect: Upward Trend of IQ Scores

An IQ (Intelligence Quotient) is measured from standard tests of intelligence. A run chart or control chart of IQ scores would reveal that they exhibit an upward trend, because IQ scores have been steadily increasing from about 1930. The trend is worldwide, and it is the same for different types of IQ tests, even those that rely heavily on abstract and nonverbal reasoning with minimal cultural influence. This upward trend has been named the *Flynn effect,* because political scientist James R. Flynn discovered it in his studies of U.S. military recruits. The amount of the increase is quite substantial: Based on a current mean IQ score of 100, it is estimated that the mean IQ in 1920 would be about 77. The typical student of today is therefore brilliant when compared to his or her great-grandparents. It is not yet clear whether the upward trend in IQ scores indicates an increasingly intelligent population or whether there are problems with the methods used for IQ testing.

Improving Quality in Cars by Reducing Variation

Ford and Mazda were producing similar transmissions that were supposed to be made with the same specifications, but it soon became apparent that the Ford transmissions required many more warranty repairs than the Japanese-made Mazda transmissions. Ford researchers investigated this and found that their transmissions were meeting the required specifications, but the *variation* in the Ford transmissions was much greater than those from Mazda. Mazda was using a better and more expensive grinder, but the increased cost was offset through fewer warranty repairs. Armed with these important results, Ford made changes and proceeded not only to meet the required specifications but also to improve quality by reducing variation. (See *Taguchi Techniques for Quality Engineering* by Phillip J. Ross.)

- **Increasing Variation:** As the run chart proceeds from left to right, the vertical variation of the points is increasing, so the corresponding data values are experiencing an increase in variation. (See Figure 14-1.) This is a common problem in quality control. The net effect is that products vary more and more until almost all of them are considered defective.

- **Upward Trend:** The points are rising from left to right, so the corresponding values are increasing over time.

- **Downward Trend:** The points are falling from left to right, so the corresponding values are decreasing over time.

- **Upward Shift:** The points near the beginning are noticeably lower than those near the end, so the corresponding values have shifted upward.

- **Downward Shift:** The points near the beginning are noticeably higher than those near the end, so the corresponding values have shifted downward.

- **Exceptional Value:** There is a single point that is exceptionally high or low.

- **Cyclical Pattern:** There is a repeating cycle.

Causes of Variation

Many different methods of quality control attempt to *reduce variation* in the product or service. Variation in a process can result from two types of causes as defined below.

> **DEFINITIONS**
>
> **Random variation** is due to chance; it is the type of variation inherent in any process that is not capable of producing every good or service exactly the same way every time.
>
> **Assignable variation** results from causes that can be identified (such as defective machinery or untrained employees).

Later in the chapter we will consider ways to distinguish between assignable variation and random variation.

The run chart is one tool for monitoring the stability of a process. We will now consider *control charts*, which are also useful for monitoring the stability of a process.

Control Charts

Because control charts were first introduced by Walter Shewhart in 1924, they are sometimes called Shewhart charts. We begin with a basic definition.

> **DEFINITION**
>
> A **control chart** (or **Shewhart chart** or **process-behavior chart**) of a process characteristic (such as mean or variation) consists of values plotted sequentially over time, and it includes a **centerline** as well as a **lower control limit** (LCL) and an **upper control limit** (UCL). The centerline represents a central value of the characteristic measurements, whereas the control limits are boundaries used to separate and identify any points considered to be *significantly high or significantly low*.

We will assume that the population standard deviation σ is not known as we now consider two of several different types of *control charts:*

1. R charts (or range charts) used to monitor variation

2. $\bar{x}$ charts used to monitor means

When using control charts to monitor a process, it is common to consider R charts and $\bar{x}$ charts together, because a statistically unstable process may be the result of increasing *variation,* changing *means,* or both.

Control Chart for Monitoring Variation: The *R* Chart

An *R* **chart** (or **range chart**) is a plot of the sample ranges instead of individual sample values, and it is used to monitor the *variation* in a process. It might make more sense to use standard deviations, but range charts are quite effective for cases in which the size of the samples (or subgroups) is 10 or fewer. If the samples all have a size greater than 10, the use of an s chart is recommended instead of an R chart. (See Exercise 13.) In addition to plotting the values of the ranges, we include a centerline located at $\bar{R}$, which denotes the mean of all sample ranges, as well as another line for the lower control limit and a third line for the upper control limit. The following is a summary of notation and the components of the R chart.

KEY ELEMENTS

Monitoring Process Variation: Control Chart for *R*

Objective

Construct a control chart for R (or an "R chart") that can be used to determine whether the *variation* of process data is within statistical control.

Requirements

1. The data are process data consisting of a sequence of samples all of the same size n.

2. The distribution of the process data is essentially normal.

3. The individual sample data values are independent.

Notation

$n =$ size of each sample or *subgroup*

$\bar{R} =$ mean of the sample ranges (the sum of the sample ranges divided by the number of samples)

Graph

Points plotted: sample ranges (each point represents the range for each subgroup)

Centerline: $\bar{R}$ (the mean of the sample ranges)

Upper control limit (UCL): $D_4 \bar{R}$ (where D_4 is a constant found in Table 14-2)

Lower control limit (LCL): $D_3 \bar{R}$ (where D_3 is a constant found in Table 14-2)

Costly Assignable Variation

The Mars Climate Orbiter was launched by NASA and sent to Mars, but it was destroyed when it flew too close to Mars. The loss was estimated at $125 million. The cause of the crash was found to be confusion over the units used for calculations. Acceleration data were provided in the English units of pounds of force, but the Jet Propulsion Laboratory assumed that those units were in metric "newtons" instead of pounds. The thrusters of the spacecraft subsequently provided wrong amounts of force in adjusting the position of the spacecraft. The errors caused by the discrepancy were fairly small at first, but the cumulative error over months of the spacecraft's journey proved to be fatal to its success.

In 1962, the rocket carrying the *Mariner 1* satellite was destroyed by ground controllers when it went off course due to a missing minus sign in a computer program.

TABLE 14-2 Control Chart Constants

n: Number of Observations in Subgroup	R Chart		$\bar{x}$ Chart		s Chart	
	D_3	D_4	A_2	A_3	B_3	B_4
2	0.000	3.267	1.880	2.659	0.000	3.267
3	0.000	2.574	1.023	1.954	0.000	2.568
4	0.000	2.282	0.729	1.628	0.000	2.266
5	0.000	2.114	0.577	1.427	0.000	2.089
6	0.000	2.004	0.483	1.287	0.030	1.970
7	0.076	1.924	0.419	1.182	0.118	1.882
8	0.136	1.864	0.373	1.099	0.185	1.815
9	0.184	1.816	0.337	1.032	0.239	1.761
10	0.223	1.777	0.308	0.975	0.284	1.716

Source: Adapted from *ASTM Manual on the Presentation of Data and Control Chart Analysis,* © 1976 ASTM, pp. 134–136. Reprinted with permission of American Society for Testing and Materials.

The values of D_4 and D_3 are constants computed by quality-control experts, and they are intended to simplify calculations. The upper and lower control limits of $D_4\bar{R}$ and $D_3\bar{R}$ are values that are roughly equivalent to 99.7% confidence interval limits. It is therefore highly unlikely that values from a statistically stable process would fall beyond those limits. If a value does fall beyond the control limits, it's very likely that the process is not statistically stable.

)─(**EXAMPLE 3**   *R* Chart of Altimeter Errors

Construct a control chart for *R* using the altimeter errors listed in Table 14-1. Use the samples of size $n = 5$ for each of the 20 days of production.

SOLUTION

Refer to Table 14-1 in the Chapter Problem on page 655 to see the column of sample ranges *R*. The value of $\bar{R}$ is the mean of those 20 sample ranges, so its value is found as follows:

$$\bar{R} = \frac{8 + 14 + \cdots + 77}{20} = 30.65 \text{ feet}$$

The centerline for our *R* chart is therefore located at $\bar{R} = 30.65$ feet. To find the upper and lower control limits, we must first find the values of D_3 and D_4. Referring to Table 14-2 for $n = 5$, we get $D_4 = 2.114$ and $D_3 = 0.000$, so the control limits are as follows:

$$\text{Upper control limit (UCL): } D_4\bar{R} = (2.114)(30.65) = 64.79$$

$$\text{Lower control limit (LCL): } D_3\bar{R} = (0.000)(30.65) = 0.0000$$

Using a centerline value of $\bar{R} = 30.65$ and control limits of 64.79 and 0.0000, we now proceed to plot the 20 sample ranges as 20 individual points. The result is shown on the next page in the Minitab display of the *R* chart.

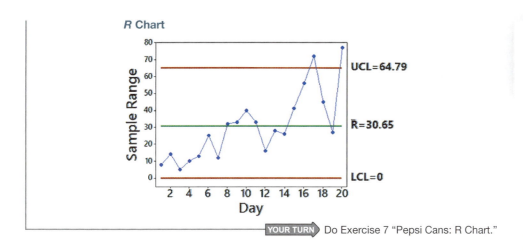

YOUR TURN Do Exercise 7 "Pepsi Cans: R Chart."

Interpreting Control Charts

When interpreting control charts, the following caution is important:

> **CAUTION** Upper and lower control limits of a control chart are based on the *actual* behavior of the process, not the *desired* behavior. Upper and lower control limits do not correspond to any process *specifications* that may have been decreed by the manufacturer.

When investigating the quality of some process, typically two key questions need to be addressed:

1. Based on the current behavior of the process, can we conclude that the process is within statistical control?

2. Do the process goods or services meet design specifications?

In this chapter we address the first question, but not the second; we are focusing on the behavior of the process with the objective of determining whether the process is within statistical control. Also, we should clearly understand the following specific criteria for determining whether a process is in statistical control (or is statistically stable).

Out-of-Control-Criteria

> **DEFINITION**
>
> A process is **not statistically stable** or is **out of statistical control** if one or more of the following out-of-control criteria are satisfied.
>
> 1. There is a pattern, trend, or cycle that is obviously not random.
>
> 2. There is at least one point above the upper control limit or at least one point below the lower control limit.
>
> 3. *Run of 8 Rule:* There are at least eight consecutive points all above or all below the centerline. (With a statistically stable process, there is a 0.5 probability that a point will be above or below the centerline, so it is very unlikely that eight consecutive points will all be above the centerline or all below it.)

Don't Tamper!

Nashua Corp. had trouble with its paper-coating machine and considered spending a million dollars to replace it. The machine was working well with a stable process, but samples were taken every so often and, based on the results, unnecessary adjustments were made. These overadjustments, called *tampering,* caused shifts away from the distribution that had been good. The effect was an increase in defects. When statistician and quality expert W. Edwards Deming studied the process, he recommended that no adjustments be made unless warranted by a signal that the process had shifted or had become unstable. The company was better off with no adjustments than with the tampering that took place.

Bribery Detected with Control Charts

Control charts were used to help convict a person who bribed Florida jai alai players to lose. (See "Using Control Charts to Corroborate Bribery in Jai Alai," by Charnes and Gitlow, *The American Statistician,* Vol. 49, No. 4.) An auditor for one jai alai facility noticed that abnormally large sums of money were wagered for certain types of bets, and some contestants didn't win as much as expected when those bets were made. *R* charts and $\bar{x}$ charts were used in court as evidence of highly unusual patterns of betting. Examination of the control charts clearly shows points well beyond the upper control limit, indicating that the process of betting was out of statistical control. The statistician was able to identify a date at which assignable variation appeared to stop, and prosecutors knew that it was the date of the suspect's arrest.

In this book we will use only the three out-of-control criteria listed above, but some companies use additional criteria such as these:

- There are at least six consecutive points all increasing or all decreasing.
- There are at least 14 consecutive points all alternating between up and down (such as up, down, up, down, and so on).
- Two out of three consecutive points are beyond control limits that are 2 standard deviations away from the centerline.
- Four out of five consecutive points are beyond control limits that are 1 standard deviation away from the centerline.

EXAMPLE 4 Interpreting the *R* Chart of Altimeter Errors

Examine the *R* chart shown in the display for Example 3 and determine whether the process variation is within statistical control.

SOLUTION

We can interpret control charts for *R* by applying the preceding three out-of-control criteria. Applying the three criteria to the display of the *R* chart, we conclude that variation in this process is *not* within statistical control. Considering the preceding three out-of-statistical-control criteria that we are using, we see that the first two criteria for being out of control are satisfied: (1) There is an obvious pattern of increasing points. (2) There are two points above the upper control limit. The third criterion is not satisfied because there are not eight consecutive points that are all above or all below the centerline.

INTERPRETATION

We conclude that the *variation* (not necessarily the mean) of the process is out of statistical control.

YOUR TURN Do Exercise 7 "Pepsi Cans: R Chart."

Control Chart for Monitoring Means: The $\bar{x}$ Chart

An $\bar{x}$ **chart** is a plot of the sample means, and it is used to monitor the *center* in a process. In addition to plotting the sample means, we include a centerline located at $\bar{\bar{x}}$, which denotes the mean of all sample means (equal to the mean of all sample values combined), as well as another line for the lower control limit and a third line for the upper control limit. Using the approach common in business and industry, the control limits are based on ranges instead of standard deviations. (See Exercise 14 for an $\bar{x}$ chart based on standard deviations.)

KEY ELEMENTS

Monitoring Process Mean: Control Chart for $\bar{x}$

Objective

Construct a control chart for $\bar{x}$ (or an $\bar{x}$ chart) that can be used to determine whether the *center* of process data is within statistical control.

Requirements

1. The data are process data consisting of a sequence of samples all of the same size n.

2. The distribution of the process data is essentially normal.

3. The individual sample data values are independent.

Notation

n = size of each sample, or *subgroup*

$\bar{\bar{x}}$ = mean of all sample means (equal to the mean of all sample values combined)

Graph

Points plotted: sample means

Centerline: $\bar{\bar{x}}$ = mean of all sample means

Upper control limit (UCL): $\bar{\bar{x}} + A_2 \bar{R}$ (where A_2 is a constant found in Table 14-2)

Lower control limit (LCL): $\bar{\bar{x}} - A_2 \bar{R}$ (where A_2 is a constant found in Table 14-2)

EXAMPLE 5 $\bar{x}$ **Chart of Altimeter Errors**

Using the altimeter errors in Table 14-1 on page 655 with samples of size $n = 5$ for each of 20 days, construct a control chart for $\bar{x}$. On the basis of the control chart for $\bar{x}$, determine whether the process mean is within statistical control.

SOLUTION

Before plotting the 20 points corresponding to the 20 values of $\bar{x}$, we must first find the values for the centerline and control limits. We get

$$\bar{\bar{x}} = \frac{2.8 + 1.6 + \cdots - 7.0}{20} = 1.19 \text{ feet}$$

$$\bar{R} = \frac{8 + 14 + \cdots + 77}{20} = 30.65 \text{ feet}$$

Referring to Table 14-2 on page 660, we find that for $n = 5$, $A_2 = 0.577$. Knowing the values of $\bar{\bar{x}}$, A_2, and $\bar{R}$, we can now evaluate the control limits.

Upper control limit (UCL): $\bar{\bar{x}} + A_2 \bar{R} = 1.19 + (0.577)(30.65) = 18.88$ feet

Lower control limit (LCL): $\bar{\bar{x}} - A_2 \bar{R} = 1.19 - (0.577)(30.65) = -16.50$ feet

The resulting control chart for $\bar{x}$ will be as shown in the accompanying Minitab display.

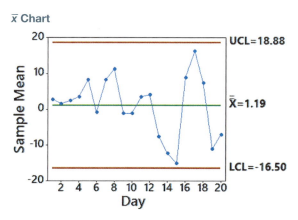

continued

High Cost of Low Quality

Microsoft announced that it would spend $1.15 billion to repair Xbox 360 game machine consoles. Industry analysts estimate that about one-third of all units were defective and in need of repair. Many users reported that the machine shut down after three flashing red lights appeared on the console.

The Federal Drug Administration reached an agreement whereby a pharmaceutical company, the Schering-Plough Corporation, would pay a record $500 million for failure to correct problems in manufacturing drugs. According to a *New York Times* article by Melody Petersen, "Some of the problems relate to the lack of controls that would identify faulty medicines, while others stem from outdated equipment. They involve some 200 medicines, including Claritin, the allergy medicine that is Schering's top-selling product."

> **INTERPRETATION**
>
> Examination of the $\bar{x}$ chart shows that the process mean is out of statistical control because at least one of the three criteria for being out of control is satisfied. Specifically, the first out-of-control criterion is satisfied because there is a pattern of increasing variation, which was also identified with the R chart.
>
> **YOUR TURN** Do Exercise 8 "Pepsi Cans: $\bar{x}$ Chart."

By analyzing the altimeter errors in Table 14-1 with a run chart, an R chart, and an $\bar{x}$ chart, we can see that the process is out of statistical control. If the manufacturing process continues without corrections, many of the altimeters will fail to satisfy FAA regulations, and the Orange Avionics Company will suffer huge losses as a result.

TECH CENTER

Control Charts

Access tech supplements, videos, and data sets at **www.TriolaStats.com**

Statdisk

Visit TriolaStats.com for detailed instructions.

Minitab

Run Chart

1. Enter all of the sample data in column *C1*.
2. Click **Stat** in the top menu.
3. Select **Quality Tools** from the dropdown menu and select **Run Chart** from the submenu.
4. In the *Single column* box enter **C1**. In *Subgroup* size enter **1**.
5. Click **OK**.

R Chart and $\bar{x}$ Chart

1. Enter sample values sequentially in column *C1* or enter the sample values in columns or rows as in Table 14-1.
2. Click **Stat** in the top menu.
3. Select **Control Charts** from the dropdown menu and select **Variables Charts for Subgroups**
4. Select **Xbar-R** from the submenu.
5. Select the format of data entered in Step 1 and then select the appropriate columns in the entry box.
6. Enter the *subgroup size* if all observations are in one column.
7. Click the **Xbar-R Options** button and then click the **Estimate** tab and select **Rbar**.
8. Click **OK** twice.

StatCrunch

1. Click **Stat** in the top menu.
2. Select **Control Charts** from the dropdown menu and **X-bar, R** from the submenu.
3. Select the columns to be used.
4. Click **Compute!**
5. Click the arrow buttons to switch between the charts and numerical results.

TI-83/84 Plus Calculator

Creating Run Charts, R Charts, and Charts using a TI-83/84 Plus calculator is possible but not recommended due to the complex procedures required and limited information displayed. Visit TriolaStats.com for detailed instructions.

Excel

XLSTAT Add-In (not available with all XLSTAT licenses)

1. Click on the **XLSTAT** tab in the Ribbon and then click **SPC**.
2. Select **Subgroup charts** from the dropdown menu.
3. For *Chart Family* click **Subgroup charts** and select **X-bar–R chart** under *Chart type*.
4. Click the **General** tab.
5. Under *Data format* select **Column** if the data are listed in separate columns or **One column** if the data are stacked in a single column.
6. If **One column** is selected, enter the sample size common to all data entries under *Common subgroup size*.
7. In the *Data* box enter the cell range of the data. If the data range includes a label, check the **Column labels** box.
8. Click the **Estimation** tab and select **R bar**.
9. Click **OK** twice.

14-1   Basic Skills and Concepts

Statistical Literacy and Critical Thinking

1. FAA Requirement Table 14-1 on page 655 lists process data consisting of the errors (ft) of aircraft altimeters when they are tested for an altitude of 2000 ft, and the Federal Aviation Administration requires that errors must be at most ± 30 ft. If $\bar{x}$ and R control charts show that the process of manufacturing altimeters is within statistical control, does that indicate that the altimeters have errors that are at most ± 30 ft? Why or why not?

2. Notation and Terminology Consider process data consisting of the amounts (oz) of Coke in randomly selected cans of regular Coke. The process is to be monitored with $\bar{x}$ and R control charts based on samples of 50 cans randomly selected each day for 20 consecutive days of production. What are $\bar{x}$ and R control charts, and what do $\bar{\bar{x}}$, $\bar{R}$, UCL, and LCL denote?

3. Lake Mead Elevations Many people in Nevada, Arizona, and California get water and electricity from Lake Mead and Hoover Dam. Shown in Exercise 4 are an $\bar{x}$ chart (top) and an R chart (bottom) obtained by using the monthly elevations (ft) of Lake Mead at Hoover Dam (based on data from the U.S. Department of the Interior). The control charts are based on the 12 monthly elevations for each of 75 consecutive and recent years. What does the $\bar{x}$ chart tell us about Lake Mead?

4. Lake Mead Elevations What does the R chart tell us about Lake Mead?

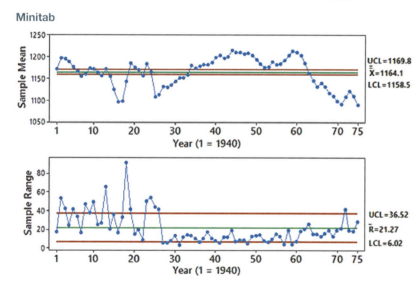

Pepsi Cans.   *In Exercises 5–8, refer to the axial loads (pounds) of aluminum Pepsi cans that are 0.0109 in. thick, as listed in Data Set 30 "Aluminum Cans" in Appendix B. An axial load of a can is the maximum weight supported by the side, and it is important to have an axial load high enough so that the can isn't crushed when the top lid is pressed onto the top. There are seven measurements from each of 25 days of production. If the 175 axial loads are in one column, the first 7 are from the first day, the next 7 are from the second day, and so on, so that the "subgroup size" is 7.*

5. Pepsi Cans: Notation After finding the sample mean and sample range for each of the 25 days, find the values of $\bar{\bar{x}}$ and $\bar{R}$. Also find the values of LCL and UCL for an R chart, then find the values of LCL and UCL for an $\bar{x}$ chart.

6. Pepsi Cans: Run Chart Treat the 175 axial loads as a string of consecutive measurements and construct a run chart. What does the result suggest?

7. Pepsi Cans: R Chart Treat the seven measurements from each day as a sample and construct an R chart. What does the result suggest?

8. Pepsi Cans: $\bar{x}$ Chart Treat the seven measurements from each day as a sample and construct an $\bar{x}$ chart. What does the result suggest?

Quarters.   *In Exercises 9–12, refer to the accompanying table of weights (grams) of quarters minted by the U.S. government. This table is available for download at www.TriolaStats.com.*

Day	Hour 1	Hour 2	Hour 3	Hour 4	Hour 5	$\bar{x}$	s	Range
1	5.543	5.698	5.605	5.653	5.668	5.6334	0.0607	0.155
2	5.585	5.692	5.771	5.718	5.720	5.6972	0.0689	0.186
3	5.752	5.636	5.660	5.680	5.565	5.6586	0.0679	0.187
4	5.697	5.613	5.575	5.615	5.646	5.6292	0.0455	0.122
5	5.630	5.770	5.713	5.649	5.650	5.6824	0.0581	0.140
6	5.807	5.647	5.756	5.677	5.761	5.7296	0.0657	0.160
7	5.686	5.691	5.715	5.748	5.688	5.7056	0.0264	0.062
8	5.681	5.699	5.767	5.736	5.752	5.7270	0.0361	0.086
9	5.552	5.659	5.770	5.594	5.607	5.6364	0.0839	0.218
10	5.818	5.655	5.660	5.662	5.700	5.6990	0.0689	0.163
11	5.693	5.692	5.625	5.750	5.757	5.7034	0.0535	0.132
12	5.637	5.628	5.646	5.667	5.603	5.6362	0.0235	0.064
13	5.634	5.778	5.638	5.689	5.702	5.6882	0.0586	0.144
14	5.664	5.655	5.727	5.637	5.667	5.6700	0.0339	0.090
15	5.664	5.695	5.677	5.689	5.757	5.6964	0.0359	0.093
16	5.707	5.890	5.598	5.724	5.635	5.7108	0.1127	0.292
17	5.697	5.593	5.780	5.745	5.470	5.6570	0.1260	0.310
18	6.002	5.898	5.669	5.957	5.583	5.8218	0.1850	0.419
19	6.017	5.613	5.596	5.534	5.795	5.7110	0.1968	0.483
20	5.671	6.223	5.621	5.783	5.787	5.8170	0.2380	0.602

9. Quarters: Notation Find the values of $\bar{\bar{x}}$ and $\bar{R}$. Also find the values of LCL and UCL for an R chart, then find the values of LCL and UCL for an $\bar{x}$ chart.

10. Quarters: R Chart Treat the five measurements from each day as a sample and construct an R chart. What does the result suggest?

11. Quarters: $\bar{x}$ Chart Treat the 5 measurements from each day as a sample and construct an $\bar{x}$ chart. What does the result suggest?

12. Quarters: Run Chart Treat the 100 consecutive measurements from the 20 days as individual values and construct a run chart. What does the result suggest?

14-1   Beyond the Basics

13. s Chart In this section we described control charts for R and $\bar{x}$ based on *ranges*. Control charts for monitoring variation and center (mean) can also be based on *standard deviations*. An *s chart* for monitoring variation is constructed by plotting sample standard deviations with a centerline at $\bar{s}$ (the mean of the sample standard deviations) and control limits at $B_4\bar{s}$ and $B_3\bar{s}$, where B_4 and B_3 are found in Table 14-2 on page 660 in this section. Construct an s chart for the data of Table 14-1 on page 655. Compare the result to the R chart given in Example 3 "R Chart of Altimeter Errors."

14. $\bar{x}$ Chart Based on Standard Deviations An $\bar{x}$ chart based on *standard deviations* (instead of ranges) is constructed by plotting sample means with a centerline at $\bar{\bar{x}}$ and control limits at $\bar{\bar{x}} + A_3\bar{s}$ and $\bar{\bar{x}} - A_3\bar{s}$, where A_3 is found in Table 14-2 on page 660 and $\bar{s}$ is the mean of the sample standard deviations. Use the data in Table 14-1 on page 655 to construct an $\bar{x}$ chart based on standard deviations. Compare the result to the $\bar{x}$ chart based on sample ranges in Example 5 "$\bar{x}$ Chart of Altimeter Errors."

Control Charts for Attributes

Key Concept This section presents a method for constructing a control chart to monitor the proportion p for some *attribute,* such as whether a service or manufactured item is defective or nonconforming. (A good or a service is nonconforming if it doesn't meet specifications or requirements. Nonconforming goods are sometimes discarded, repaired, or called "seconds" and sold at reduced prices.) The control chart is interpreted using the same three out-of-control criteria from Section 14-1 to determine whether the process is statistically stable:

Out-of-Control-Criteria

> **DEFINITION**
>
> A process is **not statistically stable** or is **out of statistical control** if one or more of the following out-of-control criteria are satisfied.
>
> **1.** There is a pattern, trend, or cycle that is obviously not random.
>
> **2.** There is at least one point above the upper control limit or at least one point below the lower control limit.
>
> **3.** *Run of 8 Rule:* There are at least eight consecutive points all above or all below the centerline. (With a statistically stable process, there is a 0.5 probability that a point will be above or below the centerline, so it is very unlikely that eight consecutive points will all be above the centerline or all below it.)

As in Section 14-1, we select samples of size n at regular time intervals and plot points in a sequential graph with a centerline and control limits. (There are ways to deal with samples of different sizes, but we don't consider them here.)

> **DEFINITION**
>
> A **control chart for p** (or **p chart**) is a graph of proportions of some attribute (such as whether products are defective) plotted sequentially over time, and it includes a centerline, a lower control limit (LCL), and an upper control limit (UCL).

The notation and control chart values are as summarized in the following Key Elements box. In this box, the attribute of "defective" can be replaced by any other relevant attribute (so that each sample item belongs to one of two distinct categories).

KEY ELEMENTS

Monitoring a Process Attribute: Control Chart for p

Objective

Construct a control chart for p (or a "p chart") that can be used to determine whether the proportion of some attribute (such as whether products are defective) from process data is within statistical control.

Requirements

1. The data are process data consisting of a sequence of samples all of the same size n.

2. Each sample item belongs to one of two categories (such as defective or not defective).

3. The individual sample data values are independent.

continued

Notation

$$\bar{p} = \text{estimate of the proportion of defective items in the process}$$

$$= \frac{\text{total number of defects found among all items sampled}}{\text{total number of items sampled}}$$

$$\bar{q} = \text{estimate of the proportion of process items that are not defective}$$

$$= 1 - \bar{p}$$

$$n = \text{size of each individual sample or subgroup}$$

Graph

Points plotted: proportions from the individual samples of size n

Centerline: $\bar{p}$

Upper control limit: $\bar{p} + 3\sqrt{\dfrac{\bar{p}\,\bar{q}}{n}}$ (Use 1 if this result is greater than 1.)

Lower control limit: $\bar{p} - 3\sqrt{\dfrac{\bar{p}\,\bar{q}}{n}}$ (Use 0 if this result is less than 0.)

CAUTION Note this distinction in the calculations: When evaluating $\bar{p}$, divide the total number of defects by the *total* number of items sampled. The computations for UCL and LCL, however, require division by n, the size of each individual sample.

We use $\bar{p}$ for the centerline because it is the best estimate of the proportion of defects from the process. The expressions for the control limits correspond to 99.7% confidence interval limits for the confidence intervals described in Section 7-1. (Section 7-1 did not include any 99.7% confidence intervals, but the z score used for a 99.7% confidence interval is $z = 2.97$, which is rounded to 3 in the expressions used for the LCL and UCL in this section.)

EXAMPLE 1 Defective Aircraft Altimeters

The Chapter Problem describes the process of manufacturing aircraft altimeters. Section 14-1 includes examples of control charts for monitoring the errors in altimeter readings. An altimeter is considered to be defective if it cannot be calibrated or corrected to give accurate readings (within 30 ft of the true altitude of 2000 ft). The Orange Avionics Company manufactures altimeters in batches of 100, and each altimeter is tested and determined to be acceptable or defective. Listed below are the numbers of defective altimeters in successive batches of 100. Construct a control chart for the proportion p of defective altimeters and determine whether the process is within statistical control. If not, identify which of the three out-of-control criteria apply.

Defects: 2   0   1   3   1   2   2   4   3   5   12   7

SOLUTION

The centerline for the control chart is located by the value of $\bar{p}$:

$$\bar{p} = \frac{\text{total number of defects from all samples combined}}{\text{total number of altimeters sampled}},$$

$$= \frac{2 + 0 + 1 + \cdots + 7}{12 \cdot 100} = \frac{42}{1200} = 0.035$$

Because $\bar{p} = 0.035$, it follows that $\bar{q} = 1 - \bar{p} = 0.965$. Using $\bar{p} = 0.035$, $\bar{q} = 0.965$, and $n = 100$, we find the control limits as follows:

Upper control limit:

$$\bar{p} + 3\sqrt{\frac{\bar{p}\,\bar{q}}{n}} = 0.035 + 3\sqrt{\frac{(0.035)(0.965)}{100}} = 0.090$$

Lower control limit:

$$\bar{p} - 3\sqrt{\frac{\bar{p}\,\bar{q}}{n}} = 0.035 - 3\sqrt{\frac{(0.035)(0.965)}{100}} = -0.020$$

Because the lower control limit is less than 0, we use 0 instead.

Having found the values for the centerline and control limits, we can proceed to plot the control chart for proportions of defective altimeters. The Minitab control chart for p is shown in the accompanying display.

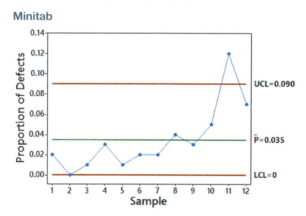

INTERPRETATION

We can interpret the control chart for p by considering the three out-of-control criteria listed earlier in this section. Using those criteria, we conclude that this process is out of statistical control for these reasons: There appears to be an upward trend, and there is point lying beyond the upper control limit. The company should take immediate corrective action.

YOUR TURN Do Exercise 5 "Euro Coins."

CAUTION Upper and lower control limits of a control chart for a proportion p are based on the *actual* behavior of the process, not the *desired* behavior. Upper and lower control limits are totally unrelated to any process *specifications* that may have been decreed by the manufacturer.

Quality Control at Perstorp

Perstorp Components, Inc., uses a computer that automatically generates control charts to monitor the thicknesses of the floor insulation the company makes for Ford Rangers and Jeep Grand Cherokees. The $20,000 cost of the computer was offset by a first-year savings of $40,000 in labor, which had been used to manually generate control charts to ensure that insulation thicknesses were between the specifications of 2.912 mm and 2.988 mm. Through the use of control charts and other quality-control methods, Perstorp reduced its waste by more than two-thirds.

TECH CENTER

Control Chart for p

Access tech supplements, videos, and data sets at **www.TriolaStats.com**

Statdisk	Minitab	StatCrunch
Visit TriolaStats.com for detailed instructions.	1. Enter the numbers of defects (or items with any particular attribute) in column **C1**. 2. Click **Stat** in the top menu. 3. Select **Control Charts** from the dropdown menu and select **Attribute charts** and **P** from the submenu. 4. In the *Variables* box enter column **C1**. 5. In *Subgroup sizes* enter the size of the individual samples. 6. Click **OK**.	1. Click **Stat** in the top menu. 2. Select **Control Charts** from the dropdown menu and **p** from the submenu. 3. Select the column to be used, select **Constant,** and enter the size of the individual samples. 4. Click **Compute!** 5. Click the arrow buttons to switch between the chart and numerical results.

TI-83/84 Plus Calculator	Excel
Not available.	**XLSTAT Add-In** *(not available with all XLSTAT licenses)* 1. Click on the **XLSTAT** tab in the Ribbon and then click **SPC**. 2. Select **Attribute charts** from the dropdown menu. 3. For *Chart Family* click **Attribute charts** and select **P chart** under *Chart type*. 4. Click the **General** tab. 5. In the *Data* box enter the cell range of the data. If the data range includes a label, check the **Column labels** box. 6. In the *Common subgroup size* box enter the sample size that is common to all data entries. 7. Click **OK**. The results include a *p* chart.

14-2   Basic Skills and Concepts

Statistical Literacy and Critical Thinking

1. Minting Quarters Specifications for a quarter require that it be 8.33% nickel and 91.67% copper; it must weigh 5.670 g and have a diameter of 24.26 mm and a thickness of 1.75 mm; and it must have 119 reeds on the edge. A quarter is considered to be defective if it deviates substantially from those specifications. A production process is monitored, defects are recorded and the accompanying control chart is obtained. Does this process appear to be within statistical control? If not, identify any out-of-control criteria that are satisfied. Is the manufacturing process deteriorating?

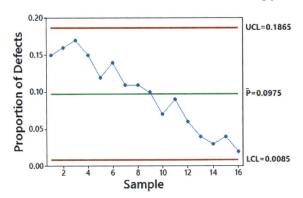

2. Notation The control chart for Exercise 1 shows a value of $\bar{p} = 0.0975$. What does that value denote, and how is it obtained? What do UCL and LCL indicate?

3. Control Limits In constructing a control chart for the proportions of defective dimes, it is found that the lower control limit is -0.00325. How should that value be adjusted?

4. Euro Coins After constructing a control chart for the proportions of defective one-euro coins, it is concluded that the process is within statistical control. Does it follow that almost all of the coins meet the desired specifications? Explain.

Control Charts for p. *In Exercises 5–12, use the given process data to construct a control chart for p. In each case, use the three out-of-control criteria listed near the beginning of this section and determine whether the process is within statistical control. If it is not, identify which of the three out-of-control criteria apply.*

5. Euro Coins Consider a process of minting coins with a value of one euro. Listed below are the numbers of defective coins in successive batches of 10,000 coins randomly selected on consecutive days of production.

32 21 25 19 35 34 27 30 26 33

6. Euro Coins Repeat Exercise 5, assuming that the size of each batch is 100 instead of 10,000. Compare the control chart to the one found for Exercise 5. Comment on the general quality of the manufacturing process described in Exercise 5 compared to the manufacturing process described in this exercise.

7. Aspirin Tablets Bottles of aspirin tablets typically have labels indicating that each tablet contains 325 mg of aspirin, but the actual amounts can vary between 315 mg and 335 mg. A tablet is defective if it has less than 315 mg of aspirin or more than 335 mg of aspirin. Listed below are numbers of defects found in batches of 1000 tablets.

16 18 13 9 10 8 6 5 5 3

8. Defibrillators At least one corporation was sued for manufacturing defective heart defibrillators. Listed below are the numbers of defective defibrillators in successive batches of 1000. Construct a control chart for the proportion p of defective defibrillators and determine whether the process is within statistical control. If not, identify which of the three out-of-control criteria apply.

8 5 6 4 9 3 12 7 8 5 22 4 9 10 11 8 7 6 8 5

9. Voting Rate In each of recent and consecutive years of presidential elections, 1000 people of voting age in the United States were randomly selected and the number who voted was determined, with the results listed below. Comment on the voting behavior of the population.

631 619 608 552 536 526 531 501 551 491 513 553 568

10. Car Batteries Defective car batteries are a nuisance because they can strand and inconvenience drivers, and drivers could be put in danger. A car battery is considered to be defective if it fails before its warranty expires. Defects are identified when the batteries are returned under the warranty program. The Powerco Battery corporation manufactures car batteries in batches of 250, and the numbers of defects are listed below for each of 12 consecutive batches. Does the manufacturing process require correction?

3 4 2 5 3 6 8 9 12 14 17 20

11. Cola Cans In each of several consecutive days of production of cola cans, 50 cans are tested and the numbers of defects each day are listed below. Do the proportions of defects appear to be acceptable? What action should be taken?

8 7 9 8 10 6 5 7 9 12 9 6 8 7 9 8 11 10 9 7

12. Smartphone Batteries The SmartBatt company manufactures batteries for smartphones. Listed below are numbers of defects in batches of 200 batteries randomly selected in each of 12 consecutive days of production. What action should be taken?

5 7 4 6 3 10 10 13 4 15 4 19

14-2 Beyond the Basics

13. *np* Chart A variation of the control chart for *p* is the ***np* chart,** in which the *actual numbers* of defects are plotted instead of the *proportions* of defects. The *np* chart has a centerline value of $n\bar{p}$, and the control limits have values of $n\bar{p} + 3\sqrt{n\bar{p}\,\bar{q}}$ and $n\bar{p} - 3\sqrt{n\bar{p}\,\bar{q}}$. The *p* chart and the *np* chart differ only in the scale of values used for the vertical axis. Construct the *np* chart for Example 1 "Defective Aircraft Altimeters" in this section. Compare the *np* chart to the control chart for *p* given in this section.

Chapter Quick Quiz

1. What are *process* data?

2. What is the difference between *random variation* and *assignable variation*?

3. Identify three specific criteria for determining when a process is out of statistical control.

4. What is the difference between an *R* chart and an $\bar{x}$ chart?

In Exercises 5–8, use the following two control charts that result from testing batches of newly manufactured aircraft altimeters, with 100 in each batch. The original sample values are errors (in feet) obtained when the altimeters are tested in a pressure chamber that simulates an altitude of 6000 ft. The Federal Aviation Administration requires an error of no more than 40 ft at that altitude.

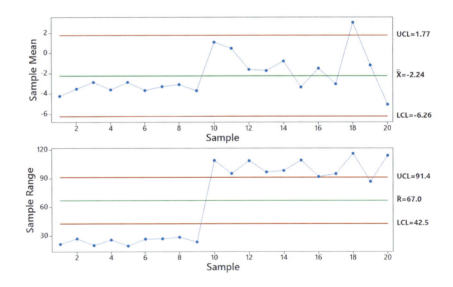

5. Is the process variation within statistical control? Why or why not?

6. What is the value of $\bar{R}$? In general, how is a value of $\bar{R}$ obtained?

7. Is the process mean within statistical control? Why or why not?

8. What is the value of $\bar{\bar{x}}$? In general, how is a value of $\bar{\bar{x}}$ found?

9. If the *R* chart and $\bar{x}$ chart both showed that the process of manufacturing aircraft altimeters is within statistical control, can we conclude that the altimeters satisfy the Federal Aviation Administration requirement of having errors of no more than 40 ft when tested at an altitude of 6000 ft?

10. Examine the following p chart for defective calculator batteries and briefly describe the action that should be taken.

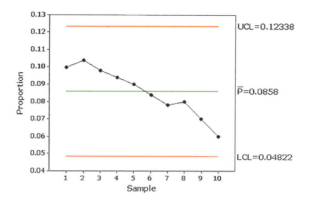

Review Exercises

Energy Consumption.   *Exercises 1–5 refer to the amounts of energy consumed in the author's home. (Most of the data are real, but some are fabricated.) Each value represents energy consumed (kWh) in a two-month period. Let each subgroup consist of the six amounts within the same year. Data are available for download at www.TriolaStats.com.*

	Jan.–Feb.	Mar.–April	May–June	July–Aug.	Sept.–Oct.	Nov.–Dec.
Year 1	3637	2888	2359	3704	3432	2446
Year 2	4463	2482	2762	2288	2423	2483
Year 3	3375	2661	2073	2579	2858	2296
Year 4	2812	2433	2266	3128	3286	2749
Year 5	3427	578	3792	3348	2937	2774
Year 6	4016	3458	3395	4249	4003	3118
Year 7	5261	2946	3063	5081	2919	3360
Year 8	3853	3174	3370	4480	3710	3327

1. Energy Consumption: Notation After finding the values of the mean and range for each year, find the values of $\bar{\bar{x}}$ and $\bar{R}$. Then find the values of LCL and UCL for an R chart and find the values of LCL and UCL for an $\bar{x}$ chart.

2. Energy Consumption: R Chart Let each subgroup consist of the 6 values within a year. Construct an R chart and determine whether the process variation is within statistical control. If it is not, identify which of the three out-of-control criteria lead to rejection of statistically stable variation.

3. Energy Consumption: $\bar{x}$ Chart Let each subgroup consist of the 6 values within a year. Construct an $\bar{x}$ chart and determine whether the process mean is within statistical control. If it is not, identify which of the three out-of-control criteria lead to rejection of a statistically stable mean.

4. Energy Consumption: Run Chart Construct a run chart for the 48 values. Does there appear to be a pattern suggesting that the process is not within statistical control?

5. Service Times The Newport Diner records the times (min) it takes before customers are asked for their order. Each day, 50 customers are randomly selected and the order is considered to be defective if it takes longer than three minutes. The numbers of defective orders are listed below for consecutive days. Construct an appropriate control chart and determine whether the process is within statistical control. If not, identify which criteria lead to rejection of statistical stability.

3   2   3   5   4   6   7   9   8   10   11   9   12   15   17

Cumulative Review Exercises

1. Internet Doctors: Confidence Interval In a survey of $n = 2015$ adults, 1108 of them said that they learn about medical symptoms more often from the Internet than from their doctor (based on a MerckManuals.com survey). Use the data to construct a 95% confidence interval estimate of the population proportion of all adults who say that they learn about medical symptoms more often from the Internet than from their doctor. Does the result suggest that the majority of adults learn about medical symptoms more often from the Internet than from their doctor?

2. Internet Doctors: Hypothesis Test Use the survey results given in Exercise 1 and use a 0.05 significance level to test the claim that the majority of adults learn about medical symptoms more often from the Internet than from their doctor.

3. Internet Doctors: Graph The accompanying graph was created to depict the results of the survey described in Exercise 1. Is the graph somehow misleading? If so, how?

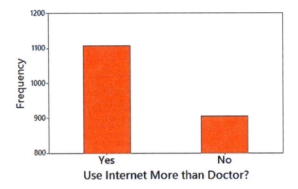

4. Internet Doctors: Probability Based on the survey data given in Exercise 1, assume that 55% of adults learn about medical symptoms more often from the Internet than from their doctor.

a. Find the probability that three randomly selected adults all learn about medical symptoms more often from the Internet than from their doctor.

b. Find the probability that when three adults are randomly selected, at least one of them learns about medical symptoms more often from the Internet than from their doctor.

5. Sunspots and the DJIA Listed below are annual sunspot numbers paired with annual high values of the Dow Jones Industrial Average (DJIA). Sunspot numbers are measures of dark spots on the sun, and the DJIA is an index that measures the value of select stocks. The data are from recent and consecutive years. Use a 0.05 significance level to test for a linear correlation between values of the DJIA and sunspot numbers. Is the result surprising?

Sunspot	45	31	46	31	50	48	56	38	65	51
DJIA	10,941	12,464	14,198	13,279	10,580	11,625	12,929	13,589	16,577	18,054

6. Sunspots and the DJIA Use the data from Exercise 5 and find the equation of the regression line. Then find the best predicted value of the DJIA in the year 2004, when the sunspot number was 61. How does the result compare to the actual DJIA value of 10,855?

7. Heights On the basis of Data Set 1 "Body Data" in Appendix B, assume that heights of men are normally distributed, with a mean of 68.6 in. and a standard deviation of 2.8 in.

a. The U.S. Coast Guard requires that men must have a height between 60 in. and 80 in. Find the percentage of men who satisfy that height requirement.

b. Find the probability that 4 randomly selected men have heights with a mean greater than 70 in.

8. Defective Child Restraint Systems The Tracolyte Manufacturing Company produces plastic frames used for child booster seats in cars. During each week of production, 120 frames are selected and tested for conformance to all regulations by the Department of Transportation. Frames are considered defective if they do not meet all requirements. Listed below are the numbers of defective frames among the 120 that are tested each week. Use a control chart for p to verify that the process is within statistical control. If it is not in control, explain why it is not.

$$3 \quad 2 \quad 4 \quad 6 \quad 5 \quad 9 \quad 7 \quad 10 \quad 12 \quad 15$$

9. Child Restraint Systems Use the numbers of defective child restraint systems given in Exercise 8. Find the mean, median, and standard deviation. What important characteristic of the sample data is missed if we explore the data using those statistics?

10. Does It Pay to Plead Guilty? The accompanying table summarizes randomly selected sample data for San Francisco defendants in burglary cases (based on data from "Does It Pay to Plead Guilty? Differential Sentencing and the Functioning of the Criminal Courts," by Brereton and Casper, *Law and Society Review,* Vol. 16, No. 1). All of the subjects had prior prison sentences. Use a 0.05 significance level to test the claim that the sentence (sent to prison or not sent to prison) is independent of the plea. If you were an attorney defending a guilty defendant, would these results suggest that you should encourage a guilty plea?

	Guilty Plea	Not Guilty Plea
Sent to Prison	392	58
Not Sent to Prison	564	14

Technology Project

a. Simulate the following process for 20 days: Each day, 200 calculators are manufactured with a 5% rate of defects, and the proportion of defects is recorded for each of the 20 days. The calculators for one day are simulated by randomly generating 200 numbers, where each number is between 1 and 100. Consider an outcome of 1, 2, 3, 4, or 5 to be a defect, with 6 through 100 being acceptable. This corresponds to a 5% rate of defects. (Hint: see the Chapter 11 Technology Project on page 563 for technology tips.)

b. Construct a p chart for the proportion of defective calculators, and determine whether the process is within statistical control. Since we know the process is actually stable with $p = 0.05$, the conclusion that it is not stable would be a type I error; that is, we would have a false positive signal, causing us to believe that the process needed to be adjusted when in fact it should be left alone.

c. The result from part (a) is a simulation of 20 days. Now simulate another 10 days of manufacturing calculators, but modify these last 10 days so that the defect rate is 10% instead of 5%.

d. Combine the data generated from parts (a) and (c) to represent a total of 30 days of sample results. Construct a p chart for this combined data set. Is the process out of statistical control? If we concluded that the process was not out of statistical control, we would be making a type II error; that is, we would believe that the process was okay when in fact it should be modified to correct the shift to the 10% rate of defects.

FROM DATA TO DECISION

Critical Thinking: Are the axial loads within statistical control? Is the process of manufacturing cans proceeding as it should?

Exercises 5–8 in Section 14-1 used process data from the manufacture of 0.0109-in.-thick aluminum cans. Refer to Data Set 30 "Aluminum Cans" in Appendix B and conduct an analysis of the process data for the cans that are 0.0111 in. thick. The values in the data set are the measured axial loads of cans, and the top lids are pressed into place with pressures that vary between 158 lb and 165 lb. The 175 axial loads are in one column, the first 7 are from the first day,

the next 7 are from the second day, and so on, so that the "subgroup size" is 7.

Analyzing the Results

Based on the given process data, should the company take any corrective action? Write a report summarizing your conclusions. Address not only the issue of statistical stability but also the ability of the cans to withstand the pressures applied when the top lids are pressed into place. Also compare the behavior of the 0.0111-in. cans to the behavior of the 0.0109-in. cans, and recommend which thickness should be used.

Cooperative Group Activities

1. Out-of-class activity Collect your own process data and analyze them using the methods of this chapter. It would be ideal to collect data from a real manufacturing process, but that might be difficult. Instead, consider using a simulation or referring to published data. Obtain a copy of computer results and write a brief report summarizing your conclusions. Here are some suggestions:

• Shoot five basketball foul shots (or shoot five crumpled sheets of paper into a wastebasket) and record the number of shots made; then repeat this procedure 20 times. Use a p chart to test for statistical stability in the proportion of shots made.

• Measure your pulse rate by counting the number of times your heart beats in 1 min. Measure your pulse rate four times each hour for several hours, then construct appropriate control charts. What factors contribute to random variation? Assignable variation?

• Search the Internet and record the closing of the Dow Jones Industrial Average (DJIA) for each business day of the past 12 weeks. Use run and control charts to explore the statistical stability of the DJIA. Identify at least one practical consequence of having this process statistically stable, and identify at least one practical consequence of having this process out of statistical control.

• Find the marriage rate per 10,000 population for several years. Assume that in each year 10,000 people were randomly selected and surveyed to determine whether they were married. Use a p chart to test for statistical stability of the marriage rate. (Other possible rates: death, accident fatality, crime.)

• Search the Internet and find the numbers of runs scored by your favorite baseball team in recent games, then use run and control charts to explore statistical control.

2. In-class activity If the instructor can distribute the numbers of absences for each class meeting, groups of three or four students can analyze them for statistical stability and make recommendations based on the conclusions.

3. Out-of-class activity Conduct research to find a description of Deming's funnel experiment, then use a funnel and marbles to collect data for the different rules for adjusting the funnel location. Construct appropriate control charts for the different rules of funnel adjustment. What does the funnel experiment illustrate? What do you conclude?

15

ETHICS IN STATISTICS

Key Concept While statistical methods give us tremendous power to better understand the world in which we live, this power also provides opportunities for uses in ways that are fundamentally unethical. It is important to consider some ethical issues in statistics, and this section considers such issues related to data collection, analysis, and reporting.

I. Data Collection

Unethical uses of statistics sometimes begin with the process of collecting data. Humans are sometimes treated in ways that are unethical, and samples can be biased in ways that are unethical.

Human Treatment When Obtaining Data

There are many research studies in which the health and safety of research subjects were compromised so that others might benefit.

Prisoner Studies In the early 1970s, it is estimated that 90% of pharmaceutical research in the United States was conducted using prisoners as human test subjects, sometimes without their knowledge or consent. The Common Rule (described later in this section) now protects human research subjects in the United States.

Milgram Experiment During the 1960s, Stanley Milgram conducted one of the most infamous psychological studies. Subjects were ordered to administer progressively stronger "electrical shocks" to actors who were in another room. These actors were not visible to the subject and they were not actually receiving shocks. Prerecorded screams were played so the subject could hear them. The actor would bang on the wall and beg for the subject to stop.

Prior to the experiment, experts estimated that only 3% of subjects would continue giving shocks after hearing the screams and the pleas to stop, but 65% of the subjects increased shocks until the maximum of 375 volts was reached. This experiment inflicted extreme emotional distress on its subjects and is now considered unethical and psychologically abusive.

Sampling Bias

Researchers can intentionally or unintentionally bias their study by using a sample that is biased. Sampling bias can seriously distort results and lead to wrong or misleading conclusions. In some cases, researchers intentionally used biased samples to ensure that desired results were obtained.

Cherry Picking The report "Are Researchers Cherry Picking Participants for Studies of Antidepressants?" by the University of Pittsburgh, Schools of the Health Sciences, described results from clinical studies conducted with the goal of gaining approval of antidepressants from the Food and Drug Administration. Only a small percentage of depressed individuals met the criteria required for participation in the clinical trial, and those who qualified had better outcomes than other depressed individuals who did not qualify. In this case, the criteria required to participate in the clinical trial created a sampling bias that favored the drug manufacturers who sponsored the research.

Nonrespondent Bias Nonrespondent bias occurs when those who do not respond to a survey differ from those who do respond. Two aspects of a survey topic that affect responses are *salience* (whether or not the topic is of interest to the respondent) and *social desirability* (whether or not the topic is threatening or embarrassing). One study found that when a topic has high interest to respondents, they were almost twice as likely to respond. Other studies have found that socially desirable behaviors like exercise and good nutrition are frequently overreported, while undesirable behaviors such as smoking are underreported. For alcohol research, several studies have shown that heavy drinkers are less likely to respond. (See "Non-response Bias in a Sample Survey on Alcohol Consumption," by V. M. Lahaut et al., *Alcohol and Alcoholism.*)

Interviewer Bias The manner in which a question is asked may affect the response. The author received a survey in the mail asking the following questions worded in a way that is intended to affect responses:

- "Do you support the creation of a national health insurance policy that would be administered by bureaucrats in Washington, D.C.?"

- "Are you in favor of creating a government-funded 'Citizen Volunteer Corps' that would pay young people to do work now being done by churches and charities, earning Corps Members the same pay and benefits given to military veterans?"

Volunteer Bias People who volunteer to participate constitute a voluntary response sample, and they often have different characteristics when compared to participants who are selected by those conducting the survey. The effects of volunteer bias have been extensively studied for the topic of human sexuality. "Volunteer Bias in Sexuality Research," by D. S. Strassberg et al., Department of Psychology, University of Utah, found that compared to non-volunteers, volunteers reported a more positive attitude toward sexuality, less sexual guilt, and more sexual experience.

II. Analysis

Research study findings are often distorted by improper analysis. Even though the sample data may be inconclusive, insufficient, or supportive of an undesirable outcome, many researchers have succumbed to pressures to report significant or "conclusive" results through negligent and unethical analysis.

Falsified Data Several surveys have attempted to determine the incidence of data falsification or fabrication among scientists. "How Many Scientists Fabricate and Falsify Research? A Systematic Review and Meta-Analysis of Survey Data," by D. Fanelli, *PLoS ONE,* Vol. 4, No. 5, found that nearly 2% of scientists admitted to having fabricated, falsified, or modified data or results at least once. This is likely to be a conservative estimate, given the sensitivity of the topic and potential for bias from the social desirability factor described earlier. A notable example of data falsification is found in the original study suggesting a link between childhood vaccines and autism.

Andrew Wakefield published a 1998 study linking the MMR vaccine to autism. In 2010, the *British Medical Journal* provided evidence that Wakefield's study relied on falsified data. The *Lancet,* which published the original study in 1998, retracted the study and Wakefield's medical license was revoked. While some autism activists still defend his actions, the scientific community has condemned his research as unethical.

Inappropriate Statistical Methods The inappropriate use of statistical methods may lead to incorrect findings and distorted results even if the data are sound. "Statistical Errors in Manuscripts Submitted to *Biochemia Medica* Journal," by Simundic and Nikolac, *Biochemia Medica,* Vol. 19, No. 3, examined 55 manuscripts that were submitted for publication and found that 48 (87%) contained at least one statistical error; 34 (62%) had an incorrect choice of the statistical test; 22% had incorrect interpretation of a *P*-value; and 75% included incorrect use of a statistical test for comparing three or more groups. Peer review is a primary defense against the use of inappropriate statistical methods, but it does not guarantee that such statistical errors will be identified and corrected.

Choosing a Significance Level It is a good practice to select a significance level for a hypothesis test *before* collecting sample data and finding a *P*-value. It is a bad practice to later adjust the significance level in order to make results appear to be significant when they did not appear to be significant based on the original significance level.

III. Ethics and Reporting

Other opportunities for unethical practices occur with the interpretation of statistical results and the decision of how and where to report findings. Personal and self-serving interests may influence final conclusions and recommendations.

Conflicts of Interest

Ghostwriting A medical ghostwriter is someone who contributes to the writing of a research study or article but is not acknowledged in the published work. Litigation over the past several years has revealed that pharmaceutical companies have hired professional writers to draft papers regarding clinical trials and they have paid physicians to accept authorship of these articles. Neither the role of the ghostwriters nor the financial compensation for the acknowledged author was mentioned in the work.

Financial Support The nonprofit organization Fair Warning reported that over the past several years, hospitals invested millions of dollars to purchase new automated heart defibrillators. A committee of the American Heart Association recommended the upgrade without the benefit of research or clinical trials showing that the new devices were better than the devices that were being used. According to Fair Warning, more than 25% of the committee members "had business ties with manufacturers of the devices" and "by one estimate, the shortcomings of the automated equipment mean that close to 1000 more hospital cardiac arrest patients die every year in the U.S." Passage in 2010 of a federal law now requires companies to disclose payments for goods and services provided to physicians or teaching hospitals. While this law does not *prohibit* financial relationships between companies and physicians or teaching hospitals, it does require disclosure of these relationships.

Reporting Nonsignificant Results A recent article appeared with the headline "In U.S., Support for Death Penalty Falls to 39 Year Low." A Gallup poll found that 61% of Americans were in favor of the death penalty and noted that this was a 3% drop from the previous year's rate. The "Survey Method" description at the end of the article indicated that "one can say with 95% confidence that the maximum margin of sampling error is 4%." With a sampling error of 4%, a 3% drop in respondents who

Anonymity and Confidentiality

A survey is conducted with *anonymity* if the identities of the respondents are not known. A survey is conducted with *confidentiality* if the identities of the respondents are not disclosed. Ideally, surveys should be both anonymous and confidential, but that isn't always practical or good. Confidentiality might be ignored if it is found that some respondents are a danger to themselves or others. In such cases, respondents should be informed with a statement such as this: "All of the information that you provide will remain confidential, unless it involves risks of serious danger to yourself or others."

The *National Observer* newspaper was discontinued, but it once hired a firm to conduct a confidential mail survey. The survey was conducted with the promise that "each individual reply will be kept confidential." One clever subscriber used an ultraviolet light to detect a unique identification code printed on the survey in invisible ink. Here, confidentiality was promised and observed, but anonymity was not promised and it was not maintained. Instead of using invisible ink, the firm should have informed the respondents that their information was not anonymous.

indicated they were in favor of the death penalty is not significant. We don't really know whether there was a decrease, increase, or no change in opinions about the death penalty.

IV. Enforcing Ethics

To help prevent unethical statistical practices, many organizations have established ethical guidelines for their members or for authors whose works they publish. The American Statistical Association, for example, ratified its "Ethical Guidelines for Statistical Practice," which provides ethical guidelines in eight general topics, including Responsibilities to Research Subjects, Responsibilities in Publications and Testimony, and Responsibilities to Other Statisticians. The complete text of the Ethical Guidelines can be found on the website www.amstat.org.

The **Common Rule** has been established in the United States as the standard of ethics for biomedical and behavioral research involving human subjects. This is the baseline standard for any government-funded research and virtually all academic institutions. Central to the Common Rule are these requirements:

1. That people who participate as subjects in covered research are selected equitably and give their fully informed, fully voluntary written consent; and

2. That proposed research be reviewed by an independent oversight group referred to as an Institutional Review Board (IRB) and approved only if risks to subjects have been minimized and are reasonable in relation to anticipated benefits, if any, to the subjects, and the importance of the knowledge that may reasonably be expected to result.

The detailed text of the Common Rule can be found on the U.S. Department of Health & Human Services website at www.hhs.gov.

As consumers of data, it is important to have a healthy skepticism when statistics are cited in support of a finding. We should ask questions such as these:

- Who is reporting about the study and do they have any financial or personal interest in the outcome?
- How were the sample data obtained? What is the potential for bias?
- Have these findings been reviewed or replicated by peers or anyone else?
- What specifics do they provide about the methods used? What is the margin of error, confidence interval, *P*-value, effect size, and so on?

Researchers must always adhere to the highest ethical standards in their work. We should all strive to behave ethically as follows:

- Be complete and honest about findings, even if the results are not what was expected or desired.
- Seek the advice of professionals when unsure about which statistical analyses are appropriate.
- Always disclose financial relationships or any other interests in the outcome of the research.
- Acknowledge only those who made a meaningful contribution to the study.
- Be ready and willing to share data, methods, analyses, and results.
- Clearly identify any assumptions, limitations, or outstanding questions in the research.

Discussion Points

1. Is it ethical to use prisoners as subjects if they understand the potential risks and have given their consent? Is it ethical to offer an incentive for consent, such as additional privileges or a reduced sentence?

2. Was the Milgram experiment described in this section unethical? Why or why not?

3. For research projects that affect human subjects, is potential psychological harm any different from potential physical harm?

4. Is it unethical to intentionally infect test subjects who are humans? What about animals?

5. Volunteer bias is well documented on the topic of human sexuality. What are some other topics that might have high risk for volunteer bias?

6. Reducing sample bias may result in a less positive outcome overall, which may delay or stop the release of a drug that a subset of the population might benefit from. Is this an acceptable tradeoff?

7. One article on the topic of statistical errors in medical research listed 47 potential sources of error. Try to identify 5 potential sources of error.

8. What is medical ghostwriting? Is the practice of medical ghostwriting unethical?

9. Should common standards exist for claiming authorship of a study? What should be included in those standards?

10. Is the law requiring disclosure of payments sufficient to protect against financial support skewing study results? Do you have any other recommendations?

APPENDIX A

Tables

TABLE A-1 Binomial Probabilities

								p								
n	*x*	.01	.05	.10	.20	.30	.40	.50	.60	.70	.80	.90	.95	.99	*x*
2	0	.980	.903	.810	.640	.490	.360	.250	.160	.090	.040	.010	.003	0+	0
	1	.020	.095	.180	.320	.420	.480	.500	.480	.420	.320	.180	.095	.020	1
	2	0+	.003	.010	.040	.090	.160	.250	.360	.490	.640	.810	.903	.980	2
3	0	.970	.857	.729	.512	.343	.216	.125	.064	.027	.008	.001	0+	0+	0
	1	.029	.135	.243	.384	.441	.432	.375	.288	.189	.096	.027	.007	0+	1
	2	0+	.007	.027	.096	.189	.288	.375	.432	.441	.384	.243	.135	.029	2
	3	0+	0+	.001	.008	.027	.064	.125	.216	.343	.512	.729	.857	.970	3
4	0	.961	.815	.656	.410	.240	.130	.063	.026	.008	.002	0+	0+	0+	0
	1	.039	.171	.292	.410	.412	.346	.250	.154	.076	.026	.004	0+	0+	1
	2	.001	.014	.049	.154	.265	.346	.375	.346	.265	.154	.049	.014	.001	2
	3	0+	0+	.004	.026	.076	.154	.250	.346	.412	.410	.292	.171	.039	3
	4	0+	0+	0+	.002	.008	.026	.063	.130	.240	.410	.656	.815	.961	4
5	0	.951	.774	.590	.328	.168	.078	.031	.010	.002	0+	0+	0+	0+	0
	1	.048	.204	.328	.410	.360	.259	.156	.077	.028	.006	0+	0+	0+	1
	2	.001	.021	.073	.205	.309	.346	.313	.230	.132	.051	.008	.001	0+	2
	3	0+	.001	.008	.051	.132	.230	.313	.346	.309	.205	.073	.021	.001	3
	4	0+	0+	0+	.006	.028	.077	.156	.259	.360	.410	.328	.204	.048	4
	5	0+	0+	0+	0+	.002	.010	.031	.078	.168	.328	.590	.774	.951	5
6	0	.941	.735	.531	.262	.118	.047	.016	.004	.001	0+	0+	0+	0+	0
	1	.057	.232	.354	.393	.303	.187	.094	.037	.010	.002	0+	0+	0+	1
	2	.001	.031	.098	.246	.324	.311	.234	.138	.060	.015	.001	0+	0+	2
	3	0+	.002	.015	.082	.185	.276	.312	.276	.185	.082	.015	.002	0+	3
	4	0+	0+	.001	.015	.060	.138	.234	.311	.324	.246	.098	.031	.001	4
	5	0+	0+	0+	.002	.010	.037	.094	.187	.303	.393	.354	.232	.057	5
	6	0+	0+	0+	0+	.001	.004	.016	.047	.118	.262	.531	.735	.941	6
7	0	.932	.698	.478	.210	.082	.028	.008	.002	0+	0+	0+	0+	0+	0
	1	.066	.257	.372	.367	.247	.131	.055	.017	.004	0+	0+	0+	0+	1
	2	.002	.041	.124	.275	.318	.261	.164	.077	.025	.004	0+	0+	0+	2
	3	0+	.004	.023	.115	.227	.290	.273	.194	.097	.029	.003	0+	0+	3
	4	0+	0+	.003	.029	.097	.194	.273	.290	.227	.115	.023	.004	0+	4
	5	0+	0+	0+	.004	.025	.077	.164	.261	.318	.275	.124	.041	.002	5
	6	0+	0+	0+	0+	.004	.017	.055	.131	.247	.367	.372	.257	.066	6
	7	0+	0+	0+	0+	0+	.002	.008	.028	.082	.210	.478	.698	.932	7
8	0	.923	.663	.430	.168	.058	.017	.004	.001	0+	0+	0+	0+	0+	0
	1	.075	.279	.383	.336	.198	.090	.031	.008	.001	0+	0+	0+	0+	1
	2	.003	.051	.149	.294	.296	.209	.109	.041	.010	.001	0+	0+	0+	2
	3	0+	.005	.033	.147	.254	.279	.219	.124	.047	.009	0+	0+	0+	3
	4	0+	0+	.005	.046	.136	.232	.273	.232	.136	.046	.005	0+	0+	4
	5	0+	0+	0+	.009	.047	.124	.219	.279	.254	.147	.033	.005	0+	5
	6	0+	0+	0+	.001	.010	.041	.109	.209	.296	.294	.149	.051	.003	6
	7	0+	0+	0+	0+	.001	.008	.031	.090	.198	.336	.383	.279	.075	7
	8	0+	0+	0+	0+	0+	.001	.004	.017	.058	.168	.430	.663	.923	8

NOTE: 0+ represents a positive probability value less than 0.0005.
From Frederick C. Mosteller, Robert E. K. Rourke, and George B. Thomas, Jr., *Probability with Statistical Applications*, 2nd ed., © 1970. Reprinted and electronically reproduced by permission of Pearson Education, Inc., Upper Saddle River, New Jersey.

NEGATIVE z Scores

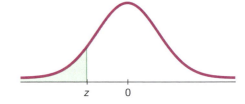

TABLE A-2 Standard Normal (z) Distribution: Cumulative Area from the LEFT

z	.00	.01	.02	.03	.04	.05	.06	.07	.08	.09
−3.50 and lower	.0001									
−3.4	.0003	.0003	.0003	.0003	.0003	.0003	.0003	.0003	.0003	.0002
−3.3	.0005	.0005	.0005	.0004	.0004	.0004	.0004	.0004	.0004	.0003
−3.2	.0007	.0007	.0006	.0006	.0006	.0006	.0006	.0005	.0005	.0005
−3.1	.0010	.0009	.0009	.0009	.0008	.0008	.0008	.0008	.0007	.0007
−3.0	.0013	.0013	.0013	.0012	.0012	.0011	.0011	.0011	.0010	.0010
−2.9	.0019	.0018	.0018	.0017	.0016	.0016	.0015	.0015	.0014	.0014
−2.8	.0026	.0025	.0024	.0023	.0023	.0022	.0021	.0021	.0020	.0019
−2.7	.0035	.0034	.0033	.0032	.0031	.0030	.0029	.0028	.0027	.0026
−2.6	.0047	.0045	.0044	.0043	.0041	.0040	.0039	.0038	.0037	.0036
−2.5	.0062	.0060	.0059	.0057	.0055	.0054	.0052	.0051 *	.0049	.0048
−2.4	.0082	.0080	.0078	.0075	.0073	.0071	.0069	.0068	.0066	.0064
−2.3	.0107	.0104	.0102	.0099	.0096	.0094	.0091	.0089	.0087	.0084
−2.2	.0139	.0136	.0132	.0129	.0125	.0122	.0119	.0116	.0113	.0110
−2.1	.0179	.0174	.0170	.0166	.0162	.0158	.0154	.0150	.0146	.0143
−2.0	.0228	.0222	.0217	.0212	.0207	.0202	.0197	.0192	.0188	.0183
−1.9	.0287	.0281	.0274	.0268	.0262	.0256	.0250	.0244	.0239	.0233
−1.8	.0359	.0351	.0344	.0336	.0329	.0322	.0314	.0307	.0301	.0294
−1.7	.0446	.0436	.0427	.0418	.0409	.0401	.0392	.0384	.0375	.0367
−1.6	.0548	.0537	.0526	.0516	.0505 *	.0495	.0485	.0475	.0465	.0455
−1.5	.0668	.0655	.0643	.0630	.0618	.0606	.0594	.0582	.0571	.0559
−1.4	.0808	.0793	.0778	.0764	.0749	.0735	.0721	.0708	.0694	.0681
−1.3	.0968	.0951	.0934	.0918	.0901	.0885	.0869	.0853	.0838	.0823
−1.2	.1151	.1131	.1112	.1093	.1075	.1056	.1038	.1020	.1003	.0985
−1.1	.1357	.1335	.1314	.1292	.1271	.1251	.1230	.1210	.1190	.1170
−1.0	.1587	.1562	.1539	.1515	.1492	.1469	.1446	.1423	.1401	.1379
−0.9	.1841	.1814	.1788	.1762	.1736	.1711	.1685	.1660	.1635	.1611
−0.8	.2119	.2090	.2061	.2033	.2005	.1977	.1949	.1922	.1894	.1867
−0.7	.2420	.2389	.2358	.2327	.2296	.2266	.2236	.2206	.2177	.2148
−0.6	.2743	.2709	.2676	.2643	.2611	.2578	.2546	.2514	.2483	.2451
−0.5	.3085	.3050	.3015	.2981	.2946	.2912	.2877	.2843	.2810	.2776
−0.4	.3446	.3409	.3372	.3336	.3300	.3264	.3228	.3192	.3156	.3121
−0.3	.3821	.3783	.3745	.3707	.3669	.3632	.3594	.3557	.3520	.3483
−0.2	.4207	.4168	.4129	.4090	.4052	.4013	.3974	.3936	.3897	.3859
−0.1	.4602	.4562	.4522	.4483	.4443	.4404	.4364	.4325	.4286	.4247
−0.0	.5000	.4960	.4920	.4880	.4840	.4801	.4761	.4721	.4681	.4641

NOTE: For values of z below −3.49, use 0.0001 for the area.

(*continued*)

*Use these common values that result from interpolation:

z Score	Area
−1.645	0.0500
−2.575	0.0050

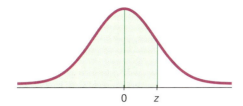

POSITIVE z Scores

0   z

TABLE A-2 *(continued)* Cumulative Area from the LEFT

z	.00	.01	.02	.03	.04	.05	.06	.07	.08	.09
0.0	.5000	.5040	.5080	.5120	.5160	.5199	.5239	.5279	.5319	.5359
0.1	.5398	.5438	.5478	.5517	.5557	.5596	.5636	.5675	.5714	.5753
0.2	.5793	.5832	.5871	.5910	.5948	.5987	.6026	.6064	.6103	.6141
0.3	.6179	.6217	.6255	.6293	.6331	.6368	.6406	.6443	.6480	.6517
0.4	.6554	.6591	.6628	.6664	.6700	.6736	.6772	.6808	.6844	.6879
0.5	.6915	.6950	.6985	.7019	.7054	.7088	.7123	.7157	.7190	.7224
0.6	.7257	.7291	.7324	.7357	.7389	.7422	.7454	.7486	.7517	.7549
0.7	.7580	.7611	.7642	.7673	.7704	.7734	.7764	.7794	.7823	.7852
0.8	.7881	.7910	.7939	.7967	.7995	.8023	.8051	.8078	.8106	.8133
0.9	.8159	.8186	.8212	.8238	.8264	.8289	.8315	.8340	.8365	.8389
1.0	.8413	.8438	.8461	.8485	.8508	.8531	.8554	.8577	.8599	.8621
1.1	.8643	.8665	.8686	.8708	.8729	.8749	.8770	.8790	.8810	.8830
1.2	.8849	.8869	.8888	.8907	.8925	.8944	.8962	.8980	.8997	.9015
1.3	.9032	.9049	.9066	.9082	.9099	.9115	.9131	.9147	.9162	.9177
1.4	.9192	.9207	.9222	.9236	.9251	.9265	.9279	.9292	.9306	.9319
1.5	.9332	.9345	.9357	.9370	.9382	.9394	.9406	.9418	.9429	.9441
1.6	.9452	.9463	.9474	.9484	.9495 *	.9505	.9515	.9525	.9535	.9545
1.7	.9554	.9564	.9573	.9582	.9591	.9599	.9608	.9616	.9625	.9633
1.8	.9641	.9649	.9656	.9664	.9671	.9678	.9686	.9693	.9699	.9706
1.9	.9713	.9719	.9726	.9732	.9738	.9744	.9750	.9756	.9761	.9767
2.0	.9772	.9778	.9783	.9788	.9793	.9798	.9803	.9808	.9812	.9817
2.1	.9821	.9826	.9830	.9834	.9838	.9842	.9846	.9850	.9854	.9857
2.2	.9861	.9864	.9868	.9871	.9875	.9878	.9881	.9884	.9887	.9890
2.3	.9893	.9896	.9898	.9901	.9904	.9906	.9909	.9911	.9913	.9916
2.4	.9918	.9920	.9922	.9925	.9927	.9929	.9931	.9932	.9934	.9936
2.5	.9938	.9940	.9941	.9943	.9945	.9946	.9948	.9949 *	.9951	.9952
2.6	.9953	.9955	.9956	.9957	.9959	.9960	.9961	.9962	.9963	.9964
2.7	.9965	.9966	.9967	.9968	.9969	.9970	.9971	.9972	.9973	.9974
2.8	.9974	.9975	.9976	.9977	.9977	.9978	.9979	.9979	.9980	.9981
2.9	.9981	.9982	.9982	.9983	.9984	.9984	.9985	.9985	.9986	.9986
3.0	.9987	.9987	.9987	.9988	.9988	.9989	.9989	.9989	.9990	.9990
3.1	.9990	.9991	.9991	.9991	.9992	.9992	.9992	.9992	.9993	.9993
3.2	.9993	.9993	.9994	.9994	.9994	.9994	.9994	.9995	.9995	.9995
3.3	.9995	.9995	.9995	.9996	.9996	.9996	.9996	.9996	.9996	.9997
3.4	.9997	.9997	.9997	.9997	.9997	.9997	.9997	.9997	.9997	.9998
3.50 and up	.9999									

NOTE: For values of z above 3.49, use 0.9999 for the area.

*Use these common values that result from interpolation:

z Score	Area
1.645	0.9500
2.575	0.9950

Common Critical Values

Confidence Level	Critical Value
0.90	1.645
0.95	1.96
0.99	2.575

Left tail

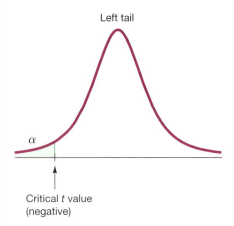

α

Critical *t* value
(negative)

Right tail

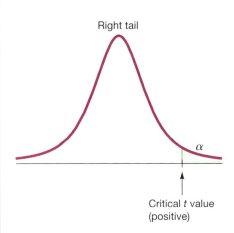

α

Critical *t* value
(positive)

Two tails

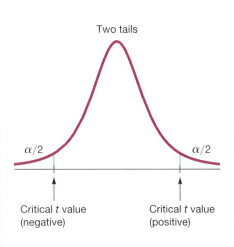

α/2   α/2

Critical *t* value   Critical *t* value
(negative)   (positive)

TABLE A-3 *t* Distribution: Critical *t* Values

Degrees of Freedom	Area in One Tail				
	0.005	0.01	0.025	0.05	0.10
	Area in Two Tails				
	0.01	0.02	0.05	0.10	0.20
1	63.657	31.821	12.706	6.314	3.078
2	9.925	6.965	4.303	2.920	1.886
3	5.841	4.541	3.182	2.353	1.638
4	4.604	3.747	2.776	2.132	1.533
5	4.032	3.365	2.571	2.015	1.476
6	3.707	3.143	2.447	1.943	1.440
7	3.499	2.998	2.365	1.895	1.415
8	3.355	2.896	2.306	1.860	1.397
9	3.250	2.821	2.262	1.833	1.383
10	3.169	2.764	2.228	1.812	1.372
11	3.106	2.718	2.201	1.796	1.363
12	3.055	2.681	2.179	1.782	1.356
13	3.012	2.650	2.160	1.771	1.350
14	2.977	2.624	2.145	1.761	1.345
15	2.947	2.602	2.131	1.753	1.341
16	2.921	2.583	2.120	1.746	1.337
17	2.898	2.567	2.110	1.740	1.333
18	2.878	2.552	2.101	1.734	1.330
19	2.861	2.539	2.093	1.729	1.328
20	2.845	2.528	2.086	1.725	1.325
21	2.831	2.518	2.080	1.721	1.323
22	2.819	2.508	2.074	1.717	1.321
23	2.807	2.500	2.069	1.714	1.319
24	2.797	2.492	2.064	1.711	1.318
25	2.787	2.485	2.060	1.708	1.316
26	2.779	2.479	2.056	1.706	1.315
27	2.771	2.473	2.052	1.703	1.314
28	2.763	2.467	2.048	1.701	1.313
29	2.756	2.462	2.045	1.699	1.311
30	2.750	2.457	2.042	1.697	1.310
31	2.744	2.453	2.040	1.696	1.309
32	2.738	2.449	2.037	1.694	1.309
33	2.733	2.445	2.035	1.692	1.308
34	2.728	2.441	2.032	1.691	1.307
35	2.724	2.438	2.030	1.690	1.306
36	2.719	2.434	2.028	1.688	1.306
37	2.715	2.431	2.026	1.687	1.305
38	2.712	2.429	2.024	1.686	1.304
39	2.708	2.426	2.023	1.685	1.304
40	2.704	2.423	2.021	1.684	1.303
45	2.690	2.412	2.014	1.679	1.301
50	2.678	2.403	2.009	1.676	1.299
60	2.660	2.390	2.000	1.671	1.296
70	2.648	2.381	1.994	1.667	1.294
80	2.639	2.374	1.990	1.664	1.292
90	2.632	2.368	1.987	1.662	1.291
100	2.626	2.364	1.984	1.660	1.290
200	2.601	2.345	1.972	1.653	1.286
300	2.592	2.339	1.968	1.650	1.284
400	2.588	2.336	1.966	1.649	1.284
500	2.586	2.334	1.965	1.648	1.283
1000	2.581	2.330	1.962	1.646	1.282
2000	2.578	2.328	1.961	1.646	1.282
Large	2.576	2.326	1.960	1.645	1.282

TABLE A-4 Chi-Square (χ^2) Distribution

Degrees of Freedom	Area to the *Right* of the Critical Value									
	0.995	0.99	0.975	0.95	0.90	0.10	0.05	0.025	0.01	0.005
1	—	—	0.001	0.004	0.016	2.706	3.841	5.024	6.635	7.879
2	0.010	0.020	0.051	0.103	0.211	4.605	5.991	7.378	9.210	10.597
3	0.072	0.115	0.216	0.352	0.584	6.251	7.815	9.348	11.345	12.838
4	0.207	0.297	0.484	0.711	1.064	7.779	9.488	11.143	13.277	14.860
5	0.412	0.554	0.831	1.145	1.610	9.236	11.071	12.833	15.086	16.750
6	0.676	0.872	1.237	1.635	2.204	10.645	12.592	14.449	16.812	18.548
7	0.989	1.239	1.690	2.167	2.833	12.017	14.067	16.013	18.475	20.278
8	1.344	1.646	2.180	2.733	3.490	13.362	15.507	17.535	20.090	21.955
9	1.735	2.088	2.700	3.325	4.168	14.684	16.919	19.023	21.666	23.589
10	2.156	2.558	3.247	3.940	4.865	15.987	18.307	20.483	23.209	25.188
11	2.603	3.053	3.816	4.575	5.578	17.275	19.675	21.920	24.725	26.757
12	3.074	3.571	4.404	5.226	6.304	18.549	21.026	23.337	26.217	28.299
13	3.565	4.107	5.009	5.892	7.042	19.812	22.362	24.736	27.688	29.819
14	4.075	4.660	5.629	6.571	7.790	21.064	23.685	26.119	29.141	31.319
15	4.601	5.229	6.262	7.261	8.547	22.307	24.996	27.488	30.578	32.801
16	5.142	5.812	6.908	7.962	9.312	23.542	26.296	28.845	32.000	34.267
17	5.697	6.408	7.564	8.672	10.085	24.769	27.587	30.191	33.409	35.718
18	6.265	7.015	8.231	9.390	10.865	25.989	28.869	31.526	34.805	37.156
19	6.844	7.633	8.907	10.117	11.651	27.204	30.144	32.852	36.191	38.582
20	7.434	8.260	9.591	10.851	12.443	28.412	31.410	34.170	37.566	39.997
21	8.034	8.897	10.283	11.591	13.240	29.615	32.671	35.479	38.932	41.401
22	8.643	9.542	10.982	12.338	14.042	30.813	33.924	36.781	40.289	42.796
23	9.260	10.196	11.689	13.091	14.848	32.007	35.172	38.076	41.638	44.181
24	9.886	10.856	12.401	13.848	15.659	33.196	36.415	39.364	42.980	45.559
25	10.520	11.524	13.120	14.611	16.473	34.382	37.652	40.646	44.314	46.928
26	11.160	12.198	13.844	15.379	17.292	35.563	38.885	41.923	45.642	48.290
27	11.808	12.879	14.573	16.151	18.114	36.741	40.113	43.194	46.963	49.645
28	12.461	13.565	15.308	16.928	18.939	37.916	41.337	44.461	48.278	50.993
29	13.121	14.257	16.047	17.708	19.768	39.087	42.557	45.722	49.588	52.336
30	13.787	14.954	16.791	18.493	20.599	40.256	43.773	46.979	50.892	53.672
40	20.707	22.164	24.433	26.509	29.051	51.805	55.758	59.342	63.691	66.766
50	27.991	29.707	32.357	34.764	37.689	63.167	67.505	71.420	76.154	79.490
60	35.534	37.485	40.482	43.188	46.459	74.397	79.082	83.298	88.379	91.952
70	43.275	45.442	48.758	51.739	55.329	85.527	90.531	95.023	100.425	104.215
80	51.172	53.540	57.153	60.391	64.278	96.578	101.879	106.629	112.329	116.321
90	59.196	61.754	65.647	69.126	73.291	107.565	113.145	118.136	124.116	128.299
100	67.328	70.065	74.222	77.929	82.358	118.498	124.342	129.561	135.807	140.169

Source: Donald B. Owen, *Handbook of Statistical Tables.*

Degrees of Freedom

$n - 1$	**Confidence interval or hypothesis test** for a standard deviation σ or variance σ^2
$k - 1$	**Goodness-of-fit test** with k different categories
$(r - 1)(c - 1)$	**Contingency table test** with r rows and c columns
$k - 1$	**Kruskal-Wallis test** with k different samples

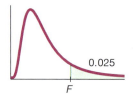

0.025

F

TABLE A-5 F Distribution ($\alpha = 0.025$ in the right tail)

					Numerator degrees of freedom (df$_1$)				
	1	2	3	4	5	6	7	8	9
1	647.79	799.50	864.16	899.58	921.85	937.11	948.22	956.66	963.28
2	38.506	39.000	39.165	39.248	39.298	39.331	39.335	39.373	39.387
3	17.443	16.044	15.439	15.101	14.885	14.735	14.624	14.540	14.473
4	12.218	10.649	9.9792	9.6045	9.3645	9.1973	9.0741	8.9796	8.9047
5	10.007	8.4336	7.7636	7.3879	7.1464	6.9777	6.8531	6.7572	6.6811
6	8.8131	7.2599	6.5988	6.2272	5.9876	5.8198	5.6955	5.5996	5.5234
7	8.0727	6.5415	5.8898	5.5226	5.2852	5.1186	4.9949	4.8993	4.8232
8	7.5709	6.0595	5.4160	5.0526	4.8173	4.6517	4.5286	4.4333	4.3572
9	7.2093	5.7147	5.0781	4.7181	4.4844	4.3197	4.1970	4.1020	4.0260
10	6.9367	5.4564	4.8256	4.4683	4.2361	4.0721	3.9498	3.8549	3.7790
11	6.7241	5.2559	4.6300	4.2751	4.0440	3.8807	3.7586	3.6638	3.5879
12	6.5538	5.0959	4.4742	4.1212	3.8911	3.7283	3.6065	3.5118	3.4358
13	6.4143	4.9653	4.3472	3.9959	3.7667	3.6043	3.4827	3.3880	3.3120
14	6.2979	4.8567	4.2417	3.8919	3.6634	3.5014	3.3799	3.2853	3.2093
15	6.1995	4.7650	4.1528	3.8043	3.5764	3.4147	3.2934	3.1987	3.1227
16	6.1151	4.6867	4.0768	3.7294	3.5021	3.3406	3.2194	3.1248	3.0488
17	6.0420	4.6189	4.0112	3.6648	3.4379	3.2767	3.1556	3.0610	2.9849
18	5.9781	4.5597	3.9539	3.6083	3.3820	3.2209	3.0999	3.0053	2.9291
19	5.9216	4.5075	3.9034	3.5587	3.3327	3.1718	3.0509	2.9563	2.8801
20	5.8715	4.4613	3.8587	3.5147	3.2891	3.1283	3.0074	2.9128	2.8365
21	5.8266	4.4199	3.8188	3.4754	3.2501	3.0895	2.9686	2.8740	2.7977
22	5.7863	4.3828	3.7829	3.4401	3.2151	3.0546	2.9338	2.8392	2.7628
23	5.7498	4.3492	3.7505	3.4083	3.1835	3.0232	2.9023	2.8077	2.7313
24	5.7166	4.3187	3.7211	3.3794	3.1548	2.9946	2.8738	2.7791	2.7027
25	5.6864	4.2909	3.6943	3.3530	3.1287	2.9685	2.8478	2.7531	2.6766
26	5.6586	4.2655	3.6697	3.3289	3.1048	2.9447	2.8240	2.7293	2.6528
27	5.6331	4.2421	3.6472	3.3067	3.0828	2.9228	2.8021	2.7074	2.6309
28	5.6096	4.2205	3.6264	3.2863	3.0626	2.9027	2.7820	2.6872	2.6106
29	5.5878	4.2006	3.6072	3.2674	3.0438	2.8840	2.7633	2.6686	2.5919
30	5.5675	4.1821	3.5894	3.2499	3.0265	2.8667	2.7460	2.6513	2.5746
40	5.4239	4.0510	3.4633	3.1261	2.9037	2.7444	2.6238	2.5289	2.4519
60	5.2856	3.9253	3.3425	3.0077	2.7863	2.6274	2.5068	2.4117	2.3344
120	5.1523	3.8046	3.2269	2.8943	2.6740	2.5154	2.3948	2.2994	2.2217
∞	5.0239	3.6889	3.1161	2.7858	2.5665	2.4082	2.2875	2.1918	2.1136

Denominator degrees of freedom (df$_2$)

(*continued*)

TABLE A-5 *(continued)* F Distribution ($\alpha = 0.025$ in the right tail)

		10	12	15	20	24	30	40	60	120	∞
					Numerator degrees of freedom (df_1)						
	1	968.63	976.71	984.87	993.10	997.25	1001.4	1005.6	1009.8	1014.0	1018.3
	2	39.398	39.415	39.431	39.448	39.456	39.465	39.473	39.481	39.490	39.498
	3	14.419	14.337	14.253	14.167	14.124	14.081	14.037	13.992	13.947	13.902
	4	8.8439	8.7512	8.6565	8.5599	8.5109	8.4613	8.4111	8.3604	8.3092	8.2573
	5	6.6192	6.5245	6.4277	6.3286	6.2780	6.2269	6.1750	6.1225	6.0693	6.0153
	6	5.4613	5.3662	5.2687	5.1684	5.1172	5.0652	5.0125	4.9589	4.9044	4.8491
	7	4.7611	4.6658	4.5678	4.4667	4.4150	4.3624	4.3089	4.2544	4.1989	4.1423
	8	4.2951	4.1997	4.1012	3.9995	3.9472	3.8940	3.8398	3.7844	3.7279	3.6702
	9	3.9639	3.8682	3.7694	3.6669	3.6142	3.5604	3.5055	3.4493	3.3918	3.3329
	10	3.7168	3.6209	3.5217	3.4185	3.3654	3.3110	3.2554	3.1984	3.1399	3.0798
	11	3.5257	3.4296	3.3299	3.2261	3.1725	3.1176	3.0613	3.0035	2.9441	2.8828
	12	3.3736	3.2773	3.1772	3.0728	3.0187	2.9633	2.9063	2.8478	2.7874	2.7249
Denominator degrees of freedom (df_2)	13	3.2497	3.1532	3.0527	2.9477	2.8932	2.8372	2.7797	2.7204	2.6590	2.5955
	14	3.1469	3.0502	2.9493	2.8437	2.7888	2.7324	2.6742	2.6142	2.5519	2.4872
	15	3.0602	2.9633	2.8621	2.7559	2.7006	2.6437	2.5850	2.5242	2.4611	2.3953
	16	2.9862	2.8890	2.7875	2.6808	2.6252	2.5678	2.5085	2.4471	2.3831	2.3163
	17	2.9222	2.8249	2.7230	2.6158	2.5598	2.5020	2.4422	2.3801	2.3153	2.2474
	18	2.8664	2.7689	2.6667	2.5590	2.5027	2.4445	2.3842	2.3214	2.2558	2.1869
	19	2.8172	2.7196	2.6171	2.5089	2.4523	2.3937	2.3329	2.2696	2.2032	2.1333
	20	2.7737	2.6758	2.5731	2.4645	2.4076	2.3486	2.2873	2.2234	2.1562	2.0853
	21	2.7348	2.6368	2.5338	2.4247	2.3675	2.3082	2.2465	2.1819	2.1141	2.0422
	22	2.6998	2.6017	2.4984	2.3890	2.3315	2.2718	2.2097	2.1446	2.0760	2.0032
	23	2.6682	2.5699	2.4665	2.3567	2.2989	2.2389	2.1763	2.1107	2.0415	1.9677
	24	2.6396	2.5411	2.4374	2.3273	2.2693	2.2090	2.1460	2.0799	2.0099	1.9353
	25	2.6135	2.5149	2.4110	2.3005	2.2422	2.1816	2.1183	2.0516	1.9811	1.9055
	26	2.5896	2.4908	2.3867	2.2759	2.2174	2.1565	2.0928	2.0257	1.9545	1.8781
	27	2.5676	2.4688	2.3644	2.2533	2.1946	2.1334	2.0693	2.0018	1.9299	1.8527
	28	2.5473	2.4484	2.3438	2.2324	2.1735	2.1121	2.0477	1.9797	1.9072	1.8291
	29	2.5286	2.4295	2.3248	2.2131	2.1540	2.0923	2.0276	1.9591	1.8861	1.8072
	30	2.5112	2.4120	2.3072	2.1952	2.1359	2.0739	2.0089	1.9400	1.8664	1.7867
	40	2.3882	2.2882	2.1819	2.0677	2.0069	1.9429	1.8752	1.8028	1.7242	1.6371
	60	2.2702	2.1692	2.0613	1.9445	1.8817	1.8152	1.7440	1.6668	1.5810	1.4821
	120	2.1570	2.0548	1.9450	1.8249	1.7597	1.6899	1.6141	1.5299	1.4327	1.3104
	∞	2.0483	1.9447	1.8326	1.7085	1.6402	1.5660	1.4835	1.3883	1.2684	1.0000

Based on data from Maxine Merrington and Catherine M. Thompson, "Tables of Percentage Points of the Inverted Beta (*F*) Distribution," *Biometrika 33* (1943): 80–84.

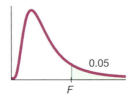

0.05

F

TABLE A-5 *(continued)* *F* Distribution ($\alpha = 0.05$ in the right tail)

					Numerator degrees of freedom (df₁)					
		1	2	3	4	5	6	7	8	9
Denominator degrees of freedom (df₂)	1	161.45	199.50	215.71	224.58	230.16	233.99	236.77	238.88	240.54
	2	18.513	19.000	19.164	19.247	19.296	19.330	19.353	19.371	19.385
	3	10.128	9.5521	9.2766	9.1172	9.0135	8.9406	8.8867	8.8452	8.8123
	4	7.7086	6.9443	6.5914	6.3882	6.2561	6.1631	6.0942	6.0410	6.9988
	5	6.6079	5.7861	5.4095	5.1922	5.0503	4.9503	4.8759	4.8183	4.7725
	6	5.9874	5.1433	4.7571	4.5337	4.3874	4.2839	4.2067	4.1468	4.0990
	7	5.5914	4.7374	4.3468	4.1203	3.9715	3.8660	3.7870	3.7257	3.6767
	8	5.3177	4.4590	4.0662	3.8379	3.6875	3.5806	3.5005	3.4381	3.3881
	9	5.1174	4.2565	3.8625	3.6331	3.4817	3.3738	3.2927	3.2296	3.1789
	10	4.9646	4.1028	3.7083	3.4780	3.3258	3.2172	3.1355	3.0717	3.0204
	11	4.8443	3.9823	3.5874	3.3567	3.2039	3.0946	3.0123	2.9480	2.8962
	12	4.7472	3.8853	3.4903	3.2592	3.1059	2.9961	2.9134	2.8486	2.7964
	13	4.6672	3.8056	3.4105	3.1791	3.0254	2.9153	2.8321	2.7669	2.7144
	14	4.6001	3.7389	3.3439	3.1122	2.9582	2.8477	2.7642	2.6987	2.6458
	15	4.5431	3.6823	3.2874	3.0556	2.9013	2.7905	2.7066	2.6408	2.5876
	16	4.4940	3.6337	3.2389	3.0069	2.8524	2.7413	2.6572	2.5911	2.5377
	17	4.4513	3.5915	3.1968	2.9647	2.8100	2.6987	2.6143	2.5480	2.4943
	18	4.4139	3.5546	3.1599	2.9277	2.7729	2.6613	2.5767	2.5102	2.4563
	19	4.3807	3.5219	3.1274	2.8951	2.7401	2.6283	2.5435	2.4768	2.4227
	20	4.3512	3.4928	3.0984	2.8661	2.7109	2.5990	2.5140	2.4471	2.3928
	21	4.3248	3.4668	3.0725	2.8401	2.6848	2.5727	2.4876	2.4205	2.3660
	22	4.3009	3.4434	3.0491	2.8167	2.6613	2.5491	2.4638	2.3965	2.3419
	23	4.2793	3.4221	3.0280	2.7955	2.6400	2.5277	2.4422	2.3748	2.3201
	24	4.2597	3.4028	3.0088	2.7763	2.6207	2.5082	2.4226	2.3551	2.3002
	25	4.2417	3.3852	2.9912	2.7587	2.6030	2.4904	2.4047	2.3371	2.2821
	26	4.2252	3.3690	2.9752	2.7426	2.5868	2.4741	2.3883	2.3205	2.2655
	27	4.2100	3.3541	2.9604	2.7278	2.5719	2.4591	2.3732	2.3053	2.2501
	28	4.1960	3.3404	2.9467	2.7141	2.5581	2.4453	2.3593	2.2913	2.2360
	29	4.1830	3.3277	2.9340	2.7014	2.5454	2.4324	2.3463	2.2783	2.2229
	30	4.1709	3.3158	2.9223	2.6896	2.5336	2.4205	2.3343	2.2662	2.2107
	40	4.0847	3.2317	2.8387	2.6060	2.4495	2.3359	2.2490	2.1802	2.1240
	60	4.0012	3.1504	2.7581	2.5252	2.3683	2.2541	2.1665	2.0970	2.0401
	120	3.9201	3.0718	2.6802	2.4472	2.2899	2.1750	2.0868	2.0164	1.9588
	∞	3.8415	2.9957	2.6049	2.3719	2.2141	2.0986	2.0096	1.9384	1.8799

(continued)

TABLE A-5 *(continued)* F Distribution ($\alpha = 0.05$ in the right tail)

	Numerator degrees of freedom (df_1)									
	10	12	15	20	24	30	40	60	120	∞
1	241.88	243.91	245.95	248.01	249.05	250.10	251.14	252.20	253.25	254.31
2	19.396	19.413	19.429	19.446	19.454	19.462	19.471	19.479	19.487	19.496
3	8.7855	8.7446	8.7029	8.6602	8.6385	8.6166	8.5944	8.5720	8.5494	8.5264
4	5.9644	5.9117	5.8578	5.8025	5.7744	5.7459	5.7170	5.6877	5.6581	5.6281
5	4.7351	4.6777	4.6188	4.5581	4.5272	4.4957	4.4638	4.4314	4.3985	4.3650
6	4.0600	3.9999	3.9381	3.8742	3.8415	3.8082	3.7743	3.7398	3.7047	3.6689
7	3.6365	3.5747	3.5107	3.4445	3.4105	3.3758	3.3404	3.3043	3.2674	3.2298
8	3.3472	3.2839	3.2184	3.1503	3.1152	3.0794	3.0428	3.0053	2.9669	2.9276
9	3.1373	3.0729	3.0061	2.9365	2.9005	2.8637	2.8259	2.7872	2.7475	2.7067
10	2.9782	2.9130	2.8450	2.7740	2.7372	2.6996	2.6609	2.6211	2.5801	2.5379
11	2.8536	2.7876	2.7186	2.6464	2.6090	2.5705	2.5309	2.4901	2.4480	2.4045
12	2.7534	2.6866	2.6169	2.5436	2.5055	2.4663	2.4259	2.3842	2.3410	2.2962
13	2.6710	2.6037	2.5331	2.4589	2.4202	2.3803	2.3392	2.2966	2.2524	2.2064
14	2.6022	2.5342	2.4630	2.3879	2.3487	2.3082	2.2664	2.2229	2.1778	2.1307
15	2.5437	2.4753	2.4034	2.3275	2.2878	2.2468	2.2043	2.1601	2.1141	2.0658
16	2.4935	2.4247	2.3522	2.2756	2.2354	2.1938	2.1507	2.1058	2.0589	2.0096
17	2.4499	2.3807	2.3077	2.2304	2.1898	2.1477	2.1040	2.0584	2.0107	1.9604
18	2.4117	2.3421	2.2686	2.1906	2.1497	2.1071	2.0629	2.0166	1.9681	1.9168
19	2.3779	2.3080	2.2341	2.1555	2.1141	2.0712	2.0264	1.9795	1.9302	1.8780
20	2.3479	2.2776	2.2033	2.1242	2.0825	2.0391	1.9938	1.9464	1.8963	1.8432
21	2.3210	2.2504	2.1757	2.0960	2.0540	2.0102	1.9645	1.9165	1.8657	1.8117
22	2.2967	2.2258	2.1508	2.0707	2.0283	1.9842	1.9380	1.8894	1.8380	1.7831
23	2.2747	2.2036	2.1282	2.0476	2.0050	1.9605	1.9139	1.8648	1.8128	1.7570
24	2.2547	2.1834	2.1077	2.0267	1.9838	1.9390	1.8920	1.8424	1.7896	1.7330
25	2.2365	2.1649	2.0889	2.0075	1.9643	1.9192	1.8718	1.8217	1.7684	1.7110
26	2.2197	2.1479	2.0716	1.9898	1.9464	1.9010	1.8533	1.8027	1.7488	1.6906
27	2.2043	2.1323	2.0558	1.9736	1.9299	1.8842	1.8361	1.7851	1.7306	1.6717
28	2.1900	2.1179	2.0411	1.9586	1.9147	1.8687	1.8203	1.7689	1.7138	1.6541
29	2.1768	2.1045	2.0275	1.9446	1.9005	1.8543	1.8055	1.7537	1.6981	1.6376
30	2.1646	2.0921	2.0148	1.9317	1.8874	1.8409	1.7918	1.7396	1.6835	1.6223
40	2.0772	2.0035	1.9245	1.8389	1.7929	1.7444	1.6928	1.6373	1.5766	1.5089
60	1.9926	1.9174	1.8364	1.7480	1.7001	1.6491	1.5943	1.5343	1.4673	1.3893
120	1.9105	1.8337	1.7505	1.6587	1.6084	1.5543	1.4952	1.4290	1.3519	1.2539
∞	1.8307	1.7522	1.6664	1.5705	1.5173	1.4591	1.3940	1.3180	1.2214	1.0000

Denominator degrees of freedom (df_2)

Based on data from Maxine Merrington and Catherine M. Thompson, "Tables of Percentage Points of the Inverted Beta (F) Distribution," *Biometrika* 33 (1943): 80–84.

TABLE A-6 Critical Values of the Pearson Correlation Coefficient r

n	$\alpha = .05$	$\alpha = .01$
4	.950	.990
5	.878	.959
6	.811	.917
7	.754	.875
8	.707	.834
9	.666	.798
10	.632	.765
11	.602	.735
12	.576	.708
13	.553	.684
14	.532	.661
15	.514	.641
16	.497	.623
17	.482	.606
18	.468	.590
19	.456	.575
20	.444	.561
25	.396	.505
30	.361	.463
35	.335	.430
40	.312	.402
45	.294	.378
50	.279	.361
60	.254	.330
70	.236	.305
80	.220	.286
90	.207	.269
100	.196	.256

NOTE: To test $H_0: \rho = 0$ (no correlation) against $H_1: \rho \neq 0$ (correlation), reject H_0 if the absolute value of r is greater than or equal to the critical value in the table.

TABLE A-7 Critical Values for the Sign Test

	α			
	.005 (one tail)	.01 (one tail)	.025 (one tail)	.05 (one tail)
n	.01 (two tails)	.02 (two tails)	.05 (two tails)	.10 (two tails)
1	*	*	*	*
2	*	*	*	*
3	*	*	*	*
4	*	*	*	*
5	*	*	*	0
6	*	*	0	0
7	*	0	0	0
8	0	0	0	1
9	0	0	1	1
10	0	0	1	1
11	0	1	1	2
12	1	1	2	2
13	1	1	2	3
14	1	2	2	3
15	2	2	3	3
16	2	2	3	4
17	2	3	4	4
18	3	3	4	5
19	3	4	4	5
20	3	4	5	5
21	4	4	5	6
22	4	5	5	6
23	4	5	6	7
24	5	5	6	7
25	5	6	7	7

NOTES:
1. *indicates that it is not possible to get a value in the critical region, so fail to reject the null hypothesis.
2. Reject the null hypothesis if the number of the less frequent sign (x) is less than or equal to the value in the table.
3. For values of n greater than 25, a normal approximation is used with

$$z = \frac{(x + 0.5) - \left(\dfrac{n}{2}\right)}{\dfrac{\sqrt{n}}{2}}$$

TABLE A-8 Critical Values of *T* for the Wilcoxon Signed-Ranks Test

	α			
	.005 (one tail)	.01 (one tail)	.025 (one tail)	.05 (one tail)
n	.01 (two tails)	.02 (two tails)	.05 (two tails)	.10 (two tails)
5	*	*	*	1
6	*	*	1	2
7	*	0	2	4
8	0	2	4	6
9	2	3	6	8
10	3	5	8	11
11	5	7	11	14
12	7	10	14	17
13	10	13	17	21
14	13	16	21	26
15	16	20	25	30
16	19	24	30	36
17	23	28	35	41
18	28	33	40	47
19	32	38	46	54
20	37	43	52	60
21	43	49	59	68
22	49	56	66	75
23	55	62	73	83
24	61	69	81	92
25	68	77	90	101
26	76	85	98	110
27	84	93	107	120
28	92	102	117	130
29	100	111	127	141
30	109	120	137	152

NOTES:

1. *indicates that it is not possible to get a value in the critical region, so fail to reject the null hypothesis.
2. Conclusions:

   Reject the null hypothesis if the test statistic *T* is less than or equal to the critical value found in this table.

   Fail to reject the null hypothesis if the test statistic *T* is greater than the critical value found in the table.

Based on data from *Some Rapid Approximate Statistical Procedures*, Copyright © 1949, 1964 Lederle Laboratories Division of American Cyanamid Company.

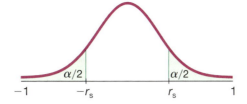

TABLE A-9 Critical Values of Spearman's Rank
Correlation Coefficient r_s

n	$\alpha = 0.10$	$\alpha = 0.05$	$\alpha = 0.02$	$\alpha = 0.01$
5	.900	—	—	—
6	.829	.886	.943	—
7	.714	.786	.893	.929
8	.643	.738	.833	.881
9	.600	.700	.783	.833
10	.564	.648	.745	.794
11	.536	.618	.709	.755
12	.503	.587	.678	.727
13	.484	.560	.648	.703
14	.464	.538	.626	.679
15	.446	.521	.604	.654
16	.429	.503	.582	.635
17	.414	.485	.566	.615
18	.401	.472	.550	.600
19	.391	.460	.535	.584
20	.380	.447	.520	.570
21	.370	.435	.508	.556
22	.361	.425	.496	.544
23	.353	.415	.486	.532
24	.344	.406	.476	.521
25	.337	.398	.466	.511
26	.331	.390	.457	.501
27	.324	.382	.448	.491
28	.317	.375	.440	.483
29	.312	.368	.433	.475
30	.306	.362	.425	.467

NOTES:

1. For $n > 30$ use $r_s \pm z/\sqrt{n-1}$, where z corresponds to the level of significance. For example, if $\alpha = 0.05$, then $z = 1.96$.

2. If the absolute value of the test statistic r_s is greater than or equal to the positive critical value, then reject H_0: $\rho_s = 0$ and conclude that there is sufficient evidence to support the claim of a correlation.

Based on data from *Biostatistical Analysis*, 4th edition © 1999, by Jerrold Zar, Prentice Hall, Inc., Upper Saddle River, New Jersey, and "Distribution of Sums of Squares of Rank Differences to Small Numbers with Individuals," *The Annals of Mathematical Statistics,* Vol. 9, No. 2.

TABLE A-10 Runs Test for Randomness: Critical Values for Number of Runs G

	Value of n_2																		
Value of n_1	2	3	4	5	6	7	8	9	10	11	12	13	14	15	16	17	18	19	20
2	1	1	1	1	1	1	1	1	1	1	2	2	2	2	2	2	2	2	2
	6	6	6	6	6	6	6	6	6	6	6	6	6	6	6	6	6	6	6
3	1	1	1	1	2	2	2	2	2	2	2	2	2	3	3	3	3	3	3
	6	8	8	8	8	8	8	8	8	8	8	8	8	8	8	8	8	8	8
4	1	1	1	2	2	2	3	3	3	3	3	3	3	3	4	4	4	4	4
	6	8	9	9	9	10	10	10	10	10	10	10	10	10	10	10	10	10	10
5	1	1	2	2	3	3	3	3	3	4	4	4	4	4	4	4	5	5	5
	6	8	9	10	10	11	11	12	12	12	12	12	12	12	12	12	12	12	12
6	1	2	2	3	3	3	3	4	4	4	4	5	5	5	5	5	5	6	6
	6	8	9	10	11	12	12	13	13	13	13	14	14	14	14	14	14	14	14
7	1	2	2	3	3	3	4	4	5	5	5	5	5	6	6	6	6	6	6
	6	8	10	11	12	13	13	14	14	14	14	15	15	15	16	16	16	16	16
8	1	2	3	3	3	4	4	5	5	5	6	6	6	6	6	7	7	7	7
	6	8	10	11	12	13	14	14	15	15	16	16	16	16	17	17	17	17	17
9	1	2	3	3	4	4	5	5	5	6	6	6	7	7	7	7	8	8	8
	6	8	10	12	13	14	14	15	16	16	16	17	17	18	18	18	18	18	18
10	1	2	3	3	4	5	5	5	6	6	7	7	7	7	8	8	8	8	9
	6	8	10	12	13	14	15	16	16	17	17	18	18	18	19	19	19	20	20
11	1	2	3	4	4	5	5	6	6	7	7	7	8	8	8	9	9	9	9
	6	8	10	12	13	14	15	16	17	17	18	19	19	19	20	20	20	21	21
12	2	2	3	4	4	5	6	6	7	7	7	8	8	8	9	9	9	10	10
	6	8	10	12	13	14	16	16	17	18	19	19	20	20	21	21	21	22	22
13	2	2	3	4	5	5	6	6	7	7	8	8	9	9	9	10	10	10	10
	6	8	10	12	14	15	16	17	18	19	19	20	20	21	21	22	22	23	23
14	2	2	3	4	5	5	6	7	7	8	8	9	9	9	10	10	10	11	11
	6	8	10	12	14	15	16	17	18	19	20	20	21	22	22	23	23	23	24
15	2	3	3	4	5	6	6	7	7	8	8	9	9	10	10	11	11	11	12
	6	8	10	12	14	15	16	18	18	19	20	21	22	22	23	23	24	24	25
16	2	3	4	4	5	6	6	7	8	8	9	9	10	10	11	11	11	12	12
	6	8	10	12	14	16	17	18	19	20	21	21	22	23	23	24	25	25	25
17	2	3	4	4	5	6	7	7	8	9	9	10	10	11	11	11	12	12	13
	6	8	10	12	14	16	17	18	19	20	21	22	23	23	24	25	25	26	26
18	2	3	4	5	5	6	7	8	8	9	9	10	10	11	11	12	12	13	13
	6	8	10	12	14	16	17	18	19	20	21	22	23	24	25	25	26	26	27
19	2	3	4	5	6	6	7	8	8	9	10	10	11	11	12	12	13	13	13
	6	8	10	12	14	16	17	18	20	21	22	23	23	24	25	26	26	27	27
20	2	3	4	5	6	6	7	8	9	9	10	10	11	12	12	13	13	13	14
	6	8	10	12	14	16	17	18	20	21	22	23	24	25	25	26	27	27	28

NOTES:

1. The entries in this table are the critical G values, assuming a two-tailed test with a significance level of $\alpha = 0.05$.

2. Reject the null hypothesis of randomness if either of these conditions is satisfied:
   - The number of runs G is less than or equal to the smaller entry in the table.
   - The number of runs G is greater than or equal to the larger entry in the table.

From "Tables for Testing Randomness of Groupings in a Sequence of Alternatives," *The Annals of Mathematical Statistics,* Vol. 14, No. 1. Reprinted with permission of the Institute of Mathematical Statistics.

Complete data sets are available at www.TriolaStats.com

This appendix lists only the first five rows of each data set. The complete data sets are available for download at www.TriolaStats.com for a variety of technologies, including Excel, TI-83/84 Plus calculators, and Minitab. These data sets are included with Statdisk, which is available for free to users of this textbook and can be downloaded at www.statdisk.org.

Data Set 1: Body Data

Data Set 2: Foot and Height

Data Set 3: Body Temperatures

Data Set 4: Births

Data Set 5: Family Heights

Data Set 6: Freshman 15

Data Set 7: IQ and Lead

Data Set 8: IQ and Brain Size

Data Set 9: Bear Measurements

Data Set 10: Manatee Deaths

Data Set 11: Alcohol and Tobacco in Movies

Data Set 12: Passive and Active Smoke

Data Set 13: Cigarette Contents

Data Set 14: Oscar Winner Age

Data Set 15: Presidents

Data Set 16: Nobel Laureates and Chocolate

Data Set 17: Course Evaluations

Data Set 18: Speed Dating

Data Set 19: Car Crash Tests

Data Set 20: Car Measurements

Data Set 21: Earthquakes

Data Set 22: Tornadoes

Data Set 23: Old Faithful

Data Set 24: Word Counts

Data Set 25: Fast Food

Data Set 26: Cola Weights and Volumes

Data Set 27: M&M Weights

Data Set 28: Chocolate Chip Cookies

Data Set 29: Coin Weights

Data Set 30: Aluminum Cans

Data Set 31: Garbage Weight

Data Set 32: Airport Data Speeds

Data Set 1: Body Data

Body and exam measurements are from 300 subjects (first five rows shown here). **AGE** is in years, for **GENDER** 1 = male and 0 = female, **PULSE** is pulse rate (beats per minute), **SYSTOLIC** is systolic blood pressure (mm Hg), **DIASTOLIC** is diastolic blood pressure (mm Hg), **HDL** is HDL cholesterol (mg/dL), **LDL** is LDL cholesterol (mg/dL), **WHITE** is white blood cell count

(1000 cells/μL), **RED** is red blood cell count (million cells/μL), **PLATE** is platelet count (1000 cells/μL), **WEIGHT** is weight (kg), **HEIGHT** is height (cm), **WAIST** is waist circumference (cm), **ARM CIRC** is arm circumference (cm), and **BMI** is body mass index (kg/m^2). Data are from the National Center for Health Statistics.

TI-83/84 list names (BODY): AGE, GENDR, PULSE, SYS, DIAS, HDL, LDL, WHITE, REDBC, PLATE, WT, HT, WAIST, ARMC, BMI

AGE	GENDER (1 = M)	PULSE	SYSTOLIC	DIASTOLIC	HDL	LDL	WHITE	RED	PLATE	WEIGHT	HEIGHT	WAIST	ARM CIRC	BMI
43	0	80	100	70	73	68	8.7	4.80	319	98.6	172.0	120.4	40.7	33.3
57	1	84	112	70	35	116	4.9	4.73	187	96.9	186.0	107.8	37.0	28.0
38	0	94	134	94	36	223	6.9	4.47	297	108.2	154.4	120.3	44.3	45.4
80	1	74	126	64	37	83	7.5	4.32	170	73.1	160.5	97.2	30.3	28.4
34	1	50	114	68	50	104	6.1	4.95	140	83.1	179.0	95.1	34.0	25.9

Data Set 2: Foot and Height

Foot and height measurements are from 40 subjects (first five rows shown here). **SEX** is gender of subject, **AGE** is age in years, **FOOT LENGTH** is length of foot (cm), **SHOE PRINT** is length of shoe (cm), **SHOE SIZE** is reported shoe size, and **HEIGHT** is height (cm) of the subject.

Data from Rohren, Brenda, "Estimation of Stature from Foot and Shoe Length: Applications in Forensic Science." Copyright © 2006. Reprinted by permission of the author. Brenda Rohren (MA, MFS, LIMHP, LADC, MAC) was a graduate student at Nebraska Wesleyan University when she conducted the research and wrote the report.

TI-83/84 list names (FOOTHT): FTSEX (1 = male), FTAGE, FTLN, SHOPT, SHOSZ, FHT

SEX	AGE	FOOT LENGTH	SHOE PRINT	SHOE SIZE	HEIGHT
M	67	27.8	31.3	11.0	180.3
M	47	25.7	29.7	9.0	175.3
M	41	26.7	31.3	11.0	184.8
M	42	25.9	31.8	10.0	177.8
M	48	26.4	31.4	10.0	182.3

Data Set 3: Body Temperatures

Body temperatures (°F) are from 107 subjects taken on two consecutive days at 8 AM and 12 AM (first five rows shown here). **SEX** is gender of subject, and **SMOKE** indicates if subject smokes (Y) or does not smoke (N). Data provided by Dr. Steven Wasserman, Dr. Philip Mackowiak, and Dr. Myron Levine of the University of Maryland.

TI-83/84 Plus list names (BODYTEMP): D1T8, D1T12, D2T8, D2T12 (no list for SEX and SMOKE). **Missing data values are represented by 9999.**

SEX	SMOKE	DAY 1—8 AM	DAY 1—12 AM	DAY 2—8 AM	DAY 2—12 AM
M	Y	98.0	98.0	98.0	98.6
M	Y	97.0	97.6	97.4	—
M	Y	98.6	98.8	97.8	98.6
M	N	97.4	98.0	97.0	98.0
M	N	98.2	98.8	97.0	98.0

(Complete data sets available at www.TriolaStats.com)

Data Set 4: Births

Data are from 400 births (first five rows shown here). For **GENDER** 1 = male and 0 = female. **LENGTH OF STAY** is in days, **BIRTH WEIGHT** is in grams, and **TOTAL CHARGES** are in dollars.

TI-83/84 list names (BIRTHS): FLOS, MLOS, FBWT, MBWT, FCHRG, MCHRG [Separate lists provided for female (F) and male (M) babies. No list for FACILITY, INSURANCE, ADMITTED, and DISCHARGED]

FACILITY	INSURANCE	GENDER (1 = M)	LENGTH OF STAY	ADMITTED	DISCHARGED	BIRTH WEIGHT	TOTAL CHARGES
Albany Medical Center Hospital	Insurance Company	0	2	FRI	SUN	3500	13986
Albany Medical Center Hospital	Blue Cross	1	2	FRI	SUN	3900	3633
Albany Medical Center Hospital	Blue Cross	0	36	WED	THU	800	359091
Albany Medical Center Hospital	Insurance Company	1	5	MON	SAT	2800	8537
Albany Medical Center Hospital	Insurance Company	1	2	FRI	SUN	3700	3633

Data Set 5: Family Heights

Height data are from 134 families (first five rows shown here). Heights are in inches. Only families with at least one child of each gender are included, and only heights of the first son and first daughter are included. The data are from a journal of Francis Galton (1822–1911), who developed the concepts of standard deviation, regression line, and correlation between two variables.

TI-83/84 list names (FAMHT): DAD, MOM, SON1, DGHT1

FATHER	MOTHER	FIRST SON	FIRST DAUGHTER
70.0	64.0	68.0	65.0
71.0	65.5	72.0	66.0
69.0	63.5	70.5	65.0
69.5	66.0	71.0	66.5
70.0	58.0	72.0	66.0

Data Set 6: Freshman 15

Weights of 67 college students are provided (first five rows shown here). **SEX** is gender of subject, **WT** is weights in kilograms, and **BMI** is measured body mass index. Measurements were made in September of freshman year and then later in April of freshman year.

Results are published in Hoffman, D. J., Policastro, P., Quick, V., and Lee, S. K.: "Changes in Body Weight and Fat Mass of Men and Women in the First Year of College: A Study of the 'Freshman 15.'" *Journal of American College Health,* July 1, 2006, Vol. 55, No. 1, p. 41. Copyright © 2006. Reprinted by permission.

TI-83/84 list names (FRESH15): WTSP, WTAPR, BMISP, BMIAP (no list for SEX)

SEX	WT SEPT	WT APRIL	BMI SEPT	BMI APRIL
M	72	59	22.02	18.14
M	97	86	19.70	17.44
M	74	69	24.09	22.43
M	93	88	26.97	25.57
F	68	64	21.51	20.10

(Complete data sets available at www.TriolaStats.com)

Data Set 7: IQ and Lead

Data are from 121 subjects (first five rows shown here). Data are measured from children in two consecutive years, and the children were living close to a lead smelter. **LEAD** is blood lead level group [1 = *low lead level* (blood lead levels < 40 micrograms/100 mL in both years), 2 = *medium lead level* (blood lead levels ≥ 40 micrograms/100 mL in exactly one of two years), 3 = *high lead level* (blood lead level ≥ 40 micrograms/100 mL in both years)]. **AGE** is age in years, **SEX** is sex of subject (1 = male; 2 = female).

YEAR1 is blood lead level in first year, and **YEAR2** is blood lead level in second year. **IQ VERB** is measured verbal IQ score. **IQ PERF** is measured performance IQ score. **IQ FULL** is measured full IQ score.

Data are from "Neuropsychological Dysfunction in Children with Chronic Low-Level Lead Absorption," by P. J. Landrigan, R. H. Whitworth, R. W. Baloh, N. W. Staehling, W. F Barthel, and B. F. Rosenblum, *Lancet,* Vol. 1, No. 7909.

TI-83/84 list names (IQLEAD):   LEAD, IQAGE, IQSEX, YEAR1, YEAR2, IQV, IQP, IQF

LEAD	AGE	SEX	YEAR1	YEAR2	IQ VERB	IQ PERF	IQ FULL
1	11	1	25	18	61	85	70
1	9	1	31	28	82	90	85
1	11	1	30	29	70	107	86
1	6	1	29	30	72	85	76
1	11	1	2	34	72	100	84

Data Set 8: IQ and Brain Size

Data are from 20 monozygotic (identical) twins (first five rows shown here). **PAIR** identifies the set of twins, **SEX** is the gender of the subject (1 = male, 2 = female), **ORDER** is the birth order, **IQ** is measured full IQ score, **VOL** is total brain volume (cm³), **AREA** is total brain surface area (cm²), **CCSA** is corpus callosum (fissure connecting left and right cerebral hemispheres) surface area (cm²), **CIRC** is head circumference (cm), and **WT** is body weight (kg).

Data provided by M. J. Tramo, W. C. Loftus, T. A. Stukel, J. B. Weaver, M. S. Gazziniga. See "Brain Size, Head Size, and IQ in Monozygotic Twins," *Neurology,* Vol. 50.

TI-83/84 list names (IQBRAIN):   PAIR, SEX, ORDER, IQ, VOL, AREA, CCSA, CIRC, BWT

PAIR	SEX (1 = M)	ORDER	IQ	VOL	AREA	CCSA	CIRC	WT
1	2	1	96	1005	1913.88	6.08	54.7	57.607
1	2	2	89	963	1684.89	5.73	54.2	58.968
2	2	1	87	1035	1902.36	6.22	53.0	64.184
2	2	2	87	1027	1860.24	5.80	52.9	58.514
3	2	1	101	1281	2264.25	7.99	57.8	63.958

Data Set 9: Bear Measurements

Data are from 54 anesthetized wild bears (first five rows shown here). **AGE** is in months, **MONTH** is the month of measurement 1 = January, **SEX** is coded with 0 = female and 1 = male, **HEADLEN** is head length (inches), **HEADWDTH** is width of head (inches), **NECK** is distance around neck (in inches), **LENGTH** is length of body (inches), **CHEST** is distance around chest (inches), and **WEIGHT** is measured in pounds. Data are from Gary Alt and Minitab, Inc.

TI-83/84 list names (BEARS):   BAGE, BSEX, BHDLN, BHDWD, BNECK, BLEN, BCHST, BWGHT (no list for MONTH)

AGE	MONTH	SEX (1 = M)	HEADLEN	HEADWDTH	NECK	LENGTH	CHEST	WEIGHT
19	7	1	11.0	5.5	16.0	53.0	26.0	80
55	7	1	16.5	9.0	28.0	67.5	45.0	344
81	9	1	15.5	8.0	31.0	72.0	54.0	416
115	7	1	17.0	10.0	31.5	72.0	49.0	348
104	8	0	15.5	6.5	22.0	62.0	35.0	166

(Complete data sets available at www.TriolaStats.com)

Data Set 10: Manatee Deaths

Annual Florida data for 24 years are provided (first five rows shown here). **DEATHS** is the annual number of Manatee deaths caused by boats, **BOATS** is the number of registered pleasure boats (tens of thousands), **POP** is the Florida population (millions), and **WATER TEMP** is the annual mean water temperature (°F).

TI-83/84 list names (MANATEE): DEATH, BOATS, POP, WTEMP (no list for YEAR)

YEAR	DEATHS	BOATS	POP	WATER TEMP
1991	53	68	13.3	71.9
1992	38	68	13.5	70.4
1993	35	67	13.7	70.5
1994	49	70	14.0	71.7
1995	42	71	14.3	70.9

Data Set 11: Alcohol and Tobacco in Movies

Data are from 50 animated children's movies (first five rows shown here). **LENGTH** is movie length in minutes, **TOBACCO** is tobacco use time in seconds, and **ALCOHOL** is alcohol use time in seconds.

The data are based on Goldstein, Adam O., Sobel, Rachel A., Newman, Glen R., "Tobacco and Alcohol Use in G-Rated Children's Animated Films." *Journal of the American Medical Association,* March 24/31, 1999, Vol. 281, No. 12, p. 1132. Copyright © 1999. All rights reserved.

TI-83/84 list names (CHMOVIE): CHLEN, CHTOB, CHALC (no list for MOVIE and STUDIO)

MOVIE	STUDIO	LENGTH (MIN)	TOBACCO (SEC)	ALCOHOL (SEC)
Snow White	Disney	83	0	0
Pinocchio	Disney	88	223	80
Fantasia	Disney	120	0	0
Dumbo	Disney	64	176	88
Bambi	Disney	69	0	0

Data Set 12: Passive and Active Smoke

Data are from 120 subjects (first five rows shown here) in three groups: **SMOKER** includes subjects who are smokers, **ETS** includes nonsmokers exposed to environmental tobacco smoke, **NOETS** includes nonsmokers not exposed to environmental tobacco smoke. All values are measured levels of serum cotinine (in ng/mL), a metabolite of nicotine. (When nicotine is absorbed by the body, cotinine is produced.) Data are from the U.S. Department of Health and Human Services, National Center for Health Statistics, Third National Health and Nutrition Examination Survey.

TI-83/84 Plus list names (SMOKE): SMKR, ETS, NOETS

SMOKER	ETS	NOETS
1	384	0
0	0	0
131	69	0
173	19	0
265	1	0

(Complete data sets available at www.TriolaStats.com)

Data Set 13: Cigarette Contents

Data are from 75 cigarettes (first five rows shown here) from three categories: **KING** includes king-sized cigarettes that are nonfiltered, nonmenthol, and nonlight; **MENTH** includes menthol cigarettes that are 100 mm long, filtered, and nonlight; and **100** includes 100-mm-long cigarettes that are filtered, nonmenthol, and nonlight. **TAR** is the amount of tar per cigarette (milligrams), **NICOTINE** is the amount of nicotine per cigarette (milligrams), and **CO** is the amount of carbon monoxide per cigarette (milligrams). Data are from the Federal Trade Commission.

TI-83/84 list names (CIGARET): KGTAR, KGNIC, KGCO, MNTAR, MNNIC, MNCO, FLTAR, FLNIC, FLCO

KING TAR	KING NICOTINE	KING CO	MENTH TAR	MENTH NICOTINE	MENTH CO	100 TAR	100 NICOTINE	100 CO
20	1.1	16	16	1.1	15	5	0.4	4
27	1.7	16	13	0.8	17	16	1.0	19
27	1.7	16	16	1.0	19	17	1.2	17
20	1.1	16	9	0.9	9	13	0.8	18
20	1.1	16	14	0.8	17	13	0.8	18

Data Set 14: Oscar Winner Age

Data are from 87 years (first five rows shown here). Data values are ages (years) of actresses and actors at the times that they won Oscars in the categories of Best Actress and Best Actor. The ages are listed in chronological order by row, so that each row has paired ages from the same year. (*Note:* In 1968 there was a tie in the Best Actress category, and the mean of the two ages is used; in 1932 there was a tie in the Best Actor category, and the mean of the two ages is used).

These data are suggested by the article "Ages of Oscar-Winning Best Actors and Actresses," by Richard Brown and Gretchen Davis, *Mathematics Teacher* magazine. In that article, the year of birth of the award winner was subtracted from the year of the awards ceremony, but the ages listed here are calculated from the birth date of the winner and the date of the awards ceremony.

TI-83/84 list names (OSCARS): OSCRF, OSCRM

ACTRESSES	ACTORS
22	44
37	41
28	62
63	52
32	41

Data Set 15: Presidents

Data are from 38 presidents of the United States (first five rows shown here). Presidents who took office as the result of an assassination or resignation are not included. **AGE** is age in years at time of inauguration. **DAYS** is the number of days served as president. **YEARS** is the number of years lived after the first inauguration. **HEIGHT** is height (cm) of the president. **HEIGHT OPP** is the height (cm) of the major opponent for the presidency.

TI-83/84 list names (POTUS): PRAGE, DAYS, YEARS, PRHT, HTOPP (no list for PRESIDENT). **Missing data values are represented by 9999.**

PRESIDENT	AGE	DAYS	YEARS	HEIGHT	HEIGHT OPP
Washington	57	2864	10	188	
J. Adams	61	1460	29	170	189
Jefferson	57	2921	26	189	170
Madison	57	2921	28	163	
Monroe	58	2921	15	183	

(Complete data sets available at www.TriolaStats.com)

Data Set 16: Nobel Laureates and Chocolate

Data are from 23 countries (first five rows shown here). **CHOCOLATE** includes chocolate consumption (kg per capita), **NOBEL** includes the numbers of Nobel Laureates (per 10 million people), **POPULATION** includes population (in millions), and **INTERNET** includes the number of Internet users per 100 people.

The data are from "The Real Secret to Genius? Reading Between the Lines," by McClintock, Stangle, and Cetinkaya-Rundel, *Chance*, Vol. 27, No. 1; and "Chocolate Consumption, Cognitive Function, and Nobel Laureates," by Franz Messerli, *New England Journal of Medicine,* Vol. 367, No. 16.

TI-83/84 list names (NOBEL): CHOC, NOBEL, POPUL, INTNT (no list for COUNTRY)

COUNTRY	CHOCOLATE	NOBEL	POPULATION	INTERNET
Australia	4.5	5.5	22	79.5
Austria	10.2	24.3	8	79.8
Belgium	4.4	8.6	11	78.0
Brazil	2.9	0.1	197	45.0
Canada	3.9	6.1	34	83.0

Data Set 17: Course Evaluations

Data are from 93 college student course evaluations (first five rows shown here). **COURSE EVAL** includes the mean course rating, **PROF EVAL** includes the mean professor rating, **PROF AGE** includes the professor age in years, **SIZE** includes the number of course evaluations per course, **PROF BEAUTY** includes the mean beauty rating based on the professor's photo.

Based on data from Andrew Gelman and Jennifer Hill, 2007, "Replication Data for Data Analysis Using Regression Multilevel/Hierarchical Models," http://hdl.handle.net/1902.1/10285.

TI-83/84 list names (EVALS): CEVAL, PEVAL, PAGE, SIZE, PBTY (no list for PROF, PROF GENDER, PROF PHOTO, and CLASS LEVEL)

PROF	COURSE EVAL	PROF EVAL	PROF GENDER	PROF AGE	SIZE	PROF BEAUTY	PROF PHOTO	CLASS LEVEL
1	4.3	4.7	female	36	24	5.000	color	upper
2	4.5	4.6	male	59	17	3.000	color	upper
3	3.7	4.1	male	51	55	3.333	color	upper
4	4.3	4.5	female	40	40	3.167	color	upper
5	4.4	4.8	female	31	42	7.333	color	upper

Data Set 18: Speed Dating

Data are from 199 dates (first five rows shown here). **DEC BY FEM** is decision (1 = yes) of female to date again, **AGE FEM** is age of female, **LIKE BY FEM** is "like" rating by female of male (scale of 1−10), **ATTRACT BY FEM** is "attractive" rating by female of male (scale of 1−10), **ATTRIB BY FEM** is sum of ratings of five attributes (sincerity, intelligence, fun, ambitious, shared interests) by female of male. Data for males use corresponding descriptors. Higher scale ratings correspond to more positive impressions.

Based on replication data from *Data Analysis Using Regression and Multilevel/Hierarchical Models,* by Andrew Gelman and Jennifer Hill, Cambridge University Press.

TI-83/84 list names (DATE): DBYF, DBYM, AGEF, AGEM, LBYF, LBYM, ABYF, ABYM, ATBYF, ATBYM

DEC BY FEM	DEC BY MALE	AGE FEM	AGE MALE	LIKE BY FEM	LIKE BY MALE	ATTRACT BY FEM	ATTRACT BY MALE	ATTRIB BY FEM	ATTRIB BY MALE
0	1	27	28	7	7	5	8	38	36
0	0	24	26	7	7	7	6	36	38
1	0	26	28	6	3	7	8	29	35
1	0	34	27	8	6	8	6	38	28
1	0	22	25	5	5	7	5	33	34

(Complete data sets available at www.TriolaStats.com)

Data Set 19: Car Crash Tests

Data are from 21 cars in crash tests (first five rows shown here). The same cars are used in Data Set 20. The data are measurements from cars crashed into a fixed barrier at 35 mi/h with a crash test dummy in the driver's seat. **HIC** is a measurement of a standard "head injury criterion," **CHEST** is chest deceleration (in g, where g is a force of gravity), **FEMUR L** is the measured load on the left femur (in lb),

FEMUR R is the measured load on the right femur (in lb), **TTI** is a measurement of the side thoracic trauma index, and **PELVIS** is pelvis deceleration (in g, where g is a force of gravity). Data are from the National Highway Traffic Safety Administration.

TI-83/84 list names (CRASH):   HIC, CHEST, FEML, FEMR, TTI, PLVS (no list for CAR and SIZE)

CAR	SIZE	HIC	CHEST	FEMUR L	FEMUR R	TTI	PELVIS
Chev Aveo	Small	371	44	1188	1261	62	71
Honda Civic	Small	356	39	289	324	63	71
Mitsubishi Lancer	Small	275	37	329	446	35	45
VW Jetta	Small	544	54	707	1048	44	66
Hyundai Elantra	Small	326	39	602	1474	58	71

Data Set 20: Car Measurements

Data are from 21 cars (first five rows shown here). The same cars are used in Data Set 19. The data are measurements from cars that have automatic transmissions. **WEIGHT** is car weight (lb), **LENGTH** is car length (inches), **BRAKING** is braking distance (feet) from 60 mi/h, **CYLINDERS** is the number of cylinders, **DISPLACEMENT** is the engine displacement (liters), **CITY** is the fuel consumption (mi/gal)

for city driving conditions, **HIGHWAY** is the fuel consumption (mi/gal) for highway driving conditions, and **GHG** is a measure of greenhouse gas emissions (in tons/year, expressed as CO_2 equivalents). Data are from the National Highway Traffic Safety Administration.

TI-83/84 list names (CARS):   CWT, CLN, CBRK, CCYL, CDISP, CCITY, CHWY, CGHG (no list for CAR and SIZE)

CAR	SIZE	WEIGHT	LENGTH	BRAKING	CYLINDERS	DISPLACEMENT	CITY	HIGHWAY	GHG
Chev Aveo	Small	2560	154	133	4	1.6	25	34	6.6
Honda Civic	Small	2740	177	136	4	1.8	25	36	6.3
Mitsubishi Lancer	Small	3610	177	126	4	2.0	22	28	7.7
VW Jetta	Small	3225	179	137	4	2.0	29	40	6.4
Hyundai Elantra	Small	2895	177	138	4	2.0	25	33	6.6

Data Set 21: Earthquakes

Data are from 600 matched pairs (first five rows shown here) of magnitude/depth measurements randomly selected from 10,594 earthquakes recorded in one year from a location in southern California. Only earthquakes with a magnitude of at least 1.00 are used.

MAGNITUDE is magnitude measured on the Richter scale and **DEPTH** is depth in km. The magnitude and depth both describe the source of the earthquake. The data are from the Southern California Earthquake Data Center.

TI-83/84 list names (QUAKE):   MAG, DEPTH

MAGNITUDE	DEPTH
2.45	0.7
3.62	6.0
3.06	7.0
3.30	5.4
1.09	0.5

(Complete data sets available at www.TriolaStats.com)

Data Set 22: Tornadoes

Data are from 500 tornadoes (first five rows shown here) arranged chronologically. **MONTH** is the month of the tornado (1 = January), **F SCALE** is the Fujita scale rating of tornado intensity, **FATALITIES** is number of deaths caused by the tornado, **LENGTH (MI)** is the

distance the tornado traveled in miles, and **WIDTH (YD)** is the tornado width in yards. Data are from the National Weather Service.

TI-83/84 list names (TORNADO): FSCAL, FATAL, TLEN, TWDTH (no list for YEAR and MONTH). The **10 missing F-scale values are represented by 9999.**

YEAR	MONTH	F SCALE	FATALITIES	LENGTH (MI)	WIDTH (YD)
1950	5	2	0	3.60	100
1950	5	4	0	34.30	150
1950	6	1	0	0.20	10
1950	8	1	0	0.10	10
1951	6	–	0	0.10	10

Data Set 23: Old Faithful

Data are from 250 eruptions (first five rows shown here) of the Old Faithful geyser in Yellowstone National Park. **INT BEFORE** is the time interval (min) before the eruption, **DURATION** is the time (sec) of the eruption, **INT AFTER** is the time interval (min) after the eruption, **HEIGHT** (ft) is the height of the eruption, and **PRED ERROR**

is the error (min) of the predicted time of eruption. Based on data from the Geyser Observation and Study Association.

TI-83/84 list names (OLDFAITH): INTBF, DUR, INTAF, OFHT, PRED

INT BEFORE (MIN)	DURATION (SEC)	INT AFTER (MIN)	HEIGHT (FT)	PRED ERROR (MIN)
82	251	83	130	4
99	243	76	125	−13
88	250	86	120	−2
92	240	82	120	−6
86	243	87	130	0

Data Set 24: Word Counts

Data are from counts of the numbers of words spoken in a day by 396 male (M) and female (F) subjects in six different sample groups (first five rows shown here). Column **M1** denotes the word counts for males in Sample 1, **F1** is the count for females in Sample 1, and so on.

Sample 1: Recruited couples ranging in age from 18 to 29
Sample 2: Students recruited in introductory psychology classes, aged 17 to 23
Sample 3: Students recruited in introductory psychology classes in Mexico, aged 17 to 25

Sample 4: Students recruited in introductory psychology classes, aged 17 to 22
Sample 5: Students recruited in introductory psychology classes, aged 18 to 26
Sample 6: Students recruited in introductory psychology classes, aged 17 to 23

Results were published in "Are Women Really More Talkative Than Men?" by Mehl, Vazire, Ramirez-Esparza, Slatcher, Pennebaker, *Science,* Vol. 317, No. 5834.

TI-83/84 list names (WORDS): M1, F1, M2, F2, M3, F3, M4, F4, M5, F5, M6, F6.

M1	F1	M2	F2	M3	F3	M4	F4	M5	F5	M6	F6
27531	20737	23871	16109	21143	6705	47016	11849	39207	15962	28408	15357
15684	24625	5180	10592	17791	21613	27308	25317	20868	16610	10084	13618
5638	5198	9951	24608	36571	11935	42709	40055	18857	22497	15931	9783
27997	18712	12460	13739	6724	15790	20565	18797	17271	5004	21688	26451
25433	12002	17155	22376	15430	17865	21034	20104		10171	37786	12151

(Complete data sets available at www.TriolaStats.com)

Data Set 25: Fast Food

Data are from 400 observations (first five rows shown here) of drive-thru service times (sec) at different fast-food restaurants. Times begin when a vehicle stops at the order window and end when the vehicle leaves the pickup window. Lunch times were measured between 11:00 AM and 2:00 PM, and dinner times were measured between 4:00 PM and 7:00 PM. Data collected by the author.

TI-83/84 list names (FASTFOOD): MCDL, MCDD, BKL, BKD, WL, WD, DDL, DDD

MCDONALDS LUNCH	MCDONALDS DINNER	BURGER KING LUNCH	BURGER KING DINNER	WENDYS LUNCH	WENDYS DINNER	DUNKIN DONUTS LUNCH	DUNKIN DONUTS DINNER
107	84	116	101	466	56	86	181
139	121	131	126	387	82	201	50
197	119	147	153	368	120	179	177
209	146	120	116	219	116	131	107
281	266	126	175	177	121	126	68

Data Set 26: Cola Weights and Volumes

Data are from 144 cans of cola (first five rows shown here). **WT** is weight in pounds and **VOL** is volume in ounces.

TI-83/84 list names (COLA): CRGWT, CRGVL, CDTWT, CDTVL, PRGWT, PRGVL, PDTWT, PDTVL

COKE REG WT	COKE REG VOL	COKE DIET WT	COKE DIET VOL	PEPSI REG WT	PEPSI REG VOL	PEPSI DIET WT	PEPSI DIET VOL
0.8192	12.3	0.7773	12.1	0.8258	12.4	0.7925	12.3
0.8150	12.1	0.7758	12.1	0.8156	12.2	0.7868	12.2
0.8163	12.2	0.7896	12.3	0.8211	12.2	0.7846	12.2
0.8211	12.3	0.7868	12.3	0.8170	12.2	0.7938	12.3
0.8181	12.2	0.7844	12.2	0.8216	12.2	0.7861	12.2

Data Set 27: M&M Weights

Data are from 100 weights (grams) of plain M&M candies (first five rows shown here). Data collected by the author.

TI-83/84 list names (MM): RED, ORNG, YLLW, BROWN, BLUE, GREEN

RED	ORANGE	YELLOW	BROWN	BLUE	GREEN
0.751	0.735	0.883	0.696	0.881	0.925
0.841	0.895	0.769	0.876	0.863	0.914
0.856	0.865	0.859	0.855	0.775	0.881
0.799	0.864	0.784	0.806	0.854	0.865
0.966	0.852	0.824	0.840	0.810	0.865

(Complete data sets available at www.TriolaStats.com)

Data Set 28: Chocolate Chip Cookies

Data are from 170 chocolate chip cookies (first five rows shown here). Brands are Chips Ahoy regular, Chips Ahoy Chewy, Chips Ahoy Reduced Fat, Keebler, and Hannaford. Values are counts of numbers of chocolate chips in each cookie. Data collected by the author.

TI-83/84 list names (CHIPS):   CAREG, CACHW, CARF, KEEB, HANNA.

CHIPS AHOY REG	CHIPS AHOY CHEWY	CHIPS AHOY RED FAT	KEEBLER	HANNAFORD
22	21	13	29	13
22	20	24	31	15
26	16	18	25	16
24	17	16	32	21
23	16	21	27	15

Data Set 29: Coin Weights

Data are from 222 coins (first five rows shown here) consisting of coin weights (grams). The "pre-1983 pennies" were made after the Indian and wheat pennies, and they are 97% copper and 3% zinc. The "post-1983 pennies" are 3% copper and 97% zinc. The "pre-1964 silver quarters" are 90% silver and 10% copper. The "post-1964 quarters" are made with a copper-nickel alloy.

TI-83/84 list names (COINS):   CPIND, CPWHT, CPPRE, CPPST, CPCAN, CQPRE, CQPST, CDOL

INDIAN PENNIES	WHEAT PENNIES	PRE-1983 PENNIES	POST-1983 PENNIES	CANADIAN PENNIES	PRE-1964 QUARTERS	POST-1964 QUARTERS	DOLLAR COINS
3.0630	3.1366	3.1582	2.5113	3.2214	6.2771	5.7027	8.1008
3.0487	3.0755	3.0406	2.4907	3.2326	6.2371	5.7495	8.1072
2.9149	3.1692	3.0762	2.5024	2.4662	6.1501	5.7050	8.0271
3.1358	3.0476	3.0398	2.5298	2.8357	6.0002	5.5941	8.0813
2.9753	3.1029	3.1043	2.4950	3.3189	6.1275	5.7247	8.0241

Data Set 30: Aluminum Cans

Data are from 350 cans (first five rows shown here) consisting of measured maximum axial loads (pounds). Axial loads are applied when the tops are pressed into place. **CANS 109** includes cans that are 0.0109 inch thick, and **CANS 111** includes cans that are 0.0111 inch thick.

TI-83/84 list names (CANS):   CN109, CN111

CANS 109	CANS 111
270	287
273	216
258	260
204	291
254	210

(Complete data sets available at www.TriolaStats.com)

Data Set 31: Garbage Weight

Data are from 62 households (first five rows shown here) consisting of weights (pounds) of discarded garbage in different categories. **HH SIZE** is household size. Data provided by Masakuza Tani, the Garbage Project, University of Arizona.

TI-83/84 list names (GARBAGE): HHSIZ, METAL, PAPER, PLAS, GLASS, FOOD, YARD, TEXT, OTHER, TOTAL

HH SIZE	METAL	PAPER	PLASTIC	GLASS	FOOD	YARD	TEXTILE	OTHER	TOTAL
2	1.09	2.41	0.27	0.86	1.04	0.38	0.05	4.66	10.76
3	1.04	7.57	1.41	3.46	3.68	0.00	0.46	2.34	19.96
3	2.57	9.55	2.19	4.52	4.43	0.24	0.50	3.60	27.60
6	3.02	8.82	2.83	4.92	2.98	0.63	2.26	12.65	38.11
4	1.50	8.72	2.19	6.31	6.30	0.15	0.55	2.18	27.90

Data Set 32: Airport Data Speeds

Data are from 50 airports (first five rows shown here) consisting of data speeds (Mbps) from four different cell phone carriers. Based on data from CNN.

TI-83/84 list names (DATASPED): VRZN, SPRNT, ATT, TMOBL (no list for AIRPORT CODE)

AIRPORT CODE	VERIZON	SPRINT	ATT	T-MOBILE
RSW	38.5	13.0	9.7	8.6
ORD	55.6	30.4	8.2	7.0
SNA	22.4	15.2	7.1	18.5
MEM	14.1	2.4	14.4	16.7
MKE	23.1	2.7	13.4	5.6

(Complete data sets available at www.TriolaStats.com)

APPENDIX C

Websites

Triola Stats: www.TriolaStats.com

Access continually updated digital resources for the Triola Statistics Series, including downloadable data sets, textbook supplements, online instructional videos, Triola Blog, and more.

Statdisk: www.Statdisk.org

Download the free Statdisk statistical software that is designed specifically for this book and contains all Appendix B data sets. Detailed information on using Statdisk is provided on the site's Help page.

StatCrunch: www.statcrunch.com

Books

*An asterisk denotes a book recommended for reading. Other books are recommended as reference texts.

Bennett, D. 1998. *Randomness.* Cambridge, Mass.: Harvard University Press.

*Best, J. 2012. *Damned Lies and Statistics.* Berkeley, Calif.: University of California Press.

*Best, J. 2004. *More Damned Lies and Statistics.* Berkeley, Calif.: University of California Press.

*Campbell, S. 2004. *Flaws and Fallacies in Statistical Thinking.* Mineola, N.Y.: Dover Publications.

*Crossen, C. 1996. *Tainted Truth: The Manipulation of Fact in America.* New York: Simon & Schuster.

*Freedman, D., R. Pisani, R. Purves, and A. Adhikari. 2007. *Statistics.* 4th ed. New York: W. W. Norton & Company.

*Gonick, L., and W. Smith. 1993. *The Cartoon Guide to Statistics.* New York: Harper Collins.

*Heyde, C., and E. Seneta, eds. 2001. *Statisticians of the Centuries.* New York: Springer-Verlag.

*Hollander, M., and F. Proschan. 1984. *The Statistical Exorcist: Dispelling Statistics Anxiety.* New York: Marcel Dekker.

*Holmes, C. 1990. *The Honest Truth About Lying with Statistics.* Springfield, Ill.: Charles C Thomas.

*Hooke, R. 1983. *How to Tell the Liars from the Statisticians.* New York: Marcel Dekker.

*Huff, D. 1993. *How to Lie with Statistics.* New York: W. W. Norton & Company.

*Jaffe, A., and H. Spirer. 1998. *Misused Statistics.* New York: Marcel Dekker.

Kaplan, M. 2007. *Chances Are.* New York: Penguin Group.

Kotz, S., and D. Stroup. 1983. *Educated Guessing—How to Cope in an Uncertain World.* New York: Marcel Dekker.

Lapp, James. 2018. *Student Solutions Manual to Accompany Elementary Statistics.* 13th ed. Boston: Pearson.

Mlodinow, L. 2009. *The Drunkard's Walk.* New York: Vintage Books.

*Moore, D., and W. Notz. 2012. *Statistics: Concepts and Controversies.* 8th ed. San Francisco: Freeman.

*Paulos, J. 2001. *Innumeracy: Mathematical Illiteracy and Its Consequences.* New York: Hill and Wang.

*Reichmann, W. 1981. *Use and Abuse of Statistics.* New York: Penguin.

*Rossman, A., and B. Chance. 2011. *Workshop Statistics: Discovery with Data.* 4th ed. Emeryville, Calif.: Key Curriculum Press.

*Salsburg, D. 2001. *The Lady Tasting Tea: How Statistics Revolutionized the Twentieth Century.* New York: W. H. Freeman.

Sheskin, D. 2011. *Handbook of Parametric and Nonparametric Statistical Procedures.* 5th ed. Boca Raton, Fla.: CRC Press.

Simon, J. 1997. *Resampling: The New Statistics.* 2nd ed. Arlington, Va.: Resampling Stats.

*Stigler, S. 1986. *The History of Statistics.* Cambridge, Mass.: Harvard University Press.

Taleb, N. 2010. *The Black Swan.* 2nd ed. New York: Random House.

Triola, M. 2018. *Minitab Student Laboratory Manual and Workbook.* 13th ed. Boston: Pearson.

Triola, M. 2018. *Statdisk 13 Student Laboratory Manual and Workbook.* 13th ed. Boston: Pearson.

Triola, M., and L. Franklin. 1994. *Business Statistics.* Boston: Addison-Wesley.

Triola, M., M. Triola, and J. Roy. 2018. *Biostatistics for the Biological and Health Sciences.* 2nd ed. Boston: Pearson.

*Tufte, E. 2001. *The Visual Display of Quantitative Information.* 2nd ed. Cheshire, Conn.: Graphics Press.

Tukey, J. 1977. *Exploratory Data Analysis.* Boston: Pearson.

Vickers, A. 2009. *What Is a P-Value Anyway?* Boston: Pearson.

Whelan, C. 2013. *Naked Statistics.* New York: W. W. Norton & Company.

Zwillinger, D., and S. Kokoska. 2000. *CRC Standard Probability and Statistics Tables and Formulae.* Boca Raton, Fla.: CRC Press.

Chapter 1 Answers

Section 1-1

1. The respondents are a voluntary response sample or a self-selected sample. Because those with strong interests in the topic are more likely to respond, it is very possible that their responses do not reflect the opinions or behavior of the general population.

3. Statistical significance is indicated when methods of statistics are used to reach a conclusion that a treatment is effective, but common sense might suggest that the treatment does not make enough of a difference to justify its use or to be practical. Yes, it is possible for a study to have statistical significance, but not practical significance.

5. Yes, there does appear to be a potential to create a bias.

7. No, there does not appear to be a potential to create a bias.

9. The sample is a voluntary response sample and has strong potential to be flawed.

11. The sampling method appears to be sound.

13. With only a 1% chance of getting such results with a program that has no effect, the program appears to have statistical significance. Also, because the average loss of 22 pounds does seem substantial, the program appears to also have practical significance.

15. Because there is a 19% chance of getting that many girls by chance, the method appears to lack statistical significance. The result of 1020 girls in 2000 births (51% girls) is above the approximately 50% rate expected by chance, but it does not appear to be high enough to have practical significance. Not many couples would bother with a procedure that raises the likelihood of a girl from 50% to 51%.

17. Yes. Each column of 8 AM and 12 AM temperatures is recorded from the same subject, so each pair is matched.

19. The data can be used to address the issue of whether there is a correlation between body temperatures at 8 AM and at 12 AM. Also, the data can be used to determine whether there are differences between body temperatures at 8 AM and at 12 AM.

21. No. The white blood cell counts measure a different quantity than the red blood cell counts, so their differences are meaningless.

23. No. The National Center for Health Statistics has no reason to collect or present the data in a way that is biased.

25. It is questionable that the sponsor is the Idaho Potato Commission and the favorite vegetable is potatoes.

27. The correlation, or association, between two variables does not mean that one of the variables is the cause of the other. Correlation does not imply causation. Clearly, sour cream consumption is not directly related in any way to motorcycle fatalities.

29. a. 700 adults   b. 55%

31. a. 559.2 respondents
    b. No. Because the result is a count of respondents among the 1165 engaged or married women who were surveyed, the result must be a whole number.
    c. 559 respondents   d. 8%

33. Because a reduction of 100% would eliminate all of the size, it is not possible to reduce the size by 100% or more.

35. Because a reduction of 100% would eliminate all plaque, it is not possible to reduce it by more than 100%.

37. The wording of the question is biased and tends to encourage negative responses. The sample size of 20 is too small. Survey respondents are self-selected instead of being randomly selected by the newspaper. If 20 readers respond, the percentages should be multiples of 5, so 87% and 13% are not possible results.

Section 1-2

1. The population consists of all adults in the United States, and the sample is the 2276 adults who were surveyed. Because the value of 33% refers to the sample, it is a statistic.

3. Only part (a) describes discrete data.

5. Statistic   7. Parameter   9. Statistic

11. Parameter   13. Continuous   15. Discrete

17. Discrete   19. Continuous   21. Ordinal

23. Nominal   25. Interval   27. Ordinal

29. The numbers are not counts or measures of anything. They are at the nominal level of measurement, and it makes no sense to compute the average (mean) of them.

31. The temperatures are at the interval level of measurement. Because there is no natural starting point with 0°F representing "no heat," ratios such as "twice" make no sense, so it is wrong to say that it is twice as warm at the author's home as it is in Auckland, New Zealand.

33. a. Continuous, because the number of possible values is infinite and not countable
    b. Discrete, because the number of possible values is finite
    c. Discrete, because the number of possible values is finite
    d. Discrete, because the number of possible values is infinite and countable

Section 1-3

1. The study is an experiment because subjects were given treatments.

3. The group sample sizes of 547, 550, and 546 are all large so that the researchers could see the effects of the paracetamol treatment.

5. The sample appears to be a convenience sample. By e-mailing the survey to a readily available group of Internet users, it was easy to obtain results. Although there is a real potential for getting a sample group that is not representative of the population, indications of which ear is used for cell phone calls and which hand is dominant do not appear to be factors that would be distorted much by a sample bias.

7. With 717 responses, the response rate is 14%, which does appear to be quite low. In general, a very low response rate creates a serious potential for getting a biased sample that consists of those with a special interest in the topic.

9. Systematic   11. Random   13. Cluster

15. Stratified   17. Random   19. Convenience

21. Observational study. The sample is a convenience sample consisting of subjects who decided themselves to respond. Such voluntary response samples have a high chance of not being representative of the larger population, so the sample may well be biased. The question was posted in an electronic edition of a newspaper, so the sample is biased from the beginning.

23. Experiment. This experiment would create an *extremely* dangerous and illegal situation that has a real potential to result in injury or death. It's difficult enough to drive in New York City while being completely sober.

25. Experiment. The biased sample created by using drivers from New York City cannot be fixed by using a larger sample. The larger sample will still be a biased sample that is not representative of drivers in the United States.

27. Observational study. Respondents who have been convicted of felonies are not likely to respond honestly to the second question. The survey will suffer from a "social desirability bias" because subjects will tend to respond in ways that will be viewed favorably by those conducting the survey.

29. Prospective study   31. Cross-sectional study

33. Matched pairs design   35. Completely randomized design

37. a. Not a simple random sample, but it is a random sample.
    b. Simple random sample and also a random sample.
    c. Not a simple random sample and not a random sample.

Chapter 1: Quick Quiz

1. No. The numbers do not measure or count anything.

2. Nominal   3. Continuous   4. Quantitative data

5. Ratio   6. Statistic   7. No

8. Observational study

9. The subjects did not know whether they were getting aspirin or the placebo.

10. Simple random sample

Chapter 1: Review Exercises

1. The survey sponsor has the potential to gain from the results, which raises doubts about the objectivity of the results.

2. a. The sample is a voluntary response sample, so the results are questionable.
   b. Statistic   c. Observational study

3. Randomized: Subjects were assigned to the different groups through a process of random selection, whereby they had the same chance of belonging to each group. Double-blind: The subjects did not know which of the three groups they were in, and the people who evaluated results did not know either.

4. No. Correlation does not imply causality.

5. Only part (c) is a simple random sample.

6. Yes. The two questions give the false impression that they are addressing very different issues. Most people would be in favor of defending marriage, so the first question is likely to receive a substantial number of "yes" responses. The second question better describes the issue and subjects are much more likely to have varied responses.

7. a. Discrete   b. Ratio
   c. The mailed responses would be a voluntary response sample, so those with strong opinions or greater interest in the topics are more likely to respond. It is very possible that the results do not reflect the true opinions of the population of all full-time college students.
   d. Stratified   e. Cluster

8. a. If they have no fat at all, they have 100% less than any other amount with fat, so the 125% figure cannot be correct.
   b. 686   c. 28%

9. a. Interval data; systematic sample
   b. Nominal data; stratified sample
   c. Ordinal data; convenience sample

10. Because there is a 15% chance of getting the results by chance, those results could easily occur by chance so the method does not appear to have statistical significance. The result of 236 girls in 450 births is a rate of 52.4%, so it is above the 50% rate expected by chance, but it does not appear to be high enough to have practical significance. The procedure does not appear to have either statistical significance or practical significance.

Chapter 1: Cumulative Review Exercises

1. 3162.5 grams. The weights all end with 00, suggesting that all of the weights are rounded so that the last two digits are always 00.

2. 0.015625   3. 16.00 is a significantly high value.

4. −6.64   5. 1067

6. 575 grams   7. 27343.75 grams^2

8. 0.20   9. 0.00065536

10. 31,381,059,609 (or about 31,381,060,000)

11. 78,364,164,096 (or about 78,364,164,000)

12. 0.000000531441

Chapter 2 Answers

Section 2-1

1. The table summarizes 50 service times. It is not possible to identify the exact values of all of the original times.

3.

Time (sec)	Relative Frequency
60–119	14%
120–179	44%
180–239	28%
240–299	4%
300–359	10%

5. Class width: 10. Class midpoints: 24.5, 34.5, 44.5, 54.5, 64.5, 74.5, 84.5. Class boundaries: 19.5, 29.5, 39.5, 49.5, 59.5, 69.5, 79.5, 89.5. Number: 87.

7. Class width: 100. Class midpoints: 49.5, 149.5, 249.5, 349.5, 449.5, 549.5, 649.5. Class boundaries: −0.5, 99.5, 199.5, 299.5, 399.5, 499.5, 599.5, 699.5. Number: 153.

9. No. The maximum frequency is in the second class instead of being near the middle, so the frequencies below the maximum do not mirror those above the maximum.

11.

Duration (sec)	Frequency
125–149	1
150–174	0
175–199	0
200–224	3
225–249	34
250–274	12

13.

Burger King Lunch Service Times (sec)	Frequency
70–109	11
110–149	23
150–189	7
190–229	6
230–269	3

15. The distribution does not appear to be a normal distribution.

Wendy's Lunch Service Times (sec)	Frequency
70–149	25
150–229	15
230–309	6
310–389	3
390–469	1

17. Because there are disproportionately more 0s and 5s, it appears that the heights were reported instead of measured. It is likely that the results are not very accurate.

Last Digit	Frequency
0	9
1	2
2	1
3	3
4	1
5	15
6	2
7	0
8	3
9	1

19. The actresses appear to be generally younger than the actors.

Age (yr)	Actresses	Actors
20–29	33.3%	1.1%
30–39	39.1%	32.2%
40–49	16.1%	41.4%
50–59	3.4%	17.2%
60–69	5.7%	6.9%
70–79	1.1%	1.1%
80–89	1.1%	

21.

Age (yr) of Best Actress When Oscar Was Won	Cumulative Frequency
Less than 30	29
Less than 40	63
Less than 50	77
Less than 60	80
Less than 70	85
Less than 80	86
Less than 90	87

23. No. The highest relative frequency of 24.8% is not much higher than the others.

Adverse Reaction	Relative Frequency
Headache	23.6%
Hypertension	8.7%
Upper Resp. Tract Infection	24.8%
Nasopharyngitis	21.1%
Diarrhea	21.9%

25. Yes, the frequency distribution appears to be a normal distribution.

Systolic Blood Pressure (mm Hg)	Frequency
80–99	11
100–119	116
120–139	131
140–159	34
160–179	7
180–199	1

27. Yes, the frequency distribution appears to be a normal distribution.

Magnitude	Frequency
1.00–1.49	19
1.50–1.99	97
2.00–2.49	187
2.50–2.99	147
3.00–3.49	100
3.50–3.99	38
4.00–4.49	8
4.50–4.99	4

29. An outlier can dramatically increase the number of classes.

Weight (lb)	With Outlier	Without Outlier
200–219	6	6
220–239	5	5
240–259	12	12
260–279	36	36
280–299	87	87
300–319	28	28
320–339	0	
340–359	0	
360–379	0	
380–399	0	
400–419	0	
420–439	0	
440–459	0	
460–479	0	
480–499	0	
500–519	1	

Section 2-2

1. The histogram should be bell-shaped.
3. With a data set that is so small, the true nature of the distribution cannot be seen with a histogram.
5. 40
7. The shape of the graph would not change. The vertical scale would be different, but the relative heights of the bars would be the same.
9. Because it is far from being bell-shaped, the histogram does not appear to depict data from a population with a normal distribution.

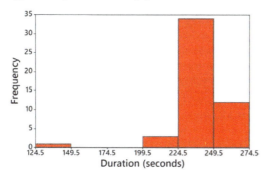

11. The histogram appears to be skewed to the right (or positively skewed).

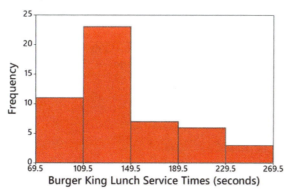

13. The histogram appears to be skewed to the right (or positively skewed).

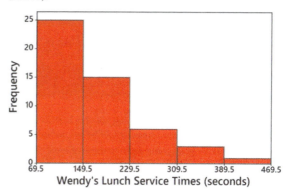

15. The digits 0 and 5 appear to occur more often than the other digits, so it appears that the heights were reported and not actually measured. This suggests that the data might not be very useful.

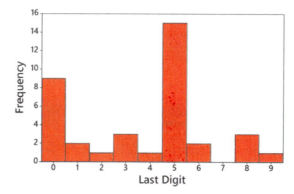

17. The ages of actresses are lower than the ages of actors.

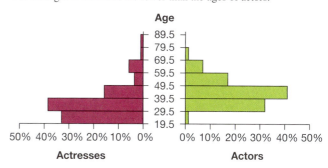

Section 2-3

1. The data set is too small for a graph to reveal important characteristics of the data. With such a small data set, it would be better to simply list the data or place them in a table.
3. No. Graphs should be constructed in a way that is fair and objective. The readers should be allowed to make their own judgments, instead of being manipulated by misleading graphs.
5. The pulse rate of 36 beats per minute appears to be an outlier.

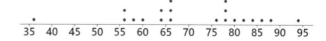

7. The data are arranged in order from lowest to highest, as 36, 56, 56, and so on.

```
3 | 6
4 |
5 | 668
6 | 044666
7 | 6888
8 | 02468
9 | 4
```

9. There is a gradual upward trend that appears to be leveling off in recent years. An upward trend would be helpful to women so that their earnings become equal to those of men.

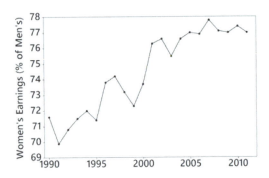

11. Misconduct includes fraud, duplication, and plagiarism, so it does appear to be a major factor.

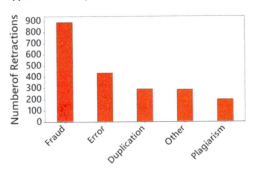

13.

15. The distribution appears to be skewed to the left (or negatively skewed).

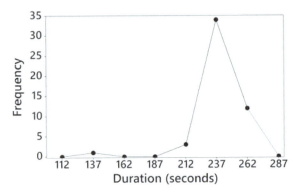

17. Because the vertical scale starts with a frequency of 200 instead of 0, the difference between the "no" and "yes" responses is greatly exaggerated. The graph makes it appear that about *five* times as many respondents said "no," when the ratio is actually a little less than 2.5 to 1.

19. The two costs are one-dimensional in nature, but the baby bottles are three-dimensional objects. The $4500 cost isn't even twice the $2600 cost, but the baby bottles make it appear that the larger cost is about five times the smaller cost.

21.
```
96. | 59
97. | 0001112333444
97. | 55666666788888999
98. | 00000000000002222233 444444444444
98. | 555566666666666666667 77777888888899
99. | 001244
99. | 56
```

Section 2-4

1. The term *linear* refers to a straight *line,* and *r* measures how well a scatterplot of the sample paired data fits a straight-line pattern.

3. A scatterplot is a graph of paired (*x, y*) quantitative data. It helps us by providing a visual image of the data plotted as points, and such an image is helpful in enabling us to see patterns in the data and to recognize that there may be a correlation between the two variables.

5. There does not appear to be a linear correlation between brain volume and IQ score.

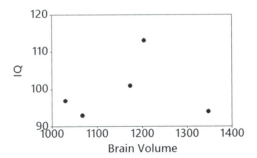

7. There does appear to be a linear correlation between weight and highway fuel consumption

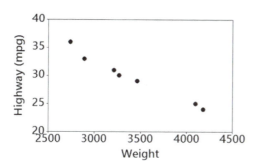

9. With $n = 5$ pairs of data, the critical values are ± 0.878. Because $r = 0.127$ is between -0.878 and 0.878, evidence is not sufficient to conclude that there is a linear correlation.

11. With $n = 7$ pairs of data, the critical values are ± 0.754. Because $r = -0.987$ is in the left tail region below -0.754, there are sufficient data to conclude that there is a linear correlation.

13. Because the P-value is not small (such as 0.05 or less), there is a high chance (83.9% chance) of getting the sample results when there is no correlation, so evidence is not sufficient to conclude that there is a linear correlation.

15. Because the P-value is small (such as 0.05 or less), there is a small chance of getting the sample results when there is no correlation, so there is sufficient evidence to conclude that there is a linear correlation.

Chapter 2: Quick Quiz

1. Class width: 3. It is not possible to identify the original data values.
2. Class boundaries: 17.5 and 20.5. Class limits: 18 and 20.
3. 40   4. 19 and 19   5. Pareto chart
6. Histogram   7. Scatterplot
8. No, the term "normal distribution" has a different meaning than the term "normal" that is used in ordinary speech. A normal distribution has a bell shape, but the randomly selected lottery digits will have a uniform or flat shape.
9. Variation
10. The bars of the histogram start relatively low, increase to some maximum, and then decrease. Also, the histogram is symmetric, with the left half being roughly a mirror image of the right half.

Chapter 2: Review Exercises

1.

Temperature (°F)	Frequency
97.0–97.4	2
97.5–97.9	4
98.0–98.4	7
98.5–98.9	5
99.0–99.4	2

2. Yes, the data appear to be from a population with a normal distribution because the bars start low and reach a maximum, then decrease, and the left half of the histogram is approximately a mirror image of the right half.

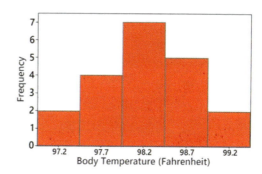

3. By using fewer classes, the histogram does a better job of illustrating the distribution.

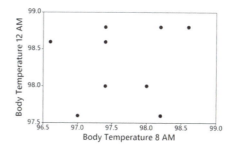

4. There are no outliers.

```
97.|125668
98.|002223466779
99.|14
```

5. No. There is no pattern suggesting that there is a relationship.

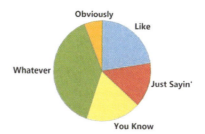

6. a. Time-series graph   b. Scatterplot   c. Pareto chart
7. A pie chart wastes ink on components that are not data; pie charts lack an appropriate scale; pie charts don't show relative sizes of different components as well as some other graphs, such as a Pareto chart.

8. The Pareto chart does a better job. It draws attention to the most annoying words or phrases and shows the relative sizes of the different categories.

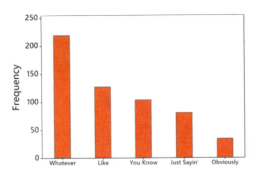

Chapter 2: Cumulative Review Exercises

1.

Total Hours	Frequency
235–239	4
240–244	3
245–249	9
250–254	8
255–259	3
260–264	3

2. a. 235 hours and 239 hours   b. 234.5 hours and 239.5 hours
   c. 237 hours
3. The distribution is closer to being a normal distribution than the others.

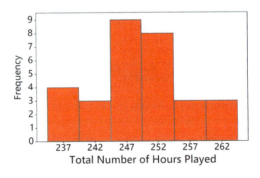

4. Start the vertical scale at a frequency of 2 instead of the frequency of 0.
5. Looking at the stemplot sideways, we can see that the distribution approximates a normal distribution.

```
23 | 6789
24 | 112555677889
25 | 00012233888
26 | 024
```

6. a. Continuous   b. Quantitative   c. Ratio
   d. Convenience sample   e. Sample

Chapter 3 Answers

Section 3-1

1. The term *average* is not used in statistics. The term *mean* should be used for the result obtained by adding all of the sample values and dividing the total by the number of sample values.
3. They use different approaches for providing a value (or values) of the center or middle of the sorted list of data.
5. $\bar{x} = 57.1$; median $= 60.0$; mode $=$ none; midrange $= 53.0$. The jersey numbers are nominal data that are just replacements for names, and they do not measure or count anything, so the resulting statistics are meaningless.
7. $\bar{x} = \$172.0$ million; median $= \$160.0$ million; mode $= \$150$ million; midrange $= \$200.0$ million. Apart from the fact that all other celebrities have amounts of net worth lower than those given, nothing meaningful can be known about the population of net worth of all celebrities. The numbers all end in 0 or 5, and they appear to be rounded estimates.
9. $\bar{x} = 7.7$ hurricanes; median $= 8.0$ hurricanes; mode $= 8$ hurricanes; midrange $= 8.5$ hurricanes. The data are time-series data, but the measures of center do not reveal anything about a trend consisting of a pattern of change over time.
11. $\bar{x} = \$1454.2$; median $= \$1500.0$; mode $= \$1500$; midrange $= \$1375.0$. The sample consists of "best buy" TVs, so it is not a random sample and is not likely to be representative of the population. The lowest price is a relevant statistic for someone planning to buy one of the TVs.
13. $\bar{x} = 32.6$ mg; median $= 39.5$ mg; mode $= 0$ mg; midrange $= 27.5$ mg. Americans consume some brands much more often than others, but the 20 brands are all weighted equally in the calculations, so the statistics are not necessarily representative of the population of all cans of the same 20 brands consumed by Americans.
15. $\bar{x} = 9.45$ in.; median $= 9.30$ in.; mode $= 9.1$ in.; midrange $= 9.50$ in. Because the measurements were made in 1988, they are not necessarily representative of the current population of all Army women.
17. $\bar{x} = \$365.3$; median $= \$200.0$; mode $= \$500$; midrange $= \$1269.5$. The amounts of $\$1500$ and $\$2500$ appear to be outliers.
19. $\bar{x} = 2.8$ cigarettes; median $= 0.0$ cigarettes; mode $= 0$ cigarettes; midrange $= 25.0$ cigarettes. Because the selected subjects *report* the number of cigarettes smoked, it is very possible that the data are not at all accurate. And what about that person who smokes 50 cigarettes (or 2.5 packs) a day? What are they thinking?
21. Systolic: $\bar{x} = 127.6$ mm Hg; median $= 124.0$ mm Hg. Diastolic: $\bar{x} = 73.6$ mm Hg; median $= 75.0$ mm Hg. Given that systolic and diastolic blood pressures measure different characteristics, a comparison of the measures of center doesn't make sense. Because the data are matched, it would make more sense to investigate whether there is an association or *correlation* between systolic blood pressure measurements and diastolic blood pressure measurements.

23. Males: $\bar{x}$ = 69.5 beats per minute; median = 66.0 beats per minute. Females: $\bar{x}$ = 82.1 beats per minute; median = 84.0 beats per minute. The pulse rates of males appear to be lower than those of females.

25. $\bar{x}$ = 0.8 and median = 1.0. Ten of the tornadoes have missing F-scale measurements.

27. $\bar{x}$ = 98.20°F; median = 98.40°F. These results suggest that the mean is less than 98.6°F.

29. $\bar{x}$ = 36.2 years, which is the same as the mean of 36.2 years obtained by using the original list of values.

31. $\bar{x}$ = 224.0 (1000 cells/μL). The mean from the frequency distribution is quite close to the mean of 224.3 (1000 cells/μL) obtained by using the original list of values.

33. 3.14; yes

35. a. 90 beats per minute   b. $n - 1$

37. 504 lb is an outlier. Median: 285.5 lb; mean: 294.4 lb; 10% trimmed mean: 285.4 lb; 20% trimmed mean: 285.8 lb. The median, 10% trimmed mean, and 20% trimmed mean are all quite close, but the untrimmed mean of 294.4 lb differs from them because it is strongly affected by the inclusion of the outlier.

39. 1.5290%

41. The median found using the given expression is 153.125 seconds, which is rounded to 153.1 seconds. This value differs by 2.6 seconds from the median of 150.5 seconds found by using the original list of service times.

Section 3-2

1. 119.0 cm³ is quite close to the exact value of the standard deviation of 124.9 cm³.

3. 401.6577 kg²

5. Range = 92.0; s^2 = 1149.5; s = 33.9. The jersey numbers are nominal data that are just replacements for names, and they do not measure or count anything, so the resulting statistics are meaningless.

7. Range = $100.0 million; s^2 = 1034.4 (million dollars)²; s = $32.2 million. Because the data are from celebrities with the highest net worth, the measures of variation are not at all typical for all celebrities.

9. Range = 13.0 hurricanes; s^2 = 10.2 hurricanes²; s = 3.2 hurricanes. Data are time-series data, but the measures of variation do not reveal anything about a trend consisting of a pattern of change over time.

11. Range = $850.0; s^2 = 61,117.4 (dollars)²; s = $247.2. The sample consists of "best buy" TVs, so it is not a random sample and is not likely to be representative of the population. The measures of variation are not likely to be typical of all TVs that are 60 inches or larger.

13. Range = 55.0 mg; s^2 = 413.4 mg²; s = 20.3 mg. Americans consume some brands much more often than others, but the 20 brands are all weighted equally in the calculations, so the statistics are not necessarily representative of the population of all cans of the same 20 brands consumed by Americans.

15. Range = 1.80 in.; s^2 = 0.27 in.²; s = 0.52 in. Because the measurements were made in 1988, they are not necessarily representative of the current population of all Army women.

17. Range = $2461.0; s^2 = 290,400.4 (dollars)²; s = $538.9. The amounts of $1500 and $2500 appear to be outliers, and it is likely that they have a large effect on the measures of variation.

19. Range = 50.0 cigarettes; s^2 = 89.7 cigarettes²; s = 9.5 cigarettes. Because the selected subjects *report* the number of cigarettes smoked, it is very possible that the data are not at all accurate, so the results might not reflect the actual smoking behavior of California adults.

21. Systolic: 14.6%. Diastolic: 16.9%. The variation is roughly about the same.

23. Males: 16.2%. Females: 11.2%. Pulse rates of males appear to vary more than pulse rates of females.

25. Range = 4.0, s^2 = 0.9, and s = 0.9

27. Range = 3.10°F; s^2 = 0.39 (°F)²; s = 0.62°F

29. 1.0, which is very close to s = 0.9 found by using all of the data.

31. 0.78°F, which is not substantially different from s = 0.62°F.

33. Significantly low values are less than or equal to 49.0 beats per minute, and significantly high values are greater than or equal to 99.0 beats per minute. A pulse rate of 44 beats per minute is significantly low.

35. Significantly low values are less than or equal to 24.74 cm, and significantly high values are greater than or equal to 29.90 cm. A foot length of 30 cm is significantly high.

37. s = 12.7 years differs from the exact value of 11.5 years by a somewhat large amount.

39. s = 68.4 is somewhat far from the exact value of 59.5.

41. a. 95%   b. 68%

43. At least 89% of women have platelet counts within 3 standard deviations of the mean. The minimum is 58.9 and the maximum is 451.3.

45. a. 24.7 cigarettes²   b. 24.7 cigarettes²   c. 12.3 cigarettes²
    d. Part (b), because repeated samples result in variances that target the same value (24.7 cigarettes²) as the population variance. Use division by $n - 1$.
    e. No. The mean of the sample variances (24.7 cigarettes²) equals the population variance (24.7 cigarettes²), but the mean of the sample standard deviations (3.5 cigarettes) does not equal the population standard deviation (5.0 cigarettes).

Section 3-3

1. James's height is 4.07 standard deviations above the mean.

3. The bottom boxplot represents weights of women, because it depicts weights that are generally lower.

5. a. 60.20 Mbps   b. 3.76 standard deviations
   c. z = 3.76
   d. The data speed of 77.8 Mbps is significantly high.

7. a. 38 beats per minute   b. 3.04 standard deviations
   c. z = -3.04
   d. The pulse rate of 36 beats per minute is significantly low.

9. Significantly low values are less than or equal to 10.9; significantly high values are greater than or equal to 31.3.

11. Significantly low weights are less than or equal to 5.51542 g, and significantly high weights are greater than or equal to 5.76318 g.

13. With z scores of 10.83 and -16.83, the z score of -16.83 is farther from the mean, so the shortest man has a height that is more extreme.

15. Male: z score $= -2.69$; female: z score $= -2.18$. The male has the more extreme weight.

17. 58th percentile   19. 34th percentile

21. 2.45 Mbps (Tech: Excel: 2.44 Mbps).

23. 3.8 Mbps (Tech: Minitab: 3.85 Mbps; Excel: 3.75 Mbps)

25. 1.6 Mbps   27. 0.5 Mbps

29. 5-number summary: 2, 6.0, 7.0, 8.0, 10.

31. 5-number summary: 128 mBq, 140.0 mBq, 150.0 mBq, 158.5 mBq, 172 mBq (Tech: Minitab yields $Q_1 = 139.0$ mBq and $Q_3 = 159.75$ mBq. Excel yields $Q_1 = 141.0$ mBq and $Q_3 = 157.25$ mBq.)

33. The top boxplot represents males. Males appear to have slightly lower pulse rates than females.

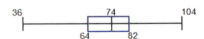

35. The top boxplot represents BMI values for males. The two boxplots do not appear to be very different, so BMI values of males and females appear to be about the same, except for a few very high BMI values for females that caused the boxplot to extend farther to the right.

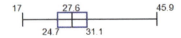

37. Top boxplot represents males. Males appear to have slightly lower pulse rates than females. Outliers for males: 40 beats per minute, 102 beats per minute, 104 beats per minute. Outliers for females: 36 beats per minute.

Chapter 3: Quick Quiz

1. 6.8 hours   2. 7.0 hours

3. Two modes: 7 hours, 8 hours   4. 1.7 hours^2

5. Yes, because 0 hours is substantially less than all of the other data values.

6. -0.93 7. 75% or 60 sleep times

8. Minimum, first quartile Q_1, second quartile Q_2 (or median), third quartile Q_3, maximum

9. 1.5 hours (from range/4)   10. $\bar{x}, \mu, s, \sigma, s^2, \sigma^2$

Chapter 3: Review Exercises

1. a. 1.0 min;   b. 0.5 min;   c. 1 min;   d. 7.5 min;
   e. 29.0 min;   f. 7.9 min;   g. 61.8 min^2;   h. -4.5 min;
   i. 2.5 min. (Tech: Minitab yields $Q_1 = -4.75$ min and $Q_3 = 3.25$ min. Excel yields $Q_1 = -4.25$ min and $Q_3 = 1.75$ min.)

2. $z = -0.13$. The prediction error of 0 min is not significant because its z score is between 2 and -2, so it is within 2 standard deviations of the mean.

3. 5-number summary: -7 min, -4.5 min, 0.5 min, 2.5 min, 22 min. (Tech: Minitab yields $Q_1 = -4.75$ min and $Q_3 = 3.25$ min. Excel yields $Q_1 = -4.25$ min and $Q_3 = 1.75$ min.)

4. 23.0. The numbers don't measure or count anything. They are used as replacements for the names of the categories, so the numbers are at the nominal level of measurement. In this case the mean is a meaningless statistic.

5. The female has the larger relative birth weight because her z score of 0.23 is larger than the z score of 0.19 for the male.

6. The outlier is 646. The mean and standard deviation with the outlier included are $\bar{x} = 267.8$ and $s = 131.6$. Those statistics with the outlier excluded are $\bar{x} = 230.0$ and $s = 42.0$. Both statistics changed by a substantial amount, so here the outlier has a very strong effect on the mean and standard deviation.

7. The minimum is 119 mm, the first quartile Q_1 is 128 mm, the second quartile Q_2 (or median) is 131 mm, the third quartile Q_3 is 135 mm, and the maximum is 141 mm.

8. With a minimum of 117 seconds and a maximum of 256 seconds, $s \approx 34.8$ seconds, which is very close to the standard deviation of 33.7 seconds found from the larger sample.

Chapter 3: Cumulative Review Exercises

1.

Arsenic (μg)	Frequency
0.0–1.9	1
2.0–3.9	0
4.0–5.9	3
6.0–7.9	7
8.0–9.9	1

2.

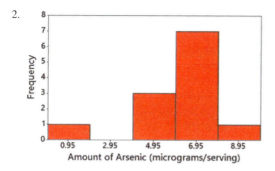

Amount of Arsenic (micrograms/serving)

3. 1. | 5
    2. |
    3. |
    4. | 9
    5. | 44
    6. | 13679
    7. | 38
    8. | 2

4. a. 6.09 μg   b. 6.45 μg   c. 1.75 μg
    d. 3.06 $(\mu g)^2$ e. 6.70 μg

5. The vertical scale does not begin at 0, so the differences among the different outcomes are exaggerated.

6. No. A normal distribution would appear in a histogram as being bell-shaped, but the histogram is not bell-shaped.

Chapter 4 Answers

Section 4-1

1. $P(A) = 1/1000$, or $0.001. P(\overline{A}) = 999/1000$, or 0.999.
3. Part (c).   5. 0, 3/5, 1, 0.135
7. 1/9 or 0.111   9. Significantly high
11. Neither significantly low nor significantly high
13. 1/2 or 0.5   15. 0.43
17. 1/10 or 0.1   19. 0
21. 5/555 or 0.00901. The employer would suffer because it would be at risk by hiring someone who uses drugs.
23. 50/555 or 0.0901. This result does appear to be a reasonable estimate of the prevalence rate.
25. 879/945 or 0.930. Yes, the technique appears to be effective.
27. 428/580 or 0.738; yes
29. 0.270. No, it is not unlikely for someone to not use social networking sites.
31. a. brown/brown, brown/blue, blue/brown, blue/blue
    b. 1/4   c. 3/4
33. 3/8 or 0.375   35. 4/16 or 1/4 or 0.25.
37. The high probability of 0.327 shows that the sample results could have easily occurred by chance. It appears that there is not sufficient evidence to conclude that pregnant women can correctly predict the gender of their baby.
39. a. In comparing the 200 mg treatment group to the placebo group, the low probability of less than 0.049 shows that the sample results could not have easily occurred by chance. It appears that 200 mg of caffeine does have an effect on memory.
    b. In comparing the 300 mg group to the 200 mg group, the high probability of 0.75 indicates the sample results could have easily occurred by chance. There is not sufficient evidence to

conclude that there are different effects from the 300 mg treatment group and the 200 mg treatment group.
41. a. 9999:1   b. 4999:1
    c. The description is not accurate. The odds against winning are 9999:1 and the odds in favor are 1:9999, not 1:10,000.
43. a. $5   b. 5:2
    c. 772:228 or 193:57 or about 3.39:1 (or roughly 17:5)
    d. Approximately $8.80 (instead of the actual payoff of $7.00)

Section 4-2

1. $P(A)$ represents the probability of selecting an adult with blue eyes, and $P(\overline{A})$ represents the probability of selecting an adult who does not have blue eyes.
3. Because the selections are made without replacement, the events are dependent. Because the sample size of 1068 is less than 5% of the population size of 15,524,971, the selections can be treated as being independent (based on the 5% guideline for cumbersome calculations).
5. 0.74
7. $P(\overline{N}) = 0.670$, where $P(\overline{N})$ is the probability of randomly selecting someone with a response different from "never."
9. 756/1118 or 0.676
11. 1020/1118 or 0.912; not disjoint.
13. a. 0.0200. Yes, the events are independent.
    b. 0.0199. The events are dependent, not independent.
15. a. 0.779. Yes, the events are independent.
    b. 0.779. The events are dependent, not independent.
17. 709/1118 or 0.634   19. 0.0156
21. a. 300   b. 154   c. 0.513
23. 0.990
25. a. 0.03   b. 0.0009   c 0.000027
    d. By using one drive without a backup, the probability of total failure is 0.03, and with three independent disk drives, the probability drops to 0.000027. By changing from one drive to three, the probability of total failure drops from 0.03 to 0.000027, and that is a very substantial improvement in reliability. Back up your data!
27. 0.838. The probability of 0.838 is high, so it is likely that the entire batch will be accepted, even though it includes many defects.
29. a. 0.299
    b. Using the 5% guideline for cumbersome calculations: 0.00239 [using the rounded result from part (a)] or 0.00238
31. a. 0.999775   b. 0.970225
    c. The series arrangement provides better protection.
33. a. $P(A \text{ or } B) = P(A) + P(B) - 2P(A \text{ and } B)$
    b. 691/1118 or 0.618

Section 4-3

1. The event of not getting at least 1 defect among the 3 iPhones, which means that all 3 iPhones are good.
3. The probability that the selected person is a high school classmate given that the selected person is female.
5. 7/8 or 0.875   7. 0.982   9. 0.344
11. 0.965. The probability is high enough so that she can be reasonably sure of getting a defect for her work.
13. a. 27/43 or 0.628   b. 16/43 or 0.372
    c. It appears that when students are given four quarters, they are more likely to spend the money than keep it.

15. a. 27/43 or 0.628   b. 12/46 or 0.261
    c. It appears that students are more likely to spend the money
       when given four quarters than when given a $1 bill.
17. 2/1155 or 0.00173. This is the probability of the test making it
    appear that the subject has hepatitis C when the subject does not
    have it, so the subject is likely to experience needless stress and
    additional testing.
19. 335/337 or 0.994. The very high result makes the test appear to
    be effective in identifying hepatitis C.
21. a. 0.9991
    b. 0.999973. The usual round-off rule for probabilities would re-
       sult in a probability of 1.00, which would incorrectly indicate
       that we are certain to have at least one working hard drive.
23. 0.490. The probability is not low, so further testing of the individ-
    ual samples will be necessary in 49% of the combined samples.
25. 0.569

Section 4-4

1. The symbol ! is the factorial symbol that represents the product
   of decreasing whole numbers, as in $6! = 6 \cdot 5 \cdot 4 \cdot 3 \cdot 2 \cdot 1 = 720$.
   Six people can stand in line 720 different ways.
3. Because repetition is allowed, numbers are selected *with
   replacement*, so the combinations rule and the two permutation rules do not apply. The multiplication counting rule can
   be used to show that the number of possible outcomes is
   $10 \cdot 10 \cdot 10 \cdot 10 = 10,000$.
5. 1/10,000   7. 1/171   9. 1/40,320
11. 1/254,251,200
13. 1/100,000,000. No, there are far too many different possibilities.
15. 1/1820 or 0.000549   17. 1/292,201,338
19. 1/100,000
21. Area codes: 800. Phone numbers: 6,400,000,000. Yes. (With
    a total population of about 400,000,000, there would be about
    16 phone numbers for every adult and child.)
23. a. 5040   b. 210   c. 1/210
25. 40,320; 1/40,320   27. 653,837,184,000
29. 1/258,890,850. There is a *much* better chance of being struck by
    lightning.
31. There are 62 different possible characters. The alphabet requires
    26 characters and there are 10 digits, so the Morse code system is
    more than adequate.
33. 128/2652 or 32/663 or 0.0483; 4.83% or about 5%
35. 12   37. 2,095,681,645,538 (about 2 trillion)

Chapter 4: Quick Quiz

1. 4/5 or 0.8   2. 0.8   3. 4/12 or 1/3   4. 0.548
5. Answer varies, but the probability should be low, such as 0.001.
6. 0.0680   7. 0.727   8. 0.00874
9. 0.00459   10. 0.0131

Chapter 4: Review Exercises

1. 0.589   2. 0.657   3. 0.592   4. 0.832
5. 0.798   6. 0.347   7. 0.198

8. $\overline{A}$ is the event of selecting a driver and getting someone who was
   not using a seatbelt. $P(\overline{A}) = 0.411$
9. $\overline{A}$ is the event of selecting a driver and getting someone who was
   killed. $P(\overline{A}) = 0.445$.
10. 0.171
11. Answer varies, but Forbes reports that about 19% of cars are
    black, so any estimate between 0.10 and 0.30 would be good.
12. a. 0.25   b. 0.316
    c. No, it is not unlikely because the probability of 0.316 shows
       that the event occurs quite often.
13. a. 1/365   b. 31/365
    c. Answer varies, but it is probably quite small, such as 0.01 or less.
    d. Yes
14. 0.0335. No.
15. a. 1/237,336   b. 1/4   c. 1/4   d. 1/3,797,376
16. 1/962,598
17. a. 999/1000 or 0.999
    b. 999,999/1,000,000 or 0.999999
18. 1/342

Chapter 4: Cumulative Review Exercises

1. a. 0.168 g/dL   b. 0.160 g/dL   c. 0.220 g/dL
   d. 0.260 g/dL   e. 0.069 g/dL   f. 0.005 (g/dL)^2
2. a. 0.09, 0.120, 0.160 0.180, 0.35 (all in units of g/dL).
      Outlier: 0.35 g/dL.
   b.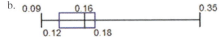
   c. .0 | 9
       .1 | 113457788
       .2 | 3
       .3 | 5
3. a. 46%   b. 0.460   c. Stratified sample
4. a. Convenience sample
   b. If the students at the college are mostly from a surrounding region
      that includes a large proportion of one ethnic group, the results
      might not reflect the general population of the United States.
   c. 0.75   d. 0.64
5. The lack of any pattern of the points in the scatterplot suggests
   that there does not appear to be an association between systolic
   blood pressure and blood platelet count.

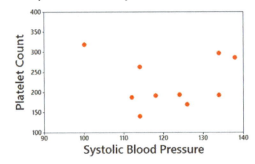

6. a. 1/2,598,960   b. 1/28   c. 1/72,770,880

Chapter 5 Answers

Section 5-1

1. The random variable is x, which is the number of girls in four births. The possible values of x are 0, 1, 2, 3, and 4. The values of the random variable x are numerical.

3. $\Sigma P(x) = 0.063 + 0.250 + 0.375 + 0.250 + 0.063 = 1.001$. The sum is not exactly 1 because of a round-off error. The sum is close enough to 1 to satisfy the requirement. Also, the variable x is a numerical random variable and its values are associated with probabilities, and each of the probabilities is between 0 and 1 inclusive, as required. The table does describe a probability distribution.

5. a. Continuous random variable   b. Not a random variable
   c. Discrete random variable   d. Continuous random variable
   e. Discrete random variable

7. Probability distribution with $\mu = 2.5$, $\sigma = 1.1$.

9. Not a probability distribution because the sum of the probabilities is 0.100, which is not 1 as required. The probabilities show that Ted needs work.

11. Probability distribution with $\mu = 1.6$, $\sigma = 0.9$. (The sum of the probabilities is 0.999, but that is due to rounding errors.)

13. Not a probability distribution because the responses are not values of a numerical random variable.

15. $\mu = 4.0$ girls, $\sigma = 1.4$ girls.

17. Significantly high numbers of girls are greater than or equal to $\mu + 2\sigma$, and $\mu + 2\sigma = 4.0 + 2(1.4) = 6.8$ girls. Because 6 girls is not greater than or equal to 6.8 girls, it is not a significantly high number of girls.

19. a. 0.109   b. 0.144   c. Part (b)
    d. No, because the probability of 0.144 is not very low (less than or equal to 0.05).

21. $\mu = 1.5$ sleepwalkers, $\sigma = 1.0$ sleepwalker

23. Significantly high numbers of sleepwalkers are greater than or equal to $\mu + 2\sigma$, and $\mu + 2\sigma = 1.5 + 2(1.0) = 3.5$ sleepwalkers. Because 3 sleepwalkers is not greater than or equal to 3.5 sleepwalkers, 3 sleepwalkers is not a significantly high number.

25. a. 0.363   b. 0.535
    c. The probability from part (b).
    d. No, because the probability of 1 or fewer sleepwalkers is 0.535, which is not low (not less than or equal to 0.05).

27. a. 1000   b. 1/1000   c. $499   d. $-50¢$
    e. Because both bets have the same expected value of $-50¢$, neither bet is better than the other.

29. a. $-$161 and $99,839   b. $-$21
    c. Yes. The expected value for the insurance company is $21, which indicates that the company can expect to make an average of $21 for each such policy.

Section 5-2

1. The given calculation assumes that the first two consumers are comfortable with the drones and the last three consumers are not comfortable with drones, but there are other arrangements consisting of two consumers who are comfortable and three who are not. The probabilities corresponding to those other arrangements should also be included in the result.

3. Because the 30 selections are made without replacement, they are dependent, not independent. Based on the 5% guideline for cumbersome calculations, the 30 selections can be treated as being independent. (The 30 selections constitute 3% of the population of 1009 responses, and 3% is not more than 5% of the population.) The probability can be found by using the binomial probability formula, but it would require application of that formula 21 times (or 10 times if we are clever), so it would be better to use technology.

5. Not binomial. Each of the weights has more than two possible outcomes.

7. Binomial.

9. Not binomial. Because the senators are selected without replacement, the selections are not independent. (The 5% guideline for cumbersome calculations should not be used because the 40 selected senators constitute 40% of the population of 100 senators, and 40% exceeds 5%.)

11. Binomial. Although the events are not independent, they can be treated as being independent by applying the 5% guideline. The sample size of 1019 is not more than 5% of the population of all adults.

13. a. 0.128   b. WWC, WCW, CWW; 0.128 for each
    c. 0.384

15. 0.0000819 (Table: 0+)   17. 0.797 (Table: 0.798)

19. 0.168   21. 0.147   23. 0.0889

25. 0.00000451. The result of 7 minorities is significantly low. The probability shows that it is very highly unlikely that a process of random selection would result in 7 or fewer minorities. (The Supreme Court rejected the claim that the process was random.)

27. a. 0.00154 (Table: 0.002)   b. 0.000064 (Table: 0+)
    c. 0.00160 (Table: 0.002)
    d. Yes, the small probability from part (c) suggests that 5 is a significantly high number.

29. a. $\mu = 18.0$ girls; $\sigma = 3.0$ girls
    b. Values of 12.0 girls or fewer are significantly low, values of 24.0 girls or greater are significantly high, and values between 12.0 girls and 24.0 girls are not significant.
    c. Yes, because the result of 26 girls is greater than or equal to 24.0 girls. A result of 26 girls would suggest that the XSORT method is effective.

31. a. $\mu = 7.5$ peas; $\sigma = 1.4$ peas
    b. Values of 4.7 or less are significantly low, values of 10.3 or greater are significantly high, and values between 4.7 and 10.3 are not significant.
    c. No, because the result of 9 peas with green pods is not greater than or equal to 10.3.

33. 0.304; no.

35. 0.662. The probability shows that about 2/3 of all shipments will be accepted. With about 1/3 of the shipments rejected, the supplier would be wise to improve quality.

37. a. 8.7 and 23.3 (or 8.6 and 23.4 if using the rounded σ). Because 19 lies between those limits, it is neither significantly low nor significantly high.
    b. 0.0736
    c. The probability of 19 or more green M&Ms is 0.242.
    d. The probability from part (c) is relevant. The result of 19 green M&Ms is not significantly high.
    e. The results do not provide strong evidence against the claim of 16% for green M&Ms.

39. a. 238.3 and 287.2 (or 287.1 if using the rounded σ). Because 308 is greater than 287.2, the value of 308 is significantly high.
  b. 0.0000369
  c. The probability of 308 or more is 0.000136.
  d. The probability from part (c) is relevant. The result of 308 voters who voted for the winner is significantly high.
  e. The results suggest that the surveyed voters either lied or had defective memory of how they voted.
41. 0.0468   43. 0.132

Section 5-3

1. $\mu = 535/576 = 0.929$, which is the mean number of hits per region. $x = 2$, because we want the probability that a randomly selected region had exactly 2 hits, and $e \approx 2.71828$, which is a constant used in all applications of Formula 5-9.
3. Possible values of x: 0, 1, 2, . . . (with no upper bound). It is not possible to have $x = 2.3$ calls in a day. x is a discrete random variable.
5. a. $P(5) = 0.158$
  b. In 55 years, the expected number of years with 5 hurricanes is 8.7.
  c. The expected value of 8.7 years is close to the actual value of 8 years, so the Poisson distribution works well here.
7. a. $P(7) = 0.140$
  b. In 55 years, the expected number of years with 7 hurricanes is 7.7.
  c. The expected value of 7.7 years is close to the actual value of 7 years, so the Poisson distribution works well here.
9. 11.6 births; 0.0643 (0.0649 using the rounded mean). There is less than a 7% chance of getting exactly 15 births on any given day.
11. a. 62.2
  b. 0.0155 (0.0156 using the rounded mean)
13. a. 0.170
  b. The expected number is between 97.9 and 98.2, depending on rounding.
  c. The expected number of regions with 2 hits is close to 93, which is the actual number of regions with 2 hits.
15. 0.9999876 or 0.9999877 (using unrounded mean). Very high chance, or "almost certain," that at least one fatality will occur.
17. 0.0000178. No, it is highly unlikely that at least one jackpot win will occur in 50 years.

Chapter 5: Quick Quiz

1. No, the sum of the probabilities is 4/3 or 1.333, which is greater than 1.
2. $\mu = 16.0; \sigma = 3.6$
3. The values are parameters because they represent the mean and standard deviation for the population of all who make random guesses for the 80 questions, not a sample of actual results.
4. No. (Using the range rule of thumb, the limit separating significantly high values is 23.2, but 20 is not greater than or equal to 23.2. Using probabilities, the probability of 20 or more correct answers is 0.163, which is not low.)
5. Yes. (Using the range rule of thumb, the limit separating significantly low values is 8.8, and 8 is less than 8.8. Using probabilities, the probability of 8 or fewer correct answers is 0.0131, which is low.)
6. Yes. (The sum of the probabilities is 0.999 and it can be considered to be 1 because of rounding errors.)
7. 4.0 flights   8. 0.8 flight^2
9. 0+ indicates that the probability is a very small positive number. It does not indicate that it is impossible for none of the five flights to arrive on time.
10. 0.057

Chapter 5: Review Exercises

1. 0.274
2. 0.999. No. The five friends are not randomly selected from the population of adults. Also, the fact that they are vacationing together suggests that their financial situations are more likely to include credit cards.
3. $\mu = 3.7$ adults and $\sigma = 1.0$ adult.
4. Using $\mu = 3.7$ adults and $\sigma = 1.0$ adult, values are significantly high if they are equal to or greater than $\mu + 2\sigma = 5.7$ adults. The result of five adults with credit cards is not significantly high because it is not equal to or greater than 5.7 adults. Also, the probability that all five adults have credit cards is 0.222, which is not low (less than or equal to 0.05).
5. Using $\mu = 3.7$ adults and $\sigma = 1.0$ adult, values are significantly low if they are equal to or less than $\mu - 2\sigma = 1.7$ adults. The result of 1 adult with a credit card is significantly low because it is less than or equal to 1.7 adults. Also, the probability that 1 or fewer adults have credit cards is 0.0181, which is low (less than or equal to 0.05).
6. No. The responses are not *numerical*.
7. No, because $\Sigma P(x) = 0.9686$, but the sum should be 1. (There is a little leeway allowed for rounding errors, but the sum of 0.9686 is too far from 1.)
8. Yes, probability distribution with $\mu = 4.6$ people, $\sigma = 1.0$ people. (The sum of the probabilities is 0.999, but that is due to rounding errors.)
9. a. 236.0 checks
  b. $\mu = 236.0$ checks and $\sigma = 12.8$ checks
  c. 210.4 checks (or 210.3 checks if using the unrounded σ)
  d. Yes, because 0 is less than or equal to 210.4 checks (or 210.3 checks).
10. a. 7/365 or 0.0192   b. 0.981
  c. 0.000182
  d. No, because the event is so rare. (But it is possible that more than one death occurs in a car crash or some other such event, so it might be wise to consider a contingency plan.)

Chapter 5: Cumulative Review Exercises

1. a. 9.6 moons   b. 5.0 moons   c. 0 moons
  d. 28.0 moons   e. 11.0 moons   f. 120.3 moons^2
  g. −12.4 moons, 31.6 moons
  h. No, because none of the planets have a number of moons less than or equal to −12.4 moons (which is impossible, anyway) and none of the planets have a number of moons equal to or greater than 31.6 moons.
  i Ratio   j. Discrete

2. a. 1/1000 or 0.001   b. 0.365
   c. 0.254   d. $-50¢$
3. a. 0.263   b. 0.342   c. 0.0688   d. 0.732   e. 0.658
4. a. 253
   b. Sample: The 347 human resource professionals who were surveyed. Population: All human resource professionals.
   c. 73% is a statistic because it is a measure based on a sample, not the entire population.
5. No vertical scale is shown, but a comparison of the numbers shows that 7,066,000 is roughly 1.2 times the number 6,000,000; however, the graph makes it appear that the goal of 7,066,000 people is roughly 3 times the number of people enrolled. The graph is misleading in the sense that it creates the false impression that actual enrollments are far below the goal, which is not the case. Fox News apologized for their graph and provided a corrected graph.
6. a. 0.254   b. 0.255 (Table: 0.256)
   c. $\mu = 5.6$ adults, $\sigma = 1.3$ adults
   d. Yes. Using the range rule of thumb, 1 is less than $\mu - 2\sigma = 3.0$; using probabilities, the probability of 1 or fewer is 0.00129 (Table: 0.001), which is low, such as less than 0.05.

Chapter 6 Answers

Section 6-1

1. The word "normal" has a special meaning in statistics. It refers to a specific bell-shaped distribution that can be described by Formula 6-1. The lottery digits do not have a normal distribution.
3. The mean is $\mu = 0$ and the standard deviation is $\sigma = 1$.
5. 0.4   7. 0.2   9. 0.6700   11. 0.6993 (Table: 0.6992)
13. 1.23   15. -1.45 17. 0.1093   19. 0.8997   21. 0.4013
23. 0.9772   25. 0.0214 (Table: 0.0215)
27. 0.0174   29. 0.9545 (Table: 0.9544)
31. 0.8413 (Table: 0.8412)   33. 0.999997 (Table: 0.9999)
35. 0.5000   37. 2.33   39. $-2.05, 2.05$
41. 1.28   43. 1.75   45. 68.27% (Table: 68.26%)
47. 99.73% (Table: 99.74%)
49. a. 2.28%   b. 2.28%   c. 95.45% (Table: 95.44%)

Section 6-2

1. a. $\mu = 0; \sigma = 1$
   b. The z scores are numbers without units of measurement.
3. The standard normal distribution has a mean of 0 and a standard deviation of 1, but a nonstandard normal distribution has a different value for one or both of those parameters.
5. 0.8849   7. 0.9053   9. 136   11. 69
13. 0.0115 (Table: 0.0116)   15. 0.6612   17. 24.9 in.
19. 20.9 in. and 26.1 in. No, 26 in. is not significantly high.
21. a. 72.11% (Table: 72.23%). Yes, about 28% of women are not qualified because of their heights.
   b. 58.2 in. to 69.2 in.
23. a. 0.92% (Table: 0.90%). Because so few men can meet the height requirement, it is likely that most Mickey Mouse characters are women.
   b. 64.0 in. to 68.6 in.

25. 0.2015 (Table: 0.2005). No, the proportion of schizophrenics is not at all likely to be as high as 0.2005, or about 20%.
27. 55.67% (Table: 55.64%). Yes, about 44% of women were excluded.
29. a. 0.1717 (Table: 0.1711)   b. 2011.5 g (Table: 2011.4 g)
   c. Birth weights are significantly low if they are 2011.4 g or less, and they are "low birth weights" if they are 2495 g or less. Birth weights between 2011.4 g and 2495 g are "low birth weights" but they are not significantly low.
31. a. 0.0038; either a very rare event occurred or the husband is not the father.
   b. 240 days
33. a. The mean is 69.5817 (69.6 rounded) beats per minute and the standard deviation is 11.3315 (11.3 rounded) beats per minute. A histogram confirms that the distribution is roughly normal.
   b. 47.4 beats per minute; 91.8 beats per minute
35. a. 75; 12
   b. No, the conversion should also account for variation.
   c. B grade: 66.3 to 75.4 (Table: 66.2 to 75.4)
   d. Use a scheme like the one given in part (c), because variation is included in the curving process.

Section 6-3

1. a. In the long run, the sample proportions will have a mean of 0.512.
   b. The sample proportions will tend to have a distribution that is approximately normal.
3. Sample mean; sample variance; sample proportion
5. No. The sample is not a simple random sample from the population of all births worldwide. The proportion of boys born in China is substantially higher than in other countries.
7. a. 4.7
   b.

Sample Variance s^2	Probability
0.0	3/9
0.5	2/9
8.0	2/9
12.5	2/9

   c. 4.7
   d. Yes. The mean of the sampling distribution of the sample variances (4.7) is equal to the value of the population variance (4.7), so the sample variances target the value of the population variance.
9. a. 5
   b.

Sample Median	Probability
4.0	1/9
4.5	2/9
5.0	1/9
6.5	2/9
7.0	2/9
9.0	1/9

   c. 6.0

d. No. The mean of the sampling distribution of the sample medians is 6.0, and it is not equal to the value of the population median (5.0), so the sample medians do not target the value of the population median.

11. a.

$\bar{x}$	Probability
34	1/16
35	2/16
36	1/16
37.5	2/16
38.5	2/16
41	1/16
42.5	2/16
43.5	2/16
46	2/16
51	1/16

b. The mean of the population is 40.5 and the mean of the sample means is also 40.5.

c. The sample means target the population mean. Sample means make good estimators of population means because they target the value of the population mean instead of systematically underestimating or overestimating it.

13. a.

Range	Probability
0	4/16
2	2/16
5	2/16
7	2/16
10	2/16
15	2/16
17	2/16

b. The range of the population is 17, but the mean of the sample ranges is 7. Those values are not equal.

c. The sample ranges do not target the population range of 17, so sample ranges do not make good estimators of population ranges.

15.

Proportion of Girls	Probability
0	0.25
0.5	0.50
1	0.25

Yes. The proportion of girls in 2 births is 0.5, and the mean of the sample proportions is 0.5. The result suggests that a sample proportion is an unbiased estimator of a population proportion.

17. a.

Proportion Correct	Probability
0	16/25
0.5	8/25
1	1/25

b. 0.2

c. Yes. The sampling distribution of the sample proportions has a mean of 0.2 and the population proportion is also 0.2 (because there is 1 correct answer among 5 choices). Yes, the mean of the sampling distribution of the sample proportions is always equal to the population proportion.

19. The formula yields $P(0) = 0.25$, $P(0.5) = 0.5$, and $P(1) = 0.25$, which does describe the sampling distribution of the sample proportions. The formula is just a different way of presenting the same information in the table that describes the sampling distribution.

Section 6-4

1. The sample must have more than 30 values, or there must be evidence that the population of grade-point averages from statistics students has a normal distribution.

3. $\mu_{\bar{x}}$ represents the mean of all sample means, and $\sigma_{\bar{x}}$ represents the standard deviation of all sample means. For the samples of 64 IQ scores, $\mu_{\bar{x}} = 100$ and $\sigma_{\bar{x}} = 15/\sqrt{64} = 1.875$.

5. a. 0.6844

b. 0.9726

c. Because the original population has a normal distribution, the distribution of sample means is a normal distribution for any sample size.

7. a. 0.1271 (Table: 0.1272)

b. 0.2510

c. Because the original population has a normal distribution, the distribution of sample means is normal for any sample size.

9. 0.7030 (Table: 0.7019). The elevator does not appear to be safe because there is about a 70% chance that it will be overloaded whenever it is carrying 27 adult males.

11. a. 131

b. 0.0000179 (Table: 0.0001)

c. No. It is possible that the 4 subjects have a mean of 132 while some of them have scores below the Mensa requirement of 131.

13. a. 140 lb

b. 0.9999999998 (Table: 0.9999)

c. 0.9458 (Table: 0.9463)

d. The new capacity of 20 passengers does not appear to be safe enough because the probability of overloading is too high.

15. a. 0.0047

b. 0.0000 (Table: 0.0001)

c. The result from part (a) is relevant because the seats are occupied by individuals.

17. a. 0.5575 (Table: 0.5564)

b. 0.9996 (Table: 0.9995)

c. Part (a) because the ejection seats will be occupied by individual women, not groups of women.

19. a. 0.8877 (Table 0.8869)

b. 1.0000 when rounded to four decimal places (Table: 0.9999)

c. The probability from part (a) is more relevant because it shows that 89% of male passengers will not need to bend. The result from part (b) gives us information about the mean for a group of 100 men, but it doesn't give us useful information about the comfort and safety of individual male passengers.

d. Because men are generally taller than women, a design that accommodates a suitable proportion of men will necessarily accommodate a greater proportion of women.

21. a. Yes. The sampling is without replacement and the sample size of $n = 50$ is greater than 5% of the finite population size of 275. $\sigma_{\bar{x}} = 2.0504584$.

   b. 0.5963 (Table: 0.5947)

Section 6-5

1. The histogram should be approximately bell-shaped, and the normal quantile plot should have points that approximate a straight-line pattern.

3. We must verify that the sample is from a population having a normal distribution. We can check for normality using a histogram, identifying the number of outliers, and constructing a normal quantile plot.

5. Normal. The points are reasonably close to a straight-line pattern, and there is no other pattern that is not a straight-line pattern.

7. Not normal. The points are not reasonably close to a straight-line pattern, and there appears to be a pattern that is not a straight-line pattern.

9. Normal   11. Not normal

13. Normal

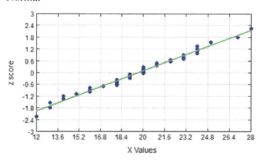

15. Not normal

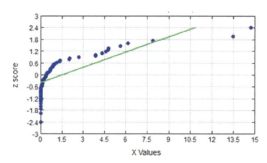

17. Normal. The points have coordinates $(32.5, -1.28)$, $(34.2, -0.52)$, $(38.5, 0)$, $(40.7, 0.52)$, $(44.3, 1.28)$.

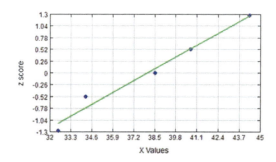

19. Not normal. The points have coordinates $(963, -1.53)$, $(1027, -0.89)$, $(1029, -0.49)$ $(1034, -0.16)$, $(1070, 0.16)$, $(1079, 0.49)$, $(1079, 0.89)$, $(1439, 1.53)$.

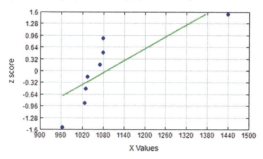

21. a. Yes   b. Yes   c. No

Section 6-6

1. a. The area below (to the left of) 502.5
   b. The area between 501.5 and 502.5
   c. The area above (to the right of) 502.5

3. $p = 0.2$; $q = 0.8$; $\mu = 20$; $\sigma = 4$. The value of $\mu = 20$ shows that for people who make random guesses for the 100 questions, the mean number of correct answers is 20. For people who make 100 random guesses, the standard deviation of $\sigma = 4$ is a measure of how much the numbers of correct responses vary.

5. 0.1102 (Table: 0.1093)

7. Normal approximation should not be used.

9. Using normal approximation: 0.2028 (Table: 0.2033); Tech using binomial: 0.2047. No, 20 is not a significantly low number of white cars.

11. Using normal approximation: 0.0549 (Table: 0.0542); Tech using binomial: 0.0513. Determination of whether 14 red cars is significantly high should be based on the probability of *14 or more* red cars, not the probability of exactly 14 red cars.

13. a. Using normal approximation: 0.0212 (Table: 0.0217); Tech using binomial: 0.0209.
   b. Using normal approximation: 0.2012 (Table: 0.2005); Tech using binomial: 0.2006). The result of 231 overturned calls is not significantly high.

15. a. Using normal approximation: 0.0114 (Table: 0.0113); Tech using binomial: 0.0113.
   b. The result of 109 is significantly low.

17. a. Using normal approximation: 0.2785 (Table: 0.2776); Tech using binomial: 0.2799.
   b. The result of 705 peas with red flowers is not significantly high.
   c. The result of 705 peas with red flowers is not strong evidence against Mendel's assumption that $3/4$ of peas will have red flowers.

19. a. Using normal approximation: 0.0000 (Table: 0.0001); Tech using binomial: 0.0000.
   b. The results suggest that the surveyed people did not respond accurately.

21. (1) 0.1723; (2) 0.1704; (3) 0.1726. No, the approximations are not off by very much.

Chapter 6: Quick Quiz

1.

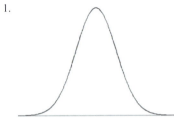

2. $z = -1.34$ 3. 0.9983   4. 0.1546 (Table: 0.1547)
5. a. $\mu = 0$ and $\sigma = 1$
   b. $\mu_{\bar{x}}$ represents the mean of all sample means, and $\sigma_{\bar{x}}$ represents the standard deviation of all sample means.
6. 0.8092 (Table: 0.8106)   7. 0.6280 (Table: 0.6292)
8. 84.6 mm Hg (Table: 84.5 mm Hg)
9. 0.9568 (Table: 0.9564)
10. The normal quantile plot suggests that diastolic blood pressure levels of women are normally distributed.

Chapter 6: Review Exercises

1. a. 0.9382   b. 0.9382   c. 0.8983
   d. -0.67 e. 0.0668
2. a. 1.13%   b. 63.8 in.
3. a. 1.42% (Table: 1.43%)   b. 59.0 in.
4. a. Normal   b. 100   c. $15/\sqrt{64} = 1.875$
5. a. An unbiased estimator is a statistic that targets the value of the population parameter in the sense that the sampling distribution of the statistic has a mean that is equal to the corresponding parameter.
   b. Mean; variance; proportion   c. True
6. a. 88.77% (Table: 88.69%). With about 11% of all men needing to bend, the design does not appear to be adequate, but the Mark VI monorail appears to be working quite well in practice.
   b. 75.1 in.
7. a. Because women are generally a little shorter than men, a doorway height that accommodates men will also accommodate women.
   b. 1, but actually a really small amount less than 1 (Table: 0.9999)
   c. Because the mean height of 60 men is less than 72 in., it does not follow that the 60 individual men all have heights less than 72 in. In determining the suitability of the door height for men, the mean of 60 heights is irrelevant, but the heights of individual men are relevant.
8. a. No. A histogram is far from bell-shaped. A normal quantile plot reveals a pattern of points that is far from a straight-line pattern.
   b. No. The sample size of $n = 13$ does not satisfy the condition of $n > 30$, and the values do not appear to be from a population having a normal distribution.
9. Using normal approximation: 0.2286 (Table: 0.2296). Tech using binomial: 0.2278. The occurrence of 787 offspring plants with long stems is not significantly low because the probability of 787 or fewer plants with long stems is not small. The results are consistent with Mendel's claimed proportion of 3/4.
10. a. 1.49% (Table: 1.50%)   b. 69.4 in.

Chapter 6: Cumulative Review Exercises

1. a. $6.02 million or $6,020,000   b. $2.80 million or $2,800,000
   c. $7.47 million or $7,470,000   d. 55.73 (million dollars)^2
   e. $z = 2.33$ f. Ratio   g. Discrete
2. a. $1.30 million, $2.80 million, $7.10 million
   b.
   c. The sample does not appear to be from a population having a normal distribution.
3. No, the distribution does not appear to be a normal distribution.

4. a. $\bar{B}$ is the event of selecting someone who does not have blue eyes.
   b. 0.65   c. 0.0429
   d. Using normal approximation: 0.1727 (Table: 0.1736); Tech using binomial: 0.1724.
   e. No
5. a. 0.7881   b. 0.9968 (Table: 0.9967)   c. 10.4 in.   d. 0.0228

Chapter 7 Answers

Section 7-1

1. The confidence level, such as 95%, was not provided.
3. $\hat{p} = 0.14$ is the sample proportion; $\hat{q} = 0.86$ (found from evaluating $1 - \hat{p}$); $n = 1000$ is the sample size; $E = 0.04$ is the margin of error; p is the population proportion, which is unknown. The value of α is 0.05.
5. 1.645   7. 2.81
9. 0.130 ± 0.087 11. $0.0169 < p < 0.143$
13. a. $\hat{p} = 0.0912$ b. $E = 0.0297$
    c. $0.0615 < p < 0.121$
    d. We have 95% confidence that the interval from 0.0615 to 0.121 actually does contain the true value of the population proportion of McDonald's drive-through orders that are not accurate.
15. a. $\hat{p} = 0.143$
    b. $E = 0.00815$
    c. $0.135 < p < 0.152$
    d. We have 90% confidence that the interval from 0.135 to 0.152 actually does contain the true value of the population proportion of returned surveys.
17. $0.462 < p < 0.529$. Because 0.512 is contained within the confidence interval, there is not strong evidence against 0.512 as the value of the proportion of boys in all births.
19. a. $11.6\% < p < 22.4\%$
    b. Because the two confidence intervals overlap, it is possible that Burger King and Wendy's have the same rate of orders that are not accurate. Neither restaurant appears to have a significantly better rate of accuracy of orders.
21. a. 0.5   b. 0.439   c. $0.363 < p < 0.516$
    d. If the touch therapists really had an ability to select the correct hand by sensing an energy field, their success rate would be

significantly greater than 0.5, but the sample success rate of 0.439 and the confidence interval suggest that they do not have the ability to select the correct hand by sensing an energy field.

23. a. 0.0276% < p < 0.0366%

(using x = 135: 0.0276% < p < 0.0367%).

b. No, because 0.0340% is included in the confidence interval.

25. Placebo group: 0.697% < p < 4.49%. Treatment group: 0.288% < p < 1.57%. Because the two confidence intervals overlap, there does not appear to be a significant difference between the rates of allergic reactions. Allergic reactions do not appear to be a concern for Lipitor users.

27. Sustained care: 77.6% < p < 88.1% (using x = 164). Standard care: 56.1% < p < 69.5% (using x = 125). The two confidence intervals do not overlap. It appears that the success rate is higher with sustained care.

29. $\hat{p}$ = 18/34, or 0.529. CI: 36.2% < p < 69.7%. Greater height does not appear to be an advantage for presidential candidates. If greater height is an advantage, then taller candidates should win substantially more than 50% of the elections, but the confidence interval shows that the percentage of elections won by taller candidates is likely to be anywhere between 36.2% and 69.7%.

31. a. 1844 (Table: 1842) b. 664 c. They don't change.

33. a. 385 b. 369

c. No, the sample size doesn't change much.

35. a. 1537 b. 1449

37. a. 752 b. 405

c. No. A sample of the people you know is a convenience sample, not a simple random sample, so it is very possible that the results would not be representative of the population.

39. 1238 (Table: 1237)

41. a. The requirement of at least 5 successes and at least 5 failures is not satisfied, so the normal distribution cannot be used.

b. 0.075

Section 7-2

1. a. 13.05 Mbps < μ < 22.15 Mbps

b. Best point estimate of μ is 17.60 Mbps. The margin of error is E = 4.55 Mbps.

c. Because the sample size of 50 is greater than 30, we can consider the sample mean to be from a population with a normal distribution.

3. We have 95% confidence that the limits of 13.05 Mbps and 22.15 Mbps contain the true value of the mean of the population of all Verizon data speeds at the airports.

5. Neither the normal nor the t distribution applies.

7. $z_{\alpha/2}$ = 2.576 (Table: 2.575)

9. 29.4 hg < μ < 31.4 hg. No, the results do not differ by much.

11. 98.08°F < μ < 98.32°F. Because the confidence interval does not contain 98.6°F, it appears that the mean body temperature is not 98.6°F, as is commonly believed.

13. 71.4 min < μ < 126.4 min. The confidence interval includes the mean of 102.8 min that was measured before the treatment, so the mean could be the same after the treatment. This result suggests that the zopiclone treatment does not have a significant effect.

15. 1.8 < μ < 3.4. The given numbers are just substitutes for the four DNA base names, so the numbers don't measure or count

anything, and they are at the nominal level of measurement. The confidence interval has no practical use.

17. 5.0 < μ < 9.0. The results tell us nothing about the population of adult females.

19. 0.284 ppm < μ < 1.153 ppm. Using the FDA guideline, the confidence interval suggests that there could be too much mercury in fish because it is possible that the mean is greater than 1 ppm. Also, one of the sample values exceeds the FDA guideline of 1 ppm, so at least some of the fish have too much mercury.

21. 143.3 million dollars < μ < 200.7 million dollars. Because the amounts are from the ten wealthiest celebrities, the confidence interval doesn't tell us anything about the population of all celebrities. The data do not appear to be from a normally distributed population, so the confidence interval might not be a good estimate of the population mean.

23. 3.67 < μ < 4.17. Because all of the students were at the University of Texas at Austin, the confidence interval doesn't tell us anything about the population of college students in Texas.

25. Females: 72.0 bpm < μ < 76.1 bpm. Males: 67.8 bpm < μ < 71.4 bpm. Adult females appear to have a mean pulse rate that is higher than the mean pulse rate of adult males. Good to know.

27. McDonald's: 161.4 sec < μ < 197.2 sec. Burger King: 139.1 sec < μ < 167.5 sec (Table: 139.2 sec < μ < 167.4 sec). There does not appear to be a significant difference between the mean dinner service times at McDonald's and Burger King.

29. The sample size is 94, and it does appear to be very practical.

31. The required sample size is 38,415 (Table: 38,416). The sample appears to be too large to be practical.

33. 4814 (Table: 4815). Yes, the assumption seems reasonable.

35. a. 425 b. 212

c. The result from part (a) is substantially larger than the result from part (b). The result from part (b) is likely to be better because it uses s instead of the estimated σ obtained from the range rule of thumb.

37. 29.4 hg < μ < 31.4 hg

39. 0.8462 g < μ < 0.8668 g; 0.8474 g < μ < 0.8656 g; the second confidence interval is narrower, indicating that we have a more accurate estimate when the relatively large sample is selected without replacement from a relatively small finite population.

Section 7-3

1. 95.0 cm³ < σ < 182.5 cm³. We have 95% confidence that the limits of 95.0 cm³ and 182.5 cm³ contain the true value of the standard deviation of brain volumes.

3. The dotplot does not appear to depict sample data from a normally distributed population. The large sample size does not justify treating the values as being from a normally distributed population. Because the normality requirement is not satisfied, the confidence interval estimate of σ should not be constructed using the methods of this section.

5. df = 24. X_L^2 = 12.401 and X_R^2 = 39.364.

CI: 0.19 mg < σ < 0.33 mg.

7. df = 146. X_L^2 = 105.741 (Table: 67.328) and X_R^2 = 193.761 (Table: 140.169). CI: 56.8 < σ < 76.8 (Table: 66.7 < σ < 96.3).

9. 0.55°F < σ < 0.72°F (Table: 0.56°F < σ < 0.74°F)

11. 29.6 min $< \sigma <$ 71.6 min. No, the confidence interval does not indicate whether the treatment is effective.

13. $1.6 < \sigma < 3.8$

15. 2.9 mi/h $< \sigma <$ 6.9 mi/h. Because traffic conditions vary considerably at different times during the day, the confidence interval is an estimate of the standard deviation of the population of speeds at 3:30 on a weekday, not other times.

17. a. Course evaluations: $0.46 < \sigma < 0.62$
    (Table: $0.47 < \sigma < 0.62$)
    b. Professor evaluations: $0.49 < \sigma < 0.66$
    c. The amounts of variation are about the same.

19. 19,205 is too large. There aren't 19,205 statistics professors in the population, and even if there were, that sample size is too large to be practical.

21. The sample size is 48. No, with many very low incomes and a few high incomes, the distribution is likely to be skewed to the right and will not satisfy the requirement of a normal distribution.

23. Using $z_{\alpha/2} = 2.575829303$: $\chi_L^2 = 109.980$ and $\chi_R^2 = 199.655$ (Table: $\chi_L^2 = 109.993$ and $\chi_R^2 = 199.638$). The values from the given approximation are quite close to the actual critical values.

Section 7-4

1. Without replacement, every sample would be identical to the original sample, so the proportions or means or standard deviations or variances would all be the same, and there would be no confidence "interval."

3. Parts b, d, e are not possible bootstrap samples.

5. $0.000 < p < 0.500$

7. a. 0.1 kg $< \mu <$ 8.6 kg   b. 1.9 kg $< \sigma <$ 6.3 kg

9. Answers vary, but here are typical answers.
    a. -0.8 kg $< \mu <$ 7.8 kg
    b. 1.2 kg $< \sigma <$ 7.0 kg

11. Answers vary, but here are typical answers.
    a. $5.3 < \mu < 8.5$. This isn't dramatically different from $5.0 < \mu < 9.0$.
    b. $1.2 < \sigma < 2.9$. This isn't dramatically different from $1.6 < \sigma < 3.8$.

13. Answers vary, but here is a typical result: $0.0608 < p < 0.123$. This is quite close to the confidence interval of $0.0615 < p < 0.121$ found in Exercise 13 from Section 7-1.

15. Answers vary, but here is a typical result: $0.135 < p < 0.152$. The result is essentially the same as the confidence interval of $0.135 < p < 0.152$ found in Exercise 15 from Section 7-1.

17. Answers vary, but here is a typical result: $3.69 < \mu < 4.15$. This result is very close to the confidence interval $3.67 < \mu < 4.17$ found in Exercise 23 in Section 7-2.

19. a. Answers vary, but here is a typical result: 233.6 sec $< \mu <$ 245.1 sec.
    b. 234.4 sec $< \mu <$ 246.0 sec
    c. The result from the bootstrap method is reasonably close to the result found using the methods of Section 7-2.

21. a. Answers vary, but here is a typical result: $2.5 < \sigma < 3.3$.
    b. $2.4 < \sigma < 3.7$
    c. The confidence interval from the bootstrap method is not very different from the confidence interval found using the methods of Section 7-3. Because a histogram or normal quantile plot shows that the sample appears to be from a population not having a normal distribution, the bootstrap confidence interval of $2.5 < \sigma < 3.3$ would be a better estimate of σ.

23. Answers vary, but here is a typical result using 10,000 bootstrap samples: $2.5 < \sigma < 3.3$. This result is very close to the confidence interval of $2.4 < \sigma < 3.3$ found using 1000 bootstrap samples. In this case, increasing the number of bootstrap samples from 1000 to 10,000 does not have much of an effect on the confidence interval.

Chapter 7: Quick Quiz

1. 0.720

2. We have 95% confidence that the limits of 0.692 and 0.748 contain the true value of the proportion of adults in the population who say that the law goes easy on celebrities.

3. $z = 2.576$ (Table: 2.575)

4. $36.9\% < p < 43.1\%$

5. 601   6. 136

7. There is a loose requirement that the sample values are from a normally distributed population.

8. The degrees of freedom is the number of sample values that can vary after restrictions have been imposed on all of the values. For the sample data described in Exercise 7, df $= 11$.

9. $t = 2.201$

10. No, the use of the χ^2 distribution has a fairly strict requirement that the data must be from a normal distribution. The bootstrap method could be used to find a 95% confidence interval estimate of σ.

Chapter 7: Review Exercises

1. $37.9\% < p < 42.1\%$. Because we have 95% confidence that the limits of 37.9% and 42.1% contain the true percentage for the population of adults, we can safely say that fewer than 50% of adults prefer to get their news online.

2. 423

3. a. 2.926
    b. $2.749 < \mu < 3.102$
    c. We have 95% confidence that the limits of 2.749 and 3.102 contain the value of the population mean μ.

4. 94

5. a. Student t distribution
    b. Normal distribution
    c. None of the three distributions is appropriate, but a confidence interval could be constructed by using bootstrap methods.
    d. χ^2 (chi-square distribution)
    e. Normal distribution

6. a. 1068
    b. 340
    c. 1068

7. -22.1 sec $< \mu <$ 308.1 sec

8. 184.0 sec $< \mu <$ 441.1 sec

9. Answers vary, but here is a typical result: 7.1 sec $< \mu <$ 293.7 sec.

10. a. $0.0113 < p < 0.0287$
    b. Answers vary, but here is a typical result: $0.0120 < p < 0.0290$.
    c. The confidence intervals are quite close.

Chapter 7: Cumulative Review Exercises

1. $\bar{x} = -3.6$ min, median $= -20.0$ min, $s = 39.9$ min, range $= 149.0$ min. These results are statistics.

2. Significantly low values are -83.4 min or lower, and significantly high values are 76.2 min or higher. Because 103 min exceeds 76.2 min, the arrival delay time of 103 min is significantly high.

3. Ratio level of measurement; continuous data.

4. -28.9 min $< \mu < 21.7$ min

5. a. 0.2553 (Table: 0.2546)
   b. 15.5 min (Table: 15.4 min)

6. 143 flights

7. $77.1\% < p < 83.5\%$

8. The graphs suggest that the population has a distribution that is skewed to the right instead of being normal. The histogram shows that some taxi-out times can be very long, and can occur with heavy traffic, but little or no traffic cannot make the taxi-out time very low. There is a minimum time required, regardless of traffic conditions. Construction of a confidence interval estimate of a population standard deviation has a fairly strict requirement that the sample data are from a normally distributed population, and the graphs show that this strict normality requirement is not satisfied.

Chapter 8 Answers

Section 8-1

1. Rejection of the claim about aspirin is more serious because it is a drug used for medical treatments. The wrong aspirin dosage could cause more serious adverse reactions than a wrong vitamin C dosage. It would be wise to use a smaller significance level for testing the claim about the aspirin.

3. a. $H_0: \mu = 174.1$ cm   b. $H_1: \mu \neq 174.1$ cm
   c. Reject the null hypothesis or fail to reject the null hypothesis.
   d. No. In this case, the original claim becomes the null hypothesis. For the claim that the mean height of men is equal to 174.1 cm, we can either reject that claim or fail to reject it, but we cannot state that there is sufficient evidence to *support* that claim.

5. a. $p > 0.5$ b. $H_0: p = 0.5; H_1: p > 0.5$

7. a. $\mu = 69$ bpm   b. $H_0: \mu = 69$ bpm; $H_1: \mu \neq 69$ bpm

9. There is sufficient evidence to support the claim that most adults would erase all of their personal information online if they could.

11. There is not sufficient evidence to warrant rejection of the claim that the mean pulse rate (in beats per minute) of adult males is 69 bpm.

13. $z = 4.28$ (or $z = 4.25$ if using $x = 333$)

15. $t = 0.657$

17. a. Right-tailed   b. P-value $= 0.1587$ c. Fail to reject H_0

19. a. Two-tailed   b. P-value $= 0.0444$ c. Reject H_0.

21. a. $z = 1.645$ b. Fail to reject H_0

23. a $z = \pm 1.96$ b. Reject H_0

25. a. Fail to reject H_0
    b. There is not sufficient evidence to support the claim that more than 58% of adults would erase all of their personal information online if they could.

27. a. Reject H_0
    b. There is sufficient evidence to warrant rejection of the claim that the mean pulse rate (in beats per minute) of adult males is 72 bpm.

29. Type I error: In reality $p = 0.1$, but we reject the claim that $p = 0.1$. Type II error: In reality $p \neq 0.1$, but we fail to reject the claim that $p = 0.1$.

31. Type I error: In reality $p = 0.87$, but we support the claim that $p > 0.87$. Type II error: In reality $p > 0.87$, but we fail to support that conclusion.

33. The power of 0.96 shows that there is a 96% chance of rejecting the null hypothesis of $p = 0.08$ when the true proportion is actually 0.18. That is, if the proportion of Chantix users who experience abdominal pain is actually 0.18, then there is a 96% chance of supporting the claim that the proportion of Chantix users who experience abdominal pain is greater than 0.08.

35. 617

Section 8-2

1. a. 270   b. $\hat{p} = 0.53$

3. The method based on a confidence interval is not equivalent to the P-value method and the critical value method.

5. a. Left-tailed   b. $z = -4.46$ c. P-value: 0.000004
   d. $H_0: p = 0.10$. Reject the null hypothesis.
   e. There is sufficient evidence to support the claim that fewer than 10% of treated subjects experience headaches.

7. a. Two-tailed   b. $z = -1.69$ c. P-value: 0.091
   d. $H_0: p = 0.92$. Fail to reject the null hypothesis.
   e. There is not sufficient evidence to warrant rejection of the claim that 92% of adults own cell phones.

9. $H_0: p = 0.10$. $H_1: p \neq 0.10$. Test statistic: $z = -0.56$. P-value $= 0.5751$ (Table: 0.5754). Critical values: $z = \pm 1.96$. Fail to reject H_0. There is not sufficient evidence to warrant rejection of the claim that the rate of inaccurate orders is equal to 10%. With 10% of the orders being inaccurate, it appears that McDonald's should work to lower that rate.

11. $H_0: p = 0.5$. $H_1: p \neq 0.5$. Test statistic: $z = 2.69$. P-value $= 0.0071$ (Table: 0.0072). Critical values: $z = \pm 2.576$ (Table: ± 2.575). Reject H_0. There is sufficient evidence to reject the claim that the proportion of those in favor is equal to 0.5. The result suggests that the politician is wrong in claiming that the responses are random guesses equivalent to a coin toss.

13. $H_0: p = 0.20$. $H_1: p > 0.20$. Test statistic: $z = 1.10$. P-value $= 0.1367$ (Table: 0.1357). Critical value: $z = 1.645$. Fail to reject H_0. There is not sufficient evidence to support the claim that more than 20% of OxyContin users develop nausea. However, with $\hat{p} = 0.229$, we see that a large percentage of OxyContin users experience nausea, so that rate does appear to be very high.

15. $H_0: p = 0.15$. $H_1: p < 0.15$. Test statistic: $z = -1.31$. P-value $= 0.0956$ (Table: 0.0951). Critical value: $z = -2.33$. Fail to reject H_0. There is not sufficient evidence to support the claim that the return rate is less than 15%.

17. $H_0: p = 0.512$. $H_1: p \neq 0.512$. Test statistic: $z = -0.98$. P-value $= 0.3286$ (Table: 0.3270). Critical values: $z = \pm 1.96$. Fail to reject H_0. There is not sufficient evidence to warrant rejection of the claim that 51.2% of newborn babies are boys. The results do not *support* the belief that 51.2% of newborn babies are boys; the results merely show that there is not strong evidence against the rate of 51.2%.

19. H_0: $p = 0.80$. H_1: $p < 0.80$. Test statistic: $z = -1.11$.
P-value $= 0.1332$ (Table: 0.1335). Critical value: $z = -1.645$. Fail to reject H_0. There is not sufficient evidence to support the claim that the polygraph results are correct less than 80% of the time. However, based on the sample proportion of correct results in 75.5% of the 98 cases, polygraph results do not appear to have the high degree of reliability that would justify the use of polygraph results in court, so polygraph test results should be prohibited as evidence in trials.

21. H_0: $p = 0.5$. H_1: $p \neq 0.5$. Test statistic: $z = -2.03$.
P-value $= 0.0422$ (Table: 0.0424). Critical values: $z = \pm 1.645$. Reject H_0. There is sufficient evidence to warrant rejection of the claim that touch therapists use a method equivalent to random guesses. However, their success rate of 123/280, or 43.9%, indicates that they performed *worse* than random guesses, so they do not appear to be effective.

23. H_0: $p = 0.000340$. H_1: $p \neq 0.000340$. Test statistic: $z = -0.66$.
P-value $= 0.5122$ (Table: 0.5092). Critical values: $z = \pm 2.81$. Fail to reject H_0. There is not sufficient evidence to support the claim that the rate is different from 0.0340%. Cell phone users should not be concerned about cancer of the brain or nervous system.

25. H_0: $p = 0.5$. H_1: $p > 0.5$. Test statistic: $z = 1.00$.
P-value: 0.1587. Critical value: $z = 1.645$. Fail to reject H_0. There is not sufficient evidence to support the claim that the probability of an NFC team Super Bowl win is greater than one-half.

27. H_0: $p = 0.5$. H_1: $p \neq 0.5$. Test statistic: $z = 2.05$.
P-value $= 0.0402$ (Table: 0.0404). Critical values: $z = \pm 1.96$. Reject H_0. There is sufficient evidence to warrant rejection of the claim that the coin toss is fair in the sense that neither team has an advantage by winning it. The coin toss rule does not appear to be fair. This helps explain why the overtime rules were changed.

29. H_0: $p = 0.5$. H_1: $p > 0.5$. Test statistic: $z = 40.91$ (using $\hat{p} = 0.64$) or $z = 40.90$ (using $x = 13,661$). P-value $= 0.0000$ (Table: 0.0001). Critical value: $z = 2.33$. Reject H_0. There is sufficient evidence to support the claim that most people believe that the Loch Ness monster exists. Because the sample is a voluntary-response sample, the conclusion about the population might not be valid.

31. H_0: $p = 0.791$. H_1: $p < 0.791$. Test statistic: $z = -29.09$ (using $\hat{p} = 0.39$) or $z = -29.11$ (using $x = 339$). P-value $= 0.0000$ (Table: 0.0001). Critical value: $z = -2.33$. Reject H_0. There is sufficient evidence to support the claim that the percentage of selected Americans of Mexican ancestry is less than 79.1%, so the jury selection process appears to be biased.

33. The P-values agree reasonably well with the large sample size of $n = 1009$. The normal approximation to the binomial distribution works better as the sample size increases. Normal approximation entries: 0.0114, 0.0012, 0.0054. Exact entries: 0.0215, 0.0034, 0.0059. Exact with simple continuity correction: 0.0117, 0.0018, 0.0054.

35. a. 0.7219 (Table: 0.7224)   b. 0.2781 (Table: 0.2776)
c. The power of 0.7219 shows that there is a reasonably good chance of making the correct decision of rejecting the false null hypothesis. It would be better if the power were even higher, such as greater than 0.8 or 0.9.

Section 8-3

1. The requirements are (1) the sample must be a simple random sample, and (2) either or both of these conditions must be satisfied: The population is normally distributed or $n > 30$. There is not enough information given to determine whether the sample is a simple random sample. Because the sample size is not greater than 30, we must check for normality, but the value of 583 sec appears to be an outlier, and a normal quantile plot or histogram suggests that the sample does not appear to be from a normally distributed population. The requirements are not satisfied.

3. A t test is a hypothesis test that uses the Student t distribution, such as the method of testing a claim about a population mean as presented in this section. The letter t is used in reference to the Student t distribution, which is used in a t test. The z test methods require a known value of σ, but it would be very rare to conduct a hypothesis test for a claim about an unknown value of μ while we somehow know the value of σ.

5. P-value $= 0.1301$ (Table: $0.10 < P$-value < 0.20)

7. P-value $= 0.2379$ (Table: P-value > 0.20)

9. H_0: $\mu = 4.00$ Mbps. H_1: $\mu < 4.00$ Mbps. Test statistic: $t = -0.366$. P-value $= 0.3579$. Critical value assuming a 0.05 significance level: $t = -1.677$ (Table: -1.676 approximately). Fail to reject H_0. There is not sufficient evidence to support the claim that the Sprint airport data speeds are from a population having a mean less than 4.00 Mbps.

11. H_0: $\mu = 0$ min. H_1: $\mu \neq 0$ min. Test statistic: $t = -8.720$. P-value: 0.0000. Critical values: $t = \pm 1.970$. Reject H_0. There is sufficient evidence to warrant rejection of the claim that the mean prediction error is equal to zero. The predictions do not appear to be very accurate.

13. H_0: $\mu = 4.00$. H_1: $\mu \neq 4.00$. Test statistic: $t = -1.638$. P-value $= 0.1049$ (Table: >0.10). Critical values: $t = \pm 1.986$ (Table: ± 1.987 approximately). Fail to reject H_0. There is not sufficient evidence to warrant rejection of the claim that the population of student course evaluations has a mean equal to 4.00.

15. H_0: $\mu = 0$. H_1: $\mu > 0$. Test statistic: $t = 0.133$. P-value $= 0.4472$ (Table: >0.10). Critical value: $t = 1.677$ (Table: 1.676 approximately). Fail to reject H_0. There is not sufficient evidence to support the claim that with garlic treatment, the mean change in cholesterol is greater than 0. There is not sufficient evidence to support a claim that the garlic treatment is not effective in reducing LDL cholesterol levels.

17. H_0: $\mu = 0$ lb. H_1: $\mu > 0$ lb. Test statistic: $t = 3.872$. P-value $= 0.0002$ (Table: <0.005). Critical value: $t = 2.426$. Reject H_0. There is sufficient evidence to support the claim that the mean weight loss is greater than 0. Although the diet appears to have statistical significance, it does not appear to have practical significance, because the mean weight loss of only 3.0 lb does not seem to be worth the effort and cost.

19. H_0: $\mu = 12.00$ oz. H_1: $\mu \neq 12.00$ oz. Test statistic: $t = 10.364$. P-value $= 0.0000$ (Table: <0.01). Critical values: $t = \pm 2.030$. Reject H_0. There is sufficient evidence to warrant rejection of the claim that the mean volume is equal to 12.00 oz. Because the mean appears to be greater than 12.00 oz, consumers are not being cheated because they are getting slightly more than 12.00 oz.

21. The sample data meet the loose requirement of having a normal distribution. $H_0: \mu = 14\ \mu g/g$. $H_1: \mu < 14\ \mu g/g$. Test statistic: $t = -1.444$. P-value = 0.0913 (Table: >0.05). Critical value: $t = -1.833$. Fail to reject H_0. There is not sufficient evidence to support the claim that the mean lead concentration for all such medicines is less than $14\ \mu g/g$.

23. $H_0: \mu = 1000$ hic. $H_1: \mu < 1000$ hic. Test statistic: $t = -2.661$. P-value $= 0.0224$ (Table: P-value is between 0.01 and 0.025). Critical value: $t = -3.365$. Fail to reject H_0. There is not sufficient evidence to support the claim that the population mean is less than 1000 hic. There is not strong evidence that the mean is less than 1000 hic, and one of the booster seats has a measurement of 1210 hic, which does not satisfy the specified requirement of being less than 1000 hic.

25. $H_0: \mu = 75$ bpm. $H_1: \mu < 75$ bpm. Test statistic: $t = -0.927$. P-value $= 0.1777$ (Table: >0.10). Critical value: $t = -1.655$ (Table: -1.660 approximately). Fail to reject H_0. There is not sufficient evidence to support the claim that the mean pulse rate of adult females is less than 75 bpm.

27. $H_0: \mu = 90$ mm Hg. $H_1: \mu < 90$ mm Hg. Test statistic: $t = -21.435$. P-value $= 0.0000$ (Table: <0.005). Critical value: $t = -1.655$ (Table: -1.660 approximately). Reject H_0. There is sufficient evidence to support the claim that the adult female population has a mean diastolic blood pressure level less than 90 mm Hg. The conclusion addresses the mean of a population, not individuals, so we cannot conclude that there are no female adults in the sample with hypertension.

29. Test statistic: $z = -1.64$. P-value $= 0.1015$ (Table: 0.1010). Critical value: $z = -1.645$. The null and alternative hypotheses are the same and the conclusions are the same. Results are not affected very much by the knowledge of σ.

31. The computed critical t score is 1.648, which is the same as the value of 1.648 found from technology. The approximation appears to work quite well.

Section 8-4

1. The sample must be a simple random sample and the sample must be from a normally distributed population. The normality requirement for a hypothesis test of a claim about a standard deviation is much more strict, meaning that the distribution of the population must be much closer to a normal distribution.

3. a. Reject H_0.
   b. Reject the claim that the new filling process results in volumes with the same standard deviation of 0.115 oz.
   c. It appears that with the new filling process, the variation among volumes has increased, so the volumes are not as consistent. The new filling process appears to be inferior to the original filling process.

5. $H_0: \sigma = 10$ bpm. $H_1: \sigma \ne 10$ bpm. Test statistic: $X^2 = 195.172$. P-value $= 0.0208$. Reject H_0. There is sufficient evidence to warrant rejection of the claim that pulse rates of men have a standard deviation equal to 10 beats per minute. Using the range rule of thumb with the normal range of 60 to 100 beats per minute is not very good for estimating σ in this case.

7. $H_0: \sigma = 2.08°F$. $H_1: \sigma < 2.08°F$. Test statistic: $X^2 = 9.329$. P-value $= 0.0000$ (Table: <0.005). Critical value: $X^2 = 74.252$ (Table: 70.065 approximately). Reject H_0. There is sufficient

evidence to support the claim that body temperatures have a standard deviation less than 2.08°F. It is very highly unlikely that the conclusion in the hypothesis test in Example 5 from Section 8-3 would change because of a standard deviation from a different sample.

9. $H_0: \sigma = 27.8$ lb. $H_1: \sigma \ne 27.8$ lb. Test statistic: $X^2 = 8.505$. P-value $= 0.0384$ (Table: <0.05). Critical values of X^2: 8.907, 32.852. Reject H_0. There is sufficient evidence to warrant rejection of the claim that cans with thickness 0.0109 in. have axial loads with the same standard deviation as the axial loads of cans that are 0.0111 in. thick. The thickness of the cans does appear to affect the variation of the axial loads.

11. $H_0: \sigma = 0.15$ oz. $H_1: \sigma > 0.15$ oz. Test statistic: $X^2 = 33.396$. P-value $= 0.1509$ (Table: >0.10). Critical value: $X^2 = 38.885$. Fail to reject H_0. There is not sufficient evidence to support the claim that the machine dispenses amounts with a standard deviation greater than the standard deviation of 0.15 oz specified in the machine design.

13. $H_0: \sigma = 32.2$ ft. $H_1: \sigma > 32.2$ ft. Test statistic: $X^2 = 29.176$. P-value: 0.0021. Critical value: $X^2 = 19.675$. Reject H_0. There is sufficient evidence to support the claim that the new production method has errors with a standard deviation greater than 32.2 ft. The variation appears to be greater than in the past, so the new method appears to be worse because there will be more altimeters that have larger errors. The company should take immediate action to reduce the variation.

15. $H_0: \sigma = 55.93$ sec. $H_1: \sigma \ne 55.93$ sec. Test statistic: $X^2 = 20.726$. P-value $= 0.0084$ (Table: <0.01). Critical values of X^2: 0.989, 20.278. Reject H_0. There is sufficient evidence to warrant rejection of the claim that service times at McDonald's have the same variation as service times at Wendy's. Drive-through service times during dinner times appear to vary more at McDonald's than those at Wendy's. Given the similar composition of the menus, McDonald's should consider methods for reducing the variation.

17. $H_0: \sigma = 55.93$ sec. $H_1: \sigma \ne 55.93$ sec. Test statistic: $X^2 = 62.049$. P-value $= 0.1996$ (Table: >0.10). Critical values of X^2: 27.249, 78.231 (Table: 27.991, 79.490 approximately). Fail to reject H_0. There is not sufficient evidence to warrant rejection of the claim that service times at McDonald's have the same variation as service times at Wendy's. Drive-through service times during dinner times appear to have about the same variation at McDonald's and Wendy's. No action is warranted.

19. Critical $X^2 = 81.540$ (or 81.494 if using $z = 2.326348$ found from technology), which is close to the value of 82.292 obtained from Statdisk and Minitab.

Chapter 8: Quick Quiz

1. a. t distribution　　b. Normal distribution
   c. Chi-square distribution
2. a. Two-tailed　　b. Left-tailed　　c. Right-tailed
3. a. $H_0: p = 0.5$. $H_1: p > 0.5$.　　b. $z = 1.39$
   c. Fail to reject H_0
   d. There is not sufficient evidence to support the claim that the majority of Internet users aged 18–29 use Instagram.
4. 0.10　　5. True　　6. False　　7. False
8. No. All critical values of X^2 are always positive.
9. The t test requires that the sample is from a normally distributed population, and the test is robust in the sense that the test works

reasonably well if the departure from normality is not too extreme. The χ^2 (chi-square) test is not robust against a departure from normality, meaning that the test does not work well if the population has a distribution that is far from normal.

10. The only true statement is the one given in part (a).

Chapter 8: Review Exercises

1. a. False   b. True   c. False   d. False   e. False
2. H_0: $p = 0.5$. H_1: $p > 0.5$. Test statistic: $z = 6.09$. P-value $= 0.0000$ (Table: 0.0001). Critical value: $z = 2.33$. Reject H_0. There is sufficient evidence to support the claim that the ballot selection method favors Democrats.
3. H_0: $\mu = 30$ years. H_1: $\mu > 30$ years. Test statistic: $t = 5.029$. P-value $= 0.0000$ (Table: <0.005). Critical value: $t = 2.370$ (Table: 2.368 approximately). Reject H_0. There is sufficient evidence to support the claim that the mean age of actresses when they win Oscars is greater than 30 years.
4. H_0: $\mu = 5.4$ million cells per microliter. H_1: $\mu < 5.4$ million cells per microliter. Test statistic: $t = -5.873$. P-value $= 0.0000$ (Table: <0.005). Critical value: $t = -2.426$. Reject H_0. There is sufficient evidence to support the claim that the sample is from a population with a mean less than 5.4 million cells per microliter. The test deals with the distribution of sample means, not individual values, so the result does not suggest that each of the 40 males has a red blood cell count below 5.4 million cells per microliter.
5. H_0: $p = 0.43$. H_1: $p \neq 0.43$. Test statistic: $z = 3.70$. P-value: 0.0002. Critical values: $z = \pm 1.96$. Reject H_0. There is sufficient evidence to warrant rejection of the claim that the percentage who believe that they voted for the winning candidate is equal to 43%. There appears to be a substantial discrepancy between how people said that they voted and how they actually did vote.
6. H_0: $\mu = 20.16$. H_1: $\mu < 20.16$. Test statistic: $t = -3.732$. P-value $= 0.0023$ (Table: <0.005). Critical value: $t = -2.821$. Reject H_0. There is sufficient evidence to support the claim that the population of recent winners has a mean BMI less than 20.16. Recent winners appear to be significantly smaller than those from the 1920s and 1930s.
7. H_0: $\sigma = 1.34$. H_1: $\sigma \neq 1.34$. Test statistic: $\chi^2 = 7.053$. P-value $= 0.7368$ (Table: >0.20). Critical values of χ^2: 1.735, 23.589. Fail to reject H_0. There is not sufficient evidence to support the claim that the recent winners have BMI values with variation different from that of the 1920s and 1930s.
8. a. A type I error is the mistake of rejecting a null hypothesis when it is actually true. A type II error is the mistake of failing to reject a null hypothesis when in reality it is false.
   b. Type I error: In reality, the mean BMI is equal to 20.16, but we support the claim that the mean BMI is less than 20.16. Type II error: In reality, the mean BMI is less than 20.16, but we fail to support that claim.

Chapter 8: Cumulative Review Exercises

1. a. 37.1 deaths   b. 36.0 deaths   c. 9.8 deaths
   d. 96.8 deaths^2 e. 28.0 deaths
   f. The pattern of the data over time is not revealed by the statistics. A time-series graph would be very helpful in understanding the pattern over time.

2. a. Ratio   b. Discrete   c. Quantitative
   d. No. The data are from recent and consecutive years, so they are not randomly selected.
3. 29.1 deaths $< \mu < 45.0$ deaths. We have 99% confidence that the limits of 29.1 deaths and 45.0 deaths contain the value of the population mean.
4. H_0: $\mu = 72.6$ deaths. H_1: $\mu < 72.6$ deaths. Test statistic: $t = -13.509$. P-value $= 0.0000$ (Table: <0.005). Critical value: $t = -2.650$. Reject H_0. There is sufficient evidence to support the claim that the mean number of annual lightning deaths is now less than the mean of 72.6 deaths from the 1980s. Possible factors: Shift in population from rural to urban areas; better lightning protection and grounding in electric and cable and phone lines; better medical treatment of people struck by lightning; fewer people use phones attached to cords; better weather predictions.
5. Because the vertical scale starts at 50 and not at 0, the difference between the number of males and the number of females is exaggerated, so the graph is deceptive by creating the false impression that males account for nearly all lightning strike deaths. A comparison of the numbers of deaths shows that the number of male deaths is roughly 4 times the number of female deaths, but the graph makes it appear that the number of male deaths is around 25 times the number of female deaths.
6. H_0: $p = 0.5$. H_1: $p > 0.5$. Test statistic: $z = 10.45$. P-value $= 0.0000$ (Table: 0.0001). Critical value: $z = 2.33$. Reject H_0. There is sufficient evidence to support the claim that the proportion of male deaths is greater than $1/2$. More males are involved in certain outdoor activities such as construction, fishing, and golf.
7. $0.763 < p < 0.854$. Because the entire confidence interval is greater than 0.5, it does not seem feasible that males and females have equal chances of being killed by lightning.
8. a. 0.512   b. 0.008   c. 0.992   d. 0.205
   e. $\mu = 40.0$ males; $\sigma = 2.8$ males
   f. Yes. Using the range rule of thumb, significantly high values are $\mu + 2\sigma$ or greater. With $\mu + 2\sigma = 45.6$, values above 45.6 are significantly high, so 46 would be a significantly high number of male victims in a group of 50.

Chapter 9 Answers

Section 9-1

1. The samples are simple random samples that are independent. For each of the two groups, the number of successes is at least 5 and the number of failures is at least 5. (Depending on what we call a success, the four numbers are 33, 115, 201,196, and 200,630 and all of those numbers are at least 5.) The requirements are satisfied.
3. a. H_0: $p_1 = p_2$. H_1: $p_1 < p_2$.
   b. There is sufficient evidence to support the claim that the rate of polio is less for children given the Salk vaccine than for children given a placebo. The Salk vaccine appears to be effective.
5. H_0: $p_1 = p_2$. H_1: $p_1 > p_2$. Test statistic: $z = 12.82$. P-value: 0.0000. Critical value: $z = 2.33$. Reject H_0. There is sufficient evidence to support the claim that vinyl gloves have a greater virus leak rate than latex gloves.

7. a. H_0: $p_1 = p_2$. H_1: $p_1 > p_2$. Test statistic: $z = -0.95$. P-value $=$ 0.8280 (Table: 0.8289). Critical value: $z = 1.645$. Fail to reject H_0. There is not sufficient evidence to support the claim that car owners violate license plate laws at a higher rate than owners of commercial trucks.

   b. 90% CI: $-0.0510 < p_1 - p_2 < 0.0148$. Because the confidence interval limits contain 0, there is not a significant difference between the two proportions. There is not sufficient evidence to support the claim that car owners violate license plate laws at a higher rate than owners of commercial trucks.

9. a. H_0: $p_1 = p_2$. H_1: $p_1 > p_2$. Test statistic: $z = 2.64$. P-value: 0.0041. Critical value: $z = 2.33$. Reject H_0. There is sufficient evidence to support the claim that the rate of success for smoking cessation is greater with the sustained care program.

   b. 98% CI: $0.0135 < p_1 - p_2 < 0.200$ (Table: $0.0134 < p_1 - p_2 < 0.200$). Because the confidence interval limits do not contain 0, there is a significant difference between the two proportions. Because the interval consists of positive numbers only, it appears that the success rate for the sustained care program is greater than the success rate for the standard care program.

   c. Based on the samples, the success rates of the programs are 25.8% (sustained care) and 15.1% (standard care). That difference does appear to be substantial, so the difference between the programs does appear to have practical significance.

11. a. H_0: $p_1 = p_2$. H_1: $p_1 > p_2$. Test statistic: $z = 6.44$. P-value $=$ 0.0000 (Table: 0.0001). Critical value: $z = 2.33$. Reject H_0. There is sufficient evidence to support the claim that the proportion of people over 55 who dream in black and white is greater than the proportion of those under 25.

   b. 98% CI: $0.117 < p_1 - p_2 < 0.240$. Because the confidence interval limits do not include 0, it appears that the two proportions are not equal. Because the confidence interval limits include only positive values, it appears that the proportion of people over 55 who dream in black and white is greater than the proportion of those under 25.

   c. The results suggest that the proportion of people over 55 who dream in black and white is greater than the proportion of those under 25, but the results cannot be used to verify the cause of that difference.

13. a. H_0: $p_1 = p_2$. H_1: $p_1 > p_2$. Test statistic: $z = 6.11$. P-value $=$ 0.0000 (Table: 0.0001). Critical value: $z = 1.645$. Reject H_0. There is sufficient evidence to support the claim that the fatality rate is higher for those not wearing seat belts.

   b. 90% CI: $0.00559 < p_1 - p_2 < 0.0123$. Because the confidence interval limits do not include 0, it appears that the two fatality rates are not equal. Because the confidence interval limits include only positive values, it appears that the fatality rate is higher for those not wearing seat belts.

   c. The results suggest that the use of seat belts is associated with fatality rates lower than those associated with not using seat belts.

15. a. H_0: $p_1 = p_2$. H_1: $p_1 \neq p_2$. Test statistic: $z = 0.57$. P-value: 0.5720 (Table: 0.5686). Critical values: $z = \pm 1.96$. Fail to reject H_0. There is not sufficient evidence to support the claim that echinacea treatment has an effect.

   b. 95% CI: $-0.0798 < p_1 - p_2 < 0.149$. Because the confidence interval limits do contain 0, there is not a significant difference between the two proportions. There is not sufficient evidence to support the claim that echinacea treatment has an effect.

   c. Echinacea does not appear to have a significant effect on the infection rate. Because it does not appear to have an effect, it should not be recommended.

17. a. H_0: $p_1 = p_2$. H_1: $p_1 < p_2$. Test statistic: $z = -7.94$. P-value: 0.0000 (Table: 0.0001). Critical value: $z = -2.33$. Reject H_0. There is sufficient evidence to support the claim that the rate of right-handedness for those who prefer to use their left ear for cell phones is less than the rate of right-handedness for those who prefer to use their right ear for cell phones.

   b. 98% CI: $-0.266 < p < -0.126$. Because the confidence interval limits do not contain 0, there is a significant difference between the two proportions. Because the interval consists of negative numbers only, it appears that the claim is supported. The difference between the populations does appear to have practical significance.

19. a. H_0: $p_1 = p_2$. H_1: $p_1 > p_2$. Test statistic: $z = 9.97$. P-value $=$ 0.0000 (Table: 0.0001). Critical value: $z = 2.33$. Reject H_0. There is sufficient evidence to support the claim that the cure rate with oxygen treatment is higher than the cure rate for those given a placebo. It appears that the oxygen treatment is effective.

   b. 98% CI: $0.467 < p_1 - p_2 < 0.687$. Because the confidence interval limits do not include 0, it appears that the two cure rates are not equal. Because the confidence interval limits include only positive values, it appears that the cure rate with oxygen treatment is higher than the cure rate for those given a placebo. It appears that the oxygen treatment is effective.

   c. The results suggest that the oxygen treatment is effective in curing cluster headaches.

21. a. H_0: $p_1 = p_2$. H_1: $p_1 < p_2$. Test statistic: $z = -1.17$. P-value $=$ 0.1214 (Table: 0.1210). Critical value: $z = -2.33$. Fail to reject H_0. There is not sufficient evidence to support the claim that the rate of left-handedness among males is less than that among females.

   b. 98% CI: $-0.0848 < p_1 - p_2 < 0.0264$ (Table: $-0.0849 < p_1 - p_2 < 0.0265$). Because the confidence interval limits include 0, there does not appear to be a significant difference between the rate of left-handedness among males and the rate among females. There is not sufficient evidence to support the claim that the rate of left-handedness among males is less than that among females.

   c. The rate of left-handedness among males does not appear to be less than the rate of left-handedness among females.

23. The samples should include 2135 men and 2135 women.

25. a. $0.0227 < p_1 - p_2 < 0.217$; because the confidence interval limits do not contain 0, it appears that $p_1 = p_2$ can be rejected.

   b. $0.491 < p_1 < 0.629$; $0.371 < p_2 < 0.509$; because the confidence intervals do overlap, it appears that $p_1 = p_2$ cannot be rejected.

   c. H_0: $p_1 = p_2$. H_1: $p_1 \neq p_2$. Test statistic: $z = 2.40$. P-value: 0.0164. Critical values: $z = \pm 1.96$. Reject H_0. There is sufficient evidence to reject $p_1 = p_2$.

   d. Reject $p_1 = p_2$. Least effective method: Using the overlap between the individual confidence intervals.

Section 9-2

1. Only part (c) describes independent samples.

3. a. Yes   b. Yes   c. 98%

5. a. $H_0: \mu_1 = \mu_2$. $H_1: \mu_1 < \mu_2$. Test statistic: $t = -22.092$. P-value $= 0.0000$ (Table: <0.005). Critical value: $t = -1.672$ (Table: -1.690). Reject H_0. There is sufficient evidence to support the claim that the contents of cans of Diet Coke have weights with a mean that is less than the mean for regular Coke.
   b. 90% CI: -0.03445 lb $< (\mu_1 - \mu_2) < -0.02961$ lb (Table: -0.03448 lb $< (\mu_1 - \mu_2) < -0.02958$ lb).
   c. The contents in cans of Diet Coke appear to weigh less, probably due to the sugar present in regular Coke but not Diet Coke.

7. a. $H_0: \mu_1 = \mu_2$. $H_1: \mu_1 < \mu_2$. Test statistic: $t = -2.979$. P-value $= 0.0021$ (Table: <0.005). Critical value: $t = -2.392$ (Table: -2.441). Reject H_0. There is sufficient evidence to support the claim that blue enhances performance on a creative task.
   b. 98% CI: $-1.05 < \mu_1 - \mu_2 < -0.11$ (Table: $-1.06 < \mu_1 - \mu_2 < -0.10$). The confidence interval consists of negative numbers only and does not include 0, so the mean creativity score with the red background appears to be less than the mean creativity score with the blue background. It appears that blue enhances performance on a creative task.

9. a. $H_0: \mu_1 = \mu_2$. $H_1: \mu_1 > \mu_2$. Test statistic: $t = 0.132$. P-value $= 0.4480$ (Table: > 0.10). Critical value: $t = 1.691$ (Table: 1.729). Fail to reject H_0. There is not sufficient evidence to support the claim that the magnets are effective in reducing pain.
   b. 90% CI: $-0.59 < \mu_1 - \mu_2 < 0.69$ (Table: $-0.61 < \mu_1 - \mu_2 < 0.71$).
   c. Magnets do not appear to be effective in treating back pain. It is valid to argue that the magnets *might* appear to be effective if the sample sizes were larger.

11. a. $H_0: \mu_1 = \mu_2$. $H_1: \mu_1 \neq \mu_2$. Test statistic: $t = 0.674$. P-value $= 0.5015$ (Table: >0.20). Critical values: $t = \pm 1.979$ (Table: ± 1.994). Fail to reject H_0. There is not sufficient evidence to warrant rejection of the claim that females and males have the same mean BMI.
    b. 95% CI: $-1.39 < (\mu_1 - \mu_2) < 2.83$ (Table: $-1.41 < (\mu_1 - \mu_2) < 2.85$). Because the confidence interval includes 0, there is not sufficient evidence to warrant rejection of the claim that the two samples are from populations with the same mean.
    c. Based on the available sample data, it appears that males and females have the same mean BMI, but we can only conclude that there isn't sufficient evidence to say that they are different.

13. a. $H_0: \mu_1 = \mu_2$. $H_1: \mu_1 \neq \mu_2$. Test statistic: $t = -2.025$. P-value $= 0.0460$ (Table: <0.05). Critical values: $t = \pm 1.988$ (Table: ± 2.023). Reject H_0. There is sufficient evidence to warrant rejection of the claim that the two samples are from populations with the same mean.
    b. 95% CI: $-0.44 < (\mu_1 - \mu_2) < 0.00$. Because the confidence interval includes negative numbers only and does not include 0, there is sufficient evidence to warrant rejection of the claim that the two samples are from populations with the same mean.

c. Yes. With the smaller samples of size 12 and 15, there was not sufficient evidence to warrant rejection of the null hypothesis, but there is sufficient evidence with the larger samples.

15. a. $H_0: \mu_1 = \mu_2$. $H_1: \mu_1 > \mu_2$. Test statistic: $t = 32.771$. P-value $= 0.0000$ (Table: <0.005). Critical value: $t = 1.667$ (Table: 1.685). Reject H_0. There is sufficient evidence to support the claim that pre-1964 quarters have a mean weight that is greater than the mean weight of post-1964 quarters.
    b. 90% CI: 0.52522 lb $< (\mu_1 - \mu_2) < 0.58152$ (Table: 0.52492 lb $< (\mu_1 - \mu_2) < 0.58182$)
    c. Yes. Vending machines are not affected very much because pre-1964 quarters are mostly out of circulation.

17. $H_0: \mu_1 = \mu_2$. $H_1: \mu_1 \neq \mu_2$. Test statistic: $t = -0.315$. P-value $= 0.7576$ (Table: >0.20). Assuming a 0.05 significance level, critical values are $t = \pm 2.159$ (Table: ± 2.262). Fail to reject H_0. There is not sufficient evidence to warrant rejection of the claim that female professors and male professors have the same mean evaluation ratings. There does not appear to be a difference between male and female professor evaluation scores.

19. $H_0: \mu_1 = \mu_2$. $H_1: \mu_1 \neq \mu_2$. Test statistic: $t = -2.385$. P-value $= 0.0244$ (Table: <0.05). Assuming a 0.05 significance level, critical values are $t = \pm 2.052$ (Table: ± 2.201). The conclusion depends on the choice of the significance level. There is a significant difference between the two population means at the 0.05 significance level, but not at the 0.01 significance level.

21. $H_0: \mu_1 = \mu_2$. $H_1: \mu_1 < \mu_2$. Test statistic: $t = -0.132$. P-value $= 0.4477$ (Table: >0.10). Critical value: $t = -1.669$ (Table: -1.688). Fail to reject H_0. There is not sufficient evidence to support the claim that men talk less than women.

23. $H_0: \mu_1 = \mu_2$. $H_1: \mu_1 < \mu_2$. Test statistic: $t = -3.450$. P-value $= 0.0003$ (Table: <0.005). Assuming a significance level of 0.05, critical value is $t = -1.649$ (Table: -1.653). Reject H_0. There is sufficient evidence to support the claim that at birth, girls have a lower mean weight than boys.

25. With pooling, df increases dramatically to 97, but the test statistic decreases from 2.282 to 1.705 (because the estimated standard deviation increases from 2.620268 to 3.507614), the P-value increases to 0.0457, and the 90% confidence interval becomes wider. With pooling, these results do not show greater significance.

27. $H_0: \mu_1 = \mu_2$. $H_1: \mu_1 \neq \mu_2$. Test statistic: $t = 15.322$. P-value $= 0.0000$ (Table: <0.01). Critical values: $t = \pm 2.080$. Reject H_0. There is sufficient evidence to warrant rejection of the claim that the two populations have the same mean.

Section 9-3

1. Only parts (a) and (c) are true.

3. The results will be the same.

5. a. $H_0: \mu_d = 0$ year. $H_1: \mu_d < 0$ year. Test statistic: $t = -2.609$. P-value $= 0.0142$ (Table: <0.025). Critical value: $t = -1.833$. Reject H_0. There is sufficient evidence to support the claim that for the population of ages of Best Actresses and Best Actors, the differences have a mean less than 0. There is sufficient evidence to conclude that Best Actresses are generally younger than Best Actors.

b. 90% CI: -16.5 years $< \mu_d < -2.9$ years. The confidence interval consists of negative numbers only and does not include 0.

7. a. H_0: $\mu_d = 0°$F. H_1: μ_d: $\neq 0°$F. Test statistic: $t = -7.499$. P-value $= 0.0003$ (Table: <0.01). Critical values: $t = \pm 2.447$. Reject H_0. There is sufficient evidence to warrant rejection of the claim that there is no difference between body temperatures measured at 8 AM and at 12 AM. There appears to be a difference.

b. $-1.97°$F $< \mu_d < -1.00°$F. The confidence interval consists of negative numbers only and does not include 0.

9. H_0: $\mu_d = 0$ in. H_1: $\mu_d \neq 0$ in. Test statistic: $t = -1.379$. P-value $= 0.2013$ (Table: >0.20). Critical values: $t = \pm 2.262$. Fail to reject H_0. There is not sufficient evidence to warrant rejection of the claim that there is no difference in heights between mothers and their first daughters.

11. H_0: $\mu_d = 0$. H_1: $\mu_d \neq 0$. Test statistic: $t = 0.793$. P-value $= 0.4509$ (Table: >0.20). Critical values: $t = \pm 2.306$. Fail to reject H_0. There is not sufficient evidence to support the claim that there is a difference between female attribute ratings and male attribute ratings.

13. -6.5 admissions $< \mu_d < -0.2$ admissions. Because the confidence interval does not include 0 admission, it appears that there is sufficient evidence to warrant rejection of the claim that when the 13th day of a month falls on a Friday, the numbers of hospital admissions from motor vehicle crashes are not affected. Hospital admissions do appear to be affected.

15. $0.69 < \mu_d < 5.56$. Because the confidence interval limits do not contain 0 and they consist of positive values only, it appears that the "before" measurements are greater than the "after" measurements, so hypnotism does appear to be effective in reducing pain.

17. a. H_0: $\mu_d = 0$ year. H_1: $\mu_d < 0$ year. Test statistic: $t = -5.185$. P-value $= 0.0000$ (Table: <0.005). Critical value: $t = -1.663$ (Table: -1.662). Reject H_0. There is sufficient evidence to support the claim that actresses are generally younger than actors.

b. 90% CI: -10.4 years $< \mu_d < -5.4$ years. The confidence interval consists of negative numbers only and does not include 0.

19. a. H_0: $\mu_d = 0°$F. H_1: $\mu_d \neq 0°$F. Test statistic: $t = -8.485$. P-value $= 0.0000$ (Table: <0.01). Critical values: $t = \pm 1.996$ (Table: ± 1.994). Reject H_0. There is sufficient evidence to warrant rejection of the claim of no difference between body temperatures measured at 8 AM and at 12 AM. There appears to be a difference.

b. 95% CI: $-1.05°$F $< \mu_d < -0.65°$F. The confidence interval consists of negative numbers only and does not include 0.

21. H_0: $\mu_d = 0$ in. H_1: $\mu_d \neq 0$ in. Test statistic: $t = -4.090$. P-value $= 0.0001$ (Table: <0.01). Critical values: $t = \pm 1.978$ (Table: ± 1.984 approximately). Reject H_0. There is sufficient evidence to warrant rejection of the claim of no difference in heights between mothers and their first daughters.

23. H_0: $\mu_d = 0$. H_1: $\mu_d \neq 0$. Test statistic: $t = 0.191$. P-value $= 0.8485$ (Table: >0.20). Critical values: ± 1.972. Fail to reject H_0. There is not sufficient evidence to support the claim that there is a difference between female attribute ratings and male attribute ratings.

25. For the temperatures in degrees Fahrenheit and the temperatures in degrees Celsius, the test statistic ($t = 0.124$) is the same, the P-value of 0.9023 is the same, the critical values ($t = \pm 2.028$) are the same, and the conclusions are the same, so the hypothesis test results are the same in both cases. The confidence intervals are $-0.25°$F $< \mu_d < 0.28°$F and $-0.14°$C $< \mu_d < 0.16°$C. The confidence interval limits of $-0.14°$C and $0.16°$C have numerical values that are $5/9$ of the numerical values of $-0.25°$F and $0.28°$F.

Section 9-4

1. a. No.   b. No.

c. The two samples have standard deviations (or variances) that are very close in value.

d. Skewed right

3. No. Unlike some other tests that have a requirement that samples must be from normally distributed populations or the samples must have more than 30 values, the F test has a requirement that the samples must be from normally distributed populations, regardless of how large the samples are.

5. H_0: $\sigma_1 = \sigma_2$. H_1: $\sigma_1 \neq \sigma_2$. Test statistic: $F = 2.3706$. P-value: 0.0129. Upper critical F value: 1.9678 (Table: Upper critical F value is between 1.8752 and 2.0739). Reject H_0. There is sufficient evidence to warrant rejection of the claim that creative task scores have the same variation with a red background and a blue background.

7. H_0: $\sigma_1 = \sigma_2$. H_1: $\sigma_1 > \sigma_2$. Test statistic: $F = 9.3364$. P-value: 0.0000. Critical F value is 2.0842 (Table: Critical F value is between 2.0540 and 2.0960). Reject H_0. There is sufficient evidence to support the claim that the treatment group has errors that vary more than the errors of the placebo group.

9. H_0: $\sigma_1 = \sigma_2$. H_1: $\sigma_1 \neq \sigma_2$. Test statistic: $F = 2.9265$. P-value: 0.0020. Upper critical F value is 1.9611 (Table: Upper critical F value is between 1.8752 and 2.0739). Reject H_0. There is sufficient evidence to warrant rejection of the claim that variation is the same for both types of Coke.

11. H_0: $\sigma_1 = \sigma_2$. H_1: $\sigma_1 > \sigma_2$. Test statistic: $F = 2.1267$. P-value: 0.0543. Critical F value is 2.1682 (Table: Critical F value is between 2.1555 and 2.2341). Fail to reject H_0. There is not sufficient evidence to support the claim that those given a sham treatment have pain reductions that vary more than the pain reductions for those treated with magnets.

13. H_0: $\sigma_1 = \sigma_2$. H_1: $\sigma_1 \neq \sigma_2$. Test statistic: $F = 4.1750$. P-value: 0.0447. Critical value: 4.0260. Reject H_0. There is sufficient evidence to warrant rejection of the claim that female professors and male professors have evaluation scores with the same variation.

15. H_0: $\sigma_1 = \sigma_2$. H_1: $\sigma_1 \neq \sigma_2$. Test statistic: $F = 2.3095$. P-value: 0.1635. Assuming a 0.05 significance level, the critical value is 3.3044 (Table: Critical value is between 3.2261 and 3.3299). Fail to reject H_0. There is not sufficient evidence to support a claim that the variation of the times between eruptions has changed.

17. $c_1 = 3$, $c_2 = 0$, critical value is 7.4569. Fail to reject H_0. There is not sufficient evidence to support a claim that the two populations of scores have different amounts of variation.

19. $F_L = 0.4103$; $F_R = 2.7006$.

Chapter 9: Quick Quiz

1. $H_0: p_1 = p_2.$ $H_1: p_1 \neq p_2.$
2. $x_1 = 258,$ $x_2 = 282,$ $\hat{p}_1 = 258/1121 = 0.230,$
   $\hat{p}_2 = 282/1084 = 0.260,$ $\bar{p} = 0.245.$
3. 0.1015 (Table: 0.1010)
4. a. $-0.0659 < p_1 - p_2 < 0.00591$
   b. The confidence interval includes the value of 0, so it is possible that the two proportions are equal. There is not a significant difference.
5. Fail to reject H_0. There is not sufficient evidence to warrant rejection of the claim that for the people who were aware of the statement, the proportion of women is equal to the proportion of men.
6. True   7. False
8. Because the data consist of matched pairs, they are dependent.
9. $H_0: \mu_d = 0$ mm Hg. $H_1: \mu_d \neq 0$ mm Hg.

10. a. $t = \dfrac{\bar{d} - \mu_d}{\dfrac{s_d}{\sqrt{n}}}$

    b. $t = \dfrac{(\bar{x}_1 - \bar{x}_2) - (\mu_1 - \mu_2)}{\sqrt{\dfrac{s_1^2}{n_1} + \dfrac{s_2^2}{n_2}}}$

    c. $z = \dfrac{(\hat{p}_1 - \hat{p}_2) - (p_1 - p_2)}{\sqrt{\dfrac{\bar{p}\bar{q}}{n_1} + \dfrac{\bar{p}\bar{q}}{n_2}}}$

    d. $F = \dfrac{s_1^2}{s_2^2}$

Chapter 9: Review Exercises

1. $H_0: p_1 = p_2.$ $H_1: p_1 < p_2.$ Test statistic: $z = -3.49$. P-value: 0.0002. Critical value: $z = -1.645$. Reject H_0. There is sufficient evidence to support the claim that money in a large denomination is less likely to be spent relative to an equivalent amount in smaller denominations.
2. 90% CI: $-0.528 < p_1 - p_2 < -0.206$. The confidence interval limits do not contain 0, so it appears that there is a significant difference between the two proportions. Because the confidence interval consists of negative values only, it appears that p_1 is less than p_2, so it appears that money in a large denomination is less likely to be spent relative to an equivalent amount in smaller denominations.
3. -25.33 cm $< (\mu_1 - \mu_2) < -7.51$ cm (Table: -25.70 cm $< (\mu_1 - \mu_2) < -7.14$ cm). With 95% confidence, we conclude that the mean height of women is less than the mean height of men by an amount that is between 7.51 cm and 25.33 cm (Table: 7.14 cm and 25.70 cm).
4. $H_0: \mu_1 = \mu_2.$ $H_1: \mu_1 < \mu_2.$ Test statistic: $t = -4.001$. P-value $= 0.0008$ (Table: <0.005). Critical value: $t = -2.666$ (Table: -2.821). Reject H_0. There is sufficient evidence to support the claim that women have heights with a mean that is less than the mean height of men. But you already knew that.
5. $H_0: \mu_d = 0.$ $H_1: \mu_d > 0.$ Test statistic: $t = 6.371$. P-value $= 0.0000$. (Table: <0.005). Critical value: $t = 2.718$. Reject H_0. There is sufficient evidence to support the claim that Captopril is effective in lowering systolic blood pressure.

6. $H_0: \mu_1 = \mu_2.$ $H_1: \mu_1 > \mu_2.$ Test statistic: $t = 2.879$. P-value $= 0.0026$ (Table: <0.005). Critical value: $t = 2.376$ (Table: 2.429). Reject H_0. There is sufficient evidence to support the claim that the mean number of details recalled is lower for the stress group. It appears that "stress decreases the amount recalled," but we should not conclude that stress is the cause of the decrease.
7. $H_0: \mu_d = 0.$ $H_1: \mu_d > 0.$ Test statistic: $t = 14.061$. P-value $= 0.0000$ (Table: <0.005). Critical value: $t = 3.365$. Reject H_0. There is sufficient evidence to support the claim that flights scheduled 1 day in advance cost more than flights scheduled 30 days in advance. Save money by scheduling flights 30 days in advance.
8. $H_0: \sigma_1 = \sigma_2.$ $H_1: \sigma_1 \neq \sigma_2.$ Test statistic: $F = 4.9933$. P-value: 0.0252. Upper critical F value: 4.0260. Reject H_0. There is sufficient evidence to warrant rejection of the claim that women and men have heights with the same variation.

Chapter 9: Cumulative Review Exercises

1. a. Because the sample data are matched with each column consisting of heights from the same family, the data are dependent.
   b. $\bar{x} = 69.7$ in.; median $= 71.0$ in.; range $= 7.7$ in.; $s = 2.6$ in.; $s^2 = 6.6$ in^2
   c. Ratio
   d. Continuous
2. There does not appear to be a correlation or association between the heights of fathers and the heights of their sons.

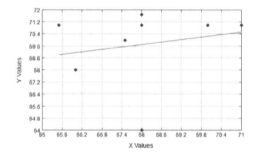

3. 67.6 in. $< \mu < 71.9$ in. We have 95% confidence that the limits of 67.6 in. and 71.9 in. actually contain the true value of the mean height of all adult sons.
4. $H_0: \mu_d = 0$ in. $H_1: \mu_d \neq 0$ in. Test statistic: $t = -1.712$. P-value $= 0.1326$ (Table: >0.10). Critical values: $t = \pm 2.365$. Fail to reject H_0. There is not sufficient evidence to warrant rejection of the claim that differences between heights of fathers and their sons have a mean of 0. There does not appear to be a difference between heights of fathers and their sons.
5. Because the points lie reasonably close to a straight-line pattern, and there is no other pattern that is not a straight-line pattern and there are no outliers, the sample data appear to be from a population with a normal distribution.
6. The shape of the histogram indicates that the sample data appear to be from a population with a distribution that is approximately normal. (Answer continued on next page.)

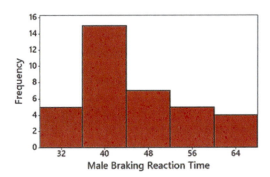

7. Because the points are reasonably close to a straight-line pattern and there is no other pattern that is not a straight-line pattern, it appears that the braking reaction times of females are from a population with a normal distribution.

8. Because the boxplots overlap, there does not appear to be a significant difference between braking reaction times of males and females, but the braking reaction times for males appear to be generally lower than the braking reaction times of females.

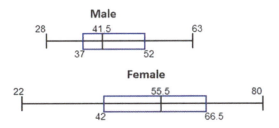

9. $H_0: \mu_1 = \mu_2$. $H_1: \mu_1 \neq \mu_2$. Test statistic: $t = -3.259$. P-value $= 0.0019$ (Table: <0.005). Critical values: $t = \pm 2.664$ (Table: ± 2.724). Reject H_0. There is sufficient evidence to warrant rejection of the claim that males and females have the same mean braking reaction time. Males appear to have lower reaction times.

10. a. Males: $40.1 < \mu < 48.7$. Females: $47.2 < \mu < 61.4$. The confidence intervals overlap, so there does not appear to be a significant difference between the mean braking reaction times of males and females.

    b. $-18.0 < \mu_1 - \mu_2 < -1.8$ (Table: $-18.2 < \mu_1 - \mu_2 < -1.6$). Because the confidence interval consists of negative numbers and does not include 0, there appears to be a significant difference between the mean braking reaction times of males and females.

    c. The results from part (b) are better.

Chapter 10 Answers

Section 10-1

1. a. r is a statistic that represents the value of the linear correlation coefficient computed from the paired sample data, and ρ is a parameter that represents the value of the linear correlation coefficient that would be computed by using all of the paired data in the population of all statistics students.

   b. The value of r is estimated to be 0, because it is likely that there is no correlation between body temperature and head circumference.

c. The value of r does not change if the body temperatures are converted to Fahrenheit degrees.

3. No. A correlation between two variables indicates that they are somehow associated, but that association does not necessarily imply that one of the variables has a direct effect on the other variable. Correlation does not imply causality.

5. Yes. $r = 0.963$. P-value $= 0.000$. Critical values: ± 0.268 (Table: ± 0.279 approximately). There is sufficient evidence to support the claim that there is a linear correlation between the weights of bears and their chest sizes. It is easier to measure the chest size of a bear than the weight, which would require lifting the bear onto a scale. It does appear that chest size could be used to predict weight.

7. No. $r = 0.117$. P-value > 0.05. Critical values: ± 0.250 (Table: ± 0.254 approximately). There is not sufficient evidence to support the claim that there is a linear correlation between weights of discarded paper and glass.

9. a.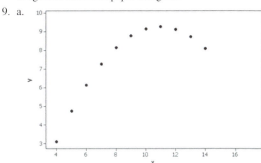

   b. $r = 0.816$. P-value $= 0.002$ (Table: <0.01). Critical values: $r = \pm 0.602$ assuming a 0.05 significance level. There is sufficient evidence to support the claim of a linear correlation between the two variables.

   c. The scatterplot reveals a distinct pattern that is not a straight-line pattern.

11. a. Answer varies. Because there appears to be an upward pattern, it is reasonable to think that there is a linear correlation.

    b. $r = 0.906$. P-value $= 0.000$ (Table: <0.01). Critical values: $r = \pm 0.632$ (for a 0.05 significance level). There is sufficient evidence to support the claim of a linear correlation.

    c. $r = 0$. P-value $= 1.000$ (Table: >0.05). Critical values: $r = \pm 0.666$ (for a 0.05 significance level). There is not sufficient evidence to support the claim of a linear correlation.

    d. The effect from a single pair of values can be very substantial, and it can change the conclusion.

13. $r = 0.799$. P-value $= 0.056$ (Table: >0.05). Critical values: $r = \pm 0.811$. There is not sufficient evidence to support the claim that there is a linear correlation between Internet users and Nobel Laureates.

15. $r = 0.992$. P-value $= 0.000$ (Table: <0.01). Critical values: $r = \pm 0.666$. There is sufficient evidence to support the claim that there is a significant linear correlation between the cost of a slice of pizza and the subway fare.

17. $r = 0.591$. P-value $= 0.294$ (Table: >0.05). Critical values: $r = \pm 0.878$. There is not sufficient evidence to support the claim that there is a linear correlation between shoe print lengths and heights of males. The given results do not suggest that police can use a shoe print length to estimate the height of a male.

19. $r = -0.959$. P-value $= 0.010$. Critical values: $r = \pm0.878$. There is sufficient evidence to support the claim that there is a linear correlation between weights of lemon imports from Mexico and U.S. car fatality rates. The results do not suggest any cause-effect relationship between the two variables.

21. $r = -0.288$. P-value $= 0.365$ (Table: >0.05). Critical values: $r = \pm0.576$. There is not sufficient evidence to support the claim that there is a significant linear correlation between the ages of Best Actresses and Best Actors. Because Best Actresses and Best Actors typically appeared in different movies, we should not expect that there would be a correlation between their ages at the time that they won the awards.

23. $r = 0.948$. P-value $= 0.004$ (Table: <0.01). Critical values: $r = \pm0.811$. There is sufficient evidence to support the claim of a linear correlation between the overhead width of a seal in a photograph and the weight of a seal.

25. $r = 0.828$. P-value $= 0.042$ (Table: <0.05). Critical values: $r = \pm0.811$. There is sufficient evidence to support the claim there is a linear correlation between the bill amounts and the tip amounts. If everyone were to tip the same percentage, r would be 1.

27. $r = 1.000$. P-value $= 0.000$. Critical values: $r = \pm0.707$. There is sufficient evidence to support the claim that there is a linear correlation between diameters and circumferences. The scatterplot confirms a linear association.

29. $r = 0.702$. P-value $= 0.000$ (Table: <0.01). Critical values: ±0.413 (Table: Critical values are between ±0.396 and ±0.444). There is sufficient evidence to support the claim that there is a linear correlation between Internet users and Nobel Laureates.

31. $r = 0.594$. P-value $= 0.007$ (Table: <0.01). Critical values: $r = \pm0.456$. There is sufficient evidence to support the claim that there is a linear correlation between shoe print lengths and heights of males. The given results do suggest that police can use a shoe print length to estimate the height of a male.

33. $r = 0.319$. P-value $= 0.017$ (Table: <0.05). Critical values: ±0.263 (Table: $r = \pm0.254$ approximately). There is sufficient evidence to support the claim of a linear correlation between the numbers of words spoken by men and women who are in couple relationships.

35. a. 0.911   b. 0.787   c. 0.9999 (largest)
    d. 0.976   e. -0.948

Section 10-2

1. a. $\hat{y} = -368 + 130x$
   b. $\hat{y}$ represents the predicted value of price.

3. a. A residual is a value of $y - \hat{y}$, which is the difference between an observed value of y and a predicted value of y.
   b. The regression line has the property that the sum of squares of the residuals is the lowest possible sum.

5. With no significant linear correlation, the best predicted value is $\bar{y} = 5.9$.

7. With a significant linear correlation, the best predicted value is 92.0 kg.

9. $\hat{y} = 3.00 + 0.500x$. The data have a pattern that is not a straight line.

11. a. $\hat{y} = 0.264 + 0.906x$
    b. $\hat{y} = 2 + 0x$ (or $\hat{y} = 2$)
    c. The results are very different, indicating that one point can dramatically affect the regression equation.

13. $\hat{y} = -8.44 + 0.203x$. Best predicted value: $\bar{y} = 5.1$ per 10 million people. The best predicted value is not at all close to the actual value.

15. $\hat{y} = -0.0111 + 1.01x$. Best predicted value: \$3.02. The best predicted subway fare of \$3.02 is not likely to be implemented because it is not a convenient value, such as \$3.00 or \$3.25.

17. $\hat{y} = 125 + 1.73x$. Best predicted value: $\bar{y} = 177$ cm. Because the best predicted value is the mean height, it would not be helpful to police in trying to obtain a description of the male.

19. $\hat{y} = 16.5 - 0.00282x$. Best predicted value: 15.1 fatalities per 100,000 population. Common sense suggests that the prediction doesn't make much sense.

21. $\hat{y} = 51.6 - 0.165x$. Best predicted value: $\bar{y} = 45$ years, which is not close to the actual age of 33 years.

23. $\hat{y} = -157 + 40.2x$. Best predicted value: -76.6 kg. The prediction is a negative weight that cannot be correct. The overhead width of 2 cm is well beyond the scope of the sample widths, so the extrapolation might be off by a considerable amount. Clearly, the predicted negative weight makes no sense.

25. $\hat{y} = -0.347 + 0.149x$. Best predicted value: \$14.55. Tipping rule: Multiply the bill by 0.149 (or 14.9%) and subtract 35 cents. A more approximate but easier rule is this: Leave a tip of 15%.

27. $\hat{y} = -0.00396 + 3.14x$. Best predicted circumference: 4.7 cm. Even though the diameter of 1.50 cm is beyond the scope of the sample diameters, the predicted value yields the actual circumference.

29. $\hat{y} = -23.2 + 0.456x$. Best predicted value: 12.9 per 10 million people. The best predicted value is not at all close to Japan's actual Nobel rate of 1.5 per 10 million people.

31. $\hat{y} = 93.5 + 2.85x$. Best predicted value: 183 cm. Although there is a linear correlation, with $r = 0.594$, we see that it is not very strong, so an estimate of the height of a male might be off by a considerable amount.

33. $\hat{y} = 13{,}400 + 0.302x$. Best predicted value: 18,200 words.

35. a. 823.64
    b. Using $\hat{y} = -3 + 2.5x$, the sum of squares of the residuals is 827.45, which is larger than 823.64, which is the sum of squares of the residuals for the regression line.

Section 10-3

1. The value of $s_e = 16.27555$ cm is the standard error of estimate, which is a measure of the differences between the observed weights and the weights predicted from the regression equation. It is a measure of the variation of the sample points about the regression line.

3. The coefficient of determination is $r^2 = 0.155$. We know that 15.5% of the variation in weight is explained by the linear correlation between height and weight, and 84.5% of the variation in weight is explained by other factors and/or random variation.

5. $r^2 = 0.764$. 76.4% of the variation in temperature is explained by the linear correlation between chirps and temperature, and 23.6% of the variation in temperature is explained by other factors and/or random variation.

7. $r^2 = 0.783$. 78.3% of the variation in waist size is explained by the linear correlation between weight and waist size, and 21.7% of the variation in waist size is explained by other factors and/or random variation.

9. $r = 0.850$. Critical values: $r = \pm 0.404$ (Table: $r = \pm 0.396$ approximately), assuming a 0.05 significance level. There is sufficient evidence to support a claim of a linear correlation between registered boats and manatee fatalities.

11. 70.5 manatees

13. 42.7 manatees $< y <$ 98.3 manatees

15. 65.1 manatees $< y <$ 106.8 manatees

17. a. 10,626.59
    b. 68.83577
    c. $38.0°F < y < 60.4°F$

19. a. 352.7278
    b. 109.3722
    c. $71.09°F < y < 88.71°F$

21. $17.1 < y < 26.0$ (values are Nobel Laureates per 10 million people)

Section 10-4

1. The response variable is weight and the predictor variables are length and chest size.

3. The unadjusted R^2 increases (or remains the same) as more variables are included, but the adjusted R^2 is adjusted for the number of variables and sample size. The unadjusted R^2 incorrectly suggests that the best multiple regression equation is obtained by including all of the available variables, but by taking into account the sample size and number of predictor variables, the adjusted R^2 is much more helpful in weeding out variables that should not be included.

5. Son $= 18.0 + 0.504$ Father $+ 0.277$ Mother

7. P-value less than 0.0001 is low, but the values of R^2 (0.3649) and adjusted R^2 (0.3552) are not high. Although the multiple regression equation fits the sample data best, it is not a good fit, so it should not be used for predicting the height of a son based on the height of his father and the height of his mother.

9. HWY (highway fuel consumption) because it has the best combination of small P-value (0.000) and highest adjusted R^2 (0.920).

11. CITY $= -3.15 + 0.819$ HWY. That equation has a low P-value of 0.000 and its adjusted R^2 value of 0.920 isn't very much less than the values of 0.928 and 0.935 that use two predictor variables, so in this case it is better to use the one predictor variable instead of two.

13. The best regression equation is $\hat{y} = 0.127 + 0.0878x_1 - 0.0250x_2$, where x_1 represents tar and x_2 represents carbon monoxide. It is best because it has the highest adjusted R^2 value of 0.927 and the lowest P-value of 0.000. It is a good regression equation for predicting nicotine content because it has a high value of adjusted R^2 and a low P-value.

15. The best regression equation is $\hat{y} = 109 - 0.00670x_1$, where x_1 represents volume. It is best because it has the highest adjusted R^2 value of -0.0513 and the lowest P-value of 0.791. The three regression equations all have adjusted values of R^2 that are very close to 0, so none of them are good for predicting IQ. It does not appear that people with larger brains have higher IQ scores.

17. For $H_0: \beta_1 = 0$, the test statistic is $t = 10.814$, the P-value is less than 0.0001, so reject H_0 and conclude that the regression coefficient of $b_1 = 0.769$ should be kept. For $H_0: \beta_2 = 0$, the test statistic is $t = 29.856$, the P-value is less than 0.0001, so reject H_0 and conclude that the regression coefficient of $b_2 = 1.01$ should

be kept. It appears that the regression equation should include both independent variables of height and waist circumference.

19. $\hat{y} = 3.06 + 82.4x_1 + 2.91x_2$, where x_1 represents sex and x_2 represents age. Female: 61 lb; male: 144 lb. The sex of the bear does appear to have an effect on its weight. The regression equation indicates that the predicted weight of a male bear is about 82 lb more than the predicted weight of a female bear with other characteristics being the same.

Section 10-5

1. $y = x^2$; quadratic; $R^2 = 1$

3. 25.5% of the variation in Super Bowl points can be explained by the quadratic model that relates the variable of year and the variable of points scored. Because such a small percentage of the variation is explained by the model, the model is not very useful.

5. Quadratic: $d = -4.88t^2 + 0.0214t + 300$

7. Exponential: $y = 1000(1.0122^x)$

9. Exponential: $y = 10(2^x)$

11. Logarithmic: $y = 3.22 + 0.293 \ln x$

13. Power: $y = 2107.9x^{0.615}$. (Result is based on 1990 coded as 1.) The projected value for 2014 is 15,271 (Using rounded coefficients: 15,261), which is considerably less than the actual value of 18,054.

15. Quadratic: $y = 0.700x^2 - 3.41x + 299$. Predicted value: 563. The decade of 2090–2099 is too far beyond the scope of the available data, so the predicted value is questionable.

17. a. Exponential: $y = 2^{\frac{2}{3}(x-1)}$ [or $y = (0.629961)(1.587401)^x$ for an initial value of 1 that doubles every 1.5 years].
    b. Exponential: $y = (1.36558)(1.42774)^x$, where 1971 is coded as 1.
    c. Moore's law does appear to be working reasonably well. With $R^2 = 0.990$, the model appears to be very good.

Chapter 10: Quick Quiz

1. Conclude that there is not sufficient evidence to support the claim of a linear correlation between enrollment and burglaries.

2. None of the given values change when the variables are switched.

3. No. The value of r does not change if all values of one of the variables are multiplied by the same constant.

4. Because r must be between -1 and 1 inclusive, the value of 1.500 is the result of an error in the calculations.

5. The best predicted number of burglaries is 92.6, which is the mean of the five sample burglary counts.

6. The best predicted number of burglaries would be 123.3, which is found by substituting 50 for x in the regression equation.

7. $r^2 = 0.249$

8. False.

9. False.

10. $r = -1$

Chapter 10: Review Exercises

1. a. $r = 0.962$. P-value $= 0.000$ (Table: <0.01). Critical values: $r = \pm 0.707$ (assuming a 0.05 significance level). There is sufficient evidence to support the claim that there is a linear correlation between the amount of tar and the amount of nicotine.

b. 92.5%

c. $\hat{y} = -0.758 + 0.0920x$

d. The predicted value is 1.358 mg or 1.4 mg rounded, which is close to the actual amount of 1.3 mg.

2. a. The scatterplot shows a pattern with nicotine and CO both increasing from left to right, but it is a very weak pattern and the points are not very close to a straight-line pattern, so it appears that there is not sufficient sample evidence to support the claim of a linear correlation between amounts of nicotine and carbon monoxide.

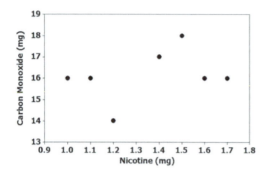

b. $r = 0.329$. P-value $= 0.427$ (Table:>0.05). Critical values: $r = \pm 0.707$ (assuming a 0.05 significance level). There is not sufficient evidence to support the claim that there is a linear correlation between amount of nicotine and amount of carbon monoxide.

c. $\hat{y} = 14.2 + 1.42x$

d. The predicted value is $\bar{y} = 16.1$ mg, which is close to the actual amount of 15 mg.

3. $r = 0.450$. P-value $= 0.192$ (Table: >0.05). Critical values: $r = \pm 0.632$ (assuming a 0.05 significance level). There is not sufficient evidence to support the claim that there is a linear correlation between time and height. Although there is no *linear* correlation between time and height, the scatterplot shows a very distinct pattern revealing that time and height are associated by some function that is not linear.

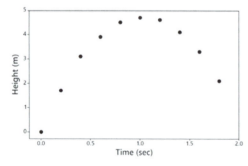

4. a. NICOTINE $= -0.443 + 0.0968$ TAR $- 0.0262$ CO, or $\hat{y} = -0.443 + 0.0968x_1 - 0.0262x_2$.

b. $R^2 = 0.936$; adjusted $R^2 = 0.910$; P-value $= 0.001$.

c. With high values of R^2 and adjusted R^2 and a small P-value of 0.001, it appears that the regression equation can be used to predict the amount of nicotine given the amounts of tar and carbon monoxide.

d. The predicted value is 1.39 mg or 1.4 mg rounded, which is close to the actual value of 1.3 mg of nicotine.

Chapter 10: Cumulative Review Exercises

1. a. $\bar{x} = 35.91$, median $= 36.10$, range $= 76.40$, $s = 31.45$, $s^2 = 989.10$.

b. Quantitative data

c. Ratio

2. $r = 0.731$. P-value $= 0.039$ (Table: <0.05) Critical values: $r = \pm 0.707$. There is sufficient evidence to support the claim of a linear correlation between the DJIA values and sunspot numbers. Because it would be reasonable to think that there is no correlation between stocks and sunspot numbers, the result is not as expected. Although there appears to be a linear correlation, a reasonable investor would be wise to ignore sunspot numbers when investing in stocks.

3. The highest sunspot number is 79.3, which converts to $z = 1.38$. The highest sunspot number is not significantly high because its z score of 1.38 shows that it is within 2 standard deviations of the mean.

4. H_0: $\mu = 49.7$. H_1: $\mu \neq 49.7$. Test statistic: $t = -1.240$. P-value $= 0.255$ (Table: >0.20). Critical values: $t = \pm 2.365$. Fail to reject H_0. There is not sufficient evidence to warrant rejection of the claim that the sample is from a population with a mean equal to 49.7.

5. $9.62 < \mu < 62.21$. We have 95% confidence that the interval limits of 9.62 and 62.21 contain the true value of the mean of the population of sunspot numbers.

6. H_0: $p = 0.10$. H_1: $p < 0.10$. Test statistic: $z = -5.25$ (using $x = 191$) or $z = -5.23$ (using $\hat{p} = 0.07$). P-value $= 0.0000$ (Table: 0.0001). Critical value: $z = -1.645$. Reject H_0. There is sufficient evidence to support the claim that fewer than 10% of police traffic stops are attributable to improper cell phone use.

7. $\bar{x} = 35.2$ years, $s = 19.7$ years, $s^2 = 389.6$ years^2

8. a. 40.13%

b. 21.5 years (Table: 21.6 years)

c. 0.1056

d. 0 + or 0.0000 (from $0.4013^{25} = 0.0000$). The audience for a particular movie and showtime is not a simple random sample. Some movies and showtimes attract very young audiences.

Chapter 11 Answers

Section 11-1

1. a. Observed values are represented by O and expected values are represented by E.

b. For the leading digit of 2, $O = 62$ and $E = (317)(0.176) = 55.792$.

c. For the leading digit of 2, $(O - E)^2/E = 0.691$.

3. There is sufficient evidence to warrant rejection of the claim that the leading digits have a distribution that fits well with Benford's law.

5. P-value $= 0.516$ (Table: >0.10). Critical value: $X^2 = 16.919$. There is not sufficient evidence to warrant rejection of the claim that the observed outcomes agree with the expected frequencies. The slot machine appears to be functioning as expected.

7. Test statistic: $X^2 = 5.860$. P-value $= 0.320$ (Table: >0.10). Critical value: $X^2 = 11.071$. There is not sufficient evidence to support the claim that the outcomes are not equally likely. The outcomes appear to be equally likely, so the loaded die does not appear to behave differently from a fair die.

9. Test statistic: $\chi^2 = 11.161$. P-value $= 0.011$ (Table: <0.025). Critical value: $\chi^2 = 7.815$. There is sufficient evidence to support the claim that the results contradict Mendel's theory.

11. Test statistic: $\chi^2 = 29.814$. P-value $= 0.000$ (Table: <0.005). Critical value: $\chi^2 = 16.812$. There is sufficient evidence to warrant rejection of the claim that the different days of the week have the same frequencies of police calls. The highest numbers of calls appear to fall on Friday and Saturday, and these are weekend days with disproportionately more partying and drinking.

13. Test statistic: $\chi^2 = 13.855$. P-value $= 0.128$ (Table: >0.10). Critical value: $\chi^2 = 16.919$. There is not sufficient evidence to warrant rejection of the claim that the likelihood of winning is the same for the different post positions. Based on these results, post position should not be considered when betting on the Kentucky Derby race.

15. Test statistic: $\chi^2 = 8.882$. P-value $= 0.031$ (Table: <0.05). Critical value: $\chi^2 = 7.815$. There is sufficient evidence to warrant rejection of the claim that the actual numbers of games fit the distribution indicated by the proportions listed in the given table.

17. Test statistic: $\chi^2 = 9.500$. P-value $= 0.147$ (Table: >0.10). Critical value: $\chi^2 = 16.812$. There is not sufficient evidence to support the claim that births do not occur on the seven different days of the week with equal frequency.

19. Test statistic: $\chi^2 = 6.682$. P-value $= 0.245$ (Table: >0.10). Critical value: $\chi^2 = 11.071$. There is not sufficient evidence to warrant rejection of the claim that the color distribution is as claimed.

21. Test statistic: $\chi^2 = 3650.251$. P-value $= 0.000$ (Table: <0.005). Critical value: $\chi^2 = 20.090$. There is sufficient evidence to warrant rejection of the claim that the leading digits are from a population with a distribution that conforms to Benford's law. It does appear that the checks are the result of fraud (although the results cannot confirm that fraud is the cause of the discrepancy between the observed results and the expected results).

23. Test statistic: $\chi^2 = 1.762$. P-value $= 0.988$ (Table: >0.10). Critical value: $\chi^2 = 15.507$. There is not sufficient evidence to warrant rejection of the claim that the leading digits are from a population with a distribution that conforms to Benford's law. The tax entries do appear to be legitimate.

25. a. 26, 46, 49, 26
   b. 0.2023, 0.3171, 0.3046, 0.1761 (Table: 0.2033, 0.3166, 0.3039, 0.1762)
   c. 29.7381, 46.6137, 44.7762, 25.8867 (Table: 29.8851, 46.5402, 44.6733, 25.9014)
   d. Test statistic: $\chi^2 = 0.877$ (Using probabilities from table: 0.931). P-value $= 0.831$ (Table: >0.10). Critical value: $\chi^2 = 11.345$. There is not sufficient evidence to warrant rejection of the claim that heights were randomly selected from a normally distributed population. The test suggests that we cannot rule out the possibility that the data are from a normally distributed population.

Section 11-2

1. a. $E = 4.173$.
   b. Because the expected frequency of a cell is less than 5, the requirements for the hypothesis test are not satisfied.

3. Test statistic: $\chi^2 = 64.517$. P-value: 0.000. Reject the null hypothesis of independence between handedness and cell phone ear preference.

5. Test statistic: $\chi^2 = 25.571$. P-value $= 0.000$ (Table: <0.005). Critical value: $\chi^2 = 3.841$. There is sufficient evidence to warrant rejection of the claim that whether a subject lies is independent of the polygraph test indication. The results suggest that polygraphs are effective in distinguishing between truths and lies, but there are many false positives and false negatives, so they are not highly reliable.

7. Test statistic: $\chi^2 = 576.224$. P-value $= 0.000$ (Table: <0.005). Critical value: $\chi^2 = 3.841$. There is sufficient evidence to warrant rejection of the claim of independence between texting while driving and driving when drinking alcohol. Those two risky behaviors appear to be somehow related.

9. Test statistic: $\chi^2 = 12.162$. P-value $= 0.001$ (Table: <0.005). Critical value: $\chi^2 = 3.841$. There is sufficient evidence to warrant rejection of the claim that whether students purchased gum or kept the money is independent of whether they were given four quarters or a \$1 bill. It appears that there is a denomination effect.

11. Test statistic: $\chi^2 = 0.064$. P-value $= 0.801$ (Table: >0.10). Critical value: $\chi^2 = 3.841$. There is not sufficient evidence to warrant rejection of the claim that the gender of the tennis player is independent of whether the call is overturned. Neither men nor women appear to be better at challenging calls.

13. Test statistic: $\chi^2 = 14.589$. P-value $= 0.0056$ (Table: <0.01). Critical value: $\chi^2 = 9.488$. There is sufficient evidence to warrant rejection of the claim that the direction of the kick is independent of the direction of the goalkeeper jump. The results do not support the theory that because the kicks are so fast, goalkeepers have no time to react. It appears that goalkeepers can choose directions based on the directions of the kicks.

15. Test statistic: $\chi^2 = 2.925$. P-value $= 0.232$ (Table: >0.10). Critical value: $\chi^2 = 5.991$. There is not sufficient evidence to warrant rejection of the claim that getting a cold is independent of the treatment group. The results suggest that echinacea is not effective for preventing colds.

17. Test statistic: $\chi^2 = 20.271$. P-value $= 0.0011$ (Table: <0.005). Critical value: $\chi^2 = 15.086$. There is sufficient evidence to warrant rejection of the claim that cooperation of the subject is independent of the age category. The age group of 60 and over appears to be particularly uncooperative.

19. Test statistic: $\chi^2 = 50.446$. P-value $= 0.000$ (Table: <0.005). Critical value: $\chi^2 = 5.991$. There is sufficient evidence to warrant rejection of the claim of independence between the state and whether a car has front and rear license plates. It does not appear that the license plate laws are followed at the same rates in the three states.

21. Test statistics: $\chi^2 = 12.1619258$ and $z = 3.487395274$, so that $z^2 = \chi^2$. Critical values: $\chi^2 = 3.841$ and $z = \pm 1.96$, so $z^2 = \chi^2$ (approximately).

Chapter 11: Quick Quiz

1. H_0: $p_0 = p_1 = \ldots = p_9$. H_1: At least one of the probabilities is different from the others.
2. $O = 27$ and $E = 30$.
3. Right-tailed.
4. df $= 9$

5. There is not sufficient evidence to warrant rejection of the claim that the last digits are equally likely. Because reported heights would likely include more last digits of 0 and 5, it appears that the heights were measured instead of reported. (Also, most U.S. residents would have difficulty reporting heights in centimeters, because the United States, Liberia, and Myanmar are the only countries that continue to use the Imperial system of measurement.)

6. H_0: Surviving the sinking is independent of whether the person is a man, woman, boy, or girl.
   H_1: Surviving the sinking and whether the person is a man, woman, boy, or girl are somehow related.

7. Chi-square distribution.

8. Right-tailed.

9. df $= 3$

10. There is sufficient evidence to warrant rejection of the claim that surviving the sinking is independent of whether the person is a man, woman, boy, or girl. Most of the women survived, 45% of the boys survived, and most girls survived, but only about 20% of the men survived, so it appears that the rule was followed quite well.

Chapter 11: Review Exercises

1. Test statistic: $X^2 = 787.018$. P-value: 0.000. Critical value: $X^2 = 16.812$. There is sufficient evidence to warrant rejection of the claim that auto fatalities occur on the different days of the week with the same frequency. Because people generally have more free time on weekends and more drinking occurs on weekends, the days of Friday, Saturday, and Sunday appear to have disproportionately more fatalities.

2. Test statistic: $X^2 = 0.751$. P-value $= 0.386$ (Table: >0.10). Critical value: $X^2 = 3.841$. There is not sufficient evidence to warrant rejection of the claim of independence between the type of filling and adverse health conditions. Fillings that contain mercury do not appear to affect health conditions.

3. Test statistic: $X^2 = 5.624$. P-value $= 0.467$ (Table: >0.10). Critical value: $X^2 = 12.592$. There is not sufficient evidence to warrant rejection of the claim that the actual eliminations agree with the expected numbers. The leadoff singers do appear to be at a disadvantage because 20 of them were eliminated compared to the expected value of 12.9 eliminations, but that result does not appear to be *significantly* high.

4. Test statistic: $X^2 = 0.773$. P-value $= 0.856$ (Table: >0.10). Critical value: $X^2 = 11.345$. There is not sufficient evidence to warrant rejection of the claim that getting an infection is independent of the treatment. The atorvastatin (Lipitor) treatment does not appear to have an effect on infections.

5. Test statistic: $X^2 = 269.147$. P-value $= 0.000$ (Table: <0.005). Critical value: $X^2 = 24.725$. There is sufficient evidence to warrant rejection of the claim that weather-related deaths occur in the different months with the same frequency. The months of May, June, and July appear to have disproportionately more weather-related deaths, and that is probably due to the fact that vacations and outdoor activities are much greater during those months.

Chapter 11: Cumulative Review Exercises

1. H_0: $p = 0.5$. H_1: $p \neq 0.5$. Test statistic: $z = 8.96$. P-value: 0.0000 (Table: 0.0002). Critical values: $z = \pm 1.96$. Reject H_0. There is sufficient evidence to warrant rejection of the claim that among those who die in weather-related deaths, the percentage of males is equal to 50%. One possible explanation is that more men participate in some outdoor activities, such as golf, fishing, and boating.

2. a. There is a possibility that the results were affected because the sponsor of the survey produces chocolate and therefore has an interest in the results.
   b. 1452

3. $82.8\% < p < 87.2\%$. We have 99% confidence that the limits of 82.8% and 87.2% contain the value of the true percentage of the population of women saying that chocolate makes them happier.

4. H_0: $p = 0.80$. H_1: $p > 0.80$. Test statistic: $z = 5.18$. P-value $= 0.0000$ (Table: 0.0001). Critical value: $z = 2.33$. Reject H_0. There is sufficient evidence to support the claim that when asked, more than 80% of women say that chocolate makes them happier.

5. Test statistic: $X^2 = 3.409$. P-value $= 0.0648$ (Table: >0.05). Critical value: $X^2 = 3.841$. There is not sufficient evidence to warrant rejection of the claim that the form of the 100-yuan gift is independent of whether the money was spent. There is not sufficient evidence to support the claim of a denomination effect. Women in China do not appear to be affected by whether 100 yuan are in the form of a single bill or several smaller bills.

6. a. $128/150 = 0.853$ b. $143/150 = 0.953$
   c. 0.727 (not 0.728)

7. $r = -0.283$. P-value $= 0.539$. Critical values: ± 0.754 (assuming a 0.05 significance level). There is not sufficient evidence to support the claim of a linear correlation between the repair costs from full-front crashes and full-rear crashes.

8. a. 630 mm
   b. 14.48% (Table: 14.46%). That percentage is too high, because too many women would not be accommodated.
   c. 0.7599 (Table: 0.7611). Groups of 16 women do not occupy a driver's seat or cockpit; because *individual* women occupy the driver's seat/cockpit, this result has no effect on the design.

Chapter 12 Answers

Section 12-1

1. a. The arrival delay times are categorized according to the one characteristic of the flight number.
   b. The terminology of *analysis of variance* refers to the method used to test for equality of the three population means. That method is based on two different estimates of a common population variance.

3. The test statistic is $F = 1.334$, and the F distribution applies.

5. Test statistic: $F = 0.39$. P-value: 0.677. Fail to reject H_0: $\mu_1 = \mu_2 = \mu_3$. There is not sufficient evidence to warrant rejection of the claim that the three categories of blood lead level have the same mean verbal IQ score. Exposure to lead does not appear to have an effect on verbal IQ scores.

7. Test statistic: $F = 5.5963$. P-value: 0.0045. Reject H_0: $\mu_1 = \mu_2 = \mu_3$. There is sufficient evidence to warrant rejection of the claim that the three samples are from populations with the same mean. It appears that at least one of the mean service times is different from the others.

9. Test statistic: $F = 7.9338$. P-value: 0.0005. Reject H_0: $\mu_1 = \mu_2 = \mu_3$. There is sufficient evidence to warrant rejection of the claim that females from the three age brackets have the same mean pulse rate. It appears that pulse rates of females are affected by age bracket.

11. Test statistic: $F = 27.2488$. P-value: 0.000. Reject H_0: $\mu_1 = \mu_2 = \mu_3$. There is sufficient evidence to warrant rejection of the claim that the three different miles have the same mean time. These data suggest that the third mile appears to take longer, and a reasonable explanation is that the third mile has a hill.

13. Test statistic: $F = 2.3163$. P-value: 0.123. Fail to reject H_0: $\mu_1 = \mu_2 = \mu_3$. There is not sufficient evidence to warrant rejection of the claim that the three different flights have the same mean departure delay time. The departure delay times from Flight 1 have very little variation, and departures of Flight 1 appear to be on time or slightly early. Departure delay times from Flight 21 appear to have considerable variation. With variances of 2.5 min^2, 709.8 min^2, and 2525.4 min^2, the ANOVA requirement of the same variance appears to be violated even for this loose requirement.

15. Test statistic: $F = 28.1666$. P-value: 0.000. Reject H_0: $\mu_1 = \mu_2 = \mu_3$. There is sufficient evidence to warrant rejection of the claim that the three different types of Chips Ahoy cookies have the same mean number of chocolate chips. The reduced fat cookies have a mean of 19.6 chocolate chips, which is slightly more than the mean of 19.1 chocolate chips for the chewy cookies, so the reduced fat does not appear to be the result of including fewer chocolate chips. Perhaps the fat content in the chocolate chips is different and/or the fat content in the cookie material is different.

17. The Tukey test results show different P-values, but they are not dramatically different. The Tukey results suggest the same conclusions as the Bonferroni test.

Section 12-2

1. The pulse rates are categorized using *two* different factors of (1) age bracket and (2) gender.

3. a. An interaction between two factors or variables occurs if the effect of one of the factors changes for different categories of the other factor.
   b. If there is an interaction effect, we should not proceed with individual tests for effects from the row factor and column factor. If there is an interaction, we should not consider the effects of one factor without considering the effects of the other factor.
   c. Because the lines are far from parallel, the two genders have very different effects for the different age brackets, so there does appear to be an interaction between gender and age bracket.

5. For interaction, the test statistic is $F = 9.58$ and the P-value is 0.0003, so there is sufficient evidence to warrant rejection of the null hypothesis of no interaction effect. Because there appears to be an interaction between age bracket and gender, we should not proceed with a test for an effect from age bracket and a test for an effect from gender. It appears an interaction between age bracket

and gender has an effect on pulse rates. (Remember, these results are based on fabricated data used in one of the cells, so this conclusion does not necessarily correspond to real data.)

7. For interaction, the test statistic is $F = 1.7970$ and the P-value is 0.1756, so there is not sufficient evidence to conclude that there is an interaction effect. For the row variable of age bracket, the test statistic is $F = 2.0403$ and the P-value is 0.1399, so there is not sufficient evidence to conclude that age bracket has an effect on height. For the column variable of gender, the test statistic is $F = 43.4607$ and the P-value is less than 0.0001, so there is sufficient evidence to support the claim that gender has an effect on height.

9. For interaction, the test statistic is $F = 1.1653$ and the P-value is 0.3289, so there is no significant interaction effect. For gender, the test statistic is $F = 1.6864$ and the P-value is 0.2064, so there is no significant effect from gender. For age, the test statistic is $F = 5.0998$ and the P-value is 0.0143, so there is a significant effect from age.

11. a. Test statistics and P-values do not change.
    b. Test statistics and P-values do not change.
    c. Test statistics and P-values do not change.
    d. An outlier can dramatically affect and change test statistics and P-values.

Chapter 12: Chapter Quick Quiz

1. H_0: $\mu_1 = \mu_2 = \mu_3 = \mu_4$. Because the displayed P-value of 0.000 is small, reject H_0. There is sufficient evidence to warrant rejection of the claim that the four samples have the same mean weight.

2. No. It appears that mean weights of Diet Coke and Diet Pepsi are lower than the mean weights of regular Coke and regular Pepsi, but the method of analysis of variance does not justify a conclusion that any particular means are significantly different from the others.

3. Right-tailed.

4. Test statistic: $F = 503.06$. Larger test statistics result in smaller P-values.

5. The four samples are categorized using only one factor: the type of cola (regular Coke, Diet Coke, regular Pepsi, Diet Pepsi).

6. One-way analysis of variance is used to test a null hypothesis that three or more samples are from populations with equal means.

7. With one-way analysis of variance, data from the different samples are categorized using only one factor, but with two-way analysis of variance, the sample data are categorized into different cells determined by two different factors.

8. Fail to reject the null hypothesis of no interaction. There does not appear to be an effect due to an interaction between sex and major.

9. There is not sufficient evidence to support a claim that the length estimates are affected by the sex of the subject.

10. There is not sufficient evidence to support a claim that the length estimates are affected by the subject's major.

Chapter 12: Review Exercises

1. Test statistic: $F = 2.7347$. P-value: 0.0829. Fail to reject H_0: $\mu_1 = \mu_2 = \mu_3$. There is not sufficient evidence to warrant rejection of the claim that males in the different age brackets give attribute ratings with the same mean. Age does not appear to be a factor in the male attribute ratings.

2. Test statistic: $F = 9.4695$. P-value: 0.0006. Reject H_0: $\mu_1 = \mu_2 = \mu_3$. There is sufficient evidence to warrant rejection of the claim that the three books have the same mean Flesch Reading Ease score. The data suggest that the books appear to have mean scores that are not all the same, so the authors do not appear to have the same level of readability.

3. For interaction, the test statistic is $F = 1.7171$ and the P-value is 0.1940, so there is not sufficient evidence to warrant rejection of no interaction effect. There does not appear to be an interaction between femur and car size. For the row variable of femur, the test statistic is $F = 1.3896$ and the P-value is 0.2462, so there is not sufficient evidence to conclude that whether the femur is right or left has an effect on load. For the column variable of car size, the test statistic is $F = 2.2296$ and the P-value is 0.1222, so there is not sufficient evidence to warrant rejection of the claim of no effect from car size. It appears that the crash test loads are not affected by an interaction between femur and car size, they are not affected by femur, and they are not affected by car size.

4. For interaction, the test statistic is $F = 0.4784$ and the P-value is 0.7513, so there is not sufficient evidence to conclude that there is an interaction effect. For the row variable of age bracket of females, the test statistic is $F = 0.3149$ and the P-value is 0.7318, so there is not sufficient evidence to conclude that the age bracket of females has an effect on the ratings. For the column variable of age bracket of males, the test statistic is $F = 1.1939$ and the P-value is 0.3148, so there is not sufficient evidence to conclude that the age bracket of males has an effect on the ratings.

Chapter 12: Cumulative Review Exercises

1. a. 2.0 min, 9.9 min, 33.4 min
   b. 10.6 min, 26.6 min, 50.3 min
   c. 112.0 min², 709.8 min², 2525.4 min²
   d. The departure delay time of 142 min is an outlier.
   e. Ratio

2. Test statistic: $t = -1.728$. P-value $= 0.1241$ (Table: >0.10). Critical values assuming a 0.05 significance level: $t = \pm 2.326$ (Table: ± 2.365). Fail to reject H_0: $\mu_1 = \mu_2$. There is not sufficient evidence to support the claim that there is a difference between the departure delay times for the two flights.

3. Because the pattern of the points is far from a straight-line pattern, the departure delay times for Flight 19 do not appear to be from a population with a normal distribution.

4. -6.8 min $< \mu < 10.8$ min. We have 95% confidence that the limits of -6.8 min and 10.8 min contain the value of the population mean for all Flight 3 departure delays.

5. a. H_0: $\mu_1 = \mu_2 = \mu_3$
   b. Because the P-value of 0.1729 is greater than the significance level of 0.05, fail to reject the null hypothesis of equal means. There is not sufficient evidence to warrant rejection of the claim that the three means are equal. The three populations do not appear to have means that are significantly different.

6. a. 0.5563 (Table: 0.5552)
   b. 0.3434 (Table: 0.3446)
   c. 1/256 or 0.00391
   d. 5.591 g

7. a. 200
   b. $0.175 < p < 0.225$
   c. Yes. The confidence interval shows us that we have 95% confidence that the true population proportion is contained within the limits of 0.175 and 0.225, and 1/4 is not included within that range.

8. a. Because the vertical scale begins at 15 instead of 0, the graph is deceptive by exaggerating the differences among the frequencies.
   b. No. A normal distribution is approximately bell-shaped, but the given histogram is far from being bell-shaped. Because the digits are supposed to be equally likely, the histogram should be flat with all bars having approximately the same height.
   c. The frequencies are 19, 21, 22, 21, 18, 23, 16, 16, 22, 22.
   d. Test statistic: $X^2 = 3.000$. P-value $= 0.964$ (Table: >0.95). Critical value: $X^2 = 16.919$ (assuming a 0.05 significance level). There is not sufficient evidence to warrant rejection of the claim that the digits are selected from a population in which the digits are all equally likely. There does not appear to be a problem with the lottery.

Chapter 13 Answers

Section 13-2

1. The only requirement for the matched pairs is that they constitute a simple random sample. There is no requirement of a normal distribution or any other specific distribution. The sign test is "distribution free" in the sense that it does not require a normal distribution or any other specific distribution.

3. H_0: There is no difference between the populations of September weights and the matching April weights. H_1: There is a difference between the populations of September weights and the matching April weights. The sample data do not contradict H_1 because the numbers of positive signs (2) and negative signs (7) are not exactly the same.

5. The test statistic of $x = 3$ is not less than or equal to the critical value of 1 (from Table A-7). There is not sufficient evidence to warrant rejection of the claim of no difference. There is not sufficient evidence to support the claim that there is a difference between female attribute ratings and male attribute ratings.

7. The test statistic of $z = -1.61$ results in a P-value of 0.1074, and it does not fall in the critical region bounded by $z = -1.96$ and 1.96. There is not sufficient evidence to warrant rejection of the claim of no difference. There is not sufficient evidence to support the claim that there is a difference between female attribute ratings and male attribute ratings.

9. The test statistic of $z = -2.66$ results in a P-value of 0.0078, and it is in the critical region bounded by $z = -2.575$ and 2.575. There is sufficient evidence to warrant rejection of the claim that there is no difference between the proportions of those opposed and those in favor.

11. The test statistic of $z = -0.24$ results in a P-value of 0.8103, and it is not in the critical region bounded by $z = -1.96$ and 1.96. There is not sufficient evidence to reject the claim that boys and girls are equally likely.

13. The test statistic of $z = -14.93$ results in a P-value of 0.0000, and it is in the critical region bounded by $z = -2.575$ and 2.575. There is sufficient evidence to warrant rejection of the claim that the median is equal to 2.00.

15. The test statistic of $z = -2.37$ results in a P-value of 0.0178 and it is not in the critical region bounded by $z = -2.575$ and 2.575. There is not sufficient evidence to warrant rejection of the claim that the median is equal to 5.670 g. The quarters appear to be minted according to specifications.

17. Second approach: The test statistic of $z = -4.29$ results in a P-value of 0.0000 and it is in the critical region bounded by $z = -1.645$, so the conclusions are the same as in Example 4. Third approach: The test statistic of $z = -2.82$ results in a P-value of 0.0048 and it is in the critical region bounded by $z = -1.645$, so the conclusions are the same as in Example 4. The different approaches can lead to very different results; see the test statistics of -4.21, -4.29, and -2.82. The conclusions are the same in this case, but they could be different in other cases.

Section 13-3

1. a. The only requirements are that the matched pairs be a simple random sample and the population of differences be approximately symmetric.
   b. There is no requirement of a normal distribution or any other specific distribution.
   c. The Wilcoxon signed-ranks test is "distribution free" in the sense that it does not require a normal distribution or any other specific distribution.

3. The sign test uses only the signs of the differences, but the Wilcoxon signed-ranks test uses ranks that are affected by the magnitudes of the differences.

5. Test statistic: $T = 16.5$. Critical value: $T = 6$. Fail to reject the null hypothesis that the population of differences has a median of 0. There is not sufficient evidence to support the claim that there is a difference between female attribute ratings and male attribute ratings.

7. Convert $T = 8323.5$ to the test statistic $z = -0.63$. P-value: 0.5287. Critical values: $z = \pm 1.96$. There is not sufficient evidence to warrant rejection of the claim of no difference. There is not sufficient evidence to support the claim that there is a difference between female attribute ratings and male attribute ratings.

9. Convert $T = 18,014$ to the test statistic $z = -16.92$. P-value: 0.0000. Critical values: $z = \pm 2.575$. There is sufficient evidence to warrant rejection of the claim that the median is equal to 2.00.

11. Convert $T = 196$ to the test statistic $z = -2.88$. P-value: 0.0040. Critical values: $z = \pm 2.575$. There is sufficient evidence to warrant rejection of the claim that the median is equal to 5.670 g. The quarters do not appear to be minted according to specifications.

13. a. 0 and 31,375
    b. 15,687.5
    c. 30,141
    d. $\dfrac{n(n + 1)}{2} - k$

Section 13-4

1. Yes. The two samples are independent. The evaluations of female professors and male professors are not matched in any way. The samples are simple random samples. Each sample has more than 10 values.

3. H_0: Evaluations of female professors and male professors have the same median. There are three different possible alternative hypotheses:
   H_1: Evaluations of female professors and male professors have different medians.
   H_1: Evaluations of female professors have a median greater than the median of male professor evaluations.
   H_1: Evaluations of female professors have a median less than the median of male professor evaluations.

5. $R_1 = 163, R_2 = 188, \mu_R = 189, \sigma_R = 19.4422$, test statistic: $z = -1.34$. P-value: 0.1802. Critical values: $z = \pm 1.96$. Fail to reject the null hypothesis that the populations have the same median. There is not sufficient evidence to warrant rejection of the claim that evaluation ratings of female professors have the same median as evaluation ratings of male professors.

7. $R_1 = 253.5, R_2 = 124.5, \mu_R = 182, \sigma_R = 20.607$, test statistic: $z = 3.47$. P-value: 0.0005. Critical values: $z = \pm 1.96$. Reject the null hypothesis that the populations have the same median. There is sufficient evidence to reject the claim that for those treated with 20 mg of Lipitor and those treated with 80 mg of Lipitor, changes in LDL cholesterol have the same median. It appears that the dosage amount does have an effect on the change in LDL cholesterol.

9. $R_1 = 1615.5, R_2 = 2755.5, \mu_R = 1880, \sigma_R = 128.8669$, test statistic: $z = -2.05$. P-value: 0.0404. Critical values: $z = \pm 1.96$. Reject the null hypothesis that the populations have the same median. There is sufficient evidence to warrant rejection of the claim that evaluation ratings of female professors have the same median as evaluation ratings male of professors.

11. $R_1 = 501, R_2 = 445, \mu_R = 484, \sigma_R = 41.15823$, test statistic: $z = 0.41$. P-value: 0.3409. Critical value: $z = 1.645$. Fail to reject the null hypothesis that the populations have the same median. There is not sufficient evidence to support the claim that subjects with medium lead levels have a higher median of the full IQ scores than subjects with high lead levels. Based on these data, it does not appear that lead level affects full IQ scores.

13. Using $U = 98.5$, we get $z = 0.41$. The test statistic is the same value with opposite sign.

Section 13-5

1. $R_1 = 164.5, R_2 = 150, R_3 = 150.5$

3. $n_1 = 10, n_2 = 10, n_3 = 10$, and $N = 30$

5. Test statistic: $H = 0.1748$. Critical value: $\chi^2 = 5.991$. (Tech: P-value $= 0.916$.) Fail to reject the null hypothesis of equal medians. There is not sufficient evidence to warrant rejection of the claim that females from the different age brackets give attribute ratings with the same median.

7. Test statistic: $H = 4.9054$. Critical value: $\chi^2 = 5.991$. (Tech: P-value $= 0.086$.) Fail to reject the null hypothesis of equal medians. The data do not suggest that larger cars are safer.

9. Test statistic: $H = 11.4704$. Critical value: $\chi^2 = 5.991$. (Tech: P-value $= 0.003$.) Reject the null hypothesis of equal medians. It appears that the three restaurants have dinner drive-through service times with different medians.

11. Test statistic: $H = 2.5999$. Critical value: $\chi^2 = 7.815$. (Tech: P-value $= 0.458$.) Fail to reject the null hypothesis of equal medians. It appears that the four hospitals have birth weights with the same median.

13. The values of t are 2, 2, 2, 2, and 4, so the values of T are 6, 6, 6, 6, and 60 and $\Sigma T = 84$. Using $\Sigma T = 84$ and $N = 19$, the corrected value of H is 0.703, which is not substantially different from the value of 0.694 found in Example 1. In this case, the large numbers of ties do not appear to have a dramatic effect on the test statistic H.

Section 13-6

1. The methods of Section 10-2 should not be used for predictions. The regression equation is based on a *linear* correlation between the two variables, but the methods of this section do not require a linear relationship. The methods of this section could suggest that there is a correlation with paired data associated by some nonlinear relationship, so the regression equation would not be a suitable model for making predictions.

3. r represents the linear correlation coefficient computed from sample paired data; ρ represents the parameter of the linear correlation coefficient computed from a population of paired data; r_s denotes the rank correlation coefficient computed from sample paired data; ρ_s represents the rank correlation coefficient computed from a population of paired data. The subscript s is used so that the rank correlation coefficient can be distinguished from the linear correlation coefficient r. The subscript does not represent the standard deviation s. It is used in recognition of Charles Spearman, who introduced the rank correlation method.

5. $r_s = 1$. Critical values are -0.886 and 0.886. Reject the null hypothesis of $\rho_s = 0$. There is sufficient evidence to support a claim of a correlation between distance and time.

7. $r_s = 0.888$. Critical values: $-0.618, 0.618$. Reject the null hypothesis of $\rho_s = 0$. There is sufficient evidence to support the claim of a correlation between chocolate consumption and the rate of Nobel Laureates. It does not make sense to think that there is a cause/effect relationship, so the correlation could be the result of a coincidence or other factors that affect the variables the same way.

9. $r_s = 1.000$. Critical values: $-0.700, 0.700$. Reject the null hypothesis of $\rho_s = 0$. There is sufficient evidence to support the claim of a correlation between the cost of a slice of pizza and the subway fare.

11. $r_s = 1$. Critical values: $-0.886, 0.886$. Reject the null hypothesis of $\rho_s = 0$. There is sufficient evidence to conclude that there is a correlation between overhead widths of seals from photographs and the weights of the seals.

13. $r_s = 0.902$. Critical values: $-0.415, 0.415$. Reject the null hypothesis of $\rho_s = 0$. There is sufficient evidence to support the claim of a correlation between chocolate consumption and the rate of Nobel Laureates. It does not make sense to think that there is a cause/effect relationship, so the correlation could be the result of a coincidence or other factors that affect the variables the same way.

15. $r_s = 0.360$. Critical values: $-0.159, 0.159$. Reject the null hypothesis of $\rho_s = 0$. There is sufficient evidence to conclude that there is a correlation between the systolic and diastolic blood pressure levels in males.

17. -0.159 and 0.159. (Use either $t = 1.975799$ from technology or use interpolation in Table A-3 with 151 degrees of freedom, so the critical value of t is approximately halfway between 1.984 and 1.972, which is 1.978.) The critical values are the same as those found by using Formula 13-1.

Section 13-7

1. No. The runs test can be used to determine whether the sequence of political parties is not random, but the runs test does not show whether the proportion of Republicans is significantly greater than the proportion of Democrats.

3. The critical values are 8 and 19. Because $G = 16$ is not less than or equal to 8 nor is $G = 16$ greater than or equal to 18, fail to reject randomness. It appears that the sequence of political parties is random.

5. $\bar{x} = 157.7$ fatalities. $n_1 = 11$, $n_2 = 9$, $G = 12$, critical values: 6, 16. Fail to reject randomness. There is not sufficient evidence to warrant rejection of the claim that there is randomness above and below the mean. There does not appear to be a trend.

7. $n_1 = 20$, $n_2 = 10$, $G = 16$, critical values: 9, 20. Fail to reject randomness. There is not sufficient evidence to reject the claim that the dates before and after July 1 are randomly selected.

9. $n_1 = 26$, $n_2 = 23$, $G = 20$, $\mu_G = 25.40816$, $\sigma_G = 3.450091$. Test statistic: $z = -1.57$. P-value: 0.1164. Critical values: $z = \pm 1.96$. Fail to reject randomness. There is not sufficient evidence to reject randomness. The runs test does not test for disproportionately more occurrences of one of the two categories, so the runs test does not suggest that either conference is superior.

11. The median is 2895.5, $n_1 = 25$, $n_2 = 25$, $G = 2$, $\mu_G = 26$, $\sigma_G = 3.49927$. Test statistic: $z = -6.86$. P-value: 0.0000. Critical values: $z = \pm 1.96$. Reject randomness. The sequence does not appear to be random when considering values above and below the median. There appears to be an upward trend, so the stock market appears to be a profitable investment for the long term.

13. b. The 84 sequences yield these results: 2 sequences have 2 runs, 7 sequences have 3 runs, 20 sequences have 4 runs, 25 sequences have 5 runs, 20 sequences have 6 runs, and 10 sequences have 7 runs.

c. With $P(2 \text{ runs}) = 2/84$, $P(3 \text{ runs}) = 7/84$, $P(4 \text{ runs}) = 20/84$, $P(5 \text{ runs}) = 25/84$, $P(6 \text{ runs}) = 20/84$, and $P(7 \text{ runs}) = 10/84$, each of the G values of 3, 4, 5, 6, 7 can easily occur by chance, whereas $G = 2$ is unlikely because $P(2 \text{ runs})$ is less than 0.025. The lower critical value of G is therefore 2, and there is no upper critical value that can be equaled or exceeded.

d. Critical value of $G = 2$ agrees with Table A-10. The table lists 8 as the upper critical value, but it is impossible to get 8 runs using the given elements.

Chapter 13: Quick Quiz

1. 1, 3, 3, 5, 3
2. The efficiency rating of 0.91 indicates that with all other factors being the same, rank correlation requires 100 pairs of sample observations to achieve the same results as 91 pairs of observations with the parametric test for linear correlation, assuming that the stricter requirements for using linear correlation are met.
3. a. Distribution-free test

   b. The term "distribution-free test" suggests correctly that the test does not require that a population must have a particular distribution, such as a normal distribution. The term "nonparametric test" incorrectly suggests that the test is not based on a parameter, but some nonparametric tests are based on the median, which is a parameter; the term "distribution-free test" is better because it does not make that incorrect suggestion.
4. Rank correlation should be used. The rank correlation test is used to investigate whether there is a correlation between foot length and height.
5. No, the P-values are almost always different, and the conclusions may or may not be the same.
6. Rank correlation can be used in a wider variety of circumstances than linear correlation. Rank correlation does not require a normal distribution for any population. Rank correlation can be used to detect some (not all) relationships that are not linear.
7. Because there are only two runs, all of the values below the mean occur at the beginning and all of the values above the mean occur at the end, or vice versa. This indicates the presence of an upward (or downward) trend.
8. a. False

   b. False
9. Because the sign test uses only *signs* of differences while the Wilcoxon signed-ranks test uses *ranks* of the differences, the Wilcoxon signed-ranks test uses more information about the data and tends to yield conclusions that better reflect the true nature of the data.
10. Kruskal-Wallis test

Chapter 13: Review Exercises

1. $r_s = 0.400$. Critical values: $-0.700, 0.700$. Fail to reject the null hypothesis of $\rho_s = 0$. There is not sufficient evidence to support the claim of a correlation between job stress and annual income. Based on the given data, it does not appear that jobs with more stress have higher salaries.
2. Test statistic: $H = 2.5288$. (Tech: P-value $= 0.2824$.) Critical value: $X^2 = 5.991$. Fail to reject the null hypothesis of equal medians. It appears that times of longevity after inauguration for presidents, popes, and British monarchs have the same median.
3. The test statistic of $z = -1.43$ results in a P-value of 0.1527 and it is not less than or equal to the critical value of $z = -1.96$. Fail to reject the null hypothesis of $p = 0.5$. There is not sufficient evidence to warrant rejection of the claim that in each World Series, the American League team has a 0.5 probability of winning.
4. $n_1 = 16$, $n_2 = 14$, $G = 11$, critical values: 10, 22. Fail to reject randomness. There is not sufficient evidence to warrant rejection of the claim that odd and even digits occur in random order. The lottery appears to be working as it should.

5. The test statistic of $x = 3$ is less than or equal to the critical value of 5 (from Table A-7). There is sufficient evidence to warrant rejection of the claim that the sample is from a population with a median equal to 5.
6. Test statistic $T = 21$ is less than or equal to the critical value of 59. There is sufficient evidence to warrant rejection of the claim that the sample is from a population with a median equal to 5.
7. $R_1 = 204.5$, $R_2 = 230.5$, $\mu_R = 255$, $\sigma_R = 22.58318$, test statistic: $z = -2.24$. Tech: P-value: 0.025. Critical values: $z = \pm 1.96$. Reject the null hypothesis that the populations have the same median. There is sufficient evidence to warrant rejection of the claim that the recent eruptions and past eruptions have the same median time interval between eruptions. The conclusion does change with a 0.01 significance level.
8. The test statistic of $x = 0$ is less than or equal to the critical value of 0. There is sufficient evidence to reject the claim of no difference. It appears that there is a difference in cost between flights scheduled 1 day in advance and those scheduled 30 days in advance. Because all of the flights scheduled 30 days in advance cost less than those scheduled 1 day in advance, it appears to be wise to schedule flights 30 days in advance.
9. The test statistic of $T = 0$ is less than or equal to the critical value of 4. There is sufficient evidence to reject the claim that differences between fares for flights scheduled 1 day in advance and those scheduled 30 days in advance have a median equal to 0. Because all of the flights scheduled 30 days in advance cost less than those scheduled 1 day in advance, it appears to be wise to schedule flights 30 days in advance.
10. $r_s = 0.714$. Critical values: ± 0.738. Fail to reject the null hypothesis of $\rho_s = 0$. There is not sufficient evidence to support the claim that there is a correlation between the student ranks and the magazine ranks. When ranking colleges, students and the magazine do not appear to agree.

Chapter 13: Cumulative Review Exercises

1. Flight 1: $\bar{x} = -1.3$ min, median is -2.0 min, $s = 1.6$ min.
   Flight 19: $\bar{x} = 9.9$ min, median is -0.5 min, $s = 26.6$ min.
   Flight 21: $\bar{x} = 33.4$ min, median is 15.5 min, $s = 50.3$ min.
   The means appear to be very different, with Flight 21 having the longest departure delay times. The medians appear to be very different, with Flight 21 having the longest departure delay times. The standard deviations appear to be very different, with Flight 21 having the greatest amount of variation. Flight 21 appears to be the least predictable flight because it has the highest variation, and it appears to have the longest departure delay times.
2. The normal quantile plot suggests that departure delay times for Flight 19 are not normally distributed.

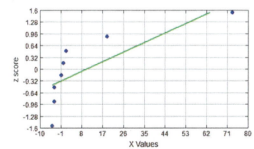

3. Kruskal-Wallis test statistic: $H = 3.2600$. Tech: P-value $= 0.1959$. Critical value: $\chi^2 = 5.991$ (assuming a 0.05 significance level). Fail to reject the null hypothesis of equal medians. There is not sufficient evidence to warrant rejection of the claim that the three samples are from populations with the same median departure delay time.

4. $3.1\% < p < 4.7\%$. We have 95% confidence that the limits of 3.1% and 4.7% actually contain the true percentage of the population of workers who test positive for drugs.

5. H_0: $p = 0.03$. H_1: $p > 0.03$. Test statistic: $z = 2.36$. P-value: 0.0092 (Table: 0.0091). Critical value: $z = 1.645$. Reject H_0. There is sufficient evidence to support the claim that the rate of positive drug test results among workers in the United States is greater than 3.0%.

6. The sample mean is 54.8 years. $n_1 = 19$, $n_2 = 19$, and the number of runs is $G = 18$. The critical values are 13 and 27. Fail to reject the null hypothesis of randomness. There is not sufficient evidence to warrant rejection of the claim that the sequence of ages is random relative to values above and below the mean. The results do not suggest that there is an upward trend or a downward trend.

7. 2401

8. There is a relatively small number of players with salaries that are substantially large, so the mean is strongly affected by those values, resulting in a large value of the mean, but the median is not affected by the small number of very large salaries.

9. H_0: $p = 0.5$. H_1: $p > 0.5$. Test statistic: $z = 1.36$. P-value: 0.0865 (Table: 0.0869). Critical value: $z = 1.645$. Fail to reject H_0. There is not sufficient evidence to support the claim that the majority of the population is not afraid of heights in tall buildings. Because respondents themselves chose to reply, the sample is a voluntary response sample, not a random sample, so the results might not be valid.

10. There must be an error, because the rates of 13.7% and 10.6% are not possible with samples of size 100.

Chapter 14 Answers

Section 14-1

1. No. If we know that the process is within statistical control, we know that none of the three out-of-control criteria are satisfied, but we know nothing about whether any specifications or requirements are satisfied. It is possible to be within statistical control while manufacturing altimeters with errors that are too large to satisfy the FAA requirements.

3. The mean is out of statistical control. The elevations have decreased substantially in recent years, so Lake Mead is becoming shallower. The decreases have been significant (and they are having a dramatic impact on the affected populations).

5. $\bar{\bar{x}} = 267.11$ lb, $\bar{R} = 54.96$ lb. For R chart: LCL $= 4.18$ lb and UCL $= 105.74$ lb. For $\bar{x}$ chart: LCL $= 244.08$ lb and UCL $= 290.14$ lb.

7. The R chart does not meet any of the three out-of-control criteria, so the variation of the process appears to be within statistical control. (Answer continued in right column)

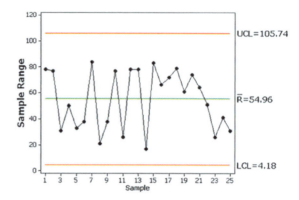

9. $\bar{\bar{x}} = 5.6955$ g, $\bar{R} = 0.2054$ g. For R chart: LCL $= 0.0000$ g and UCL $= 0.4342$ g. For $\bar{x}$ chart: LCL $= 5.5770$ g and UCL $= 5.8140$ g.

11. There are points lying beyond the upper control limit, so the process mean appears to be out of statistical control.

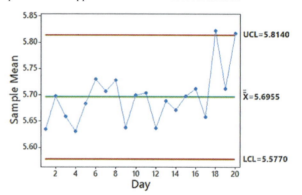

13. Except for the values on the vertical scale, the s chart is nearly identical to the R chart shown in Example 3.

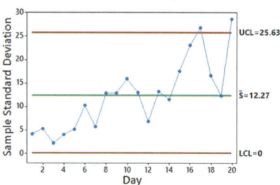

Section 14-2

1. No, the process appears to be out of statistical control. There is a downward trend and there are at least eight consecutive points all lying above the centerline. Because the proportions of defects are decreasing, the manufacturing process is not deteriorating; it is improving.

3. Because the value of -0.00325 is negative and the actual proportion of defects cannot be less than 0, we should replace that value with 0.

5. The process appears to be within statistical control. (Considering a shift up, note that the first and last points are about the same.)

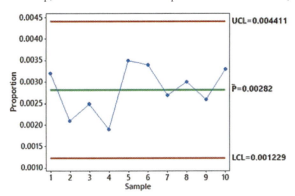

7. The process appears to be out of statistical control because of a downward trend, but the number of defects appears to be decreasing, so the process is improving. Causes for the declining number of defects should be identified so that they can be continued.

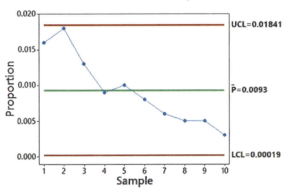

9. The process is out of statistical control because there are points lying beyond the upper control limit and there are points lying beyond the lower control limit. Also, there are eight consecutive points all lying below the centerline. The percentage of voters is increasing in recent presidential elections, and it should be much higher than any of the rates shown.

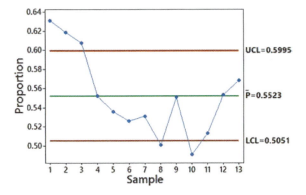

11. Although the process is within statistical control, the proportions of defects are substantially high, so immediate corrective action should be taken to substantially lower the proportions of defects.

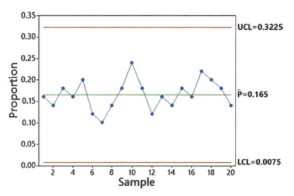

13. Except for a different vertical scale, the basic control chart is identical to the one given for Example 1.

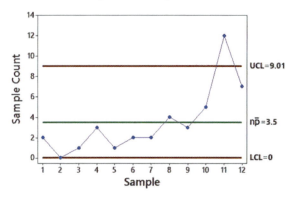

Chapter 14: Quick Quiz

1. Process data are data arranged according to some time sequence. They are measurements of a characteristic of goods or services that result from some combination of equipment, people, materials, methods, and conditions.

2. Random variation is due to chance, but assignable variation results from causes that can be identified, such as defective machinery or untrained employees.

3. There is a pattern, trend, or cycle that is obviously not random. There is a point lying outside the region between the upper and lower control limits. There are at least eight consecutive points all above or all below the centerline.

4. An R chart uses ranges to monitor variation, but an $\bar{x}$ chart uses sample means to monitor the center (mean) of a process.

5. No. The R chart has at least eight consecutive points all lying below the centerline, there are at least eight consecutive points all lying above the centerline, there are points lying beyond the upper and lower control limits, and there is a pattern showing that the ranges have jumped in value for the more recent samples. What a mess!

6. $\bar{R} = 67.0$ ft. In general, a value of $\bar{R}$ is found by first finding the range for the values within each individual subgroup; the mean of those ranges is the value of $\bar{R}$.

7. No. The $\bar{x}$ chart has a point lying beyond the upper control limit, and there are at least eight consecutive points lying below the centerline.

8. $\bar{\bar{x}} = -2.24$ ft. In general, a value of $\bar{\bar{x}}$ is found by first finding the mean of the values within each individual subgroup; the mean of those subgroup means is the value of $\bar{\bar{x}}$.

9. No. The control charts can be used to determine whether the mean and variation are within statistical control, but they do not reveal anything about specifications or requirements.

10. Because there is a downward trend, the process is out of statistical control, but the rate of defects is decreasing, so we should investigate and identify the cause of that trend so that it can be continued.

Chapter 14: Review Exercises

1. $\bar{\bar{x}} = 3157$ kWh, $\bar{R} = 1729$ kWh. R chart: LCL $= 0$ kWh, UCL $= 3465$ kWh. $\bar{x}$ chart: LCL $= 2322$ kWh, UCL $= 3992$ kWh.

2. The process variation is within statistical control.

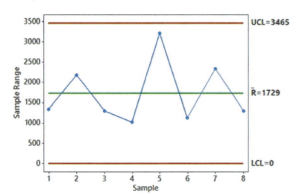

3. There appears to be a shift up in the mean values, so the process mean is out of statistical control.

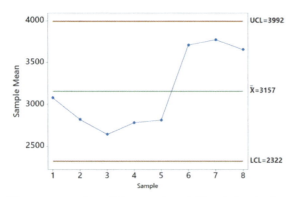

4. There appears to be a slight upward trend. There is 1 point that appears to be exceptionally low. (The author's power company

made an error in recording and reporting the energy consumption for that time period.)

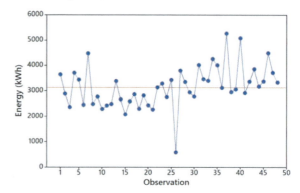

5. Because there is a distinct upward trend and there is a point beyond the upper control limit, the process is out of statistical control. Because the order times are clearly increasing, immediate corrective action should be taken.

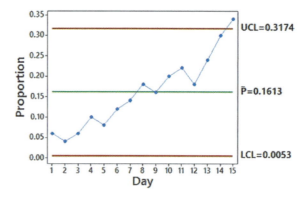

Chapter 14: Cumulative Review Exercises

1. $0.528 < p < 0.572$. Because the entire range of values in the confidence interval includes values that are all greater than 0.5, it does appear that the majority of adults learn about medical symptoms more often from the Internet than from their doctor.

2. $H_0: p = 0.5$. $H_1: p > 0.5$. Test statistic: $z = 4.48$. P-value: 0.0000 (Table: 0.0001). Critical value: $z = 1.645$.
Reject H_0. There is sufficient evidence to support the claim that the majority of adults learn about medical symptoms more often from the Internet than from their doctor.

3. The graph is misleading. The vertical scale begins with a frequency of 800 instead of 0, so the difference between the "yes" and "no" responses is greatly exaggerated.

4. a. 0.166   b. 0.909

5. $r = 0.356$. P-value: 0.313 (Table >0.05). Critical values: $r = \pm 0.632$. There is not sufficient evidence to support a claim of a linear correlation between the DJIA and sunspot numbers. Because we do not expect any relationship between sunspot numbers and the behavior of stocks, this result is not surprising.

6. $\hat{y} = 9772 + 79.2x$. The best predicted value of the DJIA in the year 2004 is $\bar{y} = 13{,}423.6$, and that value is not close to the actual 2004 value of 10,855.

7. a. 99.89% (Table: 99.88%)
   b. 0.1587

8. There is a pattern of an upward trend, so the process is out of statistical control.

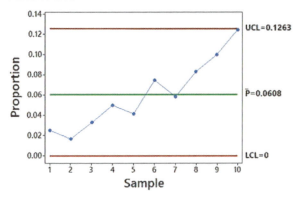

9. $\bar{x} = 7.3$, median $= 6.5$, $s = 4.2$. These statistics do not convey information about the changing pattern of the data over time.

10. Test statistic: $X^2 = 42.557$. P-value: 0.000 (Table <0.005). Critical value: $X^2 = 3.841$. There is sufficient evidence to warrant rejection of the claim that the sentence is independent of the plea. The results encourage pleas for guilty defendants.

CREDITS

Photos

Chapter 1

P1, GongTo/Shutterstock; **P1**, Stokkete/Shutterstock; **P5**, Gary Blakeley/Shutterstock; **P6**, USBFCO/Shutterstock; **P7**, Wavebreakmedia/Shutterstock; **P8**, Donskarpo/Shutterstock; **P14**, 18percentgrey/Shutterstock; **P15**, Khamidulin Sergey/Shutterstock; **P15**, Suppakij1017/Shutterstock; **P16**, Chris DeRidder/Shutterstock; **P17**, Ollyy/Shutterstock; **P18**, Allstar Picture Library/Alamy Stock Photo; **P20**, Lucky Business/Shutterstock; **P26**, Andersen Ross/Stockbyte/Getty Images; **P27**, Pixsooz/Shutterstock; **P27**, Triff/Shutterstock; **P29**, Auremar/Shutterstock; **P30**, Fujji/Shutterstock; **P31**, Dotshock/Shutterstock

Chapter 2

P40, Ministr-84/Shutterstock; **P43**, Gallofoto/Shutterstock; **P46**, Image Source; **P53**, Valua Vitaly/Shutterstock; **P58**, Stockbyte/Getty Images; **P72**, Dmitriy Eremenkov/Shutterstock

Chapter 3

P80, 06photo/Shutterstock; **P83**, Wavebreakmedia/Shutterstock; **P84**, Image Point Fr/Shutterstock; **P87**, Stefan Schurr/Shutterstock; **P98**, Kitch Bain/Shutterstock; **P99**, Alfredo Ragazzoni/Shutterstock; **P100**, Sozaijiten; **P102**, Ariwasabi/Shutterstock; **P115**, Trinacria Photo/Shutterstock; **P117**, Poleze/Shutterstock

Chapter 4

P131, Sculpies/Shutterstock; **P135**, Photomatz/Shutterstock; **P136**, Dansin/E+/Getty Images; **P136**, Pakhnyushcha/Shutterstock; **P136**, Worker/Shutterstock; **P137**, PeJo/Shutterstock; **P138**, Ssuaphotos/Shutterstock; **P139**, Bochkarev Photography/Shutterstock; **P141**, Nomad_Soul/Shutterstock; **P142**, Dgbomb/Shutterstock; **P152**, Ssuaphotos/Shutterstock; **P153**, Jim Barber/Shutterstock; **P154**, Chepe Nicoli/Shutterstock; **P160**, Brian sullivan/E+/Getty Images; **P161**, Africa Studio/Shutterstock; **P162**, Alexander Raths/Shutterstock; **P164**, Philip Gangler/Shutterstock; **P165**, Stephen VanHorn/Shutterstock; **P169**, Bioraven/Shutterstock; **P170**, Serg Shalimoff/Shutterstock; **P172**, Naki Kouyioumtzis/Pearson Education, Inc.; **P173**, Zentilia/123RF

Chapter 5

P184, Fuse/Getty Images; **P189**, JHDT Stock Images LLC/Shutterstock; **P190**, Laborant/Shutterstock; **P201**, B Calkins/Shutterstock; **P205**, Vitalinka/Shutterstock; **P225**, Alfred Eisenstaedt/The LIFE Picture Collection/Getty Images; **P225**, General Motors Cancer Research Foundation

Chapter 6

P226, Stacy L. Pearsall/Aurora/Getty Images; **P230**, Rui Santos/123RF; **P237**, Elena Stepanova/Shutterstock; **P238**, Elena Yakusheva/Shutterstock; **P269**, Ciurea Adrian/Shutterstock

Chapter 7

P297, Syda Productions/Shutterstock; **P300**, Mike Flippo/Shutterstock; **P302**, Stocksnapper/Shutterstock; **P304**, Mama_Mia/Shutterstock; **P305**, Roland Ijdema/Shutterstock; **P308**, Juan Camilo Bernal/Shutterstock; **P319**, Eric Isselée/Fotolia; **P320**, Robin W/Shutterstock; **P321**, Gary Blakeley/Shutterstock; **P322**, Rafael Ramirez Lee/Shutterstock; **P343**, Andresr/Shutterstock

Chapter 8

P356, Tiero/Fotolia; **P364**, Azuzl/Shutterstock; **P367**, Viktoria/Shutterstock; **P370**, David H.Seymour/Shutterstock; **P376**, Suravid/Shutterstock; **P379**, Alexey Burmakin/123RF; **P381**, Pete Saloutos/Shutterstock; **P392**, Zentilia/Shutterstock

Chapter 9

P414, Dorothy Alexander/Alamy Stock Photo; **P418**, Bikeriderlondon/Shutterstock; **P420**, Sergei Telegin/Shutterstock; **P421**, Alex Staroseltsev/Shutterstock; **P429**, GJS/Shutterstock; **P432**, Amy Walters/Shutterstock; **P433**, David Lee/Shutterstock; **P435**, Christy Thompson/Shutterstock; **P444**, D7INAMI7S/Shutterstock; **P445**, Andrey Arkusha/Shutterstock; **P446**, Pressmaster/Shutterstock

Chapter 10

P468, Greatandlittle/123RF; **P475**, John Roman Images/Shutterstock; **P477**, Alexander Raths/Shutterstock; **P480**, Africa Studio/Shutterstock; **P493**, Zffoto/Shutterstock; **P495**, Doglikehorse/Shutterstock; **P517**, David Madison/Photodisc/Getty Images; **P523**, Leah-Anne Thompson/Shutterstock

Chapter 11

P533, Pedro Miguel Sousa/Shutterstock; **P539**, AlexKalashnikov/Shutterstock; **P541**, Serhiy Shullye/Shutterstock; **P549**, Tatiana Popova/Shutterstock; **P551**, Lenetstan/Shutterstock; **P552**, Dundanim/Shutterstock

Chapter 12

P566, Bikeriderlondon/Shutterstock

Chapter 13

P597, 97/E+/Getty Images; **P636**, Sinisa Botas/Shutterstock; **P642**, Dotshock/Shutterstock; **P644**, Luna Vandoorne/Shutterstock

Chapter 14

P654, Maravic/Vetta/Getty Images; **P657**, Oleg Zabielin/Shutterstock; **P658**, Blaz Kure/Shutterstock; **P660**, Michael Rosskothen/Shutterstock; **P661**, Moreno Soppelsa/Shutterstock; **P662**, Age Fotostock/SuperStock; **P663**, Blend Images/Shutterstock; **P669**, Tupungato/Shutterstock

Chapter 15

P680, Umberto Shtanzman/Shutterstock

COVER

Laura A. Watt/Getty Images; Robert Essel NYC/Getty Images

Text

Chapter 1

P4, Ryan and Arnold, *American Journal of Public Health*, Vol. 101, No. 10; **P8**, From *Lies, Damn Lies, And Statistics: The Manipulation Of Public Opinion In America by 1976*. Published by Dell Publishing, © 1976; **P9**, Based on data from the Centers for Disease Control and Prevention and the Department of Energy; **P10**, *Journal of Nutrition* (Vol. 130, No. 8); **P11**, By the National Center for Health Statistics; **P14**, Based on data in "Texting While Driving and Other Risky Motor Vehicle Behaviors Among US High School Students," by Olsen, Shults, Eaton, *Pediatrics*, Vol. 131, No. 6; **P23**, Based on data from KRC Research; **P32**, Based on data from *Journal of Great Lakes Research*; **P33**, Based on data from "A Randomized Trial Comparing Acupuncture, Simulated Acupuncture, and Usual Care for Chronic Low Back Pain," by Cherkin et al., *Archives of Internal Medicine*, Vol. 169, No. 9; **P33**, Based on data from Pfizer, Inc.; **P36**, Based on data from Consumer Reports magazine

Chapter 2

P44–45, Based on data from the Centers for Disease Control; **Multiple pages**, Screenshots from Minitab. Courtesy of Minitab Corporation; **P59–60**, Based on data from "Misconduct Accounts for the Majority of Retracted Scientific Publications," by Fang, Steen, Casadevall, *Proceedings of the National Academy of Sciences of the United States of America*, Vol. 110, No. 3; **P66**, Based on data from TE Connectivity; **P66**, Based on data from *The Ladders*; **Multiple pages**, Screenshots from StatCrunch. Used by permission of StatCrunch; **Multiple pages**, Excel 2016, Windows 10, Microsoft Corporation

Chapter 3

P63–64, Based on data from *U.S. News & World Report*; **P93**, Based on data from *U.S. News & World Report*; **P112**, Based on data from the *California Health Interview Survey*; **P125**, Based on anthropometric survey data from Gordon, Clauser, et al.; **P126**, Based on data from "An Unexpected Rise in Strontium-90 in U.S. Deciduous Teeth in the1990s," by Mangano et. al., *Science of the Total Environment*

Chapter 4

P139, Based on data from "Texting While Driving … ," by Olsen, Shults, Eaton, *Pediatrics*, Vol. 131, No. 6; **P144**, Based on data from the Genetics & IVF Institute; **P146**, Based on data from "Are Women Carrying 'Basketballs'… ," by Perry, DiPietro, Constigan, *Birth*, Vol. 26, No. 3; **P146**, Based on data from "Mortality Reduction with Air Bag and Seat Belt Use in Head-on Passenger Car Collisions," by Crandall, Olson, Sklar, *American Journal of Epidemiology*, Vol. 153, No. 3; **P146**, Based on data from the Journal of the National Cancer Institute; **P148**, Based on anthropometric survey data from Gordon, Churchill, et al.; **P150**, Based on data from "Prevalence and Comorbidity of Nocturnal Wandering in the U.S. General Population" by Ohayon et al., *Neurology*, Vol. 78, No. 20; **P156**, Based on the 5% guideline for cumbersome calculations; **P156**, Data from a QSR Drive-Thru Study; **P157**, Based on data from various sources, including lifehacker.com; **P158**, Based on data from Arshad Mansoor, Senior Vice President with the Electric Power Research Institute; **P158**, Based on data from "Pacemaker and ICD Generator Malfunctions," by Maisel et al., *Journal of the American Medical Association*, Vol. 295, No. 16; **P158**, Based on data from "Association Between Helicopter vs Ground Emergency Medical Services and Survival for Adults with Major Trauma," by Galvagno et al., *Journal of the American Medical Association*, Vol. 307, No. 15; **P161**, Based on data from *USA Today*; **P164**, Based on Probabilistic Reasoning in Clinical Medicine by David Eddy, Cambridge University Press; **P167**, Based on data from "The Denomination Effect" by Priya Raghubir and Joydeep Srivastava, *Journal of Consumer Research*, Vol. 36; **P168**, Based on data from various sources including lifehacker.com; **P168**, Based on data from Arshad Mansoor, Senior Vice President with the Electric Power Research Institute; **P178**, Based on data from *USA Today*; **P178**, Data are from "Mortality Reduction with Air Bag and Seat Belt Use in Head-on Passenger Car Collisions" by Crandall Olson & Sklar, *American Journal of Epidemiology*, Vol. 153, No. 3; **P179**, Based on data from the Centers for Disease Control and Prevention; **P182**, Based on data from "Carbon Monoxide Test Can Be Used to Identify Smoker," by Patrice Wendling, *Internal Medicine News*, Vol. 40., No. 1, and Centers for Disease Control and Prevention

Chapter 5

P188, Based on data from an Adecco survey; **P196**, Based on data from the National Institutes of Health; **P196**, Based on a Microsoft Instant Messaging survey; **P196**, Based on a TE Connectivity survey; **P197**, Based on data from "Hemispheric Dominance and Cell Phone Use," by Seidman et al., *JAMA Otolaryngology—Head & Neck Surgery*, Vol. 139, No. 5; **P198**, Based on data from "Prevalence and Comorbidity of Nocturnal Wandering In the U.S. Adult General Population," by Ohayon et al., *Neurology*, Vol. 78, No. 20; **P199**, Based on data from the U.S. Department of Health and Human Services; **P200**, Based on results from a Pew Research Center survey; **Multiple pages**, ©Triola Stats. All Rights Reserved; **P208**, Based on a Pitney Bowes survey; **P209**, Based on data from a Harris Interactive survey; **P210**, Based on data from an LG Smartphone survey; **P212**, Based on data from the National Institutes of Health; **P212**, Based on test results from a sample of the items; **P213**, Based on data from ICR Survey Research Group; **P219**, Based on data from the National Highway Traffic Safety Administration; **P220**, Based on data from the Department of Transportation; **P220–221**, Based on results from an *AARP Bulletin* survey; **P221**, Based on a Coca-Cola survey; **P221**, Based on data from the National Institutes of Health; **P221**, Based on data from the U.S. National Center for Health Statistics; **P223**, HSS; **P223**, Based on data from the IBM research paper "Passenger-Based Predictive Modeling of Airline No-Show Rates," by Lawrence, Hong, and Cherrier

Chapter 6

P230, From *Statisticians and the EPA: What's the Connection Issue -371* by Barry D. Nussbaum. Published by American Statistical Association, © May 2008; **P251**, Based on anthropometric survey data from Gordon, Churchill, et al.; **P252**, Based on data from "Ethological Study of Facial Behavior in Nonparanoid and Paranoid Schizophrenic Patients," by Pittman, Olk, Orr, and Singh, *Psychiatry*, Vol. 144, No. 1; **P252**, Based on data from the National Health Survey; **P252**, Based on data from the Department of Transportation; **P270**, Based on data from University of Maryland researchers; **P273**, Based on data from the National Health and Nutrition Examination Survey; **P274**, Based on anthropometric survey data from Gordon, Churchill, et al.; **P275**, Based on data from "Neurobehavioral Outcomes of School-age Children Born Extremely Low Birth Weight or Very Preterm," by Anderson et al., *Journal of the American Medical Association*, Vol. 289, No. 24; **P289**, Based on PPG Industries; **P290**, Based on "Prevalence and Comorbidity of Nocturnal Wandering in the U.S. Adult General Population," by Ohayon et al., *Neurology*, Vol. 78, No. 20; **P290**, Based on data from ICR Research Group; **P291**, Based on data from an IBM research paper by Lawrence, Hong, and Cherrier; **P295**, Based on a study by Dr. P. Soria at Indiana University; **P295**, Based on data from the U.S. Army Anthropometry Survey (ANSUR)

Chapter 7

P312, Based on data from *QSR* magazine; **P312**, Based on data from Bristol-Myers Squibb Co.; **P312**, Based on data from the Physicians Insurers Association of America; **P312**, Based on data from *QSR* magazine; **P312**, Based on data from Purdue Pharma L.P.; **P313**, Based on data in "A Close Look at Therapeutic Touch," *Journal of the American Medical Association*, Vol. 279, No. 13; **P313**, Based on data from "Use of Prescription and

Over-the-Counter Medications and Dietary Supplements Among Older Adults in the United States," by Qato et al., *Journal of the American Medical Association*, Vol. 300, No. 24; **P313**, Based on data from ICR Research Group; **P313**, Data are from the *Journal of the National Cancer Institute*; **P314**, Based on data from "Sustained Care Intervention and Postdischarge Smoking Cessation Among Hospitalized Adults," by Rigotti et al., *Journal of the American Medical Association*, Vol. 312, No. 7; **P315**, Based on data from NetMarketShare; **P315**, Based on a 3M Privacy Filters survey; **P315**, Based on a report by Spil Games; **P329**, Based on data from "Cognitive Behavioral Therapy vs Zopiclone for Treatment of Chronic Primary Insomnia in Older Adults," by Sivertsen et al., *Journal of the American Medical Association*, Vol. 295, No. 24; **P329**, Based on data from "Effect of Raw Garlic vs Commercial Garlic Supplements on Plasma Lipid Concentrations in Adults with Moderate Hypercholesterolemia," by Gardner et al., *Archives of Internal Medicine*, Vol. 167; **P329**, Based on data from the Food and Drug Administration; **P330**, Based on data from the National Center for Education Statistics; **P340**, Based on data from "Cognitive Behavioral Therapy vs Zopiclone for Treatment of Chronic Primary Insomnia in Older Adults," by Sivertsen et al., *Journal of the American Medical Association*, Vol. 295, No. 24; **P340**, Based on data from "Effect of Raw Garlic vs Commercial Garlic Supplements on Plasma Lipid Concentrations in Adults with Moderate Hypercholesterolemia," by Gardner et al., *Archives of Internal Medicine*, Vol. 167); **P340**, Based on data from SigAlert; **P348**, Based on data from the Environmental Working Group; **P348**, Based on data from *QSR* magazine; **P348**, Based on data from Bristol-Myers Squibb Co.; **P349**, Based on data from the Physicians Insurers Association of America; **P350**, Based on data from a Rasmussen Reports survey

Chapter 8

P373, Based on data from Pfizer, Inc.; **P383**, Based on data from Pfizer; **P383**, Based on data from *QSR* magazine; **P383**, Based on data from Bristol-Myers Squibb Co.; **P384**, Based on data from Purdue Pharma L.P.; **P384**, Based on data from the Physicians Insurers Association of America; **P384**, Based on data from experiments conducted by researchers Charles R. Honts of Boise State University and Gordon H. Barland of the Department of Defense Polygraph Institute; **P385**, Based on data in "A Close Look at Therapeutic Touch," *Journal of the American Medical Association*, Vol. 279, No. 13; **P385**, Based on data from the Journal of the National Cancer Institute as reported in *USA Today*; **P385**, Based on data from "UPC Scanner Pricing Systems: Are They Accurate?" by Goodstein, *Journal of Marketing*, Vol. 58; **P385**, Based on data from "High-Dose Nicotine Patch Therapy," by Dale et al., *Journal of the American Medical Association*, Vol. 274, No. 17; **P385**, Based on data from "Holidays, Birthdays, and Postponement of Cancer Death," by Young and Hade, *Journal of the American Medical Association*, Vol. 292, No. 24; **P386**, Based on data from "Sustained Care Intervention and Postdischarge Smoking Cessation Among Hospitalized Adults," by Rigotti et al., *Journal of the American Medical Association*, Vol. 312, No. 7; **P386**, Based on data from "Use of Prescription and Over-the-Counter Medications and Dietary Supplements Among Older Adults in the United States," by Qato et al., *Journal of the American Medical Association*, Vol. 300, No. 24; **P390**, Created using JMP® software. Copyright © SAS Institute Inc. SAS and all other SAS Institute Inc. product or service names are registered trademarks or trademarks of SAS Institute Inc., Cary, NC, USA; **P396**, Based on data from "Content and Ratings of Teen-Rated Video Games," by Haninger and Thompson, *Journal of the American Medical Association*, Vol. 291, No. 7; **P397**, Based on data from "Effect of Raw Garlic vs Commercial Garlic Supplements on Plasma Lipid Concentrations in Adults with Moderate Hypercholesterolemia," by Gardner et al., *Archives of Internal Medicine*, Vol. 167; **P397**, Based on data from "Comparison of the Atkins, Ornish, Weight Watchers, and Zone Diets for Weight Loss and Heart Disease Reduction," by Dansinger et al., *Journal of the American Medical Association*, Vol. 293, No. 1; **P398**, Based on data from "Cognitive Behavioral Therapy vs Zopiclone for Treatment of Chronic Primary Insomnia in Older Adults," by Sivertsen et al., *Journal of the American Medical Association*, Vol. 295, No. 24; **P398**, Based on data from "Lead, Mercury, and Arsenic in US and Indian Manufactured Ayurvedic Medicines Sold via the Internet," by Saper et al., *Journal of the American Medical Association*, Vol. 300, No. 8; **P408**, Based on data from payscale.com; **P409**, Based on data from ICR Survey Research Group

Chapter 9

P424, Based on data from *QSR* magazine; **P424**, Based on data from "Sustained Care Intervention and Postdischarge Smoking Cessation Among Hospitalized Adults," by Rigotti et al., *Journal of the American Medical Association*, Vol. 312, No. 7; **P425**, Based on data from "Do We Dream in Color?" by Eva Murzyn, *Consciousness and Cognition*, Vol. 17, No. 4; **P425**, Based on data from Purdue Pharma L.P.; **P425**, Based on data from "Who Wants Airbags?" by Meyer and Finney, *Chance*, Vol. 18, No. 2; **P425**, Based on data from "Survival from In-Hospital Cardiac Arrest During Nights and Weekends," by Peberdy et al., *Journal of the American Medical Association*, Vol. 299, No. 7; **P426**, Based on data from "An Evaluation of Echinacea Angustifolia in Experimental Rhinovirus Infections," by Turner et al., *New England Journal of Medicine*, Vol. 353, No. 4; **P426**, Based on data from "Sustainability of Reductions in Malaria Transmission and Infant Mortality in Western Kenya with Use of Insecticide-Treated Bednets," by Lindblade et al., *Journal of the American Medical Association*, Vol. 291, No. 21; **P426**, Based on data from "Hemi-spheric Dominance and Cell Phone Use," by Seidman et al., *JAMA Otolaryngology—Head & Neck Surgery*, Vol. 139, No. 5; **P426**, Based on data from "The Denomination Effect," by Raghubir and Srivastava, *Journal of Consumer Research*, Vol. 36; **P426**, Based on data from "High-Flow Oxygen for Treatment of Cluster Headache," by Cohen, Burns, and Goadsby, *Journal of the American Medical Association*, Vol. 302, No. 22; **P427**, Based on data from "Final Report on the Aspirin Component of the Ongoing Physicians' Health Study," *New England Journal of Medicine*, Vol. 321:129–135; **P427**, Based on data from "The Left-Handed: Their Sinister History," by Elaine Fowler Costas, Education Resources Information Center, Paper 399519; **P427**, Based on data from "Association Between Helicopter vs Ground Emergency Medical Services and Survival for Adults With Major Trauma," by Galvagno et al., *Journal of the American Medical Association*, Vol. 307, No. 15; **P439**, "Based on data from "Bipolar Permanent Magnets for the Treatment of Chronic Lower Back Pain: A Pilot Study," by Collacott, Zimmerman, White, and Rindone, *Journal of the American Medical Association*, Vol. 283, No. 10."; **P440**, Based on data from "Morbidity Among Pediatric Motor Vehicle Crash Victims: The Effectiveness of Seat Belts," by Osberg and Di Scala, *American Journal of Public Health*, Vol. 82, No. 3; **P441**, Based on data from "Item Arrangement, Cognitive Entry Characteristics, Sex and Test Anxiety as Predictors of Achievement in Examination Performance," by Klimko, *Journal of Experimental Education*, Vol. 52, No. 4; **P442**, Based on data from "Effects of Alcohol Intoxication on Risk Taking, Strategy, and Error Rate in Visuomotor Performance," by Streufert et al., *Journal of Applied Psychology*, Vol. 77, No. 4; **P451**, Based on data from "Is Friday the 13th Bad for Your Health?" by Scanlon et al., *British Medical Journal*, Vol. 307, as listed in the Data and Story Line online resource of data sets; **P451**, Based on "An Analysis of Factors That Contribute to the Efficacy of Hypnotic Analgesia," by Price and Barber, *Journal of Abnormal Psychology*, Vol. 96, No. 1; **P459**, Based on data from "Effects of Alcohol Intoxication on Risk Taking, Strategy, and Error Rate in Visuomotor Performance," by Streufert et al., *Journal of Applied Psychology*, Vol. 77, No. 4; **P460**, Based on data from "Bipolar Permanent Magnets for the Treatment of Chronic Lower Back Pain: A Pilot Study," by Collacott, Zimmerman, White, and Rindone, *Journal of the American Medical Association*, Vol. 283, No. 10; **P460**, Based on data from "Item Arrangement, Cognitive Entry Characteristics, Sex and Test Anxiety as Predictors of Achievement in Examination Performance," by Klimko, *Journal of Experimental Education*, Vol. 52, No. 4; **P461**, Based on data from a Harris poll; **P462**, Based on data from "Consistency of Blood Pressure Differences Between the Left and Right Arms," by Eguchi et al., *Archives of Internal Medicine*, Vol. 167; **P463**, Based on data from "Essential Hypertension: Effect of an Oral Inhibitor of Angiotensin-Converting Enzyme," by MacGregor et al., *British Medical Journal*, Vol. 2; **P463**, Based on data from "Eyewitness Memory of Police Trainees for Realistic Role Plays," by Yuille et al., *Journal of Applied Psychology*, Vol. 79, No. 6

Chapter 10

P486, Based on data from "The Trouble with QSAR (or How I Learned to Stop Worrying and Embrace Fallacy)" by Stephen Johnson, *Journal of Chemical Information and Modeling*, Vol. 48, No. 1; **P487**, Based on data from *The Song of Insects* by George W. Pierce, Harvard University Press; **P487**, Based on "Mass Estimation of Weddell Seals Using Techniques of Photogrammetry," by R. Garrott of Montana State University; **P488**, Based on "Mass Estimation of Weddell Seals Using Techniques of Photogrammetry," by R. Garrott of Montana State University; **P491**, Created using JMP® software. Copyright © SAS Institute Inc. SAS and all other SAS Institute Inc. product or service names are registered trademarks or trademarks of SAS Institute Inc., Cary, NC, USA; **P510–511** Based on data from the Poughkeepsie Journal; **P530**, Based on data from the Motion Picture Association of America; **P531**, Based on data from Janssen Pharmaceutical Products, L.P.; **Multiple pages**, Reprint Courtesy of International Business Machines Corporation, © SPSS, Inc., and IBM Company; **Multiple pages**, Courtesy of XLSTAT™. Used with Permission; **Multiple pages**, Screenshot from Texas Instruments. Courtesy of Texas Instruments

Chapter 11

P537, Based on data from the California Department of Public Health; **P543–544** Based on data from "Participation in Cancer Clinical Trials," by Murthy, Krumholz, and Gross, *Journal of the American Medical Association*, Vol. 291, No. 22; **P544**, Based on data from "Efficacy of Hip Protector to Prevent Hip Fracture in Nursing Home Residents," by Kiel et al., *Journal of the American Medical Association*, Vol. 298, No. 4; **P549**, Based on data from "Surgery Unfounded for Tarsal Navicular Stress Fracture," by Bruce Jancin, *Internal Medicine News*, Vol. 42, No. 14; **P556**, Based on data from "Splinting vs. Surgery in the Treatment of Carpal Tunnel Syndrome," by Gerritsen et al., *Journal of the American Medical Association*, Vol. 288, No. 10; **P557**, Based on data from "Texting While Driving and Other Risky Motor Vehicle Behaviors Among U.S. High School Students" by O'Malley, Shults, and Eaton, *Pediatrics*, Vol; **P557**, Based on data from "Texting While Driving and Other Risky Motor Vehicle Behaviors Among U.S. High School Students," by O'Malley, Shults, and Eaton, *Pediatrics*, Vol. 131, No. 6; **P557**, Based on "The Denomination Effect," by Priya Raghubir and Joydeep Srivastava, *Journal of Consumer Research*, Vol. 36; **P558**, Based on data from "Action Bias Among Elite Soccer Goalkeepers: The Case of Penalty Kicks," by Bar-Eli et al., *Journal of Economic Psychology*, Vol. 28, No. 5; **P558**, Based on data from "What Kinds of People Do Not Use Seat Belts?" by Helsing and Comstock, American *Journal of Public Health*, Vol. 67, No. 11; **P558**, Based on data from "An Evaluation of Echinacea Angustifolia in Experimental Rhinovirus Infections," by Turner et al., *New England Journal of Medicine*, Vol. 353, No. 4; **P559**, Based on data from "Motorcycle Rider Conspicuity and Crash Related Injury: Case-Control Study," by Wells et al., *BMJ USA*, Vol. 4; **P559**, Based on data from "I Hear You Knocking But You Can't Come In," by Fitzgerald and Fuller, *Sociological Methods and Research*, Vol. 11, No.1; **P559**, Based on data from "Predicting Professional Sports Game Outcomes from Intermediate Game Scores," by Copper, DeNeve, and Mosteller, *Chance*, Vol. 5, No. 3–4; **P561**, Based on data from the Insurance Institute for Highway Safety; **P561**, Based on data from "Neuropsychological and Renal Effects of Dental Amalgam in Children," by Bellinger et al., *Journal of the American Medical Association*, Vol. 295, No. 15; **P562**, Based on "The Denomination Effect" by Priya Raghubir and Joydeep Srivastava, *Journal of Consumer Research*, Vol. 36; **P562**, Based on data from Parke-Davis; **P563**, Based on data from the Insurance Institute for Highway Safety; **P563**, Based on anthropometric survey data from Gordon, Churchill, et al.

Chapter 12

P566, Created using JMP® software. Copyright © SAS Institute Inc. SAS and all other SAS Institute Inc. product or service names are registered trademarks or trademarks of SAS Institute Inc., Cary, NC, USA; **P566–567**, Based on data from "Neuropsychological Dysfunction in Children with Chronic Low-Level Lead Absorption," by P. J. Landrigan, R. H. Whitworth, R. W. Baloh, N. W. Staehling, W. F. Barthel, and B. F. Rosenblum, *Lancet*, Vol. 1, Issue 7909

Chapter 13

P604, Based on data from the Genetics & IVF Institute; **P611**, Based on data from the Physicians Insurers Association of America; **P624**, Based on data from "An Unexpected Rise in Strontium-90 in U.S. Deciduous Teeth in the 1990s," by Mangano et al., *Science of the Total Environment*; **P624**, Based on data from "Item Arrangement, Cognitive Entry Characteristics, Sex and Test Anxiety as Predictors of Achievement in Examination Performance," by Klimko, *Journal of Experimental Education*, Vol. 52, No. 4; **P635**, Based on data from *Consumer Reports*; **P639**, Based on "Mass Estimation of Weddell Seals Using Techniques of Photogrammetry," by R. Garrott of Montana State University; **P640**, Based on data from *The Song of Insects* by George W. Pierce, Harvard University Press; **P648**, Based on data from "Job Rated Stress Score" from CareerCast.com; **P650**, Based on data from Quest Diagnostics

Chapter 14

P665, Based on data from the U.S. Department of the Interior; **P674**, Based on a MerckManuals.com survey; **P675**, Based on data from "Does It Pay to Plead Guilty? Differential Sentencing and the Functioning of the Criminal Courts," by Brereton and Casper, *Law and Society Review*, Vol. 16, No.1; **P660**, Source: Adapted from ASTM Manual on the Presentation of Data and Control Chart Analysis, © 1976 ASTM, pp. 134–136. Reprinted with permission of American Society for Testing and Materials

App A

P2, Mosteller, Probability with Statistical Applications, 2nd Ed., © 1970. Reprinted and Electronically reproduced by permission of Pearson Education, Inc. Upper Saddle River, New Jersey; **P6**, *Handbook of Statistical Tables*, Addison-Wesley Pub. Co.; **P7–8**, Based on data from Maxine Merrington and Catherine M. Thompson, "Tables of Percentage Points of the Inverted Beta (F)Distribution," *Biometrika* 33 (1943): 80–84; **P13**, Based on data from Some Rapid Approximate Statistical Procedures, Copyright © 1949, 1964 Lederle Laboratories Division of American Cyanamid Company; **P15**, From Tables for Testing Randomness of Grouping in a Sequence of Alternatives by Frieda S. Swed and C. Eisenhart in *The Annals of Mathematical Statistics*, Vol. 14, No.1, pp. 66–87. Copyright © Institute of Mathematical Statistics

App B

P2, Rohren, Brenda, "Estimation of Stature from Foot and Shoe Length: Applications in Forensic Science." Copyright © 2006. Reprinted by permission of the author; **P4**, Based on data from *Biostatistical Analysis*, 4th edition © 1999, by Jerrold Zar, Prentice Hall, Inc., Upper Saddle River, New Jersey, and "Distribution of Sums of Squares of Rank Differences to Small Numbers with Individuals," *The Annals of Mathematical Statistics*, Vol. 9, No. 2; **P4**, Mario. F. Triola; **P5**, Mario. F. Triola; **P6**, Mario. F. Triola; **P6**, Mario. F. Triola; **P6**, Mario. F. Triola; **P8**, Mario. F. Triola; **P9**, Based on data from Andrew Gelman, Jennifer Hill, 2007, "Replication data for Data Analysis Using Regression Multilevel/Hierarchical Models,"http://hdl.handle.net/1902.1/10285; **P9**, Based on replication data from Data Analysis Using Regression and Multilevel/Hierarchical Models, by Andrew Gelman and Jennifer Hill, Cambridge University Press; **P12**, Mario. F. Triola; **P14**, Based on data from CNN

Barrelfold

P4, *Handbook of Statistical Tables*, Addison-Wesley Pub. Co.

INDEX

A

Acceptance sampling, 158
Actual odds against, 142
Actual odds in favor, 142
Addition rule
    complementary events and, 149–150
    complements and, 159
    defining, 148
    disjoint events and, 149
    formal, 148
    intuitive, 148
    notation for, 148–149
    summary of, 155–156
Adjusted coefficient of determination, 513
Airport data speeds, 708
Alcohol and tobacco in movies, 701
α, 238, 300
Alternative hypothesis, 479
    contingency tables, 547
    defining, 359
    goodness-of-fit and, 535
    original claim for, 359–360
Aluminum cans, 707
Analysis, ethics in, 678–679
Analysis of data, 6
    potential pitfalls, 7–8
Analysis of variance (ANOVA)
    history of, 572
    interaction in, 582
    one-way, 568–582
    two-way, 582–593
Area
    between boundaries, 232
    known, 236–239
    with nonstandard normal distribution, 243
    probability and, 229, 232
    z scores and, 231, 232, 235
Arithmetic mean. *See* Mean
Aspirin, 367
Assignable variation, 658
Attributes, control charts for, 667
Average, 83

B

b_1 slope, 490
Balanced design, 584
Bar graphs, 59
Bayes' theorem, 163–165
Bear measurements, 700
bell-shaped distribution, 228
β, 228
Benford's law, 539–540

Bias
    in internet surveys, 304
    interviewer, 678
    in missing data, 22
    nonrespondent, 678
    publication, 7
    sampling, 677–678
    volunteer, 678
Biased estimators, 99, 106–107, 261
Big data, 14, 19
    applications of, 20–21
Bimodal, 85
Binomial probabilities, 683
    technology for, 288
Binomial probability distribution, 199–205
    defining, 200
    finding, 201–204
    formula, 202
    normal distribution
        approximation of, 285
    notation, 200, 284
    rationale for, 207–208
    software/calculator results, 208
Births, 699
Bivariate normal distribution, 472
Blinding, 26
Blocks, 30–31
Body data, 698
Body temperatures, 698
Bonferroni multiple comparison test
    critical values in, 576
    one-way ANOVA, 575–576
    P-values in, 576
Bootstrap resampling
    for claims testing, 393
    for dependent samples, 447
    for matched pairs, 447
Bootstrap sample
    for confidence interval, 343–344
    defining, 343
    means and, 345
    proportions and, 344
    requirements, 342
    standard deviation and, 346
Box, George E. P., 569
Boxplots, 112–120
    constructing, 119–120
    defining, 119
    for mean, 575
    modified, 121–122
    skeletal, 121
    skewness in, 119
Bribery, 662

C

Calculators. *See* Software/ calculator results
Cancer, 636
Car chases, 72
Car crash tests, 703
Car measurements, 703
Car seats, 539
Case-control study, 29
Categorical data, 14–15
    in frequency distributions, 44–45
Causation
    correlation and, 7, 68, 478
    interpretation with, 478
Censored data, 95
Census, 4
Centerline, 658
Central limit theorem, 265
    applications of, 268–270
    defining, 266
    fuzzy, 269
    sample mean and, 267
    sampling distribution and, 267
    universal truths and, 266–267
Chebyshev's theorem, 105
Cherry picking, 678
Chi-square distribution, 333–335, 687
    critical values in, 333
    degrees of freedom in, 333
    properties of, 400
Chi-square test of homogeneity, 552
Chocolate chip cookies, 707
Cigarette contents, 702
Claimed distribution
    disagreement with, 536–537
    in goodness-of-fit, 536–537
Claims testing
    with abnormal populations, 403
    bootstrap resampling for, 393
    confidence interval for, 377–378, 391–392, 402–403
    critical values for, 377, 391
    critical values in, 402
    degrees of freedom in, 400
    equivalent methods, 400
    exact methods, 379–383
    important properties of, 388
    normal approximation
        method, 374–379
    notation, 387
    number of successes, 377–378
    about population means, 387–395

about population proportion, 374–383
    P-value in, 375–376, 388–390
    requirements, 387, 399
    sign test for, 393
    about standard deviation, 399–404
    technology for, 388–389
    test statistic, 387, 394, 399
    about variance, 399–404
    Wilcoxon signed-ranks test for, 393
Class
    boundaries, 42, 43
    midpoints, 42
    size, 83
    width, 42, 43
CLI. *See* Cost of laughing index
Clinical trials, 14
    alternatives to, 551
    observational studies
        compared with, 26
    short cuts, 523
Clopper-Peason method, 309–310
Cluster sampling, 28
Coefficient
    confidence, 300
    linear correlation, 70, 472–475
    Pearson correlation, 692
    Pearson product moment
        correlation, 472
    regression, 516
    Spearman's rank correlation, 632, 695
    of variation, 106
Coefficient of determination, 507
    adjusted, 513
    multiple, 513–514
Coefficient of variation (CV)
    defining, 106
    round-off rule for, 106
Cohort study, 29
Coincidences, 165
Coin weights, 707
Cola weights, 706
Column factor, 586–587
Combinations
    defining, 170
    mnemonics for, 170
    permutations and, 173–174
    rule, 172
Commercial jet safety, 552
Common Rule, 680
Complementary event
    addition rule and, 149–150
    defining, 140

probability and, 140–141
rule of, 150
Complements
  addition rule and, 159
  multiplication rule of, 156
  probability and, 159–160
Completely randomized design, 30, 575
Composite sampling, 168, 179
Conclusions, 6
  misleading, 7
Conditional probability, 161–163
  defining, 161
  formal approach for, 161
  intuitive approach for, 161
  notation, 161
Confidence coefficient, 300
Confidence interval
  analyzing polls, 305–306
  better-performing, 309
  bootstrap sample for, 343–344
  for claims testing, 377–378, 391–392, 402–403
  Clopper-Pearson method, 309–310
  for comparisons, 336
  constructing, 304, 335–336
  critical values, 301–302
  data comparisons with, 322–323
  defining, 300, 503
  as equivalent method, 444
  for hypothesis tests, 306–307, 336, 367
  in independent samples, 433–434
  interpreting, 301, 321–322
  manual calculation, 305
  margin of error and, 322, 434–435
  for matched pairs, 443–444
  for mean, 575
  notation, 303
  one-sided, 316
  plus four method, 309
  point estimate and, 306, 322
  for population mean, 317
  for population proportion, 303–304
  for population standard deviation, 335
  population variance, 335
  procedure for, 319
  rationale for, 337–338, 421–422
  requirement check, 303, 304
  round-off rule, 303
  software/calculator results, 338, 422
  technology and, 304–305
  for two dependent samples, 443–444

in two proportions, 416, 420–421
Wald, 309
Wilson score, 309–310
Confidence level
  defining, 300
  process success rate, 301
Conflicts of interest, 679–680
Confounding, 29
Confusion of the inverse, 163
Constitution, 44
Context, 3–5
Contingency tables
  alternative hypothesis, 547
  basic concepts, 546–551
  critical thinking, 556–560
  critical values in, 548
  defining, 547
  degrees of freedom in, 548
  expected frequencies in, 548
  expected value in, 549
  notation, 547
  null hypothesis, 547
  observed frequencies in, 548
  P-values in, 547
  requirements, 547
  software/calculator results, 555
  test statistic in, 547
Continuity correction
  defining, 287
  uses of, 287–288
Continuous data, 15–16
Continuous random variable, 187
Control charts
  constants, 660
  defining, 658
  graphs, 659, 662
  interpreting, 661–662
  for mean, 656–664
  notation, 659, 662
  for p, 667
  requirements, 659, 662
  for variation, 656–664
Convenience sampling, 28
Convictions, 160
Correlation, 67–74
  basic concepts of, 470
  causation and, 7, 68, 478
  common errors involving, 478–479
  defining, 68, 470
  determining, 70
  formal hypothesis test and, 479–481
  linear, 68, 71–72, 470–473, 476–477, 477–478, 635
  negative, 471
  nonlinear relationship in, 479
  positive, 471
  rank, 632–640

regression and, 493
software/calculator results for, 481–482
Cost of laughing index (CLI), 15
Count five
  for standard deviation, 456
  for variance, 456
Course evaluations, 703
Critical region, 363
Critical thinking
  contingency tables, 556–560
  goodness-of-fit and, 543–546
  one-way ANOVA, 578–581
  sign test, 610–612
  standard deviation for, 205–208
  in two-way ANOVA, 589–591
  Wilcoxon rank-sum test, 623–625
Critical values, 686
  in Bonferroni multiple comparison test, 576
  in chi-square distribution, 333
  for claims testing, 377, 391
  in claims testing, 402
  confidence interval, 301–302
  in contingency tables, 548
  decision criteria for, 366, 641
  defining, 238, 302
  finding, 318
  for goodness-of-fit, 535
  in hypothesis testing, 362, 365
  in independent samples, 432–433
  Kruskal-Wallis test, 627
  in one-way ANOVA, 574
  in population means, 318
  rank correlation, 633
  runs test, 641, 696
  in sign test, 603, 693
  in standard deviation, 453
  in two proportions, 417, 419–420
  in variance, 453
  Wilcoxon rank-sum test, 620
  Wilcoxon signed-ranks test, 613, 694
Cross-sectional study, 29
Crowd size, 322
Cumbersome calculations, 152
Cumulative frequency distribution, 45
Curbstoning, 308
CV. *See* Coefficient of variation
Cyclical pattern, 658

D
Data
  analysis of, 6
  basic types of, 13–22
  big, 14, 19–21

body, 698
categorical, 14, 44–45
center of, 51
collection, 677
confidence intervals, 322–323
continuous, 15–16
defining, 4
discrete, 15–16
distribution of, 51
ethics, 677
falsification of, 541, 678–679
human treatment and, 677
magnitude in sets of, 20
missing, 21–22
nominal, 603, 605–606
quantitative, 14
science, 20–21
sources of, 3, 5
spread of, 51
Data transformation, 280
Death, 493
Decimals, 9
Degree of confidence, 300
Degrees of freedom (df)
  in chi-square distribution, 333
  in claims testing, 400
  in contingency tables, 548
  in independent samples, 430
  one-way ANOVA, 574
  in population means, 318
  two means and, 430
Denomination effect, 167, 462
Density curve, 229
Dependence, 151
  independence and, 201
Dependent samples
  bootstrap resampling for, 447
  confidence interval for, 443–444
  equivalent methods for, 444
  experimental design for, 442–443
  hypothesis test for, 443
  notation for, 443
  software/calculator results for, 447–448
  technology and, 446
  test statistics for, 443
Dependent variable, 489
df. *See* Degrees of freedom
Descriptive statistics, 81
Diet pills, 432
Discordant categories, 553
Discrete data, 15
  finite type, 16
  infinite type, 16
Discrete random variable, 187
Disjoint events, 149
Disobedience, 17
Dispersion. *See* Variation

Distance, 231
Distribution. *See specific types*
Distribution-free method, 342
Distribution-free tests, 599
Dotplot, 57
Double-blind, 26
Downward shift, 658
Downward trend, 658
Drake, Frank, 651
Driving tickets, 475
Drug approval, 370
Dummy variables
   logistic regression and,
      516–518
   as predictor, 516–517
Duncan test, 575

E
Earthquakes, 703
Efficiency rating, 600
Empirical rule, 104–105
Enumeration district, 28
Equally likely, 536
Error. *See* Margin of error
Estimators
   biased, 99, 106–107, 261
   defining, 261
   unbiased, 103, 106–107, 261,
      263, 299–300
Ethics
   in analysis, 678–679
   data collection, 677
   enforcing, 680
   reporting and, 679–680
   sampling bias and, 677–678
   significance level selection
      and, 679
Events
   complementary, 140–141,
      149–150
   defining, 134
   disjoint, 149
   probability of, 135–140
   rare, 141–142, 192–193, 270
   sequential, 165
   simple, 134
Exact methods
   claims testing, 379–383
   improving, 381
   simple continuity correction,
      381
Excel. *See* Software/calculator
   results
Exceptional value, 658
Exclusive or, 159
Expected frequencies
   in contingency tables, 548
   equal, 536
   finding, 536
   in goodness-of-fit, 536
   rationale for, 551

Expected value, 190–191
   in contingency tables, 549
   defining, 190
   rationale for formulas,
      194–195
Experimental units, 26
Experiment design, 25–28
   advanced, 29–31
   blinding in, 26
   completely randomized, 30,
      575
   defining, 26
   for dependent samples,
      442–443
   hypothesis testing and, 370
   matched pairs, 31
   for matched pairs, 442–443
   one-way ANOVA and, 575
   randomization in, 26
   replication in, 26
   rigorously controlled, 31
Experimenter effects, 27
Explained deviation, 506
Explained variation, 505–507
Explanatory variable, 489
Extraplotion, 493

F
Factorial rule
   defining, 169
   notation, 169
False positive/negative, 132
Falsified data, 541, 678–679
Family heights, 699
Fast food, 706
F distribution, 688–691
      one-way ANOVA, 568–569
      in standard deviation, 454
      in variance, 454
Fewer, 16
Financial support, 679
Finite population, 271
Fisher, R. A., 572
Fisher's exact test, 553
Fitted line plot, 481
5-number summary
      defining, 118
      finding, 118–119
Fluoride, 444
Flynn effect, 657
Foot and height, 698
Football kickers, 517
Formal addition rule, 148
Formal hypothesis test, 479–481
Formal multiplication rule, 150
Fractiles, 114
Fractions, 9
Frequency
      defining, 42
      expected, 536, 548, 551
      observed, 548

Frequency distribution, 44–45
   cumulative, 45
   defining, 42
   last digits in, 46
   mean calculated from, 88–89
   normal, 46
   procedure for constructing,
      43–45
   relative, 45
Frequency polygon, 61–62
Freshman 15, 699
F test
   interpreting, 454–455
   for standard deviations,
      453–456
   for variances, 453–456
Fuzzy central limit theorem, 269

G
Galton, Francis, 512
Gambling, 153
Garbage weight, 708
Gender gap, 446
Geometric distribution, 213
Geometric mean, 96
Ghostwriting, 679
Global temperatures, 346
Goodness-of-fit
   alternative hypothesis and,
      535
   Benford's law and, 539–540
   claimed distribution in,
      536–537
   critical thinking and, 543–546
   critical values for, 535
   defining, 535
   expected frequencies in, 536
   notation, 535
   null hypothesis and, 535
   P-values for, 535
   requirement check, 538
   requirements, 535
   software/calculator results,
      542
   testing for, 535
   test statistic for, 535, 541
Gosset, William, 318
Gould, Stephen Jay, 84
Graphs
   bar, 59
   control charts, 659, 662
   deceptive, 62–63
   enlightening, 57–62
   frequency polygon, 61–62
   interaction, 583–585
   with nonzero vertical axis,
      62–63
   Pareto charts, 59–60
   p chart, 668
   pictographs, 63–64
   pie charts, 60–61

R chart, 659
   regression line, 492
   software/calculator results, 64
   time-series, 58
Graunt, John, 6
Group testing, 162
Growth charts, 46

H
Hamilton, Alexander, 44
Harmonic mean, 96
Hawthorne effect, 27
Histogram, 280
Histograms, 255
   basic concepts of, 51–55
   defining, 51
   distribution shapes, 52
   important uses of, 51
   interpreting, 52
   normal distribution, 52–53,
      276
   normal quantile plots and,
      276–277
   probability, 188–189
   relative frequency, 51–52
   skewness of, 277
   uniform distribution, 53
Homogeneity, test of, 551–553
   chi-square, 552
   procedure, 552
Horvitz, Eric, 18
Hot streaks, 642
H test. *See* Kruskal-Wallis test
Hybridization, 224, 265
Hypergeometric distribution, 214
Hypothesis, 358
   alternative, 359–360, 479,
      535, 547
   null, 359–360, 479, 535, 547
Hypothesis testing, 270–271. *See
   also* Test statistic
   basic concepts of, 358–367
   big picture, 358–359
   confidence intervals for,
      306–307, 336, 367
   correlation and, 479–481
   critical region in, 363
   critical values in, 362, 365
   defining, 358
   for dependent samples, 443
   as equivalent method, 444
   experiment design and, 370
   final conclusion, 366
   formal, 479–481
   for independent samples, 430
   for matched pairs, 443
   for means, 429
   memory hints for, 367
   multiple negatives in, 366
   one-tailed, 480
   for proportions, 416, 417

P-value in, 363–364
rare event rule and, 270
rationale for, 421
rejection in, 365–366
restating, 366–367
sampling distributions in, 362
for significance, 359
significance level and,
    361–362
software/calculator results,
    381, 395, 404, 422
for standard deviation,
    453–454
test statistic in, 362, 479
type I errors, 367–369
type II errors, 367–369
for variances, 453–454

I
Identity theft, 435
Inclusive or, 148
Independence
    dependence and, 201
    key components in test of, 548
    multiplication rule and,
        151–154
    testing for, 546–551
    test statistic for tests of, 547
    in two-way ANOVA, 584
Independent events, 151
Independent samples, 428–437
    alternative methods, 434–435
    confidence interval in,
        433–434
    critical values in, 432–433
    degrees of freedom in, 430
    hypothesis testing for, 430
    notation, 430
    P-values in, 431
    requirements, 430
    software/calculator results
        for, 435–436
    standard deviation for, 435
    Wilcoxon rank-sum test for,
        619–625
Independent variable, 489
Inferences
    about matched pairs, 443
    software/calculator results,
        422
    from standard deviations,
        457–458
    about two means, 430
    about two proportions, 416
Influential points, 494
Inferential statistics, 141–142
    rare event rule for, 192–193
Institutional Review Board, 680
Interactions
    in ANOVA, 582
    defining, 583

effect, 583
graphs, 583–585
Interquartile range, 118
Interval, 300
    level, 18
    prediction, 503–504, 508,
        510
Interval estimate. *See* Confidence
    interval
Interviewer bias, 678
Intuitive addition rule, 148
Intuitive multiplication rule, 150
iPod random shuffle, 644
IQ
    brain size and, 700
    lead and, 700
    scores, 657

J
Jay, John, 44
Jet engine problems, 152
Jobs, 20

K
Known area, 236–239
Kruskal-Wallis test
    critical values, 627
    defining, 626
    key elements, 626–627
    notation, 626
    P-values, 627
    rationale, 629
    requirements, 627
    right-tailed, 627
    software/calculator results,
        630–631
    test statistic, 627

L
Law of large numbers, 137
LCL. *See* Lower control limit
Lead margin of error, 418
Least squares property, 496
Left-handedness, 237
Left-tailed test, 363
Less, 16
Levene-Brown-Forsythe test
    for standard deviation, 457
    for variance, 457
Lie detectors, 379, 392
Linear correlation
    defining, 68, 470
    determining, 71–72
    finding, 476–477
    P-values for, 71–72, 635
    strength of, 471–473
    variation in, 477–478
Linear correlation coefficient, 70
    calculation of, 472–475
    defining, 472
    formulas, 473

interpreting, 473
notation, 472
properties of, 473–474
P-values for, 473
requirements, 472–473
rounding, 473
Loaded questions, 8
Logistic regression, 516–518
Lognormal distribution, 280, 283
Longitudinal study, 29
Lottery, 165, 173
Lower class limits, 42
Lower control limit (LCL), 658
Lurking variables, 26, 478

M
MAD. *See* Mean absolute
    deviation
Madison, James, 44
Magnitudes, 17
Manatee deaths, 701
Mann-Whitney *U* test, 619, 625
Manual construction, 277–280
Margin of error
    confidence interval and, 322,
        434–435
    defining, 303
    lead, 418
    prediction interval, 504
Matched pairs
    bootstrap resampling for, 447
    claims about, 604–605,
        612–616
    confidence interval for,
        443–444
    equivalent methods for, 444
    hypothesis test for, 443
    inferences about, 443
    McNemar's test for, 553–555
    notation for, 443
    requirements, 443
    sign test for, 603
    software/calculator results
        for, 447–448
    test statistic for, 443
    Wilcoxon signed-ranks test
        for, 612–619
Matched pairs design, 31
    experimental, 442–443
Mathematical models, 522–523
McNemar's test for matched
    pairs, 553–555
Mean absolute deviation (MAD),
    103, 265
Means, 428–437. *See also*
    Population mean
    alternative methods of finding,
        434–435
    bootstrap sample and, 345
    boxplots for, 575
    calculation of, 83

calculation of, from frequency
    distribution, 88–89
confidence interval estimates
    for, 575
control charts for, 656–664
defining, 82
degrees of freedom and, 430
geometric, 96
harmonic, 96
hypothesis testing in, 429
identifying different, 572–577
inferences about, 430
informal methods of
    comparing, 575
interaction graphs for,
    583–585
notation, 83
for Poisson probability
    distribution, 215
of probability distribution,
    189
properties of, 82–83
quadratic, 96
runs test for randomness
    above or below, 644–645
sampling distributions and,
    255
software/calculator results
    for, 326, 395, 435–436
standard deviation for, 435
standard error of, 267
test statistics effected by, 573,
    574
two, 428–437
weighted, 89–90
Measurement
    interval level, 18
    levels of, 16–19
    nominal level, 17
    ordinal level, 17
    ratio level, 18
    units of, 14–15
Measures of center. *See also*
    specific types
    advanced concepts, 88–91
    basic concepts, 82–88
    critical thinking and, 87–88
    defining, 82
    rounding, 86–87
    software/calculator results,
        90–91
Median
    calculation of, 84
    class, 96
    defining, 84
    notation of, 84
    properties of, 84
    runs test for randomness
        above or below, 644–645
    of single population, 603,
        607–609

Mendel, Gregor, 224, 285–286, 288, 541
Meta-analysis, 190
Midquartile, 118
Midrange
    defining, 86
    properties of, 86
Milgram, Stanley, 17–18
Milgram experiment, 677
Minitab. *See* Software/calculator results
Misleading conclusions, 7
Misleading statistics, 16
Missing data, 21–22
    bias in, 22
    completely at random, 21
    correcting, 22
    not at random, 21–22
M&M weights, 706
Mode
    defining, 85
    important properties of, 85
Modified boxplots
    constructing, 121–122
    outliers and, 121–122
Monte Carlo method, 422
Multimodal, 85
Multinomial distribution, 214
Multiple coefficient of
    determination, 513–514
Multiple comparison tests, 575
Multiple regression equation, 518–519
    basic concepts, 511–516
    best, 514–515
    defining, 511
    finding, 512
    guidelines for, 514–515
    notation, 512
    predictions with, 516
    procedure for finding, 512
    P-values in, 514
    requirements, 512
Multiplication counting rule, 169
Multiplication rule
    of complements, 156
    formal, 150
    independence and, 151–154
    intuitive, 150
    notation for, 150–151
    rationale for, 155
    redundancy and, 154
    summary of, 155–156
Multistage sampling, 28–29

N
Negatively skewed, 53
Negative predictive value, 132
Nielsen ratings, 117
Nightingale, Florence, 59
Nobel laureates, 703

Nominal data, 603
    in sign test, 605–606
    with two categories, 605–606
Nominal level, 17
Nondistribution test, 599
Nonlinear patterns, 636–637
Nonlinear regression, 522–525
    software/calculator results, 524–525
Nonlinear relationship, 471
    in correlation, 479
Nonparametric method, 342
Nonparametric tests
    advantages of, 599
    basics of, 599–612
    defining, 599
    disadvantages of, 599
    efficiency of, 599
    misleading terminology, 599
    ranks, 600–601
Nonrandom sampling error, 31
Nonrespondent bias, 678
Nonresponse, 8
Nonsampling error, 31
Nonsignificant results, 679
Nonstandard normal distribution, 243
Nonzero vertical axis, 62–63
Normal approximation, 284–285
Normal distribution, 46
    advanced methods, 277
    binomial distribution
        approximated with, 285
    conversions of, 243
    defining, 228
    formulas for, 228–229, 232, 242
    of normal quantile plots, 276, 277
    of outliers, 276
    populations with, 276
    significance and, 247–248
    skewness and, 277
    software/calculator results, 249
    standard, 230–231
    values from known areas, 245–248
Normal distribution histograms, 52–53, 276
Normality
    assessing, 275–277
    assessment of, with normal
        quantile plots, 53–55
    claims testing and, 388
    histogram and, 280
    normal quantile plots and, 280
    outliers and, 280
    for population means, 318
    in two-way ANOVA, 584
Normality assessment, 279

Normal probability plot. *See*
    Normal quantile plots
Normal quantile plots
    assessing normality with, 53–55
    defining, 276
    histograms and, 276–277
    manual construction of, 277–280
    normal distribution of, 276, 277
    normality and, 280
    software/calculator results, 280
Not equally likely, 536
Not statistically stable, 661
Null hypothesis, 479
    contingency tables, 547
    defining, 359
    goodness-of-fit and, 535
    original claim for, 359–360
Numerical random variable, 187

O
Observational studies
    case-control, 29
    clinical trials compared with, 26
    cohort, 29
    cross-sectional, 29
    longitudinal, 29
    prospective, 29
    retrospective, 29
    types of, 29
Observed frequencies, 548
Odds, 141
    actual, against, 142
    actual, in favor, 142
    payoff, 142
Old faithful, 705
One-sided confidence interval, 316
One-tailed test, 480
One-way ANOVA, 568–582
    basics of, 569–572
    Bonferroni multiple
        comparison test, 575–576
    completely randomized
        design, 575
    critical thinking, 578–581
    critical values in, 574
    defining, 569
    degrees of freedom, 574
    experiment design and, 575
    fail to reject, 569
    F distribution, 568–569
    multiple comparison tests and, 575
    procedure for, 569
    P-value, 569
    range tests and, 575
    reject, 569

requirements, 569
    rigorously controlled design in, 575
    software/calculator results, 578
    for testing population means, 569
    test statistic in, 572–573
Online dating, 20
Orange sugar content, 320
Order of questions, 8
Ordinal level, 17
Oscar winner age, 702
Outliers, 51, 120
    modified boxplots and, 121–122
    normal distribution of, 276
    normality and, 280
    software/calculator results, 123
Out-of-control-criteria, 661, 667
Out of statistical control, 661
Out-of-town drivers, 475

P
Palm reading, 480
Parameters, 13
Parametric tests
    defining, 599
    efficiency of, 600
Pareto charts, 59–60
Passive and active smoke, 701
Payoff odds, 142
p chart, 667
    graphs, 668
    key elements, 667–668
    notation, 668
    requirements, 667
    software/calculator results, 670
Pearson correlation coefficient, 692
Pearson product moment
    correlation
    coefficient, 472
Percentages
    conversion to, from fractions, 9
    conversion to decimals, 9
    probabilities as, 137
Percentiles, 112–120, 247
    conversion of, 116–117
    defining, 115
    finding, 115
Permutations
    combinations and, 173–174
    defining, 170
    mnemonics for, 170
Permutations rule
    with different items, 171
    with some identical items, 171–172

Personal security codes, 172
Pi, 537
Pictographs, 63–64
Pie charts, 60–61
Placebo effect, 26, 465
Plus four method, 309
Point estimate, 299–300
  confidence interval and, 306, 322
  defining, 299
  population mean, 317–318
  sample proportion as, 299
Poisson probability distribution, 214
  as approximation to binomial, 215–216
  formula, 215
  mean for, 215
  parameters of, 215
  properties of, 215
  requirements, 215
  software/calculator results, 217
Polio experiment, 420
Polls, 549
Pooled estimate, 434
Pooled sample proportion, 416
Pooling, 434
Population
  abnormal, 403
  defining, 4
  finite, 271
  median of single, 603, 607–609, 616–617
  models, 523–524
  with normal distribution, 276
  standard deviation of, 102
  variance of, 102–103
Population mean
  appropriate distribution for, 325–326
  claims testing about, 387–395
  confidence interval, 317
  critical values in, 318
  degrees of freedom in, 318
  with known variables, 325–326
  normality for, 318
  notation, 317
  one-way ANOVA for, 569
  point estimate, 317–318
  requirements, 317
  round-off rule, 317
  sample size and, 323–324
  student *t* distribution, 318–319
  with unknown variables, 317–325
Population proportion
  claims testing about, 374–383
  confidence interval for, 303–304
  sample size and, 307–308

Population size, 306
  determining, 323–324
  role of, 308
Population standard deviation, 332–339
  confidence interval for, 335
  notation, 335
  requirements, 335
  round-off rule, 335
Population variance, 332–339
  confidence interval, 335
  notation, 335
  requirements, 335
Positively skewed, 53
Positive predictive value, 132
Posterior probability, 165
Power of hypothesis test, 369
Practical significance, 7
Prediction interval, 508
  defining, 503
  formulas for, 504
  margin of error, 504
  requirement, 503–504
  standard error of estimate, 504
  variation and, 510
Predictor variable, 489
Presidents, 702
Prevalence, 132
Primary sampling units (PSUs), 28
Prior probability, 165
Prisoner studies, 677
Probabilistic models, 489
Probability, 233
  area and, 229, 232
  basic concepts of, 133–141
  binomial, 288, 683
  classical approach to, 136
  complementary events and, 140–141
  complements and, 159–160
  conditional, 161–163
  convictions and, 160
  of events, 135–140
  finding values from, 246
  intuition and, 135
  notation for, 135
  as percentages, 137
  posterior, 165
  prior, 165
  relative frequency approximation of, 135
  significant results and, 270
  simulations, 137
  in statistics, 133–134
  subjective, 136
  with *z* scores, 231–232
Probability distribution
  basic concepts of, 186–193
  binomial, 199–205
  defining, 186

  mean of, 189
  parameters of, 189
  Poisson, 214–217
  requirements, 187–188
  round-off rule for, 189
  standard deviation for, 189
  variance for, 189
Probability formula, 188
Probability histogram, 188–189
Process-behavior chart. *See* Control charts
Process data, 656
Proportion, 416–423
  bootstrap sample and, 344
  confidence interval in, 416, 420–421
  critical values in, 417, 419–420
  hypothesis testing in, 416, 417
  inferences, 416
  notation for, 256, 416
  pooled sample, 416
  population, 303–304, 307–308, 374–383
  P-value in, 417–418
  requirements, 416
  rounding, 417
  sampling distributions and, 255
  software/calculator results for, 310, 381, 422
  technology and, 419
  two, 416–423
Prosecutors' fallacy, 161
Prospective study, 29
PSUs. *See* Primary sampling units
Psychologists, 549
Publication bias, 7
Push polling, 300
P-values
  in Bonferroni multiple comparison test, 576
  for claims testing, 375–376
  in claims testing, 388–390
  in contingency tables, 547
  defining, 71, 363
  for goodness-of-fit, 535
  in hypothesis testing, 363–364
  in independent samples, 431
  interpreting, 72
  Kruskal-Wallis test, 627
  for linear correlation, 71–72, 635
  for linear correlation coefficient, 473
  in multiple regression equation, 514
  in one-way ANOVA, 569
  rank correlation, 633
  in Sign test, 603
  in Standard deviation, 453

  from technology, 550
  technology and, 477
  test statistics and, 571–572
  in two proportions, 417–418
  in variance, 453
  Wilcoxon rank-sum test, 620
  Wilcoxon signed-ranks test, 613

Q

Quadratic mean, 96
Qualitative data, 14–15
Quality control, 669
Quantiles, 114
Quantitative data, 14–15
Quartiles, 112–120
  defining, 117
Questions
  loaded, 8
  order of, 8

R

Randomization, 26
Randomized block design, 30–31
Randomness. *See* Runs test for randomness
Random sample, 27
Random sampling error, 31
Random variable
  continuous, 187
  defining, 186
  discrete, 187
  numerical, 187
Random variation, 658
Range, 98
Range chart. *See* R chart
Range rule of thumb, 191, 205–206
  for significant values, 101, 191
  for standard deviation, 100–101
Range tests, 575
Rank correlation, 632–640
  advantages of, 635
  critical values, 633
  defining, 632
  disadvantage of, 635
  efficiency, 635
  nonlinear patterns, 636–637
  notation, 633
  procedure, 634
  P-value, 633
  requirements, 633
  software/calculator results, 637–638
  test statistic, 633
Ranks
  defining, 600
  handling ties among, 600–601
  in nonparametric tests, 600–601

Rare event rule, 141–142
  hypothesis testing and, 270
  for inferential statistics, 192–193
Ratio level, 18
Ratio test, 18
R chart, 659
  graphs, 659
  notation, 659
  requirements, 659
Redundancy, 153
  multiplication rule and, 154
Region, 231
Regression, 67–74
  advanced concepts, 494–498
  bad models, 492–493
  basic concepts of, 489–490
  correlation and, 493
  defining, 73
  equation, 73, 489
  fitted line plot, 481
  good models, 492–493
  logistic, 516–518
  nonlinear, 522–525
  predictions from, 492–494
  scope, 493
  software/calculator results, 498
Regression coefficients, 516
Regression equation
  slope in, 490
  technology and, 491
  y-intercept in, 490
Regression line, 489
  equation of, 490
  formulas for, 490
  graphing, 492
  notation, 490
  requirements, 490
Rejection region, 363
Relative frequency
  approximation, 135
Relative frequency distribution, 45
Relative frequency histogram, 51–52
Replacements
  sampling with, 152, 262
  sampling without, 152, 262
Replication, 26
  in experiment design, 26
Reporting, ethics and, 679–680
Residual plot, 497
Resistant statistics, 82
Response variable, 489
Retrospective study, 29
Right-tailed test, 363
Rigorously controlled design, 31
  one-way ANOVA in, 575
Root mean square (RMS), 96
Rounding
  errors, 87
  linear correlation coefficient, 473

measures of center, 86–87
slope, 490–491
two proportions, 417
y-intercept, 490
Round-off rules
  for coefficient of variation, 106
  confidence interval, 303
  notation, 307
  population mean, 317
  population standard deviation, 335
  for probability distribution, 189
  requirements, 307
  for sample size, 307, 323
  for *z* scores, 113
Rubik's cube, 174
Rule of complementary events, 150
Run, defining, 640
Run chart, 656
  defining, 657
  interpreting, 657–658
Run of 8 rule, 661, 667
Runs test for randomness, 640–647
  above and below mean, 644–645
  above and below median, 644–645
  critical values, 641, 696
  defining, 640
  fundamental principle of, 643–645
  notation for, 641
  procedure for, 642
  requirements, 641
  software/calculator results, 645
  test statistic, 641
Ryan-Joiner test, 279–280

S

Salaries, 550
Salience, 678
Salk Vaccine experiment, 25
Sample. *See also* Bootstrap sample
  defining, 4
  dependent, 443–446
  independent, 428–437, 619–625
  measurement of, 7
  reporting, 7
  self-selected, 6
  simple random, 27
  small, 230
  space, 134
  test statistic for large, 641
  test statistic for small, 641

variance of, 102–103, 573
  voluntary response, 6
Sample data
  advanced methods, 29–31
  basics of, 25–28
  collecting, 25–31
Sample mean
  behavior of, 257–258
  central limit theorem and, 267
  defining, 257–258
  sampling distributions of, 257–259
Sample proportion
  behavior of, 256–257
  as point estimate, 299
  sampling distributions of, 256
Sample size
  determining, 307, 323–324, 338
  population mean and, 323–324
  population proportion and, 307–308
  round-off rule for, 307, 323
  software/calculator results, 310
  unequal, 574–575
Sample standard deviation. *See* standard deviation
Sample variance
  behavior of, 260
  defining, 259
  sampling distributions of, 259–260, 263
Sampling. *See also* Bootstrap resampling
  bias, 677–678
  cluster, 28
  convenience, 28
  errors, 31
  multistage, 28–29
  with replacement, 152, 262
  without replacement, 152, 262
  stratified, 28
  systematic, 28
  in two-way ANOVA, 584
Sampling bias, 677
Sampling distribution, 267
  determining, 362
  general behavior of, 255
  in hypothesis testing, 362
  mean and, 255
  notation for, 267
  proportions and, 255
  of sample mean, 257–259
  of sample proportion, 256
  of sample variance, 259–260, 263
  of statistics, 256
  variance and, 255
Sampling method, 3, 5
  common, 27

Scatterplots, 67–74
  defining, 68
  interpreting, 471
  of *z* scores, 481
Scheffé test, 575
Self-selected samples, 6
Semi-interquartile range, 118
Sequential events, 165
Shakespeare's vocabulary, 302
Shewhart chart. *See* Control charts
Shuffles, 170
Significance, 251
  hypothesis testing for, 359
  normal distribution and, 247–248
  practical, 7
  statistical, 7
Significance level
  defining, 361
  ethics and, 679
  hypothesis testing and, 361–362
Significantly high, 247
Significantly low, 247
Significant results
  identifying, 192
  probability and, 270
Significant values, 101, 114, 191
Sign test, 601–612
  basic concept of, 601–602
  for claims testing, 393
  critical thinking, 610–612
  critical values in, 603, 693
  defining, 601
  matched pairs for, 603
  nominal data in, 605–606
  notation, 603
  procedure, 602
  P-values in, 603
  requirements, 603
  software/calculator results, 609–610
  test statistic in, 608–609
Simple continuity correction, 381
Simple events, 134
Simple random sample, 27, 430
Simulations, 137
Single population
  median of, 603, 607–609, 616–617
  Wilcoxon signed-ranks test and, 616–617
Six degrees of separation, 18
Skeletal boxplots, 121
Skewness
  in boxplots, 119
  of histograms, 277
  to left, 53
  normal distribution and, 277
  to right, 53

Slope
    formulas for, 490
    in regression equation, 490
    rounding, 490–491
Small samples, 230
Smoking, 575
    cancer and, 636
SNK test, 575
Snowden, Eric, 19
Social desirability, 678
Software/calculator results
    binomial probability
        distribution, 208
    confidence interval, 338
    contingency tables, 555
    for correlation, 481–482
    for dependent samples,
        447–448
    goodness-of-fit, 542
    graphing, 64
    hypothesis testing, 381, 395,
        404, 422
    for independent samples,
        435–436
    inferences, 422
    Kruskal-Wallis test, 630–631
    for matched pairs, 447–448
    for means, 326, 395
    measures of center, 90–91
    measures of variation, 107
    multiple regression equations
        and, 518–519
    nonlinear regression, 524–525
    normal distribution, 249
    normal quantile plots, 280
    one-way ANOVA, 578
    outliers, 123
    p chart, 670
    Poisson probability
        distribution, 217
    prediction intervals, 508
    for proportions, 310, 381, 422
    rank correlation, 637–638
    regression, 498
    runs test, 645
    sample size, 310
    sign test, 609–610
    standard deviation, 338, 404
    for standard deviations,
        457–458
    for two means, 435–436
    in two-way ANOVA, 587–588
    for variance, 457–458
    Wilcoxon rank-sum test, 623
    Wilcoxon signed-ranks test,
        617
    z scores, 239
Sort data, 38
Spearman's rank correlation
    coefficient, 632, 695
Speed dating, 703

Standard deviation
    bootstrap sample and, 346
    Chebyshev's theorem, 105
    claims testing about, 399–404
    count five for, 456
    for critical thinking, 205–208
    critical values in, 453
    defining, 103
    F distribution in, 454
    hypothesis test for, 453–454
    for independent samples, 435
    Levene-Brown-Forsythe test
        for, 457
    notation, 453
    population, 332–339
    of population, 102
    for probability distribution, 189
    P-values in, 453
    range rule of thumb for,
        100–101
    requirements in, 453
    software/calculator results,
        338, 404
    technology for, 456
    test statistic for, 453
    for two means, 435
Standard deviations
    F test for, 453–456
    inferences from, 457–458
    software/calculator results
        for, 457–458
Standard error of estimate, 504
Standard error of the mean, 267
Standardized score. See z scores
Standard normal distribution,
    230–231
StatCrunch. See Software/
    calculator results
Statdisk. See Software/calculator
    results
Statistically stable, 657
Statistical methods, 6
    inappropriate, 679
Statistical significance, 7
Statistics. See also specific types
    in data science, 20–21
    defining, 4, 13
    inferential, 141–142
    misleading, 16
    origins of, 6
    probability in, 133–134
    resistant, 82
    sampling distributions of, 256
Stemplots, 58
Stigler, Stephen, 572
Stocks, 99
Stratified sampling, 28
Student t distribution, 318–319,
    686
Subgroup, 659
Subjective probabilities, 136

Surveys, 31
Survivorship bias, 5
Systematic sampling, 28

T
Tank serial numbers, 321
t distribution, 318–319, 686
    student, 318–319
Teacher evaluations, 477
Technology. See also Software/
    calculator results
    for claims testing, 388–389
    confidence interval and,
        304–305
    dependent samples and, 446
    projects, 651
    P-values and, 477, 550
    regression equation and, 491
    for standard deviation, 456
    two proportions and, 419
    for variances, 456
Test of homogeneity, 551
Test of significance. See
    Hypothesis testing
Test sensitivity, 132
Test specificity, 132
Test statistic. See also Hypothesis
    testing
    calculation of, 573
    claims testing, 387, 394, 399
    in contingency tables, 547
    effect of mean on, 573
    for goodness-of-fit, 535, 541
    in hypothesis testing, 362, 479
    Kruskal-Wallis test, 627
    for large samples, 641
    for matched pairs, 443
    notation, 453
    for one-way ANOVA, 572
    in one-way ANOVA, 572–573
    P-value and, 571–572
    rank correlation, 633
    runs test, 641
    in sign test, 608–609
    for small samples, 641
    for standard deviation, 453
    for tests of independence, 547
    for two dependent samples, 443
    value of, 627
    for variances, 453
    Wilcoxon rank-sum test, 620
    Wilcoxon signed-ranks test,
        613
TI-83/84 Plus Calculator. See
    Software/calculator results
Time-series graph, 58
Total deviation, 506
Total variation, 506
Tree diagram, 155
True positive/negative, 132
True zero, 18

t test. See Student t distribution
Tukey test, 575
Twins, 445
Two-tailed test, 363
Two-way ANOVA, 582–593
    balanced design in, 584
    column factor, 585
    column factor in, 586–587
    critical thinking, 589–591
    fail to reject, 584
    independence in, 584
    normality in, 584
    procedure for, 584, 585
    rejection, 584
    requirements, 584
    sampling in, 584
    software/calculator results in,
        587–588
    variation in, 584
Two-way frequency table. See
    Contingency tables
Type I errors, 367
Type II errors, 367

U
UCL. See Upper control limit
Unbiased estimators, 103,
    106–107, 263
    defining, 261
    in point estimates, 299–300
Unequal sample size, 574–575
Unexplained deviation, 506
Unexplained variation, 505–507
Uniform, 277
Uniform distribution, 229–230,
    277
Uniform distribution histograms,
    53
Upper class limits, 42
Upper control limit (UCL), 658
Upward shift, 658
Upward trend, 658

V
Value of a statistical life (VSL), 30
Variables
    advanced concepts, 103–107
    continuous random, 187
    dependent, 489
    discrete random, 187
    dummy, 516–518
    explanatory, 489
    independent, 489
    lurking, 26, 478
    numerical random, 187
    in population mean, 317–326
    predictor, 489
    random, 186–187
    response, 489
    software/calculator results,
        107

Variance. *See also* Analysis of
   variance
   claims testing about,
     399–404
   count five for, 456
   critical values in, 453
   F distribution in, 454
   F test for, 453–456
   hypothesis test for,
     453–454
   Levene-Brown-Forsythe test
     for, 457
   population, 332–339
   for probability distribution,
     189
   requirements in, 453
   of samples, 102–103, 573
   sampling distributions and,
     255
   software/calculator results
     for, 457–458
   technology for, 456
   test statistic for, 453
Variation
   assignable, 658
   basic concepts of, 97–103
   coefficient of, 106
   comparing, 105–106

   control charts for, 656–664
   explained, 505–507
   increasing, 658
   in linear correlation,
     477–478
   measures of, 97–107
   of population, 102–103
   prediction interval and, 510
   properties of, 103
   random, 658
   reduction of, 658
   total, 506
   in two-way ANOVA, 584
   unexplained, 505–507
Voluntary response sample, 6
Volunteer bias, 678
VSL. *See* Value of a statistical
   life

W
Wald confidence interval, 309
Weighted means, 89–90
Wilcoxon rank-sum test
   critical thinking, 623–625
   critical values, 620
   defining, 619
   for independent samples,
     619–625

   key elements, 620
   notation, 620
   P-values, 620
   requirements, 620
   software/calculator results,
     623
   test statistic, 620
Wilcoxon signed-ranks test
   for claims testing, 393
   critical values, 613, 694
   defining, 612
   for matched pairs, 612–619
   notation, 613
   procedure, 613–614
   P-value, 613
   rationale for, 616–617
   requirement, 613
   single population and,
     616–617
   software/calculator results,
     617
   test statistic, 613
Wildlife, 318
Wilson score confidence interval,
   309–310
Within statistical control, 657
Witnesses, 421
Word counts, 705

X
$\bar{x}$, 83
$\bar{x}$, sampling distribution, 257
$\bar{x}$ chart, 662–664

Y
Yates's correction for continuity,
   560
y-intercept
   formulas for, 490
   in regression equation, 490
   rounding, 490

Z
$0+$, 188
Zeros, true, 18
z scores, 112–120
   area and, 231, 232, 235
   defining, 113
   known area and, 236–239
   negative, 684
   positive, 685
   probability with, 231–232
   properties of, 113
   round-off rule for, 113
   scatterplots of, 481
   significant values and, 114
   software/calculator results, 239

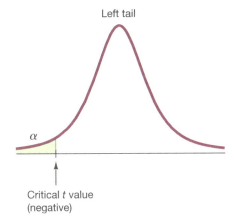

Left tail

α

Critical *t* value
(negative)

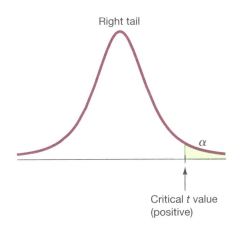

Right tail

α

Critical *t* value
(positive)

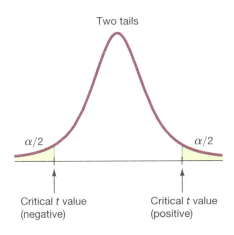

Two tails

$\alpha/2$ $\alpha/2$

Critical *t* value   Critical *t* value
(negative)   (positive)

TABLE A-3 *t* Distribution: Critical *t* Values

Degrees of Freedom	Area in One Tail				
	0.005	0.01	0.025	0.05	0.10
	Area in Two Tails				
	0.01	0.02	0.05	0.10	0.20
1	63.657	31.821	12.706	6.314	3.078
2	9.925	6.965	4.303	2.920	1.886
3	5.841	4.541	3.182	2.353	1.638
4	4.604	3.747	2.776	2.132	1.533
5	4.032	3.365	2.571	2.015	1.476
6	3.707	3.143	2.447	1.943	1.440
7	3.499	2.998	2.365	1.895	1.415
8	3.355	2.896	2.306	1.860	1.397
9	3.250	2.821	2.262	1.833	1.383
10	3.169	2.764	2.228	1.812	1.372
11	3.106	2.718	2.201	1.796	1.363
12	3.055	2.681	2.179	1.782	1.356
13	3.012	2.650	2.160	1.771	1.350
14	2.977	2.624	2.145	1.761	1.345
15	2.947	2.602	2.131	1.753	1.341
16	2.921	2.583	2.120	1.746	1.337
17	2.898	2.567	2.110	1.740	1.333
18	2.878	2.552	2.101	1.734	1.330
19	2.861	2.539	2.093	1.729	1.328
20	2.845	2.528	2.086	1.725	1.325
21	2.831	2.518	2.080	1.721	1.323
22	2.819	2.508	2.074	1.717	1.321
23	2.807	2.500	2.069	1.714	1.319
24	2.797	2.492	2.064	1.711	1.318
25	2.787	2.485	2.060	1.708	1.316
26	2.779	2.479	2.056	1.706	1.315
27	2.771	2.473	2.052	1.703	1.314
28	2.763	2.467	2.048	1.701	1.313
29	2.756	2.462	2.045	1.699	1.311
30	2.750	2.457	2.042	1.697	1.310
31	2.744	2.453	2.040	1.696	1.309
32	2.738	2.449	2.037	1.694	1.309
33	2.733	2.445	2.035	1.692	1.308
34	2.728	2.441	2.032	1.691	1.307
35	2.724	2.438	2.030	1.690	1.306
36	2.719	2.434	2.028	1.688	1.306
37	2.715	2.431	2.026	1.687	1.305
38	2.712	2.429	2.024	1.686	1.304
39	2.708	2.426	2.023	1.685	1.304
40	2.704	2.423	2.021	1.684	1.303
45	2.690	2.412	2.014	1.679	1.301
50	2.678	2.403	2.009	1.676	1.299
60	2.660	2.390	2.000	1.671	1.296
70	2.648	2.381	1.994	1.667	1.294
80	2.639	2.374	1.990	1.664	1.292
90	2.632	2.368	1.987	1.662	1.291
100	2.626	2.364	1.984	1.660	1.290
200	2.601	2.345	1.972	1.653	1.286
300	2.592	2.339	1.968	1.650	1.284
400	2.588	2.336	1.966	1.649	1.284
500	2.586	2.334	1.965	1.648	1.283
1000	2.581	2.330	1.962	1.646	1.282
2000	2.578	2.328	1.961	1.646	1.282
Large	2.576	2.326	1.960	1.645	1.282

NEGATIVE z Scores

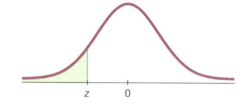

TABLE A-2 Standard Normal (z) Distribution: Cumulative Area from the LEFT

z	.00	.01	.02	.03	.04	.05	.06	.07	.08	.09
−3.50 and lower	.0001									
−3.4	.0003	.0003	.0003	.0003	.0003	.0003	.0003	.0003	.0003	.0002
−3.3	.0005	.0005	.0005	.0004	.0004	.0004	.0004	.0004	.0004	.0003
−3.2	.0007	.0007	.0006	.0006	.0006	.0006	.0006	.0005	.0005	.0005
−3.1	.0010	.0009	.0009	.0009	.0008	.0008	.0008	.0008	.0007	.0007
−3.0	.0013	.0013	.0013	.0012	.0012	.0011	.0011	.0011	.0010	.0010
−2.9	.0019	.0018	.0018	.0017	.0016	.0016	.0015	.0015	.0014	.0014
−2.8	.0026	.0025	.0024	.0023	.0023	.0022	.0021	.0021	.0020	.0019
−2.7	.0035	.0034	.0033	.0032	.0031	.0030	.0029	.0028	.0027	.0026
−2.6	.0047	.0045	.0044	.0043	.0041	.0040	.0039	.0038	.0037	.0036
−2.5	.0062	.0060	.0059	.0057	.0055	.0054	.0052	.0051 *	.0049	.0048
−2.4	.0082	.0080	.0078	.0075	.0073	.0071	.0069	.0068	.0066	.0064
−2.3	.0107	.0104	.0102	.0099	.0096	.0094	.0091	.0089	.0087	.0084
−2.2	.0139	.0136	.0132	.0129	.0125	.0122	.0119	.0116	.0113	.0110
−2.1	.0179	.0174	.0170	.0166	.0162	.0158	.0154	.0150	.0146	.0143
−2.0	.0228	.0222	.0217	.0212	.0207	.0202	.0197	.0192	.0188	.0183
−1.9	.0287	.0281	.0274	.0268	.0262	.0256	.0250	.0244	.0239	.0233
−1.8	.0359	.0351	.0344	.0336	.0329	.0322	.0314	.0307	.0301	.0294
−1.7	.0446	.0436	.0427	.0418	.0409	.0401	.0392	.0384	.0375	.0367
−1.6	.0548	.0537	.0526	.0516	.0505 *	.0495	.0485	.0475	.0465	.0455
−1.5	.0668	.0655	.0643	.0630	.0618	.0606	.0594	.0582	.0571	.0559
−1.4	.0808	.0793	.0778	.0764	.0749	.0735	.0721	.0708	.0694	.0681
−1.3	.0968	.0951	.0934	.0918	.0901	.0885	.0869	.0853	.0838	.0823
−1.2	.1151	.1131	.1112	.1093	.1075	.1056	.1038	.1020	.1003	.0985
−1.1	.1357	.1335	.1314	.1292	.1271	.1251	.1230	.1210	.1190	.1170
−1.0	.1587	.1562	.1539	.1515	.1492	.1469	.1446	.1423	.1401	.1379
−0.9	.1841	.1814	.1788	.1762	.1736	.1711	.1685	.1660	.1635	.1611
−0.8	.2119	.2090	.2061	.2033	.2005	.1977	.1949	.1922	.1894	.1867
−0.7	.2420	.2389	.2358	.2327	.2296	.2266	.2236	.2206	.2177	.2148
−0.6	.2743	.2709	.2676	.2643	.2611	.2578	.2546	.2514	.2483	.2451
−0.5	.3085	.3050	.3015	.2981	.2946	.2912	.2877	.2843	.2810	.2776
−0.4	.3446	.3409	.3372	.3336	.3300	.3264	.3228	.3192	.3156	.3121
−0.3	.3821	.3783	.3745	.3707	.3669	.3632	.3594	.3557	.3520	.3483
−0.2	.4207	.4168	.4129	.4090	.4052	.4013	.3974	.3936	.3897	.3859
−0.1	.4602	.4562	.4522	.4483	.4443	.4404	.4364	.4325	.4286	.4247
−0.0	.5000	.4960	.4920	.4880	.4840	.4801	.4761	.4721	.4681	.4641

NOTE: For values of z below −3.49, use 0.0001 for the area.

(*continued*)

*Use these common values that result from interpolation:

z Score	Area
−1.645	0.0500
−2.575	0.0050

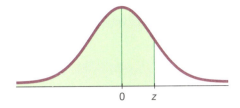

POSITIVE z Scores

TABLE A-2 *(continued)* Cumulative Area from the LEFT

z	.00	.01	.02	.03	.04	.05	.06	.07	.08	.09
0.0	.5000	.5040	.5080	.5120	.5160	.5199	.5239	.5279	.5319	.5359
0.1	.5398	.5438	.5478	.5517	.5557	.5596	.5636	.5675	.5714	.5753
0.2	.5793	.5832	.5871	.5910	.5948	.5987	.6026	.6064	.6103	.6141
0.3	.6179	.6217	.6255	.6293	.6331	.6368	.6406	.6443	.6480	.6517
0.4	.6554	.6591	.6628	.6664	.6700	.6736	.6772	.6808	.6844	.6879
0.5	.6915	.6950	.6985	.7019	.7054	.7088	.7123	.7157	.7190	.7224
0.6	.7257	.7291	.7324	.7357	.7389	.7422	.7454	.7486	.7517	.7549
0.7	.7580	.7611	.7642	.7673	.7704	.7734	.7764	.7794	.7823	.7852
0.8	.7881	.7910	.7939	.7967	.7995	.8023	.8051	.8078	.8106	.8133
0.9	.8159	.8186	.8212	.8238	.8264	.8289	.8315	.8340	.8365	.8389
1.0	.8413	.8438	.8461	.8485	.8508	.8531	.8554	.8577	.8599	.8621
1.1	.8643	.8665	.8686	.8708	.8729	.8749	.8770	.8790	.8810	.8830
1.2	.8849	.8869	.8888	.8907	.8925	.8944	.8962	.8980	.8997	.9015
1.3	.9032	.9049	.9066	.9082	.9099	.9115	.9131	.9147	.9162	.9177
1.4	.9192	.9207	.9222	.9236	.9251	.9265	.9279	.9292	.9306	.9319
1.5	.9332	.9345	.9357	.9370	.9382	.9394	.9406	.9418	.9429	.9441
1.6	.9452	.9463	.9474	.9484	.9495 *	.9505	.9515	.9525	.9535	.9545
1.7	.9554	.9564	.9573	.9582	.9591	.9599	.9608	.9616	.9625	.9633
1.8	.9641	.9649	.9656	.9664	.9671	.9678	.9686	.9693	.9699	.9706
1.9	.9713	.9719	.9726	.9732	.9738	.9744	.9750	.9756	.9761	.9767
2.0	.9772	.9778	.9783	.9788	.9793	.9798	.9803	.9808	.9812	.9817
2.1	.9821	.9826	.9830	.9834	.9838	.9842	.9846	.9850	.9854	.9857
2.2	.9861	.9864	.9868	.9871	.9875	.9878	.9881	.9884	.9887	.9890
2.3	.9893	.9896	.9898	.9901	.9904	.9906	.9909	.9911	.9913	.9916
2.4	.9918	.9920	.9922	.9925	.9927	.9929	.9931	.9932	.9934	.9936
2.5	.9938	.9940	.9941	.9943	.9945	.9946	.9948	.9949 *	.9951	.9952
2.6	.9953	.9955	.9956	.9957	.9959	.9960	.9961	.9962	.9963	.9964
2.7	.9965	.9966	.9967	.9968	.9969	.9970	.9971	.9972	.9973	.9974
2.8	.9974	.9975	.9976	.9977	.9977	.9978	.9979	.9979	.9980	.9981
2.9	.9981	.9982	.9982	.9983	.9984	.9984	.9985	.9985	.9986	.9986
3.0	.9987	.9987	.9987	.9988	.9988	.9989	.9989	.9989	.9990	.9990
3.1	.9990	.9991	.9991	.9991	.9992	.9992	.9992	.9992	.9993	.9993
3.2	.9993	.9993	.9994	.9994	.9994	.9994	.9994	.9995	.9995	.9995
3.3	.9995	.9995	.9995	.9996	.9996	.9996	.9996	.9996	.9996	.9997
3.4	.9997	.9997	.9997	.9997	.9997	.9997	.9997	.9997	.9997	.9998
3.50 and up	.9999									

NOTE: For values of z above 3.49, use 0.9999 for the area.

*Use these common values that result from interpolation:

z Score	Area
1.645	0.9500
2.575	0.9950

Common Critical Values

Confidence Level	Critical Value
0.90	1.645
0.95	1.96
0.99	2.575

GUIDED WORKBOOK

LAURA IOSSI

Broward College

ELEMENTARY STATISTICS
WITH INTEGRATED REVIEW
TWELFTH EDITION

Mario F. Triola

Dutchess Community College

Pearson

Boston Columbus Indianapolis New York San Francisco
Amsterdam Cape Town Dubai London Madrid Milan Munich Paris Montreal Toronto
Delhi Mexico City Sao Paulo Sydney Hong Kong Seoul Singapore Taipei Tokyo

Copyright © 2017 Pearson Education, Inc.
Publishing as Pearson, 501 Boylston Street, Boston, MA 02116.

ISBN-13: 978-0-13-449543-9
ISBN-10: 0-13-449543-8

2 2019

www.pearsonhighered.com

Contents

Integrated Review 1: Introduction to Integrated Review   1

Integrated Review 2: Decimals, Fractions, Percentages, and Graphs   13

Integrated Review 3: Getting Ready to Evaluate Statistical Formulas   29

Integrated Review 4: Preparing for Probability   41

Integrated Review 5: Evaluating Formulas for Probability Distributions   59

Integrated Review 6: The Building Blocks for Working with Normal Distributions   69

Integrated Review 7: Intervals of Numbers   79

Integrated Review 8: Formulas for Hypothesis Testing   91

Integrated Review 9: Formulas for Two-Sample Hypothesis Testing   97

Integrated Review 10: Linear Equations   109

Appendix A: Statistics Symbols   127

Appendix B: Units of Measurement   128

Appendix C: Conversions between Units of Measurement   129

Answers to Practice Problems   131

Integrated Review 1: Introduction to Integrated Review

Elementary Statistics **Chapter 1: Introduction to Statistics**

Objectives:

1. Learn how statistics is different.
2. Construct a plan of action for success.
3. Identify variables in context.
4. Recognize and convert between units of measurement.
5. Classify data as a type of number.
6. Evaluate exponents.
7. Translate numbers between scientific notation and standard form and vice versa.
8. Identify the number of significant digits.
9. Interpret technology output that is in scientific notation.

Welcome to statistics! In the corresponding chapter of the Triola text, you will begin the course with an introduction to the basic terminology used in statistics. Most students are not as familiar with statistics as they are with algebra. So, you may not be sure what to expect in this course. We will begin the *Guided Workbook for Elementary Statistics with Integrated Review* with a brief discussion of how statistics is different.

Since you have found your way to the *Guided Workbook with Integrated Review*, let me begin by explaining what this resource is and how you can use it.

This workbook parallels Triola, *Elementary Statistics,* 13th edition. The 10 integrated review chapters correlate with the first 10 chapters in the *Elementary Statistics* textbook. This workbook will review the algebra and arithmetic that you may need for the corresponding chapter in the text. For example, we will review how to evaluate a formula by substituting and simplifying in this workbook, and you will learn the statistical utility of the formula in the text. So, you may want to complete the integrated review for each chapter before beginning your work in the textbook. You may also wish to come back to this resource as you encounter difficulties with the algebraic and arithmetic calculations within the text.

The format of this resource is that of a workbook. The objectives that we will cover in each chapter will be clearly stated at the beginning of each unit. We will review each objective with an explanation or example(s). After each example, a My Turn! Problem will be provided. You will find space to work out these problems directly in the workbook. The answers to the My Turn! problems are located at the end of each chapter before the Practice Problems. The answers to all the Practice Problems are located at the end of the Workbook. In addition, you can find unlimited practice problems that correspond to this integrated review on MyStatLab. The Integrated Review problems on MyStatLab include corresponding help features such as videos, additional examples, and interactive Help Me Solve This experiences.

I hope that you will find this integrated review to be helpful, and I wish you success in this course. Let's start our journey!

Objective 1: Learn how statistics is different.

Most students join a statistics course not exactly sure what they have signed up for. It isn't algebra, is it? It definitely isn't calculus. So, what is it? I highly recommend that you open your textbook (either the paper version or etext) and peruse the chapter titles. You won't see terminology such as *factor* or *solve*. What you will see are such terms as *data*, *summarizing*, *graphing*, *estimating*, *inference*, and *probability*. Statistics involves making decisions based on information. You will learn how to gather, organize, and glean meaning from data. Yes, there will be calculations involved, and these calculations often involve large sets of data. Therefore, much of what you would encounter in statistics would best be handled with appropriate technology.

You may notice that many of the problems in the textbook appear to be "word problems." I find that this often makes students anxious. However, look closer at the problems. You are not going to be solving for *x*. The application problems that you typically encounter in the textbook are more interesting to read. The words are describing a real-world scenario. That is, they describe where the data came from and even provide the data we are dealing with. You may want to view the problems as a sort of information source from which you need to extract the desired information and numbers, rather like a dictionary or telephone book. (Do those exist anymore?)

Many students find that they enjoy statistics a lot more than a basic algebra class, as it feels more relevant to them. Statistics is everywhere. Once you delve into the course, you will be able to apply what you learn almost immediately.

Objective 2: Construct a plan of action for success.

As much as you may end up loving statistics, you probably want to have to take this course only once. As such, you should form a plan of action so that you succeed.

1. Determine your weaknesses and develop a strategy to overcome each of them.
    If you tend to lose focus or procrastinate, find a study buddy to motivate you.
    If you tend to leave materials for class at home, perhaps you can locate a locker on campus or have a prepacked statistics bag that contains your calculator, pencils, etc.
    If you tend to get anxious during test, find ways to simulate the testing environment as much as possible BEFORE you take a test.
2. Determine at least three ways to get help.
    What are you going to do when you get stuck on a topic? It is best to come up with several ways to get help before you are frazzled.
    Some ideas to consider:
    Watching the videos that accompany this text.
    Going to the math lab (It might be called something different at your college.).
    Visiting your professor during office hours.
3. Use appropriate technology.
    Statisticians use technology in the real world. If your professor allows it and recommends a particular technology tool, make the investment and gain access to it. Get your hands on

the technology as soon as you can. Practice with it. Use the same technology when you practice as you would during testing situations.

4. Determine an appropriate study schedule.

   It is generally recommended that you spend 2–3 hours studying/practicing outside of class for every hour that you spend in class. If you are taking this course online or in hybrid format, you will want to increase this recommended time. If you tend to struggle with math, you will also want to increase this time allotment.

   Also, remember to break the time into manageable pieces. The material will stick with you better in the long term if you look at it every day than if you have a marathon session the night before an exam.

If you need more study tips, there are entire books devoted to success strategies and you may want to consider finding one you like. For now, let's focus on our first objective of the integrated review!

Objective 3: Identify variables in context.

A variable represents an unknown quantity. In statistics, this could be a data value. You may want to consider using a variable name that is helpful in identifying what it represents (e.g., w for weight, h for height).

Note that within this course, you will feel as though you are reading Greek at times because you will be! A lot of Greek letters are used as variables in this course. The good news is that if you ever travel to Greece, you will be able to read some of the signs there. To see some of the Greek alphabet that you may encounter in this course, please see Appendix A.

Example 1 Let n represent the number of respondents to a survey about exercise. What does $n = 50$ mean?

It means that 50 people responded to the survey about exercise.

My Turn!

Let h represent the height of a wave in feet at "Jaws" in Maui. What does $h = 30$ mean?

Example 2 A six-sided die is rolled.

    a) Let d represent the side that faces up. What are the possible values for d?
       A six sided die has sides with 1, 2, 3, 4, 5, and 6 dots. So, d has the possible values of 1, 2, 3, 4, 5, or 6.

    b) Let t represent the sides that face up that are multiples of 3. What are the possible values for t?
       The faces of a die that are multiples of 3 are 3 and 6. Therefore, t has the possible values of 3 or 6.

    c) Now, the die is rolled 4 times. Let f represent the number of times that the die lands with a 5 facing up. What are the possible values for f?
       The die could land on a five 0, 1, 2, 3, or 4 times. So, f has the possible values of 0, 1, 2, 3, and 4.

My Turn!

Let b represent the number of boys in a family of 5. What are the possible values for b?

Objective 4: Recognize and convert between units of measurement

Since statistics deals with measurements, you should feel comfortable recognizing both metric and English units of measurement along with their abbreviations.

The metric system uses the following basic prefixes.

Prefix	Meaning
kilo	1000
hecto	100
deca	10
deci	0.1
centi	0.01
milli	0.001

The following are some common units of measurement in the metric system along with their abbreviations.

Metric	
Unit of Measurement	Abbreviation
gram	g
meter	m
liter	l

The following are some common units of measurement in the English system.

English	
Unit of Measurement	Abbreviation
foot	ft.
mile	mi.
ounce	oz.
inch	in.
pound	lb.

The above tables are not comprehensive lists, and if you are unfamiliar with a unit of measurement, you may want to look it up in one of our more comprehensive tables in Appendix B.

Occasionally, you may have to convert your data from one unit of measurement to another unit of measurement for the sake of consistency. That is, you generally want your data to be in the same unit of measurement before you perform any further calculations.

Here are a few of the most commonly used conversions.

12 inches = 1 foot
5280 feet = 1 mile
1 meter ≈ 3.28 feet
1 kilogram ≈ 2.2 pounds
1 mile ≈ 1.62 kilometers

Additional commonly used conversions can be found in Appendix C.

Example 3 Convert 72 inches to feet.

We will use the conversion factor that there are 12 inches in 1 foot. We write this as a fraction and multiply by it. We can do this, since the numerator is equal to the denominator; so, we are essentially multiplying by 1.

$$72 \; \cancel{\text{inches}} \times \frac{1 \text{ foot}}{12 \; \cancel{\text{inches}}} = 6 \text{ feet}$$

Answer 6 feet

My Turn!

Convert 5 miles to feet.

Example 4 Convert 8960 meters to kilometers.

There are 1000 meters in 1 kilometer.

$$8960 \ \cancel{\text{meters}} \times \frac{1 \, \text{kilometer}}{1000 \ \cancel{\text{meters}}} = 8.960 \ \text{kilometers}$$

Answer 8.96 kilometers

My Turn!

Convert 56 grams to milligrams.

Objective 5: Classify data as a type of number.

You will encounter several classifications for data in the text. Although we will not go over those here, we will indirectly practice classifying, using some concepts you probably know from prerequisite material.

Example 5 Select which of the following types of numbers the number 6 belongs to.

Whole numbers, integers, real numbers

Whole numbers = {0, 1, 2, 3, 4, ...}

Integers = {..., −3, −2, −1, 0, 1, 2, 3,...}

Real numbers = all the numbers on a number line; this includes rational and irrational numbers.

The number 6 belongs to each of these sets of numbers.

My Turn!

Select which of the following types of numbers the number 2.5 belongs to.

Whole numbers, integers, real numbers

Objective 6: Evaluate exponents.

Recall that exponents are shorthand for repeated multiplication. The value of the exponent tells you how many times the factor repeats.

For instance, $3^4 = 3 \cdot 3 \cdot 3 \cdot 3$.

Example 6 Evaluate 8^3.

We expand 8^3, and then calculate the repeated multiplication. Most technology can calculate exponents, and you may want to explore how to do this on your calculator now.

$$8^3 = 8 \cdot 8 \cdot 8 = 512$$

Answer 512

My Turn!

Evaluate 7^4.

Objective 7: Translate numbers between scientific notation and standard form and vice versa.

Scientific notation is a version of shorthand for very large or very small numbers.

A number is in **scientific notation** when it is written in the form $a \times 10^b$, where a is a number greater than or equal to 1 and less than 10.

Example 7 Translate 0.000478 to scientific notation.

We have to move the decimal point until there is one digit to the left of it. (This is how you make a into a value greater than or equal to 1 and less than 10.) This would move the digit immediately to the right of 4. Count the number of places that we moved the decimal point and note the direction we moved it. We moved the decimal point 4 places to the right. Since we moved it to the right, the exponent on the 10 will be negative.

$$0.000478 = 4.78 \times 10^{-4}$$

Answer 4.78×10^{-4}

My Turn!

Write 0.0392 in scientific notation.

Example 8 Translate 760,000 to scientific notation.

For scientific notation, we have to move the decimal point until there is one digit to the left of it. This would move the digit immediately to the right of 7. Count the number of places that we moved the decimal. We moved the decimal 5 places to the left. Since we moved it to the left, the exponent on the 10 will be positive.

$$760,000 = 7.6 \times 10^5$$

Answer 7.6×10^5

My Turn!

Write 9,670,000 in scientific notation.

Example 9 Write 3.79×10^3 in standard form.

To translate from scientific notation to standard form, we are essentially doing the reverse of the above examples. Whenever the exponent is positive, we will move the decimal point to the right. Since the exponent in this example is positive 3, the decimal point will move right three places. (This is because we are multiplying by 1000.)

$3.79 \times 10^3 = 3790$

Answer 3790

My Turn!

Write 8.1×10^{-7} in standard form.

Objective 8: Identify the number of significant digits

The **significant digits (or significant figures)** of a number are the digits that carry meaning.

The following are considered significant digits:

1. All nonzero digits are significant
2. Zeros between nonzero digits are significant.
3. Leading zeros are never significant.
4. In a number with a decimal point, trailing zeros, those to the right of the last non-zero digit, are significant (they indicate that the number represents something that was measured to that level of precision).
5. In a number without a decimal point, trailing zeros may or may not be significant, and more information is needed to make that determination. (For purposes of our text, we will assume that they are not significant unless otherwise stated.)

Example 10 Identify the number of significant digits in the following numbers.

a) 938,000
   9, 3, and 8 are each significant digits, since they are nonzero digits. The trailing zeros will be considered as insignificant for purposes of this text.
   There are 3 significant figures.
b) 0.00012
   The leading zeros are insignificant. The 1 and 2 are significant digits. So, there are 2 significant digits.
c) 0.000103
   The leading zeros are insignificant. The 1 is significant. The 0 to the right of the 1 is significant, since zeros between nonzero digits are always significant. The 3 is significant, since it is a nonzero digit. So, there are 3 significant digits.

My Turn!

Identify the number of significant digits in the following numbers.

    a) 1,030,000

    b) 101

    c) 0.000002

Objective 9: Interpret technology output that is in scientific notation.

Depending on the technology that you use, you may encounter scientific notation that looks slightly different on your screen than how we have been writing it thus far. You will want to try calculating something like 2^{50} using technology. The result is probably too large to write in standard form on your screen. On a TI-84 calculator, it may show up as 1.125899907E15. This output indicates that the number has been expressed in scientific notation. That is,1.125899907E15 is equivalent to $1.125899907 \times 10^{15}$.

Example 11 Round 2.058911321E14 (which is 3^{30}) to 3 significant digits.

2.058911321E14 is equivalent to $2.058911321 \times 10^{14}$.

If we round to 3 significant digits, we begin looking at the digits from left to right. The 2 is significant since it all nonzero digits are significant. The 0 is significant, since it is between two nonzero digits. The 5 is significant, since it is nonzero. This would be our 3 significant digit positions. However, for the 5 we have to decide whether to keep it or to round up. Since the digit immediately to the right of it is an 8, we round the 5 up to a 6.

So, our rounded value is 2.06×10^{14}.

Answer 2.06×10^{14}

My Turn!

The Excel output for 4^{30} is 1.15292E+18. Round this number to 3 significant digits.

Answers to My Turn!

1. A wave that is 30 feet high at "Jaws"
2. The possible values for b are 0, 1, 2, 3, 4, and 5.
3. 26,400 feet
4. 56,000 mg
5. Real numbers
6. 2401
7. 3.92×10^{-2}
8. 9.67×10^{6}
9. 0.00000081
10. a) 3 significant digits b) 3 significant digits c) 1 significant digit
11. 1.15×10^{18}

Practice Problems

1. Let d represent the days since a baby was born. If a baby was born on January 3 and today is January 15, what does d equal?
2. If t represents the time in years since the last U.S. presidential election, what are the possible values for t?
3. Convert 5 meters to feet.
4. Convert 35 kilometers to meters.
5. Select which of the following types of numbers the number -4 belongs to: whole numbers, integers, real numbers.
6. Evaluate 9^{3}.
7. Write 0.000706 in scientific notation.
8. Write 270,000 in scientific notation.
9. Write 9.71×10^{7} in standard form.
10. Identify the number of significant digits in the following numbers.
a) 503,000.2
b) 909
c) 3000
11. Excel output for $\dfrac{1}{4^{30}}$ is 8.67362E-19. Round this number to 3 significant digits.

Integrated Review 2: Decimals, Fractions, Percentages, and Graphs

Elementary Statistics Chapter 2: Exploring Data with Tables and Graphs

Objectives:

1. Round decimals.
2. Write a fraction in lowest terms.
3. Convert between decimals, fractions, and percentages.
4. Calculate relative frequencies.
5. Find the percentage of a number.
6. Plot points.
7. Review the skills for graphing.

In this chapter of the Triola text, you will be summarizing data using tables and graphs. In order to create these tables and graphs, you will have to be comfortable with dealing with numbers in several formats: decimals, percentages, and fractions. We will review how to convert from one number format to another in this *Guided Workbook with Integrated Review*. Also, throughout the Triola text, you will be expected to round answers according to various rounding rules. So, we will review the basics of rounding in this workbook. Finally, in the related chapter of the textbook you will learn about several graphs. To help you with this, we will review the basics of plotting points and the basic skills for graphing.

Objective 1: Round decimals.

Throughout the text, you will be asked to round your final answer. You may be asked to round in one of two ways: (1) to a particular place value or (2) to a certain number of significant figures.

Let's begin with a refresher on place value.

The table below indicates the place value for each digit in the number 1234.5678.

1	2	3	4	.	5	6	7	8
Thousands	Hundreds	Tens	Ones	Decimal Point	Tenths	Hundredths	Thousandths	Ten thousandths

You will often be asked to round your final answer to a particular place value. So, we will review the rounding basics in this first example.

Example 1 Round 3456.789 inches to the nearest tenth of an inch.

Step 1: Locate the digit in the desired place value. Let's call this digit R for purposes of our discussion.

The digit 7 is in the tenth place.

We are trying to decide whether the given number is closer to 3456.7 or 3456.8

Step 2: If the digit in the place value immediately to the right of R is a 0, 1, 2, 3, or 4, then keep R as it is and all digits to the right of R are zeros. If the zeros are not placeholders (that is, removing them does not result in other digits losing their place value), then they can be removed. If the digit in the place value immediately to the right of R is a 5, 6, 7, 8, or 9, then increase R by 1 (round up) and all digits to the right of R are zeros. Again, if these zeros are not placeholders, then they can be removed.

Since the digit immediately to the right of 7 is an 8, we must round up to 3,456.800. As the zeros are not placeholders, we can remove them.

Answer 3456.789 inches rounded to the nearest tenth of an inch is 3456.8 inches.

My Turn!

Round 7654.32198 miles to the nearest hundredth of a mile.

Example 2 Round 0.050748 cm to 3 significant digits.

We went over significant digits in the first chapter of this integrated review. Now, we will round numbers to a particular number of significant figures.

First, leading zeros are never significant digits. So, we will begin counting significant figures from left to right, beginning with the 5. The 0 immediately to the right of 5 is also a significant digit, since it is between two nonzero digits. The 7 to the right of the 0 is also significant, since all nonzero digits are always significant. So, by rounding to 3 significant digits, we are trying to decide whether 0.050748 is closer to 0.0507 or 0.0508. Since the digit immediately to the right of 7 is 4, we will keep the third significant digit at 7.

Answer 0.050748 rounded to three significant digits is 0.0507.

My Turn!

Round 0.36409 cm to 3 significant digits.

Objective 2: Write a fraction in lowest terms.

There are times that your result may be a fraction, especially when we work on the chapter on probability. Whenever you give a final result as a fraction, it must be fully reduced. In order for a fraction to be in lowest terms, there should be no common factors in the numerator and denominator.

Example 3 Write $\dfrac{48}{72}$ in lowest terms.

We may not all arrive at a reduced fraction by taking the same route. What is key is that we all check to make sure that there are no common factors left in the numerator and denominator. If there are any, cancel them out, since $\dfrac{a}{a} = 1$.

$$\frac{48}{92} = \frac{2 \cdot 24}{2 \cdot 46} = \frac{24}{46} = \frac{2 \cdot 12}{2 \cdot 23} = \frac{12}{23}$$

Note that many calculators (including the TI-83/84) can rewrite fractions in lowest form for you. You may want to become comfortable using this feature, especially for fractions with large numerators and denominators.

My Turn!

Write $\dfrac{36}{102}$ in lowest terms.

Objective 3: Convert between decimals, fractions, and percentages.

There will be many times in this course that you will want to convert between decimals, fractions, and percentages. We will review how to do this for the most commonly used conversions in statistics.

Example 4 Convert 8% to a decimal.

To convert from a percentage to a decimal, you will remove the percent symbol and move the decimal point two places to the left. (This is because the percent symbol means per hundred; so, you are essentially dividing by 100.)

Currently, the decimal point is to the right of 8. (An invisible decimal point is always to the right of the rightmost digit.) So, we will move it two places to the left. Since there is only one digit to the left, we will have to insert a placeholder of 0. Also, it is good practice to write a 0 to the left of the decimal point, as it results in less confusion to the reader. (Decimal points are pretty tiny for reading.)

$$8\% = 0.08$$

Answer 0.08

My Turn!

Convert 0.3% to a decimal.

Example 5 Convert 0.006 to a percent.

Now, we will do the reverse and convert a decimal to a percent. To do this you will move the decimal point two places to the right and insert the percent symbol.

$$0.006 = 0.6\%$$

Answer 0.6%

My Turn!

Convert 2.3 to a percent.

Example 6 Convert $\frac{7}{8}$ to a percent.

First, we will convert the fraction to a decimal. Then, we will follow the steps from Example 5 to change the decimal to a percentage.

To change a fraction to a decimal, you divide the numerator by the denominator.

$$\frac{7}{8} = 7 \div 8 = 0.875$$

Now, we will convert the decimal 0.875 to a percentage by moving the decimal point two places to the right and inserting the percent symbol.

$$0.875 = 87.5\%$$

Answer 87.5%

My Turn!

Convert $\frac{9}{16}$ to a percent.

Example 7 Convert 0.06 to a fraction.

To convert a decimal to a fraction, you write the number with the decimal point and leading zeros removed as the numerator of a fraction over a denominator that is a power of 10. The exponent on the 10 is the number of digits to the right of the decimal point in the given number. For 0.06, there are 2 digits to the right of the decimal point. So, the denominator will be 10^2, which can be rewritten as 100. Once you have a fraction, be sure to simplify it fully.

$$0.06 = \frac{6}{10^2} = \frac{6}{100} = \frac{3}{50}$$

Answer $\frac{3}{50}$

My Turn!

Convert 0.0028 to a fraction.

```
```

Objective 4: Calculate relative frequencies.

As you construct tables and graphs in the text, you will be calculating relative frequencies, which are the ratio of the amount of times an event occurs to the number of times the event could occur. In other words, it is the number of times something happens divided by the total number of times it could happen. Relative frequencies are often represented as percentages.

Example 8 If 30 out of 80 people responded that they skip flossing at least one day a week, calculate the percentage of respondents that skip flossing at least one day a week.

So, the relative frequency for this problem is 30 out of 80, which can be written as $\frac{30}{80}$. We can then turn this fraction into a decimal. Finally, we convert our answer to a percentage by moving the decimal point two places to the right and inserting %.

$$\frac{30}{80} = 30 \div 80 = 0.375 = 37.5\%$$

Answer 37.5% of respondents skip flossing at least one day a week.

My Turn!

If 450 out of 800 people responded that they own more than one computer, calculate the percentage of respondents who own more than one computer.

```
```

Objective 5: Find the percentage of a number.

Sometimes, you may need to determine what the percentage of a number is.

What is $p\%$ of x?

To do this, you will multiply the percentage written as a decimal times the number.

That is,

$$p\% \text{ as a decimal} \cdot x$$

Example 9 What is 30% of 150?

Write 30% as a decimal and multiply it by 150.

$$30\% \text{ of } 150 = 0.3 \cdot 150 = 45$$

Answer 30% of 150 is 45.

My Turn!

What is 65% of 600?

Example 10 There were 900 respondents to a survey. If 40% of respondents indicated that they live in an apartment, how many respondents live in an apartment?

This amounts to answering the question, "What is 40% of 900?"

To find the percentage of a number, you multiply the percent written as a decimal times the number.

$$x = 0.4 \cdot 900 = 360$$

Answer 360 respondents live in an apartment.

My Turn!

There were 80 respondents to a survey. If 30% of respondents indicated that they own a DVD player, how many respondents own a DVD player?

Objective 6: Plot points.

In statistics, we will often study data that come in pairs. It can be helpful to see the paired data on a graph. Each set of paired data can be represented by a point that can be written as (x, y). The points can be plotted on a Cartesian coordinate plane. The origin, or the place where the horizontal (x) and vertical (y) axes intersect, is the starting point The first value in the paired data (the x value) will tell you to move left (negative) or right (positive). The second value (the y value) will place the point up (positive) or down (negative). For example, you can plot the set of paired data $(3, 5)$ by counting three units to the right of the origin and up five units.

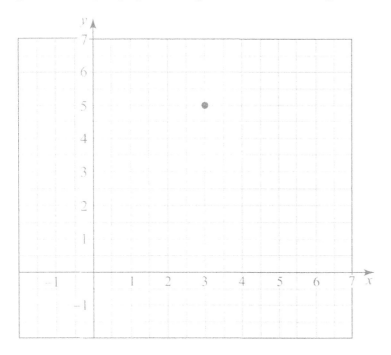

Example 11 Plot the following paired data.

Number of People in a Household (x)	Number of TVs in a Household (y)
1	2
2	4
3	5

These data can be written as three points (1, 2), (2, 4), and (3, 5). Then, we can plot these three points on the Cartesian plane. As you learn statistics, you will find that most of the data that you will encounter will be positive values. As a result, many times you can limit your Cartesian plane to the first quadrant (the upper right-hand section). Also, there may be times that you will want to change the scale on the axes. Be sure to label your axes with the variables that are being used.

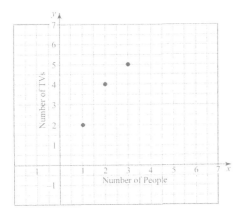

My Turn!

Plot the following paired data.

Number of Rooms in a House	Number of Paintings in a House
5	15
7	20
8	26

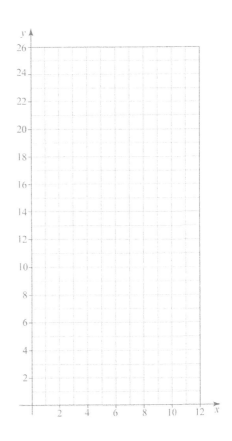

Objective 7: Review the skills for graphing.

In the corresponding chapter of the text, you will begin interpreting graphs. Some of the graphs will be in the Cartesian plane, as in the last example. Other graphs may include some that you may have seen before in the media, such as bar charts, pie charts, and time series graphs. We will do a sampling of two here. Notice how the graphs that follow are carefully labeled so that the information is clear to the reader. Whenever you create your own graphical representations of data, label everything clearly so that the graph conveys all the necessary information to the reader.

Example 12 The following is called a bar chart. It represents the types of patients that a veterinarian had in one day. What percentage of the vet's patients that day were cats? Round the answer to the nearest percent.

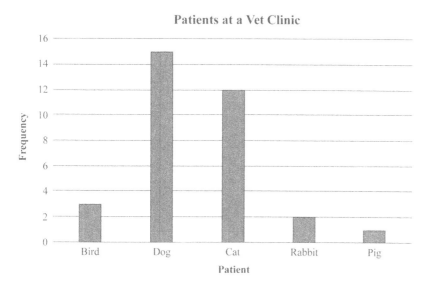

In order to answer this, we need to figure out how many cats were patients that day and how many animals were seen that day at the veterinarian's clinic.

The height of the bar labeled "Cat" will tell us that there were 12 cats seen that day.

To find the total number of animals seen that day, we have to sum up all of the heights (frequencies) of the bars.

Total number of animals seen that day $= 3 + 15 + 12 + 2 + 1 = 33$.

So, to find the relative frequency, we have to divide 12 by 33 to change it to a decimal: $\frac{12}{33} = 0.\overline{36}$. (The 36 has a bar above it since those digits are repeating.) We will now convert this to a percentage and round to the nearest percent.

$$0.\overline{36} = 36.\overline{36}\% \approx 36\%$$

Answer 36% of the patients were cats.

My Turn!

The bar chart represents the frequencies for the types of rooms that are in a home. What percentage of the house's rooms are bathrooms? Round the answer to the nearest percent.

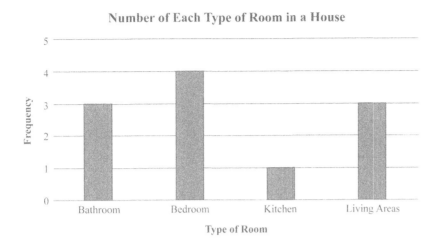

Number of Each Type of Room in a House

Example 13 The following is called a pie chart. It represents the college majors for a sample of students. What percentage of the college students were math majors? Round the answer to the nearest percent.

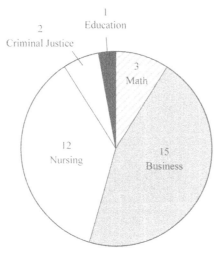

College Major

First, we need to figure out how many students have math as their college major. There were 3 math majors, as indicated by the number 3 in the appropriate slice of the pie chart. Next, we need to figure out the total number of students who were included in the sample. We can do that by summing up the frequencies for the 5 college majors.

Total number of students sampled $= 3 + 15 + 12 + 2 + 1 = 33$.

So in order to find the relative frequency, we have to divide 3 by 33 to change it to a decimal: $\frac{3}{33} = 0.\overline{09}$.

We will now convert this to a percentage and round to the nearest percent.

$$0.\overline{09} = 9.\overline{09}\% \approx 9\%$$

Answer About 9% of college students sampled were math majors.

My Turn!

The pie chart represents the frequencies for various types of toys in a pediatrician's waiting room. What percentage of the toys were dolls? Round the answer to the nearest percent.

Toys in a Pediatricians's Waiting Room

Answers to My Turn!

1. 7654.32 miles

2. 0.364 cm

3. $\dfrac{6}{17}$

4. 0.003
5. 230%
6. 56.25%

7. $\dfrac{7}{2500}$

8. 56.25%

9. 390

10. 24 respondents

11.

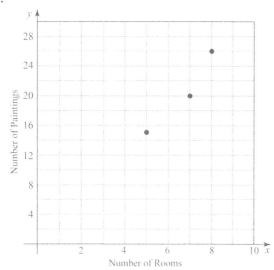

12. 27%
13. 19%

Practice Problems

1. Round 24,967.0784 kilograms to the nearest tenth of a kilogram.
2. Round 1.04708 pounds to 3 significant digits.
3. Write $\dfrac{84}{196}$ in lowest terms.
4. Convert 2.5% to a decimal.
5. Convert 0.007 to a percent.
6. Convert $\dfrac{2}{25}$ to a percent.
7. Convert 0.018 to a fraction.
8. If 30 out of 600 people tested positive for a cat allergy, calculate the percentage of participants who tested positive for a cat allergy.
9. What is 32% of 700?
10. A researcher surveyed 300 people. If 22% of those surveyed indicated that they are vegan, how many subjects responded that they are vegan?
11. Plot the following paired data.

Number of Times Exercising in a Week	Number of Meals Eaten in a Week
5	20
2	17
0	15

12. The bar chart represents the frequency of various colored suitcases on an airport baggage claim conveyor belt. What percentage of suitcases were red? Round the answer to the nearest percent.

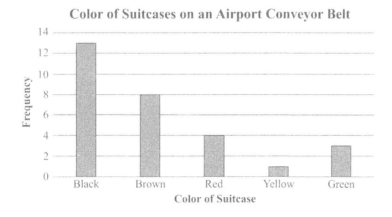

13. The pie chart represents the frequencies of various state license plates for cars on a Florida highway. What percentage of the license plates were from New York? Round the answer to the nearest percent.

License Plates for Cars on a Florida Highway

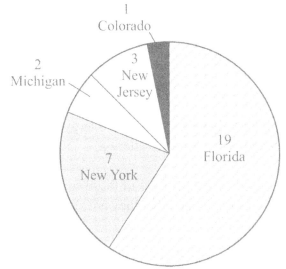

Integrated Review 3: Getting Ready to Evaluate Statistical Formulas

Elementary Statistics Chapter 3: Statistics for Describing, Exploring, and Comparing Data

Objectives:

1. Apply the order of operations.
2. Evaluate square roots.
3. Evaluate expressions and formulas.
4. Apply summation notation.

In this chapter of the Triola text, you will begin working with some basic statistical formulas that will be used to help describe data. In this integrated review chapter, we will review the concepts that will help you evaluate these statistical formulas.

Objective 1: Apply the order of operations.

We use the order of operations to simplify a numerical expression. You will have many numerical expressions that you will have to simplify in statistics, especially if you are evaluating formulas.

It is important that we do the operations in the correct order to arrive at a correct answer. It is critical that each step is completed from left to right.

Below you will find a listing of the order of operations. Always calculate the operations from left to right. This is especially important to remember for steps 3 and 4. This means that if there is a division to the left of multiplication in an expression, you would perform the division first.

1. **Parentheses**

2. **Exponents**

3. **Multiply** or **Divide**

4. **Add** or **Subtract**

Example 1 Perform the indicated operations: $5 \cdot 8^3$.

First we must calculate the exponent, since exponents should be done before multiplication according to the order of operations. (We reviewed calculating exponents in *Integrated Review* Chapter 1.)	$5 \cdot \boxed{8^3} = 5 \cdot \boxed{512}$
Next, we multiply.	$5 \boxed{\cdot} 8^3 = 5 \boxed{\cdot} 512 = 2560$

Answer $5 \cdot 8^3 = 2560$

My Turn!

Perform the indicated operations: $6^4 \div 4$.

Example 2 Perform the indicated operations: $15 \div 3 \cdot 2 + 9$.

First, we must multiply or divide from left to right. Since the division is to the left of the multiplication, we must do that first. It is a common misconception to think that you always multiply before dividing. This is not the case; you complete the operations of multiplication and division as they appear from left to right.	$15 \boxed{\div} 3 \cdot 2 + 9 = 5 \cdot 2 + 9$
Now, we multiply.	$5 \boxed{\cdot} 2 + 9 = 10 + 9$
Finally, we perform addition.	$10 \boxed{+} 9 = 19$

Answer $15 \div 3 \cdot 2 + 9 = 19$

My Turn!

Perform the indicated operations.

$5 + 2 \cdot 8$

Example 3: Perform the indicated operations: $\dfrac{(5-3)^2 + (7-3)^2 + (9-3)^2}{7}$

First, we must simplify within each set of parentheses.	$\dfrac{(5-3)^2 + (7-3)^2 + (9-3)^2}{7} = \dfrac{2^2 + 4^2 + 6^2}{7}$
Now, calculate the exponents.	$= \dfrac{4+16+36}{7}$
You can think of the entire numerator as being within an invisible set of parentheses. So, you would simplify that next.	$= \dfrac{56}{7}$
Finally, you can perform the division (or reduce the fraction).	$= 8$

Answer $\dfrac{(5-3)^2 + (7-3)^2 + (9-3)^2}{7} = 8$

My Turn!

Perform the indicated operations: $\dfrac{10}{(5+3)^2 + (8+3)^2}$.

Round the answer to the nearest tenth.

Objective 2: Evaluate square roots.

Recall the meaning of square root.

The principal square root of a number a is denoted by $\sqrt{a}$. Furthermore, $x = \sqrt{a}$ if $x^2 = a$. So, $3 = \sqrt{9}$, since $3^2 = 9$.

Note that for the order of operations a square root can be treated as an exponent.

Example 4 Evaluate $\sqrt{64}$.

We are essentially trying to find a number such that the number times itself is 64.

$$\sqrt{64} = \sqrt{8 \cdot 8} = 8$$

Answer $\sqrt{64} = 8$

My Turn!

Evaluate $\sqrt{25}$.

If the value under the radical symbol (i.e. the radicand) is not a perfect square, you are going to have to try a different strategy. You may recall having to simplify radicals for algebra. However, for the way that you will be using square roots for statistics, you won't have to simplify square roots as you would in an algebra course. You will simply use a calculator for radicands that are not perfect squares. However, you may want to come up with an estimate of which two integers the square root is between in order to catch careless calculator entry errors.

Example 5 Which two integers is $\sqrt{88}$ between?

You need figure out which two perfect squares 88 is between.

$$81 < 88 < 100$$

$$9^2 < 88 < 10^2$$

So,

$$9 < \sqrt{88} < 10$$

Answer: $\sqrt{88}$ is between 9 and 10.

My Turn!

Which two integers is $\sqrt{65}$ between?

Example 6 Approximate $\sqrt{88}$ to the nearest hundredth.

Use a calculator to evaluate this, since 88 is not a perfect square.

For the TI-83/84 calculators, the keystrokes are 2nd $\sqrt{}$ 88 enter. Note that the $\sqrt{}$ is the second function for the x^2 button. If you need to find other roots, press MATH.

Your calculator output is approximately 9.3808. This rounds to 9.38, which agrees with our findings from Example 5, as it is between 9 and 10.

Answer $\sqrt{88} \approx 9.38$. The symbol $\approx$ means "is approximately equal to." We use this symbol to indicate that our value is not an exact answer, since it is a rounded value.

My Turn!

Approximate $\sqrt{130}$ to the nearest hundredth.

Objective 3: Evaluate expressions and formulas.

Example 7 Perform the indicated operations:

$$\frac{18-11}{\dfrac{3}{\sqrt{9}}}$$

In order to simplify this expression, we will apply the order of operations to an expression that contains a square root. We treat a radical symbol as an exponent in the order of operations. We can do this because radicals can be rewritten as rational exponents (e.g. $\sqrt{9}=9^{\frac{1}{2}}$).

We can begin by viewing the numerator 18 – 11 as being contained within invisible parentheses and the denominator $\dfrac{3}{\sqrt{9}}$ as being contained within another set of parentheses. We can work on simplifying each of these. The simplification of the numerator is shown first.	$\dfrac{(18-11)}{\left(\dfrac{3}{\sqrt{9}}\right)}=\dfrac{7}{\left(\dfrac{3}{\sqrt{9}}\right)}$
To simplify the denominator $\dfrac{3}{\sqrt{9}}$, we must find $\sqrt{9}$.	$\dfrac{7}{\left(\dfrac{3}{\sqrt{9}}\right)}=\dfrac{7}{\left(\dfrac{3}{3}\right)}=\dfrac{7}{1}$
We find our final answer by dividing (or simplifying the fraction).	$\dfrac{7}{1}=7$

Answer $\dfrac{18-11}{\dfrac{3}{\sqrt{9}}}=7$

My Turn!

Perform the indicated operations:

$$\frac{6.8 - 6.1}{\dfrac{1.1}{\sqrt{16}}}$$

Round the answer to the nearest hundredth.

A formula is a special type of equation that expresses the relationship between several variables. On one side of the equation (usually the left side), you have the dependent variable. On the other side you have the independent variables. To evaluate a formula, we substitute known values for the dependent variables and follow the order of operations to simplify. We will include one example here, and you will see numerous examples in the chapters that follow.

Example 8 Evaluate the formula

$$E = z \cdot \frac{\sigma}{\sqrt{n}}$$

for $z = 1.96$, $\sigma = 2.1$, and $n = 100$. Round the answer to the nearest thousandth.

First, we must substitute the given values for the independent variables.

$$E = z \cdot \frac{\sigma}{\sqrt{n}} = 1.96 \cdot \frac{2.1}{\sqrt{100}}$$

In order to simplify the right-hand side, we will apply the order of operations to an expression that contains a square root.

We treat a radical symbol as an exponent in the order of operations. So, we will begin there. The square root of 100 is 10.	$E = 1.96 \cdot \dfrac{2.1}{\sqrt{100}} = 1.96 \cdot \dfrac{2.1}{10}$
We can simplify the fraction by dividing by 10.	$E = 1.96 \cdot \dfrac{2.1}{10} = 1.96 \cdot 0.21$
We find our value for the dependent variable, E, by multiplying.	$E = 1.96 \cdot 0.21 = 0.4116$
Now, we round our final answer to the nearest thousandth. Since the value in the ten thousandth place is a 6, we round to 0.412.	$E \approx 0.412$

Answer $E \approx 0.412$ when $E = z \cdot \dfrac{\sigma}{\sqrt{n}}$ and $z = 1.96$, $\sigma = 2.1$, and $n = 100$.

My Turn!

Evaluate the formula

$$n = \frac{z^2 \cdot \hat{p} \cdot \hat{q}}{E^2}$$

For $z = 2.575$, $\hat{p} = 0.3$, $\hat{q} = 0.7$, and $E = 0.05$. Round the answer to the nearest whole number.

Objective 4: Apply summation notation.

You will often see the summation symbol, $\sum$, in statistical formulas. It is the capital Greek letter "sigma." Whenever you see this symbol in a formula, it is telling you to add up values.

For instance, $\bar{x} = \dfrac{\sum x}{n}$ is the formula for finding the mean (average), $\bar{x}$, where x represents the data values and n is the number of data values.

Example 9 Find $\bar{x} = \dfrac{\sum x}{n}$ for the values in the accompanying table.

x
34
48
25
76
22
29

There are 6 values for x, so $n = 6$.

The formula is telling us to add up the values of x and divide by n.

$$\bar{x} = \frac{\sum x}{n} = \frac{34 + 48 + 25 + 76 + 22 + 29}{6} = \frac{234}{6} = 39$$

Note that the above formula could also be written as

$$\sum_{i=1}^{6} \frac{x_i}{n}$$

Answer $\bar{x} = 39$

My Turn!

Find $\bar{x} = \dfrac{\sum x}{n}$ for the values in the accompanying table, where n is the number of given values. Round to the nearest tenth, as needed.

x
5
4
8
7
2

Example 10 Find

$$\frac{\Sigma(x-39)^2}{n-1}$$

for the accompanying table and $n = 6$.

x
34
48
25
76
22
29

This notation is telling us to subtract 39 from each x value first, since we work out parentheses first in the order of operations. Next we would calculate exponents per the order of operations. Then, we sum up these squares of these differences. Once we have this sum, then we can perform division by the denominator.

$$\frac{\Sigma(x-39)^2}{n-1} = \frac{(34-39)^2+(48-39)^2+(25-39)^2+(76-39)^2+(22-39)^2+(29-39)^2}{6-1}$$

$$= \frac{(-5)^2+(9)^2+(-14)^2+(37)^2+(-17)^2+(-10)^2}{6-1}$$

$$= \frac{25+81+196+1369+289+100}{5}$$

$$= \frac{2060}{5}$$

$$= 412$$

Answer $\dfrac{\Sigma(x-39)^2}{n-1} = 412$ for the given table of values.

My Turn!

Find

$$\frac{\Sigma(x-2)^2}{n-1}$$ for the accompanying table and $n = 4$. Round to the nearest tenth, as needed.

x
4
3
2
1

Answer to My Turn!

1. 324
2. 21
3. 0.1
4. 5
5. 8 and 9
6. 11.4
7. 2.55
8. $n \approx 557$
9. $\bar{x} = 5.2$
10. 2

Practice Problems

1. Perform the indicated operations: $17 - 3 \cdot 4$.
2. Perform the indicated operations: $5 + 3^4 \div 9$.
3. Perform the indicated operations: $\dfrac{4^2}{(9-2)^2 + (8-2)^2}$. Round the answer to the nearest hundredth.
4. Evaluate $\sqrt{36} + \sqrt{100}$.
5. Which two integers is $\sqrt{101}$ between?
6. Approximate $\sqrt{255}$ to the nearest hundredth.
7. Perform the indicated operations. Round the answer to the nearest hundredth.
$$\frac{0.9 - 0.8}{\sqrt{\dfrac{0.8 \cdot 0.2}{50}}}$$
8. Evaluate the formula $\sigma = \sqrt{n \cdot p \cdot (1 - p)}$ for $n = 60$ and $p = 0.25$. Round the answer to the nearest tenth.
9. Find $\dfrac{\Sigma x}{n}$ for the values in the given table, where n is the number of given values. Round to answer to the nearest tenth.
10. Find $\dfrac{\Sigma(x-2)^2}{n-1}$ for the values in the given table and $n = 4$. Round the answer to the nearest tenth.

x
8
7
2
9

Integrated Review 4: Preparing for Probability

Elementary Statistics Chapter 4: Probability

Objectives:

1. Apply operations to fractions.
2. Apply operations to decimals.
3. Evaluate factorials.
4. Evaluate $_nP_r$ and $_nC_r$.
5. Determine the intersection, union, and complement of two sets.

In this chapter of the Triola text, you will learn the basics of probability. The probability, or likelihood, that some event will occur can be written either as a decimal or a fraction. For this reason, when you are evaluating probability formulas you will be dealing with operations on decimals and fractions. We will begin this integrated review chapter by reviewing operations on fractions and decimals. Then we will review some basic calculations that you will use for calculating other probability formulas. Finally, we will review the common operations of intersection, union, and complement of sets, as these concepts tie in with some other probability calculations. Once again, we will not explain the probability concepts in this integrated review. We will review only the basic calculation techniques here, and you will see the probability applications in the textbook itself.

Objective 1: Apply operations to fractions.

We will review the basic operations of addition, subtraction, multiplication, division, and exponentiation of fractions. Although you may typically calculate these using a calculator within a statistics course, we will briefly review how to do them without technology. However, you may also want to try each operation using a calculator to make sure you know how to work with fractions on your calculator.

Adding Fractions

Example 1 Perform the indicated operation:

$$\frac{3}{16}+\frac{9}{16}$$

In order to add fractions, you need a common denominator. The two given fractions already have the common denominator of 16. So, all you need to do is add the two numerators together and keep the common denominator. Always write your answers in reduced form.

$$\frac{3}{16}+\frac{9}{16}=\frac{(3+9)}{16}=\frac{12}{16}=\frac{3}{4}\cdot\frac{4}{4}=\frac{3}{4}\cdot 1=\frac{3}{4}$$

Answer $\dfrac{3}{16}+\dfrac{9}{16}=\dfrac{3}{4}$

My Turn!

Perform the indicated operation:

$$\frac{5}{8}+\frac{1}{8}$$

Example 2 Perform the indicated operation:

$$\frac{1}{9}+\frac{2}{15}$$

Since we need a common denominator in order to add or subtract fractions, our first step will be to rewrite each fraction over the lowest common denominator. The lowest common denominator is 45. Then, you follow the same procedure as in Example 1 and add the two numerators together and keep the common denominator.

$$\frac{1}{9}+\frac{2}{15}=\frac{1}{9}\cdot\frac{5}{5}+\frac{2}{15}\cdot\frac{3}{3}$$

$$=\frac{5}{45}+\frac{6}{45}$$

$$=\frac{(5+6)}{45}$$

$$=\frac{11}{45}$$

Answer $\dfrac{1}{9}+\dfrac{2}{15}=\dfrac{11}{45}$

My Turn!

Perform the indicated operation:

$$\frac{3}{10} + \frac{5}{12}$$

Subtracting Fractions

Example 3 Perform the indicated operation:

$$\frac{5}{6} - \frac{1}{3}$$

You also need a common denominator when subtracting fractions. So, our first step will be to rewrite each fraction over the lowest common denominator. The lowest common denominator is 6. Once you have rewritten each fraction over the lowest common denominator, subtract the two numerators and keep the common denominator. Be sure that your final answer is reduced!

$$\frac{5}{6} - \frac{1}{3} = \frac{5}{6} - \frac{1}{3} \cdot \frac{2}{2}$$

$$= \frac{5}{6} - \frac{2}{6}$$

$$= \frac{(5-2)}{6}$$

$$= \frac{3}{6}$$

$$= \frac{1}{2}$$

Answer $\dfrac{5}{6} - \dfrac{1}{3} = \dfrac{1}{2}$

My Turn!

Perform the indicated operation:

$$\frac{11}{15} - \frac{5}{12}$$

Since there are several topics in statistics that require subtracting fractions from 1, we will cover that type of subtraction problem here.

Example 4 Perform the indicated operation:

$$1 - \frac{3}{8}$$

The number 1 can always be rewritten as $\frac{a}{a}, a \neq 0$. We will select a as the value of the common denominator. So, we set $a = 8$ for this problem. Then

$$1 - \frac{3}{8} = \frac{8}{8} - \frac{3}{8} = \frac{5}{8}$$

Answer $1 - \dfrac{3}{8} = \dfrac{5}{8}$

My Turn!

Perform the indicated operation:

$$1 - \frac{16}{19}$$

Multiplying Fractions

Example 5 Perform the indicated operation:

$$\frac{5}{6} \cdot \frac{1}{3}$$

You do *not* need a common denominator in order to multiply fractions.

One way to multiply fractions is to multiply all numerators together and make that product the numerator, multiply all denominators together and make that product the denominator, and then reduce the fraction. Another way to do it is to "cancel out" any common factors that are in any numerator with any denominator (recall that you can do this because any number divided by itself is equal to 1) and then multiply across numerators and denominators. However, you do not have to decide on your preferred method for this particular example as there are no common factors in the numerators and denominators.

$$\frac{5}{6} \cdot \frac{1}{3} = \frac{5 \cdot 1}{6 \cdot 3} = \frac{5}{18}$$

Answer $\dfrac{5}{6} \cdot \dfrac{1}{3} = \dfrac{5}{18}$

My Turn!

Perform the indicated operation:

$$\frac{4}{15} \cdot \frac{5}{12}$$

Dividing Fractions

Example 6 Perform the indicated operation:

$$\frac{5}{6} \div \frac{1}{3}$$

You do *not* need a common denominator in order to divide fractions.

To divide, multiply by the reciprocal of the second fraction.

(Perhaps, you have heard the expression, "Keep. Change. Flip." This pneumonic device helps some individuals remember to keep the first fraction the same, to change the division to multiplication, and to flip the divisor.)

Then, you follow the steps for multiplication. That is, multiply across the numerators, multiply across the denominators, and simplify.

$$\frac{5}{6} \div \frac{1}{3} = \frac{5}{6} \cdot \frac{3}{1} = \frac{5 \cdot 3}{6 \cdot 1} = \frac{15}{6} = \frac{5}{2}$$

You do not have to rewrite as a mixed number. Mixed numbers are generally not used in statistics.

Also, notice that you could have canceled out the common factor of 3 from the 3 in the second numerator and from the 6 in the first denominator before multiplying.

$$\frac{5}{6} \cdot \frac{3}{1} = \frac{5 \cdot 1}{2 \cdot 1} = \frac{5}{2}$$

Answer $\dfrac{5}{6} \div \dfrac{1}{3} = \dfrac{5}{2}$

My Turn!

Perform the indicated operation:

$$\frac{4}{9} \div \frac{7}{12}$$

Example 7 Perform the indicated operation:

$$\left(\frac{5}{6}\right)^3$$

Recall that an exponent is simply shorthand for repeated multiplication.

$$\left(\frac{5}{6}\right)^3 = \frac{5}{6} \cdot \frac{5}{6} \cdot \frac{5}{6} = \frac{125}{216}$$

Answer $\left(\frac{5}{6}\right)^3 = \frac{125}{216}$

My Turn!

Perform the indicated operation:

$$\left(\frac{2}{5}\right)^4$$

Objective 2: Apply operations to decimals.

We will review the basic operations of addition, subtraction, multiplication, division, and exponentiation of decimals. Although you may typically compute these using a calculator within a statistics course, we will briefly review how to do them without technology. Then, you may want to try these examples again using your calculator.

Adding Decimals

Example 8 Perform the indicated operation:

$$1.26 + 0.8$$

In order to add decimals, you need to line up the decimal points vertically. Then, add as you would normally and vertically line up the decimal point in the sum with those in the addends.

$$
\begin{array}{r}
1.26 \\
+0.80 \\
\hline
2.06
\end{array}
$$

Answer $1.26 + 0.8 = 2.06$

My Turn!

Perform the indicated operation: $6.005 + 1.34$

Subtracting Decimals

Example 9 Perform the indicated operation:

$$1.26 - 0.8$$

You also need to line up the decimals points vertically when you are subtracting decimals.

$$
\begin{array}{r}
1.26 \\
-0.80 \\
\hline
0.46
\end{array}
$$

Borrowing was involved in the subtraction. Since you will be using technology in this course, I encourage you to familiarize yourself with the procedure for performing basic operations on your preferred technology.

Answer $1.26 - 0.8 = 0.46$

My Turn!

Perform the indicated operation:

$$9 - 1.004$$

Multiplying Decimals

Example 10 Perform the indicated operation:

$$3.628 \cdot 0.2$$

You do *not* line up decimal points for multiplication.

Count the number of places to the right of each decimal point in each factor. After completing standard multiplication, you will have to insert the decimal point in the answer. The product will have the same number of digits to the right of the decimal point as the sum of the number of places to the right of the decimal point from the factors.

3.628 has three digits to the right of the decimal point.

0.2 has one digit to the right of the decimal point.

The product will have $3+1=4$ digits to the right of the decimal point.

$$3.628$$
$$\times\ 0.2$$
$$\overline{0.7256}$$

Answer $3.628 \cdot 0.2 = 0.7256$

My Turn!

Perform the indicated operation:

$$5.2 \cdot 31.57$$

Dividing Decimals

Example 11 Perform the indicated operation:

$$3.628 \div 0.2$$

First, we will set up the values for long division.

$$0.2\overline{)3.628}$$

Then we must move the decimal point in the divisor so that it is at the right-hand side of the divisor. In this case, we would need to move it one place value to the right (equivalent to multiplying by 10). We also need to do this same movement to the decimal point in the dividend. (You can think about why we are allowed to move the decimal point the same number of places in the divisor and dividend.)

So, we now have

$$2\overline{)36.28}$$

Perform long division.

$$
\begin{array}{r}
18.14 \\
2\overline{)36.28} \\
-2 \\
\hline
16 \\
-16 \\
\hline
02 \\
-2 \\
\hline
08 \\
-8 \\
\hline
0
\end{array}
$$

Make sure you line up the decimal point in your quotient directly above the one in the dividend.

Even though we have briefly reviewed the procedure here, you are strongly encouraged to calculate these operations with technology.

Answer $3.628 \div 0.2 = 18.14$

My Turn!

Perform the indicated operation:

$$34.6953 \div 5.1$$

Objective 3: Evaluate factorials.

Now that we have reviewed operations on fractions and decimals, let's go over some concepts that we use in counting and probability. The first notation that we will look at is called factorial notation. The symbol $n!$ is read as "n factorial" and is telling us to multiply the factors beginning with n all the way down to 1.

$$n! = n \cdot (n-1) \cdot (n-2) \cdot \ldots \cdot 3 \cdot 2 \cdot 1$$

Example 12 Evaluate $7!$

The $7!$ (7 factorial) is telling us to multiply the factors beginning with 7 all the way down to 1.
$$7! = 7 \cdot 6 \cdot 5 \cdot 4 \cdot 3 \cdot 2 \cdot 1 = 5040$$

Answer $7! = 5040$

My Turn!

Perform the indicated operation:

$$6!$$

Objective 4: Evaluate $_nP_r$ and $_nC_r$.

You will learn the contextual meaning of $_nP_r$ (permutation) within the Triola text. Here you will just practice the calculation.

$$_nP_r = \frac{n!}{(n-r)!}$$

Example 13 Evaluate $_{10}P_3$.

We begin by substituting 10 and 3 into the formula. Then, we will perform the subtraction within the parentheses in the denominator. If you think about it, this makes sense, since simplifying parentheses come first in the order of operations. We will then expand the factorials. Once this is written out, you will quickly see that many factors cancel out. Our answer is ultimately determined by multiplying together the three decreasing factors beginning with 10. Note that you are always able to compute $_nP_r$ by multiplying the r decreasing factors starting with n.

$$_{10}P_3 = \frac{10!}{(10-3)!} = \frac{10!}{7!} = \frac{10 \cdot 9 \cdot 8 \cdot 7 \cdot 6 \cdot 5 \cdot 4 \cdot 3 \cdot 2 \cdot 1}{7 \cdot 6 \cdot 5 \cdot 4 \cdot 3 \cdot 2 \cdot 1} = 10 \cdot 9 \cdot 8 = 720$$

Answer $_{10}P_3 = 720$

My Turn!

Evaluate $_8P_2$.

Now we will practice the calculating of combinations, which are denoted by $_nC_r$.

$$_nC_r = \frac{n!}{r!\,(n-r)!}$$

Example 14 Evaluate $_{10}C_3$.

We begin by substituting 10 and 3 into the formula. Then, we will perform the subtraction within the parentheses in the denominator. Once this is written out, you will quickly see that many factors cancel out. Our answer is ultimately determined by multiplying together the three decreasing factors beginning with 10 and dividing by the three decreasing factors beginning with 3. Note that you are always able to compute $_nC_r$ by dividing the product of the r decreasing factors starting with n by $r!$.

$$_{10}C_3 = \frac{10!}{3!\,(10-3)!} = \frac{10!}{3!\,7!} = \frac{10\cdot9\cdot8\cdot7\cdot6\cdot5\cdot4\cdot3\cdot2\cdot1}{3\cdot2\cdot1\cdot7\cdot6\cdot5\cdot4\cdot3\cdot2\cdot1} = \frac{10\cdot9\cdot8}{3\cdot2\cdot1} = \frac{720}{6} = 120$$

Answer $_{10}C_3 = 120$

My Turn!

Evaluate $_8C_2$.

Objective 5: Determine the intersection, union, and complement of two sets.

A set is simply a collection of items. We will review three operations on sets.

The **intersection** of two sets consists of all the elements that are simultaneously in both sets. The symbol that we use to indicate intersection is $\cap$.

Let A be a set and let B be a set; then $A \cap B$ consists of the elements that are in A **and** in B.

The **union** of two sets consists of all the elements that are in at least one of the sets. The symbol that we use to indicate union is $\cup$.

Let A be a set and let B be a set; then $A \cup B$ consists of the elements that are in A **or** in B.

The **complement** of a set A consists of the items that are in the universal set, U (the set that contains everything or all elements under consideration), that are **not** in A. Several symbols are commonly used to denote the complement; however, we will use $\overline{A}$ (read "A-bar").

Example 15 Let $U = \{1,2,3,4,5,6,7,8,9\}$, $A = \{2,4,6,8\}$, and $B = \{1,2,3,4,5\}$.

a) Find $A \cap B$.
    To find the intersection, we must look for the elements in A and in B.
    $A = \{\boxed{2},\boxed{4},6,8\}$, and $B = \{1,\boxed{2},3,\boxed{4},5\}$ Those elements are 2 and 4.
    $$A \cap B = \{2,4\}$$

b) Find $A \cup B$.
    To find the union, we must look for the elements that are in A or in B (That is, join them all together). We only need to write each item once, though, in our solution set.

    $$A \cup B = \{1,2,3,4,5,6,8\}$$

c) Find $\overline{A}$.

To find the complement of A, we must identify the elements in the universal set that are not in A. The elements in A are $A = \{2, 4, 6, 8\}$. The elements that are in the universal set that are not in A are boxed: $U = \{\boxed{1}, 2, \boxed{3}, 4, \boxed{5}, 6, \boxed{7}, 8, \boxed{9}\}$. So, the complement of A consists of the elements 1, 3, 5, 7, and 9.

$$\overline{A} = \{1, 3, 5, 7, 9\}$$

Answer $A \cap B = \{2, 4\}$, $A \cup B = \{1, 2, 3, 4, 5, 6, 8\}$, and $\overline{A} = \{1, 3, 5, 7, 9\}$

My Turn!

Let $U = \{1, 2, 3, 4, 5, 6, 7, 8, 9\}$, $A = \{1, 2, 3, 4, 5, 6\}$, and $B = \{3, 6\}$.

Find $A \cap B$, $A \cup B$, and $\overline{A}$.

Answers to My Turn!

1. $\dfrac{3}{4}$

2. $\dfrac{43}{60}$

3. $\dfrac{19}{60}$

4. $\dfrac{3}{19}$

5. $\dfrac{1}{9}$

6. $\dfrac{16}{21}$

7. $\dfrac{16}{625}$

8. 7.345
9. 7.996
10. 164.164
11. 6.803
12. 720
13. 56
14. 28
15. $A \cap B = \{3,6\}$, $A \cup B = \{1,2,3,4,5,6\}$, and $\overline{A} = \{7,8,9\}$

Practice Problems

1. Perform the indicated operation: $\dfrac{5}{18} + \dfrac{1}{18}$.

2. Perform the indicated operation: $\dfrac{3}{7} + \dfrac{7}{10}$.

3. Perform the indicated operation: $\dfrac{8}{15} - \dfrac{2}{5}$.

4. Perform the indicated operation: $1 - \dfrac{9}{10}$.

5. Perform the indicated operation: $\dfrac{3}{16} \cdot \dfrac{2}{9}$.

6. Perform the indicated operation: $\dfrac{4}{9} \div \dfrac{9}{16}$.

7. Perform the indicated operation: $\left(\dfrac{1}{4}\right)^{3}$.

8. Perform the indicated operation: $3.576 + 2.3$.

9. Perform the indicated operation: $5.6 - 2.354$.

10. Perform the indicated operation: $1.05 \cdot 2.5$.

11. Perform the indicated operation: $7.05 \div 0.5$.

12. Evaluate $9!$.

13. Evaluate $_9P_4$.

14. Evaluate $_9C_4$.

15. Let $U = \{$red, blue, orange, yellow, green, purple, pink, white, black$\}$, $A = \{$red, blue, yellow$\}$, and $B = \{$red, pink, purple$\}$.
    Find $A \cap B$, $A \cup B$, and $\overline{A}$.

Integrated Review 5: Evaluating Formulas for Probability Distributions

Elementary Statistics Chapter 5: Discrete Probability Distributions

Objectives:

1. Evaluate formulas for probability distributions.

2. Evaluate the binomial probability formula.

3. Evaluate expressions with *e*.

4. Evaluate the Poisson formula.

Chapter 5 of the Triola text will continue to discuss probability, and several related formulas will be introduced. In this integrated review chapter, we will go over how to evaluate some of those formulas. You will learn the meaning of the formulas in the corresponding chapter of the textbook.

Objective 1: Evaluate formulas for probability distributions.

We reviewed summation notation in *Integrated Review* Chapter 3. As you may recall, the summation symbol, Σ, tells you to add up values.

In the related chapter from the text, we will often check to see if the sum of a table of values equals 1.

Example 1 Determine whether $\Sigma P(x) = 1$.

P(x)
0.2
0.15
0.05
0.3
0.2
0.01

So, we will add up the six values that were given in the table and check to see if the sum is equal to 1.

$$\Sigma P(x) = 0.2 + 0.15 + 0.05 + 0.3 + 0.2 + 0.01 = 0.91$$

$$0.91 \neq 1$$

Answer $\Sigma P(x)$ does not equal 1.

My Turn!

Determine whether $\Sigma P(x) = 1$.

P(x)
0.03
0.22
0.15
0.2
0.25
0.15

Sometimes, the summation may require that you use values from two columns of a table.

Example 2 Find $\Sigma\left[x \cdot P(x)\right]$.

x	P(x)
0	0.2
1	0.15
2	0.05
3	0.3
4	0.2
5	0.1

This notation is telling us to add up the products of the first column of x values with the second column, $P(x)$. You can elect to find the sum in a horizontal fashion, or you can elect to insert a third column to the right with the products and then total that column.

$$\Sigma\left[x \cdot P(x)\right] = (0 \cdot 0.2) + (1 \cdot 0.15) + (2 \cdot 0.05) + (3 \cdot 0.3) + (4 \cdot 0.2) + (5 \cdot 0.1)$$
$$= 0 + 0.15 + 0.1 + 0.9 + 0.8 + 0.5$$
$$= 2.45$$

Now, we will perform the same calculations in a table format.

x	$P(x)$	$x \cdot P(x)$
0	0.2	0
1	0.15	0.15
2	0.05	0.1
3	0.3	0.9
4	0.2	0.8
5	0.1	0.5
$\Sigma\left[x \cdot P(x)\right] =$		2.45

Answer $\Sigma\left[x \cdot P(x)\right] = 2.45$

My Turn!

Find $\Sigma\left[x \cdot P(x)\right]$.

x	$P(x)$
0	0.3
1	0.35
2	0.15
3	0.2

Example 3 Find $\Sigma\left[x^2 \cdot P(x)\right]$ using the same given table as in Example 2.

x	$P(x)$
0	0.2
1	0.15
2	0.05
3	0.3
4	0.2
5	0.1

This notation is telling us to add up the products of the squares of the first column of x values with the second column, $P(x)$. Again, this could be done horizontally or by using extra columns in a table. Note that we are following the order of operations and will be squaring before multiplying and then finally adding.

$$\Sigma\left[x^2 \cdot P(x)\right] = (0^2 \cdot 0.2) + (1^2 \cdot 0.15) + (2^2 \cdot 0.05) + (3^2 \cdot 0.3) + (4^2 \cdot 0.2) + (5^2 \cdot 0.1)$$
$$= (0 \cdot 0.2) + (1 \cdot 0.15) + (4 \cdot 0.05) + (9 \cdot 0.3) + (16 \cdot 0.2) + (25 \cdot 0.1)$$
$$= 0 + 0.15 + 0.2 + 2.7 + 3.2 + 2.5$$
$$= 8.75$$

Or, by expanding on the given table, we get the following:

x	x^2	$P(x)$	$x^2 \cdot P(x)$
0	0	0.2	0
1	1	0.15	0.15
2	4	0.05	0.2
3	9	0.3	2.7
4	16	0.2	3.2
5	25	0.1	0.25
	$\Sigma\left[x^2 \cdot P(x)\right] =$		8.75

Answer $\Sigma\left[x^2 \cdot P(x)\right] = 6.5$

My Turn!

Find $\Sigma\left[x^2 \cdot P(x)\right]$.

x	$P(x)$
0	0.3
1	0.35
2	0.15
3	0.2

Example 4 Evaluate $\mu + 2\sigma$ when $\mu = 25.2$ and $\sigma = 2.3$.

We need to substitute 25.2 for μ (the lowercase Greek letter mu) in the beginning of the expression and 1.3 for σ (the lowercase Greek letter sigma) at the end of the expression.

$$\mu + 2\sigma = 25.2 + 2(2.3)$$

Following the order of operations, we need to multiply before we add.

$$25.2 + 2(2.3) = 25.2 + 4.6 = 29.8$$

Answer $\mu + 2\sigma = 29.8$ when $\mu = 25.2$ and $\sigma = 2.3$.

My Turn!

Evaluate $\mu + 2\sigma$ when $\mu = 9.9$ and $\sigma = 0.4$.

Example 5 Evaluate $\mu = n \cdot p$ given $n = 150$ and $p = 0.35$.

We are evaluating a formula given the values for all but one variable, μ.

$$\mu = n \cdot p = 150 \cdot 0.35 = 52.5$$

Answer $\mu = 52.5$

My Turn!

Evaluate $\mu = n \cdot p$ when $n = 2000$ and $p = 0.04$.

62 : Integrated Review 5

Example 6 Evaluate $\sigma = \sqrt{np(1-p)}$ $n = 50$ and $n = 0.2$. Round the answer to the nearest tenth.

Begin by replacing n and p with the given values. You will be substituting 0.2 for p in two places. After you have done this, follow the order of operations.

$$\sigma = \sqrt{np(1-p)} = \sqrt{50 \cdot 0.2 \cdot (1-0.2)} = \sqrt{50 \cdot 0.2 \cdot (0.8)} = \sqrt{10 \cdot 0.8} = \sqrt{8} \approx 2.82 \approx 2.8$$

Answer $\sigma \approx 2.8$

My Turn!

Evaluate $\sigma = \sqrt{np(1-p)}$ for $n = 100$ and $p = 0.38$. Round the answer to the nearest tenth.

Objective 2: Evaluate the binomial probability formula.

Example 7 Evaluate $_nC_x \cdot p^x \cdot (1-p)^{n-x}$ for $n = 10$, $x = 2$, and $p = 0.3$. Round to three significant digits.

First, we must substitute the given values in for the given variables.

$$_nC_x \cdot p^x \cdot (1-p)^{n-x} = {_{10}C_2} \cdot 0.3^2 \cdot (1-0.3)^{10-2}$$

Next, we will clean up the third factor.

$$= {_{10}C_2} \cdot 0.3^2 \cdot (0.7)^8$$

Then, we will evaluate the combination. Even though we are showing the detailed calculations here, you may want to practice evaluating combinations using technology.

$$_{10}C_2 = \frac{10!}{2!(10-2)!} = \frac{10!}{2! \, 8!} = \frac{10 \cdot 9 \cdot 8 \cdot 7 \cdot 6 \cdot 5 \cdot 4 \cdot 3 \cdot 2 \cdot 1}{2 \cdot 1 \cdot 8 \cdot 7 \cdot 6 \cdot 5 \cdot 4 \cdot 3 \cdot 2 \cdot 1} = \frac{10 \cdot 9}{2 \cdot 1} = \frac{90}{2} = 45$$

Copyright © 2017 Pearson Education, Inc.

We now have $_{10}C_2 \cdot 0.3^2 \cdot (0.7)^8 = 45 \cdot 0.3^2 \cdot (0.7)^8$. Per the order of operations, we will evaluate exponents next.

$$= 45 \cdot 0.09 \cdot 0.05764801$$

Next, we multiply.

$$= 0.2334744405$$

Finally, we round to 3 significant figures.

$$\approx 0.233$$

Answer $_nC_x \cdot p^x \cdot (1-p)^{n-x} \approx 0.233$ for $n = 10$, $x = 2$, and $p = 0.3$.

My Turn!

Evaluate $_nC_x \cdot p^x \cdot (1-p)^{n-x}$ for $n = 20$, $x = 5$, and $p = 0.4$. Round to 3 significant digits.

Objective 3: Evaluate expressions with *e*.

The mathematical constant *e* is the number $e \approx 2.71828$.

Example 8 Evaluate e^5. Round the answer to the nearest tenth.

You perform calculations with base *e* as you would with any other base. You will want to find powers of *e* using technology. (Most calculators have an e^x button.)

$$e^5 \approx 148.413 \approx 148.4$$

Answer $e^5 \approx 148.4$

My Turn!

Evaluate $3e^5$. Round the answer to the nearest tenth.

Objective 4: Evaluate the Poisson formula.

Example 9 Let $P(x) = \dfrac{\mu^x \cdot e^{-\mu}}{x!}$ and let $\mu = 2$. Find $P(3)$. Round your answer to three significant figures.

You may notice that this formula is written in function notation. We are being asked to evaluate the function when $x = 3$. So, we will substitute 2 in for μ and 3 in for x. Once we have done that, we will follow the order of operations. We will finish by rounding.

$$P(3) = \frac{2^3 \cdot e^{-2}}{3!} \approx \frac{8 \cdot 0.13533528}{3!} = \frac{1.08268224}{6} = 0.18044704 \approx 0.180$$

Answer $P(3) \approx 0.180$

My Turn!

Let $P(x) = \dfrac{\mu^x \cdot e^{-\mu}}{x!}$ and let $\mu = 4$. Find $P(3)$. Round your answer to 3 significant figures.

Answers to My Turn!

1. Yes
2. 1.25
3. 2.75
4. 10.7
5. 80
6. 4.9
7. 0.0746
8. 445.2
9. 0.195

Practice Problems

1. Determine whether $\Sigma P(x) = 1$.

$P(x)$
0.18
0.02
0.35
0.2
0.05
0.15

2. Find $\Sigma\left[x \cdot P(x)\right]$.

x	$P(x)$
0	0.2
1	0.45
2	0.15
3	0.2

3. Find $\Sigma\left[x^2 \cdot P(x)\right]$.

x	$P(x)$
0	0.1
1	0.25
2	0.45
3	0.2

4. Evaluate $\mu - 2\sigma$ when $\mu = 265.9$ and $\sigma = 3.8$.
5. Evaluate $\mu = n \cdot p$ given $n = 1125$ and $p = 0.86$.
6. Evaluate $\sigma = \sqrt{np(1-p)}$ for $n = 200$ and $p = 0.48$. Round the answer to the nearest tenth.
7. Evaluate $_nC_x \cdot p^x \cdot (1-p)^{n-x}$ for $n = 15$, $x = 5$, and $p = 0.7$. Round to 3 significant digits.

8. Evaluate $6e^{-3}$. Round the answer to the nearest tenth.

9. Let $P(x) = \dfrac{\mu^x \cdot e^{-\mu}}{x!}$ and let $\mu = 5$. Find $P(4)$. Round to 3 significant digits.

Integrated Review 6: The Building Blocks for Working with Normal Distributions

Elementary Statistics Chapter 6: Normal Probability Distributions

Objectives:

1. Find area.

2. Interpret inequality notation.

3. Evaluate formulas for normal probability distributions.

In this chapter of the Triola text, you will begin working with normal distributions (bell-shaped curves). You will be finding the area under portions of the normal curve. We will start this chapter of the integrated review by briefly looking at the area of a rectangle. Then, we will go over some of the basics for inequality notation. Finally, we will practice evaluating the formulas you will be using in the related chapter.

Objective 1: Find area.

In the chapter in the text on normal distributions, we will be finding area and associating it with probability. For the topic of uniform distributions, you will be finding area under rectangles. So, we will take a moment to review finding the area of a rectangle.

Example 1 Find the area of the rectangle shown.

0.5 cm

2 cm

The area of a rectangle is given by $base \times height$ (or $length \times width$). We can write this as a formula.

$$A = b \cdot h$$

For the above rectangle, the base is 2 cm and the height is 0.5 cm.

We can plug these values into the formula.

$$A = b \cdot h = 2\,\text{cm} \cdot 0.5\,\text{cm} = 1\,\text{cm}^2$$

Answer The area is 1 cm^2.

My Turn!

Find the area of the shaded portion of the rectangle.

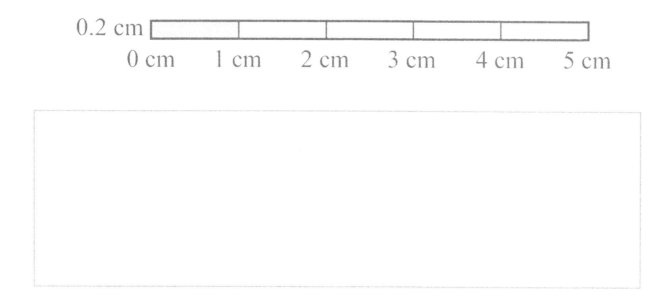

Objective 2: Interpret inequality notation.

In the related section of the textbook, we will be dealing with translating intervals in context to notation that we can use for calculating purposes. Also, you will want to be comfortable with indicating which integers would be located within an interval. The following table includes a column for inequality notation, a column with the inequality expressed in words, and a third column that indicates which integers from 0 to 10 satisfy the inequality from the first column.

Notation	Translation	$X = \{0, 1, 2, 3, 4, 5, 6, 7, 8, 9, 10\}$
$x < 5$	x is less than 5	0, 1, 2, 3, 4
$x > 7$	x is greater than 7	8, 9, 10
$2 < x < 6$	x is greater than 2 and less than 6	3, 4, 5
$x \leq 4$	x is less than or equal to 4	0, 1, 2, 3, 4
$x \geq 9$	x is greater than or equal to 9	9, 10

Example 2 Which integers from 0 to 10 satisfy the following inequality?

$$x < 7$$

This translates to "x is less than 7." The integers that are less than 7 include 0, 1, 2, 3, 4, 5, and 6.

Answer The integers from 0 to 10 that satisfy $x < 7$ are 0, 1, 2, 3, 4, 5, and 6.

My Turn!

Which integers from 5 to 15 satisfy the following inequality?

x is greater than or equal to 10.

There are some phrases that you may encounter as part of the text that will translate into inequalities. The following table includes a sampling of these.

Phrase	Inequality
At least 5 puppies	$x \geq 5$
No more than 5 puppies	$x \leq 5$
Fewer than 5 puppies	$x < 5$
More than 5 puppies	$x > 5$
Exactly 5 puppies	$x = 5$

There are other possibilities beyond the phrases used in the above table. The more problems you practice, the more comfortable you will be become with the various ways to describe inequalities.

Example 3 Translate the following phrases into inequalities.

    a) A heart rate of at least 80 bpm

        We can locate the phrase of *at least* in the table and see that it implies $\geq$. If a heart rate is at least 80 bpm, it means that it can be 80, 81, 82, …bpm.
        Letting h represent heart rate, we can translate the phrase into $h \geq 80$.

    b) A speed of no more than 55 mph

        We can locate the phrase *no more than* in the second row of the table and see that it translates to $\leq$. If the speed is no more than 55 mph, it means that it can be anywhere from 0 mph up to and including 55 mph.
        If we let s represent speed, the phrase can be rewritten as $s \leq 55$.

    c) A temperature below 98.6 degrees Fahrenheit

        The phrase *below* is not contained within the table. However, if we think about it, a temperature below 98.6 degrees means that it can be any value less than 98.6, not including 98.6.
        Therefore, using t for temperature, our translation becomes $t < 98.6$.

Answer a) $h \geq 80$ b) $s \leq 55$ c) $t < 98.6$

My Turn!

Translate the following phrases into inequalities.

    a) A weight of no more than 180 pounds.
    b) A height of at least 60 inches
    c) Exactly 9 doors in a house

Objective 3: Evaluating formulas for normal probability distributions.

You will be learning about several formulas related to the normal distribution in the corresponding section of the textbook. Here, we will practice evaluating these formulas. You will learn the meaning of the formulas and how to apply them in context within the text section itself.

Example 4 Evaluate $z = \dfrac{x - \mu}{\sigma}$ if $x = 21.3$, $\mu = 23.5$, and $\sigma = 2.1$. Round the answer to the nearest hundredth.

First, we must substitute for the variables in the correct location. Then, we must simplify the numerator. (This is because you can envision the entire numerator within parentheses, and the order of operations dictates that parentheses must be simplified first.) Next, we will divide, and then we will round our final answer.

$$z = \frac{21.3 - 23.5}{2.1} = \frac{-2.2}{2.1} \approx -1.047 \approx -1.05$$

Answer $z \approx -1.05$

My Turn!

Evaluate $z = \dfrac{x - \mu}{\sigma}$ if $x = 10.3$, $\mu = 10.2$, and $\sigma = 1.3$. Round the answer to the nearest hundredth.

Example 5 Evaluate $x = \mu + z \cdot \sigma$ if $\mu = 6.7$, $z = 1.96$, and $\sigma = 0.9$. Round the answer to the nearest tenth.

First, we must substitute for the variables in the correct location. Then, we will follow the order of operations for the right-hand side. So, we will have to multiply before adding.

$$x = \mu + z \cdot \sigma = 6.7 + 1.96 \cdot 0.9 = 6.7 + 1.764 = 8.464 \approx 8.5$$

Answer $x \approx 8.5$

My Turn!

Evaluate $x = \mu + z \cdot \sigma$ if $\mu = 9.8$, $z = 2.575$, and $\sigma = 0.3$. Round the answer to the nearest tenth.

Example 6 Solve for x when $z = \dfrac{x - \mu}{\sigma}$ and $z = 2.15$, $\mu = 3.5$, and $\sigma = 0.3$. Round the answer to the nearest tenth.

We need to substitute 2.15 for z to the left of the equal sign, and 3.5 for μ and 0.3 for σ on the right-hand side.

$$2.15 = \frac{x - 3.5}{0.3}$$

Now, we need to solve for x. The first step is to multiply both sides of the equation by 0.3 to eliminate the fraction.

$$2.15(0.3) = \frac{(x - 3.5)}{0.3}(0.3)$$
$$0.645 = x - 3.5$$

Add 3.5 to both sides to isolate x.

$$0.645 + 3.5 = x - 3.5 + 3.5$$
$$4.145 = x$$

Now, we must round our final answer to the nearest tenth and $x \approx 4.1$.

Answer $x \approx 4.1$

My Turn!

Solve for x when $z = \dfrac{x - \mu}{\sigma}$ and $z = -0.34$, $\mu = 21.5$, and $\sigma = 0.8$. Round the answer to the nearest tenth.

Example 7 Evaluate

$$z = \dfrac{\bar{x} - \mu}{\dfrac{\sigma}{\sqrt{n}}}$$

if $\bar{x} = 9.8$, $\mu = 25.6$, $\sigma = 1.9$, and $n = 49$. Round your answer to the nearest hundredth.

We need to substitute 9.8 for $\bar{x}$ and 25.6 for μ in the numerator, and 1.9 for σ and 49 for n in the denominator.

$$z = \frac{9.8 - 25.6}{\dfrac{1.9}{\sqrt{49}}}$$

We can then simplify the right-hand side using the order of operations. You may want to visualize an invisible set of parentheses around the numerator and another invisible set of parentheses around the denominator.

$$z = \frac{9.8 - 25.6}{\dfrac{1.9}{\sqrt{49}}}$$

$$= \frac{(9.8 - 25.6)}{\left(\dfrac{1.9}{\sqrt{49}}\right)}$$

$$= \frac{-15.8}{\left(\dfrac{1.9}{\sqrt{49}}\right)}$$

$$= \frac{-15.8}{\left(\dfrac{1.9}{7}\right)}$$

$$= \frac{-15.8}{0.27142857}$$

$$= -58.2105266$$

Finally, we must round our final answer to the hundredth, which gives us $z \approx -58.21$.

Answer $z \approx -58.21$

My Turn!

Evaluate

$$z = \frac{\overline{x} - \mu}{\dfrac{\sigma}{\sqrt{n}}}$$

if $\overline{x} = 7.2, \mu = 5.8, \sigma = 1.2,$ and $n = 81.$ Round the answer to the nearest hundredth.

Answers to My Turn!

1. 0.2 in^2
2. $10, 11, 12, 13, 14, 15$
3. a) $w \leq 180$ b) $h \geq 60$ c) $d = 9$
4. $z \approx 0.08$
5. $x \approx 10.6$
6. $x \approx 21.2$
7. $z = 10.5$

Practice Problems

1. Find the area of the shaded region of the following rectangle.

2. Which integers from 10 to 20 satisfy the following inequality? $x < 12$
3. Translate the following phrases into inequalities.
    A length of more than 18 inches
    A volume of less than 60 ft^3
4. Evaluate $z = \dfrac{x - \mu}{\sigma}$ if $x = 67.8$, $\mu = 70.2$, and $\sigma = 2.3$. Round the answer to the nearest hundredth.
5. Evaluate $x = \mu + z \cdot \sigma$ if $\mu = 78$, $z = 1.96$ and $\sigma = 10.3$. Round the answer to the nearest tenth
6. Solve for x when $z = \dfrac{x - \mu}{\sigma}$ and $z = 1.47$, $\mu = 99.5$ and $\sigma = 4.8$. Round the answer to the nearest tenth.
7. Evaluate
    $z = \dfrac{\bar{x} - \mu}{\dfrac{\sigma}{\sqrt{n}}}$ if $\bar{x} = 54.2, \mu = 52.6, \sigma = 3.2 \ and \ n = 64$. Round the answer to the nearest hundredth.

Integrated Review 7: Intervals of Numbers

Elementary Statistics Chapter 7: Estimating Parameters and Determining Sample Sizes

Objectives:

1. Find the middle value for an interval written either as an inequality or in interval notation.

2. Find the distance from the middle value of an interval to its endpoints.

3. Write and interpret three different forms of intervals (as used for confidence intervals).

4. Evaluate formulas used for confidence intervals.

In the chapter in the text on estimating parameters and determining sample sizes, we will be utilizing intervals a lot. So, in this integrated review chapter, we will spend some time discussing several topics related to intervals.

Objective 1: Find the middle value for an interval written either as an inequality or in interval notation.

One topic that you will have to understand in the text is how to find the middle value for a given interval. The given interval can be written either as an inequality or in interval notation. We will review how to handle both of these formats.

Example 1 Let $3.8 < \mu < 6.2$ represent an interval on the number line. Find the value that is in the middle of the interval.

From an algebra class, you may recall that this is the same thing as finding the midpoint for two points on a number line. In order to find the middle value, you have to add the two values of the endpoints and divide that sum by 2.

The middle value of $a < \mu < b$ is $\dfrac{a+b}{2}$.

We have $3.8 < \mu < 6.2$. So, substitute 3.8 for a and 6.2 for b. Then, the middle value of the given interval of numbers is found by following the order of operations. (Recall that the entire numerator can be thought of as being within parentheses and is therefore simplified first.)

left endpoint   midpoint   right endpoint

$a = 3.8$ $\dfrac{a+b}{2}$ $b = 6.2$

$$\frac{a+b}{2} = \frac{3.8+6.2}{2} = \frac{10}{2} = 5$$

Answer The value of 5 is the middle value of the interval $3.8 < \mu < 6.2$.

My Turn!

Let $105.67 < \mu < 118.59$ represent an interval on the number line. Find the value that is in the middle of the interval.

Example 2 If $(0.124, 0.498)$ is an interval on the number line, find the value that is in the middle of the interval.

This problem is completed in the same way as Example 1. It simply differs in how the given interval is written. The interval $(0.124, 0.498)$ could be rewritten as $0.124 < \mu < 0.498$. In order to find the middle value, you have to add the two values of the endpoints and divide that sum by 2. That is,

$$\text{the middle value of } (a,b) = \frac{a+b}{2}$$

0.124   midpoint   0.498

We have $(0.124, 0.498) = (a,b)$. So, substitute 0.124 for a and 0.498 for b and simplify.

$$\frac{a+b}{2} = \frac{0.124+0.498}{2} = \frac{0.622}{2} = 0.311$$

Answer 0.311 is the middle value of the interval $(0.124, 0.498)$.

My Turn!

If $(6.48, 15.84)$ is an interval on the number line, find the value that is in the middle of the interval.

Objective 2: Find the distance from the middle value of an interval to its endpoints.

Example 3 Let $(0.368, 0.549)$ represent an interval on the number line. Find the distance from the middle of the interval to either endpoint.

Let E represent the distance from the midpoint to an endpoint.

There are two ways to solve this problem:

Method 1: Find the middle value of the interval first and then find the distance from the midpoint to the endpoint.

Step 1: Follow the process for Objective 1 to find the middle value of the interval.

The middle value of $(0.368, 0.549)$ is

$$\frac{0.368 + 0.549}{2} = \frac{0.917}{2} = 0.4585$$

Step 2: Find the distance from the middle value that you found in step 1 to the endpoint.

You can find the distance from the middle value to an endpoint by subtracting the left endpoint from the midpoint of the interval.

$$\text{distance from middle value to endpoint} = \text{middle value} - \text{left endpoint}$$

(You can also calculate this distance by subtracting the middle value from the right endpoint.)

For this example, the distance from the middle value to either endpoint is

$$E = 0.4585 - 0.368 = 0.0905$$

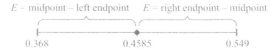

Method 2: Find the distance directly without actually finding the midpoint.

The distance from the middle value to the endpoint of an interval (a,b) can be given by

$$E = \frac{b-a}{2}$$

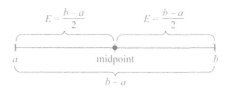

So, the distance from the middle value to the endpoint of the interval $(0.368, 0.549)$ can be found as follows:

$$E = \frac{0.549 - 0.368}{2} = \frac{0.181}{2} = 0.0905$$

You can feel free to use the method that is more intuitive for you. Sometimes you may need to know the middle value anyway. In that case, method 1 is definitely the way to go. If you don't need to know the midpoint, then you may want to use method 2.

Answer The distance is 0.0905.

My Turn!

Let $(26.8, 32.4)$ represent an interval on the number line. Find the distance from the middle of the interval to either endpoint.

Objective 3: Write and interpret three different forms of intervals (as used for confidence intervals).

As you may have gathered from the examples above, there are several different forms of intervals. We will review the various forms you will see in the textbook. Often, technology will give you the output in a form that is different from how you may want to write it. So, it is important to understand how to translate from one format to another.

The key point that we have to remember is that $a < \mu < b$ and (a, b) represent the same interval for μ.

Example 4: Express $2.5 < \mu < 5.3$ in interval notation.

The endpoints of the interval are 2.5 and 5.3. So, we can rewrite $2.5 < \mu < 5.3$ as $(2.5, 5.3)$.

Answer $(2.5, 5.3)$

My Turn!

Express $-3.5 < \mu < 1.9$ in interval notation.

Example 5 Express $(0.88, 0.96)$ in inequality notation, assuming that the variable under study is μ.

The endpoints of the interval are 0.88 and 0.96. So, we can rewrite $(0.88, 0.96)$ as $0.88 < \mu < 0.96$.

Answer $0.88 < \mu < 0.96$

My Turn!

Express $(0.63, 0.92)$ in inequality notation, assuming that the variable under study is p.

Sometimes, you will want to write your interval in the following form:

$$\text{middle value}\,(m) \pm \text{distance from middle value to endpoint}\,(E) = m \pm E$$

The notation $\pm$ translates to *plus or minus*.

Example 6 Let $4.0 < \mu < 4.8$ represent an interval on the number line. Write the given interval in the format $m \pm E$.

First, we must find the value that is in the middle of the interval and let the variable m represent that.

$$m = (4.0, 4.8) = \frac{4.0 + 4.8}{2} = \frac{8.8}{2} = 4.4$$

Next, we must find the distance from the middle of the interval to either endpoint and let E represent that.

$$E = 4.4 - 4.0 = 0.4$$

Finally, we must write the given interval in the desired format $m \pm E$.

So, $4.0 < \mu < 4.8$ can be rewritten in the desired form of 4.4 ± 0.4.

Answer 4.4 ± 0.4

My Turn!

Let $0.18 < \mu < 0.36$ represent an interval on the number line. Write the given interval in the format $m \pm E$.

Objective 4: Evaluate formulas used for confidence intervals.

You will be learning about several formulas in the related section of the textbook on confidence intervals. Here, we will practice evaluating these formulas. You will learn the meaning of the formulas and how to apply them in context within the text section itself.

Example 7 Evaluate the formula

$$E = z \cdot \sqrt{\frac{\hat{p}(1 - \hat{p})}{n}}$$

when $z = 1.96$, $\hat{p} = 0.516$, and $n = 1000$. Round the answer to the thousandths.

First, we must substitute for the variables in the correct location. Then, we will follow the order of operations for the right-hand side.

$$E = z \cdot \sqrt{\frac{\hat{p}(1 - \hat{p})}{n}}$$

$$= 1.96 \cdot \sqrt{\frac{0.516(1 - 0.516)}{1000}}$$

$$= 1.96 \cdot \sqrt{\frac{0.516(0.484)}{1000}}$$

$$= 1.96 \cdot \sqrt{\frac{0.249744}{1000}}$$

$$= 1.96 \cdot \sqrt{0.000249744}$$

$$\approx 1.96 \cdot 0.0158032908$$

$$= 0.03097445$$

$$\approx 0.0310$$

Answer $E \approx 0.0310$

My Turn!

Evaluate the formula $E = z \cdot \sqrt{\dfrac{\hat{p}(1-\hat{p})}{n}}$ when $z = 1.645$, $\hat{p} = 0.874$, and $n = 500$. Round the answer to the thousandths.

Example 8 Evaluate the formula

$$E = t \cdot \frac{s}{\sqrt{n}}$$

when $t = 2.626$, $s = 0.93$, and $n = 200$. Round your answer to the nearest hundredth.

By now, you probably have the idea that we need to substitute and simplify!

$$E = t \cdot \frac{s}{\sqrt{n}} = 2.626 \cdot \frac{0.93}{\sqrt{200}} \approx 2.626 \cdot \frac{0.93}{14.14213562} \approx 0.1726882039 \approx 0.17$$

Answer $E \approx 0.17$

My Turn!

Evaluate the formula

$$E = t \cdot \frac{s}{\sqrt{n}}$$

when $t = 2.756$, $s = 0.64$, and $n = 30$. Round your answer to the nearest hundredth.

Example 9 Evaluate the formula

$$n = \frac{z^2 \cdot p(1-p)}{E^2}$$

when $z = 1.645$, $p = 0.38$, and $E = 0.04$. Round the answer up to the next whole number.

Once again, substitute and simplify. If you are doing this on your calculator, be really careful with the order of operations.

$$
\begin{aligned}
n &= \frac{z^2 \cdot p(1-p)}{E^2} \\
&= \frac{1.645^2 \cdot 0.38(1-0.38)}{0.04^2} \\
&= \frac{1.645^2 \cdot 0.38(0.62)}{0.04^2} \\
&= \frac{2.706025 \cdot 0.38 \cdot 0.62}{0.0016} \\
&= \frac{0.63753949}{0.0016} \\
&\approx 398.4621813
\end{aligned}
$$

Note the special rounding instructions for this problem. We are not following traditional rounding rules. If you don't get an exact whole number, you want to automatically go up to the next higher whole number. (You will find out why we do this when you learn about this formula in the textbook.)

So, for this problem we would round up to 399.

Answer $n = 399$

My Turn!

Evaluate the formula

$$n = \frac{z^2 \cdot p(1-p)}{E^2}$$

when $z = 1.96, p = 0.24$, and $E = 0.09$. Round the answer up to the next whole number.

Answers to My Turn!

1. 112.13
2. 11.16
3. 2.8
4. $(-3.5, 1.9)$
5. $0.63 < p < 0.92$
6. 0.27 ± 0.09
7. 0.024
8. 0.32
9. 87

Practice Problems

1. Let $0.45 < p < 0.85$ represent an interval on the number line. Find the value that is in the middle of the interval.
2. If $(2.57, 17.64)$ is an interval on the number line, find the value that is in the middle of the interval.
3. Let $(36.9, 42.4)$ represent an interval on the number line. Find the distance from the middle of the interval to either endpoint.
4. Express $3.5 < \mu < 11.9$ in interval notation.
5. Express $(0.63, 0.92)$ in inequality notation, assuming that the variable under study is μ.
6. Let $24.1 < \mu < 26.3$ represent an interval on the number line. Write the given interval in the format $m \pm E$.
7. Evaluate the formula

$$E = z \cdot \sqrt{\frac{\hat{p}(1-\hat{p})}{n}}$$

when $z = 1.96$, $\hat{p} = 0.526$, and $n = 300$. Round the answer to the thousandths.

8. Evaluate the formula

$$E = t \cdot \frac{s}{\sqrt{n}}$$

when $t = 2.678$, $s = 0.64$, and $n = 51$. Round your answer to the nearest hundredth.

9. Evaluate the formula

$$n = \frac{z^2 \cdot p(1-p)}{E^2}$$

when $z = 1.96$, $p = 0.37$, and $E = 0.08$. Round the answer up to the next whole number.

Integrated Review 8: Formulas for Hypothesis Testing

Elementary Statistics Chapter 8: Hypothesis Testing

Objective:

1. Evaluate formulas for hypothesis testing.

You will be learning about several formulas for hypothesis testing in the related section of the textbook. Here, we will practice evaluating these formulas. You will learn the meaning of the formulas and how to apply them in context within the text section itself.

Objective 1: Evaluate formulas for hypothesis testing.

Example 1 Evaluate the formula

$$z = \frac{\hat{p} - p}{\sqrt{\dfrac{p \cdot q}{n}}}$$

when $\hat{p} = \dfrac{x}{n}$, $x = 80$, $n = 100$, $p = 0.75$, and $q = 1 - p$. Round your answer to the nearest hundredth.

First we need to figure out the value for $\hat{p}$ so that we can substitute in that value. Since, $\hat{p} = \dfrac{x}{n}$, with $x = 80$ and $n = 100$, we get

$$\hat{p} = \frac{x}{n} = \frac{80}{100} = 0.8$$

We may also want to find the value of q. Since $q = 1 - p$ and $p = 0.75$ (Be really careful with p and $\hat{p}$, as they are indeed different), we get
$q = 1 - p = 1 - 0.75 = 0.25$

Now, we can substitute in values for all of the variables and solve for z.

$$z = \frac{\hat{p} - p}{\sqrt{\dfrac{p \cdot q}{n}}} = \frac{0.8 - 0.75}{\sqrt{\dfrac{0.75 \cdot 0.25}{100}}} = \frac{0.05}{\sqrt{\dfrac{0.1875}{100}}} = \frac{0.05}{\sqrt{0.001875}} \approx \frac{0.05}{0.0433} \approx 1.1547$$

Now, we will round our value for *z* to the nearest hundredth.

$$z \approx 1.15$$

Answer $z \approx 1.15$

My Turn!

Evaluate the formula

$$z = \frac{\hat{p} - p}{\sqrt{\dfrac{p \cdot q}{n}}}$$

when $\hat{p} = \dfrac{x}{n}$, $x = 90$, $n = 120$, $p = 0.35$ and $q = 1 - p$. Round your answer to the nearest hundredth.

Example 2 Evaluate the formula

$$t = \frac{\overline{x} - \mu}{\dfrac{s}{\sqrt{n}}}$$

when $\mu = 99.2$, $\overline{x} = 99.5$, $n = 100$, and $s = 1.3$. Round your answer to the nearest hundredth.

Begin by substituting the given values for the variables in the formula. Then, follow the order of operations to simplify the right hand side.

$$t = \frac{\overline{x} - \mu}{\dfrac{s}{\sqrt{n}}} = \frac{99.5 - 99.2}{\dfrac{1.3}{\sqrt{100}}} = \frac{0.3}{\dfrac{1.3}{\sqrt{100}}} = \frac{0.3}{\dfrac{1.3}{10}} = \frac{0.3}{0.13} \approx 2.3077 \approx 2.31$$

Note that if 100 were not a perfect square, you would want to use your calculator to find the value of the square root.

Answer $t \approx 2.31$

My Turn!

Evaluate the formula

$$t = \frac{\bar{x} - \mu}{\frac{s}{\sqrt{n}}}$$

when $\mu = 16$, $\bar{x} = 17.5$, $n = 80$, and $s = 2.4$. Round your answer to the nearest hundredth.

Example 3 Evaluate the formula

$$\chi^2 = \frac{(n-1) \cdot s^2}{\sigma^2}$$

when $\sigma = 3.9$, $n = 50$, and $s = 3.67$.

Round your answer to the nearest tenth.

Note that χ^2 (chi-squared) is considered as an entity. You do not approach this problem trying to solve for χ. You strive to get a value for χ^2. That is, $\chi^2 = \square$.

We can go straight to substituting the given values in for the variables and simplifying.

$$\chi^2 = \frac{(n-1) \cdot s^2}{\sigma^2}$$

$$= \frac{(50-1) \cdot 3.67^2}{3.9^2}$$

$$= \frac{(49) \cdot 3.67^2}{3.9^2}$$

$$= \frac{(49) \cdot 13.4689}{15.21}$$

$$= \frac{659.9761}{15.21}$$

$$\approx 43.3909$$

$$\approx 43.4$$

Answer $\chi^2 \approx 43.4$

My Turn!

Evaluate the formula

$$\chi^2 = \frac{(n-1) \cdot s^2}{\sigma^2}$$

when $\sigma = 4.5$, $n = 60$, and $s = 4.6$. Round your answer to the nearest tenth.

Answers to My Turn!

1. $z \approx 9.19$
2. $t \approx 5.59$
3. $\chi^2 \approx 61.7$

Practice Problems

1. Evaluate the formula

$$z = \frac{\hat{p} - p}{\sqrt{\dfrac{p \cdot q}{n}}}$$

when $\hat{p} = \dfrac{x}{n}$, $x = 90$, $n = 150$, $p = 0.82$, and $q = 1 - p$. Round your answer to the nearest hundredth.

2. Evaluate the formula

$$t = \frac{\bar{x} - \mu}{\dfrac{s}{\sqrt{n}}}$$

when $\mu = 44$, $\bar{x} = 45.2$, $n = 80$, and $s = 3.1$. Round your answer to the nearest hundredth.

3. Evaluate the formula

$$\chi^2 = \frac{(n-1) \cdot s^2}{\sigma^2}$$

when $\sigma = 11.3$, $n = 160$, and $s = 11.4$. Round your answer to the nearest tenth.

Integrated Review 9: Formulas for Two-Sample Hypothesis Testing

Elementary Statistics **Chapter 9: Inferences from Two-Samples**

Objective:

1. Evaluate formulas for two-sample hypothesis testing.

You will be learning more formulas for hypothesis testing in the related section of the textbook. Once again, we will practice evaluating these formulas here. You will learn the meaning of the formulas and how to apply them in context within the text section itself.

Objective 1: Evaluate formulas for two-sample hypothesis testing.

Example 1 Evaluate the formula

$$\bar{p} = \frac{x_1 + x_2}{n_1 + n_2}$$

for $x_1 = 15$, $x_2 = 18$, $n_1 = 35$, and $n_2 = 25$. Round your answer to the nearest ten thousandth.

First, we substitute the given values in for the variables in the formula. Notice that the variables with x and n have subscripts: The variables x_1 and x_2 represent two different unknowns. So, it is critical that we substitute the appropriate values into the variables with the correct subscripts.

$$\bar{p} = \frac{x_1 + x_2}{n_1 + n_2} = \frac{15 + 18}{35 + 25}$$

Once we have completed substituting, we apply the order of operations to the right-hand side. We can group the numerator and the denominator in two sets of parentheses to remind us to individually simplify each of them first by adding. Then, we divide the numerator by the denominator to change the fraction to a decimal. No rounding of the final answer is necessary in this problem.

$$\bar{p} = \frac{(15 + 18)}{(35 + 25)} = \frac{33}{60} = 0.55$$

Answer $\bar{p} = 0.55$

My Turn!

Evaluate the formula

$$\overline{p} = \frac{x_1 + x_2}{n_1 + n_2}$$

for $x_1 = 22$, $x_2 = 18$, $n_1 = 40$, and $n_2 = 30$. Round your answer to the nearest ten thousandth.

Example 2 Evaluate the formula

$$z = \frac{(\hat{p}_1 - \hat{p}_2) - (p_1 - p_2)}{\sqrt{\dfrac{\overline{p} \cdot \overline{q}}{n_1} + \dfrac{\overline{p} \cdot \overline{q}}{n_2}}}$$

if $p_1 - p_2 = 0$, $x_1 = 40$, $x_2 = 60$, $n_1 = 80$, $n_2 = 100$, $\hat{p}_1 = \dfrac{x_1}{n_1}$, $\hat{p}_2 = \dfrac{x_2}{n_2}$, $\overline{p} = \dfrac{x_1 + x_2}{n_1 + n_2}$, and $\overline{q} = 1 - \overline{p}$.

Round your answer to the nearest hundredth.

We have a lot of variables to substitute for in the formulas. We even have formulas to find out the values for some of these variables! Let's begin by using the formulas for the variables that we need to plug into the formula for z. We will find numerical values for $\hat{p}_1$, $\hat{p}_2$, $\overline{p}$, and $\overline{q} = 1 - \overline{p}$.

$$\hat{p}_1 = \frac{x_1}{n_1} = \frac{40}{80} = 0.5$$

$$\hat{p}_2 = \frac{x_2}{n_2} = \frac{60}{100} = 0.6$$

$$\overline{p} = \frac{x_1 + x_2}{n_1 + n_2} = \frac{40 + 60}{80 + 100} = \frac{100}{180} = 0.5556$$

$$\overline{q} = 1 - \overline{p} = 1 - 0.5556 = 0.4444$$

As you can see, we round $\overline{p}$ and $\overline{q}$ to the nearest ten thousandth even though we are rounding our final answer to the nearest hundredth. We keep a few extra places in intermediate steps to minimize rounding error. Ideally, we would try to complete all the calculations at one time using technology so that there is no intermediate rounding. However, when that is not possible, an extra two places should generally be sufficient.

Now, we are ready to substitute into the formula for z and simplify. Note that entire expression under the radical symbol can be thought of as being inside parentheses. This means we cannot take a square root until we have simplified the entire expression under the $\sqrt{}$.

$$z = \frac{(\hat{p}_1 - \hat{p}_2) - (p_1 - p_2)}{\sqrt{\dfrac{\overline{p} \cdot \overline{q}}{n_1} + \dfrac{\overline{p} \cdot \overline{q}}{n_2}}}$$

$$= \frac{(0.5 - 0.6) - 0}{\sqrt{\dfrac{0.5556 \cdot 0.4444}{80} + \dfrac{0.5556 \cdot 0.4444}{100}}}$$

$$= \frac{-0.1}{\sqrt{\dfrac{0.5556 \cdot 0.4444}{80} + \dfrac{0.5556 \cdot 0.4444}{100}}}$$

$$\approx \frac{-0.1}{\sqrt{\dfrac{0.2469}{80} + \dfrac{0.2469}{100}}}$$

$$\approx \frac{-0.1}{\sqrt{0.0031 + 0.0025}}$$

$$= \frac{-0.1}{\sqrt{0.0056}}$$

$$\approx \frac{-0.1}{0.0748}$$

$$\approx -1.3369$$

$$\approx -1.34$$

Answer $z \approx -1.34$

My Turn!

Evaluate the formula

$$z = \frac{(\hat{p}_1 - \hat{p}_2) - (p_1 - p_2)}{\sqrt{\dfrac{\overline{p} \cdot \overline{q}}{n_1} + \dfrac{\overline{p} \cdot \overline{q}}{n_2}}}$$

if $p_1 - p_2 = 0$, $x_1 = 40$, $x_2 = 30$, $n_1 = 60$, $n_2 = 80$, $\hat{p}_1 = \dfrac{x_1}{n_1}$, $\hat{p}_2 = \dfrac{x_2}{n_2}$, $\overline{p} = \dfrac{x_1 + x_2}{n_1 + n_2}$, and $\overline{q} = 1 - \overline{p}$.

Round your answer to the nearest hundredth.

Example 3 Evaluate

$$E = z \cdot \sqrt{\frac{\hat{p}_1 \cdot \hat{q}_1}{n_1} + \frac{\hat{p}_2 \cdot \hat{q}_2}{n_2}}$$

if $z = 1.96$, $x_1 = 50$, $x_2 = 60$, $n_1 = 70$, $n_2 = 100$, $\hat{p}_1 = \dfrac{x_1}{n_1}$, $\hat{p}_2 = \dfrac{x_2}{n_2}$, $\hat{q}_1 = 1 - \hat{p}_1$, and $\hat{q}_2 = 1 - \hat{p}_2$.

Round your answer to 3 significant digits.

Let's begin by using the formulas for the variables that we need to plug into the formula for E. Let's find numerical values for $\hat{p}_1$, $\hat{p}_2$, $\hat{q}_1$, and $\hat{q}_2$. Throughout this solution, we will round to five significant digits (two more significant digits than what is desired in our final answer is normally adequate) in intermediate steps to minimize rounding error. Recall that the leading zeros don't count as significant digits.

$$\hat{p}_1 = \frac{x_1}{n_1} = \frac{50}{70} \approx 0.71429$$

$$\hat{p}_2 = \frac{x_2}{n_2} = \frac{60}{100} = 0.6$$

$$\hat{q}_1 = 1 - \hat{p}_1 \approx 1 - 0.71429 = 0.28571$$

$$\hat{q}_2 = 1 - \hat{p}_2 = 1 - 0.6 = 0.4$$

Since we now have numbers for each of the variables on the right-hand side, we can plug them into the formula for E. Then, we follow the order of operations on the right-hand side to simplify it.

$$E = z \cdot \sqrt{\frac{\hat{p}_1 \cdot \hat{q}_1}{n_1} + \frac{\hat{p}_2 \cdot \hat{q}_2}{n_2}}$$

$$\approx 1.96 \sqrt{\frac{0.71429 \cdot 0.28571}{70} + \frac{0.6 \cdot 0.4}{100}}$$

$$\approx 1.96 \sqrt{\frac{0.20408}{70} + \frac{0.24}{100}}$$

$$\approx 1.96 \sqrt{0.0029154 + 0.0024}$$

$$= 1.96 \sqrt{0.0053154}$$

$$\approx 1.96 \cdot 0.072907$$

$$\approx 0.14290$$

$$\approx 0.143$$

Answer $E \approx 0.143$

My Turn!

Evaluate

$$E = z \cdot \sqrt{\frac{\hat{p}_1 \cdot \hat{q}_1}{n_1} + \frac{\hat{p}_2 \cdot \hat{q}_2}{n_2}}$$

if $z = 2.575$, $x_1 = 32$, $x_2 = 33$, $n_1 = 50$, $n_2 = 60$, $\hat{p}_1 = \frac{x_1}{n_1}$, $\hat{p}_2 = \frac{x_2}{n_2}$, $\hat{q}_1 = 1 - \hat{p}_1$, and $\hat{q}_2 = 1 - \hat{p}_2$.

Round your answer to 3 significant digits.

Example 4 Evaluate the formula

$$t = \frac{(\bar{x}_1 - \bar{x}_2) - (\mu_1 - \mu_2)}{\sqrt{\dfrac{s_1^2}{n_1} + \dfrac{s_2^2}{n_2}}}$$

for $\bar{x}_1 = 210$, $\bar{x}_2 = 190$, $\mu_1 - \mu_2 = 0$, $s_1 = 11$, $s_2 = 7$, $n_1 = 45$, and $n_2 = 60$. Round your answer to the nearest hundredth.

We can begin this problem by immediately substituting the given values for the variables. Once again, we use the order of operations to guide us in simplifying the right-hand side. Be careful not to evaluate the square root until you have simplified the entire expression under the $\sqrt{}$.

$$t = \frac{(\bar{x}_1 - \bar{x}_2) - (\mu_1 - \mu_2)}{\sqrt{\dfrac{s_1^2}{n_1} + \dfrac{s_2^2}{n_2}}}$$

$$= \frac{(210 - 190) - (0)}{\sqrt{\dfrac{11^2}{45} + \dfrac{7^2}{60}}}$$

$$= \frac{20}{\sqrt{\dfrac{11^2}{45} + \dfrac{7^2}{60}}}$$

$$= \frac{20}{\sqrt{\dfrac{121}{45} + \dfrac{49}{60}}}$$

$$\approx \frac{20}{\sqrt{2.6889 + 0.8167}}$$

$$= \frac{20}{\sqrt{3.5056}}$$

$$\approx \frac{20}{1.8723}$$

$$\approx 10.6820$$

$$\approx 10.68$$

Answer $t \approx 10.68$

My Turn!

Evaluate the formula

$$t = \frac{(\bar{x}_1 - \bar{x}_2) - (\mu_1 - \mu_2)}{\sqrt{\dfrac{s_1^{\,2}}{n_1} + \dfrac{s_2^{\,2}}{n_2}}}$$

for $\bar{x}_1 = 68$, $\bar{x}_2 = 64$, $\mu_1 - \mu_2 = 0$, $s_1 = 4$, $s_2 = 3$, $n_1 = 80$, and $n_2 = 90$.

Round your answer to the nearest hundredth.

Example 5 Evaluate

$$E = t \cdot \sqrt{\frac{s_1^{\,2}}{n_1} + \frac{s_2^{\,2}}{n_2}}$$

if $t = 2.449, s_1 = 1.5, s_2 = 1.8, n_1 = 50,$ and $n_2 = 40$. Round your answer to the nearest tenth.

By now, you know the routine; substitute and simplify!

$$E = t \cdot \sqrt{\frac{s_1^{\,2}}{n_1} + \frac{s_2^{\,2}}{n_2}}$$

$$= 2.449 \cdot \sqrt{\frac{1.5^2}{50} + \frac{1.8^2}{40}}$$

$$= 2.449 \cdot \sqrt{\frac{2.25}{50} + \frac{3.24}{40}}$$

$$= 2.449 \cdot \sqrt{0.045 + 0.081}$$

$$= 2.449 \cdot \sqrt{0.126}$$

$$\approx 2.449 \cdot 0.3550$$

$$\approx 0.8694$$

$$\approx 0.9$$

Answer $E \approx 0.9$

My Turn!

Evaluate $E = t \cdot \sqrt{\dfrac{s_1^{\,2}}{n_1} + \dfrac{s_2^{\,2}}{n_2}}$ if $t = 1.99, s_1 = 0.6, s_2 = 0.8, n_1 = 30,$ and $n_2 = 50$. Round your answer to the nearest tenth.

Example 6 Evaluate

$$t = \dfrac{\overline{d} - \mu_d}{\dfrac{s_d}{\sqrt{n}}}$$

for $\overline{d} = 1.2, \mu_d = 0,$ $s_d = 0.14,$ and $n = 49.$ Round your answer to the nearest hundredth.

Let's substitute the four given values in for the appropriate variables. Be careful that you follow the order of operations carefully, especially for the denominator of the complex fraction. If you are using your calculator, you may have to insert parentheses.

$$t = \dfrac{\overline{d} - \mu_d}{\dfrac{s_d}{\sqrt{n}}} = \dfrac{1.2 - 0}{\dfrac{0.14}{\sqrt{49}}} = \dfrac{1.2}{\dfrac{0.14}{\sqrt{49}}} = \dfrac{1.2}{\dfrac{0.14}{7}} = \dfrac{1.2}{0.02} = 60$$

Answer $t = 60$

My Turn!

Evaluate

$$t = \dfrac{\overline{d} - \mu_d}{\dfrac{s_d}{\sqrt{n}}}$$

for $\overline{d} = 0.9, \mu_d = 0,$ $s_d = 0.4,$ and $n = 64.$ Round your answer to the nearest hundredth.

Answers to My Turn!

1. $\bar{p} \approx 0.5714$
2. $z \approx 3.42$
3. $E \approx 0.241$
4. $t \approx 7.30$
5. $E \approx 0.3$
6. $t = 18$

Practice Problems

1. Evaluate the formula

$$\bar{p} = \frac{x_1 + x_2}{n_1 + n_2}$$

for $x_1 = 46$, $x_2 = 37$, $n_1 = 80$, and $n_2 = 70$. Round your answer to the nearest ten thousandth.

2. Evaluate the formula

$$z = \frac{(\hat{p}_1 - \hat{p}_2) - (p_1 - p_2)}{\sqrt{\dfrac{\bar{p} \cdot \bar{q}}{n_1} + \dfrac{\bar{p} \cdot \bar{q}}{n_2}}}$$

if

$p_1 - p_2 = 0, x_1 = 96, x_2 = 83, n_1 = 100, n_2 = 100, \hat{p}_1 = \dfrac{x_1}{n_1}, \hat{p}_2 = \dfrac{x_2}{n_2}, \bar{p} = \dfrac{x_1 + x_2}{n_1 + n_2}$, and $\bar{q} = 1 - \bar{p}$.

Round your answer to the nearest hundredth.

3. Evaluate

$$E = z \cdot \sqrt{\frac{\hat{p}_1 \cdot \hat{q}_1}{n_1} + \frac{\hat{p}_2 \cdot \hat{q}_2}{n_2}}$$

if $z = 1.96$, $x_1 = 45$, $x_2 = 76, n_1 = 50, n_2 = 80, \hat{p}_1 = \dfrac{x_1}{n_1}, \hat{p}_2 = \dfrac{x_2}{n_2}, \hat{q}_1 = 1 - \hat{p}_1$, and

$\hat{q}_2 = 1 - \hat{p}_2$.

Round your answer to 3 significant digits.

4. Evaluate the formula

$$t = \frac{(\bar{x}_1 - \bar{x}_2) - (\mu_1 - \mu_2)}{\sqrt{\dfrac{s_1^{\,2}}{n_1} + \dfrac{s_2^{\,2}}{n_2}}}$$

for $\bar{x}_1 = 230, \bar{x}_2 = 242.5, \mu_1 - \mu_2 = 0, s_1 = 4.5, s_2 = 3.9, n_1 = 40,$ and $n_2 = 45$. Round your answer to the nearest hundredth.

5. Evaluate

$$E = t \cdot \sqrt{\dfrac{s_1^{\,2}}{n_1} + \dfrac{s_2^{\,2}}{n_2}}$$

if $t = 1.664, s_1 = 0.9, s_2 = 0.7, n_1 = 80,$ and $n_2 = 70$. Round your answer to the nearest tenth.

6. Evaluate

$$t = \frac{\bar{d} - \mu_d}{\dfrac{s_d}{\sqrt{n}}}$$

for $\bar{d} = 1.2, \mu_d = 0, s_d = 0.2,$ and $n = 100$. Round your answer to the nearest hundredth.

Integrated Review 10: Linear Equations

Elementary Statistics Chapter 10: Correlation and Regression

Objectives:

1. Find and interpret slope.
2. Find and interpret the *y*-intercept of a line.
3. Find values from a linear equation or graph.
4. Graph a linear equation.
5. Find and interpret a linear model ($y = mx + b$).

In this chapter of the Triola text, you will be modeling data with linear equations. In order to understand the statistical concepts, it is imperative that you are comfortable with lines in the coordinate plane. We will review slope, *y*-intercept, linear equations and their graphs, and linear modeling.

Objective 1: Find and interpret slope.

The slope of the line is its tilt or steepness.

Line with positive slope   Line with negative slope   Line with slope of 0

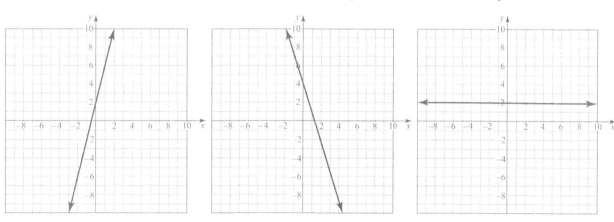

Here are some phrases related to slope that you may have heard:

$$\text{Slope} = \frac{\text{rise}}{\text{run}} = \frac{\text{change in y}}{\text{change in x}}$$

The slope of the line m between the points (x_1, y_1) and (x_2, y_2) can be given by

$$m = \frac{y_2 - y_1}{x_2 - x_1}$$

Example 1 Find the slope of the line that contains the points (3, 5) and (6, 9).

We can label the points as follows $(x_1, y_1) = (3,5)$ and $(x_2, y_2) = (6,9)$.

Now, we will substitute the given points into the formula for slope:

$$m = \frac{y_2 - y_1}{x_2 - x_1} = \frac{9-5}{6-3} = \frac{4}{3}$$

Answer The slope of the line that passes through the points (3, 5) and (6, 9) is $\frac{4}{3}$.

My Turn!

Find the slope of the line that contains the points (0.5, 2) and (0.7, 5). Round the answer to the nearest hundredth.

Example 2 Find the slope of the line given in the following graph.

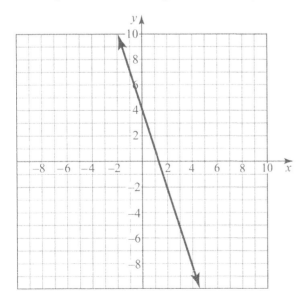

Begin by locating two points on the line. We will use the points $(1, 1)$ and $(2, -2)$, although you could use any two points on the line.

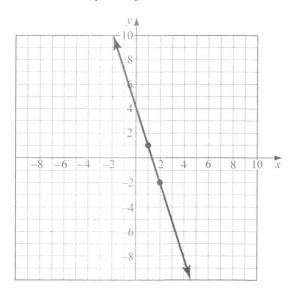

You could then either use the formula or map out the rise over run from the graph.

Formula Method:
$$m = \frac{y_2 - y_1}{x_2 - x_1} = \frac{-2 - 1}{2 - 1} = \frac{-3}{1} = -3$$

Rise-Over-Run Method: Let's look at this problem as if we are travelling from $(2, -2)$ to $(1, 1)$. The rise from -2 to 1 is 3 units. The run from 2 to 1 is -1. The run is negative because we went to the left, which is the direction of the negatives on the number line. (You could have also traveled from $(1, 1)$ to $(2, -2)$, which has a rise of -3 and a run of 1, resulting in the same slope.)

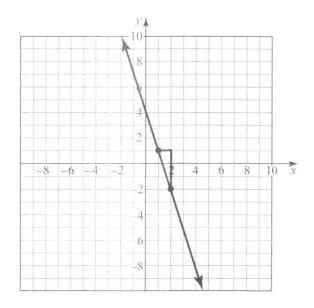

$$m = \frac{\text{rise}}{\text{run}} = \frac{3}{-1} = -3$$

Answer The slope is -3.

My Turn!

Find the slope of the line given in the following graph.

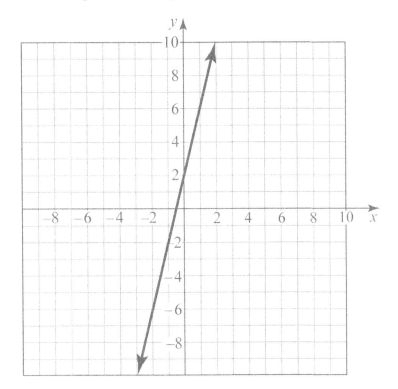

Objective 2: Find and interpret the *y*-intercept of a line.

The *y*-intercept(s) of a graph is (are) the point(s) where a graph crosses the *y*-axis. For a line, there is at most one *y*-intercept. Any *y*-intercept should be written as an ordered pair in the format $(0, b)$, where *b* is the *y*-coordinate of the *y*-intercept.

Example 3 Find the *y*-intercept of the following graph.

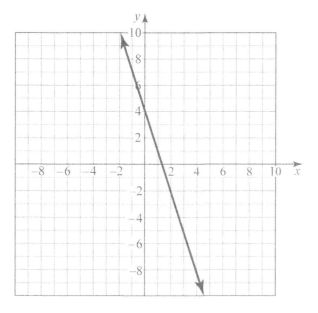

Looking at the graph, we can see that the line clearly crosses through the *y*-axis at the point (0,4).

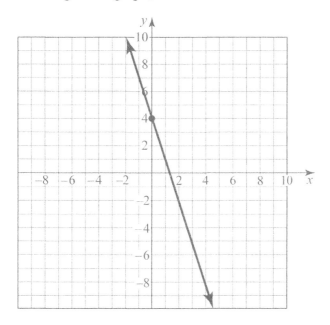

Answer The *y*-intercept of the line is (0, 4).

My Turn!

Find the *y*-intercept of the following graph.

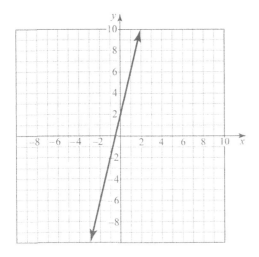

For any linear equation written in slope-intercept form, $y = mx + b$. The *y*-intercept can be written as $(0, b)$. That is, the constant value *b* is the *y*-coordinate of the *y*-intercept.

Example 4 Find the *y*-intercept of the line $y = 7x - 3$.

The equation is written in the form $y = mx + b$. The value of *b* is -3. Since the *y*-intercept should be written as an ordered pair $(0, b)$, the *y*-intercept for this line is $(0, -3)$.

Answer The *y*-intercept is $(0, -3)$.

My Turn!

Find the *y*-intercept of the line.

$$\hat{y} = \frac{2}{3}x + \frac{1}{2}$$

Objective 3: Find values from a linear equation or graph.

Example 5 Find three pairs of data values that would be on the line $\hat{y} = 5x + 7$.

There are actually an infinite number of points on any line. So, answers can vary.

If you plug in any x value, you will get a corresponding $\hat{y}$ (y-hat) value.

Let's plug in 0, 1, and 2 for x.

x	$\hat{y} = 5x + 7$	**Paired data values**
0	$\hat{y} = 5(0) + 7 = 0 + 7 = 7$	(0, 7)
1	$\hat{y} = 5(1) + 7 = 5 + 7 = 12$	(1, 12)
2	$\hat{y} = 5(2) + 7 = 10 + 7 = 17$	(2, 17)

Answer The points (0, 7), (1, 12), and (2, 17) are on the line $\hat{y} = 5x + 7$.

My Turn!

Find three pairs of data values that would be on the line $\hat{y} = -2x + 3$.

Example 6 In the following graph, the horizontal axis represents the height of human beings and the vertical axis represents the weight of human beings. Find the weight (in pounds) for somebody who is 60 inches tall.

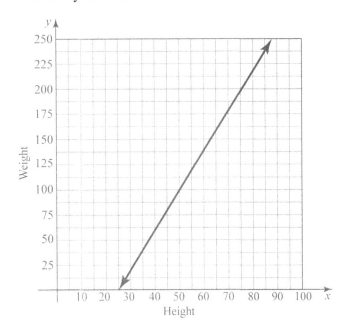

The horizontal axis represents height and the vertical axis represents weight. So, we begin by locating the given value of height = 60 on the horizontal axis and then move up to locate the point on the given line. Then, we look over to the vertical axis and find the corresponding *y* value.

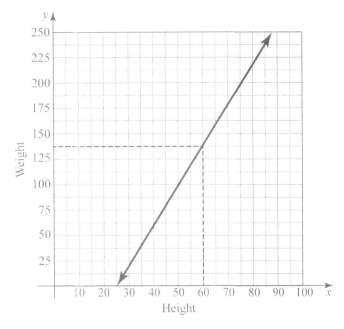

The corresponding weight for a height of 60 inches looks to be about 137.5 pounds, since it halfway between 125 and 150 pounds.

Answer 137.5 pounds

My Turn!

In the following graph, the horizontal axis represents the weight (in pounds) of dogs and the vertical axis represents the weight (in pounds) that the dog eats in a week. How many pounds of dog food does a dog who weighs 48 pounds eat in a week?

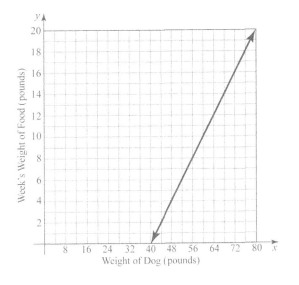

Objective 4: Graph a linear equation.

Example 7 Graph $y = -\dfrac{2}{3}x + 5$.

There are quite a few strategies for graphing lines from an equation. For this example, we will use the slope and the y-intercept to graph the line.

First, extract the y-intercept from the linear equation. Since $b = 5$, the y-intercept for this line is $(0, 5)$. Plot the y-intercept.

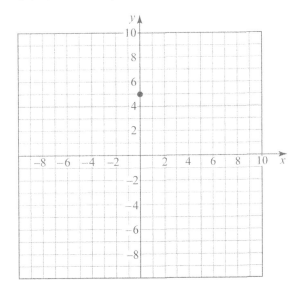

Once you have the *y*-intercept, you can use the slope to locate another point. The slope in this problem is $-\dfrac{2}{3}$. So, another point will be down 2 units and to the right 3 units. (Remember rise/run.)

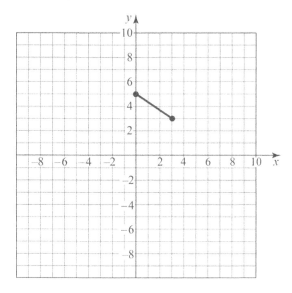

You can now extend this line using a straightedge.

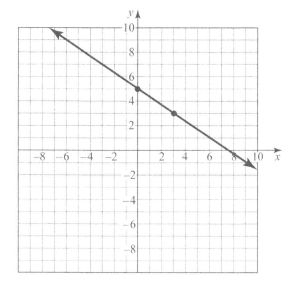

My Turn!

Graph $y = \dfrac{1}{4}x + 3$.

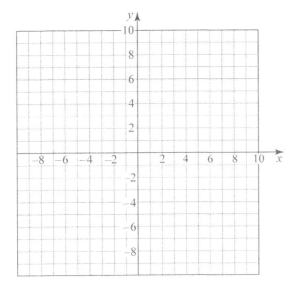

Example 8 Graph $y = 4x - 10$.

You could graph this equation using the method of Example 6 by using the y-intercept of $(0, -10)$ and the slope of 4. However, we will review another method in this example. You can find two points on the line and then connect them with a straightedge. Finding a third point is recommended as a check (or you could use the y-intercept as a check).

x	$y = 4x - 10$	Paired data values
1	$y = 4(1) - 10 = 4 - 10 = -6$	$(1, -6)$
2	$y = 4(2) - 10 = 8 - 10 = -2$	$(2, -2)$

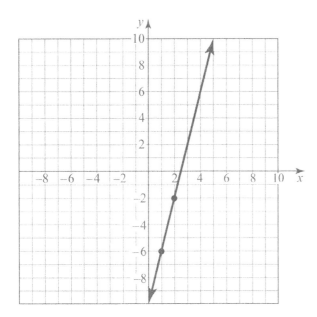

We can see that the line also passes through the y-intercept $(0, -10)$, which serves as our check.

My Turn!

Graph $y = \frac{1}{2}x - 5$. (Hint: You may want to substitute values for x that are multiples of the denominator.)

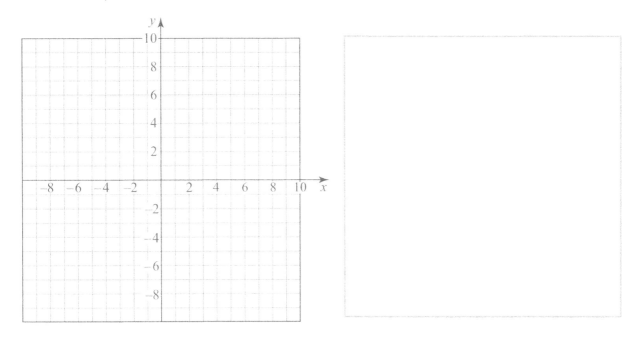

Objective 5: Find and interpret a linear model ($\hat{y} = mx + b$).

You will be studying linear models in great detail in the textbook. For those problems, you will be looking at real data that can sometimes be messy and may best be handled with a calculator. In this workbook, we will use a simplistic model to help prepare you.

Example 9 Find a linear model for the following data for the number of pages in a book and its weight in ounces.

Number of Pages	Weight (oz.)
150	6
200	8
250	10
300	12

The first thing you may want to do is plot the points to determine whether they fall on a line.

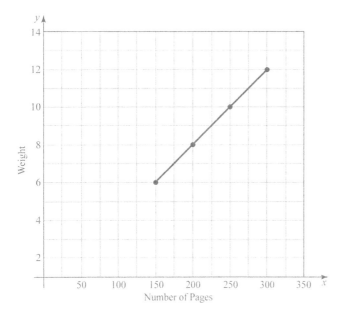

It does appear that the data fall on a line. So, we will use two points to find the slope.

$$m = \frac{8-6}{200-150} = \frac{2}{50} = \frac{1}{25}$$

We now need to find the y- intercept. Since it was not one of the given points, we can substitute any given point for x and $\hat{y}$ into $\hat{y} = \dfrac{1}{25}x + b$ and solve for b.

$$\hat{y} = \frac{1}{25}x + b$$

$$6 = \frac{1}{25}(150) + b$$

$$6 = 6 + b$$

$$0 = b$$

So, the linear model that fits the data is $\hat{y} = \frac{1}{25}x + 0$.

Answer $\hat{y} = \frac{1}{25}x$

My Turn!

Find a linear model for the following data for the number of hours spent studying and the final exam score.

Number of Hours Spent Studying	Final Exam Score
4	40
6	50
12	80
18	100

Answers to My Turn!

1. 15
2. 4
3. (0, 2)
4. (0, ½)
5. Answers will vary. (0,3), (1,1), (2,−1)
6. 4 pounds
7.

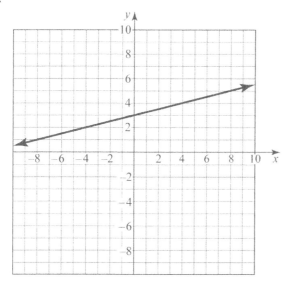

8.

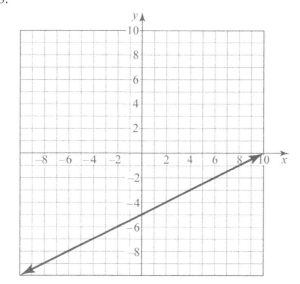

9. $\hat{y} = 5x + 20$

Practice Problems

1. Find the slope of the line that contains the points (3.5, 2.4) and (4.5, 1.6).
2. Find the slope of the line in the following graph.

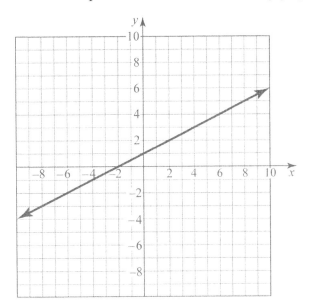

3. Find the y-intercept of the graph shown for Practice Problem 2.
4. Find the y-intercept of the line $\hat{y} = -0.28x + 1.5$.
5. Find three pairs of data values that would be on the line $\hat{y} = -x + 0.2$.
6. The horizontal axis in the following graph represents the number of days that a car is rented, and the vertical axis represents the cost in dollars. How much does it cost to rent a car for 2 days?

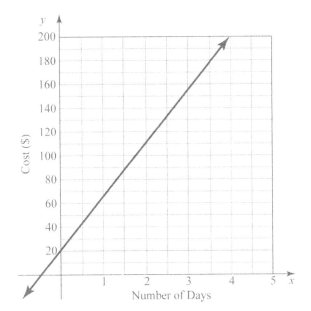

7. Graph $y = -x + 1$ using the slope and y-intercept.

8. Graph $y = \dfrac{4}{5}x - 3$ by plotting points.

9. Find a linear model for the following data for the number of courses that a student is enrolled in and the number of hours spent watching TV on a Wednesday night.

Number of Courses	Number of Hours Watching TV
2	4
3	3
4	2
5	1

Appendix A: Statistics Symbols

Symbol	Read as
α	Alpha ('ælfə)
β	Beta ('biːtə)
μ	Mu ('mjuː)
$\bar{x}$	x-bar
σ	Sigma ('sɪgmə)
ρ	Rho ('rəʊ)
χ	Chi ('kaɪ)
$\hat{p}$	p-hat
$\hat{q}$	q-hat
$\hat{y}$	y-hat
$\bar{d}$	d-bar
μ_d	Mu-sub-d
s_d	s-sub-d
$\bar{A}$	A-bar, the complement of event A (used for probability)
$\approx$	Is approximately equal to
$<$	Is less than
$>$	Is greater than
$\geq$	Is greater than or equal to
$\leq$	Is less than or equal to
df	Degrees of freedom

Appendix B: Units of Measurement

Metric		
Unit of Measurement	**Abbreviation**	**Measures**
Gram	g	Mass (weight)
Meter	m	Length
Liter	l	Volume
Degree Celsius	° C	Temperature

English		
Unit of Measurement	**Abbreviation**	**Measures**
Ounce	oz.	Mass (weight)
Pound	lb.	Mass (weight)
Ton	tn.	Mass (weight)
Inch	in.	Length
Foot (plural feet)	ft.	Length
Yard	yd.	Length
Mile	mi.	Length
Pint	pt.	Volume
Quart	qt.	Volume
Gallon	gal.	Volume
Degree Fahrenheit	° F	Temperature

Appendix C: Conversions between Units of Measurement

Within English System	
16 ounces	1 pound
2240 pounds	1 ton
12 inches	1 foot
3 feet	1 yard
5280 feet	1 mile
2 pints	1 quart
4 quarts	1 gallon

Within Metric System	
Kilo-	1000 basic units (meters, liters, or grams)
Hecto-	100 basic units (meters, liters, or grams)
Deka- (or deca-)	10 basic units (meters, liters, or grams)
Deci-	0.1 basic units (meters, liters, or grams)
Centi-	0.01 basic units (meters, liters, or grams)
Milli-	0.001 basic units (meters, liters, or grams)

Between English and Metric Systems	
1 pound	0.454 kilograms
1 inch	2.54 centimeters
1 mile	1.61 kilometers
1 pint	0.473 liters
1 fluid ounce	0.0296 liters

The conversion from degrees Fahrenheit (F) to degrees Celsius (C) requires the use of the following formula:

$$C = \frac{5}{9}(F - 32)$$

The conversion from degrees Celsius (C) to degrees Fahrenheit (F) requires the use of the following formula:

$$F = \frac{9}{5}C + 32$$

Answers to Practice Problems

Integrated Review 1

1. $d = 12$
2. In the United States, we have presidential elections every four years. So, the possible values for t are 0 (election year), 1, 2, 3, or 4.
3. 16.4 feet
4. 35,000 meters
5. Integers, real numbers
6. 729
7. 7.06×10^{-4}
8. 2.7×10^{5}
9. 97,100,000
10. a) 7 significant digits
    b) 3 significant digits
    c) 1 significant digit
11. 8.67×10^{-19}

Integrated Review 2

1. 24,967.1 kilograms
2. 1.05 pounds
3. $\dfrac{3}{7}$
4. 0.025
5. 0.7%
6. 8%
7. $\dfrac{9}{500}$
8. 5%
9. 224
10. 66 subjects

11.

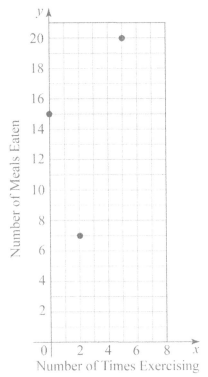

12. 14%
13. 22%

Integrated Review 3

1. 5
2. 14
3. $\dfrac{16}{85} \approx 0.19$
4. 16
5. 10 and 11
6. 15.97
7. 1.77
8. $\sigma \approx 3.4$
9. 6.5
10. 36.7

Integrated Review 4

1. $\dfrac{1}{3}$
2. $\dfrac{79}{70}$

3. $\dfrac{2}{15}$

4. $\dfrac{1}{10}$

5. $\dfrac{1}{24}$

6. $\dfrac{64}{81}$

7. $\dfrac{1}{64}$

8. 5.876

9. 3.246

10. 2.625

11. 14.1

12. 362,880

13. 3024

14. 126

15. $A \cap B = \{\text{red}\}$, $A \cup B = \{\text{red, blue, yellow, pink, purple}\}$ and

   $\overline{A} = \{\text{orange, green, purple, pink, white, black}\}$

Integrated Review 5

1. No
2. 1.35
3. 3.85
4. 258.3
5. 967.5
6. 7.1
7. 0.00298
8. 0.3
9. 0.175

Integrated Review 6

1. 0.3 in.^{2}
2. 10, 11
3. $l > 18$, $v < 60$
4. $z \approx -1.04$
5. $x \approx 98.2$
6. $x \approx 106.6$
7. $z = 4$

Integrated Review 7

1. 0.65
2. 10.105
3. 2.75
4. (3.5, 11.9)
5. $0.63 < \mu < 0.92$
6. 25.2 ± 1.1
7. $E \approx 0.057$
8. $E \approx 0.24$
9. $n \approx 140$

Integrated Review 8

1. $z \approx -7.01$
2. $t \approx 3.46$
3. $\chi^2 \approx 161.8$

Integrated Review 9

1. $\bar{p} \approx 0.5533$
2. $z \approx 3.00$
3. $E \approx 0.0959$
4. $t \approx -13.6$
5. $E \approx 0.2$
6. $t = 60$

Integrated Review 10

1. -0.8
2. $\dfrac{1}{2}$
3. $(0, 1)$
4. $(0, 1.5)$
5. Answers will vary. $(0, 0.2)$, $(1, -0.8)$, $(2, -1.8)$
6. $110

7.

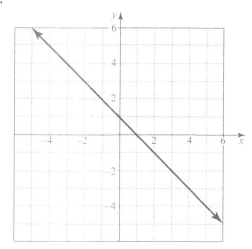

8.

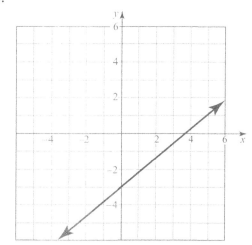

9. $\hat{y} = -x + 6$